3rd International Symposium on Architecture Research Frontiers and Ecological Environment (ARFEE 2020)

E3S Web of Conferences Volume 237 (2021)

Zhangjiajie, China
18 - 20 December 2020

Editors:

R. Weerasinghe
C. Fang

ISBN: 978-1-7138-2717-7

Printed from e-media with permission by:

Curran Associates, Inc.
57 Morehouse Lane
Red Hook, NY 12571

Some format issues inherent in the e-media version may also appear in this print version.

For additional information, please contact EDP Sciences – Web of Conferences
at the address below.

EDP Sciences – Web of Conferences
17, Avenue du Hoggar
Parc d'Activité de Courtabœuf
BP 112
F-91944 Les Ulis Cedex A
France

Phone: +33 (0) 1 69 18 75 75
Fax: +33 (0) 1 69 28 84 91

contact-edps@webofconferences.org

Additional copies of this publication are available from:

Curran Associates, Inc.
57 Morehouse Lane
Red Hook, NY 12571 USA
Phone: 845-758-0400
Fax: 845-758-2633
Email: curran@proceedings.com
Web: www.proceedings.com

TABLE OF CONTENTS

Preface

The 2020 3rd International Symposium on Architecture Research Frontiers and Ecological Environment (ARFEE 2020) was successfully held in the form of online conference on December 18-20, 2020, which should be held in Zhangjiajie, China as planned. Due to the impact of COVID-19 and travel restrictions, ARFEE 2020 was held via online collaboration tool Zoom, which is different from the traditional way. This can not only effectively reduce people gathering and prevent COVID-19, but also enable the scholars to make academic communications as scheduled. The conference mainly focuses on the research fields of architectural science, ecological environment and sustainable development, and is dedicated to provide the experts, scholars, engineers, etc. from different colleges and universities, research institutes, enterprises and institutions from home and abroad, with an academic platform for sharing of academic research findings, exploration of the cutting-edge engineering issues and discussion on the current opportunities and challenges, in a concerted effort to promote international cooperation and communication and the industrialization of scientific research results.

In the opening ceremony of the conference, Prof. Congqi Fang, Shanghai Jiaotong University/ Shanghai Normal University, China presided over the opening ceremony and introduced the experts and guests. Afterwards, Prof. Congqi Fang made opening speeches for the conference and announced the formal opening of the conference. During the conference, the conference model was divided into three sessions, including keynote speeches, oral presentations and online Q&A discussion.

In the keynote speeches session, keynote speakers were each allocated 30-40 minutes to hold their speeches. We were honored to invite six distinguished experts as keynote speakers to present their insightful speeches. Dr.Rohitha Weerasinghe from University of the West of England, England, made a speech about Do Green Buildings Really Work?-A Technical Insight into UK Building Sector. Prof. Congqi Fang from Shanghai Jiaotong University/Shanghai Normal University, China, shared his study on Double Skin Semi-Base-Isolated Structures. Vice Dean/Prof. Yubo Cui from Dalian Minzu University, China and his speech title is: New Contaminants, Antibiotics and Antibiotic Resistance Genes in Sludge Treatment Wetlands. Vice Dean/Prof. Zheng Chen from Guangxi University, China, held a speech on the topic of Coupled models for the combined diffusion-reaction behaviour of chloride and sulphate ions in concrete. Prof. Antonio Caggiano from Darmstadt University of Technology/ CONICET and University of Buenos Aires, Germany and Argentina, his speech is about Energy Storage in Porous Cementitious systems: recent developments and future trends. Prof. Tao Zhang from China Agricultural University, China, made a speech about Biomass waste valorization to generate modified biochar to recover phosphorus from animal manure wastewater.

In the oral presentation's session, some scholars were given about 10-15 minutes to perform their oral presentations one by one. We were honored to invite Prof. Ying Wang (Shanghai Normal University, China) and Prof. Yi Wang (China University of Geosciences, China) as Session Chair of the Oral Presentation's Sessions. The scholars presented their research results on the topics of architectural science, environmental engineering on the sessions. In the Q&A session, all participants were invited to join in academic groups to discuss and explore the academic issues after the presentations.

The proceedings collected the latest research and results in the field of Architecture, Environment and Energy. The topic of this collection is related to architecture science, environment and energy science, sustainable development and so on. All manuscripts published in the proceeding have been through rigorous review and process to meet the requirements of International publication standard.

Finally, we would like to express our sincere gratitude to the Chairman, the distinguished keynote speakers, as well as all the participants. Because of your enthusiastic participation, the conference was successfully held as planned. Thanks to the information technology, the online conference made it easier for more people join in our conference. We had about 180 attendees this time, who are from China, UK, Italy, Spain and more countries and districts. We are expecting more and more experts and scholars from all over the world to join this international event next year. We also want to thank the publisher for publishing the proceedings. We are grateful to all those who have contributed to the success of ARFEE 2020. We hope that all participants and other interested readers benefit scientifically from the proceedings.

The Organizing Committee of ARFEE 2020

 https://doi.org/10.1051/e3sconf/202123700001

Conference Committees

Synthesis and characterization of thiol-polystyrene microsphere and its adsorption properties of thallium

Youze Xu[1], Yuehui Chen[2*], Yuanyuan Zhao[1*], Shuang Zhou[1], Guangyi Fu[1], Yu Zhong[1]
[1]. Key Laboratory of Water Pollution Control Technology, Hunan Research Academy of Environmental Science, Changsha Hunan 410004
[2]. Department of Environmental Science and Engineering, College of Environment and Resources, Xiangtan University, Xiangtan, 411105

Abstract: Thallium (Tl) is highly toxic in the environment. In order to solve the harm of Tl pollution to the environment, thiol-polystyrene microsphere (TPM) with good adsorption properties for Tl were prepared in this paper. The GAUSSIAN software was used to calculate the adsorption performance of the TPM on Tl^+, and the adsorption mechanism was discussed. The thermogravimetric analysis show that the TPM prepared with chloromethyl polystyrene microspheres (CPM) and thiourea as the main raw materials has good chemical thermal stability. In the Tl^+ solution of 1mg/L, the adsorption and removal rate of the TPMs in was 99%, with a significant adsorption of Tl^+ properties. When the dosage of TPM is 0.4L/g, pH is 6-10, temperature is 5-20 °C, and adsorption time reaches 640min, thallium-containing water with Tl^+ concentration of 5mg/L can be treated to within the national discharge standard (5μg/L). The analytical effect of 0.5mol/L dilute sulfuric acid on TPM can reach 99.96%, and the adsorption capacity of TPM on Tl^+ is reduced by only 0.1% after repeated use 4 times, which has good regeneration performance. The modified polystyrene microspheres were characterized by EDS, FT-IR, Raman spectra and XPS, and it was proved that the TPM were ligated with Tl^+ through the grafted mercapto group, so as to adsorb and remove Tl^+ from the solution. According to the isotherm and kinetic model of TPM adsorption Tl^+, the maximum saturated adsorption amount of TPM to Tl^+ can reach 3.47mg/g, mainly chemical adsorption, and the membrane diffusion stage is the main control step of TPM adsorption Tl^+. The results showed that the TPM could be a promising adsorbent with good adsorption capacity and reusability for Tl^+ removal from wastewater

1. Introduction

In the natural environment, the Tl exists mainly in various sulfides (such as silver, antimony, arsenic and other sulfides) ore, chalcopyrite, pyrite, silicate and other minerals[1]. Tl is highly toxic and its toxicity to mammals is higher than arsenic, lead, mercury, copper and zinc[2-4]. It has also serious health effects on humans by the food chain[5]. Tl poisoning can cause damage or change of nerve cells and mitochondria in animals, which can cause gastroenteritis, multiple neuropathy and hair loss, and even can pose a threat to human life under high dose intake[6, 7]. The World Health Organization defined Tl poisoning when the concentration of Tl in human urine is at the range of 4.5-6μg/L[8], and the minimum lethal dose to the human is at the range of 10-15 mg/kg[9].

Nowadays, treatment methods for thallium-containing wastewater include oxidation-flocculation[10, 11], chemical precipitation[12], solvent extraction[13, 14], ion exchange [15, 16], adsorption[17-20] and some composite methods based on the above ways. Tang et al.[21] synthesized a novel titanium ferromagnetic nano-adsorbent for the treatment of thorium-containing wastewater, and the removal rate of Tl^+ could reach 83% in 30 minutes. Mueller et al.[22] investigated sulphuric acid-reducing bacteria to produce cerium sulfide (Tl_2S) precipitates, thereby reducing the concentration of Tl from wastewater. Reddy et al.[23] recycled the Tl from the lead smelting dust by solvent extraction and reduction conversion. The results showed that the extraction rate of Tl reached 99%. Quintas et al.[24] treated thallium-containing wastewater in the environment with a raw and L-cysteine chemically modified yerba mate derived from domestic waste, and the results showed that the adsorption material had the largest biosorption capacity at 328K. Amin et al.[25] fixed the Tl on the styrene divinylbenzene anion exchange resin, then measured the absorbance of Tl at 636 nm and 830 nm. The method has been successfully application to test Tl in water. However, there are still many problems in the treatment of thallium-containing wastewater by the above methods. For example, the chemical precipitation method is difficult to meet the discharge standard after the thallium-containing wastewater treatment, the biological method is demanding on the treatment conditions, and the solution extraction method is only effective for the treatment of specific thallium-containing wastewater, and the composite method is difficult to operate. Therefore, it

*Corresponding author. E-mail: zyyhjgc0601@163.com (Yuanyuan Zhao) and 61080510@qq.com (Yuehui Chen)

is a major demand to develop a simple and efficient method for treating wastewater containing Tl for promoting the sustainable development of the involved enterprises and ensuring the ecological environment and personal safety.

Previous reports indicate that cysteine and glutathione containing thiol groups are involved in the synthesis of some proteins and enzymes in vivo, and thus their special structure makes the sulfhydryl groups easy to combine with heavy metals[26]. Tl and thiol group also have a high affinity[27], resulting in the inability to synthesize proteins and enzymes in the organism and showing pathological phenomena. Since the thiol group contains unpaired lone pairs of electrons, a coordination bond can be formed with the heavy metal ion. Yang et al.[28] reported that mercapto-modified resin has certain adsorption to gold, and the recovery rate reached 90%-100%. Utkelov et al.[29] used CPM as raw materials to graft amino propylene to obtain a sulfhydryl chelating resin having three donor atom reactive groups, and the α-position contains a sulfhydryl group, which can form a strong bond with the metal ion.

Therefore, this thesis intends to use polystyrene microspheres for the treatment of thorium-containing wastewater by utilizing the high affinity between thiol groups and thallium. Moreover, the thiol-modified polystyrene microspheres has certain reproducibility, the synthesis method is simple, and the particle size is controllable[30], which has a high application prospect in the selection of the thallium-containing wastewater.

2. Materials and Method

2.1 Materials

CPM (Nankai University Chemical Plant), thiourea, sodium hydroxide, hydrochloric acid, nitric acid, absolute ethanol, Pb^{2+}, Zn^{2+}, Tl^{+} standard solution.

2.2 Preparation and Characterization of TPM

7.5 g of a mixture of CPM and thiourea in a mass ratio of 2:1 was added in 30mL of absolute ethanol, and then heated in a water bath at 50°C for 6 h. After the reaction, the obtained microspheres were rinsed with pure water, and the washed microspheres were poured into a 50% sodium hydroxide solution and hydrolyzed at 70°C for 10 hours to obtain TPM. The sample was washed with pure water and then rinsed with 5% dilute hydrochloric acid. The final product was washed several times and vacuum dried at 50°C for 6h.

The physical and chemical properties of the TPM were tested by optical microscopy, specific surface area analysis, elemental analysis, Raman spectroscopy, and thermogravimetric analysis.

2.3 Adsorption ability of TPM

Gaussian view5 software was used to construct the model of TPM combined with different heavy metal ions, and the stable configuration and frontier orbital energy of the structure were calculated by the Ground State module DFT algorithm in Gaussian09W. The calculation of the structure of sulfhydryl and benzene rings was based on 6-311G+(d,p) group, heavy metal ions adopt the LanL2DZ basis group.

10 mL of Tl^{+} (4 mg/L), Zn^{2+} (4 mg/L) and Pb^{2+} (4 mg/L) simulated wastewater were added to a series of 50 mL centrifuge tubes, respectively, and 0.1 g of TPM was added. The initial pH of the solution was adjusted to 7, and the reaction conditions were also adjusted. The mixture was shaken under the temperature of 25°C and a rotation speed of 180 rpm, and the adsorption reaction time was 640 min. After the reaction was completed, the microspheres were allowed to settle for 30s. The supernatant was taken to measure the remaining concentrations of Tl^{+}, Zn^{2+} and Pb^{2+}, and the adsorption removal rate was calculated.

2.4 Single factor influence experiment

The thallium-containing wastewater with a concentration of 5 mg/L was placed in a series of conical flasks in varying volume in the presence of 0.1 g TPM. The pH was adjusted to neutral, and the reaction kept for 600 min at the temperature of 25°C. Firstly, for the optimum dosage of TPM during the process, the effects of different dosages of TPM on the adsorption of Tl^{+} were investigated. Secondly, the optimum adsorption time was determined with the unparalleled dosage of TPM at the same pH value and temperature. Finally, under the optimum adsorption time and the optimum dosage, the influence of the pH and temperature on adsorption was investigated.

2.5 Adsorption isotherms and kinetics

Three groups of 30mL of thallium-containing wastewater with varying concentration of Tl^{+} (from 8 to 20 mg/L) were placed in 50mL of centrifuge tubes, respectively. After adjusting the initial pH to 7, each group in the presence of 0.1g TPM was stirred at 180rpm for 12h at 5, 15, 25°C, respectively. Then, the concentration of Tl^{+} in the supernatant was measured, and the amount of adsorption of TPM was calculated.

2.6 Recycling experiment

The microspheres were desorbed with different concentrations of dilute sulfuric acid. The details are as follows: adding 30 mL of dilute sulfuric acid with concentration of 0.1, 0.2, 0.3, 0.4 and 0.5 mol/L in a 50mL centrifuge tube in the presence of 0.3g adsorbed TPM, respectively. Next, these tubes were stirred for 10 h in a constant temperature of 25°C shaking incubator. After standing for 30s, the microspheres were allowed to sink to the bottom, and the concentration of Tl^{+} in the solution was measured to calculate the desorption efficiency.

https://doi.org/10.1051/e3sconf/202123701001

For recycling experiment, the used TPM in the suspension was gathered from after each run by vacuum filtration, and washed several times by the double distilled water until neutral pH. Then the washed catalyst dried overnight in oven at 65°C.

The "adsorption-desorption-adsorption" experiment was repeated 4 times to determine the regeneration performance of TPM.

3. Result and Discussion

3.1 Physical and chemical properties of TPM

Fig.1 shows that the TPM are regular spheres with an average particle size of 0.5mm and a smooth surface, which is a pale yellow microsphere. The specific surface area of TPM was investigated. As shown in Table 1, the modified TPM has a lower specific surface area than the unmodified ones, and the resins have larger particle sizes[31]. The microspheres with a larger particle size are easier to perform solid-liquid separation in the process than the smaller particle size microspheres, and the larger specific surface area is beneficial to the adsorption site of Tl^+ on the microspheres in the wastewater.

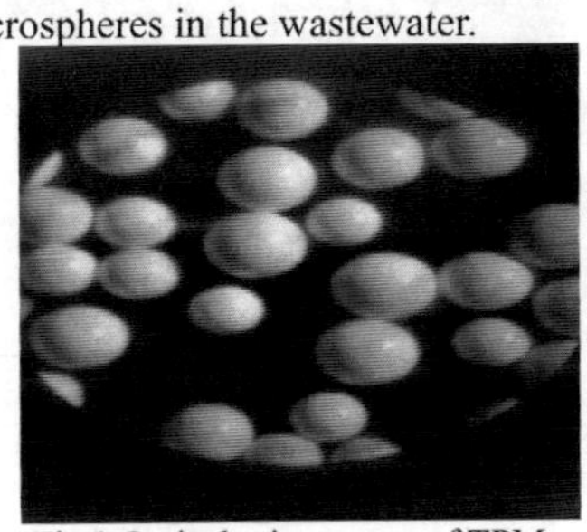

Fig.1 Optical microscopy of TPM

To understand the chemical properties of TPM, energy dispersive spectrometer was applied and the analysis results are shown in Table 1.

Table.1 Pore and surface characteristics of resins

Items	CPM	TPM
S_{BET} (m^2/g)	59.7	48.9
V_t (cm^3/g)	50.6	37.1
R_m (nm)	0.614	0.503

It can be seen from the table that TPM has a lower level of chlorine than that before modification, and the amount of sulfur is significantly improved, which may indicate that the thio-substituted chloromethyl group is supported on the microspheres. The sodium appeared after modification may attribute to the washing process using NaOH. In order to further prove the chemical structure of the TPM, the functional group structure is analyzed by Raman spectroscopy, which could promote the thiol group to easily visualize in the infrared spectrum[27]. The results are shown in Fig. 2. As shown in Fig 2, the C-Cl peak at 678 cm^{-1} after modification disappears, and a new peak appears at 2560 cm^{-1}, corresponding to -SH group[32]. This trend indicates that the thiol group replaced the chloromethyl group successfully and loaded on the microspheres. After adsorption, the -SH peak disappeared since Tl^+ binds to the sulfhydryl group[33].

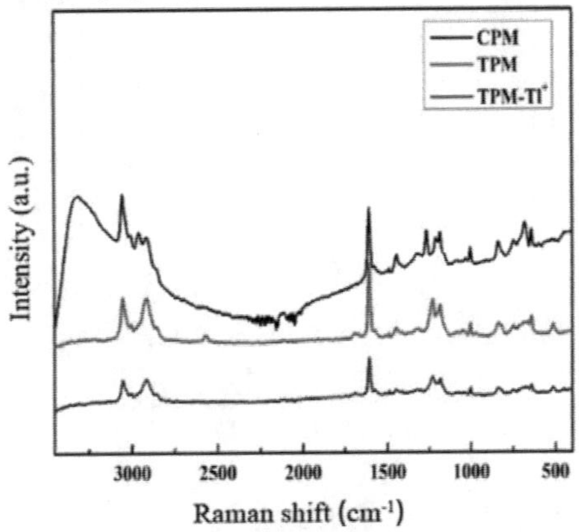

Fig.2 Raman spectrum of TPM and CPM

The thermogravimetric analysis was displayed in in Fig. 3 for determining the chemical stability of the TPM. As can be seen from the Fig.3, the microspheres have been experienced three mass losses from 0 to 800°C. The first mass loss occurred around 150°C. This is because the water contained the sample at which evaporates. The second mass loss occurs at 320°C because the sulfhydryl groups on the microspheres are thermally decomposed. The third time is 410°C, at which temperature the structure of carbon chains is destroyed[31]. It can also be seen from the results that the thermal stability of the TPM is good, and the chemical structure of the TPM is not changed less than 300°C. Therefore, the TPM are available for the treatment of thallium-containing wastewater in the mild environment.

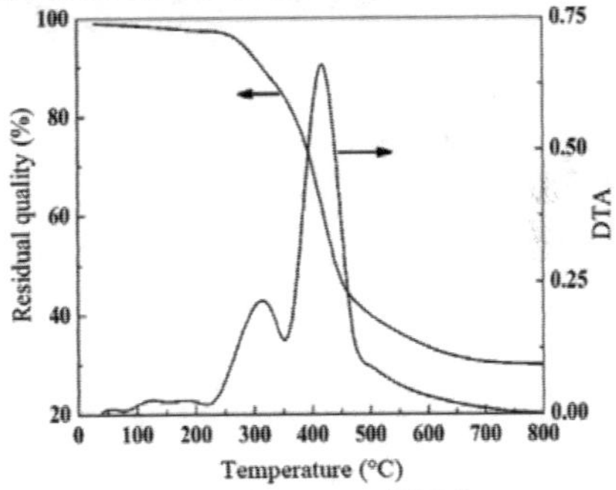

Fig.3 TGA curve of TPM

In order to determine the adsorption performance of the prepared TPM on Tl^+, the adsorption effect of microspheres on Tl^+ before and after modification was measured, and the results are shown in Fig.4. The CPM before modification have basically no adsorption capacity for Tl^+, indicating that the chemical groups and physical structures existing in them do not have any effect on the adsorption; Adsorption removal rate of modified tpm for Tl + increased from 5% to 99.9%, indicating that the modified polystyrene microspheres had a significant adsorption effect on Tl^+, and the adsorption effect was mainly caused by the supported thiol groups. This result is consistent with previous research[34].

https://doi.org/10.1051/e3sconf/202123701001

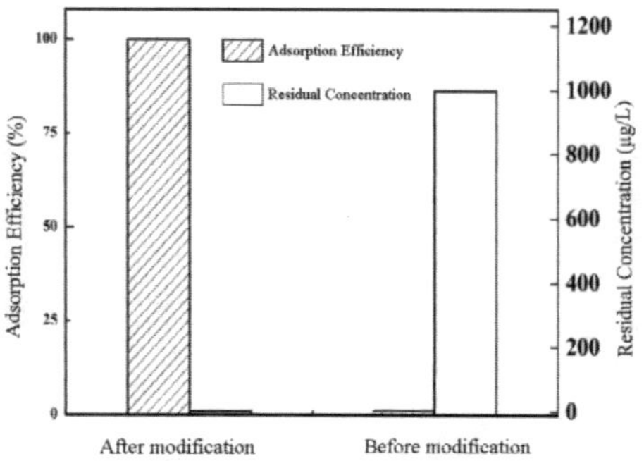

Fig. 4 Adsorption removal rate of Tl^{+} before and after the modification of TPM

3.2 Adsorption ability of TPM

The stable configuration and frontier orbital energy of between the TPM and heavy metal ions were calculated according to GAUSSIAN software[31] (Fig.5 and Table 2).

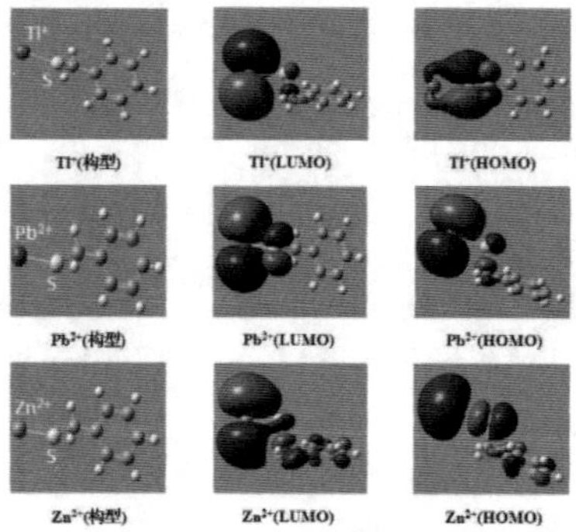

Fig.5 The stable configuration of sulfhydryl with heavy metal ions, and LUMO, HOMO

According to the Fig.5, it can be seen that the distribution of electron cloud between LUMO and HOMO of TPM and Tl^{+} is more symmetric, indicating that the coordination ability is better. In the LUMO and HOMO diagrams, the electron cloud is mainly distributed on the 6p orbit of Tl^{+}, and the bonding mode is π bond. In addition, the frontier orbital energy of TPM and each heavy metal ion was calculated. The results are shown in the Table 2. The frontier orbital energy of TPM and Tl^{+} is 0.1043, and the frontier orbital energy of Pb^{2+} and Zn^{2+} is 0.07945 and 0.06425, respectively. The energy is related to the strength of the chemical activity of the complex. The chemical activity increased as the energy fell, and the stability of the structure decreased, and vice versa. In this way, the binding ability of TPM and Tl^{+} is the strongest, followed by Pb^{2+} and Zn^{2+}.

Table 2 Sulfhydryl is associated with heavy metal ions in the front-line orbital energy

Metal ions	E_{LUMO}	E_{HOMO}	ΔE
Tl^{+}	-0.0719	-0.1962	0.1043
Pb^{2+}	-0.06487	-0.14432	0.07945
Zn^{2+}	-0.09286	-0.15711	0.06425

In order to verify the calculation results, an adsorption competition experiment was carried out. As shown in Fig.6, the adsorption rate of TPM to Tl^{+}, Pb^{2+}, Zn^{2+} heavy metal ions are faster. Within 5 min of the test, the concentration of these heavy metal ions decreased by 85.84%, 53.17% and 65.75%, respectively, and then gradually became stable, and the final concentration of Tl^{+} was 3μg/L. The experimental results demonstrate that the metal ions bind to the active sites on the surface very quickly when they are in contact with the TPM. After the surface active sites are combined with the heavy metal ions, the ions adsorbed on the surface diffuse into the deep layers of the TPM and enter the slow phase. Thus, the adsorption rate of metal ions in the solution gradually stabilizes, and finally reaches the adsorption equilibrium. Therefore, the presence of other heavy metal ions will compete with Tl^{+} for the adsorption site of TPM, which will have a certain influence on the adsorption performance of TPM to Tl^{+}.

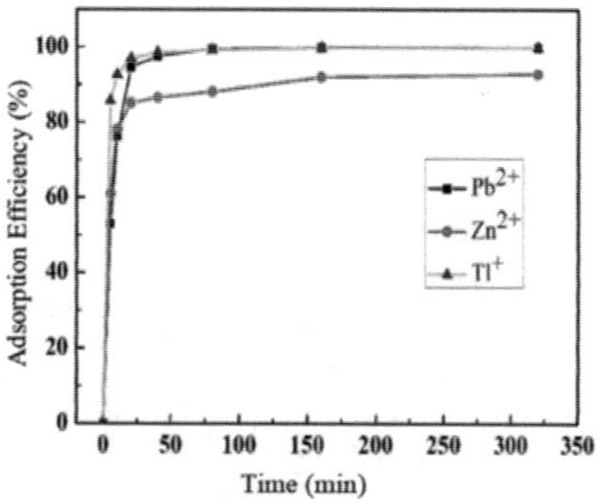

Fig.6 Effect of time on absorption capacity of Pb^{2+}、Zn^{2+}、Tl^{+} by TPM

3.3 Effect of adsorption conditions

The effects of different dosage, treatment time, environmental pH and ambient temperature on the adsorption of by TPM were investigated. The variations in the removal efficiency of Tl^{+} at different dosage are shown in Fig.7. The liquid-solid ratios are 0.2, 0.4, 0.6, 0.8, and 1 L/g, respectively. It can be seen from the figure that when the concentration of Tl^{+} is 5mg/L and the liquid-solid ratio is 0.6~1.0L/g, the concentration of Tl^{+} was higher than 180μg/L in remaining solution, which was much higher than the national emission standard of 5μg/L. This means that the concentration of Tl^{+} in the wastewater after the reaction is lower than the national standard of 5μg/L. Under these conditions, the adsorption capacity of TPM to Tl^{+} was calculated to be about 2.73mg/g. Compared with other adsorbent materials, the TPM prepared in this study has a higher adsorption capacity[35].

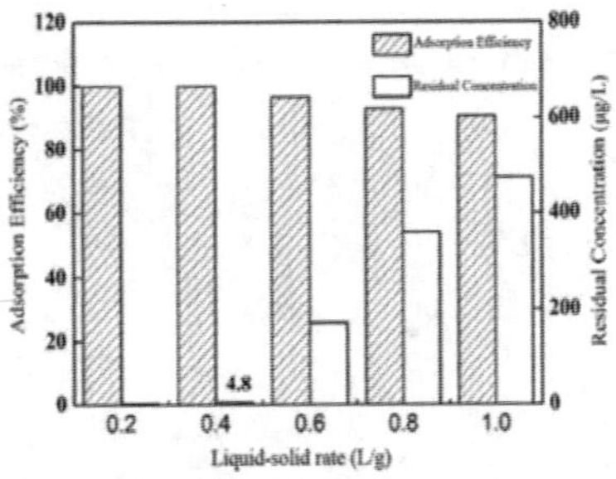

Fig.7 Effect of dosage on absorption capacity of Tl by TPM

Fig.8 shows the effect of reaction time on Tl^+ adsorption by TPM. The process of Tl^+ adsorption by TPM is segmental. In the initial 40min, the concentration of Tl^+ in the solution decreased rapidly, and the rate of adsorption reaction was fast. The rate of adsorption slowed down in the subsequent 120 min, and then began to decrease rapidly until the adsorption equilibrium was reached at 640 min, and the Tl^+ concentration was lowered to 3μg/L.

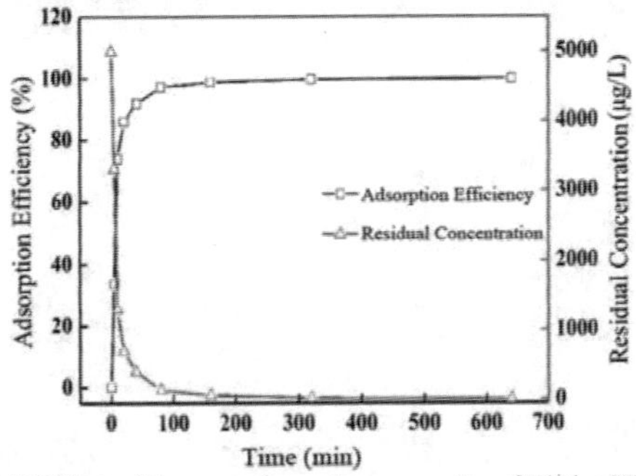

Fig.8 Effect of time on absorption capacity of Tl^+ by TPM

The effect of different pH on the adsorption was also investigated (Fig.9). At the Tl^+ concentration of 5mg/L, the adsorption removal rate of Tl^+ by TPM is above 99.7% in the range of pH 2-10, indicating that the pH value has a weak influence on the adsorption process. After 640 min, the residual concentration of Tl^+ was 12.34μg/L at pH=2, which did not reach the national standard of 5μg/L. However, with the increase of pH, the residual concentration of Tl^+ decreased continuously. When pH=6, the remaining concentration is 4.41μg/L, which is lower than the national standard. At pH=10, the residual concentration is 0.78μg/L, which is far lower than the national standard, indicating that the alkaline condition is beneficial to the adsorption of TPM to Tl^+.

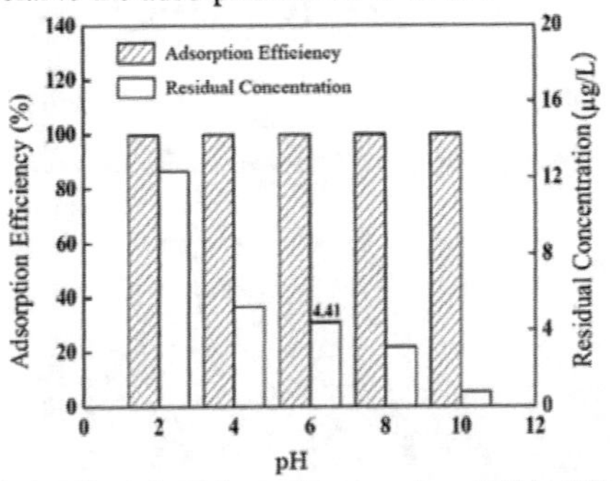

Fig.9 Effect of pH absorption capacity of Tl^+ by TPM

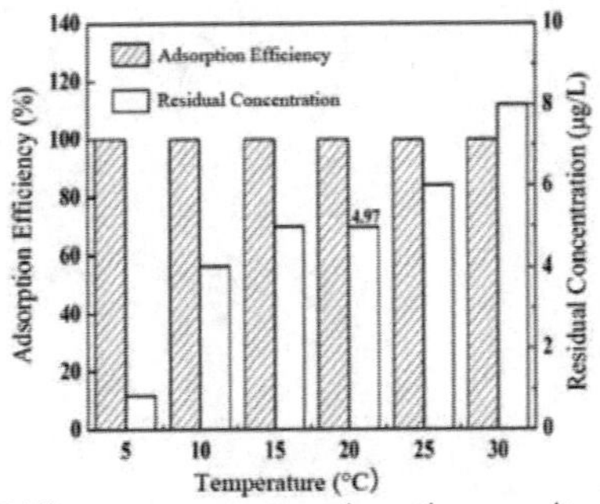

Fig.10 Different temperature on absorption capacity of Tl^+ by TPM

The adsorption performance of TPM on Tl^+ under different temperature conditions is shown in Fig.10. The removal rate of Tl^+ by TPM is above 99.7% when the temperature is in the range of 5-30 °C, indicating that the temperature has little effects on the adsorption process in this study. When the reaction temperature is in the range of 5~20°C, the residual concentration of Tl^+ is reduced from 4.97μg/L to 0.83μg/L, which is lower than the national emission standard, suggesting that the lower temperature is beneficial to the adsorption of Tl^+ by TPM.

3.4 Adsorption isotherm

The adsorption isotherms were measured at 5, 15, and 25°C and were fitted by Langmuir, Freundlich and Dubinin-Radushkevich equations respectively. The fitting equation is as follows:

Langmuir[36, 37]:

$$\frac{C_e}{q_e} = \frac{C_e}{q_m} + \frac{1}{bq_m}$$

Freundlich[38]:

$$\ln q_e = \frac{1}{n} \ln C_e + \ln K_f$$

Dubinin-Radushkevich[39-41]:

$$\ln q_e = \ln q_m + K_D \varepsilon^2$$

$$\varepsilon = \mathrm{RT} \ln\left(1 + \frac{1}{C_e}\right)$$

$$\mathrm{E} = \frac{1}{\sqrt{-2K_D}}$$

Where q_e is the theoretical adsorption amount, mg/g; q_m is the empirically calculated unit mass adsorbent adsorption amount, mg / g; C_e is the concentration of the adsorbed substance in the solution, mg/L; K_f, n is the Freundlich empirical constant; K_D is related to the adsorption free energy, mol^2/kJ^2; ε is the Polanyi adsorption potential energy; E is the average adsorption energy, $kJ \cdot mol^{-1}$. As shown in Fig.11, the isothermal fitting curve of TPM adsorption T1+ has a good linear relationship, and the relevant fitting parameters are all above 0.95, indicating that TPM adsorption of Tl^+ is monolayer adsorption, which is preferential adsorption, and the adsorption process is exothermic.

https://doi.org/10.1051/e3sconf/202123701001

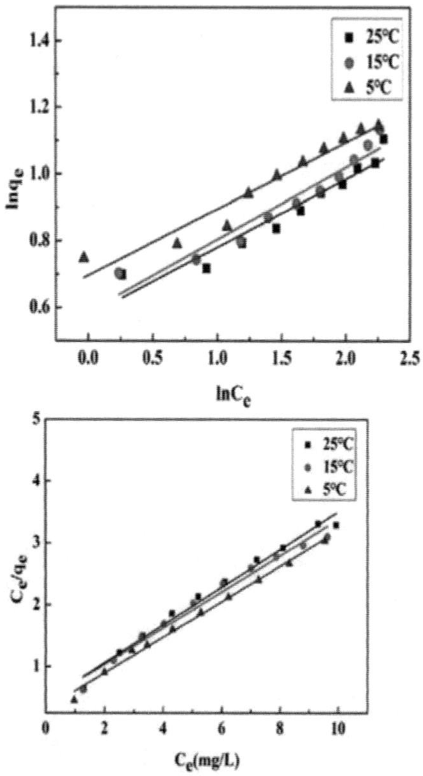

Fig.11 Langmuir isotherems for adsorption of Tl^+ on TPM

In addition, the free energy of adsorption can be calculated by fitting the results of DR isotherm. The values are shown in Table 3. The average adsorption energy obtained is between 15.63-18.14kJ•mol-1. According to the literature, the average adsorption energy is physical adsorption between 1.0-8.0, ion exchange adsorption between 9.0-16.0, and chemical adsorption above 16.0[42, 43]. Therefore, the process of TPM adsorption by TPM is dominated by chemical reactions, supplemented by ion exchange.

Table 3 D-R isotherems parameters for adsorption of Tl^+ on TPM

Temperature（℃）	25	15	5
E（kJ/mol）	18.14	17.08	15.63

3.5 Adsorption kinetics

Fig.12 shows the kinetic curves of the adsorption process under different concentrations of Tl^+ and different pH conditions. The kinetics curves at different pH conditions can be divided into three stages. The first stage is 0-10 min, which is the surface mass transfer stage and the adsorption rate is the fastest. The second stage occurs at 10-80 min, followed by the third stage, which reflects the diffusion of the adsorbate in the pore structure of the material during the adsorption process. Under different concentration conditions, it can also be divided into three stages, but the membrane diffusion stage time at a lower concentration is longer than that at a high concentration.

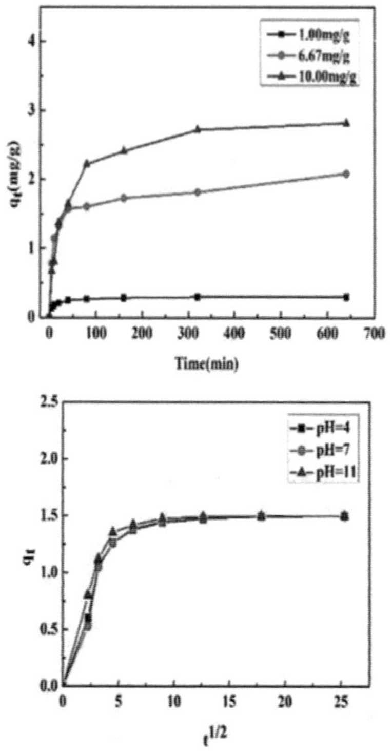

Fig.12 Particle diffusion model plots for the adsorption of Tl^+ on TPM, pH=4,7,11; 1,6.67and10mg/L

The results were fitted by quasi-second-order kinetic equation[44-46], and the fitting results are shown in Fig.13. The correlation coefficients are all above 0.999, which indicates that the adsorption process of Tl^+ by TPM can be used a good description of the stage dynamics under the experimental conditions.

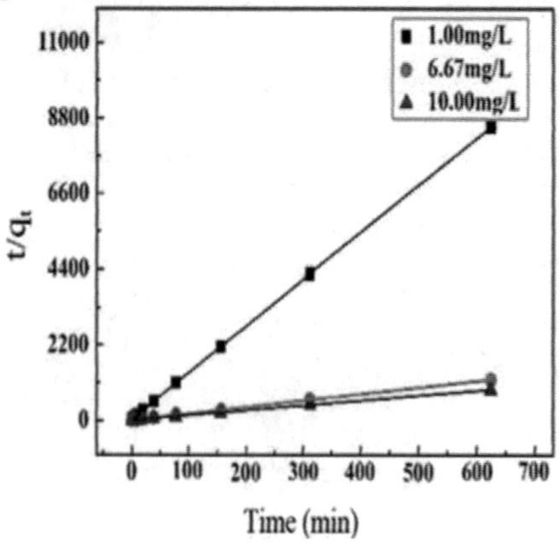

https://doi.org/10.1051/e3sconf/202123701001

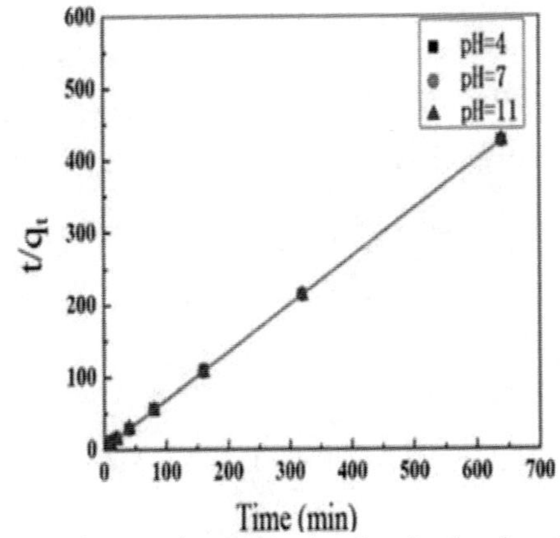

Fig.13 Pseudo-second-order kinetics plots for the adsorption of Tl^+ on TPM, pH=4,7,11; 1,6.67and10mg/L

3.6 Desorption and regeneration of TPM

Desorption experiment was carried out in the presence of 0.1, 0.2, 0.3, 0.4 and 0.5mol/L of dilute sulfuric acid to investigate the desorption effect of different concentrations of reaction solution. As shown in Fig.14, the desorption rate of TPM increased gradually as the increase of sulfuric acid concentration, and the desorption efficiency increased from 89.38% to 99.96%. When the sulfuric acid concentration is 0.5mol/L, it is basically possible to desorb all Tl^+ on the TPM. The TPM was recycled in a 0.5mol/L sulfuric acid solution to investigate the regeneration performance of Tl^+ adsorption by TPM. As displayed in Fig.15, the results suggested that the adsorption rate of TPM to Tl^+ decreased from 99.93% to 99.83% after four cycles. This means that TPM has good cycle regeneration performance. Therefore, it can be proved that TPM is a highly efficient material for adsorbing Tl^+ and the regeneration performance is better.

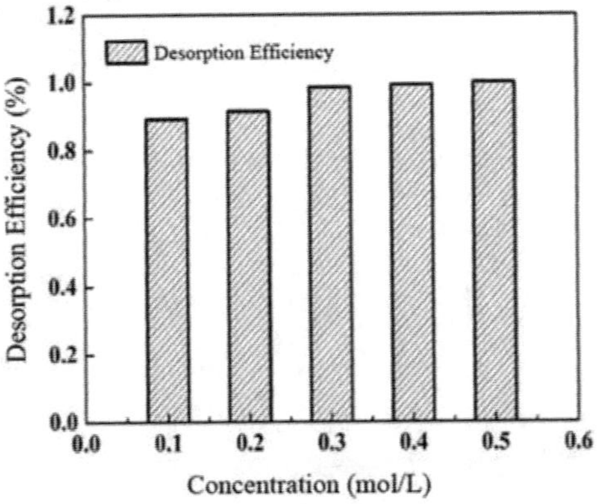

Fig.14 The efficiency of desorption with different concentration of desorption

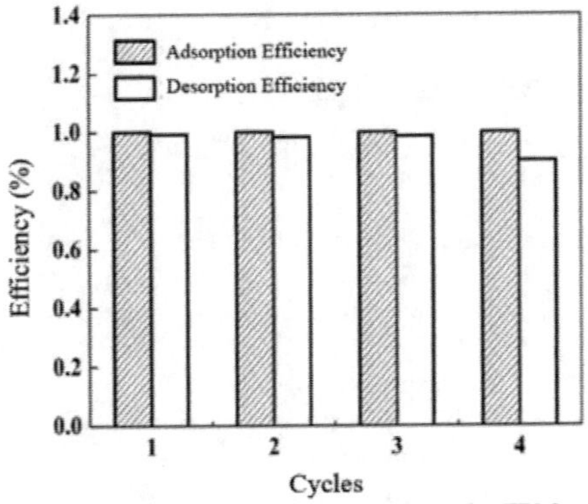

Fig.15 Adsorption-desorption cycles for TPM

4. Conclusion

In this study, chloromethylated polystyrene microspheres were used as raw materials to prepare efficient Tl^+ adsorbents through thiol modification. The adsorption characteristics and mechanism of Tl^+ in solution under different conditions were studied. The main conclusions of the research in this paper are summarized as follows: Elemental analysis, Fourier Transform Infrared Spectrometer, and thermogravimetric analysis proved that the thiol groups were successfully grafted on the polystyrene microspheres surface and the TPM had good thermal stability. When the dosage of TPM is 0.4L/g, pH is 6-10, temperature is 5-20 °C, and adsorption time reaches 640min, thallium-containing water with Tl^+ concentration of 5mg/L can be treated to within the national discharge standard (5μg/L). The analytical effect of 0.5mol/L dilute sulfuric acid on TPM can reach 99.96%, and the adsorption capacity of TPM on Tl^+ is reduced by only 0.1% after repeated use 4 times. Raman analysis showed that the thiol peak disappeared after Tl^+ was adsorbed by TPM, indicating that the functionality of the thiol group disappeared after the thiol group was successfully coordinated with Tl^+. Langmuir fitting results showed that the maximum saturated adsorption capacity of the TPM was 3.47mg/g, and the adsorption process was exothermic. According to the results of kinetics studies, the adsorption experiment is dominated by chemical adsorption and supplemented by ion exchange. In addition, the intra-particle diffusion model show that the average value of the membrane diffusion coefficient and the pore diffusion coefficient is 1.68×10^{-7} and 5.71×10^{-7}, indicating that the adsorption process is mainly controlled by membrane diffusion.

Acknowledgements

Thanks Feng Cong for completing part of the experimental operation and preliminary data collection.

Funding

The study was financially supported by Research and Development plan of key areas in Hunan Province: Key

Technologies for Controlling Heavy Metal Pollution of Inlet River Water and Deep Purification of Antimony Micro-pollution (2019SK2281).

References

1. R. Tyagi, P. Rana, A. R. Khan, D. Bhatnagar, M. M. Devi, S. Chaturvedi, R. P. Tripathi and S. Khushu, *Journal of Applied Toxicology*, 2011, **31**, 663-670.
2. A. R. Despić, D. M. Dražić, M. M. Purenović and N. Ciković, *Journal of Applied Electrochemistry*, 1976, **6**, 527-542.
3. C. J. Rickwood, M. King and P. Huntsman-Mapila, *Ecotoxicology and Environmental Safety*, 2015, **115**, 300-308.
4. A. K. Das, R. Chakraborty, M. L. Cervera and M. d. l. Guardia, *Analytical & Bioanalytical Chemistry*, 2006, **385**, 665-670.
5. C. Easton, A. Turner and G. Sewell, *Environmental Pollution*, 2011, **159**, 3504-3508.
6. A. L. J. Peter and T. Viraraghavan, *Environment International*, 2005, **31**, 493-501.
7. M. J P, *Veterinary and human toxicology*, 1993, **35**, 445-453.
8. T. Xiao, J. Guha, D. Boyle, C.-Q. Liu, B. Zheng, G. C. Wilson, A. Rouleau and J. Chen, *Environment International*, 2004, **30**, 501-507.
9. R. Riyaz, S. L. Pandalai, M. Schwartz and Z. N. Kazzi, *Journal of Medical Toxicology*, 2013, **9**, 75-78.
10. Xiong and Yongliang, *Environmental Chemistry*, 2009, **6**, 441.
11. C. Casiot, M. Egal, O. Bruneel, N. Verma and F. Elbaz-Poulichet, *Environmental Science & Technology*, 2011, **45**, 2056-2064.
12. J. E. Dutrizac, T. T. Chen and S. Beauchemin, *Hydrometallurgy*, 2005, **79**, 138-153.
13. T. Sato, *Hydrometallurgy*, 1992, **30**, 367-383.
14. S. Shirazian, F. Fadaei and S. N. Ashrafizadeh, *Solvent Extraction & Ion Exchange*, 2012, **30**, 490-506.
15. T. Vincent, J.-M. Taulemesse, A. Dauvergne, T. Chanut, F. Testa and E. Guibal, *Carbohydrate Polymers*, 2014, **99**, 517-526.
16. P. J. Faustino, Y. Yang, J. J. Progar, C. R. Brownell, N. Sadrieh, J. C. May, E. Leutzinger, D. A. Place, E. P. Duffy, F. Houn, S. A. Loewke, V. J. Mecozzi, C. D. Ellison, M. A. Khan, A. S. Hussain and R. C. Lyon, *Journal of Pharmaceutical and Biomedical Analysis*, 2008, **47**, 114-125.
17. L. Zhang, T. Huang, M. Zhang, X. Guo and Z. Yuan, *Journal of Hazardous Materials*, 2008, **157**, 352-357.
18. S. G. Nielsen, L. E. Wasylenki, M. Rehkämper, C. L. Peacock, Z. Xue and E. M. Moon, *Geochimica et Cosmochimica Acta*, 2013, **117**, 252-265.
19. J. Zolgharnein, N. Asanjarani and T. Shariatmanesh, *Toxicological & Environmental Chemistry Reviews*, 2011, **93**, 207-214.
20. H. Dashti and H. Aghaie, *CLEAN - Soil*, 2013, **41**, 673-679.
21. J. Tang, W. Wu, L. Yu, X. Fan, G. Liu and Y. Yu, *Science of The Total Environment*, 2019, **694**, 133625.
22. R. F. Mueller, *Mine Water and the Environment*, **20**, 17-29.
23. B. R. Reddy, S. V. Rao and K. H. Park, **22**, 500-505.
24. Q. P. Y, D. G. L, D. S. S. Maris and E. L. B, *Environmental science and pollution research international*, 2019, **26**, 32285-32297.
25. A. S. Amin, A.-A. M. El-Sharjawy and M. A. Kassem, *Spectrochimica Acta Part A: Molecular and Biomolecular Spectroscopy*, 2013, **110**, 262-268.
26. D. Pelclova, P. Urban, P. Ridzon, Z. Senholdova, E. Lukas, P. Diblik and L. Lacina, *Human & Experimental Toxicology*, 2009, **28**, 263-272.
27. G. A. Kilic and M. Kutlu, *Food & Chemical Toxicology*, 2010, **48**, 980-987.
28. Y. Jin-hui, W. Jin-song, Z. Shu-kui, R. Lin and D. Qin-wen, *Bioinformatics and Biomedical Engineering*, 2010, DOI: 10.1109/ICBBE.2010.5517179, 1-4.
29. B. A. Utkelov, E. E. Ergozhin and R. K. Ashkeeva, *Reactive Polymers*, 1991, **14**, 187-191.
30. M. Node, K. Kumar, K. Nishide, S.-i. Ohsugi and T. Miyamoto, *Tetrahedron Letters*, 2001, **42**, 9207-9210.
31. D. Chen, L. Wang, Y. Ma and W. Yang, *NPG Asia Materials*, 2016, **8**, 301.
32. G. S. Mandair, Z. Yu, N. Galaffu, M. Bradley and A. E. Russell, *Applied Spectroscopy*, 2004, **58**, 1282-1287.
33. R. Sanghi and P. Verma, *Bioresource Technology*, 2009, **100**, 501-504.
34. T. Saegusa, S. Kobayashi, K. Hayashi and A. Yamada, *Polymer Journal*, 1978, **10**, 403-408.
35. L. Zhang, T. Huang, N. Liu, X. Liu and H. Li, *Microchimica Acta*, 2009, **165**, 73-78.
36. T. Powell, G. M. Brion, M. Jagtoyen and F. Derbyshire, *Environmental Science & Technology*, 2000, **34**, 2779-2783.
37. S. Luo, X. Xu, G. Zhou, C. Liu, Y. Tang and Y. Liu, *Journal of Hazardous Materials*, 2014, **274**, 145-155.
38. B. K. Nandi, A. Goswami and M. K. Purkait, *Journal of Hazardous Materials*, 2009, **161**, 387-395.
39. M. M. Dubinin, E. D. Zaverina and L. V. Radushkevich, *Zhurnal Fizicheskoi Khimii*, 1947, **21**, 1351-1362.
40. G. Thirupathi, C. P. Krishnamoorthy and S. Pushpavanam, *Chemical Engineering Journal*, 2007, **125**, 177-186.
41. Y. Seki, S. Seyhan and M. Yurdakoc, *Journal of Hazardous Materials*, 2006, **138**, 60-66.
42. M. S. Onyango, Y. Kojima, O. Aoyi, E. C.

Bernardo and H. Matsuda, *Journal of Colloid and Interface Science*, 2004, **279**, 341-350.
43. I. Kiran, T. Akar, A. S. Ozcan, A. Ozcan and S. Tunali, *Biochemical Engineering Journal*, 2006, **31**, 197-203.
44. D. Duranoğlu, A. W. Trochimczuk and U. Beker, *Chemical Engineering Journal*, 2012, **187**, 193-202.
45. A. R. Tehrani-Bagha, H. Nikkar, N. M. Mahmoodi, M. Markazi and F. M. Menger, *Desalination*, 2011, **266**, 274-280.
46. Z. Yin, Y. Liu, S. Liu, L. Jiang, X. Tan, G. Zeng, M. Li, S. Liu, S. Tian and Y. Fang, *Science of The Total Environment*, 2018, **639**, 1530-1542.

E3S Web of Conferences **237**, 01002 (2021)
ARFEE 2020 https://doi.org/10.1051/e3sconf/202123701002

Enzyme degradation mechanism of white rot fungi and its research progress on Refractory Wastewater

Huang ShiYuan*, Li Sheng, Wang ZhenYu, Lin SenHuan, Deng Jian

University of South China,School of Civil Engineering,Hengyang,Hunan 421001,China

Abstract. The lignin-degrading enzyme system of white rot fungi is highly efficient and non-specific, and can degrade a variety of pollutants, including dyes, phenolic compounds and pesticides.The article presents an overview of the mechanism of enzymatic degradation of white rot fungi and its research status in several refractory wastewater were described.

1 Introduction

The development of printing and dyeing, tanning, synthetic resin and other chemical industries has produced a class of refractory wastewater, which has the characteristics of complex composition, refractory and high biotoxicity.Pollutants discharged into the water environment will cause great harm to the ecological environment and human health.Traditional treatment methods are far to reach ideal treatment effect, so it is urgent to explore new treatment technologies for this kind of wastewater.

White rot fungi(WRF) is a kind of filamentous fungus which grows on wood. Its hyphae can penetrate into wood, release lignin degrading enzyme to degrade lignin into white sponge substance, and completely oxidize lignin to CO_2 and H_2O[1].The oxidability of ligninolytic enzymes has attracted much attention in wastewater treatment.Its ability to oxidize organics in liquid media has been verified and recognized, and has been widely used in the treatment of refractory organic pollutants[2].The catalytic mechanism of ligninase degradation system and the application of WRF in several refractory wastewater were reviewed.

2 Enzyme degradation mechanism

The degradation of pollutants by white rot fungi is mainly completed by the lignin degrading enzyme system,including lignin peroxidase (LiP), manganese peroxidase (MnP) and Laccase (Lac).They can attach to the cell wall or secreted outside to form a synergistic degradation system.

2.1 Catalytic Mechanism of LiP&MnP

LiP has a high redox potential, and can oxidize many substances directly or indirectly with the aid of free radical catalysis. The degradation mechanism of the direct oxidation of lignin peroxidase is shown in the figure[3]: LiP(Fe^{3+}) is excited by H_2O_2 and loses electrons, generating compound I ($[Fe^{4+}]^{+}\bullet$); compound I gains electrons and transforms into compound II(Fe^{4+}), and a free radical (R•) is generated at the same time; Compound II continues to lose electrons to form Fe^{3+} and a free radical (R•) is generated. The free radical is unstable and easy to undergo a reduction reaction. Indirect oxidation of lignin peroxidase can be assisted by substances that are easily oxidized to generate free radicals.

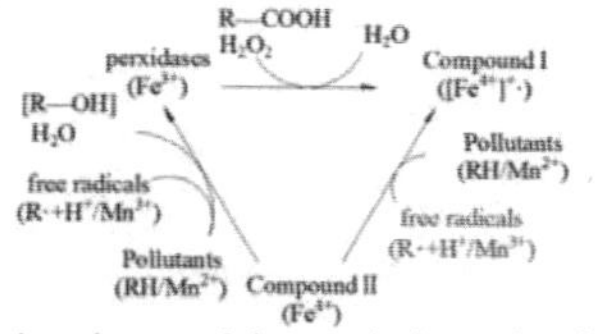

Fig1. Generic scheme of the catalytic cycle of perxidases

2.2 Catalytic mechanism of Lac

Laccase is an oxidase protease containing copper ions. Its catalytic mechanism mainly depends on the synergistic action of four copper ions in laccase molecules and the production of free radicals.The oxidative degradation mechanism of laccase is shown in the figure[4]:The white-rot fungal laccase type I Cu^{+} reacts with O_2 to form the intermediate oxide type II Cu^{+}. It reacts with H_2O to form type III Cu^{2+}, which can continue to produce inactivated laccase type III Cu^{2+}.Inactivated laccase type III Cu^{2+} loses 4 electron reduce to laccase type I Cu^{+}, which converts the substrate molecule into free radicals and simultaneously generates water. Substrate free radicals are generally unstable and prone to redox.

* Corresponding author: 550903597@qq.com

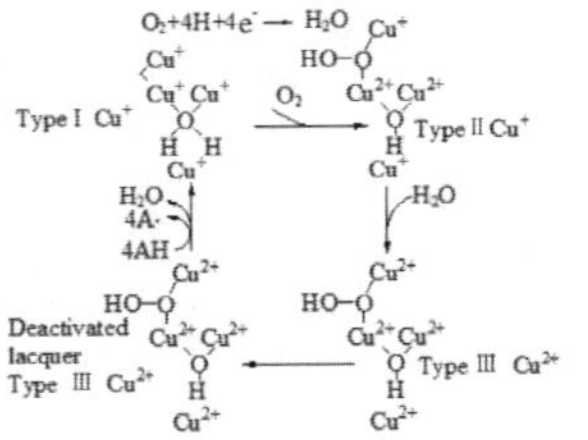

Fig2. The catalytic cycle of laccases

3 Research on dyeing wastewater

Dyeing wastewater has high concentration, high COD, high biotoxicity, poor biodegradability and other characteristics, is an important kind of refractory wastewater.Nowadays, dyes are developing towards good light stability, hard oxidation and high temperature resistance, and conventional treatment methods cannot meet the requirements of degradation[5].The degradation of pollutants by the extracellular enzymes of WRF is nonspecific and can transform substrates without distinction.Its degradation characteristics can adapt to the water quality characteristics of printing and dyeing wastewater.

Si Jing et al.[6]studied the decolorization effect of the azo dye Congo red by Trametes pubescens under non-nutrient conditions.The removal was up to 80.52%(pH2.0,30°C,dye concentration 80mg/L).The phytotoxicity test of the treated dye solution showed that the toxicity of degradation products and dyes was greatly reduced.This study shows that white rot fungi can degrade and detoxify dyes, and have great application potential in dye wastewater treatment.Bugra et al.[7]studied the decolorization effect of Trametes versicolor on RB220 and methyl red Dyes in soil medium.The degradation of methyl red and RB220 were 91%-50%,80% -49%. The removal decreases with increasing concentration, and the decolorization rate is high at low concentration, which may be caused by dye toxicity.Wu Yi et al.[8] used Porodaedalea laricis as the research object to deal with a variety of dyes, including reactive bright blue X-BR, rema bright blue R, Congo red,etc.The decolorization rate reached 95.64%, 97.21%, 91.63% after 7 days. Experimental results show that this strain has a good degradation effect on synthetic dyes.

According to the above, it can be recognized that WRF have a degrading and detoxifying effect on dyes.However,the poor biodegradability of dye wastewater is difficult to ensure the rapid propagation of fungi and the secretion of lignin extracellular enzymes, resulting in low treatment efficiency.This problem is the first to be solved before white rot fungi are applied in actual wastewater treatment.

4 Research on heavy metal wastewater

Unlike organic pollutants, heavy metal ions are non-biodegradable and easy to accumulate through the biological chain.The accumulation of heavy metals in organisms can cause many kinds of damages through thiol binding, protein denaturation, essential metal displacement and oxidative stress and other side effects, which is extremely harmful to human health and ecological environment safety.The fungal cell wall has a high content of polysaccharides and proteins, and contains a large number of functional groups that can be combined with heavy metal ions. In recent years, the removal of heavy metals by biosorption has attracted the attention of many scholars.

Muhammad et al.[9]studied the potential of immobilized Pleutrotus sajor-caju, Agaricus bitorquis, and Ganoderma lucidum to repair wastewater containing Cu(II), Pb(II), Cr(III) and Cr(VI) under the influence of strain dosage, initial concentration, time and other factors.Adsorption experiments were carried out at a pH of 4.5, 30°C and 120r/min. The results showed that the maximum potential of immobilized fungi to repair Cu(II), Pb(II), Cr(VI) and Cr(III) was 205.1, 208.5, 226.6 and 207.3mg/g, respectively.Chen et al.[10] found that Ganoderma Sinensis could enrich Cd(II) in aqueous solution, and the removal rate of Cd(II) by this strain ranged from 21.9% to 6.78% when the concentration of Cd(II) increased (10~30mg/L).Anwei Chen et al.[11] removed Cd(II) from aqueous solution with Phanerochaete chrysosporium.At a pH of 6.5, the removal rate reached the maximum of 83.93%.The highest removal rate of Cd(II) at different initial concentrations (2, 5, 10, 15, 20, 30 mg/L) is 63.62%(5mg/L).

In conclusion, WRF can adsorb and enrich heavy metals to a certain extent. However, the adsorption capacity of the same strain to different heavy metal ions and different strains to the same heavy metal ion is significantly different, and high concentrations of heavy metal ions can inhibit the biological activity of WRF.

5 Research on wastewater containing phenol

Phenolic substances are highly toxic, and even at very low concentrations, they will pose a huge threat to environmental safety and human health.Due to the cell wall rich in chitin and the production of non-specific extracellular enzymes, WRF are more resistant to toxic compounds and have the potential to treat high-concentration phenolic wastewater.

Fan et al.[12]used different carriers to immobilize Trametes to treat phenol-containing wastewater under different factors(pH, concentration and temperature), and analyzed its treatment effect and optimal treatment conditions.The treatment effect is best when wood chips are used as the carrier.Under the conditions of pH 6.5, temperature 30°C, and initial concentration of phenolic compounds 250mg/L, the removal rate of phenolic compounds by the fungus was up to 89.97%.Kietkwanboot et al.[13] investigated the ability of Trametes Hirsuta AK04 to utilize phenolic compounds as a single and mixed substrate in mineral media and palm oil effluent.The initial concentrations of phenolic compounds tested were 100, 200, 500, and 1,000 mg/L for the single substrate systems. The removal ranged from 86-95% for phenol, 78-98% for benzoic acid derivatives, 78-92% for cinnamic acid derivatives, and 87-90% for phenylacetic acid derivatives after 7 days.The mixed substrate medium was prepared to have equal concentration of each phenolic compound at 100, 200, 400, and 600 mg/L to be used in the mixed substrate experiment.The results showed that the removal rate of phenolic substances reached more than 87% after 7 days of treatment(100 and 200mg/L), and the biodegradation efficiency of phenolic substances exceeded 200mg/L decreased.When the concentration was 400 and 600mg/L, the removal rates were 72-83% and 51-69%, respectively.Bernats et al.[14] studied the degradation of phenol by a single white-rot fungus under different factors (biological inoculation amount, pH and temperature). The degradation rate of total phenol in batch flasks by four fungal monocultures of Trametes versicolor, Phanerochaete chrysosporium, Gloeophyllum trabeum and Irpex lacteus in synthetic medium was compared.The batch study of four fungi showed that the degradation effect of Ganoderma lucidum was the best. The optimum conditions were pH 5~6, temperature 25°C, biomass 10%(V/V), and the total phenol degradation rate reached 93% after 7 days.

In summary, WRF can effectively degrade phenolic compounds, immobilization can improve the degradation efficiency.Relatively speaking, WRF have higher tolerance to single phenolic substances than mixed phenolic substances.

6 Conclusion and prospect

The research on white rot fungi is still at the basic experimental level.The application research of white rot fungi needs the following research: a) To conduct in-depth research on the secretion conditions and degradation mechanism of lignin-degrading enzyme systems, so as to provide theoretical guidance for actual wastewater treatment. b) Free microorganisms are not resistant to impact load, and changes in water quality have a greater impact on microbial activity and affect the solid-liquid separation effect. Immobilization can solve the above problems.It is necessary to develop new immobilized carrier with high economy, high mechanical strength and good mass transfer performance. c)Research on inactivated white rot fungus body adsorption.The deactivated white rot fungus is not restricted by culture environment and water quality and can be used as adsorptive material in heavy metal wastewater treatment.

References

1. Li.Hui-rong. Research progress of White-rot Fungi[J].Progress in Environmental Science,1996, 4(6):69-77.
2. Lucas D, Barceló D, Rodriguez-Mozaz S. Removal of pharmaceuticals from wastewater by fungal treatment and reduction of hazard quotients[J].Science of The Total Environment, 2016, 571: 909-915.
3. Xu W D, Wen X H. Application of microorganism in the treatment of dye-containing wastewater [J].Environmental Pollution Treatment Technology and Equipment,2000,1(2): 9-16. (in Chinese)
4. Dirk W., Irene K., Spiros N A. et al. White-rot fungi and their enzymes for the treatment of industrial dye effluents [J].Biotechnology Advances,2003,22(1/2):161-187.
5. Paramaguru G, Kathiravan A, Selvaraj S, et al. Interaction of anthraquinone dyes with lysozyme: Evidences from spectroscopic and docing studies[J].Journal of Hazardous Materials,2010,175(1/3): 985-991.
6. Si Jing, YAN Zhi-hui, CUI Bao-kai, et al. Decolorization of the azo dye gangue red by mycelium of Villi Bolus under nutrient-free conditions [J].Chinese Journal of Microbiology, 2014, 41(002):218-228.
7. Bugra Dayi, Aidai Duishemambet Kyzy, Hatice Ardag Akdogan. Characterization of recuperating talent of white-rot fungi cells to dye-contaminated soil/water[J]. Chinese Journal of Chemical Engineering, 2019, 27 (3): 634-638.
8. Wu Y, MA H F, CAO Y J, et al. Optimization of lassase producing liquid medium for White rot fungus Larch rust and its decolorization effect on dyes [J].Bulletin of Biotechnology, 2020,36(01): 45-59.
9. Hanif M A, Bhatti H N. Remediation of heavy metals using easily cultivable, fast growing, and highly accumulating white rot fungi from hazardous aqueous streams[J]. Desalination and Water Treatment, 2015, 53(1): 238-248.
10. Chen J Y, Shen Y Y, Li X Z.The ability of white rot fungus Ganoderma Sinense to remove cadmium and anthracene[J].Journal of environmental engineering, 2016,10(2): 787-791.
11. Chen A, Zeng G, Chen G, et al. Simultaneous cadmium removal and 2, 4-dichlorophenol degradation from aqueous solutions by Phanerochaete chrysosporium[J].Applied microbiology and biotechnology, 2011, 91(3):811-821.
12. Fan P Y, Cui J G, Jia H. Treatment of phenolic wastewater by immobilized white rot fungi [J].Journal of Environmental Engineering, 2014 (5):1977-1981.
13. Kietkwanboot A, Chaiprapat S, Müller R, et al. Biodegradation of phenolic compounds present in palm oil mill effluent as single and mixed substrates by Trametes hirsuta AK04[J].Journal of Environmental Science and Health, Part A, 2020:1-14.
14. Juhna T,Bernats M. Factors Governing Degradation of Phenol in Pharmaceutical Wastewater by White-rot Fungi: A Batch Study[J].Open Biotechnology Journal, 2015, 9(1):93-99.

https://doi.org/10.1051/e3sconf/202123701003

Study on Diversity and Protection Countermeasures of Zhongshan Park in Wuhan City

Shijing He*, Tingting Dou

Wuhan Institute of Design and Sciences, Wuhan 430205, China

Abstract. Park green space is an important part of garden green space, concentrated expression of the diversity of urban plants. In order to explore the diversity of plants in the park, a typical site in Zhongshan Park in Wuhan was selected for this survey. The species importance, richness index, diversity index and evenness index were used to analyze the characteristics of plant diversity.There are a total of 184 plants of 114 genera and 64 families in Wuhan Zhongshan Park. Among them, there are 86 species of plants of 40 families and 64 genera, 60 species of plants of 26 families and 38 genera, and 26 species of herbs, 11 families and 15 genera.The richness was herb layer > tree layer > shrub layer.Pielou uniformity was herb layer > tree layer > shrub layer;Shannon-wiener diversity was herb layer > tree layer > shrub layer.Simpson diversity was herb layer > tree layer > shrub layer.The diversity of herbaceous plants in the park is high, and the index of shrub layer is generally low, with few species. Plants in the park grow well, some plants are not evenly distributed.

1 preface

Plant diversity is the basis of urban ecosystem, and the protection of plant diversity is the key point of urban biodiversity protection, and plant diversity is also an important index for the evaluation of China's national ecological garden cities[1].The improvement of the stability and healthy development of urban ecosystem functions are often closely related to urban plant diversity[2].Therefore, with the gradual improvement of urbanization level and the rapid increase of population, ecological problems, especially the plant diversity under the influence of urbanization, have become a hot issue for people to study[3].As an important part of urban green space system, park green space has large patch area, rich species, complex plant community structure and good landscaping effect, which are the main manifestations of urban plant diversity[4]. This paper analyzes the plant diversity and discusses its garden plant configuration characteristics, and puts forward corresponding suggestions for the diversity protection and green space construction of the park as well as the effective protection and sustainable utilization of plant landscape resources.

2 Overview and methods of research

2.1 Research Site Overview

Wuhan is located in the east of Jianghan plain, the Yangtze river middle reaches.Wuhan has a typical subtropical monsoon climate, cold in winter and hot in summer, with plenty of rainfall and sunshine. Zhongshan Park in Wuhan is a famous historical park, formally established in 1928,which is a national key park. It is a comprehensive park with many service functions, and the annual visitor volume is as high as 10 million person-times.

2.2 The research methods

2.2.1 Survey method

The diversity of Zhongshan Park in Wuhan was investigated, and representative typical places were selected in the park. Each sample plot was 20m×20m, and the species, name, DBH, plant height, crown width, canopy density and growth status of the trees in the sample plot were recorded.A 5m×5m shrub quadrate was set in each plot, and the species name, plant height, number of plants, coverage and growth status of the shrub were recorded by the quadrate method. Meanwhile, a 1m×1m herb quadrate was also set to record the species name, coverage and growth status of the herb.

2.2.2 Data calculation

Importance value of trees and shrubs = (relative density + relative coverage + relative frequency) / 3.

Herb importance value = (relative coverage + relative frequency) / 2 [5].

Shannon-wiener diversity index is calculated as formula (1) ;Simpson index of diversity is calculated as formula (2); Pielou uniformity index is calculated as formula (3); Margalef richness index is calculated as

* Corresponding author: shijing0221@qq.com

formula (4):

$$H=-\Sigma Pi\ln Pi \quad (1)$$

$$D=1-\Sigma pi^2 \quad (2)$$

$$E=H/\ln S \quad (3)$$

$$R=(S-1)/\ln N \quad (4)$$

In the formula, Ni is number of individuals of the ith species; N is Total number of individuals of all species, S is Number of plant species in sample plot.

3 Results and analysis

3.1 Park plant species composition

According to the survey, there are 184 species of plants in 114 genera and 64 families in Zhongshan Park in Wuhan, among which 86 species of trees in 40 families and 64 genera account for 46.74% of the plant species in the park (Table 1).There are 60 kinds of shrubs, 26 families and 38 genera in the park, accounting for 32.61% of the whole plant species in the park. There are 26 herbaceous species, 11 families and 15 genera in the park, accounting for 14.13% of the plant species in the park. The ratio of arbor to shrub species in the garden was about 1.43:1, and the appropriate ratio of arbor to shrub was 1 (1-1.5) [6], indicating that the arbor and shrub ratio planning in Zhongshan Park was reasonable. There are 48 native species in the park and 136 alien species, accounting for 73.91% of the plant species in the park. In Wuhan zhongshan park which contains the advantage of more than 8 kinds of tree species have two kinds, the first is rosaceae have 11 genera and 13 species, the second is a oleaceae has 9 kinds of 8 genera.

Table1. Composition of plant community in Zhongshan Park

Life-type	Families	Genus	Species
Trees	40	64	86
Shrubs	26	38	60
The herb	11	15	26
In total	64	114	184

3.2 Diversity analysis of plants in the park

3.2.1 Species significance analysis

The important value is a comprehensive quantitative index of the status and function of a species in a population, which can reflect the degree of dominance of each species. Among the 8 quadrates selected from Zhongshan Park in Wuhan, there are 45 species of plants, 30 families and 45 genera.There were 17 species, 15 families and 17 genera of Joe, 9 species, 8 families and 9 genera of Irrigation, and 19 species, 11 families and 19 genera of herb.In the arbor layer, *Magnolia grandiflora, Lagerstroemia indica, Cinnamomum bodinieri*, *Fraxinus chinensis*, *Taxodium ascndens* important value is relatively high(Table 2). *Lagerstroemia indica*, *Cinnamomum bodinieri*, *Fraxinus chinensis*, *Taxodium ascndens* are all native species, *Magnolia grandiflora* for foreign species.

Table2. Ranking of important values of arbor layers in Zhongshan Park, Wuhan (top 5)

Number	Plant	Important value
1	*Taxodium ascndens*	0.0846
2	*Fraxinus chinensis*	0.0719
3	*Cinnamomum bodinieri*	0.0482
4	*Lagerstroemia indica*	0.0447
5	*Magnolia grandiflora*	0.0236

In the shrub layer, the most important value in the top five are Osmanthus fragrans, Loropetsalum chinense, Mahonia fortune, Hypericum monogynum, Aucuba chinensis (Table 3). Among them Osmanthus fragrans, Mahonia fortune, Aucuba chinensis. are native tree species, and Loropetsalum chinense, Hypericum monogynum are alien tree species. All indexes of Osmanthus fragrans rank the first. As a native plant in Wuhan, Osmanthus fragrans has a good growth and a wide planting area. Meanwhile, as an evergreen shrub, Osmanthus fragrans has a good landscape effect. Loropetsalum chinense is often arranged as a low shrub, although the frequency of the garden is not high, but planted in pieces, forming a good landscape.Mahonia fortune, which is in the middle of the index, grows well as a native species. The indexes of Hypericum monogynum and Aucuba chinensis are relatively low, and the planting in the garden is sparse, with general growth condition and poor landscape effect. From the overall index of the shrub layer, the garden shrub planting range is small, few varieties, uneven distribution, and landscape effect is relatively general.

Table3. Ranking of important values of shrub layer in Zhongshan Park, Wuhan (top 5)

Number	Plant	Important value
1	*Osmanthus fragrans*	0.0657
2	*Loropetsalum chinense*	0.0294
3	*Mahonia fortune*	0.0226
4	*Hypericum monogynum*	0.0158
5	*Aucuba chinensis*	0.0151

In the ranking of important values for the herb layer (Table 4), the top five values are Ophiopogon bodinieri, Duchesnea indica, Cynodon dactylon, Petunia hybrida, Eleusine indica. Among them, Duchesnea indica, Cynodon dactylon, Eleusine indica are native species, and Ophiopogon bodinieri, Petunia hybrida are alien species. Ophiopogon bodinieri is among the best in all the indexes. It has strong adaptability and can endure heat and cold as well as humidity. Its root system is well developed, and although it is an alien species, it has good covering effect and is A common ground cover plant. The indexes of Duchesnea indica and Cynodon dactylon are similar. As common native plants, they spread fast, grow well and have good landscape effect. Petunia hybrida, as an ornamental plant, is rich in landscaping and frequently appears in the garden. However, Petunia hybrida prefers warm environment and is not resistant to frost and waterlogging. As an alien species, its relative coverage is not high.

Table4. Ranking of important values of herb layer in Zhongshan Park, Wuhan (top 5)

Number	Plant	Important value
1	*Ophiopogon bodinieri*	0.1667
2	*Duchesnea indica*	0.0883
3	*Cynodon dactylon*	0.0832
4	*Petunia hybrida*	0.0772
5	*Eleusine indica*	0.0624

3.2.2 Species richness index analysis

The richness of species can measure the richness of garden plants in the garden green space, reflecting the adaptability of species to the environment and the utilization of resources. The more abundant the plants, the higher the value will be[7]. It can be seen that the Margalef richness index of arbor and shrub is arranged as: herb layer > tree layer > shrub layer（Table 5）. The herbaceous layer in the sample land is the most abundant, because of the human disturbance, through the fence and other obstacles to visitors to trample herbaceous plants, so that all kinds of herbaceous plants can grow naturally.

Table5. Measurement results of plant diversity index in park

Level	Richness	Uniformity	Shannon-Wiener	Simpson
Tree Layer	0.7072	1.0301	1.1645	0.4712
shrub Layer	0.5403	0.7219	0.5701	0.4720
Herb Layer	1.5702	1.4425	2.3567	0.8007

Species richness is a measure of the richness of garden plants in the garden green space. The more uniform the population distribution in the plant community, the higher the community uniformity index. On the contrary, the more concentrated the population distribution, the lower the community evenness index[8].The Pielou evenness index of arbor and shrub was arranged in the order of herb layer > tree layer > shrub layer.Compared with shrub layer, the uniformity index of tree layer and herb layer is higher. In the park, the distribution of tree layer and herb layer is more uniform, with low ecological dominance, while the distribution of shrub layer is not uniform, with high ecological dominance.

Species diversity is an important index to measure the richness and evenness of community structure. The higher the plant diversity in the community, the more complex the structure and function will be[9]. Shannon-winner diversity index is used to reflect the degree of individual disorder and uncertainty of community species. The diversity index of shannon-winner in the overall layout of the park is arranged as herb layer > tree layer > shrub layer（Table 5）. At some of the landscape nodes of the park, the management has constructed a landscape layer with high coverage and rich colors, with native species as the main structure for good landscape effect. Among all the plots, shrub layer has the simplest structure and the worst ecological function. Only a few shrub species such as *Osmanthus fragrans*, *Loropetsalum chinense* and *Mahonia fortune* are widely used in the garden, while others are planted sparsely and have a single ecological function. Herbaceous layer shannon-Winner diversity index was almost greater than 2, herbaceous layer species richness was high. The extensive management of herbaceous plants by the park managers shows that many wild herbaceous plants can grow naturally, resulting in high diversity of herbaceous layers.

Simpson diversity index refers to the probability that two individuals belonging to the same species are randomly selected from a community[10].The Simpson diversity index was the herb layer > shrub layer > tree layer（Table 5）.On the whole, the vegetation in the tree layer and herb layer is more abundant, and the landscape effect is better than that in the shrub layer.

4 Discussion and Suggestions

4.1 Discuss

The selection and arrangement of plant species is often an important factor that determines the style of garden plant landscape and affects garden art and environmental greening level [11].Through this survey, it is found that WuHan ZhongShan Park green space is relatively rich, a total of 64 families, 114 genera and 184 species of plants.Among them, there are 48 native plants, accounting for 26.09% of the total plant species in the garden. It can be seen that the proportion of native species planted in the garden is relatively low, and most of them are imported species.The reasonable collocation of native species and alien species should be paid attention to when plant configuration is carried out in the park. The herbaceous plants in the park are abundant, and the shrub species are the least, among which the species of *Rosaceae* and *Luteaceae* are more dominant, which are outstanding in the greening of the park. It was found that some areas of the garden had unreasonable arrangement of greening, relatively simple plant community and relatively simple plant arrangement. In the park as a whole, the richness index, diversity index and evenness index of the tree layer, shrub layer and herb layer in wuhan zhongshan park were mostly as follows: herb layer >tree layer > shrub layer. The stability of had strong stability, while shrub layer had the weakest stability.

4.2 Suggest

The diversity of garden plants is an important part of the urban green space system, and native plants are in a key position among garden plants[12]. Add native plants appropriately in the garden, and use native species to build urban landscapes with obvious regional characteristics.Native species are those with low cost, high survival rate and quick effect, with lax requirements on conservation and management[13].It is suggested to enrich the variety of garden plants, a variety of evergreen plants and an appropriate amount of some excellent native plants to view flowers and leaves, so as to adorn the color of the park in four seasons. We suggest designers to optimize the structure of shrub layer, increase the variety and quantity of shrub, and enrich the plant species. We need to combine the ornamental value and function through different landscape features, which can not only achieve good landscape effect, but also increase the stability of the park green space system. At the same time we should pay attention to increase the variety and number of shrubs. Due to the low shrub diversity in the park, we should try our best to choose some evergreen shrubs. At the same time, we should also pay attention to the arrangement of some flower viewing, leaf viewing or fruit viewing shrubs in the park to increase the seasonal variation of the park landscape. For example, *Cycas revoluta, Sabina procumbens, Distylium chinense, Fatsia japonica* and other evergreen native shrubs can be planted in the garden, and then some flowers such as *Gardenia jasminoides*、*Malus halliana*、*Michelia figo*、*Serissa japonica* as shrubs. There are many lawns in Zhongshan Park in Wuhan, and the herb planting in some areas is relatively simple, so some plants can be added for flower viewing.

When we construct the plant community in the park, we should try to form a multi-layered plant community. The plant community should be dominated by trees, supplemented by shrub layer and herb layer, which can enrich the diversity of landscape structure. [14]. Secondly, we should make full use of the limited space and combine different planting forms to improve the number of garden plant varieties, green rate of parks and plant landscape quantity .

Acknowledgements

This research was financially supported by the Humanities and Social Science Research Program of Hubei Provincial Education Department (16G216) and School-level quality course construction project (201513).

References

1. L. Shuo, M. X. Wang, L. Dong, etc. LA, 26,106-110(2019).
2. Crane K R . Kinzig A . Science, 308(2005).
3. Mc Kinney . BC, 127(2006).
4. X. Li, X. Wen. LNFSAT, 5, 24-27(2013).
5. Y. S. Wang, T. L. ShangGuan. JSXU (NSE), 02, 312-316(2010).
6. Z. Y. Zhu, S. Qing. LA, 01, 39-46(1995).
7. Q. S. Ren. CJOE, 21, 67-70(2002).
8. C. Zhu, Y. H.Pan, Y. M. Feng. etc. LA, 3, 83-86(2009).
9. J. Ma, S.J. Zheng, K. Zhou. etc. ZJAS, 9(2016).
10. K. P. Ma, J. H. Hang, S. L.Yu, etc. AES, 15, 268-277(1995).
11. B.Wang, S. L. Dai, X.Z. Yang. CASB, 27, 290-295(2011).
12. Y. S. Xu, L. J. Wu. HBAS, 57, 80-85(2018).
13. J. L. He, Y. N. Li, J. B. Xi, SXA, 42, 178-179(2016).
14. J. R. Lei, X. Q. Song, R.X. He, CJOE, 35, 118-124(2016).

Distribution of Triarrhena lutarioriparia and its reserve characteristics of nitrogen and phosphorus in Dongting Lake

Youze Xu[1], Guangyi Fu[1],Nan Tang[1], Zhonghao He [2], Lincheng Jian[1], and Yuanyuan Zhao[1*]

[1]. Hunan Research Academy of Environmental Science, Hunan Provincial Key Laboratory of Water Pollution Control Technology, 410000 Changsha Hunan, China
[2]. Beijing Normal University-Hong Kong Baptist University United International College, 519000 Zhuhai Guangdong, China

Abstract. *Triarrhena lutarioriparia*, a typical and most abundant macrophyte in Dongting lake wetland, was in the state of abandonment following the papermaking industry revocation in the lake basin. In order to provide scientific basis for precise management of *T. lutarioriparia*, the *T. lutarioriparia* distribution charateristics in Dongting Lake and its storage characteristics of nutrients were investigated in this study. Remote sensing interpretation results showed that the total area of *T. lutarioriparia* in Dongting Lake wetland was 58,450 ha, 48.31% of which distributed in South Doting Lake wetlands. The nutrients contents were significantly different in *T. lutarioriparia* tissues, ranking in the descending order of spikes (TN 27.90 mg/g, TP 3.46 mg/g)>leaves (TN 16.38 mg/g, TP 2.11 mg/g)>stems (TN 5.38 mg/g, TP 0.85 mg/g). The total P quantities in each *T. lutarioriparia* tissue were ranked in the order: stems (560.26 t)>leaves (396.52 t)>spikes (284.67 t), while the total N quantities were within the range of 2170.02~2801.3 t. It was estimated that about 7712.99 t of TN and 1241.45 t of TP were annually removed from Dongting Lake by reaping *T. lutarioriparia*. The nutrients stored in the dead tissues of *T. lutarioriparia* might possess non-negligible impact on the water quality of Doting Lake.

1 Introduction

Dongting Lake is known as the "Kidney of the Yangtze River", and its runoff is about 1/3 of the total runoff of the Yangtze River [1]. Dongting Lake is a very important natural flood storage reservoir and internationally important wetlands, bearing major responsibilities for water ecology safety in the middle and lower reaches of the Yangtze River Basin [1]. Aquatic plants are indispensable parts both for both the structure and function of lake ecosystems [2]. *Triarrhena lutarioriparia*, a unique fibrous perennial herbaceous macrophyte in China, widely distributed in Dongting lake wetland [3-5]. The ecological functions of *Triarrhena lutarioriparia* such as purification of water bodies are extremely important to the ecosystem of Dongting Lake wetland. Meanwhile, compared with Phragmites australis, *T. lutarioriparia* has higher biomass per unit area and excellent cellulose quality [3, 6]. Thus, *T. lutarioriparia* has been applied as the main raw material for pulp production in Dongting Lake area since the early 1960s [7-10]. In order to protect Dongting Lake from pulping and papermaking wastewater pollution, an implementation plan was issued by Hunan Provincial People's Government for guiding exit of papermaking enterprises in Dongting Lake Area (Government Office Letter No. 36, 2018), requesting paper-making enterprises around Dongting Lake to fully withdraw from pulp production capacity and to withdraw from paper-making production capacity in 2019. The overall withdrawal of papermaking makes it difficult to develop alternative resource utilization methods for *T. lutarioriparia* in the lake area within a short time. Therefore, a large amount of *T. lutarioriparia* will stay uncut in the lake wetland, causing problems in how to manage and protect them from exerting non-negligible impact on the water quality of Doting Lake [11-13].

Mastering the distribution status and nutrient characteristics of *T. lutarioriparia* is highly essential to formulate an effective management for it [14]. However, as the artificial planting development of *T. lutarioriparia* in recent years, the current distribution of *T. lutarioriparia* in Dongting Lake is still unclear. Moreover, the nutrient characteristics in T. lutarioriparia have been given less concern. The lack of research on *T. lutarioriparia* greatly increases the difficulty in the precise management of *T. lutarioriparia*. Therefore, remote sensing interpretation methods were used in this study to explore the distribution characteristics of *T. lutarioriparia* in Dongting Lake wetland. The dry-weight biomass and nutrient contents of each *T. lutarioriparia* tissue in different areas of Dongting Lake were investigated, and the total quantities of nutrients stored in *T. lutarioriparia* were further calculated. Results obtained in this study can provide scientific basis and strong supports for management of *T. lutarioriparia* after the withdrawal of the paper industry in Dongting Lake district.

*Corresponding author. E-mail: zyyhjgc0601@163.com

https://doi.org/10.1051/e3sconf/202123701004

2 Method and Material

2.1 Sample Collection and Pretreatment

After preliminary investigation and research in November 2018, a global positioning system was used to locate *T. lutarioriparia* in the whole lake area of Dongting Lake, and 6 sampling points were selected to collect *T. lutarioriparia* samples. The stems, leaves, and spikes of each plant of *T. lutarioriparia* are separately processed and marked on site. Then the collected *T. lutarioriparia* samples are washed with ultrapure water, dried and weighed. After green removing at 105 °C for 30 min, the plant samples were dried to constant weights at 80 °C. The dried samples were then placed in a desiccator, cooled to room temperature and weighed, and water contents of the plant samples were measured.

2.2 Remote sensing interpretation of T. lutarioriparia resources in Dongting Lake

The geological conditions and plant species of the Dongting Lake area were investigated. Furthermore, the variations in remote sensing pictures of *T. lutarioriparia* due to seasonal changes were analyzed. Based on the cloud analysis of difference in water levels due to seasonal changes and the remote sensing pictures, the distribution of *T. lutarioriparia* in Dongting Lake wetland was interpreted by applying the Landsat 8 OLI-TIRS image data [15, 16]. The remote sensing images of the Dongting Lake area in early spring of 2018 were selected for interpretation, and ENVI 5.3 and ArcMap 10.2 software were used to process the images. The maximum likelihood method is used to supervise and classify the images through the visual method. In order to improve the accuracy of the interpretation work, the unclear distribution areas of *T. lutarioriparia* in Dongting Lake wetlands were clarified by on-site survey and sampling.

2.3 Data analysis

The total nitrogen (TN) and phosphorus (TP) contents in the remaining litter (leaves, stems and spikes) were measured using Kjeldahl digestion followed by colorimetric analysis. Data shown in the figures herein are averages. Analysis of the variance was performed by one-way ANOVA and two-tailed paired t test using the PASW statistics software package version 18 (SPSS, Inc., Chicago, IL). Differences were considered significant when the significance level was <0.05 (e.g. α=0.05).

3 Result and Discussion

3.1 The distribution characteristic of *T. lutarioriparia* in Dongting Lake

Comprehensively considering the three factors including *T. lutarioriparia* distribution, natural conditions and geographic location, six sampling points (D1-D6 seen in Figure 1) were set up in the field investigation for the precise interpretation of the distribution of *T. lutarioriparia* in the Dongting Lake area. Figure 1 illustrated the interpretation results of *T. lutarioriparia* distribution in Doting Lake wetland. The specify areas of *T. lutarioriparia* in different districts were shown in Table 1. The total distribution area of *T. lutarioriparia* in early 2018 was nearly 58,467 ha (Table 1). As shown in Table 1, the specify areas of T. lutarioriparia in different districts were ranked in the descending order: South Dongting Lake (50.40%) > East Dongting Lake (31.39%) > West Dongting Lake (18.21%). In the basin of South Dongting Lake, *T. lutarioriparia* with the area of 29,457.78 ha was mainly distributed in district of Yuanjiang city, accounting for 95.8% of the total *T. lutarioriparia* in South Dongting Lake. In the basin of East Dongting Lake, the distribution area of *T. lutarioriparia* in Yueyang City and Xiangyin city has an area of 9,707.68 ha and 7,948.50 ha, accounting for 96.40% of the total distribution area in this basin. In the basin of West Dongting Lake, *Triarrhena lutarioriparia* was mainly distributed in Hanshou County followed by Anxiang County. Among all the districts in Dongting Lake basin, Yuanjiang City has the widest distribution area with 28,235.89 ha of *T. lutarioriparia*, followed by Yueyang County (9,707.68 ha) and Xiangyin County (7,948.5 ha).

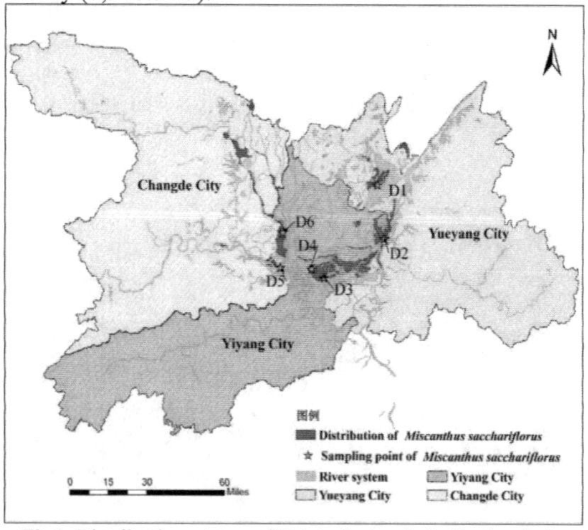

Fig1. Distribution status of T. lutarioriparia in Doting Lake wetland

Table1. Distribution area of *T. lutarioriparia* in different Doting Lake regions

Lake	District name	Distribution area (ha)	Total area (ha)
South Dongting Lake	Ziyang	28.07	29457.78
	Nanxian	1193.82	
	Ruanjiang	28235.89	
West Dongting Lake	Anxiang	2951.36	10646.45
	Jinshi	32.15	
	Hanshou	7252.43	
	Dingcheng	410.51	
East Dongting Lake	Junshan	132.78	18345.86
	Yueyang	9707.68	

https://doi.org/10.1051/e3sconf/202123701004

Xiangyin	7948.50
Huarong	557.00

3.2 Nitrogen and Phosphorus distribution in *T. lutarioriparia*

Ten copies of each *T. lutarioriparia* tissue were collected from sampling sites (D1-D6). Basic properties including biomass and water content were statistically calculated and showed in Table 2. The proportion for dry-weight biomass of over-ground *T. lutarioriparia* tissues was ranked in the descending order of stems (70.73%) > leaves (20.98%) > spikes (8.28%). The TN and TP contents of *T. lutarioriparia* in D1-D6 sampling sites of Doting Lake wetlands were demonstrated in Figure 2 and Figure 3, respectively. The TN (5.76 ~11.07 mg/g) and TP contents (1.14 ~2.11 mg/g) varied siginificantly with different *T. lutarioriparia* tissues ($P<0.05$), both exhibiting the sequence of spikes > leaves > stems. The average contents of TP in spikes, leaves, stems and roots of *T. lutarioriparia* tissues were 3.76 mg/g, 2.07 mg/g, 0.85 mg/g and 2.08 mg/g, respectively. Meanwhile, the average contents of TN in spikes, leaves, stems and roots of *T. lutarioriparia* tissues were 27.58 mg/g, 15.37 mg/g, 4.42 mg/g and 6.23 mg/g, respectively. No significant differences were observed in the nutrient content of *T. lutarioriparia* from different area of Dongting Lake wetlands ($P>0.05$). The nutrient contents in rhizosphere soil where *T. lutarioriparia* grew were shown in Table 3. The TN and TP contents in rhizosphere soils were within the range of 11.8~13.4 mg/g and 6.13-11.1 mg/g. There is a positive correlation ($R2> 0.99$) between the nutrient contents in *T. lutarioriparia* and its rhizosphere soil.

Table2. Basic properties of *T. lutarioriparia* in Doting Lake wetlands

Tissues	Moisture content/%	Dry-weight biomass per plant /g
Spikes	12.96	10.10
Leaves	12.79	25.58
Stems	28.69	86.23
Roots	37.06	72.88

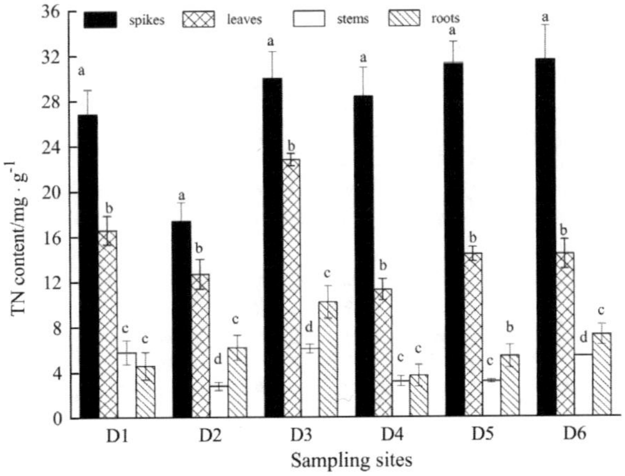

Fig2. TN content of T. lutarioriparia tissues in Dongting Lake wetland

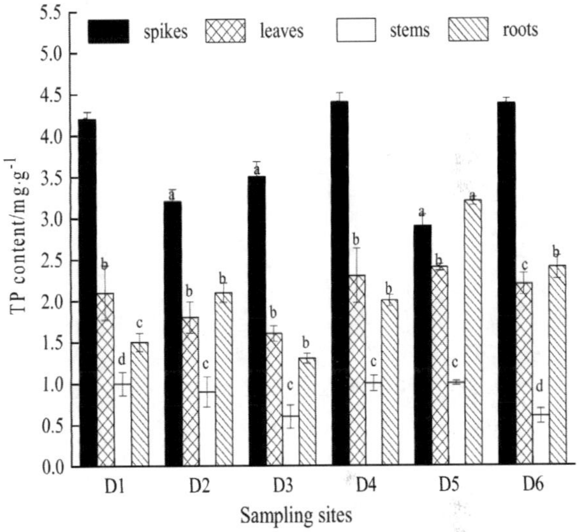

Fig3. TP content of T. lutarioriparia tissues in Dongting Lake wetland

Table3. Nitrogen and phosphorus content of rhizosphere soil in T. lutarioriparia growing area

Sampling sites	TN /mg·g^{-1}	TP / mg·g^{-1}
D1	12.18	6.7
D3	13.4	6.46
D5	11.8	7.7

3.3 Quantities of Nitrogen and Phosphorus stored in *T. lutarioriparia*

According to statistics, the total output of *T. lutarioriparia* in Dongting Lake is about 900,000 tons/year (dry weight). Based on the biomass proportion and the nutrient (N and P) content of each *T. lutarioriparia* tissue, the annual yield and nutrient reserves of each *T. lutarioriparia* tissue were calculated and shown in Table 4. The annual yields of the over-ground parts of *T. lutarioriparia* tissues in Dongting Lake wetland were 74,500 tons, 188,800 tons, and 636,700 tons for spikes, leaves and stems, respectively. The quantities of nitrogen and phosphorus that can be annually removed from Doting Lake by cutting the *T. lutarioriparia* were 7712.99 tons and 1241.45 tons, respectively. There was no significant difference in the total amount of nitrogen (2,170-2,801 tons/year) stored in different tissues of *T. lutarioriparia* in Dongting Lake. On the contrary, the phosphorus reserves of different *T. lutarioriparia* tissues were quite different, exhibiting the descending order of stems (560.26 tons/year) > leaves (396.52 tons/year) > spikes (284.67 tons/year). In the absence of harvesting, the dead tissues of the uncut *T. lutarioriparia* with large scale will lie down and decay in the lake water, and the release of *T. lutarioriparia* nutrients during decomposing process possesses non-negligible impact on the water quality of Doting Lake.

Table4. Annual output and nutrient reserves of T. lutarioriparia tissues

Tissues	Biomass/(*10^4 t)	N Storage/t	P Storage/t
Spikes	7.45	2170.02	284.67
Leaves	18.88	2741.67	396.52
Stems	63.67	2801.3	560.26
Total	90	7712.99	1241.45

4 Conclusion

The total growing area of *T. lutarioriparia* in Dongting Lake wetland was about 58,450 ha, and *T. lutarioriparia* mainly distributed in South Doting Lake wetlands, accounting for 50.40% of the total growing area. The annual yields (calculated in dry weight) of the over-ground parts of *T. lutarioriparia* tissues in Dongting Lake wetland were 74,500 tons, 188,800 tons, and 636,700 tons for spikes, leaves and stems, respectively. The nutrients showed similar distributing characteristics in over-ground tissues of *T. lutarioriparia*. Both the average contents of TN and TP exhibited the high-to-low sequence of spikes (TN 27.58 mg/g, TP 3.76 mg/g) > leaves (TN 15.37 mg/g, TP 2.07 mg/g) > stems (TN 4.42 mg/g, TP 0.85 mg/g). The quantities of nitrogen and phosphorus that can be annually removed from Doting Lake by cutting the *T. lutarioriparia* were 7712.99 tons and 1241.45 tons, respectively. Thus, the dead tissues of the uncut *T. lutarioriparia* with large amount of nutrient reserves may possess non-negligible impact on the water quality of Doting Lake.

References

1. S.M. Wang, H. S. Dou, *China Lake Records. Science ress*, Beijing, China (in Chinese)
2. P.E. O'Sullivan, C.S. Reynolds. *Chapter 11. Aquatic Plants and Lake Ecosystems* [M]//*The Lakes Handbook*, *Limnology* and Limnetic Ecology, **1**, Blackwell Science Ltd (2007)
3. S.L. Chen, S.A. Renvoize, *A new species and a new combination from China of Miscanthus (Poaceae)*, Kew Bulletin, **60(4)**,605-607 (2011)
4. G. Han, G. Zhang, *On the feedaimed potentialities and further development exploitation of Phragmites communis and Miscanthus sacchariflorus in the region of DongTing lake*, *Resources and Environment in the Yangtze Basin*, 232–236 (1998)
5. Liou L, Cao X, *Studies on the Autotetraploid of Triarrhena lutarioriparia*, *Acta Genetica Sinica*, **24**, 544-549 (1997)
6. N. Dang Z. Huang, H. Li, *Advances in Biology and Applications of Fiber Plant Triarrhena lutarioriparia*,[J], *Tropical and Subtropical Botany*, **20(4)**, 418-424 (2012, in Chinese)
7. X. Chen, L. Zhu, Z. Deng, et al., *Effects of Clipping on Bud Bank and Population Regeneration of Triarrhena lutarioriparia in Dongting Lake Wetland, China*[J], Wetlands, 1-8 (2020)
8. L. Liu, *Unique parper-making raw materials in China: Triarrhena lutarioriparia*, *Plants*, **17**, 6-7 (1990)
9. Xianjun W. Study on the Comprehensive Utilization of Reeds in the South Dongting Lake Area, *Agricultural Technology and Equipment*, 2019. (in Chinese)
10. L. Liu, *Chinese endemic papermaking materials-Miscathus lutarioriparius*[J], *Plant magazine*, **5**, 6-7 (1990, in Chinese)
11. Y. He, N. Song, H. L. Jiang, *Effects of dissolved organic matter leaching from macrophyte litter on black water events in shallow lakes*, *Environmental Science and Pollution Research* (2018)
12. E.M. Eid, K. H. Shaltout, Y.M. Al-Sodany, *Decomposition dynamics of Phragmites australis litter in Lake Burullus, Egypt*, *Plant Species Biology*, **29(1)**, 47-56 (2014)
13. H. P. van Dokkum, D. M. E. Slijkerman, L. Rossi, M. L. Costantini, *Variation in the decomposition of Phragmites australis litter in a monomictic lake: the role of gammarids*, *Hydrobiologia*, **482**, 69–77. (2002)
14. Y. Xie, Y. Xie, H. Xiao, et al., *Controls on Litter Decomposition of Emergent Macrophyte in Dongting Lake Wetlands*[J], *Ecosystems* (2017)
15. R. C. Estoque, Y. Murayama, *Classification and change detection of built-up lands from Landsat-7 ETM+ and Landsat-8 OLI/TIRS imageries: A comparative assessment of various spectral indices*[J], *Ecological Indicators*, **56**, 205-217 (2015)
16. H. Huang, L. Zhang, Z. Gao, *The vegetation resource at the intertidal zone in Shanghai using remote sensing*[J], *Acta Ecologica Sinica*, **25(10)**, 2686-2693. (2005, in Chinese)

Factors influencing electrochemically enhanced sacrificial anode phosphate recovery

Xiaoyan ZHOU[1], and Gang ZHANG[1*], Haijian SHI[2]

[1]School of Civil Engineering and Architecture, University of Jinan, 336 Nanxinzhuang West Road, Jinan, 250022, China
[2]Weihai ecological environment bureau wendeng branch,95 Fengshan Road Wendeng District, Weihai, 264400, China

Abstract. The recovery of phosphate from wastewater has been attracting increasing attention. In this study, the technology of an electrochemically enhanced sacrificial anode was applied to investigate the phosphate recovery process from sewage. The experimental results indicated that, when residence times ranged from 20 to 120 s, and phosphate removal efficiencies ranged from 24.65% to 95.27%. When the current density ranged from 10 to 70 A/m^2, the phosphate removal efficiencies ranged from 56.73% to 95.17%. When the plate spacing ranged from 1 to 5 cm, the phosphate removal efficiencies ranged from 83.48% to 9.48%. The experimental results confirm the conclusions of this study about controlling operational factors in practical applications to increase phosphate removal efficiencies.

1 Introduction

As we all know, phosphate resources are gradually decreasing on the earth[1]. To effectively recover phosphate resources, several techniques of recycling phosphate (e.g., adsorption and precipitation) have been explored[2]. Phosphate can be removed by using coagulants and adsorbents[3]. Iron salts and aluminum salts are commonly used coagulants for the removal of phosphate; these include ferrous sulfate, ferric chloride, and polyaluminum chloride. Adsorption has the advantage of fast adsorption speed and simple operation, but there are problems such as dissolution loss of adsorbent and difficulties in regeneration. Precipitation not only entails adding chemical agents but uses chemical reactions to generate precipitation and thus removes the phosphate in the wastewater. The work of Kruk et al[4] shows that struvite precipitation using a magnesium sacrificial anode for phosphate recovery is technically feasible. They find that, when high-purity magnesium alloy is used as an anode, not only the removal of phosphate but the recovery of struvite can be highly efficient in the wastewater treatment of a supernatant of fermented waste activated sludge from wastewater. Precipitation has the advantages of simple operation and strong impact load resistance, but the operational cost is high. Some researchers have been concerned with the electrochemical reduction of nitrates by using different cathodes such as Cu, Ti, Ru, and Fe [5]. In the electrochemical process, the cathode can produce hydrogen, which is an electron donor, from nutrient denitrifying bacteria. Iron cathodes have been shown to be relatively effective in promoting nitrate reduction[6]. On this basis, the electrochemical phosphate removal process can be added, and it can treat not only nitrates but also phosphates[7]. The electrochemical phosphate removal process has been used on domestic sewage and it now finds increased application in a range of small- and medium-sized companies, and it also has been received more and more attention[8].

Therefore, in the current study, the phosphate removal effect of different operating conditions, such as plate spacings, current densities, and residence time were studied by using an electrochemically phosphate recovery process.

2 Materials and methods

2.1 Materials

The secondary treatment water in a community water station was used as the treatment object. The raw water of the reclaimed water station was domestic sewage, and we adopted aeration regulation + hydrolysis acidification + anoxic treatment + biochemical treatment + precipitation + filtration + disinfection as the main treatment process. The typical sewage has a pH of 7.0–8.5 and a phosphate content of 3–5 mg/L.

2.2 Experimental system

As shown in Fig. 1, the electrochemical phosphate removal reaction takes place in a rectangular container. The rectangeular container was made of plexigalss. The dimensions of the electrode plates used in these experiments (Fe) were 10 × 100 × 100 mm. The water temperature and voltage were measured during all the experiments at a sampling interval of 5–40 min. The pH values of the solution were measured before and after each experiment. All experiments were at least duplicated, and

*Corresponding author: email address: cea_zhangg@ujn.edu.cn

the results showed good repeatability.

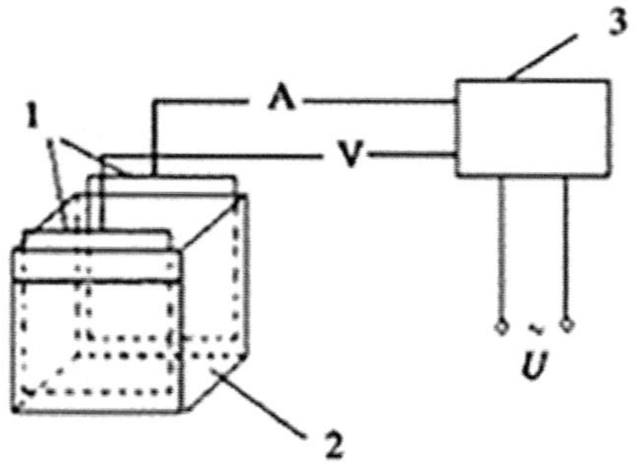

1: electrode plate material
2: electrochemical reactor
3: DC power

Fig1. Laboratory-scale electrochemical system used in the experiments.

2.3 Analytical method

All glass materials that came in contact with the water samples were rinsed with HCl acid and pure water in advance. Water samples were filtered through a φ15 cm filter paper. Phosphate removal was measured using the UV spectrophotometer UV-5800 (Yuanxi, China). The pH value was determined with a pH meter (FE20K, Mettler Toledo, USA). The control of the current was supplied by DC power (DPS-3005D, Zhaoxin, China).

3 Results and discussion

3.1 Effect of residence time

Residence time has an important effect on the electrochemically enhanced sacrificial anode phosphate recovery process, because the parameter is directly dependent on the dissolved anode metal. Therefore, the effect of residence time on the phosphate removal efficiency was studied within the range of 20–120 s at a current density of 10 A/m^2 (i = 0.1 A), with a plate spacing of 2 cm, an iron anode, and an initial neutral pH condition. Fig. 2 shows the relationship between phosphate removal efficiency and residence time.

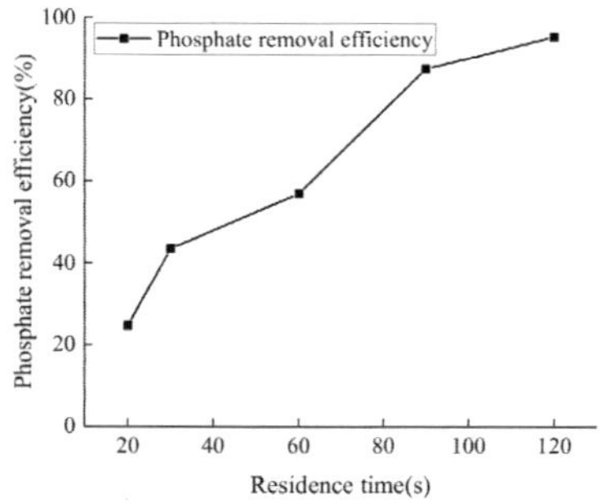

Fig2. Effect of residence time on phosphate removal efficiency.

As observed from Fig. 2, longer residence time increased the effect on the efficiency of phosphate removal in the electrochemically enhanced sacrificial anode phosphate recovery process. When the residence time was from 20 to 60 s, the rate of phosphate removal increased from 24.65% to 56.93%. In this residence time interval, the phosphate removal rate increased rapidly with the increase of residence time. When the residence time increased from 60 to 120 s, the rate of phosphate removal increased from 56.93% to 95.27%. In this residence time interval, the rate of phosphate removal increases with the increase of residence time gradually slowed. To save on power cost, in subsequent experiments a residence time of 60 s was used.

Because of the constant current density, the amounts of ferrous ions and iron ions both gradually increased with the extension of residence time, and the precipitation of ferrous phosphate and iron phosphate was greatly increased by ferrous ions and iron ions combining with phosphate. Ferrous ions and iron ions would hydrolyze to form flocculants such as ferric hydroxide and ferrous hydroxide, which could adsorb phosphate. Hence, the removal rate of phosphate increased with the residence time. However, the anode was likely to become passivated as the residence time increased. A dense oxide film was formed on the surface of the iron electrode that inhibited the anode from continuing to dissolve and release ferrous ions and iron ions. Consequently, the residence time has little effect on the phosphate removal efficiencies.

3.2 Effect of plate spacing

Plate spacing is one of the operational parameters that greatly influences the removal efficiencies of phosphate from reclaimed water using the electrochemically enhanced sacrificial anode phosphate recovery process. The effect of plate spacing on the removal of phosphate was explored within the range of 1–5 cm at a current density of 10 A/m^2 (i = 0.1 A), with an initial neutral pH condition, using an iron anode, and with a residence time of 1 min. The optimum plate spacing and corresponding phosphate removal percentage in this study are presented in Fig. 3.

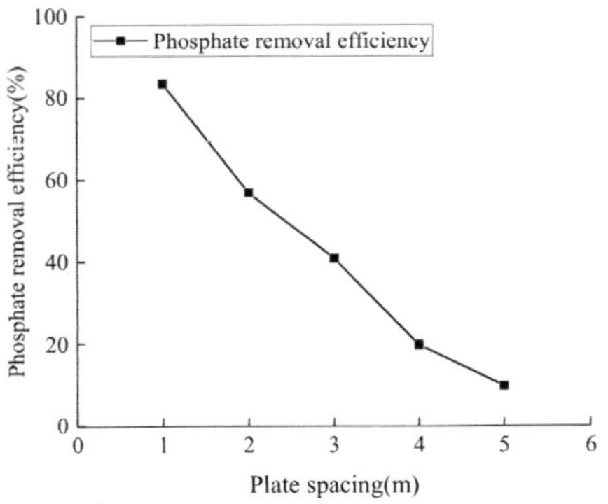

Fig3. Effect of plate spacing on phosphate removal efficiency.

Fig. 3 shows that plate spacing can affect the phosphate removal efficiency. When the plate spacing was increased from 1 to 5 cm, the phosphate removal efficiency decreased from 83.48% to 9.48%. Therefore, the longer the plate spacing, the more the phosphate removal rate would gradually decrease.

As a result of the smaller plate spacing, the current density flowing between the plates was higher, which produced more ferrous ions and iron ions[9].Ferrous ions and iron ions could combine with more phosphate to precipitate ferrous phosphate and iron phosphate. Based on this study, because the plate spacing affected the phosphate removal efficiencies, it is necessary to choose a suitable plate spacing for the electrochemically enhanced sacrificial anode phosphate recovery process. Given the influence of actual processing and other factors, a plate spacing of 2 cm was chosen in subsequent experiments.

3.3 Effect of current density

Current density is a crucial parameter for controlling the reaction rate in the electrochemically enhanced sacrificial anode phosphate recovery process. The operational current density is critical to the electrochemically enhanced sacrificial anode phosphate recovery process as it is the only operational factor that can be controlled directly[10]. The effect of current density on the removal of phosphate was explored within the range of 10–70A/m^2 at a plate spacing of 2 cm, with an initial neutral pH condition, using an iron anode, and with a residence time of 1 min. Fig. 4 demonstrates the effect of current density on the phosphate removal percentage.

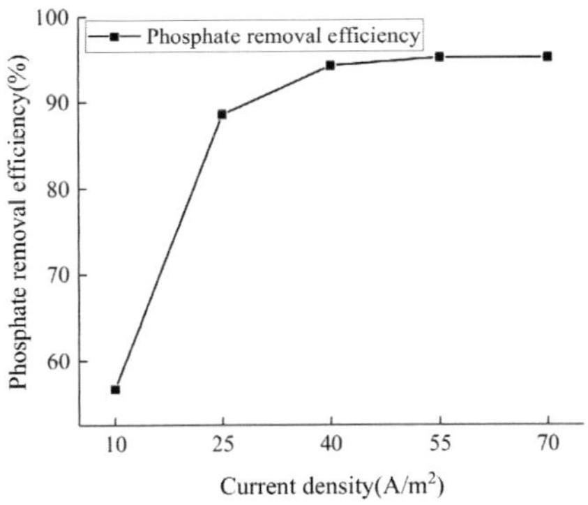

Fig4. Effect of current density on phosphate removal efficiency.

As can be seen in Fig. 4, the results indicate that current density can make a difference in the phosphate removal efficiency. When the current density increased from 10 A/m^2 to 40 A/m^2, the phosphate removal efficiency increased from 56.73% to 94.23% and the phosphate removal rate was faster. When the current density increased from 40 A/m^2 to 70 A/m^2, the phosphate removal efficiency increased from 94.23% to 95.17% and the phosphate removal rate slowed down. Thus, the optimal current density is 40A/m^2.

On the basis of Faraday's law, the current density can attribute to the increase in dissolution rate of the sacrificial iron ele0ctrode[11]. The amounts of dissolved ferrous ions and iron ions were greatly increased. As the current density increased, the amounts of ferrous ions and iron ions increased in the anode plate per unit time, and the chance of ferrous ions and iron ions colliding with phosphate to form precipitates of ferrous phosphate and iron phosphate was greatly increased. In addition, the increase of ferrous ions and iron ions also enhanced the flocculation of the electrochemical reaction, and many bubbles were produced during the high-current-density electrochemical reaction, so that these effects were both beneficial for high phosphate removal efficiencies[12].

4 Conclusion

The phosphate removal effects of residence time, plate spacing, and current density by using a process of electrochemically phosphate recovery have been explored. A range of electrochemically phosphate recovery process options were thoroughly investigated to assess the performance and maintenance requirements of various actual scenarios. Under the same experimental conditions, our results reveals that the optimum operating conditions are plate spacing is 2 cm, residence time of 60s and current density of 40A/m^2 should be considered. During operation, the phosphate removal percentage was found to increase with both longer residence time and higher current density. Therefore, the electrochemical process ensures that the process is very efficient, reduces overall energy, and shortens retention time. However, in practice, the optimum operating conditions need to be a harmonious combination among these factors for the greatest efficiency and cost savings.

Acknowledgment

This study was financially supported by the Natural Science Foundation of Shandong Province (ZR2017MEE061) and the Key Research and Development Project of Shandong Province (2016GSF117007). The authors would like to thank reviewers and the manuscript editors for their constructive comments.

Author Disclosure Statement

No competing financial interests exist.

References

1. L.L.E Bayon. F.C. Ballesteros Jr., S.G Segura, M.C Lu, Water reuse nexus with resource recovery: On the fluidized-bed homogeneous crystallization of copper and phosphate from semiconductor wastewater. J. Clean. Prod, 236, 117705 (2019)
2. J.S Wei, X.G Meng, X.H Wen, Y.H Song, Adsorption and recovery of phosphate from water by amine fiber, effects of co-existing ions and column filtration. J Environ Sci-China, 87, 123 (2020).
3. Y.B Chu, M Li, J.W Liu, W Xu, S.H Cheng, Molecular insights into the mechanism and the efficiency-structure relationship of phosphorus removal by coagulation. Sci. Water Research, 147, 195 (2018).
4. D.J Kruk, M Elektorowicz, J.A Oleszkiewicz, Struvite precipitation and phosphorus removal using magnesium sacrificial anode. Sci. Chemosphere, 101, 28 (2014).
5. Y Gao, Y.W Xie, Q Zhang, Y.X Wang, L.Y Yang, Intensified nitrate and phosphorus removal in an electrolysis-integrated horizontal subsurface flow constructed wetland. Sci. Water Research, 108, 39 (2017).
6. X.X Ju, S.B Wu, Y.S Zhang, R.J Dong, Intensified nitrogen and phosphorus removal in a novel electrolysis-integrated tidal flow constructed wetland system. Sci. Water Research, 108, 37 (2017).
7. D.D Nguyen, H.H Ngo, W Guo, T.T Nguyen, S.W Chang, A Jang, Y.S Yoon, Can electrocoagulation process be an appropriate technology for phosphorus removal from municipal wastewater. Sci. Sci Total Environ, 563, 549 (2016).
8. V Kuokkanen, T Kuokkanen, J Ramo, U Lassi, J Roininen, Removal of phosphate from wastewaters for further utilization using electrocoagulation with hybrid electrodes-Techno-economic studies. Journal of Water Process Engineering, 8, 50 (2015).
9. A Shalaby, E Nassef, A Mubark, M Hussein, Phosphate removal from wastewater by electrocoagulation using aluminium electrodes. Sci. Environ Eng Sci, 1, 90 (2014).
10. X.Y Zheng, H.N Kong, D.Y Wu, C Wang, Y Li, H.R Ye, Phosphate removal from source separated urine by electrocoagulation using iron plate electrodes. Water Sci Technol, 60, 2929 (2009).
11. A Attour., G.N Ben, T.M Mouldi, F Lapicque, J.P Leclerc, Intensification of phosphate removal using electrocoagulation treatment by continuous pH adjustment and optimal electrode connection mode. Sci. Desalin Water Treat. 57,13255 (2016).

Quantitative Evaluation of Plant Priority Protection in Cao Lake Wetland Nature Reserve

Xinyang Zhang*

Wuhan Institute of Design and Sciences, 430205 Wuhan, China

Abstract. Based on the survey data of plant resources in Cao Lake Wetland Nature Reserve, using analytic hierarchy process (AHP) and square root method, the evaluation system of plant priority protection was constructed. From three aspects of endangered coefficient, genetic coefficient and species coefficient, the plant species in the reserve were evaluated quantitatively and comprehensively, and the protection levels were divided. The results showed that there were 179 plant species in nature reserve, including 5 endangered species, 9 vulnerable species, 41 near threatened species and 124 least concerned species. The priority protection levels of plants were determined, including 1 species of the primary protection, 12 species of the secondary protection, 20 species of the tertiary protection and 146 species of the delayed protection, and corresponding protection suggestions were put forward. The purpose is to provide more effective background information and data support for the plant protection work and to help protect biodiversity.

1 Preface

With the rapid development of global economy and the rapid growth of population, the research and protection of biodiversity had became a global concern. As one of the important resources for human survival, wetland plays an important role in the global climate environment, ecosystem and biodiversity. In order to protect and restore the wetland ecological environment, a large number of wetland nature reserves had been established in large and medium-sized lakes and important rivers in China [1].

As the background composition of wetland nature reserve, plants are also the primary productivity of the ecosystem. Plant diversity is the basis of biodiversity. The key to the protection of plant diversity is the protection of rare and endangered plants. The establishment of plant protection area and the priority of plant protection can improve the efficiency of plant protection. Therefore, it is of great practical significance to slow down or contain the loss of plant diversity for the protection of regional ecosystem diversity [2].

2 Research area and methods

2.1 Overview of research area

Cao Lake Wetland Nature Reserve is located in Huangpi district, Wuhan city, Hubei province. The geographical location is 30°44′01.5″~ 30°46′29.2″ N, 114° 26′53.07″ ~ 114°29′13.3″ E. The total area of the reserve is 1148.2 hectares. The core area of the reserve is 387.8 hectares, the buffer area is 3.9 hectares, and the experimental area is 756.5 hectares.

The key protected objects are white crane, Oriental White Stork, cygnet and other rare water birds, wetland ecosystem and biodiversity. The nature reserve is a combination of natural and constructed wetlands. The terrain in the reserve is flat and belongs to the geomorphic type of lacustrine plain. It is mainly composed of yellow brown soil and fluvo aquic soil and herbaceous plants. The climate has four distinct seasons, the annual average sunshine hours are 1540~ 2180 hours, the average annual precipitation is 1104.6 mm, and the average temperature is 16.4°C. It belongs to the subtropical monsoon climate [3].

2.2 Survey methods

From July to August 2017, the team visited the reserve for many times to conduct plant resource survey and plant priority protection level study. The investigation team used the route survey method and ecological typical sample method.

The survey route was formulated according to the current situation of the wetland nature reserve. Multiple lines covered the whole reserve as much as possible, and the survey sample line perpendicular to the center of the lake was set to find out the ecological environment with different characteristics in the reserve. GPS was used to locate the whole survey, and the plant species along the survey route were recorded.

At the same time, according to the distribution of vegetation and hydrogeological conditions, a total of 10 sample plots with different habitats and vegetation types were set up. The names of plant species, longitude and latitude, altitude, soil condition and human disturbance

* Corresponding author: xinyangyy@qq.com

were recorded. Combined with relevant literature and previous investigation reports, the plant list of Cao Lake Wetland Nature Reserve was determined [4].

2.3 Methods of evaluation system construction

2.3.1 Evaluation model

The evaluation model of plant priority protection level in Cao Lake Wetland Nature Reserve was constructed based on analytic hierarchy process (AHP) [5-6], which included three levels: evaluation target layer, evaluation criteria layer and evaluation index layer (Table 1) .

Table1. Evaluation model of plant priority protection level

Evaluation target layer (A)	Evaluation criteria layer (B)	Evaluation index layer (C)
Species priority protection value (A)	Endangered coefficient (B_1)	Domestic distribution frequency (C_1)
		Provincial distribution frequency (C_2)
		Sample plot distribution frequency (C_3)
	Genetic value coefficient (B_2)	Species type (C_4)
		Endemic condition (C_5)
	Species value coefficient (B_3)	Ecological value (C_6)
		Medicinal value (C_7)
		Other values (C_8)

2.3.2 Evaluation model weight

Using the analytic hierarchy process (AHP), on the basis of the ordered hierarchical structure model, the weight coefficient of the index was comprehensively calculated by comparing the relative importance of each index at the same level. The weight of evaluation index system of plant priority protection level in Cao Lake Wetland Nature Reserve was determined. The result of composite ranking of index layer and target layer of evaluation system was obtained (Table 2). From the consistency test results, the consistency ratio is less than 0.1, so the results of hierarchical total ranking have satisfactory consistency.

Table2. Composite ranking of evaluation index layer and target layer*

Evaluation index layer	B_1 (0.6370)	B_2 (0.2583)	B_3 (0.1047)	Composite weight (ω_i)
C_1	0.637			0.4058
C_2	0.2583			0.1645
C_3	0.1047			0.0667
C_4		0.75		0.1937
C_5		0.25		0.0646
C_6			0.1047	0.0110
C_7			0.2583	0.0270
C_8			0.637	0.0667

*i=1, 2, 3… 7, 8.

2.3.3 Evaluation index assignment and calculation of endangered coefficient

The endangered coefficient refers to the endangered degree of a plant population in natural distribution. According to the existing living conditions of plants in Cao Lake Wetland Nature Reserve and relevant literature, the evaluation criteria of endangered coefficient were determined (Table 3).

Table3. Evaluation index assignment of endangered coefficient

Index subdivision	Qualitative description	Score
Domestic distribution frequency (C_1)	1 province	5
	2 to 3 provinces	4
	4 to 6 provinces	3
	7 to 10provinces	2
	≥11 provinces	1
Provincial distribution frequency (C_2)	1 county	5
	2~3 counties	4
	4~6 counties	3
	7~10 counties	2
	≥11 counties	1
Sample plot distribution frequency (C_3)	1 plot	5
	2 to 3 plots	4
	4 to 5 plots	3
	6 to 7 plots	2
	≥8 plots	1

According to the indexes in Table 3, the plant species in the reserve were graded, and the endangered coefficient of plant species in the reserve was calculated according to Formula 1.

$$V_1 = \sum_{i=1}^{3}\left(\omega_i \frac{X_i}{X_{i\max}}\right) \quad (1)$$

In the formula, ω_i is the composite weight of index Ci, X_i is the actual score of index Ci, and X_{imax} is the highest score of index Ci, (i=1,2,3).

According to the calculated values, the endangered grades of plants were classified in different value ranges.

Table4. Standards of plant endangered grading

Score of endangered coefficient	Endangered grading
$V_1 \geq 0.7$	Endangered
$0.6001 \leq V_1 < 0.7$	Vulnerable
$0.4 \leq V_1 < 0.6$	Near threatened
$V_1 < 0.4$	Least concern

2.3.4 Evaluation index assignment and calculation of genetic coefficient

According to the genetic loss caused by the extinctions of the investigated plant species to the existing plant diversity, the evaluation criteria of the genetic value coefficient were determined (Table 5).

Table5. Evaluation index assignment of genetic coefficient

Index subdivision	Qualitative description	Score
Species type (C_4)	Only 1 species in family	5
	2 to 3 species in family	4
	Only 1 species in genus	3
	2 to 3 species in genus	2
	≥4 species in genus	1
Endemic condition (C_5)	Only in here	5
	Only in Hubei	4
	In region specific	3
	Only in China	2
	In China and other countries	1

The species type was determined according to the classification of families and genera in Flora of China, and the number of families and genera containing species of the investigated plant species was evaluated. The endemic situation was evaluated according to the records of the Flora of China and the Flora of Hubei.

The plant species in the reserve were graded by various indexes in Table 5, and the genetic value coefficient of plant species in the reserve was calculated according to Formula 2.

$$V_2 = \sum_{i=4}^{5}\left(\omega_i \frac{X_i}{X_{i\max}}\right) \quad (2)$$

In the formula, ω_i is the composite weight of index Ci, X_i is the actual score of index Ci, and X_{imax} is the highest score of index Ci, (i=4,5).

2.3.5 Evaluation index assignment and calculation of species coefficient

According to the value of the investigated plant species recognized and exploited by human beings, the evaluation index of plant utilization value coefficient in Cao Lake Wetland Nature Reserve was determined mainly in the aspects of medicine, food, industry, raw materials and Ecology (Table 6).

The ecological value was determined according to the community structure in the sample plots. The medicinal value was determined according to the records of New Chinese Materia Medica and Chinese Pharmacopoeia. The value of other plants was determined according to the records of China Resource Plants.

According to the indexes in Table 6, the plant species in the reserve were graded, and the species value coefficient of the plants in the reserve was calculated according to Formula 3.

Table6. Evaluation index assignment of species coefficient

Index subdivision	Qualitative description	Score
Ecological value (C_6)	Edificators	5
	Co-species of a community	4
	Dominant species other than edificators	3
	Subdominant	2
	Others	1
Medicinal value (C_7)	Included in the Chinese Pharmacopoeia	4
	Included in the New Chinese Materia Medica	3
	General folk medicinal plants	2
	Others	1
Other values (C_8)	Having 3 kinds and above	4
	Having 2 kinds	3
	Having only 1 kind	2
	None	1

$$V_3 = \sum_{i=6}^{8}\left(\omega_i \frac{X_i}{X_{i\max}}\right) \quad (3)$$

In the formula, ω_i is the composite weight of index Ci, X_i is the actual score of index Ci, and X_{imax} is the highest score of index Ci, (i=6,7,8).

2.3.6 Calculation of priority protection coefficient

The endangered coefficient, genetic value coefficient and species value coefficient of each plant species in Cao Lake Wetland Nature Reserve were calculated based on the established plant priority protection level model. According to the proportion weight of each coefficient, the species priority protection coefficient (V) was calculated according to Formula 4.

$$V = B_1 \times V_1 + B_2 \times V_2 + B_3 \times V_3 \quad (4)$$

According to the score of species priority protection coefficient, the standards of plant priority protection grading were developed (Table 7).

Table7. Standards of plant priority protection grading

Score of species priority protection coefficient	Protection grading
$V \geq 0.6$	Primary protection
$0.5 \leq V < 0.6$	Secondary protection
$0.4 \leq V < 0.5$	Tertiary protection
$V < 0.4$	Delayed protection

3 Research results and analysis

3.1 Plant species composition

There were 179 species of plants belonging to 143 genera and 62 families in Cao Lake Wetland Nature Reserve. There were 1 family, 1 genus and 1 species of gymnosperms and 61 families, 142 genera and 178 species of angiosperms. Plant species were not abundant.

The above plant composition characteristics were related to the wetland type of the reserve and the protection object was rare waterfowl. Although the number of plant species was not much, but the proportion of herbs was rich. There were 165 species of herbs, accounting for 92.18% of the total number of seed plants in this area. These characteristics were determined by the environmental factors such as soil and hydrology in Cao Lake Wetland Nature Reserve.

3.2 Endangered coefficient evaluation

According to the data obtained from the survey, combined with the classification standard of plant endangered grade and the classification standard of plant priority protection level, the endangered situation of plants in the reserve was counted.

It can be seen from Table 8 that most of the plants with high endangered coefficient in the reserve were distributed less in the whole country and only exist in some provinces in China. For example, *Ceratopteris pterioides*, *Polygonum bistorta*, etc. There were also a few plants widely distributed in China, but they were rare in Hubei province and had low frequency in the sample plots. For example, *Dichondra repens*, *Daucus carota*, *Pluchea indica*, etc.

It can be seen from Table 9 that among the 179 plant species in Cao Lake Wetland Nature Reserve, 5 species were endangered species, accounting for 2.79% of the total. They were scattered in the reserve and their individual volume was rare. There were 9 species of vulnerable species, accounting for 5.03% of the total. Some species had special growth environment and few individuals. There were 41 species of near threatened species, accounting for 22.91% of the total. They had a large number of individuals, scattered scattered or dominated by small patches and zonal distribution. There were 124 least concerned species, accounting for 69.27% of the total. On the whole, the species with $V_1 \geqslant 0.4$ and above account for more than one third of the total, indicating the necessity of establishing the reserve. The existing endangered species, vulnerable species and near threatened species should be monitored and protected in time to make their living environment stable and the population number stable.

Table8. Evaluation of plant endangered coefficient*

Species name	C_1	C_2	C_3	V_1
Ceratopteris pterioides	4	5	5	0.8715
Oxalis pes-caprae	3	5	5	0.7441
Solanum pseudocapsicum	3	5	5	0.7441
Alnus cremastogyne	3	5	5	0.7441
Leersia hexandra	3	5	5	0.7441
Setaria faberii	3	4	5	0.6927
Geranium carolinianum	3	5	2	0.6813
Commelina communis	4	1	5	0.6657
Polygonum bistorta	4	1	5	0.6657
Artemisia lavandulaefolia	3	4	3	0.6508
Daucus carota	3	3	5	0.6412
Euphorbia hypericifolia	2	5	4	0.6376
Pluchea indica	2	5	5	0.6167
Zoysia japonica	2	5	5	0.6167

*$V_1 \geqslant 0.6$ are listed.

Table9. Evaluation of plant endangered grading

Endangered grading	Number of species
Endangered	5
Vulnerable	9
Near threatened	41
Least concerned	124

3.3 Genetic coefficient evaluation

According to the data obtained from the survey, combined with the evaluation index of plant genetic value coefficient, the scores of plant genetic value coefficient were calculated.

Table10. Evaluation of plant genetic coefficient*

Species name	C_4	C_5	V_2
Myosoton aquaticum	3	1	0.50
Houttuynia cordata	3	1	0.50
Aeschynomene indica	3	1	0.50
Nelumbo nucifera	2	1	0.35
Nymphaea tetragona	2	1	0.35
Hydrocharis dubia	2	1	0.35
Humulus scandens	2	1	0.35
Gelsemium elegans	2	1	0.35
Melia azedarach	2	1	0.35
Cirsium setosum	2	1	0.35
Zizania latifolia	2	1	0.35
Polygonum bistorta	1	2	0.25
Alnus cremastogyne	1	2	0.25

*$V_2 \geqslant 0.25$ are listed.

It can be seen from Table 10 that there were 13 species with genetic value coefficient of 0.25 and above in the reserve. Among them, *Houttuynia cordata*, *Aeschynomene indica* and *Myosoton aquaticum* were the highest (V_2=0.5). There were 8 species with genetic value coefficient of 0.35. Such as *Nelumbo nucifera*, *Nymphaea tetragona*, *Hydrocharis dubia* and so on. There were 2 species with genetic value coefficient of 0.25. They were *Polygonum bistorta* and *Alnus cremastogyne*. There were no rare and ancient relic plants in the reserve, so the overall genetic value coefficient was not high.

3.4 Species coefficient evaluation

According to the survey data, combined with the evaluation index of species value coefficient, the species value coefficient scores were calculated.

It can be seen from Table 11 that there were 14 species with value coefficient above 0.8 in the reserve. Some of them were important constructive species or dominant species in the reserve, such as *Cynodon dactylon*, *Vicia sepium*, *Imperata cylindrica*, etc. Some of them were widely exploited and utilized by human beings, such as *Morus alba*, *Typha orientalis*, *Zizania latifolia*, etc. There were also some plants with great medicinal value, such as *Celosia argentea*, *Broussonetia papyrifera*, *Phragmites australis*, etc.

Table11. Evaluation of plant species coefficient*

Species name	C_6	C_7	C_8	V_3
Broussonetia papyrifera	4	4	4	0.9791
Phragmites communis	4	4	4	0.9791
Typha orientalis	1	4	4	0.9162
Celosia argentea	1	4	4	0.9162
Eichhornia crassipes	5	2	4	0.8709
Humulus scandens	5	2	4	0.8709
Cynodon dactylon	5	2	4	0.8709
Imperata cylindrica	5	2	4	0.8709
Vicia sepium	5	2	4	0.8709
Sapium sebiferum	4	2	4	0.8499
Cinnamomum camphora	3	2	4	0.8290
Melochia corchorifolia	3	2	4	0.8290
Morus alba	3	2	4	0.8290
Zizania latifolia	2	2	4	0.8081

*$V_3 \geqslant 0.8$ are listed.

3.5 Priority protection evaluation

According to the quantitative evaluation indexes and calculation methods of species priority protection value, the endangered coefficient, genetic coefficient, species coefficient and species priority protection value of 179 species in Cao Lake Wetland Nature Reserve were calculated. Then the priority protection order of plant species was analyzed.

It can be seen from Table 12 that there were 13 species with $V \geqslant 0.5$, accounting for 7.26% of the total. The highest priority protection coefficient (V=0.6646) was *Ceratopteris pterioides*. Plants with high endangered coefficient generally had higher priority protection coefficient. Plants with $V \geqslant 0.5$, their endangered coefficient was greater than 0.6. For example, *Oxalis pes-caprae*, *Alnus cremastogyne*, *Alternanthera philoxeroides*, etc. Although the weight ratio of genetic coefficient and species coefficient asw not high, it also played a key role in plant evaluation system. For example, the endangered coefficient of *Setaria faberi* was not in the top five, but its genetic coefficient was the highest, so its priority protection coefficient ranks fourth.

It can be seen from Table 13 that among the 179 plant species in Cao Lake Wetland Nature Reserve, 1 species was of the primary protection, accounting for 0.56% of the total. They were rare in the nature reserve and have weak anti-interference ability, so it is urgent to protect and stabilize the population. There were 11 species of secondary protection, accounting for 6.70% of the total. The number of them in the reserve was small. So they need to monitor the status of the population in time, maintain and feel the living environment of the population, and prevent the population from greatly reducing. There were 20 species of tertiary protection, accounting for 11.17% of the total. Most of them were dominant species or sub dominant species in the reserve. They can be used reasonably without special protection measures. There were 147 species of delayed protection, accounting for 81.56% of the total. Most of them were constructive species in the nature reserve, and their self recovery ability is strong after disturbance.

Table12. Evaluation of plant priority protection*

Species name	V_1	V_2	V_3	V
Ceratopteris pterioides	0.8715	0.20	0.4686	0.6646
Oxalis pes-caprae	0.7441	0.20	0.5633	0.5921
Alnus cremastogyne	0.7441	0.25	0.4040	0.5883
Setaria faberii	0.6927	0.20	0.6278	0.5655
Alternanthera philoxeroides	0.7441	0.20	0.3093	0.5655
Solanum pseudocapsicum	0.7441	0.20	0.3093	0.5655
Leersia hexandra	0.7441	0.20	0.2448	0.5587
Geranium carolinianum	0.6813	0.20	0.4594	0.5406
Commelina communis	0.6657	0.20	0.4385	0.5283
Polygonum bistorta	0.6657	0.25	0.2448	0.5209
Daucus carota	0.6412	0.20	0.4686	0.5156
Artemisia lavandulaefolia	0.6508	0.20	0.3076	0.5049
Zoysia japonica	0.6167	0.20	0.4878	0.5017

*$V \geqslant 0.5$ are listed.

Table13. Evaluation of plant priority protection grading

Protection grading	Number of species	Percentage
Primary protection	1	0.56%
Secondary protection	12	6.70%
Tertiary protection	20	11.17%
Delayed protection	146	81.56%

4 Conclusion and discussion

The evaluation results of the endangered coefficient of 179 plants species in Cao Lake Wetland Nature Reserve showed that there were 5 endangered species, 9 vulnerable species, 41 near threatened species and 124 Least concerned species, accounting for 2.79%, 5.03%, 22.91% and 69.27% of the total respectively. Among them, there were 14 endangered and vulnerable species, accounting for 7.82% of the total. The assessment results of the priority protection level of plants were as follows: 1 species under primary protection, 12 species under secondary protection, 20 species under tertiary protection, and 146 species under delayed protection, accounting for 0.56%, 6.70%, 11.17% and 81.56% of the total respectively.

The protection of rare and endangered plants depends on the overall natural environment and reasonable human intervention. Corresponding protection measures should be taken according to different priority protection levels of plants. For the primary protected plants, artificial propagation should be adopted to increase the population number while preventing human destruction. For the secondary protected plants, it is necessary to prevent man-made destruction. In general, it is not necessary to take special protection measures for them. If the population continues to decline, corresponding measures should be taken in time. The tertiary protected plants can be slightly developed and utilized in the reserve. Avoid the irreversible consequences of over exploitation. Strengthen management and strictly control the development intensity. The number of the population was kept at a relatively stable level. Generally, there is no need to pay too much attention to the delayed protection plants and get corresponding safety protection in the reserve.

Rare and endangered plants, due to their special habitat and rare number, are the strong evidence of plant origin, flora evolution and paleogeological changes. Therefore, the research on the protection of rare and endangered plants plays a key role in the protection of plant diversity and has irreplaceable scientific significance. According to the actual situation of Cao Lake Wetland Nature Reserve, the plants priority protection grading was determined. Compared with the national list of endangered plants, it can provide more effective background information and data support for the reserve to formulate targeted protection plan.

Acknowledgments

This research was financially supported by the Humanities and Social Science Research Program of Hubei Provincial Education Department (17G119), and the Teaching Research Project of Hubei Provincial Universities (2017505).

References

1. R. R. Yan, X. B. Cai, X. L. Wang, Y. Chen, X. Han, WS, 11, 1, 136-144 (2013)
2. Z. R. Liu. IMAU, 101-134 (2017)
3. X. Y. Zhang, S. Y. Tan, H. D. Pang, JAU(NSE), 2, 103-108 (2020)

https://doi.org/10.1051/e3sconf/202123701006

4. X. Y. Zhang, J. H. Shang, S. J. He, STE, 21, 92-98, (2019)
5. X. Y. Zhang, X. Y. Cai, E3S web of conferences (EDP Sciences, 79, 03010, 2019)
6. Q. Q. Peng, Z. X. Wang, T. T. Li, B. M. Xiong, Q. C. Yang, GUI, 37, 7, 859-867 (2017)

Methodology study on voluntary emission reduction of greenhouse gases in port off road mobile machinery hybrid power transformation project

Guo xu *

Key Laboratory of Environmental Protection in Water Transport Engineering Ministry of Communications,

Tianjin Research Institute of Water Transport Engineering, Tianjin 300456, China

Abstract. The greenhouse gas emissions of ports mainly come from the combustion of fossil fuels produced in the loading and unloading production. With the continuous advancement of port equipment electrification, non road mobile machinery has gradually become the main source of fossil fuel emissions. On the basis of the existing voluntary emission reduction methodology, this paper puts forward the greenhouse gas voluntary emission reduction methodology of port non road mobile machinery hybrid power transformation project, and verifies the correctness of the methodology through examples. In the future, it can be used as the accounting method of trading volume in voluntary carbon emission trading, and effectively promote the energy conservation and emission reduction of port industry and green port construction.

1 Introduction

The greenhouse gas emissions of ports mainly come from the combustion of fossil fuels produced in the loading and unloading production. With the continuous advancement of port equipment electrification, non road mobile machinery has gradually become the main source of fossil fuel emissions. With the increasingly stringent requirements of greenhouse gas emissions, it is necessary to carry out the hybrid transformation of various machinery to reduce the use of fossil fuels.

On this basis, the market approach to achieve voluntary emission reduction is an important means to improve the initiative of enterprises and improve the effect of emission reduction. At present, China has carried out various carbon trading practices[1-4].For example, Shanghai has incorporated aviation into the carbon trading system, and Shenzhen has adopted the carbon trading mechanism to promote new energy vehicles. At the same time, various experts and scholars have also carried out relevant research, Qian Zhenglin et al. conducted a comparative study on seven international voluntary emission reduction standards, including the gold standard and the Chicago Climate Exchange Standard, from five aspects of basic information, project information, accounting methods, sustainability requirements, verification and registration, and put forward relevant countermeasures and suggestions for the development and standardization of China's voluntary emission reduction market [5]. In view of the problems existing in China's voluntary carbon emission reduction market, Ma Qiujun and others put forward countermeasures and suggestions such as establishing emission reduction awareness, building institutional environment, improving the decomposition principle of emission reduction indicators, unifying trading standards, and improving market structure[6].

2 Introduction of voluntary emission reduction methodology

2.1 Definition

Port non road mobile machinery： all kinds of port operation equipment, such as forklift truck, tire crane, empty container stacker, etc., which are mainly powered by fossil fuel and used for loading, unloading and stacking operations.

Hybrid power transformation: in the case of the original functions of port non road mobile machinery, the energy supply mode is changed from the original high-power diesel generator set to the power supply by the municipal power grid, and only the small power diesel generator set is equipped to provide the power needed for short-distance movement.

Workload: Port non road mobile machinery completes a unit weight of goods handling or overturning and other complete operation process.

* Corresponding author: 18630836305@163.com

https://doi.org/10.1051/e3sconf/202123701007

2.2 Applicable conditions

This methodology is applicable to the oil-electricity hybrid transformation of various existing port non road mobile machinery, so as to achieve emission reduction projects (without considering the impact of equipment type and high-voltage and low-voltage frequency conversion in power grid).

2.3 Baseline scenario

Baseline emission refers to the multiplication of energy consumption per unit operation volume of port non road mobile machinery, annual average operation volume of single machine and emission coefficient of fuel used by equipment under normal operation conditions. The calculation formula is as follows.

$$BE_y = \sum_i EF_{BL,TEU,i} \times DD_{i,y} \times N_{i,y} \times 10^{-6} \quad (1)$$

BE_y: Represents the total emissions from the baseline in year y (tCO2) .

$EF_{BL,TEU,i}$: Emission factors for baseline RTG type i (gCO_2/t) .

$DD_{i,y}$: Average annual work load (t) .

N_y: *Number of RTG operations.*

$$EF_{BL,TEU,i} = SFC_i \times NCV_{BL,i} \times EF_{BL,i} \times IR^t \quad (2)$$

SFC_i: Fuel consumption rate (g/t)

$NCV_{BL,i}$: Net calorific value of fossil fuel consumption (J/g)

$EF_{BL,i}$: Emission factors of fossil fuel consumption (gCO_2/J)

IR^t: The technical improvement factor of baseline RTG in t year. The technical improvement rate is applied to each calendar year, and the default value of technical improvement factor is 0.99 for all RTG types of baseline.

t: Years of technical improvement.

2.4 Emission calculation

Project emissions: including emissions from fossil fuel consumption of port non road mobile machinery, calculated as follows:

$$PE_y = \sum_i EF_{PJ,TEU,i,y} \times DD_{i,y} \times N_{i,y} \quad (3)$$

PE_y: Total project emissions (tCO_2)

$EF_{PJ,TEU,i,y}$: Emission factor of unit work volume (tCO_2/t)

$N_{i,y}$: Number of RTGs

$DD_{i,y}$: Average annual work load

The emission factor is calculated as follows:

$$EF_{PJ,TEU,i,y} = \sum_i SEC_{PJ,TEU,i,y} \times \frac{EF_{elect,y}}{1-TDL_y} \times 10^{-3} + \sum_i SFC_{PJ,TEU,i,y} \times NCV_{PJ,i} \times EF_{PJ,i} \times 10^{-6} \quad (4)$$

$SEC_{PJ,TEU,i,y}$: Power consumption rate per unit work load (kWh/t) .

$EF_{elect,y}$: CO_2 emission factor of power consumption ($kgCO_2$/ kWh) .

$SFC_{PJ,TEU,i,y}$: Fossil fuel consumption rate per unit operation volume (g/t) .

$EF_{PJ,i}$: CO2 emission factors of fossil fuels (gCO_2/J) .

$NCV_{PJ,i}$: Net calorific value of fossil fuel consumption (J/g) .

TDL_y: Average loss of technical transmission and distribution of electricity.

3 Example calculation

3.1 Baseline emission calculation

The emission factors of fossil fuels are as follows:
20.2t-C/TJ*44/12=74.067g-CO_2/MJ
=74.067*10^{-6} g-CO_2/J

The net calorific value of fossil fuel consumption is 42652j / g, the technical improvement coefficient is 0.93, and the unit consumption of crane operation is 116g/ t, then the baseline emission factor is:

$$EF_{BL,TEU,i} = SFC_i \times NCV_{BL,i} \times EF_{BL,i} \times IR^t$$
$$= 116 * 74.067 * 10 - 6 * 42652 * 0.93$$
$$= 341g - CO_2/t$$

If the annual operation capacity of the company's revamped equipment is 41588030t, the baseline emission is:

$$BE_y = F_{BL,TEU,i} \times DD_{i,y} \times N_{i,y} \times 10^{-6}$$
$$= 341g - CO_2/t * 41588030\ t$$
$$= 14185.59t - CO_2$$

3.2 Emission calculation

The annual operation capacity of the company's crane is 41588030t, the fuel consumption is 372350l, and the diesel density is 0.86, which is equivalent to 320221kg, and the power consumption is 4687830kwh. Then the unit consumption is:

(320221kg+4687830kWh) /41588030t
=(0.077kg+1.13kWh) /t

The power emission factor of the port area is 0.7598 t-co2 / MWh, equivalent to 759.8kg- CO_2 / kwh; the average loss of power technology transmission and distribution is 20%, and the net calorific value of fossil

https://doi.org/10.1051/e3sconf/202123701007

fuel consumption is 42652j / g; the emission factor of fossil fuel is: 0.7598 t-CO_2 / MWh, equivalent to 759.8kg- CO_2/kwh

20.2t-C/TJ*44/12=74.067g-CO_2/MJ=74.067*10^{-6} g-CO_2/J

Therefore, the emission factor of the equipment is as follows:

$$F_{PJ,TEU,i,y} = \sum_i SEC_{PJ,TEU,i,y} \times \frac{EF_{elect,y}}{1-TDL_y} \times 10^{-3}$$

$$+\sum_i SFC_{PJ,TEU,i,y} \times NCV_{PJ,i} \times EF_{PJ,i} \times 10^{-6}$$

=1.13Kwh/t×(759.8kg-CO_2/Kwh)/(1-0.2)×10^{-3}

+0.077kg/t×（42652J/g×74.067×10^{-6} g

=1.03kg-CO_2

Then the total annual emission of the project is:

$$PE_y = \sum_i EF_{PJ,TEU,i,y} \times DD_{i,y} \times N_{i,y}$$

=1.073×41588030=4462t

3.3 Emission calculation

ER_y=BE_y-PE_y

=14185.59-4462.40

=9723.19t-CO_2

4 Conclusion

Port off road mobile machinery is the main source of port greenhouse gas emissions. Based on the existing voluntary emission reduction methodology, this study proposes the greenhouse gas voluntary emission reduction methodology of port off road mobile machinery hybrid power transformation project, and verifies the correctness of the methodology through examples. It can be used as the accounting method of trading volume in voluntary carbon emission trading in the future, and can be effectively promoted Energy saving and emission reduction of port industry and green port construction.

Table 1. Parameters to be monitored

Parameter	description	monitoring method
$DD_{i,y}$	Average operation volume of equipment *i* in the year *y* (t/year)	**Mode A:** monitor all operating equipment. **Method B:** Sampling survey was used for each type. The sampling of this project shall be randomly selected according to the latest version of "general provisions for sampling and investigation of small-scale CDM project activities", and the sample size shall be determined with 90% confidence interval and ± 10% error. The lower limit of 95% confidence interval should be used for the driving distance of annual draw.
$SEC_{PJ,i,y}$	By monitoring the oil consumption se of the equipment and the electricity SF after transformation, the fossil fuel consumption / power consumption (g / t and kwh / t) per unit operation volume of RTG type I project in *y* year are obtained. The corresponding formula is as follows: $SEC_{PJ,i,y}$= SE/ $DD_{i,y}$; $SFC_{PJ,\ i,y}$=SF/ $DD_{i,y}$;	**Mode A:** monitor the consumption of all non road mobile machinery. **Method B:** measure the power / fossil fuel consumption per unit operating capacity of each type of non road mobile machinery under typical working conditions; sampling shall be selected randomly according to the latest version of "general provisions for sampling and investigation of small-scale CDM project activities", and the sample size shall be determined with 90% confidence interval and ± 10% error. The upper limit of 95% confidence interval should be used for power / fossil fuel consumption rate.
$SFC_{PJ,I,y}$	By monitoring the oil consumption se of equipment and the electricity SF after transformation, the fossil fuel consumption / power consumption (g / T and kwh / T) per unit operation volume of non road mobile machinery of project type *i* was obtained. The corresponding formula is: $SEC_{PJ,\ i,y}$=SE/ $DD_{i,y}$; $SFC_{PJ,\ i,y}$=SF/ $DD_{i,y}$.	**Mode A:** monitoring the consumption of non road mobile machinery in all projects **Method B:** measure the power / fossil fuel consumption per unit operating capacity of each type of non road mobile machinery under typical working conditions; the sample size shall be determined by 90% confidence interval and ± 10% error according to the latest version of "general provisions for sampling and investigation of small-scale CDM project activities". The upper limit of 95% confidence interval should be used for power / fossil fuel consumption rate.
EF_{elec}	Grid CO_2 emission factors for type *i* non road mobile machinery	According to the methodology cms-002-v01 / cms-003-v01

https://doi.org/10.1051/e3sconf/202123701007

References

1. Liu L L etc, Study on methodology of voluntary emission reduction of greenhouse gases from shore power projects (journal of waterway and harbor), 445-449,2019,40(4).
2. Guo X etc, Research on construction technology of energy consumption onlinemonitoring system for port energy equipment. (journal of waterway and harbor), 746-752,2018,39(6).
3. WANG C Y etc,Components of regional emergency linkage system for ship pollution accidents ,(journal of waterway and harbor), 340-346,2020,41(3).
4. XIONG H X etc. Enlightenments of Tokyo bay revitalization action plan (2003 ~ 2012) in Janpan to Bohai bay in China.(journal of waterway and harbor) 119-124 ,2020,41(1).
5. Qian Z L,Ma X M. A comparative study of international voluntary emission reduction standards (Eco economy) 39-42,48,2012,(5).
6. Ma Q J,Liu X. Countermeasures for the development of China's voluntary carbon emission reduction market(economic review) 58-61.2011,(11).

Establishment and research of combined forecasting model of carbon

Zuxu Zou, Jiaojiao Huang, and Chenlu Li

(Wuhan Polytechnic University, College of Civil Engineering and architecture, Hubei, Wuhan 430000)

Abstract. This paper analyzes the influencing factors of carbon dioxide emissions from four aspects: Population, economy, industrial structure and energy, then from the carbon emissions, economic development, industrial structure, energy consumption structure to show the status quo of carbon emissions in Hubei Province. Based on the analysis of the influencing factors, the main influencing factors of carbon emission are population, regional gross product and coal consumption The multivariate linear regression model and the polynomial curve model are established and the error analysis is carried out. The combination weight coefficients of two single models are obtained through the linear programming model and the combination forecasting model is established, finally, the corresponding countermeasures to reduce carbon emissions are put forward.

1 Introduction

In March 2011, Hubei Province was selected as one of the country's first seven carbon trading pilot provinces and cities. In August 2013, Hubei Province formulated and promulgated the " measures for the administration of carbon emissions trading in Hubei Province", to complete the system construction including management system, quota allocation, trading platform, Verification Report, etc. . The effective use of energy is closely related to the great economic development. Under the extensive economic development pattern of many years, Hubei Province has developed bottlenecks: insufficient industrial resources, low energy utilization efficiency, and worsening environmental problems, it can be seen that the development of intensive and low-carbon economy is indispensable.The study and comparative analysis of the current situation of carbon emissions in Hubei Province and the whole country can not only provide better decision-making recommendations for the environmental protection and low-carbon economic transformation in Hubei Province, it also provides the effective reference for the National Energy Utilization and the carbon emission control, and has certain practical significance. In this paper, the data of carbon emission in Hubei Province is analyzed and compared with the national carbon emission situation, combining the influence factors of population, regional GDP, coal, oil, natural gas and hydropower consumption, the main influencing factors are screened through the grey relational grade evaluation model, and the multiple linear regression and polynomial curve combination model are established to forecast the carbon emission of Hubei Province, based on the assessment of the achievement of the carbon emission reduction target and the analysis of the emission reduction potential in Hubei Province, the corresponding countermeasures for reducing carbon emissions are put forward.

2 The establishment of combination forecasting model

Combining the historical forecast data with the actual value, we can give an optimal weight combination of various forecast models, so that the total error of these historical forecast data is minimum after weighting. There are many ways to build a combination forecasting model, the typical ones are:

(1)standard deviation

Suppose there are n models to be combined. Suppose the standard deviation of Model i is σ_i (i = 1,2,3... n)(the standard deviation is the square root of the average of the sum of squares of the differences between each data and its average) , $\sigma = \sum_{i=1}^{n} \sigma_i$ (i= 1,2,3... n),to minimize σ ,then

$$\omega_i = \frac{\sigma - \sigma_i}{\sigma} \bullet \frac{1}{n-1} \text{ (i=1,2,3...n)} \quad (1)$$

This method first needs to calculate the standard deviation of n models.

(2)sum of squares of errors

Suppose that two single models need to be combined, let them be model a and model b. The errors of the two models are one e_a,e_b respectively, and let the error sum $e = \omega_a \cdot e_a + \omega_b \cdot e_b$, ω_a 、ω_b are the weight values of models a and b. Because there are positive errors and negative errors, in order to find the minimum value of e, the square of the error is

$$Var(e) = \omega_a^2 Var(e_a) + \omega_b^2 Var(e_b) + 2\omega_a\omega_b cov(e_a, e_b) ,$$

 https://doi.org/10.1051/e3sconf/202123701008

Minimizing w_a to e is worth finding

$$\omega_a = \frac{Var(e_b) - cov(e_a, e_b)}{Var(e_a) + Var(e_b) - 2cov(e_a, e_b)} \quad (2)$$

The principle is that $Var(e_a) \to \infty, w_a = 0; Var(e_b) \to \infty, w_a = 1$ the more reliable model a is, the smaller the prediction error,the larger the model a's weight coefficient, and vice versa.

The above two methods, the principle is similar, is based on the error minimum establishment function solution. Then we introduce the third method, the principle is the same as the second method, but we need to solve the weight value with the help of the established linear programming model, the calculation method is simpler and the results are more intuitive. By comparing the errors of the combination forecasting model and the single forecasting model, the rationality of the weight value and the superiority of the combination forecasting model can be tested.

Because there are large and small errors in the prediction model, and the single model can only contain local carbon emission information, in order to make full use of the effective information reflected by the single model and reduce the randomness of the single model prediction, consider building a combined forecasting model. The combination forecasting model can effectively use the information of multiple samples and determine the contribution value of single model to the combination model by the weight value, so the error can be minimized.

There are m kinds of methods to predict the same object $m \geq 2$, and. If the actual value of the t period is , the forecast value is , and the forecast error is e_t $(e_t = \hat{y}_t - y_t)$, then for the m method, the corresponding t period actual value, the forecast value and the forecast error are respectively $\hat{y}_{mt}, y_{mt}, e_{mt}$. Suppose the weight of the m method is: ω_i $(i=1,2,...m; \sum_{i=1}^{m} \omega_i = 1)$, then the combination forecast value and the combination forecast error of the t period are respectively $\hat{y}_{ct} = \sum_{i=1}^{m} \omega_i \hat{y}_{mt}$ $e_{zt} = \sum_{i=1}^{m} \omega_i e_{mt}$, the combination forecast Error Square sum is e^2_{zt}, so

$$e_{zt}^2 = \sum_{i=1}^{m} e_{zt}^2 = \sum_{i=1}^{m} \sum_{j=1}^{m} \omega_i \omega_j \sum_{t=1}^{n} e_{mt}^2, t = 1,2,....n \quad (3)$$

Based on the principle of the combination model, that is, the sum of the squares of the prediction error is the minimum, the linear programming equation is established to obtain the weight of each prediction method, and the linear programming model is as follows:

$$\begin{cases} f = mine_{zt}^2 \\ \sum \omega_i = 1 \\ \omega_i > 0; i = 1,2....m \end{cases} \quad (4)$$

By using the error and variance between the predicted value and the actual value of the multivariate linear equation and the polynomial curve model, the optimal solution of the linear programming equation can be obtained, and then the weight value can be obtained.

3 Analysis of combinatorial forecasting models

Using the multiple linear regression equation and the quadratic curve model, the carbon dioxide emissions of Hubei Province from 2018 to 2022 can be measured. The actual value of carbon emissions from 2018 to 2022 can be calculated through the data found in the statistical yearbook. The value is compared with the actual value, and then the error percentage can be calculated, as shown in Table 1:

Table1. Comparison of predicted value with actual value and error of prediction model

Year	Actual value	Multivariate linear regression equation		POLYNOMIAL quadratic equation	
		Predicted value	Percentage error (%)	Predicted value	Percentage error (%)
2018	29360	28071.9	4.39	28267	3.72
2019	28193	30101.7	6.77	29455.6	4.48
2020	30736	31086.9	1.14	30556	0.59
2021	30125	30469.7	1.14	31574.6	4.81
2022	33183	31932.3	3.77	32508.5	2.03

Combining the data analysis of Table 1 with the formula (1) ~ (2) , we get $w_1 = 0.096; w_2 = 0.904$, so the combined forecasting model can be expressed as:

$$\begin{aligned} Y = {} & 0.096 \times (-197843.9 + 34.991x1 + 1.676x2) \\ & + 0.904 \times [-4144204 + 2079t - 42.4(t-1999)2] \end{aligned} \quad (5)$$

4 Test of combination forecasting model

4.1 the combination forecasting model is compared with the multiple LINEAR REGRESSION MODEL

Based on the consideration of population and raw coal consumption, the combined forecasting model also considers the Change Law of carbon emission with time (year) in Hubei Province, which is equivalent to giving more influence factors to the sequence, the dependence of the predicted results on the data of population and raw coal consumption is reduced. Using the indexes MAD, MSE and Mape, the errors of the combined forecasting model and the multiple linear regression equation were compared and analyzed:

Table2. Comparison of errors between multiple linear regression model and combined forecasting model

Indicators \ Model	MULTIPLE LINEAR REGRESSION MODEL	Combined Forecasting Model
MAD	1028.62	927.75
MSE	1421707	1068985
MAPE	3.44	3.11

Obviously, through comparative analysis, the three

error index values of the combined forecasting model are smaller than those of the multiple linear regression model, which proves that the combined forecasting model is superior to the single multiple linear regression model and can achieve the goal of reducing the error.

4.2 the combination forecasting model is compared with the polynomial curve model

The combined forecasting model retains the polynomial curve model and gives higher weight, but does not exclude the multivariate linear regression and gives up its use. It shows that both models are suitable for fitting and predicting CO_2 emissions and that each of the two single models has its own focus. While affirming that the carbon emission in Hubei Province has obvious characteristics of changing with time, the effect of population and raw coal consumption on carbon emission can be incorporated into the combined forecasting model to reduce the fluctuation of data. Using the indexes MAD, MSE and MAPE, the errors of combined forecasting model and polynomial curve model are compared and analyzed as follows: Table 3:

Table3. Comparison of errors between polynomial curve model and combined forecasting model

Indicators \ Model	Polynomial curve model	Combined Forecasting Model
Mad	931.94	927.75
MSE	1075500	1068985
MAPE	3.13	3.11

Through comparative analysis, although the error index value of the polynomial curve model is close to that of the combination forecasting model, it is not difficult to see that the index value of the combination forecasting model is smaller, which proves that the combination forecasting model is better than the single polynomial curve model, the prediction accuracy is improved when the multivariate linear regression equation is used, so the combined prediction model can be used to predict the CO_2 emission in Hubei Province.

5 Summary

On the basis of factor analysis, the model combination and prediction and analysis are carried out in this paper. The contents include:

(1) the linear programming model is established on the principle of minimum variance, the combination weights of multivariate linear regression model and polynomial curve model are 0.096 and 0.904 respectively. The composite model can be expressed as:

$$Y=0.096\times(-197843.9+34.991x_1+1.676x_2) +0.904\times[-4144204+2079t-42.4(t-1999)^2]$$

(2) using MAD, MSE and MAPE to test the superiority of the combination forecasting model, the combination forecasting model is superior to the single forecasting model.

(3) the forecast value of CO_2 emission in Hubei Province from 2018 to 2022 is calculated by using the combined forecasting model, and it is found that the growth rate is 1%~2% , and the carbon emission is increasing year by year.

(4) to analyze the result of carbon emission prediction, and to calculate the emission reduction potential of Hubei Province by sensitivity analysis method. The population is more sensitive to carbon emission.

Fund

This paper is based on the Philosophy and Social Science Research Project of the Hubei Provincial Department of Education (19Y078) "Influencing factors of Hubei Province's carbon emissions under the new economic normal and the establishment and research of their combined forecasting model"

References

1. Ehrlich P. Holdren J. P. Impact of population [J] . Science, 171: 1212-1217(1971)
2. Thomas Dietz, Eugene a Rosa. Effects of population and influence on CO2 emissions [J] . Ecology, 94 : 175-179(1997)
3. Ratnakar Pani, Ujjaini Mukhopadhyay. Identifying The major players behind increasing global divergences emission decomposition analysis [J] . The Environmentalist, 30(2) : 183-205(2010)
4. Brant Liddle Sidney Lung. Age-structure, urbanization, and climate change in developed countries: revisiting STIRPAT for disaggregate population and p-consumption related Impacts. Popul Environ, (1)007-031.(2010)
5. Impact of Carbon Dioxide Emission on GNP Growth: Interpretation of Proposed Scenarios [R] . Presentation to the Energy and Industry Subgroup, Response Strategies Working Group, IPCC, Paris, (1989)
6. Hulten C.DEVISA Index numbers[J].Econometrical, 41(6) : 1017-1025(1973)
7. Howarth R. B. Manufacturing Energy use in eight OECD countries [J].Energy Economics,13(2) : 135-42(1991)
8. Ang B. W. , Zhang F. Energy, 23(6) : 489-495(1998)
9. Potential futures for road transportation CO2 emissions in the Asia Pacific [J] . Asia Pacific Viewpoint,48(3) : 355-377(2007)
10. CHANGE IN CO2 emissions and its transmission between Korea and using Japan International input-output analysis [J].Ecological Economics,,58(4): 788-800(2006)
11. Bin Shui, Robert C. Harriss. The role of CO2 granditent in US-China Trade [J] . Energy Policy, 30(6) : 4062-4070.(2006)
12. Druckman A, Jackson T. The carbon footprint of UK households 1990-2004: Asocioeconomic ally

segregated, quasi-multi-regional input-output model [J] . Ecological Economics, 30(8) : 2066-2077(2009)

Application of a novel a core-shell microstructured nanocomposites as a fire extinguishant using seawater

Liu Hao[1], Zhang Tianwei[1,2*], Zhang Cunwei[1,2], Xiao Lei[1,2], Liang Qiang[1]

[1] China People's Police University, School of Rescue and Command, 065000 Langfang, China
[2] China People's Police University, National Engineering Laboratory for Fire and Emergency Rescue, 065000 Langfang, China

Abstract: Water has unsurpassed capabilities in fire extinguishing. However, water scarcity is rapidly increasing in many regions and water shortage problems have led to find either new water resources or improve seawater treatment technologies. In this study, a new environment-friendly and high-efficiency powdered fire extinguishing agent with a core-shell structure is designed. The designed extinguishing agent is a colloidal powder with a 92% liquid content and is prepared through physical modification and high-speed shearing. Fumed silica is used for the external shell due to its chemical inertness, whereas seawater is used as the core due to its high extinguishing efficiency. Results show that the size of the prepared agent ranged between 100–200μm. The two fire extinguishing experiments of different scales demonstrate that the employing inorganic salts with seawater as the core can significantly increase the efficiency of the new powder in extinguishing gasoline and diesel fire compared with ultrafine and common ABC dry powders. The suppression mechanism of action is caused by the combined flame chemical and pyrolysis theories of the potassium compounds. This study would be of great theoretical significance and application value for improving the application scope of the seawater.

1 Introduction

Petroleum and petroleum products, especially gasoline and diesel oil, are widely used in various applications in daily life[1]. Different types of finished gasoline and diesel products, which are highly combustible, can be stored in oil tanks[2]. However, casualties and property losses could arise should an oil tank combust and explode. Furthermore, oil tank combustion generates abundant thermal radiation, which hinders close-distance extinguishment[3]. The fire can affect adjacent oil tanks and cause a domino effect that will expand the reach of the accident and intensify the damage[4].

In fire outbreaks caused by oil tank combustion, traditional extinguishing methods apply water to cool the tank walls. When the temperature of the oil products have declined to a certain point, fire-extinguishing foam is applied to cover the burning liquid's surface until the fire is fully extinguished[5]. However, if the flame and tank wall temperature is 1050–1400 °C and > 1000 °C, respectively, during the oil tank fire disaster, increased water will be needed to cool down the tank wall due to thermal radiation. These factors determine the amount of water and time that will be consumed before the oil surface sufficiently cools for foam application. These conditions cause water pollution and influence the firefighting opportunity. Moreover, vapors resulting from the contact between the hot oil surface and water can push the upper oil products out of the tank, causing boiling over and spraying. This occurrence may cause secondary damages[2]. In contrast, foams evaporate and disappear upon contact with the combusting liquid surface. After extinguishment, a thick foam layer is maintained on the oil surface to prevent after-combustions. Hence, a considerable amount of foam is needed to completely extinguish an oil tank fire. The toxic substances generated from the heating of the foam can damage the atmosphere to a certain extent. In brief, developing an environment-friendly, clean, and highly efficient fire extinguishing agent for oil tank fires is necessary and urgent.

Compared with water and foam, dry powder extinguishing agents cover and extinguish fire through chemical reactions and can be applied onto the center of the fire without the need for special nozzles. However, dry powder has no cooling effect and has poor mobility and these properties limit its applications[6]. The new powder extinguishing agent with a core-shell structure combines the coverage capacity of the dry powder and the cooling effect of water and was prepared through high-speed shearing method. The core-shell particle(CSP) can successfully extinguish oil tank fire accidents and has advantages over the traditional extinguishing agents.

The designed extinguishing agent used hydrophobic particles for the shell and contained chemical extinguishing components comprising large amounts of water and some white powder[7]. Chen[8] developed a dry water extinguishing agent that used hydrophobic SiO_2 for the shell and NaCl solution as the core. The core water content was > 60% and the addition of poval and

*Corresponding author: zhangtianwei_119@outlook.com

ammonium sulfate significantly improved the distribution of the particle size. This dry water extinguishing agent was superior to the commercial ABC and BC dry powders in terms of octane fire extinguishing efficiency. Ni[7] developed an extinguishing agent that employed hydrophobic SiO_2 for the shell and distilled water as the core and used it against gasoline fire. This agent developed a synergistic effect during extinguishment and was significantly superior to SiO_2 or water. Furthermore, the same extinguishing agent, which had been stored for one year, was used in the comparison experiment. The extinguishing efficiency remained the same and demonstrated good storage performance of the powder. Han[9] studied a dry water extinguishing agent that employed hydrophobic SiO_2 for the shell and phosphoric acid solution as the core and found that the acid solution can exist as a stable core. Moreover, he improved the traditional spraying device; hence, significantly saving extinguishing agents from spraying loss.

In the aforementioned studies on dry powder extinguishing agents, SiO_2 particles were used in the production of rigid droplet-shaped shells and the extinguishing components served as the core. The studies mainly focused on the physical and chemical properties of the extinguishing agents. The agents' uniqueness and efficiency in extinguishing gasoline or diesel oil fire were noted. Further studies are still needed to elucidate the materials' structural strength. Zhang[10] discovered that dissolving some potassium compounds into water can increase the chemical extinguishing efficiency of pure water. Moreover, appropriate proportions of potassium compounds can react with Gellan gum to form a gel that possesses structural strength and high elasticity. These materials can increase the strength of the core-shell structure.

In this paper, a powder extinguishing agent with a gel core-shell structure was prepared using SiO_2 for the shell and seawater as the core through high-speed shearing method. This extinguishing agent was added with Gellan gum and methyl hydrogen silicone oil to increase its structural strength. The preparation technique, morphology, pressure resistance, and extinguishing efficiency against gasoline and diesel oil fire of the retardant were studied. The extinguishing mechanism was discussed based on theory and experimental results.

2 Experimental Method

2.1 Sample Preparation and Characterization

The raw materials, including fumed SiO_2, the inorganic salts prepared for artificial seawater, Gellan gum, and methyl hydrogen silicone oil were purchased from Aladdin® and required no further purification. The powder extinguishing agent was prepared using a 966B homogenizer. The power, rotating speed range, and volume of the homogenizer were 800 W, 10000–30000 r/min, and 2.0 L, respectively.

The method of seawater configuration was as follows: 25g sodium chloride (NaCl), 11g magnesium chloride hexahydrate ($MgCl_2 \cdot 6H_2O$), 1.6g calcium chloride dihydrate ($CaCl_2 \cdot 2H_2O$) and 4g sodium sulphate (Na_2SO_4) were separately dissolved in 1L distilled water (< 0.06μS/cm).

A certain quantity of hydrophobic SiO_2, seawater, Gellan gum, and methyl hydrogen silicone oil were weighed on electronic scales. Gellan gum was added to the boiling pure water and mixed evenly through stirring. Next, the mixture was poured into the stirring container of the homogenizer and certain amounts of SiO_2, inorganic salts required in seawater and methyl hydrogen silicone oil were added and blended continuously for 30 s at a rate of 17000 r/min. The final products were left to cool at room temperature. After the temperature cooled to 30°C to 45 °C, the powder extinguishing agent with a gel-type core-shell structure was prepared.

Morphology of the prepared extinguishing agent was observed under a PH50-3A43L-EP/PL ternary TV optical microscope with the total magnification power of the microscope was 1600×, the microscope of which was equipped with objective lens with four different magnification powers (4, 10, 40, and 100×).

2.2 Fire Extinguishing Tests

Many methods exist for testing extinguishing efficiency, such as cup-burner, micro-nozzle, chemical analysis, portable fire extinguisher, and pressure tank spraying methods. Each method has its advantages, disadvantages, and specific application. The fire model, which is used in the portable fire extinguisher and pressure tank spraying method, resembles actual fire accidents. These methods are applied by many international standards to test the extinguishing efficiency of condensed-phase extinguishing agents. However, international standard experiments cover a large scale and require considerable manpower and materials. Therefore, our laboratory study assessed the extinguishing efficiency of the prepared retardant on a reduced scale according to the GA499.1-2010, GB 4066.2-2004, and GB50898-2013 designs[11-13].

The small-scale extinguishing experiment was carried out in a limited space of 1 m^3 (1.2 m × 0.7 m × 1.2 m). In the experiment, a portable 0.5 kg extinguisher was used. The inner diameter and the cone angle of the nozzle were 1.8 mm and 60°, respectively. To address the effects of empirical factors on the portable fire extinguisher, we placed a nozzle with an external steel tube at the spraying entrance. The spray angle was limited to 0°-5° to decrease effects of operational variability on the experimental results. The experimental set-up is shown in Fig. 1a.

The spray entrance was located 91 cm from the fire source and the nozzle was 110 cm above the ground. In the experiment, a 20 cm × 20 cm × 5 cm square oil pan was used. Four pieces of K-type nichrome alloy thermocouple were employed to measure the temperature at different heights of the flame at 10 cm intervals. The bottom thermocouple was 6 cm above the bottom of the oil pan. The experiment consumed 200 mL of gasoline and diesel fuels. Before the experiment, the net weight of the extinguisher cylinder (M_1) was measured and a certain amount (M_2) of the prepared extinguishing agent was loaded into the extinguisher. The tank was filled with nitrogen at 1 MPa filling pressure. At the beginning of experiment, the gasoline in the oil pan was ignited and pre-burned for 30 s under uncovered conditions. The box gate was closed after the flame has stabilized and the extinguishing agent was sprayed. The time from spraying to extinguishing was noted with a mechanical stopwatch. Application of the extinguishing agent was stopped after the flame was fully extinguished. The weight of the extinguisher (M_3) was measured, and the actual consumption of the extinguishing agent was calculated. Each experiment was repeated for at least five times and the mean was calculated. The experiment was recorded with a digital camera.

The medium-scale extinguishing experiment was carried out in a limited space of 27 m^3 (3 m × 3 m × 3 m). In this experiment, a 5 kg tank extinguisher was used and the nozzle was set at the exit of the powder transmission pipe. The inner diameter and the cone angle of the nozzle were 1.8 mm and 60°, respectively. The experimental set-up is shown in Fig. 1b.

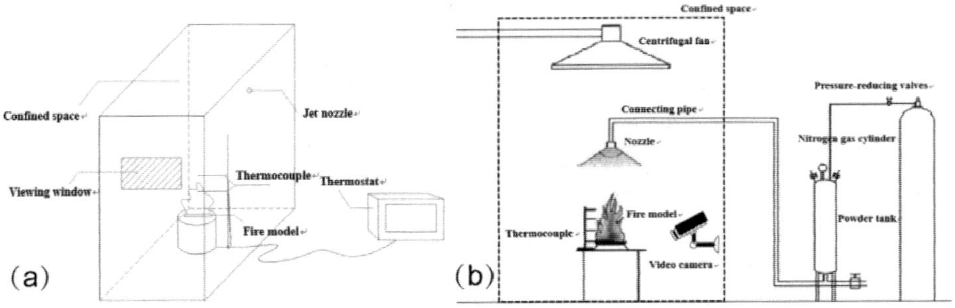

Fig1. Diagrammatic sketch of the fire extinguishing test (a) small-scale and (b) medium-scale.

The nozzle was installed 150 cm above the bottom of the oil pan. In the experiment, the diameter and height of the oil pan were 50 cm and 10 cm, respectively. Four K-type nichrome alloy thermocouples were installed at 20 cm intervals. The bottom thermocouple was placed 3 cm below the oil surface. The experiment consumed 500 mL of gasoline and diesel fuels. The bottom of the oil pan was 15 cm above from the ground. The oil pan's diameter and height were 50 and 10 cm, respectively, and was set for pre-burning. The pan was filled with 200 mL n-heptane to ignite the diesel. Before the experiment, the net weight of the extinguisher cylinder (M_1) was measured. A certain amount (M_2) of the prepared extinguishing agent was loaded into the tank, which was then filled with nitrogen at 1 MPa. The gasoline or n-heptane was ignited and the fireproof door was closed. The timing was simultaneously started with the ignition of the gasoline or diesel. The extinguishing agent was applied after 30 s, and the application was stopped immediately after the flame was extinguished. The time from spraying to extinguishment was noted with a mechanical stopwatch. The weight of the extinguisher (M_3) was measured and the actual consumption of the extinguishing agent was calculated. Considering the uncertainty in the timing and the quality of the extinguishing experiment, the margin of error was set to ± 10%.

3 Result and discussion

3.1 Sample Morphology

According to previous experimental results, the water content of CSP could reach as high as 95%. However, the overly high water content may cause adverse effects on moisture retention and structural strength. A comparison of the particle sizes of the powders which is prepared using different liquid/solid ratios (seawater/SiO_2 ratio) is shown in Fig.2.

The particle size gradually decreases with increasing liquid/solid ratio. Chen[8] concluded that the powder extinguishing agent achieved optimal mobility and performance at 100-300μm. According to the particle size test results, the 100-300μm particles account for the highest proportion (72.6%) in comparison with the other sizes at 92:8 liquid/solid ratio. Further increasing the SiO_2 content (liquid/solid ratio = 91:9) results in the production of a considerable amount of non-enveloped powder, and the raw materials are not completely able to form a core-shell structure. As shown in Fig.2, all of the particles are spheres or spheroid. Particle size is positively related with the liquid/solid ratio which is minimal at 92:8 liquid/solid ratio with most particles range between 100-300μm.

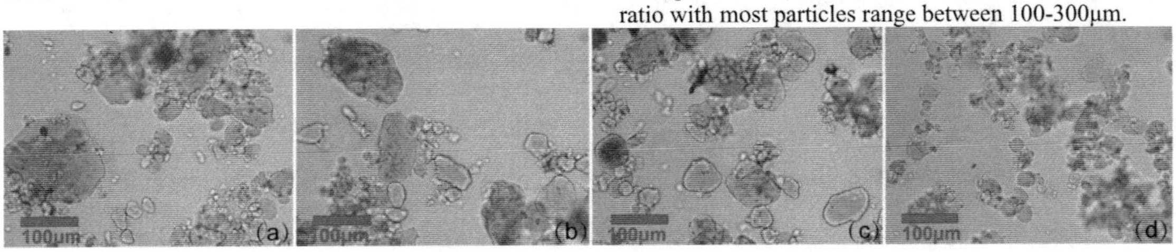

Fig2. Images of the particles morphology at different seawater/SiO_2 ratios: (a) 95:5, (b) 94:6, (c) 93:7, and (d) 92:8.

In this paper, the mass fractions of the Gellan gum and methyl hydrogen silicone oil are kept constant during the preparation. The Gellan gum could reduce the water loss and increase the internal structural strength and moisture retention of the agent. Methyl hydrogen silicone oil could enhance the resistance of powders to agglomeration and increase its mobility. Hence, adding a certain mass fraction of methyl hydrogen silicone oil could increase the adsorption capacity of the SiO_2 solution and offset the influence of the extinguishing components and heating process on the surface tension of the liquid phase[9]. Gellan gum and silicone oil could also influence the surface tension of the liquid phase[14], and the relationship between the variation in mass fraction and types of the additives on the surface tension of the liquid phase will be introduced in future studies.

3.2 Fire Extinguishing Efficiency

In the extinguishing efficiency experiment, ultrafine and common ABC dry powders are used for comparison with the loading contents in the storage tank were 100 and 400 g, respectively. Experimental results of the two experimental scales are shown in Table 1, including the composition, particle size, amount consumed, and extinguishing time.

Table1. Suppression efficiency of the samples in extinguishing gasoline and diesel fires.

Sample composition	Particle size /μm	Fire model/size	Mass of consumed sample /g	Extinguishing time /s
CSP (8 *wt*% SiO_2 + 92 *wt*% water)	100-300	Gasoline/Small	54.8/48.3/52.1/49.6/55.7	6.5/4.7/4.3/5.9/6.7
CSP (8 *wt*% SiO_2 + 92 *wt*% seawater)	100-300	Gasoline/Small	24.6/21.9/23.9/22.6/24.4	2.7/1.4/2.4/2.0/2.4
Ultrafine ABC powder (> 50 *wt*% $NH_4H_2PO_4$)	9.2-18.7	Gasoline/Small	36.7/36.1/34.8/33.2/35.4	3.4/4.8/5.2/3.8/4.7
CSP (8 *wt*% SiO_2 + 92 *wt*% seawater)	100-300	Gasoline/Medium	145.4/168.2/155.3/147.7/168.7	8.4/10.2/9.0/11.1/9.2
		Diesel/Medium	111.5/82.7/100.9/113.4/95.6	8.2/7.7/8.1/7.9/7.1
CSP (8 *wt*% SiO_2 + 92 *wt*% water)	100-300	Gasoline/Medium	153.7/166.4/175.5/169.2/188.2	9.2/9.5/8.4/11.3/10.4
		Diesel/Medium	122.1/117.8/113.5/120.8/107.4	9.1/8.8/8.5/8.9/9.5
Commercial ABC powder (> 50 *wt*% $NH_4H_2PO_4$)	90-150	Gasoline/Medium	339.7/364.2/352.5/382.6/377.7	12.0/12.4/14.1/-/-
		Diesel/Medium	280.4/254.2/268.9/310.5/277.4	10.4/9.8/11.4/10.5/9.7

As shown in Table 1, the average consumption of the as-prepared samples is 23.48-39.8 g with the average extinguishing time is 2.18-5.08 s. No re-burn is observed after the extinguishment. The extinguishing efficiency of as-prepared samples is equivalent to the ultrafine ABC dry powder (35.24 g at 4.38 s). The seawater-CSP shows the lowest consumption and the shortest extinguishing time among all the agents, which is ~1.8 times higher than the ultrafine ABC dry powder. The extinguishing efficiency of the powders that used solutions as the core is better than the powders that used pure water (52.1 g at 5.62s). The extinguishing process in the small-scale experiment is shown in Fig.3.

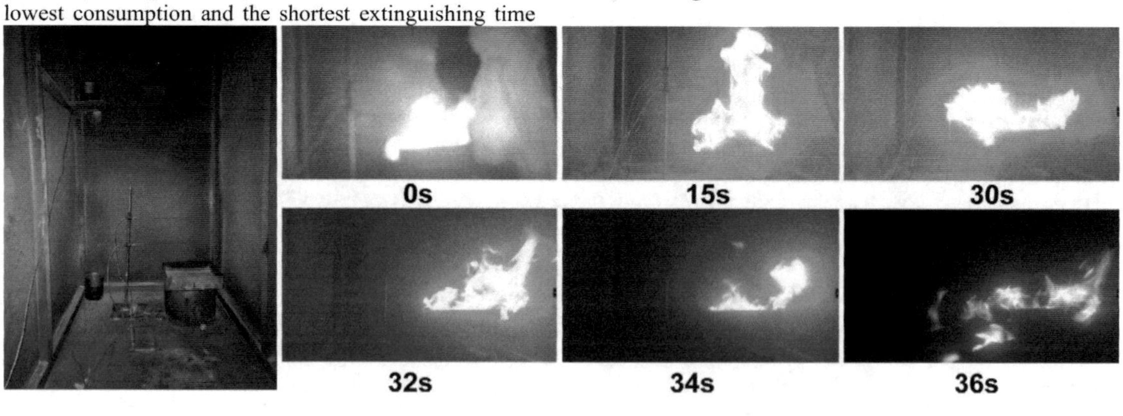

Fig3. Images of the fire extinguishing process in the small-scale test using seawater/SiO_2 particles.

As shown in Table 1, both seawater-CSP and water-CSP can extinguish gasoline and diesel fires in the medium-scale experiment, while an extinguishing failure is observed in the gasoline pool fire experiment using common ABC dry powder. The average consumption of the agents and the extinguishing times of seawater-CSP are 157.06g/9.58s for gasoline fire and 100.82g/7.8s for diesel fire, respectively. And for the water-CSP, the number of which are 170.61g/9.76s and 116.32 g/8.96s, respectively. The average consumption of commercial ABC powder in suppressing gasoline and diesel fires is 363.34g and 278.28 g with the extinguishing times of 12.83 and 10.36 s, respectively. The results of which are significantly higher than that of seawater-CSP and water-CSP. Ni[7] noted that the extinguishing efficiency of water-CSP in the gasoline fire experiment was higher than the common ABC dry powder due to the particle size of water-CSP was 10-30μm, which was lower than that of common ABC dry powder (70-100μm). The particle dispersion and specific surface area increases with the particle size decreases, which are conducive for the particle in the flame free radicals (FFRs) adsorption and strengthen the homogeneous phase inhibition by increasing the endothermic decomposition and reaction rate between the inorganic salts in the seawater and FFRs. On this basis, decreasing the particle size of the seawater-CSP might increase the extinguishing efficiency. The typical extinguishing process of the medium-scale experiment using seawater-CSP is shown in Fig. 4.

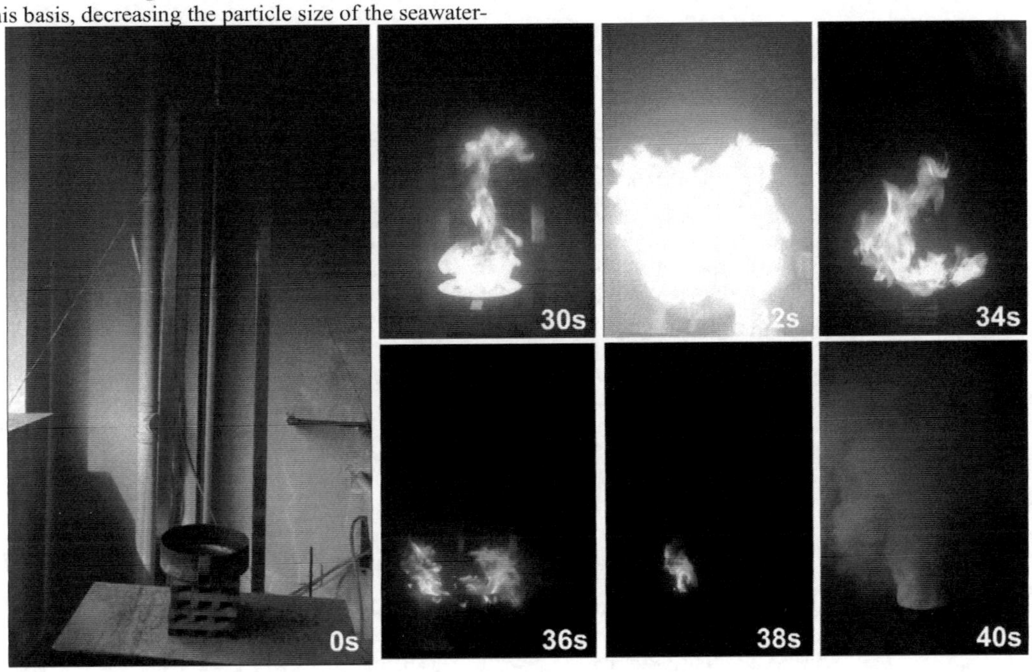

Fig4. Images of the gasoline pool fire extinguishing process in the medium-scale test using seawater-CSP.

3.3 Fire Extinguishing Mechanism

The homogeneous chemical effects of seawater-CSP in fire extinguishing is the elimination effects of gaseous active species on FFRs, which is the decomposition products of solutes under the flame temperature [15]. The species Cl which is produced by NaCl, $MgCl_2 \cdot 6H_2O$ and $CaCl_2 \cdot 2H_2O$ of seawater under high temperature conditions could act as an inhibitor in promoting the conversion of H and OH to H_2 and H_2O through the reaction (1)-(4). The concentration of FFRs is reduced, which clearly influences the key chain reaction (i.e., (5)) of propane combustion, limiting the degree of chain reaction or catalyzing the chain termination process [16-18].

$$Cl+Cl+M \rightarrow Cl_2+M \quad (1)$$
$$Cl_2+H \rightarrow HCl+Cl \quad (2)$$
$$HCl+H \rightarrow H_2+Cl \quad (3)$$
$$H+OH+Cl \rightarrow H_2O+Cl \quad (4)$$
$$O_2+H \rightarrow OH+O \quad (5)$$

The homogeneous chemical effects of Na-containing species in flame suppression is the FFR scavenging effects by gaseous decomposition products through the reaction (6)-(9), which promotes the transformation of H, O and OH to a stable molecule [19-21].

$$Na+OH+M \rightarrow NaOH+M \quad (6)$$
$$NaOH+H \rightarrow Na+H_2O \quad (7)$$
$$NaOH+OH \rightarrow NaO+H_2O \quad (8)$$
$$NaO+O \rightarrow Na+O_2 \quad (9)$$

Ca-containing species can also exert homogeneous chemical effects through the reaction (10)-(12) to reduce FFR concentrations, and $Ca(OH)_2$, CaOH and CaO are the main active species [22,23].

$$CaOH + H \rightarrow CaO \quad (10)$$
$$CaO + H_2O + M \rightarrow Ca(OH)_2 \quad (11)$$
$$Ca(OH)_2 + H \rightarrow CaOH \quad (12)$$

The inert SiO_2 particles have high thermal stability, which could neither absorb heat by themselves nor decompose to form gaseous active species, and thus the flame cannot be extinguished [24,25]. However, SiO_2 particles suspended around the flame can be regarded as an "inert curtain wall" which can both absorb partly energy of FFRs to reduce temperature and reduce the number of FFRs by heterogeneous collision effect [26,27]. When SiO_2 particles are combined with liquid into seawater-CSP, the fire extinguishing efficiency is greatly improved based on the synergy effects.The temperature changes of the three types of particles used in small-scale experiment are shown in Fig. 5.

 https://doi.org/10.1051/e3sconf/202123701009

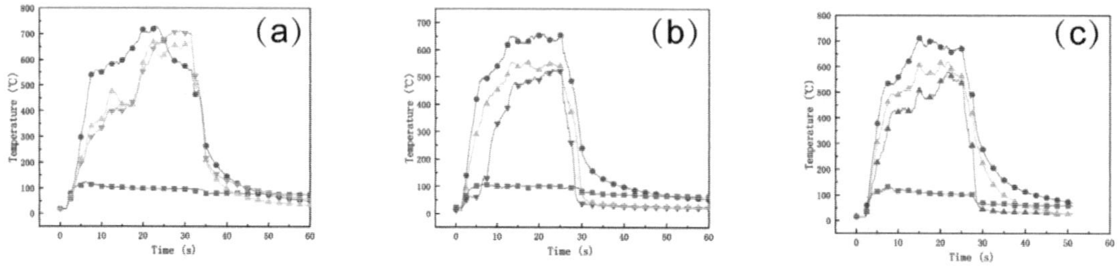

Fig5. Flame temperature in small-scale tests: (a) water-CSP, (b) seawater-CSP, (c)ultrafine ABC powder.

Based on the total flooding mechanism, the fire extinguished by ultrafine ABC powder is mainly depended on the physical and chemical interactions when powders and flames are mixed. The amount of FFRs are reduced by the recombination reactions with gaseous active substances generated by thermal decomposition, and the oxygen is isolated by a glassy-like cover layer formed on the surface of the combustion material. However, as seen from Table 1, little difference in fire extinguishing efficiency between seawater-CSP and ultrafine ABC powder, although the particle size of seawater-CSP is nearly 10 times than that of ultrafine ABC powder. This indicates that seawater-CSP has good suspension and high dispersion, the effect of endothermic cooling of inner liquid and dilution by seawater-CSP is equivalent to the gas-phase chemical inhibition of ultrafine ABC powder.

Temperature changes of the particles used in the medium-scale experiment are shown in Fig. 6. However, the oxygen dilution is not the main extinguishing mechanism due to the adequate oxygen content in the medium-scale extinguishing tests. The extinguishment mechanism of seawater-CSP is the result of the combined effects of coverage, dilution, heat and thermal radiation absorption. More heat absorption by the phase change occurs in the flame under the protection of the heat-resistant shell(i.e. SiO_2), which could reduce the evaporation loss of the fine water droplets during the ejecting process. Reflected in the flame temperature, that is, the flame temperature drop rate under the action of seawater-CSP is fast, the re-ignition phenomenon occurs after the common ABC powder is completely applied due to the poor cooling effect, causing the flame temperature increasing.

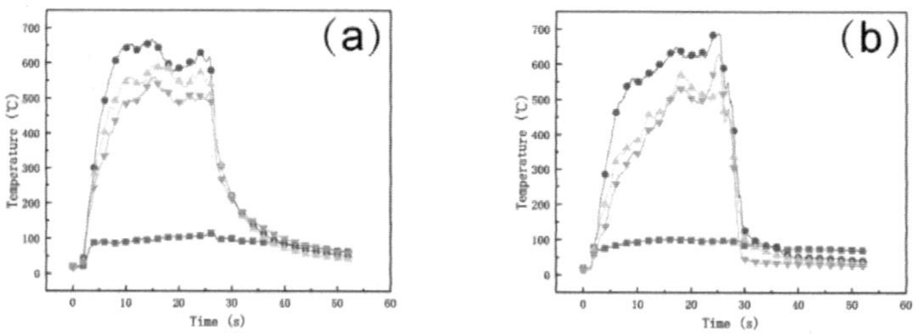

Fig6. Flame temperatures in the medium-scale tests: (a) seawater-CSP and (b) Common-ABC

The efficiency improvement of seawater-CSP relative to common ABC powder in the diesel fire is more significant than in the gasoline fire, as shown in Fig. 6, which is because of the diesel demonstrates a higher surface temperature (~300 °C) compared to gasoline (< 100 °C) during combustion. Liu[28] indicated that surface cooling was the main extinguishing mechanism in liquid fuels with low combustion surface temperature, and the extinguishment only occurs for burning liquid fuels with high temperatures under the synergistic effect of surface and flame cooling. The seawater-CSP has good suspension characteristics, and a heat-resistant shell which leads most of the heat absorption of water by phase change in diesel flames. Therefore, GDW has higher efficiency and is more suitable for extinguishing the fires with higher temperature of fuel surface during combustion, lower consumption and a shorter dousing time of the extinguishing agents is observed in diesel fire compared to that of gasoline fire.

4 Conclusion

Gasoline and diesel tanks have high risks for fire accidents. Explosive combustion of gasoline and diesel tanks is a great threat to personnel safety and to the surrounding structures. Traditional extinguishing agents (e.g., water and dry powder) have limited effects on the extinguishment of gasoline or diesel fires. Therefore, a novel core-shell particle with inner seawater as a fire extinguishant is designed and developed, which can combine the advantages of the traditional extinguishing agents and offset their disadvantages, and several extinguishing mechanisms could be responsible for the extinguishing process.

At present, water resources are becoming increasingly strained due to the decrease of the total amount of water resources caused by the uncertainty of global climate change, and the application of seawater (accounting for 97.5% of global water resources) as the liquid core is one of the important researches and development aspects of CSP materials. The strong corrosion behavior on the storage and injection equipment is likely the main reason why seawater has not yet been applied in fire-fighting practice at the full scale, although the inorganic salts from seawater exhibit a certain chemical effect in the process of fire extinguishing. The physical barrier formed by nanoscale particles between seawater and equipment can exactly reduce the corrosion rate of metal theoretically, which makes the seawater has more potential than fresh water as liquid core of CSP materials for fire extinguishants.

Acknowledgment

Financial support by the National Natural Science Foundation of China (No. 51804314), Natural Science Foundation of Hebei (No. E2019507007), Key R&D Program of Hebei Province(No. 20375503D) and Key Scientific Research Projects in China People's Police University (No. 2019zdgg006)

References

1. L. Kong, Y. Gao, Q. Zhou, Z. Sun. J.Hazard.Mater., **343**(2017)
2. R.M. Leite, F. R. Centeno. J.Hazard.Mater., **342**(2017)
3. X. Liu, L.Tong, H.Wang, Z. Zheng, M. Yao. Appl. Energ., **189**(2017)
4. V. Suchá, M. Mihaljevič, V. Ettler, L. Stmad. J. Environ. Manage., **171**(2016)
5. L. H. Hu, S. Liu, W. Peng, R. Huo. J.Hazard.Mater., **169**(2009)
6. X. Ni, X. Wang, S. Zhang, M. Zhao. J. Fire Sci., **32**(2014)
7. X. Ni, S. Zhang, Z. Zheng, X Wang. J.Hazard.Mater., **341**(2017)
8. X. Chen, A. Fan, B. Yuan, Y. Sun, Y. Zhang. J. Loss Prevent Proc., **59**(2019)
9. Z. Han, Y. Zhang, Z. Du, F. Xu, S. Li, J. Zhang. J. Clean. Prod. **166**(2017)
10. T. Zhang, Z. Han, Z. Du, Z. Zhang, K. Liu. Int. J. Hydrogen Energ. **41**(2016)
11. T. Zhang, Z. Han, Z. Du, K. Liu, Z. Zhang. Appl. Therm. Eng., **107**(2016)
12. Y. Yan, Z. Du, Z. Han. Fire Mater., **43**(2019)
13. T. Zhang, Z. Du, Z. Han, K. Liu. Appl. Therm. Eng., **102**(2016)
14. T. Zhang, H. Liu, Z. Han, Z. Du. Appl. Therm. Eng., **122**(2017)
15. V. Babushok, W. Tsang, G. T. Linteris, D. Reinelt. Combust. Flame., **115**(1998)
16. R. Zheng, K. N. C. Bray, B. Rogg. Combust Sci. Technol., **126**(1997)
17. X. Cao, J. Ren, Y. Zhou, Q. Wang, X. Gao, M. Bi. J.Hazard.Mater., **285**(2015)
18. W. E. Wilson, J. T. O'donovan, R. M. Fristrom. Symposium (International) on Combustion, **12**(1969)
19. . A. J. Hynes, M. Steinberg, K. Schofield. J. Chem. Phys., **80**(1984)
20. B. A. Williams, J. W. Fleming. HOTWC, **4**(1999)
21. O. Dounia, O. Vermorel, T. Poinsot. Combust. Flame., **193**(2018)
22. Y. Koshiba, T. Haga, H. Ohtani. Fire Safety J., **109**(2019)
23. D. H. Cotton, D. R. Jenkins. Trans. Faraday Society, **67**(1971)
24. A. Hamins. Symposium (International) on Combustion, **27**(1998)
25. Jr. W. A. Rosser, S. H. Inami, H. Wise. Combust. Flame., **7**(1963)
26. K. Kuang, W. K. Chow, X. Ni, D. Yang. Fire Mater., **32**(2008)
27. M. Krasnyansky. Fire Mater., **35**(2011)
28. Z. Liu, A. K. Kim, D. Carpenter. Fire Safety J., **42**(2007)

Study on the Rhizosphere Soil Microbial Community Structure Associated with Five Land Use Types in Jinchuan Mining Area

Tian-peng GAO[1,5*], Jing-wen FU[2], Ming-bo ZUO[2], Yu-Bing LIU[3], Dang-hui XU[4], Guo-hua Chang[1], Xi-sheng Tai[1], Bing Yue[1], Zhuo-xin Yin[1], and Qing Zhang[1]

[1]College of Geography and Environmental Engineering, Engineering Center for Pollution Control and Ecological Restoration in Mining of Gansu Province, Lanzhou City University, Lanzhou 730070, China
[2]College of Geography and Environmental science, Northwest Normal University, Lanzhou 730070, China
[3]Key Laboratory of Stress Physiology and Ecology in Cold and Arid Regions of Gansu Province, Northwest Institute of Eco-Environment and Resources, Chinese Academy of Sciences, Lanzhou, Gansu 730000, China
[4]State Key Laboratory of Grassland Agro Ecosystems/School of Life Science, Lanzhou University, Lanzhou 730000, Gansu, China
[5]School of Biological and Environmental Engineering, Xi'an University, Xi'an710065, China

Abstract. Five different land use types (desert, farmland, mining park, slag heap and tailing dam) were selected as variables around the Jinchuan Cu-Ni mining area in Jinchang, Gansu Province in the present study. The *Atriplex canescens (Pursh) Nutt.'s* rhizosphere bacterial abundance, diversity and community composition were examined taking advantage of High-throughput sequencing technology to discuss the effect of soil physicochemical properties on soil microbial community structure. The result indicated that the phylum *Proteobacteria* and *Firmicutes* was the most dominant taxon in desert, farmland and mining park, with a high abundance more than 30%. The phylum *Proteobacteria* was the most dominant taxon in slag heap and tailing dam, with a high abundance more than 40%. The tailing dam had the highest bacterial Chao indexes and the farmland had the highest bacterial Observed species indexes, Shannon indexes and Simpson indexes. Observed species indexes and Shannon indexes between the five sites were significantly different. The redundancy analysis and principal component analysis showed that the main environmental factors caused the different of rhizosphere bacterial community structure in five land use types were Mg, Ca, Cu, TN and moisture, followed by Ni, Cr, K, Pb, Zn content and pH. Hence, the result indicates that land use and soil environmental factors had significant impact on the diversity of soil microbial community structure.

1 Introduction

Soil microbial community is the most significant part of ecosystem and important participant of material circulation and energy flow. Soil quality and nutrient cycling were affected by human activities such as industry, forestry, and agriculture, resulting in the reconstruction of soil microbial communities. The Jinchuan mining area was one of the famous large Cu-Ni sulfide deposits with multimetal symbiosis in the world. Various land use types around the mining area might be affected by different levels of heavy metal pollution. Heavy metals not only change soil fertility, but also interfere with microbial communities, leading to the loss of biodiversity.

Research on soil microbial communities currently focused on land use types, vegetation types, heavy metal stress and nutrient elements which affected their diversity and structure. High-throughput sequencing was widely used to study the impact of environmental factors on microbial communities due to a large amount of biological information resources acquired quickly and a low cost. Previous studies used high-throughput sequencing to analyze combined effects of green manure returning and addition of sewage sludge compost on plant growth and microorganism communities in golden tailings, effects of inoculation with lignocellulose-degrading microorganisms on nitrogen conversion and denitrifying bacterial community during aerobic composting and microbial community diversity of nitrogen-contaminated groundwater[1-3].Therefore, high-throughput sequencing technology was the most reliable method for exploring the microbial community structural characteristics changed by various types of land use types with complex soil environments and heavy metal pollution near mining areas.

Changes in the composition of contaminated soil microbial communities depended on species replacement. Including, tolerant species replaced sensitive species, resulting in a decrease in biodiversity, which was considered to be one of the main threats to soil. Studies had investigated that exposure to heavy metals could give rise to the proliferation of metal-resistant microbial community, such as *Proteobacteria* and *Firmicutes*.

* Corresponding author: zkgtp@163.com; liuyb@lzb.ac.cn

Their community not only depends on the chemical properties of the soil, including pH, organic matter and soil nutrients, but also affects by the toxic heavy metals, K, Ca, Na, and Mg in the soil. Therefore, it was necessary to combine them to investigate and evaluate the sensitivity of soil microorganisms to soil ecosystem protection, but so far, there had been few such studies. Hence, five different land use types (desert, farmland, mining park, slag heap and tailing dam) were selected as variables around the Jinchuan Cu-Ni mining area in Jinchang, Gansu Province in the present study. The *Atriplex canescens (Pursh) Nutt.'s* rhizosphere bacterial abundance, diversity and community composition were examined taking advantage of high-throughput sequencing technology to discuss the effect of soil physicochemical properties on soil microbial community structure.

Table 1. Site code and latitude and longitude information

Site number	Distinction	Latitude(N)	Latitude(N)
A	Desert	38° 28' 52.14"	38° 28' 52.14"
B	Farmland	38° 39' 16.23"	38° 39' 16.23"
C	Mining park	38° 37' 01.71"	38° 37' 01.71"
D	Slag heap	38° 30' 57.5"	38° 30' 57.5"
E	Tailing dam	38° 29' 27.6"	38° 29' 27.6"

2 Methods and Materials

2.1 Soil and plants

The test soil was collected from the Jinchuan mining area and its surroundings (Table 1). A site was located in the desert surrounding the Jinchuan mining area, which was not disturbed by human activities. The soils of C, D, and E site were taken from mining park, slag heap and tailing dam that had been polluted by heavy metals for a long time. The soil of B site was taken from farmland. As one of the dominant species, the *Atriplex canescens (Pursh) Nutt.* is widely distributed in five sites. Three plants were collected from each site for analysis. For soil sampling, the top soil was shovel off and the soil below 5cm with the root system was collected into a sterile numbered bags. Rhizosphere soil was obtained by the method of shaking root, and the remaining soil was kept as the bulk soil. Each soil site was divided into two parts, the first part was stored at 4℃ for microbial detection and analysis, while, and the others was naturally air-dried for the analysis of soil physical and chemical properties.

2.2 Determination of soil physical and chemical properties

Soil moisture was detected by infrared moisture meter (MS-100, Yuda Electronics Co., Ltd., Harbin, China). Electrical conductivity (EC) and pH were detected by conductivity meter (DDSJ-308A, Leici Electronics Co., Ltd., Shanghai, China) and pH meter (PHS-3E, Leici Electronics Co., Ltd., Shanghai, China), respectively, with a 5:1 water-soil ratio. Total nitrogen (TN) was detected by an automatic azotometer (SKD-100, Peiou Analytical Instrument Factory, Shanghai, China) with Kieldahl's method, while total phosphorus (TP) was detected by ultraviolet and visible spectrophotometer (TU-1810, Beijing Purkinje General Instrument Co., Ltd., Beijing, China) with molybdenum blue colorimetry. Calcium (Ca) and magnesium (Mg) were detected by titration and the other metallic elements were detected by a flame atomic absorption spectrophotometer (TAS-990F, Beijing Purkinje General Instrument Co., Ltd., Beijing, China).

2.3 DNA extraction, PCR amplification and sequence analysis

Genomic DNA of each soil site was extracted by PowerSoil DNA Isolation Kit (MoBio Co., Ltd.) following the provided procedure. The extracted genomic DNA was used as a template for PCR amplification of the V4 region (515F and 806R) of the microbial 16S rRNA gene. PCR amplification conditions were adjusted to: denaturation at 94 ℃ for 5 min, then 30 cycles (94℃/30s, 52℃/30s, 72℃/30s), and finally extended at 72 ℃ for 10 min. The fragment length and concentration of PCR products were detected by 1% agarose gel electrophoresis. Sites with bright main strip in between can be used for further experiments. PCR products were mixed in equidensity ratios according to the GeneTools Analysis Software (Version4.03.05.0, SynGene). Then, mixture PCR products were purified with EZNA Gel Extraction Kit (Omega, USA). Sequencing libraries were generated using NEBNext® Ultra™ DNA Library Prep Kit for Illumina® (New England Biolabs, USA) following manufacturer's recommendations and index codes were added. The library quality was assessed on the Qubit@ 2.0 Fluorometer (Thermo Scientific) and Agilent Bioanalyzer 2100 system. At last, the library was sequenced on an IlluminaHiseq2500 platform and 250 bp paired-end reads were generated.

2.4 Statistical analysis

Statistical analyses and visualization were performed in SPSS 23.0, Origin 2019b for Windows. Redundancy analysis (RDA) was performed with CANOCO 5.0 to reveal the relationships between bacterial community diversity and environmental properties. The data were mean ± standard error of three repeats.

3 Results and Discussions

3.1 Soil chemical properties and heavy metal concentrations

The soil moisture content of five sites was all at a relatively low level, and belong to alkaline soil (8.34-8.74, Table 2). The overall pollution of the five metals was evaluated using the Nemerow Pollution Index (PN). The content of TN, EC, TK, Ca, Mg, and Na was the highest in B site. E site had the highest PN, achieving moderate pollution, in which the content of Cu and Ni exceeded Soil environmental quality Risk control standard for soil contamination of agricultural land (GB15618-2018). The Cu content in B and D sites exceeded the national standard, and the Zn content was close to the national standard. The other three sites were not seriously polluted by heavy metals. Moreover, the content of TP, K, Ca, Mg, five heavy metals, pH, moisture and EC showed significant variation among these sites.

Table 2. Soil chemical properties and heavy metal concentrations

Land use types	A	B	C	D	E
Moisture(%)	0.79±0.03e	7.19±0.05a	2.64±0.09c	1.18±0.11d	2.96±0.04b
pH	8.65±0.10b	8.34±0.02e	8.52±0.03c	8.74±0.12a	8.37±0.08d
TN(mg/kg)	0.03±0.01c	0.08±0.00a	0.06±0.01b	0.04±0.01bc	0.05±0.00b
TP(mg/kg)	22.72±0.21a	16.95±0.12c	8.81±0.07e	18.88±0.15b	10.50±0.13d
EC(μS/cm)	85.09±2.56e	539.47±17.21a	370.51±11.21b	146.14±7.28c	121.98±5.36d
K(mg/kg)	33.24±1.25e	60.55±1.56d	89.66±2.01b	63.01±1.99c	109.53±2.38a
Ca(mg/kg)	0.65±0.01e	4.07±0.08a	1.32±0.03c	2.72±0.05b	0.74±0.01d
Na(mg/kg)	0.05±0.00d	0.49±0.01a	0.05±0.00cd	0.06±0.00cd	0.12±0.00b
Mg(mg/kg)	0.08±0.01e	0.95±0.02a	0.10±0.01d	0.13±0.01c	0.28±0.01b
Cu(mg/kg)	10.63±0.12e	17.72±0.17d	113.12±9.51c	211.66±8.37b	222.31±7.62a
Ni(mg/kg)	6.52±0.35e	9.85±0.41d	156.32±8.65b	52.43±2.46c	591.20±15.37a
Cr(mg/kg)	7.58±0.08e	11.62±0.16d	31.66±0.31b	15.54±0.0.15c	130.42±3.58a
Pb(mg/kg)	30.22±0.74e	33.43±0.96d	42.35±1.27b	70.47±1.37a	35.50±1.88c
Zn(mg/kg)	47.93±2.39e	85.12±1.47c	104.56±2.18b	295.69±7.15a	91.17±3.24d
PN	0.14	0.23	0.89	1.59	2.38

[a]Different small letters indicate significant differences between sites for that parameter.
[b]The grading standards: PN≤1, warning limit; 1≤PN≤2, slight pollution; 2≤PN≤3, moderate pollution; PN≥3, heavy pollution.

3.2 Analysis of microbial diversity and composition of rhizosphere soil

Soil microbial biomass and diversity were influenced by various soil environmental variables[4].The soil microbial community compositions at the phylum level in the five sites were shown in Fig. 1, and a total of 15 bacterial phyla were detected. *Proteobacteria* were dominant bacteria in many types of soils, such as mining area soil, wetland soil, karst soil and farmland soil. The predominant bacterial phyla (relative abundance>9%) were *Proteobacteria*, *Actinobacteria*, *Firmicutes* and *Bacteroidetes* in A, B, E sites, which accounted for 77%, 80%, and 87% of the total soil bacteria. The predominant bacterial phyla were *Proteobacteria* and *Actinobacteria* in C, D sites, which accounted for 99% and 97% of the total soil bacteria. The dominant microorganisms were relatively abundant in the soil and played an important role in the regulation of ecological functions. Because the envelope of *Proteobacteria* was mainly composed of lipopoly saccharide, it could protect internal genetic material from external interference, so that it could survive and reproduce in most environments as the dominant bacteria phyla[5-6]. Additionally, some species were highlighted even though their relative abundances were low, such as *Planctomycetes*, *Verrucomicrobia* and *Chloroflexi*. However, the relative abundances of these phyla varied greatly at the five sites. Interestingly, the relative abundance of *Actinobacteria*, *Bacteroidetes*, *Acidobacteria*, *Chloroflexi* in the C, D sites were significantly lower than in the other three sites. When heavy metals entered the soil in large quantities, tolerant species increased, microbial diversity decreased and community structure substantially changed[7]. In present study, compared with the other three land types, the relative abundance of *Proteobacteria*, the dominant phyla in plots C and D, increased significantly. The relative abundances of *Firmicutes*, *Actinomycetes* and *Acidobacteria* were significantly reduced.

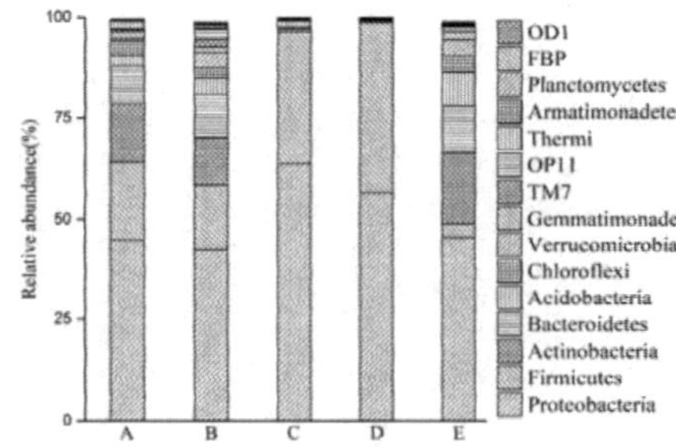

Fig. 1. Soil bacterial relative abundance at phylum level

Four alpha diversity indexes (Chao1, Observed species Shannon and Simpson) were used to evaluate the soil microbial community richness, diversity, coverage, evenness and dominant species status, respectively (Table 3). Chao index ranged from 527.79 to 1987.65. Observed species index ranged from 242.56 to 1068.67. Shannon index ranged from 3.06 to 7.86. Shannon index ranged from 0.77 to 0.98. The soil at E site had the highest Chao indexes, while that at B site had the highest

Observed species, Shannon and Shannon indexes. Observed species and Shannon indexes were significantly different among the five sites.

Table 3. Community diversity analysis of soil rhizosphere microorganisms at phylum level

Sites	Chao1	Observed species	Shannon	Simpson
A	527.79±188.79c	242.56±102.04d	3.06±0.29d	0.77±0.01d
B	1859.93±58.10a	1068.67±13.50a	7.86±0.08a	0.98±0.00a
C	721.82±170.02bc	347.56±39.64c	3.56±0.18c	0.81±0.03c
D	1686.95±58.15b	919.60±17.46b	6.95±0.09b	0.96±0.01ab
E	1987.65±47.55a	1047.11±14.38a	7.05±0.12b	0.96±0.00b

[a]Different small letters indicate significant differences between sites for that parameter.

3.3 Redundancy analysis of bacterial community structure and soil properties

The composition of the microbial community in the long-term contaminated soil had changed. However, this change both depended on the soil chemical properties and heavy metals. For this study, the main contributions of heavy metal fractions and soil chemical properties to bacterial communities were analyzed using redundancy analysis (RDA) and principal component analysis (PCA) (Figure 2., Table 4.). The first two comprehensive indicators were retained (52.80%, 36.60%), in accordance with the principle that the contribution rate was greater than 80%. The cumulative contribution rate was 89.40%, indicating that fourteen environmental factors could explain the differences in the composition of microbial communities among the sites. The first main component mainly includes the content of Mg, Ca, Cu, TN and moisture. They were the main environmental factors which caused differences in the rhizosphere bacterial community structure of five land use types. Next, the second main component mainly includes the content of Ni, Cr, K, Pb, Zn content and pH. According to the contribution rate, the relative importance of each comprehensive index could be known.

Table 4. Coefficient and contribution of comprehensive indexes

Principal component	Each comprehensive indexes														Contribution/%
	moisture	pH	TN	TP	EC	K	Ca	Na	Mg	Cu	Ni	Cr	Pb	Zn	
1	.90	-.58	.87	.06	.76	.55	.91	.86	.92	.87	.51	.533	.68	.66	52.81
2	-.41	.69	-.46	-.85	-.36	.79	-.30	-.49	-.37	.43	.82	.80	.63	.65	36.60

The content of K, Ca, Na, Mg and so on, would affect the toxicity of heavy metals in the different land use types[8-9]. Figure 2(a) showed that the relationship between soil bacterial community structure and the content of K, Ca, Na, Mg. The characteristic values of axis 1 and 2 were 0.5837 and 0.0728, respectively. The cumulative interpretation of changes in all microbial community data is 65.65%. The cumulative interpretation of soil environmental factors by change was 100.00%. It contained most information of the microbial community and environmental, which not only explained the factors' relationship, but also expressed the relationship between the soil bacterial community structure and K, Ca, Na, Mg content. We could easily find that Ca, Na, Mg content had significant effects on rhizosphere bacterial community structure. K content had a negative effect on soil bacterial community structure. Among the dominant bacteria phyla, the relative abundance of *Proteobacteria* and *Firmicutes* was significantly negatively correlated with the content of Ca, Na, Mg in the soil ($P<0.05$). The relative abundances of *Actinobacteria* and *Bacteroidetes* were significantly positively correlated with the Na content in the soil ($P<0.01$). Combined Table 2. and Figure 1., as the first principal components, it could be seen that the Ca and Mg content was positively correlated with *Actinobacteria*, *Bacteroidetes*, *Acidobacteria* and *Chloroflexi* relative abundance. Interestingly, the relative abundance was the higher in B and E sites with higher Ca content. The relative abundance of *Chloroflexi* and *Actinobacteria* was the highest in the E site with the highest Ca content. The relative abundance of Actinobacteria and Bacteroidetes was the higher in A, B and E sites with higher Mg content. It could be seen that the Ca and Mg content was positively correlated with Planctomycetes relative abundance. Moreover, the relative abundance of Planctomycetes was the lowest in the C site with the lowest Ca content. This was similar to the results of previous studies[10-11].

Figure 2(b) showed that the relationship between soil bacterial community structure and the content of moisture, pH, EC, TP and TN. The characteristic values of axis 1 and 2 were 0.7992 and 0.0726, respectively. The cumulative interpretation of changes in all microbial community data is 87.18%. The cumulative interpretation of soil environmental factors by change was 99.99%. We could easily find that pH had significantly positive effects on the relative abundance of *Proteobacteria* and *Firmicutes*, and had significantly negative effects on the relative abundance of *Actinobacteria*, *Bacteroidetes*, *Acidobacteria* and *Chloroflexi*. It was the same as TP, without significantly correlation with the phylum bacteria. Moisture was negatively related to the relative abundance of *Firmicutes* ($P<0.05$). TN was negatively related to the relative abundance of *Firmicutes*, and was positively related to the relative abundance of *Chloroflexi* ($P<0.05$).Combined Table 2 and Figure 1., as the first principal components, it could be seen that the relative abundance of *Proteobacteria* and *Firmicutes* was the highest in C sites with lowest moisture, meeting the negative correlation between them. Moisture was significantly positively related to the relative abundance of *Actinobacteria* and *Chloroflexi* ($P<0.05$). Although it had no significant correlation with *Acidobacteria*, *Bacteroides*, *Chloroflexi* and *Verrucomicrobia*, they were positively correlated. In addition, their relative abundance was also the highest in the E example with the highest moisture.

Figure 2(c) showed that the relationship between soil bacterial community structure and the content of Cu, Cr, Ni, Zn, Pb. The characteristic values of axis 1 and 2 were 0.7949 and 0.0729, respectively. The cumulative interpretation of changes in all microbial community data is 86.78%. The cumulative interpretation of soil environmental factors by change was 100.00%. We could easily find that among the dominant bacterial, the relative abundance of *Proteobacteria* was significantly negatively related to the content of Cu ($P<0.05$), and the relative abundance of *Proteobacteria* was negatively related to the content of Ni, Cr, Pb, Zn. The relative abundance of *Firmicutes* was negatively related to the content of Cu ($P<0.05$), and the relative abundance of *Firmicutes* was positively related to the content of Ni, Cr, Pb, Zn. This was the opposite of the condition of *Bacteroidetes*. The relative abundance of *Actinobacteria* was positively related to the content of Cu, this is consistent with previous studies[12]. The relative abundance of *Planctomycetes* was significantly positively related to the content of Cu ($P<0.01$). In addition, the relative abundance of *Planctomycetes* was also the highest in the C site with the highest Cu content, indicating that it had a better adaptability to Cu pollution.

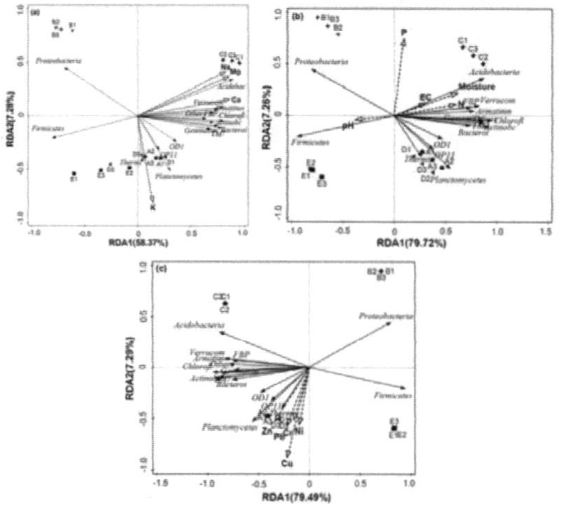

Fig. 2. Redundancy analysis for bacterial and environmental factors at phylum level

4 Conclusions

(1) *Proteobacteria* and *Firmicutes* were dominant bacteria in the five land use types, accounting for more than 40% and 30%, respectively.

(2) The tailing dam had the highest *Atriplex canescens (Pursh) Nutt.'s* rhizosphere bacterial richness, and the farmland had the highest bacterial diversity, coverage, evenness and dominant species status. Observed species indexes and Shannon indexes between the five sites were significantly different.

(3) The main environmental factors which caused differences in the rhizosphere bacterial community structure of five land use types, were the content of Mg, Ca, Cu, TN and moisture, followed by the content of Ni, Cr, K, Pb, Zn and pH.

Acknowledgements

The research was funded by the National Natural Science Foundation of China (31860176, 41977204), the Key research and Development Program of Gansu Province (20YF3FA037), the Key Research and Development Program of Shanxi Province (2020ZDLLSF06-06). We are grateful to all anonymous reviewers whose comments improved the quality of the manuscript.

References

1. Y.J. Ai, F.P. Li, H.H. Gu. Environ. Sci. Pollut. R. 2-4 (2020)
2. J. Yuan, J. Gu, X. J. Wang. Bioresource Technolo. **313**, 123644-123654 (2020)
3. S. Li, M.X. Yang, H. Wang. Environmental Pollut. **267**, 114875-114889 (2020)
4. D. Geisseler, K.M. Scow. Soil Biol. Biochem. **75**, 54–63 (2014)
5. H.G. Song, O.S. Kim, J. J. Yoo. Joral. Microbiol. **42**, 285-291 (2004)
6. W.R. Whalley, B. Riseley, P.B. Leeds Harrison. Eur. J. Soil Sci. **56**,353-360 (2005)
7. C.L. Chen, M. Liao, L.S. Zeng. Acta Ecologica Sinica. **26**, 3404-3412 (2006).
8. A. Kenarova, G. Radeva, I. Traykov. Ecotox. Environ. Safe. **100**, 226-232 (2014)
9. M. Stemmer, A. Watzinger, K. Blochberger. Soil Biol. and Biochem. **39**,3177-3186 (2007)
10. T. Narihiro, H. Tamaki, A. Akiba. Plos One.**9**,104752 (2014)
11. J.M. Kim, A.S. Roh, S.C. Choi. Joral Microbiol. **54**, 838-845 (2016)
12. D.P.H. Lejon, J.M.F. Martins, J. Lévêque. Environ. Sci. Technol. **42**, 2819-2825 (2008)

Green belt's effect on fine particle matter（PM2.5）in residential cluster: based on CFD simulation

Xiong changwei[1], Chen qingchang[1,*]

[1] School of Urban Construction and Safety Engineering, Shanghai Institute of Technology, 201418 Shanghai Haiquan Road No.100, China

Abstract. In the area of residential green belt planning, most planners pay attention to the landscape function of green belts, while few researchers consider the impact of green belt on the concentration of fine particulate matter in the air. Based on site investigation, information about plants, buildings and weather in the selected area were collected, combined with air pollution measurement, four CFD models with different green belt composition were built and simulated. The results showed that at the residential cluster scale, green belts had two effects on fine particles: blocking and agglomeration. Under the two effects, the role of green belts in reducing fine particulate pollution was not always positive, improper green belts could even aggravate air pollution. This study discussed the impact of different greenbelt composition on PM2.5 concentration in residential clusters by CFD simulation, providing theoretical and methodological support for green belt planning and healthy city planning.

1 Introduction

In recent years, the concentration of particulate matter with aerodynamic equivalent diameters of less than 2.5 microns (PM2.5) caused by industrial development has become higher and higher in urban areas, posing problem to urban residents' health[1]. In 2016, the State Council of China promulgated the "healthy 2030" planning outline, which clearly declared health problem had become one of the most important problems affect social development[2]. In this context, urban green space, which can eliminate air pollution and maintain residents' health, has attracted more attention of planners. Previous studies have shown that plants can reduce the concentration of PM2.5 in the air, and use urban green space to reduce air pollution is very economical. Nowak et al.,[3] find that shrubs and trees in the city can remove $7.11*10^5$ tons of air pollutants each year in US, creating about \$38 billion economic value.

However, most planners focus on landscape design and functional renovation of large and centralized urban park green space, few pay attention to small and scattered residential green belt. Mcdonald et al.,[4] and Dai Fei et al.,[5] find a negative correlation between green space area and air pollutant concentration at the urban scale, but this does not mean that their research results are still valid at the street or cluster scale. In order to provide reference and guidance for residential green belt planning, more research based on residential cluster should be conducted.

Although the scale of residential green belt is very small, it has confirmed impact on street's PM2.5 distribution. Guo xiaohua et al.,[6] study the effect of different green belts on PM2.5 concentrations in street by using Computational Fluid Dynamics (CFD) simulation software. They find green belts can significantly reduce the concentration of PM 2.5 in street sidewalks by blocking PM2.5 on motorway. Li xiangrong[7] finds green belts composed of different plants have different resistance to PM2.5. Besides, he also points out green belts with shrubs reduce the accumulation of PM2.5 in front of buildings, which could be beneficial to optimize indoor air quality of buildings in residential area.

Since green belt can affects the distribution of PM2.5 in street canyons, exploring green belt's effect on fine particle concentration in residential area will make sense. In order to achieve this goal, one control group and three experimental groups with different kinds of green blet will be set. Based on field measurement equipment and CFD simulation software ENVI-met, the effect of green belts on the concentration and distribution of PM2.5 in residential clusters will be explored.

2 Research objects and methods

2.1 Overview of the study area and subjects

Shanghai (120°52E–122°12E, 30°40N–31°53N), a coastal city, located in eastern China. In Shanghai, green belt is widely used in residential areas as green landscape infrastructure. In most residential clusters, green belts ware often planted with shrubs, trees or both shrubs and trees. A representative residential cluster in Fengxian district with a greening rate of 25% were chosen for the study. The residential cluster consists of 4 residential buildings face north and south, each building had six

* Corresponding author: Cqc@sit.edu.cn

floors, each floor about three meters high. The green belts ware mainly distributed in three areas: the front of the house, between the two houses and behind the house (figure 1). The experiment was taken place in November, the time when the northwest wind became the dominant wind direction, sending air pollutants from industrial zone to the experimental area.

2.2 Parameters and meteorological data of CFD models

Based on ZHU Lingling et al.,[8] and Guo xiaohua et al.,[6]'s research, parameters required to establish residential cluster CFD model, including grid accuracy, building size and green belt size of the simulated area, were determined and listed (table 1). PM2.5 concentration at the upwind part of the residential cluster required by the simulation experiment was measured by air quality transmitter. The measurement equipment's accuracy is 1μ g/m^3, and it was installed 1.5 meters above the ground. Besides, meteorological data, which could affect PM2.5 concentration in CFD models, were also measured and listed (table 2).

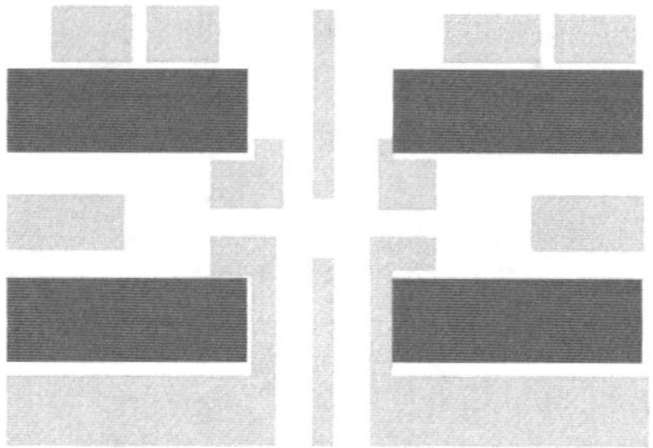

Fig1. Distribution of buildings and green belts of the experimental area

Table1. CFD model parameters

Parameter	Value
Grid accuracy	1 m × 1 m × 2 m
Building height	18 m
Percentage of greenbelt	25%
Building size	18 m × 64 m

Table2. Meteorological data and PM2.5 concentration

Data	Value
temperature	11.5℃
Relative humidity	80%
Wind speed	2.39m/s
Wind direction	307.5°
PM2.5 concentration	31μg/m^3

2.3 Green belt setting

After a survey of residential areas in Fengxian District, three widely used forms of green belts were selected. They were respectively applied to three sets of CFD models (Table 3,4). In addition, a CFD model with no green belt was set as control group.

2.4 CFD simulation

The simulation experiment would calculate the concentration of PM2.5 after 24 hours' spread in residential cluster. The meteorological data and PM2.5 emission rate used in the simulation experiment were set according to the previous measurement data. Each CFD model would be simulated separately. after the simulation, data calculated by the CFD software would be used for analysis.

Table3. Three different forms of green belt

Type	Description
Shrub	Only shrubs in the green belt
Shrub and tree	1-meters-wide shrub wall on the outside with trees 6 meters apart from each other on the inside
tree	trees 6 meters apart from each other

Table4. Plant parameters in CFD models

	Shrub	Tree
Species	Photinia	Camphor tree
Hight	1.5m	6m
Leaf area profile	2.5	0.1(0-2.4m); 0.6(2.4-3m); 1(3-5.4m); 0.6(5.4-6m)

3 Results and discussion

3.1 Results and analysis

The simulation results confirm that green belts do have impacts on the concentration of PM2.5 in residential cluster, and clusters with different greenbelts have different PM2.5 distribution characters (figure 2). Comparing with the control group, it could be found that fine particles in the shrub group were more likely to gather in a strip at the central area of the cluster. Besides, the shrub group had the highest overall PM2.5 concentration with the maximum value occurred in the area between green belt and the south side of the building (building d). On the contrast, the PM2.5 distribution of the tree group was much smoother. It had higher PM2.5 concentration in both inner and downwind areas for about 3.3μg/m^3 than control group. As for shrub and tree group, its PM2.5 distribution had characteristics of the above two groups. In the shrub and tree group, the highest concentration of PM2.5 had reached 22.6μg/m^3, compared with 26.4μg/m^3 in the shrub group and 21.6μg/m^3 in the tree group. In terms of overall PM2.5 concentration rank in the cluster, control group was the lowest, tree group was the third, while the group of shrub and tree came second, and the shrub group became the highest.

https://doi.org/10.1051/e3sconf/202123701011

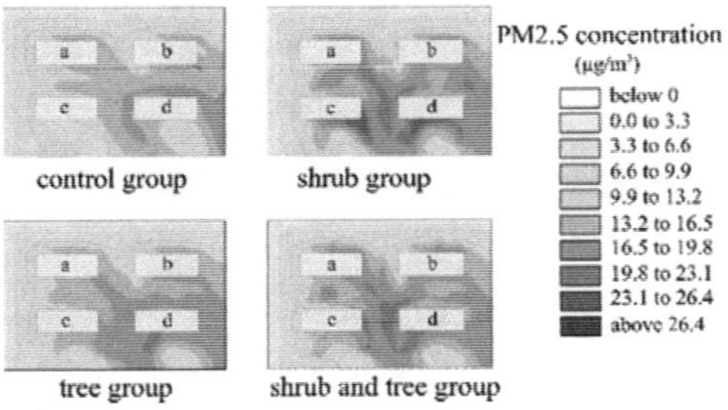

Fig2. PM2.5 concentration of the four groups.

Through the analysis of PM2.5 concentration on the north and south sides of the building, the impact of different green belts on residents' health could be seen (figure 3 and 4). For buildings in the upwind area (building a, b and c), all kinds of green belt in this experiment led to higher PM2.5 concentration in building's north and south side; for the building in downwind area (building d), green belt made the concentration rise in south side and down in north side.

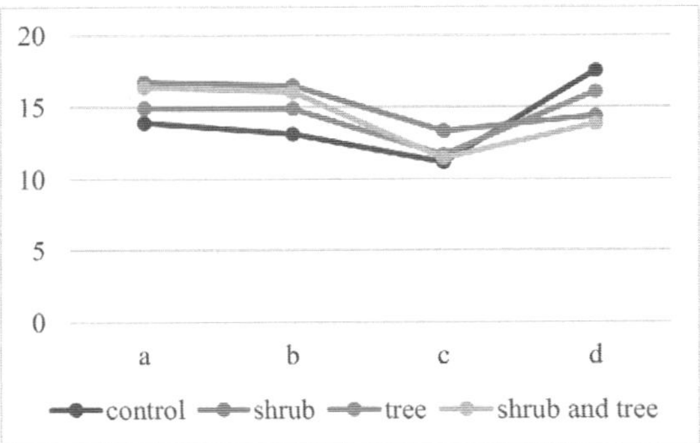

Fig3. Average PM2.5 concentration on the north side of the buildings.

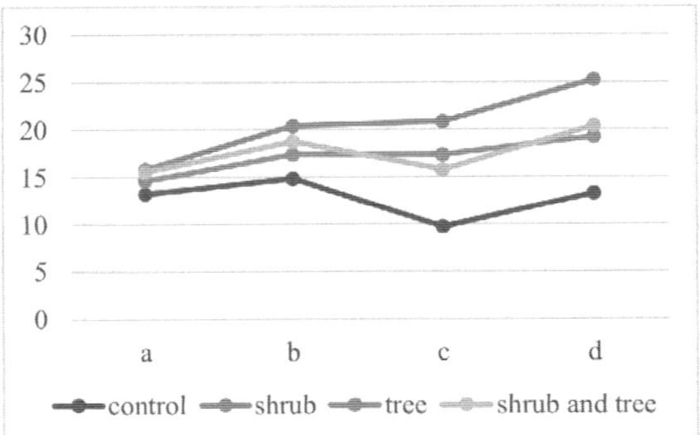

Fig4. Average PM2.5 concentration on the south side of the buildings.

3.2 discussion

According to Prusty et al.,[9] and Tallis et al.,[10]'s research, plants have two effects on fine particles in the air, one called the adsorption effect, and the other called retention effect. Adsorption effect allows plants to remove part of the fine particles in the air permanently, and is one of the main reasons why green spaces reduce the PM2.5 concentration in the air. As for the retention effect, the amount of particulate pollutants retained by retention effect is related to the concentration of particulate pollutants in the air. Retention effect is unstable, particles can easily return to the air due to increased wind speed or other external forces if there is no rainfall. Combine their research with Sanchez and Mccollin[11]'s study , it can be concluded that in the absence of rainfall, a higher concentration of PM2.5 will accumulate near the green belt then affect downwind areas, Causing PM2.5 concentration rise in residential areas. This conclusion can help explain the simulation experiment results. For instance, shrub group's PM2.5 mainly gathers in the centre road of the residential cluster where there were three green belts arranged approximately in parallel. Because of the CFD model did not set rainfall factor and particles could not be deposited to the ground, a high concentration of particles had accumulated. Similar things happened in shrub and tree group. Meanwhile, the experimental results also prove that it is feasible to reduce the PM2.5 concentration in the downwind area by setting up multiple green belts in the upwind area.

As for the tree group, at the height of the canopy, the distribution of fine particles was just consistent with that of the shrub groups; at the height of the shrub layer, the fine particles were evenly distributed. Combined with previous research, the most believable explanation was that the canopy reduced air velocity and caused PM2.5 accumulation.

4 Conclusions

This project studied the impact of different green belts on PM2.5 concentration in residential clusters, so as to provide new methods and ideas for urban green belt landscape planning in Shanghai.

Based on the field measurement data collected in Fengxian district, three different CFD models of green belts and a control model without green belts were established. According to the simulation and analysis results, this study proves that the types of plants in the green belt and the location of green belt are decisive factors affecting the concentration of fine particulate pollutants in residential cluster. Moreover, this experiment found that green belts did not necessarily reduce PM2.5 concentrations at the cluster scale. On the contrary, under the condition of high external PM2.5 concentration and lack of rainfall, improper planned green belts could even aggravate air pollution.

This research can provide useful data and methods for the planning and renovation of green belts in residential areas in Shanghai. In addition, this experiment is based on CFD simulation. Results calculated by computers inevitably have some different from actual situation. In the future, taking more factors into consideration will make the calculation results more realistic.

https://doi.org/10.1051/e3sconf/202123701011

References

1. Cohen A J, Brauer M, Burnett R, et al. Estimates and 25-year trends of the global burden of disease attributable to ambient air pollution: an analysis of data from the Global Burden of Diseases Study 2015[J]. Lancet: S0140673617305056 (2017)
2. WANG Zhi-peng, WANG Wei, SHEN Meng, et al. Impacts of green space on residents' health in urban residential areas: A study in two communities in Hefei [J]. Journal of Environmental and Occupational Medicine, 37(8): 772-776 (2020)
3. Nowak D J, Crane D E, Stevens J C. Air pollution removal by urban trees and shrubs in the United States[J]. Urban Forestry & Urban Greening, 4(3-4):115-123 (2006)
4. Mcdonald A G, Bealey W J, Fowler D, et al. Quantifying the effect of urban tree planting on concentrations and depositions of PM10 in two UK conurbations[J]. Atmospheric Environment, 41(38):8455-8467 (2007)
5. Dai Fei, Chen Ming, Zhu Shengwei, Chen Hong, Fu Fan. Effects of Different Green Coverage in Block Scale on PM 10 and PM 2.5 Removal—A Case Study of the Main City of Wuhan[J]. Chinese Landscape Architecture, 34(03):105-110 (2018)
6. Guo Xiaohua, Dai Fei, Yin Lihua Simulation Study on reduction effect of PM2.5 by Road Green Belt Planning and Design Based on ENVI-MET. Landscape Architecture, 25 (12) :75-80 (2018)
7. Li xiangrong. Study on PM2.5 Pollution and Numerical Simulation of Street Canyon in Chongqing [D]. ChongQing university (2019)
8. ZHU Lingling, GU Kangkang, FANG Yunhao Study residential areas and PM 2.5 concentration based on Envi-met. Ecology and Environmental Sciences 28(8): 1613-1621 (2019)
9. Prusty BAK , Mishra PC , Azeez PA. Dust accumulation and leaf pigment content in vegetation near the national highway at Sambalpur, Orissa, India. Ecotoxicology and Environmental Safety, 60: 228-235 (2005)
10. Tallis M, Taylor G, Sinnett D, et al. Estimating the re-moval of atmospheric particulate pollution by the urban tree canopy of London, under current and future environments. Landscape and Urban Planning, 103: 129-138 (2011)
11. SANCHEZ I A, MCCOLLIN D, A comparison of microclimate and environmental modification produced by hedgerows and dehesa in the Mediterranean region: A study in the Guadarrama region, Spain [J]. Landscape and Urban Planning, 143(1): 230-237 (2015)

Research on Water Restoration System in Wuhan City Based on Pocket Park Concept

Li Hongmin*, Fan Luping, and Zou Zuxu

School of Civil Engineering and Architecture, Wuhan Polytechnic University, Wuhan, China

Abstract. Wuhan city in Hubei Province is endowed with unique natural water resources and numerous lakes. It has a good framework of natural "green sponge", which provides a broad space for stormwater regulation and storage. However, with the increase of infrastructure construction in Wuhan, many lakes have been buried and the city is becoming a "hard shell city" made of steel and cement. This article combines the natural landform of Wuhan and the pocket park system theory to make an organic combination of the ongoing urban infrastructure construction and the restorative construction of the municipal water supply system in Wuhan, so as to make an exploratory research for the early construction of Wuhan into a sponge city.

1 INTRODUCTION

Along with the speeding up of urbanization, the city's impervious area gradually increases, the city is turning into concrete built in "city scale", the rain cannot get effective discharge or use. Meanwhile, water circulation system and ecological environment deteriorate even further.

In the three years from 2012 to 2015, more than 360 cities in China experienced waterlogging. In order to improve the urban ecological environment, to eliminate urban waterlogging, the department of housing and urban-rural development issued a "sponge urban construction technology guide - low impact development of rainwater system build (trial)" in October 2014. In January 2015, the government launched a pilot program to construct sponge urban cities and there were 16 cities in it including Wuhan.

In foreign countries such as the United States, Japan, Australia and other countries in Europe, they studied the practices of urban water resource utilization management much early. They put forward "BMPs (Best Management Practices) water resources comprehensive management", "LID low impact development technology", "WSUD water sensitive urban design" concepts which obtained the very good applications in urban plannings and designs, and those can provide a good reference for our country.

Best Management Practices (BMPs) originated in the United States. The initial focus was on the reduction and control of non-point source pollution sources. In 1987, the United States published the Amendments to the Clean Water Act, and the acts stipulate the provisions of non-point source pollution control and incorporating the control requirements of stormwater runoff pollution into the national pollution discharge permit system. In 2003, BMP was expanded to include integrated stormwater management decision-making systems such as stormwater flood control, soil erosion control and reduction and control of non-point source pollution sources, with more emphasis on ecological design and non-engineering management measures that combine with natural ecology.

Low impact Development (LID) is a new concept of urban stormwater management developed on the basis of BMPs. LID emphasizes that in the case of rainfall, various rainwater control technologies such as storage, infiltration, evaporation, filtration, purification and retention should be used to restore the drainage state of rainwater after urban development close to the state before urban development.

In the late 1990s, water sensitive urban design (WSUD) is put forward in Australia, aimed at the problem of traditional drainage system and it put forward a rainwater management mode and method. The theory regards the urban water cycle as a whole, that is to say the tap water, sewage, drainage are all the components of the urban water cycle, and they should be considered comprehensive and thoroughly.

Since the 1990s, the concept of green infrastructure has been expanded in Europe and the United States. Although the meaning of green infrastructure construction is different in different countries and fields, in a broad sense, it all refers to the effective use of green space, wetland and rainwater storage and infiltration facilities. To be specific, in urban development and land use, it is devoted to the construction of rainwater storage and infiltration facilities, low-lying land, constructed wetland and other infrastructure by taking advantage of the functions of natural environment such as prevention of vegetation and water quality purification.

* Corresponding author: 313833354@qq.com

2 Sponge City and Pocket Park

2.1 Overview of Sponge City

Sponge city means that cities, like sponges, have good resilience in adapting to environmental changes and responding to natural disasters brought by rain. The concept of sponge city is put forward under the concept of green and environmental protection. However, the sponge transformation of the residential area needs to combine the sponge technology with the terrain environment of the residential area, and in this process, there are many fields to explore and study. As one of the sponge city pilot cities, Wuhan also has the task of urban renovation, and some of the renovation problems faced by the residential areas are representative in the whole Yangtze River Basin and even the whole country.

The first goal of building a sponge city is to protect the original water ecosystem. Secondly, to restore the damaged water ecology is another purpose. For the urban green space, water body and wetland that have been damaged under the traditional extensive urban construction mode, the comprehensive application of physical, biological and ecological technology ways gradually restore the hydrologic cycle characteristics and ecological functions, maintain a certain proportion of urban ecological space, and promote the improvement of urban ecological diversity. Many places in China combine the point source wastewater treatment with the implementation of "river length system". It does great help in water pollution control, and it improves the water ecology as well. Third, low impact development. Fourth, various low-impact measures and system combinations are adopted to effectively reduce surface water runoff and reduce the impact of heavy rain on urban operation.

2.2 Pocket park

Under the guidance of the traditional "drainage oriented" stormwater management concept, most cities and towns quickly and directly discharge the collected rainwater into the receiving water body through the municipal rainwater pipe network, with little consideration given to infiltration into the soil, resource utilization and other ways of rainwater. The traditional municipal facilities are unable to cope with the sudden increase in rainfall during heavy rainfall or in the case of continuous rainfall for several days, which causes the city waterlogging problem.

However, the pocket park is another alternative in the system of sponge city. Figure 1 shows us the water drainage system of sponge city.

Pocket parks are characterized by flexible site selection, small area and discrete distribution. They can appear in large quantities in cities, which are like oases in the desert for cities crowded with tall buildings. They can change the urban environment to a large extent and partially meet the needs of people for parks in high-density urban centre.

The concept of pocket Park was first proposed at an exhibition organized by the New York Parks Association in 1963 as "a new Park for New York". Its prototype was to establish patches of Midtown Park, or pocket Park system, scattered in the high-density urban centre.

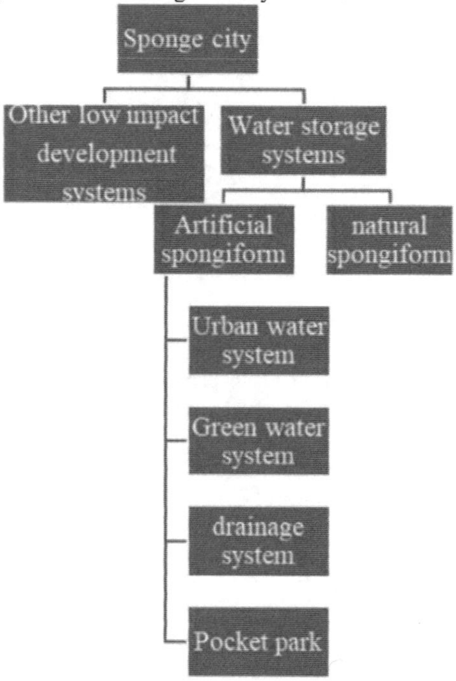

Fig1. . Water drainage system of sponge city

On May 23, 1967, Paley Park on 53rd Street in New York, the United States officially opened. This new form of urban public space came into being, marking the official birth of pocket Park. Peregrina Park responds well to Manhattan's conditions in terms of size and function, and it is as significant as Central Park.

3 Based on the pocket park concept of the water system

3.1 Analysis of main causes of waterlogging

3.1.1 The hardened area of the underlying surface

As can be seen from the rain drainage path diagram shown in Figure 2, a large number of hardened-pavement is adopted on the land, which results in the infiltration path of rain water being blocked, and the rain can only form runoff on the surface and then be drained away. This causes great pressure to the drainage of municipal pipes and drains. At the same time, many residential areas in Wuhan have the problem of poor construction of hardened-pavement, resulting in uneven roads, resulting in the destruction of the originally designed drainage slope, so that rain water is often concentrated in uneven potholes in rainy days.

3.1.2 The green space is not set properly

Green land is one of the underlying surface. Its surface runoff coefficient is not so high, but in the design, it is considered more about its landscape functions, usually set higher than the road after the saturated soil infiltration degrees. And that will be formed on the surface of green land runoff, go through the pipe line again. Nevertheless, soil in Wuhan is cohesive soil, the permeability of the soil is low, so, most of the rainfall can only pass through drainage discharge.

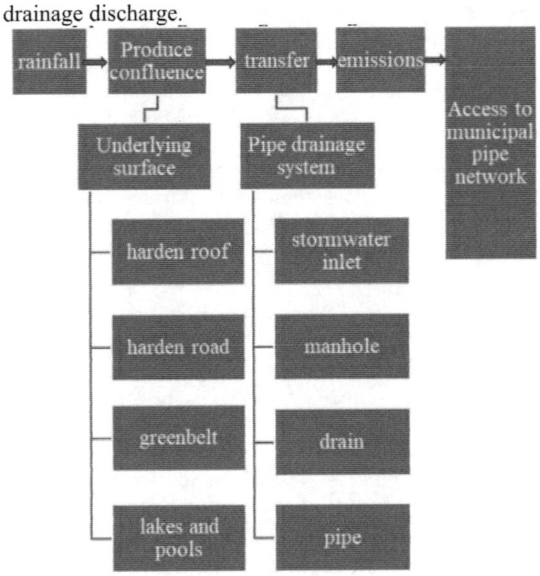

Fig2. rain drainage path

Another problem is the choose of vegetation. They were selected at will. Some plots chose vegetation that required a lot of water in the process of growth, resulting in excessive absorption of moisture in the soil by vegetation. Some communities chose some exotic vegetation, but as a result, the vegetation could not adapt to the climate of Wuhan or the protection situation was not ideal, and that lead to their death, which still could not play a good role in intercepting the runoff.

3.1.3 The drainage capacity of the pipeline is insufficient

After the formation of runoff, rainwater is gradually collected and flows into rainwater inlet and other water collection facilities, and then flows into the pipe and canal system for transfer. However, some rainwater outlets or other water collection facilities in some communities become inefficient due to poor maintenance. At the same time, the drainage pipe and canal are not enough. The drainage pipe diameter of many early-built communities in Wuhan is only 300mm, and the designed flow capacity can only meet the basic drainage. However, as long as there is heavy rainfall, it will be overloaded, resulting in the failure of timely and effective discharge of rainfall, resulting in water accumulation in the community.

3.2 Analysis of current situation of rainwater and flood control in residential areas in Wuhan

The total water area of Wuhan is 2,217.6 square kilometres, with a coverage rate of 26.10%. There are 166 lakes in Wuhan, covering a total area of 867 square kilometres, accounting for 10.2% of the total land area of Wuhan. The wetland covers an area of 3,195.85 square kilometres. Wuhan is a large city in China with abundant fresh water resources, so it provides a strong foundation for the construction of Wuhan sponge city. Affected by the quasi-stationary front of the Yangtze River and Huai River. Wuhan is very representative among the cities in the Yangtze River basin. Wuhan is located at the northern edge of the subtropical humid monsoon climate zone, with abundant rainfall. The average annual rainfall is 1304.3mm, the maximum annual rainfall is 2105.9mm, and the average annual evaporation is 949.8mm. Heavy rain is mostly concentrated in May to August every year, with the maximum daily rainfall of 473mm and maximum hourly rainfall of 120mm. The occurrence frequency of summer and autumn rainstorm is relatively high, especially mesoscale rainstorm, which usually has the characteristics of small range, large intensity, short duration and strong suddenness, etc.. And that can easily lead to the formation of waterlogging in the residential area of Wuhan in a short time. It is a typical rainy city in the south.

However, at present, most of the rainwater control methods in existing communities in Wuhan are based on traditional drainage facilities, such as wells, pipes and channels. The rainwater runoff process can be divided into three stages: generating confluence, transferring and discharging. The confluence stage refers to the process in which the rainwater gradually converges to form runoff after a series of process such as depression, infiltration and evaporation on the underlying surface of the site after the occurrence of rainfall, and finally drains to the stormwater inlet and other water collection facilities. The transfer stage is the process in which stormwater runoff is collected in the pipeline and transferred to the outlet. The discharge stage is the way to treat rainwater in the final stage of rainwater. The usual way to dispose of rainwater in the community is to drain it into the municipal pipeline network.

3.3 Pocket Park Solution

The "pocket park" covers an area of 300 ~ 5,000 square meters, and many of them are built by using the "spare parts" in the urban construction land. The site selection is flexible and the construction period is short, so they have advantages in creating the personality and characteristics of the park. The Table 1 shown below is an overview forms of pocket parks.

Through the investigation of urban street corner green space in the old city, it is found that most street corners in the city are mainly hardened with a large area of the ground, lacking the necessary green shady areas and leisure places. The large area of hardened ground intensifies the reflection of light and heat, and does not have the function of dustproof and noise reduction, so people are not willing to go to this kind of space for leisure and activities, causing a waste of social public resources. Therefore, a new form of "pocket park" should be created to meet the needs in Wuhan style.

Table1. Overview of pocket parks

	type	Scale /m²	Construction forms
1	Small open Spaces in non-urban green Spaces	10~400 ≥400	Shrubs+ stools shrubs+ stools+ entourage tree array
2	greenbelt	10~100 ≥100	Shrubs+ stools Green space+ public facilities
3	Street green space	10~400 ≥400	Small green space+ public facilities; public facilities+ entourage green space+ sport space+ public facilities
4	Green space for public facilities	10~400 ≥400	Small green space+ stools Green space+ sport space
5	Residential green space	≥100	Green space+ public facilities
6	Community park	≥4000	Green space+ sport space+ public facilities

Pocket park is mainly a public space with small scale, close to people's life and flexible use function, which is made of unbuilt urban green space, idle land and urban micro-space. For example, in the event of a major natural disaster, it can be used to set up temporary relief tents and distribute relief supplies.

It is reported that an average of about 40 "street parks" have been built in Wuhan every year since 2017. And a green highlight is the construction of 100 "pocket parks" in Wuhan in 2020 greening plan.

3.4 The Pocket Park application case

Lingang district Home Service Station and Green Leisure Square is a typical pilot project, it have been transformed into a "pocket park" with amazing feedback. This is one of the projects of Lingang district in Shanghai to promote sponge city construction.

Before transformation, rainfall was often accumulated in the area on rainy days. And that causes a lot of problems.

In view of the problems existing in the original, and combining with problems such as parking, recreation, sports and other requirements, the designer chose "permeable pavement + flat drain + rain garden" technology combination. The permeable brick can let the rain infiltration, excessive amounts of rainwater drainage collected through the cover board in the rainwater garden. In addition, "green roof" not only realizes the roof rainwater runoff absorption, but also solves the perennial water leakage problem; "Ecological parking space" has green plants and it is permeable, which can also solve the parking problem of residents. The construction of "permeable circular track" provides residents with fitness trails.

The improved "pocket park" also upgraded its waterlogging capacity. It is reported that the green area of Lingang Home service station and green leisure square has reached 1.3 hectares, accounting for 66% of the total area. The pocket park is able to absorb 158 cubic meters of rainwater runoff in the local block and has a surplus capacity of 190 cubic meters.

4 Conclusion and prospect

The green work plan for 2020 in Wuhan was released on May 26. and it declared that the city will build 700 hectares of new urban green space by combining the establishment of a national ecological garden city and an international wetland city with the revitalization after the epidemic. In addition, 100 pocket parks will be built.

Wuhan city will start the construction of "5-minute Service Circle of urban Park green space" for two years, and plans to carry out the construction of urban park and "pocket park" for two consecutive years, so that the service radius of urban park green space covers more than 90% of the residential land, and citizens are expected to enter the park green space within 5 minutes walking out of their homes.

The construction of pocket park can not only improve the life of residents, but also change the urban waterlogging situation. Pocket park is an important part of sponge city construction.

References

1. Liu H.M., Sun W.J., Shen S.L., et al. Flood risk assessment in metro systems of mega-cities using a GIS-based modeling approach. Science of The Total Environment, **626,** 1012-1025 (2018)
2. Technical guide for sponge city construction (Trial), Ministry of housing and urban-rural development (2014)
3. Li J., Deng C., Li Y., et al. Comprehensive Benefit Evaluation System for Low-Impact Development of Urban Stormwater Management Measure. Water Resources Management, **31**(14), 1-14(2017)
4. Park J., Kim J. H, Dong K L., et al. The influence of small green space type and structure at the street level on urban heat island mitigation. Urban Forestry & Urban Greening, **21,** 203-212 (2016)
5. Li J, Deng C, Li H, et al. Hydrological Environmental Responses of LID and Approach for Rainfall Pattern Selection in Precipitation Data- Lacked Region. Water Resources Management, **32,** 4(2018)
6. Wu Lianfeng, A case study of the sponge city construction plan in built-up areas, Water& Wastewater engineering, **45,** 11(2019)
7. Jennifer R. Wolch, Jason Byrne, Joshua P. Newell. Urban green space, public health, and environmental justice: The challenge of making cities 'just green enough'. Landscape and Urban Planning, **125**, 234-244 (2014)

https://doi.org/10.1051/e3sconf/202123701013

Lumbar shape factors extraction influencing the ratio of waist-darts' volumes for young female

Xue Zhang, Hongshu Jin

Wuhan Textile University, Wuhan, Hubei, 430073, China

Abstract. To improve the fitness of bottoms of individuals, this study is proposed to extract lumbar shape factors influencing the ratio of waist dart's volumes of young females. A total of 27 measurement items including lumbar shapes and waist dart's volumes were collected through overlapped graphics of lumbar feature sections obtained from 30 female college students aged 18 and 24via 3D body scanning method. Results of correlation analysis showed that the total waist dart's volume was significantly correlated with the lateral lumbar bending level (ytc), and the ratio of each waist dart's volume was significantly correlated with lumbar anteversion angle (β), lumbar anteverted amount (ytq) and buttock projected volume (ab) respectively with a correlation coefficient above r=0.7. Furthermore, a total of 2 factors with a characteristic value that was larger than 1 were extracted through principle component analysis, whose cumulative contribution rate reached 80.241%. Principle Component 1 represented factors of lumbar anteversion, which was related to the ratio of anterior/posterior waist dart's volume; and Principle Component 2 represented the lateral lumbar shape characteristics affecting the total waist dart's volume.

1 Introduction

Fitness is a core evaluating indicator of the quality of garment customisation [1]. In the field of bottom pattern-making for individuals, it is effective if the ratio of waist dart's volumes is matched with different lumbar shapes, so as to improve the fitness of bottoms. However, the waist-hip difference is a standard and the only parameter related to the total waist dart's volume for clothing mass production without taking the features of individual lumbar shapes into account.

There are many theoretical researches on parametric descriptions of somatotype differences and human body shape subdivisions [2] were oriented to garment customisations, according to statistical analyses based on multiple body shape variables collected via three-dimensional human body scanning method. Meanwhile, it is a widespread and potential approach that was applied in the field of basic researches, such as body type classification and garment customisation, etc.

There are some researches related to pattern making, such as parametrises of pants back piece corresponding to 4 types of gluteal groove morphologies by draping [3]; the clustering of upper crotch length of man's trousers based on linear regression relationships with lower body height variables [4] as well as the linear regression model of garment shoulder angle based on BMI and shoulder angle [5], through which the relationships between human body parameters and garment patterns were discussed based on statistical analyses considering differences in human body shapes.

For the dart's volume determination, studies on the waist-darts volume defined by standard young women mannequins [6] represented the average ratio of waist-darts, which was inadaptable to performing differences in human body shapes. Mathematical predicting models of breast and the back dart volumes were respectively established with characteristic variables of breast and acromion based on geometric features of chest and back of female bodies [7], and discussions on body shape characteristics were limited to specified geometric parameters that showed standardized human body shapes. Otherwise, Keiko Hiraragi [8] analysed relationships between morphological characteristics of female upper bodies and the ratio of each waist-dart's volume in a trunk basic pattern according to classifications of upper body shapes. This study provided theoretical clues for darts volume assignments related to garment pattern-making, which took differences in human body shapes into consideration.

Therefore, the purpose of this study is to extract lumbar shape factors influencing the ratio of waist dart's volumes based on characteristic analyses about lumbar shapes of young females. In this paper, lumbar shapes and dart's volumes measured through three-dimensional human body cross sections are main data sources for lumbar shape feature analyses.

2 Method of Data Acquisition of Lumbar

The subjects 72 young women aged 18 to 25 years old, were asked to wear light-coloured fitting shorts and vests.

2.1 Measurement Points of Lumbar

According to the definition of human body surface measurement point [9], there are 4 measurement points manually confirmed on lumbar surface, including lateral waist point, abdomen projected point, buttock projected point and maximum thigh point, all of which were marked with labels initially.

2.2. Method of Three-dimensional Human Body Scanning and Overlapping of Cross-sections

Body surfaces were scanned via an OKIO-Body Scan Tian Yuan human three-dimensional scanner. During the scanning, arms of subjects were spread out for about 30°, kept upright and looking ahead.

Characteristic cross-sections of lumbar belonging to aforementioned lumbar measurement points were selected and input into AutoCAD; then, these characteristic cross-sections were overlapped as a measuring graphics.

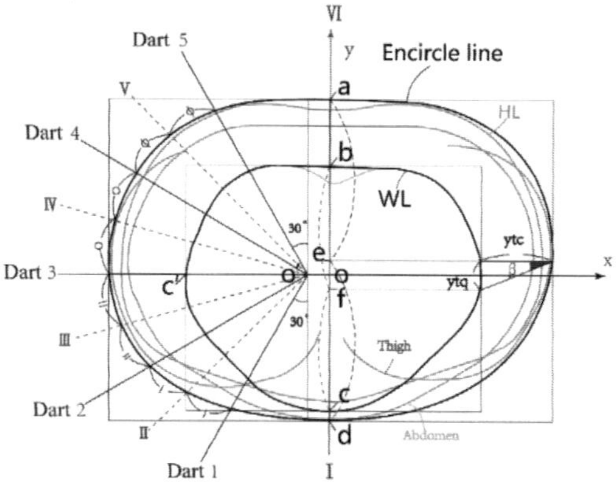

Figure 1. Waist overlapping sections diagram.

The measuring method of overlapping graphic of lumbar cross-sections is shown in Figure 1, and measurement procedures are as follows: 1) Via the waist dart measuring method [9], encircling lines of characteristic cross-sections of lumbar are drawn through connecting the measurement feature points; 2) The x-axis (1/2 point of the distance between central point 'f' of the waist thickness and central point 'e' of the outer circumferential thickness) and the y-axis (1/2 point of the waist breadth) intersect at point 'o', then take the length of 1/2 waist thickness (c'o') on y-axis to determine the focal point 'o'' ; 3) Take the focal point 'o'' as central point, draw the rays at an interval of 30°as waist dart lines and the midlines between each two adjacent waist dart lines (dotted line, II-V).

2.3. Measurement Items of Overlapping Graphic of Lumbar Cross-sections

A total of 27 items were measured on the overlapping graphic of lumbar cross-sections, including 21 for the measurement of lumbar shapes and 6 for that of waist-darts.

Measurement items of lumbar shapes include circumferences, widths and thicknesses of lumbar, and calculation items include waist flattening rate, etc.

A difference between encircling and waist lines is the total waist dart volume; and a difference among midlines of each two adjacent waist-dart lines is each waist-dart's volume. A part of definitions of lumbar shape measurement items and waist-dart's ratios are shown in Table 1.

Table 1. Measurement items in the graphic of partial waist overlapping sections.

Ordina	Measuring project	Definition
1	Buttock projected volume (ab)	The distance between the encircling and the WL line at the front center.
2	Abdomen projected volume (cd)	The distance between the outer perimeter and the WL line at the back center.
3	Lumbar anteverted amount (ytq)	The distance between the outer perimeter thick coronal line 'e' and the WL thick coronal line 'f'.
4	Lateral lumbar bending level (ytc)	The distance between outer perimeter line of lateral lumbar line and WL line.
5	Lumbar anteversion angle (β)	The angle of a right triangle with `ytq' and `ytc' as right-angled edges.
6	Total waist dart's volume (s)	Difference between the encircling line and the WL line.

7	Dart 1 (%)	The difference between the encircling line and the WL line between Front Center I and Dart Midline II, and its percentage to the total dart's volume.
8	Dart 2 (%)	The size difference between the outer perimeter and the WL perimeter between Dart Midline II and Dart Midline III accounting for the percentage of the total dart's volume.
9	Dart 3 (%)	The size difference between the outer perimeter and the WL perimeter between Dart Midline III and Dart Midline IV accounting for the percentage of the total dart's volume.
10	Dart 4 (%)	The size difference between the outer perimeter and the WL perimeter between Dart Midline IV and Dart Midline V accounting for the percentage of the total dart's volume.
11	Dart 5 (%)	The size difference between the outer perimeter and the WL perimeter between Dart Midline V and Posterior Central VI accounting for the percentage of the total dart's volume.

3 Basic Statistics of Lumbar Shape

According to results of Kolmogorov-Smirnov test through SPSS statistics 23, all lumbar feature variables follow a normal distribution.

Taking A-type upper body size in GB/T+1335.2-2008--"standard sizing for garment-women" as a test value, one-sample t-tests were done, whose results are shown in Table 2. The average waist circumference of subjects taken in this study was significantly different from the national standard of A-type, which showed the basic characteristics of subjects taken in this study.

Tab 2. Results of single-sample t-tests at the control site compared with national A-type women.
n=30

Control variable	Average value (cm)	Standard deviation	Inspection value (cm) 160/84A	*t*	Significance
Height	160.00	4.70	160.00	-0.008	N.S. [a]
WL	66.42	4.20	68.00	-2.067	* [b]
HL	90.84	4.88	90.00	0.948	N.S. [a]

[a] $p > 0.05$: No significant difference;
[b] $0.01 < p < 0.05$: difference;
[c] $p < 0.01$: Significant difference.

4 Correlative Relationships between Waist-darts Volume and Lumbar Shape Features

The results of correlation analyses between waist-darts and the measurement of lumbar morphological features on overlapped diagrams are shown in Table 3. Abdomen projected volumes (cd) respectively showed a highly significant correlation over r=0.8 with the waist darts.

Tab 3. Results of correlation analyses between waist shape variables and the allocation rate of dart's volume.

Waist dart	Lumbar anteversion angle (β)	Lumbar anteverted amount (ytq)	Buttock projected volume (ab)	Abdomen projected volume (cd)	Lateral lumbar bending level (ytc)
Total waist dart's volume	0.031	0.223	.389*	0.151	.830**
Dart 1 (%)	-.917**	-.910**	-.780**	.865**	0.15
Dart 2 (%)	-.841**	-.814**	-.863**	.494**	0.238
Dart 3 (%)	-.367*	-0.312	-.509**	-0.092	0.244
Dart 4 (%)	.796**	.813**	.774**	-.625**	0.015
Dart 5 (%)	.949**	.911**	.896**	-.685**	-0.302

** At level 0.01 (two-tailed), the correlation is significant;
* At level 0.05 (two-tailed), the correlation is significant.

A significant correlation of r=0.815 was shown between the total waist-dart's volume and the lateral lumbar bending level (ytc), which indicated a close relationship between the side waist shape and the total waist dart's volume.

Morphological variables of lumbar anteversion angle (β), lumbar anteversion amount (ytq), lateral lumbar bending level (ytc) and buttock projected volume (ab) were negatively correlated with the ratio of front waist dart's volumes over r=0.7 (Dart 1%, Dart 2%), which were positively correlated to the ratio of back waist-dart's volumes (Dart 4, Dart 5) respectively. It is shown that the ratio of front waist-dart's volumes decreased with increases in the degree of lumbar anteversions, and the ratio of back waist-dart's volumes increased with increases in the degree of lumbar anteversions.

5 Extraction of Lumbar Shape Feature Factors Affecting the Ratio of Waist Dart's Volumes

Based on the relationship shown between the ratio of waist-dart's volume and the lumbar morphological feature variables, principle component analyses were carried out to extract lumbar shape feature factors affecting the ratio of waist-dart's volumes. 11 variables featured lumbar shapes were selected after eliminating variables uncorrelated with waist-dart's volumes and verified non-collinear. Results of principle component analyses through maximum variance method are shown in Table 4.

A total of 2 factors with a characteristic root that was larger than 1 were extracted, whose cumulative contribution rate reached 80.241%. In Principle Component 1, ratios of front and back waist dart's volumes (Dart 1, Dart 2, Dart 4 and Dart 5) and characteristic variables (β, ytq, and ab) showed a high loading coefficient that was above 0.7, thus, this principle component could be interpreted as a lumbar anteversion factor; in Principle Component 2, the total waist-dart's volume and the lateral lumbar bending level (ytc) showed a high loading coefficient that was above 0.8, which was interpreted as a lateral lumbar shape factor.

In summary, two factors influencing the ratio of waist-dart's volumes were extracted, of which Feature Factor 1 affected the ratio of front and back waist-dart's volumes, and Feature Factor 2 affected the total waist-dart's volume.

Tab 4. Results of factor analyses.

Variable	Factor	
	1	2
Lumbar anteverted amount (ytq)	0.972	0.099
Lumbar anteversion angle (β)	0.971	-0.081
Dart 5(%)	0.961	-0.124
Dart 1 (%)	-0.932	0.137
Buttock projected volume (ab)	0.901	0.257
Dart 2 (%)	-0.888	0.023
Dart 4 (%)	0.853	0.133
Abdomen projected volume (cd)	-0.755	0.239
Lateral lumbar bending level (ytc)	-0.017	0.888
Total waist-dart's volume	0.264	0.885
Wide outer perimeter	-0.222	0.602
Characteristic root	6.700	2.127
Variance contribution rate (%)	60.906	19.335
Cumulative contribution rate (%)	60.906	80.241

6 Conclusion

Waist morphological features that affected the ratios of waist-dart's volumes of young women were extracted in this study. Main conclusions are as follows:

1)In the results of correlation analyses, the total waist-dart's volume significantly showed a correlative relationship with the lateral lumbar bending level (ytc), and the ratio of waist-dart's volumes was significantly correlated with variables representing the degree of lumbar anteversions.

2)Two principle component factors influencing the ratios of waist-dart's volumes were extracted through principle component analysis. Principle Component 1 was interpreted as a lumbar anteversion factor that affected the ratio of each waist-dart's volumes; and Principle Component 2 was interpreted as a lateral lumbar factor that affected the total waist-dart's volume.

References

1. Hwa K S and Susan P A 2012 Development of Automated Custom-Made Pants Driven by Body Shape *J. Clothing and Textiles Research Journal.* vol 30, no 4. 315-329
2. Hwa K S and Susan P A 2011 Categorization of lower body shapes for adult females based on multiple view analysis *J. Textile Research Journal.* vol 81, no 9. 315-329
3. Xia Y, Shi X Q and Wang H F 2015 Analysis on young female's hip convex curve and trousers'

https://doi.org/10.1051/e3sconf/202123701013

back part structure *J. Journal of Textile Research*. vol 36, no 11. 115-120

4. Wang Y B, Wang H F and Ke Y 2018 Influential factors of upper crotch length in young male's pants *J. Journal of Textile Research*. vol 39, no 9. 115-119
5. Li X H, Zhang S Y, Zhao C T and Bian F 2018 Relationship between garment shoulder angle and characteristics of human body *J. Journal of Textile Research*. vol 39, no 11. 111-115
6. Zhao X G, Pan L, Wang J, Yu Z J, Yao T, Zhou B X and Wang X N 2015 Method of chest-waist difference distribution for clothing prototype of young women in northeast China *J. Journal of Dalian Polytechnic University*. vol 34, no 5. 370-373
7. Jiang L J and Wang W L 2014 Elationship between Female Bust and Back Characteristics and the Individual Prototype Darts *J. Journal of Dong hu a University (Natural Scence)*. vol 40, no 5. 571-574
8. Keiko H 2013 A Study on the Allocation Ratio of Women's Upper Half Body Prototype Waist Darts on Different Body Types-Body Classification Using a Polymerization Horizontal Section Figure- *J. Bunka Gakuen University*. vol 44. 1-11
9. Milyoshi M 2000 *The Theory of Fashion Modeling, 1st ed* (Beijin: China Textile & Apparel Press) p 116,178

Status and inspiration on the development of the air monitoring system ANITA for European Space Agency

Zhengyang Yuan[1*], Jiayong Zhou[1], Cuncun Qian[2], Shizhong Wang[1], Juntao Zhao[1], Hongqi Shi[1]

[1]Wuhan Second Ship Design and Research Institute, 430205, Wuhan, China
[2]Huazhong University of Science and Technology, Audit Office, 430074, Wuhan, China

Abstract. This paper describes the technical characteristics and application status of ANITAII (Analysing Interferometer for Ambient Air), an atmospheric monitoring system currently used by ESA (European Space Agency), reveals the characteristics and advantages of FTIR (Fourier transform infrared spectroscopy) technology in multi-component gas monitoring, and analyzes the application prospect of FTIR technology in the field of environmental monitoring.

1 Introduction

ANITA is a permanent continuous trace gas monitoring system on the International Space Station (ISS), which is jointly developed by ESA and NASA (National Aeronautics and Space Administration)[1]. Based on FTIR technology, the air analyser can simultaneously monitor 32 trace gases with ppm or sub-ppm detection limits[2]. The system can measure with a time resolution in the order of minutes, which allows to trace the dynamic changes in the concentrations of trace gases in the ISS for the first time. In the view of the successful application of FTIR technology in ANITA project, this paper analyzes the application prospect of FTIR technology in the field of environmental monitoring.

2 Basic principles of FTIR

FTIR had been evaluated by ESA as the best technology for continuous monitoring of cabin air, and its technology maturity had been successfully confirmed in a large number of Breadboard (BB) experiments and blind sample tests of NASA[3].

The principles of infrared spectrum measurement is shown in Fig.1.

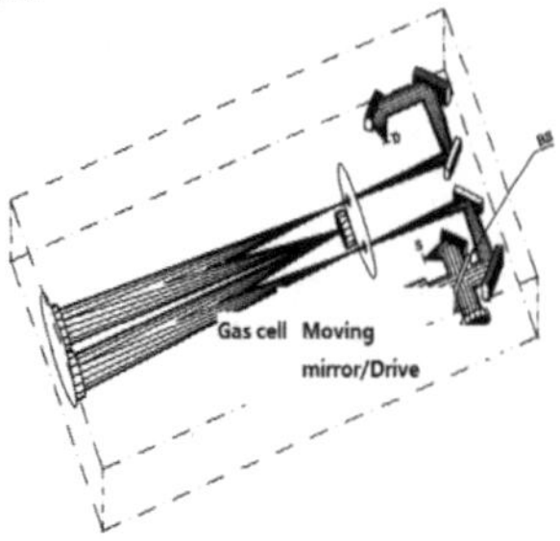

Fig1. Operation principle of the FTIR measurement system, S means light source.

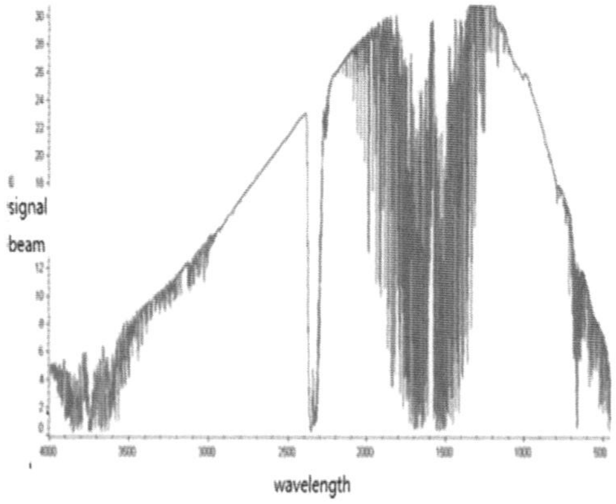

Fig2. Typical FTIR spectrum detected by ANITA.

The modulated infrared radiation from the broadband light source is guided through the air chamber where the air sample (target gas) exists, and then the system detects the radiation, and transforms the measured signal by the mathematical Fourier transform method, finally obtaining the characteristic spectrum as shown in Fig. 2. The wave number range is selected among 600~3500cm-1, in order to monitor the characteristic absorption line of each gas component in parallel. The position and details of absorption band are used for qualitative identification of gas, and the absorption depth is used for corresponding quantitative analysis.

Infrared (IR) technology has several advantages, including high wavelength stability, high optical resolution and high efficiency, which make IR spectrometer suitable for gas measurement. At the same time, the complex situation in multi-gas analysis requires complex data analysis. Therefore, by the time ESA developed the monitoring technology based on FTIR principle in the TGM (Trace Gas Monitoring) scheme, it

* Zhengyang Yuan: 869077438@qq.com

also developed the data analysis technology based on PLS (partial least square) method.

The FTIR modulator is the core of a FTIR spectrometer, which represents Michelson Interferometer as shown in Fig.3. It has a beam splitter, a fixed and a movable mirror. The incoming ray signal is modulated by the movable mirror, and the interference figure produced by coherent ray is detected by the detector. By applying the mathematical Fourier transform method to the detection signal which is a function of time, the measurement signal is converted into a spectrogram for performing the corresponding gas analysis. The micro-vibration injected into the system during the measurement process will cause the moving mirror to tilt and move, thus reducing the system performance.

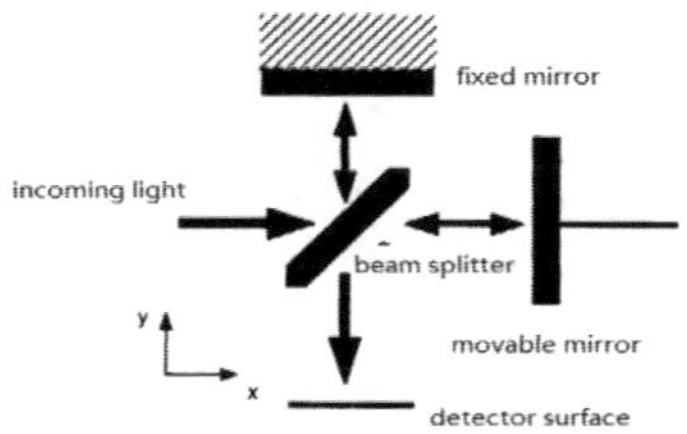

Fig3. FTIR modulator.spectral ghosts

3 ANITA & its technical characteristics

In Fig.4, ANITA I measurement is shown on the left, and ANITA II on the right.

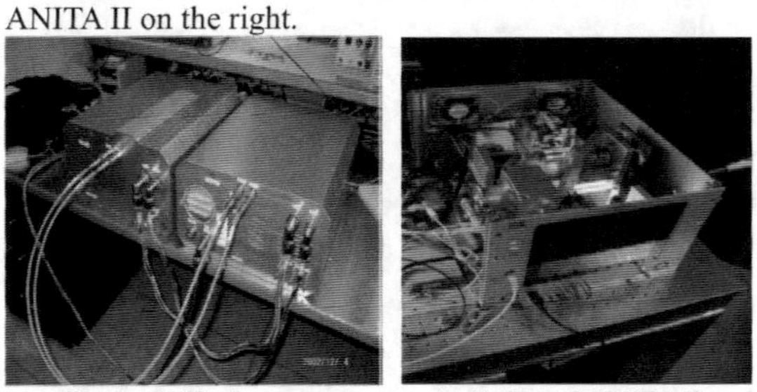

Fig4. ANITA I measurement units (left) and ANITA II measurement units (right).

ANITA I program came from ESA's TGM scheme, it began in 1991, and was launched in August in 2007. It has been running on the ISS for 11 months and produced outstanding data sets[4].

ANITA can automatically detect and quantify various gases in complex mixtures, even if the background gas such as H2O and CO2 vary greatly. The total signal-to-noise ratio (SNR) in the absorption spectrum depends on the SNR of the foreground and background spectra. In each measurement, the absorption spectrum is calculated by a foreground spectrum (the spectrum measured on the air sample) and a background spectrum as a reference measurement. Based on the stability of ANITA instrument and the effective baseline drift compensation in analysis software, the background spectrum is set to be updated every 12 hours in order to minimize the cycle time. As a compromise between the need for short cycle time on the hand and the allocation of time for signal integration and air sample update on the other hand, the cycle time of each measurement includes one minute of signal integration time and two minutes of response time[5].

For long-term health problems, such as cancer risk caused by air pollutants, system cycle is not important. However, for possible short-term problems, such as direct danger of crew poisoning, detection of leakage or overheating of equipment, and inspection of effect of corrective measures, low cycle time may be necessary. The cycle time is set to 5 minutes.

In the laboratory, we usually measure the background spectrum in a gas chamber filled with pure N2. Provided the gas chamber strong enough, high vacuum can be used instead of N2. ANITA relies on another solution, that is, obtaining low pressure through internal pump for sampling. The pump can reduce the air chamber pressure to 1500Pa (about 1.5% of the measured pressure). This process provides a clean background[6].

ANITAI can monitor 32 gases online at the same time with sub-ppm accuracy, the gas varieties are as follows in Tab.2[7].

Table1. Gas scenario from ANITAI

Compound		CUCL(ppm)	Estimated detection limit (ppm)
No.	Name		
1	methanol	70	0.10
2	ethanol	30	0.3
3	2-propanol	30	0.5
4	1-butanol	50	1.0
5	formaldehyde	5	0.10
6	acetaldehyde	30	0.4
7	propionaldehyde	50	0.4
8	butyraldehyde	40	0.3
9	toluene	6	0.9
10	meta-xylene	15	0.4
11	ortho-xylene	6	0.4
12	para-xylene	20	0.2
13	ethyl benzene	20	0.9
14	Ethyl acetate	5	0.3
15	n-butyl acetate	4	0.2
16	dichloro methane	12	0.3
17	Freon 11	2	0.013

 https://doi.org/10.1051/e3sconf/202123701014

18	Freon 12	1	0.04
19	Halon 1301	1.25	0.05
20	Freon 113	3	0.05
21	perfluoro propane	100	0.03
22	hexane (n-)	12	0.3
23	acetone	10	0.4
24	2-butanone	30	0.3
25	hexamethyl cyclo-trisiloxane	2	0.003
26	octamethyl cyclo-tetrasiloxane	1.2	0.003
27	Decamethyl cyclo-pentasiloxane	1	0.012
28	ammonia	4	0.06
29	carbon monoxide	10	0.05
30	methane	500	-
31	carbon dioxide	10000	-
32	water	25028	-

Essential characteristics of the calibration methods include[8]:

- Synthetic calibration exploiting the advantages of the FTIR technique
- Control of optical spectral interference (the main problem)
- Baseline drift compensation (solves a major problem)
- Optimal noise handling utilizing information from numerous spectral data points
- Control of possible optical saturation effects (too high spectral absorption)
- Simple, fast, and fully automatic runtime calculations
- Easy adaptations by recalibration, ever after deployment in space (without astronaut involvement)

The concentration of each gas refers to 25 and standard atmospheric pressure, that is, 101.3kPa. The temperature and pressure of air samples will lead to the change of infrared spectral line structure. Tests show that ANITA is robust enough to handle relatively small temperature changes (for example ±2) and moderate pressure drops (10kPa) with high quality. Large changes may lead to slight or significant quality degradation, but will not lead to failure.

During the ANITA I mission, strong artifacts were observed in many spectra. In the "wrong" spectral region, these artifacts appear as signals similar to CO2 and H2O, that is why they are called "ghosts" [9]. Fig.5 shows spectral ghosts after eliminating all real features from the spectrum. This effect was investigated after ANITA I mission. The researchers specially developed a mathematical model and successfully simulated the results quite close to reality. The result shows that there is vibration acting on the modulator of ANITA I instrument, and finally found that the hardware with problems is the movable mirror (or its whole base), fixed mirror and/or beam splitter[10].

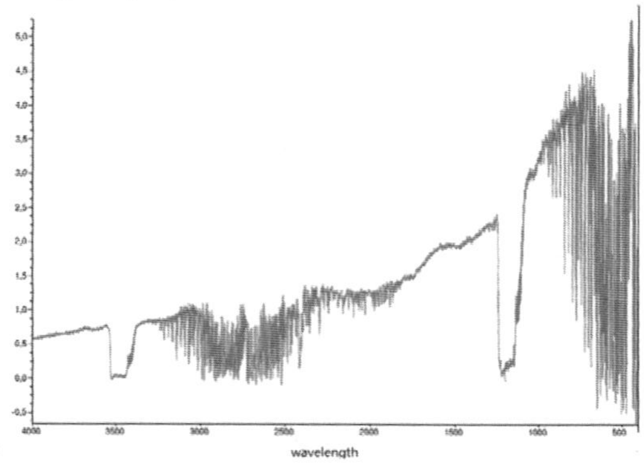

Fig5. Spectral ghosts.

ANITA II is mainly characterized by miniaturization, and its performance is mainly to eliminate the hardware-related interference and instability as well as increase the detection accuracy. Tab.2 compares the instrument characteristics of ANITA I and ANITA II [11].

Table2. Comparison of ANITA I and ANITA II instrument

System property	ANITA I	ANITA II
Mass	54kg	27kg
Elements	Two mid-deck locker inserts, connecting gas tubes and data cables, laptop, breakout board	One mid-deck locker insert
Power	150W (peak)	100W (peak)
Sensitivity	Good	Very good
Response time (full set of data with 33 gases and potential outliers)	3-5 minutes	3-5 minutes

The sensitivity of each gas is an independent value, which to a great extent depends on the gas itself. ANITA II is usually an order of magnitude more sensitive than ANITA I in the operation of the ISS.

Base on the experience of ANITAI and the previous research of ANITAII, a new modulator driving device has been developed, which greatly reduces the micro-vibration of modulator and the displacement tilt of mirror.

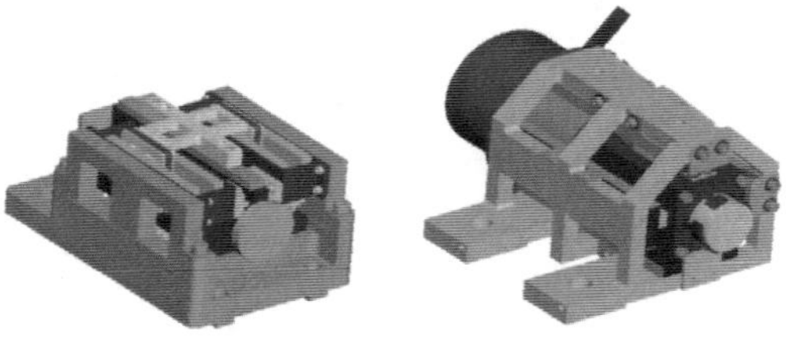

Fig6. Design sketch of the new modulator (left) compared to previous version.

For the purpose to reduce the required volume, mass and power consumption, it is necessary to develop a laser diode system with stable temperature to replace the heavy He-Ne laser. The requirement of frequency stability of diode laser leads to the development and manufacture of customized analog electronic board suitable for DFB laser diode. The total mass and volume of the circuit board are about 10% of the mass and volume of the He-Ne laser[12].

4 Prospect

ANITAII may be the most outstanding achievement in the application of FTIR technology. It is small in size, light in weight, stable in operation and powerful in function, and can monitor 32 gases online at the same time with sub-ppm accuracy. In addition to short-term manned spacecraft and medium-to-long term manned space stations, ESA will also develop ANITAIII to apply in projects such as Moon or Mars base, which will greatly promote the development of human spaceflight. At the same time, ANITAII can also be applied to other confined spaces such as submersibles, large-scale warehousing and cold storage, and underground hedging facilities to ensure the safety and reliability of scientific research or hedging environment.

The successful application of ANITA will also strengthen our confidence in FTIR technology. In terms of the breadth of gas monitoring, FTIR technology has advantages that other monitoring technologies do not have, such as TDLAS and DOAS. In the field of complex gas monitoring, such as atmospheric environment monitoring and toxic harmful gases in industrial areas, FTIR technology can continue to shine brilliantly.

References

1. T. Stuffler, D. Kampf, H. Mosebach, et al. The flight experiment ANITA-a high performance air analyser for manned space cabins, Acta Astronautica 55 (2004) 573-579.
2. T. Stuffler, H. Mosebach, D. Kampf, et al. ANITA and ANITAII Optimised Analysers for Advanced Air Quality Monitoring in Crewed Space Craft. SAE 2004-01-2265, ICES 2004, Colorado.
3. T. Stuffler, D. Kampf, H. Mosebach, et al. ANITA- the optimised spacecraft cabin air monitor on its way to orbit, ICES 2003, Vancouver.
4. T. Stuffler, H. Mosebach, D. Kampf, et al. The Air Quality Monitor ANITA- Going into Operation on the International Space Station, SAE 2007-01-3148, ICES 2007, Chicago.
5. T. Stuffler, H. Mosebach, D. Kampf, et al. Advanced ISS Air Monitoring-The ANITA and ANITA 2 Missions. SAE International 2009-01-2523, ICES 2009, Savannah.
6. T. Stuffler, H. Mosebach, D. Kampf, et al. Evaluation of ANITA Air Monitoring on the International Space Station. SAE 2009-02-2520, ICES 2009, Savannah.
7. T. Stuffler, H. Mosebach, D. Kampf, et al. The ANITA Air Monitoring Programme and Instrumentation-ISS and Other Applications. SAE 2006-01-2176, ICES 2006, Virginia.
8. T. Stuffler, H. Mosebach, D. Kampf, et al. ANITA2-The Future High Performance Air Monitor for ISS Trace Gas Contamination and Gas Concentration Dynamics Monitoring, AIAA 2010-6093, ICES 2010.
9. Honne A., Schumann-Olsen H., Kampf D., et al. ANITA air monitoring on the International Space Station Part2: Air Analyses, SAE 2008-01-2043, ICES 2009.
10. T. Stuffler, S. Gutruf, H. Mosebach, et al. Status Report on ANITA2-The European Programme on High Performance Air Monitoring of ISS Trace Gas Contamination and Gas Concentration Dynamics. AIAA 2011-5211, ICES 2011, Oregon.
11. T. Stuffler, S. Gutruf, K. Nader, et al. Progress in ANITA2, the Upcoming High Performance ISS Air Monitor for Continuous In-Orbit Operation, ICES-2014-68, Arizona.
12. T. Stuffler, P. Hofmann, V. Klein, et al. Progrress in the Development of the Advanced ISS Air Monitor ANITA2-BB results and alternative ways for an implementation of the program. ICES-2015-91, Washington.

Characteristics of Physical and Chemical Properties of Soil in Different Reclamation Years of Opencast Coal Mine Dump in Grassland

Guowang Wang [1], Jiancheng Yan [2,3,*], Junhai Gao [2] and Changshun Wang [4]

[1]Shenhua Baori Hiller Energy Co., Ltd.,Hulunbuir 021500,China;
[2]Tangshan Branch, China Coal Science and Technology Ecological Environment Technology Co, Ltd. ,Tangshan 063012,China
[3]Tiandi(Tangshan) Mining Science and Technology Company Ltd.,Tangshan 063000,China;
[4]Hulunbuir University, Hulunbuir 021000, China.

Abstract. Taking the reclamation waste dump of the open-pit mining area in the east of Inner Mongolia Steppe as our study area, we sampled field investigation and laboratory measurements to study the changes of soil physical and chemical properties in the waste dump of the open-pit mining area during its recovery years. The objective of this study was aiming to provide theoretical support for the ecosystem restoration and reconstruction in the open-pit mining area of grasslands in the Northwest of China. The results showed that: 1) the mechanical composition of soil all layers in different reclamation years was mainly composed of silt, except for 20-30cm; the silt content was significantly higher than the control area (CK), and the silt content of surface soil (0-5cm) in reclamation 3 years (2016) was significantly higher than CK, but the sand content was significantly lower than CK; 2) In the early stage of vegetation restoration and succession, the physical and chemical properties of soil in different reclamation years were significantly different. With the increase of reclamation years, TOC and TN content decreased, but TP content did not change significantly during reclamation years. Based on the above results, the change characteristics of soil physical and chemical properties in the early stage of vegetation restoration succession were discussed. The research results have certain theoretical significance and practical guidance value for the ecological restoration and reconstruction of open-pit mining areas in the north of Northwest China.

1 Introduction

Open pit mining makes the mining area become a fragile ecosystem with strong erosion, which seriously threatens the sustainable development of the mining area. The waste dump of open pit mine is a special deposit of rock and soil mixed dump, which not only occupies a lot of land, but also has a great impact on the surrounding ecological environment. Moreover, the quality of reconstructed soil is poor, and it is not easy to be reclaimed. Therefore, the reclamation of waste dump has always been an important content of Open-pit Mine Reclamation [1,2]. The domestic research on the reclamation of waste dump began in the 1980s. Since then, the research has been further accelerated. The suitability evaluation and economic benefit analysis of land reclamation of waste dump have been carried out, [3,4]. After the 1990s, the reclamation of open-pit mine waste dump began to move towards practical research. Scholars focused on Antaibao [8-10], Anjialing [11] and other open-pit mine dumps with long mining time. The research on land reclamation of open-pit mine waste dump in grassland area was carried out late [12], but it made rapid progress. Scholars focused on Soil and water conservation [13], soil quality change [1, 14-16] and other directions.

Soil reconstruction is an effective way to solve the problem of insufficient surface soil resources of disturbed soil in mine engineering. The results of soil reconstruction will ultimately affect the restoration of ecosystem and biodiversity in mining area. The study on the change of soil quality after soil reconstruction with different reclamation years will provide theoretical support for the restoration and reconstruction of grassland and even the whole open pit mining area in Northwest China.

2 Study Area

Shenhua Baorixile open-pit coal mine is located in Baorixile Town, Chenbalhu Canner, Hulunbuir City, in the east of Inner Mongolia grassland area. The geographical coordinates are between E119°15'and 119°36', N49°14'-49°30', and the altitude is 667m-684m. It is a semi-arid continental climate in the middle temperate zone, with an average annual temperature of - 2.4 ℃, an average annual precipitation of 350 mm, an average annual evaporation of 1247 mm, and an average annual wind speed of 3.5 m / s.

The natural grassland type in the study area is *Leymus chinensis* + *Stipa baikalensis* steppe. The main plants are

*Corresponding author: jc_yan@163.com

Leymus chinensis, *Stipa baicalensis*, *pocokia ruthenica*, *Cleistogenes squarrosa*, etc. The zonal soil type is dark chestnut soil. In 2015, the traditional 33°slope was further cut to 14 ~ 16 °and the reclaimed plant was *Elymus dahuricus* with a sowing rate of 40g per m^2.

3 Research Methods

3.1 Sample Collection and Oreparation

In July 2019, the reconstruction and reclamation time of different parts of the dump (2016, 2017 and 2018) was determined by consulting relevant records and visiting the staff of mine greening section. After comprehensively considering the similarity of slope, aspect and irrigation conditions, the south slope of North dump was selected as the research object, and the natural grassland without grazing around the study area was taken as the control (CK).

The depth of each sample is 30cm ~ 10cm, and the soil is sampled at a depth of 20cm, and the soil is sampled at a depth of 5cm. We taked all samples back to the laboratory to dry, grind and sift through 2 mm sieve for use.

3.2 Soil Sample Analysis

Ls13320mw laser particle size analyzer (Beckman Coulter) diffraction analyzer was used to determine the mechanical composition of soil [17]. The soil texture classification system [18] was used to determine the clay particle size of 0-0.002mm, the powder particle size of 0.002-0.05mm, and the sand particle size of 0.05-2 mm. Organic matter (TOC) was determined by potassium dichromate sulfate oxidation capacity method; total nitrogen (TN) was determined by potassium sulfate copper sulfate selenium powder digestion and determined by nitrogen analyzer automatic analysis method; total phosphorus (TP) was determined by sulfuric acid perchloric acid digestion and molybdenum antimony resistance colorimetry. The above analysis methods are shown in soil agrochemical analysis method [19].

3.3 Aata Processing and Analysis

WPS software was used for drawing, SPSS16.0 software was used for multiple comparison (Turkey test), and the confidence interval was 95%.

4 Results and Analysis

4.1 Effect of Different Reclamation Years on Soil Mechanical Composition of Waste Dump

Soil mechanical composition is one of the most basic data for studying soil. The temporal variability of soil mechanical composition can be used to evaluate the degree of soil erosion.The difference of soil composition in 30cm depth is shown in the table. It can be seen from Table 1 that the soil mechanical composition of each soil layer in different reclamation years is mainly composed of silt, which is more than 60%, and the content of soil clay is the least, less than 10%; the proportion of the three is roughly sand: silt: Clay = 6:72:22. However, CK's sand: Powder: Clay = 8:65:26.

From the perspective of soil sand content, the sand content of 0 ~ 5cm soil increased with the increase of reclamation years, and the sand content of 5 ~ 10cm, 10 ~ 20cm and 20 ~ 30cm soil decreased with the increase of reclamation years. Except that the difference of 10 ~ 20cm reclamation for 3 years and reclamation for 2 years was not significant, the sand content of other layers had significant difference. Compared with CK, there were significant differences, but no obvious regularity. From the perspective of soil silt content, the silt content of 0 ~ 5cm soil decreased with the increase of reclamation years, and the silt content of the other three layers increased with the increase of reclamation years. Except for 5 ~ 10cm, there was no significant difference among other layers. Compared with CK, there were significant differences except 20-30cm. From the perspective of soil clay content, the clay content of 0 ~ 5cm soil decreased with the increase of reclamation years, and the clay content of the other three layers increased with the increase of reclamation years, but there was no significant difference. The content of clay in 0 ~ 5cm soil was significantly lower than that in CK, except that there was no significant difference between 0 ~ 5cm soil clay and CK.

The silt content of 0 ~ 5cm soil was significantly higher than that of the control, the content of sand was significantly lower than that of the control, and there was no significant difference between the soil clay content and the control, indicating that soil and water loss would still occur in the surface soil after 3 years of reclamation.

Table1. Particle size distribution of soil in different reclamation years

layer	year	Sand (%)	Silt (%)	Clay (%)
0~5cm	2016	20.92±0.03a	73.13±0.22a	5.95±0.23a
	2017	21.73±0.19b	72.42±0.4a	5.85±0.4a
	2018	22.36±0.16c	71.87±0.4a	5.78±0.31a
	CK	44.08±0.27d	50.56±0.16c	5.36±0.12a
5~10cm	2016	24.61±0.32a	70.08±0.36a	5.31±0.55a
	2017	22.36±0.19b	71.74±0.31b	5.90±0.54a
	2018	14.23±0.27c	79.63±0.4c	6.13±0.57a
	CK	16.98±0.34d	74.71±0.24d	8.31±0.1b
10~20cm	2016	24.78±0.07a	70.8±0.49a	4.42±0.6a
	2017	24.07±0.11a	71.45±0.6a	4.48±0.66a
	2018	12.75±0.30b	80.77±0.09b	6.48±0.34a
	CK	22.91±0.21c	68.06±0.39c	9.02±0.23b
20~30cm	2016	27.37±0.19a	68.51±0.61a	4.12±0.89a
	2017	24.03±0.32b	70.97±0.57a	5.00±0.36ab
	2018	20.28±0.31c	72.61±0.77a	7.11±0.79b
	CK	21.45±0.36c	68.47±0.18a	10.08±0.2c

Note: Different small letters in the same column indicate significant difference among different years in the same soil layer ($P < 0.05$). The same below.

4.2 Relationship Between Species Siversity and Soil Chemical Properties

4.2.1 Characteristics of Soil TOC Content

It can be seen from table 2 that the TOC content in 0-5cm, 5-10cm and 10-20cm soil layers decreases with the

 https://doi.org/10.1051/e3sconf/202123701015

increase of reclamation years, and there is a significant difference between them; however, there is no obvious rule of organic carbon content in 20-30cm soil layer. The content of soil organic carbon in 0 ~ 5cm, 5 ~ 10cm and 10 ~ 20cm in three years and two years of reclamation were significantly lower than that of the control, but the difference was not significant between one year reclamation and the control, but it was also slightly lower than the control.

Table2. Total organic carbon content of soil in different reclamation years

layer（cm）	2016	2017	2018	CK
0~5	21.26±0.4a	23.97±0.49b	26.17±0.57c	27.31±0.4c
5~10	16.93±1.19a	21.49±0.47b	31.43±0.98c	31.52±1.19c
10~20	16.69±1.73a	23.88±1.29b	30.31±0.22c	24.98±1.73bc
20~30	18.78±1.45a	22.52±0.75a	21.69±0.3a	21.5±1.45a
Average	18.42±1.19a	22.97±0.75b	27.4±0.52c	25.9±1.19c

4.2.2 Characteristics of Soil TN Content

With the increase of reclamation years, the content of TN in 0-5cm, 5-10cm and 10-20cm soil layers decreased significantly, while the content of total nitrogen in 20-30cm soil layer did not change significantly. Compared with CK, the content of TN in 0 ~ 5cm soil layer was lower than that of CK, but the difference was not significant; the TN content of 5 ~ 10cm and 10 ~ 20cm soil was significantly lower than that of CK; the TN content of 20 ~ 30cm soil layer was slightly higher than that of CK, but the difference was not significant.

Table3. Total nitrogen content of soil in in different reclamation years

layer（cm）	2016	2017	2018	CK
0~5	1.693±0.033a	1.757±0.024a	1.767±0.08a	1.897±0.118a
5~10	1.387±0.055a	1.687±0.081b	1.843±0.019c	2.417±0.147d
10~20	1.497±0.069a	1.64±0.029b	1.70±0.049c	2.033±0.192d
20~30	1.617±0.07a	1.96±0.057a	1.907±0.084a	1.583±0.2a
Average	1.55±0.06	1.79±0.05	1.8±0.06	1.98±0.16

4.2.3 Characteristics of Coil TP Content

The content of TP in soil layers increased with the increase of reclamation years, but the difference was not significant. Compared with the control, the content of TP in each layer of soil was higher than that of the control, but the difference was not significant.

Table4. Total phosphorous content of soil in in different reclamation years

layer	2016	2017	2018	CK
0~5cm	0.403±0.024a	0.367±0.022a	0.343±0.014a	0.350±0.012a
5~10cm	0.378±0.021a	0.367±0.009a	0.356±0.02a	0.363±0.044a
10~20cm	0.346±0.017a	0.346±0.008a	0.357±0.017a	0.317±0.045a
20~30cm	0.389±0.029a	0.390±0.019a	0.362±0.019a	0.280±0.038a
Average	0.379±0.023a	0.368±0.015a	0.355±0.018a	0.328±0.035a

5 Discussion

Soil mechanical composition is one of the most basic data for soil research, which can reflect the size and quantity of soil mineral particles, affect soil and water conservation and the supply of plant nutrients [20]. The results show that, after enclosure of degraded grassland, with the restoration of vegetation, the clay content in soil significantly increases [21], which is different from the results of this study. With the increase of reclamation years, the content of sand increases, while the content of clay decreases, which is consistent with the research results of Liu Jun on the waste dump of Huolin Henan open-pit coal mine [22], which may be due to the natural restoration of soil Even if the upper layer is covered with the original topsoil, the original sequence of occurrence layer will be destroyed, and eventually the soil properties will be changed.

Organic matter is an important part of soil and plays an important role in soil fertility and environmental protection [23]. The input of soil organic matter mainly comes from the accumulation of surface biomass and the contribution of human activities, while the output is mainly from the decomposition of various life activities and the output with soil and water loss . Generally speaking, soil organic matter content increases with the increase of reclamation years, but this study shows that with the increase of reclamation years, soil organic carbon content decreases. The lowest content of organic matter was found in the soil after 3 years of reclamation, which may be due to the loss of original accumulated nutrients with the loss of soil clay after soil disturbance. At the same time, due to the short recovery time, the vegetation coverage is not as good as the surrounding grassland, and the surface litter remains less, which can't effectively supplement soil nutrients, which is basically consistent with the research results of Su Zhongfei et al. The content of TN in different reclamation years was consistent with that of TOC, and decreased with the increase of reclamation years.

Most of the phosphorus in soil comes from soil parent material except for a small part from dry wet deposition. In addition to the loss of surface runoff, the phosphorus in soil is almost negligible [23]. Some studies have shown that, in the natural grassland vegetation restoration succession, the soil total P content also increases with the increase of the restoration year, and has a significant correlation with the restoration year. In this study, although there are similar characteristics, the difference is not significant. It is speculated that the reason is due to the implementation of reclamation project, the original 0 ~ 30 cm soil layer is mixed and then covered again.

In this study, the soil quality of three years of reclamation is obviously inferior to the surrounding natural grassland, but the soil quality of one-year reclamation is similar to that of surrounding natural grassland, which is the same as the dynamic evolution law of soil quality of Yimin mining area in Inner Mongolia: "the growth rate of soil quality in the initial stage of reclamation is fast, and with the increase of reclamation time, the recovery speed of soil quality is gradually slow, close to the original landform". It has been reported in some studies that after full enclosure of severely degraded grassland, good restoration effect can be achieved about 8-12 years; for waste dump restored by artificial intervention, it is reported that the surface soil quality can reach the level of undisturbed soil after 5 years of Reclamation [14]. It can be seen that the artificial intervention will accelerate the reclamation speed of the waste dump in the mining area, but the specific recovery time of the ecological system structure and function of the dump still needs to strengthen the long-term positioning monitoring.

Acknowledgment

This paper is supported by National Key R&D Program of China(2016YFC0501105) and Special Youth Project of Science and Technology Innovation and Entrepreneurship Fund of Tiandi Science and Technology Co., Ltd (2018-TD-QN041).

References

1. J. Wang , Z.Bai, R.Yang . Succession Law and Model of Reclaimed Soil Quality of Opencast Coal Mine Dump in Grassland. Transactions of the Chinese Society of Agricultural Engineering, **28**,229-235(2012).
2. Z. Hu. The 30 Years Land Reclamation and Ecological Restoration in China: Review, Rthinking and Prospect. Coal Science and Technoloy, **47**,25-35(2019).
3. HU Zhen-qi. Spatial Variability of Physical Properties of Reclaimed Soil in Surface Coal Mine[J]. Journal of China University of Mining & Technology, 34-40(1992).
4. W. Wang. Preliminary Studies on the Evaluation Method of the Suitability of Land Reclamation in Open Cast Mines. Opencast Mining Technology, 72-75(2013).
5. G.Liu, R. Bi, Y. Cao. On the Economic Benefits of Land Reclamation in Open Cast Coal Mine Dump Site .China Coal, 109-111(2008).
6. J. Huang, J.Sun. Present Situation and Assumption of Open Pit Reclamation in Masteel [J]. Metal Mine Design and Construction, 47-50(1994).
7. S. Chai, Z. Bai. Study on Lnad Reclamation-growing and Runoff Characteristic of Surface Rock Soil on Soil-discharged Fields in Antaibao Opencut Coal Mine. Coal Mine Environmental Protection, 22-26 (1999).
8. S. Chai, Z. Bai. Relationship between Surface Runoff Characteristics and Plantation in reclaimed land on dumps of Antaibao open-Pit mine. Metal Mine Design and Construction, 49-53(1995).
9. G. Wang, X.Wang, D. Li, et al. Correlation between Nutrients of Reclaimed Soil in Antaibao Opencast Mine.Coal Mine Environmental Protection, 25-27 (2001).
10. X. Li, Z. Zhang. Analysis of Land Reclamation in Mine Dump of Fuxin. Journal of Liaoning Technical University(Natural Science),**S2**,251-253(2009).
11. P. Tai, T. Sun , H.Jia , et al. Restoration for Refuse Dump of Open-Cast Mine in Steppe Region. Journal of Soil and Water Conservation, 90-93 (2002).
12. J. Wei. Discussion of Land Rehabilitation Pattern After Mining in Open Pit Mine . Coal Science and Technoloy,**S2**,384-385(2013) .
13. H.Sun, X. Li ,Z. Hu, et al. Variance of reclamation soil quality in Majiata opencast mine region., 205-209,**24**(2008).
14. X. LIU, Q. Zhu, Z. Hu, et al. Reclamation Soil Quality Change Characteristics of Open-pit Coal Mine Dump Site Slope in Weakened Ecological Grassland Region .JiangSu Agricultural Sciences,**45**, 242-244 (2017).
15. S. Wang, Y. Cao, G. Luo, et al. Effects of Reconstructing Soil in Open Cast Mining Areas of the Loess Plateau on Soil Physical Properties and Vegetative Growth. Journal of Agricultural Resources and Environment, **36,** 344-354 (2019).
16. J.Wang, R. Yang, Z. Bai. Succession Law and Model of Reclaimed Soil Quality of Opencast Coal Mine Dump in Grassland. Transactions of the Chinese Society of Agricultural Engineering, **28**, 229-235(2012).
17. Y. Chen, Y. Zhang, Z. Tan, et al. Comparison of Particle-size Results of Sediments Measures by the MS2000 and LS13320Laster Diffraction Particle-size Analyzers. Acta Scientiarum Naturalium Universitatis Sunyatseni,**57**,48-55(2018).
18. K. Wu, R. Zhao. Soil Texture Classification and Its Application in China. Acta Pedologica Sinica , **56**,227-241 (2019).
19. S. Bao. *Agrochemical Analysis of Soil.* (Agriculture Press ,Beijing,2000).
20. L. Wu, X. Jia, H. Wang. Effects of Maize Growth Durations on Soil Physical and Chemical Characteristics After a Slope Hill to a Terrace Landform.Land Development and Engineering Reaearch, **5**,38-42(2020).
21. H. FU, Y. Chen, Z. Zhou, et al. Changes of Vegetation and Soil Environment of Desert Grasslandin the Early Period of Restoration in Alxa , Inner Mongolia. Journal of Desert Research,58-61(2003).
22. J. Liu, X. Dang. Effect of Planting Years of *Hippophae rhamnoides* on Soil Fertility Quality of Dump Slope in Surface Mine. Journal of Zhejiang Forestry Science and Technology ,**37**,9-14(2017).
23. C. Huang. *Pedology* . (China Agriculture Press, Beijing, 2002).

Research on InSAR Monitoring Precision Analysis of Coal Mining Influence Based on Long Time Sequence

Junhai Gao[1,2], Rongjiu Gao[2], Lisheng Chen[2], Yongbin Han[2], Jiancheng Yan [1,2,*] and Xingli Li[2]

[1]Tangshan Branch, China Coal Science and Technology Ecological Environment Technology Co, Ltd, ,Tangshan 063012,China;
[2]Tiandi(Tangshan) Mining Science and Technology Company Ltd.,Tangshan 063000,China

Abstract. In this paper, in view of the scope of coal mining subsidence, the magnitude of subsidence deformation and the impact on buildings (structures), InSAR technology is selected for monitoring, time series analysis method combined with D-InSAR method for processing, and Sentinel- 1 Satellite IW data and RADARSAT-2 satellite XF data have been monitored for four years in the mining-affected area and its surrounding areas. The results show that there are obvious subsidence ranges on the surface of the current mining area and the unstable settlement area. The Shen District has been in a stable state in recent years, and the time sequence analysis for more than 3 years has verified the high accuracy and reliability of the monitoring results. The analysis of the monitoring results shows that the InSAR technology can be used for mining subsidence monitoring, especially when determining the subsidence boundary and whether it is stable or not, it has its technical superiority.

1 Introduction

The surface of coal mining produces subsidence and deformation. The scope, degree of influence, development trend and deformation law of mining subsidence are the current research hotspots in the industry. The research methods mainly include the establishment of observation stations, comprehensive analysis and computer numerical simulation[1], Laboratory simulation of similar materials, etc. In recent years, the application of technology such as total station robots, laser scanners [2,3], and drones [4,5], especially the development and application of technologies such as remote sensing and InSAR [6-10], it provides a new method for obtaining three-dimensional data information of mining subsidence. The advantages of InSAR technology applied to deformation monitoring are as follows: the analysis of the monitoring dimension and deformation law of traditional monitoring methods stays at the "point line" level, which can not realize regional monitoring, while InSAR can improve the mining surface deformation monitoring from the "point line" level to the "area" level.Traditional monitoring methods can not ensure the acquisition of deformation information in the key deformation period, while InSAR can extract the deformation information of the previous surface affected by mining using historical data, so as to trace the past deformation; the surface deformation characteristics obtained by traditional monitoring methods are not really "the same state". A monitoring method that can instantly obtain the surface state information of the entire area of interest is needed. InSAR can quickly scan the entire monitoring area through satellites, and achieve simultaneous acquisition of the monitoring target area. result.

This paper applies InSAR technology and adopts time series processing methods to monitor the building group and its surrounding area at the boundary of the mining-affected area, and analyze in detail to determine whether the building group is located in the subsidence deformation area and the degree of influence, and solve the local government and villagers, concerns with scientific methods Security issues. The key monitoring area this time is about 30 km^2. While implementing InSAR monitoring, leveling of the monitoring area was carried out, which can be compared and analyzed to verify the accuracy and reliability of the results.

2 Analysis and Selection of Interferometric Methods

According to the analysis of monitoring purposes and site conditions, the D-InSAR differential interferometry monitoring and analysis method is selected to carry out coherent target timing analysis, which is convenient for grasping the impact of long-term coal mining and determining the impact range and degree changes in recent years.

D-InSAR interferometric measurement method: Synthetic Aperture Radar Differential InSAR (D-InSAR) technology refers to the use of two interferometric images in the same area, one of which is the interferometric phase image before deformation, and the other is after deformation The acquired interferometric phase image is then processed by differential processing (removing the influence of the earth's curved surface and topographical undulations) to obtain the measurement

*Corresponding author: jc_yan@163.com

technology of surface deformation.

With the help of the collected high-precision DEM in the monitoring area, the 2-pass differential interference is selected this time. The principle of the two-pass method is as follows: It is assumed that there is a deformation in the SAR observation slope direction during the time period when the two SAR images are acquired. The interference phase ϕ_1 formed by the interference of two SAR images can be expressed as:

$$\phi_1 = -\frac{4\pi}{\lambda}[(R_1 + \Delta r) - R_2] \quad (1)$$

$$\phi_0 = -\frac{4\pi}{\lambda}[R_1 - R_2] \quad (2)$$

$$\Delta\phi = \phi_1 - \phi_0 = -\frac{4\pi}{\lambda}\Delta r \quad (3)$$

R_1 and R_2 are respectively the slant distance of the main and auxiliary images, λ is the wavelength, and ϕ is the phase simulated by the existing DEM according to the interference baseline and incident angle. The differential interference result is only related to the wavelength, not to the baseline

Coherent target timing analysis method: The project uses the coherent target timing analysis combined with Small Baseline Subset (SBAS) and Persistent Scatterer InSAR (PS-InSAR) technology to combine all SAR images covering the same area. Several sets are formed, and the coherent target points are calculated based on the small baseline set combination principle. The principle is: the SAR image interference baseline distance (including time baseline distance and spatial baseline distance) in the collection is small, and the baseline distance between subsets is large. Then after simple and effective merging, all available small baseline interferograms are obtained. The merging here is based on the minimum deformation rate standard, which is easily obtained by using the singular value decomposition (SVD) method. The SBAS-InSAR method limits the geometric decoherence caused by the long baseline, and makes more SAR images participate in the deformation calculation, increasing the time sampling.

Assuming that there are M (M>2) scene SAR images, according to the imaging time sequence, set the cumulative shape of the same resolution unit *(x, r)* at imaging time i (i=0, 1,..., *M*-1) The variable phase is $\phi(t_i, x, r)$. If the critical baseline condition is met, there are N different interference image pairs, which can generate N scene differential interferograms:

$$\frac{M}{2} \le N \le \frac{M(M-1)}{2} \quad (4)$$

3 Data selection and processing

3.1 Data selection and collection

The research work collected necessary InSAR data, DEM data and basic geographic information data for cartographic analysis.

3.1.1 Overview of the project research area

The research project area is located in the Kailuan mining area, and the southwest of the monitoring area is the mining area. The location of the project area is shown in Figure 1.

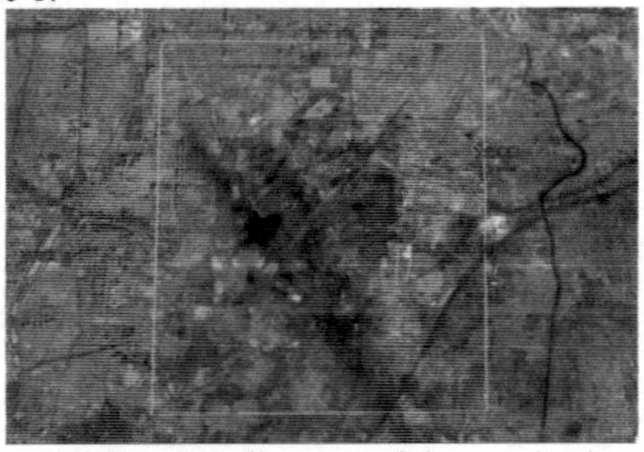

Figure 1 Schematic diagram of the monitoring area

3.1.2 Radar data

Sentinel-1 satellite IW mode radar data and RADARSAT-2 satellite XF mode radar data are used this time, of which RADARSAT-2 satellite XF mode radar data is the main data source. The Sentinel-1 satellite has a replay cycle of 12 days, C-band, and IW data in the interference wide mode is selected. The single-scene data coverage is 250km×180km. The data cycle is from April 2017 to November 2018, with 48 periods. The RADARSAT-2 satellite has a return period of 24 days, C-band, and XF data in scan mode. The single-scene data coverage is 125km×125km. The data period is from November 2016 to December 2019, with a total of 34 periods.

3.1.3 SRTM DEM data in the work area

The horizontal resolution is 90 meters (3 seconds), the elevation accuracy of the monitoring area is about 10 meters, and the terrain of the work area is flat, which meets the requirements of terrain compensation in InSAR processing

3.2 Data processing analysis

3.2.1 Radar data registration

The registration of SAR image is to accurately fit two or more images of the same target area together in space. Select one scene from the Sentinel-1 sequence radar data and the RADARSAT-2 sequence radar data as the main image (for example, RADARSAT-2 selects the radar imaging time of 20170515 as the main image), and register the remaining scene images as the secondary images. Correct the secondary image to the spatial position of the main image. Orthogonal correlation technology is used to achieve precise registration of radar data. The registration accuracy of Sentinel-1 data must be higher than 0.002 pixels, and the registration accuracy of RADARSAT-2 data must be higher than 0.2 pixels.

According to Sentinel-1/RADARSAT-2 main image registration DEM data, through DEM data information and radar imaging geometric information, remove the terrain phase in the subsequent interferometry. Apply the SRTM DEM data covering the test area to transform the coordinate system of the external DEM based on the imaging geometry of the SAR image, that is, geocoding, establish a conversion lookup table between the coordinate system between the SAR image and the external DEM, and simulate the SAR image with the DEM. The texture matching of SAR images has a matching accuracy higher than 0.2 pixels.

3.2.2 Selection of interference image pair and D-InSAR processing

According to the baseline distribution characteristics of the acquired SAR data sequence, combined with the basic characteristics of regional ground deformation, select the data with a spatial baseline less than 300m and a time interval less than 360 days to form an interferometric image pair sequence; interferometric image pairs that meet the baseline requirements are subjected to interference processing. Generate interferogram sequence; introduce precise orbital parameters for baseline estimation, remove flat ground based on geometric imaging model, use DEM sampled into radar coordinate system, simulate terrain phase through geometric imaging model, and obtain the phases including deformation phase, atmospheric phase and noise phase Differential interferogram: Screen the differential interferogram, and eliminate the interferogram with obvious influence of strong convection and the interferogram with obvious residual interference phase fringes.

3.2.3Timing analysis of coherent targets

Perform time series analysis on the interference phase of coherent point targets, estimate atmospheric fluctuations, DEM errors, and noise according to the temporal and spatial characteristics of each phase component, separate them from the differential interference phase one by one, and finally obtain the deformation rate of each coherent point target .

Selection of coherent point targets: Coherent targets are radar targets that have stable backscatter characteristics in time series and maintain a stable phase. These targets are generally point targets or continuous targets. The solution of the surface deformation variables is based on the phase changes of these targets. Generally, to obtain correct statistical results, more than 20 scenes of SAR data are required. The point target detection algorithm considers a point target to be a coherent target, and determines whether it is a relevant target according to the degree of correlation between the sub-views.

Time series differential phase analysis: use the differential interference phase obtained in the previous period to extract the interference phase of the coherent point target, and use the spatial phase unwrapping for the point target phase, construct the Delaunay triangulation network according to the position of these points, and determine the residual in the triangulation network For the handicap, the minimum cost flow algorithm is used to connect the positive and negative residual point pairs, and the unwrapping result of each point is obtained. For the differential interference phase sequence of the coherent point target, the two-dimensional periodogram is used for parameter inversion to estimate the movement rate of adjacent points and the elevation error correction. Through multiple error correction iterations, the correlation coefficient of the two-dimensional periodic function is 0.8-0.85 , accurately extract the linear deformation rate.

4 InSAR Monitoring Results and Analysis

Through two processing methods of conventional D-InSAR and coherent target timing analysis, the RADARSAT-2 radar data is used to obtain the average land subsidence rate of the Nanhu area from November 2016 to December 2019 and the cumulative land surface of the three years Subsidence, to obtain the distribution of land subsidence and the development trend of subsidence; use Sentinel-1 radar data to obtain the average deformation rate, cumulative subsidence and subsidence contour map of the Nanhu area from April 2017 to November 2018

4.1 Monitoring results and analysis of land subsidence in Nanhu area of Tangshan City

4.1.1 D-InSAR results based on RADARSAT-2 data

For the Nanhu area of Tangshan City, we used RADARSAT-2 data to perform multiple phases of interference processing from November 2016 to February 2019, and obtained 6 phases of interference patterns, as shown in Figure 4-1~6. In the figure, a color circle represents a 28 mm deformation.

https://doi.org/10.1051/e3sconf/202123701016

Figure 4-1 Deformation interference pattern of Nanhu area (20161128~20170304)

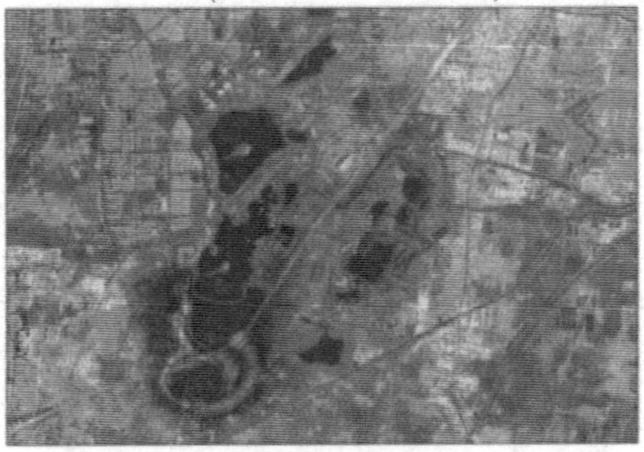

Figure 4-2 Deformation interference pattern of Nanhu area (20161222~20170115)

Figure 4-3 Deformation interference pattern of Nanhu area (20171123~20180227)

Figure 4-4 Deformation interference pattern of Nanhu area (20180203~20180227)

Figure 4-5 Deformation interference pattern of Nanhu area (20181212~20190222)

Figure 4-6 Deformation interference pattern of Nanhu area (20191020_20191207)

As can be seen from Figure 4-2, from December 22, 2016 to January 15, 2017, the ground subsidence rate around the mining center in the Yuexu District of Tangshan Mine was rapid, and the deformation center interference fringes exceeded 6 in only 24 days. Color week, both the settlement amount exceeds 160 mm. Other places are uniform in color, and the ground is stable.

It can be seen from Figure 4-4 and Figure 4-5 that in different time periods, with the change of mining activities, the settlement center, settlement range and settlement speed will change accordingly, reflecting the changes induced by coal mining during the monitoring period. Deformation development status.

4.1.2 Timing analysis results of coherent targets based on Sentinel-1 data

For the Nanhu area of Tangshan City, using the coherent target time series analysis method and using Sentinel-1 data to obtain the InSAR monitoring results of the settlement of the Nanhu area of Tangshan City (centered on the area affected by mining in Tangshan) from April 2017 to November 2018 The sequence has 5 periods, see Figure 4-7~11.

https://doi.org/10.1051/e3sconf/202123701016

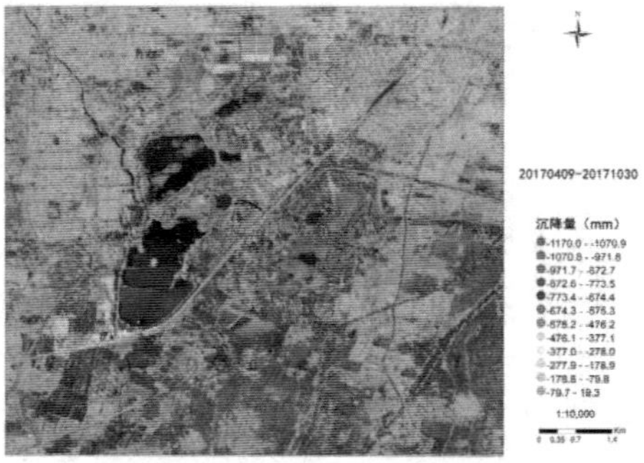

Figure 4-7 InSAR monitoring results of land subsidence in the study area (20170409~20171030)

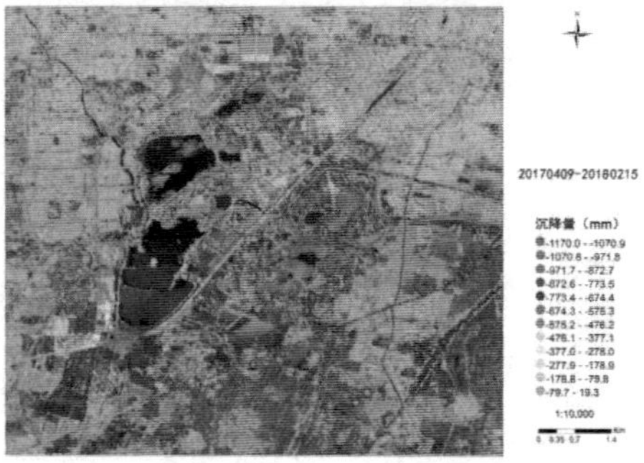

Figure 4-8 InSAR monitoring results of land subsidence in the study area (20170409~20180215)

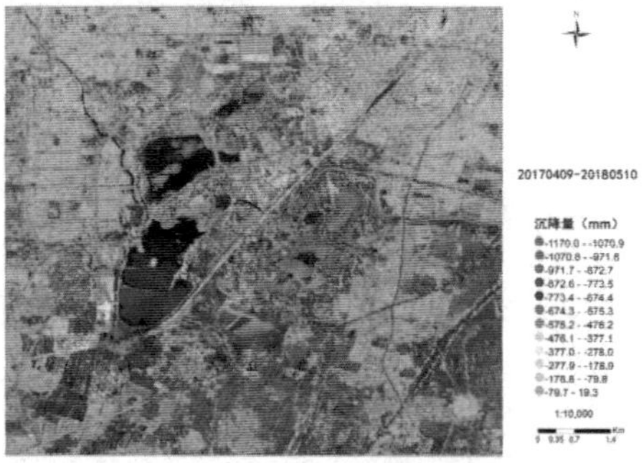

Figure 4-9 InSAR monitoring results of land subsidence in the study area (20170409-20180510)

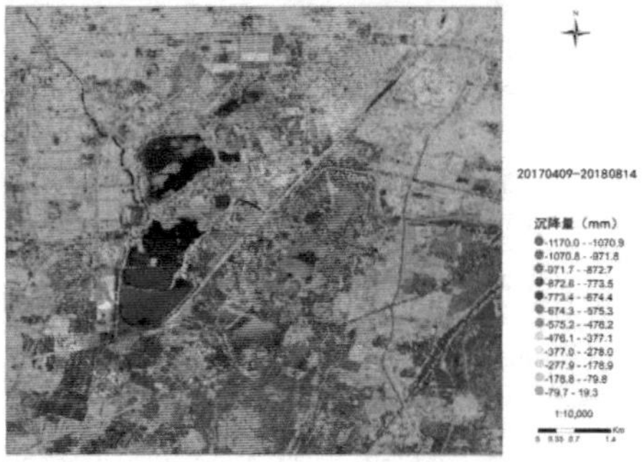

Figure 4-10 InSAR monitoring results of land subsidence in the study area (20170409~20180814)

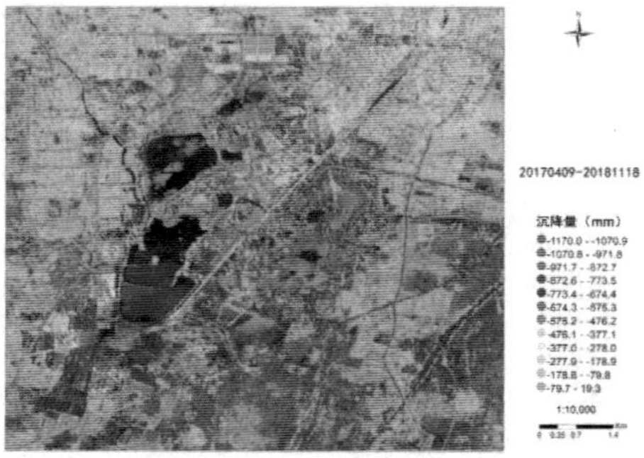

Figure 4-11 InSAR monitoring results of ground subsidence in the study area (20170409~20181118)

Figure 4-12 Contour map of the subsidence in the study area (20170409~20181118)

It can be clearly seen from the Sentinel-1 monitoring results that the changes in the settlement of the Nanhu area mainly occurred in the southwestern part of the Nanhu Lake, which was the main mining area of coal mines from 2016 to 2018. The monitoring results can also accurately reflect the intensity and scope of underground mining. During this monitoring period, the maximum settlement in this area reached more than 1100mm. Based on the Sentinel-1 settlement monitoring results, the settlement contour map of the Nanhu area from April 2017 to November 2018 was drawn, as shown in Figure 4-12. It can be clearly seen from the figure that the eastern part of the Nanhu area is generally stable and there is no significant land subsidence.

4.2 Time series analysis of coherent targets based on RADARSAT-2 in land subsidence in the eastern part of Nanhu Lake

Based on the 34 phases of RADARSAT-2 satellite XF mode radar data from November 2016 to December 2019, and using the coherent target timing analysis method, the ground subsidence rate map of the eastern region of the monitoring area from November 2016 to December 2019 was obtained , As shown in Figure 4-13. It can be seen from the figure that the impact of land subsidence basically disappears to the east side of Nanhu Lake, and the maximum subsidence rate in the figure reaches 28 mm/year, which is located in the southwest corner of Figure 4-13; the location of the monitoring point marked in the figure (RP Location) The surrounding area is outside the scope of influence of coal mining, and the area is generally stable without significant ground subsidence.

The time series of land subsidence at typical coherent target points (RP points) are analyzed. Figure 4-14 is a time series diagram of land subsidence at a typical target point. It can be concluded from the figure that there is no significant seasonal land subsidence at this point, showing the characteristics of ground stability.

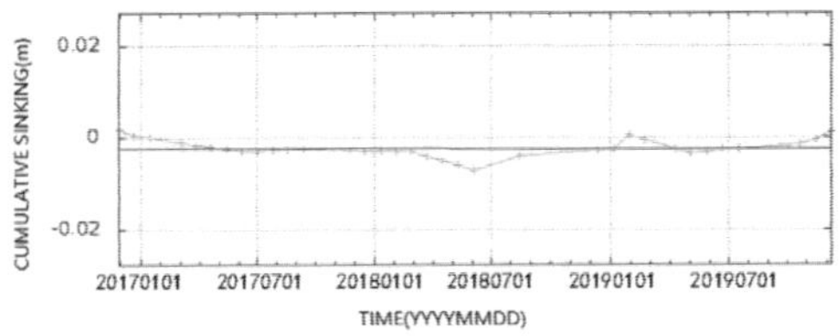

4-13 Time series diagram of land subsidence at typical coherent target points

At the same time, the cumulative subsidence map for the three years starting from November 2016 was obtained. The overall land subsidence development trend in the eastern part of Nanhu Lake is shown in Figure 4-15~17.

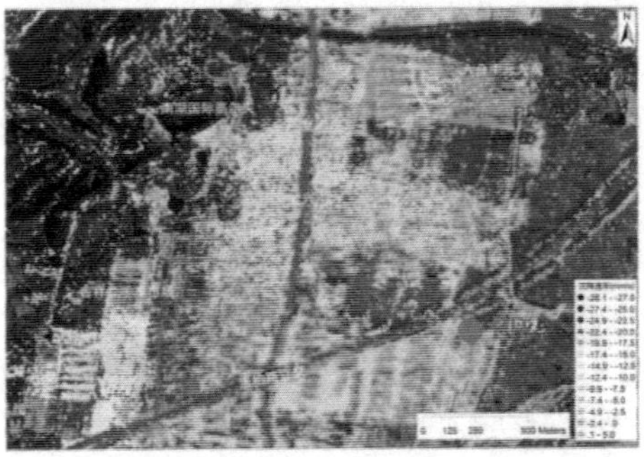

Figure 4-14 Map of land subsidence rate in the eastern part of Nanhu Lake (2016.11~2019.12)

Figure 4-15 Map of cumulative land subsidence in the eastern part of Nanhu Lake (2016.11~2017.11)

Figure 4-16 Map of cumulative land subsidence in the eastern part of Nanhu Lake (2016.11~2018.12)

Figure 4-17 Map of cumulative land subsidence in the eastern part of Nanhu Lake (2016.11~2019.12)

The time series of land subsidence at typical coherent target points and the cumulative land subsidence map show that the eastern area of Nanhu Lake is in a relatively stable condition, with no significant accumulation of land subsidence and no significant uneven settlement.

5 Conclusion

Based on the space-borne InSAR technology, the land subsidence monitoring of the Nanhu area from 2016 to 2019 was completed, and information such as the cumulative amount of land subsidence and the average annual subsidence rate in the monitoring area were extracted. Through the long time sequence analysis of the monitoring results for more than 3 years, the high accuracy and reliability of the monitoring results are verified, which can be used for large-scale regional monitoring of coal mining areas.

Through the analysis of the ground subsidence monitoring results, it is concluded that the changes mainly occur in the current mining affected area, and the eastern area of Nanhu Lake has become stable and has stabilized, and there is no significant uneven subsidence.

InSAR technology can be used for mining subsidence monitoring, especially when determining the boundary of subsidence and whether it is stable or not, it has its technological advantages.

For the current mining area of the mining area, Sentinel-1 data increases the time sampling rate and reduces the impact of time incoherence, which can enhance the description of the deformation process of the mining center of the mining area. For the slower ground subsidence, the RADARSAT-2 satellite can Describe the details of the deformation field more finely in space. For large deformation areas, that is, InSAR decoherence areas, it can also be supplemented with conventional measurements to improve the current deficiencies of InSAR technology.

Acknowledgement

This paper is supported by General Project of Science and Technology Innovation and Entrepreneurship Fund of Tiandi Science and Technology Co., Ltd (2018-TD-MS052),National Key R&D Program of China(2016YFC0501105) and Hebei Province"333 Talent Project " (A202001096).

Referneces

1. X.Yuan,W.Lyu,D.Yuanet al.Application of Numerical Fitting in Mine Subsidence Monitoring. MINE SURVEYING. **47**,6-9(2019).
2. J. Bai, C. Li. Application of Three -Dimensional Laser Scanning Technology in Mines.MINE SURVEYING. **47**, 116-119 (2019).
3. C.Guo, J. Li ,H. Feng,et al.The Application of 3D Laser Scanning in Dam Subsidence Monitoring in Mining Area.MINE SURVEYING. **42**, 70-72(2014).
4. E.Hou, J. Zhang, X. Xie, et al.Contrast Application of Unmanned Aerial Vehicle Remote Sensing and Satellite Remote Rensing Technology Relating to Ground Surface Cracks Recognition in Coal Mining Area.Geological Bulletin of China. 443-448, **38**(2019).
5. F.Tang, L.Li, X.Li, et al. Research on Characteristice of Minging-induced Surface Cracks Based on UAV images.Coal Science and Technology.**48**, 130-136(2020).
6. F.Tang, L.Dong, Z.Wang, et al.A 3-D inversion model for InSAR detected displacements based on ground subsidence symmetry induced by horizontal coal mining.Journal of China Coal Society.**44**,210-220(2019).
7. Y. Luan, Y. Liang, Z.Ji, et al.Monitoring and Analysis of Surface Subsidence Caused by SBAS-InSAR Technology. Coal Science and Technology.**14**,198-204 (2020).
8. T. Liang .Monitoring Surface Deformation in Mining Area Using Short Baseline Set InSAR Technology. Bull Surv Map.**S1**,82-84 (2014).
9. D. Li, Z. Deng, X. Gao,et al.Monitoring and Analysis of Surface Subsidence in Mining Area Based on SBAS-InSAR. Geomatics and Information Science of Wuhan University.**43**, 1531-1537(2018).
10. S. Han, B.Zhao, Y.Bai,el al. Mining Subsidence Research Based on Sbas-insar in Yaojie Caol Mine.MINE SURVEYING. **47**, 1-5(2019).

Review of research on high-temperature behavior of geopolymer paste

Yang Hang[1], Zhu Mei-chun[1], Fang Cong-qi[1,*]

[1]School of Civil Engineering, Shanghai Normal University, 201418 Shanghai, China

Abstract. Geopolymer composites have been widely researched during recent years as an alternative to sustainable construction materials, which can minimize CO_2 emission for its application of industry by-products. Past researches on geopolymer show that it has comparable strength and better high temperature stability compared to ordinary Portland cement. In this paper, the high-temperature behavior of geopolymer paste has been discussed through the last data regarding geopolymer, mainly including its bonding performance with steel, stress-strain characteristics, structural analysis of different observation scales and the performance of special geopolymer paste. In summary, some problems that need to be studied in future researches are put forward.

1 Introduction

The term 'geopolymer' was invented in 1970's by Prof. Joseph Davidovits and the French researchers, which is a new eco-friendly cementitious material, also termed as alkali-activated cement, and could be a superior alternative to ordinary Portland cement (OPC) (Shill 2019)[1]. Due to its simple production process, environmental character, and better engineering properties than OPC, in order to enable the civil engineering industry to get faster and better development, geopolymers have also been favored by the majority of researchers. Geopolymer composites is a new type of cementitious material that conforms to the development of the times. The commonly used geopolymers include kaolin based geopolymer, metakaolin based geopolymer and fly-ash based geopolymer.

Comparing to the OPC which degrades in strength irreversibly starting at 200 ℃ , geopolymer has been regarded as a potential fire-resisting material in civil engineering, which is especially important in high-rise buildings, underground buildings and large-span buildings. The research of geopolymer on the fire resistance is mainly to study the influence of the engineering characteristics of geopolymer when the temperature rises gradually, and the performance is whether the compressive strength changes, or whether there are obvious cracks and so on. About the research on geopolymer is mainly divided into three aspects of the characteristics of geopolymer paste, geopolymer mortar and geopolymer concrete. This paper presents the experiments and conclusions on the fire resistance performance of geopolymer paste in recent years.

2 Bonding behavior during and after exposure to elevated temperatures

The bonding behavior of steel to fly ash based geopolymer paste at fire condition was evaluated, and an extensive experimental study was carried out to investigate the effect of high temperature on the thermal-physical behaviors and mechanical properties of fly ash based geopolymer paste by Jiang (Jiang 2020)[2] According to the visual observation of OPC and GPC specimens subjected to high temperatures, there was a color change in GPC when the temperature increased to 800 C. There were fewer visible cracks and no spalling phenomenon in GPC specimens at very high temperatures (800-1200℃) compared with OPC, indicating that the GPC has better fire-resistance. While OPC completely disintegrated at 800C, the GPC still retained some residual bond strength even after exposure to 1200 ℃ , which indicated that GPC paste had an excellent bond performance at extremely severe temperatures. A positive linear relationship between the residual bond strength (RBS) and residual compressive strength (RCS) indicated that higher compressive strength could result in better bond performance both for GPC and OPC specimens, as shown in Fig1.The XRD results showed that the phases in GPC were much more stable than those in OPC at high temperatures. Furthermore, SEM micrographs demonstrated that the porous system in GPC provided escape routes for water. This characteristic led to better mechanical properties of GPC than that of OPC at elevated temperatures.

*Corresponding author: congqifang@msn.com

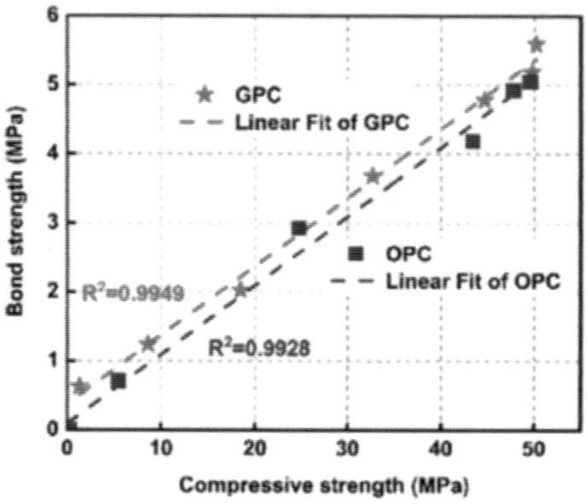

Fig1. Relationship between RBC and RCS(Jiang 2020).

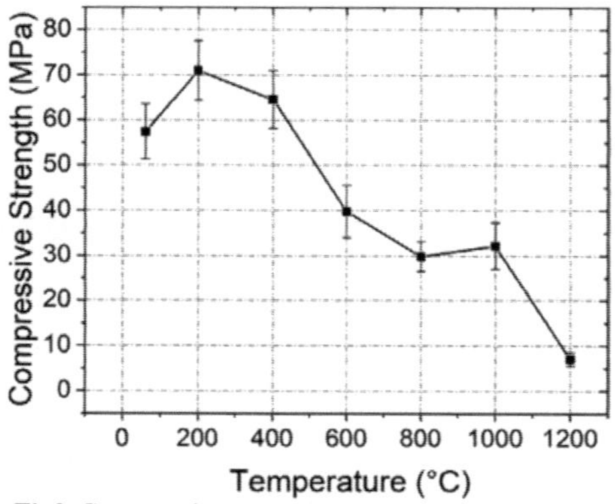

Fig2. Compressive strength of geopolymer after heat treatment(Panjasila 2020).

3 Specific geopolymer paste

Jiang (Jiang 2020)[3] researched the properties of fly ash-based gepolymer paste containing waste glass powder (WPG) at high temperature The addition of WPG excited reactive silicon and alumina, changed the Si/Al ratio of geopolymers, and thus changed the mechanical properties of the materials, which is similar to (Peem 2020)[4] and (Khan 2020)[5]. It was found that the GPC paste had the best optimum compressive and bonding strengths when the content of WPG was 20%. After exposure to high temperatures (up to 1200℃), the unreacted WGP melted to fill the porous microstructures of the geopolymer matrix, which mitigated the increase of pores, and improved the fire resistance. It was found that Si: Al = 2.5 effectively improved the high temperature resistance of GPC materials, but it was found that Si: Al = 4.17 was better (Peem 2020)[4], the difference of the best Si/Al developed by the two researches may be explained that the latter used a high-calcium fly ash-based geopolymer. For this, it can be verified by the effect of different calcium content on the high temperature resistance of geopolymer.

The changes of compressive strength, microstructure and magnetic properties of high-calcium fly ash based geopolymer at high temperature were studied(Panjasila 2020)[6]. The residual strength of the calcined geopolymer was shown in Fig2. The results showed that the compressive strength was increased at 200℃, and started to adversely affect the properties of specimens at temperatures around 400℃. But at 1000℃, because of the formation of new iron oxide phase, the crystallinity of the paste was increased, the rate of strength decline was actually decreasing. Aziz (Aziz 2019)[7] also suggested that the rearrangement of some amorphous phases to form temperature-stable crystalline phases contributed to the strength of geopolymers. Panjasila also believed that the compressive strength may be related to the magnetic properties of the specimens and that the measurement of magnetic properties can be used as a non-destructive method for structural health monitoring. As for this hypothesis, there should be more experiments to test and refine it in future studies, because it is leading the development of this industry, it is also necessary and innovative to solve practical engineering problems.

4 Stress-strain characteristics at high-temperature

Choi (Choi 2020)[8] used surface fractal dimension analysis method to study the change of mechanics and microstructure of fly ash based geopolymer paste at high temperature, the different dosages of Na_2O geopolymer pastes were exposed to different temperatures by the control variable method. The results showed that the fractal dimension increased with the increase of temperature and Na_2O content, and the compressive strength was proportional to the surface fractal dimension. It showed that the content of Na_2O and the temperature were proportional to the compressive strength in a certain range, as was shown in Fig3. As for the effect of temperature, it can be seen from (Panjasila 2020)[6], (Aziz 2019)[7] and (Zhang 2020)[11] that the intensity will increase in a certain temperature range. Through the fractal geometry method used in this experiment contrast with other methods, the results are much more convincing. It can be supposed that the content of Na_2O on the impact of compressive strength should also exist a maximum, the related tests are worth examining. In connection with the experiments of (Jiang 2020)[3] (Peem 2020)[4], it is conceivable to design an experiment in which Na_2O and reactive SiO_2 and other related chemicals were taken as independent variables, the strength of fly ash-based GPC paste and GPC concrete at high temperature was tested under the best combination.

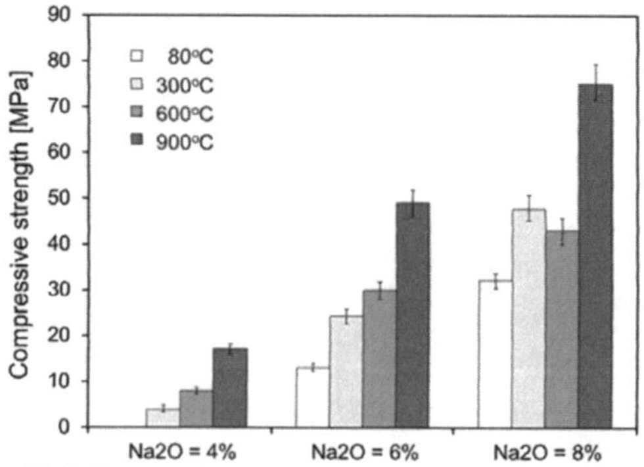

Fig3. Compressive strengths of the exposed specimens according to the Na2O content (Choi 2020).

 https://doi.org/10.1051/e3sconf/202123701017

Performance change of GGBFS based geopolymer at high temperatures was carried out (Aziz 2019)[7]. It was found that the unexposed geopolymers had a compressive strength of 95 MPa. The compressive strength decreased to 72 MPa at 800℃ but increased to 152 MPa at 1000℃. As can be seen in Fig4 (red-coloured region represents cavern), the crack of geopolymer treated at 1000℃ was reduced and partially melted, which meaned the crack was healed at high temperature then the compressive strength was improved. Moosavi (Moosavi 2019)[9] also believed that the presence of liquid phase in the composite was an important factor to affect the microstructure and mechanical properties of the composite. At 1000 ℃, gehlenite (Ca_2Al_2SiO), mayenite ($Ca_{12}Al_{14}O_{33}$) and Larnite (Ca_2SiO_4) were also formed in geopolymer paste, and their structures were relatively ordered. It was supposed that the strength increase was mainly due to the rearrangement of amorphous phase to form temperature-stable crystalline phase.

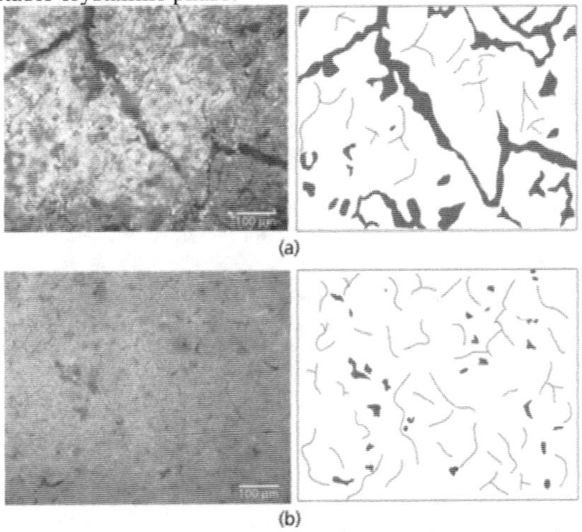

Fig4. Optical microscopy images and schematic representation of crack appearance of GGBFS exposed at (a) 800°C and (b) 1000°C(Aziz 2019).

5 Mixed design and control factors for high temperature applications

The fire resistance of geopolymer was classified into three scales: micro-scale (including crack formation, pore structure changes, densification, sintering, melting), meso-scale (phase composition change: Crystal growth or failure, geopolymer paste transformation) and macro-scale (geopolymer paste thermal deformation) by Lahoti (Lahoti 2019)[10]. Interrelation between micro-, meso-, andmacro-scale properties are discussed related to the performance of geopolymer at high temperature exposure.

Researches on the performance of geopolymers at micro-scale show that geopolymers are chemically stable and do not undergo breakdown of chemical structure in contrary to OPC hydration products which undergo severe deterioration on thermal exposure. The factors such as Si/Al ratio, alkali cation type and alkali content are important to determine the chemical structure of geopolymers before and after exposure to elevated temperatures.

High-temperature exposure causes thermal shrinkage and expansion as well in geopolymers, which in turn causes macro-cracking. It is importance to control the water content in geopolymer mix to regulate the meso-scale thermal deformations. Due to the fact that metakaolin-based geopolymers require a significantly higher amount of water as compared to fly ash-based geopolymers for workability purpose, fly ash-based geopolymers might be preferred for structural fire-resistance applications Thermal deformations are also influenced by the composition of alkali activating solution as well as the type of alkali cation. Using potassium as alkali cation, thermal shrinkage decreases when compared to using sodium as alkali cation.

Compressive strength of geopolymers can be affected by microstructural changes and phase composition changes,so factors affecting the compressive strength of geopolymers on exposure to elevated temperatures may include the choice of precursor, use of aggregates, total alkali content in geopolymer It should be noticed that not all geopolymer formulations perform soundly at ambient as well as on high-temperature exposure, tailoring of mix design to produce geopolymers with good thermal stability is essential. With careful mix design, remarkable chemical stability, volume changes, strength endurance, and spalling resistance can be achieved for geopolymers. Mix design factors which need to be controlled include Si/Al ratio, Al/Na ratio, and water/solids ratio other than the factors such as precursor type, alkali cation type, and curing condition. The performance of geopolymer paste as well as the selection of compatible aggregates, fibers and/or fillers, are important considerations for high temperature applications.

In short, to play the best fire resistance of geopolymer paste, it is necessary to strictly control the mixed design and control factors, such as the choice of precursor, use of aggregates, total alkali content in geopolymer and water content, to achieve ideal chemical stability, volume stability, strength endurance and spalling resistance.

6 Conclusions and prospect

From the finding of the recent studies that investigated the high-temperature behavior of geopolymer, the author's views can be summarized in Table1. Furthermore, some problems that need to be studied in future researches are presented as follows:

• More experiments need to be carried out to research the high-temperature bond behaviour of different types of bars (for example Prestressed steel wire and parallel steel wire bundles) and geopolymer, which can provide fundamental information for use geopolymer in prestressed structures and long span cable structures.

• It is necessary to find an optimal mix proportion to realize the best fire resistance of geopolymer. Experiments can be carried out in which Na_2O and reactive SiO_2 and other related chemicals are taken as independent variables, the strength of fly ash-based GPC paste and GPC concrete at high temperature is tested under the best combination. These experiments may have a great breakthrough in solving some practical engineering problems.

• According to the construction and use of safety requirements, geopolymer components now need to meet certain high-temperature duration requirements. Under this requirement, the geopolymer component can be subjected to a long period of high temperature resistance test according to the international temperature-rise curve and the relatively practical temperature-rise curve, and is verified by the finite element analysis software, it is necessary to simulate the performance change of the component in a reasonable and real heating process.

Acknowledgements

This work was financially supported by foundations from Science and Technology Commission of Shanghai Municipality (18070502900; 18ZR1427900). The financial support is appreciated.

Table1. Summary of the literature

Author	Date	Conclusion and opinion
Jiang	2020	Compared with the OPC, the GPC has less visible cracks and no spalling phenomenon, less mass loss and good bonding property at 800,100,1200℃.
Jiang	2020	It is found that the addition of waste glass powder (WPG) excites more active silicon and alumina, thus changes the Si/Al ratio of Geopolymer, and then changes the mechanical properties of the material at high temperature.
Panjasila	2020	The changes of compressive strength, microstructure and magnetic properties of high-calcium fly ash based geopolymer at high temperature were studied. When the geopolymer was heated to 600℃, the strength began to decrease significantly, and the magnetic properties changed from ferromagnetic to paramagnetic at 1200℃,so he hypothesized that compressive strength might be related to the magnetic properties of high calcium fly ash based geopolymer.
Choi	2020	The changes of mechanical and microstructure characteristics of fly ash based geopolymer materials at high temperature were studied by using fractal geometry method. With the increase of temperature and Na_2O content, the fractal dimension of the surface of macropores increased, the compressive strength is proportional to the surface fractal dimension.
Aziz	2019	The relative reduction of cracks and partial melting of crushed slag base polymer treated at 1000 °C means the crack healed and the strength increased at high temperature, the strength increase is mainly due to the rearrangement of amorphous phase to form temperature-stable crystalline phase.
Lahoti	2019	The fire resistance of geopolymers was studied on micro scale, meso scale and macro scale.

References

1. Shill S K, Al-Deen S, Ashraf M, et al. Fly A GPC Tem.J. Con Bui Mat, **239**,117886(2019)
2. Jiang X, Xiao R, Zhang M M, et al. Lab Fly A GPC Tem.J. Con Bui Mat, **254**,119267(2020)
3. Jiang X, Xiao R, Ma Y T, et al.WGP Fly A GPC Tem.J. Con Bui Mat, **262**,120579(2020)
4. Peem N, Pitcha J, et al. R H Fly A GPCC F.J. Cleaner Production , **252**, 119797(2020)
5. Khan M N N, Sarker P K. H Tem GGBFS.J. Con Bui Mat, **263**,120177(2020)
6. Panjasila P, Nattapong C, Rattakarn Y, et al. H-Ca Fly A GPC Tem.J. Con Bui Mat, **265**,120650(2020)
7. Aziz I H, Abdullah M, Heah C Y, Liew Y M,et al.BFS GPC Tem.J. Cement Research, **18**,162(2019)
8. Choi Y C, Park B.H-Tem Fly A GPC.J.Mat Tec, **4**,7655 (2020)
9. Moosavi A, Asadi S, Shoraki H J. Mic Mec GPC Tem .J. Ceramics Inter, **1**,246(2019)
10. Lahoti M, Tan K H, Yang E H. GPC F-resistance.J. Con Bui Mat, **221,**514 (2019)
11. Zhang H G, Lang L, Wang Q G, et al. Heat-Cured Fly A GPCC H- Tem.J. Con Bui Mat, **262**,120924(2020)

Spatio-temporal variation analysis of snow cover in China-Pakistan Economic Corridor based on remote sensing technology

Fan ZHANG[1,2,3], ZhiHua ZHANG[1,2,3*]

[1]Faculty of Geomatics, Lanzhou Jiaotong University; 730070 Lanzhou, China
[2]National-Local Joint Engineering Research Center of Technologies and Applications for National Geographic State Monitoring; 730070 Lanzhou, China
[3]Gansu Provincial Engineering Laboratory for National Geographic State Monitoring; 730070 Lanzhou, China

Abstract. Based on the MODIS10A2 snow product data from 2001 to 2019, the characteristics of annual variation, interannual variation and spatial distribution of snow cover in China Pakistan Economic Corridor from 2001 to 2019 are analyzed by using remote sensing technology. The result shows that: the snow-cover in a year generally starts from the middle of October, and the snow cover area reaches the maximum value in January of the next year, and reaches the minimum value in July. From 2001 to 2019, the snow area of China-Pakistan Economic Corridor generally showed a decreasing trend. The distribution of snow in the China-Pakistan Economic Corridor is extremely uneven. The northern area is obviously more than that in the south. The mountainous and plateau areas are with high frequency of snow cover, and the plains are areas with low frequency of snow-cover. Permanent snow-cover is relatively low. Few, mainly concentrated in the Karakoram Mountains, in terms of distribution range, mainly distributed in 4452 ~ 8378 m.

1 Introduction

The snow, as one of the natural factors of the earth's surface is active, and sensitive to climate change, it reflected the weak by the strong absorption characteristics of important influence on water cycle of the earth's radiation balance, is the indicator of global climate change, 98% of the world's seasonal snow is located in the northern hemisphere [1-3] China-Pakistan economic corridor area belongs to arid half arid areas, mountain snow melt water is one of the important ways of water supply area rivers.

At present, there have been studies on the temporal and spatial variation of snow-cover in East Asia [4], Central Asia [5], the northern hemisphere [6] and other regions, but there are few studies on the temporal and spatial variation of snow-cover in the China-Pakistan Economic Corridor. Tang Z G [7] use MODIS10A2 snow products such as spatial and temporal distribution of Tianshan snow in 2001-2015 was studied, the results showed that 2001-2015, Tianshan snow-area is a downward trend overall. Hen W Q [8], such as combination of MODIS data and the DEM elevation data, analyzes the different heights with central Asian snow coverage rate and snow days, the result shows that the snow coverage in different altitude difference is very big, but permanent snow season change of time and space difference is not obvious.

China-Pakistan economic corridor as the region's flagship project all the way, as the use of Pakistan's location advantage, established by the cooperation and mutual benefit, in Kashgar, eventually to Gwadar port trade corridors, on the one hand, it can promote the Pakistani region along the corridor and the economic development of Xinjiang province of China, on the other hand, it has a positive role in improving our homeland [11] and openness with expanded in china-Pakistan economic corridor construction in China in recent years, the snow became one of the main restricting factors, especially the construction of the highway railway, therefore, accurate monitoring of China-Pakistan economic corridor, snow space-time change is of great significance.

2 Study area

China-Pakistan economic corridor (Fig. 1) throughout the pamirs and adjacent areas, along the northern mountainous area the highest elevation of 8553 m, the maximum elevation of 3848 m, is the world's highest areas which is affected by the Indian monsoon and westerly winds, China-Pakistan economic corridor to the warm temperate zone continental arid climate in China, Pakistan for more than three-quarters of the study area is subtropical desert grassland climate rainfall less than 200 mm [12-14].

The data in this paper are MODIS 10a2 snow product data set, which is composed of MODIS / Terra snow cover for 8 days provided by the National Snow and Ice Data Center of the United States. Different pixel values of each image represent different ground features, 25 for no snow, 38 for lakes, 50 for clouds, 200 for snow [15-16]. Using MRT software to splice modis10a2 products, the

* ZhiHua ZHANG: 43447077@qq.com

coordinate system is transformed into WGS84 coordinates, and the projection is transformed into Albert coordinates. The elevation data is the digital elevation model of strm, and the spatial resolution is 90 M. after resampling, it has the same spatial resolution as MODIS data (500 m).

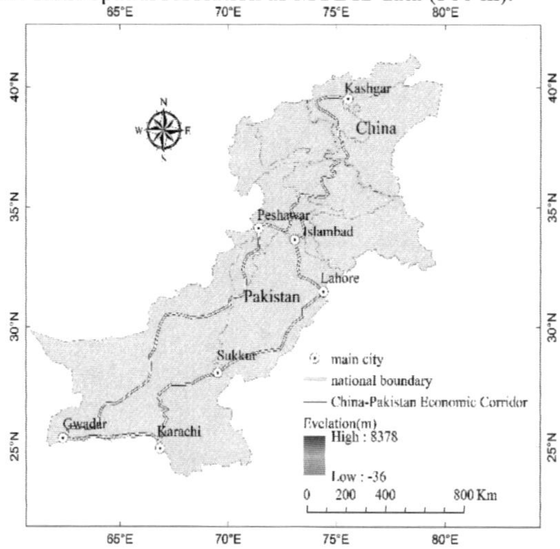

Fig. 1. Study area

3 Temporal distribution characteristics of snow-cover in China-Pakistan Economic Corridor

3.1 Interannual variation of snow-cover

The maximum area image of snow-cover in a year is obtained by superposition operation of 46 periods of data in a year. The judgment rule of snow pixel is: if there is snow in the image of any 8 days in a year, the pixel is determined as the annual snow pixel. Through the analysis of the annual snow-cover of the China-Pakistan Economic Corridor from 2001 to 2019 (Fig. 2), it can be seen that the snow-cover rate of the China-Pakistan Economic Corridor shows a decreasing trend from 2001 to 2019. From 2001 to 2005, the snow-cover rate increased obviously. The snow-cover rate fluctuated greatly from 2005 to 2009. In 2007, the minimum snow- cover rate was 9.44% in the past 19 years. However, in 2009, the snow-cover rate reached the maximum value of 11.59%, which was the maximum value of 19 years.

Compared with 2001, the annual snow-cover rate in 2009 increased significantly, reaching 1.3%. By comparing and analyzing the annual snow-cover maps of 2001 and 2009, we can see that the change of snow cover area in the north of China Pakistan Economic Corridor is small, while the increase of snow-cover area mainly occurs in the southern mountain area.

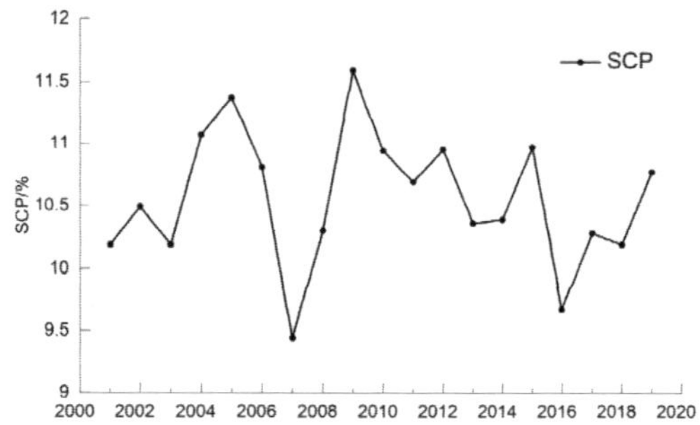

Fig. 2. Interannual snow-cover percent

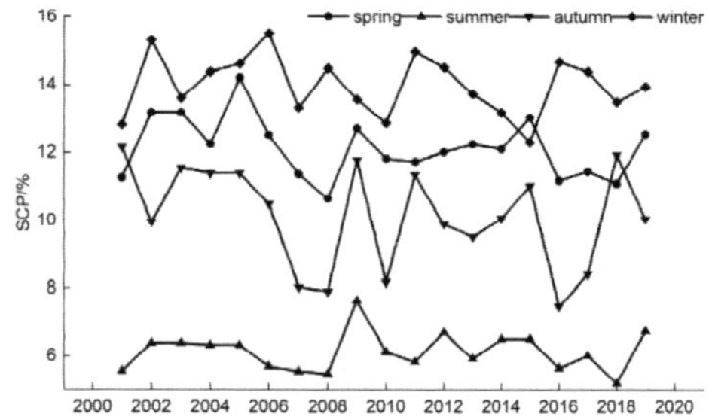

Fig. 3. Seasonal snow-cover percent

3.2 Seasonal variation of snow-cover

The snow-cover of China-Pakistan Economic Corridor has obvious seasonal variation. The maximum area of snow-cover in each season is selected as the snow area of that season. The seasons are divided into March, April and may as spring, June, July and August as summer, September, October and November as autumn, December, January and February of the next year as winter. The interannual variation trend of the four periods was obtained (Fig. 3). According to the analysis within the year, the difference of snow-cover rate is obvious in different seasons, summer is far lower than spring, autumn and winter, the highest snow-cover rate is in winter in most years, but the highest snow-cover rate is in spring in 2009. According to the interannual analysis, the snow-cover rate in spring and winter is more than 10%, and the interannual oscillation amplitude of the change of snow-cover rate in autumn is large, and the snow-cover in summer is relatively stable and the change trend is not obvious because it is permanent snow at high altitude.

3.3 Monthly mean change of snow-cover

In this paper, the monthly snow image in 19 years is calculated by grid, and the operation rule is: if a pixel shows snow in any scene image in the current month, it is determined as the monthly snow pixel. The snow cover image of China Pakistan Economic Corridor is as shown in figure(fig. 4). For a year, the month with the largest snow-cover rate was January, reaching 14.64%; the month with the smallest snow-cover rate was July, only 5.37%. Winter is the month with the largest snow coverage rate, with snow coverage rate exceeding 13.5%, followed by March and November, with snow coverage rate of 13.18%

https://doi.org/10.1051/e3sconf/202123701018

and 12.41%. The snow-cover rate in other months is less than 12%. For interannual, the change of snow-cover rate is obvious in the same month, especially in the month with high snow-cover, the interannual fluctuation is greater, and the main interannual fluctuation months are January February and October December.

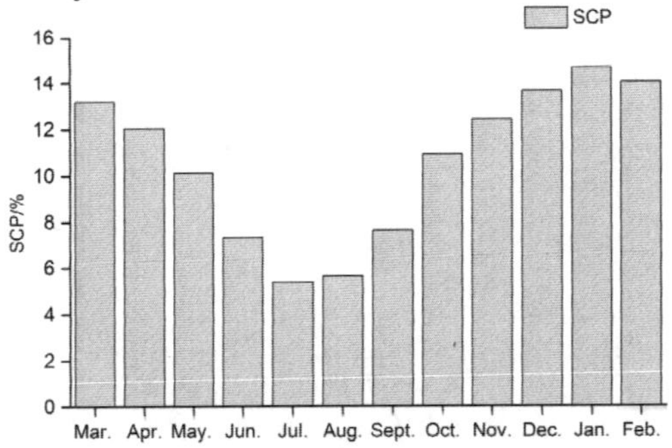

Fig. 4. Monthly snow-cover percent

4 Spatial distribution characteristics of snow-cover in China-Pakistan Economic Corridor

4.1 Altitude distribution of snow-cover

The altitude span of the China-Pakistan Economic Corridor is large, and the environment of each altitude is obviously different.Therefore,the natural breakpoint classification method is used to re classify the altitude based on the digital elevation data. This method uses statistical formula to determine the natural clustering of attribute values. Its function is to reduce the differences at the same level and increase the differences of different levels, so as to maximize the differences [17-19]. According to the principle of the natural breakpoint method, some continuous but not concentrated data are classified into grades. These grades will set boundaries at the positions where the data values are relatively different, so that the variance within each level is the minimum, and the variance between adjacent levels is the largest.The results relatively eliminate the subjectivity of artificial differentiation. The calculation formula is as follows:

$$ssd_{i-j} = \sum_{k=1}^{j}(a[k] - M_{i-j})^2 \qquad (1 \le i \le j \le n) \qquad (1)$$

Where, ssd is the boundary value of risk to different degrees; a is an array (array length is k); M is the average value in each level.

According to the above rules, the annual snow cover ratios of five sub zones from 2001 to 2019 are obtained (Table 1).

Table 1. Zonation based on DEM

Zonation	Elevation(m)	Area ratio (%)
1	-36 ~ 707	38.81%
2	707 ~ 1772	23.53%
3	1772 ~ 3106	11.32%
4	3106 ~ 4452	13.85%
5	4452 ~ 8378	12.50%

4.2 Spatial distribution of permanent snow-cover

There is permanent snow-cover in the study period of China Pakistan Economic Corridor. The image data of June, July and August in summer in the study area were selected, with a total of 12 periods per year and 228 periods in 19 years. Considering the above two factors, several periods of images with snow cover rate less than 5% are selected, and then the data with cloud pixel number less than 1 × 10^4 are selected. The intersection of snow pixels is taken to extract the permanent snow, and the spatial distribution map of permanent snow is obtained (Fig. 5).

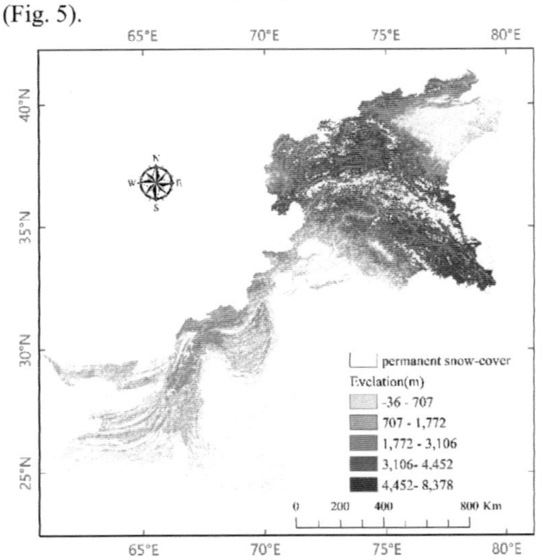

Fig. 5. Permanent snow-cover

The figure shows the spatial distribution of permanent snow-cover in the China Pakistan Economic Corridor. It can be seen from the figure that the permanent snow-cover is less distributed in the southern mountains and plains, with a total area of only 150.32 km^2, while it is widely distributed in the Himalayas, Hindukush mountains and Karakoram mountains, with an area of 33822 km^2.

4.3 Spatial distribution of snow-cover frequency

The number of times a pixel is covered by snow is the snow-cover frequency of that pixel. A total of 874 images from 19 years in the study area were analyzed and processed, and the annual snow-cover frequency map of China-Pakistan Economic Corridor from 2001 to 2019 (Fig.6) was synthesized by grid operation. It can be seen from the figure that the high value areas of snow- cover frequency are mainly Pamir Plateau, Karakoram mountains, Himalayas and Xingjiang mountains from north to south Hindukush mountains, the snow frequency in these areas is more than 450 times; the sub high value areas are mainly distributed near the Pamir Plateau, and the snow frequency in the plain areas of the Indus River Basin is low, ranging from 0 to 150 times.

According to the snow-cover ratio of short-term snow-

 https://doi.org/10.1051/e3sconf/202123701018

cover and long-term snow-cover [23-25], the snow-cover frequency is divided into two categories: short-term snow-cover and long-term snow cover. It can be seen from the table that there are great differences in the snow cover of each sub zone. Zone1 is mainly composed of no snow-cover and short-term snow-cover, zone2 is more than zone1 without snow-cover and short-term snow-cover, and the ratio of medium and long-term snow-cover area of sub zone3, 4 and 5 is the largest.

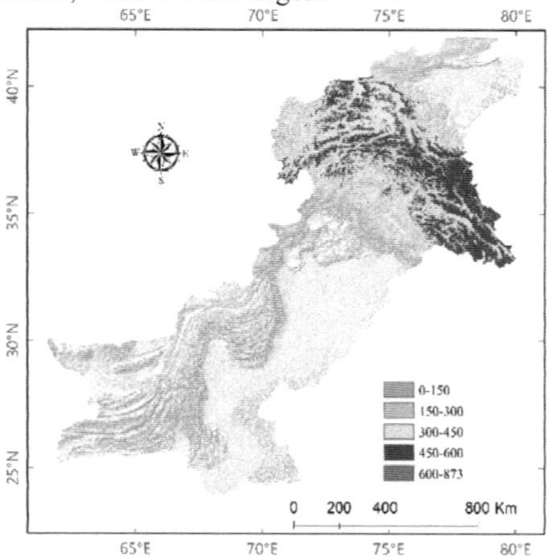

Fig. 6. Snow-cover frequency

5 Results and discussion

Based on the analysis of the spatial-temporal situation of snow cover in China Pakistan Economic Corridor from 2001 to 2019, the following conclusions can be drawn:

(1) The snow-cover rate of China-Pakistan Economic Corridor showed a slow decrease trend in the whole 19 years. The snow-cover rate showed a slight upward trend from 2001 to 2005, and fluctuated greatly from 2005 to 2009. In 2007, the snow cover rate was the lowest value in 19 years, but in 2009, the snow-cover rate increased sharply and reached the highest value in 19 years. The snow-cover rate will decrease slowly from 2009 to 2019. The change of snow-cover rate in the northern part of the China -Pakistan Economic Corridor is small, while that in the south is obvious, mainly distributed in mountains and plains.

(2) During the 19 years, the snow accumulated in October every year and reached the maximum value of snow-cover rate in January of the next year. The snow began to melt from March to May and reached the minimum value in July. Compared with the proportion of snow-cover in four seasons, the maximum snow-cover rate is 15.5% in winter and 5.2% in summer.

(3) The snow-cover in summer from 2001 to 2019 is mainly permanent snow-cover, mainly concentrated in the range of 4452-8378 m, and the change trend is not obvious. The snow-cover rate in autumn showed a decreasing trend. The frequency of snow-cover in mountain area was higher (450-873), while that in plain and watershed was lower (0-150).

Table 2. Snow frequency area ratio table of each dividing zone

	Zone1	Zone2	Zone3	Zone4	Zone5
None	18.426	14.524	8.576	4.352	1.024
Short-term	88.182	54.356	38.452	22.124	11.389
Medium-term	6.826	10.547	30.455	45.245	14.214
Medium and long-term	0.000	1.128	12.154	32.741	38.147
Long-term	0.000	0.000	0.000	13.356	44.354

Acknowledgments

The work was supported by the National Key R&D Program of China (2017YFB0504201, 2017YFB05042 03), the National Natural Science Foundation of China (Project No. 41861059, 41761082, 61862039), and partially supported by LZJTU EP 201806. The authors are grateful to the editors and reviewers more enough for their self-giving works.

References

1. R.L. Armstrong,M.J. Brodzik, Recent northern hemisphere snow extent: A comparison of data derived from visible and microwave satellite sensors 28(19):3673-3676(2001).
2. S.S Mashtayeva, Dai L, Che T, et al, Spatial and Temporal Variability of Snow Depth Derived from Passive Microwave Remote Sensing Data in Kazakhstan. JMR,30(06):1033-1043(2016).
3. Liu J F,Chen R S,Song Y X, Distribution and Variation of Snow Cover in China. Climate Change Research, 8(05):55-62(2012).(in Chinese)
4. Wang Y F,Sun X G,Yang X Q, Role of snow depth in the influence of EL Nino on the summer climate anomalies over East Asia. Chinese Journal of Geophysics. 60(09):3325-3337(2017).(in Chinese)
5. Brown R. D., Robinson D. A., Northern Hemisphere spring snow cover variability and change over 1922–2010 including an assessment of uncertainty. The Cryosphere. 5(12):219-229. (2011)
6. Zhang N L,Fan X T,Zhu J J, Spatial and temporal distribution characteristics of snow cover in northern hemisphere based on MODIS Snow products. Remote Sensing Information, 27(06):28-34(2012).(in Chinese)
7. Tang Z G,Wang J,Wang X, et al, Remote Sensing Study on temporal and spatial variation of snow cover in Tianshan area in recent 15 years. Remote Sensing Technology and Application. 32(03):556-563(2017).(in Chinese)
8. Zhong Z T,Li X,Xu X C, et al, Remote sensing analysis of spatial and temporal variation of snow cover in arid region of Central Asia. Chinese Science Bulletin, 63(25):2641-2654(2018).(in Chinese)
9. Wang Y Z,Li H,Liang P, "One belt, one road" construction and China's economic development

https://doi.org/10.1051/e3sconf/202123701018

strategy: a case study of the Sino Pakistani Economic Corridor. Qinghai Social Sciences ,2020(03):83-88(2020).(in Chinese)

10. Zhao Y X, Yao L K,Wei Y X, et al, Characteristics of platforms along the river of China-Pakistan Economic Corridor and key points of railway route selection. Railway Engineering, 60(07):102-106(2020).(in Chinese)
11. Yin X,Hu X, Achievements, challenges and Countermeasures of infrastructure interconnection project of China-Pakistan Economic Corrido. South Asian Studies Quarterly,2019(03):32-41+5(2019).(in Chinese)
12. Zhang J,Han T,Wang J, Characteristics of platforms along the river of China-Pakistan Economic Corridor and key points of railway route selection. Journal of Glaciology and Geocryology, 2005(05):649-654(2005).(in Chinese)
13. Zhou B,Liu B,Zhan L, Study on monitoring of important geographical elements in China-Pakistan Economic Corridor. Geomatics and Spatial Information Technology, 43(06):160-164+171(2020).(in Chinese)
14. Yao C,Wang X,Zhao X R, et al, Temporal and spatial variation characteristics of ice lake in China- Pakistan Economic Corridor from 1990 to 2018. Journal of Glaciology and Geocryology ,42(01):33-42(2020).(in Chinese)
15. Lei X C,Song K S,Wang Z M,et al, Accuracy analysis of MODIS and AMSR-E snow cover products in Heilongjiang Basin. Journal of University of Chinese Academy of Sciences, 28(01):43-50(2011).(in Chinese)
16. Wang L F,Zhang S C,Zhang C L, Application of ground based GPS in snow depth inversion in Altay. Desert and Oasis Meteorology,13(01):93-98(2019).(in Chinese)
17. Chen X N,Bao A M,Liu P, Accuracy evaluation of mod10a1 classification in Tianshan Mountains based on multi-scale statistical samples. Remote Sensing for Land & Resources, (03):80-85(2010).(in Chinese)
18. Zhou S Y,Xiao P F,Feng X Z, et al, Snow recognition in SAR image based on Markov random field model.Journal of Nanjing University(Natural Science),51(05):976-986(2015).(in Chinese)
19. Yuan L,Li C E,Chu S L, Influence of cold season deepening on AMSR-E monitoring snow-cover in Inner Mongolia. Pratacultural Science. (08):26-30(2018).(in Chinese)

Review on remediation of contaminated groundwater by Fe^0-PRB technology

Xiaohui Zhang [1,*], Xiaodong Wang[1], and Xianfeng Sun [1]

[1]School of Architecture and Civil Engineering, University of Jinan, Jinan 250022, China

Abstract. PRB technology is a new groundwater remediation technology, which is currently in the pilot test stage in China. Zero-valent iron is one of the most commonly used fillers in permeable reaction walls. In this paper, the principle and structure of PRB technology, the working principle of zero-valent iron in PRB, the preparation method and the optimization and improvement of application technology, the application examples at home and abroad and the prospect of PRB technology are introduced.

1 Introduction

Groundwater resources are important living resources for human beings. The pollution of surface water and the scarcity of resources make groundwater resources more precious. As a result of long-term non-standard operation of enterprises and factories in industrial activities and leakage of storage equipment and pipelines, pollutants seep down and cause groundwater pollution, which has a negative impact on human life and health. With the promulgation of 《the Action Plan for The Prevention and Control of Soil Pollution》 in 2016, the remediation of soil and groundwater has gradually attracted the attention of people at all levels and of all kinds in China.

At present, groundwater remediation technologies include natural attenuation monitoring technology, groundwater ectopic remediation technology, groundwater in-situ remediation technology, phytoremediation technology, electrodynamic remediation technology, permeable reactive barrier remediation technology, etc. [1]. Among them, the permeable reactive barrier (PRB) technology is favored by researchers because of its simple engineering facilities, long duration, low cost and good effect [2]. PRB technology, which uses zero-valent iron as filler, has been widely used and continuously improved and optimized by researchers.

PRB technology uses the wall filled with reaction medium to block the pollutants in the pollution plume, and uses the filling medium to have physical, chemical, biological and other reactions with the pollutants to degrade the pollutants, so as to achieve the purpose of groundwater remediation. In China, PRB technology has not been commercialized yet, and is currently in the laboratory research and pilot test stage.

Liu Jun et al. [3] took quartz sand as the carrier and made iron oxide crystallize on it to treat organic silicon wastewater and synthetic pharmaceutical wastewater. Under optimized operating conditions, the removal rate of COD and TOC could reach 80% and 85%, and the reduction amount of total iron (Fe^{3+}) could reach 26%. Gao Fang et al [4] to hydroxyapatite as the carrier, the simple and feasible through chemical solution of the hydroxyapatite prepared nZVI composite material removal of uranium (Ⅵ), the effect is remarkable.

This paper introduces the principle and structure type of PRB technology, the working principle of zero-valent iron in PRB, the preparation method and the optimization and improvement of application technology, the application examples at home and abroad and the prospect of PRB technology.

2 Permeable reaction wall technology

2.1 The principle of PRB

PRB technology is a kind of filling technology, the permeable wall grid reaction medium permeable response, in the process of groundwater repair, placed in the downstream of groundwater, reaction gate on the vertical of plume in the natural hydraulic gradient through the response under the action of grid and adsorption, precipitation with fillings, biodegradation and redox reaction, remove the pollutants in the groundwater, and the groundwater meets the standards [5].

Common treatment forms and principles of PRB technology are as follows:

(1) Adsorption type: The permeable reaction wall uses the adsorbent as the filling medium. When the pollution plume flows through the reaction wall, adsorption, ion exchange, surface complexation and other reactions occur. Common adsorption materials include zeolite [6], activated carbon, silicoaluminate, etc.

(2) Precipitation type: The permeable reaction wall uses the precipitant as the filling medium. When the pollution plume flows through the reaction wall, the

* Corresponding author: 13853189137@163.com

precipitation reaction occurs and precipitates are generated, so it is trapped. Common precipitators include calcium carbonate and hydroxyapatite [7].

(3) Oxidizing reduction prototype: the permeable reaction wall uses oxidizing agent or reducing agent as the filling medium. When the pollution plume flows through the reaction wall, it is oxidized or reduced into non-toxic valence or precipitation and removed. Common filling media include hydrogen peroxide, calcium peroxide, ozone, sodium persulfate, zero-valent iron [8], etc.

(4) Biodegradable: The filling medium in the Bio-PRB[9] mainly provides microorganisms with carbon source, oxygen, electron acceptor and other substances suitable for growth, so that microorganisms can degrade pollutants better.

2.2 Positioning

Single-processing system PRB and multi-unit processing system PRB are currently applied PRB structure types. Among them, the single treatment system PRB includes two basic types: continuous wall PRB, funnel-guide door PRB, and improved types such as water-barrier wall-in-situ reactor PRB and siphon PRB. The multi-unit processing system PRB consists of series and parallel structures.

2.2.1 Continuous wall PRB

Continuous wall PRB is the most basic PRB structure. In its application, pervious grille should be placed vertically in the downstream of groundwater, and the pollution plume should flow through the reaction wall and react with the filling medium under the action of natural hydraulic gradient, so as to finally get the outlet meeting the requirements. The action process does not change the flow direction of groundwater, and has a small impact on geology. The overall structure is relatively simple, and this structure is suitable for the contaminated site with a relatively small range of pollution plume [10].

2.2.2 Funnel - guide door type PRB

The funnel-gate type PRB is composed of a water barrier funnel, a water barrier gate and a treatment unit. The water barrier funnel is embedded in the water barrier layer to prevent seepage into the surrounding clean groundwater area due to the large pollution plume area. The water barrier wall is used to control and guide the underground water flow to collect, and the filled reaction medium is used to remove pollutants. This structure type is suitable for the contaminated site with relatively shallow landfill depth and large pollution plume range. Its advantage is that the filling media are all used in the reaction funnel, so the reaction is concentrated and the filling medium quality is less required. The disadvantage is that it will change the groundwater flow direction to some extent and interfere with the natural groundwater flow field [11].

2.2.3 The modified PRB

The water barrier wall - in-situ reactor PRB uses the water barrier wall to divert contaminated groundwater into the inlet of the upper end of the in-situ reactor. The contaminated water enters the reactor and flows downward under the action of gravity and fully reacts with the filling material. Finally, the water flows out through the drain pipe.

The siphon PRB uses the natural hydraulic gradient to introduce the polluted groundwater into the reaction chamber [12]. After the reaction is completed, the groundwater in the reaction chamber will be discharged by siphoning. The advantages of this structure type are low operating cost and small floor space. Disadvantages are small scope of application and high requirements for installation technology [13].

2.2.4 The modified PRB

The multi-unit processing system includes series and parallel structure types. The series multi-unit treatment system is composed of multiple permeable reaction walls in series. The filling medium in the permeable reaction wall can be the same or different, depending on the pollution component contained in the pollution plume, so as to achieve the purpose of removing multiple pollutants simultaneously. The parallel multi-unit treatment system is composed of several permeable reaction walls in parallel. The system is mainly used in the case that the pollution plume is relatively wide and the pollution component is relatively single.

3 Fe^0-PRB

3.1 Technical introduction to zero valent iron

Zero-valent iron can react with organic and inorganic pollutants in groundwater [14], and reduce high-priced inorganic pollutants to low-priced non-toxic or precipitated ones. It provides electrons for organic pollutants, converts them into non-toxic substances, and eventually achieves the purpose of water treatment.

3.2 Preparation of nanometer zero valent iron

At present, the common preparation method of nZVI is liquid phase reduction. Strong reducing agent was added to Fe^{2+} or Fe^{3+} water solution to reduce Fe^{2+} or Fe^{3+} to Fe^0. The most common reducing agent is $NaBH_4$. The preparation process of nZVI is simple and easy to operate, and the product has a good effect. However, it is easy to be oxidized by air and other oxides in the process of operation, usually under the protection of nitrogen. The reaction equation is as follows:

$$4Fe^{3+} + 3BH_4^- + 9H_2O \rightarrow 4Fe^0(s) + 3H_2BO_3^- + 12H^+ + 6H_2(g) \quad (1)$$

$$Fe^{2+} + 2BH_4^- + 6H_2O \rightarrow Fe^0(s) + 2B(OH)_3 + 7H_2(g) \quad (2)$$

3.3 Optimization and improvement of zero valent iron application

Zero-valent iron has strong reductibility and large specific surface area. In the process of removing pollutants in water, it is easy to be oxidized to form oxides to cover its surface, which reduces the specific surface area of the reaction, hindering the further progress of the reaction and unable to mineralize pollutants in water. At present, there are two common improvement methods. One is the carrier fixation method, in which nZVI is loaded on the carrier to make its particles disperse and reduce agglomeration. Second, surface modification can reduce agglomeration of particles and increase stability.

3.3.1 Carrier fixation method

Using porous media material as the carrier, the nZVI is fixed on the carrier, and the carrier characteristics are effectively utilized to improve the stability and dispersion of the nZVI particles and overcome the defects of the nanoparticles themselves. The commonly used carriers are quartz sand, attapulgite clay, chitosan, zeolite, etc.

CHEN et al. [15] prepared nZVI using activated carbon as the carrier and applied it to dechlorination reaction of hexachlorobenzene (HCB). The experimental results showed that the iron-loaded method could significantly increase the removal efficiency and dechlorination performance of HCB. Wang Hong etc. [16] by chemical liquid phase reduction of attapulgite prepared load zero-valent iron composite material, and optimize the preparation technology of composite materials, the degradation of the experimental results show that the composite material has higher stability and dispersion, improve the processing efficiency of methylene blue.

3.3.2 The surface modification

The surface modification of nZVI is mainly made by adding polymeric polymer electrolyte or surfactant, which can reduce or prevent the charge and dipole suction between nZVI particles, thus preventing the agglomeration between nZVI particles and increasing its stability. The commonly used modification methods include physical assistant method and chemical additive method.

Sun et al. [17] selected polyelectrolyte mixture for surface modification of nZVI particles, and found that the average particle size was only about 15 nm, and that the adsorption capacity and migration capacity were significantly improved. Jing Li et al. [18] modified nZVI simultaneously with non-ionic surfactant polyvinylpyrrolidone and anionic surfactant sodium oleate. The results showed that the working efficiency of zero-valent iron was optimized, the reaction was balanced for 2min, and the removal efficiency was more than 99%.

4 Application cases at home and abroad

4.1 Domestic application cases

The research on PRB technology in China started relatively late and is still in the laboratory research and pilot test stage, which still needs further research and exploration to provide theoretical basis for the realization of commercial application.

Some pilot tests have also been conducted in China. Tian Lei [19] carried out a pilot test of PRB technology for remediation of TCE and toluene contaminated groundwater in an engineering demonstration site in Henan Province with zero-valent iron composite material as filler, with good results. However, the service life of zero-valent iron will be affected by the content of other substances in groundwater. Hou, etc.[20] the construction of underground water in Shenyang in Liaoning province PRB demonstration project, with zeolite as the filling medium pollution treatment of NH_4^+-N, after a period of time, water concentration in accordance with GB/T 14848 《underground water quality standard》 Ⅲ class water quality requirements.

4.2 Overseas application cases

Zero-valent iron PRB technology has been widely used in groundwater remediation abroad [21], with a high success rate, and has been commercialized in European and American countries [22].There have been more than one hundred successful cases of the reaction wall built with zero-valent iron as the filling medium.

A continuous reaction wall with length of 5.5m, depth of 9.7m and thickness of 1.5m was constructed in a contaminated site of Boden Base in Ontario, Canada, using zero-price iron and concrete as fillers to treat TCE and PCE. The final removal rate of TCE and PCE was 90% and 88%[23].A north Carolina Elizabeth city pollution sites using zero-valent iron electroplating workshop for packing construction has a length of 45.7 m, 7.3 m deep, thick 0.6 m of continuous reaction wall removal of TCE and hexavalent chromium, Cr (Ⅵ) concentration of 5 mg/L to did not check out, concentration of TCE by 7 mg/L dropped to less than 0.005 mg/L [24].

5 Outlook

PRB technology is a simple and efficient groundwater remediation technology. After the installation of the reaction wall, almost no additional costs will be incurred except for the necessary maintenance costs. Although PRB technology in China has a late start and is still in the pilot test stage, it has a broad prospect and will be commercialized soon. The problem of groundwater remediation in China has attracted the attention of many experts and scholars, and the widespread application of PRB technology will also become a trend.

References

1. China Environmental Protection Industry Association . Technical Catalogue of Contaminated Sites Remediation (First batch) [A] . Beijing: Report on the Development of China's Environmental Protection Industry (2014), 2015.
2. BAI Y H, ZHANG S J . Permeable reactive barrier as a remedy technology of contaminated groundwater [J] . Yunnan Environmental Science, 2005, 24(4) : 51-54.
3. Liu Jun, Lou Yuefeng, Li Jun. Method and characteristics of loading iron oxide on quartz sand surface[J] . Chemical progress, 2016,35 (2) : 624-628.
4. Gao Fang, Zhang Weimin, Guo Yadan, et al. Hydroxyapatite nano zero-valent iron removal of uranium (Ⅵ) in aqueous solution [J] . Chinese ceramics, 2015,51 (8) : 10-15.
5. UESPA.A citizens guide to permeable reactive barriers[M]. Washington D.C.:US Environmental Protection Agency Press, 2001:1-2.
6. Zuo Simin, JingZhaoqian, Tao Mengni, et al. Study on the application of natural zeolite and modified zeolite in wastewater treatment [J] Application of chemical, 2019,48(5): 1136-1145.
7. Zeng Yunzheng, Zhang Weimin, Wang Huidong, et al. Study on removal of Manganese from aqueous solution with quartz sand - loaded hydroxyapatite [J] Water treatment technology, 2017,43(5): 72-79.
8. Xu Haiyu, Zhang Mingqing, Chen Yiyu. Removal of hexavalent chromium from water by organic attapulgite loaded nano zero valent iron [J] Environmental Science in China, 2019,39(12) : 5079-5084.
9. Liu N, Ding F, Wang L,et al. Coupling of bio-PRB and enclosed in-well aeration system for remediation of nitrobenzene and aniline in groundwater [J] Environmental Science & Pollution Research , 2016,23(19): 9972-9983.
10. Ministry of Environmental Protection, PRC. Technical Report on remediation of contaminated Sites [R]. Beijing: Ministry of Environmental Protection, PRC, 2014.
11. Zhang Xiaohui, Ge Fangzhou, Dong Yujing, et al. Research progress of permeable reactive wall in situ remediation of contaminated groundwater [J]. Industrial water and wastewater, 2015,46 (3) : 1-5.
12. Cun Junfang, Zheng Xilai, Lin Guoqing. Permeable reactive wall technology for groundwater organic pollution treatment [J]. Advances in Water Science, 2003,14 (3) .
13. JEEN S W, GILLHAM R W, PRZEPIORA A . Predictions of long- term performance of granular iron permeable reactive barriers: field- scale evaluation[J] . Journal of Contaminant Hydrology, 2011, 123 (1/2) : 50-64 .
14. Wang Yan, Gu Xiaogang, Miu Zhouwei, et al. Review on the screening methods of groundwater remediation in contaminated sites [J]. China Municipal Engineering, 2018,5 (10) : 25-27.
15. Wang Hong, Zhu Liangliang, Yuan Lingwei. Preparation and degradation of methylene blue from modified attapulgite loaded nano - zero valent iron [J]. Sichuan chemical, 2020,23 (1) : 6-10.
16. CHEN W, PAN L, CHEN L, et al. Dechlorination of hexachlorobenzene by nanozero-valent iron/ activated carbon composite: iron loading, kinetics and pathway[J]. RSC Advances, 2014,4(87): 46689-46696.
17. Sun Y P, Li X Q, Zhang W X. A method for the preparation of stable dispersion of zero-valent iron nanoparticles[J]. Colloids and Surfaces A: Physicochemical And Engineering Aspects, 2013,308(1-3):60-66.
18. Jing Li, Mingjie Fan, Miao Li, et al. Cr(VI) removal from groundwater using double surfactant-modified nanoscale zero-valent iron (nZVI): Effects of materials in different status[J]. Science of the Total Environment. 2020.
19. Tian Lei . Study on the Removal of mixed pollution of chlorinated hydrocarbon and benzene Series from groundwater by PRB[D] . Beijing: China University of Geosciences, Beijing, 2014.
20. HOU G, LIU F, LIU M, et al . Performance of a permeable reactive barrier for in situ removal of ammonium in groundwater[J] . Water Science and Technology: Water Supply, 2014, 14(4): 585-592.
21. LUDWIG R D, SMYTH D J, BLOWES D W, et al . Treatment of arsenic , heavy metals , and acidity using a mixed ZVI-compost PRB[J] . Environmental Science and Technology, 2009, 43(6) :1970-1976.
22. VESELA L, NEMECEK J, SIGLOVA M, et al . The biofiltration permeable reactive barrier: practical experience from Synthesia [J] . International Biodeterioration & Biodegradation, 2006, 58(3/4): 224-230.
23. O'HANNESIN S F, GILLHAM R W, VOGAN J L . TCE degradation in groundwater using zero-valent iron [R] . Washington DC: American Chemical Society Extended Abstract, Industrial and Engineering Chemistry Division, 1995: 55-58.
24. USCG . In situ permeable reactive barrier for treatment of contaminated groundwater at the US Coast Guard Support Center, Elizabeth City, North Carolina[R] . Elizabeth City, NC: US Coast Guard Support Center, 1998.

 https://doi.org/10.1051/e3sconf/202123701020

Research on Sustainable Development and green and low-carbon clothing design

Dong Fanglin

College of Textiles and Clothing, Shanghai University of Engineering Sciences, Shanghai, China

Abstract: With the rapid development of the world economy, science and technology, more and more people are realizing that the serious situation caused by the environmental crisis affects people's lives. In the new era, people's awareness of green and low-carbon has been continuously raised, especially the daily use of clothing as a way of life and the flooding of new and old clothes, green and low-carbon concept of clothing design has become the trend of sustainable development in the future, around the sustainable development, green and low-carbon concept of clothing design and integration for the clothing industry to do a stable innovative development direction.

1 Introduction

Under the background of the early ancient times, the research found that people already have the green and low-carbon environmental consciousness, which imperceptibly influences the continuation and development of the future developed economy and science and technology, the main fabric of early people's clothing to animal fur, natural things: linen, wool, silk, bark and other fabrics made of clothing. With the continuous improvement of modern people's quality of life, people's awareness of green and low-carbon has become more and more intense. People live in harmony with nature. The awareness of green, low-carbon and environmental protection has become a mainstream trend in contemporary society, however, people neglect the transfer of clothes from closets to landfills and incinerators, and the world is facing the direction of old clothes shelving, slowly becoming a mainstream issue, green, low-carbon, recycled, re-used waste extraction material, the garment field of recycled fabric has attracted wide attention. The integration of green and low-carbon Clothing Sustainable Development Research, fully let people live in harmony with the natural environment, and clothing design products are inextricably linked.

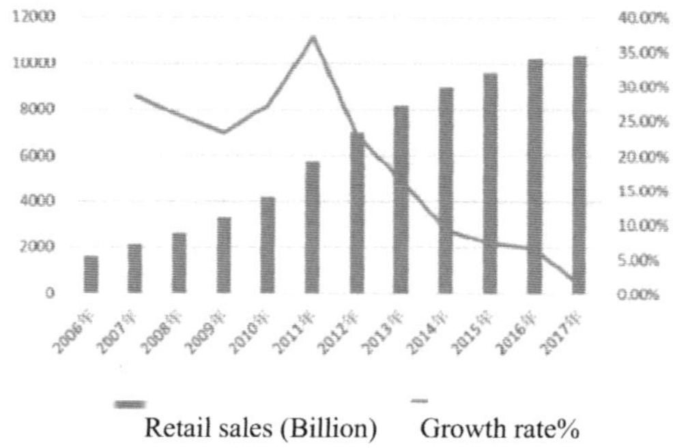

Figure 1. The recovery rate of textile products

1.1 Trends in sustainable clothing

In China's traditional garment manufacturing industry, cotton, linen, are commonly used materials, such materials in the traditional garment manufacturing process is difficult to be recovered and recycled. Because of the characteristics of its mode of production and clothing products, people are more and more aware of the problem of green and low carbon environment, and keep the clothing industry in line with the transformation form of green and low carbon, the importance of green and low-carbon fashion design, culture, science and technology and other factors complement each other. It has a deep influence on the research of green and low-carbon fashion design.

The whole life of green eco-low-carbon clothing from production to consumption hasn't caused pollution to the ecological environment and harm to human body, providing people with health, comfort and fashion to use while also saving natural resources to the greatest extent, guarding the home, to achieve the goal of harmonious coexistence between man and nature, to promote the clothing of the green ecological low-carbon. Compared with the traditional fashion design, the green eco-low-carbon fashion design uses the recycling of used materials to design, in line with the current and future direction, can effectively ensure the practicality of clothing and green low-carbon; Advocated the idea of resource conservation. To solve the low-carbon environment to create a new green design methods and ideas.

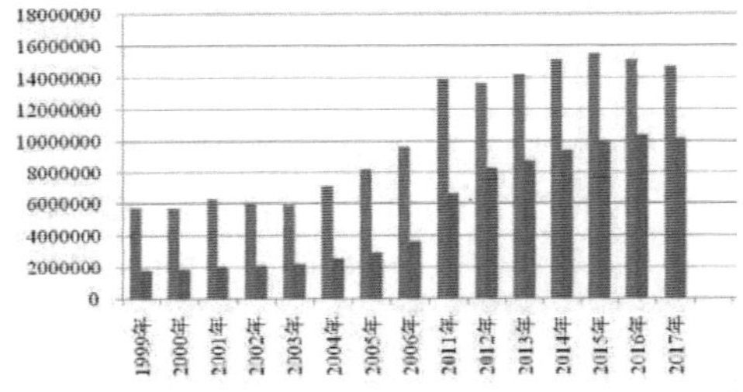

Chian1990-2017Textiles over the years Cumulative finished products (Ten thousand yuan)

Chian1999-2017Cumulative finished garments over the years (ten thousand yuan)

Figure2.Cumulative finished products of textiles and clothing in 1999-2017

1.2 Purpose and characteristics of green low-carbon clothing

Green low-carbon environmental protection clothing design to a certain extent in line with the design of green, low-carbon, safety, health, comfort. Now the global trend of green and low-carbon fashion design is unstoppable. All countries in the world are warming up to study the green and low-carbon fashion design to restrict the market circulation and consumption of non-green and low-carbon products, study the standard of Overall Comfort, health and harmlessness of environment.

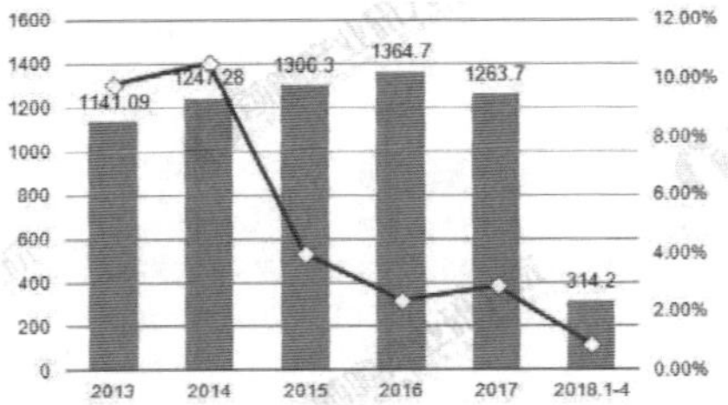

Total profit of textile and garment enterprises above De signated Size:Billion

Year-on-year growth

Figure 3 Profit chart of textile and garment enterprises

When the Green low-carbon clothing design carries on the product development, first introduces the low-carbon idea and with the traditional design factor overall consideration balanced ecology need as well as the profit space, grasp the factors such as green and low-carbon-based, investment, product performance and other aspects. Secondly, on the premise of being able to guarantee the profit space, on the premise of being able to effectively ensure the maximum profit growth space of the enterprise, on the premise of being able to effectively ensure the safety and health of the enterprise's products and the green and low-carbon character, to avoid the ecological imbalance caused by negative effects. Again, the product isn't the low-carbon factor external reflection, internal and external factors from a variety of expectations, otherwise it is difficult to occupy a favorable market, it is difficult to play out the green low-carbon factors.

2 The embodiment of sustainable development fashion design

2.1 Current market situation of sustainable fashion

2.1.1 Current situation of sustainable development of apparel in China.

The rate at which people change clothes in our country is increasing. In the case of most people do not want to buy second-hand clothing, there is no reasonable and disposal of used clothing is growing every year. Direct disposal, on-the-spot burial and burning of open fire garbage will cause serious environmental pollution to the natural environment, ecology and water resources. It is conceivable that such a large number of old clothes will bring about environmental pressure. At present, green and low-carbon sustainable development is the direction of national and even global development.

Li Ning, a leading domestic sportswear company that has been responding to low carbon fashion earlier this year, has formed a partnership with empire. He invented ECO circle, ECO durable fabrics, and launched a recycling program for used clothing. Li Ning's stores recycle used clothing before sending it to the Teijin polyester fiber group, based in Tokyo, Japan, China, where it is broken down by natural chemicals and converted into natural polyester fibers, developed and manufactured from recycled materials for the production of new green sustainable recycling products; September 20,2009, during the special meeting on the research of low-carbon Environmental Protection Clothing, which was jointly conducted by the Environmental Protection Research Institute, the Goodall Center for Environmental Knowledge and culture in Jurchen people, Li Ning Company and Teijin fiber group, Teijin was the only one who gave the idea of "clothes ring" design in a constructive way, advocates that people should adhere to the green low-carbon environmental protection.

Puma has launched a free-of-charge recycling program for used sportswear in Shanghai and Beijing, and H & M, which has announced the launch of a year-round recycling program for used clothing, has adopted a similar traditional approach to Puma's. H & M and S & Puma have both chosen I: Co as strategic partners.

Some large-scale garment and textile industry is also helpful to systematize and standardize the pollution control after treating the most advanced garment and textile pollution equipment

In order to fully facilitate more citizens to recycle and dispose of used clothes by themselves in first-and second-tier cities, starting from May 2010, a city-wide pilot scheme was launched in Shanghai to recycle used clothes by sorting them, bringing them into residential communities like used clothes collection bins, and through the used clothes collection bins, the collection and disposal of used clothes is carried out by specialized companies of used clothes, and many urban areas have adopted the method of classified management and recycling of used clothes.

Taiwan's clothing brand SCAFE, introduced a lower carbon environmental protection clothing manufacturing process. They have reached agreements with some coffee retailers to recycle coffee grounds, paper cups and other coffee "waste" into trousers, shirts and suits that not only look the same as ordinary clothes, the utility model has the functions of coffee odor diffusion control, ultraviolet radiation protection, quick-freezing drying and other protection functions.

Based on a series of mature application cases in China, many green, natural and low-carbon fabric concepts have been used at different technical levels, such as green, natural and low-carbon fabric, primary energy recycling and "year ring of clothing" , especially in the use of natural low-carbon raw materials and renewable resources of low-carbon cloth technology has in-depth study. Some key areas of China's national research working group for the Biodegradable plastic have further research findings, breakthrough currently facing the major issue of green environmental protection and low carbon. The domestic clothing brand speeds up the regeneration, the second use, provides the transformation low carbon green clothing design unceasingly.

2.1.2 Status quo of su stainable de velopment of apparel abroad.

Green low-carbon awareness of foreign countries are also very concerned about the development of the foreign used clothes market, the awareness of old clothes transformation is also used by many foreign design.

Jake Danaher, New York, USA. The "plastic cloth" is popular with many environmentalism. The products in the shop use garbage to turn into treasure. A large number of disposable plastic bottles are processed in a series to make swimsuits, changing the fashion world's linear use of re-materials to a system that has a circulation of goods is also what green low-carbon clothing design would like to see.

MIYAKE's clothes focus on simple science and technology, he runs through the concept of Oriental plane composition in three-dimensional cutting methods, reduce artificial traces, keep comfort and versatility, walking in a green and low-carbon fashion. More attention is paid to the harmony between man and nature.

Countries such as the United States and Japan, which are the main examples, are all gradually promoting green, low-carbon environmental protection and regeneration. The transformation model of the circular economy is gradually speeding up. Due to the increasing shelving of household clothes and garbage, the threat to the environment and ecology is becoming greater and greater, the imbalance between man and nature. In order to get rid of "make-then-use-and-throw-away" , the fashion pattern needs to be updated, so that the major clothing brands to create a green and low-carbon fashion design trend mainstream wave sustainable progress.

3 design strategies for sustainable development

3.1 Research on sustainable design

3.1.1 Sustainable design materials.

Material is one of the basic elements of clothing, and the progress of science and technology also provides us with the sustainable development design concept. The green clothing low-carbon and environmental protection clothing design provides a new type of using material, the fabric of the clothing design transformation method twice, that is to take all or part of the original low-carbon clothing for reorganization, transformation of a brand-new clothing design style, the old clothes through disassembly, reorganization and transformation into both fashionable and personalized clothing, to create and design new and more environmentally friendly low-carbon clothing. Designers can also use their own expertise to transform the material to take on a variety of styles of appearance.

In the development and utilization of fabric recycling flow pattern, common fabrics are cotton, hemp, silk, wool, chemical fiber materials, as well as synthetic materials and recycled materials. Low-carbon green fabric all over the world are developing new types of textiles, divided into natural fiber, renewable environmental protection fiber and biodegradable synthetic fiber class. Natural Green fabrics are also environmentally friendly and green is also an important embodiment of low-carbon environmental protection. This is also a continuous production line. This kind of fabric technology, which is refined by natural vegetable oil and pure manual process, although complicated, but it is very environmental protection, in the numerous international fashion show hall clothing, many are also the application of natural environmental protection fabric products. With the new type of pulp environment-friendly cloth as raw material, in the surface coating mortar plays the role of wind and rain. The water-based fabric burying is usually convenient and can absorb water and decompose by itself, which reduces the pollution to the environment. With the rapid development of science and technology in the garment

https://doi.org/10.1051/e3sconf/202123701020

and fabric manufacturing industry in China, it's also believed that carbon fibers in dyed coffee can be obtained by directly using the charcoal residue of dyed coffee filtered at high temperature as the raw material for dyeing, the fabric has the functions of producing negative ions, bacteriostasis, Deodorization and anti-ultraviolet

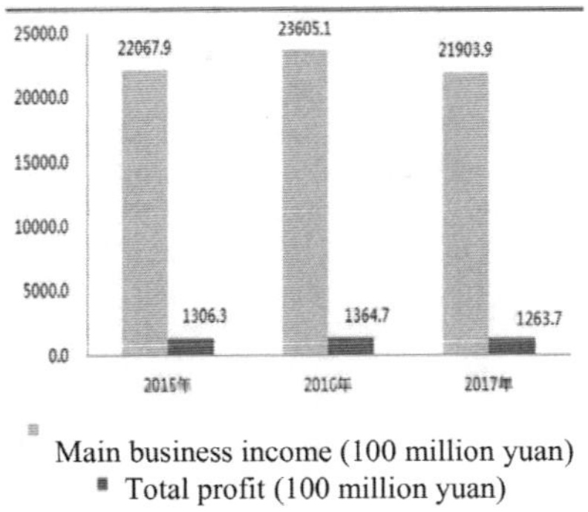

Figure 4 Profit chart of textile and clothing

For the secondary processing design of recycled fabric, the green and low-carbon clothing design is embodied through the design of addition and subtraction, deformation design and combination design. The design of adding, subtracting and sending is mainly based on the original foundation through traditional means such as embroidery, filling and decoration. The design of "old" and "regeneration" is also the direction of development combined with science and technology, and many of them are designed with the techniques of handicraft, recycling of the weaving process also reflects the green low-carbon clothing design, showing a different level of beauty, one of the performance of green low-carbon fabrics. Many brands are increasingly interested in recycled and alternative fabric.

3.1.2 Sustainable color design.

Sustainable Development type of clothing color is one of the indispensable, color is also the first visual perception of people, different color collocation and proportion give people different senses. For the Green low-carbon clothing, in the return to the natural ecological natural color similar to the visual organs of the people and Comfort. For the second design of color, the degree of color in the recycled old clothes is almost low, and then on the basis of the second processing, through the extraction of natural plant dyeing, tie-dye, Batik, pattern design to make the whole look brand-new. Therefore, the combination of the traditional way and technology means, blend with each other to better reflect the green low-carbon clothing design.

3.1.3 Sustainable design.

Green low-carbon clothing style design, the characteristics of human body structure, design simple and comfortable as well as functional and practical style clothing, completely green low-carbon clothing to do the foundation. In today's clothing styles mostly use three-dimensional cutting, the waste and consumption of fabrics also have an impact on the design of styles using plane cutting and three-dimensional cutting plus the use of a small number of fabrics, using diversified design, can use splicing, replacement, etc. , clothing parts such as sleeves, collars and pockets can be designed into a variety of combinations of old clothing fabric, different from the traditional clothing design, the pursuit of Natural Comfort Green low-carbon clothing and highlight personalized display.

3.2 Sustainable development of clothing in China

Green low-carbon clothing in the design, production, use of the entire life cycle should be achieved circular economic flow model, a new round of development model to update the circulation system, to promote rapid economic development, the brand effect follows the principle of "3R1D" which can reduce, repeat, recycle and degrade. For example, actively develop "zero waste" manual plate-making and worn-out clothing processing technology, use worn-out leftover materials method processing worn-out clothing or other accessories, will be worn-out leftover materials and worn-out processed clothing pre-processing classification, multi-cycle production mode of manufacturing.

3.3 Sustainable and innovative application of modern technology in garment design

China has a long history of using garment dyes for printing and dyeing, and it's difficult to understand the ways of printing and dyeing commonly used in the manufacture of dyes, fabrics and garments in cultural tradition, including China's traditional Dyes Batik, tie-dye, weaving and other printing and dyeing materials used are considered to be natural low-carbon. Under the background of the high-tech era, people advocate the green and low-carbon clothing design in the clothing industry and society, pay attention to the regeneration and alternative cycle design, and design production process by hand, the secondary recycling of materials using green low carbon products reduces and avoids the secondary pollution damage to the environment produced by complex industrial processes and the excessive waste of resources, thus reducing carbon emissions, for the low-carbon Green clothing design provides innovation and new ideas forward.

3.4 Development and collision between traditional culture and modern technology

3.4.1 Theories and methods of sustainable design.

Design is like a person's life, with a cycle, the end of the cycle, the meaning of design also began to disappear.

But the truly classic design can always withstand the test of time, unlimited extension and expansion of their own cycle. The wind direction of art and design is changing, so is the aesthetic of human beings. Works that are truly enduring have prolonged their existence indefinitely because their designs conform to the most basic aesthetic rules and the aesthetic laws of human beings, this is the enduring power of the rule. In the concept of sustainable design, it is necessary to fully apply the concept of design, many of the classic, long-lasting design works, even after many years have their infinite vitality, although we can not replicate such success, but we can deeply carry out such thought and theory in the design, through the reasonable design, creates one conforms to the present esthetic, has thousands of years vitality excellent design.

The promotion of the Green and low-carbon concept stimulates the development of various industries, renews and rotates the Global Model, promotes the design of low-carbon and green clothing, and also makes people aware of the innovation of the green and low-carbon concept, let the traditional pattern and technology combine to create a green and low-carbon clothing design, fully understand the sustainable development of green and low-carbon clothing, so that the whole becomes a more healthy, comfortable, pure and fashionable environment, let people really understand the concept of green low-carbon.

3.4.2 Inherit ance and in novation of traditional culture in fashion art.

In recent years, a new voice has appeared on the stage of the fashion industry. The Chinese traditional culture, represented by the Chinese style, is gradually entering the international fashion industry and forming the fashion of the Chinese style to a certain extent. Exploring the reasons for the rise, we can easily find that with the development of China's economy and social progress, people's spiritual needs are more prominent, and the accelerated development of the Internet has provided a vast space for the cultivation of personality, the richness of Chinese culture also provides ample space for discovering one's hobbies and cultivating one's own interests. In such a climate, many people who try to combine the traditional culture with the modern fashion art try to innovate and transform the traditional culture to form a cultural trend with more modern artistic characteristics and to form a more individual flavor, the elements of the traditional culture of fashion, in the information age, are amplified by the spread of the Internet and social media to deepen its identity on a larger scale.

The inheritance and innovation of traditional culture in fashion art means that the tolerance of cultural ideas should be further expanded. Attention should be paid to the development and prevalence of ancient customs in the role of youth. The combination of traditional culture and modern fashion is undoubtedly giving new vitality to the traditional culture and restoring its vitality and vitality in a modern way.

3.4.3 Application of traditional handicraft in fashion.

In the age-old project is gradually expanded, the decline of contemporary, there are always some old and pure skills, in the information age development of today by the flood submerged, we forget. But fortunately, there are more and more people around us to explore the ancient elements of China into their own design and ideas. The traditional handicraft is one of the important inspiration sources of contemporary design, the true non-abandoned handicraft emanates the natural, humanistic, natural beauty, and the characteristics it displays, advocated the idea is that we now people can not imagine the exquisite and ingenuity. A piece of non-abandoned handicraft embodies the hard work and wisdom of craftsmen. It shows US unique folk customs and contains the essence and strength of Chinese traditional culture. An artifact has stood the test of time and the vicissitudes of things for thousands of years.

In today's fashion world, the traditional hand-made embroidery on the runway often comes from the cross-border association between the hand-made inheritors and the designers, so that many non-heritage handicrafts can be brought back to life, and the works can be innovatively manufactured and put on the market platform, so as to promote the development and spread of ich.

3.4.4 Sustainable and innovative app lication of modern technology in fashion art.

To create better sustainable products, modern technology needs to be incorporated into the design. First of all, to incorporate readable content into the design, with good readable fonts, pictures are the foundation, design is the connotation, typography should be creative and innovative, no matter when, any device that looks at the content of a post should be attractive enough to be recognized by the user in any situation. Fancy Fonts designed to appeal to the reader can be eye catching in certain situations, but may not always be able to understand what they are trying to convey. In the design of visual communication, should be in place, the most important direct and clear expression to express the meaning of the message, the design of the text, pictures, text should be directly linked with precision, create a comprehensive sense of experience and reading sense. In the visual communication should also be applied to some interactive design, speaking of different elements interact together, users will have a combination of interaction with the sensory experience, small interaction details can make the design more authentic, give the project a different meaning. When designing interactive elements, the USABILITY and practicability of design have to be considered in the category of products. The interaction brought by more competitive works should be unique, interactivity should be unique and irreplaceable, used in visual communication to tell a story, act as a guide to the client, and delight the client with subtle effects.

https://doi.org/10.1051/e3sconf/202123701020

4 Conclusion

Green and low-carbon is an important trend in the future. The concepts of Chinese garment production and consumers are becoming more and more mature. They have a good grasp of the changing factors of people, nature, time and space background, industrial ecological environment, etc. , make the clothing with Chinese characteristics green and low carbon industry. In recent years, more and more attention has been paid to the issue of environmental protection, through the green and low-carbon ecological design, people's ecological civilization level has been continuously improved, and the research and application of clothing design have been brought into play to realize the sustainable development of green and low-carbon. In the design of green low-carbon clothing, not only consider "serving people" , "people-oriented" , zero waste fashion is equal to not make garbage fashion green low-carbon clothing brand, but also consider the balance of the environment. The key to the development of green and low-carbon clothing lies in ecological, natural, environmental protection, health and so on. With the development of the new era, the low-carbon Green clothing is gradually penetrating into people's life. The daily recycling pattern of clothing plays an important role in the clothing industry.

References

1. Zhou Jie. Discussion on the application of green design in Garment Design J. Western leather, 201714:99. 2.
2. Cao Yu, Zheng Qiyun. Study on the decision of recycling dual perishable products under stochastic market demand [J] . Journal of Anhui Normal University (humanities and Social Sciences Edition) , 2016(6).
3. Meng Qiang, Li Fangyi, Li Ching, Zhou Lirong, Ji Qin. Rapid Life Cycle assessment method for scheme design based on Green FeaturesJ.Computer-integrated manufacturing, 21,20153:626-633.
4. Yao Weixin. Research on the mode of recycling and reuse of waste clothing based on enterprise-oriented -- taking as example J. International Textile Review, March 2014.
5. Guo Yan. Research on the concept, principle and hierarchical structure management of Zero Abandonment J. Textile reporting, October 2014(10).
6. Wang Chen. Research on the application of three-dimensional modeling technique in Clothing Regeneration D. Zhejiang: Zhejiang Sci-Tech University, 2014.
7. Bai Yuling. The present situation and strategic choice of low-carbon Textile clothing in China J. Foreign Trade and economic practice, July 2012,(7).
8. Hao Shuli. RESEARCH ON LOW-CARBON PATTERN OF GARMENT INDUSTRY J. Shandong Textile Economy, 2011(08).
9. Zhu Yuansheng. Fabric and Costume Design M. SHANGHAI: China Textile University Press, 2008.11

Effect and mechanism of C_2HF_5 on premixed flame of runaway gas (syngas) in lithium-ion batteries

Chenyu Zhao[1], Zidong Guo[2,*]Guiyun Zhang[1], Tianwei Zhang[3], Zhaoqing Liu[4] and, Hao Liu[3]

[1]China People's Police University, Graduate School, 065000 Langfang Hebei, China
[2]China People's Police University, School of Fire Protection Engineering, 065000 Langfang Hebei, China
[3]China People's Police University, School of Rescue and Command, 065000 Langfang Hebei, China;
[4]Fire and Recuse Department of Hebei Province, 050000 Hebei, China;

Abstract. Lithium-ion battery is a new type of clean energy. Due to manufacturing defects, improper use, unexpected accidents and other reasons, lithium-ion battery thermal runaway fire and explosion accidents are frequently re-ported, and its safety has become an important issue to be solved. In this paper, the effect of second-generation halon substitute C_2HF_5 on runaway gas premix flame of lithium-ion battery are studied by experiments and numerical simulations.The bunsen burner was used to measure the changing trend of the premixed flame velocity of the runaway gas/air under different equivalent ratios and different concentrations of C_2HF_5. The results show that the flame velocity is in good consistency with the model under the conditions of lean flame and low inhibitor addition.The promotion effect of inhibitors on combustion has been verified by bunsen flame. Compared with CH_4, C_2HF_5 is more suitable for inhibiting syngas flame of lithium-ion batteries, and the effect on the rich flames is better than lean flames.

1 Introduction

With the development of industrialization worldwide, human society has become increasingly dependent on fossil fuels. The resulting problems such as excessive greenhouse gas emissions, environmental pollution, and energy shortages have seriously threatened the sustainable development of human society. As a new type of renewable and clean energy, lithium-ion batteries has received widespread attention due to its environmental protection and high efficiency characteristics, and has been applied in many fields such as power source of battery powered vehicles and hybrid electric vehicles [1]. However, in the actual application process, the storage and use of lithium-ion batteries are dangerous. Factors such as manufacturing defects, overcharge, impact, and short circuit can easily cause thermal runaway of lithium-ion batteries and release high heat [2], and generate a large amount of toxic and combustible gas mixture[3]. Therefore, it is necessary to propose corresponding solutions to improve the safety and reliability of lithium-ion batteries.

The fire extinguishing mechanism of halon extinguishing agent (1211,1301,2402) widely used in the early days is chemical fire extinguishing. However, at the same time, halon was banned in 1987 due to its destructive effect on the ozone layer [4].

C_2HF_5 is a new halon replacement product developed by dupont of the United States, with zero ODP and low GWP and ATL values, listed in NFPA-2001(2018)[5] and ISO 14520 (2015)[6] standards. At room temperature and pressure, C_2HF_5 is a colorless, almost odorless, non-conductive, no residual clean gaseous chemical extinguishing agent that is compatible with rubber and plastic sealing materials. Due to the advantages of low boiling point and low potential greenhouse effect, short atmospheric survival time and superior fire extinguishing performance, the U.S. Department of Defence chose C_2HF_5 as the fire protection for aircraft engines.

The FAA conducted aerosol tank tests on C_2HF_5 found that when the amount of added C_2HF_5 was lower than its inert concentration (8.9%), the explosion pressure was 2.6 times higher than that in the dry test[7,8], and the flame combustion was strengthened. This phenomenon of enhanced combustion has attracted the attention of scholars. Pagliaro et al. [9] measured the flame velocity and pressure changes of methane/air premixed laminar flow with different amounts of added C_2HF_5 using a constant volume burner. Babushok et al. [10] established a model and found through simulation calculation that the addition of C_2HF_5 promoted combustion in the lean combustion side. NIST's Takahash et al. [11] tested with a cup burner and found that the addition of a certain amount of C_2HF_5 would increase the heat release rate, thus promoting combustion.

The syngas fuel used to simulate the thermal runaway of lithium-ion batteries is mainly composed of H_2 and CO, and may also contain N_2, CO_2, water vapor (H_2O), hydrocarbons and other substances, such as CH_4. The effect of syngas combustion composition ratio on the minimum ignition energy, flame velocity and flame stability is the current research hotspot, but there are few studies on the effect of halon replacement on the

* Corresponding author: 804777429@qq.com

combustion rate of syngas. In particular, the composition ratio of syngas used in this paper is based on the composition of the thermal runway gas of lithium ion battery measured in the experiment.

Based on the Bunsen burner method, this article studies the influence of the amount of C_2HF_5 on the adiabatic temperature and flame velocity in the mixed gas/air laminar premix flames with different equivalence ratios. The first purpose of current work is to reproduce the experimental phenomenon of FAA-ACT in Bunsen burner experiment. To test the performance of C_2HF_5 as an inhibitor, and to discuss the feasibility of application. The second purpose of this paper is to enrich the experimental values of the laminar flame velocity of syngas in lithium-ion batteries.

2 Experimental methodology

2.1 Experimental setup

The content and composition of the fuel was the same as in our previous research [12], i.e., H_2-29.6%, CO-22.3%, CO_2-35.5%, CH_4-6.9%, and C_2H_4-5.7%.

Four combustion diameters (diameter 8, 10, 12, and 15mm) are designed for the nozzle of the tubular burner, and the pipe length is designed to be 900mm, all of which are made of stainless steel. At the same time, the smoothness of the nozzle edge of the burner is improved to ensure that the air flow at the outlet of the nozzle of the burner is within the laminar flow state range. In order to reduce the influence of flame stretch and curvature on the measurement of flame velocity, different nozzle diameters are set to ensure flame stability at low equivalence ratio. The experimental data in this paper are mainly obtained at 10mm and 12mm diameters (8mm flame has poor stability, and 15mm is limited by gas velocity). When $\varphi \geq 1$, the measurement uncertainty of 12mm and 10mm nozzles is consistent [13]. The system diagram of the experimental device is shown in Fig. 1.

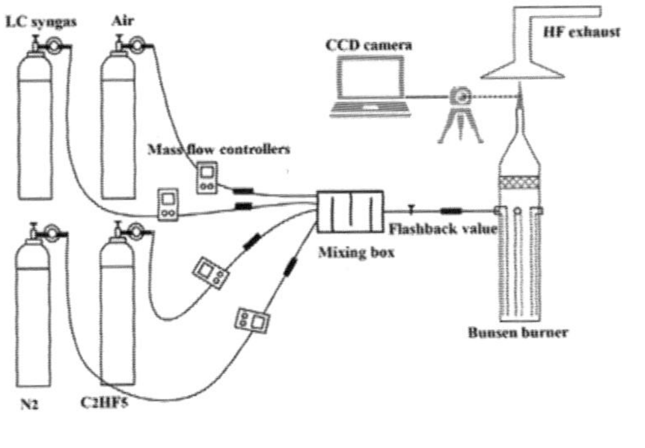

Fig. 1. Schematic diagrams of the experimental apparatus.

Mass flowmeter should be used in gas pipeline control (FLOWS INSTRUMENT company Dmass mass flowmeter, the entire range accuracy of ±1.0% in the range, repeatability of ±0.1%). Before the trial of the syngas are collected and used to use GC-MS INSTRUMENT analysis, ensure that each group of experimental mixed gas composition error less than 0.1% (21% of synthetic air is 79% N_2, O_2). In the experiment, the flame height maintained was 2 to 3 times of the nozzle diameter of the burner. Excessively low flame would increase the heat loss of the burner [13], and excessively high flame would have an effect on the elongation rate. The flame image was captured by a 105mm, f / 4.5 UV lens with a resolution of 2320 × 1720 and a pixel size of 7μm × 7μm with a shooting speed of 500FPS with a high-speed CCD camera. 500 images were taken in each experimental group and averaged to reduce the influence of disturbance caused by the surrounding environment. All experiments were carried out at standard atmospheric pressure, and the temperature of the unburned mixture was 298±2K.

In the experiment, the uncertainty of gas mix ratio comes from the uncertainty of fuel component measurement and gas flow velocity measurement, which is about 0.002-0.004. The total uncertainty of S_L obtained by the test is about 1.5-3.0cm/s.

2.2 Laminar flame velocity measurement method and flame stability analysis

Laminar flame velocity was measured using a Cone Angle Method (Fig. 2.) in which the flame propagated to unburned mixture at an Angle α/2. The calculation formula of laminar flame velocity is as follows:

$$Su = U\, sin\alpha/2 \qquad (1)$$

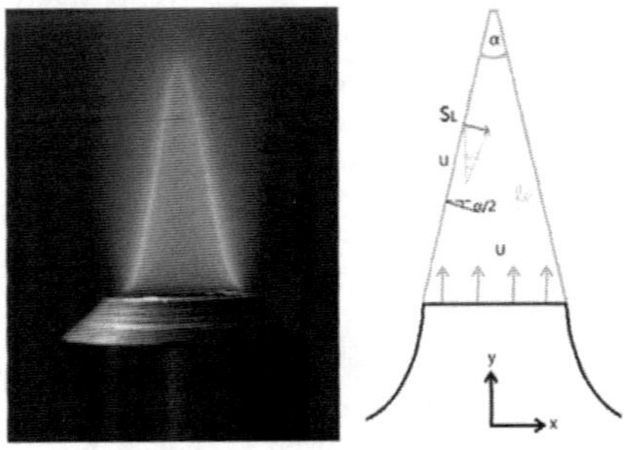

Fig. 2. Measure laminar flame velocity (angle method).

The stable flame morphology is related to the nozzle velocity, combustion velocity, effective Lewis number and environmental conditions. The instability of gas flow and the stretching phenomenon of flame will greatly affect the flame stability of tubular burner. In the process of premixed gas combustion, the ratio of heat transfer to mass diffusion plays a decisive role in flame morphology and stability, and the effective Lewis number is the key parameter of flame stability. In order to reduce the influence of these factors, the method of quantifying the effective Lewis number at different equivalence ratios is adopted. In the study of this paper, the effective Lewis number is first calculated for any working condition to ensure that it is within the stable range. When the effective Lewis number is approximately 1, the flame can be considered stable.

Based on Lapalme's [14] method for determining the

effective Lewis number of multi-component mixed-gas fuel, according to the composition of syngas, the effective Lewis number is calculated at different equivalence ratios [15,16]:

$$Le_{f/O_2} = 1 + \frac{(Le_{exc}-1)+(Le_{def}-1)A_1}{1+A_1} \quad (2)$$

Where, $Le_{f/O2}$ is the effective Lewis Numbers of syngas/oxygen, Le_{exc} is the effective Lewis Numbers of reactant excess and Le_{def} is the effective Lewis Numbers of reactant shortage.Parameters $A_1 = 1 + Ze(\Phi- 1)$, Φ as reactant under the condition of excess and deficiency relative to the quality of the chemical equivalence ratio (when $\varphi \leqq 1$, $\Phi = 1/\varphi$;When the $\varphi > 1$, $\Phi = \varphi$), Ze for Zel' dovich number, expression is:

$$Ze = E_a(T_{ad} - T_u)/(RT_{ad}^2) \quad (3)$$

Where R is the general gas constant, E_a is the activation energy, and the expression is:

$$E_a = -2R\partial\, ln(rS_L)/\partial(1/T_{ad}) \quad (4)$$

Based on the method of Sun [17], the sensitivity of laminar flame velocitys to flame temperature was obtained to determine the value of E_a under the condition of not changing the mixture equivalence ratio by making small adjustments to the content of CO_2.

2.3 Modeling

The simulation was carried out by CHEMKIN-PRO program group [18-20]. The Back Diffusion ffect, Heat Diffusion Soret effect and Molecular Diffusion effect of gas inlet boundary were taken into account in the setting. The mixture of syngas and air was calculated in the Mixture-averaged method. The simulated environmental conditions were normal temperature and pressure (298K, 0.101Mpa), and the air composition was the same as that of the synthetic air. The design parameters CURV were set as 0.05 and 0.05 respectively, and the generated mesh quantity ranged from 350 to 500. The mesh quantity met the requirements of grid independence, and the absolute error of laminar flame velocitys was < 1%.

3 Results and discussion

3.1 Experimental validation

In order to evaluate the reliability of the experimental device, the premixed flame velocity of CH_4/air and syngas/air within a certain equivalent range (298K, 0.1 MPa) was measured, and compared with the literature and model values. The results are shown in the Fig. 3.

For the CH_4/air premixed flame, as shown in Fig. 3., due to the different research methods of scholars, different mixing ratios and measurement errors, the data points are relatively scattered. But in general, each data point is consistent with the laminar flame velocity variation trend simulated by the model. The maximum value of CH_4/air laminar flame velocity occurs between 1.0 and 1.2, with a velocity distribution of 35~41cm/s. The maximum laminar flame velocity obtained by the author is 40.3cm/s (φ=1.1), and the value is within a reasonable range. Between mixed gas φ=0.8~1.1 and other scholars use different methods to measure the data is relatively close, especially in the lean flame ($\varphi \leq 1.0$).

The study of Burbano [21] showed that the prediction ability of different hydrocarbon combustion mechanism models was consistent when the equivalence ratio was less than 1.5, which was also the focus of this paper. At present, the obtained experimental data is slightly higher than the research results of other scholars (only lower than the Lintercis data), which is highly consistent with the model used in the whole range.

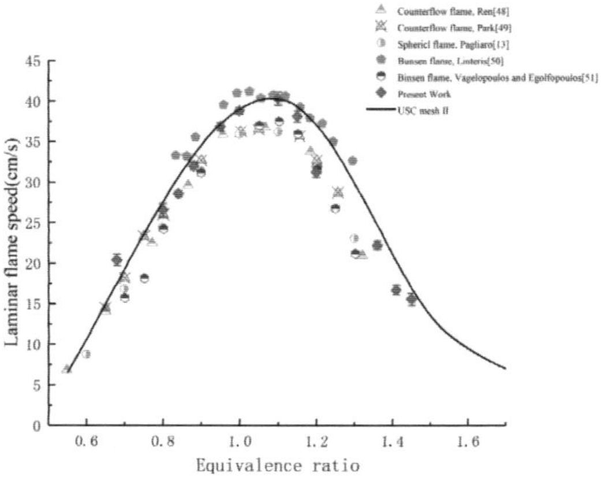

Fig. 3. Flame velocity and calculated adiabatic flame temperature of premixed CH_4/air flames with added C_2HF_5 for different initial equivalence ratios; line: simulations using detailed mechanism.

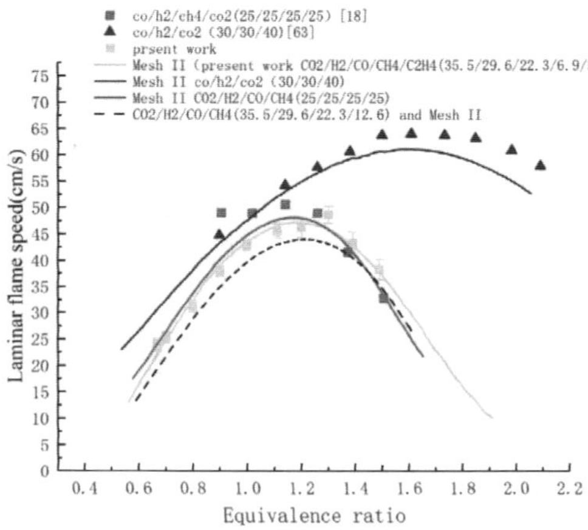

Fig. 4. Laminar flame velocity of syngas/air (298K, 0.1MPa) and previously corresponding published results.

No measurements have been made for the combustion of syngas (component) in lithium ion batteries. According to Le-Chatelier formula [22], the flammable limit calculated in normal conditions (298K, 0.1mpa) syngas/ air in a certain equivalence ratio range is 7.6%~52.5%. The corresponding equivalence ratio is φ=0.22~2.99. Experimental measurement and model simulation were carried out within the range of equivalence ratio, and the results were compared with the research conclusions of other scholars on syngas with similar components, Fig. 4.

As shown in Fig. 4, the experimental value of wang [23] has a large error in the rich combustion area of φ>1.4, and the experimental value has a good consistency with the simulation results under other working conditions. Lapalme [24] showed that the increase of CO_2 and CH_4 content in syngas would reduce the flame velocity and lead to the reduction of the equivalent ratio of the maximum flame velocity. The flame velocity curve moved to the lean flame side, which was more sensitive to CH_4 content. The CO_2 content in the syngas component of the lithium ion battery used is higher than 25%, the CO content is lower than 25%, and the CH_4 content is between 0 and 25%. As a result of the component change, the inhibition effect on flame velocity is enhanced and the promotion effect is weakened, making the experimental and simulated values lower than Lapalme experimental conditions. While in the case of φ>1.4, a situation occurred in which the experimental speed of synthetic gas used was higher than Lapalme experimental condition, which can be explained by the use of synthetic gas H_2:CO>1:1, H_2 in the rich combustion region has a greater impact on the flame velocity[16].

In conclusion, the simulation and experimental results of CH_4/air and syngas/air premixed laminar flames are in good agreement with the research conclusions of other scholars. The verification experiment shows that the device and model can get reasonable experimental results. Although the verification experiment was carried out before adding C_2HF_5, it was inserted into the overall experiment process to ensure the reliability of the device and the repeatability of the experiment.

3.2 Analysis of laminar flame velocitys variation

The USC Mech Version II model has achieved good consistency in the CH_4/ air premix flame process without adding C_2HF_5, but its application in the synthesis gas of lithium ion batteries is unknown. In the current study, the equivalence ratio (not adding inhibitor) within the range of 0.6 ~1.2 add C_2HF_5 (X_a = 0%~5%, within the scope of the published data is more, at the same time, this paper mainly focus on lean burn condition add trace inhibitor changes in the velocity of the flame) of CH_4/air. Premixed flame of synthetic gas/air has carried on the experiment and simulation, C_2HF_5 mechanism research needs based on the experiment the Bunsen burner flame. Inhibitor addition is carried out under the condition of determining the ratio of fuel to air. The initial conditions are T_0=298±2K and P_0= 0.1Mpa. The addition of inhibitors is carried out in the form of the volume proportion in the mixture:

$$X_a = X_{(agent)}/\left[X_{(fuel)} + X_{(air)} + X_{(agent)}\right] \quad (5)$$

Where X (agent) is the mole number of inhibitor components in the mixture. Novec 1230 was selected as the research baseline because its experimental fire-extinguishing concentration was low, and it was also the second generation of halon substitute product, so the data was of certain comparative significance. At the same time, the simulated and experimental values of N_2 only playing a thermal role were set as the background baseline, as shown in Fig. 5.

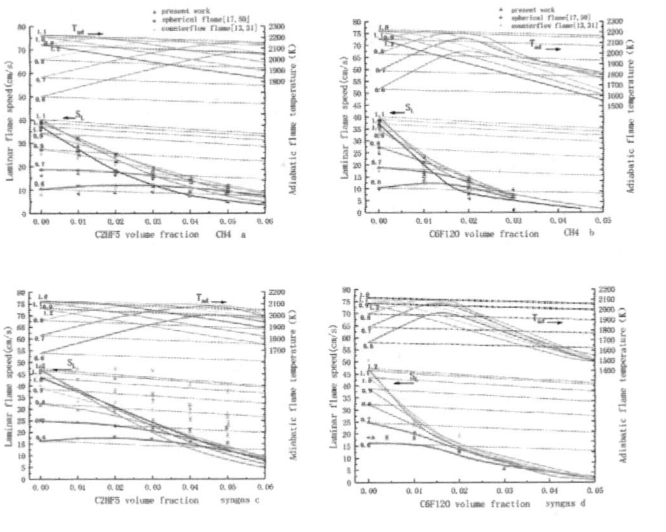

Fig. 5. Flame velocity and calculated adiabatic flame temperature of premixed CH_4/air (a,b) syngas/air (c,d) flames with added $C_6F_{12}O$ (Novec 1230) for different initial equivalence ratios; lines: simulations using detailed mechanism; symbols: experimental data.

Compare methane/air premixed flame velocitys (C_2HF_5 Fig. 5a and Novec 1230 Fig. 5b) measurement value, the predicted values and the published experimental measurement data hedge flame compare, it can be found that the data is very close. That further prove that adopted by the numerical model of equivalence ratio in range of 0.6 ~ 1.2 can better predict inhibitors added case flame properties change. At the equivalent of close to the combustion limit (φ=0.6), the model predicted value is slightly higher than the experimental value. For example, when C_2HF_5 X_a=0.02, the simulated value was 2.6% higher than the experimental value (Bunsen burner flame), and when Novec 1230 X_a=0.02, the simulated value was 30% higher than the experimental value. This is due to the limitation of flame combustion experiment in lean flame side. Compared with other equivalent experiments, flame velocity is lower and flame radiant heat loss is greater. With the increase of gas equivalence ratio, the gap between experimental value and simulated value is further narrowed. From the overall trend, the burnt-out side has a good fitting degree in the range of φ=0.7~1.0. In the condition of rich combustion-side (φ>1), the simulated value showed a trend of being lower than the experimental value, and the inhibitory effect of the inhibitor was overestimated. When φ=1.2, C_2HF_5 X_a=0.02, the simulated value is lower than the experimental value 3.38%, 1230 X_a=0.02, the simulated value is lower than the experimental value 12.25%. This is caused by the entrainment of ambient air by the Bunsen burner flame in an open environment (further burning of excess fuel). However, with the further increase of inhibitor addition, the consistency between the simulated and experimental values became better.

As for premixed syngas/air lean flame at φ= 0.6 (C_2HF_5 Fig .5c, Novec1230 Fig. 5d). The flame velocity increases first and then decreases with the increase of the addition amount of inhibitor (experimental and simulation). Such as relative to add C_2HF_5 flame velocity, the size of the X_a is respectively 1%, 1.5%, 2%, 2.5%, 3% of the cases, the flame velocity increased by 6.32%, 7.38%, 7.13%, 5.48%, 1.99%. When X_a=0.001 and 0.004, the combustion rate increased by 1.8% and 3.9%. The corresponding promotion effect was more obvious in methane flame (ab). Subsequently, the flame velocity decreased with the increase of addition amount. Previous studies have involved the analysis of the reasons for promoting flame combustion [9, 25-28], which is caused by the competitive relationship between the thermal effect of the inhibitor itself as a fuel combustion and the flame inhibition effect of the contained F element. And in the equivalence ratio of φ>0.67 showed inhibitory effect, chemical equivalent and more sensitive to the addition of inhibitors, in the case of φ=0.7, 1.0, 1.2, add C_2HF_5 X_a=0.01, flame velocity reduced by 4.73%, 6.17%, 27.89%, this is because at this time inhibitor containing F inhibition is stronger than its thermal effect as fuel.

However, at the range of 0.6~1.0 equivalence ratio, the consistency between the simulation and the experimental values becomes worse with the increase of inhibitor addition. Respectively, when the equivalence ratio was 0.6, 0.7, 0.8, 0.9, 1.0, and the addition amount of C_2HF_5 X_a=2%, the difference between the experimental and simulated values was 3.42%, 0.9%, 7.13%, 9.22%, and 28.44%. This is due to the limitations of the Bunsen flame and the nature of the syngas itself. The reaction rates of combustible gas components in syngas and oxygen in ambient gas are different, furthermore, under the action of extinguishing agent, the combustion equilibrium of syngas was broken. Nevertheless, the numerical model ignores the large error caused by the flame disturbance caused by the high addition of extinguishing agent. The addition of Novec 1230 also has similar conclusions, but the simulation conclusion of the model is more consistent than that of C_2HF_5.

Further discussion on temperature change. At the lean flame and chemical equivalent flame (φ=0.6~1.0), laminar adiabatic flame temperature with low addition of inhibitor will lead to an increase in temperature. Under the condition of the same proportion of inhibitor addition, the temperature rise of adiabatic flame temperature increases with the lower the fuel equivalent ratio (0.6, X_a (C_2HF_5, Novec 1230)=0.01). Compared with C_2HF_5, the rise of flame temperature is more sensitive to the addition of Novec 1230. Novec 1230 can make the adiabatic flame temperature reach the peak faster at the same low amount of premixed gas. This provides a certain basis for the explanation of the increase of fire speed. Temperature has a particularly significant influence on the fire speed. Previously, some scholars have put forward relevant conclusions about the relationship between temperature and flame velocity [29].

In the process of experimental simulation, it was found that if the flame velocity of 5cm/s (premixed gas φ=0.55) as the laminar flame limit flame velocity[30,31], when mixed gas φ=0.5, using the model is not able to calculate. When adding C_2HF_5 X_a=0.01, laminar flame velocity is 10cm/s. From the perspective of simulation, this indicates that the addition of inhibitors with fuel properties will expand the combustion limit of mixed gas [9,12,32,33]. Previously, some scholars used constant volume incendiary bombs to conduct experiments and obtained similar conclusions[10], which further supported the conclusion of promoting combustion with low equivalence ratio and low addition, and verified the numerical simulation results in this article.

For the premixed syngas used, USC Mech Version II model can obtain better simulation results in the case of mixed gas in the lean combustion side and adding low inhibitory dose. Even in the case of poor consistency in the rich combustion side simulation, but in the case of φ>1, inhibitors show inhibitory effect, which is not the focus of this paper. Compared with previous applications of CH_4 and C_3H_8 flame[9,10], C_2HF_5 and Novec 1230 are more suitable for the flame generated by uncontrolled heat gas in lithium ion batteries. In fact, from the comparison of data in ac and bd graph, we can see that under the condition of φ=0.6 and the addition of inhibitor X_a=0.01, the combustion rate of methane flame increased by 10.93% (C_2HF_5) and 17.43% (Novec 1230). For the gas distribution flame, the flame velocity increased by 6.32% (C_2HF_5) and decreased by 1.51% (Novec 1230) with the same amount of additive. The different flame promotion effect is related to the fuel properties of the inhibitors used. With the same addition amount, Novec 1230 can make the flame temperature reach the peak earlier, and the peak value moves to the rich combustion side. In terms of the flame promotion effect on the poor combustion side, Novec 1230 promotes the peak flame velocity more than C_2HF_5 (methane flame).

For the condition of rich combustion (as shown in Fig. 6), the simulation results can be compared. When the same amount of inhibitor was added, the inhibitory effect of Novec 1230 was better than that of C_2HF_5 (as can be seen from the tangent slope of the curve in the Fig. 6.).

For φ =1.0~1.6, the initial flame velocity and temperature is still at a high value, when adding a small number of inhibitors can quickly get a better control. With the gradual increase of inhibitor addition, the slope of laminar flame velocitys and temperature change gradually decreases, and finally reaches a stable value at a high dosage, which is consistent with the description of inhibitor saturation effect in literature [34,35]. Therefore, the focus of this paper is still a discussion on the demarcation point of promoting and inhibiting the burnt-out side.

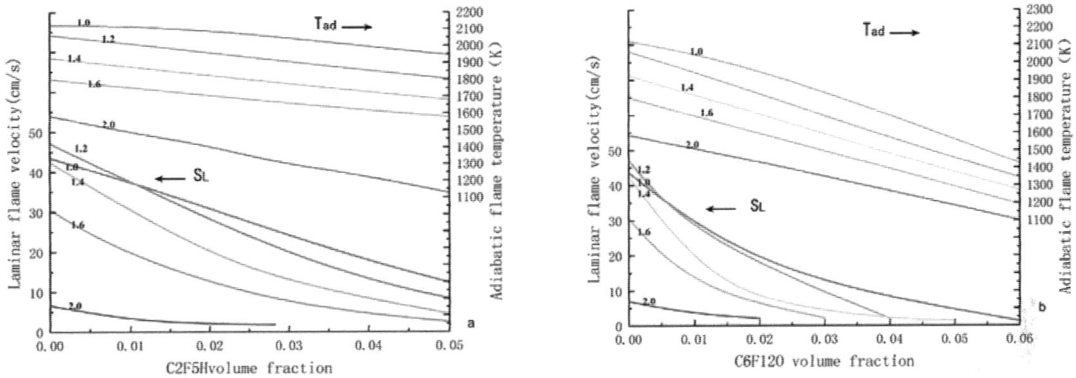

Fig. 6. Burning velocity and calculated adiabatic flame temperature of premixed syngas/air (a,b) flames(φ≥1) with added C_2HF_5 (Novec 1230) for different initial equivalence ratios; lines: simulations using detailed mechanism.

3.3 Changes of flame morphological photographs

In previous studies on the application of halon substitutes, many scholars used pool fire experiments or cup-burner experiments loaded on the oxidant end, which would generate similar diffusion flames similar to the actual flames. Therefore, the experimental conclusions are often used to guide practical engineering applications. However, the mechanism and efficiency of action of inhibitor and fuel gas in premixed condition are different, so it is necessary to conduct Bunsen burner flame experiment. For the measurement of laminar flow velocity in premixed combustion, strict experimental conditions are required for the analysis of instantaneous schlieren images obtained from constant volume incendiary bomb experiments. However, the structure of the opposed flame is different from the actual flame because it is a one-dimensional plane flame.

With the inhibition of syngas/ air flame as an example, Fig. 7. shows the equivalence ratio of φ=0.6, added C_2HF_5 (added amount X_a=0~0.04) and $C_6F_{12}O$ (added amount X_a=0~0.02) syngas/air premixed flame photos (inhibitor involved in CH_4/air premixed flame combustion process chemical thermodynamics and kinetic mechanism has been confirmed).

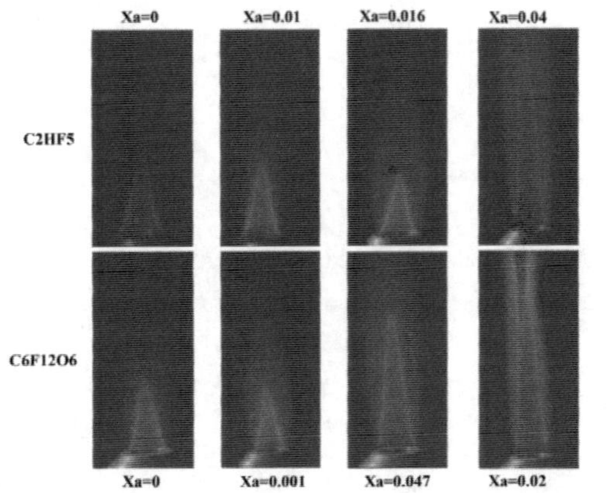

Fig. 7. Morphological photographs of the premixed syngas/air (φ=0.6)Bunsen flames with the addition of 0~0.02 $C_6F_{12}O$(upper flames) and 0~0.04 C_2HF_5(lower flames); conditions: (298K, 0.1MPa).

When no inhibitor was added, the conical flame was light blue with obvious contour. When the trace inhibitor was added (C_2HF_5 X_a=0.005, Novec 1230 X_a=0.001, which showed the effect of promoting flame combustion), the flame turned bright blue and the reaction area was thickened. With the increase of the amount of inhibitor, the flame length was further shortened (the combustion rate was increased). When the C_2HF_5 was X_a=0.016 and the $C_6F_{12}O$ was X_a=0.0047, the flame combustion rate reached the extreme value of the promotion effect under this equivalent. At this time, the flame was the most active and brighter, reflecting a significantly larger area (compared with the condition without inhibitor). This is because the inhibitor C_2HF_5, Novec 1230) itself, as part of the fuel, participates in the combustion process, consumes oxygen and produces a heating reaction, which promotes combustion and expands the flame reaction area. When the amount of inhibitor was further increased, the flame brightness decreased while the length was lengthened. As the condition for the existence of the conical stable flame of Bunsen burner is that the flame combustion velocity is equal to the airflow velocity, when the flame velocity decreases, the root of the flame tends to break away from the edge of the burner under the action of no gas (on the contrary, it is tempering). In fact, at this time, the premixed flame has reached the point where it can be extinguished at random under certain concentration of inhibitor (that is, small changes in air flow may cause the flame to blow off). In addition, as the density of combustion gas decreases under the action of heat, the flame will be aggravated by the influence of buoyancy, which will also lead to the extinction of the flame.

Bunsen burner flame (low yield, low inhibitory concentration to promote combustion) can clearly reproduce the phenomenon of flame intensification in the FAA-ACT experiment, and can intuitively observe the stable and bright conical premixed flame. At the same time, under the action of inhibitor, the flame burning speed, thermal diffusion and frontal position will change. Therefore, the application of premixed flame test equipment of tubular burner can also be used to deeply understand the phenomenon of flame intensification in FAA-ACT experiments.

4 Conclusions

The promotion/inhibition properties of the second generation halon replacement extinguishing agent C_2HF_5 (Novec 1230 and N_2 as pairs) on the heat runaway gas syngas flame (CH_4/ air premix flame as contrast) of the lithium ion battery were studied by Bunsen burner experiment. The parameters of laminar flame velocity and adiabatic flame temperature with addition of C_2HF_5 (X_a=0~0.06), Novec 1230 (X_a=0~0.05) and N_2 were studied and simulated in the range of equivalence ratio φ =0.6~1.2. The cone Angle method was used to measure the flame velocity, and the USC Mech Version II model was experimentally verified. With a low additive amount on the lean side (φ=0.6~1.0, the inhibitor additive amount range was related to the nature of the inhibitor used), a good consistency was obtained, and a large deviation was observed in the simulation on the rich combustion side.

In terms of laminar combustion velocity, compared with previous applications of CH_4 and C_3H_8 flame, C_2HF_5 and Novec 1230 are more suitable for the flame generated by thermal runaway gas in lithium ion batteries. With φ =0.6 and inhibitor addition X_a=0.01, the combustion rate of methane flame increased by 10.93% (C_2HF_5) and 17.43% (Novec 1230). For the gas distribution flame, the flame velocity increased by 6.32% (C_2HF_5) and decreased by 1.51% (Novec 1230) with the same amount of additive. According to the kinetic model analysis, the inhibitor will promote the reactivity under the condition of low equivalent and low addition, and then increase the flame burning speed.

Bunsen flame reappeared the flame intensification phenomenon in the FAA-ACT experiment with low equivalent and low inhibitory concentration, and the flame intensification effect could be clearly observed through the change of flame shape, position and color. The results show that the flame height decreases with the increase of laminar flame velocity. When the flame is promoted by the inhibitor, the flame is brighter and a lot of heat is released. The success of Bunsen flame experiment provides a convenient experimental approach for further analysis of combustion products. In practical engineering application, the dosage of inhibitor is usually higher than the designed concentration. However, in the case of low concentration addition, the role of promoting combustion cannot be ignored, and this characteristic also exists in the case of premix of other fluorocarbon inhibitors and hydrocarbon fuels. Current experiments show that it is feasible to use Bunsen burner flame to study the flame promotion/inhibition effect of inhibitors.

Acknowledgement

This work was sponsored by the National Natural Science Foundation of China (51804314), Natural Science Foundation of Hebei Province (E2019507007) and Key R&D Program of Hebei Province (No. 20375503D) and Key Scientific Research Projects in China People's Police University (No. 2019zdgg006).

References

1. Martínez C.M, Hu X, Cao D, Velenis E, et al. IEEE Trans. Veh. Technol **66**, 1-16 (2016)
2. Feng X, Fang M, He X, et al. J. Power Sources **255**, 294-301 (2014)
3. Koch S, Fill A, Birke KP. J. Power Sources **398**, 106-112 (2018)
4. Benedick. Int. N. **1**, 231-246 (1987)
5. Association NFP. NFPA 2001 (2018 Edition). MA 021692018.
6. Standardization IOf. ISO14520-1(Third edition 2015-12-01). 2015.
7. Reinhardt J.W. Behavior of bromotrifluoropropene and pentafluoroethane when subjected to a simulated aerosol can explosion. In: Administration FA, editor. Washington,D.C. USA2004.
8. Reinhardt J.W. Aircraft cargo MPS test of FK-5-1-12. In: Administration FA, editor. International Aircraft Systems Fire Protection Working GroupOctober 25-26, 2006.
9. Pagliaro J.L, Linteris G.T, Sunderland P.B, et al. Combust. Flame **162**, 41-49 (2015)

10. Babushok V.I, Linteris G.T, Burgess Jr. D.R, et al. Combust. Flame **162**, 1104-1112 (2015)
11. Takahashi F, Katta V.R, Linteris G.T, et al. Proc. Combust. Inst. **35**, 2741-2748 (2015)
12. Liu L, Du Z, Zhang T, et al. Int. J. Hydrogen Energy **44**, 22282-22300 (2019)
13. Bouvet N, Chauveau C, Gkalp I, et al. Int. J. Hydrogen Energy **36**, 992-1005 (2011)
14. Linteris G.T, Babushok V.I, Sunderland P.B, et al. Proc. Combust. Inst. **34**, 2683-2690 (2013)
15. Addabbo R, Bechtold J.K, Matalon M. Proc. Combust. Inst. **29.2**, 1527-1535 (2002)
16. Bechtold J.K, Matalon M. Combust. Flame **127**, 1906-1913 (2001)
17. Sun C.J, Sung C.J, He L, et al. Combust. Flame **118**, 108-128 (1999)
18. Kee RJ, Rupley FM, Miller JA. Chemkin-II: A Fortran chemical kinetics package for the analysis of gas-phase chemical kinetics. United States1989.
19. CHEMKIN-PRO 15101. Reaction design: San Diego, CA; 2013.
20. Kee RJ, Warnatz J, Miller JA. Fortran computer-code package for the evaluation of gas-phase viscosities, conductivities, and diffusion coefficients. [CHEMKIN]. United States1983.
21. Burbano H.J, Pareja J, Amell A.A. Int. J. Hydrogen Energy **36**, 3232-3242 (2011)
22. Lau L.J. J Econ Theory. **12**, 131-163 (1976)
23. Wang Z.H, Weng W.B, He Y, et al. Fuel. **141**, 285-292 (2015)
24. Lapalme D, Seers P. Int. J. Hydrogen Energy **39**, 3477-3486 (2014)
25. Moffat R.J. Exp. Therm Fluid Sci. **1**,3-17(1988)
26. Pagliaro J.L, Linteris G.T. Fire Saf. J. **87**,10-17 (2017)
27. Wu, Xu, Yong, et al. J. Fire Sci. **34**, 289-304 (2016)
28. Ren X, Jiang Y, Xu W. J. Fire Sci. **34**,416-430(2016)
29. Stephen R. An Introduction to Combustion: Concepts and Applications. Boston2000.
30. Egerton A, Thabet S.K. Proc. R. Soc. London, ser. A. **211**, 445-471 (1952)
31. Westbrook C.K. Combust. Sci. Technol. **34**, 201-225 (1983)
32. Shebeko Y.N, Azatyan V.V, Bolodian I.A, et al. Combust. Flame **121**, 542-547 (2000)
33. Azatyan V.V, Shebeko Y.N, Shebeko A.Y, et al. Russ. J. Phys. Chem. A **4**, 760-768 (2010)
34. Noto T,Babushok V,Hamins A, et al. Combust. Flame **112**, 147-160 (1998)
35. G.T. Linteris AM, Tsang W. Halon Replacements ACS Symposium Series 611: American Chemical Society; p. 260-274(1995).

An Empirical Study on the Relationship between Atmospheric Environmental Quality and Economic Growth in Shaanxi Province

Rong Guo[1*]

[1]Xi'an Jiaotong University City College, Management department, Shaanxi province, 710018, China

Abstract. According to the panel data of Shaanxi province from 2005 to 2018, the simultaneous equations were constructed through cubic polynomials to study the relationship between the atmospheric environment and economic growth in Shaanxi. The results show that the relationship between air pollution and GDP per capita in Shaanxi province presents an inverted n-shaped curve, which indicates that the economic growth in Shaanxi province causes the fluctuation of air quality, and reflects the relationship that the air pollution in Shaanxi improves first, then worsens and finally improves further with the economic growth. At the same time, the research results show that adjusting the industrial structure, increasing investment in environmental management, reducing the emission of atmospheric pollutants, and reducing energy consumption can promote the sustainable development of the Shaanxi economy and improve the quality of the atmospheric environment.

1 The introduction

China implements the strategy of developing the western region to promote the coordinated development of the regions in 2000. General secretary Xi Jinping has issued the "One Belt And One Road" initiative in 2013. Shaanxi plays the role of bridgehead in western development and is also the core area of "One Belt And One Road", which is the important fulcrum of the Eurasian land bridge. At the same time, High and frequent air pollution is a threat to people's health, Stable and rapid economic development affects people's life. How to find a balance between atmospheric environment and economy has become the focus of the government and many scholars. Because of this, based on the time-series data from 2005 to 2018 and panel data from 10 cities and 1 demonstration area in Shaanxi, this paper uses empirical analysis to study the relationship between economic growth and the atmospheric environment in Shaanxi, to provide a decision-making basis for sustainable economic development and atmospheric environment improvement.

2 Index selection and model setting

The relationship between economic development and environmental quality began at the end of the 20th century. The representative scholars were Grossman and Krueger (1991), who concluded that there was an "inverted U-shaped" curve between them through the analysis of economic and environmental data of the North American Free Trade Area. David I. Stern (1994), Shaw (2001), and other scholars supported the existence of an "inverted U-shaped" curve between economy and environment through research. Shafik (1994), Hannes (2001), and other scholars concluded through empirical research that there are other curve characteristics between economy and environment, such as "positive U-shaped", "positive N-shaped" and "inverted N-shaped".

Domestic scholars began to study the relationship between economic development and environmental quality at the beginning of the 21st century. For example, Yuping Wu (2002) conducted an empirical study to simulate the curve relationship between environment and economy by collecting data from Beijing from 1985 to 1999. According to the research of domestic and foreign scholars, the indexes selected in this paper are as follows:

2.1 Index selection

Since environmental data have been published since 2005, the original data collected in this paper are from The Statistical Yearbook of Shaanxi province, The Statistical Yearbook of Shaanxi Province, and the Statistical Yearbook of Xi'an from 2005 to 2018. The indicators are obtained through collation and corresponding calculation.

Also, since the emission indicators of Xi Xian New Area were not counted until 2017, to maintain consistency, this paper selected 10 cities and 1 demonstration area that were continuously counted in the statistical yearbook of Shaanxi province. The indicators selected are as follows:

- Economic growth, measured by *Per capita GDP*.

* Corresponding author: 121415940@qq.com

- Atmospheric environment quality (*Q*). According to the statistical indicators of the statistical yearbook, three indicators were adopted in this paper, namely, per capita sulfur dioxide emissions (*SO_2, ton / 10,000 people*), per capita NOX emissions (*NO, ton / 10,000 people*), and per capita carbon emission (*S, ton / 10,000 people*).
- Air pollution regulation (*F*) is represented by three indicators, namely, the number of exhaust gas treatment facilities per capita (*F_a, set / 10,000 people*), the treatment capacity of exhaust gas treatment facilities per capita (*F_b, cubic meter/hour/person*) and the operating cost of exhaust gas treatment facilities per capita (*F_c, yuan/person*).
- The industrial structure is measured by two indicators, namely, the per capita gross industrial output value (*IOV, yuan/person*) and the per capita energy consumption of industrial enterprises above the scale (*C, ton standard coal/person*).
- Control variables, other indicators that affect economic development, and atmospheric pollutant emissions. Population density (*P, people/ square km*) is obtained by dividing the number of permanent residents in each city by its land area at the end of each year. Annual precipitation (*TP, ml*) is represented by the annual precipitation of each city.

2.2 Model building

This paper constructs simultaneous equations based on cubic polynomials to investigate the relationship between atmospheric environment and economic growth in Shaanxi province.

The air quality is represented by three indicators, namely, per capita sulfur dioxide emissions, per capita NOX emissions, and per capita carbon emission.

$$LnGDP_{it}=\alpha_0+\alpha 1lnQ_{it}+\alpha_2 lnIOV_{it}+\varepsilon_{it} \quad (1)$$

$$LnQ_{it}=\beta_0+\beta_1 lnGDP_{it}++\beta_2(lnGDP_{it})^2+\beta_3(lnGDP_{it})^3 +\beta_4 lnF_{it}+\beta_5 lnC_{it}+\beta_6 lnP_{it}+\beta_7 lnTP+\delta_i \quad (2)$$

Equation (1) is the output equation, indicating that in the form of Cobb-Douglas production function, GDP per capita is atmospheric environment quality (*Q*) and industrial output per capita above the scale (*IOV*).

Equation (2) is the environmental equation, where air pollution (*Q*) contains three proxy variables, namely, per capita carbon emission (*S*), per capita sulfur dioxide emission (*SO_2*) and per capita nitrogen oxide emission (*NO*).

By the regression coefficient of the above equation β, the relationship curve between the atmospheric environment and economic growth can be judged. The curve may present the following situations:

Table 1. Relationship between parameter symbols and curve morphology.

Coefficient values	**The relationship between economic growth and atmospheric environment**
$\beta_1=\beta_2=\beta_3=0$	No correlation
$\beta_1>0$, $\beta_2=\beta_3=0$	Linear relation: Monotone increasing
$\beta_1<0$, $\beta_2=\beta_3=0$	Linear relation: Monotone decreasing
$\beta_1>0$, $\beta_2<0$, $\beta_3=0$	Inverted U-shaped curve
$\beta_1<0$, $\beta_2>0$, $\beta_3=0$	U curve
$\beta_1>0$, $\beta_2<0$, $\beta_3>0$	N curve
$\beta_1<0$, $\beta_2>0$, $\beta_3<0$	Inverted N-shaped curve

From 2005 to 2010, the exhaust emission index of Shaanxi Statistical Yearbook did not count the per capita NOX emission index. In 2018, the emission targets of Xi'an city were counted in Xi ' a Statistical Yearbook, but none of the other cities were counted. Therefore, the original data of this paper include 10 cities and 1 district in Shaanxi province, with a total of N = 143 observed values. The non-equilibrium panel is used for empirical study.

The main statistical values of each variable are shown in Table 2.

Table 2. Main statistics of variables.

Variable code	**The mean**	**Standard error**	**The minimum value**	**The maximum**
GDP	32928.02	21703.49	4800	112845
S	262.35	613.34	0.96	4050.57
SO_2	164.07	184.68	3.06	974.35
NO	137.78	168.89	3.68	655.46
F_a	1.44	1.40	0.29	11.26
F_b	47.75	221.55	0.19	1707.43
F_c	74.55	83.39	2.17	471.21
IOV	19168.10	13733.70	1850.86	71038.40
C	5.50	9.96	0.00	48.65
P	372.55	408.06	57.25	1550.37
TP	669.65	362.62	248.70	4318.00

Before the parameter estimation of the antithetical equation model, according to the results in Table 2, the parameter identification problem is first investigated. According to the order conditions of the identifiability, the model set in this paper has many exogenous variables, and both the output equation and the environmental equation belong to over-identification, which is similar to the research results of Ying Feng. Secondly, through the Hausmann test, the model established in this paper has a simultaneous relation, and the simultaneous equation model of the atmospheric environment and economic growth is established.

3 Empirical results and analysis

3.1 The empirical results

This paper USES 3 SLS STATA15.0 software for data processing, at the same time, due to the atmospheric environmental quality by the per capita emissions of sulfur dioxide (SO2), Per capita NOX emissions (NO), Carbon emission per capita (S), three atmospheric pollution indicators to measure, so the regression results show that the results of the three equations, the results are shown in Table 3.

Table 3. Regression results.

Equation	The output equation			Environmental equation		
Result	lnGDP	lnGDP	lnGDP	lnS	$LnSO^2$	lnNO
lnGDP				-1.467	-1803.0	-1.318
				(-0.32)	(-0.82)	(-0.29)
$(lnGDP)^2$				0.621	179.3	0.569
				(0.71)	(0.83)	(0.69)
$(lnGDP)^3$				-0.0386	-5.929	-0.0351
				(-0.90)	(-0.83)	(-0.90)
lnS	-0.0050***					
	(-4.50)					
$lnSO^2$		-0.0018*				
		(-1.89)				
lnNO			-0.0008**			
			(-2.38)			
lnF_a				0.104	0.886	0.817***
				(0.24)	(0.83)	(3.70)
lnF_b				-0.0126	0.184	0.326**
				(-0.06)	(0.62)	(2.42)
lnF_c				0.181	0.351	0.348**
				(0.62)	(0.98)	(2.42)
lnC				0.333	0.906	0.341***
				(1.32)	(1.58)	(2.94)
lnIOV	0.995***	0.998***	1.002***			
	(630.92)	(754.50)	(1430.65)			
lnP				-0.111	0.351	0.0571
				(-0.33)	(0.67)	(0.41)
lnTP				-0.704	-0.184	-1.058***
				(-0.93)	(-0.16)	(-2.78)
Constant	0.0982***	0.0418	-0.0032	0.0001***	6026.387	0.0001
	(3.30)	(1.60)	(-0.24)	(3.30)	(0.82)	(0.30)
N	110	110	60	110	110	60
R^2	0.9999	0.9999	0.9999	0.2260	0.2706	0.6938

Annotation : ***, p<0.01, **, p<0.05, *, p<0.1.

3.2 Results analysis

3.2.1 The result analysis of output equation

According to the empirical results of the output equation, firstly, the impact coefficients of the three air pollution indicators in the output equation on per capita GDP were all significantly negative, indicating that the economic development level of Shaanxi would decrease with the increase of the emission of air pollutants.

A 1% increase in soot emissions will lead to a 0.005% reduction in per capita production, a 1% increase in SO2 emissions will lead to a 0.002% reduction in per capita production, and a 1% increase in NO emissions will lead to a 0.0008% reduction in per capita production, that is, the emission of air pollutants will lead to a reduction in per capita output, affecting per capita income.

Second, the per capita industrial output value of the city above the scale of 99% confidence, the impact coefficient on per capita output is a significantly positive correlation, indicating that this factor is an important factor affecting per capita output.

3.2.2 The result analysis of the environmental equation

First, analyze the result of atmospheric environment quality and economic growth. result analysis. From the perspective of the environmental equation, the calculation results of the estimated coefficients of the three atmospheric environmental pollution indicators selected in this paper show that: β1 the results for -1.467, -1803.0 and -1.318, namely β1<0; β2 the results for 0.621, 179.3 and 0.569, namely β2>0; β3 the results for -0.0386, -5.929 and -0.0351, namely β3<0. According to the parameter symbol and curve shape in Table 1, β1<0, β2>0, β3<0, that is, the relationship between atmospheric environmental pollution and per capita GDP shows an "Inverted N-shaped" curve, and economic growth causes fluctuations in atmospheric environmental quality. The inverted N-shaped curve reflects that atmospheric environmental pollution first improves the environmental problems with economic growth, then enters the stage of environmental deterioration, and finally improves the environmental problems. This indicates that Shaanxi's economy grows slowly in the early stage and its consumption of resources and energy is relatively low. In the middle of the statistical fourteen years, rapid economic growth accelerated the consumption of resources and energy, which led to the aggravation of air pollution. In recent years, the government and the public have realized the importance of environmental issues and reduced the emission of air pollution by adjusting the industrial structure and upgrading the industry.

Secondly, the results of air pollution regulation and air environment quality are analyzed. There are three indicators of air pollution regulation, among which the number of exhaust gas treatment sets per capita is positively related to the air environment quality. The capacity of per capita exhaust gas treatment facilities has a positive relationship with per capita sulfur dioxide emissions and per capita nitrogen oxide emissions, and a negative relationship with per capita smoke (dust) emissions. The operating cost of exhaust gas facilities per capita is positively related to the three indexes of atmospheric environment quality. The three indicators of air pollution regulation and the three indicators of air environment quality are only positively correlated with per capita NOX emissions, and the coefficient is significantly correlated. The results are similar to the conclusions of Feng Ying et al., Tongsheng Xu, and other scholars. This indicates that air pollution control has not improved environmental pollution. The reasons may be as follows: the facilities invested in air pollution control and the cost are less. According to the environmental governance data of developed countries, to improve the environment, environmental protection funds need to be invested in 1% to 1.5% of the gross domestic product. However, in 2017, for example, Weinan city and Lin Yu city, which are heavily polluted, invested 0.66% and 0.29% of their GDP respectively. Secondly, the facilities and costs of governance are not stable and continue to increase in accordance with the emission of pollution. Xi'an increased 206 governance facilities in 2017 over the previous year, but the cost decreased by 29.7 percent. The lack of stability of government facilities and costs may be the reason for this empirical result.

Third, the results of the analysis of the industrial structure and atmospheric environment quality. Energy consumption of industrial enterprises above the per capita size is positively correlated with atmospheric environmental quality, and only significantly correlated with nitrogen oxide emissions per capita. This result shows that per capita NOX emissions increase with the increase of energy consumption of industrial enterprises above the size. From 2005 to 2010, the energy consumption of large-scale industrial enterprises continued to increase, and the emission of air pollutants increased, reaching the highest level in 14 years of statistical data. In the three years from 2011 to 2013, the energy consumption of industrial enterprises above a certain size decreased, but the decline was small. From 2013 to 2018, with the upgrading and optimization of industrial structure, the energy consumption of industrial enterprises above a certain scale has been greatly reduced, and the emission of air pollutants has been reduced accordingly.

Fourth, analyze the results of population density, total annual precipitation, and atmospheric environment quality. The indicators of population density on the atmospheric environment are positively correlated with per capita sulfur dioxide emissions and per capita nitrogen oxide emissions, but negatively correlated with per capita smoke (dust) emissions, and are not significant with the empirical results of the three, indicating that the increase of population density does not have an impact on atmospheric environment quality. The total annual precipitation has a negative relationship with the three indicators of atmospheric environment, which is only significant for the per capita NOX emissions. According to the empirical results, precipitation has little influence on the two indicators of SULFUR dioxide emissions and NOX emissions, and precipitation has little effect on their improvement. The reason may be that Shaanxi is a long and narrow region with high terrain from north to south and low terrain in the middle, which is not conducive to the outward diffusion of pollutants. It may even cause the transport of external pollutants to the central region due to wind direction, humidity, and other problems, leading to the accumulation of atmospheric pollutants.

4 Conclusions

In this paper, according to the Shaanxi province in 2005-2018 panel data for empirical research, according to the results of empirical analysis, Shaanxi atmospheric environment quality and the strong correlation between the economic development.

There is an inverted U-shaped curve relationship between atmospheric environmental quality and economic development, which reflects that atmospheric environmental pollution first improves the environmental problems with economic growth, then enters the stage of environmental deterioration, and finally improves the environmental problems.

Second, the emission of air pollutants will lead to a decrease in per capita output, affecting per capita income. An increase of 1% in the emissions of air pollutants for three indicators would result in a reduction of about 0.003% in per capita production.

Finally, with the increase in energy consumption per capita of industrial enterprises above a certain size, the emission of air pollution increases.

Acknowledgment

Fund Items: 2019 Scientific Research Plan of The Education Department of Shaanxi Province, An empirical study on the relationship between atmospheric environmental quality and economic growth in Shaanxi province (Project Approval No. :19JK0493)

References

1. Hongyun Bu, Jie HUANG, Research on the Harmonious Relationship between Industrial Structure and Environmental Quality in Beijing-Tianjin-Hebei Region, Commercial Economics Research, **5** (2018).
2. Ying Feng, Xiaoning LI, Guojun QU, Mei Xu, Research on the relationship between water environmental pollution and economic growth in China, Journal of northwest a&f university (social science edition), **17** (2017).
3. Danlin Yan, Lu GAO, Research on energy Structure Adjustment of Urban Haze Control in China, Commercial Economics Research, **13** (2015).
4. Feng He, Dongdong Ma, Research on the correlation between haze and industrialization development, Soft Science, **29** (2015).

Research on Magnetic Property of Environmental Friendly Material SnO_2: Mn, S

Su Lei[1], Huang Yi[1,*]

[1] School of Electronic Information, Southwest Minzu University, 610041Chengdu, China

Abstract. Tin dioxide (SnO_2) is a commonly known material with the rutile structure of wide band gap n-type semiconductor which is widely used like ZnO common oxide materials in daily life. But comparing with ZnO, it has a wider band gap (about 3.6 eV), and a higher exciton binding energy 130 meV. Because of its excellent optical, electrical and other excellent physical and chemical characteristics, SnO_2 has been widely adapted in thermoelectric film, gas sensor, photovoltaic devices, magnetic materials, and other related fields. A large number of theories and experiments illustrate that, after the proper doping, the remarkable improvements can be achieved. Based on the first principle, we investigated the photoelectric properties and magnetic properties when the Mn and S were doped in SnO_2. It was shown by calculation that a Mn atom provides 1.52 μB magnetic moment and a S atom provides 0.06 μB, while O and Sn atoms rarely contribute to the system. In the system the magnetism is mainly derived from the Mn-3d electronic spin polarization.

1 Introduction

As an environmental friendly material, SnO2 has a rutile structure and is also an n-type wide bandgap semiconductor with a band gap of 3.6 eV. In an experiment, due to the successful preparation of TiO2 diluted magnetic semiconductor [1], people began to broaden their horizons to other oxide semiconductor materials, such as tin dioxide we are going to talk about here. Since then, people began to study SnO2 dilute magnetic semiconductors both experimentally and theoretically. Dietl and his collaborators used pulsed laser deposition (PLD) technology to grow Co-doped SnO2 thin film samples [2-4]. They found that when the Co doping concentration is less than 3%, the sample film will show ferromagnetism and be highly transparent. The Curie temperature of the sample is close to 650 K, the saturation magnetization is about 7.5±0.5 μB/Co, and the magnetic moments in the Co and Co clusters are 1.67 μB/Co and 2.1 μB/Co, respectively. It can be seen that the film The cobalt magnetic moment of the sample is much larger than that of cobalt and cobalt clusters and that of cobalt in other cobalt oxides, but there is a little regret that its magnetization will rapidly weaken as the Co concentration increases.

Fukumura et al. also successfully prepared iron-doped SnO2 semiconductor materials by pulsed laser deposition technology, and the prepared samples showed ferromagnetic and highly transparent characteristics [5]. Fitzgerald and his collaborators took the work one step further and found room temperature ferromagnetism in Sn1-xMxO2 (M= Mn, Fe, Co, x=0.05) semiconductor materials [6]. Fitzgerald and his collaborators later studied the use of PLD to prepare transition metal doped SnO2 [7-9].

Menzel and his collaborators used magnetron sputtering to prepare Co-doped SnO2 thin film samples on a silicon substrate. They doped the thin film samples at a concentration of 3%, 5%, and 7%, respectively. Among the three samples prepared All have room temperature ferromagnetism, and the cluster size of Co in the sample is less than 1 nm [10]. Mishra et al. studied Fe-doped SnO2 nanoparticles, and the experimental results showed that the sample showed paramagnetism, and the Fe3+ in the sample was in a paramagnetic state [11-15].

Mn is also often used as a transition metal cation to dope SnO2 crystals to study its magnetic modification effect. Yu Li, Zheng Guang and others mentioned that the magnetic moment is mainly derived from the 3d orbital of the doped element in the calculation of Mn doped SnO2. Spin polarization of electrons [16]. The doping of S can introduce oxygen vacancies, and the effect of the presence of oxygen vacancies on the magnetic properties of the system cannot be ignored. As there is no report about the co-doping of Mn and S with SnO2, this article will explore the co-doping of Mn and S with SnO2.

2 Calculation method and model

In this paper, we use the software package to calculate. The specific calculation parameters are set as follows: select the PBE exchange correlation potential under the generalized gradient approximation (GGA), set the cut off energy to 300 eV, set the k-point grid to 4×4×6, and set the initial Spin is set to 3. When the total energy change of the system stabilizes within 1.0×10^{-5} eV, the self-consistent field (SCF) of the electronic structure stops calculating. In the calculation process, the selected valence electron

* Corresponding author: soliderhy@126.com

configurations of tin, oxygen, manganese and sulfur atoms are ($5s^25p^2$), ($2s^22p^2$), ($3d^54s^2$) and ($3s^23p^2$) respectively. The calculation model is shown in Figure 1. This 2×2×2 supercell has 48 atoms in total (15 Sn atoms, 31 O atoms, 1 Mn atom and 1 S atom). The doping method is that one Mn atom replaces one Sn atom, and one S atom replaces one O atom. The doping concentrations of Mn and S are 6.25% and 3.125%, respectively.

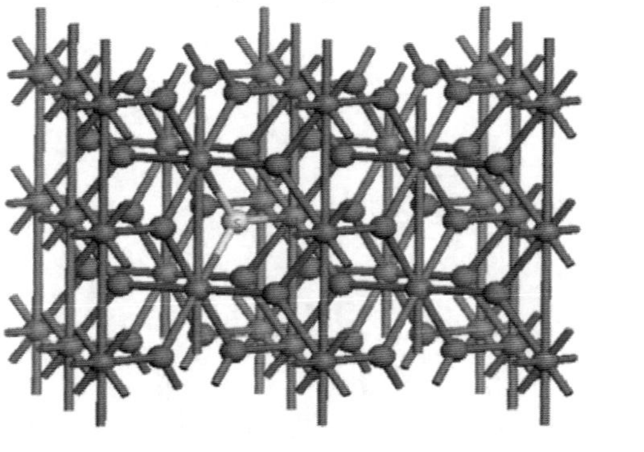

Fig. 1. The model graph Cu, S co-doping SnO_2.

3 Results and discussion

3.1 Magnetic properties

According to the above model and calculation parameters, this paper calculates the Mn and S co-doped SnO_2 system. After optimizing the supercell, the lattice constant is divided into a=4.847 Å, c=3.316 Å, and the magnetic moment distribution of the system is calculated. The results show that the main magnetic moment of the system comes from Mn atoms. Of course, the interaction of Mn atoms with S atoms and the nearest neighbor O atoms also provides part of the magnetic moment. The magnetic moment of Mn atom is 1.52 μB, the magnetic moment of S atom is -0.06 μB, and the magnetic moment of O atom ranges from 0 to 0.02 μB. Sn atoms basically have no contribution to the magnetic moment. The total magnetic moment is 1.5 μB. The integer magnetic moment shows that the doped SnO_2 is not a semi-metal. We will further discuss this issue through the density of states. Although O atom makes a small contribution to the magnetic moment, we can still find that O atom's contribution to the magnetic moment depends on its position. It is O atom, Mn's nearest neighbour, that produces part of the magnetic moment, while other O atoms make little contribution to the magnetic moment. As we all know, the physical and chemical properties of materials mainly depend on the electronic structure of the materials. Therefore, it is extremely important to study the electronic structure of doped SnO_2 samples. At the same time, in order to further study the origin of the magnetism of the system, we calculated the relevant electronic density of states diagram, as shown in Figure 2. It can be seen from the figure that the doped SnO_2 undergoes spin cleavage. In the total density of states, the number of spin-up electrons is more than the spin-down electrons. It can be found from the partial density of states that the magnetism of the system is mainly caused by the spin splitting of Mn atoms. S atoms and O atoms also have a small contribution to the magnetic moment, while the Sn atom spin-up and spin-down electronic density of states is basically symmetrical, which makes little contribution to magnetism. It can be seen from the figure that there is no obvious exchange effect between Mn atoms and O atoms and S atoms. The magnetic moment of the system is mainly caused by the internal structure of the Mn atom itself.

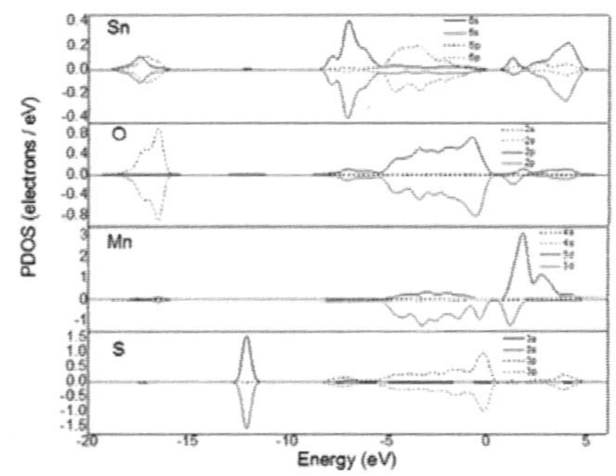

Fig. 2. The DOS of Mn, S co-doping SnO_2.

In order to further explore its magnetic origin, let us consider the partial wave state density of each atom, as shown in Figure 3. The results show that O-2p and S-3p electrons have a small contribution to the magnetic moment, and most of the magnetic moment is provided by Mn-3d electrons. At the same time, we can also see that although it is relatively weak, there is still a coupling between the Mn-3d and S-3p orbitals near the Fermi level, and there is also an interaction between Mn-3d and O-2p. But as we have seen, the interaction is weak, and the main magnetic moment of the system may be due to the spin splitting of Mn-3d electrons.

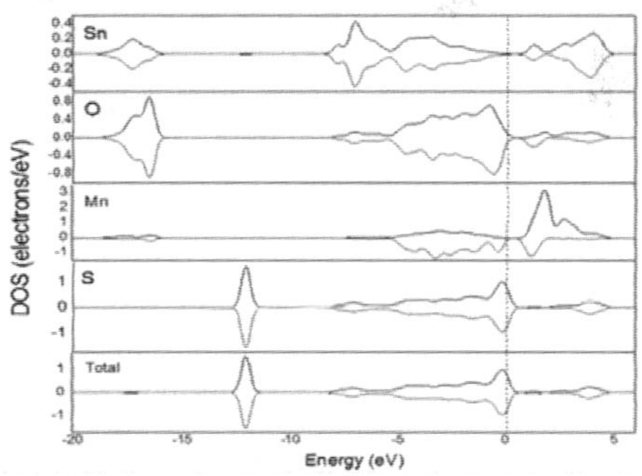

Fig. 3. The PDOS of Mn, S co-doping SnO_2.

3.2 Dielectric function

The complex dielectric function is a physical quantity describing the propagation of electromagnetic waves in a medium, and it is also a bridge connecting the external electric field and the polarization of the medium. From the knowledge of physics, the polarization current can be divided into two parts. One part is generated by the real part of the dielectric, which lags the electric field by $\pi/2$. This part of the current does not lose energy, and the other part is caused by the imaginary part of the dielectric. The current and the electric field are in phase, which will cause the absorption of electromagnetic waves, thereby consuming energy. We can even calculate that the energy absorbed by the medium per unit time is proportional to the imaginary part of the dielectric. Next, we will analyze the dielectric function of Mn and S co-doped SnO_2. The calculation results are shown in Figure 4.

The calculation results show that resonance excitation occurs at three points A, B and C, and the corresponding energy values of A, B and C are 7.349 eV, 8.485 eV and 9.859 eV respectively. These points have high energy loss due to the resonance excitation. Let's look at the dielectric imaginary part spectrum again. Its peak corresponds to the inter-band transition and will also cause significant energy loss. As shown in Figure 4, this system will undergo inter-band transition at D and E, and the corresponding energy values are 2.822 eV and 5.029 eV respectively.

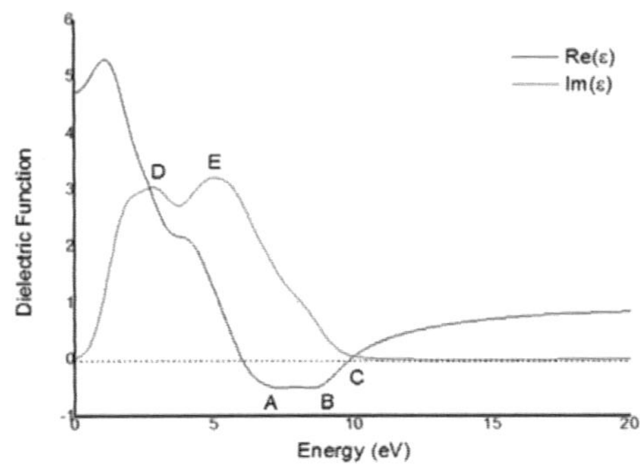

Fig. 4. The dielectric function of Mn, S common doping SnO_2.

3.3 Complex refractive index

The dispersion relationship between refractive index and frequency, not only has important applications in various fields of optics, but even in particle physics. It is also used to study strong interactions, which shows its importance. The extinction coefficient also has important applications in many fields. After doping, we calculated the complex refractive index which was shown in Figure 5.

The calculation results show that 0-1.17 eV and 9.65-20 eV are the normal dispersion regions. At this time, the phase velocity of light decreases with the increase of frequency, and 1.17-9.65 eV is the abnormal dispersion region. At this time, the phase velocity of light was increased with frequency. It can even exceed the speed of light in vacuum. It is not the propagation speed of signals (or energy). It does not violate the theory of relativity. We can also see that at 3.16 eV and 6.01 eV, the extinction coefficient curve has two obvious peaks, and the peaks are 0.83 and 1.18 respectively. The results indicated that resonance absorption occurs here and the light wave energy corresponding to the natural frequency of the system also nearby.

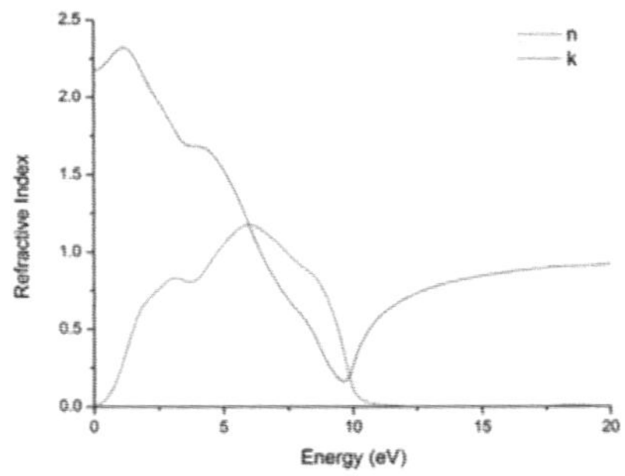

Fig. 5. The complex refractive index curve of Mn, S common doping SnO_2.

4 Conclusion

(1) When Mn and S are doped into SnO_2, compared with pure SnO_2, the lattice constant is slightly increased, but the change is not big. It is indicating that the doping of Mn and S does not affect the crystal structure of SnO_2.

(2) In the Mn and S co-doped SnO_2 system, the phenomenon of spin splitting appears, and the system exhibits a ferromagnetic order. Regarding the origin of magnetism, we believe that the main magnetic moment of the system may be due to the spin splitting of Mn-3d electrons. At the same time, the coupling of the Mn-3d orbital and the S-3p orbital also makes a significant contribution to the magnetic properties of the system. One Mn atom provides a magnetic moment of 1.52 μB, and an S atom provides a magnetic moment of -0.06 μB. The adjacent O atoms also have a weak contribution to the magnetic moment, while Sn has almost no contribution to the magnetic moment. The total magnetic moment of the system is 1.5 μB, indicating that the doped system is not semi-metallic.

(3) The doping system in the 0-10 eV band is mainly anomalous dispersion, and the 10-20 eV band is normal dispersion; the absorption of light is also concentrated in the 0-10 eV band, and the 10-20 eV band is basically light No absorption.

Acknowledgements

This work is supported by the innovative research projects for graduate (CX2020SZ85), Southwest Minzu University.

References

1. J. Hassoun, S. Panero, P. Simon, P. L. Taberna, B. Scrosati, Adv. Mater. J. **19** (2007)
2. Dietl T, Ohno H, Science, 2000, **287**(2000)
3. T. Jungwirth, W. A. Atkinson, B. H. Lee, A. H. MacDonald, Physical Review B, **59**(1999)
4. H. Shimizu, T. Hayashi, T. Nishinaga, M. Tanaka, Applied Physics Letters, **74**(1999)
5. Fukumura T, Jin Z, Ohtomo A, AppL Phys. Lett, **75**(1999)

6. Matsumoto Y, Murakami M, Shono T, Science, **291**(2001)
7. C. B. Fitzgerald, M. Venkatesan, A. P. Douvalis, S. Huber, and J. M. D. Coey, J. Appl. Phys, **95**, (2004)
8. T. Fukumura, Z. W. Jin, M. Kawasaki, T. Shono, T. Hasegawa, S. Koshihara, H. Koinuma, Appl. Phys. Lett. **78**, 7 (2001)
9. X. W. Lou, L. A. Archer, Adv. Mater, **20**, (2008)
10. R. Menzel, A. Duerrbeck, E. Liberti, H. C. Yau, D. McComb, M. S. P. Shaffer, Chem. Mater, **25** (2013)
11. Y. K. Mishra, S. Kaps, A. Schuchardt, I. Paulowicz, J. Kona Powder and Particle, **31** (2014)
12. J. M. D. Coey, A. P. Douvalis, C. B. Fitzgerald, M. Venkatesan, Appl. Phys. Lett, **84** (2004)
13. H. Hori, S. Sonoda, T. Sasaki, Y. Yamamoto, S. Shimizu, K. Suga, K. Kindo, Physica B, **324**, 142 (2002)
14. K. Ueda, H. Tabata, T. Kawai, Appl. Phys. Lett, **79**, 7 (2001)
15. Z. H. Wen, Q. Wang, Q. Zhang, J. H. Li, Adv. Funct. Mater, **17** (2007)
16. Y. Li, Z. Guang, H. K. Hua, J. Phys. Chem, **26** (2010)

Structural feature factors extraction of upper body shape for young female

Yujie Ren, Hongshu Jin*

Wuhan textile university, Wuhan, Hubei 430073, China

Abstract. Human body shape feature points are the key information and basic unit for human body model we are constructing, which performance the difference of body shapes. The purpose of the study is to extract the structural feature factors related to the upper body surface feature points for young females. The 12 feature points of upper body surface were manually confirmed from anthropometric expertise. A total of 31 measurements items, including 3 body surface measurement and 28 photo measurement, were collected for 33 females college students According to the results of correlation analysis, the feature variables of the width, thickness and height dimension based on 12 feature points significantly respectively correlated to the variables of their coordinate orientation, furthermore, the correlated relationship which reflected the width and height features of neck and shoulders shape mainly affected by local skeletal structures. Then, four principle component factors account for upper body shapes of young females, such as width, thickness, height and shoulder shape with characteristic value all over 1, were extracted by the principal component analysis, and the cumulative contribution rate reached 87.387%. Therefore, a total of 8 feature variables sifted from each principle component factor with a loading coefficient over 0.7 as fundamental typical indicators represent the three-dimensional characteristics of body surface feature points reflecting the divergence of body shapes, and it is useful structural information for individual human body modelling.

1 Introduction

Under the background of the intelligent development of the clothing industry and the trend of personalized clothing consumption, virtual human modelling has become a hotspot in the field of multidisciplinary cross-application research. The human body surface feature points (measurement points) are the key information reflecting the differences of human body shapes, and also the basic parametric information for subdividing and identifying the human body shape.

In the field of theoretical research, the body shape classification based on diverse analysis of human body parameters, such as the square ratio of cross-section's area and girth on BL, WL and HL[1], and the descriptions of the differences of trunk longitudinal section[2], supplied and refined the national standard body type classifications. Furthermore, the studies on the body local morphology subdivision [3-4] proposed diversified human body morphological parameters based on feature points, as well as the method of morphological information combination for representing the difference of body shape.

On the method of human body shape measurement, non-contact 3D human body scanning and 2D photo measurement were commonly used for collecting the data by the human body surface features points labelled manually currently. However, the original technical problems of 3D human body scanning, such as the high cost of labor, the local data missing, and the deformation of automatic local shape renovation, limit its application in individual human body modelling accurately, as well as in the field of clothing mass customization[5]. In addition, the two-dimensional human body image is used as a reliable data source for obtaining human body shape feature values [6], describing and identifying important information of the human body shape differences.

In this study, it is aiming to extract the structural feature factors related to the upper body surface feature points for young female. For unified body measurement standard, the identification of body surface feature points was confirmed from manual measuring methods. And the extracted structural feature factors were the important references for typical indices selecting represent the three-dimensional characteristics of body surface feature points reflecting the divergence of body shapes, and it is useful structural information for individual human body modelling.

2 Method of measurement data acquisition

* Corresponding author: 2013025@wtu.edu.cn

2.1 Measurement object

The measurement subjects were 33 female college students aged 18-23.

2.2 Definition of body surface feature points

The 12 upper body surface feature points were confirmed from anthropometric expertise manually and showed in Figure 1.

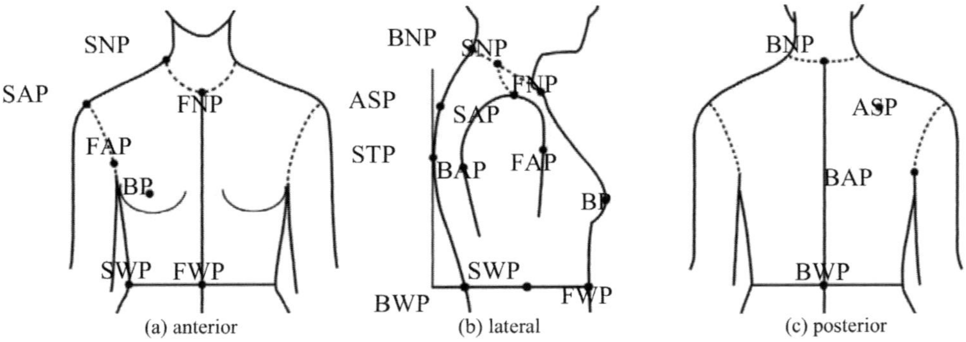

Figure 1. Feature points on body surface.

Among them, the feature points such as SNP (side neck point), FNP (front neck point), BNP (back neck point), SAP (shoulder acromion point), BP (bust point) adopt the general definition of national standard GB/T 16160-2008 "Anthropometric definition and methods for garment". For the definitions, the other feature points are referred to the ergonomic measurement point definition by Machiko MIYOSHI[7], including the armpit front/back point is the extreme point of skin fold in front of and behind the trunk and upper arm which is simple and easy to confirm.

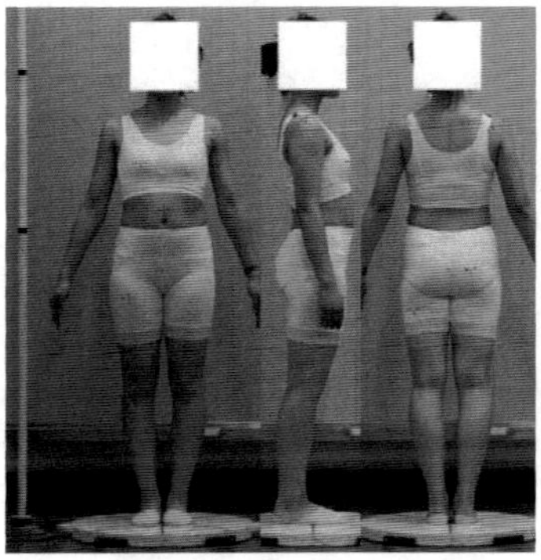

Figure 2. Photo used for measuring.

2.3 Photography and measurement methods

The subjects were required to wear fitted underwear, keep upright and look ahead.

The 12 feature points and related reference lines were marked using triangular markers on the body surface.

The anterior, lateral and posterior photos for each subject were shot using Canon EOS 5D Mark II SLR camera. The shooting distance was 10 m, and the height of centre of lens was 1 m, which was roughly the same as the height of waist. At the same time, a pole marked 50 cm was placed on the right side of subject horizontally as a reference for converting and recording the actual measurement. The photo for measuring is shown in Figure 2.

2.4 Upper body measurement items

There is a total of 31 items for the upper body measurement, including 28 photo measurement items and 3 body surface measurements (upper body control area, bust, waist, height).

The photo measurement items show as Figure 3, including 7 items of front/back width measurement (Figure 3(a)), 12 items of thickness measurement (Figure 3(b)), and 9 items of height measurement (Figure 3(c)).

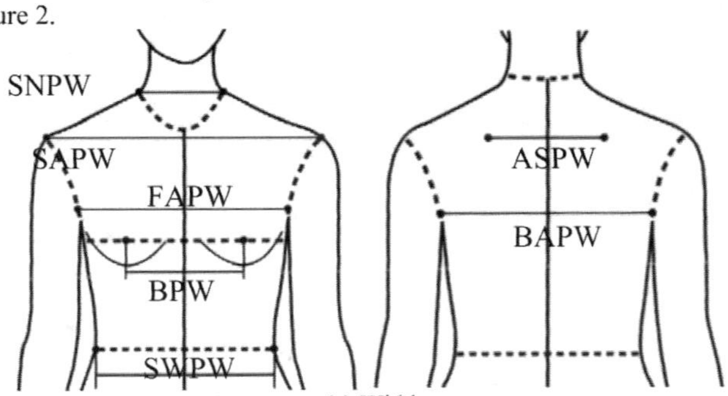

(a) Width

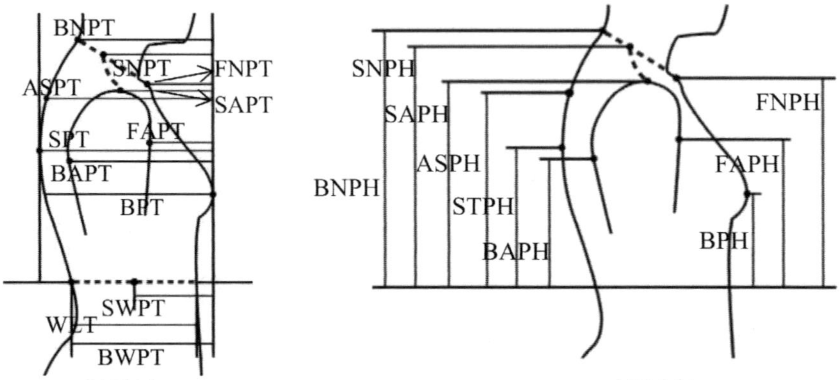

(b) Thickness (c)Height

Figure 3. Schematic diagram of measurement items.

3 Extraction of upper body features of young women

For extracting the morphological feature factors of the upper body of young women, the statistical analysis method of Kolmogorov-Smirnov test, correlation analysis and principle component analysis were utilized by IBM SPSS Statistics 21 data analysis software.

3.1. Upper body data distribution test

According to the results of Kolmogorov-Smirnov test, the upper body morphologic feature variables all follow the normal distribution.

Taking the size of A's upper body control part (height, bust and waist) in GB/T+1335.2-2008 "Standard sizing for garment-Woman" as the test value, the One-Samples T-Test were done and the results shown in Table 1.

Table 1. Mean test of sample female youth and national standard female data.

Project	Mean (cm)	Standard deviation(cm)	Test value(cm)	Sig.(2-tailed)
height	162.67	4.75	160	**[a]
bust	82.17	5.21	84	**[a]
Waist	65.25	5.80	68	N.S.[b]

[a] **P<0.01. n=33

[b] No significant

The average height and bust of the subjects taken in this study showed significant differences with standard size. Compared with the national average, the average height and bust girth were slightly higher lower respectively.

3.2. Relationship between upper body feature variables

3.2.1. Correlation relationship between surface morphological characteristics variables.

The results of correlation analysis of upper body surface morphological feature variables showed that: 1) For the width variables, SWPW(WL width) showed a highly significant correlation with FAPW(armpit front fold point width) and BAPW(armpit back fold point width), respectively; SAPW（shoulder acromion point width） and SNPW(side neck point)showed a correlation between r=0.6 and above, reflecting the characteristics of the shoulder and neck width; The correlation coefficient between ASPW(suprascapular point width) and other width characteristic variables was lower than r=0.6, showing the width characteristic based on the bone structure of the shoulder.

2) Among height variables, the correlation coefficient between SNPH(side neck point height), BNPH(back neck point height) and FNPH(front neck point height) reached more than r=0.85.SAPH(shoulder acromion point height); BAPH(armpit back fold point height) and ASPH (suprascapular point height) showed a high correlation above r=0.8, showing the height characteristics of neck and shoulder respectively.

3) Among the thickness variables, the correlation coefficient between SNPT(side neck point thickness), FNPT(front neck thickness) and BNPT(back neck point thickness) were above r=0.9;The correlation coefficients between SAPT(shoulder acromion point thickness), FAPT(armpit front fold point thickness), BAPT (armpit back fold point thickness) and ASPT (suprascapular point width) were more than r=0.8; The correlation coefficient between waist thickness and BPT(chest thickness) and STPT(body thickness) is close to r=0.9, which describes the thickness characteristics of neck-shoulder and bust-waist respectively.

To sum up, the width, height and thickness variables of upper body surface morphology showed high significant correlations respectively, and the main morphological features of shoulder, neck and trunk cadres were also shown.

3.2.2. Relationship between basic size and characteristic variables.

Bust and waist were significantly correlated with SWPW(WL width), BPT(bust thickness), WLT(WL thickness) and other variables over r=0.8, indicating the close relationship between the shape characteristics of upper body bust and waist and the size of circumference.

In addition, FAPW(armpit front fold point width), BAPW(armpit back fold point width) and SWPW(point width) were significantly correlated with BPT(bust thickness), STPT(body thickness) and WLT(WL thickness) respectively, and the relationship between upper body morphology and local characteristics was described.

The correlation coefficient between BPW (bust point width) and other morphological characteristic variables is lower than r=0.5, and the position of bust point is greatly affected by the shape and material of bra cup, showing no correlation with morphological characteristics.

3.3 Extraction of principal components of surface morphological features

Based on the results of correlation analysis, excluded variables unrelated and repetitive, 19 morphological characteristic variables are screened out and used for the principle component analysis. The results of principal component analysis are shown in Table 2. four principal components with eigenvalues above 1 were extracted by the maximum variance factor rotation method, and the cumulative variance reached 87.387%.

Table 2. Extraction of morphological features of upper body after rotation.

Characteristics of the variables	Ingredients			
	1	2	3	4
FAPT	.958			
SAPT	.912			
ASPT	.863			
BAPT	.846			
BNPT	.839			
FNPT	.831			
STPT	.791			
BPT	.683			
BAPH		.950		
ASPH		.933		
SAPH		.933		
BNPH		.887		
FNPH		.871		
BAPW			.842	
WLT			.821	
BWPT			.807	
SWPW	.		.782	
SAPW				.891
ASPW				.761
Eigenvalue	6.283	4.346	4.217	1.758
Explain the variance (%)	33.069	22.873	22.195	9.251
Cumulative variance (%)	33.069	55.942	78.136	87.387

1) According to the 8 variables related to the thickness of upper body show over 0.6 high loading coefficient, the first principal component could be interpreted as trunk and shoulder-neck thickness factor. Inside, BWPT (WL back centre point thickness) and BNPT (back neck point thickness) and STPT (body thickness) in the first principal component expressed lateral dorsal curve morphology. In particular, BNPT (back neck point thickness) and SAPT(shoulder acromion point thickness), expressed the lateral morphological characteristics of neck and shoulder characteristic, and STPT (body thickness) expressed the lateral morphological characteristics of trunk, which were the typical indictors of thickness factor.

2) In consideration of height variables with over 0.7 high loading coefficient in the second principal component, it could be interpreted as upper body height factor. Inside, BNPH (back neck point) and SAPH(shoulder acromion point height) explained the height characteristics of neck and shoulder. Meanwhile, BNPH (back neck point height) also expresses the height characteristics of trunk. These variables were selected as typical indictors of the second principle component.

3) The third and fourth principal component includes 4 and 2 variables with above 0.7 high loading coefficient variables which described the trunk shape characterises and shoulder-width characterises, it could be interpreted as waist shape factor and shoulder-width

https://doi.org/10.1051/e3sconf/202123701024

factor, respectively. SAPW (shoulder acromion point width), SWPW (WL width) and BWPT(WL back centre point thickness) of the third and fourth principal components are respectively typical features to express shoulder width and waist shape.

The above principal components include the thickness, height and width of the trunk, shoulders and neck, and comprehensively express the relationship between the characteristic variables of three azimuths on the body surface. In particular, the anthropometry definition of BNP(back neck point) and SAP(shoulder acromion point) is clear, and the influence of manual operation error is small. The related factors reflect the main characteristics of the shape of human shoulder and neck.

And then, a total of 8 typical indexes describing the overall and local morphological characteristics of human body were extracted from the three azimuths of thickness, height and width. Collinearity detection results of typical characteristic variables in each principal component showed that VIF (Variance Inflation Factor) among typical indicators was less than 10. Therefore, statistically, the extracted 8 typical indicators were reasonable.

4 Conclusion and future work

The aim of this study is to extract the structural feature factors related to the upper body surface feature points for young females. Based on 12 surface feature points of the human body, human body feature measurements were collected. The main results of this study summarized as follows:

1) the relationship between each feature size was obtained through correlation analysis.

2) principal component analysis was used to extract 8 typical feature indexes that can describe the basic feature information of the human body.

In future work, the author plans to increase the sample size and continue to use the approach for morphological subdivision and body modelling to promote the development of clothing customization.

References

1. Wang Q M 2016 Female body classification in Jiangsu and Zhejiang based on cross-sectional area of body J. Journal of Textile Research. vol 37, no 05. 131-136
2. Xia F Q, Wu G, Xie H Y and Zhong Y 2017 Classification of body shape based on longitudinal section curve J. Journal of Textile Research. vol 38, no 06. 86-91
3. Hamad M, Thomassey S and Bruniaux P 2017 A new sizing system based on 3D shape descriptor for morphology clustering J. Computers & Industrial Engineering. vol 113. 683-692
4. Sun N L and Song X X 2014 Research on chest classification of young women in Shanghai J. Journal of Silk. vol 51, no 09. 34-39
5. Ni S M, Jin J F and Pang C F 2014 Research of young female somatotype based on longitudinal profile J. Textile Research. vol 35, no 08. 87-93
6. Luo S H and Wang J P 2013 Research on 2-D image-based non-contact anthropometric technology J. Journal of Textile Research. vol 34, no 08. 151-155
7. Milyoshi M 2000 The Theory of Fashion Modeling, 1st ed (Beijin: China Textile & Apparel Press) p67

Strategy Analysis of Promoting Pollution Control of SMEs

Zhang Yanying[1], *Jiang Gui*[1*], *Yu Ziwei*[2]

[1]Wuhan University of Science and Technology, Wuhan, 430065, China
[2]Hubei Research Center for SMEs

Abstract. The pollution problem of SMEs is an important problem to be solved in the process of China's economic development. Based on the game theory, this paper takes the government and SMEs as the two sides of the game, constructing the game theory model of pollution control of SMEs, putting forward the strategies to promote the pollution control of SMEs according to the analysis results of the model.

1 Introduction

The report of the 19th National Congress of the Communist Party of China pointed out that the construction of ecological civilization is a great event for the development of the Chinese nation for thousands of years. As Xi Jinping said, lucid waters and lush mountains are invaluable assets. It is necessary to integrate the economic and social development and ecological civilization construction organically, so as to pursue good environmental benefits as well as good economic and social benefits.

According to the Ministry of Industry and Information Technology, by the end of 2018, the number of small and medium-sized enterprises (SMEs) in China has exceeded 30 million, accounting for 99% of the total number of enterprises. The booming SMEs play an important role in promoting the sustainable development of the national economy. However, at the same time, they also produce huge production resources consumption and environmental pollution problems during the production process. According to the investigation of the State Bureau of Statistics and the State Environmental Protection Administration, more than 80% of the industrial production has pollution problems, accounting for 60% of the pollution sources in China.

As far as the supervision of SMEs is concerned, the current regulatory measures are applicable to all enterprises, and they are also applied to the majority of SMEs at the same time. Due to the large number and wide distribution of SMEs and the huge cost of government supervision, the pollution control of SMEs in China has been very little. How to deal with the pollution problem of SMEs has become an important problem to be solved in the process of economic development in China.

In my opinion, the pollution control of SMEs is mainly a game between the government supervision department and the pollutant discharging enterprises. Therefore, this paper studies the game between them in the pollution supervision, trying to promote the possibility of government supervision and SMEs pollution control, and provide assistance for environmental protection.

2. Analysis of pollution control of SMEs the perspective of game theory

2.1. Model hypothesis

In this paper, we take SMEs and the government as the two sides of the pollution supervision game. SMEs in the market produce products at the same time discharge pollutants to the external environment. The government formulates pollution policies according to the bearing capacity of the environment.

Fistly, assuming that both sides of the game are completely rational economic men, their goals are to maximize their own interests. At the same time, assuming that the information is complete, both parties know the other party's choice of action, but the other party does not know the specific behavior.

Secondly, the strategic space of both players in the game is as follows: SMEs choose to control pollution or not; government (environmental regulatory agency) chooses to inspect or not inspect SMEs. Assuming that the probability of SMEs to control pollution is x, the probability of not treating pollution is $(1-x)$; the probability of government inspection is y, and the probability of not checking is $(1-y)$.

Thirdly, assuming that the regulatory cost required by the government is constant, that is, K. When SMEs register, they need to apply to the government for pollutant discharge permit and pay a license fee for this. Assuming that the amount of pollutant discharged by the enterprise is n when applying for the license, the license fee for handling the permit is $M=a \cdot n\ (a>0)$, where a is the rate of emission permit. Assuming that the actual pollution discharge of an enterprise is $n'(n>0)$, in the case of pollution control, the enterprise needs to pay the cost of $M' = B>n\ (b>0)$, where B is the cost rate of pollution control

* Corresponding author: 1012894113@qq.com

of the enterprise, and the government will give the enterprise a total of *T* subsidies and tax incentives after checking the pollution control achievements of the enterprise. If the enterprise does not manage, it will cause adverse consequences to the environment. In order to make up for the damage of environmental resources, the government needs to pay this part of the governance cost, i.e. *M*. If the enterprise fails to pay the corresponding pollution control cost, but the government finds out the behavior of the enterprise due to inspection, it will impose a fine of $F=C \cdot n$ $(c>0)$, where $C>>b$, the government will extract the part of $M' = B>n$ $(b>0)$ from the fine for pollution control.

2.2 Model construction

As mentioned above, the income function matrix of this model can be obtained, as shown in Table 1.

Table 1. income matrix of SMEs and governments

	pollute (x)	Not pollute (1-x)
inspect (y)	an-K-T, -an-bn'+T	an-K+cn'-bn', -an-cn'
not inspect (1-y)	an, -an-bn'	an-bn', -an

It can be seen from the above table that the expected returns of government departments are as follows:

$$E1=y[x(an-K-T)+(1-x)(an-K+cn'-bn')]$$
$$+(1-y)[x \cdot an+(1-x)(an-bn')] \quad (1)$$

The expected returns of SMEs are as follows:

$$E2=x[y(-an-bn'+T)+(1-y)(-an-bn')]$$
$$+(1-x)[y \cdot (-an-cn')+(1-y)(-an)] \quad (2)$$

As mentioned above, it is assumed that the government and SMEs are fully rational and pursue the maximization of their own interests. When deriving E1, let its reciprocal be zero, that is, if $\frac{\partial E1}{\partial y} = 0$, formula (3) can be obtained.

$$x = \frac{cn'-K}{cn'+T} = 1-\frac{K+T}{cn'+T} \quad (3)$$

In the same way, if we take the derivative of E2 and make its reciprocal be zero, that is, if $\frac{\partial E2}{\partial x} = 0$, we can get the formula (4).

$$y = \frac{cn'}{cn'+T} = 1-\frac{T}{cn'+T} = \frac{c}{c+T/n} \quad (4)$$

2.3 Result analysis of game model

According to formula (3), to improve the probability of government supervision, the following measures can be taken: 1) reducing the government inspection cost *K;* 2) reducing the pollution control reward *T* paid by the government to enterprises; 3) increasing the fine rate *c*.

Rationality analysis:

When the income is fixed, the less the cost is, the easier the action subject will take the action. At the same time, when the cost is determined, the higher the income, the more willing the action subject is to take the action. For the government, the lower the inspection cost and the less rewards paid to the enterprises, the more likely the government is to act. Therefore, it is of positive significance to improve the inspection probability of the government, and measures 1) and 2) are reasonable. In the same way, increasing the penalty rate means the increase of revenue for the government, and can also improve the probability of government inspection, and the measures are reasonable.

According to formula (4), to improve the probability of pollution control of enterprises, there are the following situations: 1) increasing the fine rate *C*; 2) increasing the government incentive subsidy *T*; 3) increasing the emission volume *n'*.

Rationality analysis:

The penalty and reward measures adopted by the government are negative and positive incentives respectively. When the fine rate is increased, that is, when the negative incentive is enhanced, the cost of non pollution control SMEs will increase accordingly, which can promote the SMEs to avoid the action of no pollution control. When the government subsidies increase, that is to say, the positive incentive is enhanced. The enterprises can get more benefits from the pollution control behavior, which can prompt the SMEs to take pollution control actions. The conclusion 1) and 2) is reasonable. The cost of pollution control of enterprises is affected by many factors, such as the technical conditions of enterprises, the amount of pollutants discharged, the pollution treatment equipment and other factors. For a specific enterprise, the cost of the purchased sewage equipment is certain. When the maintenance cost is ignored, the larger the enterprise's emission volume is, the lower the marginal emission cost the enterprise will pay for each additional unit of pollution discharge. As the specified rate is fixed, in the case of not treating pollution, the larger the amount of pollution, one has to pay the higher the penalty cost. Therefore, the larger the discharge volume of the enterprise itself, the more likely the enterprise is to take pollution control action. The conclusion 3) is reasonable.

3. Strategies to strengthen the supervision pollution control of SMEs

3.1 A normal inspection mechanism should be established to reduce the cost of government inspection

The frequency of the opportunistic behavior of the SMEs in violating the regulations varies with the difficulty of supervision by the supervision department. When the supervision department does not supervise effectively, or the frequency of supervision is low, the probability of opportunism is high, and the probability of illegal discharge of pollutant discharging enterprises is high. If the monitoring frequency of supervision departments increases, the probability of enterprises' illegal emission

 https://doi.org/10.1051/e3sconf/202123701025

detection will increase. Under the condition of fixed income, the higher the cost of opportunism, and the emission enterprises may take the initiative to reduce opportunistic behavior. SMEs have a large number and wide distribution, so their illegal discharge has the characteristics of "difficult to discover", "short of time", "difficult to obtain evidence".

Moreover, the life cycle of SMEs is relatively short, with an average of 7 years. In the process of supervision, the change and update of the enterprise list is also a big problem. The government can not be on duty to supervise the enterprises every day. Therefore, on the one hand, environmental supervision departments should improve their own monitoring technology level. On the other hand, they should pay attention to the use of scientific supervision methods, optimize the monitoring route, strengthen the links in the whole supervision work, and reduce the cost of information communication.

Firstly, in terms of the inspection mode, the working mode of regular inspection and reinspection after receiving reports is widely used in the traditional pollution control field. This work mode is too passive and easy for SMEs to form a routine to avoid inspection. The government can use the combination of regular inspection and random inspection to form a dynamic supervision on the pollutant emission of SMEs, so as to better grasp the pollution discharge situation of SMEs.

Secondly, we can reduce the cost by transferring the cost of government inspection to enterprises. SMEs should be required to install continuous monitoring devices and provide pollution treatment reports regularly when they apply for discharge permits. For those who refuse to install equipment, the government departments have the right to postpone the issuance of emission permits and refuse their production requests. For those who refuse to provide pollution control reports or provide false information in violating regulations in the process of regular inspection and random inspection, they shall be given corresponding administrative punishment.

Thirdly, we can selectively improve the inspection probability of some enterprises, especially those with severe pollution. Due to the lack of perfect pollutant emission reduction facilities for SMEs, there are often many heavy polluters. In order to alleviate the contradiction between the economic development of SMEs and the natural environment pollution, and to promote the development of circular economy, we should focus on the treatment of enterprises with heavy pollution. There are a lot of heavy pollution enterprises in the whole country. By focusing on improving the prevention and control concept of environmental pollution, it can play an environmental demonstration role for others. On the one hand, they can strengthen the recognition of the importance of environmental protection in the field of social and economic production through the treatment of key pollution enterprises, and on the other hand, they can apply key treatment to reflect the determination of pollution control, so as to lay the foundation for the future development of management work.

Fourthly, in the supervision of other social subjects, we can take public supervision to improve the detection probability of illegal enterprises. The number of enterprises that the government departments can pay attention to is limited, but the residents around the enterprises have a full understanding of the enterprises, and the community residents have the most say on whether the enterprises have related violations. The government can rely on the residents around the enterprise to transmit relevant information. Measures can be taken to let the residents actively supervise, such as increasing publicity intensity and frequency, so that residents can understand the importance of environmental protection. At the same time, measures should be taken to encourage reporting. If residents report illegal discharge of pollutants by SMEs, they will be given certain material rewards after investigated to be true.

3.2 The cost of pollution control should be reduced ; market mechanism should be introduced

The reason why SMEs are not willing to carry out pollution control is mainly due to the high cost of pollution control, which is affected by the technical level and scale of SMEs. In terms of technical level, SMEs in China generally have low technical level and inferior production techniques, especially township enterprises often adopting technologies eliminated by urban industrial departments. Due to the low level of technology and the lack of human capital, SMEs not only pollute seriously, but also are difficult to reduce pollution through technological progress. In terms of enterprise scale, the economic scale of SMEs is too small. For example, according to national regulations, the economic operation scale of alkali recovery process, the key link in straw pulp papermaking wastewater treatment, is at least 1000 tons/day. There is a large gap between the emission scale of most SMEs and the above-mentioned economic operation scale, resulting in higher pollution treatment costs. From the perspective of enterprise pollution control facilities, the average life span of SMEs in China is 7 years, which is lower than the depreciation period of pollution control facilities, which also increases the cost of pollution control and reduces the willingness and ability of enterprises to control pollution.

The policy of "who pollutes, who governs" is applicable to enterprises with a certain scale. But it is not realistic for SMEs. Market mechanism can be adopted to solve the dilemma. As long as the government gives correct guidance, it is completely possible to concentrate SMEs in certain professional areas or industrial parks. Meanwhile, professional companies with environmental protection facilities operation qualification are responsible for centralized treatment, so as to solve the problem that pollution control facilities cannot meet the requirements of economic operation scale due to the small scale of enterprises.

On the basis of carrying out the pilot project of third-party environmental services, we can further promote the in-depth cooperation among environmental protection departments, industrial parks and enterprises, provide environmental protection contract management services for the government or enterprises through "environmental stewards", and charge fees according to the final pollution control effect or income. We should establish a service quality evaluation system, as well as supervision and restraint mechanism, and formulate supporting policies with clear responsibilities and orderly competition.it is necessary to carry out the operation of large-scale environmental protection facilities and pollution control services by a third party, or entrust environmental protection experts to monitor, supervise the construction and operation of environmental protection facilities, and other integrated services in the form of regions or groups.

3.3 Information linkage should be strengthened; policy publicity and implementations should be promoted

In the game hypothesis of this paper, the information between the government and SMEs is complete, but in reality, the information gap between the government and enterprises exists. Marked by the launch of the new environmental protection law in 2015, China's major industries and enterprises have ushered in the era of "cleaner production" since the central environmental supervision in 2016. Environmental protection storm roars past, but there are also many SMEs that are shut down because they do not understand the government policies. Therefore, the publicity and implementation of government policies is also very important for enterprises to choose pollution control.

Firstly, we should improve the service system of enterprises, take districts and counties as units, smooth the daily communication channels with enterprises, and establish regular contact and exchange mechanism. Many measures should be taken, such as supporting the industry to build private enterprise information service platform, guiding small and medium-sized enterprises practitioners to learn and understand national environmental protection policies, regulations, standards, as well as mastering the development trends of advanced green technology at home and abroad. We should strengthen intellectual support for green development of SMEs, and provide effective guidance and services for enterprises of different industries and sizes. Also，we should increase the training frequency of environmental protection laws and regulations, standards, technology and management knowledge of private enterprises, in order to help them to understand relevant laws and regulations.

Secondly, a platform should be built for government enterprise communication through a third party. Enterprises and the government can contact through environmental protection industry, society and other industrial organizations. They entrust associations and societies to participate in the formulation of environmental protection policies, plans and standards by purchasing services. Through industry organizations, the development demands of enterprises are fully expressed, and various policy requirements and information are timely transmitted to member enterprises to enhance their confidence and motivation in environmental governance. The government should keep track of the latest trend of the development of industry technology, publish economic and applicable technology and relevant information in real time, and help small and medium-sized enterprises to choose appropriate management technology or a third-party professional intermediary organization. At the same time, the third-party organizations should also give full play to the self-discipline role of industry organizations, regulate the behavior of members, and improve the quality of service.

4. Conclusion

SMEs are large in number and widely distributed. No matter how to strengthen supervision and reduce supervision cost, supervision organizations will always face the problem of cost. As a profit-making organization, the primary consideration of enterprises is their own interests. In the process of government enterprise game, the conflict of interests between the two sides will continue to exist. I believe that in order to make enterprises consciously control pollution and the government can effectively supervise, we can formulate strategies from the aspects of information communication between government and enterprise and the cost reduction of government and enterprise to promote the pollution control of SMEs.

References

1. Jin GR, Zhang L. China soft science, **1**, 126-133, (2014)
2. Long WB, Li SH, Song XZ, Public Administration Review, **6**, 25-56, (2015)
3. Long WB, Li SH, Xing DR. China Administration, **12**, 66-73, (2017)
4. Wang GH, Innovation and Entrepreneurship Management, **2**, 25-33 (2015)
5. Mei CM, Enterprise Economy, **12,** 33-37, (2017)
6. Duan JJ. Charming China, **29**, 39, (2017)
7. Zhou CG., Digital Users, **8**, 222, (2018)
8. Chen Weiheng. Shenzhen Stock Exchange, **6**, 59-64 (2018)
9. Guo T, Environmental Science and Management, **9**, 10-14, (2019)
10. Chen H. Environment and Development, **4**, 53, (2019)

Study on soil and water conservation regionalization in Changji Prefecture

Xiaozhuang Zhang[1], *Yajun Gao*[2,*]

[1] College of Conservancy Engineering,Zhengzhou University, 450001,Zhengzhou, Henan , China;Tel:18203662619(Xiaozhuang Zhang)
[2] Yellow River Institute of Hydrology and Water Resources, Hydrographic Bureau of Yellow River Conservancy Commission, 450003,Zhengzhou, Henan, China

Abstract: Taking once during as the study area, the course by using the DPSIR model and a lot of literatures screening of high frequency index, and quantitative analysis combined with the expert appraisal method, on once during regionalization index system establishment, course planning method was studied, determining the topographical features, water loss and soil erosion prevention and control system, land exploitation and utilization of direction, after years of average rainfall, population density, coverage of trees, soil and water loss area total area percentage as indicators of dividing the partition of soil and water conservation, such as in the villages and towns as the basic unit. Through principal component analysis, cluster analysis, the GIS space superposition method for regionalization, will once during divided into course will once during is divided into six course four areas: the southern mountain freeze-thaw district, Zhongshan in southern hilly water conservation ecological reserve, with low hills in southern soil conservation area, central oasis environment farmland protection area in northern desert ecological maintenance, low mountain in northern wind sand area, ecological maintenance area 6 level 4 area of water conservation, determine the various functional areas of soil erosion prevention and control of system layout, the direction of land use, for once during planning of soil and water conservation and soil erosion prevention and control system of the course layout provides the scientific basis.

1 Overview of the research area

Changji Hui Autonomous Prefecture (hereinafter referred to as "Changji Prefecture") is one of the administrative regions under the jurisdiction of the Xinjiang Uygur Autonomous Region.Is located in tianshan mountain, the southeast margin of junggar basin, with east, Shihezi city west, south and Turpan city, general Mongolian autonomous prefecture of adjacent, north border with Tuscaloosa, Altay region, the northeast border with Mongolia, border 227.3 km, is a new north road, the ancient "silk road" to the path to the countries of central Asia and Europe. Geographical coordinates between east longitude 85 ° 17 '50 "15" ~ 91 ° 32', north latitude 43 ° 31 '06' ~ 45 ° 38 '50 ", a state of things is 541 km long, north and south 285 kilometers wide, land area of 73485 km2, during is course construction is an important part of the silk road of Xinjiang economic belt core, and Urumqi-Changji stone city in the Northern Slope economic zone is take the lead in the development of the core area, from something north surrounded on three sides of Urumqi, with the development of integration in Urumqi city location advantage, state Changji is 35 km away from Urumqi city center,18 km away from Urumqi International Airport. The Northern Xinjiang Railway, 312 National Highway, 216 national Highway, 115 provincial highway and Tu-Wu-Da highway run through the whole territory, serving as a bridgehead and golden passage connecting the eastern China and the central Asian and European markets in the west. The terrain of Changji Autonomous Prefecture is high in the south and low in the north, and slopes from the southeast to the northwest, with the Tianshan mountains in the south, the vast alluvial plain in the middle and the vast desert basin in the north. This area is the Northern Slope of The Tianshan Mountains stretching across the south, which is commonly known as the "Northern Slope of Tianshan Mountains". Changji city belongs to the temperate zone and has a typical continental arid climate, characterized by cold winter, hot summer and large temperature difference between day and night. Due to the influence of topographic conditions, there is a great difference in climate from south to north, and there is more precipitation in summer in the south. The desert climate in the north is remarkable. The sunshine in Changji Prefecture is abundant, with 2700 hours of sunshine per year. The annual cumulative temperature ≥10°C is 3450°C, and the annual average temperature is 6.8°C. The average temperature in January is -15.6°C, and the average temperature in July is 24.5°C. The average annual precipitation is 190 mm, and the precipitation in summer is significantly higher than that in winter. The annual frost-free period is 160-190 days.

*Corresponding author: 1018950743@qq.com

The soil erosion types in Changji prefecture are complex and diverse, which can be divided into wind erosion, hydraulic erosion and other main types. Wind erosion is the most important and typical type of erosion in Changji Prefecture. Although hydraulic erosion has a small area, it is highly destructive and has a large influence area. It is also one of the main forms of soil erosion in Changji Prefecture. The soil erosion area of the study area is 46,300 km2, accounting for 5.2% (4.63/88.54) of the soil erosion area of Xinjiang, and 63.0% (4.63/7.35) of the territorial land of the whole prefecture, among which the hydraulic erosion area is 4,300 km2 and the wind erosion area is 42,000 km2.According to the grade of soil erosion intensity, the areas of mild, moderate, intense, extreme and violent erosion are 26,900 km2, 710,000 km2, 74,400 km2, 39,900 km2 and 100,000 km2, respectively, accounting for 58.16%, 15.29%, 15.95%, 8.42% and 2.19% of the total soil erosion area, among which the proportion of mild erosion area is the largest, reaching 58.16%.Soil erosion is a serious threat to the living environment of local people.

2 Index selection method and regionalization method

By reading a large number of soil and water conservation districts literature, most of these studies are some indicators of regional soil and water loss is analyzed, and does not take into account different indicators relate to each other and influence each other, the question of causality as the social economy and environment, caused by human activities and environmental status of the feedback effect of the society, cannot reflect the various aspects of different factors, therefore, the selection of index for soil and water loss can only have a vague guidance, is not specific to every aspect of society, economy and nature. Based on this, this paper puts forward a method of selecting soil erosion index by DPSIR model. Based on DPSIR model can be more comprehensive analysis of the related factors of soil erosion, more scientific to select indicators for soil and water loss, high frequency division indexes, a lot of literatures and tertiary thinning area zoning index, based on the DPSIR model, selection of high frequency index, level 3 area after refining division index, and then take the expert evaluation method, the topographical features, grass coverage, years of average rainfall, soil erosion type and thinning intensity, population density, In addition, the percentage of soil and water loss in the total area, the soil loss control system, the direction of land development and utilization, and the thickness of soil layer are the regionalization indexes of Changji Prefecture.

The principal component analysis (PCA) with the function of dimensionality reduction and aggregation combined with system clustering and GIS spatial superposition were used to determine the results of regionalization according to the principle of centralized concatenation and regional conjugation

3 Data Sources

To avoid the interannual fluctuation of data and the error of data statistics, so that the characteristics is more representative, this article are based on the original data from 2003-2018 ".During water course bulletin" and "Changji hui autonomous prefecture, statistical yearbook (Changji hui autonomous prefecture, statistics bureau, 2018)," .During the social and economic statistical yearbook course, China's economic and social development statistical database, etc., including literature data access, etc. At the same time because of too many indicators, makes analysis of the complexity of the increase, so the dimension reduction function and the polymerization of the combination of principal component analysis and system clustering method.

4 Zoning results

At once, during the four area, six pieces of course based on the spatial pattern of soil and water loss prevention, according to the national soil and water conservation districts guidelines (trial) ", this division USES 4 class partition system, primary area for the overall pattern of area, the secondary area for regional coordination area, level 3 area as the basic function, level 4 area (state level partition) as the partition of soil and water conservation.

In the national soil and water conservation zoning, Changji Prefecture is located in the northern sandstorm area level I subdivision, the northern Xinjiang mountain basin area level II subdivision, the North Slope of Tianshan Mountain habitat environment farmland protection area level III.

In accordance with the requirements of the overall direction, the soil and water conservation contents related to natural forest protection, grassland protection construction, land consolidation, desert management and overall urban and rural development shall be comprehensively coordinated, and on the basis of soil and water conservation regionalization, the first-level zoning layout of soil and water conservation in Changji Prefecture shall be formulated. For ease of reflecting the different regional location and the characteristics of soil and water loss and soil erosion prevention and control of the direction. During area is divided into six level course: Zhongshan south mountain freeze-thaw zones, ecological protection areas, with low hills in southern hilly water conservation soil conservation area, central oasis environment farmland protection area in northern desert ecological maintenance of wind sand area, water conservation ecological maintenance, low mountain in northern area.

Table 1.Results of soil and water conservation zoning in Changji Prefecture

Level 3 area	State - level division name	Area（km^2）

Tianshan North Slope living environment farmland protection area	Alpine freeze-thaw zone in the south	3507
	South Zhongshan Hill Water conservation ecological protection zone	9071
	Southern low mountain foothill soil retention area	8865
	Central oasis habitat environment farmland protection area	18472
	Northern desert ecological maintenance windbreak sand control area	30111
	North, low mountain water conservation and ecological conservation area	3459

4.1 Layout of regional prevention and control system

The southern part of Changji prefecture is a mountainous area in the north slope of Tianshan Mountain, which accounts for about 20% of the land area of the whole prefecture and is the main tourist landscape belt of Changji Prefecture.The beautiful Jiangbulake scenic spot, the majestic Bogda Peak (5445m), the picturesque alpine lake Tianchi, the mysterious Kangjia Shi-gate rock painting, the mysterious Fenghuang Mountain scenic spot and other high quality tourist attractions are located in this area. This part of the tourism area is mostly prohibited from development, so special attention should be paid to environmental issues in the development process. Scenic spots need to set up core area, buffer zone and tourist area, the core area is forbidden to tourists, is the key protection area of the ecological environment, scenic area development is mainly concentrated in the tourist area.

4.1.1 Southern alpine freezing-thawing area

This area is located at the top of the Tianshan Mountain in the south, the high mountain area is covered with snow all the year round, there are a lot of modern glaciers, the altitude is about 3000-5445m, from the west to the east there are roughly parallel To the Iran-Habiergat Mountain, Tiangeer Mountain, Bogda Mountain composition, is the main source of the river in the area. Freezing-thawing erosion is dominant in this area. This area is far away from residential areas and all located within the national prevention zone. It should be sealed off and protected to reduce human interference.

4.1.2 South Zhongshan Hill Water Conservation Ecological Protection Zone

This area is the Tianshan Mountain in the south, with undulations in the Zhongshan belt. Within the range of 1150-3000m above sea level, the vertical zone of mountain vegetation is relatively complete, the mountain is precipitous, the maximum height difference is over 100 meters, the mountain area is cut strongly, and the valley is developed. Its natural characteristics are abundant rainfall, fertile soil, rich water and grass, dense virgin forests, rich in various plants and animals and mineral resources. Moderate hydraulic, mild wind and gravity combined erosion are dominant in this area.
Problems: Grassland overgrazing, grass quantity decreased, grass quality deteriorated; Artificial soil erosion caused by tourism, water conservancy projects, mining and other development and construction projects has not been effectively controlled.

4.1.3 Southern low mountain and hilly soil conservation area

The southern hilly region is flat and belongs to the original denudation tectonic landform. In the low hills and hills with an altitude of 950-1150m, the thick tertiary strata are mostly exposed. Due to the scouring action of rivers and floods, the topography is crisscross with gullies and terraces developed on both sides of the valleys. Every year from winter to autumn, snow cover and glacial meltwater in high mountain areas combine with seasonal snowmelt water, rainfall and rainstorm in low mountain areas to provide water for rivers and become runoff forming areas of rivers. The intensity of hydraulic and moderate wind erosion is dominant in this area.

4.1.4 Farmland Protection zone of central Oasis human settlement environment

The central area of Changji prefecture is the piedmont plain area at the northern foot of Tianshan Mountain, which accounts for about 25% of the total land area of the prefecture. This region is the main agricultural area, but also Changji prefecture politics, economy, culture more developed region. Since ancient times, it has been an important place for people of all ethnic groups to develop agriculture and animal husbandry, and to live and reproduce. Historical sites and scenic spots left by various social and economic activities are very common. At the same time, it has formed a recreational belt around the city with large and medium-sized cities as the center and urban residents' needs as the core. This area belongs to the restricted development area. Although it also belongs to the ecologically fragile area, compared with the natural ecological area in the south, it is more suitable for tourism development. Tourism can be combined with the construction of agriculture, culture and characteristic towns to create characteristic tourism towns.

The farmland protection area in the central part of the living environment is an alluvial and alluvial inclined plain and fine soil plain area formed by the diluvial and alluvial action of various rivers. The topography here is undulating, and the terrain slopes from southeast to northwest, with lacustrine deposition distribution in some parts. The elevation is 400 ~ 1000m, and the topographic gradient is 1% ~ 15%. On the alluvial and alluvial fans, the scouches with a depth of 2 ~ 10m are formed on the fan surface due to the erosion and cutting action of running water. The alluvial plain area is flat, and due to the erosion of running water, pinniform and branchlike gullies are developed with a depth of 3 ~ 5m. In some areas, due to the flood or the exposure of groundwater, there are puddles and reeds growing. The area is the core of human life and production activities, as well as an important production base of petroleum, chemical industry, agriculture, forestry and animal husbandry in Xinjiang. Oases form wherever rivers go. Most of the river runoff from the mountains is used up in the oasis, and only a few rivers have rain water into the desert. Therefore, this area is the utilization and loss area of river runoff. Moderate wind force and slight hydraulic erosion are dominant in this area.

4.1.5 Northern desert ecological maintenance windbreak and sand control area

The northern part of Changji Prefecture is mostly covered by deserts, which are 400 ~ 600m above sea level. This area has a dry climate and little rain. Rich deserts and sandy vegetation form different types of desert landscapes, which have become an important tourist resource in this area. General Temple, multicoloured Bay, silicified wood .All the national geoparks are in this area. From the perspective of resource quality, some of them belong to the unique resources in this area and have strong development value. However, this region also belongs to a region with extremely fragile ecological environment and is a prohibited development zone, so it needs to pay attention to environmental protection in the development process just like the southern natural ecological tourism zone.

It mainly refers to the Gurbantungu Desert and The General Gobi Desert, with an altitude of 280 ~ 1100m.The desert area is 280 ~ 850m above sea level.

Problems: Lack of drinking water, poor combination of water and grass, weak disaster resistance of livestock, often affected by strong wind and low temperature; The exploitation of mineral resources will cause man-made disturbance, which will easily cause additional soil and water loss.

4.1.6 North water conservation and Ecological Conservation area in middle and low Mountains

This area refers to the mountains and hills to the north of the desert. The mountains are low and low, with an altitude of 800 ~ 3000m. They are mainly composed of North Tashan, large and small Hafutik Mountains and Kekse Mountains. The surface is mostly covered by weathering and semi-weathering, and rainfall is rare, with annual rainfall of 100 ~ 200mm.The vegetation coverage rate is low, except for larch scattered in the shade slope of Beita Mountain, most of the low xerophytic desert plants only grow in the mountain gully area. This area is the winter pasture of Mulei County, due to the lack of water, the pasture utilization rate is only 30%.Moderate wind force and moderate hydraulic erosion dominated the area.

References

1. Hu J.B. (2018) Discussion on soil erosion and ecological management in Karst area of Guizhou Province. China Resources Comprehensive Utilization, 36(02):124-125+134.
2. Fu W.B.,Dai Q.H.,Yan YJ. (2015)The response of soil erosion in Karst slope and its shallow underground crevasse ratios. Journal of Soil and Water Conservation, 29(02):11 — 16+22.Yue K.Q. (2016)
3. Lu G.Y.,Li S.,Wei X.H., et al. (2013) The processes of soil degradation in Rock Desertification area of northern Guangdong. Journal of Soil and Water Conservation, 27(02):20-25.
4. Li Y, Bai X, Zhou Y, et al. (2016)Spatial-Temporal Evolution of Soil Erosion in a Typical Mountainous Karst Basin in SW China, Based on GIS and RUSLE.
5. Arabian Journal for Science and Engineering, 2016, 41(1):209-221.Chen G.Y.(2018)
6. M.A. Busari, S.S. Kukal, A. Kaur, R. Bhatt, A.A. D ulaziConservation tillage impacts on soil, crop and the environment Int Soil Water Conserv Res, 3 (2015),pp. 119129, 10.1016/j.iswcr.2015.05.0 02

Analysis of the water quality condition of the Zijiang River in summer and the applicability of evaluation methods

Tao Zhang[1,*], Peng Xiang[2], Shijiang Zhu[2,*], Yang Liu[3],Wen Xu[2],Xiaoling Li[2],Aihua Sun[2],Wan Meng[2], Shubiao Gui[2],Jin Zhu[2]

[1]Yichang Dongfengqu Irrigation Area Administration,China
[2]College of Hydraulic and Environmental Engineering, China Three Gorges University
[3]China Institute of Water Resources and Hydropower Research
* Author:T.Zhang, mainly engaged in the management and research of water conservancy and water ecology.

Abstract: The single-factor evaluation method, improved Nemerow index method, fuzzy mathematical synthesis evaluation method and comprehensive nutrient status index method were used to evaluate the water quality of the Zijiang River in summer, and to analyze the applicability of the evaluation methods. The results show that: (1) The Zijiang River is heavily polluted with nitrogen and TN is the main pollutant in the summer. The overall nutrient status index is 34.33, which is not eutrophic. (2) The single-factor evaluation method is simple and suitable for rapid identification of excessive pollutants, but cannot reflect the water quality. the Nemerow index method takes into account the weight of each evaluation factor in the river water quality, and is suitable for making a comprehensive evaluation of the river water quality, but has a certain degree of subjectivity when classifying the water quality. The fuzzy comprehensive evaluation method is suitable for reflecting the influence of different evaluation factors on water quality by calculating the weight and membership degree of each evaluation factor, but the calculation is complicated.The Integrated Nutrient Status Index method is suitable for determining whether eutrophication is occurring in a water body and cannot determine the water quality category.

1. Introduction

Comprehensive research and application of water quality evaluation methods at home and abroad, there is currently no unified evaluation method[1].It has been found that there are seasonal characteristics of water quality in rivers and lakes[2,3,4], this paper uses the widely used single-factor evaluation method, improved Nemerow index method, fuzzy mathematical evaluation method, integrated nutrient status index method, four methods to evaluate the water quality of the Zijiang River in summer, and the applicability of the four evaluation methods analysis.

2. Materials and methods

2.1 Overview of Zijiang

Zijiang River is the third largest tributary of the Dongting Lake in the central part of Hunan province.With a total length of 653km and a watershed area of 28142km^2, the annual precipitation is 1200mm-1800mm, and it rains a lot from April to August, and the runoff of the Zijiang River increases significantly in July [5].

2.2 Water samples taken

Water withdrawals along the main stem of the Zijiang River from June 27 to July 1, 2019. A total of 36 intake sections, each taking surface, middle, and bottom water at the left, middle, and right banks, respectively.From July 2 to 4, 2019, five indicators including total nitrogen, total phosphorus, ammonia nitrogen, permanganate index and chlorophyll a were measured in the laboratory of Three Gorges University. The sample point layout is shown in Fig 1.

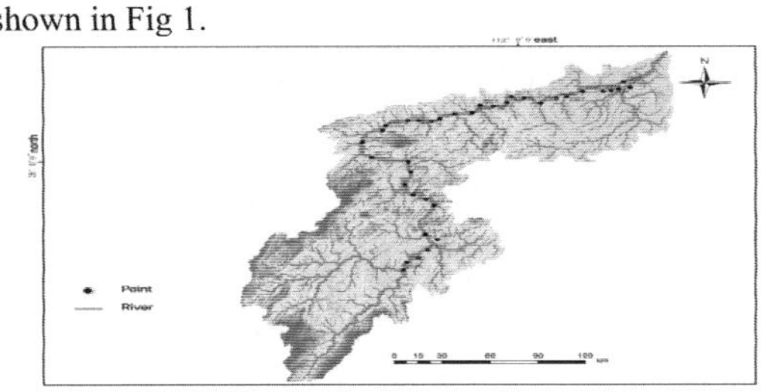

Fig 1.Distribution of sampling points

* Corresponding author:S.J.Zhu, PhD, associate professor, Email:46212465@qq.com.

2.3 Water quality assessment methods

2.3.1 Single factor evaluation method

Single-factor evaluation is a method to determine the water quality level by comparing the extreme or average value of the evaluation factor with the magnitude of the standard value. The formula is [6,7]:

$$P_i = \frac{C_i}{C_0} \quad (1)$$

Pi is the relative pollution value of an indicator,Ci is the maximum measured concentration value of an indicator,and C_0 is the maximum allowable standard value of an indicator.

2.3.2 Fuzzy mathematical synthesis evaluation method

The specific steps are as follows [8,9]:

(1)Establish the evaluation object factor set U={U1,U2,U3,U4,U5}, where U is the measured mass concentration value of the n pollution indicators involved in the evaluation.

(2)Establish the evaluation set v={I, II, III, IV, V}.

(3)Determine the weight coefficients and the normalization process. Each evaluation factor Ui is given a weight ai according to the principle of "the more the measured concentration exceeds the standard limit, the greater the weight", and the weights of the n evaluation factors constitute the weight set A={a1,a2,a3,a4,a5}. The formula for determining the weights according to the superscalar multiplier method is:

$$S_i' = \frac{1}{m}\sum_{j=1}^{n} S_{ij} \quad (2)$$

$$I_i = \frac{X_i}{S_i} \quad (3)$$

$$W_i = \frac{I_i}{\sum_{j=1}^{n} I_i} \quad (4)$$

X_i is the measured value of factor i, S_{ij} is the j_{th} criterion of factor i, m is the number of levels of evaluation criteria, Si′ is the average value of the various levels of criteria of factor i, W_i is the weight of the i-th evaluation factor, and the normalization process yields the weight set, W={W1, W2, ... , Wn}.

(4)Establishment of affiliation functions:

$$r_{ij} = \begin{cases} 1 & , X_i < S_j \\ \dfrac{S_{j+1} - X_i}{S_{j+1} - S_j}, & S_j < X_i < S_{j+1} \\ 0 & , X_i \geq S_{j+1} \end{cases} \quad (5)$$

rij indicates the affiliation of evaluation factor i to level j (i=1, 2, ..., n; j=1, 2, ..., m), Xi is the measured value of evaluation factor i, and Sj indicates the water quality criterion for level j of evaluation factor i. The water quality criterion for level j of evaluation factor i is indicated.

(5)The fuzzy relationship between water quality evaluation factors and evaluation criteria is established, and the m×n fuzzy evaluation matrix R:

$$R = \begin{bmatrix} r_{11} & r_{12} & \cdots & r_{1m} \\ \vdots & & & \\ r_{n1} & r_{n2} & \cdots & r_{nm} \end{bmatrix} \quad (6)$$

(6)Matrix Compounding, Development of Integrated Evaluation Models:

$$B = W * R$$

Based on the principle of affiliation, the quality of the water body is rated j.

2.3.3 Improvement of the Nemerow index method

The formula is as follows [10]:

$$F_i = \frac{C_i}{S_{ij}} \quad (7)$$

$$P = \sqrt{\frac{F_{Max}^{\,2} + F'^{2}}{2}} \quad (8)$$

C_i is the measured concentration of the category i evaluation factor, S_{ij} is the category j standard concentration of the category i evaluation factor, and F_{Max} is the maximum value of F_i, which is the value corresponding to the pollution factor with the greatest weight.

$$r_i = \frac{s_{Max}}{s_i} \quad (9)$$

$$w_i = \frac{r_i}{\sum_{i=1}^{m} r_i} \quad (10)$$

r_i is the correlation ratio for the i-th pollutant, s_i is the standard value for each pollutant, s_{Max} is the maximum standard value for the i-th pollutant, and m is the number of pollutants evaluated.

2.3.4 Comprehensive nutrient status index method

Four detection indicators of chlorophyll a (Chl-a), TP, TN, and COD_{Mn} were chosen for the study, which were calculated as follows [11,12]:

$$TLI(Chl\text{-}a)=10\times(2.5+1.086\ln Chl\text{-}a) \quad (11)$$

$$TLI(TP)=10\times(9.436+1.624\ln TP) \quad (12)$$

$$TLI(TN)=10\times(5.453+1.694\ln TN) \quad (13)$$

$$TLI(COD_{Mn}) = 10\times(0.109+2.661\ln COD_{Mn} \quad (14)$$

TLI(Chl-a) is the chlorophyll a nutrient index, TLI(TP) is the total phosphorus nutrient index, TLI(TN) is the total nitrogen nutrient index, and TLI(CODMn is

 https://doi.org/10.1051/e3sconf/202123701027

the potassium permanganate nutrient index.

$$TLI(\Sigma)=\sum_{j=1}^{m} w_j \times TLI(j) \tag{15}$$

$$Wj=\frac{r_{ij}^{2}}{\sum_{j=1}^{m} r_{ij}^{2}} \tag{16}$$

TLI(∑) is the composite nutritional status index, Wj is the relevant weight of the nutritional status index for the jth parameter, TLI(j) is the nutritional status index for the jth parameter, and m is the number of parameters evaluated.

Table 1. The value of each parameter

parameter	r_{ij}	r_{ij}^2
Chl-a	1	1
TP	0.84	0.7056
TN	0.82	0.6724
COD_{Mn}	0.83	0.6889

Table 2. Nutritional Status Index Rating

No	classification	TLI(∑)
1	Poor nutrition	TLI(∑)<30
2	Medium nutrition	30≦TLI(∑)≦50
3	Eutrophication	TLI(∑)>50
4	Mild eutrophication	<50TLI(∑)≦60
5	Moderate eutrophication	<60TLI(∑)≦70
6	Severe eutrophication	TLI(∑)>70

3. Results and analysis

3.1 Single factor evaluation method

The relative pollution value P_i was calculated from the single-factor evaluation formula, and the Pi values for TP, NH3-N, and COD_{Mn} were all less than 1. TN had 14 sections with P_i values greater than 1, namely ZS02, ZS03, ZS06, ZS07, ZS08, ZS10, ZS15, ZS16, ZS21, ZS24, ZS25, ZS29, ZS33, and ZS35. The proportion is 39%. Fig 2 shows the relative pollution value of TN, and the measured TN concentrations in the 14 sections with P_i values greater than 1 are all greater than 2.0 $mg.L^{-1}$ in the Surface Water Quality Standard V, and TN can be judged as a pollutant exceeding the standard.

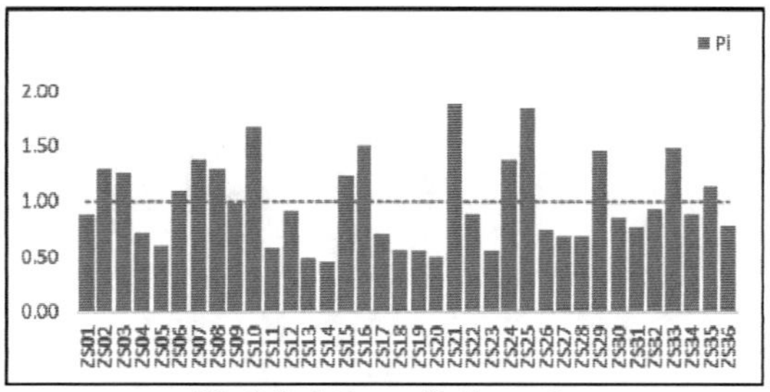

Fig 2. Relative pollution value

3.2 Fuzzy mathematical synthesis evaluation method

The fuzzy mathematical comprehensive evaluation method was applied to evaluate 36 cross sections, most of which are class III and have good water quality, with 10 cross sections in ZS10, ZS16, ZS17, ZS19, ZS21, ZS24, ZS25, ZS29, ZS32, ZS36 having poor water quality.

Through the analysis of weights W_i, the 10 sections of poor water quality TN accounted for the largest weight W_i, such as Fig 3, in order of 0.707, 0.763, 0.605, 0.632, 0.828, 0.764, 0.817, 0.764, 0.588, 0.692, indicating that TN is the main pollution factor, ZS10, ZS17, ZS24, ZS32 section NH3-N weights W_i were 0.190, 0.230, 0.104, 0.249, second only to TN weights, indicating that NH_3-N is a minor pollution factor, ZS16, ZS19, ZS21, ZS25, ZS29, ZS36 section COD_{Mn} weights Wi were 0.101, 0.206, 0.072, 0.111 0.105 and 0.172, second only to the TN weight, indicating a high degree of contamination by organic pollutants and reducing substances; among the 10 sections, the TP weight W_i was smaller than the other three indicators, indicating that TP has the lowest degree of influence on water pollution.

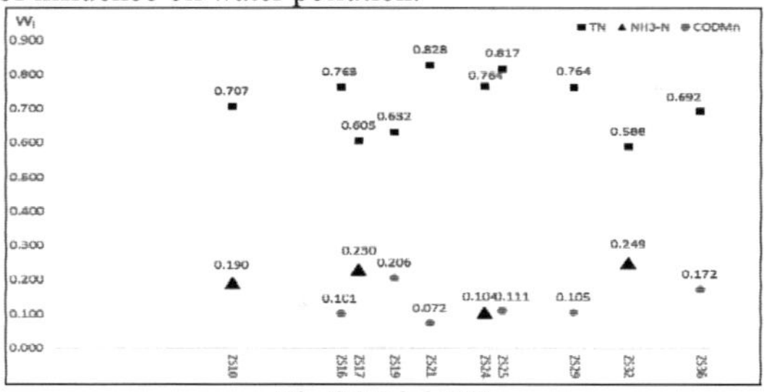

Fig 3. Weight W_i

3.3 Improvement of the Nemerow index method

The improved Nemerow index method was applied to evaluate the water quality of 36 sections, and the improved Nemerow index P is shown in Fig 4.

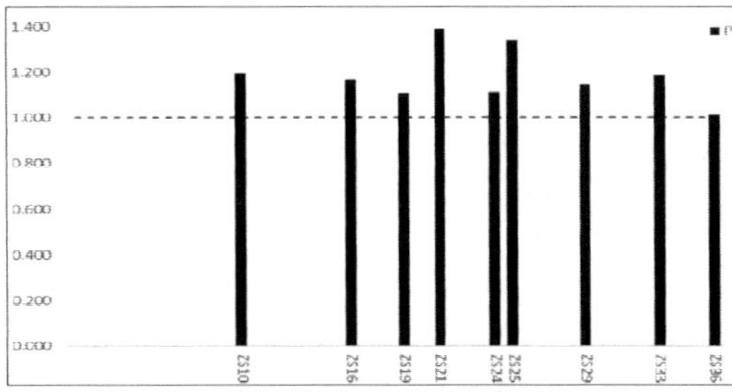

Fig 4.Nemerow Index

3.4 Comprehensive nutrient status index method

Using the integrated nutrient status index method, the nutrient status index TLI<50 was calculated for 36 sections, with a maximum TLI value of 43.12 and a mean value of 34.33, which is a medium nutrient level and the water body is not eutrophic.

4. Evaluate the results and analyse the applicability of the four methodologies

The results of the single-factor evaluation method differ greatly from the other three methods, with only ZS13, ZS14 and ZS20 having water quality of Type III. The results of the Nemerow index method and the fuzzy mathematical method are similar, with 90% coincidence in the sections with poor water quality and TN as the main polluting factor. The results of the Nemerow index method show that 11% of the sections are of Type II and 77.8% of the sections are of Type III class. The results of the fuzzy mathematical method evaluation showed that 72% of the sections were class III.In the overall analysis of the Zijiang River, there is no obvious pattern in the water quality evaluation results. This phenomenon may be related to the increased flow and the uneven distribution of aquatic plants in the Zijiang River in summer. There are seasonal changes in the biological community, and the ability of aquatic plants to remove pollutants is different in different seasons.There are seasonal changes in the biological community in the water, the ability of aquatic plants to purify pollutants is different in different seasons, the runoff increases significantly in July, and there is uncertainty in the water quality due to the complex composition of river water sources during the high water period[13].Through investigations, it was found that Zijiang aquatic plants are widely distributed, with 22 species. Studies have shown that aquatic plants have the highest removal efficiency of total nitrogen and ammonia nitrogen in summer [14]. Wang Xiaoyan and others found that the total nitrogen content in the Taihu Lake Basin is the lowest in summer and autumn[15] .

The single-factor evaluation method judges the water quality grade based on the comparison between the detection index and the water quality standard. Its advantage is that the evaluation is simple and fast, and it can directly reflect the pollution factor exceeding the standard. The disadvantage is that it cannot fully reflect the overall water quality of the river section.The advantage of the improved Nemerow index method is that the weights of TN, TP, NH3-N, and COD_{Mn} in the water quality evaluation are considered, and the worst index is used to judge the water quality of the section, avoiding ignoring some harms due to small measured concentrations.The impact of pollution factors with large coefficients on water quality evaluation results is suitable for judging toxic and harmful substances in rivers, and can more comprehensively evaluate river water quality. Its disadvantage is that there is a certain subjectivity in the classification of water quality.The advantage of fuzzy mathematics comprehensive evaluation method is that it takes into account the ambiguity of water quality classification boundaries and the weight of pollution factors, and can objectively and comprehensively evaluate the water quality status. By calculating the weight of evaluation factors and the degree of membership, the impact of each evaluation factor on water quality can be known.The advantage of the comprehensive nutrient status index method is that the eutrophication status of the water body can be directly judged through calculation. If the water body is eutrophication, combined with the improved Nemerow index method, the pollution factors that lead to the eutrophication of the water body can be determined, and the two are correlated, Analysis can verify the evaluation results. The disadvantage is that only the nutritional status can be determined in a single application, and the pollution factor cannot be judged.

5. Conclusion

(1) The water quality of 36 sections in Zijiang was evaluated. Zijiang River has serious nitrogen pollution in summer and TN is the main pollution factor.

(2) The single-factor evaluation method is simple to calculate and is suitable for quickly identifying excessive pollutants, but cannot comprehensively reflect the water quality of the water body. The Nemerow index method takes into account the weight of each evaluation factor in the river water quality, and is suitable for making a comprehensive assessment of the river water quality Evaluation, but it is subjective in the classification of water quality. The fuzzy comprehensive evaluation method calculates the weight and membership degree of each evaluation factor, which is suitable for reflecting the impact of different evaluation factors on water quality, but the calculation is complicated. The comprehensive nutritional status index method is applicable.To determine whether the water body has eutrophication, it is impossible to determine the water quality category, and it can be combined with other methods to determine the pollution factors that lead to the eutrophication of the water body.

References:

1. Y.L.Zhang,Y.Zhou, et al.*Application Comparison of Different Water Quality Evaluation Methods i n Danjiangkou Basin Water Quality Evaluation,*

Environ.Monit.China,**31**(2015)

2. Kimura, Masaomi,Noda.*Seasonal characteristics of surface water quality in the wastewater catc hment system of an urbanizing basin*.Paddy & Water Environment,(2018)
3. M.A.Soares Cruz,A.D.A.Goncalves,et al.*Spatial a nd seasonal variability of the water quality cha racteristics of a river in Northeast Brazil,*Enviro nmental Earth ences,**78**(2019)
4. Peters M,Guo Q,Strauss H,et al.*Seasonal effects on contamination characteristics of tap water fr om rural Beijing: A multiple isotope approach*.J ournal of Hydrology,(2020)
5. T.J. Hou,Y.Gao,*Analysis on the Characteristics a nd Causes of Water and Sediment Variation in Zishui, Hunan*,Water Resources and Hydropower Express,**39**(2018)
6. F.Luo,G.R.Wu,C.Wang,*Application of Nemerow p ollution index method and Single factor evaluati on method in water quality evaluation*,Environm ent and Sustainable Development,(2016)
7. Z.Wang,S.J.Zhu,Y.Liu, *Comparison of Water Qua lity Assessment Methods in the Lower Luan Riv er Region*,Water-saving irrigation,**10**(2019)
8. L.J.Wang,C.H.Li,*Groundwater quality evaluation of Shenyang section of Liaohe River based on i mproved fuzzy mathematics*,Ground Water, (2019)
9. R.X.Huang,*The evaluation of groundwater enviro nmental quality in bozhou based on fuzzy math ematics comprehensive evaluation*,Ground Water, (2018)
10. Y.M.Ning,F.N.Yin,*Water Quality Evaluation Base d on Improved Nemerow Pollution Index Metho d and Grey Clustering Method*,J.Cent. China No rm. Univ.Nat.Sci,**54**(2020)
11. M.C.Wang, X.Q.Liu,J.H.Zhang,*Evaluation method and classification standard of lake eutrophicati on*,Environ. Monit.China,**5**(2002)
12. G.Q.Bao ,L.Yin ,J.L. Yu ,C. Liu,X.C.Qiu,*Eutrop hication evaluation of Heihe based on compreh ensive nutritional status index and BP neural n etwork,*Bulletin of Soil and Water Conservation, **38**(2018)
13. K.E.Boyer,P.Fong,R.R.Vance,R.F.Ambrose,Wetland s,**21**(2001)
14. C.Wang, Y.Q.Chen,L.J.Cai, Z.Zuo,B.X.Cheng, *Pu rification effects of large-scale ecological purific ation projects in different seasons on nitrogen i n raw water,*Chin. J. Environ.Eng,**9**(2015)
15. X.Y.Wang,S.Zhong,*Study on Changes of Ammoni a Nitrogen and Total Nitrogen Content in Taihu Lake Basin in Different Seasons*,Anhui Agric.S ci,(2014)

Increase the rate of plasma-assisted synthesis of silver nanoparticles through additives

Zhai Limin[1], Lu Ye[2], Chen Dong[3], Chen Xinyi[3], Liu Linsheng[4] and Li Chuanqi[5]

[1]Zhai Limin, College of Electronic Engineering, Guangxi Normal University, China
[2] Lu Ye, College of Electronic Engineering, Guangxi Normal University, China
[3]Chen Dong, College of Physics and Electronics, Nanning Normal University, China
[3]Chen Xinyi, College of Electronic Engineering , Guangxi Normal University, China
[4]Liu Linsheng, College of Electronic Engineering, Guangxi Normal University, China
[5]Li Chuanqi, College of Physics and Electronics , Nanning Normal University, China

Abstract. The plasma-assisted method was used to synthesize silver nanoparticles, and the growth process of silver nanoparticles (Ag-NPs) was monitored in real time by localized surface plasmon resonance (LSPR) absorption spectroscopy. The effect of additives on the synthesis of Ag-NPs was verified. It is found that the addition of isopropanol and glucose can increase the plasmon resonance absorption intensity of the reaction solution, and promote the synthesis of Ag-NPs. In the plasma-assisted method, the additives can effectively improve the synthesis efficiency of Ag-NPs ,which has great inspiration for the synthesis of other metal nanoparticles.

1 Introduction

In recent years, the research and application of nanotechnology has become more and more extensive, especially the noble metal nanomaterials have gradually been applied to the fields of biomedicine [1], environmental health [2], photocatalysis [3] and new energy[4]. Among them, silver nanoparticles (Ag-NPs) have antibacterial and bactericidal properties [5], which are particularly prominent in biomedical applications. The current methods for synthesizing Ag-NPs are mostly chemical methods [6], but these methods require the introduction of toxic and harmful reducing agents during the experiments, which will pollute the environment. Plasma assisted method uses part of the reducing substances generated after the discharge plasma contacts the liquid surface to replace the reducing agent added in the chemical method to reduce silver ions [7]. This method has the advantage of being more environmentally friendly. And the experimental equipment used is simpler to synthesize metal nanoparticles on a large scale.

So far, there are few researches on the synthesis of silver nanoparticles by plasma electrochemical method at domestic and abroad. Chen Q and others have carried out some related mechanism and application research [8-10]. They studied plasma-liquid reaction system by changing the current and adding different active materials. In this paper, a control group experiment was set up. Isopropanol and glucose were added and the concentration was changed. The growth process of Ag-NPs was monitored in real time by localized surface plasmon resonance (LSPR) absorption spectroscopy, so as to discuss the influence of these two additives on the growth rate of Ag-NPs.

2 Experimental Setup

When the discharge plasma contacts the liquid surface, reducing and oxidizing substances are generated. [10] The plasma-assisted method for synthesizing Ag-NPs uses these reducing substances (mainly hydrated electrons, atomic H, etc.) in the precursor solution to reduce Ag^+ to Ag^0. The reducing substances and oxidizing substances in the solution coexist, so the reduction reaction and the oxidation reaction in the reaction system proceed simultaneously. The main reductions in the solution are as follows:

$$e + H_2O \rightarrow e_{aq}^- \quad (1)$$

$$e_{aq}^- + Ag^+ \rightarrow Ag \quad (2)$$

To prevent the Ag-NPs generated in the solution from agglomerating during the synthesis process, polyvinyl pyrrolidone (PVP) was added to the solution.

The setup used in the experiment is shown in Figure 1. In the experiment, a negative voltage (high voltage DC power supply LAS-20KV-50mA) was applied to the tungsten steel electrode 3mm above the liquid surface to excite plasma, and argon passed through the gas flow controller (D08-3F, Qixing Huachuang).Flow rate of the electrode nozzle is controlled at 20 sccm. The circulation process of the reaction is as follows: First, a peristaltic pump (RZ1030 box type, Runze fluid CO.Ltd) pumps

*Liu Linsheng: linshengliu@163.com *Li Chuanqi:lcq@gxnu.edu.

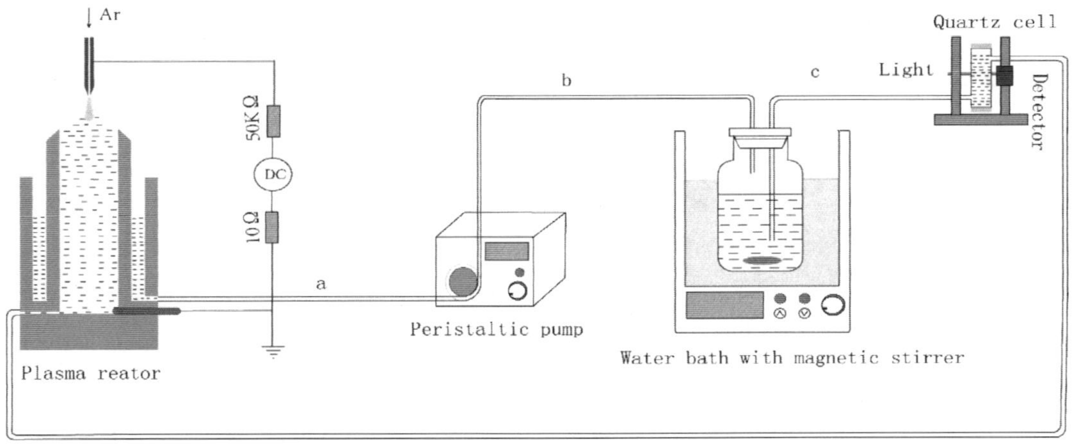

Fig1. Schematic diagram of the experimental set.

the reaction solution into a gas scrubber in a water bath and mixes with magnetic stirring (DFY-5L/10, Yu Hua, the water temperature is controlled at 25℃) and then enter the cuvette (optical path length 10mm) to monitor the absorption spectrum of the solution in real time. Then the solution flows from the cuvette to the plasma circulating reactor, and the solution flowing from the bottom to the top of the reactor reacts with the plasma. Finally, the reacted solution returns to the peristaltic pump to enter the next cycle. The light source used in the experiment is a deuterium halide lamp (DH2000, Wenyi Optoelectronics); the spectrometer used for real-time detection of absorption spectra is (USB6500-Pro, Jingyi Optoelectronics).

The reagents used in the experiment are as follows: $AgNO_3$ (analytical grade, Sinopharm Group); Polyvinyl Pyrrolidone $((C_6H_9NO)_n$, excellent grade pure, Sinopharm Group); isopropanol (C_3H_8O, analytical grade, Sinopharm Group)), Glucose($C_6H_{12}O_6$, analytical pure, Sinopharm Group). The ratio of $AgNO_3$ to PVP concentration used in the experiment is 1:2. The ratio of $AgNO_3$ to Glucose concentration used in the experiment is also 1:2.

3 Results

The OH radicals generated in the reaction system will directly or indirectly affect the reactivity of the solution and the synthesis process of Ag-NPs. Figure 2(a) is a graph of the plasmon resonance absorption intensity of the $AgNO_3$/PVP mixed solution discharged for 20 minutes. It is found that the peak of the solution is around 400 nm, indicating that the spherical silver nanoparticles have been successfully synthesized. Compared with Figure 2(a), it can be found from Figure 2(b) that the OH radical scavenger isopropanol has a great influence on the LSPR of Ag nanoparticles. Although Ag nanoparticles have a resonance absorption peak at 400 nm. However, the intensity of the local plasmon resonance absorption peak is significantly enhanced compared with not using isopropanol. It can be seen from Figure 2(c) that as the concentration of isopropanol added increases, the reduction reaction becomes more intense, and the intensity of the absorption peak increases more obviously.

As the reaction progresses, the hydroxide radicals in the solution will indirectly generate hydrogen peroxide through direct passage. The oxidation reaction and the reduction reaction in the solution will proceed simultaneously, so there is a competitive relationship between the two reactions. The oxidizing substance in the solution is mainly hydrogen peroxide, and the electrons generated during the reaction of isopropanol with the discharge plasma will quickly react with OH radicals, so isopropanol can scavenge OH radicals. As the content of OH radicals in the solution decreases, the concentration of hydrogen peroxide also decreases. Therefore, the oxidation reaction in the solution is suppressed and the reduction reaction is promoted.

As the concentration of isopropanol increases, the rate of reduction increases. But when the reduction rate is too fast, the concentration of silver ions in the solution will be consumed in a short time. Therefore, the absorption peak intensity will begin to flatten and no longer increase.

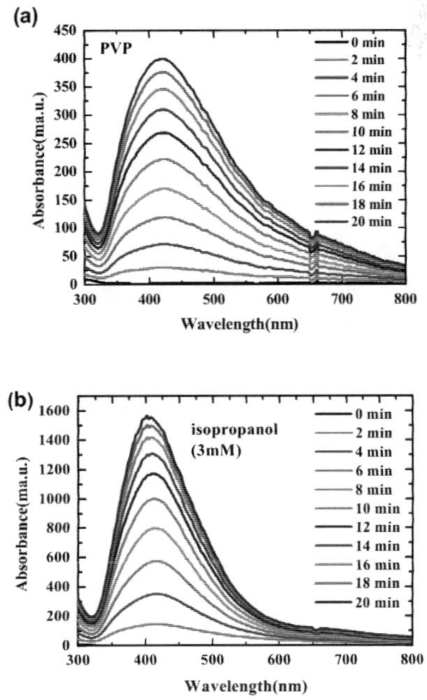

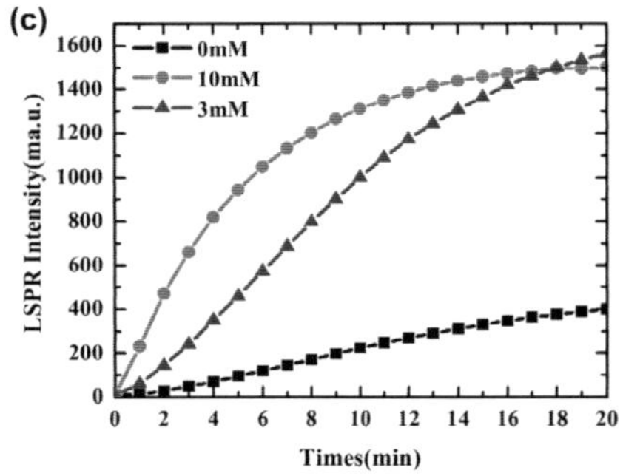

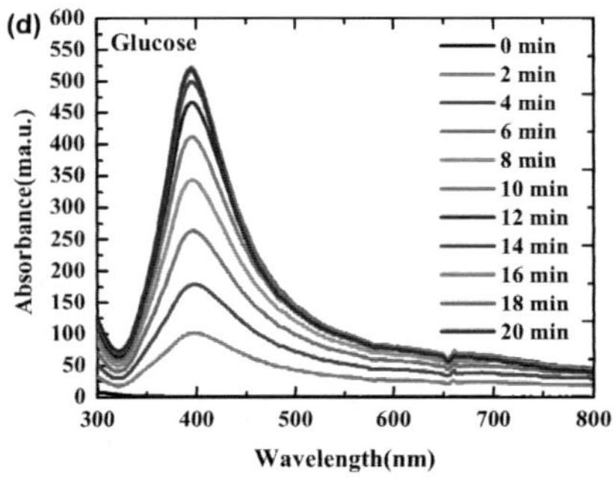

Fig2. (a) Time dependence of LSPR on Ag nanoparticles synthesized from $AgNO_3$/PVP solution,(b) Time dependence of LSPR on Ag nanoparticles synthesized from $AgNO_3$/PVP/ isopropanol solution,(c) The dependence of the absorption peak intensity on Ag nanoparticles synthesized from different concentrations of isopropanol solution,(d) Time dependence of LSPR on Ag nanoparticles synthesized from $AgNO_3$/glucose solution.

The reducibility of glucose will change with the difference of PH. In the plasma-liquid reaction system, the pH of the solution is less than 8, so the reducibility of glucose is weak [11]. From the comparison of Figure2 (a) and (d), it is found that the peak position of the resonance absorption peak of the solution added with glucose has blue-shifted, and the half-width of the absorption peak is smaller.The absorption peak curve starts to overlap at 20 minutes of reaction, and the absorption peak intensity has reached saturation.

In the plasma-liquid reaction system, the PH of the solution is less than 8. At this time, there are two main reactions that can reduce Ag^+ to Ag: one is that e_{aq}^- and H react with Ag^+; the other is that glucose directly reduces Ag^+ to Ag. In the reaction system, glucose can directly react with Ag^+ to generate Ag elemental substance. Compared with the solution with PVP as a stabilizer, the introduction of glucose brings a new way to generate Ag for the reaction system, which greatly promotes Ag-NPs.

In our synthesis, glucose plays two roles: one is to reduce Ag^+ to Ag^0, and the other is to act as a stabilizer for Ag atoms. In the process of reducing the metal salt to the corresponding metal nanoparticle, part of the glucose is oxidized to carboxylic acid. This carboxylic acid can produce a dense coating on the surface of Ag nanoparticles to counteract the van der Waals force responsible for particle coalescence, thereby stabilizing Ag nanoparticles [12].

In addition, the biomedical field requires high product safety. Glucose is a non-toxic reagent. Using glucose as a stabilizer can integrate the synthesized particles into various application systems. Glucose, as a small molecule stabilizer, can be easily removed when attached to the surface of nanoparticles. Therefore, the nanomaterials synthesized by this method have better biocompatibility.

4 Conclusion

Plasma-assisted synthesis of silver nanoparticles has the advantages of high efficiency and environmental protection, and the growth process of silver nanoparticles can be monitored in real time through local plasmon resonance absorption spectroscopy. It is found that the addition of isopropanol can greatly increase the absorption peak intensity of the solution and reduce its half-height width; greatly improve the synthesis efficiency and product uniformity. Compared with PVP, adding glucose has a dual effect, not only as a stabilizer, but also to promote the synthesis of silver nanoparticles. It has a more efficient synthesis effect, can complete the synthesis of nanoparticles in a short time, and its synthesis efficiency is significantly improved.

Acknowledgments

This work was supported by the doctoral scientific research fund of Guangxi Normal University (Grant No.2017BQ019).

References

1. Azharuddin M, Zhu G H, Das D, et al. A repertoire of biomedical applications of noble metal nanoparticles[J]. Chemical Communications, 2019,55:6964-6996
2. Kowalska E, Endo M, Wei Z, et al. Noble Metal Nanoparticles for Water Purification[M]//Nanoscale Materials in Water Purification. Elsevier, 2019: 553-579.
3. Krysiak O A, Barczuk P J, Bienkowski K, et al. The photocatalytic activity of rutile and anatase TiO2 electrodes modified with plasmonic metal nanoparticles followed by photoelectrochemical measurements[J]. Catalysis Today, 2019, 321: 52-58.
4. Ağbulut Ü, Sarıdemir S. A general view to converting fossil fuels to cleaner energy source by adding nanoparticles[J]. International Journal of Ambient Energy, 2019: 1-6.
5. Wu J, Li F, Hu X, et al. Responsive Assembly of Silver Nanoclusters with a Biofilm Locally Amplified Bactericidal Effect to Enhance Treatments against Multi-Drug-Resistant Bacterial Infections[J]. ACS Central Science, 2019:1366-1376.
6. Xia Y, Xiong Y, Lim B, et al. Shape - controlled synthesis of metal nanocrystals: simple chemistry meets complex physics?[J]. Angewandte Chemie

International Edition, 2009, 48(1): 60-103.

7. Tochikubo F, Shimokawa Y, Shirai N, et al. Chemical reactions in liquid induced by atmospheric-pressure dc glow discharge in contact with liquid[J]. Japanese Journal of Applied Physics, 2014, 53(12): 126201.
8. Chen Q, Li J, Li Y. A review of plasma–liquid interactions for nanomaterial synthesis[J]. Journal of Physics D: Applied Physics, 2015, 48(42): 424005.
9. Chen Q, Kaneko T, Hatakeyama R. Rapid synthesis of water-soluble gold nanoparticles with control of size and assembly using gas–liquid interfacial discharge plasma[J]. Chemical Physics Letters, 2012, 521: 113-117.
10. Gong X, Ma Y, Lin J, et al. Tuning the Formation Process of Silver Nanoparticles in a Plasma Electrochemical System by Additives[J]. Journal of The Electrochemical Society, 2018, 165(11): E540 E545.
11. T. V. Bukreeva, I. V. Marchenko, B. V. Parakhonskiy et al. "Formation of silver nanoparticles on shells of polyelectrolyte capsules using silver-mirror reaction". Colloid Journal, 2009, 71(5): 596.
12. M. Giersig, T. Ung, L. M. Liz-Marzan et al. "ChemInform Abstract: Direct Observation of Chemical Reactions in Silica-Coated Gold and Silver Nanoparticles." Cheminform, 2010, 28(31): no–no.

https://doi.org/10.1051/e3sconf/202123701029

Research on evolution game of clothing enterprises' carbon emission reduction investment considering free riding behavior

Hangxin Guo[1], Zheng Liu[1,2*], Mengmeng Zhao[3] and Huihui Wang[4]

1. School of Management, Shanghai University of Engineering Science, Shanghai, 201620;
2. Antai College of Economics and Management, Shanghai Jiao Tong University, Shanghai, 200240
3. School of textile and clothing, Shanghai University of Engineering Science, Shanghai, 201620;
4. Business school, University of Shanghai for Science and Technology, Shanghai 200093

Abstract: "low carbon economy" puts forward new requirements for carbon emission reduction in all links of the supply chain. This paper takes the three-level clothing supply chain composed of government, supplier and manufacturer as the research object, considering the "free riding behavior" of clothing supplier and manufacturer, by constructing the three-party evolutionary game model, using the replication dynamic equation to analyze the evolution process of the three actors, and studies the influence of parameter change on the strategy selection of each agent. Finally, the simulation is carried out by Matlab. The results show that the strategies are affected by the government's punishment, the "free ride" revenue of suppliers and manufacturers, and the cost of carbon emission reduction investment. Finally, based on the analysis results, some suggestions are put forward for enterprises to invest in carbon emission reduction.

1 Introduction

With the increasingly prominent environmental problems, low carbon is regarded as the basis of national economic development in the world. As the largest emitter of carbon dioxide, China has the responsibility and obligation to contribute to carbon emission reduction, which requires all-round social cooperation and participation [1]. The investment of carbon emission reduction leads to the decrease of enterprise profits, so usually enterprises are not willing to actively invest in carbon emission reduction, which requires the supervision of government departments. At the same time, the supervision of government departments needs cost, which leaves opportunities for enterprises not to carry out carbon emission reduction or "hitchhiking". Therefore, using evolutionary game theory to analyze the input behavior of carbon emission reduction has certain significance. In recent years, many scholars have used evolutionary game theory to study carbon emission reduction investment.

Z Liu et al. used edge computing to intelligently optimize the supply chain under the constraint of carbon tax, so as to achieve the dual goals of economic benefits and emission reduction [2]. J Z Xu et al. studied the factors that affect the main strategies of the three-party game by building the three-party evolutionary game model [3]. S C Tang et al. studied the carbon emission reduction strategy of supply chain under the leadership of manufacturer and retailer based on social responsibility [4]. J P Xu et al. studied the dynamic equilibrium strategy of the integration of procurement, coal blending and distribution in thermal power enterprises, and proved that the modified method can achieve low-carbon economy through examples [5]. S P Sun et al constructed a tripartite evolutionary game model and analyzed the local stability of the model by using the dynamic replication equation [6].

J L Jiao et al. studied the influence of carbon quota, supervision cost and other factors on evolutionary stability strategy under dynamic and static reward and punishment system [7]. F J Wang et al. studied the impact of government pressure on the carbon emission reduction performance of enterprises, and the research shows that the impact is positive [8]. D D Fan et al. used the Nash model and the Steinberg model to analyze the impact of government subsidies on the carbon emission reduction decision-making of enterprises under different power structures [9]. X Y Cao et al. studied the influence of carbon tax policies and government subsidies on the optimal carbon emission reduction and order quantity of enterprises [10]. R Zhao et al. analyzed how the government and the enterprise choose the strategy to achieve the emission reduction target [11];

The existing literature mainly studies the impact of government on the decision-making behavior of all parties and the emission reduction decision-making behavior of upstream and downstream enterprises in the supply chain, but few literatures take "free riding behavior" into account. Based on this, this paper uses evolutionary game theory and numerical simulation method to study the the asymmetric dynamic evolution process of the three parties in the game. Finally, based on the analysis results, some suggestions are put forward for enterprises to invest in carbon emission reduction.

*Corresponding author. Email: 03140011@sues.edu.cn

2 The construction of tripartite game model

This paper takes the three-level clothing supply chain composed of government, supplier and manufacturer as the research object, and studies the evolution process of the tripartite game. The emission reduction investment refers to the behavior of investment to reduce carbon dioxide emissions in the process of production, processing, transportation, retail or use [16]. It's raw products are low-carbon products, and the products without carbon emission reduction investment are ordinary products.

In order to facilitate the analysis, the following assumptions are made for this paper:

Hypothesis 1: The probability of government choosing the "regulatory" strategy is x, then the probability of choosing the "non regulatory" strategy is $1-x$; the probability of suppliers and manufacturers choosing the "carbon emission reduction investment" strategy is y and z.

Hypothesis 2: the cost of government regulation is C; when both the supplier and the manufacturer invest in carbon emission reduction, the environmental benefit obtained by the government is W_1; when only the supplier does it, it is W_2; when only the manufacturer does it, it is W_3; when neither party does it, it is W_4, and the fines imposed by the government are F_s and F_m respectively.

Hypothesis 3: when no carbon emission reduction investment is made, the cost of the supplier and the manufacturer is C_s and C_m, and the revenue is R_s and R_m. when the supplier is making the investment, the revenue increase rate is α_0, and when both parties are making the investment, it is α_1; similarly, when the manufacturer is making the investment, the revenue increase rate is β_0, and when both parties are making the investment, it is β_1.

Hypothesis 4: when faced with the opportunity of carbon emission reduction investment, when one enterprise invests in emission reduction, the other enterprise can "hitchhike". As a result, the manufacturer's revenue from hitchhiking is R'_m and the supplier's revenue is R'_s.

3 Model solution

Based on the assumptions and parameters, eight strategic combinations of the tripartite evolutionary game are constructed. Government regulation is recorded as A_1, no regulation as A_2, suppliers and manufacturers' carbon emission reduction investment is recorded as A_3 and A_5 respectively, and no carbon emission reduction investment is recorded as A_4 and A_6 respectively. Then the three-party income after the game is shown in table 1.

Table 1. Benefits of tripartite game

Event	A_1		A_2	
	A_3	A_4	A_3	A_4
A_5	W_1-C $(1+\alpha_1)R_s-C_s$ $(1+\beta_1)R_m-C_m$	W_3-C+F_s R'_s-F_s $(1+\beta_0)R_m-C_m$	W_1 $(1+\alpha_1)R_s-C_s$ $(1+\beta_1)R_m-C_m$	W_3 R'_s $(1+\beta_0)R_m-C_m$
A_6	W_2-C+F_m $(1+\alpha_0)R_s-C_s$ R'_m-F_m	$W_4-C+F_s+F_m$ R_s-F_s R_m-F_m	W_2 $(1+\alpha_0)R_s-C_s$ R'_m	W_4 R_s R_m

4 Model analysis

4.1 Stability strategy of government departments

Assuming that the expected benefit of the government's choice of "regulatory" strategy is U_{gy}, that of "non regulation" is U_{gn}, and the average expected revenue of the government is $\overline{U_g}$, then:

$$U_{gy} = yz(W_1 - W_2 - W_3 - W_4) + z(W_3 - W_4 - F_m) + y(W_2 - W_4 - F_s) + W_4 - C + F_s + F_m \quad (1)$$

$$U_{gn} = yza_3 + y(1-z)a_7 + (1-y)za_4 + (1-y)(1-z)a_8 = yz(W_1 - W_2 - W_3 + W_4) + y(W_2 - W_4) + z(W_3 - W_4) + W_4 \quad (2)$$

$$\overline{U_g} = xU_{gy} + (1-x)U_{gn} = -xzF_m - xyF_s + x(F_s + F_m - C) + yz(W_1 - W_2 - W_3 + W_4) + y(W_2 - W_4) + z(W_3 - W_4) + W_4 \quad (3)$$

The replication dynamic equation of the government is as follows:

$$F(x) = \frac{dx}{dt} = x(U_{gy} - \overline{U_g}) = x(1-x)(F_m + F_s - C - zF_m - yF_s)$$

(4)

If $F(x)=0$, then $x_1 = 0$, $x_2 = 1$, $y_0 = \frac{F_m + F_s - C - zF_m}{F_s}$. When $y = y_0$, $F(x) \equiv 0$, all levels are stable points; When $y \neq y_0$, then $x_1 = 0$

and $x_2 = 1$ are two stable points.
The derivation of equation (4) shows that:

$$\frac{dF(x)}{dx} = (1-2x)(F_m + F_s - C - zF_m - yF_s) \quad (5)$$

①When $F_m + F_s \le C$, $\left.\frac{dF(x)}{dx}\right|_{x_1=0} < 0$, $\left.\frac{dF(x)}{dx}\right|_{x_2=1} > 0$,

so $x_1 = 0$ is the evolutionary stability strategy;

②When $F_m + F_s > C$:

When $y > \frac{F_m + F_s - C}{F_m + F_s}$, $\left.\frac{dF(x)}{dx}\right|_{x_1=0} < 0$, $\left.\frac{dF(x)}{dx}\right|_{x_2=1} > 0$,

so $x_1 = 0$ is the evolutionary stability strategy;

When $y < \frac{F_m + F_s - C}{F_m + F_s}$, $\left.\frac{dF(x)}{dx}\right|_{x_1=0} > 0$, $\left.\frac{dF(x)}{dx}\right|_{x_2=1} < 0$,

so $x_1 = 1$ is the evolutionary stability strategy;

Through the analysis, the government's evolution stability strategy is shown in figure 1.

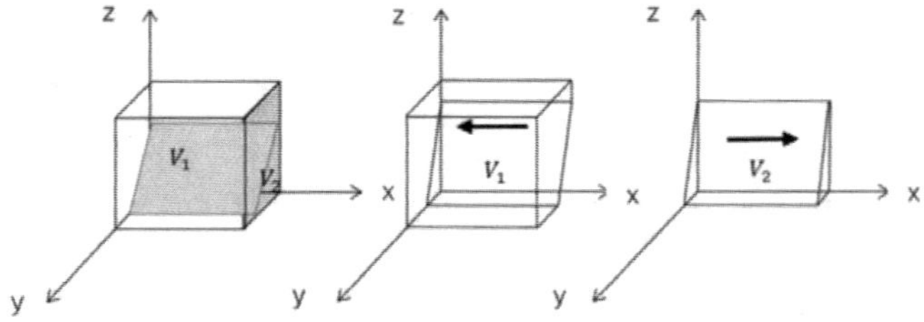

Figure 1. Supplier's impact on government strategy

It can be seen from figure 1 that when the initial state is in space V_1, $y > y_0$ and $x_1 = 0$ is the equilibrium point, because the cost of government supervision is greater than the benefit, the "no regulatory " strategy is chosen; When the initial state is in space V_2, $y < y_0$ and $x_2 = 1$ is the equilibrium point, and the government chooses the " regulatory " strategy.

Due to the cost pressure of government, some enterprises are opportunistic and pursue economic development unilaterally; If the government supervises and punishes the enterprises' extensive production, enterprises can be encouraged to actively invest in carbon emission reduction. Therefore, the government should play an active role to promote enterprises to do this.

4.2 Stability strategy of supplier Department

Assuming that the expected return of the supplier's choice of "carbon emission reduction input" strategy is U_{sy}, the expected return of the choice of "no carbon emission reduction input" strategy is U_{sn}, and the average expected return of the supplier is $\overline{U_s}$, then:

$$\overline{U_s} = yU_{sy} + (1-y)U_{sn} = (\alpha_1 - \alpha_0)yzR_s + \alpha_0 yR_s - yC_s - xF_s + zR_s' + (1-z)R_s + xyF_s - zyR_s' + zyR_s \quad (6)$$

The replication dynamic equation of the supplier is as follows:

$$F(y) = \frac{dy}{dt} = y\left(U_{sy} - \overline{U_s}\right) = y(1-y)\left\{\alpha_0 R_s - C_s - \left[R_s' - (\alpha_1 - \alpha_0 + 1)R_s\right]z + xF_s\right\} \quad (7)$$

If $F(y) = 0$, then $y_1 = 0$, $y_2 = 1$, $z_0 = \frac{\alpha_0 R_s - C_s + xF_s}{R_s' - (\alpha_1 - \alpha_0 + 1)R_s}$, and the derivation of equation (6) shows that:

$$\frac{dF(y)}{dy} = (1-2y)\left\{\alpha_0 R_s - C_s - \left[R_s' - (\alpha_1 - \alpha_0 + 1)R_s\right]z + xF_s\right\} \quad (8)$$

(1)When $z = z_0$, $F(y) \equiv 0$, all levels are stable points;

(2)When $z < z_0$, $\left.\frac{dF(y)}{dy}\right|_{y_1=0} > 0$, $\left.\frac{dF(y)}{dy}\right|_{y_2=1} < 0$,

so $y_2 = 1$ is the evolutionary stability strategy;

(3)When $z > z_0$, $\left.\frac{dF(y)}{dy}\right|_{y_1=0} < 0$, $\left.\frac{dF(y)}{dy}\right|_{y_2=1} > 0$,

so $y_1 = 0$ is the evolutionary stability strategy;

Through the analysis, the evolution stability strategy of suppliers is shown in figure 2.

 https://doi.org/10.1051/e3sconf/202123701029

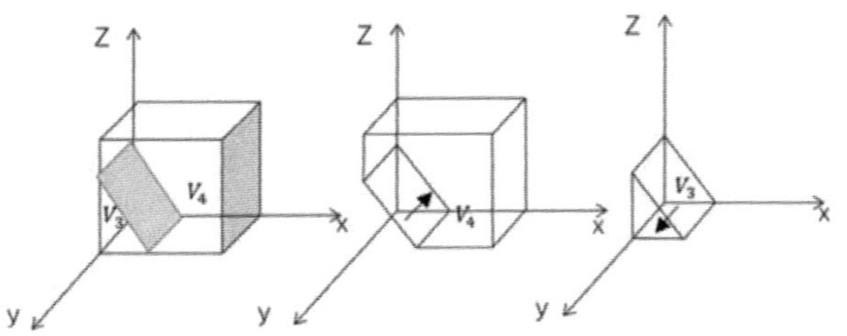

Figure 2. Manufacturer's impact on supplier strategy

It can be seen from figure 2 that when the initial state is in space V_3, there are $z < z_0$ and $y_2 = 1$ as the equilibrium point, and supplier chooses to invest in carbon emission reduction. At this time, enterprises tend to choose the investment strategy of carbon emission reduction in consideration of the fine mechanism existing in government supervision and their own net income. On the contrary, when in V_4, there is $z > z_0$, $y_1 = 0$ as the equilibrium point, and they choose the strategy of "no investment ".

4.3 Manufacturer's stabilization strategy

Suppose that the manufacturer's expected return for choosing "carbon emission reduction input" strategy is U_{my}, the expected return for choosing "no carbon emission reduction input" strategy is U_{mn}, and the manufacturer's average expected return is $\overline{U_m}$, then:

$$\overline{U_m} = zU_{my} + (1-z)U_{mn} = (\beta_1\text{-}\beta_0)yzR_m + \beta_0 zR_m - zC_m - xF_m + yR'_m + (1-y)R_m + xzF_m - yz(R'_m - R_m) \quad (9)$$

The replication dynamic equation of the manufacturer is as follows:

$$F(z) = \frac{dz}{dt} = z\left(U_{my}\text{-}\overline{U_m}\right) = z(1-z)\left\{\beta_0 R_m - C_m + xF_m - y\left[R'_m - R_m(\beta_1\text{-}\beta_0+1)\right]\right\} \quad (10)$$

If $F(z) = 0$, then $z_1 = 0$, $z_2 = 1$, $y_0 = \dfrac{\beta_0 R_m - C_m + xF_m}{R'_m - R_m(\beta_1\text{-}\beta_0+1)}$, and the derivation of equation (9) shows that:

$$\frac{dF(z)}{dz} = (1-2z)\left\{\beta_0 R_m - C_m + xF_m - y\left[R'_m - R_m(\beta_1\text{-}\beta_0+1)\right]\right\} \quad (11)$$

(1)When $y = y_0$, $F(z) \equiv 0$, all levels are stable points;

(2)When $y < y_0$, $\left.\frac{dF(z)}{dz}\right|_{z_1=0} > 0$, $\left.\frac{dF(z)}{dz}\right|_{z_2=1} < 0$, so $z_2 = 1$ is the evolutionary stability strategy;

(3)When $y > y_0$, $\left.\frac{dF(z)}{dz}\right|_{z_1=0} < 0$, $\left.\frac{dF(z)}{dz}\right|_{z_2=1} > 0$, so $z_1 = 0$ is the evolutionary stability strategy;

Through the above analysis, the evolution stability strategy of the manufacturer is shown in figure 3.

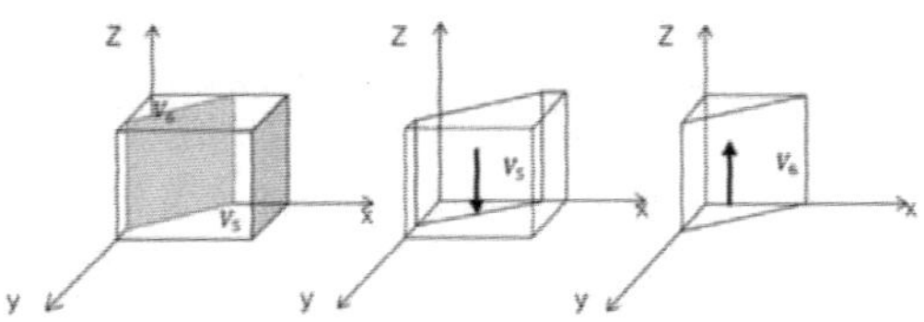

Figure 3. Supplier's impact on manufacturer strategy

It can be seen from Figure 3 that when the manufacturer's initial state is in space V_6, there is $y < y_0$, $z_2 = 1$ as the equilibrium point, and the manufacturer chooses the " investment" strategy. At this time, the enterprise will tend to choose the "investment" strategy considering the fine mechanism and its own net income situation existing in the government supervision; when in V_5, there is $y > y_0$, as the equilibrium point, which tends to choose“ no investment "strategy.

Which strategy supplier and manufacturer will ultimately choose is related to the government's punishment mechanism and the free riding behavior of upstream and downstream enterprises. In order to make enterprises choose the strategy of " investment", it's necessary to properly use the government to supervise and guide the market regulation.

4.4 Comprehensive analysis of game equilibrium strategy of fireworks among government, supplier and manufacturer

Combined with figure 1, 2 and 3, the results of evolutionary game among government, supplier and manufacturer are summarized, as shown in table 2. Because the following equilibrium states are not robust to small disturbances, there is no evolutionary stability policy (ESS) in the game process.

Table 2. Equilibrium results of tripartite evolutionary game

strategy combination	final strategy selection	strategy combination	final strategy selection
(V_1, V_3, V_5)	(no regulation, investment, no investment)	(V_2, V_3, V_5)	(regulation, investment, no investment)
(V_1, V_3, V_6)	(no regulation, investment, investment)	(V_2, V_3, V_6)	(regulation, investment, investment)
(V_1, V_4, V_5)	(no regulation, no investment, no investment)	(V_2, V_4, V_5)	(regulation, no investment, no investment)
(V_1, V_4, V_6)	(no regulation, investment, investment)	(V_2, V_4, V_6)	(regulation, investment, investment)

5 Numerical simulation

This paper only considers the situation of "free riding" between supplier and manufacturer. Suppose the specific parameter settings are shown in table 3. Matlab is used to simulate the stability strategy results of three actors under the change of specific parameters.

Table 3. Specific parameter values

Parameter symbols	α_0	α_1	β_0	β_1	C	C_m	C_s	R_s	R_s'	R_m	R_m'	F_s	F_m
Numerical value	0.235	0.455	0.245	0.375	7	13.5	8	35	52.5	57	71.5	9	6

5.1 The influence of government punishment on the equilibrium strategy of enterprises

Figure 4 shows the impact of the change of government's punishment on the evolution results. Change the value of F_s without changing any other parameters. As shown in the figure, the initial proportion of each entity is the same. After the reduction of the government's penalty for manufacturing supplier. The government's strategy has changed from regulatory to non regulatory, and the supplier choose "no investment" strategy due to the small punishment. At this time, the manufacturer choose "investment" strategy to get more benefits, so the supplier can get more benefits through free riding.

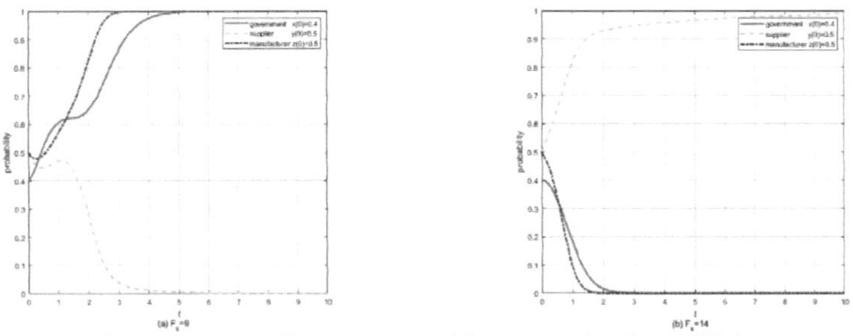

Figure 4. Impact of government punishment on tripartite equilibrium strategy

5.2 The influence of cost on the equilibrium strategy of three parties

Figure 5 shows the impact of cost input on the evolution results of the three parties. Other parameters remain unchanged, changing C and C_m respectively, as shown in the figure, from $C=7$ to $C=10$, the government tends to choose non regulatory strategy due to the increase of regulatory cost; when the manufacturer's cost C_m changes from 13.5 to 18, the supplier tends to choose "investment" strategy, and the manufacturer eventually tends to choose "no investment" strategy. Because at this time, when the supplier is faced with greater punishment, the benefit generated by free riding is less than that generated by carbon emission reduction investment, so the supplier will finally adopt the carbon emission reduction investment strategy. Therefore, the government can change the equilibrium strategy of enterprises by increasing the punishment to enterprises and reducing the supervision cost.

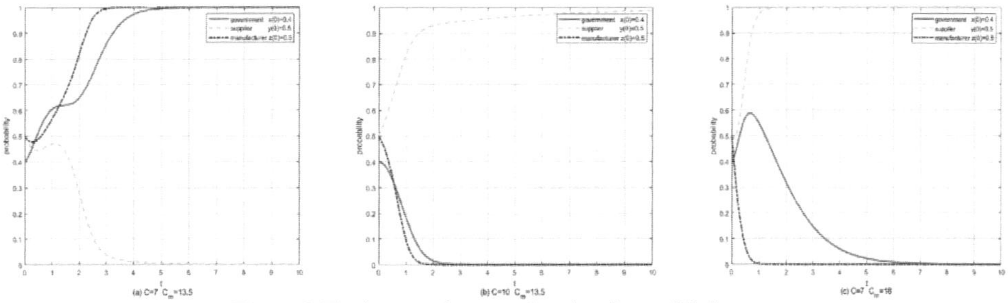

Figure 5. The impact of cost on the tripartite equilibrium strategy

6 Conclusion and prospect

Based on the existing research on the evolutionary game of carbon emission reduction investment, this paper considers the "free riding" behavior among clothing enterprises. By constructing a tripartite evolutionary game model, the evolution process of the three actors is analyzed by using the replication dynamic equation, and the influence of the strategy selection between the actors and the parameter change on the strategy selection of each actor is studied. Finally, Matlab is used for simulation. The results show that when the government chooses the regulatory strategy and the manufacturer (retailer) chooses to invest in carbon emission reduction, the retailer (manufacturer) will adopt the "free riding" strategy; At the same time, when the government punishment is low and the regulatory cost is high, the enterprise will also adopt this strategy.

In this regard, the government should cooperate with various departments, strengthen supervision, formulate reasonable punishment system, and set up carbon trading policies and carbon emission rights, so as to promote enterprises to carry out low-carbon emission reduction independently. Enterprises should increase efforts to independently develop carbon emission reduction equipment or technical means, and at the same time, by means of cooperation and coordination or revenue sharing contract, avoid "free riding" behavior.

Acknowledgments

The finding is sponsored by the National Social Science Fund of China (Grant No. 18CGL015).

References

1 Liu Z, Hu B and Zhao Y J 2020 Research on Intelligent Decision of Low Carbon Supply Chain Based on Carbon Tax Constraints in Human-Driven Edge Computing *IEEE* **08** 48264-48273

2 Wang Q P, Zhao Z G and He F L 2014 Strategies Selection and Behavioral Evolution of Carbon-Emission-Reduction Investment in Supply Chain Firms *IEEM* **28** 180-189

3 Xu J Z and Lv X C 2014 Evolutionary Analysis Between Governments、Manufacting Enterprises and Consumers on Low-Carbon Decision-making Behavior under the Background of Low-Carbon Economy *Operations Research and Management Science* **23** 81-91

4 Tang S C, Liu Y X and Xiao T J 2020 Pricing and carbon emission reduction decision in a socially responsible supply chain Chinese Journal of Management Science 28 99-108

5 Xu J P, Feng Q, Lv C W and Huang. Q 2019 Low-carbon electricity generation-based dynamic equilibrium strategy for carbon dioxide emissions reduction in the coal-fired power enterprise Enviro Sci Pollut Res 26 36732-53

6 Sun S P and Sun X Y 2016 Evolutionary Analysis Between Governments、Manufacting Enterprises and Consumers on Low-Carbon Decision-making Behavior under the Background of Low-Carbon Economy Operations Research and Management Science 25 113-9

7 Jiao J L, Chen J, Li L L and Li F Y 2017 A Study of Local Governments' and Enterprises' Actions in the Carbon Emission Mechanism of Subsidy or Punishment Based on the Evolutionary Game J Manag Sci China 25 140-150

8 Wang F J and Sun J Q 2019 Research on the impact of government pressure on the performance of carbon emission reduction of enterprises Forecasting 38 37-44

9 Fan D D and Xu Q 2018 Decisions of Enterprises Carbon Emission Reduction and Government Subsidy in Different Power Structures Soft science 32 64-70

10 Cao X Y and Zhang J F 2018 Carbon Reduction Optimization and Coordination of Supply Chain under Carbon Reduction Subsidy and Carbon Tax Operations Research and Management Science 27 57-61

11 Zhao R, Neighbour G, Han J J, McGuire M and Deutz P 2012 Using game theory to describe strategy selection for environment risk and carbon emissions reduction in the green supply chain Journal of Loss Prevention in the Process Industries 25 927-36

12 Jones G, Plassmann K and York EH 2009 Vulnerability of exporting nations to the development of a carbon label in the United Kingdom Environmental Science & Ploicy 12 479-490

Study on the dilemma and Countermeasures of rural environmental pollution control under the background of Rural Revitalization Strategy

Xiaojuan Han[1]

[1] Shangdong women s university, No.2399 Daxue Road, Changqing District, Jinan City, Shandong Province,China

Abstract. Environmental pollution has been the focus of attention in recent years. The 19th National Congress of the Communist Party of China put forward the strategy of rural revitalization, which pointed out that we should adhere to the harmonious coexistence of man and nature. Under the background of Rural Revitalization Strategy, we should improve the rural living environment and accelerate the transformation and upgrading of agriculture. We should establish and improve the rural environmental governance system, realize the construction of ecological civilization, and adhere to the implementation of green development and innovative development. Based on the background of Rural Revitalization Strategy, taking YZ town as an example, this paper analyzes the difficulties of rural environmental pollution control, such as weak awareness of environmental pollution control, low efficiency of environmental protection funds, and puts forward corresponding countermeasures.

1 Research background and significance

1.1 Research background

In recent years, environmental problems have attracted the attention of all sectors of society. In view of the problem of environmental pollution, the government has issued a series of laws and regulations in order to realize the construction of ecological civilization. We should adhere to the implementation of green development and build a beautiful China. The 19th National Congress of the Communist Party of China put forward the strategy of rural revitalization, which pointed out that we should adhere to the three steps of Rural Revitalization. At present, the implementation of Rural Revitalization Strategy has achieved certain results, but it still needs to continue to work hard. In order to build ecological and livable villages proposed in the strategic plan of rural revitalization, we should control the rural environmental pollution, adhere to the harmonious coexistence of human and nature, and take the road of rural green development.

1.2 Research significance

1.2.1 Theoretical significance

Under the background of Rural Revitalization Strategy, we should do a good job in rural human settlements, establish and improve the rural environmental pollution control system, and adhere to the harmonious coexistence of human and nature. Rural environmental pollution control work is complex, we should plan scientifically and implement it effectively. Taking YZ Town, Jining City, Shandong Province as an example, this paper investigates the current situation of rural environmental pollution control, analyzes the existing problems of rural environmental pollution control, and puts forward targeted solutions on this basis. These studies are expected to play a complementary role in the theoretical study of environmental pollution control.

1.2.2 Practical significance

The quality of rural environment is not only related to the quality of life of rural residents, but also affects the development of rural economy. Based on the background of Rural Revitalization Strategy, this paper analyzes the problems of rural environmental pollution, and enhances the environmental awareness of rural residents. These studies hope to play a certain reference role in the construction of ecological livable villages.

2 The current situation of rural environmental pollution under the background of Rural Revitalization Strategy

In this paper, Jining City, Shandong Province, YZ Town, as an example, we conduct field research to understand the rural environmental pollution under the strategy of Rural

https://doi.org/10.1051/e3sconf/202123701030

Revitalization. Since the implementation of the Rural Revitalization Strategy, YZ town has taken a series of measures, such as the government of YZ town to encourage villagers to establish ecological agricultural parks, introduce strawberry industry, increase aquaculture, etc., and increase the economic sources of villagers. In recent years, great changes have taken place in YZ Town, and the implementation of Rural Revitalization Strategy has achieved obvious results. But there are also some problems. For example, employees in YZ town have low educational background, as shown in Figure 1.

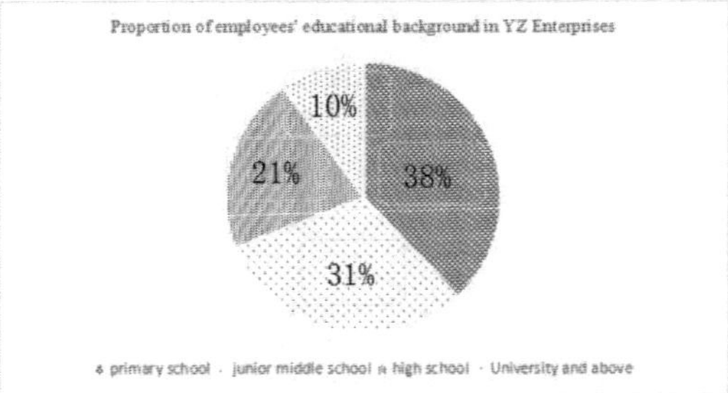

Fig1. Proportion of employees' educational background in YZ town.

Most enterprises in YZ town are garment production and processing enterprises. According to the survey on the educational background of employees in YZ Town, we can see that nearly 70% of employees have junior high school or primary school education and only 10% of them have university degree or above. These facts tell us that the educational level of employees in YZ town is relatively low and the knowledge of environmental protection is relatively limited.

In the process of toilet transformation, some villagers didn't actively cooperate. Through the investigation of toilet renovation in YM village of YZ Town, we can get Figure 2. As can be seen from Figure 2, about 25% of the villagers didn't cooperate with the renovation of the toilet and couldn't understand the event; about 57% of the villagers actively cooperated with the renovation of the toilet; and about 23% of the villagers cooperated with the renovation of the toilet.

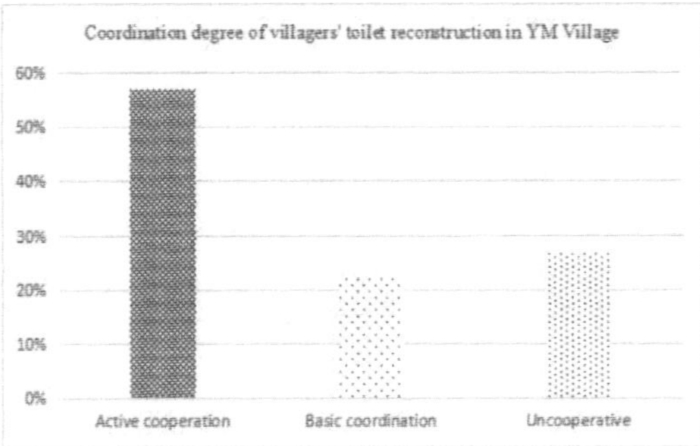

Fig2. Coordination degree of villagers' toilet reconstruction in YM village, YZ town

3 The dilemma of rural environmental pollution control under the background of Rural Revitalization Strategy

Based on the investigation of the environmental situation of YZ Town, combined with the current situation of rural environmental pollution control in YZ Town, this paper analyzes the existing problems of rural environmental pollution control in YZ town from the aspects of government, enterprises and the masses.

3.1 The Government level

3.1.1 The grass roots environmental protection governance mechanism isn't perfect

The grass-roots government departments are the front-line departments to implement the relevant national environmental protection policies. They have a lot of contact with enterprises, the masses and the media. However, the grass-roots environmental protection governance mechanism isn't perfect. On the one hand, the government lacks communication with enterprises and the masses, and has not established a perfect communication mechanism. Enterprises, the masses and other participants aren't aware of environmental protection, and they can't attract enough attention. On the other hand, the environmental protection facilities of the government are insufficient.The government of YZ town is lack of high-tech monitoring equipment, which can't accurately monitor and control environmental pollution in advance. In addition, the grass-roots environmental protection supervision personnel are limited, considering the complexity of environmental protection issues, grass-roots law enforcement departments are often unable to do their best in law enforcement and supervision. They tend to reconcile the situation of rural environmental pollution, and it is difficult to solve environmental protection problems from the source.

3.1.2 The information of environmental pollution control isn't timely grasped

The source of rural environmental pollution is extensive, the means are various, and the randomness is strong. According to the characteristics of rural environmental pollution, YZ town government has set up a special patrol team to patrol and inspect the rural environmental pollution. However, the timeliness of the government and other parts of the understanding of environmental pollution information is still poor. They often grasp the information after the occurrence of environmental pollution, unable to do a good job in precontrol. For example, for the behavior of villagers burning straw, the government usually finds a lot of smoke and then stops extinguishing the fire. It is difficult to find the perpetrators of the burning behavior.

3.1.3 The assessment and incentive mechanism isn't perfect

For example, under the guidance of the grass-roots government departments, YZ town has basically completed the transformation of rural toilets. However, the evaluation and incentive mechanism of environmental pollution control work isn't perfect, which leads to the low enthusiasm of relevant participants and limited treatment effect. Environmental protection work involves all aspects of life, the evaluation incentive mechanism isn't perfect, the evaluation content guidance isn't clear, which can not timely grasp the plight of environmental governance, resulting in low efficiency of environmental pollution control.

3.1.4 Lack of long-term mechanism for environmental governance

YZ town is equipped with garbage collection, clothing collection and other facilities, and set up a special garbage disposal place. However, rural residents still have the phenomenon of randomly discarding garbage and burning straw at will, and the utilization efficiency of infrastructure is low. At the same time, the investment of environmental pollution control facilities is often sufficient in the early stage, and the utilization rate is low after the investment, which will lead to the phenomenon of idle and waste in the later stage. When the facilities are damaged, they don't maintain or manage them.

3.2 The Enterprise level

3.2.1 Lack of legal concept of environmental protection

The enterprises in YZ town are mainly garment production enterprises, and the environmental pollution mainly involves dust pollution. Most of the employees in YZ town are residents of YZ Town, which lacks the concept of legal management of environmental protection. On the one hand, the educational level of employees in YZ town is generally low, about 60% of them are junior high school graduates. Their environmental awareness is weak and they can't pay enough attention to environmental protection and environmental governance. On the other hand, their learning effect on environmental protection policy isn't ideal.

3.2.2 Low level of industrial structure and insufficient participation in corporate governance

As one of the main bodies of environmental pollution control, enterprises play an important role in environmental pollution control. Most enterprises in YZ town are small and micro enterprises. According to the statistical data, we can see that most enterprises have not installed relevant sewage facilities. Due to the limited funds, enterprises pay more attention to how to reduce production costs and improve economic benefits, but less investment in environmental protection. The government have brought many preferential policies to enterprises through tax reduction and fee reduction. However, in order to pursue economic benefits, enterprises ignore environmental protection and governance, and lack of intensity to participate in environmental pollution control.

3.2.3 Low utilization rate of funds

YZ town enterprise environmental pollution control fund source is mainly from the government, the enterprise investment is less. Based on the perspective of pursuing economic interests, the funds invested by the government aren't fully and effectively used in the environmental pollution control of enterprises, and the utilization efficiency of funds is low.

3.3 The Mass level

3.3.1 Lack of awareness of environmental protection and low awareness of participation

The villagers are lack of understanding of environmental pollution control and can't fully realize the importance of environmental pollution control.Most residents think that environmental protection has nothing to do with themselves. Through investigation, we found that about 40% of rural residents do not think that burning straw will cause serious impact on environmental pollution. About 20% of the rural residents do not support the renovation of toilets in the process of toilet renovation by YZ town government. These villagers believe that traditional toilets will not cause serious pollution to the environment. Rural residents are lack of awareness of environmental pollution control, and the enthusiasm and initiative of rural residents to participate in environmental governance are low.

3.3.2 Decrease of rural resident population

The rural living environment involves many aspects. The number of villagers in YZ town who go out to work and study is increasing, and the rural permanent population is decreasing, and most of them are the elderly, women and children, which will cause a lot of idle and waste of infrastructure. In terms of environmental pollution control, the government has invested funds in the construction of infrastructure. With the loss of young villagers in YZ Town, the problems of idle infrastructure and low utilization efficiency will affect the implementation of Rural Revitalization Strategy to a certain extent.

3.3.3 Insufficient third party or public supervision

The environmental atmosphere of villages and towns depends on the living habits of villagers. Influenced by traditional ideas, most villagers can't attach importance to environmental protection and environmental pollution control. Rural areas lack the supervision of third parties or public media, such as villagers burning garbage. The supervision of the third party or the public media isn't enough, and the villagers are lack of consciousness.

4 Countermeasures of rural environmental pollution control under the background of Rural Revitalization Strategy

4.1 The Government level

4.1.1 Establish and improve the grass-roots environmental protection mechanism and strengthen supervision

On the one hand, the government can establish a sound communication mechanism to communicate with enterprises, the masses and the public in a timely manner. Through communication, the government can obtain sufficient environmental information to deal with or prevent environmental pollution incidents in time. On the other hand, the government can increase the investment of governance funds and provide sufficient financial guarantee. For example, the government can optimize the allocation of infrastructure, purchase high-tech environmental protection equipment, establish environmental protection big data platform, and improve work efficiency. Third, the government should increase the proportion of environmental supervision personnel, strengthen the government's understanding of environmental pollution, and achieve prevention and control in advance.

4.1.2 Timely grasp information and play a leading role of the government

The government should understand the rural environmental conditions through various channels and ways. For example, the government can build an environmental protection big data platform to timely understand the latest environmental pollution situation, actively mobilize multiple parties to participate in environmental pollution control and protection work, and play the leading role of the government on the basis of equal communication and exchange. The government should establish and improve the environmental protection accountability system and establish an effective information sharing platform.

4.1.3 Establish and improve the assessment and incentive mechanism of environmental protection governance

The government should establish and improve the assessment mechanism of environmental protection governance in combination with the specific situation of villages and towns. The government should set up the content of the assessment and incentive mechanism, and give appropriate rewards and punishments according to the specific situation of environmental pollution control in villages and towns and the environmental protection supervision. Enterprises or villages with better environmental pollution control should be given appropriate rewards to encourage them to continue to actively carry out environmental protection work; for enterprises or villages and towns with poor environmental pollution control, the government should give appropriate punishment and warn them to conscientiously abide by and implement relevant environmental protection policies and regulations.

4.1.4 Establish long-term environmental governance mechanism

The government should increase investment in high technology in environmental pollution control and environmental protection. In the traditional supervision mode, the government supervision is often based on human resources. With the development of information technology, the government can strengthen the supervision and management of environmental conditions through database and other information platforms or monitoring facilities, establish a long-term environmental pollution control system, and timely maintain and protect infrastructure. At the same time, the government should make clear the responsibilities of the main participants in environmental protection work, so as to achieve the unity of rights and responsibilities and long-term effectiveness.

4.2 The Enterprise level

4.2.1 Enhance the legal responsibility consciousness of enterprise environmental protection

Under the background of Rural Revitalization Strategy, we should improve rural living environment and establish ecological livable villages. The most important thing is that we should improve the legal awareness of environmental protection. On the one hand, enterprises should improve the educational background of employees and strengthen the environmental awareness of enterprise management and employees. On the other hand, enterprises can regularly invite environmental experts to train their employees, strengthen the study of environmental pollution control knowledge, and deepen the understanding of relevant national environmental protection policies. In addition, enterprises can strengthen the environmental knowledge education of employees in the corporate culture.

4.2.2 Adjust industrial structure and increase participation in environmental protection

First, enterprises can upgrade and transform traditional industries. For example, we can develop new processes, use new technologies to produce products, and control environmental pollution from the source by means of information technology such as big data. Secondly, enterprises should develop modern agriculture, attract college students to work in their hometown and village, and increase the participation of each subject in environmental protection.Third, enterprises should establish large-scale green agricultural production base and strengthen the upgrading and transformation of industrial structure. Enterprises should introduce professional talents and improve environmental protection technology. For example, enterprises can select college students village officials, seek technical support for environmental protection planning, combine internal and external, train local talents and increase environmental protection participation.

4.2.3 Improve the utilization efficiency of environmental protection funds

The construction of ecological livable villages is inseparable from the financial guarantee. Under the current policy, the government have given a lot of special funds for environmental protection. Enterprises should use the special funds for special purposes and use them in the most appropriate places to improve the efficiency of funds. We should give full play to the proper use of government special funds and stimulate the enthusiasm of enterprises to participate in environmental protection.

4.3 The Mass level

4.3.1 Strengthen villagers' awareness of environmental protection and enhance their awareness of environmental protection participation

Under the background of rural strategy, as one of the main participants in environmental protection, we must strengthen villagers' awareness of environmental protection and enhance their awareness of environmental protection participation. First, we should strengthen the publicity and education of villagers to make them realize the importance of environmental protection and environmental pollution control. Second, we should pay attention to multi-channel and multi-way publicity, interpret environmental protection policies and strengthen environmental protection publicity. Third, we should encourage villagers to actively participate in environmental and economic pollution control work, protect the environment, and establish ecological and livable villages.

4.3.2 Attracting college students to return home

The grass-roots departments of villages and towns can set up corresponding policies to attract talents, attract rural college students to return home, and strengthen the work of rural environmental pollution control. The grass-roots departments should support the young people to return home, increase the permanent rural population and improve the efficiency of environmental protection infrastructure. Villagers should actively participate in the implementation of Rural Revitalization Strategy and strive to implement various policies and regulations.

4.3.3 Enhance the supervision of the third party or the public

Rural grass-roots departments should build information-based big data platform to improve the supervision of the third party or the public. Villagers timely understand the relevant information of environmental pollution control and actively participate in environmental pollution control and environmental protection work. Through the supervision of the third party or the public, we can strengthen the villagers' awareness of environmental protection and highlight the main role of the villagers. Village cadres should strengthen the sense of leadership, lead villagers to correctly understand the importance of environmental protection, make villagers actively participate in environmental pollution control, and make villagers make certain contributions in the process of implementing the strategy of Rural Revitalization.

References

1. Yucan Huo,Study on the dilemma and solution path of feed enterprises supporting rural ecological environment governance under the background of Rural Revitalization ,China feed, **15** (2020)
2. Wenbo Qu, Shuling Li. Public participation in environmental pollution control in China: An Empirical Study Based on dynamic spatial panel model,Journal of Beijing University of Technology (SOCIAL SCIENCE EDITION), **22** (2020)
3. Chun Jin. Research on the impact of fiscal decentralization on environmental pollution -- Based on the threshold effect of policy environment,Friends of accounting, **06** (2020)
4. Zhen Cao, Zhanhao Gu,Discussion on rural ecological livable construction under the background of Rural Revitalization -- Based on the investigation of Zhejiang Province ,Chinese Youth Social Sciences, **38**(2019)
5. Jixi Gao, Qinfang Sun, Zhu Lin, Implementing Rural Revitalization Strategy and promoting rural ecological civilization construction, Environmental protection, **46**(2018)
6. Maoyou Deng,Ecological environment problems of rural tourism in China and countermeasures, Agricultural economy, **04** (2017)

https://doi.org/10.1051/e3sconf/202123701031

Degradation of methylene blue organic wastewater by plasma assisted method

Zhuansun Menglin[1], Liu Linsheng [2*], Wang Yanting [3], Zhai Limin [4], Chen Xinyi [5], Tian Chunsuo [6]

[1]Zhuansun Menglin, College of Electronic Engineering, Guangxi Normal University, China
[2*]Liu Linsheng, College of Electronic Engineering, Guangxi Normal University,China
[3]Wang Yanting, College of Electronic Engineering, Guangxi Normal University, China
[4]Zhai Limin, College of Electronic Engineering, Guangxi Normal University, China
[5]Chen Xinyi, College of Electronic Engineering, Guangxi Normal University, China
[6]Tian Chunsuo, College of Electronic Engineering, Guangxi Normal University, China

Abstract: In the current situation of water shortage and serious pollution, plasma discharged technology has a useful application prospect in sewage treatment. In order to control the degradation rate of pollutants, the effects of different gas environment, OH scavenger ethanol and H_2O_2 on the degradation of methylene blue simulated organic wastewater by plasma assisted method were studied. The experimental results show that the effect of air plasma discharge is slightly better than that of atmospheric pressure argon discharge. Ethanol can remove OH, thus reducing the degradation rate of methylene blue, while H_2O_2 can improve the degradation rate.

1 Introduction

At present, human beings are facing the problem of shortage of fresh water resources. With the rapid development of modern industry, water pollution has become a new reason for the lack of water resources. In recent years, the situation of sewage treatment in China is more severe, such as large amount of sewage disposal, low treatment efficiency and disorderly discharge. Water resources is polluted or damaged. The problem of dye wastewater is extremely serious. It is difficult to decolorize because of its deep color, high content of organic pollutants, complex components and high biological toxicity. In addition, the dyes have anti-photolysis and anti-oxidation properites, which are difficult to be effectively degraded by conventional methods, causing serious pollution of the water environment.

The traditional water treatment methods are needed to renew materials, the cost is high, and sometimes chemical reagents are used, which may cause water pollution again[1]. With the demand for new methods of water treatment, advanced oxidation technology began to appear. When advanced oxidation technology is applied to water treatment, it can produce strong oxidizing substances in water, and can effectively purify sewage, so it has a good application prospect[2]. There are many types of advanced oxidation methods, such as Fenton reaction, photocatalytic oxidation and electrochemical oxidation. Fenton reaction and photocatalytic oxidation need chemical reagents, which inevitably lead to secondary pollution and expensive reagents. Electrochemical advanced oxidation technique is a green and efficient water treatment method, which is based on the generation of hydroxyl radicals by electron transfer. The advantage of electrochemical oxidation method is that only a little or no chemical reagents are needed, and organic pollutants can be treated more quickly[3].

Plasma discharge technique combines the advantages of advanced oxidation methods. The inelastic collision between high-energy electrons and water in the process of plasma discharge leads to a series of complicated processes such as excitation, dissociation and ionization, and a large number of active free radicals and strong oxidizing molecules are produced. Among them, the most important one is OH, which is second only to fluorine in oxidizing properties. It can destroy the C-C bond of organic matters and form C-OH bond in the process of reaction with organic matters, which will destroy the structures of organic matters[4].

In this study, atmospheric pressure glow discharge plasma was used to degrade methylene blue to simulate the degradation process of organic wastewater. The degradation rate of methylene blue by plasma was investigated in detail by adding additives and changing discharge gas environment.

2 Experimental Setup And Principle

2.1. Experimental setup and degradation principle

Plasma is an ionized gaseous substance composed of atoms and positive and negative particles, which can be produced by different types of high voltage discharge.

* xxx: author@e-mail.org Liu Linsheng: linshengliu@163.com

When the plasma acts on the aqueous solution, a large number of active substances, such as OH, O_3, O and e_{aq}^-, will be produced in the liquid[5]. The process of water treatment by plasma technique includes many types of advanced oxidation processes, such as ozone oxidation, UV-photocatalytic oxidation and pyrolysis. It is a comprehensive and environmentally friendly water treatment technique. Organic pollutants in the water can be rapidly degraded by the oxidation and reduction of the substances, or in the synergistic effect, so as to achieve the purpose of sewage purification[6].

The experimental cycle setup is shown in Figure 1. The high voltage DC power supply(LAS 20kV, 50mA) was used to discharge at the tungsten steel electrode, which located 3 mm above the flowing solution surface on the top of the plasma reactor, and plasma was generated between the liquid surface and the electrode. Argon (Ar) is sprayed from the conical nozzle of tungsten steel electrode at a flow rate of 20 SCCM through the gas mass flow controller(D07-19B).

The circulation process of the reaction solution is as follows: firstly, the solution is pumped into the blending bottle in the water bath by the peristaltic pump (RZ1030) through the silicone tube, mixed by magnetic stirring (DFY-5L/10, water temperature controlled at 25 ℃), and then flows into the quartz cuvette (10 mm in optical lengthth) to monitor the absorption spectrum in real time. Then the solution flows from the cuvette into the plasma circulation reactor, and flows from the bottom of the reactor to the top to react with the plasma. Finally, the solution reacts with the plasma, and then returns to the peristaltic pump for the next cycle. The flow of liquid and magnetic stirring can promote the diffusion of organic dyes, so that organic dyes can be uniformly degraded.

The light source used for the experiment is tungsten halogen lamp (DH-2000), and a micro fiber spectrometer (USB6500 Pro) is used for real-time monitoring. The experimental setup is simple and the experimental cost is low.

The volume of the solution was 80 ml, the concentration was 5 mg/L, the flow rate of the peristaltic pump was 50 ml/min, and the reagent used was methylene blue (1%, Aladdin).

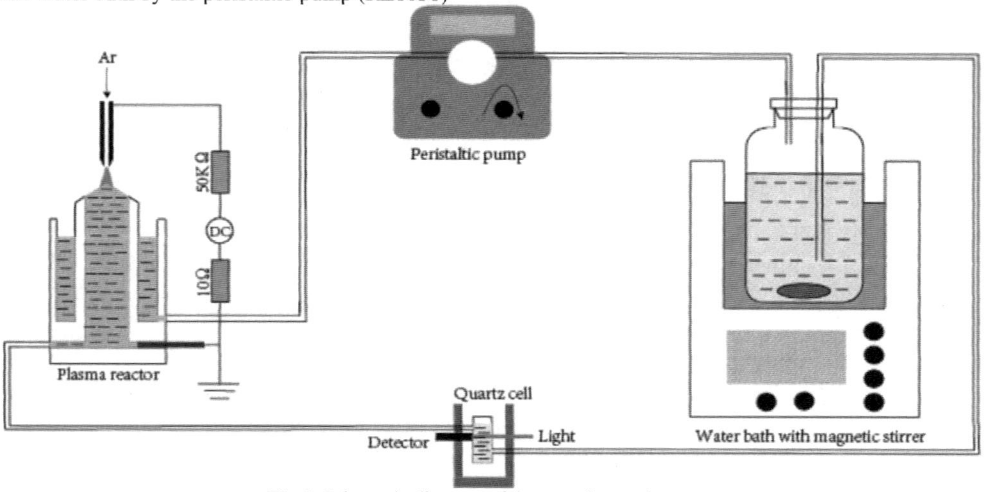

Fig.1. Schematic diagram of the experimental setup.

According to the bond dissociation energy(BDE) theory, the lower the BDE is, the more active the chemical bond is and the easier it is for old bonds to break and new bonds to form.The molecular structure of methylene blue is shown in the Figure 2, the main chromogenic group is '-S-'. '9C-S' and '10C-S' are easily broken by oxidation[7] .

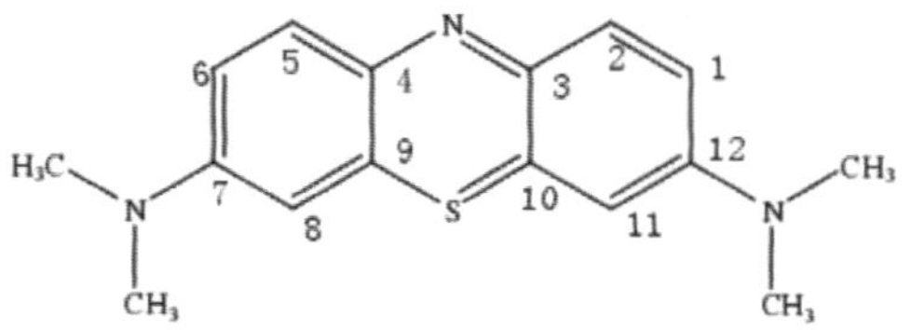

Fig.2. Molecular diagram of methylene blue

2.2 Methylene blue concentration monitoring

Methylene blue solution has a maximum absorption peak at about 660 nm, and this is an isochromatic absorption point. The absorption intensity at the isochromatic absorption point is only related to the concentration of methylene blue in the solution. Therefore, the absorbance at this point can represent the relative concentration of methylene blue in solution. The change of absorbance with time can reflect the decomposition rate of methylene blue. The UV Vis absorption spectra of the solution was measured by spectrometer(USB6500 Pro), and the light source was tungsten halogen lamp(DH-2000). All the absorption spectra were normalized.

2.3 Decolorization rate of methylene blue

The absorbance of methylene blue after degradation was measured by optical fiber spectrometer, and the corresponding concentration value was calculated. The degradation rate of methylene blue was calculated according to the following formula.

$$\eta = \frac{C_0 - C_1}{C_0} \times 100\% \quad (1)$$

η represents the removal rate of methylene blue.

C_0 represents the initial concentration, mg/L.

C_1 represents the degraded concentration, mg/L.

Through this formula, we can directly show the removal rate of methylene blue.

3 Results and discussion

3.1 Gas environment

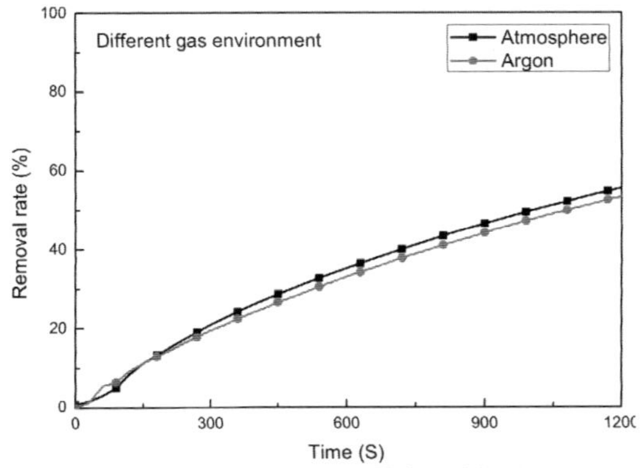

Fig.3.1 Decolorization rate of methylene blue in air and argon gas environment,30 mA current, 20 min reaction time

In this experiment, the gas environment of plasma is changed. Atmospheric pressure argon discharge and air discharge were used. As shown in Fig. 3.1, the decomposition rate of air discharge was slightly higher than that of atmospheric pressure argon discharge. In the case of air discharge, HNO_3 and HNO_2 were formed in the solution due to the presence of a large amount of N_2 and O_2. As a result, the PH value of the solution decreases and $ONOO^-$ was formed, which plays an auxiliary role in oxidation. Therefore, the decomposition rate was slightly faster than that of argon discharge at atmospheric pressure[8].

3.2 Adding ethanol

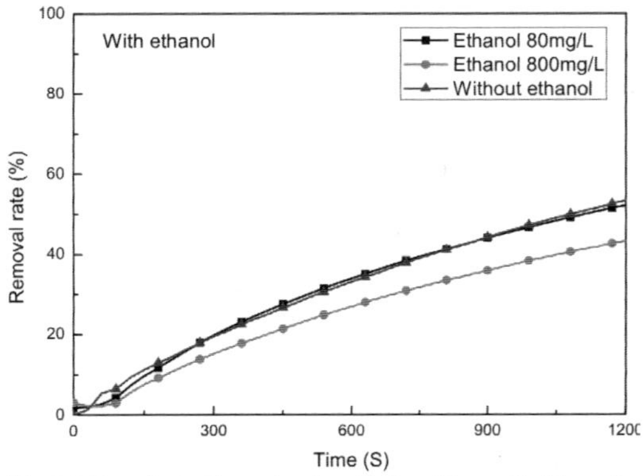

Fig.3.2 Decolorization rate of methylene blue with 80 mg/L and 800 mg/L concentration of ethanol, 30 mA current,20 min reaction time

Among a series of active substances generated in the contact reaction between plasma and liquid surface, OH as a strong oxidant, played an important role in the process of oxidative degradation of methylene blue. Ethanol can provide electrons to react with OH to act as OH scavenger. The addition of ethanol decreased the concentration of OH and the decomposition rate of methylene blue. As shown in Figure 3.2, when low concentration ethanol was added, the degradation rate decreased slightly after 15 minutes. When the concentration of ethanol improved, the degradation rate of methylene blue decreased significantly[9].

3.3 Adding H_2O_2

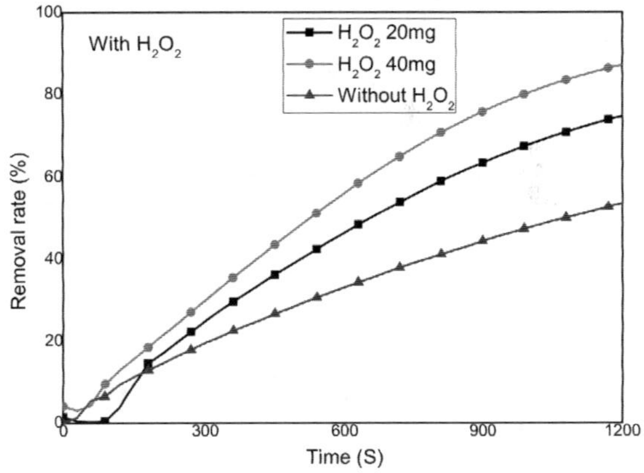

Fig.3.3 Decolorization rate of methylene blue with 20 mg/L and 40 mg/Lconcentration of H_2O_2,30 mA current, 20 min reaction time

When hydrogen peroxide was added, hydrogen peroxide and most organic compounds will not react significantly, but because of its existence, it can react with other substances in different ways to produce more OH. Thus, the degradation of methylene blue was accelerated[10].

$$H_2O_2 \rightarrow H^+ + HO_2^- \quad (2)$$

$$O_3 + HO_2^- \rightarrow \cdot O_2^- + \cdot OH + O_2 \quad (3)$$

It can be concluded from this group of experiments that H_2O_2 is a very important material in the experiment of methylene blue degradation by plasma method.

4 Conclusion

The degradation rate of methylene blue in air plasma discharge was slightly higher than that in argon plasma discharge. When ethanol was added into the solution, the OH concentration in the solution could be reduced, and the degradation rate of methylene blue could be reduced. When H_2O_2 was added, H_2O_2 generates more OH through the reaction to accelerate the degradation rate of methylene blue. Plasma degradation of organic pollutants in-depth study, can further improve the treatment efficiency, has a good application prospect, and provides an important safeguard for the treatment of sewage.

Acknowledgments

This work was supported by the doctoral scientific research fund of Guangxi Normal University (Grant No. 2017BQ019); Innovation Project of Guangxi Graduate Education(Grant No. JGY2020024); Special project of Guangxi Normal University for cultivating students' scientific research ability(Natural Science): (Grant No.2019YR007).

References

1. SONUNE A, GHATE R. Developments in wastewater treatment methods[J]. Desalination, 2004, 167(1):55-63.
2. Glaze W H, Kang J W, Chapin D H. The Chemistry of Water Treatment Processes Involving Ozone, Hydrogen Peroxide and Ultraviolet Radiation[J]. Ozone ence & Engineering, 1987, 9(4):335-352.
3. Nidheesh P V, Gandhimathi R. Trends in electro-Fenton process for water and wastewater treatment: An overview[J]. Desalination, 2012, 299:1-15.
4. Lukes P. Water treatment by pulsed streamer corona discharge[D]. Prague: Institue of Chemical Technology, 2001.
5. Sarangapani C, Dixit Y, Milosavljevic V, et al. Optimization of atmospheric air plasma for degradation of organic dyes in wastewater[J]. Water Science & Technology, 2017, 75(1):207-219.
6. S. Ma, K. Kim, S. Chun, S. Y. Moon, and Y. Hong, Plasma-assisted advanced oxidation process by a multi-hole dielectric barrier discharge in water and its application to wastewater treatment[J], Chemosphere, 2020, 243:125377.
7. Huang F, Chen L, Wang H, et al. Analysis of the degradation mechanism of methylene blue by atmospheric pressure dielectric barrier discharge plasma[J]. Chemical Engineering Journal, 2010, 162(1):250-256.
8. Bangbang H, Xinning G, Xin W, et al. What Are the Effective Reactants in the Plasma-Induced Wastewater Treatment[J]. Journal of the Electrochemical Society, 2018, 165(10):E454-E459.
9. Z. Shen, Y. Zhou, and J. Wang, Bioresour. Technol., 2013, 131:33.
10. Lukes P, Dolezalova E, Sisrova I, et al. Aqueous-phase chemistry and bactericidal effects from an air discharge plasma in contact with water: evidence for the formation of peroxynitrite through a pseudo-second-order post-discharge reaction of H_2O_2 and HNO_2[J]. Plasma Sources ence & Technology, 2014, 23(1):015019.

Morphological factors extraction of upper limb for young female

Xin He, Hongshu Jin*

Wuhan textile university, Wuhan, Hubei 430073, China

Abstract. This paper is proposed to extract the morphological factors of upper limb shapes related to the personalized sleeve design. The 36 items of lateral upper limb morphology were measured for 50 young female aged 19 to 22 years old by the photo measurement method. Based on the correlation relationships of morphological variables of upper limb, there are 4 morphological factors of upper limb shapes with eigenvalues above 1 were extracted by the principal component analysis, and the cumulative variance reached 83.504%. Among them, the factors of girth and height of upper limb were relevance to the girths of torso and stature respectively, and the factor of upper limb oblique angle is influenced by the upper body axis inclination, and the arm root height is interpreted as an independent factor describing the arm root shape. These morphological factors provided the references in key feature indicator sifting for upper limb shape subdivisions and the critical parameters in personalized sleeve structure designing.

1 Introduction

The arm full length is the only parameter about upper limb size for sleeve structure designing and grading in GB/T 1335-2008 "Standard sizing systems for garment-Woman". However, the more structural parameters reflecting the upper limb shape features were needed for garment design responding to the demands of personalized costume consummations.

In the theoretical research field of body type classification, the feature variables of body surface angles of torso [1-2], girth differences, girth ratios [3] are mostly used as the typical indicators for classifications of human body shape. For the morphological feature subdivision of human body [4-5], the statistical analysis methods of the feature extraction, clustering, and combination of typical indicators for describing the morphological differences of human body, which provides a direction on this research of upper limb morphological factors extraction.

For the methods of human body data acquisition, the photo measurement method has the advantages of multi types of human contour feature data acquiring, such as low cost, fast and accurately [6-7]. Besides, the lateral contour of human body surface performances the main morphological features [8], which is beneficial to acquiring multiple morphological data of upper limbs. Therefore, the research of upper limb shape subdivision using three upper limb angle variables by the method of photo measurement [9] is uncompleted in description of upper limb morphological characteristics.

In view of these, this paper is aiming to extract the morphological factors of upper limbs by the photo measurement method. The morphological factors are accounting for the differences of upper limb shape, and the typical indicators of each morphological factor are the key variables for upper limb shape subdivisions and the critical parameters in personalized sleeve structure designing.

2 Upper limb morphological data acquisition

2.1. Upper limb morphological measurement method

The morphological data of upper limb are acquired by the photo measurement method. In addition, the body surface measurement method was aided for the size of upper torso control parts acquiring.

2.1.1. Measurement object. The measurement objects are 50 young female college students aged 19-22 wearing fitted underwear and standing upright naturally, with the upper limbs hanging down naturally.

2.1.2. Photo shooting and measurement requirements. The photos for measurements are shot from 10 m away using the camera model of Canon EOS 5D Mark III and the lens EF24-105 mm. And the center of the lens is on the same horizontal line as the waistline. The photo of right half body for upper limb morphological measuring is shown in Figure 1. At the same time, a benchmark of 50 cm is placed on the back of the subject as the reference for converting and recording the actual measurement.

* 2013025@wtu.edu.cn

https://doi.org/10.1051/e3sconf/202123701032

Figure 1. Photo of right half body used for measuring.

2.2. Measurement items of upper limb shape

A total of 36 items are measured including 28 items for upper limb and 8 items for upper torso. The photo measurement items are shown in Figure 2.

The measurement items of upper limb include 23 photo measurements such as heights, widths, lengths, and oblique angles; and 5 body surface measurements include arm full length (al), upper-arm girth (uag), elbow girth (e), wrist girth (wr) and armscye girth (ag).

The measurement items of upper torso include 5 photo measurements of bust depth (bd), waist depth (wd), upper back angle (∠m), lower back angle (∠n), and upper body axial angle (∠o); and 3 body surface measurements of stature (h), bust girth (b), and waist girth (w).

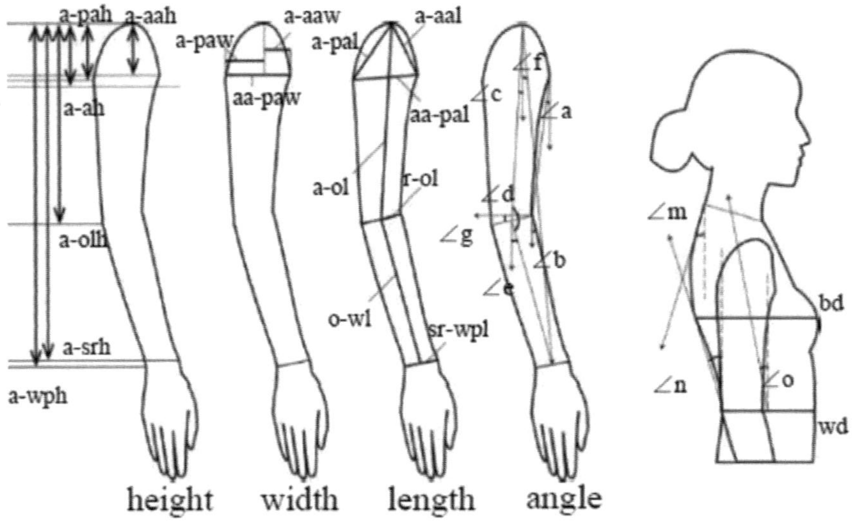

Figure 2. Photo measurement items of upper limb and upper torso.

3 Basic statistical analysis of upper limb morphological variables

The results of basic statistics and the single-sample T test of the sizes of control parts were acquired by IBM SPSS Statistics 23 and shown in Table 1. When the test values is the standard measurements of Y body type in the GB/T 1335.2-2008 "Standard sizing systems for garment-Woman", the average values of control parts showed significant differences except waist girth (w), which indicated that the stature (h) and arm full length (al) are slightly larger, bust girth (b) is slightly smaller in average than the standard size respectively.

According to the results of Kolmogorov-Smirnov test, the sizes of upper torso control parts and morphological variables of upper limb all follow the normal distribution.

Table 1. Mean test of sample female youth and national standard female data.

Project	Mean (cm)	Standard deviation (cm)	Test value (cm)	Sig. (2-tailed)
h	161.8	4.6	160.0	**[a]
al	52.4	1.9	50.5	**[a]
b	82.1	5.6	84.0	*[b]
w	65.2	5.8	64.0	N.S.[c]

[a]**P<0.01; [b]*P<0.05; [c] No significant n=50

4 Morphological factors extraction of upper limb shape

The morphological factors of upper limbs were extracted by principal component analysis based on the correlation coefficients between the morphological variables of upper limbs. Furthermore, the correlated relationships between the variables of upper limbs in each morphological factor and the sizes of upper torso control parts are confirmed.

4.1. Relationships in upper limb morphological variables

The heights and lengths of upper limb showed the high significant correlation above r=0.7, which indicated the heights and lengths were approximately in described the longitudinal features of upper limbs.

The height of acromion to posterior armpit point (a-pah) and the height of acromion to armpit horizontal position (a-ah) showed a significant correlation coefficient above r=0.7, which described the relationship between the morphological variables of the arm root heights. The width between anterior and posterior armpit points (aa-paw) and armscye girth (ag), upper-arm girth (uag) showed a highly significant correlation above r=0.8 respectively, which showed that the arm root widths closely related to the arm root girths. The arm full inclined angle (∠f) is significantly correlated with the angle of upper arm (∠c) and angle of lower arm (∠e) with correlation coefficient above r=0.7, which indicated oblique angle of upper limbs entirely influenced by the full arm oblique angle.

4.2. Extraction of principal components of upper limb morphological variables

For extracting the morphological factors of upper limb, the principal component analysis was conducted based on the results of correlation analysis. After removing the variables with weak correlations and showed collinearity between each other, 14 morphological variables of upper limb were sifted and used for analyzing finally.

The results of principal component analysis are shown in Table 2. There are 4 principal components with eigenvalues above 1 were extracted by the maximum variance factor rotation method, and the cumulative variance reached 83.504%. The principle components are explained as follows:

1) In principal component 1, the variables of the width between anterior and posterior armpit points (aa-paw), upper-arm girth (uag), armscye girth (ag), elbow girth (e) showed loading coefficient above 0.8, which described the features of upper limb girth and width, and it can be summarized as the upper limb girth factor.

2) In principal component 2, the variables with larger loading coefficient included the height of acromion to wrist point (a-wph), arm full length (al), and the length of olecranon to wrist (o-wl), that showed the features of upper limb height, which can be summarized as the upper limb height factor.

3) In principal component 3, the variables with loading coefficient above 0.7 included arm full inclined angle (∠f) and angle of upper arm (∠c), that showed the features of upper limb oblique angle, which can be summarized as the upper limb oblique angle factor.

4) In principal component 4, the variables with larger loading coefficient included the height of acromion to armpit horizontal position (a-ah) and the height of acromion to posterior armpit point (a-pah), which showed the features of arm root height and summarized as the arm root height factor.

Table 2. Extraction of morphological factors of upper limb after rotation.

Characteristics of the variables	Ingredients			
	1	2	3	4
aa-paw	0.930			
uag	0.920			
ag	0.892			
e	0.891			
a-wph		0.962		
al		0.954		
o-wl		0.831		
a-olh		0.746		
∠f			0.988	
∠c			0.819	
∠e			0.758	
a-ah				0.842
a-pah				0.832
a-aah				0.708

Eigenvalue	3.588	3.369	2.481	2.252
Explain the variance (%)	25.632	24.065	17.722	16.082
Cumulative variance (%)	25.632	46.697	67.419	83.504

Therefore, the morphological differences of upper limb could be described as 4 factors of upper limb girth, upper limb height, upper limb oblique angle, and arm root height.

4.3. The relationship between the sizes of upper torso control parts and the morphological variables of upper limb

For confirming the sizes of upper torso control parts related to morphological variables of upper limbs, the correlation analysis was conducted between the sizes of upper torso control parts and the variables of each upper limb morphological factors.

The bust girth (b) /waist girth (w) showed a significant correlation coefficient with the width between anterior and posterior armpit points (aa-paw) and armscye girth (ag) above r=0.7 respectively. In addition, the bust depth (bd) /waist depth (wd) also showed a significant correlation with the variables of the upper limb girth factor. It could be explained that the girth/width of upper torso control parts were closely related to the size of upper limb.

The stature (h) showed a significant correlation coefficient with the height of acromion to wrist (a-wph) and arm full length (al) above r=0.7 in the upper limb height factor, which explained the upper limb height (length) dimensions influenced by stature obviously.

The upper torso axial angle (∠o) significantly correlated with arm full inclined angle (∠f) in upper limb oblique angle factor above r=0.6, which explained the correlation relationship between the oblique angles of upper limb and upper torso for keeping the balance of upper body.

The arm full length (al) showed a significant correlation coefficient above r=0.7 with the height of acromion to olecranon (a-olh), which indicated that the arm full length is closely related to the heights of upper limb.

To sum up, the factors of girth and height of upper limb were closely related to the sizes of upper torso control parts, which is the references of sleeve size grading. And the factor of upper limb oblique angle is influenced by the upper body axis inclination. In addition, the arm root height factor was the independent factor for arm root shape describing.

5 Conclusions

The aim of this paper is to extract the morphological factors of upper limbs by the photo measurement method for young females. The main results summarized as follows:

1) Based on the results of correlation relationships of morphological variables of upper limb, the upper limb shape features interpreted by 4 morphological factors which extracted by principal component analysis.

2) The factors of girth and height of upper limb were relevance to the girths of torso and stature respectively, and the factor of upper limb oblique angle is influenced by and the inclination of upper body axis, and the factor of arm root height is interpreted as an independent factor describing the arm root shape.

These results provided the references in key feature indicator sifting for upper limb shape subdivisions and the critical parameters in personalized sleeve structure designing.

References

1. Young L, Yun J 2010 Classification of upper lateral body shapes for the apparel industry J. Human Factors and Ergonomics in Manufacturing & Service Industries. vol 20, no 05. 378-390
2. Wookyung L, Haruki I 2010 Classification of body shape characteristics of women's torsos using angles J. International Journal of Clothing Science and Technology. vol 22, no 04. 297-311
3. He T H, Liu Y N, Zhao X 2018 Body classification and prototype amendment of female college students in Tianjin J. Wool Textile Journal. vol 46, no 07. 53-57
4. Xu J Y, Kuang C Y, Liu G L 2008 Research on classification of youth male body based on body surface morphology J. Silk Textile Technology Overseas. no 03. 15-17
5. Sun N L, Song X X 2014 Research on chest classification of young women in Shanghai J. Journal of Silk. vol 51, no 09. 34-39
6. Wang Z J, Dai H 2007 Research on the body parameters influencing the sleeve pattern design of young women garments J. Journal of Xi'an University of Engineering Science and Technology. vol 21, no 01. 48-51
7. Ge Y, Liu G L 2007 Study on photogrammetric measurement of body in apparel industry J. Journal of Textile Research. vol 28, no 10. 78-81
8. Satoru NAKAZAWA 2000 The human body and clothing M. Yuan G L translation, 1st ed (Beijing: China Textile & Apparel Press) p 176
9. Chen Q, Tuo W, Zhang D 2009 Sleeve structure of men's western-style garment based on arm shape J. Journal of Textile Research. vol 30, no 07. 99-102

Research on the Impact Test and Control Countermeasures of Residential Elevator Operation Noise

Peng Hong[1,*], Lv Zhong[2], and Chen Mingfeng[3]

[1]Chongqing Jian Zhu College, Chongqing 400070, China;
[2]Chongqing Construction Science Research Institute, Chongqing 400050, China;
[3]Chongqing University, Chongqing 400030, China

Abstract. The influence of the residential elevator operating noise on the indoor personnel of the adjacent elevator room is becoming more and more prominent. Through on-site testing of a household in Chongqing affected by elevator noise, the noise distribution in key areas affected by elevator noise was experimentally studied, and compared with the current standard limits. Combined with the noise spectrum characteristics, the experimental analysis obtained the noise spectrum diagrams of different areas such as elevator machine rooms, corridors and bedrooms. It is believed that low-frequency noise has a greater impact on the residential acoustic environment. Combined with the noise propagation path of elevators, the technical measures and improvement schemes for noise reduction of elevator equipment operation noise are proposed, which can provide reference for the design and improvement of residential acoustic environment.

1 Preface

With the rise of high-rise buildings, the use of elevators has become increasingly widespread and has become one of the most frequent special equipment in various buildings. Because of its manned function, elevator operation has put forward stricter requirements on its reliability, vibration, noise and comfort. The "Elevator Technical Conditions" GB/T 10058-2009 clearly stipulates that for residential elevators with a rated speed less than or equal to 2.5m/s, the maximum noise value in the elevator car during operation should not exceed 55dB; for residential elevators with speeds between 2.5 and 6m/s, the maximum noise value in the elevator car during operation should not exceed 60dB; during the door opening and closing process of all residential elevators, the maximum noise value should not exceed 65dB. Due to the unreasonable design, construction or use of the elevator, the vibration and noise generated by the elevator during operation will diffuse to the surrounding rooms through the walls and floors of the building in a solid-borne sound transmission mode. The resulting elevator noise will affect the indoor acoustic environment quality of the building and endanger people's physical and mental health[1-3]. In recent years, there have been more and more complaints about elevator noise in high-rise residential buildings. Therefore, the prevention and control of residential elevator noise has caused widespread concern in the society[4]. In this paper, combined with the actual measurement of specific engineering cases, the characteristics of elevator noise are analyzed, and its control methods and technical measures are discussed, so as to provide reference for the management of elevator noise in high-rise residential buildings.

2 The source and characteristics of elevator noise

2.1 The source of elevator noise

During the operation of the elevator, noise will be generated due to the rotation of the traction machine, the opening and closing of the door, the attraction and release of the relay and the contactor, the friction between the guide rail and the guide shoe, etc. This kind of noise is transmitted in the building through the components supporting the elevator and directly affects people's life and work. According to the different nature of the noise generated by elevator operation, it is mainly divided into three types: mechanical noise, aerodynamic noise and electromagnetic noise.

(1)Mechanical noise. There are three main sources of mechanical noise: First, friction and impact occur on the rotating part of traction motor in the machine room, centrifugal force caused by excessive dynamic balance deviation of brake, friction between meshing surfaces of worm gear of reducer and sliding friction between traction wheel and steel wire rope, all of which will produce mechanical noise. Second, mechanical noise is also caused by the following situations[5-6]: high-speed rotation of the rope sheaves on the top of the elevator car in the hoistway, loose elevator car fasteners or unreasonable installation positions and quantities, vibration caused by the vibration frequency of the elevator car wall being close to the

* Corresponding author: penghong2952@163.com

 https://doi.org/10.1051/e3sconf/202123701033

system vibration frequency, friction between the elevator car and the counterweight when running up and down along the guide rail, due to the problem of guide rail installation, the vibration of the elevator car and the counterweight during operation, the friction between the guide shoe and the guide rail, the insufficient preload of the guide shoe spring or the unreasonable intermittent between the guide shoe and the guide rail surface, etc. Third, when the elevator hall door and car door are opened and closed, the action sound of the interlocking mechanism and the collision sound of the safety contact plate will also produce mechanical noise.

(2) Aerodynamic noise. When the elevator car is running in the hoistway, as the speed of the box body increases, the gas is compressed sharply, and the hoistway air is disturbed up and down and airflow collision occurs. At this time, a lot of aerodynamic noise will be generated, and even pressure waves will be generated. At the same time, the flow velocity of the gas in the gap between the outer wall of the elevator car and the wall of the hoistway increases sharply due to the sudden reduction of the flow area, which will produce a lot of aerodynamic resistance, and even separation and vortexes may occur at the upstream and downstream of the box. This tunnel-like aerodynamic effect also has a certain impact on the aerodynamic noise[7].

(3) Electromagnetic noise. The electromagnetic force that changes in time and space between the various parts of the elevator motor, as well as the unbalanced unilateral magnetic pulling force generated by the eccentric effect of the motor stator and rotor, will cause electromagnetic noise. At the same time, the generation of electromagnetic noise is closely related to the slot fit, slot inclination and the stiffness of the motor structure[8].

2.2 The propagation characteristics and hazards of elevator noise

Because the elevator unit is based on a steel frame, the vibration generated by the motor and brake cannot be sufficiently attenuated. The vibration is transmitted to the floor and wall through the pedestal, and then propagated through the rigid structure such as concrete floor, column and beam, and then radiates the noise when it reaches the residence. Among the sources of elevator noise, mechanical noise caused by vibration during elevator operation is the most obvious. Relevant studies have shown that this type of noise transmission is mainly in the form of low-medium frequency vibration, which is transmitted through solid media such as walls and floors[9]. Different from high-frequency noise, the low-frequency noise with frequency between 20-200Hz attenuates slowly, and the sound wave is very long, which can easily pass through obstacles and enter the human ear. Because low-frequency noise is mainly transmitted through the structure, it is easy to cause people's sensory resonance, so people's tolerance to low-frequency noise is relatively low.

Although the direct impact of low-frequency noise on human physiology is not as obvious as that of high-frequency noise, the impact on the overall physical and mental health of the human body is more significant. Due to the frequent operation of residential elevators, people will be exposed to the pollution of low-frequency elevator noise for a long time. Their auditory system, nervous system, cardiovascular system, and digestive system will be significantly damaged, mainly as tinnitus, earache, tympanic membrane damage, sympathetic nervousness, upset, grumpy, blurred vision, etc. At the same time, the low-frequency noise of elevators can cause disorders of internal secretion and decreased immune function, and increase the consumption of various vitamins, amino acids, glutamic acid, lysine and other nutrients in the human body. This has a great impact on the growth and development of the human body, especially children. Secondly, low-frequency noise is particularly harmful to pregnant women and fetuses, causing uterine contractions in pregnant women, affecting the blood supply of the fetus, and then affecting the development of the fetal nervous system. Generally speaking, pregnant women, children, the elderly, people with chronic diseases, and people with sensitive physique are more susceptible to low-frequency noise, and the damage is more serious[10-12].

3 Noise measurement of residential elevator

3.1 Case overview

A resident on the 24th floor of a community in Chongqing reported that the elevators produced significant noise during the start, stop and operation, and they were more prominent at night, which seriously affected the daily life of the residents. This paper takes this residence as the test object, and conducts on-site measurement of residential elevator operation noise. The residence is a 2+1 split-level on the top floor, and the unit type is shown in Figure 1. The black boxed area in the figure was originally an outdoor area, and the resident was transformed on the basis of the original construction drawings. The 1F-3F black-framed areas were transformed into balconies, kitchens, bedrooms and corridors, and the transformed areas were close to the elevator shaft and elevator machine room. According to residents' descriptions, this area is significantly affected by elevator noise, and the 2F bedroom is the most affected.

The reasons for the excessive indoor noise of the residents were preliminarily judged on the spot. One is that the elevator machine room is separated from the bedroom by a wall, and the wall is made of blocks with poor sound insulation; the other is the rigid connection between the elevator load-bearing steel frame and the building structure. The solid noise generated by the elevator is transmitted to the lower floor through the main engine base, and then transmitted from the floor to the adjacent bedroom walls and rooms, causing solid and metal sound transmission, and finally causing the indoor noise of the residents to exceed the standard.

3.2 Testing method

According to the relevant standards, the testers

respectively selected four locations as inspection points after site surveys, as shown in Figure 1 (b) and (c). The daytime when the elevator runs frequently is selected as the test time period, and the CESVA building sound insulation test system with a test range of 23dB-140dB and an accuracy of ±0.7dB is used as the test tool to continuously detect elevator noise. The test points layout description are shown in Table 1.

Table1. Description of test points layout

test point 1	test point 2	test point 3	test point 4
living room	transformed bedroom (abbreviated as "bedroom")	renovated corridor (abbreviated as " corridor ")	elevator machine room

(a) 1F plane graph (b) 2F plane graph (c) 3F plane graph

Fig1. The plane graph of the residential in the case

4 Results and analysis

4.1 Comparative analysis of test results and standard values

It can be seen from Figure 1 that the test point in the living room area is closer to the outdoors. Normally, this area should be more affected by the outdoor acoustic environment. The test point in the corridor area is closer to the indoor (elevator shaft) and should be less affected by the outdoor acoustic environment. However, according to the test data in Table 2, the noise of the corridor test point is higher than that of the living room test point, so it can be determined that the noise mainly comes from the elevator machine room. The bedroom and the corridor are close to the elevator shaft and the elevator machine room respectively, and the test data shows that the noise value of the corridor is significantly higher than the noise value of the bedroom. This phenomenon shows that the mechanical noise generated by the vibration of the equipment in the elevator machine room has a significantly higher impact on the adjacent rooms than the aerodynamic noise in the elevator shaft.

At the same time, it can be seen from Table 2 that the noise value of the 250Hz frequency band in the octave band sound pressure level in the daytime corridor area exceeds the standard. The specific performance is: the standard value is 44dB, and the test value is 48.3dB (non-mandatory index), which exceeds the standard by 4.3dB. Although the night noise test has not been carried out, the noise standard value at night is lower than that during the day, so it can be determined that when the elevator is operating at night, there should be a number of low frequency noise values exceeding the standard. It can be seen from Table 2 that the influence of background noise is also significant, and the background noise values of the sound pressure level in octave bands are close to the noise test value. Among them, the background noise value of the corridor area in the 31.5Hz frequency band is significantly higher than the noise test value. After residents fall asleep at night, the impact of background noise on human sleep will be more obvious.

Table2. Comparison table of field test data and standard value

test points		test period	the type of the room	≤equivalent continuous A – weighted sound pressure level (dB)	≤sound pressure level in octave bands (dB)				
					31.5Hz	63Hz	125Hz	250Hz	500Hz
standard value		daytime	bedroom	45.0	76.0	59.0	48.0	39.0	34.0
		daytime	living room	45.0	79.0	63.0	52.0	44.0	38.0
test point 1— living room	test noise values	daytime	living room	31.0	42.8	41.7	43.0	38.5	27.0
	background			27.4	37.5	35.3	34.0	27.8	20.6

	noise values								
	correction values			28.0	40.8	38.7	42.0	38.5	26.0
test point 2—bedroom	test noise values	daytime	bedroom	31.0	42.8	41.7	43.0	38.5	27.0
	background noise values			27.4	37.5	35.3	34.0	27.8	20.6
	correction values			28.0	40.8	38.7	42.0	38.5	26.0
test point 3—corridor	test noise values	daytime	living room	41.1	52.6	42.6	45.8	48.3	37.5
	background noise values			27.4	54.0	38.1	34.0	27.8	23.0
	correction values			41.1	52.6	42.6	45.8	48.3	37.5
test point 4—elevator machine room	test noise values	daytime	—	70.8	48.0	57.2	65.2	73.8	71.4

Note: The standard value comes from "Social Life Environment Noise Emission Standard" GB 22337-2008.

4.2 Spectrum Energy Analysis of Elevator Operation Noise

Only by deeply mastering the spectrum characteristics of noise sources can noise control be carried out more effectively. This paper analyzes the noise spectrum energy of the elevator machine room, corridor and bedroom area during the three periods of elevator start, operation, and stop. The results are shown in Figure 2-Figure 4.

From Figure 2-Figure 4, we can see that the noise spectrum energy intensity of the elevator machine room is mainly concentrated in the 200-500Hz region, and the frequency is divided in the intermediate frequency range; the noise spectrum energy intensity of the corridor area is mainly concentrated in the 125-250Hz region, and the frequency is divided in the low frequency range; the energy intensity of the noise spectrum in the bedroom area is mainly concentrated in the 60-150 Hz area, and the frequency is divided in the low frequency range. The noise spectrum energy intensity of the corridor and bedroom area is relatively low, and a small amount of frequency intensity below 60Hz (ultra low frequency) appears. Therefore, from the perspective of spectral energy intensity, the residence is mainly affected by the low-medium frequency (60-250Hz) noise generated during elevator operation. The resident is exposed to the low-medium frequency noise environment caused by elevator operation for a long time, and his physical and mental health will inevitably be greatly damaged.

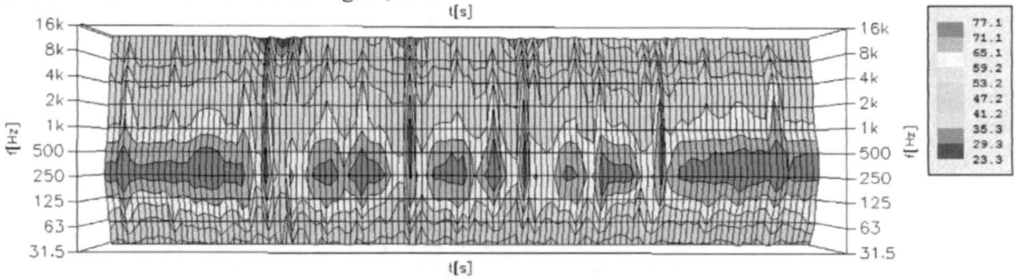

Fig2. Noise spectrum energy diagram of the elevator machine room

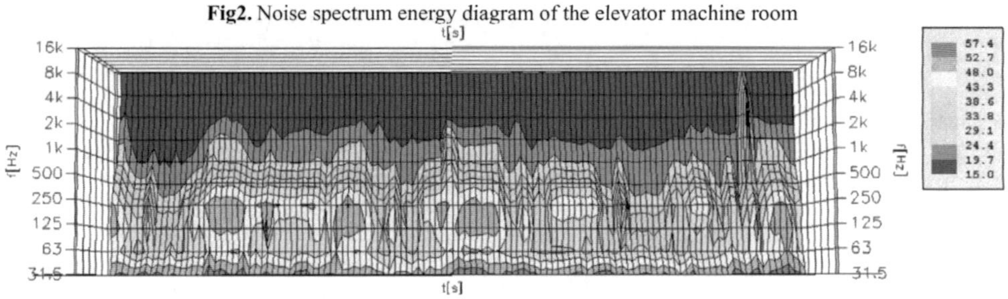

Fig3. Noise spectrum energy diagram of the corridor

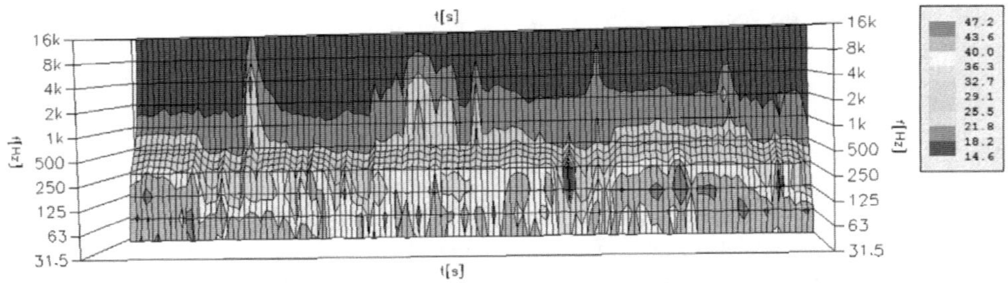

Fig4. Noise spectrum energy diagram of the bedroom

However, the current existing noise evaluation methods underestimate the actual impact of low-frequency noise components on people. At present, our country's current acoustic environmental quality standard system mainly uses equivalent A sound level to test and evaluate environmental noise. The frequency characteristic of A weighting is that the low frequency and high frequency of noise are greatly attenuated. Therefore, when the equivalent A sound level is used to test the elevator noise, the sound level of low-frequency noise has been greatly attenuated, resulting in the detection instrument not being able to accurately display the noise in this frequency band. As a result, when most of rooms which are adjacent to elevator machine rooms or elevator shaft use conventional instruments to test the indoor noise environment, the test results do meet the national standard values, but in fact, residents still suffer from elevator noise pollution for a long time. This is also true of the feedback from the households tested in this paper. For this reason, the authors believe that as people's requirements for the quality of the living environment increase, the equivalent A sound level evaluation method has certain limitations, and environmental testing standards and testing methods for low-frequency noise should be further formulated.

5 Noise reduction measures

The methods of controlling noise mainly include controlling the noise source, cutting off the transmission route and protecting the spreader. The medium and low frequency noise of the elevator is mainly generated by the vibration of the equipment, and then spreads through the building structure, especially through the walls and floors with poor sound insulation. In order to reduce the influence of elevator noise, the authors propose the following technical measures.

(1) The installation quality of the elevator traction machine will directly affect the noise decibel value of the elevator operation. The part where the traction machine base is connected to the I-beam load-bearing beam or concrete foundation should adopt spring shock absorbers, rubber shock absorbers or sound-absorbing sponges to achieve vibration reduction and adjustment. The load end of the traction sheave shaft is only allowed to have an upward inclination angle. At the same time, the channel formed by the traction rope through the floor will further enhance the noise propagation. Elevator noise is introduced into the elevator shaft through this channel, and then affects the rooms around the elevator shaft and the elevator car. It is recommended to install a muffler or sound isolator at a suitable position under the floor[14].

(2) Improving the sound insulation performance of the wall in the elevator machine room is an effective way to reduce elevator noise. It can be achieved by using light steel keel to fix sound-absorbing cotton or spraying new sound-absorbing materials on the surrounding walls of the elevator machine room, and then installing soft and porous sound-absorbing panels or asbestos sound-proof panels on the outer layer. In particular, it is necessary to install a board with both sound insulation and sound absorption functions on the wall of the elevator machine room near the side of the residential room, and fill the wall with dense sound insulation material. This can not only increase the sound insulation effect of intermediate frequency noise, but also absorb the energy of low frequency noise and reduce intermediate frequency reverberation noise.

(3) The control of aerodynamic noise in the elevator shaft should not be underestimated. The distance between the elevator hoistway wall and the outer wall of the elevator car should be fully considered, and vents or ventilation equipment should be arranged reasonably. At the same time, sound absorption and sound insulation processing devices can be installed in the elevator hall and car to reduce the aerodynamic noise caused by wind pressure in the elevator shaft during operation.

6 Conclusion

Through the analysis of the characteristics of elevator noise and the actual measurement of elevator operating noise, the following conclusions are obtained from three aspects: building design, detection and control of low-frequency noise generated by elevator vibration, and daily operation management.

(1) Elevator noise mainly comes from mechanical noise caused by equipment vibration during operation. This kind of noise is low-medium frequency noise, and the physical and mental health of people exposed to elevator noise for a long time will be seriously affected.

(2) Elevator noise has a particularly serious impact on the rooms which are adjacent elevator machine rooms or elevator shaft rooms. From the noise spectrum point of view, the noise frequencies affecting households are mainly concentrated in 60-250Hz. There are still shortcomings in noise detection and evaluation, and it is necessary to further develop acoustic environmental quality standards for low-frequency noise.

(3) Strengthening the maintenance of elevators is also conducive to controlling elevator noise. For example, add lubricating oil to all mechanical parts in time to reduce

friction, and tighten loose parts and screws to reduce vibration. This can not only ensure the safe operation of the elevator, but also reduce the influence of elevator noise.

Acknowledgments

Chongqing Education Commission Science and Technology Research Project (KJQN202004302) "Research on Water Resistance and Strength Performance of Large Volume Cement-Desulfurized Gypsum Composite Cement System"

References

1. Tianshui Guo. Vibration and noise analysis of high-speed traction elevator. Master's thesis, Guizhou University, China, 2019 (in Chinese).
2. Xiaowei Wang, Yanjie Yu, Ruijun Zhang, et al. Noise and Vibration Control, 2014, 34(3), 1(in Chinese).
3. Baek, Kwang-Hyun, Kim,et al. Transactions of the Korean Society for Noise and Vibration Engineering, 2011, 21(1).
4. Wenhuan Qiu. China Elevator, 2019, 30(12), 49(in Chinese).
5. Jicun Wang, Menghua Chen. Safety Technology of Special Equipment, 2020, (5), 41(in Chinese).
6. Zhi Xu. China Elevator, 2018, 29(18), 45(in Chinese).
7. Xiaolin Zhang , Jingwei Ji . China Elevator, 2018, 29(18), 43(in Chinese).
8. Yanbin Yin, Hongliang Zhang. Mechanical & Electrical Engineering Technology, 2019, 48(11), 240(in Chinese).
9. Zhengxin Cai, Xuhong Guo. Value Engineering, 2019, (30), 161(in Chinese).
10. Xiaohua Teng, Yongxin Yao, Da Duan, et al. Military Medical Journal of South China, 2015, 29(9), 660(in Chinese).
11. Jie Yin. Environmental Study and Monitoring, 2017, 30(2), 56(in Chinese).
12. Bin Liu. Science and Technology & Innovation, 2016, (21), 104(in Chinese).
13. Shiguang Cui. Environmental Protection and Circular Economy, 2019, (7), 61(in Chinese).
14. Yanqiu Chen, Lizhong Yang, Zhijian Fu, et al. Building Simulation, 2018, 11(4).

Preliminary study on the integration control platform of construction waste based on "BIM+GIS" technology

Aoping Wang[1], *Ning* Wang[2], *Kuanqu* Li[2], and *Fumin* Ren[2,*]

[1] Beijing Jiaotong University School of Software Engineering, Email:19126317@bjtu.edu.cn, China
[2] Beijing Jiaotong University School of Civil Engineering, environmental engineering, Email:fmren@bjtu.edu.cn, China

Abstract. In recent years, the output of construction waste in China has been increasing. 70% of more than 600 large and medium-sized cities in China are surrounded by waste, of which the contribution rate of construction waste reaches 40%. Compared with some developed countries, China lacks real-time monitoring and intelligent control of the whole process of construction waste. In this paper, the popular "BIM + GIS" technology is introduced into the field of construction waste supervision. Based on this technology, a real-time monitoring and intelligent management and control information management platform for construction waste is established, so as to achieve accurate management and control of construction waste.

1 Introduction

According to statistics, the total amount of construction waste produced in recent years in China is about 15.5 billion tons to 24 billion tons. The huge amount of construction waste disposal is a challenge[1] faces in urban management. Construction waste occupies a lot of land resources, affects air quality, pollutes water area, destroys city appearance and environmental health, and has potential safety hazards. Therefore, the research of construction waste precise control technology system has important practical significance[2].

2 "BIM + GIS" and construction waste

2.1 Overview of "BIM + GIS"

BIM, namely "building information model", is an information model that can integrate engineering information, process and resources in different stages of the whole life cycle of the project, and can be easily used by all parties involved in the project[3]. Through 3D digital technology to simulate the real information of the building, it can provide a mutually coordinated and internally consistent information model for engineering design and construction[4].The model is a digital representation of the physical and functional characteristics of a facility. Building information model is a new tool for architecture, engineering and civil engineering , which is based on the information data of construction projects, builds the building model, and simulates the real information of buildings through digital information simulation. It has five characteristics of visualization, coordination, simulation, optimization and printability[5].

Geographic Information System (GIS) is a technical system which collects, stores, manages, calculates, analyzes, displays and describes the geographic distribution data in the whole or part of the earth's surface (including atmosphere) space with the support of computer hardware and software systems. As an important tool, technology and discipline to acquire, process, manage and analyze geospatial data, GIS has been widely concerned and developed rapidly in recent years[6]. From the perspective of technology and application, GIS is a tool, method and technology to solve spatial problems. From the perspective of discipline, GIS is a discipline developed on the basis of geography, cartography, surveying and computer science, with an independent discipline system. From the perspective of function, GIS has the acquisition, storage, display, editing, processing, analysis, output and response of spatial data. From the perspective of systematology, GIS has certain structure and function and is a complete system. GIS is a comprehensive subject, which combines Geography with cartography, remote sensing and computer science. It has been widely used in different fields. It is a computer system for inputting, storing, querying, analyzing and displaying geographic data[7].

2.2 "BIM + GIS" promotes precise control of construction waste

BIM is the carrier of building internal spatial layout and related management information, while GIS can display the overall layout of the city and integrate BIM model

* Corresponding author: fmren@bjtu.edu.cn

with GIS platform, which not only represents the equal display of BIM model in GIS, but also brings a lot of engineering construction information to the GIS map, providing accurate positioning and information for future construction waste supervision and treatment personnel complete and high simulation BIM construction waste visualized operation and maintenance management and emergency response service[8]. Authorized personnel can overview the geographical location of the source and disposal sites of construction waste, query the type and output of the building, review the real-time monitoring image of the construction waste disposal site, view the traceability information of recycled products of construction waste, intuitively understand the construction waste information generated by the whole city or a single building, at the same time, they can also supervise and manage the subsequent cleaning, transportation and disposal of the construction waste online[9].

"BIM + GIS" is a professional application system. The system is based on BIM + GIS platform for application development, which is used to manage urban buildings and underground facilities with a certain type of standard form, including topology information and asset information, covering multi-dimensional dynamic spatial information such as surface, ground and underground. The system meets the management and application requirements of the construction waste supervision department. To provide real-time monitoring and optimal allocation and transportation decision-making scheme for the construction waste clearing and transportation unit, early warning of various possible accidents and decision support for accident handling, so as to achieve the goal of strengthening facility management and creating a good internal environment for accurate control of construction waste[10].

3 Research content

3.1 Research and development of BIM + GIS integration and intelligent collaborative platform

(1) Theoretical research and model construction of BIM based decision management system for the whole process of construction waste including: site investigation, research the construction waste classification system, research on the prediction theory of construction waste production based on BIM, research and formulate a theoretical system for construction waste disposal, research on the decision-making model of construction waste disposal based on the above theoretical content.

(2) Demand analysis and design the system's function. Preliminarily designed the functional modules of the construction waste integrated management system based on BIM+GIS: construction waste information management, BIM model application and system management. The construction waste information management module includes the management of various information in the construction waste disposal process. Among them, the realization of the production forecast and disposal decision-making functions depends on the research theory in (1). The construction waste information management module and the BIM application module are mainly aimed at the users of all parties involved in the project. After the users register and log in, they can manage and apply the project information through BIM model application and other functional modules. The specific functional module of system is shown in figure 1.

(3) Database selection and design. The plan is to use MySQL database as the main storage and Redis as the cache database. Use Redis where high performance is needed and MySQL where high performance is not needed, to speed up access. Research and design the entity relationships of the system, design E-R diagrams, and design specific data tables.

(4) Function design of BIM information integrated management system platform software development By providing a unified API (Application Programming Interface), users can directly program IFC data, avoid dealing with the complex entity relationship of IFC Standard, and develop programs according to professional application requirements[11].

(5) Provide model view display and BIM software application function for BIM information integrated management system platform.Users can directly browse the information model, select different information model display options, view the components that make up the information model and obtain the relevant properties of the components, and then through the professional software to carry out various performance simulation analysis of the building information model, complete the data exchange and sharing between the various disciplines, and realize the collaborative work between the project participants.

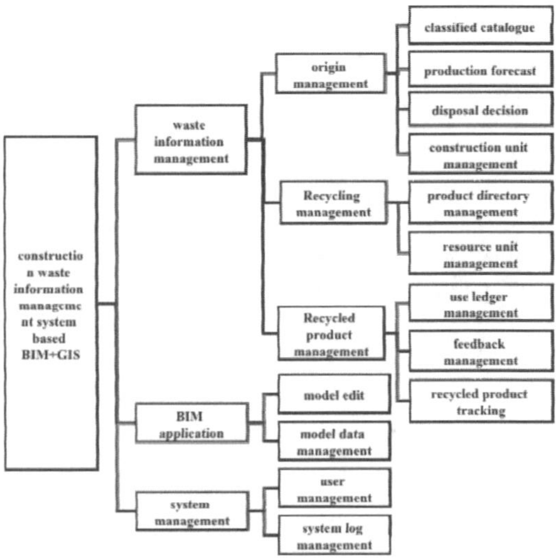

Fig. 1. BIM+GIS information integrated management system

(6) The application of GIS technology in the construction waste management platform. In order to

https://doi.org/10.1051/e3sconf/202123701034

realize the visual management of construction sites and construction waste disposal sites in the city, it is planned to use open source GIS services such as ArcGIS to draw dynamic spatial location distribution maps.

3.2 Based on the integration of BIM + GIS and real-time monitoring technology of the whole process of construction waste, a platform for real-time monitoring and intelligent management and control of the whole process of construction waste is built

(1) Real time monitoring and intelligent control platform for the whole process of construction waste.

Integrate the construction waste management system based on BIM+GIS to realize the intelligent comprehensive management of the whole process of construction waste from generation, transportation, disposal, recycling to the application of recycled products. The platform integrates construction waste generation source characteristics information, transportation supervision and distribution information, stacking, landfill, final disposal information and resource recovery information, and comprehensively uses technologies of The Internet of Things (IOT) and Artificial Intelligence (AI) to realize the path tracking and intelligent identification of transportation vehicles, safety monitoring and warning of the construction environment, etc.

(2) Related theory and technology research.

In order to truly realize intelligent management, in addition to the BIM-based accurate prediction model and construction waste disposal decision-making model mentioned above, it is necessary to study rapid identification technology based on remote sensing data, vehicle identification technology based on surveillance video, and construction waste transportation vehicles. Path planning algorithm, environmental risk assessment technology, etc.

(3) System development and integration.

Research the development of the platform, the general agreement in the integration process, the integration interface form of each subsystem, the communication protocol with the monitoring equipment, as well as the call interface, data dictionary library, data format, data collection frequency, data output form and frequency, coding method, data encryption and verification method, etc., and formulate the construction of the real-time monitoring and intelligent control platform for the whole process of construction waste Standard, guide platform integration[12]. The system is planned to adopt microservice architecture. Each microservice can be developed independently, with low coupling between services, free choice of languages and tools, and easy expansion of requirements. The system architecture diagram of the preliminary design is shown in Figure2.

(4) Real time monitoring and intelligent control platform project demonstration of the whole process of construction waste.

In the project demonstration project, organize training and promotion application, in the promotion and application stage, constantly improve the integrated platform function, keep close to the user's use habits, and become an effective tool for precise management and control of construction waste.

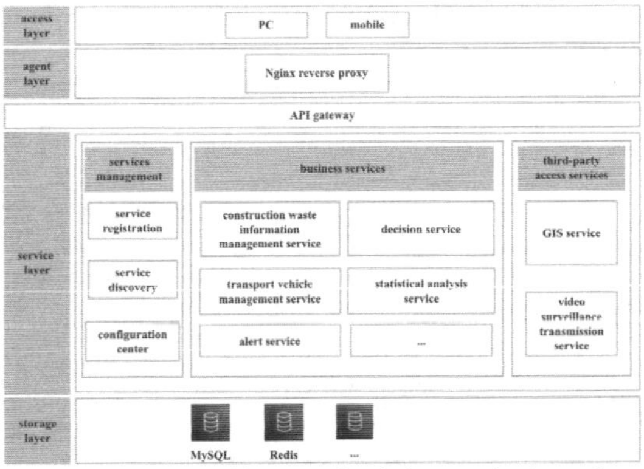

Fig. 2. System architecture of real-time monitoring and intelligent control platform for the whole process of construction waste

3.3 Based on the integration of BIM + GIS and real-time monitoring technology of the whole process of construction waste, a platform for digital city construction waste comprehensive management platform is built

The construction of digital urban management system can realize the integration of system, location, function, resource, information and location resources, integrate the existing urban emergency management, electronic video monitoring, mayor's hotline and other systems and functions, share video, data and services, and directly integrate the decentralized construction of computer room to the municipal emergency linkage center, and build the information exchange hub. "Unified supervision" refers to the establishment of an independent digital city management and supervision center to effectively supervise the city's urban management. The digital urban management information system is to divide the city into several management grids by digitizing the urban management objects, taking 1km^2 as a unit; in view of the particularity of the management of three links (generation, transportation, consumption and disposal) of urban construction waste disposal, the division principles of the grid, components and events of the three links are determined respectively[13].

(1) Division principle of grid, component and event

In the process of construction waste source generation, the grid is divided into urban area as the unit; the components are divided into the red line map of the construction project as the unit; the event management is mainly for the approval and coordination of construction waste disposal. For the construction waste transportation link, the grid is divided into urban area as a unit; the components are divided into transportation enterprises as

https://doi.org/10.1051/e3sconf/202123701034

a unit; the event management mainly monitors and manages the spilling, omission and illegal spoil in the process of construction waste transportation.

In the process of construction waste absorption and disposal, the grid is divided into urban area as a unit; the components are divided into construction waste absorption yard as a unit; the event management mainly focuses on the monitoring of the unauthorized establishment of waste disposal site, the emergency of waste soil landslide and the supervision of temporary waste disposal yard .

(2) Information exchange mechanism of grid, component and event management.

The integration of grid, component and event management information includes three blocks: the approval information of disposal Emission Declaration, the information of vehicle transportation market and the information of coordination, absorption and disposal

1) The information exchange mechanism between the application procedures of the construction in progress department and the application and approval of the construction waste disposal and discharge is conducive to the approval and supervision of the construction waste discharge amount and disposal plan during the construction of the construction project;

2) The communication mechanism of vehicle information, construction waste transportation administrative license information and administrative law enforcement information of public security and transportation departments is conducive to the monitoring and joint law enforcement management of construction waste transportation license approval, spilling omission and illegal waste soil in the transportation process;

3) The information exchange mechanism between land ownership information of land department and urban construction planning information of planning department and construction waste adjustment, absorption and disposal information is conducive to the reasonable establishment of temporary disposal site and the management of illegal spoil.

(3) Operation and management mechanism of construction waste management system.

Digital construction waste management needs both technical support and mechanism guarantee. According to the current management practice of urban construction waste in each city, the management measures such as the implementation measures of digital urban construction waste management, the standards and time limit of construction waste management events and components disposal, and the management and disposal process of digital urban construction waste are formulated. Based on above-mentioned management measures,establish a working mechanism of "unified acceptance at the municipal level, coordinated command of the city and departments, and disposal of problems by relevant units".

4 Prediction and analysis of platform effectiveness

4.1 Preparation of unified specifications and standards

In the process of building the "BIM + GIS" integrated platform for precise management and control of construction waste, it is necessary to establish a standardized integrated platform to ensure the unity, stability and scalability of the platform.

4.2 Research and development of BIM + GIS fusion technology platform based on Web3D

Integrate multi-source heterogeneous data into a unified platform of GIS and BIM integration, and realize the Internet visual control service of real-time monitoring data of all kinds of construction waste in the whole process with lightweight Web3D technology. In terms of function, the platform can realize virtual space display, information release, data analysis, risk assessment, early warning and prediction, emergency response, comprehensive management, statistical analysis and other functions; in terms of performance, it can ensure the stability, reliability, fast response and strong operability of the software platform; in terms of security, the software platform can prevent the leakage of important data and information and ensure the operation of the software platform Safety of the line.

4.3 Preparation of unified specifications and standards

Establish a real-time monitoring and intelligent management and control platform for the whole process of construction waste; realize the allocation and control amount accounting for more than 95% of the construction waste production. Relying on the digital urban management information system, gradually establish the urban construction waste information management platform, and take the information platform construction as the starting point to build the construction waste supply and demand information system, and improve the socialized and standardized level of the comprehensive utilization of construction waste.

5 Conclusion

Based on the "BIM + GIS" technology, the information management platform for urban construction waste is gradually established, and the information platform construction is taken as the starting point to build the construction waste supply and demand information system, so as to improve the socialized and standardized level of comprehensive utilization of construction waste. The whole process real-time monitoring and intelligent management and control platform of construction waste is oriented to service and data sharing in the design, establishing and providing comprehensive services based on the resource utilization of the construction industry, and providing standard open data interface to realize data sharing with other systems. The data of the management and control platform can be collected in multiple dimensions, and the relationship between data can be established, and multi-dimensional data verification and analysis can be provided to realize closed-loop data management. Combined with the construction and application of digital city management platform, it provides effective support for government supervision.

The platform provides users with an information platform to improve the operation efficiency of the clearing and transportation enterprises. With the standardized process of the platform, the urban management department simultaneously establishes the source approval and the issuance of franchise qualification. Continuously promote the organic integration and co governance of government, society and enterprises. Through online business activities and market factors, enterprises form a dual effect governance ecology of self-management and government supervision.

This research is closely related to the development of social economy and society, the national ecological environment protection strategy, the people's life and health, and the environmental security. It is an urgent strategic demand of the country, and has important scientific value and social, economic and ecological benefits.

The research was conducted with in financial support from the National Key Research and Development Program (2018YFC0706000), China National Railway Group Science and Technology Project (N2019Z003) and Beijing Jiaotong University basic scientific research business fee project (2020JBZ112).

References

[1] Wei wei. The road to a smart city based on BIM+GIS technology[C]. China Smart City Construction Technology Seminar, 2015.

[2] Gu Jiazhen, Qian junyu, Wang Shan, et al. Research on the application of BIM+GIS integrated technology in construction engineering[J]. China's real estate industry, 2017(6).

[3] Cheng J C P , Ma L Y H . A BIM-based system for demolition and renovation waste estimation and planning[J]. Waste Management, 2013, 33(6):1539-1551.

[4] Yuan H. Critical Management Measures Contributing to Construction Waste Management: Evidence From Construction Projects in China[J]. Project Management Journal, 2013, 44(4):101-112.

[5] Kim Y C , Hong W H , Park J W , et al. An estimation framework for building information modeling (BIM)-based demolition waste by type[J]. Waste Management & Research, 2017:0734242X1773638.

[6] Osmani M , Glass J , Price A D F . Architects' perspectives on construction waste reduction by design[J]. Waste Management, 2008, 28(7):1147-1158.

[7] Maier S, Oliveira L B. Economic feasibility of energy recovery from solid waste in the light of Brazil's waste policy: The case of Rio de Janeiro[J]. Renewable & Sustainable Energy Reviews, 2014, 35(35):484-498.

[8] Diamond J. Bower Building and Decoration by the Bowerbird Amblyornis inornatus[J]. Ethology, 2010, 74(3):177-204.

[9] Zhao X F , Wang Q. Analysis on Structural Problems in Building Decoration[J]. Applied Mechanics and Materials, 2014, 578-579:536-539.

[10] XIAO Ping-xi. Investigation of Creative Inheritance of Building Decoration[J]. Journal of Chongqing Architecture University, 2005, 27(4):15-19.

[11] Zeng Hong, Mao Xiaoli. Research on Application of BIM in Construction Waste Management[J]. Shanxi Architecture, 2013, 39(19):184-185.

[12] Gao Jiayong. Research on Construction Waste Reduction Management under the Framework of Green Construction[J]. Engineering Technology: Digest Edition:0018400184.

[13] J SolísGuzmán, Marrero M, Montesdelgado M V, etc. A Spanish model for quantification and management of construction waste.[J]. Waste Manag, 2009, 29(9):2542-2548.

Status quo and development potential of full life cycle evaluation of construction waste

Jingming Jia[1],Fumin Ren[1,*]and Jintong Zhu[1]

[1]School of Civil Engineering, Beijing Jiao Tong University, Beijing, China

Abstract. It is critically important to synthesize the life-cycle assessment (LCA) and construction waste content, especially in term in recycling of construction waste resources[1] and optimizing critria in the life-cycle evaluation. That is, the development of construction waste resources is entitled to be boosted, as same as criteria and elements in periodic evaluation concentrating on the module of construction waste . This paper analyzes the disadvantages of construction waste and potential promotion in each stage, which is of great significance for improving the evaluation system and management of construction waste.

1 Introduction

The construction industry, as a pillar industry in China, has been in a state of rapid development. The problems exposed are: the exhaustion of building material resources and the surge of construction waste.According to relevant research, the production of construction waste in China accounts for 40% of the total urban waste, in comparison to the global construction industry accounts for 40% of the energy consumption, which has only generated 47% of the global waste. Moreover, there are many kinds of complex components, and the pollution degree caused by them is also different. All these mentioned factors have sharpened the conflicts between reusing and disposing of waste resources in Chinese construction industry.According to the Development Report on the Industrialization of Construction Waste Resources in China, from 2011 to 2019, the amount of construction waste generated by housing construction in China was as high as 977 to 2133 million tons, taking up about 30% to 40% of the total amount of municipal solid waste.However, the recycling rate of construction waste in China is still less than 5%[2]In the whole life cycle of buildings, waste of building materials and consumption of energy lead to serious environmental problems such as acidification, global warming, haze as well as ozone hole. In order to deter the environmental deterioration, it is estimated that, by 2030, 65% of building materials in China and 75% of decoration waste are supposed to be recycled. Therefore, a full-developed evaluation covering the whole life cycle of construction waste is demanded to provide information support and approaching guidance for the sustainable development of construction waste.

The LCA of construction waste can summarize several kinds of factors that have a great impact on the environment from different urban cases, and at the same time, the hazard of construction waste can be quantitatively assessed according to the corresponding model. In addition, it can be used as one of the key evaluation standards of urban green construction.

2 Full life cycle evaluation of construction waste

2.1 Definition of full life cycle evaluation of construction waste

The whole life cycle impact assessment of construction waste is based on the life cycle assessment, which is an assessment method for the whole process of construction waste.The full life cycle impact assessment of construction waste can be regarded as the product of combining LCA with practical engineering and specialization in the field of construction waste. Life cycle assessment (LCA) is a globally recognized environmental load quantification standard[3], which plays an important role in evaluating the environmental impact and resource consumption of product life cycle[4]. Foreign research on LCA has been half a century, mainly in the impact assessment of breakthroughs. Since 1990s, Life cycle assessment has become a research hotspot in domestic academic circles[5]. A large number of researchers have carried out a large number of studies in life cycle methodology research and specific materials and product applications. Yang Jianxin et al. established a method system to evaluate China's environmental impact and resource consumption, which provided a theoretical basis for domestic life cycle assessment research[6].

So far, there have been several executions of whole life cycle impact assessment stagnated in some specific cases, for example. Therefore, Webber and his team, by using LCA evaluation model and practical application of the construction waste processing, after compareing direct landfill technology with the shenzhen city recycling

* Fuming Rem:fmren@bjtu.edu.cn

program , figured out two approaches to evaluate the economic costs associated with the quantitative evaluation and environmental load.

With regard to some defects in the whole life impact assessment of construction waste, In order to improve the relevant knowledge system and enhance the reference value of its practical application, this paper analyzes the deficiency and potential developing direction of the full life cycle impact assessment of construction waste from two aspects[7].

2.2 Indicators of full life cycle evaluation of construction waste

According to the overall process of construction waste,the indexes of the impact assessment of the full life cycle of construction waste can be divided into the following parts:

2.2.1 Evaluation on the source of construction waste

The source of construction waste determines the outputing quantity of the entire building garbage and its initial impacts on the environment. The impact evaluation on the source includes the aseessments on the influences of building materials, soil quality in ocuupied constructing land, the overall regional planning, and the the surrounding water quality. In addition to assess the potential hazards of the materials used in buildings, the purpose of the evaluation stage is to find out the impacts of the buildings at the initial stage of construction on the surrounding environment.

2.2.2 Transport process impact evaluation index

According to relevant statistics, nearly 15% of the garbage in the source area is scattered on the way when it is transported to the disposal site or landfill site, which brings additional pollution burden to the nearby residents and the surrounding environment. Therefore, the reduction of garbage in the process of transportation has become an important index for the evaluation of the whole life cycle, which is mainly about the variation of garbage in the process of origin and disposal, the problem of too long transportation and the deterioration and oxidation of garbage.

2.2.3 Impact evaluation index of disposal end

Disposal sites and landfills have been the main "destination" of construction waste in the past two decades, and the final site's capacity, leaching toxicity, and disposal methods have become typical indicators of environmental impact. Excessive amount of one factor will harm the surrounding environment. For example, heavy metals in construction waste migrate to the deep soil, causing soil compaction and saline-alkali land production. Therefore, it is necessary to strictly control all elements at the disposal end.

2.2.4 Impact evaluation index of remanufactured products

Since the 13th Five-Year Plan, recycled products have been the main products of the national emphasis on the recycling of construction waste. Recycled products, such as recycled bricks and recycled concrete, have become the key evaluation indicators of recycled products in terms of their promotion scope and practical effect after application. Although it does not involve environmental pollution factors, it is of great significance to improve the recycling rate of construction waste, and it is necessary to become an evaluation project of the whole life cycle of construction waste[8].

3 Review of the full life cycle evaluation of construction waste

3.1 Source evaluation of construction waste

At present, the sources of construction waste in China are widely distributed and the components are complex. According to the investigation of domestic demonstration cities, there are generally three classification methods: according to the source, according to the physical composition, and according to the reuse performance. Traditional construction waste will be divided into land excavation, the excavation of roads, the old building demolition, construction and building materials production and waste five categories, mainly by the sediment, crushed stone, waste mortar, brick and tile fragments, concrete, asphalt, plastic, scrap metal, waste wood and so on, but it is not clear for construction waste of scientific evaluation system. The lack of life cycle evaluation system leads to the lack of factual basis for the evaluation at the source of the life cycle of construction waste, which leads to the lack of accurate and scientific information in the whole evaluation system.

In order to improve this situation, typical evaluation methods of construction waste from abroad can be used for reference, such as construction waste classification assessment, reuse product quality control and other technologies, to constantly improve China's construction waste classification system, and further promote the maturity of the evaluation system, the Fig.1 shows the growing trend of standards related to construction waste.

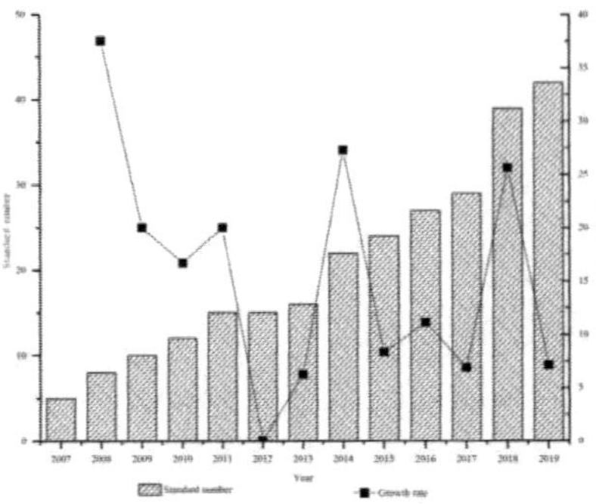

Fig1. The growing trend of standards related to construction waste

At the same timebased on the whole industry chain to carry out the evaluation of the life cycle of construction waste.

For different types of construction waste (take decoration waste as an example), different classification methods and information acquisition methods can be used. Through the analysis of decoration waste in Huzhou, Zhou Jijiao used the charging mechanism to obtain detailed information about the quantity and category of decoration waste on the one hand, and on the other hand, to promote the development of prefabricated building technology to achieve the reduction of decoration waste. Through the analysis of concrete examples to constantly improve the evaluation system of garbage, and finally build a construction waste in China in accordance with the actual life cycle evaluation system.

3.2 Construction waste transport evaluation

The construction waste transport process is in the position of connecting the past and connecting the future in the whole life cycle evaluation system. Safe and complete transport is the important premise to guarantee the origin of garbage and disposal site garbage, whose properties are unified. At the same time, it is also one of the important means to avoid the environmental pollution around the transportation. Therefore, the dynamic characteristics of construction waste in the transportation process should be included in the whole evaluation system.

After practical investigation, it is found that in most cities, transportation units control the disposal direction of construction waste, especially playing an important role in the optimization and allocation of construction waste in the aspect of "redundancy balance" resource recovery.For example, in the comprehensive utilization of earthworks, excavation is carried out in the early stage of A project, and the transport party is entrusted to carry out external transportation. The transport party will allocate some earthworks for backfilling in the later stage of the surrounding projects according to the construction conditions of surrounding projects, and send the rest to landfill or temporary storage yard. When backfilling is required from the completion of Project A to the later stage of the project, the transporter will also transfer the earthwork required by Project A from other projects or temporary landfill sites. This has played an important role in reducing the landfill volume of construction waste terminal. However, because transportation party itself is

chasing interests in the process of construction waste disposal, so also is the important factor of construction waste control problem is one of the main performance is: the vast majority of urban construction waste transportation cost is the "front end settlement", namely the construction waste generated will use each way to transport for disposal, construction waste, and synchronization according to the capacity and settlement. Since this mode of operation fixed the total revenue of the transporter, it was inevitable that he would try to reduce the cost to increase the profit, thus leading to the phenomenon of "construction waste discarded at random nearby and leaking out".

In view in the process of transportation due to theeconomic benefits, the environment for the construction waste pollution problem occurred frequently, to alleviate the phenomenon, and the actual situation can be accurate records, and shall put the specific indicators such as litter lose balance, attribute changes, transportation vehicle control indexes into the evaluation system of the life cycle.

3.3 Evaluation of construction waste disposal end

Construction waste recycling is the best way to achieve green site waste disposal. The main difficulties in recycling construction waste are material pollution, economic benefits and disposal methods. According to the investigation from different demonstration cities, the disposal methods of construction waste in China are mainly classified disposal :(1) waste soil disposal in projects; (2) mud disposal in projects; (3) waste metal disposal in projects; (4) inorganic nonmetal disposal in projects; (5) concrete disposal; (6) waste bricks disposal;

(7) waste mortar disposal, etc. This method is mostly designed to promote resource recovery. However, there is still a gap in the evaluation of the disposal methods of toxic and harmful construction waste. The disposal methods of toxic and harmful construction waste are generally landfill and incineration. The vast majority of disposal sites choose incineration as the main means of solution. The method uses high temperature to decompose and remove toxic substances, and the high heat generated by the combustion process can also be used to generate electricity, but the overall cost is high. As for landfill, its low cost will cause secondary pollution to the environment and occupy land resources. Today, however, disposal landfills are equipped with impervious layers to prevent leachate from entering the environment. However, foreign research found that regardless of the nature of the impermeable layer, leachate will break through from the impermeable layer sooner or later, contaminating groundwater and critical soil.

Paying attention to the leaching characteristics of toxic substances from construction waste has become an important index for the full life cycle assessment of construction waste. Investigation includes not only toxic

 https://doi.org/10.1051/e3sconf/202123701035

substances leaching law of vertical and horizontal, more should be to focus on its borders on soil water leaching behavior, due to the influence of factors such as surface tension, toxic substances in the junction of sports is different from pure medium, this to slow construction waste environmental impact, promote restoration of the soil and water conservation behavior is of great significance, the present evaluation is not to be classified as assessment.

3.4 Evaluation of construction waste recycling products

The field of renewable products has been developing slowly over the past few decades, with most renewable resources ending up in landfills. With the introduction of waste recycling, green construction and other concepts, the development of recycled products gradually put on the agenda. At present, the attention of the government of China on recycled products is focus on construction waste in terms of recycled fine aggregate, recycled coarse aggregate and recycled concrete, but there is a shortage of relevant standards on the technology, quality and comprehensive properties (physical and chemical properties) of recycled products.

Applicable tracking of recycled products is the fundamental guarantee of their existence. Only when the recycled materials prove to be useful and usable can they be accepted by the market. Applicable facts of recycled products, such as service life, application breadth and variation frequency, these attributes control the root of recycled products and should be included in the evaluation system. Through the evaluation system to further standardize the properties of recycled products, enhance the application value of recycled products, and develop related technologies.

In the demonstration cities involved in this paper, through the optimization and upgrading of the recycled products, the mature products include square bricks, permeable bricks, roadbed, aggregate, etc., to realize the deep recycling of construction waste. The recycling rate of this city has reached 95%, which is a typical example of green city construction and implementation of green construction. However, after in-depth investigation, it was found that there were still loopholes in the feedback and evaluations of recycled products, and follow-up evaluations were not implemented. Therefore, in general, in the stage of regenerated products, the evaluation system still has a large shortcoming.

4 Development potential of full life cycle evaluation of construction waste

In order to change the current situation of construction waste disposal and fundamentally improve the technical value of construction waste full life cycle evaluation, the potential value of cycle evaluation in different stages is explored through the analysis of the whole process of construction waste, so as to improve the full life cycle evaluation system.

4.1 Development direction of construction waste generation source evaluation

Looking back at the past construction cases, the whole construction is often a mess, soil pollution, extreme waste of resources, building information chaos, has brought a great burden on the environment and economic costs. The traditional method of collecting construction waste information is relatively complicated, and its process is shown in Figure 2.

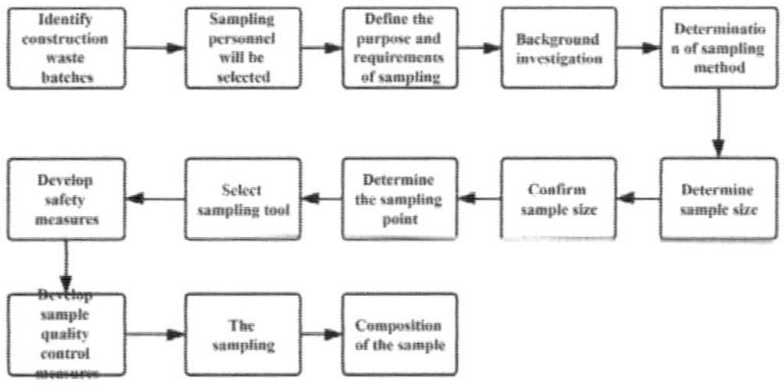

Fig2. The traditional method of collecting construction waste process

With the advent of the information age, BIM technology and green construction have injected new vitality into the whole life cycle evaluation of construction waste. BIM technology can perfectly integrate the overall information of construction projects[13], and the establishment of relevant information model and the integration of the concept of "green construction", can deeply dig the precise point of full life cycle evaluation, to help achieve refined management.

The advantages of BIM technology are in the planning and initial stages of buildings. The technology can be used to demonstrate the design scheme in the planning stage with the full life cycle evaluation, including scheme design, cost analysis and sustainability evaluation[9], comprehensive utilization of the site, etc., which implements the green construction with environmental protection and resource conservation as the core. Meanwhile, it is also the concrete embodiment of the theme of "sustainable development" in the project construction. Wang Ning et al. studied the technical system combining BIM and GIS, explored the precise control of construction waste, and established a real-time monitoring and intelligent control information platform for construction waste. This technology broadens the field of construction waste supervision, and combines life cycle evaluation with information technology to improve the socialized and standardized level of comprehensive utilization of construction waste, and also provides effective support for the government's monitoring and management.

With the gradual popularization of BIM and the improvement of application maturity, the application of BIM technology in building renovation projects will gradually increase, and the breadth, depth and accuracy of information provided by BIM technology will gradually increase, which will further promote the application of BIM in building renovation projects LCA. It can be predicted that there will be more and more researches on the application of BIM in building renovation LCA in the future, and the collaborative interaction between BIM and LCA tools will be better and better[10].

To sum up, the visualization and simulation function of BIM shows the characteristics of green construction and green building, so that BIM can be widely used in the practice of building projects, which will also make outstanding contributions to the construction of ecological civilization in China.

4.2 Evaluation potential of construction waste transportation

In order to realize accurate management and accurate information evaluation of the transported wastes in the process of construction waste transportation, a series of safeguard measures, such as increasing the number of monitoring probes of transport vehicles, should be established to ensure real-time monitoring of garbage and reduce pollution to the road environment.

At present, to improve the information evaluation in the process of construction waste, the following aspects can be used to strengthen the cleaning and transportation of construction waste:

4.2.1 Unified construction garbage collection and transportation charge system

Unified transportation cost of construction garbage is an important link to improve the construction garbage full life cycle evaluation system and realize construction garbage collection and transportation at the same time.

4.2.2 Strictly manage the cleaning and transportation process

On the one hand, use the special sealing performance of the vehicle is very good, in order to avoid leakage, dust and other problems. In addition, the establishment of receipt system is a key factor to ensure the fixed-point transportation of transport vehicles, and is also an important reference information for periodic evaluation. In addition, transportation routes should be strictly inspected, such as setting UP GPS and Beidou terminals, and adding remote sensing equipment, etc., so as to real-time monitor and control the behavior of candid and disorderly shooting, so as to achieve accurate control.

Dynamic control of temporary storage points:

In order to realize the construction waste transportation can be fixed, requires the dynamic control for temporary storage of construction waste, more convenient to follow-up the vehicles for garbage pickup precision, is beneficial to jointly supervise each department, using information technology means to improve the evaluation system is to break the information barriers between departments, the greatest degree of fine evaluation details.

4.3 Development direction of construction waste disposal evaluation

The disposal site is the area where construction waste frequently pollutes the environment, so it is necessary to accurately evaluate and control the disposal site such as landfill.

4.3.1 Explore scientific disposal techniques

Based on the status quo of comprehensive classification of construction waste, different evaluation systems are established for different types of waste. Starting from the whole life cycle, we will study through technological reform and management policy, focusing on reducing quantitative information and supplementing classified information. Under the premise of normal guidance, actively promote the nearby reuse of construction waste, which not only effectively promotes the resource recovery of construction waste disposal, but also further improves the evaluation system. For example, the separation and disposal of waste plastics are shown in the table1 below.

Table 1-1 Recycling products and uses of waste plastics

Waste plastics	Recycled product	use
Abandoned PVC	drain	Renovate the railway base bed to turn slurry and remove mud
	It is made into small spheres with a diameter of 20-30 mm	Instead of stone slag as foundation, subgrade cushion material
	Polymer alloy	Materials make boats, border walls and decorate parks
Waste plastics	Plastic "wood" products are resistant to moisture and corrosion	Make fences, road signs
	Production of carpet fiber, clothing insulation lining raw materials	Production of carpet fiber, clothing insulation lining
Waste mineral water bottles	Polyvinyl chloride chemical fiber	Mix it with wool to make a sweater
bottles	Fine plastic	Long lace fibers

4.3.2 Planning a reasonable disposal site

Construction waste disposal site is not only an important urban infrastructure, but also an important source of basic data for the whole life cycle evaluation system. In the initial stage of building formation, the total amount of possible production should be predicted and strictly evaluated according to the amount of resource recovery and disposal that should be undertaken and the land scale, access system and enterprise qualification involved in the corresponding technological process[11].

4.3.3 Promote the recycling of construction waste

The biggest characteristic of construction waste is its high added value, and the best way to dispose of it is to make use of resources. Therefore, all localities and departments should actively explore ways to recycle construction waste[11], continuously improve the full life cycle evaluation of different approaches, increase the proportion of its resource utilization, and actively explore new technologies, new models and new evaluation system for construction waste recycling.

4.3.4 Strengthen the evaluation system of the whole life cycle

Accurate control of construction waste is one of the important means to improve the evaluation system, and the "Intelligent Supervision System of Construction Waste in Shenzhen" set up in Shenzhen is a typical case of accurate control and mature full life cycle evaluation system. This sharing of data breaks down existing barriers across functions[11], making the law enforcement process no longer exist blind areas, the evaluation system no longer exist lack of loopholes, so that the overall difficulty of supervision is reduced. Many cities should learn from its advanced experience.

4.4 Tracking and evaluation direction of construction waste recycling products

The development of recycled products is an important channel for the recycling of construction waste, and the tracking of recycled products is the survey feedback after its practical application. The evaluation of the feedback results is conducive to expanding the development direction of the recycled products and opening up new ideas for the reclamation of construction waste. Moreover, the problems arising from the feedback results also provide opportunities for the development of the recycling technology. Therefore, the evaluation of recycled products is an important link in the whole life cycle evaluation. The current practical evaluation direction can be divided into the following parts:

4.4.1 Evaluation of material utilization of recycled products

Limited by the regenerative technology, the amount of recycled materials added in most of the regenerative products is inversely proportional to the performance of the final product. So how to allocate the proportion of recycled materials in recycled products has become a hot topic at present. This ratio not only affects the properties of finished products, but also relates to whether the recycling technology has been used. It is an important evaluation index to evaluate whether a resource chemical plant carries out the recycling technology. Therefore, in the future, it should be an important link to evaluate the utilization rate of renewable resources of different renewable products on the premise of not affecting their normal service life and use properties.

4.4.2 Deformation evaluation of recycled products

In the practical application process, the physical and chemical properties of the regenerated products will be denaturated to a certain extent with the different application scenarios. At present, there are kinds of research in this field, such as compressive strength and tensile strength indexes of recycled concrete, water absorption rate and water permeability indexes of recycled aggregate, etc. However, a large number of studies have only focused on the degeneration of recycled products in different cases. However, the denaturation evaluation of commonly used recycled products in a certain field is still in the blank stage, and most of the performance evaluation is still in the fragmentation stage. In the future, the performance evaluation of recycled products can focus on the commonly used recycled products in an industry, so as to form a relatively complete performance evaluation system of recycled products.

4.4.3 Market evaluation of recycled products

The market evaluation is mainly divided into the application scope evaluation and the application depth evaluation of the recycled products. As for the application scope of the recycled product, it shows the market potential of the recycled product and whether it has the future development space. The evaluation result directly affects the corresponding production technology and the recycling rate of some waste. In addition, the in-depth evaluation of application lies in whether the regenerated product can be used in high-end industries to meet the needs of some large markets, which is conducive to the expansion of the application of the regenerated product.

5 Summary

The whole life cycle evaluation of construction waste not only aims at improving the whole construction waste industry chain, but also is an important measure to promote the development of green construction and green construction in China. Through the in-depth analysis of the four links of source, transportation, terminal and product, new requirements and ideas are proposed for the whole cycle evaluation, and the shortage and development direction of construction waste recycling are further summarized to constantly improve its recyclable products. The evaluation of life cycle implements the development strategy of sustainable development, conforms to the new development concept of "clear water and green mountains are gold and silver mountains" proposed by President Xi jinping, and provides a technical guarantee for the sustainable and healthy development of national economy, as well as a new evaluation system for the construction of national ecological civilization.

Acknowledgement

This work was financially supported by the National Key Research and Development Project (2018YFC0706000) and the Fundamental Research Funds of Beijing Jiaotong

University (2020JBZ112).

References

1. Huang Zhong. Game Analysis and Development Strategy Research on sales link of construction waste Recycling Products [D].Shaanxi: Xi 'an University of Architecture and Technology, (2018).
2. Liu Xia, HU Xi. Research on the Quantitative Method of construction waste based on the whole life cycle [J]. Science and Technology Innovation,(2020),(2):175-176.
3. Liu Hua. Theory and Method of green Product evaluation and its application in powder metallurgy Guangdong: South China University of Technology,2005.
4. Tang L L. Life Cycle environmental Impact Assessment of construction waste Treatment [J]. Renewable Resources and Circular Economy, 209 (6):29-31.
5. Wang Ning, Lou Dai, Chen Daqing, et al.Preliminary Study on the Construction waste Accurate control Information Management Platform based on "BIM+GIS" technology [J].Environmental engineering, 2020, 20 (3) : 46-50.
6. Bai Lu, SUN Qihong, Qiao Qi.Review on the progress of life cycle evaluation in China [J]. Anhui agricultural science,(2010),38(5):2553-2555.
7. Chen Yamin. Preliminary study on land use Ecological Impact Assessment method in LCA [D]. Shanghai: Fudan University,(2012).
8. Wang Bo. Study on the treatment Model of construction waste in Shenzhen Based on life cycle Evaluation [D]. Hubei: Huazhong University of Science and Technology, 2012. DOI:10.7666/d.D230636.
9. Yang Keren. Application Research of Architectural Design Quality Evaluation based on BIM Technology [D]. Shaanxi: Xi 'an University of Architecture and Technology,2018.
10. Yang Xining. Study on environmental Impact Assessment of the Life cycle of Building Renovation Project based on BIM [D].Chongqing: Chongqing University,(2018).
11. Xie Zhifeng, ZHAO Jinzhu.Treatment, Disposal, utilization and Countermeasures of Construction waste in China [J]. Urban Construction Theory Research (Electronic Version),(2011),(27).
12. Hu Chunwei. Application of BIM Technology in Construction Engineering Management [J]. Construction Engineering Technology and Design,2018,(11):3301.
13. Hua Wei. Study on construction waste Management in Henan Province [D]. Henan: Zhengzhou University,(2017).

Study on the effect of Fe and N co-doped supported TiO_2/GF photocatalytic oxidation of nitrobenzene wastewater

Yafeng Li[1], Qun Fu[1] and Jiahui Yu[1]

[1]Municipal and environmental engineering, Shenyang Jianzhu University, Shenyang, Liaoning, 110000, China

Abstract. This research aims to treat the refractory nitrobenzene wastewater in the industrial water treatment industry with modified supported TiO_2, and to study its influence on the treatment effect.Experiments have found that by preparing Fe and N co-doped TiO_2/GF supported catalysts, it has better catalytic performance. Under the same conditions, Fe and N co-doped TiO_2/GF can degrade nitrobenzene wastewater better than single Doped with Fe-TiO_2/GF and N-TiO_2/GF, the removal rates of nitrobenzene and COD are as high as 97.2% and 92.5%, respectively.The Fe and N co-doped TiO_2/GF catalyst overcomes the agglomeration of traditional powdered TiO_2 in solution, and has a good removal effect on nitrobenzene wastewater.

1 Research background and significance

Photocatalysis is a new advanced oxidation technology with simple equipment, mild conditions, and obvious effects of oxidizing and degrading organic matter. It can convert organic matter into water and carbon dioxide by using the energy of sunlight, which can meet environmental protection requirements.[1]Inorganic metal oxide semiconductor materials are commonly used photocatalysts, and TiO_2 has become an ideal photocatalyst due to its stable chemical properties, non-toxicity, and low cost. [2]There are three main crystal types of TiO_2: anatase , rutile and brookite. A large number of studies have shown that anatase is the most active crystal type, and its energy band gap energy (Eg) is 3.2eV, which is equivalent to the photon energy with a wavelength of 387.5nm[3].Frank and Bard[4] pioneered the application of semiconductor materials to sewage treatment, and cyanate oxide used TiO_2 as a photocatalyst.TiO_2 photocatalytic oxidation technology is considered to be one of the most promising photocatalytic materials at this stage [5].

Nitrobenzene wastewater is a key and difficult point in the industrial water treatment industry. In order to solve the harmful effects of nitrobenzene wastewater on people's production and life, domestic and foreign scholars have conducted a lot of research on the treatment methods of nitrobenzene wastewater.Shang Xiaoyuan [6] used the biodegradation method to study the granulation of aerobic sludge and the removal effect of granular sludge on nitrobenzene wastewater. The experimental results showed that the removal rate of TD was not ideal.Yang Pengfei[7] conducted experiments with physical treatment methods. The experimental research showed that the removal rate of wastewater is not affected by the pH of the solution. At the same time, the extraction phase can be regenerated and used for multiple extractions. However, some industrial extractants are organic solvents, which are easy to be used in wastewater. Cause secondary pollution, and the treatment effect is not thorough enough.

This study mainly studies the preparation conditions of Fe and N co-doped TiO_2/GF photocatalysts and the process conditions for the treatment of nitrobenzene wastewater, which provides a basis for accelerating the industrial use of photocatalytic oxidation technology.

2 Materials and methods

2.1 Test device and test carrier

The photocatalytic oxidation device used in the experiment is self-made, which is composed of light source, reaction device and aeration device.The glass fiber used in the experiment is linear AR glass fiber, also known as alkali-resistant glass fiber, which is 100% inorganic fiber.

2.2 Experiment method

In this experiment, the Fe and N co-doped TiO_2/GF photocatalysts were prepared by the sol-gel/dipping-climbing method. At the same time, the ultraviolet mercury lamp with a wavelength of 253.7nm and 45W was used as the light source to investigate the different ion doping in the preparation process. The impurity amount and doping type affect the performance of TiO_2/GF supported photocatalyst, and the optimal doping amount is determined.

3 Results and discussion

Test standard conditions: 45W ultraviolet lamp, 0.15g supported photocatalyst sample, 1000ml nitrobenzene concentration is about 78.86mg/L, COD concentration is about 1582mg/L.The supernatant was taken to determine the absorbance at 545nm and 465nm.

3.1 The effect of nitrogen doping on the treatment effect

Keep other benchmark conditions unchanged, weigh 0.15g nitrogen-doped TiO_2 containing n(N): n(Ti) ratios of 0.05, 0.1, 0.2, 0.4, 0.6, and the benchmark conditions of the photocatalysis test remain unchanged, calculate The removal rate of nitrobenzene and COD, draw a curve, and investigate the effect of different nitrogen doping on the catalytic performance.The test results are shown in Figure 1 and 2:

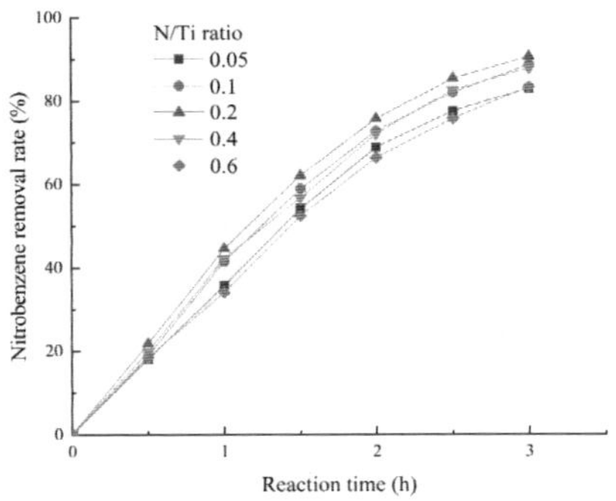

Fig1. Influence of nitrogen content on removal of nitrobenzene in wastewater

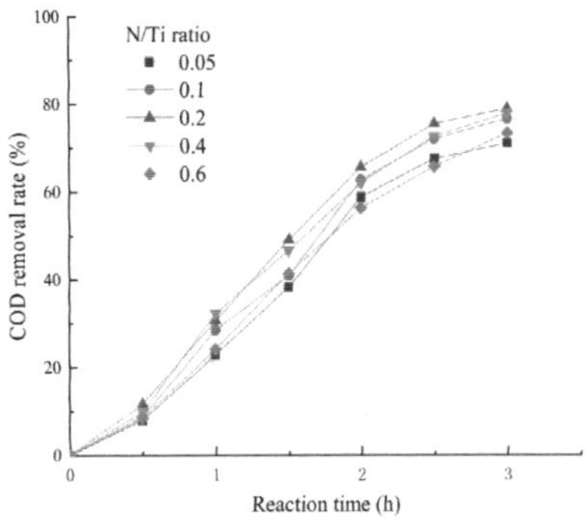

Fig2. Influence of nitrogen content on COD removal in wastewater

The test results show that under the same reaction conditions and within the same reaction time, different doping amounts of N have little effect on the catalytic removal of wastewater.When the ratio of n(N):n(Ti) is 0.2, the catalytic performance of N-TiO_2/GF photocatalyst is the best, and the removal rate of nitrobenzene and COD reaches the maximum of 90.6% and 78.9%, respectively.

3.2 The influence of iron content on the treatment effect

The benchmark conditions of the photocatalysis test remain unchanged. Weigh 0.15g of iron-doped TiO_2 with the ratio of n(Fe):n(Ti) to 0.3%, 0.5%, 0.7%, 0.9%, 1.1%, and calculate the nitro group. The removal rate of benzene and COD, draw a curve, and investigate the influence of different iron doping amounts on the catalytic performance.The test results are shown in Figure 3 and 4:

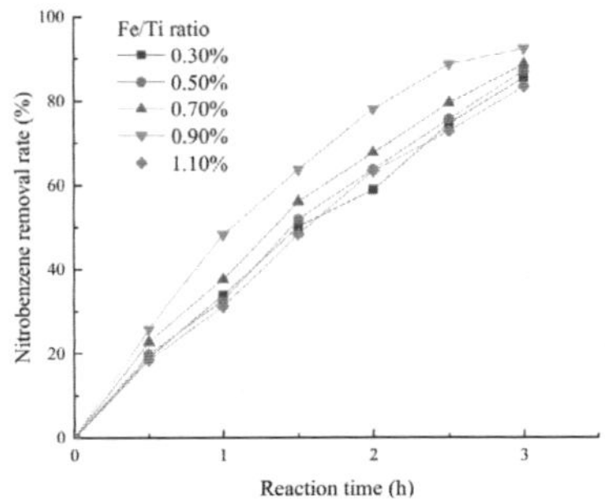

Fig3. Influence of iron content on removal of nitrobenzene in wastewater

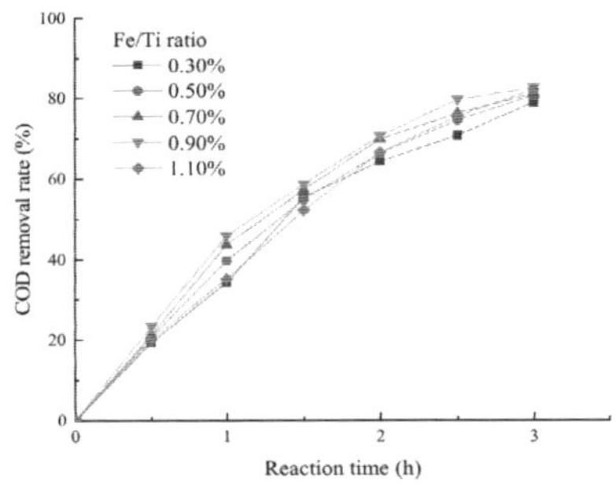

Fig4. Influence of iron content on COD removal in wastewater

The test results show that the catalytic effect of Fe-TiO_2/GF photocatalyst is better than that of N-TiO_2/GF photocatalyst.With the increase of Fe doping concentration, the catalytic removal effect of wastewater first increases and then decreases.When the doping concentration reaches the ratio of n(Fe):n(Ti) to 0.9%, the catalytic effect is the best, and then when the concentration increases again, the catalytic removal effect decreases.

3.3 Orthogonal test

The optimal preparation conditions of Fe and N co-doped TiO_2/GF photocatalyst and its treatment effect on nitrobenzene wastewater were determined by four-factor three-level orthogonal experiment.Orthogonal test design: The experiment selects four influencing factors: Fe doping amount, N doping amount, calcination time and calcination temperature, each of which takes 3 level conditions, and the design adopts $L_9(3^4)$ orthogonal test.

Table1. Orthogonal design table of factor conditions.

Level	Value of each factor			
	Calcination time（h）	Calcination temperature（℃）	$n_{(Fe)}$： $n_{(Ti)}$	$n_{(N)}$： $n_{(Ti)}$
1	2.0	400	0.4	0.25
2	2.5	450	0.5	0.45
3	3.0	500	0.6	0.65

Table2. Orthogonal test results.

Test number	factor				Removal rate（%）
	Calcination time（h）	Calcination temperature（℃）	$n_{(N)}$： $n_{(Ti)}$	$n_{(Fe)}$： $n_{(Ti)}$	
1	2	400	0.1	0.70%	89.8
2	2	450	0.2	0.90%	94.2
3	2	500	0.4	1.10%	79.7
4	2.5	400	0.2	1.10%	81.2
5	2.5	450	0.4	0.70%	97.5
6	2.5	500	0.1	0.90%	84.3
7	3	400	0.4	0.90%	80.8
8	3	450	0.1	1.10%	91.4
9	3	500	0.2	0.70%	86.5

Table3. Data processing results table.

factor	Calcination time（h）	Calcination temperature（℃）	Fe doping ratio（%）	N doping ratio（%）
K1	266.7	251.8	265.6	270.8
K2	260	283.1	264.9	262.3
K3	258.7	250.5	255	252.3
k1	88.9	83.9	88.5	90.26
k2	86.67	93.37	88.3	87.43
k3	86.23	83.5	85	84.1
R	1.67	9.73	3.5	6.16
best combination	Calcination time2.5h，Calcination temperature450℃，Fe doping amount n (Fe): n (Ti) ratio is 0.7% (mole fraction) and N doping amount n (N): n (Ti) ratio is 0.4 (mole fraction)			

It can be seen from the orthogonal test that the doping ratio of iron and nitrogen has different effects on the performance of the photocatalyst.The order of the influence of each influencing factor on the photocatalytic performance of the catalyst is: calcination temperature>Fe doping amount>N doping amount>calcination time.There are optimal preparation conditions for each influencing factor: calcination time 2h, calcination temperature 450℃, Fe doping amount n(Fe): n(Ti) ratio is 0.7% (molar fraction) and N doping amount n(N) : The ratio of n(Ti) is 0.4 (molar fraction), and the removal rate of nitrobenzene wastewater reaches 97.5%.

4 In conclusion

（1）Under the same reaction conditions and the same reaction time, different doping amounts of N have little effect on the catalytic removal of wastewater.When the ratio of n(N):n(Ti) is 0.2, the catalytic performance of N-TiO_2/GF photocatalyst is the best, and the removal rate of nitrobenzene and COD reaches the maximum of 90.6% and 78.9%, respectively.

（2）The catalytic effect of Fe-TiO_2/GF photocatalyst is better than that of N-TiO_2/GF photocatalyst.When the doping concentration reaches 0.9% of n(Fe):n(Ti), the catalytic effect is the best.

（3）The doping ratio of iron and nitrogen has different effects on the performance of the photocatalyst.There are optimal preparation conditions for each influencing factor: the calcination time is 2h, the calcination temperature is 450℃, the ratio of Fe doping amount n (Fe): n (Ti) is 0.7% (molar fraction) and N doping amount n (N) ： The ratio of n(Ti) is 0.4 (molar fraction), and the removal rate of nitrobenzene wastewater reaches 97.5%.

Subject number:2018ZX07601001.Integration and Application of Water Pollution Control and Water Environment Management Technology in Liaohe River Basin

References

1. J. Zhu， Y. Xiong ， R. Guo . Research Progress of Modification of Titanium Dioxide Photocatalyst[J].Inorganic Salt Industry,**52**， 23-54 （2020）
2. Y. Zhang ， W. Wu . Preparation of iron-doped modified titanium dioxide and its photocatalytic treatment of wastewater[J].Water treatment technology,**46**， 60-63 （2020）
3. Wen Y, Ding H. Preparation and Photocatalytic Activity of Ag@AgCl Modified natase TiO_2 Nanotubes[J]. Chinese Journal of Catalysis， **32**， 36-45 （2011）
4. Frank SN,Bard AJ.Journal of the American Chmical Society[J].**99**， 303-304 （1977）
5. Y. Huang . Study on Preparation of New Type TiO_2 Composite Photocatalyst and Its Performance in

Treatment of Organic Wastewater[D]. Nanchang Aeronautical University (2012) .

6. X. Shang . Study on the degradation of nitrobenzene by aerobic granular sludge in SBR[D].Nanjing Agricultural College (2016) .
7. P. Yang. Super-gravity enhanced extraction-homogeneous catalytic ozonation to treat high-concentration nitrobenzene wastewater[D].North University of China (2018) .

Analysis of Air Pollutant concentration change trend in Urumqi city under the condition of city closed in winter

Zhang Haizhen[1], Wei Jiang[2*],
[1]College of Continuing Education, Chongqing Vocational Institute of Engineering, Chongqing 402260, China;
[2]Tourism School of Xinjiang University of Finance and Economics,830021,China

Abstract. During the epidemic period，Urumqi has been sealed off from the city's management, just as"suspended" state.From an environmental point of view, the reduction of energy consumption during the closure of the city can be considered as an energy control to study the resulting reduction of atmospheric pollutant concentration changes.In this paper, the monitoring data of air pollutant concentration in the same period of city closure and normal years are compared, and the results show that the air pollutant concentration has decreased in different degrees during the period of city closure.The largest decrease was-44.66% for NO_2,-40.13% for CO,-36.44% for $PM_{2.5}$, and the smallest was-2.06% for SO_2.Multivariate analysis of variance showed that energy control had a significant effect on the concentration of pollutants during the city closure, for example NO_2（F=128.96，Sig.=0.000)，PM10（F=29.58，Sig=0.000）,$PM_{2.5}$（F=13.98，Sig.=0.000），CO(F=46.34；Sig.=0.000）. Through the analysis of the data, it can be concluded that the air quality of Urumqi in winter is poor and the concentration of pollutants is high. The energy control during the closing period played a positive role in pollutant emission reduction and effectively improved the quality of atmospheric environment.

1 Introduction

The formation and development of a city is a process in which all kinds of factors, such as human flow and material flow, gather together in a certain region, and it is also a dense area of human life and production.When the concentration of pollutants discharged into the atmosphere in the process of human life and production exceeds the threshold of the atmospheric environmental capacity in the region, the problem of atmospheric environmental pollution arises.This has become a common problem in many large and medium-sized cities, especially in developing countries with relatively backward economies. The status of air pollution is usually closely related to the city's geographical location, the city's development stage and the structure of energy consumption[1].

Urumqi is the capital of Xinjiang,it is also the Center of the political, economic and Cultural of Xinjiang. In the course of urban development, it has also experienced serious atmospheric environmental problems, which are mainly related to the typical "valley-type" geographical position and energy structure[2].From the point of view of pollution status, the concentration of air pollutants in the city presents a "U"-shaped change trend, that is, the concentration of pollutants in the heating period is much higher than that in the non-heating period, and the difference of pollutant content in different regions shows that there are some differences of pollution sources in different regions, the quality of surrounding environment is related to industrial production and transportation, while the urban area is mainly related to motor vehicle exhaust and social life[3].How much does the motor vehicle exhaust and social life in the city center have to do with air pollution? This is a question that environmental researchers have been thinking about, but it seems to be an incalculable problem, and social management during an epidemic provides an opportunity to study it.

From January 29 to March 4,2020, the city of Urumqi was closed down due to the outbreak of the new crown disease. Other social production and living activities were "suspended" except those related to the national economy and the people's livelihood.This is an extremely disadvantageous state for economic and social development, but it also provides a rare opportunity for environmental research.In this paper, we compare the monitoring data of air pollutant concentration in the same period of epidemic time (energy control) and in the normal state, and discuss the air pollutant that the city life releases to the air environment.

2 Data acquisition and processing

In this paper, the daily average data of air pollutants in the same period in 2019 and 2020 from state-controlled environmental quality monitoring sites in Urumqi are selected as the analysis objects.Air Pollutants include (PM_{10}, $PM_{2.5}$, SO_2, NO_2, CO and O_3-8h) .For ease of

* Corresponding author: 16826657@qq.com

https://doi.org/10.1051/e3sconf/202123701037

analysis, the areas covered by the monitoring points are divided into southern, central and northern urban areas (Figure 1).The samples were collected by Thermo Scientific1405 Particle Monitor. The data is the average value of 0-23 points per day. Using WPS and SPSS19.0 to analyze the data.

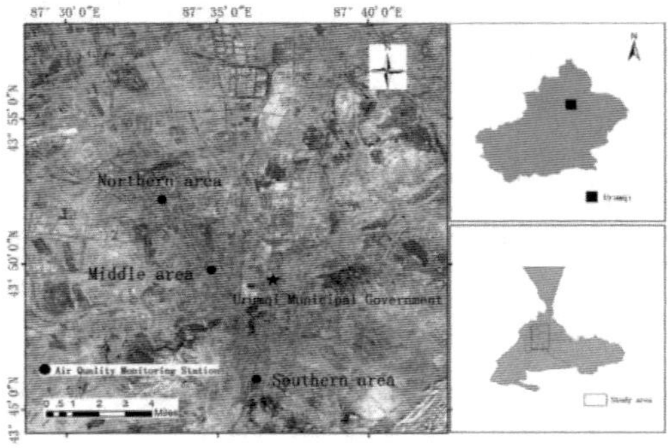

Figure 1 Monitoring point location area

3 Results

3.1 Changes in environmental quality

The statistics show that the environmental quality has improved significantly during the period of the city closure, with the I-class days appearing in different areas, moreover, the number of days of II environmental quality has also increased significantly, with 142% in the south area, 75% in the middle area, and 44.44% at the lowest in the north area. The days of III to VI levels of environmental quality showed a decreasing trend. Specific trends are shown in Table 1.

Table 1 Changes in air quality

Air Quality Rating	2019			2020			Rate of change(%)		
	Southern	Central	Northern	Southern	Central	Northern	Southern	Central	Northern
I				3	5	2			
II	7	8	9	17	14	13	142.86	75.00	44.44
III	12	11	8	8	8	6	-33.33	-27.2	-25.0
IV	5	7	5	4	4	6	-20.00	-42.86	20.00
V	10	7	10	3	3	6	-70.00	-57.14	-40.00
VI	1	2	3		1	2	-100.00	-50.00	-33.3

3.2 Variation of pollutant concentrations in different regions

During the city's closure, all motor vehicles (except special vehicles) , restaurants and small industrial enterprises have been closed, resulting in a reduction in emissions of pollutants (Table2).

The change of pollutant concentration is very obvious, the concentration of PM_{10}, in the particulate matter, showed a trend of low in the middle and high in the southern and north, with an average decrease of about 35% , showing as the northern area>southern>middle area, that is |-35.72% |>|-34.40% | >|-34.33% |.The spatial distribution of $PM_{2.5}$ is the same as that of PM_{10}, but the decreasing range shows the trend from south to north, which is southern>middle area>northern area , that is |-36.44% | > |-28.48% | > |-16.38% | ; Among the gaseous pollutants, the concentration of NO2 decreased the most, and also showed a trend of decreasing from south to north, that is |-44.66%|>|-37.90%|>|-33.69%|; The trend of CO change was basically the same as that of NO_2, but the decrease of CO in the southern region was larger than that in the middle and north regions, that is |-40.13% | >|-20.31% | > |-20.69% |. SO_2 decreases differently in different regions, but the overall magnitude is not large, ranging from-15% to-25% . However, O_3-8H is different from other pollutants, but shows an increase trend, the increase range is relatively large, all in 50% , from the spatial variation, the Southern >the Northern >the Central Zone, namely 81.64% >52.91%> 51.93% .

Table2variation of pollutant concentration in different regions

Unit:μg/m³

Pollutants	Years	Southern	Central	Northern
PM_{10}	2019	127.51	125.54	134.23
	2020	83.65	82.44	86.29
	(%)	-34.40%	-34.33%	-35.72%
$PM_{2.5}$	2019	124.20	117.37	134.29
	2020	78.94	83.94	112.29
	(%)	-36.44%	-28.48%	-16.38%
SO_2	2019	6.37	10.94	11.31
	2020	4.83	8.26	9.57
	(%)	-24.22%	-24.54%	-15.40%
NO_2	2019	80.54	84.06	71.40
	2020	44.57	52.20	47.34
	(%)	-44.66%	-37.90%	-33.69%
CO	2019	2.29	2.01	1.82
	2020	1.37	1.60	1.45
	(%)	-40.13%	-20.31%	-20.69%
O_3-8H	2019	36.26	48.14	50.06
	2020	65.86	73.14	76.54
	(%)	81.64%	51.93%	52.91%

3.3 Results of multivariate analysis of variance

The statistical data show that the concentration of pollutants mainly showed a downward trend during the period of closure of the city, the environmental quality improved obviously, and the rate of decrease of the concentration of pollutants in different regions was different. Multiple factor difference test results, as shown in Table3, the most significant impact of energy control was NO_2(F = 128.96, Sig.= 0.000) , followed by CO (F = 46.34; Sig. = 0.000) , PM_{10}(F = 29.58, Sig. = 0.000) , $PM_{2.5}$(F = 13.98.65, Sig. =0.000) , SO2(F =

48.38, Sig. = 0.000) and O_3-8H (F = 31.52,Sig. =0. 000) were the significant differences among the regions.

Table 3 Multivariate analysis of variance

Tests of Between-Subjects Effects										
Dependent Variable: PM_{10}						Dependent Variable: $PM_{2.5}$				
Source	Type III Sum of Squares	df	Mean Square	F	Sig.	Type III Sum of Squares	df	Mean Square	F	Sig.
Corrected Model	115984.952a	5	23196.99	6.03	0	86838.667a	5	17367.733	4.10	0.001
Intercept	2351510.876	1	2351510.88	611.3	0	2472389.505	1	2472389.505	584.27	0
Energy control	113774.019	1	113774.02	29.577	0	59136.076	1	59136.076	13.98	0
areas	2154.581	2	1077.29	0.28	0.756	22969.295	2	11484.648	2.71	0.069
Energy control * areas	56.352	2	28.18	0.007	0.993	4733.295	2	2366.648	0.56	0.572
Error	784734.171	204	3846.74			863239.829	204	4231.568		
Total	3252230	210				3422468	210			
Corrected Total	900719.124	209				950078.495	209			
a.R Squared = .122 (Adjusted R Squared = .100)						a.R Squared = .091 (Adjusted R Squared = .069)				
Dependent Variable: SO_2						Dependent Variable: NO_2				
Corrected Model	1158.195a	5	231.639	23.916	0	53282.895a	5	10656.579	27.90	0
Intercept	15342.976	1	15342.976	1584.129	0	842840.076	1	842840.076	2206.98	0
Energy control	2.1	1	2.1	0.217	0.642	49250.743	1	49250.743	128.96	0
areas	937.152	2	468.576	48.379	0	2750.467	2	1375.233	3.60	0.029
Energy control * areas	218.943	2	109.471	11.303	0	1281.686	2	640.843	1.68	0.189
Error	1975.829	204	9.685			77907.029	204	381.897		
Total	18477	210				974030	210			
Corrected Total	3134.024	209				131189.924	209			
a.R Squared = .37 (Adjusted R Squared = .345)						a.R Squared = .406 (Adjusted R Squared = .392)				
Dependent Variable:CO						Dependent Variable: O_3-8H				
Corrected Model	21.705a	5	4.341	11.89	0	68923.498a	5	13784.7	21.10	0
Intercept	647.683	1	647.683	1774.19	0	642528.052	1	642528.052	983.62	0
Energy control	16.915	1	16.915	46.34	0	15180.36	1	15180.36	23.24	0
areas	1.574	2	0.787	2.16	0.118	41181.349	2	20590.674	31.52	0
Energy control * areas	3.216	2	1.608	4.40	0.013	11300.933	2	5650.466	8.65	0
Error	74.472	204	0.365			156121.187	239	653.227		
Total	743.86	210				842344.05	245			
Corrected Total	96.177	209				225044.685	244			
a.R Squared = .226 (Adjusted R Squared = .207)						a.R Squared = .306 (Adjusted R Squared = .292)				

4 Analysis and Discussion

4.1 The change of PM_{10} and $PM_{2.5}$

In winter, the air pollution in Urumqi is serious, mainly because the heating consumes a lot of fossil energy, at the same time, the motor vehicle hot-running time is long, the driving speed is slow, which causes a lot of tail gas emissions.Under normal conditions, the concentration of particulate pollutants appears to be high in the north and south area, and low in the middle area, which is directly related to the surrounding environment of the monitoring point. The North and south points are at both ends of the city, and on the one hand they are easily affected by the urban heat island effect, on the other hand, the main factors that affect the concentration of pollutants are the exposed ground surface in the suburbs and the dust produced by vehicles entering the city[4].

During the period of the city closure, PM_{10} decreased considerably. From the point of view of the extent of the decrease, namely the north >the south>the central, mainly after the closure of the city, the most important change was the total suspension of motor vehicles (except working vehicles and special vehicles) , therefore, the dust generated by motor vehicle operation has been contained and reduced to the lowest point, and the large drop in the northern area is due to its high concentration in the early stage, and the concentration of PM_{10} in the three monitoring sites after the closure of the city is very close, you could say it represents the background value of the city in winter.

The concentration of $PM_{2.5}$ decreased obviously, but the average decrease was smaller than that of PM10. The main reason was that the ambient concentration of PM2.5 was almost the same as that of PM_{10} under normal conditions. Although the emission of motor vehicle exhaust was greatly reduced after the closure of the city, it might be affected by the inversion layer in winter, the ability of reducing pollutants is still limited, and the trend of $PM_{2.5}$ is the same as that of PM_{10}, which is consistent with the conclusion of the former study[5].

The results of variance analysis also showed that city closure had significant effect on controlling PM_{10} and $PM_{2.5}$, and the effect on PM_{10} was greater than that on $PM_{2.5}$.

4.2 The change of SO_2 and NO_2.

The decrease rate of SO_2 is obviously lower than that of NO_2, which is mainly due to the adjustment of energy structure such as "coal to gas" , the production of SO_2 emissions is almost reduced to a minimum.

The concentration of NO_2, the main pollutant of motor vehicle emission, has the biggest reduction. Under normal conditions, vehicle exhaust is the main pollution source of NO_2 in urban area. After the closure of the city, motor vehicle exhaust emissions will almost reach zero, and the concentration of pollutants will also decrease, with the largest decrease in the southern area, mainly due to the complete shutdown of the passenger stations near the monitoring points, combined with a sharp decline in the number of vehicles on the nearby city's high-speed roads, the central district followed, with the northern district having the lowest rate of decline, also reaching-33.69% . The results of variance analysis showed that the effect of urban closure on NO_2 was the most significant, which indicated that the concentration of NO_2 in Urumqi urban area in winter was about 40% .

4.3The change of CO and O_3-8H

During the period of city closure, the CO reduction from south to North showed a decreasing trend. From the concentration point of view, the concentration in the south and the north of city was close to each other. But from the pollutant concentration point of view, the concentration value itself was low, and it was difficult to have a higher reduction.The analysis of variance showed that the effect of city closure on CO reduction was significant.O_3-8H is the main air pollutant, and it is also the only one whose concentration increased after the closure of the city, with an average increase of over 50% .It can be concluded that the atmospheric environmental quality is improved, the atmospheric radiation is enhanced and the concentration of O_3-8H is increased due to the reduction of the overall pollutant emission.This is basically consistent with the conclusion that particles in the atmosphere during the heating period help to increase the diffraction and scattering of light.

5 Conclusion

The air pollution of Urumqi is serious in winter, and the concentration of pollutants is high.After the closure of the city, the concentration of pollutants showed a decreasing trend.

On the premise of constant consumption of conventional energy sources such as industrial production and heating, the concentration of pollutants caused by motor vehicles has dropped remarkably, it is shown that proper traffic diversion and control of the number of motor vehicles are effective means to improve the atmospheric environment quality.

The change range of different pollutants in the atmospheric environment is quite different from that of energy control and regional types. For example, the reduction of motor vehicles has the biggest effect on NO_2 and the second effect on the concentration of atmospheric particulate matter, SO_2 and O_3-8H require regional energy regulation.

City closed is an effective means to control the spread of the epidemic, which is extremely disadvantageous to economic and social development, but it is a rare opportunity to study the state of air pollution. However, due to the lack of previous experimental preparation and experience, there are some scientific problems, didn't take advantage of the opportunity to figure it out.

References

1. Noel de Nevers.Air Pollution Control Engineering(second edition) [M]. McGraw-Hill.（2000）.
2. Wei jiang, Wangguohua, Renquan etc al.Study on Quantitative Model of Concentration of Air Pollutants over Urumqi[J]. ARID ZONE RESEARCH,28,5(2011).
3. Chen xuegang, Wei Jiang, Ren quan,etc al. Effects of the urban expansion on the spatially varying trends of distribution in air pollutants concentration: a case study of Urumqi[J].ecology and environmental sciences,22,6(2013).
4. Juan Li, Guoshun Zhuang, Kan Huangect. Characteristics and sources of air-borne particulate in Urumqi,China, the upstream area of Asia dust [J]. Atmospheric Environment, 42（2008）.
5. Binggan Wei, Fengqing Jiang , Xuemei Li.Spatial distribution and contamination assessment of heavy metals in urban road dusts from Urumqi, NW China[J]. Microchemical Journal, (2009).

Study on the Design of Avocado-fiber Skincare Knitted Underwear

Wang Shi*, Li Junjia and Zhao Fu

Dalian Polytechnic University

Abstract: With the continuous changes taken place in people's lifestyles, functional underwear has drawn attention from more consumer groups. Skincare, comfortable knitted underwear fabrics can be made by adding refined avocado oil into viscose to yield fibers and then blending it with other fibers such as acrylic and wormwood fiber. This article first introduces the moisture-retaining principles, raw material selection, and skincare property of avocado-fiber knitted fabric, and discusses the comfortability of the fabric, the requirements of underwear fitness, the style design optimization, and the functional area partitioning, aiming to provide design ideas for the development of the new skincare functional knitted underwear.

1. Background

With the rise of living standards, consumers are seeing a diversified demand for clothing and become increasingly demanding of functional fibers. Over the past few years, the beauty industry has developed rapidly, followed by the emergence of skincare fibers. By adding additives with cosmetic and skincare ingredients into fibers, researchers develop skincare fabrics and achieve the purpose of fulfilling both the wearability and beauty care results in fabrics. Due to its rich nutrients, avocado has been gradually gaining favor and popularity among health seekers. The extracted avocado oil boasts low acidity, non-irritation, and fine transdermal absorption. Therefore, apart from its edibility, avocado oil can also be used as a natural raw material for advanced skincare products and to make moisturizing skincare fibers that can adjust the skin conditions. Knitted fabric made of such fibers can meet the demand of female consumers for comfortability and skin beauty, and hence enjoys great market potential.

2. The Development of Avocado-fiber Functional Knitted Fabric

Traditional fall/winter underwear is often thermal solely, without taking into account the moisturizing need of skin. The dry and cold environment, especially in north China, can easily generate static electricity from frictions between the skin and clothes and between clothes and result in discomfort in wearing. On the other hand, wrapped in poorly ventilated, water-deficient underwear for a long while, human body will suffer burdens on surface skin and dry skin peeling, and harbor bacteria. So, the moisturizing effect of fall/winter underwear fabric turns out to be a key factor to improve the wearing experience.

2.1. The Moisture-retaining Principles of Avocado Fiber

The main constituent elements of human sebum are fatty acids, one of which is oleic acid. A major cause for dry skin is the poorly-moisturizing stratum corneum and the moisture imbalances. Oleic acid is a great moisturizer that can soften keratin and maintain skin sebum resilience. Avocado oil is rich in oleic acid and palmitoleic acid that can condition the skin environment. It also contains rich anti-aging vitamin E, which can gently protect the skin, make it smoother, and effectively relieve skin dryness in winter.

2.2. Selection of Raw Materials and Blending Composition

The development of fall/winter knitted underwear fabrics should be based on the consideration of good thermal protection, so acrylic fiber is chosen as the main blending material. Acrylic fiber, also known as "artificial wool", shares similar properties with wool, but its warmth retention, elasticity and strength are all better than wool, and it is also quite soft. Second, the fabric should be somewhat antibacterial, so wormwood fiber is added. Wormwood fiber, made by adding nano-sized wormwood leaf powder into viscose fiber, is antibacterial and deodorant, and can improve the skin environment and promote the skin surface health. The third fiber is avocado fiber, which can provide important moisturizing and skincare effects. Finally, a small amount of modal and spandex blends is added to increase moisture absorption and elasticity of the fabric and improve its comfortability and durability. The specific composition

*wangshi@dlpu.edu.cn

is: acrylic fiber 50.6%, wormwood fiber 18.4%, avocado fiber 13.8%, modal 9.2%, and spandex 8%; and the fabric weight is 220g/m2.

2.3. Fabric Performance

The fabric is basically weaved with flat weft-knitted, single-sided stitches by circular knitting machines. Besides its long-lasting moisture retention and smooth skin-friendly feel, it also boasts great moisture absorption and breathability, with a moisture absorption rate that is 1.7 times that of ordinary like products. It is also thermal and antimicrobial (see Table 1). With such an excellent comprehensive performance, this fabric is ideally suited for the development of fall/winter underwear.

Table 1. Test Results of The Fabric's Moisture Absorption Rate & Comfortability

Condition	Time/h	Moisture Absorption Rate(%)	
		Original Sample	Contrast Sample
40°C ×90%RH	0	0	0
	1	7.4	4.4
	3	9.9	5.7
	5	10.6	6.1
	7	10.9	6.4
	24	11.4	6.6
20°C ×65%RH	25	10.2	5.7
	27	8.8	5.4
	29	8.5	5.1
	31	8.4	4.9
	48	8	4.5

No.	Test Item	Standard	Test Result	Conclusion
		$220g/m^2$		
1	Pilling (Level)	≥3	3-4	Qualified
2	Surface Charge Density ($\mu C/m^2$)	≤7	3.2	Qualified
3	Moisture Absorption and Elevated Temperature in 30 mins (°C)	Maximum Elevated Temperature ≥4	7	Qualified
4		Average Elevated Temperature ≥3	2.5	——
5	Moisture Retention Rate	≥30	28	——

3. Analysis of Design Factors of Skincare Functional Underwear

In developing functional clothing, it is necessary to position the product accurately, that is, which consumer group the product is targeted at and what functional and performance requirements it should meet. The development of functional clothing should adapt to the current economy and people's demand for material culture. Different consumer groups have diverse ideas on consumption and functionality.

3.1. Target Customer

Avocado-fiber functional knitted underwear features great care for the skin, which coincides naturally with females' consumption preferences. With the improvement of living standards, women attach importance to not only facial care but also body care. Body lotions containing whitening and scrubbing agents have become a must-have for most women, revealing females' enthusiasm for body beauty. Therefore, it's very important to fully display the product's functionality. On the other hand, female consumers who prefer skincare functional knitted underwear are not only concerned about the product quality, but also tend to set higher expectations for beauty, which not only entails the exterior design of product but also its skincare effect.

3.2. Fabric Comfortability

Underwear is the clothing that directly touches human skin, so its comfortability turns out to be fairly important. The knitted fabric is a suitably resilient material that can adapt to the flexing, bending and twisting of human body. Its fine breathability can also keep fluffy and soft even when the skin is well moisturized.

3.3. Fitness

The knitted underwear fabric developed based on avocado fiber features skin moisturizing, so the product should fit the skin as much as possible to give full play to the fabric's moisturizing effect. Female body curves are mainly shown in their breast, waist and hips and the undulation of muscles. Traditional knitted underwear rarely has split lines, so it is quite loose and fails to fit the body well, especially under the breast and arms, and at wrists and legs (see Figure 1). Skintight sportswear usually fits the body ideally (see Figure 2). The fitness of knitted underwear developed based on the moisture retention of avocado fiber should lie somewhere between traditional knitted underwear and skintight sportswear, so that it can serve better moisture retention and skincare.

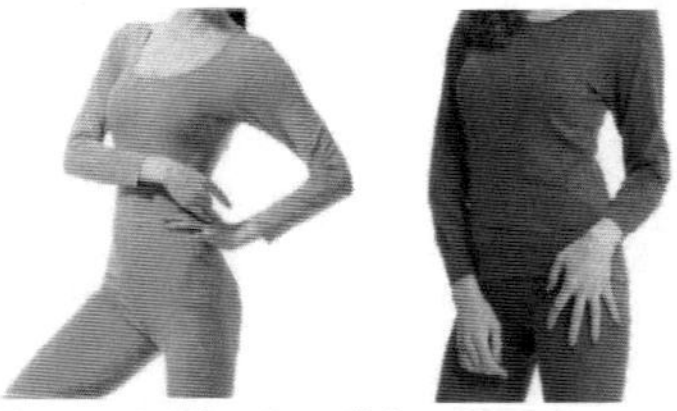

Figure 1 Fit of traditional underwear

Figure 2 Ideal fit of functional underwear

3.4. Factors of Style Design

Ordinary knitted underwear on the market is often linear and lacks a targeted body-fitting structure. It is plane while draped for hanging, so both its wearing effect and fitness are far from satisfactory. To improve moisture retention, the product should comprise targeted splits and splices at the breast, waist, hips and the undulating parts of muscles to showcase the overall curve shape and maintain its modelling when it is not worn, so that it can fit the female body better during wear.

For example, as shown in Figure 3, Style A contains an arc-shaped plane splits at breast and hips to highlight its solid structure. Its V-shaped split line on the abdomen not only exhibits a balance of the body volume, but also visually makes the waist look slenderer. Furthermore, the split lines on the back and limbs also fit well with the muscle undulations to enhance the wearing comfortability and visually lift the body up. In Style B, proper pleats are added on the breast to highlight its modeling and well fit the body at the lower part of breast; the split lines at waist and back are also functional and nice-looking; and for the underpants, the small pieces of splices at the raised parts on the legs have further enhanced the fitness.

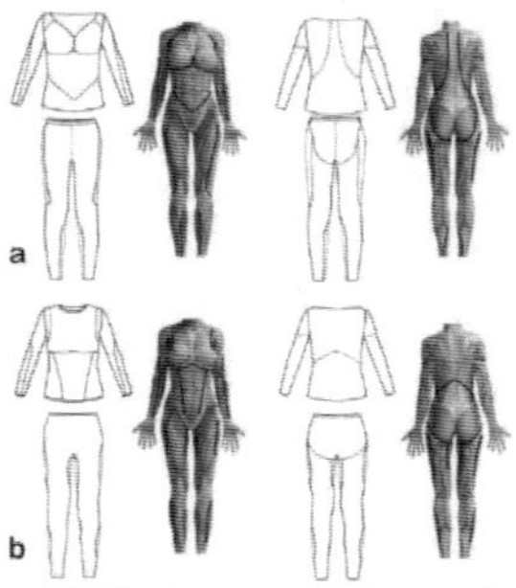

Figure 3 Underwear design reference

3.5. Functional Area Partitioning

The development of functional underwear should be based on reasonable structural division and design targeted at different parts' functional requirements. For example, for underarm areas, moisture retention is not needed, but smell-removing and germicidal functional fabrics are much preferred. Second, the warm-keeping demand of the waist and abdomen. Both men and woman are likely to get a chill on these areas, so the thermal effect should be accordingly strengthened, and pile stitches with better heat preservation should be selected. In addition, considering the functionality of body beauty, fabrics on the inner thigh and the outer calf should be thicker to put a certain pressure on the skin, tighten the figure, and thus achieve body-shaping effect.

4. Conclusion

What counts most in the design of avocado-fiber knitted underwear is to realize its moisturizing function, wearing comfortability and overall aesthetics. The factor that influences the moisturizing effect is the underwear's fitness with the skin. To improve such fitness, it's necessary to analyze thoroughly the relationship between clothing and human body and design "people-oriented" products. Innovative design of style and pattern is an important solution. On the other hand, comfortability is an essential requirement of underwear. The selection of fabrics and the partitioning of human body's functional areas are major factors to improve comfortability. Ultimately, with functionality and comfortability fulfilled, aesthetics should also be integrated into product design to meet female consumers' propensity to beautify themselves and cater to the market demand.

References

1. Lv Ning, Gao Wenjuan. Research Progress of New Skin Care Fiber[J]. Melliand China, 2016, 44(05): 16-18 & 20.
2. Matcha, Mustard. A Popular Food-grade Cosmetic Raw Material: The Application of Shea Butter in Cosmetics[J]. China Cosmetics Review, 2018 (12): 56-59.
3. Wang Yuhong, Wang Sujuan, Zhao Lina, & Liu Xingkun. Research and Development of Functional Clothing[J]. Shandong Textile Economy, 2020 (02): 10-12.
4. Zhang Yingli. The Clothing for Cosmetic Mask A New Experience of Skin Care[J]. Progress in Textile Science & Technology, 2010 (04): 4-5 & 9.
5. Zhou Yi, Chen Fanliang. The Design Ethics and Development Trends of Female Underwear[J]. West Leather, 2020, 42 (01): 82.
6. Li Yuanyuan. Aesthetic Culture Research on the Current Female Body Consumption in China[D]. Northwest University for Nationalities, 2019.
7. Hu Jun. Children's Smart Watch is not Omnipotent for Anti-Abduction [N]. China Consumer Journal, 2015-07-30 (001).
8. Bai Rui, Chen Linyun, & Wu Yufeng, & Huang Xiaoyun. Development of Camellia Oil Fiber Knitted Functional Underwear Fabric[J]. Knitting Industries, 2020 (04): 5-8.

https://doi.org/10.1051/e3sconf/202123701039

Green Development Assessment of a Stainless-steel Industrial Park Based on Material Flow Analysis (MFA)

*Zhijie Jia, Mengfan He, Junjie Mou, Geng Ge, Quanhui Yang, Wei Liu**

College of resources and environment, Chengdu University of Information Technology, 610225, China

Abstract. Based on the material flow analysis method, the green development index system of industrial park was constructed. By using the PSR model, 13 indexes were divided into pressure layer, state layer and response layer, the indexes were standardized, weight distribution and synthesis. The green development level of industrial park was divided into 5 grades combined with the pressure state response factors. The results show that the energy consumption per unit industrial output of a stainless-steel industrial park is 0.7116 tons of standard coal per million yuan, the water consumption per unit industrial output is 6.1466 cubic meters per million yuan, the wastewater discharge per unit industrial added value is 3.12 tons per million yuan, the reduction rate of carbon emissions per unit industrial added value is 19.53%, and the green development index is 0.65, which belongs to the third level of green development. The problems are reflected in particulate emission, resource recycling, energy efficiency, industrial structure, and park greening. In the future, the stainless-steel industrial park should focus on industrial restructuring and transformation, environmental protection infrastructure construction, new policy guidance, and ecological environment construction of the park.

1 Introduction

Industrial parks are the most direct manifestation of mature industrial systems and will be generally established and developed in the process of industrial structure adjustment and upgrading throughout the country. With the upsurge of industrial park construction, if not properly managed, it will be easy to cause serious resource consumption and environmental pollution problems caused by industrial park production activities, and then restrict the future development of industrial park [1-2].

Material flow analysis is an effective method to study the industrial metabolism of specific substances in a country or region [3]. According to the principle of industrial ecology, through the inter-enterprise material, energy and information integration, the formation of inter-industry metabolism and symbiosis coupling relationship, improve the output rate and ecological level of industrial parks are improved, and the overall competitive advantage and sustainable development ability of industrial parks are enhanced [4-5].

The core of green development of industrial park is the construction of ecological industry, and ultimately win-win of ecological environment and economy. The green development level of industrial park is to measure the impact of economic development on resource consumption and environment under the comprehensive consideration of economic development, resource utilization and ecological environment protection. Through the establishment of industrial park green development level evaluation index system, reflect the influence of industrial park on regional environmental quality and improvement potential [4-5].

The material flow of industrial parks is analysed to understand the flow of resources and energy in industrial parks and the emission of pollutants in production activities. Based on the material flow analysis, the evaluation index system of green development of industrial parks are constructed. The index system is used to systematically evaluate the industrial parks, and the ways to save resources and energy and improve environmental quality are obtained. Finally, the transformation of industrial parks to green sustainable development is achieved.

2 Research method

2.1 Material Flow Analysis of Industrial Park

MFA of industrial park is to describe the source, confluence and path of material flow in specific industrial park within a given time span, revealing the correlation characteristics and laws between economic development of industrial park and resource input and pollution output, so as to serve the circular transformation and ecological transformation and upgrading of industrial park. The fundamental frame of MFA about the industrial park is made up of three blocks which include material input, park system and material output[4-5]. MFA indicators generally include system

* Wei Liu: weling9@163.com

 https://doi.org/10.1051/e3sconf/202123701039

scale, structure, level and the indicators of efficiency, such as resource output rate, pollution emission intensity, direct material input, direct environmental emission and material stock changes.

2.2 Model

The model of Pressure-State-Response (PSR) uses the logical thinking of "Pressure-State-Response" so as to reflect the interaction between human beings and the environment[6-7]. See Figure 1 for the PSR model framework of green development of industrial park.

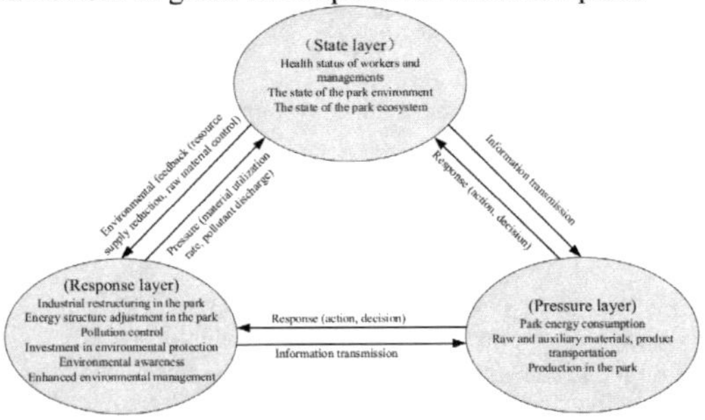

Fig. 1. The PSR model framework of green development of industrial park.

2.3 Assessment of green development level of industrial park

2.3.1 Evaluation index system for green development of industrial park

The construction process of evaluation indexing system for green development of industrial parks includes the steps: to build the evaluation indexing system; analyze and determine the connotation of the industrial park, collect design indicators in terms of economic development, environmental governance, industrial structure, recycling, park management and ecological livability, and select indicators with high frequency. Finally, the evaluation indexing system of green development of industrial parks is determined[8-13].

2.3.2 Evaluation method for green development of industrial parks

Reference value determined. Reference values are determined after considering the following principles, such as: considering the country or industry, the standard shall adopt the specified values in the standard as far as possible; consistent with or better than the target value of relevant policy research; considering the actual development of the park; taking the development planning as the target value; shortage of significant statistical, using expert consultation, qualitative description and other methods to determine.

Standardized treatment. Due to the different properties and dimensions of each index, it cannot be used for direct calculation, so it is necessary to standardize the index, which is equal to relative values can be obtained by comparing the selected reference values and index values. The processing mode is shown in Table 1:

Table 1. Reference value standardization processing method.

<table>
<tr><th colspan="2">Target classification</th><th>Standardized results</th><th>Remarks</th></tr>
<tr><td rowspan="3">An index which reference value is superior to the index value</td><td>The larger the index data, the better:</td><td>$C_i = 1 - \frac{(S_i - S)}{(S_i - S_{min}{}^1)}$</td><td rowspan="3">In the formula, C_i is the standardized processing value of the index, S_i is the reference value of the index, S is the value of the index, S_{max} is the maximum value of the index, and S_{min} is the maximum value of the current status of the index. Among: $S_{max}{}^1 = S_{max} * 1.05$ $S_{min}{}^1 = S_{min}/1.05$</td></tr>
<tr><td>The smaller the indicator data, the better:</td><td>$C_i = 1 - \frac{(S - S_i)}{(S_{max}{}^1 - S_i)}$</td></tr>
<tr><td>For other indicators:</td><td>$C_i = 1 - \frac{\lvert S_i - \rvert}{(S_{max}{}^1 - S_{min}{}^1)}$</td></tr>
<tr><td colspan="2">An indicator whose value is superior to a reference value</td><td>The standardized treatment value is 1</td><td>/</td></tr>
</table>

Index weight is determined. In this study, analytic hierarchical process (AHP) is used to assess the indexes of green developing level of industrial parks to determine the weight. Build the hierarchical structure about the evaluation of green development level in industrial parks the target layer is green development level of industrial parks, the criterion layer is pressure, state and response factors, and the project hierarchy is various indexes. The relative importance of each layer is obtained by comparing the factors of each layer in pairs with the scale method. If n factors of a certain level are compared, B1, B2,.... The influence of Bn on A factor A in the upper layer is identified by Kij as the ratio of Bi and Bj to the importance of A, and the judgment matrix K is obtained through the pair comparison:

$$K = \begin{bmatrix} k_{11} & k_{12} & \cdots & k_{1n} \\ k_{21} & k_{22} & \cdots & k_{2n} \\ \vdots & \vdots & \ddots & \vdots \\ k_{m1} & k_{m2} & \cdots & k_{mn} \end{bmatrix}$$

Scaling methods 1-9 of 1, 3, 5, 7, 9, respectively two factors are equally important, the former is slightly more important than the latter, the former is more important than the latter obviously, the former is more important than the latter strongly, the former is more important than the latter extreme, 2, 4, 6, 8 identification before the latter than in the middle of the values, the degree of the former than the latter is not important with reciprocal $(K_{ij} = 1/K_{ij})$ said.

By calculating the eigenvalue and vector of the judgment matrix, the priority is completed. The second part concludes that the judgment calculates the maximum eigenvalue and eigenvector X, and then normalizes the eigenvector X. The vector component X obtained is the weight of the factor applied to the upper layer. This research is realized by MATLAB program.

Table 2. The calculation of weight ranking.

Steps	Formulas
Matrix K per row index product T_i	$T_i = \prod_j^n k_{ij}$
The eigenvector is divided into the vector $\overline{X_i}$	$\overline{X_i} = \sqrt[n]{T_i}$
The normalization is going to give me the weight X_i	$X_i = \frac{\overline{X_i}}{\sum_{j=1}^{n} \overline{X_i}}$
The largest eigenvalue λ	$\lambda = \frac{1}{n}\sum_{i=1}^{n}\frac{(KX)_i}{X_i}$

In order to check the consistency of matrix, consistency index CI value and average random consistency index RI value are introduced, as shown in the following table, CI= (λ-n) / (n-1) , and CI is compared with RI, denoted as CR, CR=CI/RI. When CR≤0.1, the judgment matrix has acceptable consistency; otherwise, the judgment matrix needs to be adjusted until the requirements are met.

Table 3. RI value of random consistency index.

n	1	2	3	4	5	6	7	8	9
RI	0	0	0.58	0.90	1.12	1.24	1.32	1.41	1.45

Index value synthesis method. The index synthesis method is shown in Table 4.

Table 4. Index synthesis method.

Steps	Formulas	Remarks
Primary index value	$A_i = \sum_{i=1}^{n} W_i C_i$	In the formula: A_i is the first-level index value, W_i is the weight value corresponding to the second-level index, C_i is the standard value of the index, n is the number of subordinate indicators.
Composite development index	$Z = \sum_{i=1}^{m} X_i A_i$	In the formula: Z is the first-level index value, X_i is the weight value corresponding to the first–level index, A_i is the first-level index value, m is the number of items of the first-level index.

The grade of green development level of industrial parks is divided. The green development level of industrial parks is affected by pressure, state and response factors. Stress index represents the effect of human economic and social activities on the environment, mainly describing the impact and stress brought by human activities on the environment. It can reflect the intensity of resource utilization and its changing trend in a certain period from the aspects of energy, transportation, construction and industry. The status index represents the environmental status and environmental changes at a specific time stage. It reflects the changes of environmental factors, mainly considering the biological, physical and chemical characteristics and ecological functions in the environment or ecosystem, namely land, air, suitability and satisfaction. Response index refers to that when human activities change the natural and ecological environment and bring about adverse impacts on regional environmental quality, the society will respond through environmental, economic and other policies and changes in consciousness and behavior to mitigate or eliminate the possible adverse impacts, such as a legal system, economic means, technological change and international obligations. The green development level of the industrial park is divided into 5 levels according to the response factors of pressuring state.

Table 5. Evaluation grade of green development level of industrial parks.

Grades	Scores	Fundamental states
First stage	0.9~1	**Pressure**: Low carbon emission intensity, lo emission coefficient, low energy consumptic intensity, low energy flow density, very go development of the park
		Status: The air environment quality is very good, the comprehensive utilization degree of pollutants is high, and the public is satisfied with the environment
		Response: The distribution of the three industries is very reasonable, with high proportion of clean energy, high investment in environmental protection, good energy-saving technology and strong awareness of environmental protection
Second stage	0.8~0.9	**Pressure**: Low carbon emission intensity, low emission coefficient, low energy consumption intensity, low energy flow density, and good development of the park
		Status: The air environment quality is good, the comprehensive utilization degree of pollutants is high, and the public is satisfied with the environment
		Response: Rational distribution of the three industries, relatively high proportion of clean energy, high investment in environmental protection, better energy-saving technology, strong awareness of environmental protection
Third stage	0.6~0.8	**Pressure**: moderate carbon emission intensity, moderate emission coefficient, moderate energy consumption intensity, moderate energy flow density, general development of the park
		Status: Good air quality, moderate comprehensive utilization of pollutants, moderate public satisfaction with the environment
		Response: The distribution of the three industries is reasonable, the proportion of clean energy is moderate, the investment in environmental protection is moderate, the technology of energy saving is general, and the awareness of environmental protection of enterprises and citizens is general
Fourth stage	0.4~0.6	**Pressure**: high carbon emission intensity, high emission coefficient, high energy consumption intensity, high energy flow density, and poor development of the park
		Status: Poor air quality, low comprehensive utilization of pollutants, low public satisfaction with the environment

		Response: Unreasonable distribution of the three industries, low proportion of clean energy, low investment in environmental protection, backward energy-saving technology, low awareness of enterprise environmental protection
Fifth stage	0~0.4	**Pressure**: The carbon emission intensity is very high, the emission coefficient is very high, the energy consumption intensity is very high, the energy flow density is very high, and the development of the park is very poor
		Status: The air environment quality is very poor, the comprehensive utilization degree of pollutants is very low, and the public's satisfaction degree to the environment is very low

3 Results and discussion

3.1 Research data collection and processing

Based on the data of a stainless-steel industrial park, this research summarized 37 key enterprises and several other enterprises, with a total annual output value of 13,974.225 million yuan, they are mainly in the stainless-steel, steel, cement, and forging industries. The index data used basically comes from official authoritative data, including the second national pollution source survey, environmental quality bulletin, and water resource bulletin. Most of the evaluation index data can be obtained directly, and some data need to be processed. The preliminary data processing results are shown in Figure 2, the total amount of energy equivalent to standard coal is 994,440 tons, the total amount of water taken is 8,589,460 cubic meters, the total amount of pollutants discharged is 199,220 tons, and the total input material is 18,1733.39 tons.

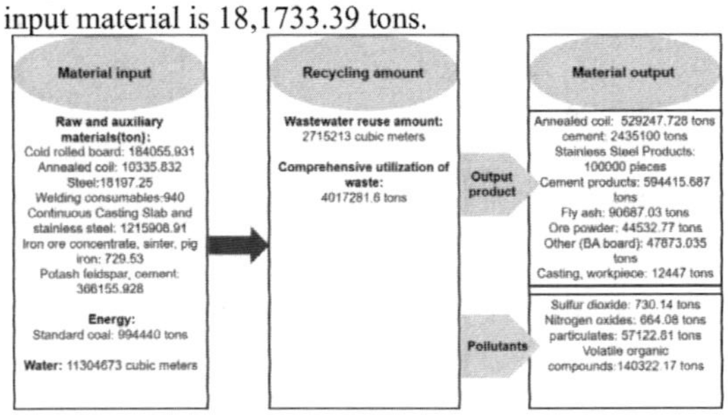

Fig. 2. Panorama of MFA about the industrial park.

3.2 Results of MFA of the industrial park

The evaluation indicators of MFA about the industrial park select resource output rate, pollutant emission intensity, and material input.

Table 6. Evaluation Index of MFA of the Industrial park.

Indicator name	The formula	The formula	unit
Substance input	Direct material input add indirect material input	1817323.39	ton
resource output rate	Annual industrial output value divided by substance input	0.7689	Ten thousand yuan/ton
Energy consumption intensity per unit industrial output value	Total energy consumption divided by annual industrial output value	0.7116	tec/ten thousand yuan
Water consumption intensity per unit industrial output value	Industrial water divided by the annual output value of industry	6.1466	Cubic meter/ten thousand yuan
Pollutant emission intensity per unit industrial output value	Pollutant emissions divided by annual industrial output value	0.1426	Tons/ten thousand yuan

Due to the analysis of the evaluation index results of the MFA of the industrial park, the enterprises in the industrial park have overcapacity, insufficient innovation and development capabilities, unsustainable economic development, and new difficulties in business operations; The industrial park is small in scale, the products are relatively single and the added value is low, the advantages of industrial development are not obvious, and the development experience is insufficient; The ferrous metal smelting and rolling processing industry in the industrial park, as well as the cement industry's resource consumption and pollution reduction, are under pressure; New environmental problems such as heavy metals, persistent organic pollutants, and particulate matter have become increasingly prominent, and pollution reduction and environmental pressure have increased.

3.3 Results of the evaluation of the green development level of the industrial park

Refer to Table 7 for the reference value, current value and corresponding standardized value of the green development level evaluation index of the industrial park.

Table 7. The current value and standardization of the green development level indicators of the industrial park.

Target layer	First level indicator	Serial number	Secondary indicators	Current value	unit	Reference	standardization
Green development	Stress indicator	1	Energy consumption per unit of industrial output value	0.7116	Tons of standard coal/ten thousand yuan	≤0.9 Tons of standard coal/ten thousand yuan	1

level of industrial park		2	Water consumption per unit industrial output value	6.1466	Cubic meter/ten thousand yuan	≤9 Cubic meter/ten thousand yuan	1
		3	Wastewater discharge per unit of industrial added value	3.12	Tons/ten thousand yuan	≤8Tons/ten thousand yuan	1
		4	Reduction rate of carbon emissions per unit of industrial added value	19.53	%	≥22%	0.274
	Status indicator	5	Air quality good rate	67.39	%	Meet the functional area standard	1
		6	Comprehensive utilization rate of industrial solid waste	99.39	%	≥85%	1
		7	Industrial water reuse rate	52.83	%	≥75%	0.176
		8	Centralized treatment rate of domestic sewage	90	%	≥85%	1
		9	Green coverage	23.8	%	≥40%	0.065
		10	Public satisfaction with the environment	80%	%	≥90%	0.571
	Response indicator	11	The proportion of high-tech industries in the industrial output value of the whole district	59.56	%	50%~80%	1
		12	Percentage of the added value of green industry in the industrial added value of the park	6.27	%	≥15%	0.033
		13	Clean energy usage rate	20.37	%	≥50%	0.032
		14	Percentage of green buildings in new industrial buildings	0	%	≥50%	0
		15	Environmental report prepared by the park	1	Period/year	1Period/year	1
		16	Implementation rate of cleaner production audit of enterprises above designated size	100	%	100%	1

The calculation of the weights of the evaluation indicators for the green development level of the industrial park is as follows, where A represents the target level, that is, the green development level of the industrial park.

Table 8. First-level index judgment matrix.

A	**pressure**	**status**	**response**
pressure	1	3	1/2
status	1/3	1	1/5
response	2	5	1

Note: The consistency ratio of the judgment matrix is 0.0032.

The consistency ratio of pressure, state and response index judgment matrix is 0, 0.0562 and 0.0241, respectively.

See Table 9 for the first-level indicator values of the green development index of the industrial park, the green development index of the industrial park is 0.65, and the green development index belongs to the third level.

Table 9. The first-level indicator value of the green development level of the industrial park.

	Stress indicator	**Status indicator**	**Response indicator**
Numerical	0.72775	0.62998	0.61372

Green Industrial park development level at the middle level, there is some discrepancy from the target green industrial park, compared with the PSR model, the main problems restricting the green development of the industrial park are analysed as follows:

The regional ambient air quality exceeds the standard, which is specifically manifested in the excessive inhalable particulate matter. The main reason is that the industrial park is located on the edge of the basin and belongs to the mountainous area of the basin. The landform is mainly mountainous, which is not conducive to the diffusion of pollutants; The second is that the industrial park has two large steel companies and two cement companies, and the total amount of pollutant emissions is large. At the same time, the informative construction of the industrial park is poor, the ecological industrial information platform lacks much information, and the construction of the platform is not perfect. The public cannot have a deep understanding of the process of Ecological degree, greening, and sustainable development of the industrial park. In summary, residents around the industrial park are less satisfied with the environmental construction of the industrial park;

The industrial park is dominated by steel, cement, and forging enterprises. There are very few green industries, resulting in a very low proportion of green industry added value. At the same time, the construction of environmental protection infrastructure in the industrial park is lagging behind and no industrial solid waste disposal site has been built, and solid waste recycling, treatment and disposal are in a blank state; The sewage pipe network is not sound and the sewage collection rate is low, which makes little contribution to sewage treatment. The concept of circular economy in industrial park enterprises is poor, and only large-scale enterprises have carried out technological transformation. Most small and medium-sized enterprises value short-term economic benefits, with low resource recycling rate, low use rate of clean energy, and low investment in environmental protection facilities;

Compared with the construction area, the green area of the industrial park is relatively small, the green coverage rate is low, and there are no green buildings in the newly-built industrial buildings, and the ecological environment of the industrial park is poor.

4 Conclusion

Based on the MFA method, a stainless-steel industrial park output of a stainless-steel industrial park in 2017 were studied. At the same time, combined with the evaluation of the green development level of the industrial park, 17 indexes were established to evaluate the industrial park, and 8 indexes were not up to the standard: reduction of carbon emissions per unit of industrial value added, reuse of industrial water, green coverage, public satisfaction with the environment, the proportion of green industry added value to the industrial added value of the park, the utilization rate of clean energy, the proportion of green buildings in newly built industrial buildings, the perfection of ecological industry information platform.

The energy consumption of the industrial park enterprises in coal, oil and other oil and other resource-based materials, high in proportion to the proportion of iron and steel, cement, casting and other resource-based industries, the industrial structure of the industrial park needs to be optimized, the following recommendations are made:

Industrial category adjustment and distribution: the transformation and upgrading of traditional heavy industries, the development of high-tech green industries and modern service industries, and the active introduction of new materials, new energy, advanced equipment and other "high-tech light and clean" emerging industries. Metallurgical building materials industry upgrade transformation, orderly withdrawal from cement, steel primary, low-end ceramics and other traditional industries;

The industrial park should strengthen the construction of environmental protection infrastructure, especially sewage treatment and solid waste comprehensive utilization and disposal facilities. Environmental access to construction projects also needs to be adjusted accordingly: strict control of high pollution, high energy consumption, high water consumption projects into the park; require existing steel, cement, and other high energy consumption enterprises to implement ultra-low emission transformation and strictly limit production capacity; to improve and optimize the industrial structure of the three types of parks, improve new materials, new energy sources, green building materials, the proportion of modern logistics and other industries; enterprises that do not meet the current national industrial policy and their government environmental access requirements are prohibited from entering the park; new construction, reconstruction and expansion projects must adopt advanced production technology, equipment and pollution control technology, clean production level is not lower than the second level; the existing and new projects in the park strictly implement national and local pollutant emission standards;

The industrial park should adhere to the combination of fast-growing tree species and slow-growing tree species, evergreen trees and deciduous trees, scientific allocation of public green space and greening tree species. To actively build green corridors of industrial parks, with emphasis on strengthening the greening of ecological corridors such as main roads, secondary trunk roads and branch roads;

Through policy guidance, we can arouse the enthusiasm of enterprises in the application of energy saving and emission reduction technology, promote the development of energy saving and environmental protection industry, and promote the protective development of enterprises to the environment. Enterprises are encouraged to carry out scientific technical reform, improve the repeated use of industrial water, increase the recycling of industrial solid waste, and establish the circular economy of the park.

Acknowledgement

This research was supported by the undergraduate innovation and entrepreneurship training program: study on the green development of typical industrial park in Sichuan Province based on MFA (No. S202010621120).

References

1. R.N. Zhao, Z. Ma, Q. Qiao, D.H. Hu, Y. Zhang, M.H. Xie, J. Guo. J. RES. **33 (02)**. 511 (2020)
2. Y.X. Xu, C.H. Cui, K. Chen, Y.D. Tang. J. EP. **47 (21)**. 69 (2019)
3. H. Cheng, X.C. Peng, Z.L. Chen, J.H. Dong. J. ESM. **36 (10)**. 142 (2011)
4. P.H. Gu, Z.C. Liu. J. EG. **37 (04)**. 142 (2017)
5. C.X. Zhang, Y. Wang. J. SM. **06**. 51 (2019)
6. H. Zhao, X. Yan, F.Q. Wang, P.P. Kang. J. WRP. **36 (04)**. 21 (2020)
7. S. Das, B. Pradhan, P.K. Shit, A.M. Alamri. J. S. **12 (15)**. (2020)
8. Y.L. Xing, B. Feng. J. EPS. **45 (06)**. 39 (2019)
9. Y. Zhang, Q. Qiao, Y. Yao, L. Fang, J. Guo, W.N. Bai. J. CPRE. **25 (06)**. 12 (2015)
10. X.M. Sun, Z.J. Cui, L. Zhu, L. Liu. J. CPRE. **20 (01)**. 124 (2010)
11. N.Z. Xu, W.H. Zeng, P.L. Xue, F. Dong, G.M. Zhou. J. CPRE. **20 (03)**. 44 (2010)
12. Q. Qiao. J. JEET. **1 (01)**. 82 (2011)
13. B.J. Lai, J.P. Tian, W. Liu, T. Liu, L.J. Chen. J. AES. **34 (22)**. 6745 (2014)

Effect of temperature on pyrolysis of sewage sludge: biochar properties and environmental risks from heavy metals

Zhipu Wang[1,3]*, *Shun* Liu[1], *Kai* Liu[2]**, *Shibo* Ji[3], *Mingming* Wang[4], and *Xinqian* Shu[3]

[1]State Key Laboratory of Heavy Oil Processing, China University of Petroleum–Beijing at Karamay, 834000 Karamay, China
[2]School of Municipal and Environmental Engineering, Henan University of Urban Construction, 467036 Pingdingshan, China
[3]School of Chemical and Environmental Engineering, China University of Mining & Technology (Beijing), 100083 Beijing, China
[4]PetroChina Xinjiang Oilfield Company, 834000 Karamay, China

Abstract. Biochars were prepared via the pyrolysis of sewage sludge at temperatures ranging from 350–550 °C. The properties and behaviors of heavy metals in the biochars were investigated. The results indicated that the pH values and ash contents of the biochars increased, while biochar yield and C, H, and N contents decreased with the increasing temperature. A high pyrolysis temperature contributed to a developed biochar pore structure. The specific surface area and pore volume of the biochars increased, while the average pore width decreased, with the increasing temperature. Heavy metals in the biochars were further enriched with the increasing temperature. TCLP tests demonstrated that the leaching potential of heavy metals from the biochars significantly decreased with the increasing temperature, indicating the decrease of potential ecological risks of heavy metals to the environment. Additionally, BCR tests confirmed the transformation of heavy metals from mobile fractions (F1 and F2) to stable fractions (F3 and F4). The evaluation results showed that a high pyrolysis temperature can effectively inhibit the ecological risks of heavy metals in the biochars. Thus, the conversion of sewage sludge into biochar via pyrolysis is a promising method for the safe disposal of sewage sludge.

1 Introduction

Sewage sludge is a by-product of municipal wastewater treatment plants. With continuous urbanization worldwide, many municipal sewage treatment plants have been constructed, and the amount of discharged sewage sludge is increasing dramatically. Sewage sludge contains a large amount of heavy metals and pathogenic microbes that can cause serious environmental pollution and even threaten human health once directly discharged into the environment [1]. The proper disposal of sewage sludge has attracted great interest, and the organic composition in sewage sludge is also considered as an important biological resource. Therefore, it is necessary to develop appropriate disposal technology to both reuse resources and safely dispose of sewage sludge.

Pyrolysis may provide a feasible and attractive strategy for sludge recycling [2], since it can reduce sludge volume, kill pathogens and parasites [3], and convert the organic matter in sewage sludge into bio-oil, pyrolysis gas, and biochar [4]. Sludge-derived biochar is a kind of porous material with an abundance of nutrients including N, P, and K [5]. It is also a potential amendment to contaminated soil, and its application to soils may decrease soil bulk density [6], increase soil porosity [7], enhance soil water holding capacity [8-10], increase soil nutrient retention capacity [10], adsorb soil pollutants [11], improve the nutrient supply for plants [12], and enhance plant growth [13].

The characteristics of sludge-derived biochars have an important influence on their improvement to degraded soils. Detailed research is required on the elemental compositions, physicochemical properties, and pore structures of sludge-derived biochars. Moreover, sludge-derived biochars are enriched in heavy metals [14], and the behaviors of those heavy metals require further research to minimize the risks of their release into the environment. Pyrolysis temperature is the key factor that determines the characteristics of biochar [15-16]. Therefore, the aims of this study are (1) to investigate the effect of pyrolysis temperature on the properties of sludge-derived biochar, (2) to explore the effects of pyrolysis temperature on the transformations of heavy metals during the pyrolysis process, and (3) to reduce the potential ecological risks of heavy metals in biochar.

2 Materials and methods

*Corresponding author mail: wzpcumt@163.com
**Corresponding author mail: liukaicumtb@163.com

2.1 Pyrolysis feedstocks

The sewage sludge used in this study was collected from the sludge dehydrating unit of a municipal wastewater treatment plant in Urumqi, Xinjiang Province, China, and was air dried outdoors. The sludge was dried to constant weight at 85 °C in a laboratory oven, ground through a 60-mesh sieve, and packed in sealed plastic bags for further use. The sludge is labeled as SS. The properties of the sludge are listed in Table 1.

2.2 Pyrolysis experiments

The pyrolysis experiments were conducted at a pyrolysis temperature ranging from 350 to 550 °C [17] and a heating rate of 20 °C/min in an electric heating furnace. Figure 1 presents a schematic diagram of the furnace consisting of a nitrogen conveyor, a pyrolysis reactor, and a liquid separator.

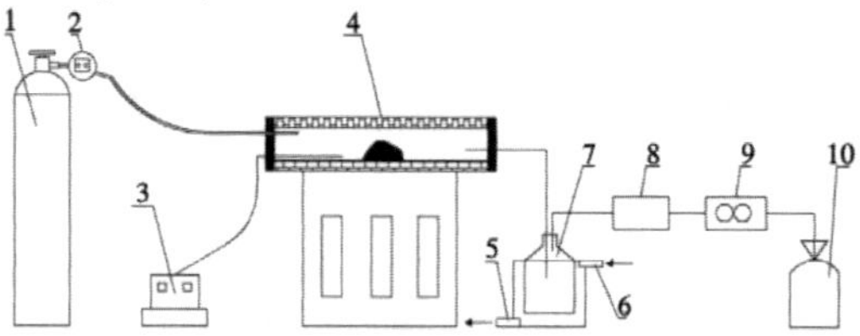

1. nitrogen gas, 2. pressure gage, 3. temperature controller, 4. electrical heating furnace, 5. water outlet tube, 6. water inlet tube, 7. water bath, 8. desiccator, 9. active carbon filter, 10. gas collector.

Fig. 1. Schematic diagram of the pyrolysis furnace.

Before and during the pyrolysis process, a continuous inflow of N_2 at 0.5 L/min was sent to the furnace to achieve an oxygen-free environment. A total of 30 g of dry SS was placed in the quartz tube, fed into the middle portion of the furnace, and heated for 2 h after reaching the target temperature. The solid products (biochars) were cooled to room temperature and weighed, and the biochar yield was determined according to equation (1). The biochar prepared by the pyrolysis of the sludge at X (°C) was labeled as SSBX.

$$Y_b = \frac{M_b}{M_s} \quad (1)$$

where Y_b represents the yield of the biochar, and M_b and M_s represent the dry weights of the biochar and sludge, respectively.

2.3 Characterization of the sludge and biochars

The C, H, and N contents of the sludge and biochars were determined simultaneously by using an elemental analyzer (Flash, Ea11121, USA). The pH values of the samples (sample/water, 1:20, w/v) were measured with a digital pH meter (PHS-3C, China). The BET (Brunauer–Emmett–Teller) surface area, pore volume, and pore size of the samples were determined with an automated surface area and pore size analyzer (ASAP2020; Micrometrics, USA). The ash contents of the samples were measured according to the methods listed in the standard procedures of the Chinese standard GB/T212-2008 [18].

Table 1. General properties of the sludge and biochars.

Samples	Yield (%)	pH	Element contents (%)			H/C	Ash[b] (%)
			C	H	N		
SS	-[a]	7.03 ± 0.05	26.15	3.86	4.57	0.15	46.66 ± 2.15
SSB350	73.33 ± 1.20	7.42 ± 0.03	25.32	2.16	3.15	0.09	55.14 ± 2.39
SSB450	67.13 ± 1.13	8.07 ± 0.02	24.28	1.71	2.98	0.07	64.36 ± 3.11
SSB550	61.28 ± 0.92	8.54 ± 0.03	19.89	1.09	2.35	0.05	72.42 ± 2.84

[a]Not determined.
[b]On a dried base.

2.4 Analytical methods for heavy metals

Before the total contents of heavy metals in the sludge and biochars were tested, these samples were first digested according to Wang et al [19]. The BCR (the European Community Bureau of Reference) sequential extraction method was used to analyze the chemical form distributions of heavy metals in the samples [20]. The TCLP (toxicity characteristic leaching procedure) was used to analyze the leachable toxicities of heavy metals in the samples [21]. The concentrations of heavy metals in the solutions were determined by using inductively coupled plasma optical emission spectrometry (ICP-OES, Optima5300, Perkin Elmer, USA). The ecological risks of heavy metals in the sludge and biochars were assessed according to Wang et al [19].

2.5 Statistical analyses

Statistical analyses were performed by using the SPSS statistical package (v. 19.0, IBM Corp., Armonk, NY, USA). The data were subjected to variance analysis. The means were separated by using the protected least significant difference test at $p < 0.05$.

3 Results and Discussions

3.1 General Properties

The general properties of the samples (the sludge and biochars) are listed in Table 1. The biochar yield decreased from 73.33% to 61.28% with the increase of pyrolysis temperature from 350 to 550 °C. Moreover, compared with the sludge, the contents of C, H, and N of biochar SSB350 obviously decreased after pyrolysis at 350 °C. The decomposition of organic matter into volatile substances during the pyrolysis process was the reason for these decreases [22]. Furthermore, the contents of these elements gradually decreased with the increase of pyrolysis temperature, indicating the increased thermal decomposition of organic matter at higher temperatures. These results were consistent with the research of Jin et al [23]. The H/C atomic ratio of the biochars obviously decreased with the increase of temperature, indicating that an elevated temperature can increase the aromatization degree of biochar [20], [24-25].

The ash content of biochar was positively correlated with pyrolysis temperature. When the pyrolysis temperature increased from 350 to 550 °C, the ash content of the biochar increased from 55.14% to 72.42%. In this study, higher pyrolysis temperatures resulted in higher alkalinity of the biochar. There were two main mechanisms to explain this phenomenon. First, more alkali salts were released from the organic structure at higher temperatures. Second, more organic amines were converted to pyridines with the increase of pyrolysis temperature, resulting in the decrease of acid functional groups on the biochar surface.

3.2 Characterization of the biochars

The pore structure characteristics of the sludge and biochars are exhibited in Table 2. The results suggest that pyrolysis of the sludge at higher temperatures promoted the development of the pore structures of biochars. When the temperature increased from 350 to 550 °C, the specific surface area and total pore volume of the biochars increased from 8.01 m^2/g and 0.041 cm^3/g to 10.23 m^2/g and 0.062 cm^3/g, respectively, while the average pore size decreased from 23.50 nm to 16.40 nm. The volatilization of organic matter in biochar is helpful to the development of pore structure [20].

Table 2. Pore structures of the sludge and biochars.

Pore structure	Samples			
	SS	SSB350	SSB450	SSB550
Surface area (m^2/g)	4.35	8.01	8.78	10.23
Total pore volume (cm^3/g)	0.013	0.041	0.049	0.062
Average pore width (nm)	55.41	23.50	22.10	16.40

In this study, the biochars prepared at different temperatures were mesoporous materials. As shown in Figure 2, the N_2 adsorption isotherms of these biochars corresponded to similar type IV adsorption isotherms according to IUPAC (International Union of Pure and Applied Chemistry) classification. As shown in Figure 3, the pore size distributions of the biochars were similar, and had similar peak horizontal positions.

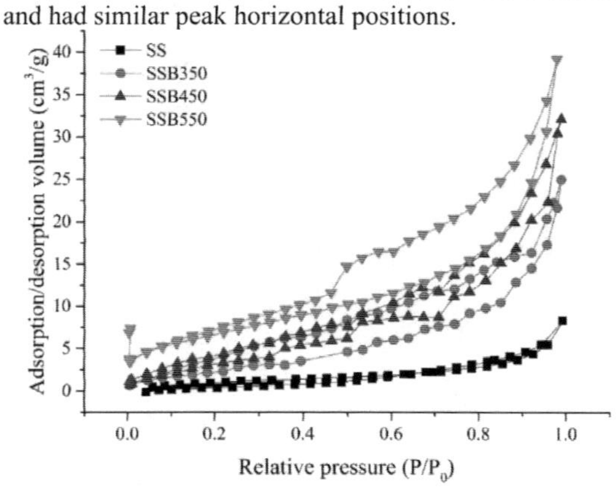

Fig. 2. Nitrogen adsorption-desorption isotherms of biochars obtained at different pyrolysis temperatures.

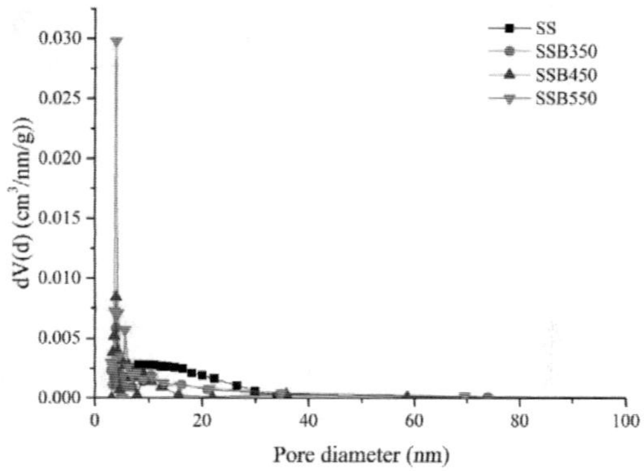

Fig. 3. The pore size distribution of biochar obtained at different pyrolysis temperatures.

3.3 Metal analysis

3.3.1 Total concentrations of heavy metals

The total concentrations of metals in the sludge and biochars are presented in Figure 4. Of the six heavy metals, the total concentration of Zn in the sludge was the highest and reached 587.04 mg/kg, which may be due to the widespread use of galvanized drainage pipes in municipal drainage networks.

The total concentrations of the six heavy metals in biochar SSB350 increased significantly after pyrolysis at 350 °C, and, except for Cd, the higher the pyrolysis temperature was, the greater the total concentration of heavy metal was. This result indicates that higher pyrolysis temperatures caused the enrichment of heavy metals in the biochars.

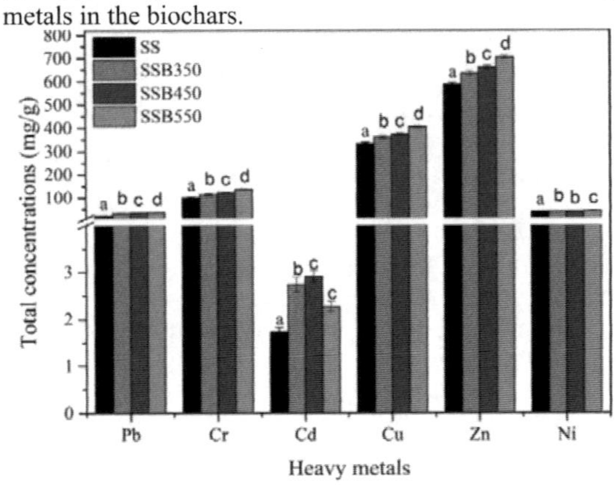

Fig. 4. Total concentrations of heavy metals in the sludge and biochars.

3.3.2 Concentrations of BCR-extractable fractions

The BCR (European Community Bureau of Reference) method is always used to quantitatively analyze the chemical form distributions of heavy metals in sediment, soil, and biochars [4,26]. According to the BCR method, heavy metals are classified into four chemical forms: (1) acid soluble/exchangeable fraction (F1), (2) reducible fraction (F2), (3) oxidizable fraction (F3), and (4) residual fraction (F4), including residual and silicate-bound heavy metals [27]. The chemical form distributions of the six heavy metals in the sludge and biochars are presented in Figure 5.

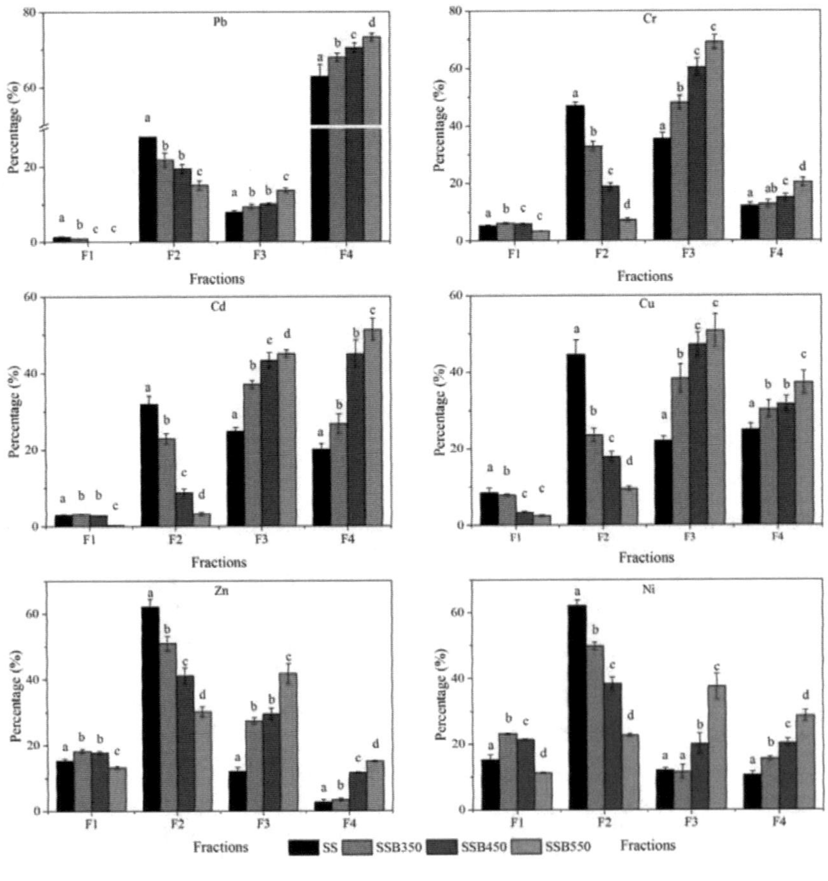

Fig. 5. Chemical form distributions of heavy metals in the sludge and biochars.

For Pb, the percentages of F1, F2, F3, and F4 in the sludge were 1.36%, 27.90%, 7.92%, and 62.82%, respectively. Compared with the sludge, the percentages of Pb in F1 and F2 in biochar SSB350 decreased, while those in F3 and F4 obviously increased. Furthermore, when the temperature increased from 350 °C to 550 °C, the percentage of Pb in F1 decreased from 0.90% to 0%, and that in F2 decreased from 27.90% to 21.88%. In contrast, the percentages of Pb in F3 and F4 increased from 7.92% to 9.31% and 62.82% to 67.91%, respectively. These results suggest that pyrolysis at a high temperature effectively promoted the transformation of Pb from F1 and F2 to F3 and F4. Meanwhile, the increase of temperature had similar effects on the chemical form distributions of the other five heavy metals (Cd, Cu, Cr, Zn, and Ni).

During the pyrolysis process, organic compounds in the sludge were decomposed into phenols, acids, and aromatic compounds, and heavy metals combined with organic functional groups, such as hydroxyl groups, to form stable coordination compounds [28]. Moreover, some metals combined with Si-O bonds [26]. These factors caused the transformations of those heavy metals from F1 and F2 to F3 to F4. Previous research has confirmed that metals presenting in F3 and F4 are more stable in the environment than those in F1 and F2 [29]. Therefore, pyrolysis can effectively stabilize heavy metals in sludge and reduce the possibility of their release into the environment.

Table 3. Leaching toxicities of the heavy metals from the sludge and biochars.

Samples	Heavy metals (mg/g)					
	Pb	Cr	Cd	Cu	Zn	Ni
SS	0.53(2.55[a])	3.21(3.20)	0.16(9.34)	30.80(9.22)	34.99(5.96)	1.34(3.53)
SSB350	0.31(1.21)	1.73(1.52)	0.14(4.78)	20.13(5.56)	22.83(2.34)	0.51(1.75)
SSB450	0.29(1.07)	1.50(1.22)	0.13(4.44)	12.11(3.24)	22.19(2.18)	0.46(1.28)
SSB550	0.25(0.83)	1.05(0.78)	0.10(3.27)	10.65(2.63)	15.07(1.39)	0.43(1.03)
Standard[b]	5.0	5.0	1.0	Not enlisted	Not enlisted	5.0

[a]Leaching rate (%): the ratio of the leaching content of heavy metals to the total concentration of each heavy metal.
[b]USEPA test methods for evaluating solid waste; physical/chemical methods of total concentration of each heavy metal (SW-846) (Huang & Yuan, 2016).

3.3.3 Leachable toxicity of heavy metals

The leaching characteristics of the heavy metals in the sludge and biochars are exhibited in Table 3. In this study, the leaching contents of all six heavy metals from the sludge and biochars were below the permissible limits [18]. Compared with the sludge, the leaching contents of the six heavy metals in biochar SSB350 decreased significantly, and the increase in temperature further reduced the leaching contents of these heavy metals. For Pb, the leaching content decreased from 0.31 mg/g to 0.25 mg/g, and the leaching rate decreased from 1.21% to 0.83%. The leaching characteristics of the other five heavy metals exhibited similar trends. Thus, pyrolysis at a high temperature had a noticeable effect on the alleviation of the leaching toxicities of the six heavy metals in the sludge. This observation was in accordance with that obtained by Jin et al [23].

3.3.4 Ecological risk assessment of heavy metals

The ecological risk index (GRI) is always used to evaluate the ecological risks of single and multiple metals in water, sediment, soil, and biochars [18], [20]. In this study, this method was used to assess the ecological risks of heavy metals in the sludge and biochars prepared at different pyrolysis temperatures, and the values were computed by using the following formulas:

$$ICF = \frac{C_{F1+F2+F3}}{C_{F4}} \quad (2)$$

$$GCF = \sum_{i=1}^{n} ICF_i \quad (3)$$

$$GRI = \sum_{i=1}^{n} Tr^i ICF_i \quad (4)$$

where $C_{F1+F2+F3}$ is the sum of the total concentrations of a single heavy metal presenting in fractions Fl + F2 + F3, C_{F4} is the total concentration of a single heavy metal in fraction F4, and Tr^i is the toxic response factor, which is the response factor for the ecological toxicity of a single heavy metal [30]. Furthermore, ICF and GCF are the contamination factors of a single heavy metal and multi-metals, respectively. GRI represents the ecological risks of heavy metals (Pb, Zn, Cd, Cr, Cu, and Ni) in the samples. The ICF, GCF, and GRI values of the samples are listed in Table 4.

Table 4. Assessment of ecological risks of heavy metals in the sludge and biochars.

Samples	ICF[a]						GCF[b]	GRI[c]
	Pb	Cr	Cd	Cu	Zn	Ni		
SS	0.59	7.28	1.49	3.02	8.49	8.43	29.20	137.43
SSB350	0.47	6.84	2.36	2.30	6.47	5.45	23.89	137.25
SSB450	0.42	5.68	1.22	2.16	7.62	3.95	21.05	94.20
SSB550	0.39	3.94	0.95	1.69	5.66	2.49	15.12	69.13

[a]*ICF*: ≤1, 1–3, 3–6, and >6 denote low, moderate, considerable, and high contamination risks, respectively.

[b]*GCF*: <6, 6–12, 12–24, and >24 denote low, moderate, considerable, and high contamination risks, respectively.
[c]*GRI*: <150, 150–300, 300–600, and >600 denote low, moderate, considerable, and high ecological risks, respectively.

The assessment results reveal that the ICF value of Zn was the highest of the six heavy metals for both the sludge and the biochars. The GCF value of the sludge was 29.20. After pyrolysis at 350 °C, the GCF value of the biochar SCB350 decreased to 23.89. Moreover, the GCF values of the biochars obviously decreased with the increase of pyrolysis temperature (Table 4). In the GRI analysis, the GRI values of the sludge and SSB350 were 137.43 and 137.25, respectively, which are both at the low risk level. Additionally, the GRI values of the biochars decreased with the increase of pyrolysis temperature. These results suggest that the evaluated pyrolysis temperatures effectively reduced the ecological risks of both single and multiple heavy metals in the biochars.

This can be explained as follows. During the pyrolysis process, metals presenting in mobile fractions F1 and F2 were released from the organic structure of the sludge and transferred into stable fractions F3 and F4, resulting in the decrease of ecological risks of those heavy metals. Furthermore, the increase of pyrolysis temperature enhanced this process, as shown in Figure 5.

4 Conclusions

With the increase of pyrolysis temperature, the biochar yield decreased obviously, but the pH value and ash content increased. Additionally, the increase of pyrolysis temperature reduced the H/C ratio of the biochar, increased its specific surface area and total pore volume, and reduced its pore size. the total concentration of heavy metals in the biochars further increased with the increase of pyrolysis temperature. Elevated pyrolysis temperatures effectively inhibited the leaching potential of heavy metals from the biochars, contributed to the transformation of heavy metals from mobile fractions (F1 and F2) to stable ones (F3 and F4), and effectively reduced the potential ecological risks of heavy metals in the biochars. These results suggest that biochar prepared via the pyrolysis of sewage sludge at a high temperature is more suitable for soil improvement. However, the effects of pyrolysis residence time and heating rate on the characteristics of sewage sludge-derived biochar require further investigation.

Acknowledgements

This work was supported by the Science and Technology Foundation of Karamay (No. 2020CXRC0013), Joint Foundation of China University of Petroleum-Beijing at Karamay (No. XQZX20200014), and the Natural Science Foundation of China (No. 51074170).

References

1. Y. Zhou, Y. Liu, W. Jiang, L. Shao, L. Zhang, L. Feng, Effects of pyrolysis temperature and addition proportions of corncob on the distribution of products and potential energy recovery during the preparation of sludge activated carbon, Chemosphere **221,** 175–183 (2019).
2. Shao, W. Jiang, L. Feng, L. Zhang, Co-production of activated carbon, fuel-gas, and oil from the pyrolysis of corncob mixtures with wet and dried sewage sludge, Waste Manag Res **32**, 519–526 (2014).
3. Z. Li, H. Deng, L. Yang, G. Zhang, Y. Li, Y. Ren, Influence of potassium hydroxide activation on characteristics and environmental risk of heavy metals in chars derived from municipal sewage sludge, Bioresour Technol **256**, 216–223 (2018).
4. X. Wang, Q. Chi, X. Liu, Y. Wang, Influence of pyrolysis temperature on characteristics and environmental risk of heavy metals in pyrolyzed biochar made from hydrothermally treated sewage sludge, Chemosphere **216,** 698–706 (2019).
5. K. Wang, Y. Zheng, X. Zhu, C.E. Brewer, R.C. Brown, Ex-situ catalytic pyrolysis of wastewater sewage sludge – A micro-pyrolysis study, Bioresour Technol **232,** 229–234 (2017).
6. W. Buss, M.C. Graham, J.G. Shepherd, O. Mašek, Suitability of marginal biomass-derived biochars for soil amendment, Sci Total Environ **547**, 314–322 (2016).
7. S. Van Wesenbeeck, W. Prins, F. Ronsse, M.J. Antal, Sewage Sludge Carbonization for Biochar Applications. Fate of Heavy Metals, Energy & Fuels **28**, 5318–5326 (2014).
8. S. Bruun, S.L. Harmer, G. Bekiaris, W. Christel, L. Zuin, Y. Hu, L.S. Jensen, E. Lombi, The effect of different pyrolysis temperatures on the speciation and availability in soil of P in biochar produced from the solid fraction of manure, Chemosphere **169**, 377–386 (2017).
9. A. Méndez, A.M. Tarquis, A. Saa-Requejo, F. Guerrero, G. Gascó, Influence of pyrolysis temperature on composted sewage sludge biochar priming effect in a loamy soil, Chemosphere **93**, 668–676 (2013).
10. Y. Yue, L. Cui, Q. Lin, G. Li, X. Zhao, Efficiency of sewage sludge biochar in improving urban soil properties and promoting grass growth, Chemosphere **173**, 551–556 (2017).
11. D. Zhou, D. Liu, F. Gao, M. Li, X. Luo, Effects of Biochar-Derived Sewage Sludge on Heavy Metal Adsorption and Immobilization in Soils, Int J Environ Res Public Health **14**, 7 (2017).
12. H. Yuan, T. Lu, Y. Wang, Y. Chen, T. Lei, Sewage sludge biochar: Nutrient composition and its effect on the leaching of soil nutrients, Geoderma **267**, 17–23 (2016).
13. M.K. Hossain, V. Strezov, P.F. Nelson, Comparative Assessment of the Effect of Wastewater Sludge Biochar on Growth, Yield and Metal

Bioaccumulation of Cherry Tomato, Pedosphere **25**, 680–685 (2015).

14. Z. Khanmohammadi, M. Afyuni, M.R. Mosaddeghi, Effect of pyrolysis temperature on chemical and physical properties of sewage sludge biochar, Waste Manag Res **33**, 275–283 (2015).
15. H. Chen, Y. Zhai, B. Xu, B. Xiang, L. Zhu, L. Qiu, X. Liu, C. Li, G. Zeng, Characterization of bio-oil and biochar from high-temperature pyrolysis of sewage sludge, Environ Technol **36**, 470–478 (2015).
16. L.J. Leng, X.Z. Yuan, H.J. Huang, H. Wang, Z.B. Wu, L.H. Fu, X. Peng, X.H. Chen, G.M. Zeng, Characterization and application of bio-chars from liquefaction of microalgae, lignocellulosic biomass and sewage sludge, Fuel Processing Technology **129**, 8–14 (2015).
17. X.D. Song, X.Y. Xue, D.Z. Chen, P.J. He, X.H. Dai, Application of biochar from sewage sludge to plant cultivation: Influence of pyrolysis temperature and biochar-to-soil ratio on yield and heavy metal accumulation, Chemosphere **109**, 213–220 (2014).
18. H.J. Huang, X.Z. Yuan, The migration and transformation behaviors of heavy metals during the hydrothermal treatment of sewage sludge, Bioresour Technol **200**, 991–998 (2016).
19. Z. Wang, J. Wang, L. Xie, H. Zhu, X. Shu, Influence of the Addition of Cotton Stalk during Co-pyrolysis with Sewage Sludge on the Properties, Surface Characteristics, and Ecological Risks of Biochars, J. Therm. Sci. **15,** 551–556 (2019).
20. J. Jin, M. Wang, Y. Cao, S. Wu, P. Liang, Y. Li, J. Zhang, J. Zhang, M.H. Wong, S. Shan, P. Christie, Cumulative effects of bamboo sawdust addition on pyrolysis of sewage sludge: Biochar properties and environmental risk from metals, Bioresour Technol **228**, 218–226 (2017).
21. Z. Wang, X. Shu, H. Zhu, L. Xie, S. Cheng, Y. Zhang, Characteristics of biochars prepared by co-pyrolysis of sewage sludge and cotton stalk intended for use as soil amendments, Environ. Technol. **41**, 1347–1357 (2020).
22. S. Wei, M. Zhu, X. Fan, J. Song, P.a. Peng, K. Li, W. Jia, H. Song, Influence of pyrolysis temperature and feedstock on carbon fractions of biochar produced from pyrolysis of rice straw, pine wood, pig manure and sewage sludge, Chemosphere **218**, 624–631 (2019).
23. J. Jin, Y. Li, J. Zhang, S. Wu, Y. Cao, P. Liang, J. Zhang, M.H. Wong, M. Wang, S. Shan, P. Christie, Influence of pyrolysis temperature on properties and environmental safety of heavy metals in biochars derived from municipal sewage sludge, J Hazard Mater 320, 417–426 (2016).
24. M.I. Al-Wabel, A. Al-Omran, A.H. El-Naggar, M. Nadeem, A.R. Usman, Pyrolysis temperature induced changes in characteristics and chemical composition of biochar produced from conocarpus wastes, Bioresour Technol **131**, 374–379 (2013).
25. Y.K. Choi, E. Kan, Effects of pyrolysis temperature on the physicochemical properties of alfalfa-derived biochar for the adsorption of bisphenol A and sulfamethoxazole in water, Chemosphere **218**, 741–748 (2019).
26. B. Zhao, X. Xu, S. Xu, X. Chen, H. Li, F. Zeng, Surface characteristics and potential ecological risk evaluation of heavy metals in the bio-char produced by co-pyrolysis from municipal sewage sludge and hazelnut shell with zinc chloride, Bioresour Technol **243**, 375–383 (2017).
27. B.S. Sagagi, C.M. Davidson, A.S. Hursthouse, Adaptation of the BCR sequential extraction procedure for fractionation of potentially toxic elements in airborne particulate matter collected during routine air quality monitoring, Int. J. Environ. Anal. Chem. **10**, 1–13 (2019).
28. N. Gao, C. Quan, B. Liu, Z. Li, C. Wu, A. Li, Continuous Pyrolysis of Sewage Sludge in a Screw-Feeding Reactor: Products Characterization and Ecological Risk Assessment of Heavy Metals, Energy & Fuels. **31**, 5063–5072 (2017).
29. G. Rauret, Extraction procedures for the determination of heavy metals in contaminated soil and sediment, Talanta **46**, 449–455 (1998).
30. P. Devi, A.K. Saroha, Risk analysis of pyrolyzed biochar made from paper mill effluent treatment plant sludge for bioavailability and eco-toxicity of heavy metals, Bioresour Technol **162**, 308–315 (2014).

Study on Sewage Sludge Drying System With Built-in Solar Drying Bed

Zhipan Gu[1], *Jichun* Yang[2], *Jing* Liu[2], *Leren* Tao [1,3,*], *Ye* Zhang[2], and *Lihao* Huang[1,3]

[1] Instiute of Refrigeration and Cryogenics, School of Energy and Power; University of Shanghai for Science and Technology; Shanghai 200093; PR China
[2] School of Architecture and Civil Engineering; Jiaxing University; Jiaxing; Zhejiang 314001, PR China
[3] Shanghai Key Laboratory of Multiphase Flow and Heat Transfer in Power Engineering, University of Shanghai for Science and Technology, Shanghai 200093, PR China

Abstract. In this paper, the renewable energy solar energy is used as the heat source. The combination of solar drying bed and traditional hot air drying can effectively reduce energy consumption and operation cost. The drying chamber is divided into three layers. The top air supply outlet supplies hot air, the middle layer places wet sludge, and the bottom layer uses hot water coil to dry the sludge. The whole drying process is a heat and mass transfer process with convective heat transfer and radiation heat transfer. After analysis and comparison with traditional energy drying, it is found that drying 97.5kg of sewage sludge will save 79% energy, save 12.84 kg of standard coal, reduce 32 kg of carbon dioxide and 1.284 kg of sulphur dioxide.

1 Introduction

Sewage sludge is produced as a by-product in the process of sewage treatment. It contains a lot of water and refractory organics, heavy metals and salts, pathogenic bacteria, parasite eggs, etc[1]. Sewage sludge is a complex mixture of water, inorganic compounds, microorganism and certain undigested substances [2]. It may cause secondary pollution, and even harmful components may eventually enter the food chain[3]. With the development of global economy, sewage sludge is produced in large quantity worldwide. In 2013, China's total dry sludge output was 6.25 million tons. Canada and the United States produce approximately 660,000 tons and 6.5 million tons of dry sludge annually[4]. Due to the large volumes of sludge, the focus has shifted towards reduction, stabilization, harmlessness, and resource utilization of sludge. The thermal drying is an essential intermediate stage for sludge management. Traditional thermal drying needs more fossil energy. Access to renewable, clean and cheap energy is a major issue in socio-economic development. In recent years, sustainable and renewable energy resource like solar energy is widely applied in drying process.

There are many examples in the literature on using solar energy in agriculture, food production as an energy source. Khanlari et al. [5]experimentally and numerically analyzed the effect of simple tube-type solar air heater in drying applications. Hatami et al. [6] experimentally and numerically analyzed an indirect solar dryer to be used in drying agricultural products. Elmustapha et al. [7] conducted a cross case analysis of solar thermal and solar photovoltaic systems. Kasaeian et al. [8] analyzed the solar radiation, storing heat, extracting, and the application of solar ponds in the recent decades. Tong et al. [9] investigated different nanofluids to enhance thermal performance of a flat-plate solar collector. Vera-Medina et al. [10] experimentally and numerically developed a new freeze protection system by using flexible silicone peroxide tubes inside the absorber of the solar collector. Poblete et al.[11] pointed out that drying air flow rate is the most important factor that affect the performance of a closed solar dryer. Karthikeyan et al.[12] indicated that drying air temperature has significant effect on dried agricultural product quality through research on a solar tunnel dryer.

This work adopts the combination of solar drying bed and traditional solar drying technology. Solar collector and solar drying bed are used to convert solar energy into heat energy, and the sewage sludge is heated and dried, so that the moisture in the sludge can be transported and vaporized to achieve the effect of sludge reducing moisture content. This provides a new research idea for the exploration and development of an efficient, environmental friendly and energy saving sludge resource treatment system. Sewage

2 Sewage sludge drying System

2.1 Formatting the title, authors and affiliations

* Corresponding author: cryo307@usst.edu.cn (Leren Tao)

The schematic diagram of sludge drying system of solar drying bed is shown in Fig. 5. The system components are shown in Table 1.

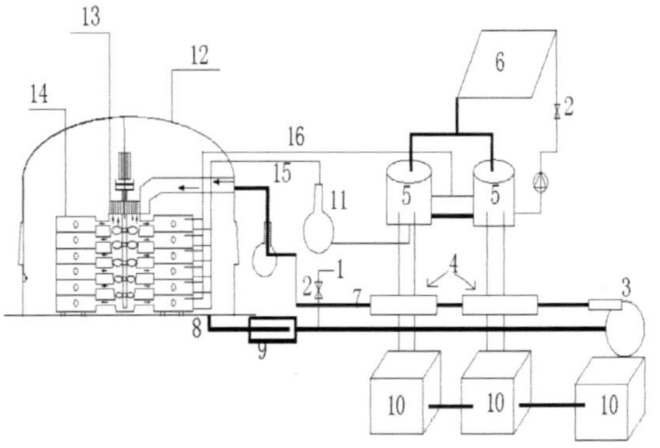

Fig. 1. Sludge drying system of solar drying bed

Table 1. System components.

Serial number	Expression	Serial number	Expression
1	Air	9	Dehumidifier
2	Valve	10	Underground Hot Water Tank
3	Blower	11	Pump
4	Heat Exchanger	12	Drying Room
5	Heat Preservation Water Tank	13	Air Distributor
6	Solar Collector	14	Drying chamber
7	Air Pipe	15	Water Supply Pipe
8	Return Air pipe	16	Water Return Pipe

2.2 System operation

The whole sludge drying system is composed of two parts: one is solar sludge drying bed, the other is hot air convection drying. The heat of both systems comes from the solar collector, which produces hot water stored in a hot tank. Part of the hot water enters the heat exchanger to heat the air, and the hot air is sent to the drying chamber through the fan. It passes through the heat transfer between the sludge surface and the sludge interior, so that the water in the sludge gradually vaporizes and enters the return air pipe together with the wet air. The other part of hot water in the hot water tank is fed into the inner coil of the drying bed through the water supply pipe, and flows into the hot water tank through the return pipe after heat exchange.

2.3 Sludge drying chamber

In this system, the sludge drying chamber is a combination of hot air convection drying and drying bed drying. The sewage sludge drying chamber is shown in Fig. 2. In the drying chamber, the drying box is used to dry the sludge layer by layer, which improves the drying efficiency. The drying chamber is divided into three layers. The top air supply outlet supplies hot air, the middle layer places wet sludge, and the bottom layer uses hot water coil to dry the sludge. The whole drying process is a heat and mass transfer process with convective heat transfer and radiation heat transfer. The whole process of the system is simple, which reduces the loss of heat energy and improves the utilization rate of heat energy. Compared with the traditional thermoelectric drying system, the energy consumption and pollutant emission are greatly reduced.

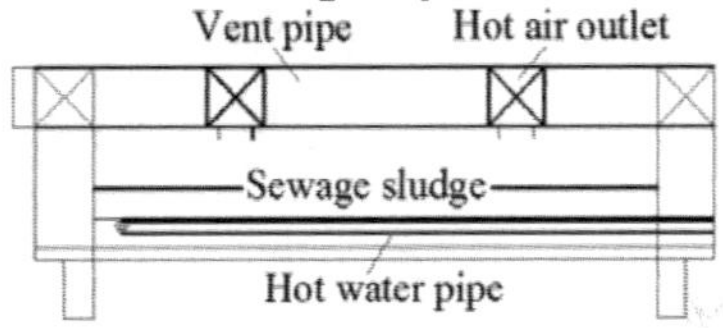

Fig. 2. Sludge drying chamber

3 Theoretical design calculation

3.1 Thermal calculation

According to the law of conservation of energy, the thermodynamic balance calculation of sludge drying system has the following energy conservation relations.

$$Q_{ss} + Q_w + Q_L = Q_{so} \quad (1)$$

Where Q_{ss} is the heat required for drying sludge(KJ), Q_w is heat required for evaporation of water (KJ). Q_{so} is the heat provided by solar drying system. Q_L is the heat loss during drying (KJ).

$$Q_{ss} = M_2 (T_2 - T_1)[(1 - R_2) C_s + R_2 C_i] \quad (2)$$

$$M_2 = M_1 (1 - R_1)/(1 - R_2) \quad (3)$$

Where M_2 is the sludge mass at a certain moisture content after drying (kg). T_1 and T_2 are the temperature of sludge before and after drying (℃). R_2 is the moisture content of dried sewage sludge (%).C_s is the specific heat of the sewage sludge (KJ/(Kg· ℃)). C_s is the specific heat of the water (KJ/(Kg·℃)). M_1 is the sludge mass at a certain moisture content before drying (kg). R_1 moisture content of sludge before drying (%).

$$Q_w = W [r_i + (T_2 - T_1)C_i] \quad (4)$$

$$W = M_2 (1 - R_2)/(x_1 - x_2) \quad (5)$$

$$x = R /(1 - R) \quad (6)$$

Where W is evaporation water (kg). r_i is the latent heat of vaporization of water (kJ/kg). x_1 and x_2 are moisture content of dry basis before and after drying.

$$Q_L = \alpha(Q_{ss} + Q_w) \quad (7)$$

Where α is Heat loss coefficient in drying process.

The thickness of sludge in each layer is 20cm, and the moisture content of each layer of sludge is 80%. After weighing, the weight of each layer of sludge is 97.5kg.

It takes 1 hour to dry the first layer of sludge. The 97.5kg sludge with 80% moisture content is treated to make the moisture content of sewage after drying reach 50%, the specific heat capacity of sludge C_s = 1.244 KJ/(Kg·℃), and the density $\rho = 1.3 \times 10\text{Kg/m}^3$.

$$M_2 = 97.5 \times (1 - 80\%)/(1 - 50\%) = 39 \text{ kg} \quad (8)$$

$$W = 39 \times (1 - 50\%)/(4 - 1) = 58.5 \text{ kg}$$

$$Q_w = 58.5 \times [2257.2 + (40 - 20) \times 4.1868] = 136968.15\text{KJ}$$

$$Q_{ss} = 39 \times (40 - 20)[(1 - 50\%) \times 1.244 + 0.5 \times 4.1868] = 2118.02\text{KJ}$$

$$Q_L = 0.05 \times (136968.15 + 2118.02) = 6954.3\text{KJ}$$

$$Q_{so} = 136968.15 + 2118.02 + 6954.3 = 146010.5\text{KJ}$$

3.2 Calculation of solar collector area

The calculation formula of solar collector area is as follows

$$A_c = (Q_{so} - Q_0)f / J_T\eta_{cd}(1 - \eta_L) \quad (9)$$

Where A_c is the collection area of the solar collector (m^2). Q_0 is heat stored in drying chamber (KJ). f is the guarantee rate of solar energy is 30%–80%, according to the solar radiation during the service life of the system, system economy and user requirements. J_T is the annual average daily solar radiation on the day lighting surface of the local collector (KJ/m^2). η_{cd} is the annual average collection efficiency of the collector, according to the experience, the value is 0.25-0.50, and the specific value should be determined according to the actual test results of the collector. η_L is the heat loss rate of the collector pipeline. According to experience, the value(η_L) is 0.20-0.30.

When calculating the heat collection area of solar energy drying system, it is considered that the sludge drying room can provide part of the heat needed for sludge drying, so the heat collection area of solar collector can be reduced. In summer, the temperature in the solar greenhouse can reach about 60 ℃, and the air quantity needed for drying per hour is 6541.25m3/h after measurement. The heat provided by hot air is 180997.56kJ. The heat stored in drying chamber (Q_0) is set to 40% of the heat provided by the hot air.

$$A_c = (146010.5 - 180997.56 \times 40\%) \times 0.475 / (2268 \times 0.5(1 - 0.3)) \approx 45\text{m}^2$$

The sludge drying system is installed in Jiaxing City, Zhejiang Province. The solar collector is located in the same direction as the building, facing due south, and the inclination angle is the same as the local latitude. 50 pieces of horizontal plug-in vacuum tube collector with the specification of 47×1500mm are selected as a unit collector. The unit collector size is 3100 mm×1647 mm in width. The heat collection area of unit collector is 5.11m^2. According to the total area of the collector and the heat collection area of the unit collector, the number of collectors required is calculated to be 9 groups.

4 Energy saving analysis

Solar sludge drying equipment can convert solar energy into heat energy through solar energy conversion equipment, effectively reducing the use of traditional energy.

According to the amount of 97.5kg sludge per hour, the energy consumption of traditional electric heating per hour is 40.6KWh. The energy consumption of electrical equipment such as pumps and fans in the solar sludge drying system is estimated to be 8.5kw/h. Compared with the traditional electric heating, the energy consumption is reduced by 79%. The use of solar energy sludge drying equipment can effectively reduce energy consumption. At the same time, the command and control system of water pump and fan can further reduce the power consumption of system electrical equipment, and the energy saving effect is remarkable.

5 Emission reduction analysis

According to the research, saving 1kwh electricity can save 0.4kg standard coal and reduce pollution emission of 0.997kg carbon dioxide and 0.03kg sulfur dioxide. According to the previous calculation results, drying 97.5kg sludge per hour can save 32.1kwh of electric energy. It can save 12.84 kg of standard coal, reduce 32 kg of carbon dioxide and 1.284 kg of sulfur dioxide.

The energy saving and emission reduction calculated above is only one layer of sludge energy saving and emission reduction in drying chamber. The whole drying chamber can be composed of four to five layers, which can save more energy and reduce more greenhouse gas emissions.

6 Application prospect analysis

The research technology of this system can effectively solve the drying problem of sludge in urban sewage treatment plant, and reduce the energy consumption in the process of harmless, reducing and resource treatment of urban sludge.

This system is expected to be promoted and used in BOT mode or as the technical transformation project of sewage treatment plant, which is suitable for sludge treatment project of sewage treatment plant in small and medium-sized cities. The technical field is in development. Due to the small amount of single project, most domestic large and medium-sized enterprises have not set foot in it. At present, it is the best entry period for solar energy to apply dry sludge market. This system can take advantage of the technical advantages of independent research and development, and will become a market leader after industrialization.

7 Conclusion

In this system, the solar collector is used to convert solar energy into heat energy to heat and dry the sludge, so as to transfer and evaporate the water in the sludge, so as to reduce the water content of the sludge. Compared with traditional electric heating, drying 97.5kg of sewage sludge will save 79% energy, save 12.84 kg of standard coal, and reduce 32 kg of carbon dioxide and 1.284 kg of sulphur dioxide. This provides a new research idea for the exploration and development of efficient, environmental protection, energy-saving sludge treatment system.

The system uses renewable energy solar energy as heat source, and its operation and management are simple. It can effectively reduce the energy consumption and operation cost of drying. It has a good application prospect.

Acknowledgments

This work was supported by Zhejiang Provincial Natural Science Foundation of China [No. LQ19E080014], Public Welfare Technology Research Plan of Zhejiang Province [No. LGG19E060005], Science and Technology Program of Ministry of Housing and Urban-Rural Development China [2015-K7-026] and Project of Key Laboratory of Shanghai [1N-15-301-101].

References

1. G.Yang, G. Zhang, H.Wang, *Water Res*, 2015, 78, 60-73.
2. E. Eriksson, N. Christensen, J.E. Schmidt, A. Ledin, *Desalination*, 2008, 226, 371-388.
3. A. Pathak, M.G. Dastidar, T.R. Sreekrishnan, *J. Environ. Manage*,2009, 90, 2343-2353.
4. A.K. Wahidunnabi, C. Eskicioglu, *Water. Res*, 2014, 66, 430-446.
5. A. Khanlari, H.O. Güler, A.D. Tuncer, C. Sirin, Y.C. Bilge, Y. Yılmaz, A. Güngor, *Renew. Energy*, 2020, 145, 1677-1692.
6. S.Hatami, G.Payganeh, A.Mehrpanahi, *J. Clean. Prod.* 2020, 244, 118809.
7. H. Elmustapha, T. Hoppe, H. Bressers, *J. Clean. Prod.* 2018, 186, 155-167.
8. A. Kasaeian, S. Sharifi, W.M. Yan, *Sol. Energy*, 2018, 174, 189-206.
9. Y. Tong, H. Lee, W. Kang, H. Cho, *Appl. Therm. Eng*. 2019, 159, 113959.
10. J. Vera-Medina, I. Lillo-Bravo, J.Hernandez, M. Larraneta, *Appl. Therm. Eng*, 2018, 135, 446-453.
11. R. Poblete, E. Cortes, J. Macchiavello, J. Bakit, *Renew. Energy*, 2018, 126, 978-986.
12. A.K. Karthikeyan, S. Murugavelh, *Renew. Energy*, 2018, 128, 305-312.

Effect of nitrogen fertilization on maize yield responses to soil microbial activity and root length density in the North China Plain

Qiuhua Li[1,2], *Jingjing* Sun[1], *Jun* Yao[3], and *Qunhui* Wang[1*]

[1]School of Energy and Environmental Engineering, University of Science and Technology Beijing, Beijing 100083, People's Republic of China
[2]School of Biotechnology and Health Sciences, Wuyi University, Jiangmen 529020, People's Republic of China
[3]School of Water Resources and Environment, China University of Geosciences, Beijing 100083, People's Republic of China

Abstract. A maize field experiment in the North China Plain was conducted to understand the effect of different N fertilizer rate on the yield of maize, using soil microbial activity and root length density (RLD) as performance parameters, due to their possibility to enhance productivity. The four N fertilizer rates were 0 (N0), 120 (N120), 210 (N210) and 300 (N300) kg N hm^{-2}. The results indicated that nitrogen (N) fertilizer had a significant influence not only on yield (p<0.05), but also on root length density (p<0.05) and soil microbial activity (p<0.05). In addition, the soil microbial activity and RLD were significantly related with maize yield. RLD differences were generally evident within the 100 cm soil layer, whereas there was no difference in the deeper soil under different N treatments. The most RLD concentrated in 0-60cm soil layer under N0, N120 and in 0-90cm soil layer under N210, N300. The microbial growth rate constant (k) was greater in N210 than other treatments. Generally, N fertilizer application can stimulate root growth and microbial activity, meanwhile, they can interact with each other, heighten the availability of N fertilizer in soil, thus enhanced yield of maize. According to our study, 210 kg N hm^{-2} was the optimum N fertilizer rate to achieve maximum yield and sustain the soil productivity.

1 Introduction

Maize (*Zea mays* L.) is one of the major crops in China, and occupies 17% of global maize production. The North China Plain is the largest agricultural production region in China ([45])accounts for more than 30% of China's maize production ([33]). Therefore, increasing maize yield in this region plays an important role in the national food security ([42]). Fertilization by especially nitrogen (N) inputs is a key management tool for maintaining crop productivity. Optimal N fertilizer applied into soil can increase nutrient of crop, soil fertility ([46]) and thus yield. However, high N fertilization is common in China ([8]) to ensure high yield, and local farmers have applied N fertilizer at a rate of 266 kg N hm^{-2} from 2010 to 2012 ([43]). Excessive N fertilizer would cause negative impacts on the environment ([10,16,37]) and in this region in particularly.

In agricultural production, the root systems are the primary part of the crop that determines the acquisition water and nutrients from the soil ([17]). Root length density (RLD) is an important parameter which plays a crucial role in crop growth, nutrient uptake and yield ([11]). For example, increasing maize root depth that the yield of maize increased from 6.0 to 7.8 t hm^{-2} ([3]) and increasing RLD of maize by 13.2% that can be raised grain yield by 17.3% and enhanced post-silking N uptake by 26.7% ([26]).

Soil micro-organisms in the agricultural soil environment have a great contribution. They are involved in the cycling of nitrogen, carbon, and sulfur ([22, 40]), decomposition of organic matter ([36]), and improvement of soil structure ([6]).They also serve as indicators of soil quality and fertility ([7]). In addition, N fertilizer can have a positive ([24]) or no effect ([39])on soil microbial activity. Furthermore, optimum N fertilizer was relative to soil N availability, improved soil microbial activity subsequently ([21]). Many of these studies of the soil microbial activity by micro-calorimetry focused on laboratory, field data have been rarely studied in maize soil.

It is essential to reduce excessive N fertilizer applied for development of sustainable agriculture in the North China Plain. Previous studies focus on improving the N use efficiency([8, 47]) and N management ([1]) based on biomass production and N accumulation in the soil. Few of these studies used root length density and soil microbial activity to study the efficiency of N and how these properties impacted yield of maize.

The primary objective of this study was to (1) investigate the effects of different N fertilizer rate on soil

* Corresponding author: Wangqh@163.com

microbial activity through microcalorimetry, RLD and yield, (2) evaluate whether soil microbial activity, RLD and yield were correlated. We hypothesized that enhanced maize yield would be related to soil microbial activity and RLD through N fertilizer rate stimulation. To explore this effect, this work was based on a maize field experiment in the North China Plain.

2 Materials and methods

2.1 Experimental site description

The field experiment was conducted in a typical maize-wheat rotation cropping system at Baoding Farming Experimental Station ($38^0$5'N, $115^0$30'E), Hebei Province, China. This region has a temperate continental monsoon climate. The annual mean air temperature is 12°C. The average annual precipitation is 550mm, mainly falls from July to September. The soil type was a moisture loam containing soil organic matter 16.1 g kg^{-1}, total N 0.96 g kg^{-1}, Olsen-P 16.6 mg kg^{-1} and Olsen-K99.9mg kg^{-1} respectively in the top soil layer (0-20cm) before the experiment.

2.2 Experimental design

The experiment was established in 2009 during the maize growing season. Summer maize (Zhengdan 958), a commonly used variety in North China Plain was planted after wheat harvest without tilling the soil. The row spacing was 60 cm and the density was 60000 plants hm^{-2}. The experiment comprised four treatments: 0 (N0), 120(N120), 210(N210), 300 kg N hm^{-2}(N300).Fertilizers used were urea (46%), phosphorus pentoxide(12%) and potassium sulfate(60%). Urea was applied three times during the growth stage, 40% of urea was applied as basal fertilizer, 40% of urea was applied at huge bellbottom stage, and 20% of urea was applied at tasselling. Total P and K fertilizer was applied both as basal fertilizer, which comprised 90 kg (P_2O_5)·hm^{-2} and 180kg (K_2O) ·hm^{-2}. Depending on rainfall period, the experiment was irrigated 60 mm during the early growing stage after planting. All the straw was chopped and returned to the soil after maize harvest.

2.3 Grain yield and root collection

After harvest in 2013, grain yield (kg hm^{-2}) was measured for all maize cobs in a $20m^2$ area in each plot at moisture content of 14%. Root length density (RLD) (cm cm^{-3}) was calculated from the volume (cm^3) of the sampling core of the root length (cm) ([12]).

2.4 Soil sampling and analysis

Soil samples were taken after maize harvest at three depths: 0-30, 30-60, 60-90 cm. Fresh soil samples were passed immediately through a 2 mm sieve, and then air-dried for soil microbial activity analyses.

Soil microbial activity was determined by the microcalorimetry method ([13]). Soils (1.0g) were placed in ampoules (4.5ml) at 28°C for several days. The heat flow of the samples was recorded continuously by the computer automatically. The heat generated by biological processed in living cells is directly recorded by microcalorimetry over time. Data were collected as different heat flow power-time until the heat signal returned to baseline indicating no further measurable metabolic activity.

2.5 Data analyses

The difference of yield of maize was determined by significant difference ($p< 0.05$). Data were analyzed using correlation analysis and analysis of variance (ANOVA) using the SPSS 18. RLD was analyzed and calculated using Win Rhizo Pro Vision 5.0a. Microbial activity analyzed using Origin 8.0.3. Correlation analysis was used to analyze the relationship between micro-calorimetric parameters, RLD, yield and N fertilizer. ANOVA was used to detect N fertilizer influencing soil parameters. Regression analysis was applied to calculate the optimum maize yield.

3 Results

3.1 Maize yield

Yields were affected by N fertilizer are shown in Table1. After four years of N fertilizer experiment, the yields of maize were significantly ($p<0.05$, Table 2) affected by N fertilizer.

Table 1. Yield of maize under different N fertilizer rate in the four seasons from 2010 to 2013.

Treat-ment	2010	2011	2012	2013
N0	8947±322a	7470±117b	6794±284c	6264±57c
N120	9014±238a	7603±146b	8954±159ab	9473±193ab
N210	9276±166a	9513±98a	9704±86a	9987±72a
N300	8999±286a	8729±201ab	9172±143ab	9627±a111b

Note: Maize yields of treatments followed by different letters are significantly different at $p<0.05$, ±: Standard Deviation.

The first maize yields in 2010 had no difference among N treatments. The second maize yields in 2011, the highest yield was observed in N210 which was significantly higher than N0 and N120, but it was no significant difference between N210 and N300. The maize yields in 2012 and 2013, the difference of yield varied with the rate of N fertilizer compared to the control, but there were no significant differences between N120 and N300. The results suggested that increased N fertilizer to 300kg hm^{-2} had no positive effect on yield. Generally, the yield of N0 from 2010 to 2013 always decreased and had the lowest yield which was significantly lower compared to other treatments in 2012 and 2013.

 https://doi.org/10.1051/e3sconf/202123701042

Table 2. The analysis of the Correlation between micro-calorimetric parameters, RLD, yield and N fertilizer and the analysis of variance.

	Nitrogen	Yield	RLD	Q_{total}	P_{max}	k
Correlation analysis						
Nitrogen	1					
Yield	0.770**	1				
RLD	0.899**	0.827**	1			
Q_{total}	0.038	0.063	-0.038	1		
P_{max}	0.255	0.230	0.299	-0.632*	1	
k	0.520	0.824**	0.612*	0.249	0.556	1
Analysis of variance						
Treatments		*	*	ns	ns	*

RLD: root length density, Q_{total}: the total thermal, P_{max}: peak height values, k: growth rate constant.
*: significant at $P<0.05$, **: significant at $P<0.01$, ns: not significant at $P<0.05$.

3.2 Root length density

Root length density (RLD) was found to be one of the most important parameters influencing grain yield ([23]). The effects of N fertilizer on RLD was investigated (Fig.1). Overall RLD in the 0-100cm soil layer was changed more obvious than in deeper soil layer.At N0, 49.8% in the 0-30 cm soil layer, 24.8% in the 30-60 cm soil layer, 9.6% in the 60-90 cm soil layer, 13.4% in the 90-120 cm soil layer 8.8%, in the 120-150 cm soil layer and 1.0% in the 150-180 cm soil layer, respectively, most RLD in the 0-60cm soil layer and about 50% RLD in the 0-30 cm soil layer. At N120, 58.6% in the 0-30 cm soil layer, 27.1% in the 30-60 cm soil layer, 1.6% in the 60-90 cm soil layer, 11.2% in the 90-120 cm soil layer, 0.8% in the 120-150 cm soil layer and 0.8% in the 150-180 cm soil layer, respectively, most RLD in the 0-60cm soil layer while 60% RLD in the 0-30 cm soil layer.At N210, 38.2% in the 0-30 cm soil layer, 26.7% in the 30-60 cm soil layer, 22.7% in the 60-90 cm soil layer, 5.9% in the 90-120 cm soil layer, 4.6% in the 120-150 cm soil layer and 1.8% in the 150-180 cm soil layer, respectively, most RLD in the 0-90cm soil layer, probably 50% RLD in the 30-90 cm soil layer.At N300, 38.8% in the 0-30 cm soil layer, 23.1% in the 30-60 cm soil layer, 18.6% in the 60-90 cm soil layer, 7.1% in the 90-120 cm soil layer, 8.8% in the 120-150 cm soil layer and 3.6% in the 150-180 cm soil layer, respectively, most RLD in the 0-90cm soil layer. Significant differences were found among the four treatments.

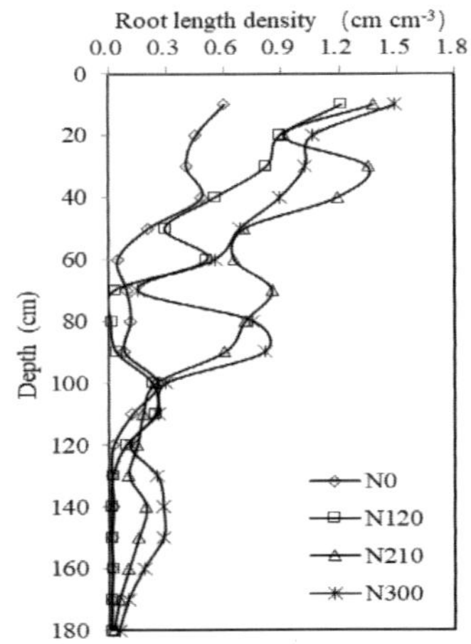

Fig.1 Root length in the 0-180 cm soil layer under different N fertilizer rate.

3.3 Microbial activity

The power-time curves under different N rates showed a huge difference (Fig. 2). The curves of surface soil layer presented a higher activity than deeper soil layer. The parameters of microcalorimetry can be observed in Table 3 and Fig.3. In the 0-30 cm soil layer, N210 had the highest P_{max}, k and the biggest Q_{total}. In the 30-60cm soil layer, N300 had a higher P_{max} and Q_{total} than other treatments. In the 60-90cm soil layer, the curves showed a flat curve with less k and P_{max}, all these indicated low microbial activity in deeper soil layer. It could be seen from these curves and parameter, the microbial growth was influenced by N fertilizer rate and soil layer.

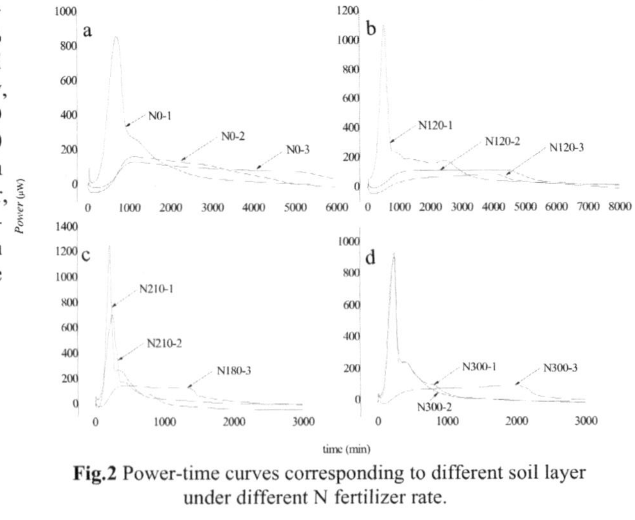

Fig.2 Power-time curves corresponding to different soil layer under different N fertilizer rate.

Table 3. Growth rate constant (k) and correlation coefficients (r^2) of microbial growth

Treatment	Soil layer (cm)	k (min^{-1})×10^{-3}	r^2
N0	0-30	0.0052±0.00009	0.9973
	30-60	0.0013±0.00006	0.9952
	60-90	0.0015±0.00004	0.9908
N120	0-30	0.00915±0.00003	0.9998
	30-60	0.00912±0.00005	0.9908
	60-90	0.00708±0.00008	0.9934
N210	0-30	0.00965±0.00008	0.9998
	30-60	0.00925±0.00003	0.9906
	60-90	0.00828±0.00001	0.9904
N300	0-30	0.00926±0.00006	0.9985
	30-60	0.00868±0.00005	0.9986
	60-90	0.00392±0.00006	0.9992

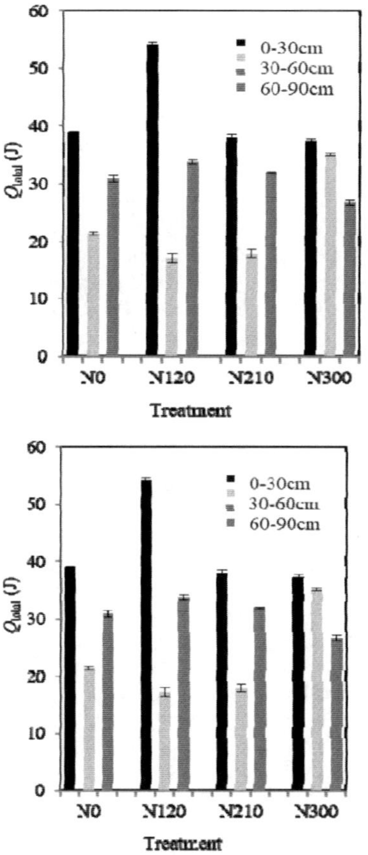

Fig.3. Microcalorimetric parameters from power–time curves. Error bars indicate the standard error of the mean.

3.4 Correlation analysis and Analysis of variance

N fertilizer application affected yield, RLD and k ($P<0.05$) according to analysis of variance. Meanwhile, correlation analysis showed that N fertilizer was positively correlated with yield, RLD and Q_{total}, P_{max}, k(Table2) as well, and there was a significant positively correlation ($P<0.01$) between yield and RLD, k. This is the reason why there was a perfect linear correlation between yield and RLD (Fig. 4).

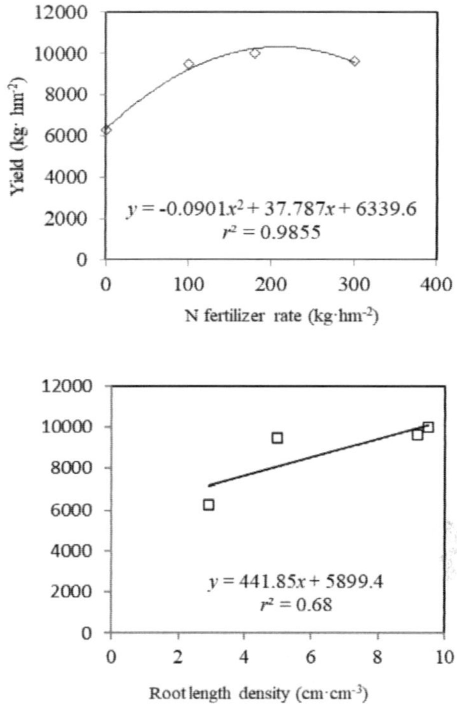

Fig.4. The relationship between N fertilizer rate and yield was expressed by fitting model and the relationship between root length density and yield was expressed by linear model (in the experiment of 2013).

3.5 Yield and root length density in response to N fertilizer rate.

The correlation between N fertilizer and yield (Fig.4) can be expressed by the regression equation: $Y_{yield} = -0.0901X_N^2 + 37.787X_N + 6339.6$to calculate the optimum N fertilizer rate, the result showed that when N fertilizer rate exceeded 210 kg hm^{-2} there was no further yield. Compared with N300 and N120, N210 had higher RLD and microbial activity than other treatments which the two parameters also had a significantly positive ($p<0.05$) correlation between RLD and k (Table2) can mostly take up N from the soil and the correlation between root length density and yield (Fig. 4) both can be explained why N210 had a greater yield. In this view, the optimal range of N fertilizer rate in this region was ≤210 kg hm^{-2}.

4 Discussion

Root length density (RLD) and soil microbial activity in maize fields were differences under different N fertilizer and soil layers.Half RLD was found in the 0-60cm soil layer under N210 and N300, whereas half RLD was concentrated in the 0-30cm soil layer under N0 and

N120. These results are consistent with previous studies, a rainfed filed experiment in Samaru showed that about 64% of root length in the surface 0-15cm soil layer ([31]) and 80% of the total root length accumulated in the top 40cm ([34]) and RLD decreased with soil depths ([41, 44]). Maize RLD depended on nutrients which was reported by Peng ([32]).

With the soil deepen, the power-time curves show a huge difference (Fig. 2). Because microbial activity was significantly affected by soil depth ([18]), so we studied an experiment on the soil depth. In the 0-30 cm soil layer, N210 showed a stronger microbial activity with the highest P_{max}, k and the biggest Q_{total}. In the 30-60cm soil layer, N300 showed a higher P_{max} and Q_{total}, perhaps due to apply large N fertilizer in the soil. In the 60-90cm soil layer, the curves showed a flat curve with less k and P_{max}, this could be considered as poor soil microbial activity correspond with low heat ([28]).

Yield is an important criterion for evaluating the maize performance. The results showed that fertilizer is related to maize yield closely ([35]), especially depended on N fertilizer ([23, 49]). These findings were consistent with previous studies conducted on the N application which increased from 250 to 380 kg hm^{-2} while yield had no further increased ([20]), and by using a statistical method demonstrated reduced N fertilizer would not reduce yield ([5]), even better decrease the negative impact on the environment, also reported that higher N supply led to a non-significant increase in yield and changing the rate of N application had no a significant effect on the yield([25]).

Many studies have shown that maize yields are influenced by various factors, such as fertilizer ([3, 29]), water ([19]), tillage ([4]), agricultural management ([38]). In this study we suggest that high soil microbial activity and more RLD caused increased yield of maize. Certainly, the correlation between yield and soil microbial activity and RLD are consistent with this hypothesis (Table 2). Moreover, highest yield under 210kgN hm^{-2} was observed (Fig. 4). Because optimum N fertilizer management which can greatly sustain soil productivity, thus high crop yield ([48]).

Higher RLD development is related to high grain yield production([15]). Indeed, according to Table1, N210 had the highest yield. Correlations (Table 2) between yield and RLD suggest that 210 kg(N)·hm^{-2} in soil increased the root growth ([27]), more RLD may cause more water and nutrient could be used by crop. Microbial activity is related to RLD (Table 2) because root can generate root exudate that was known to have a multitude of functions in ecological interaction with the microbial soil activity ([2]), then high microbial activity can involve in the cycling of nitrogen to crop. Thus resulted in high yield ultimately.

Soil microbial activity as an biological indicator that evaluate the efficiency of soil fertility which directly influenced the yield of crop. For example, maize yield in Kano State positively correlated with soil organic C, and soil organic C accounted for 75% of the variation in maize yield ([9]). The microbial activity is related to soil organic matter decomposition while the soil cycles of soil C and N dominated by microbial activity are beneficial for the improvement in soil fertility ([6]), further increase the crop biomass and yield ([4]).

5 Conclusions

In conclusion, this study demonstrated that N fertilizer can increase maize yield by influenced soil microbial activity and root length density. High N fertilizer rate cannot be increased yield of maize, soil microbial activity and root length density. Compare to high N fertilizer rate, within this region the optimum rate of N fertilizer for great yield was$\leq$210 kg hm^{-2} while it kept the high maize yields.

Acknowledgements

This work was supported in part by grants from the International Joint Key Project from Chinese Ministry of Science and Technology (2010DFB23160), National Natural Science Foundation of China (41273092), Public welfare project of Chinese Ministry of Environmental Protection (201409042), and Overseas, HongKong and Macau Young Scholars Collaborative Research Fund (41328005), and Basic and Theoretical Science and technology Research Project of Jiangmen city (2016030100270007378, 2017030100620016213).

References

1. Allen, D.E., Kingston, G., Rennenberg, H., Dalal, R.C., Schmidt, S., 2010. Effect of nitrogen fertilizer management and water logging on nitrous oxide emission from subtropical sugarcane soils. Agriculture, *Ecosystems & Environment* 136, 209-217.
2. Baetz, U., Martinoia, E., 2014. Root exudates: the hidden part of plant defense. *Trends in plant science* 19, 90-98.
3. Bedada, W., Karltun, E., Lemenih, M., Tolera, M., 2014. Long-term addition of compost and NP fertilizer increases crop yield and improves soil quality in experiments on small holder farms. Agriculture, *Ecosystems & Environment* 195, 193-201.
4. Botha, J.J., Anderson, J.J., Van Staden, P.P., 2015. Rainwater harvesting and conservation tillage increase maize yields in South Africa. *Water Resources and Rural Development.*
5. Chen, J., Huang, Y., Tang, Y., 2011. Quantifying economically and ecologically optimum nitrogen rates for rice production in south-eastern China. Agriculture, *Ecosystems & Environment* 142, 195-204.
6. Cui, J., Holden, N.M., 2015. The relationship between soil microbial activity and microbial biomass, soil structure and grassland management. Soil and Tillage Research 146, 32-38.
7. Dinesh, R., Ghoshal Chaudhuri, S., 2013. Soil biochemical/microbial indices as ecological indicators of land use change in mangrove forests.

Ecological Indicators 32, 253-258.
8. Duan, Y., Xu, M., Gao, S., Yang, X., Huang, S., Liu, H., Wang, B., 2014. Nitrogen use efficiency in a wheat–corn cropping system from 15 years of manure and fertilizer applications. *Field Crops Research* 157, 47-56.
9. Ekeleme, F., Jibrin, J.M., Kamara, A.Y., Oluoch, M., Samndi, A.M., Fagge, A.A., 2014. Assessment of the relationship between soil properties, Striga hermonthica infestation and the on-farm yields of maize in the dry Savannas of Nigeria. Crop Protection 66, 90-97.
10. Erisman, J.W., Bleeker, A., Galloway, J., Sutton, M.S., 2007. Reduced nitrogen in ecology and the environment. Environ Pollut 150, 140-149.
11. Gao, Y., Duan, A., Qiu, X., Liu, Z., Sun, J., Zhang, J., Wang, H., 2010. Distribution of roots and root length density in a maize/soybean strip intercropping system. Agricultural Water Management 98, 199-212.
12. Gao, Y., Xie, Y., Jiang, H., Wu, B., Niu, J., 2014. Soil water status and root distribution across the rooting zone in maize with plastic film mulching. Field Crops Research 156, 40-47.
13. Guo, H., Yao, J., Cai, M., Qian, Y., Guo, Y., Richnow, H.H., Blake, R.E., Doni, S., Ceccanti, B., 2012. Effects of petroleum contamination on soil microbial numbers, metabolic activity and urease activity. Chemosphere 87, 1273-1280.
14. Haque, M.M., Kim, S.Y., Kim, G.W., Kim, P.J., 2015. Optimization of removal and recycling ratio of cover crop biomass using carbon balance to sustain soil organic carbon stocks in a mono-rice paddy system. Agriculture, Ecosystems & Environment 207, 119-125.
15. Hayashi, T., Yoshida, T., Fujii, K., Mitsuya, S., Tsuji, T., Okada, Y., Hayashi, E., Yamauchi, A., 2013. Maintained root length density contributes to the waterlogging tolerance in common wheat (Triticum aestivum L.). Field Crops Research 152, 27-35.
16. Kern, J., Hellebrand, H.J., Scholz, V., Linke, B., 2010. Assessment of nitrogen fertilization for the CO2 balance during the production of poplar and rye. Renewable and Sustainable Energy Reviews 14, 1453-1460.
17. Li, C., Sun, J., Li, F., Zhou, X., Li, Z., Qiang, X., Guo, D., 2011. Response of root morphology and distribution in maize to alternate furrow irrigation. Agricultural Water Management 98, 1789-1798.
18. Li, F., Liu, M., Li, Z., Jiang, C., Han, F., Che, Y., 2013. Changes in soil microbial biomass and functional diversity with a nitrogen gradient in soil columns. Applied Soil Ecology 64, 1-6.
19. Liu, C.-A., Zhou, L.-M., Jia, J.-J., Wang, L.-J., Si, J.-T., Li, X., Pan, C.-C., Siddique, K.H.M., Li, F.-M., 2014a. Maize yield and water balance is affected by nitrogen application in a film-mulching ridge–furrow system in a semiarid region of China. European Journal of Agronomy 52, 103-111.
20. Liu, J., Zhu, L., Luo, S., Bu, L., Chen, X., Yue, S., Li, S., 2014b. Response of nitrous oxide emission to soil mulching and nitrogen fertilization in semi-arid farmland. Agriculture, Ecosystems & Environment 188, 20-28.
21. Liu, Y., Dell, E., Yao, H., Rufty, T., Shi, W., 2011. Microbial and soil properties in bentgrass putting greens: Impacts of nitrogen fertilization rates. Geoderma 162, 215-221.
22. Lucas, R.W., Casper, B.B., Jackson, J.K., Balser, T.C., 2007. Soil microbial communities and extracellular enzyme activity in the New Jersey Pinelands. Soil Biology and Biochemistry 39, 2508-2519.
23. Mahanta, D., Rai, R.K., Mishra, S.D., Raja, A., Purakayastha, T.J., Varghese, E., 2014. Influence of phosphorus and biofertilizers on soybean and wheat root growth and properties. Field Crops Research 166, 1-9.
24. Mandal, A., Patra, A.K., Singh, D., Swarup, A., Ebhin Masto, R., 2007. Effect of long-term application of manure and fertilizer on biological and biochemical activities in soil during crop development stages. Bioresource technology 98, 3585-3592.
25. Mansouri-Far, C., Modarres Sanavy, S.A.M., Saberali, S.F., 2010. Maize yield response to deficit irrigation during low-sensitive growth stages and nitrogen rate under semi-arid climatic conditions. Agricultural Water Management 97, 12-22.
26. Mu, X., Chen, F., Wu, Q., Chen, Q., Wang, J., Yuan, L., Mi, G., 2015. Genetic improvement of root growth increases maize yield via enhanced post-silking nitrogen uptake. European Journal of Agronomy 63, 55-61.
27. Munt, O., Arias, M., Hernandez, M., Ritter, E., Schulze Gronover, C., Prüfer, D., 2012. Fertilizer and planting strategies to increase biomass and improve root morphology in the natural rubber producer Taraxacum brevicorniculatum. Industrial Crops and Products 36, 289-293.
28. Núñez-Regueira, L., Rodríguez-Añón, J.A., Proupín-Castiñeiras, J., Núñez-Fernández, O., 2005. Influence of the agricultural exploitation processed on the productivity capacity control of soils. Journal of Thermal Analysis and Calorimetry 80, 35-41.
29. Ngome, A.F., Becker, M., Mtei, K.M., Mussgnug, F., 2011. Fertility management for maize cultivation in some soils of Western Kenya. Soil and Tillage Research 117, 69-75.
30. Nyakudya, I.W., Stroosnijder, L., 2014. Effect of rooting depth, plant density and planting date on maize (Zea mays L.) yield and water use efficiency in semi-arid Zimbabwe: Modelling with AquaCrop. Agricultural Water Management 146, 280-296.
31. Oikeh, S.O., Kling, J.G., Horst, W.J., Chude, V.O., Carsky, R.J., A, A.L.-J., 1999. Growth and distribution of maize roots under nitrogen fertilization in plinthite soil. Field Crops Research 62, 1-13.
32. Peng, Y., Yu, P., Zhang, Y., Sun, G., Ning, P., Li, X., Li, C., 2012. Temporal and spatial dynamics in root length density of field-grown maize and NPK

in the soil profile. Field Crops Research 131, 9-16.

33. PRC, M.o.A., 2012. China Agriculture Statistical Report. China Agriculture Press, Beijing (in Chinese).
34. Qin, R., Stamp, P., Richner, W., 2006. Impact of tillage on maize rooting in a Cambisol and Luvisol in Switzerland. Soil and Tillage Research 85, 50-61.
35. Rasool, R., Kukal, S., Hira, G., 2008. Soil organic carbon and physical properties as affected by long-term application of FYM and inorganic fertilizers in maize–wheat system. Soil and Tillage Research 101, 31-36.
36. Sotomayor-Ramírez, D., Espinoza, Y., Acosta-Martínez, V., 2009. Land use effects on microbial biomass C, β-glucosidase and β-glucosaminidase activities, and availability, storage, and age of organic C in soil. Biology and Fertility of Soils 45, 487-497.
37. Stuart, D., Schewe, R.L., McDermott, M., 2014. Reducing nitrogen fertilizer application as a climate change mitigation strategy: Understanding farmer decision-making and potential barriers to change in the US. Land Use Policy 36, 210-218.
38. Thierfelder, C., Matemba-Mutasa, R., Rusinamhodzi, L., 2015. Yield response of maize (Zea mays L.) to conservation agriculture cropping system in Southern Africa. Soil and Tillage Research 146, 230-242.
39. Wei-Dong, K., Yong-Guan, Z., Bo-Jie, F., Xiao-Zeng, H., Lei, Z., Ji-Zheng, H., 2008. Effect of Long-Term Application of Chemical Fertilizers on Microbial Biomass and Functional Diversity of a Black Soil. Pedosphere 18, 801-808.
40. Wichern, F., Mayer, J., Joergensen, R.G., Müller, T., 2007. Release of C and N fron roots of peas and oats and their availability to soil microorganisms. Soil Biology and Biochemistry 39, 2829-2839.
41. Worku, M., Bänziger, M., Schulte auf'm Erley, G., Friesen, D., Diallo, A.O., Horst, W.J., 2012. Nitrogen efficiency as related to dry matter partitioning and root system size in tropical mid-altitude maize hybrids under different levels of nitrogen stress. Field Crops Research 130, 57-67.
42. Xu, X., He, P., Pampolino, M.F., Chuan, L., Johnston, A.M., Qiu, S., Zhao, S., Zhou, W., 2013. Nutrient requirements for maize in China based on QUEFTS analysis. Field Crops Research 150, 115-125.
43. Xu, X., He, P., Qiu, S., Pampolino, M.F., Zhao, S., Johnston, A.M., Zhou, W., 2014. Estimating a new approach of fertilizer recommendation across small-holder farms in China. Field Crops Research 163, 10-17.
44. Xue, Y.-F., Zhang, W., Liu, D.-Y., Yue, S.-C., Cui, Z.-L., Chen, X.-P., Zou, C.-Q., 2014. Effects of nitrogen management on root morphology and zinc translocation from root to shoot of winter wheat in the field. Field Crops Research 161, 38-45.
45. Yu, Q., Saseendran, S.A., Ma, L., Flerchinger, G.N., Green, T.R., Ahuja, L.R., 2006. Modeling a wheat–maize double cropping system in China using two plant growth modules in RZWQM. Agricultural Systems 89, 457-477.
46. Zhang, K., Yang, D., Greenwood, D.J., Rahn, C.R., Thorup-Kristensen, K., 2009. Development and critical evaluation of a generic 2-D agro-hydrological model (SMCR_N) for the responses of crop yield and nitrogen composition to nitrogen fertilizer. Agriculture, Ecosystems & Environment 132, 160-172.
47. Zhang, M., Fan, C.H., Li, Q.L., Li, B., Zhu, Y.Y., Xiong, Z.Q., 2015. A 2-yr field assessment of the effects of chemical and biological nitrification inhibitors on nitrous oxide emissions and nitrogen use efficiency in an intensively managed vegetable cropping system. Agriculture, Ecosystems & Environment 201, 43-50.
48. Zhao, S., Qiu, S., Cao, C., Zheng, C., Zhou, W., He, P., 2014. Responses of soil properties, microbial community and crop yields to various rates of nitrogen fertilization in a wheat–maize cropping system in north-central China. Agriculture, Ecosystems & Environment 194, 29-37.
49. Ziadi, N., Cambouris, A.N., Nyiraneza, J., Nolin, M.C., 2013. Across a landscape, soil texture controls the optimum rate of N fertilizer for maize production. Field Crops Research 148, 78-85.

Research on saving-energy air-conditioning systems for the heavy truck cab

Chunxue Gao[1,*], Huayang Pan[1], Qiuxin Liu[1,2]

[1]Wuhan University of Science and Technology, Wuhan 430065,China
[2]City college, Wuhan university of science and technology, Wuhan 430083,China

Abstract: This study analyzed the cooling demand of different cab structures and types of heavy trucks under the most unfavorable conditions. The thermal environment of the cab under different air supply schemes is also studied. The utilization of waste heat the heavy truck exhaust as the heat source of adsorption refrigeration systems makes the thermal comfort of local areas of the cab guaranteed while the input cooling energy is reduced.

1 Introduction

Air-conditioning systems in heavy truck cabs are necessary to provide a comfortable driving environment, thus reducing accidents caused by emotional irritability and anxiety of automobile vehicle drivers[1]. Research shows that at least 51% of the energy consumed by automobile engines is converted into waste heat[2]. If the waste heat is recycled and used as the heat source for adsorption refrigeration systems to supply cooling air for cabs, the driving environment for truck drivers can be ensured, the energy efficiency of heavy trucks improved and emissions reduced.

The refrigeration system adopted for heavy trucks in this study combines basic solid single-effect intermittent adsorption refrigeration and energy storage device. According to the different driving conditions of trucks, the characteristics of uneven matching between waste heat and cooling load are analyzed to obtain the refrigerating capacity of the system under the most unfavorable conditions. Besides, distributions of temperature field, velocity field, PMV and PPD in the cab are obtained by simulating and studying the operation of refrigeration systems under the most unfavorable working conditions. Then an optimalscheme is found out.

2 Air-conditioning load and truck waste heat

When calculating the load of truck air conditioners, there are many uncertain factors and there is no uniform standard at home and abroad. In view of this, the calculation is made with reference to *Design Code for Heating Ventilation and Air-Conditioning of Civil Buildings* and *Application of Refrigeration and Air-Conditioning Technology in Transportation Engineering*[3]. The indoor calculation parameters of the truck (Wuhan) and the main performance parameters of the truck are shown in Table 1 and Table 2, and the structural parameters of the truck cab are shown in Table 3.

Table 1. Indoor air calculation parameters of trucks in summer

Calculation parameters	Parameter values
Temperature (°C)	26~28
Relative humidity (%)	50~60
Air velocity (m/s)	0.25~0.5
Fresh air volume [m^3/(person· h)]	20~30
Vertical temperature difference (°C)	<1.5

* Corresponding author: gaochunxue@wust.edu.cn

Table 2. Main performance parameters of the truck

Performance parameters	Parameter value
Internal dimensions of cab (mm): Length × width × height	2500×2300×1400
Number of passengers allowed (person)	3
Number of rows (row)	1
Rated power (kw)	280
Rated load (ton)	17
Horsepower	350

Table 3. Structural parameters of truck cab

Structure	Materials	Thickness (mm)	Area (m^2)	Thermal conductivity ($w/m{\cdot}k^{-1}$)
Front windshield	Class a laminated glass	4.8	2	0.754
Front wall	Steel plate	1	1.33	48
Gate glass	Tempered glass	3.5	0.57×2	0.754
Side wall	Steel plate	1	2.08×2	48
Rear windshield	Tempered glass	3.2	0.375	0.754
Back wall	Steel plate	1	1.81	48
Roof	Glass steel	0.7	5.75	0.52
Air contact floor	Steel plate	0.7	1.25	48
Engine upper floor	Steel plate	0.7	2.19	48

Based on the above-mentioned indoor calculation parameters of trucks in summer in Table 1, it is assumed that the indoor design temperature is 26°C, the relative humidity 50%, the air velocity 0.3m/s, the fresh air volume 30m^3/h and the number of indoor passengers 2 to calculate the cooling load of the cab. Besides, the direction of the truck is from south to north. On July 20, a sunny day, there are 2 people in the truck.

The calculation result of the daytime cooling load of the truck cab in summer is 7590W. At night, the radiation heat transfer of enclosure structure, glass, etc. is not included in the calculation since there is no solar radiation. The night cooling load in summer is 5672W at driving condition, and 5476W at rest condition where the engine is in a flameout state.

Residual heat is an important factor that affects the refrigerating capacity of adsorption refrigeration system. Truck residual heat mainly includes two aspects: engine residual heat and exhaust residual heat. Due to the large difference between the exhaust temperature (which can reach 500°C) and the cooling air temperature, the exhaust residual heat is more conducive to the adsorption and desorption process than the engine residual heat. Therefore, the exhaust heat is used as the driving heat source of adsorption refrigeration air-conditioning system in this paper.

In order to further analyze the adsorption refrigeration system based on waste heat utilization, the exhaust volume and exhaust temperature of a brand diesel engine under different rotational speeds and loads have been tested, and the exhaust heat has been calculated by Formula (1).

$$Q_e = M_e \cdot C_f \cdot (t_1 - t_2) \tag{1}$$

In the formula:

M_e—— exhaust mass flow (kg/s);

C_f——average specific heat of flue gas at constant pressure (kJ/kg·k);

t_1—— exhaust gas temperature (°C);

t_2—— exhaust outlet temperature (°C);

It is assumed that the flue gas flow is steady, and the automobile exhaust is misty gas with complex composition, the majority of which are CO, HC, and NO compounds. At about 450°C, the density of exhaust gas is similar to the air density, so the average specific heat of flue gas at constant pressure (C_f) is taken as 1.06 kJ/(kg.k). Because the low exhaust temperature will cause dew point corrosion of acidic oxides, the exhaust outlet temperature (t_2) is taken as 180°C.

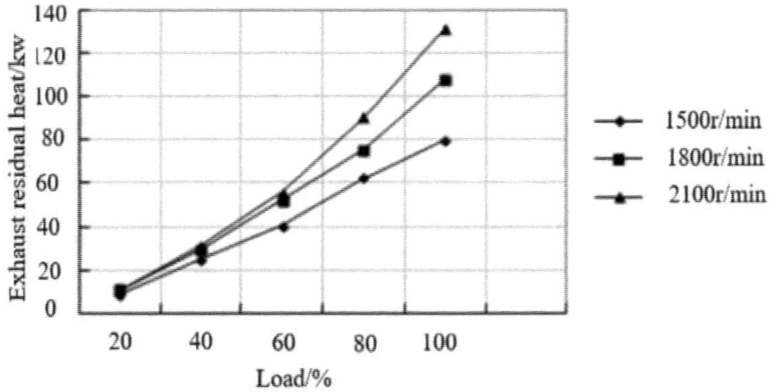

Fig. 1. Exhaust residual heat of diesel engine

According to the Formula (1), the exhaust residual heat of a diesel engine can be calculated and analyzed. As shown in Fig. 1, the exhaust residual heat increases with the increase of rotational speed and load,. When the diesel engine is running at full load (the rotational speed is 2100r/min), the exhaust residual heat is maximum, reaching 131.3 kW. It can be concluded from the figure that the exhaust residual heat is mainly affected by the load. Besides, under a large load condition, the rotational speed has a considerable impact on the exhaust residual heat. A heater speed will result in greater amount of exhaust residual heat. As the operating conditions of diesel engines are changeable and the residual heat is changing, it is necessary to carry out energy analysis on the exhaust residual heat.

Considering that trucks are always driving under changeable road conditions such as high speed and urban roads, and the driving conditions are complex and changeable, it is difficult to determine the residual heat changing with time and road conditions, and the accurate calculation is more difficult. Therefore, this study takes one day as the running period to analyze and calculate the residual heat when there are one and two people in the cab respectively.

When there is only one driver in the cab, it is assumed that the truck is running at 1800r/min and at full load for 8h, the exhaust temperature is 463°C, the exhaust volume is 0.36kg/s, and the exhaust heat is 108kW. Besides, assuming that the truck is running at 1500r/min at full load for 3 hours, the exhaust temperature is 459°C, the exhaust volume is 0.27kg/s, and the exhaust heat is 79.85 kW. For the remaining 13 hours, the diesel engine is in a state of shutdown.

When two people take turns driving within 24 hours, it is assumed that the truck is running at 1800r/min at full load for 13 hours, the exhaust heat is 108kW. Moreover, assuming that the truck is running at 1500r/min at full load for 11 hours, the exhaust heat is 79.85kW.

Based on the study of the basic single adsorption bed refrigeration system, the coefficient of performance (COP) of the system is 0.1 ~ 0.3 when zeolite molecular sieve-water is adopted as the adsorption working pair, and the heat source temperature is 180 ~ 463°C, the cooling temperature 32°C, and the desorption/adsorption cycle time 20 ~ 40min. When one day is taken as an example, the total refrigerating capacity throughout the day is 228kW and the average refrigerating power is 9.5 kW (calculated in 24 hours) under the most unfavorable situation where the COP is 0.1 and the cycle time is 1h, and two people take turns driving within 24 hours, The refrigerating capacity of 228kW is sufficient to meet the indoor cooling demand of 161kW of the whole day. When only one driver is driving, the total refrigerating capacity throughout the day is 110kW and the average refrigerating power is 4.58kW (calculated by 24 hours). The total refrigerating capacity in this case is equal to the full-day cooling load of 110kW (calculated by working for 8 hours during the day, 2 hours in the evening and resting in the car for 7 hours at night).

Considering that the calculation is carried out under the most unfavorable conditions, and the cooling load varies greatly with the influence of time, geographical location, working time, and truck running state. Therefore, it can be considered that the balance between supply and demand can be basically guaranteed under practical circumstances. However, due to the unstable refrigeration power affected by the running state of the truck, there is still the possibility that the refrigeration capacity of the system is not sufficient to meet the indoor comfort requirements within a certain period of time. The setting of the cold accumulator can play a certain role in buffering the occurrence of this kind of situation. Considering it is essential to reasonably set up the terminals of the system and the indoor air distribution to improve the thermal confort of drivers in the heavy truck cab,[2,4-5] this paper further studies the local air-conditioning system of the cab under the most unfavorable condition where only one driver is in the cab (calculated by working for 8 hours during the day, 2 hours in the evening and resting in the car for 7 hours at night).

3 Numerical simulation

3.1 Physical model

Detailed physical model information is described as below.

1) Design the air-conditioning system in the cab

A terminal with four-sided air outlet and return air inlet in the middle is equipped above the driver and the co-pilot respectively. Size of each side of supply air outlet is 0.4 m × 0.05 m, and the size of the return air inlet 0.5 m × 0.5 m. The rear side wall of the driver is equipped with a rectangular slit supply air outlet with a size of 0.4 m × 0.1 m. There are supply air outlets at the feet of the main driver and the co-pilot respectively with a size of 0.15 m × 0.1 m. There is an return air inlet under the driver's seat side, with a size of 0.5 m × 0.2 m.

2) Inlet boundary

Inlet boundaries of the researched space are supply air outlets. According to the cooling load, the air supply temperature is determined to be 16°C. The air supply speed of the top four-sided outlets 3m/s, and the air supply angle is 30 degrees from the X axis as shown in Figure 2. The air supply speed of the rear side wall is 2m/s and that of the foot tuyere is 2m/s. The wind direction is perpendicular to the tuyere plane.

3) Outlet boundary

Outlet boundaries are return air inlets. The air speed of the middle return air inlet is 0.96 m/s, and that of the lower return air outlet is 1.4 m/s. The pressure is taken from the external environment pressure.

4) Wall boundary

The front, side, rear enclosure and roof of the truck are the second boundary conditions. The floor and seats are assumed to be heat insulation. Glass is calculated based on heat flux density, and sunshade is not considered.

5) Cooling load in the cab

the heat dissipation rate of the driver is set as 224W, and that of the instrument equipment as 196W.

6) Convergence error

The convergence residual is set to be 0.1%.

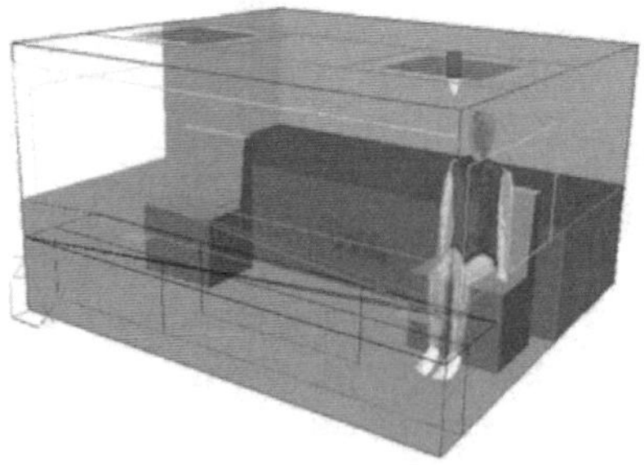

Fig. 2. Cab model

3.2 Grid division

Considering the complex internal structure of the cab, the grid is locally encrypted around the human body, tuyere, etc. to make the indoor temperature field, velocity field, PMV, PPD, etc. to be well simulated, and satisfy sufficient accuracy requirements. The grid number is well controlled at the same time, so as to improve convergence stability and reduce the amount of calculation. Grid division is shown in Figure 3 and 4.

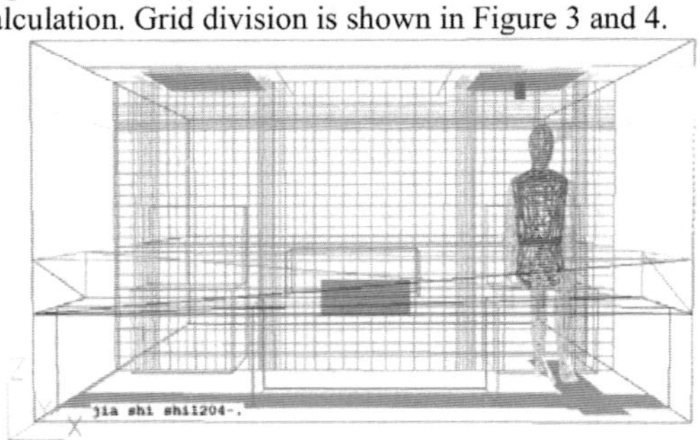

Fig.3. Y=0. 975m cross-sectional grid

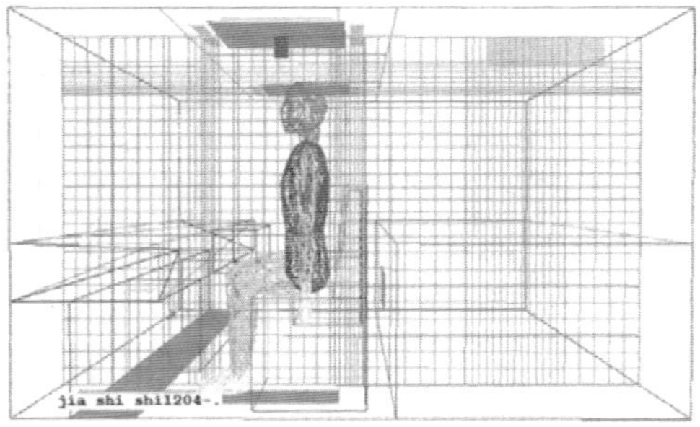

Fig.4. X=2. 155m cross-sectional grid

3.3 Simulation study

By changing the air supply temperature, air supply angle and air supply speed, the indoor temperature field and speed field are simulated and analyzed. The distribution of predicted mean vote (PMV) and predicted percentage of dissatisfied(PPD) are forecasted.

3.3. 1 Influence of air supply temperature

The air supply temperature of 16 °C, 18 °C and 20 °C is selected for the simulation research respectively. The air supply angle is 30 °C and the air supply speed is 3m/s, which remains unchanged. The simulation results at 18°C are as follows:

https://doi.org/10.1051/e3sconf/202123702001

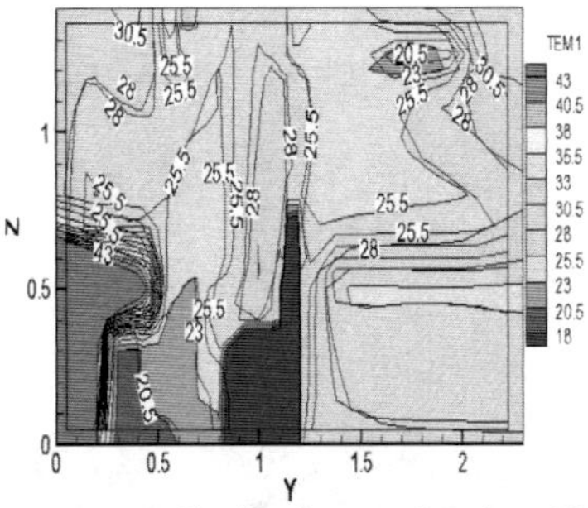

Fig.5. Temperature distribution on the plane X=2. 155m at the air supply temperature of 18°C

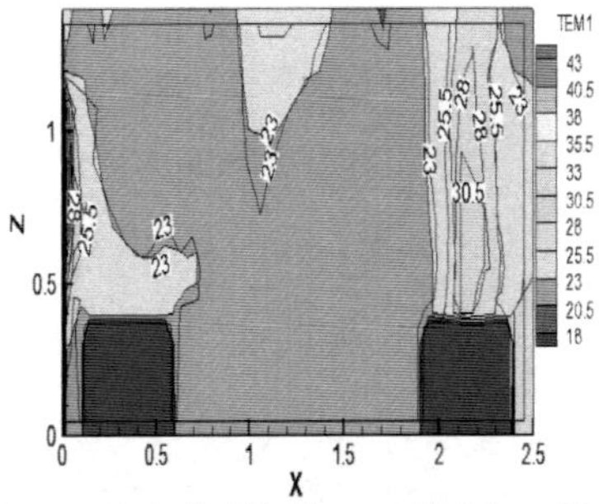

Fig.6. Temperature distribution on the plane Y=0.975m at the air supply temperature of 18°C

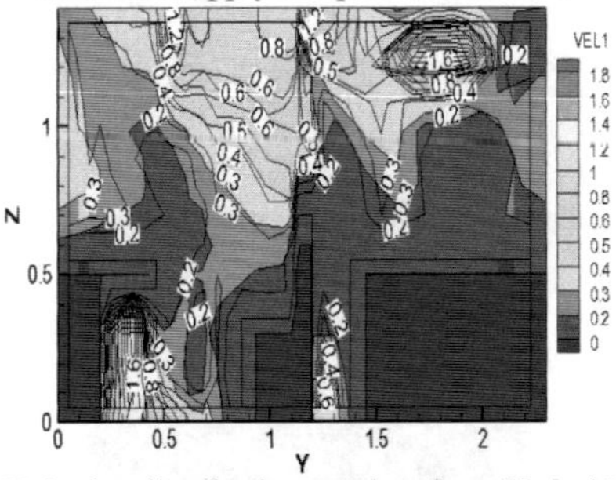

Fig.7. Velocity distribution on the plane X=2. 155 m at the air supply temperature of 18°C

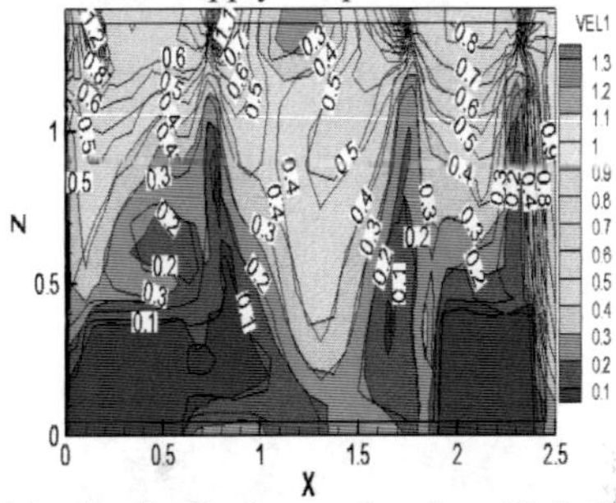

Fig.8. Velocity distribution on the plane Y=0. 975m at the air supply temperature of 18°C

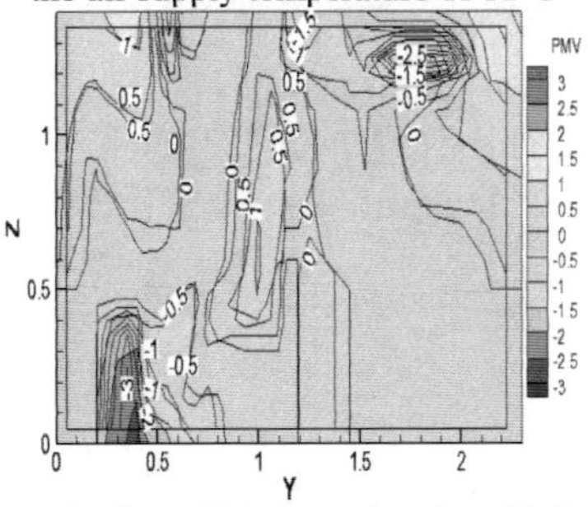

Fig.9. Distribution of PMV on the plane X=2. 155mat the air supply temperature of 18°C

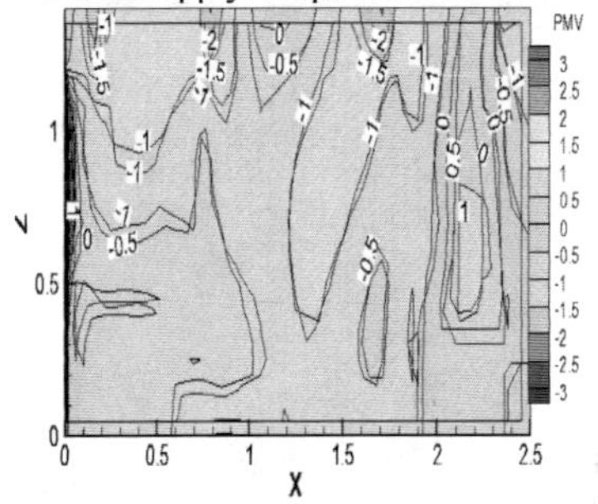

Fig.10. Distribution of PMV on the plane Y=0. 975m at the air supply temperature of 18°C

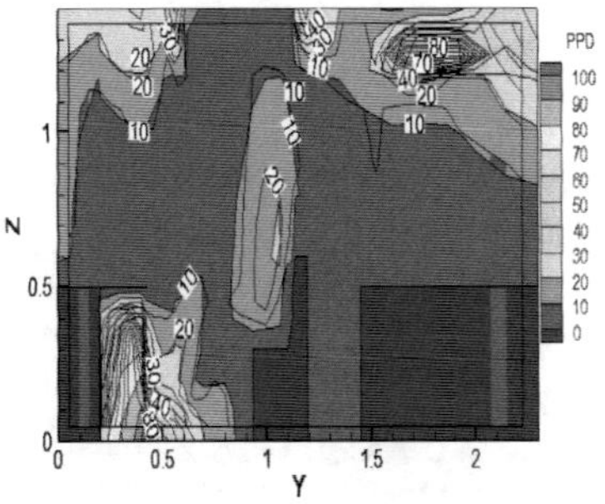

Fig.11. Distribution of PPD on the plane X=2. 155m at the air supply temperature of 18°C

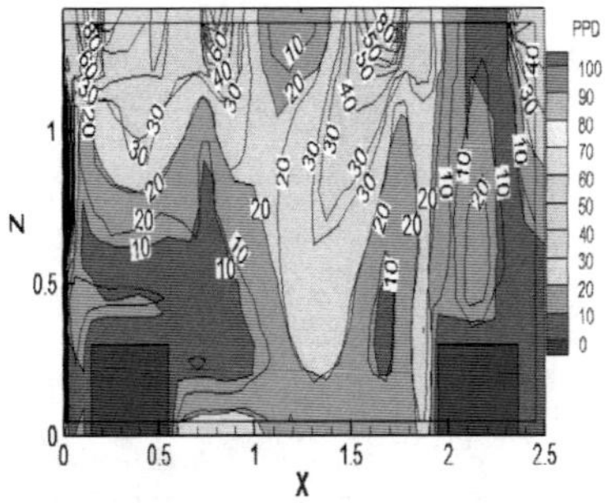

Fig.12. Distribution of PPD on the plane Y=0. 975m at the air supply temperature of 18°C

According to the simulation results, when the air supply temperature rises from 16°C to 20°C, the distribution of temperature field basically remains unchanged, but the temperature stratification phenomenon gradually appears, and the indoor wind speed basically remains at 0.4 m/s. According to the change trend of PMV and PPD, it can be seen that when the air supply temperature is 18 °C, the comfort is the best. The reduction of air supply temperature within a certain range is conducive to improving the comfort index of the cab environment.

3.3. 2 Influence of air supply angle

Under the air supply temperature of 18°C and the air supply speed of 3m/s, the three modes of air supply angles of 30°, 45° and 60° are simulated and studied respectively. The results of simulating the air supply angle of 30° are as follows:

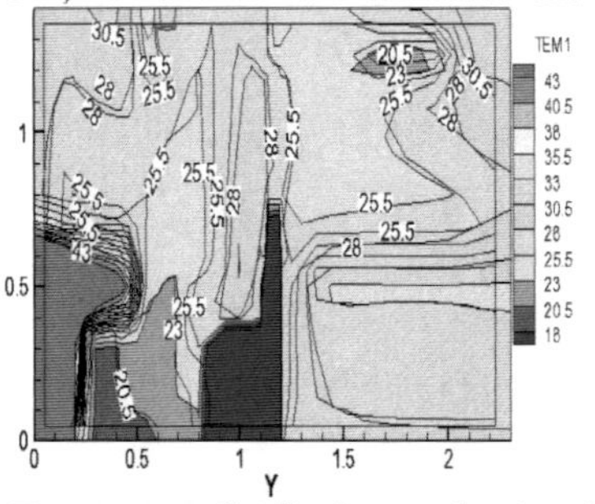

Fig.13. Temperature distribution on the plane X=2. 155m at the air supply angle of 30°

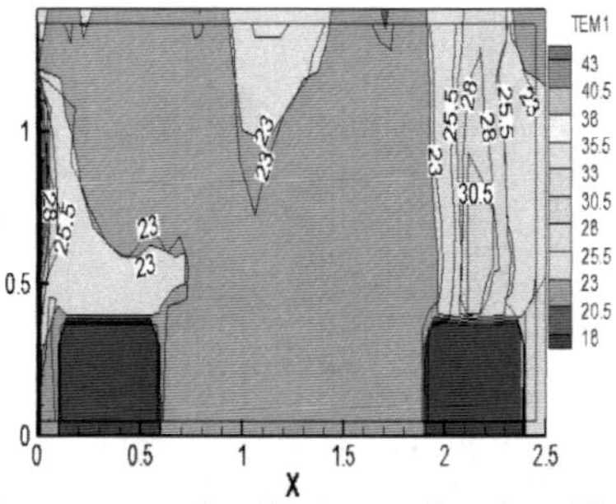

Fig.14. Temperature distribution on the plane Y=0.975m at the air supply angle of 30°

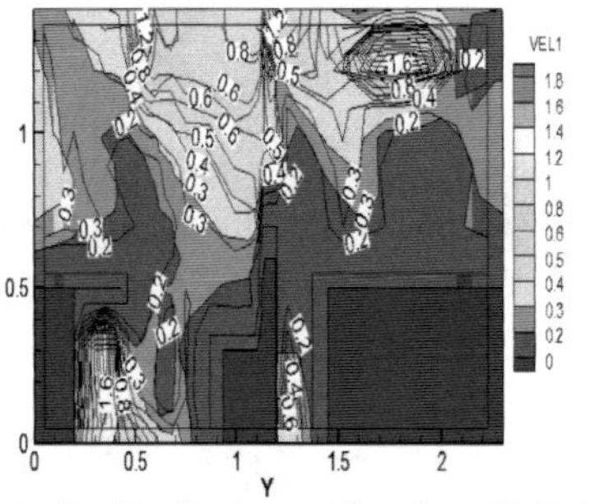

Fig.15. Velocity distribution on the plane X=2. 155m at the air supply angle of 30°

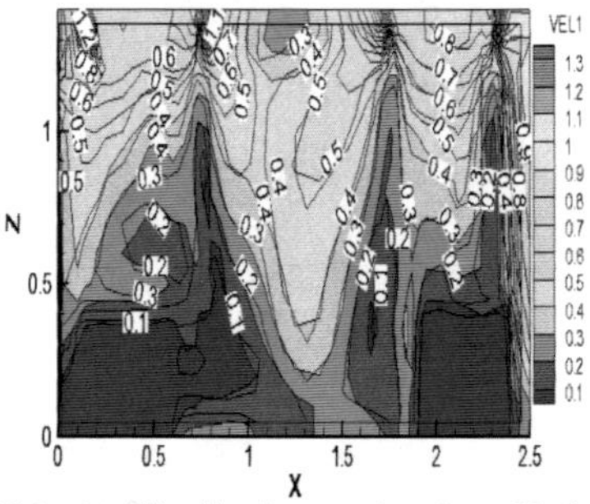

Fig.16. Velocity distribution on the plane Y=0. 975m at the air supply angle of 30°

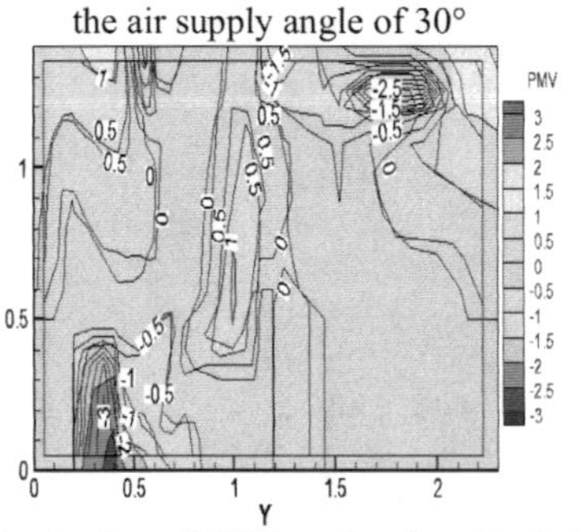

Fig.17. Distribution of PMV on the plane X=2. 155m at the air supply angle of 30°

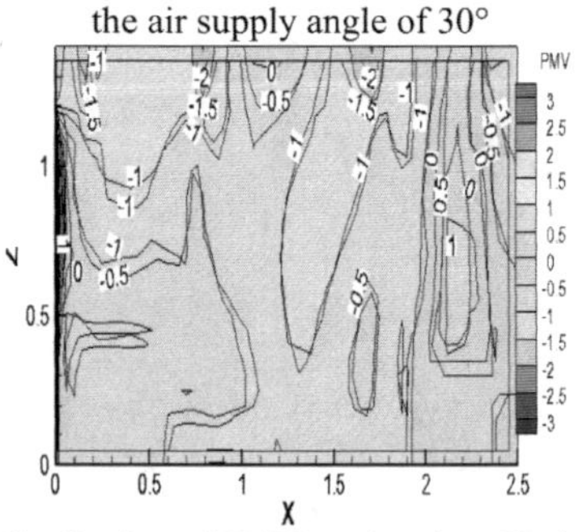

Fig.18. Distribution of PMV on the plane Y=0. 975m at the air supply angle of 30°

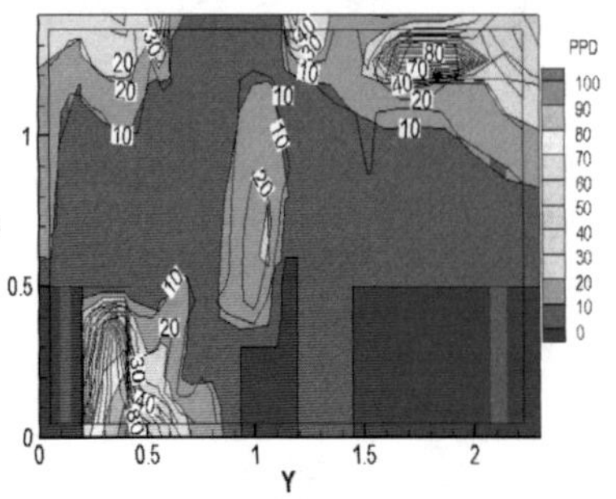

Fig.19. Distribution of PPD on the plane X=2. 155m at the air supply angle of 30°

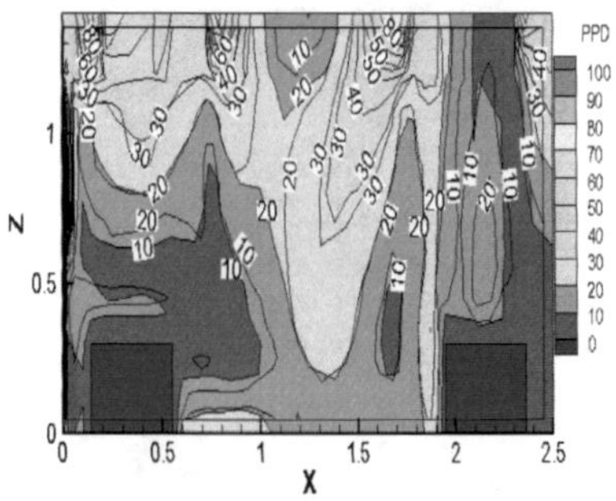

Fig.20. Distribution of PPD on the plane Y=0.975m at the air supply angle of 30°

According to the analysis of simulation results, it can be seen that changing the air supply angle has a great influence on the environmental comfort in the cab. Through comparison of PMV and PPD maps, it can be found that with the increase of air supply angle, the environmental comfort in the cab gradually decreases. The main reason is the combined effect of the driver's increased blowing feeling and temperature reduction. It

 https://doi.org/10.1051/e3sconf/202123702001

can be seen from the figure that reducing the air supply angle is helpful to improve the thermal comfort environment in the cab. When the air supply angle is 30°, the indoor environment comfort is the best.

3.3.3 Influence of air supply speed

The space in the truck cab is narrow, and the air supply outlets on all sides are located above the head. To satisfy the large cooling load, the air flow rate is taken at a large value. The simulation study compares the indoor thermal comfort under three air supply speeds (2.5 m/s, 3m/s, 3.5 m/s) when the air supply temperature is 18°C and the air supply angle is 30°. The simulation results for 2.5 m/s are as follows:

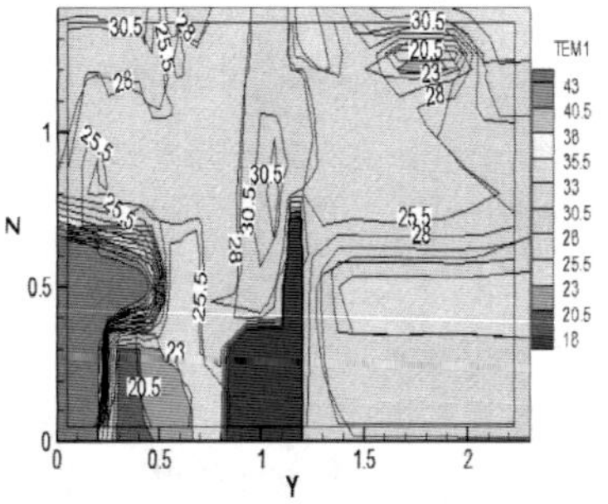

Fig.21. Temperature distribution on the plane Y=0. 975m at the air supply velocity of 2.5 m/s

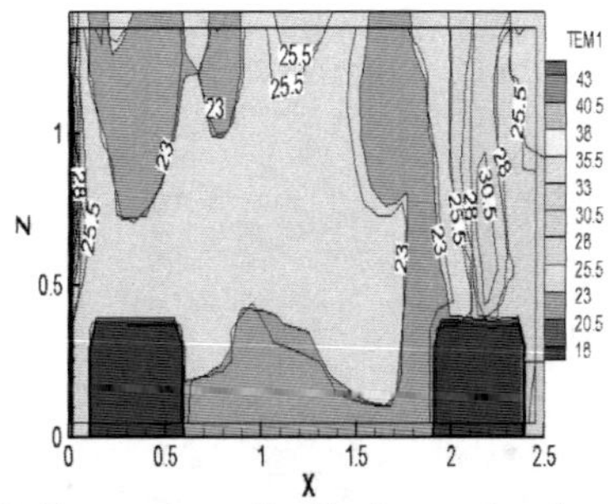

Fig.22. Temperature distribution on the plane X=2. 155m at the air supply velocity of 2.5 m/s

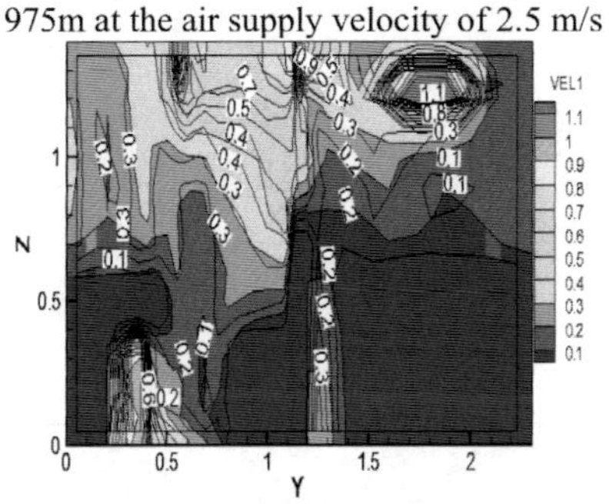

Fig.23. Velocity distribution on the plane Y=0. 975m at the air supply velocity of 2.5 m/s

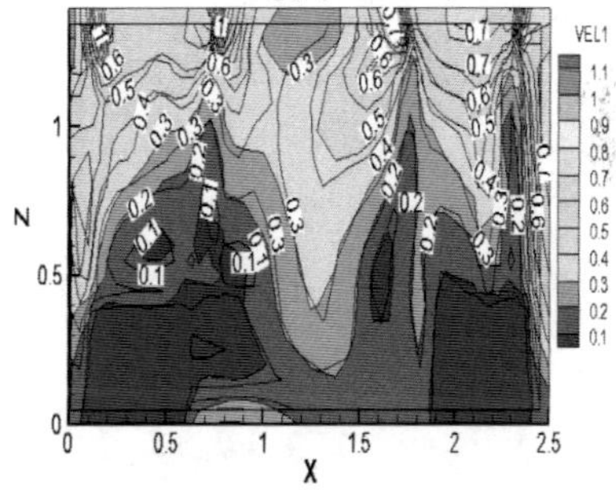

Fig.24. Velocity distribution on the plane X=2. 155m at the air supply velocity of 2.5 m/s

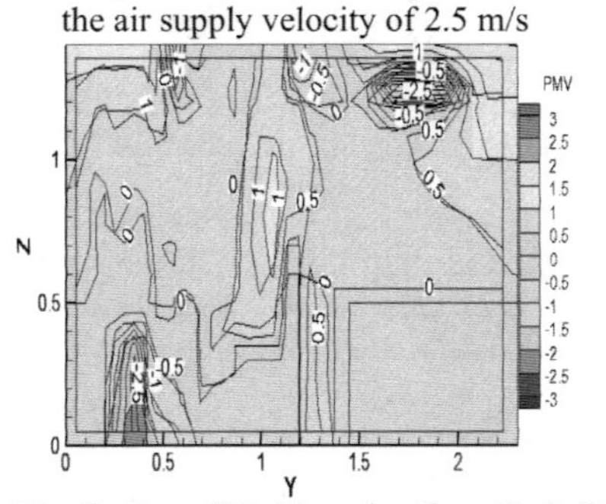

Fig.25. Distribution of PMV on the plane Y=0. 975m at the air supply velocity of 2.5 m/s

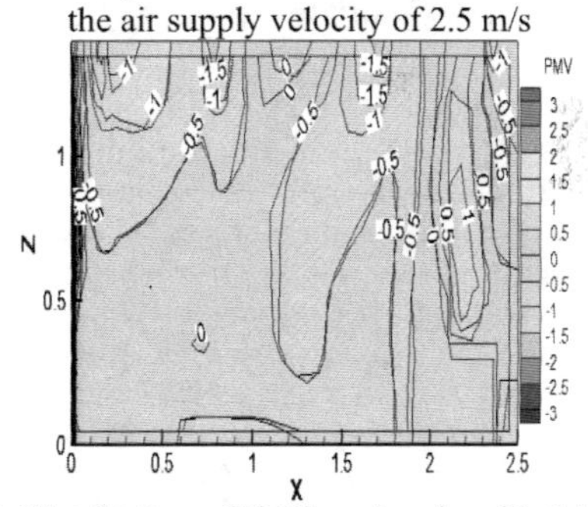

Fig.26. Distribution of PMV on the plane X=2.155m at the air supply velocity of 2.5 m/s

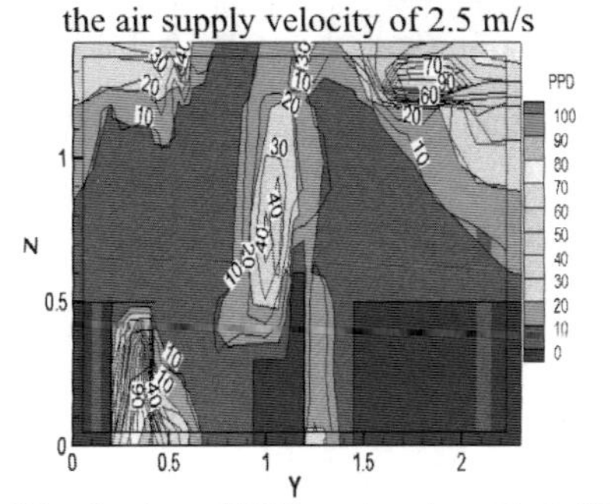

Fig.27. Distribution of PPD on the plane Y=0. 975m at the air supply velocity of 2.5 m/s

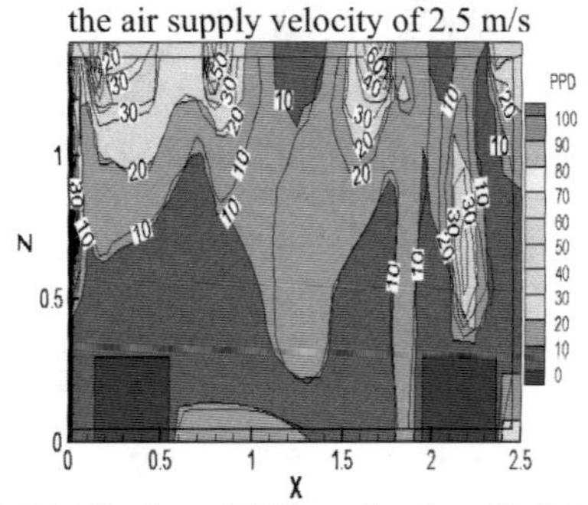

Fig.28. Distribution of PPD on the plane X=2. 155m at the air supply velocity of 2.5 m/s

 https://doi.org/10.1051/e3sconf/202123702001

From the simulation results, it can be seen that with the increase of air supply speed, the indoor average temperature decreases from 27°C to 24°C, and the temperature in the area where the human body is located decreases from 26°C to 23°C. This is because the air supply outlet is located above the driver. With the increase of air supply speed, the input of cold energy also increases.

When only the wind speed is changed, the distribution of air velocity in the cab remains basically unchanged. By comparing the distribution maps of the three air supply speeds, it can be found that the lower the wind speed is, the lower the probability of cold air blowing on the head is, which is conducive to improving human comfort.

From the distribution maps of PMV and PPD, it can be seen that when the wind speed is 2.5 m/s, the overall indoor comfort is the best, and the values of PPD and PMV are basically kept within the standard comfort range, while with the increase of wind speed, uncomfortable areas begin to appear locally in the cab. Therefore, by reducing the air supply speed within a certain range, a better comfortable environment can be obtained.

4 Economic analysis of the systems

The economic comparison of the three kinds of truck air conditioners is shown in the following table.

Table 4. Cost comparison of three types of air conditioning

Types of air conditioning	**Purchase cost/ yuan**	**Life/ year**	**Average annual depreciation cost/ yuan**	**Annual operating cost/ yuan**	**Annual maintenance cost/ yuan**	**Average annual cost/ yuan**
Compression air conditioner	4000	7	570	200	50	820
Adsorption refrigeration air conditioner	15920	25	400	0	50	450
Solar semiconductor refrigeration air conditioner	85560	25	386	0	0	386

A comprehensive comparison of the three air-conditioning systems shows that the initial investment of adsorption refrigeration air conditioning is relatively high, but its service life is long and it has no pollution to the environment. In addition, it can make full use of low-grade heat energy and produce certain economic and social benefits.

5 Conclusion

The refrigeration system combining basic solid single-effect intermittent adsorption refrigeration and energy storage device is selected in this study, because the system can produce certain economic and social benefits. For the local air-conditioning system of the truck cab, the study is carried out through theoretical analysis and numerical simulation. And the results show that under the one-person driving condition, the air supply scheme of "opening the upper tuyere of the main driver, the air supply temperature is 16°C, the air supply angle is 45°, and the air supply velocity is 3m/s" can ensure a good comfortable environment in the local area where the driver moves.

References

1. T.T. Huang, X.B. Zhao, Y.W. Guo, Journal of Nanjing Normal University, *Theoretical study of mass recovery adsorption refrigeration driving by vehicle waste heat*, **12**,27-31(2015)
2. Meunier F. Applied Thermal Engineering, *Solid Sorption Heat Powered Cycles for Cooling and Heat Pumping Applications*, **18**, 715-729 (1998)
3. Y.M. Sun, *Application of Refrigeration and Air-Conditioning Technology in Transportation Engineering* (2012)
4. R.Z. Wang, L.W Wang, J.Y. Wu, *Theory and Application of Adsorption Refrigeration*, Science Press, (2007)
5. Y.M. Sun, *Application of Refrigeration and Air-Conditioning Technology in Transportation Engineering*, Mechanical Industry Press, (2012)

The relationship between management and control indexes of distribution network construction based on Apriori algorithm

YANG Zhuo[1], YU Jinxiong[2,*], LI Huitao[3], ZHENG Zhicong[4], WANG Lili[1], WANG Kehui[2],PAN Zhaolun[2], SUN Huijun[2]

[1] State Grid Henan Electric Power Company, Henan Zhengzhou, China
[2] North China Electric Power University, Beijing Changping, China
[3] State Grid Henan Electric Power Company Economic and Technology Research Institute, Henan Zhengzhou, China
[4] State Grid Pingdingshan Eletric Power Company, Henan Pingdingshan, China

Abstract. There are a large number of distribution network construction projects with different types, and a large number of complex data relationships are generated in the construction process. This paper combs the key management and control indicators of distribution network construction projects, and constructs the management and control index system of each stage of the project. Based on Apriori algorithm, this paper studies the relationship between the key management and control indicators, and provides reference for the management and control of distribution network construction.

1 Introduction

Distribution network is an important public infrastructure for national economic and social development, which is the key link to connect the transmission network and users and ensure the reliability of power supply. However, the distribution network has the characteristics of large-scale construction, various types of equipment, wide distribution range and large regional differences. In addition, the distribution network has many engineering stages, complex control objects, large amount of data involved, strict management requirements, close connection between the project stage and the control object, and complex data relationship, which makes the management and control of the whole process of the distribution network project more difficult. Therefore, it is necessary to study the key factors of the whole process management and control of the distribution network engineering construction.

At present, the management research of distribution network construction projects mainly focuses on single distribution network projects, and few studies on key control indexes in the whole process of project group[1-3]. This paper starts with the construction of the key management and control index system of distribution network, studies the relationship between the control indexes, finds out the relationship among them, and provides reference for the management and control of the construction of distribution network.

2 Key index system for whole process control of distribution network construction

This chapter searches for the key control indicators of the whole process of distribution network construction project, and divides them into decision-making, design, construction and completion phases in chronological order, and each phase is divided into cost, quality and progress. With reference to Literature and practical construction experience, the construction of key index system for the whole process control of distribution network construction is analyzed as follows.

2.1 Decision stage indicators

In terms of cost, the research fee mainly includes two major costs: tool use fee and survey labor cost; in terms of quality, the site selection rationality of relevant projects is considered; in terms of progress, the survey progress is mainly composed of two progress indicators: Inspection progress and decision-making progress.

2.2 Design stage indicators

In terms of cost, the design input cost is considered. In terms of quality, the rationality of design and the adequacy rate of capital preparation are considered. In terms of progress, the design progress and project approval progress are considered.

* Corresponding author: jinxiong.yu@163.com

2.3 Construction stage indicators

In terms of distribution network construction cost, four indicators are selected: labor cost, material cost, machinery cost and measure cost. The measure cost is subdivided into construction increase cost and temporary facility cost. In terms of quality, construction method, material qualification rate and construction qualification rate are considered. In terms of progress, the five indicators of equipment operation efficiency, construction personnel efficiency, material supply rate, social and economic factors and adverse weather delay rate are considered for control.

2.4 Completion stage indicators

In terms of cost at the completion stage, the test tool usage fee and acceptance labor fee are used as the key control indicators at this stage. In terms of quality, it is believed that the acceptance criteria will affect the quality of the project, so it is listed as an indicator. In terms of progress, the acceptance progress is introduced as an indicator.

In summary, the key index system for the whole process of distribution network construction is shown in Figure 1.

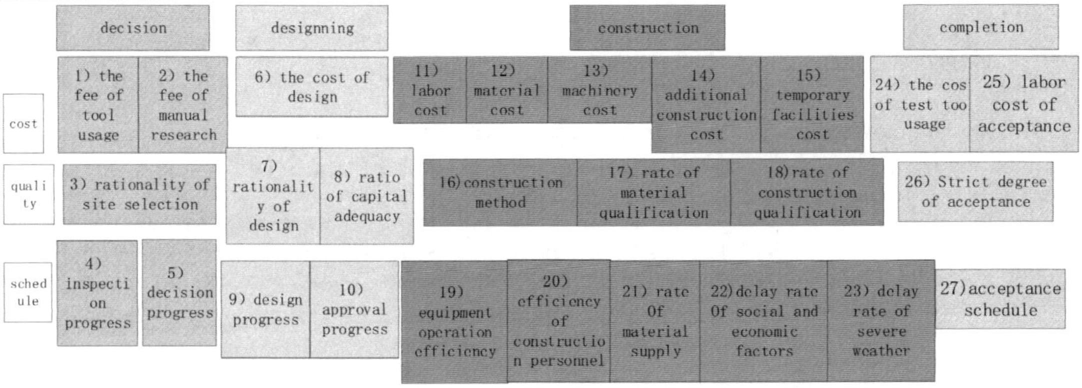

Fig1. Key Index System for the Whole Process Control of Distribution Network Construction

3 Association relationship mining model of key management and control indicators for distribution network construction projects

3.1 Association rules for key management and control indicators of distribution network construction projects

Assuming that m key indicators are set during the construction process, these indicators need to be evaluated. Collected *n* index records related to the distribution network construction project, each record contains the completion of the set m key indicators. Each distribution network construction project index record has at least one key management and control index deviation.

$$x_{ij}=\begin{cases}0, \text{There is no deviation in the } j\text{-th index of the } i\text{-th project}\\ 1, \text{There is deviation in the } j\text{-th index of the } i\text{-th project}\end{cases} \quad (1)$$

Where: x_{ij} is the implementation of the *j*-th index of the *i*-th distribution network construction project; $i=1,2,...,n$; $j=1,2,...,m$.

Calculate the $support(j \Rightarrow k)$, $confident(j \Rightarrow k)$, $lift(j \Rightarrow k)$ between the *j*-th and *k*-th indicators.

$$support(j \Rightarrow k)=\frac{\sigma_{jk}}{n} \quad (2)$$

$$confidence(j \Rightarrow k)=\frac{\sigma_{jk}}{\sigma_j} \quad (3)$$

$$\sigma_j=\sum_{i=1}^{n} x_{ij} \quad (4)$$

$$lift(j \Rightarrow k)=\frac{support(j \Rightarrow k)}{support(j)*support(k)} \quad (5)$$

$$support(j)=\frac{\sigma_j}{n} \quad (6)$$

In the formula: σ_{jk} is the number of projects with deviations in the *k*-th index among n distribution network construction projects when the *j*-th index has a deviation. σ_k and σ_j respectively represent the number of projects with deviations in the *k*-th and *j*-th indicators among the n distribution network construction projects. *support(j)* represents the support of the *j*-th indicator.

If the $support\ (j \Rightarrow k)$, $confidence\ (j \Rightarrow k)$ and $lift\ (j \Rightarrow k)$ between the *j*-th and *k*-th indexes meet the three conditions of equation (7-9), it is considered that the possibility of deviation of the *k*-th index will increase when the *j*-th index deviates. Therefore, it is considered that the *k*-th index is related to the *j*-th index; otherwise, there is no correlation between the *k*-th index and the *j*-th index.

$$support(j \Rightarrow k) \geq T_1 \quad (7)$$

$$confidence(j \Rightarrow k) \geq T_2 \quad (8)$$

$$lift(j \Rightarrow k) \geq T_3 \quad (9)$$

Where T_1, T_2 and T3 are the thresholds of support, confidence and lift respectively.

3.2 Association relationship mining based on Apriori algorithm

The mining of frequent itemsets in the Apriori algorithm is gradually realized through multiple iterations. In the first iteration, calculate the support, confidence and lift of each item in the indicator system and record all the frequent items found as *k*-. In the subsequent rounds of iterations, the *k*-item set is used as the seed itemset of this round of iteration to generate a candidate $(k+1)$-item set. In this iteration, in order to mine all $(k+1)$-frequent itemsets as the seed itemsets of the next iteration, it is necessary to calculate the actual support, confidence and lift of each candidate $(k+1)$-itemsets . The iterative process continues until no new frequent itemsets can be generated.

According to the definition of frequent itemsets, in order to find out all frequent itemsets and determine whether various combinations are frequent items, it is necessary to exhaust all combinations of indicators in a distribution network construction project, and calculate the support, confidence and lift of each combination. Apriori algorithm uses the following two attributes to reduce the computation of item set combination.

Property 1: any nonempty subset of a frequent itemset is frequent.

Property 2: any superset of an infrequent itemset is non frequent.

4 Empirical analysis

The support, confidence, and lift threshold settings will have a greater impact on the results of the association relationship. If the threshold is too small, some less closely related associations will be generated; if the threshold is too large, the closely related associations will be deleted. According to the principle of not missing important associations, the thresholds of support, confidence, and lift are set to 0.3, 0.7, and 1, respectively.

In this study, 27 key indicators are determined temporarily. According to the data records of 178 distribution network construction projects collected, the support degree, confidence degree and lift degree between the indicators are calculated, and compared with the threshold value, and some correlation relations are selected to describe and interpret. Taking the rationality of site selection as an example, Table 1 shows the calculation results of the indicators related to the rationality of site selection.

Table1. Location Rationality Association

No.	Association	Support	Confidence	Lift
1	Rationality of site selection→ design progress	0.38	0.79	1.12
2	Rationality of site selection→ approval progress	0.35	0.75	1.05
3	Rationality of site selection→ labor cost	0.62	0.86	1.05
4	Rationality of site selection→ machinery cost	0.63	0.76	1.21
5	Rationality of site selection→ equipment operation efficiency	0.53	0.84	1.02
6	Rationality of site selection→ efficiency of construction personnel	0.50	0.78	1.17

It can be seen that in the first association, the support degree of rationality of site selection→decision progress is 0.38, which indicates that 38% of the records in the project concentration are biased in the rationality of site selection and the design progress. The confidence level of 0.79 indicates that there is a 79% probability that the design progress will also deviate in the project records with deviation in site selection rationality. The increase degree is greater than 1, which indicates that the two are positively correlated, and the higher the value, the higher the correlation.

Approval schedule, labor cost, machinery cost, equipment operation efficiency and construction personnel efficiency are also related to the rationality of site selection. It shows that when there is deviation in the rationality of site selection, the indicators related to the rationality of site selection are more likely to be biased. In the process of management and control, when the rationality of site selection is found to be deviated, the related indicators can be controlled in advance.

The material qualification rate, construction qualified rate, construction personnel efficiency, material supply rate are related to the capital adequacy rate, which indicates that when the capital adequacy rate deviates, the indicators related to the capital adequacy rate tend to deviate. The shortage of funds may lead to the low qualified rate and supply rate due to insufficient material purchase fee, and the working enthusiasm and efficiency of construction personnel will be reduced, so the qualified rate will also decline.

The efficiency of construction personnel is related to labor cost, which indicates that the deviation of labor cost will lead to the decline of construction personnel efficiency.

The qualified rate of construction is related to the construction method. Improper construction method is easy to lead to unqualified construction.

The qualified rate of construction and the progress of acceptance are related to the qualified rate of materials, which indicates that when the qualified rate of materials has problems, the two indicators of construction qualified rate and acceptance progress are more likely to have deviation.

The acceptance progress is associated with the construction pass rate, which indicates that the construction acceptance rate is not up to standard, which is more likely to lead to the deviation of the acceptance progress.

Labor cost and construction increase cost are related to the delay rate of social and economic factors, which indicates that the frequent occurrence of delay caused by social and economic factors is more likely to lead to excessive labor cost and increased construction cost.

Labor cost, construction increase cost and temporary facilities cost are related to the delay rate of bad weather, which indicates that when the delay caused by bad weather is more, it is easier to lead to deviation of labor cost, construction increase cost and temporary facility cost.

By integrating the above relationships, the key management and control indicators association diagram of distribution network construction project is obtained as shown in Figure 2.

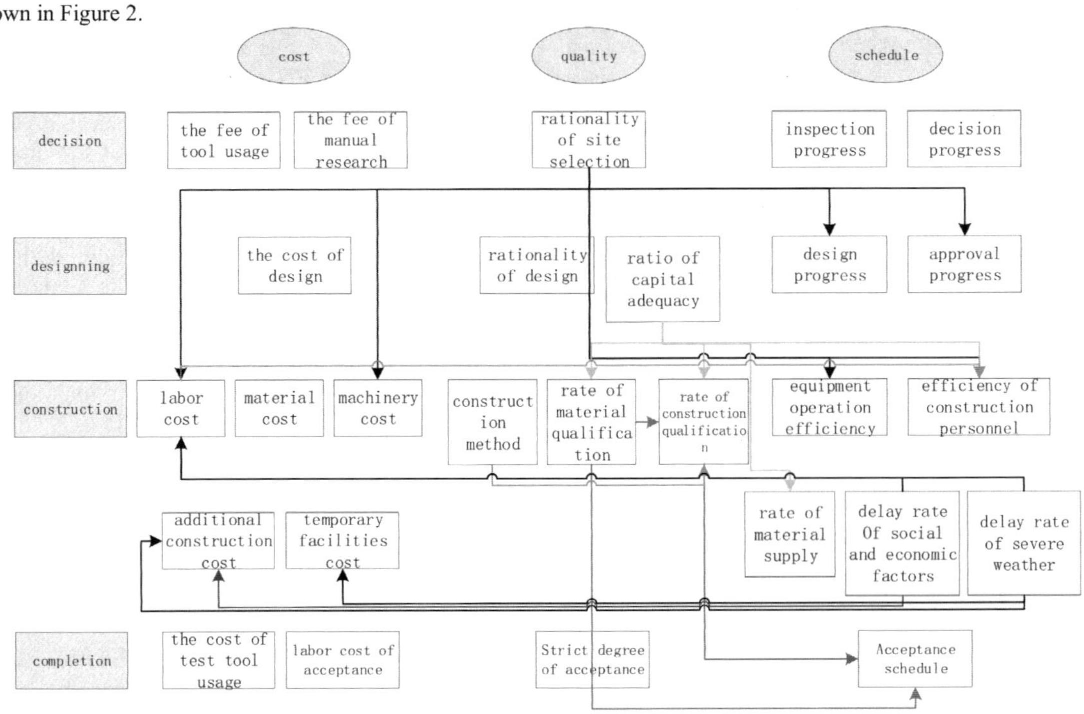

Fig2. Correlation diagram of key management and control indicators of distribution network construction projects

5 Conclusion

In this paper, the key management and control index system is established according to the three dimensions of cost, quality and progress in each stage of distribution network construction project. The association algorithm is first applied to the research on the association relationship of key management and control indicators, and the mining model of association relationship is constructed. According to the collected data of distribution network construction project, the support degree, confidence degree and promotion degree between the key management and control indicators are calculated, and compared with the set value The correlation between the key control indicators is obtained by comparing the thresholds. In the future management of distribution network construction projects, if a certain index deviates, the related indicators can be controlled in advance to reduce the impact on the project implementation and achieve the effect of prior control.

Acknowledgment

This work is supported by Research on Big Data Correlation Analysis of Key Nodes in Power Distribution Network Engineering Project Management (SGTYHT/18-JS-209).

References

1. 1.LI Weiguo, XU Wenwen, QIAO Zhenyu, WANG Xuguang. Distribution section fault location method based on transient zero sequence current bump feature[J]. Power System Protection and Control,2020,48(10):164-173.
2. YANG Feiran, YU Yongjin. Fault diagnosis of distribution network based on time-constrained hierarchical fuzzy Petri net[J]. Power System Protection and Control,2020,48(02):99-106.
3. Z. Q. Bo,X. N. Lin,Q. P. Wang,Y. H. Yi,F. Q. Zhou. Developments of power system protection and control[J]. Springer Singapore,2016,1(1):1-8.
4. Winters Tobey. Retail electricity markets require marginal cost real-time pricing[J]. The Electricity Journal.2001,14(9):74-81.

Research on Optical Energy Storage System Based on Rule Logic Control

Lu Yan[1,*], Rui Xu[1], and Xin Zhang[1]

[1]School of Electrical Engineering, Shanghai Dianji University, 201306 Shanghai, China

Abstract. Photovoltaic energy storage system is composed of photovoltaic power generation, energy storage battery, load and power grid, it has researched on different modes of bi-directional DC/AC converter, taking into account a variety of photovoltaic output fluctuations, energy storage battery charging and discharging management, load increasing and decreasing, etc. Aiming at the energy management strategy, using DC bus voltage fluctuation as the basis of energy layer mobilization, the power balance control based on regular control is proposed. Compared with the DC bus voltage level energy management, the power tracking accuracy is further improved and energy loss is reduced. Finally, through MATLAB/Simulink being in off/connected model and the charging/discharging model, the effectiveness of the energy control strategy was verified separately by simulation. The simulation results show that: the proposed control strategy can improve the power tracking accuracy and reduce the energy loss of the system; the proposed strategy guarantees the bidirectional exchange of system energy and maintains the power balance in the network, the DC bus voltage is stable, and the energy storage battery achieves smooth power output and peak load shifting.

1 Introduction

In the field of new energy power generation, such as photovoltaic power generation and wind power generation, there are volatility, intermittentness and randomness. These unavoidable factors will cause output power fluctuations and bring non-negligible negative effects on the grid and load. The integrated application of energy and energy storage devices can help improve the local consumption of renewable energy and solve the power fluctuations in the grid caused by intermittent energy output [1-3]. How to solve the time and space mismatch between production capacity and energy use through energy storage, and realize the efficient use of large-scale renewable energy is the focus of the energy Internet research. At the same time, because it contains a variety of energy flow conditions, how to achieve different units Inter-energy management and coordinated control are also a major research focus. In response to this hot spot, a lot of research has been done in related literature [4-6].

Aiming at the problem that photovoltaic power generation is affected by external environmental factors such as light intensity and temperature, the output power fluctuates, and the stability is affected when it is connected to the grid. An energy storage battery is designed as an energy storage device to achieve complementary solar storage systems. The energy of each part is the flow takes the DC bus as the connection point. It is proposed to use the DC bus voltage fluctuation as the basis for energy layer transfer, and the power balance control based on rule logic control as the corresponding energy management control strategy to realize the load, energy storage battery, photovoltaic power generation and power grid. Energy flow between. The charging and discharging effect of the energy storage battery realizes smooth photovoltaic output power fluctuations, so that the power of photovoltaic power input to the grid remains stable for a certain period of time, realizes the function of peak shaving and valley filling, and reduces the impact of intermittent power output fluctuations on power quality.

2 Optical storage system topology

The composition diagram of the photovoltaic energy storage system is shown in Figure 1. The entire system is output by photovoltaic modules through maximum power tracking (MPPT) to ensure the maximum utilization of photovoltaic energy. The energy storage converter (PCS) is used to realize the two-way exchange of energy with the grid, and the energy storage battery is used to solve the fluctuation of photovoltaic output. The energy required by the load is first met by photovoltaic output, and the shortfall is provided by the grid or energy storage batteries. Energy storage batteries mainly ensure that the energy flow within the grid is in a balanced state. Finally, a complementary system model for optical storage is established to achieve smooth output.

* Corresponding author: author@e-mail.org

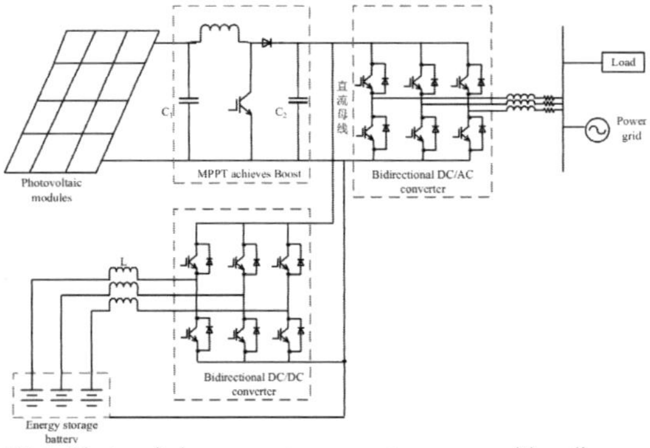

Fig1. Photovoltaic energy storage system composition diagram

3 Optical storage system rules control operation mode

3.1 System energy management solution

During the operation of the entire optical storage system, its control principle is shown in Figure 2. Among them, the DC bus voltage calculation module calculates the real-time value of the DC bus voltage according to the output voltage of different units; the working state selection module compares the calculated value of the DC bus with the set threshold to determine the working state of the energy storage converter; the power tracking calculation module is based on At this moment, the photovoltaic output and the power required by the load calculate the charge and discharge power of the energy storage battery; the mode selection module performs power distribution according to the working status and the power calculation result, and controls the bidirectional DC/DC converter for power throughput.

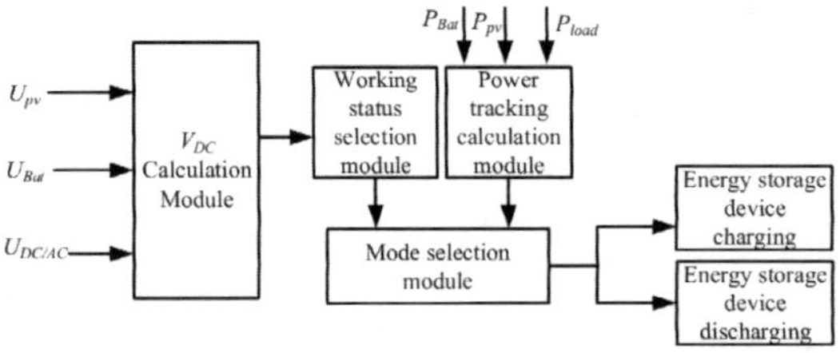

Fig2. System composition diagram of system energy management

In the figure: U_{PV}, U_{Bat}, $U_{DC/AC}$ are the photovoltaic output voltage, energy storage battery output voltage, and grid-side converter transmission power converted to the DC side voltage respectively. P_{PV}, P_{Bat}, and P_{load} are the output of the photovoltaic module, the output power of the energy storage battery, and the power required by the load, respectively.

3.2 Regular control mode and energy flow state

The flow of power in the network is ultimately directly reflected in the increase and decrease of the DC bus voltage. Maintaining the power balance in the network is equivalent to stabilizing the DC bus voltage. Therefore, the system operating mode switching is based on the DC bus voltage. In order to ensure the optimization of mode selection and improve the operating efficiency of the proposed smooth control strategy, a rule-based control strategy is designed according to mode selection and power balance [7]. Its core content is: the collected data is allocated according to the established principles for rule configuration, and when the corresponding conditions are met, it enters the corresponding mode to work [8]. The research of rule control is based on the DC bus voltage and power characteristics in the network under different working conditions. The specific control process is shown in Figure 3. U_{dc} is the measured voltage value, U_{H2} represents the DC bus voltage is higher due to power surplus on the DC side, close to 1.1 U_{dc}^*; UH1 represents the DC bus has a slightly higher fluctuation than the normal voltage, approximately 1.05 U_{dc}^*; UL1 represents the DC bus is normal Slightly lower fluctuations, approximately 0.95 U_{dc}^*; U_{L2} means that the DC bus is low due to power shortage on the DC side, approximately 0.9 U_{dc}^*[9]. The whole system is based on the DC bus voltage adjustment to realize power balance control to ensure voltage balance and power balance within the network.

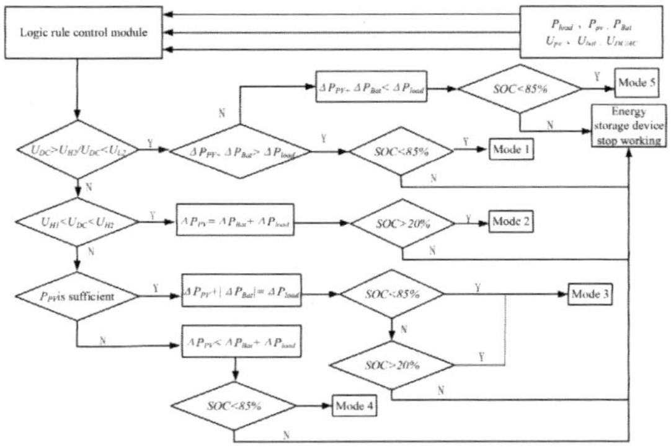

Fig3. Regular control flow chart

4 Simulation and experiment

4.1 Comparative analysis of energy management strategies

Using different energy management strategies to track the actual load demand and photovoltaic output, the results are shown in Figure 6 and Figure 7. The results show that the energy management strategy based on rule logic control reduces the tracking loss of photovoltaic output and load forecasting, and improves the power tracking accuracy.

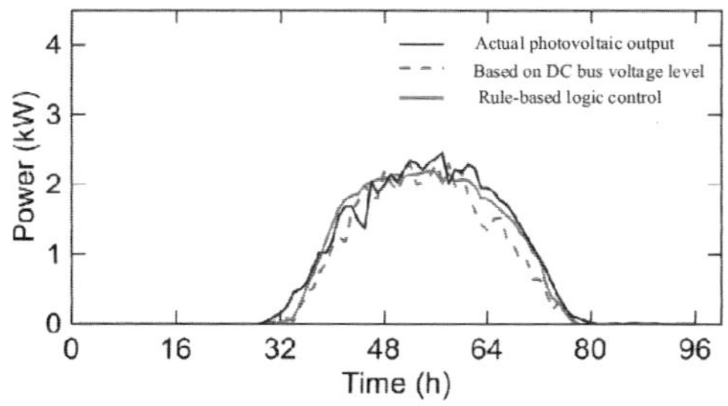

Fig4. Load demand tracking

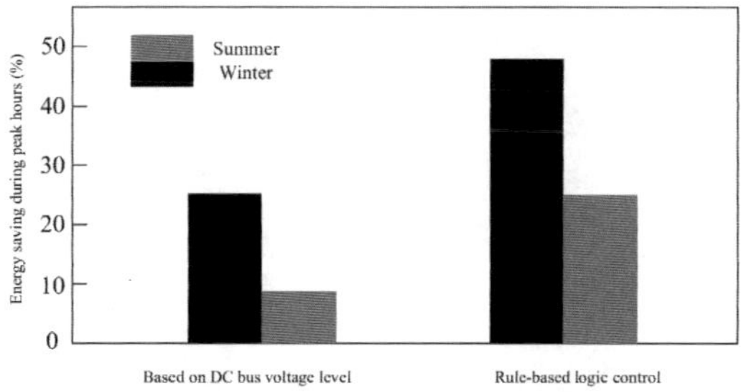

Fig5. Save energy during peak hours(%)

4.2 Mode switching simulation

4.2.1 Grid-connected state mode switching simulation

The initial operating light intensity of the system is $600W/m^2$. The lack of power on the DC side causes a large drop in the DC side bus voltage, and the energy storage battery power supply energy is insufficient. The system working mode is 5, the PCS works in the rectifying state, and the AC side delivers energy to charge the battery. The light intensity rises to $1000W/m^2$ by 0.8s, the photovoltaic output is increased from the original 196kW to 498.9kW, the system working mode is switched from 5 to 1, the photovoltaic output is sufficient and the energy storage battery still has remaining power transmitted to the grid after working at the upper limit of charge In the whole process, the DC bus voltage is basically stable at about 715V.

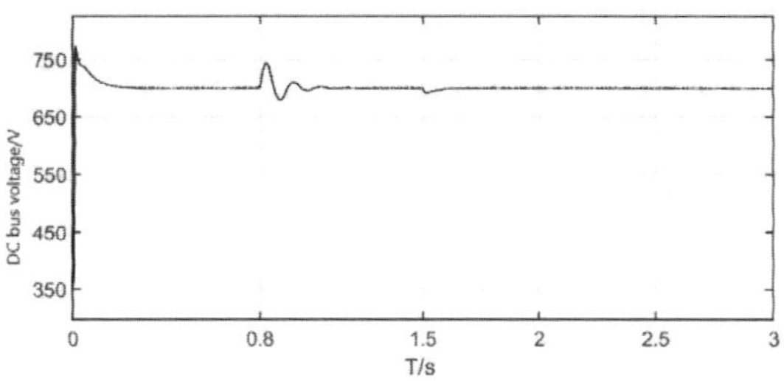

Fig6. DC voltage bus

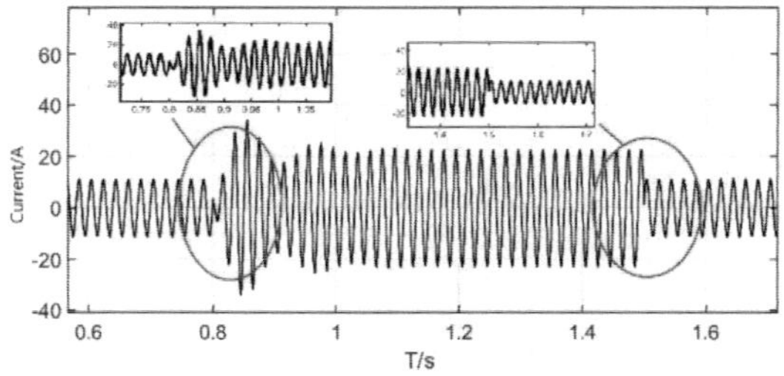

Fig7. Rectifier and inverter mode switching

4.2.2 Simulation analysis of off-grid state mode switching

When operating off-grid, photovoltaic output and energy storage battery output jointly maintain power balance while ensuring that photovoltaics are used to the maximum.

When the photovoltaic output is 498.9kW in 0-0.8s, when the operation reaches 0.8s, due to the change of the external environment, the light intensity suddenly drops from $1000W/m^2$ to $600W/m^2$ and the output power drops to 196kW. Figure 8 shows the power required by the load, which is 350kW for 0-1.5s and 450kW for 1.5s-3s.

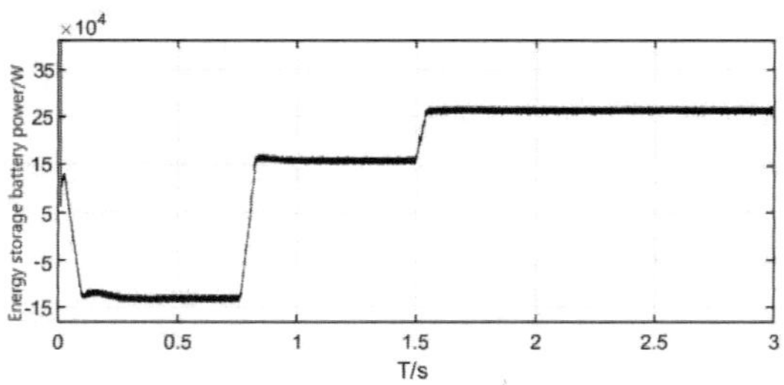

Fig8. Energy storage battery power

5 Conclusion

This article aims to stabilize the DC bus voltage and accurately track the actual demand of photovoltaic output and load by the energy storage battery to achieve peak shaving and valley filling. The power balance control based on rule logic control is designed to achieve 5 modes of optical storage system operation, and then the two-way DC/ Control strategies for DC converters and energy storage converters.

The simulation results show that the energy management strategy based on rule-based logic control proposed in this paper corresponds the working mode of the energy storage converter and the charging and discharging state of the energy storage battery to the specific working conditions, and solves the problem that may be caused by the fluctuation of photovoltaic output when the external environment changes. System power imbalance problem, and stabilize the DC bus voltage, use energy storage battery to achieve smooth output power, and achieve the goal of peak clipping and valley filling. At the same time, it achieves accurate tracking of photovoltaic output and load demand, and improves power tracking accuracy, which proves the effectiveness of this strategy in reducing system energy loss and improving energy utilization.

References

1. LC Wang, YP Gao. Electrical Measurement & Instrumentation, [J].52(14): 49-54.(2015).
2. M Ding, JX Shi, PP Han. China Electric Power[J/OL]: 1-7[2020-04-24] .
3. Kai Sun,Li Zhang,Yan Xing. IEEE Transactions on Power Electronics. 2011
4. Yao Bo, Fan Yanfang. Renewable Energy[J]. 2017, 35(02): 232-239.
5. Wang Yi, Zhang Lirong, Li Heming, Liu Junpeng. Proceedings of the Chinese Society of Electrical Engineering[J]. 2013, 33(04): 16-24+4.
6. Xue Zhang, Pei Wei, Deng Wei, Qu Hui, Shen Ziqi, Zhao Zhenxing. Proceedings of the Chinese Society of Electrical Engineering, 2014, 34(31) :5553-5562.
7. BANVAIT H,ANWAR S. 2009 American control conference. 2009
8. Zhendong Zhang. Hubei University of Technology, 2018.
9. Maoliang Hu. University of Electronic Science and Technology of China, 2014.

Characterization of interaction between electric vehicles and smart grid

ZHANG Yan[2,*], HE Yongxiu[2], SU Fengyu[2], WANG Xiaoqing[2] and ZHANG Diling[1]

[1]Beijing Key Laboratory of Demand Side Multi-Energy Carriers Optimization and Interaction Technique , Beijing Haidian District 100085, China
[2]North China Electric Power University, Beijing Changping District 102206, China

Abstract. With the development of smart distribution technology in the future, electric vehicle users can not only charge reasonably based on peak-valley price, they can also discharge electricity into the power grid to realize their economic benefit when it's necessary and thus promote peak load shifting. According to the characteristic that future electric vehicles can discharge, this paper studies the interactive characteristics between electric vehicles and smart grid. In this paper, the example shows that the charging and discharging behaviour of EV users will bring significant fluctuation effect to the power grid load, and the reasonable TOU strategy can stimulate EV users to conduct reasonable charging and discharging, so as to smooth the peak and valley difference of some power grid.

1 Introduction

In the future, with the development of smart power technology, the coupling and interaction characteristics between the load end and the grid will become more and more significant[1]. For electric vehicles, they can carry out the real time electric energy exchange with the grid based on V2G (Vehicle-to-Grid) technology. This technology can use the price signal to guide users to charge and discharge so that users can realize the target of optimizing the power consumption, and the power grid can realize the economic dispatch and alleviate the pressure in peak periods [2-4]. At present, many scholars have done relevant research, such as discussing the benefit of different charging modes and combining the EVs with energy storage technologies[5], but the research which integrates electric vehicle and real-time electricity price is less. Balijepalli et al. analysed the benefits of demand response and reviewed the correlational research on it, including research on EVs and this study has referential value for our research to a certain extent. This paper analyses the interaction effect of electric vehicles and grid based on the real time electricity price under different scenarios.

2 Analysis on characterization of interaction of electric vehicle based on V2G technology

V2G (Vehicle-to-Grid) technology is a new type of grid technology, which can realize the reverse charging of electric vehicles to the grid. Electric vehicles with V2G can not only serve as consumers of electricity, but also serve as power sources to provide power to the power grid, so as to realize two-way interaction and electrical energy exchange between electric steam and the power grid. Charging of electric vehicles has a strong randomness in time distribution. Under the situation of widespread access of electric vehicles, it is likely to have a certain negative impact on operation management, scheduling planning and other aspects of power grid. This paper takes electric private cars as an example to analyse their electricity characteristics and load curve.

2.1 Initial charging time of electric vehicle

Assuming that the travel time, daily mileage and other data of private electric vehicles are the same as those of fuel vehicles, and referring to the results of the National Household Travel Survey (NHTS) conducted by the US Department of Transportation in all 50 states, the most probable likelihood estimation method is adopted to approximate the normal distribution at the beginning of charging for private electric vehicles. Its probability density function is shown in the following equatio[1]n.

$$f_s(t)=\begin{cases}\dfrac{1}{\sigma_s\sqrt{2\pi}}e^{-\frac{(t-\mu_s)^2}{2\sigma_s^2}} & \mu_s-12<t\le 24\\ \dfrac{1}{\sigma_s\sqrt{2\pi}}e^{-\frac{(t+24-\mu_s)^2}{2\sigma_s^2}} & 0<t\le\mu_s-12\end{cases} \quad (1)$$

Where, t is the charging start time of EV,and $\mu_s=17.6$, $\sigma_s=3.6$.

2.2 Battery capacity and initial charge

In this paper, it is assumed that the battery capacity of

* Corresponding author: hsezzhangyan@163.com

 https://doi.org/10.1051/e3sconf/202123702004

electric private cars is gamma distribution.Some battery parameters of electric private cars are shown in Table 1:

Table1. Battery parameters of electric private car

category	parameter α	parameter β	battery capacity /kWh	maximum capacity for quick charging /kWh
parameter	10.8	0.8	15	5

Note: The maximum capacity of fast charging indicates that fast charging is possible only if the capacity is below this set value

The distribution of battery capacity of electric private cars is shown in Equations (2) ~ (3) :

$$f(cap;\alpha,\beta)=\frac{1}{\beta^{\alpha}\Gamma(\alpha)}cap^{\alpha-1}e^{-\frac{cap}{\beta}}\ (\alpha,\beta>0) \tag{2}$$

$$\Gamma(\alpha)=\int_0^{+\infty}x^{\alpha-1}e^{-x}dx \tag{3}$$

Assuming that the decrease of battery charge is in direct proportion to the driving distance of the electric vehicle, the real-time charge SOC_t of the electric vehicle at time t is expressed as:

$$SOC_t=\eta\times SOC_{t-1}\times(1-D/Ran) \tag{4}$$

Where, η represents the energy conversion coefficient, which is used to correct the energy loss caused by the acceleration and braking of electric vehicles, and the value range is [0.9,1.0]; SOC_{t-1} Represents the charge capacity of the electric vehicle at the time point t-1; D represents the distance travelled between time periods T-1 and T ;Ran represents the maximum mileage; SOC_0 represents the battery charge at the first departure time of that day. Assuming that each charge is full, the battery charge at the departure time is 1.

2.3 Charging power

In the previous EV models, ev charging power was treated as a constant value for the sake of simple calculation. However, in the actual process, the power change curve of lithium battery charging process will also change due to different charging ratio and initial charge. The specific change rule is shown in Figure 1[1]. It can be seen from the figure that at the beginning of charging, the charging power increases gradually, and then decreases gradually after reaching the peak value.

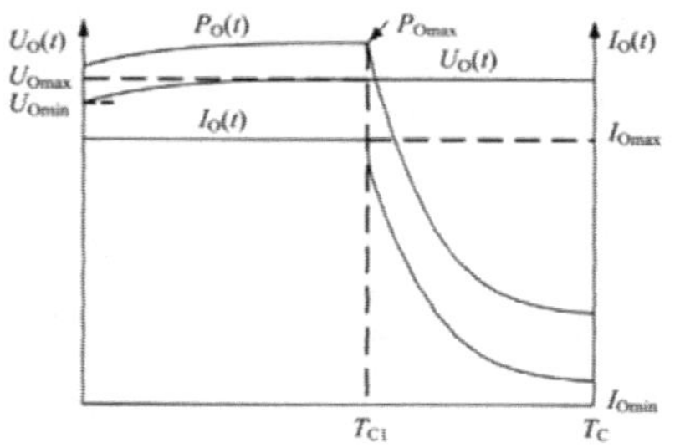

Fig1. Power curve of lithium battery charging

To simplify the analysis, this paper takes the constant current stage as a linear function and the constant voltage stage as an exponential function to obtain the function model of power and time as follows:

$$P_O(t)=\begin{cases}I_{O\max}U_O(t) & 0\le t<T_{C1}\\ U_{O\max}I_O(t) & T_{C1}\le t<T_C\end{cases} \tag{5}$$

Set $U_{O\min}=k\times U_{O\max}$, he formula can be deformed as :

$$P_O(t)=\begin{cases}U_{O\max}I_{O\max}[(1-k)\frac{t}{T_{C1}}+k] & 0\le t<T_{C1}\\ U_{O\max}I_{O\max}e^{-a(t-T_{C1})} & T_{C1}\le t<T_C\end{cases} \tag{6}$$

3 Description of interactive characteristics of electric vehicles under time-of-use price

With the implementation of time-of-use tariff policy, EV users will choose charging time according to the different electricity price, and try to charge when the electricity price is low. Charging decisions made by users at different electricity price levels in different time periods can be described by the demand price elasticity, that is, the ratio between the rate of change of EV charging load and the rate of change of price:

$$\varepsilon_{uv}=\frac{\Delta Q_u/Q_u^{(0)}}{\Delta\rho_v/\rho_v^{(0)}} \tag{7}$$

$$\Delta Q_u=Q_u-Q_u^{(0)} \tag{8}$$

$$\Delta\rho_u=\rho_u-\rho_u^{(0)} \tag{9}$$

Where, $Q_u^{(0)}$ and Q_u are respectively charging loads in u period before and after the implementation of TOU, $\rho_v^{(0)}$ and ρ_v are respectively the time price before and after the implementation of TOU。When $T_u=T_v$, ε_{uv} is the self-elasticity coefficient ($\varepsilon_{uv}<0$), representing the charging demand elasticity of the current period ; when $T_u\neq T_v$, ε_{uv} is the cross elasticity coefficient ($\varepsilon_{uv}>0$) , representing the charging demand elasticity of the electricity price in time period T_u to time period T_v .

After the implementation of TOU, as rational consumers, users can transfer the charging load from the period of high electricity price to the period of low electricity price by flexible charging arrangement (adjustable charging time and power) without changing the charging demand and charging position, thus reducing the charging cost.

The charging load in the trough period is

$$Q_o=Q_o^{(0)}-\Delta Q_{op}+\Delta Q_{po}-\Delta Q_{os}+\Delta Q_{so}+\Delta Q_{v2g} \tag{10}$$

The charging load in the flat period is

$$Q_s=Q_s^{(0)}-\Delta Q_{sp}+\Delta Q_{ps}-\Delta Q_{so}+\Delta Q_{os}+\Delta Q_{v2g} \tag{11}$$

The charging load during peak hours is

$$Q_p = Q_p^{(0)} - \Delta Q_{ps} + \Delta Q_{sp} - \Delta Q_{po} + \Delta Q_{op} - \Delta Q_{v2g} \qquad (12)$$

Where, $Q_i^{(0)}$ represents the initial value of the load in each period , in which i represents different time periods (O represents the trough, S represents the flat segment, P represents the peak) ; ΔQ_{ij} represents the load transferred from period I to period J, ΔQ_{v2g} represents the load transferred by electric vehicles to the grid through V2G technology.

4 Example analysis

In this section, the coupling interaction between electric private cars and the power grid is calculated, and the change of power grid load under the scenarios of 40% discharge, 60% discharge and 80% discharge is respectively calculated under the difference of electricity price between high, medium and low peak and valley.

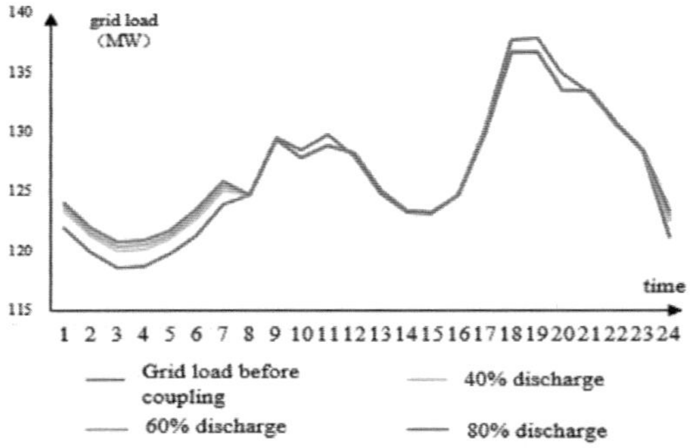

Fig2. Changes of grid load with different discharge ratios of EV under low electricity price

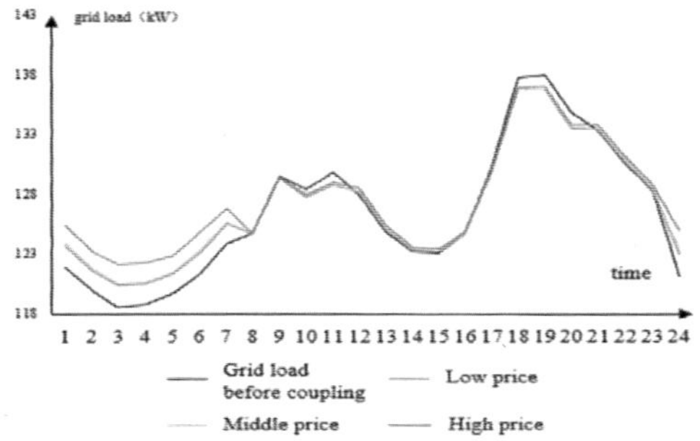

Fig3. Changes of grid load under different electricity prices at 60% EV discharge

The two figures respectively show the coupling between electric vehicles and the power grid under the condition of low electricity price with different discharge ratio and the change of power grid load under the condition of different electricity price level under the condition of 60% discharge ratio of electric vehicles. The results show that in the case of 60% discharge, the higher the electricity price level is, the more obvious the effect of EVs participating in peak load cutting is.

Then the peak-to-valley ratio of grid load is introduced to analyze the coupling effect of EV. Peak-to-valley ratio of power grid load refers to the ratio between the maximum load and the minimum load of power grid, which can reflect the peak-to-valley difference of power grid load in a day. The larger the peak-to-valley ratio is, the more unstable the power grid load is, and the more measures are needed to control the load. The peak-valley ratio of grid load before and after EV interaction is calculated, and the calculation results are shown in Figure 4 below.

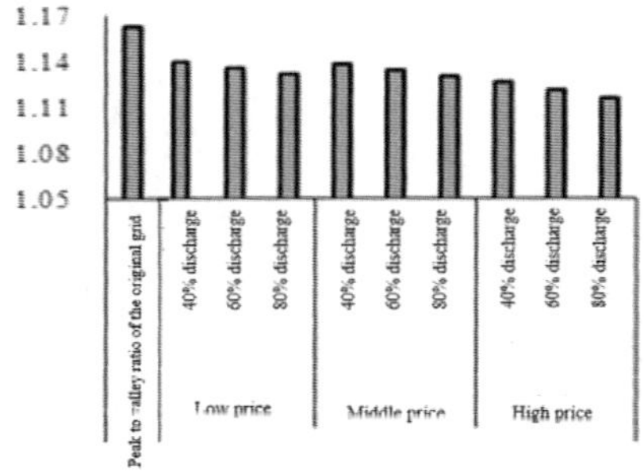

Fig4. Peak-valley ratio of grid load under different scenarios

As can be seen from the figure, before the participation of electric vehicles in coupling, the peak and valley ratio of the power grid is relatively large. Based on the above research, it can be seen that the charging and discharging behaviors of EV users will bring significant fluctuation effect to the power grid load, and a reasonable TOU strategy can motivate EV users to conduct reasonable charging and discharging, so as to smooth the peak and valley difference of part of the power grid.

Therefore, in the future, with the continuous popularization of V2G technology of electric vehicles, the grid load will be confronted with great fluctuations. In order to suppress the fluctuations caused by large-scale charging and discharging behaviors of electric vehicles on the grid, EV users need to be guided to participate in the coupling interaction with the grid in a reasonable and orderly manner.

5 Conclusion

V2G technology effectively realizes the two-way electric energy interaction between electric vehicles and smart grid, which greatly promotes the participation of electric vehicles in the process of demand response. However, the interaction between electric vehicles and the power grid is affected by the peak-valley price difference and the charging and discharging behaviour of electric vehicles. In the future, electric vehicles' participation in the power grid interaction must have a more perfect mechanism to guide the charging and discharging behaviour of electric vehicle users, so as to prevent more fluctuations in the power grid load curve, which is not conducive to the safe and stable operation of the power grid.

Acknowledgments

This work is supported by Open Fund of Beijing Key Laboratory of Demand Side Multi-Energy Carriers Optimization and Interaction Technique(No.YDB51202001975)

https://doi.org/10.1051/e3sconf/202123702004

References

1. LIU Qing, QI Zhongyi. The Load Forecast Model for Power Grid with the Accessing of Large-scale Electric Vehicles by Considering Spatial Motion Characteristics[J], Modern Electric Power, 2015, **32**(1):76-82..
2. Nguyen H K, Ju B S, Zhu H. Demand side management to reduce Peak-to-Average Ratio using game theory in smart grid. *Computer Communications Workshops* (INFOCOM WKSHPS), 2012:91-96.
3. Shao S, Pipattanasomporn M, Rahman S. Challenges of PHEV penetration to the residential distribution network. *Power & Energy Society General Meeting*, 2009:1-8.
4. Li X R, Hui J H, Qian J, et al. Influence of wind power generation on side Load modeling of distribution network. *Power system automation*, 2009, **33**(13):89-94 .
5. Shang Y L,Sun B,Cui N X, et al. Study on charging Load of large-scale electric vehicles under different charging modes and its influence on power grid. *Proceedings of the 33rd China Control Conference*. Nanjing :2013.

Application of Edge Computing and SCADA System in Second Class Substation

Zhenyu Gao[1], Pan Chen[1], Zhonghua Chen[1], Guochao Shi, Zhoncai Liu[2*], Shiwei Feng[2]

[1]Hangzhou Electric Power Design Institute Co., Ltd, Hangzhou, China
[2]Shanghai Makesens EnergyStorage Technology Co.,Ltd, Shanghai, China

Abstract. This article adopts the way of progressive transformation of edge calculation and the combination of the substation, this paper makes a brief introduction to the calculation of edge, and describes its development situation at home and abroad, and then stand for intelligent SCADA system on 750 kV substation briefly, the architecture will eventually margin calculation and combination of SCADA system, discuss the practicability of EC - SCADA system.

1 Introduce

With the increasing demand of electric load in all parts of the country year by year, the contradiction between China load Centre and primary energy centre is becoming more and more obvious, and the optimal allocation of energy resources in China is gradually paid attention to. In the current transmission mode of power system, UHV transmission reduces the engineering cost of unit power transmission with large transmission capacity and long transmission distance, reduces line loss, and saves the area of line corridor [1]. Improving comprehensive economic and social benefits plays an increasingly important role. As an important part of the intelligent substation of the second-class power transmission system, the SCADA (supervisory control and data acquisition) control and data acquisition system are responsible for many important uses, such as telecontrol function, automatic control function, meter function, interface and system use, relay protection and related functions. Its operation state can directly endanger the safe and stable operation of the power grid system.

2 Edge Computing

2.1 Edge Computing Brief

Edge computing refers to a new network architecture and open platform that integrates network, computing, storage and application of core capabilities at the edge of the network near the source of human, object or data acquisition.EC, the traditional centralized cloud computing processing mode is changed to move the storage computing power down to the edge of the network, that is the terminal oriented mode, which provides the edge intelligent service nearby. The edge computing mode greatly reduces the burden and blockage of the network caused by data return, realizes the relatively high transmission speed with low delay, and can quickly respond to the terminal request and improve the system performance. Figure 1 is a schematic diagram of cloud computing and edge computing.

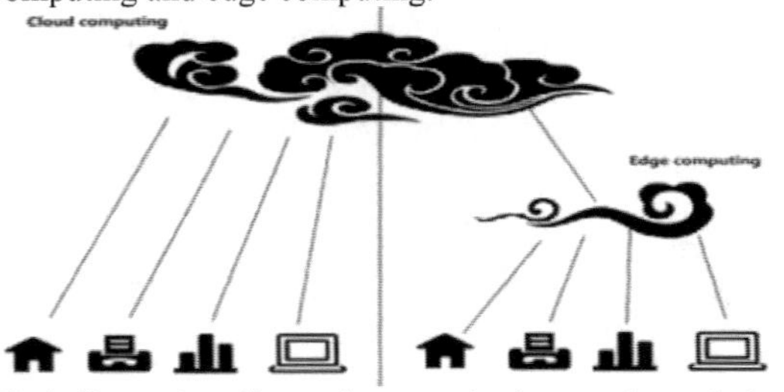

Fig.1. Comparison diagram between cloud computing and edge computing.

EC concept comes from 5G technology based on EC technology can realize the mutual operation function of different manufacturers and platform systems. The concept of EC can be extended to the Internet of things, and some experts are exploring EC application prospects in service manufacturing and other fields [2].

The previous network cloud computing model is based on centralized core nodes. With the rapid development of intelligent substation, the research and development of various online monitoring systems, the number of process layer devices and systems is increasing day by day. Higher requirements are given for intelligent substation centralized data processing mode. For economic reasons, in order to prevent the centralization node from becoming a potential risk point and inhibit the development factor, the EC adopts a closer mode to the terminal side to respond to the data at the edge of the network, which not only improves the response speed of the system, but also reduce the system transmission requirements.

* Corresponding author: liuzhongcai0@163.com

2.2 Edge Computing Domestic and Foreign Development

The edge computing is still in its early stages, Amazon, Google and Microsoft are still in the leading position in the edge cloud computing world, Amazon and AWS Greengrass jointly developed the edge computing field in 2017, and Microsoft plans to invest $5 billion in emerging technology research in the next four years. Google also released two new products in 2017, hardware chip Edge TPU and software stack Cloud loT Edge [3]. Google said it Cloud loT Edg extend the powerful data processing and machine learning capabilities of Google Cloud to billions of edge devices.

Huawei Technology Co., Ltd., Shenyang Institute of Automation, Chinese Academy of Sciences, China Information and Communication Research Institute, Intel Corporation, ARM and Soft-com Power Information Technology Co., Ltd. jointly established in Beijing Edge Computing Consortium (ECC) [4]. A total of 62 members, covering scientific research institutions, industrial manufacturing, energy and electricity and other different fields. Version 1.0 and 2.0 of the Edge Computing Reference Architecture were published in 2016 and 2017, respectively, combing the test bed of edge computing, the solutions of edge computing in industrial manufacturing, power energy, smart city, transportation and other industries are proposed. Edge computing is one of G five core functions, Is the key to achieve 5 G performance improvement. China Communication Standardization Association (CCSA) launched the Edge Computing Research Project in 2017. The CCSA Wireless Communications Commission (TC5) and the Industrial Internet Task Force (ST8) have established projects on edge computing, respectively. The three major operators have carried out extensive exploration in computing edge computing. In 2018, China Unicom announced the official launch of Edge-Cloud scale pilot projects and planning and construction of thousands of edge data centers in 15 provinces and cities across the country; China Mobile diverted to CDN edge nodes through core sinking gateways in Jiangsu and Zhejiang, and explored some commercial scenarios.

3 Intelligent Station SCADA System Introduction and Architecture Model

3.1 SCADA System presentation

SCADA system is a comprehensive application of modern automation, computer, signal processing and modern communication technology in the field of power substation, that is, the secondary equipment such as measurement and control, fault recording, protection and telecontrol device of substation are combined through function to realize automatic monitoring, measurement, control and information protection interaction of primary and secondary equipment of substation, as well as a series of comprehensive automation performance such as communication with dispatching center. SCADA system can complete the functions of monitoring, communication and information uploading by protecting, detecting, collecting, analyzing and calculating the information of the whole station.

3.2 SCADA System Architecture Model in 750 kV Intelligent Substation

For example, 750kV intelligent substation, SCADA system uses "Three-tier two-network [5]" hierarchical distributed architecture with 61850 protocol as the basic communication specification; "Three-tier" means the station control layer, the spacer layer and the process layer, it mainly includes backstage monitor, telecontrol, microcomputer five prevention, fault recording and other intelligent devices in the station; intelligent terminal, merging unit. The two networks are the station control layer network and the process layer network, corresponding to the MMS,GOOSE and SV network respectively. The three-tier two-net architecture is shown in Figure 2.

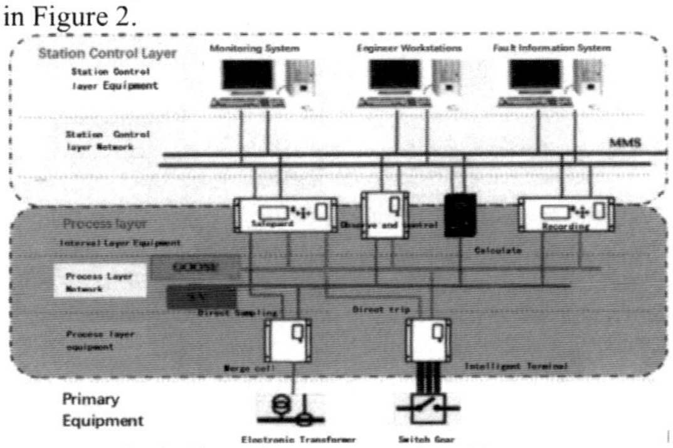

Fig.2. Three-tier two-network architecture.

3.3 Failure analysis of SCADA system of intelligent station

The normal and reliable operation of the intelligent station SCADA system is a necessary condition for the stable operation of the ultra-high voltage power transmission system. The normal and reliable operation of the SCADA system of the intelligent station not only depends on the functional design of the system itself, but is also closely related to the transmission channel, the status of the background server and the network status. Defects in any of these links may cause abnormalities in the entire intelligent station SCADA system. This section focuses on the shortcomings of the SCADA system of the smart station, namely communication interruption, server downtime and network storm.

3.3.1 Communication interruption

Communication interruption is one of the common factors of abnormal SCADA system of smart station. The communication transmission media of intelligent stations include optical cables, pigtail cables, jumpers and twisted-pair cables. Small animal infringements, ground subsidence and excessive bending of man-made optical fibres are the main reasons for the damage of communication media. In addition to the communication interruption caused by the destruction of the communication medium and the destruction of the physical link, there are also communication interruptions caused by the terminal or background hardware and programs.

In general, communication interruption will not cause protection to directly affect the operation of primary equipment [6]. However, communication interruption can cause the background and dispatch centre to lose the remote information data such as the actual position of the primary equipment and the actual state of the secondary equipment, and the remote measurement data such as the current and voltage amplitude and phase, resulting in the loss of remote control of the primary and secondary equipment, and even The failure of automatic voltage and reactive power regulation (AVC) and other remote adjustments can endanger the reliable operation of the grid system in severe cases.

3.3.2 Server down

Due to the long running time, large amount of processed data and environmental factors, server downtime is also one of the reasons for the failure of the SCADA system of the smart station [7].

On May 20, 2018, the operator of a 750kV substation found that the monitoring screen of the background monitoring machine No. 2 was black and could not start normally, and the background monitoring machine 2 server alarm light was always on. Initially, it is judged that the server program of the background monitoring machine 2 is dead, and the abnormality is still restored after restarting the server and the monitoring machine. After on-site inspection by maintenance personnel, it was judged that the hard disk of the background monitoring machine 2 server was damaged. After the hard disk was replaced, the system was restored to normal. It took about 19 hours from the discovery of the defect to the elimination of the defect.

Interactive output layer: Provide interactive data for the cloud background monitoring machine and dispatch centre, and provide man-machine dialogue.

The fast response layer is composed of a rule engine and a fast gateway, allowing core data to quickly communicate with the core processing layer and the interactive input layer.

Core processing layer data can be accessed by edge devices

The SCADA system server of the intelligent station is configured with double sets, and the operator station can be used as a monitoring machine. Therefore, restarting the server will generally not cause the station to lose monitoring.

3.3.3 Network storm

Due to the network topology and connection reasons, UDP packets are replicated on a large scale on the network and data frames are sent, which makes the network communication function drop sharply, and in severe cases, the network is completely paralyzed. The causes of network storms usually include: communication hardware reasons, equipment communication module abnormalities, network structure abnormalities and the use of hacker viruses [8].

As one of the reasons for the failure of the SCADA system of the smart station, the network storm can lead to the overall failure of the SCADA system of the smart station, that is, the complete paralysis of the station control layer network.

4 EC-SCADA System Application Inquiry

EC-SCADA system can provide edge intelligent service nearby, realize the information exchange of automatic monitoring, measurement, control and protection of primary and secondary equipment of substation, and a series of comprehensive automation performance with dispatching center. According to the "three layers and two networks" architecture of intelligent substation EC-SCADA the system architecture can realize the parallel calculation and analysis of various protection, measurement and control, stability control, measurement and accident record multiple data, so as to quickly draw the operation conclusion to meet the requirements of intelligent substation intelligent monitoring service.

4.1 EC—SCADA System architecture

EC—SCADA system architecture is shown in Figure 3, which is divided into four levels: interactive output layer, fast response layer, core processing layer and edge device interface layer. The interactive output layer provides interactive data for cloud background monitor and dispatch center, provides man-machine dialogue; fast response consists of rule engine and fast gateway to enable core data to communicate quickly with interactive output layer; core processing layer data can be integrated with terminal equipment, including core data center, metadata, registry and configuration [9].

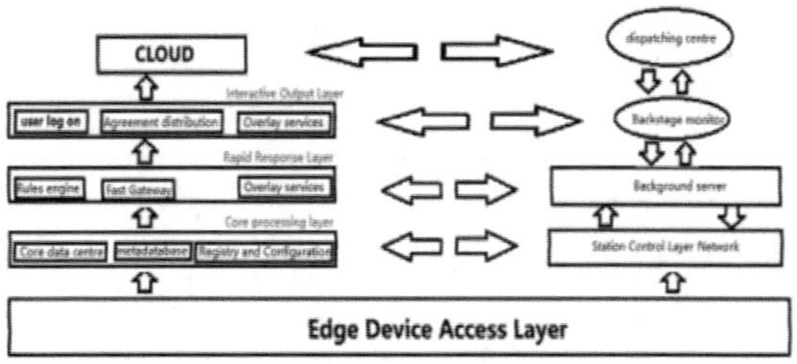

Fig.3. EC—SCADA system architecture.

4.2 Comparison of EC -SCADA and SCADA Systems Scheduling Remote Control

SCADA the remote control of the dispatching system, the remote control is first selected by the dispatching master station through telecontrol [10]. After success, the terminal equipment carries on the remote-control reverse calibration to the dispatching master station through the telecontrol. After receiving the Return Correction, the dispatching master station performs remote control to the terminal equipment [11].

EC-SCADA system scheduling remote control, through the dispatching master station at the same time to the cloud and telecontrol remote control selection, due to EC-SCADA system characteristics, communication speed reasons, cloud command priority through the core processing layer sent to the terminal equipment to select, after success, the terminal equipment Return Correction, the core processing layer directly controls the terminal equipment after receiving the return [12]. EC-SCADA system diagram is shown in Figure 4.

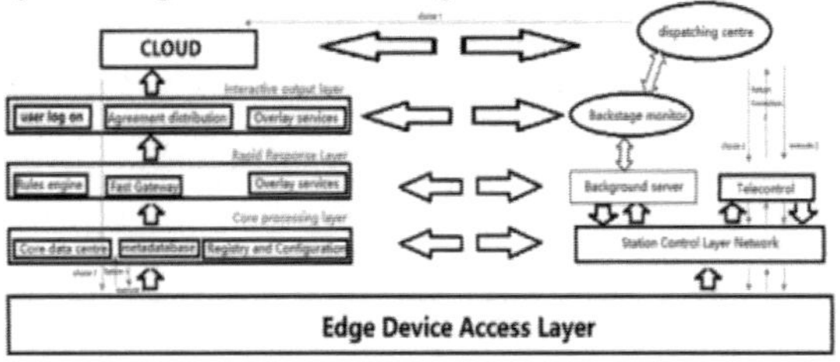

Fig.4. EC-SCADA system diagram.

Through simulation and comparison EC-SCADA the system performs faster and more reliably than the SCADA system. Meanwhile EC-SCADA the system can avoid the problem of partial function locking and system paralysis caused by SCADA system because of link selection.

5 Significance of EC-SCADA systems

EC-SCADA system improves the intelligence of the system, reduces the dependence of centralized data center, reduces the processing cost of data transmission, can complement each other with telecontrol and existing background monitoring server, and EC-SCADA the system to improve the terminal data processing and decision-making ability of intelligent substation. Moreover, the system improves the processing ability of heterogeneous data in remote monitoring disk of intelligent substation, and provides a reliable guarantee for the safe and stable operation of UHV transmission.

6 Conclusion

EC provide intelligent services in the network edge measurement fusion network, computing, storage and so on, especially suitable for the application of SCADA systems with a large number of different manufacturers and different principles of intelligent terminals. This paper has made the framework design of the EC-SCADA system for the existing situation, but there are still some key technical problems in practice. With the continuous development of the technology and the improvement of the relevant standards, under the background of the development of global energy interconnection EC knitting can promote the technological innovation of the next generation intelligent substation.

References

1. Huangzhang Lv, Chen Dan, Fan Bin, et al. *Standardization progress and case analysis of edge computing* (JCRD, Guangzhou, 2018)
2. Bin Li , Bincheng Jia, Wangzhang Cao, et al. *Application prospect of edge computing in power demand response business* (Phys. S.T, Shanghai, 2018)
3. Zhang Ning, Jingku Liu. *The ultra-high voltage grid risk analysis and its assessment framework* (Phys. S.P.C, Beijing, 2013).
4. Chao Chen, Shouwei Zhang, et al. *Research on power industry data application based on big data.* (MET, Shanghai, 2014).
5. Wang Xun, Zhao Meng, et al. *Discussion on big data analysis and application of power transmission and transformation equipment* (Phys.SBD, Shengzhen, 2018)
6. Ziming Zhao, Zhiping Cai, et al. *Edge computing: plating, application and challenge* （JCRD, Xi'an, 2018）
7. Yanzhen Lin, Qingwu Gong. *Relay protection risk assessment based on fuzzy information granulation algorithm* (EPC, Beijing, 2016)
8. Yue Yu, Chunxia Zhou, Rongrong Zhan, et al.*reaserch on network stress cause and test method of relay protection in smart substation*(PSPC, Changsha,2017)
9. Haitao Yang, Ping Ji, Yuan Ren, et al. *Analysis of safety monitor function and technical requirements for UHV power grid* (PST, Shanghai, 2016)
10. Miao Miao,Ping Ji et al. *analysis on interrelationship between future UHV power grid structural form and power source composition in china* (AEPS, Shanghai,2018)
11. Zhaoyang Xue, Yening Lai. *Integration of macro energy thinking and big data thinking part one big data and power big data* (AEPS, shanghai, 2016)
12. Zhaoyang Qu, Shuai Chen, Fan Yang, et al. *An attribute reducing method for electric power system* (AEPS, Shanghai, 2014)

https://doi.org/10.1051/e3sconf/202123702006

Research on hot air ice melt system for cable sheaths of Er'qi Yangtze River Bridge

Liu Qianqian[1*], Han yongtao[1] ,and Fan xueliang[1]
([1]China Railway Wuhan Survey and Design Co., Ltd., Wuhan 430070, China)

Abstract. To prevent the cable sheaths of Er'qi Yangtze River Bridge from freezing under bad weather in winter, which may lead to ice fall and bring traffic safety accidents, a theoretical model and technical scheme of hot air ice melt system for Er'qi Yangtze River Bridge has been proposed. Based on the established heat transfer model for cable sheaths, the characteristics of heat transfer between internal hot air and outdoor environment have been analyzed. Under the condition when the temperature of the end of cable sheath's outwall is more than 3℃, the range of temperature and speed of input hot air were estimated, which provides a theoretical basis for hot air melt system for cable-stayed bridge sheaths.

1 Introduction

In recent years, our country's bridge construction has been developed rapidly. The cable-stayed bridge is widely built in crossing the river due to its beautiful shape and convenient construction. However, in the middle and lower reaches of the Yangtze River, there is continuous low temperature rain and snow in winter, and recent years, some cable stayed bridges appeared freeze and ice fall. In 2015, the ice fell from Tianxingzhou Bridge in Wuhan. In January of 2018, the cables of the Poyang Lake Bridge has been frozen, and in February, the ice of Anqing Yangtze River Bridge's tower cables melted and fell to the cars on the bridge; In January of 2019, the Er'qi Yangtze River Bridge and the Baishazhou Bridge in Wuhan also appeared freeze and then ice melt with fall. The cable sheaths of cable-stayed bridge were frozen with ice fall posed detrimental effect on traffic safety accidents and caused certain social negative influence.

2 Research review of ice melt system for cable sheaths

The cable sheaths of cable-stayed bridge froze was a relatively low probability phenomenon under bad weather in winter, only when multiple negative meteorological factors appear at the same time. The solution of this problem is still in the experimental research stage at home and abroad, and no mature engineering technical measures have been applied to practical projects. In previous studies on the cable deicing and melting system mainly has the following directions: the First is to use the thermal resistance material to heat the cable sheaths to deicing[1];The second is to use high temperature air to heat the cable sheaths[2]. The third is to brush hydrophobic paint on the outside of the cable sheaths. The Fourth is use of physical and mechanical deicing devices [3]. This paper attempts to take Er'qi Yangtze River bridge as the research object to demonstrate the feasibility of the method of heating cable sheath with hot air.

The Er'qi Yangtze River Bridge is a cable-stayed bridge, has a main span of 2×616m from three concrete towers, the width of deck is 30.5m and the height of the main tower is 205m with two opposing 8 lanes of traffic. The cable of the bridge was built by parallel steel strand, the cable is composed of multiple unbonded high strength parallel galvanized steel strand, and the outer layer is equipped with HDPE sheath. There are 264 cables in total, the maximum length of a single cable is 300m, and each cable is equipped with HDPE De250 protective sheath.

3 Parameter of cable sheaths model

In this paper, the model of conveying the hot air to heat cable sheath is analyzed, that is deliverying the hot air from one end of the cable sheath to the other end to keep the temperature of the cable sheath out wall higher than the freezing point of water, and prevent the sheath from freezing. The parameters of single cable sheath model is presented in Tab.1.

Tab.1 The parameters of single cable sheath model

Project	Symbol	Value	Unit
External diameter of cable sheath	D_2	0.25	m
Thicknesses of cable sheath	σ	0.008	m
Internal diameter of cable sheath	D_1	0.234	m
Net area of cable sheath (steel strand is deducted)	A	0.032	m^2

* Corresponding author: queenaliu0608@163.com

Project	Symbol	Value	Unit
Thermal conductivity of cable sheath	λ_1	0.5	$W/(m \cdot K)$
Length of cable	l	300	m

The meteorological parameter on February 11, 2019, which the weather of the day that ice fall occurred on the Er'qi yangtze river bridge in Wuhan, was used as a reference. The weather on February 11, 2019 was sleet and rainy, during this period, the maximum wind force was level 3-4 from the north, and the wind speed was about 3.4-7.9m/s. The minimum temperature at night was -1℃, and the maximum temperature in the day was 3℃.Considering the changes of river wind speed and air temperature comprehensively, the outdoor meteorological parameters and conditions are shown in Tab.2.

Tab.2 The outdoor meteorological parameter

Project	Symbol	Value	Unit
Air temperature	T_w	-3	℃
Wind speed	v_2	5	m/s
Density	ρ_2	1.293	kg/m^3
Specific heat capacity	c_{p2}	1005	$J/(kg \cdot K)$
Thermal conductivity	λ_2	0.0244	$W/(m \cdot K)$

4 Thermodynamic calculation of heat transfer model for cable sheath

Based on the calculation of Reynolds number, the heat transfer of the cable sheath for cable-stayed bridge is divided in three processes, that is the turbulent convection heat transfer between the hot air inside the sheath and the inner wall of the sheath, thermal conduction heat transfer between the inner wall and the outer wall of the sheath, and forced convection heat transfer between the outer wall of the sheath and the air outside the sheath. The heat transfer capacity above-mentioned are φ_1、φ_2、φ_3 respectively. The total heat transfer between the hot air inside the sheath and the outdoor air is Q_1, and the heat transfer Q_2, sent by the hot air into the sheath. According to the law of energy conservation, $\varphi_1 = \varphi_2 = \varphi_3 = Q_1 = Q_2$。

To simplify the model, when the steel cable inside the sheath and the hot air reach thermal balance, there is no more heat transfer between the two, so the heat exchange between the steel cable and the hot air can be ignored.

To prevent the cable sheath from freezing, the end temperature of the outer wall of the sheath t_b' should be kept at $t_b' \geq 3°C$,so the hot air's inlet temperature t_1 and wind speed v_1 of entering the sheath was calculated and analyzed. The thermally physical properties of the fluid (specific heat capacity C_p、thermal conductivity λ、density ρ) are related to the fluid temperature, due to the cable structure and melting point of the cable sheath on the cable-stayed bridge, and the hot air inlet temperature should be considered at $t_1 \leq 60°C$.

(1) When the outer wall temperature of the end of sheath$t_b' \geq 3°C$, the outlet temperature of hot air input from the sheath is calculated as t_2.

$$Re = \frac{\rho_2 v_2 D_2}{\mu} \quad (1)$$

$Re = 92248$,the flow state is turbulent;

In the above formula, ρ_2、$v_2$are presented in Tab. 2; D_2 is presented in Tab. 1; μ is defined as Aerodynamic viscosity, Pa.s。

$$Pr = \mu c_{p2}/\lambda_2 \quad (2)$$

In the above formula,c_{p2}、$\lambda_2$are presented in Tab.2;

$$Nu = 0.266 R_e^{0.805} Pr^{1/3} \quad (3)$$

Convection coefficient between the out wall of sheath and the air is h_2.

$$h_2 = Nu\lambda_2/D_2 \quad (4)$$

The convection heat transfer of cold air across the outer wall of sheath [4].

$$\varphi_3 = Ah_2\Delta t_2 = 32.9kW \quad (5)$$

Based on the law of conservation of energy,$\varphi_3 = \varphi_2$, so the heat of thermal conductivity between inner and outer wall of sheath $\varphi_2 = 32.9kW$。

$$\varphi_2 = 2\pi l\lambda_1(t_a - t_b)/ln\,(r_2 - r_1) \quad (6)$$

In the above formula, t_a、t_b are the inner and outer wall temperature of sheath,℃; r_1、r_2 are the inside and outside radius, m; $\lambda_1$is thermal conductivity of sheath, $W/(m \cdot K)$。

The end temperature of the inner wall of the sheath $t_a' = 5.31°C$.consequently, considered the temperature difference in heat transfer, the outlet temperature of the hot air input in the sheath $t_2 \geq 6°C$。

(2) There are three heat transfer processes between the hot air inside the sheath and the outdoor air, namely the turbulent convection heat transfer of the hot air inside the sheath, the thermal conductivity between inner and outer wall of sheath and the force convection heat transfer of cold air across the outer wall of sheath. The total heat transfer coefficient K is calculated according to the following formula.

$$\frac{1}{K} = \frac{1}{\alpha_0} + \frac{d_0}{2\lambda_w} ln\frac{d_0}{d_i} + \frac{1}{\alpha_i}\frac{d_0}{d_i} \quad (7)$$

In the above formula,K is the total heat transfer coefficient,$W/(m^2 \cdot K)$; α_i、$\alpha_0$are the fluid heat transfer efficient of internal and external sheath, $W/(m^2 \cdot K)$; d_i、d_0 are the internal and external diameters, m; λ_w is the thermal conductivity of sheath, $W/(m \cdot K)$; α_i、α_0 were calculated as following:

$$\alpha_0 = Nu_o\lambda_2/d_0,\ \alpha_i = Nu_i\lambda_1/d_i \quad (8)$$

In the above formula, Nu_i、Nu_o are the Nusselt number of internal and external fluid.

(3) The heat transfer temperature difference Δt between hot air and the outer wall of the sheath is the logarithmic temperature difference.

$$\Delta t = [(t_1 - t_w) - (t_2 - t_w)]/ln\left[\frac{t_1 - t_w}{t_2 - t_w}\right] \quad (9)$$

In the above formula,t_1、t_2 are the inlet and outlet temperature of hot air, ℃; t_w is the air temperature outside the sheath, ℃; the value is -3;

(4) The heat exchange between the hot air inside the sheath and the cold air outside the sheath Q_1 can be converted into function related to t_1、t_2 and R_e、P_r, namely the equation as following:

$$Q_1 = \frac{A[(t_1-t_w)-(t_2-t_w)]/\ln[\frac{t_1-t_w}{t_2-t_w}]}{\frac{1}{\frac{1}{(0.0266Re_o^{0.805}Pr_o^{1/3})\lambda_2/d_o}+\frac{d_o}{2\lambda_w}\ln\frac{d_o}{d_i}+\frac{1}{(0.023Re_i^{0.8}Pr_i^{0.3})\lambda_1/d_i}\frac{d_o}{d_i}}} \quad (10)$$

In the above formula, Re_i、Pr_i are the functions of t_1、v_1 ,which are the hot air inlet temperature and velocity. Therefore the Q_1 can be converted into function of t_1、v_1,$Q_1 = f(t_1, v_1)$。

(5) Based on the principle of energy conservation and the continuity equation, the heat input into the sheath Q_2 is calculated as follows:

$$Q_2 = c_{p1}\left(\frac{\pi d_i^2}{4}\right)v_1\rho_1(t_1 - t_2) \quad (11)$$

In the above formula, c_{p1} is specific heat capacity of hot air, $J/(kg \cdot K)$; ρ_1 is density of hot air, kg/m^3。The Q_2 can be converted into the function related to t_1、v_1, $Q_2 = g(t_1, v_1)$.

5 Thermodynamic solution of the heat transfer model of the cable sheath

(1) Under the following assumptions:

considered the ice melt and energy save, the outer wall temperature of the end of the sheath is $t_b' \geq 3°C$ and outlet temperature of hot air is $6°C \leq t_2 \leq 10°C$. Under the premise of ensuring the safety of bridge structure and sheath melting point, $t_1 \leq 60°C$、$v_1 \leq 50m/s$.

(2) The approximate value of equation $Q_1 = Q_2$ was calculated by the MATLAB compiler in the way of discrete iteration. The unknown value t_1、v_1 were dispersed equably within the range of values, $t_1 = [1:1:60]$, $v_1 = [1:1:50]$. We plug the t_1、v_1 into equation $Q_1 = f(t_1, v_1)$, $Q_2 = g(t_1, v_1)$, and calculate in loop iteration solution.

(3) Conditions of solution: when the modulus of Q_1 and Q_2 meet the following requirement, which is $|Q_1 - Q_2| \leq 0.5kW$, we pick up the values of t_1、v_1、$Q_1(Q)$ as a group of approximate solutions of the equation. The evaluation results are as following.

(4)when $t_2 = 6°C$, $30°C \leq t_1 \leq 58°C$, $34m/s \leq v_1 \leq 50m/s$, $63.62kW \leq Q \leq 88.93kW$。As shown in Fig.1, the points marked in red in the figure are the approximate solution of $Q_1 = Q_2$ with $t_2 = 6°C$.

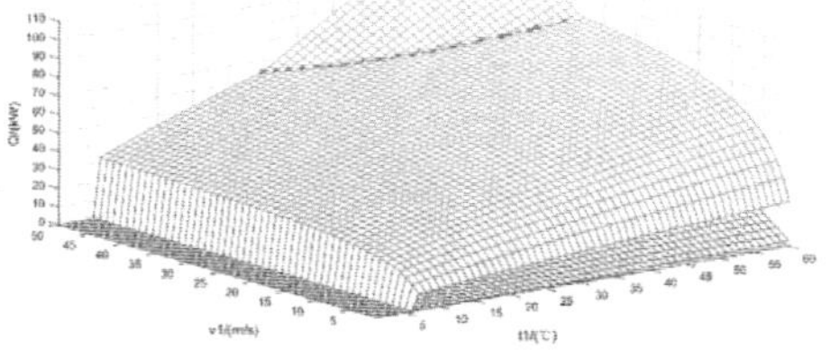

Fig.1. The value of t_1、v_1、Q with $t_2 = 6°C$

(5)when $t_2 = 8°C$, $41°C \leq t_1 \leq 60°C$, $38m/s \leq v_1 \leq 48m/s$, $81.49kW \leq Q \leq 98.91kW$ .As shown in Fig.2, the points marked in red in the figure are the approximate solution of $Q_1 = Q_2$ with $t_2 = 8°C$.

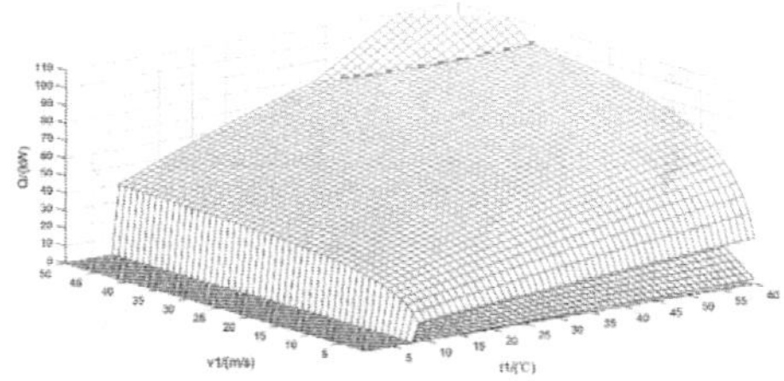

Fig.2. The value of t_1、v_1、Q with $t_2 = 8°C$

(6)when $t_2 = 10°C$, $47°C \leq t_1 \leq 60°C$, $43m/s \leq v_1 \leq 50m/s$, $94.34kW \leq Q \leq 106.77kW$ .As shown in Fig.3, the points marked in red in the figure are the approximate solution of $Q_1 = Q_2$ with $t_2 = 10°C$.

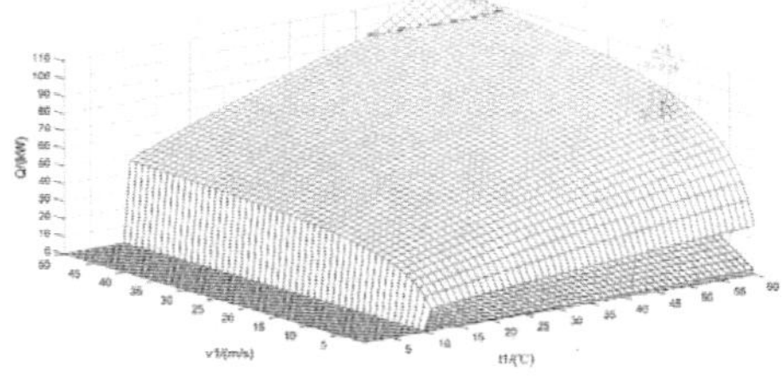

Fig.3. The value of t_1、v_1、Q with $t_2 = 10°C$

(7)when $6°C \leq t_2 \leq 10°C$, the value of t_1、v_1、Q was shown in Fig.3. The data shown in gray is the design recommendation value in Tab.3.

Tab.3 The value of t_1、v_1、Q

t_2=6℃			t_2=7℃			t_2=8℃			t_2=9℃			t_2=10℃		
Q kW	t_1 °C	v_1 m/s	Q kW	t_1 °C	v_1 m/s	Q kW	t_1 °C	v_1 m/s	Q kW	t_1 °C	v_1 m/s	Q kW	t_1 °C	v_1 m/s
88.93	58	34	93.97	59	36	98.91	60	38	102.05	59	41	106.77	60	43
85.51	54	35	91.40	56	37	96.25	57	39	99.25	56	42	103.90	57	44
86.46	55	35	88.76	53	38	97.25	58	39	97.44	54	43	102.03	55	45
82.94	51	36	87.05	51	39	94.54	55	40	95.58	52	44	100.12	53	46
83.91	52	36	84.26	48	40	91.75	52	41	93.68	50	45	99.27	52	47
81.28	49	37	85.29	49	40	89.93	50	42	91.73	48	46	97.29	50	48
78.57	46	38	82.45	46	41	88.07	48	43	90.85	47	47	94.34	47	50

79.57	47	38	83.49	47	41	87.24	47	44	88.83	45	48
76.80	44	39	81.64	45	42	85.31	45	45	87.91	44	49
77.82	45	39	79.75	43	43	83.33	43	46	86.96	43	50
74.98	42	40	77.81	41	44	82.42	42	47			
76.01	43	40	76.92	40	45	81.49	41	48			
74.17	41	41	76.01	39	46						
72.27	39	42	73.01	36	48						
71.40	38	43	72.03	35	49						
69.43	36	44	71.03	34	50						
68.51	35	45									
67.57	34	46									
66.61	33	47									
65.64	32	48									
64.64	31	49									
63.62	30	50									

6 Hydraulic calculation of cable sheath model

When $t_1 = 54°C$, $v_1 = 35m/s$, the net area of the sheath A=0.032m^2, and the hydraulic calculation of the sheath is shown in Tab.4.

Tab.4 The hydraulic calculation of the sheath

Air quantity	Diameter	Length of sheath	V	R	ΔPy	ζ	Dynamic pressure	ΔPj	Total resistance
(kg/h)	(mm)	(m)	(m/s)	(Pa/m)	(Pa)		(Pa)	(Pa)	(Pa)
4990	205	300	35.06	68.35	20503.91	2	736.18	1472.36	21976.28

As shown in Tab.4, the frictional resistance $\Delta p_y =$ 20503.91Pa ,the total resistance $\Delta p = \Delta p_y + \Delta p_j =$ 34436.78Pa. The pressure value meets the pressure value of PVC material.

7 Conclusion

Based on 300m of a single sheath the temperature of input hot air supply $t_1 = 54°C$, the supply air speed$v_1 = 35m/s$ and $Q = 85.51kW$. The total number of the bridge cables is 264, the maximum length of a single cable is 300m, and the minimum length of a single cable is 210m. To simplify calculation, the cable lengths were all considered as 300m, so the total heat input into the sheath is 22,574.64kW.

Based on the establishment of heat transfer model for the cable-stayed bridge's single cable sheath, in order to prevent the cable sheath from freezing in winter under the bad weather, and ensure the temperature of the end of sheath's outwall is more than 3℃, according to the law of conservation of energy, the range of temperature and speed of input hot air were estimated. The calculation of the model provides a theoretical basis and design reference for the application of hot air to the cable sheath melt system for cable-stayed bridge.

On the premise of keeping the safety of the bridge and maintaining the existing facade, and aiming at the structural characteristics of the existing cable-stayed bridge, the use of the hot air ice melt system of cable sheath for the existing cable-stayed bridge has a reference significance for the application in reconstruction of the existing cable-stayed bridge to prevent freezing.

However, the huge energy consumption demand, the choice of heat source and equipment installation are all technical difficulties, the design scheme become used still need to be further demonstrated, Such as the relief valve is added to both ends of the sheath, which can not only guarantee the hot air not to lose instantly, reduce energy consumption, but also ensure the pressure relief timely when the pressure reaches a certain value.

References

1. China Railway Bridge Scientific Research Institute Co., LTD. A system and Method of cable deicing and anti-icing for cable-stayed bridge: China, 201810141060[P].2018-08-17
2. China Railway Bridge Scientific Research Institute Co., LTD. A system and method of intelligent dehumidification and deicing for cable-stayed bridge: China, 201810073182[P].2018-07-20
3. China Railway Bridge Scientific Research Institute Co., LTD. A device of deicing and snow removal for cable-stayed bridge: China, 201810140798[P].2018.08.10
4. Zhang ximing, Ren zepei. Heat transfer theory[M].4 versions. Beijing:China building industry press:2001:4-5

Market clearing price forecast for power peak shaving auxiliary service

Dunnan Liu[1], Mengjiao Zou[1,*], Yue Zhang[1],Lingxiang Wang[1],Tingting Zhang[1],and Mingguang Liu[1]

[1]School of Economics and Management, North China Electric Power University, Changping District, Beijing 102206, China

Abstract. The use of new energy to generate electricity in the power system and the large-scale increase of new energy grid connection has led to increasingly insufficient power system regulation,in order to solve this problem, the peak shaving auxiliary service market came into being.This article comprehensively analyzes the factors those affect the market clearing price of power peak shaving auxiliary services:The macro factors include energy economic policies (renewable energy and electric energy substitution), technological innovation, market operation rules, etc., and the micro factors include the quotation and demand of thermal power plants and wind power generation.The power peak shaving auxiliary service market is an important part of the power market. Its appearance makes the grid operation safer and more reliable, and the reasonable fluctuation of clearing prices and total market costs reflects the market's sensitivity to peak shaving resource demand.This paper uses the BP neural network model to select 31 consecutive days of peak shaving auxiliary service clearing price data in North China for prediction.

1 Introduction

In order to ensure the safe and stable operation of the North China Power Grid and the Beijing-Tianjin-Tangshan Power Grid, it is necessary to promote the optimization of the power auxiliary service resources of the North China Power Grid on a larger scale. Therefore, the relevant entities in the power system predict the market clearing price of auxiliary services, which is more conducive to the understanding of the new market situation for all power entities and facilitates efficient participation in market competition.

With the progress of the electricity reform, the electricity market has gradually been divided into the main energy market and the auxiliary service market, and the two markets need to be balanced to achieve the optimal configuration [1],The current energy consumption pattern is developing in the direction of "electricity substitution" and "clean substitution". The method of replacing coal with electricity and oil with electricity has optimized the energy consumption structure [2],however, the volatility of renewable energy will reduce the stability of the power system. As its installed capacity increases, the demand for power peak shaving auxiliary services will also increase [3].The price mechanism can be used to guide the peak of thermal power units to obtain profits [4]. Some scholars have proposed a two-part economic compensation mechanism and specific algorithm for peak shaving that combines peak shaving cost compensation and excess power generation revenue sharing [5].In foreign countries, Northern Europe uses the "ancillary service market + imbalance + settlement+ regulation power market" model, the United Kingdom adopts the "integrated transaction" model, and the United States adopts the PJM. These models provide reference for my country's peak shaving auxiliary service market transactions. [6].The system marginal price, capacity cost and grid auxiliary service cost determine the electricity price level of the entire grid[8], the difference caused by the volatility of wind power output[10], coal consumption, loss, environment, fuel consumption, etc. will all have an impact on the cost of power peak shaving[7].

The market clearing price of peak shaving has the function of market guidance. A clear price signal attracts more units to provide peak shaving capacity to promote the effective consumption of clean energy in the receiving-end regional power grid [9].The importance of ancillary services has gradually increased in recent years and has not received much attention before. Therefore, there are relatively few domestic literatures on the clearing price prediction of the power peak shaving ancillary service market;There is no power peak shaving service in mature ancillary service markets abroad, and the time-sharing adjustment of the spot market electricity price has solved the peak shaving problem. Therefore, foreign research on this aspect is rare. Most of the forecasting models of electricity prices in the current literature predict short-term or medium- and long-term electricity transaction prices in the electricity market, such as:Using random forest regression to predict the price of spot market clearing [15], combined with support vector machine and autoregressive moving average and external input module hybrid mid-term power mcp prediction [12], applying generalized autoregressive sliding model to predict electricity prices [14], based on multiple support vector machines to build mid-term power mcp forecasting model [11], based on wavelet transform, chaotic least square support vector machines and exponential generalized autoregressive conditional

* Corresponding author: zoumengjiao_98@163.com

heteroscedasticity model based day-ahead electricity price forecasting new technology [13].

Based on the above background, this article will analyze the factors that affect the clearing price of the peak shaving market from both the macro and micro aspects. Because the BP neural network has self-learning function and associative storage function, it can find the optimal solution at a high speed and has good forecast stability.This paper mainly uses the Bp neural network model to predict the clearing price of the peak shaving auxiliary service market.

2 Macro factors

2.1 Energy Economic Policy

In recent years, people's consumption demand has been increasing, and the upgrading of the consumption structure has promoted the continuous development of China's economy. However, because my country's current energy structure is mainly based on coal, the rapid energy consumption means that pollutant emissions are increasing year by year. The transformation of energy consumption patterns can start from both energy production and terminal consumption.

2.1.1 Renewable Energy

Affected by national policies, the installed capacity of renewable energy power generation has grown rapidly, but there are problems such as the high cost of renewable energy compared with traditional energy generation, the low flexibility of the power system, and the large-scale grid integration of renewable energy power generation. The intermittent, volatility and other characteristics of renewable energy itself have resulted in inefficient utilization of renewable energy.When renewable energy continues to develop, the market size increases, and the subsidy gap is further expanded, and the quota system emerges at the historic moment. At this time, in the absence of green certificate trading, the government compensates for corporate losses through subsidies; If green certificate trading is implemented, the price of renewable energy power is determined by the price of green certificates and the price of conventional energy power.In the peak shaving ancillary service market, the strong support of national policies and increased government subsidies are conducive to the increase of the supply-side renewable energy installed capacity and the reduction of electricity prices. Among them, the feed-in tariff subsidy policy has a significant impact on the installed capacity of wind power.

2.1.2 Electricity alternative

The State Grid proposed the development strategy of electric energy substitution in 2013. Compared with coal, oil and other energy sources, electric energy has the advantages of being safe, convenient, clean and efficient, and stable in supply. The current areas of electric energy substitution are very wide: electric heating, electric refrigeration, electric vehicles, households Electrification, electric kilns in industrial production, electrified railways, etc.However, it should be pointed out that the development of electric heating is not intended to replace central heating with combined heat and power. Electric heating is a supplementary heating method in the entire heating system.The vigorous implementation of the electric energy substitution policy has also promoted the continuous development of my country's electrification level. People's dependence on coal, gas and kerosene has weakened, and electric energy has met people's demand for coal and oil as a substitute. At the same time, the promotion of national policies, rewards, subsidies and other methods have stimulated the public's electricity demand, which has increased the demand-side electricity in the peak shaving auxiliary service market, and the electricity price may fall under the intervention of national policies.

2.2 Technological innovation

With the continuous expansion of the development and utilization of renewable energy in my country, the level of its technical equipment has been significantly improved. my country has mature large-scale hydropower design, construction, management and operation capabilities; low wind speed wind power technology has made breakthroughs and has been widely used in the Middle East and South; and established an internationally competitive photovoltaic power generation industry chain.Due to the influence of weather, season and other factors in the power generation process of clean energy, it is difficult for us to systematically control and predict the power generation and time of clean energy. At present, the problem of wind curtailment in my country's "Three Norths" area during the winter heating period is serious. This is because the proportion of combined heat and power units in the "Three Norths" area is high, which leaves less room for wind power. The flexibility of thermal power must be improved. At present, the flexibility of thermal power can be modified from two aspects: enhancing the flexibility of unit operation and enhancing the flexibility of boiler fuel. In the power peak shaving ancillary service market, innovations in power generation technology will reduce the cost of power generation, and thus electricity prices will fall to a certain extent, which will further stimulate demand and promote economic growth.

2.3 Technological innovation

The operating rules of the power peak-shaving auxiliary service market are an invisible hand used by the government to monitor the power peak-shaving auxiliary service market, which will affect the market demand and price of the peak-shaving service to a certain extent. Considering that the rapid development of photovoltaics in the Northeast region has a significant impact on the power system, on January 14, 2019, the new rules issued on the official website of the Northeast Energy

Regulatory Bureau officially included photovoltaics in the power auxiliary service market. The Northwest region has a high penetration rate of new energy. According to its regional characteristics, the virtual energy storage mode and the adoption of the settlement method of "deep adjustment power generation rights substitution" were first established, which not only promoted the development of the market but also made the process more streamlined. In different regions, time periods, and unit types are different, there will be slight gaps in the compensation benchmark value and segmented quotation method of paid peak shaving.

3 Micro factors

3.1 Quotation of thermal power plants and market demand

The real-time electricity market is to encourage generators to participate in the regulation of power imbalances (under-generation management, over-generation management, congestion management). The quotation of thermal power plants sometimes refers to the quotation of the previous day's market (but whether the real-time market's quotation will be The market is high or the quotation deviation will be relatively large, it still depends on the supply and demand relationship), so the determination of the price mainly depends on the effectiveness of the real-time market price, that is, whether the price signal of the real-time market can adjust the supply and demand relationship. The market price is jointly determined by the supply side and the demand side: the quotations of thermal power plants are often relatively subjective and not very predictable; however, the peak shaving demand in the market is relatively objective. There are many factors that affect the demand for peak shaving in the market, such as: unit maintenance plan, cross-regional and cross-province power purchase, wind power generation, etc. However, the most important factor that will affect short-term price changes is the power of wind power, because wind resources are volatile and intermittent. In wind power generation, we need to pay attention to the conditions of wind resources, mainly wind speed and its stability. This is the determinant of the availability of wind resources, as well as the most basic factor that determines the amount of wind power and the cost of wind energy, and these factors will also have a negative effect on the final clearing price.

3.2 Wind power

The price of the power peak-shaving auxiliary service market fluctuates greatly, but because the fuel efficiency of thermal power units has not changed much in the short term and the current level of equipment in the market is relatively stable,I think thermal power units have little effect on prices, and the reason for the large price fluctuations is the use of new energy.Wind power and solar energy are all intermittent energy sources, and their power and quality of power generation are unpredictable in different seasons, time periods or due to changes in weather factors. The production, supply and consumption of wind power products are carried out almost at the same time, so the production and consumption of wind power must maintain a real-time balance.China's wind power resources are mainly concentrated in the "Three North" area, and its load has obvious reverse distribution characteristics. Wind power output has the characteristics of uncertainty, randomness and intermittent. The power grid considers its own ability to accept wind power and the safe operation of grid-connected power, and will limit part of the wind power output at a certain moment.

In recent years, the total installed capacity of wind power has steadily increased, and the wind power market has gradually become an important emerging market. However, whether wind power can be connected to the grid for power generation depends on whether the power system has sufficient peak shaving capacity.Due to the operation mode of cogeneration units "using heat to determine electricity", a large number of wind curtailments will occur during the heating season in northern my country.In the peak shaving ancillary service market, the efficiency of thermal power units will not change in the short term, and the cost of power generation will not change much in a short period of time. The results obtained by predicting short-periodical and highly volatile wind power generation power will affect the market peak shaving demand to a certain extent, thereby affecting the quotation of thermal power plants. If a large amount of wind power resources flood into the market during a certain period of time, the cost of wind power generation will be lower at this time because of the large supply, and the clearing price will also decrease.

3.3 Selection of price influence factors

As shown in Figure 1:analyze the factors that affect prices from both macro and micro aspects: once the macro policy of a region is determined, it is long-term and will not change much in the short term, Therefore, when predicting short-term prices, you can temporarily ignore the macro factors in this area. What needs to be considered is the micro factors that may change in the short term. Through the above analysis, it can be concluded that the wind power generation power is mainly considered among the micro factors. Short-term power prediction of wind power can improve the economic performance of system operation and increase the capacity of wind power absorption. For the power grid, accurate prediction is conducive to real-time adjustment of the power grid dispatching plan, reducing operating costs. For the electric field, it helps to ensure the grid connection rate of wind power and maximize the benefits.

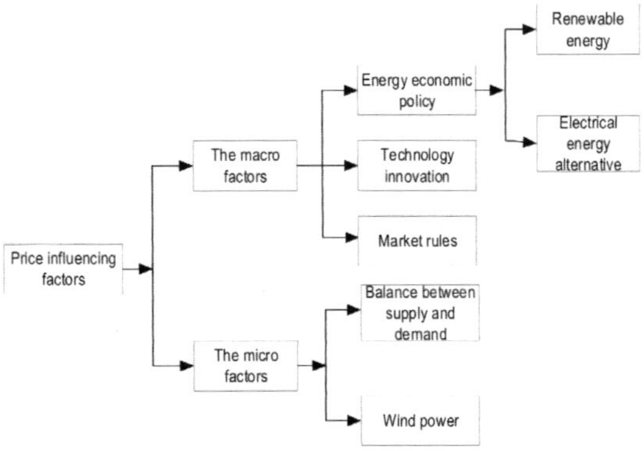

Fig. 1. Analysis of price influence factors

4 Price Modeling of Peak-shaving Auxiliary Service Based on BP Neural Network

The BP neural network includes input layer i, hidden layer j, and output layer k. The upper and lower layers are connected to each other but there is no connection between the same layers. There are network weights between the upper and lower neurons that indicate the connection strength of the two neurons ω ,the learning rate is η ,The input vector is $\vec{X}=[x1,x2\ldots xi\ldots xn],i=1,2,\ldots,n$; the output vector is $\vec{Y}=[y1,y2\ldots yk\ldots ym],k=1,2,\ldots,m$; the output of each neuron in the t-th hidden layer is $\vec{a}^{(t)}=[a_1^{(t)},a_2^{(t)}\ldots a_j^{(t)}\ldots a_{st}^{t}],j=1,2,\ldots,st$.

Suppose the number of nodes in the input layer is n, the number of nodes in the hidden layer is l, and the number of nodes in the output layer is m. The weight from the input layer to the hidden layer is ω_{ij}, the weight from the hidden layer to the output layer is ω_{jk}, the bias from the input layer to the hidden layer is a_j, and the bias from the hidden layer to the output layer is b_k .Let $\omega_{ij}^{(t)}$ be the weight between the i-th neuron in the input layer and the j-th neuron in the t-th hidden layer, and $b_i^{(t)}$ is the bias of the i-th neuron in the t-th layer, and there is a calculation formula:

$$\mathrm{net}_i^{(t)}=\sum_{j=1}^{st-1}W_{ij}^{(t)}a_{ij}^{t-1}+b_i^{(t)} \tag{1}$$

$$a_i^{(t)}=f(net_i^t) \tag{2}$$

Where $\mathrm{net}_i^{(t)}$ is the input of the i-th neuron in layer t, and $f(\bullet)$ is the activation function.

The output of the hidden layer is:

$$H_j=g(\sum_{i=1}^{n}\omega_{ij}x_i+a_j) \tag{3}$$

The output of the output layer is:

$$O_K=\sum_{j=1}^{1}H_j\omega_{jk}+b_k \tag{4}$$

If there are m training samples:
$$\begin{matrix} x_1 & r_1 & \cdots & y_1 \\ x_2 & r_2 & \cdots & y_2 \\ \cdots & \cdots & \cdots & \cdots \\ x_m & r_m & \cdots & y_m \end{matrix}$$

d_i is the expected output of x_i ,define the error as:

$$E=\frac{1}{m}\sum_{i=1}^{m}E(i) \tag{5}$$

Where $E(i)$ is the training error of a single sample:

$$E(i)=\frac{1}{2}\sum_{k=1}^{n}(d_k(i)-y_k(i))^2 \tag{6}$$

When the error is X, the error is fed back from the back neuron level by level, and the Bp neural network will iteratively update the weight and bias during the feedback process,the way is:

$$W_{ij}^{(t)}=W_{ij}^{(t)}-\alpha\frac{\partial E}{\partial W_{ij}^{(t)}} \tag{7}$$

$$b_i^{(t)}=b_i^{(t)}-\alpha\frac{\partial E}{\partial b_i^{(t)}} \tag{8}$$

5 Case analysis

In order to verify the effectiveness of the peak shaving auxiliary service clearing price model based on BP neural network proposed in this paper, this paper uses the peak shaving auxiliary service clearing price in North China to verify.Select electricity price data for 31 consecutive days, with 15 minutes as a step, a total of 96 points per day, but there are only 40 time points for the implementation of the transaction, that is, 26 time points in the 0:45-7:00 time period and 12:45 -16: 00 time period 14 points. The data obtained are: North China. Jingjintang load (MW), North China. Jingjintang wind power total active power (MW), North China. Jingjintang photovoltaic output (MW), clearing price (yuan/MWH).

The 40 time points of each day are cycled from 1-40 as the time points of the corresponding data.Then determine the week attribute of the corresponding data day and whether it is a holiday, but through analysis, it is found that the correlation between the weekly attribute of the corresponding data and whether it is a holiday is not very relevant to the clearing price, Therefore, these two types of data were not considered when constructing the prediction model.In the Bp neural network prediction model, the time point, North China. Jingjintang load (MW), North China. Jingjintang total active power (MW), and North China. Jingjintang photovoltaic output (MW) are used as input layers. The clearing price are used as the final output layer. The data of the first 30 days are training examples, and the clearing price on the 31st day is a test example that is predicted. The predicted results are shown in Figure 2 and Figure 3:

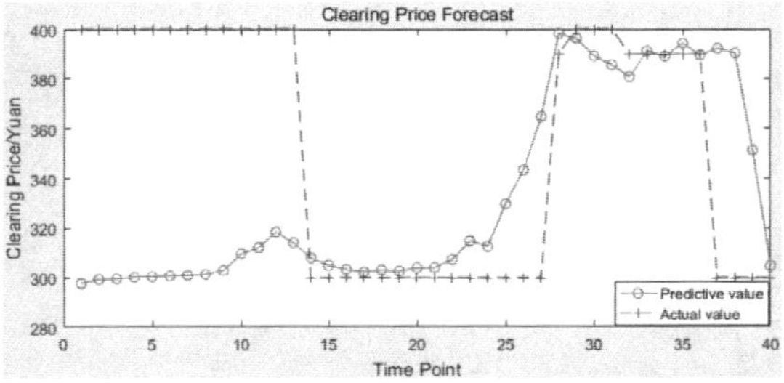

Fig. 2. Clearing price forecast

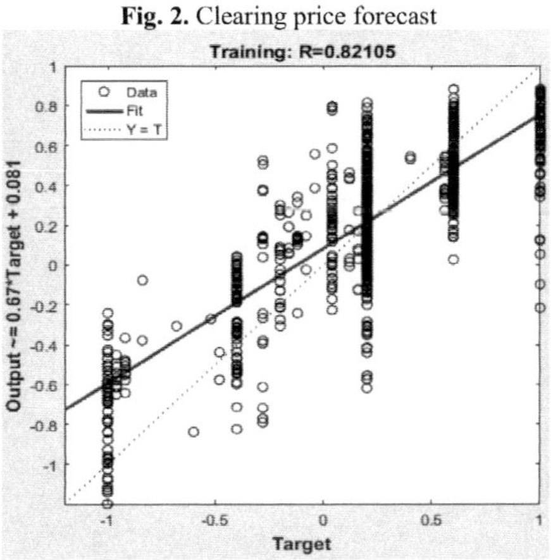

Fig. 3. Correlation coefficient

From the prediction result in Figure 2, we can see that the average absolute percentage error (MAPE) between the actual value of the clearing price on the 31st day and the predicted value is 11.8016%. The image shows that from the 14th time point onwards, the change trend of the actual value curve and the predicted value curve is roughly the same, the values are also relatively similar, and the prediction effect is generally satisfactory.

6 Conclusion and Outlook

In combination with this article, the following conclusions are obtained: the emergence of the power peak shaving auxiliary service market is to solve the problem of the imbalance between the supply and demand of electrical energy in the power system during a certain period of time. Compared with the US PJM model, which uses changes in real-time electricity prices to guide power generation companies to actively participate in peak shaving: the income of generator sets with different load rates is reflected and adjusted by changes in real-time electricity prices in the energy market during different periods.Most areas of China have not yet established a spot market based on time-of-use electricity prices, so China's peak shaving is achieved through the power peak shaving auxiliary service market.There are many factors that affect the clearing price. The macro factors include energy economic policies, technological innovation, market operation rules, etc., and the micro factors include the quotation and demand of thermal power plants and the power of wind power.After a comprehensive analysis, it is found that the most important factor affecting the market clearing price of peak shaving auxiliary services is wind power.In the process of predicting the clearing price through the BP neural network, it is found that the clearing price sequence of the power peak shaving auxiliary service market is highly volatile, and it is difficult to find the periodic law of the sequence.

This article analyzes the factors that affect the market clearing price of peak shaving auxiliary services, and uses BP neural network to predict clearing prices. However, due to time constraints, there is still room for improvement in the following aspects: the analysis of influencing factors in this article is qualitative analysis, it is also possible to further analyze wind power generation power from climatic factors such as temperature, humidity, wind direction, wind speed, and in-depth analysis of the correlation between various influencing factors and clearing prices;the current input data uses full-time data, which is difficult to reflect the error of market prices; there is a value of 0 in the data, and its processing method needs to be further studied; the prediction level achieved by the model used in this article is not high enough, and further optimization or exploration of more suitable models is needed.

Reference

1. G. Yang. *Research on Joint Optimization Decision of Electric Energy Main Market and Auxiliary Service Market*[D],SJTU(2011)
2. G. Huang .*Research on the Development Potential of Electric Energy Substitution and Its Economic and Environmental Benefit*[D],NCEPU(2019)
3. X.L.Zhao, M.Wang, Y.Zhao, Q.B.Wu. *Improved model of compensation mechanism for peak shaving auxiliary service based on capacity difference of thermal power units*[J]. Automat Electron Power Sys.**04**:57-61(2013)
4. H.P.Zhang, D.B.Gao, J.N.Zhang, Z.Xu. *The Development Road of Northeast Electric Power Peak Shaving Auxiliary Service Market*[J].China Power Enterprise Management,**28**:26-29(2018).
5. Y.S. Zhang, J.Zhang, T.T.Feng. *Research on th -e Economic Compensation Mechanism of Chin-a's Nuclear Power Plants' Peak Shaving Auxiliary Service*[J].Power Syst Technol.**07**:2131-2138(2017)
6. Y.H. He, M.Zhou, Z.Y.Long, J.Xu.*The operation mode of typical foreign electricity balance market and its enlightenment to China*[J]. Power Sys Techno.**11**:3520-3528(2018)
7. L.Qi, B.Cheng, R.Zhao, X.J.Gao.*Cost and benefit analysis of providing peak shaving auxiliary services for coal-fired thermal power units*[J].Power Syst and Big Data,**10**:23-29(2019)
8. P.S.Dai.*Research on the Price Formation Mechanism of China's Electricity Market*[D].XMU(2004)
9. F.Xu, C.Q.Ge, X.Wu, M.J.Zhu, M.F.Tu.*The market mechanism and clearing model of peak shaving auxiliary services in regional power*

grids[J],Automat Electron Power Sys.**16**:109-120(2019).

10. J.F.hu.*Research on Cost Allocation Mechanism and Model of Auxiliary Services for Wind Power Access*[D].NCEPU(2014)
11. Xing Y,Nurul A. Chowdhury. *Mid-term electricity market clearing price forecasting: A multiple SVM approach*[J].Int J Electr Power Energy Syst.**58**(2014).
12. Xing Yan,Nurul A. Chowdhury. *Mid-term electricity market clearing price forecasting utilizing hybrid support vector machine and auto-regressive moving average with external input*[J]. Int J Electr Power Energy Syst.**63**(2014).
13. Zhang JL, Tan ZF. *Day-ahead electricity price forecasting using WT, CLSSVM and EGARCH model.* Int J Electr Power Energy Syst.**45**:362–8(2013).
14. X.C.Liu, J.Shen, Y.G.Li. *A Generalized Auto-Re gressive Conditional Heteroscedasticity Model fo r System Marginal Price Forecasting Based on Weighted Double Gaussian Distribution*[J].Pow-e r Sys Techno „**1**:139-144(2010).
15. Q.Wei,S.J.Cheng,W.B.Huang,G.W.Ma,C.H.Tao.*Fo recasting method of spot market clearing price based on random forest regression*,Proc Chin Soc Electrical Eng,1-10(2020).

A power trading mode based on blockchain for prosumers

Zhichao Ren[1], Wei Wang[1], Bo Chen[1], Xin Li[1], Yuhong Zhang[1], Yu Hu[2,*], and Huaqiang Li[2]

[1]State Grid Sichuan Economic Research Institute, 610000 Chengdu, China
[2]College of Electrical Engineering Sichuan University, 610000 Chengdu, China

Abstract. With the increase of the penetration rate of distributed generation on the distribution network side, the access of a large number of prosumers makes the trading information massive, and the demand of prosumers for more flexible power trading mechanism is also strengthened. Therefore, a weak-centralized power trading mode based on blockchain is proposed in this paper. Trading information is automatically stored in the blockchain in the form of smart contracts. The centralized organization only manages congestion and does not participate in the process of trading matching and settlement. In the distributed security verification, the successive over relaxation (SOR) iterative method is improved in this paper, which improves the iterative efficiency and convergence stability of the distributed algorithm. Finally, a case consisting of six nodes is presented to verify the feasibility of the method.

1 Introduction

In recent years, with the development of distributed generation technology, users gradually change from one-way consumers to prosumers. The emergence of a large number of prosumers will produce new business opportunities [1]. The increase of the number of market participants makes the trading information massive. However, the traditional centralized power trading mode has some problems, such as high operating cost, difficult to guarantee the fairness and transparency of the transaction. It is urgent to seek a safe, efficient and fair power trading mode.

Therefore, this paper proposes to use blockchain technology to manage the power trading between prosumers. As a distributed accounting system with asymmetric encryption, blockchain has the characteristics of decentralization, unforgeability and anonymity, which can well adapt to the decentralized system structure of distribution network side, and is in line with the demands of power trading between prosumers [2]. The introduction of blockchain technology into the electricity trading of prosumers can achieve the transaction in a decentralized way, and ensure the efficient and reliable power trading between prosumers.

However, for the present stage, it is difficult to realize the fully decentralized multilateral electricity trading, because the power trading is different from the ordinary commodity trading, and the network constraints should also be considered. In the fully decentralized environment, it is still difficult for each node to reduce and adjust the trading volume spontaneously and eliminate the line congestion. Therefore, this paper proposes a weakly-centralized electricity trading mode for prosumers. In this mode, the centralized organization still exists, but only manages congestion, and does not directly participate in the matching and settlement process of electricity retail trading between prosumers, which limits the power of the central organization, and also satisfies the safe and reliable operation of the distribution network.

2 Distributed trading and blockchain technology

Blockchain is a chained data structure that connects the blocks of stored data in chronological order. It is a decentralized database and can be regarded as a distributed ledger [3-4]. The simplified structure of block chaining is shown in

. It links the blocks with certain associations in the time sequence generated by blocks. Each block is composed of a block header and a block body. The block header contains the version number, the hash value of the previous block, Merkle root and time stamp. The block body is equivalent to a database, which records the specific trading information. It takes Merkle function as the root and stores each trading information in the form of an inverted tree.

* Corresponding author: huyu1129@126.com

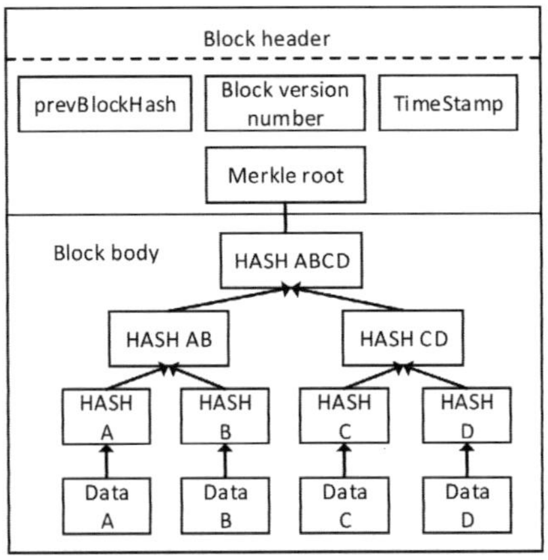

Fig1. Blockchain structure diagram

The features of blockchain, such as high security and decentralization, are just in line with the demands of distributed power trading. The specific analysis are as follows:

(1) The decentralized characteristics of blockchain makes trading between prosumers without the assistance of a third-party central organization, and the trading cost between prosumers decreases with the reduction of trust cost.

(2) The tamper resistant characteristics of blockchain ensures the information security of distributed trading, and maintains the fairness of distributed trading.

(3) Due to the data traceability characteristics of blockchain, users can know the types of electric energy purchased by each transaction, such as traditional energy, wind energy, photovoltaic power generation, etc., which enriches the selection of energy purchase types for users.

3 Overall structure

The specific process of weakly-centralized distributed trading based on blockchain is shown in Fig 2:

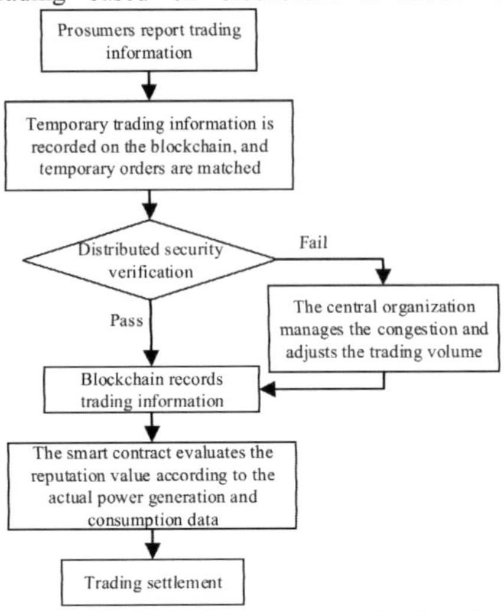

Fig2. Distributed power trading flow chart

(1) When the deviation occurs in the real-time phase, the user reports the trading price and the deviation quantity, and requests to eliminate the deviation.

(2) The blockchain platform collects trading requests reported by all users and matches trading orders to form temporary trading orders.

(3) Each user conducts distributed security verification according to temporary trading orders. If there is no power flow overrun, the temporary order will be regarded as the final trading result, and the trading data will be recorded in the blockchain. Otherwise, the central organization will manage the congestion and record the adjusted trading result in the blockchain.

(4) In the whole trading process, the energy data of each prosumer will be recorded by smart meters. In the trading settlement stage, the smart contract will confirm the actual output or power consumption of users, so as to evaluate the transaction reputation value of each user, and impose a certain liquidated damage on the default user.

The details of bidding strategy, trade matching mechanism, distributed security verification, congestion management, evaluation of transaction reputation value, etc. are described in detail below.

4 Bidding strategy and matching mechanism of prosumers

First of all, the following market parameters are set: $p_{\text{sell.w}}$ is the unit price of electricity sold to the grid by prosumers, that is, the grid connected price; $p_{\text{buy.w}}$ is the price for market participants to purchase electricity directly from the grid; The smooth operation of the distributed power trading market should meet the basic premise that the income obtained by the users participating in the distributed power trading should not be less than the income obtained by trading directly with the power grid. Therefore, the quote range of prosumers should be expressed as $\left[p_{\text{sell.w}}, p_{\text{buy.w}}\right]$.

Let n be the price interval parameter, which indicates that there are n bidding prices $(n \geq 2)$: $p_{\text{sell.w}}$, $p_{\text{sell.w}}+\frac{p_{\text{buy.w}}-p_{\text{sell.w}}}{n-1}$, $\cdots$, $p_{\text{buy.w}}$; ΔT is the step size of trading price change, $\Delta T=\frac{p_{\text{buy.w}}-p_{\text{sell.w}}}{n-1}$.

For the seller, there are n bidding strategies: $S(1)$, $S(2)$, $\cdots$, $S(n)$. The quotation of $S(r)$ can be expressed as $p_{\text{buy.w}}-(r-1)\Delta T$; For the buyer, there are n bidding strategies: $B(1)$, $B(2)$, $\cdots$, $B(n)$. The quotation of $B(r)$ can be expressed as $p_{\text{sell.w}}+(r-1)\Delta T$; It can be seen that the higher the r selected by the seller, the lower the price, indicating that the user has stronger demand for electricity sales; the higher the r selected by the buyer, the higher the price, indicating that the user has stronger demand for electricity purchase. In this paper, the trading willingness degree is introduced to measure the intensity of prosumers' willingness to trade electric energy. The trading willingness degree of user i is shown in formula

(1), where ε_{tw} is the factor of willingness to trade.

$$\chi_i(r) = \varepsilon_{tw} \cdot r \quad (1)$$

Temporary trading orders are matched according to the priority of trading price, reputation value and reporting time. The trading price of temporary orders is shown in formula (2):

$$G_{ij}^{\Delta t} = \frac{S(r_i) + B(r_j)}{2} \quad (2)$$

Where $G_{ij}^{\Delta t}$ is the trading price of temporary order formed by seller i and buyer j; $S(r_i)$ is the seller's willing selling price; $B(r_j)$ is the buyer's willing purchase price.

5 Security verification

In the distributed security verification method, reference [5] calculates the phase angle of each node iteratively, so as to find out the power flow of all lines. In reference [6], a distributed power flow calculation method based on successive over relaxation iteration (SOR) is proposed. Compared with the traditional distributed iteration method, the algorithm has faster convergence speed and stronger convergence stability. In this paper, based on the SOR iterative method, an improved SOR iterative algorithm is proposed, which makes the convergence speed faster and the convergence stability further enhanced. The specific calculation steps of the improved SOR iterative algorithm are as follows:

Firstly, let's select a reference node in the network, set its phase angle $\theta_0=0$, and set the initial phase angle of other nodes as 0. Secondly, calculate the injection power $\mathbf{P}_{inj}^{\Delta t}$ of each node to the network. When $P_{inj.i}^{\Delta t}>0$, it means that node i sells electric energy to the other nodes, otherwise it means that node i purchases electric energy from the other nodes. In the iteration process, each node broadcasts the results of each phase angle iteration to the network for the other nodes connected with itself. The iterative formula is shown in formula (3):

$$\theta_i^{\Delta t}(k+1)=\theta_i^{\Delta t}(k)-\lambda_i(-P_{inj.i}^{\Delta t}+\sum_{j\in\Omega_i}[\theta_i^{\Delta t}(k)-\theta_j^{\Delta t}(\alpha)]) \quad (3)$$

Where $\theta_i^{\Delta t}(k)$ is the phase angle value of node i in the k-th iteration; Ω_i is the set of all nodes connected with node i; $\lambda_i=\omega/\sum_{j\in\Omega_i}\gamma_{ij}$, ω is the relaxation factor and satisfies $0<\omega<2$; $\gamma_{ij}=1/X_{ij}$, which is the reciprocal of impedance between node i and j; $\alpha=\begin{cases} k & j>i \\ k+1 & j<i \end{cases}$.

The convergence condition is shown in formula (4):

$$|-P_{inj.i}^{\Delta t}+\sum_{j\in\Omega_i}\gamma_{ij}[\theta_i^{\Delta t}(k)-\theta_j^{\Delta t}(\alpha)]|<\varepsilon \quad (4)$$

Where ε is a positive number as small as possible.

The power flow distribution of each branch can be calculated from the final phase angle of each node. The power flow calculation formula of branch from node i to node j is shown in formula (5):

$$P_{ij}^{\Delta t} = (\theta_i^{\Delta t} - \theta_j^{\Delta t}) \cdot \gamma_{ij} \quad (5)$$

6 Congestion management

In the aspect of congestion management, the trading weight factor is introduced in this paper, so that the trading volume can be reduced while the user's willingness to trade can be fully considered, so as to make targeted reduction. The calculation of the trading weight factor is shown in formula (6):

$$\sigma_{ij}=\chi_i+\chi_j \quad (6)$$

Where χ_i and χ_j are the trading willingness degree corresponding to the bidding strategy selected by the buyer i and the seller j respectively.

For convenience, the following parameters are defined: let m be the number of network nodes; τ is the number of network branches; w is the number of distributed trading participants in the network. Let $\mathbf{Y}$ be the diagonal matrix, and the diagonal element is $1/x_{ij}$, where x_{ij} is the reactance from node i to node j, and $\mathbf{A}$ is the reduced order node incidence matrix of the corresponding row with the slack nodes removed; $\mathbf{T}_{temp}$ is the temporary trading matrix formed by temporary trading orders, the element t_{ij} in the matrix represents the electric energy provided by the prosumer i to j:

$$\mathbf{T}_{temp} = \begin{bmatrix} 0 & t_{12} & \cdots & t_{1\tau} \\ t_{21} & 0 & \cdots & t_{2\tau} \\ \vdots & \vdots & 0 & \vdots \\ t_{\tau 1} & t_{\tau 2} & \cdots & 0 \end{bmatrix} \quad (7)$$

$\mathbf{u}$ is the w-dimensional unit column vector; $\mathbf{M}$ is the network trading association matrix, where:

$$m_{ij} = \begin{cases} 0 & \text{Prosumer } j \text{ is not on node } i \\ 1 & \text{Prosumer } j \text{ is on node } i \end{cases} \quad (8)$$

Let $\mathbf{q}_{inj}$ be the injection power of each node, then $\mathbf{q}_{inj} = \mathbf{M}\times\mathbf{T}_{temp}\times\mathbf{u}$, which is the m-dimensional vector. Let $\mathbf{q'}_{inj}$ be the injection power of other nodes except the relaxed node. Let $\mathbf{S}$ be a set of security domains, which is composed of all trading matrices $\mathbf{T'}$ meeting the security verification. The essence of congestion management is to find a matrix in $\mathbf{S}$ with the smallest change compared with the original trading matrix by reducing the trading volume. Let the adjusted trading matrix be $\mathbf{T}_L$, it can be transformed into the optimization problem shown in formula (9):

$$min \sum_{i,j=1,i\neq j}^{n} \sigma_{ij}\left(t_{ij}-t_{Lij}\right)^2 \quad (9)$$

The constraint condition is shown in formula (10):

$$-\mathbf{P}_{lmax}^{\Delta t} \leq \mathbf{YA}(\mathbf{A}^T\mathbf{YA})^{-1}\mathbf{q'}_{inj.L}=\mathbf{H}_{PTDF}\mathbf{q'}_{inj.L} \leq \mathbf{P}_{lmax}^{\Delta t} \quad (10)$$

Where $\mathbf{P}_{lmax}^{\Delta t}$ is the power flow margin vector of the line in the current period; $\mathbf{H}_{PTDF}$ is the power transmission distribution factor; $\mathbf{q}'_{\text{inj.L}}$ is the node injection power vector corresponding to the optimized trading matrix $\mathbf{T}_{\text{L}}$ after removing the slack node;

7 Evaluation of transaction reputation value and settlement

In the distributed power trading, users may not generate and use electricity according to the contract. Therefore, it is necessary to punish the defaulting users, deduct the transaction reputation value and charge the liquidated damages.

7.1 Calculation of reputation value

When the actual power generation or consumption deviation is within a certain range, the transaction reputation value of the user will not be deducted; otherwise, the transaction reputation value will be deducted. The reputation value is shown in formula (11):

$$H_i^{\Delta t} = \begin{cases} 100, & \dfrac{D_i^{\Delta t}}{T_i^{\Delta t}} \le \mu_{\text{pdtv}} \\ 100(1-\alpha_{\text{rvc}}\dfrac{D_i^{\Delta t}}{T_i^{\Delta t}}), & \dfrac{D_i^{\Delta t}}{T_i^{\Delta t}} > \mu_{\text{pdtv}} \end{cases} \quad (11)$$

Where $H_i^{\Delta t}$ is the reputation value of user i in the t-th trading cycle; $T_i^{\Delta t}$ is the electricity quantity that user i should trade in the t-th trading cycle according to the contract; $D_i^{\Delta t}$ is the deviation electricity quantity of user i in the t-th trading cycle; α_{rvc} is the reputation value coefficient; μ_{pdtv} is the power deviation threshold value;

The transaction reputation value curve is shown in Fig 3:

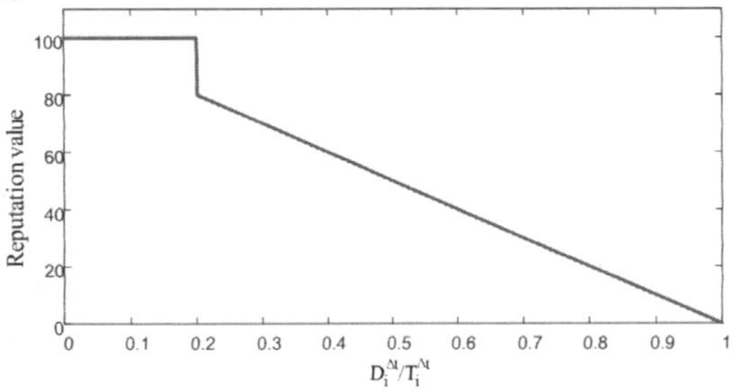

Fig3. Reputation value curve diagram

7.2 Liquidated damages

If the reputation value of the user is deducted, a certain liquidated damage will be added on the basis of the trading price.

For the buyer, the liquidated damage is:

$$G_{\text{LD}.i}^{\Delta t} = \beta_{\text{pc}}(1-\frac{H_i^{\Delta t}}{100})G_{ij}^{\Delta t} \quad (12)$$

For the seller, the liquidated damage is:

$$G_{\text{LD}.i}^{\Delta t} = \beta_{\text{pc}}(1-\frac{100}{H_i^{\Delta t}})G_{ij}^{\Delta t} \quad (13)$$

Where $G_{\text{LD}.i}^{\Delta t}$ is the liquidated damage of user i in the t-th trading cycle; $G_{ij}^{\Delta t}$ is the trading price of user i and j in the t-th trading cycle; and β_{pc} is the penalty coefficient.

The final settlement price of user i is the sum of trading price and the liquidated damage, that is:

$$G_{\text{fsp}.i}^{\Delta t} = G_{\text{LD}.i}^{\Delta t} + \text{G}_{ij}^{\Delta t} \quad (14)$$

According to the final settlement price formula, only when the reputation value is not deducted in the current trading cycle, the final settlement price of the user ($G_{\text{fsp}.i}^{\Delta t}$) is equal to the trading price ($G_{ij}^{\Delta t}$). Otherwise, the users whose reputation value is deducted will bear some losses, and the matching priority in the next trading cycle will be reduced [7]. Therefore, the introduction of reputation value can supervise the user to complete the trading contract in strict accordance with the contract content, which can play a certain role in restraining the user's illegal behavior.

8 Example analysis

8.1 Example setting

In order to verify the economic benefits of distributed trading between prosumers, the 6-node system shown in Fig4 is used to simulate the trading.

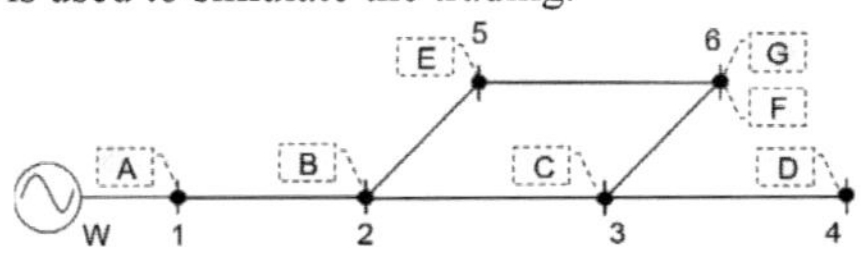

Fig4. Six node distribution network diagram

In the network, A-G are the prosumers with distributed generation equipment, and W is the distribution network operator, which can balance the fluctuating power of the distribution network. Because it is connected to the power grid, it can be equivalent to a generator with infinite capacity. The grid-connected price of distributed generation equipment is 0.4 yuan/kWh ($p_{\text{sell.w}}$=0.4 yuan/kWh), and the unit price of electricity purchased directly from the seller is 1.2 yuan/kWh ($p_{\text{buy.w}}$=1.2 yuan/kWh).

Assume that the distributed power trading is a seller's market, that is, the total amount of electricity sold by the prosumers is less than the amount of electricity purchased. If the price interval parameter n=21 , the trading information reported by each prosumer according to the bidding strategy in section 4 at time t is shown in Table 1, where $H_i^{\Delta t-1}$ represents initial reputation value of each user, $W_i^{\Delta t}$ represents residual electricity of prosumers, $T_i^{\Delta t}$ represents the trading information reporting time of each user; WP represents the willingness price corresponding to the bidding strategy selected by each user.

Table1. Trading information reported by prosumers

Users	$H_i^{\Delta t-1}$	$W_i^{\Delta t}$ /kWh	$T_i^{\Delta t}$ /S	Bidding strategy	WP (yuan/kWh)
A	100	30	0.01	S(16)	0.60
B	100	20	0.02	S(10)	0.84
C	80	-15	0.01	B(13)	0.88
D	100	16	0.01	S(10)	0.84
E	100	-18	0.03	B(7)	0.64
F	100	-32	0.02	B(13)	0.88
G	100	-12	0.02	B(15)	0.96

8.2 Trade matching

According to the trading reporting information of prosumers, 6 temporary trading orders are successfully matched according to the priority of "trading price, reputation value and reporting time [8]". Electricity quantity that failed to match was deemed to have been purchased from or sold to the distribution network operator W. The trade matching results are shown in Table 2, where $t_{\text{temp.}ij}$ represents the amount of electricity temporarily traded between user i and j ; $G_{ij}^{\Delta t}$ represents the trading price of temporary order formed by user i and j .

Table2. Trade matching result of prosumers

Buyer	Seller	$t_{\text{temp.}ij}$ /kWh	$G_{ij}^{\Delta t}$ /(yuan/kWh)
G	A	12	0.78
F	A	18	0.74
	D	14	0.86
C	D	2	0.86
	B	13	0.86
E	B	7	0.74
	W	11	1.20

8.3 Comparison of security verification algorithms

The improved SOR iterative algorithm described in section 5 is used to calculate the distributed power flow of the temporary trading orders, and the power of each branch is shown in Table 3.

Table3. Calculation results of distributed power flow

Branch	Actual power/kW	Transmission capacity margin/kW
1-2	41.000	50
2-3	25.142	28
3-4	-16.000	28
2-5	35.857	28
3-6	26.142	28
5-6	17.857	28

From the security verification results, the actual power of branch 2-5 exceeds the transmission capacity margin, and congestion management is needed to eliminate power flow overrun.

The performance comparison of the three distributed algorithms is shown in Table 4, where TDA represents traditional distributed algorithm.

Table4. Performance comparison of three distributed algorithms

TDA	SOR algorithm	Improved SOR algorithm
When λ=0.06, the number of iterations is 209. When λ>0.061, the algorithm can not converge.	When ω= 1, the number of iterations is 171. When ω>1.1, the algorithm can not converge.	When ω=1, the number of iterations is 94, while when ω=1.46, the minimum number of iterations is only 24. When 0<ω<2, the distributed algorithm can converge reliably.

The phase angle convergence process of node 5 in the three iterative algorithms is shown in Fig 5

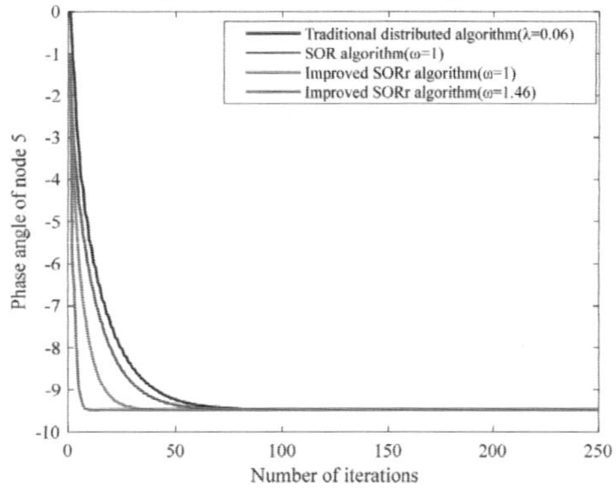

Fig5. Phase angle convergence process of node 5

From the above performance comparison, the improved SOR iterative algorithm proposed in this paper has higher convergence speed than the distributed iterative algorithm in reference [5] and reference [6], and has stronger convergence stability.

8.4 Congestion management

The trading contract formed by the central organization after congestion management in combination with formula (9) and formula (10) is shown in Fig 6. In Fig 6, the second column of each transaction is the optimization result without considering the user's trading willingness (i.e., the weight factor $\sigma_{ij}=1$ for all trade), and the third column is the optimization result after considering the trading weight factor of formula (6). ($\varepsilon_{\text{tw}}=1$)

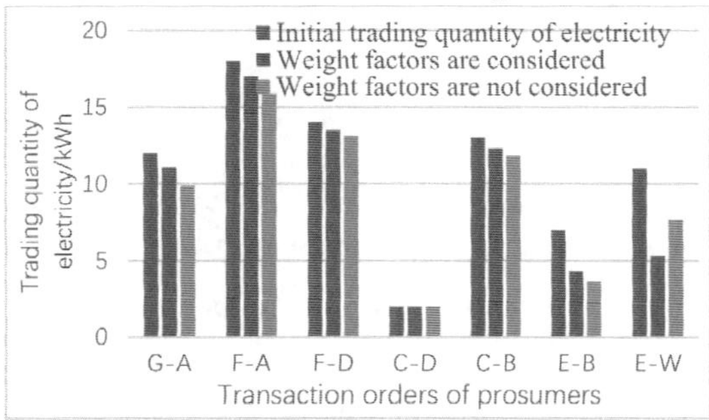

Fig6. Congestion management optimization results

The trading reduction of each prosumer after congestion management is shown in

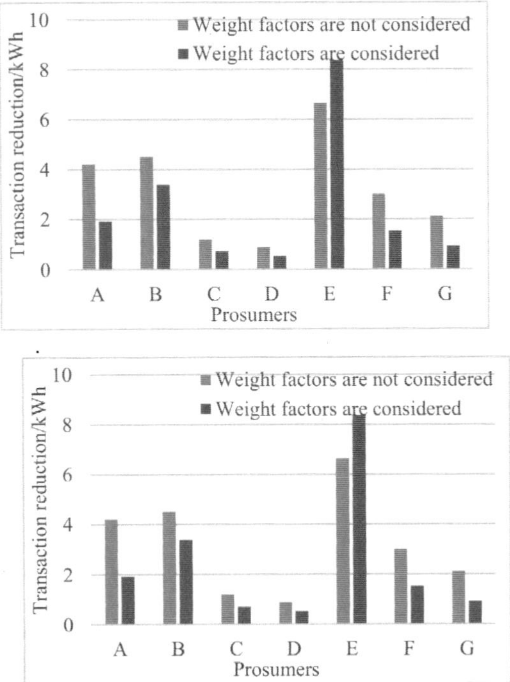

Fig7. Trading reduction of prosumers

Comparing two kinds of congestion management results, we can see that when the trading weight factor is not considered, the congestion management is to reduce the trading according to the minimum square sum of the adjustment volume, and treat all temporary transaction orders equally. When considering the trading weight factor, because A, G, F and some other users choose the bidding strategy with higher trading willingness degree, the trading volume is reduced relatively less in the congestion management. On the contrary, because the buyer E chooses the bidding strategy with lower trading willingness degree, it is located at the end of the matching queue, and its trading reduction is relatively increased during congestion management. Therefore, the introduction of trading weight factor can reflect the trading intention of users to a certain extent, enhance the enthusiasm of prosumers to participate in distributed trading, and make users with stronger trading intention can purchase or sell more electric energy.

8.5 Recording and settlement of transactions

When the final trading contract is reached, all prosumers and the distribution network operator who participate in the transaction will use the private key to authenticate the trading contract, and then broadcast the contract to other nodes in the network for consensus authentication. The trading information is packed into a set in the form of Merkle tree as shown in Fig 8. The hash value of each trading data is calculated in turn, and recursion is continued until the root node data value is obtained. The blockchain only needs to save the data of the root node, so that the recorded trading information has good tamper resistance. If a malicious node attempts to tamper with the trading information, such as tampering the trading contract between user W and user E into the trading between user A and user E (as shown in Fig 8), the data value of the root node of the Merkle tree will also change, indicating that the trading information has been tampered with. At the same time, we can find the location of the tampered trading information by comparing the root node value of each subtree step by step from the root node.

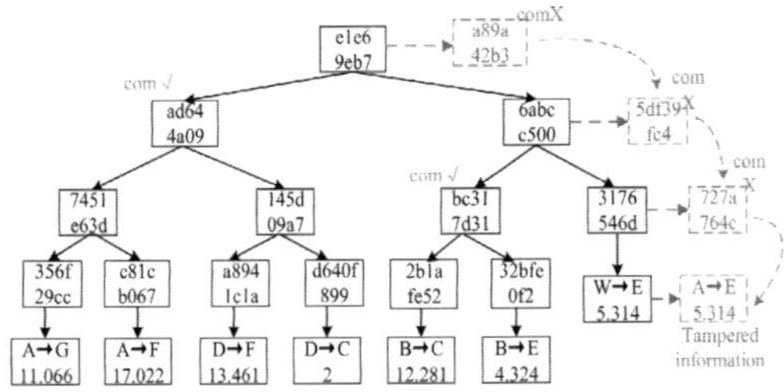

Fig8. Merkle tree records trading information

When the trading settlement phase is reached, the smart contract will automatically read the energy data collected by smart meters, and clear the actual power delivery of each prosumer. According to formula (11-13), the smart contract will change the reputation value of the defaulting users and punish them with liquidated damages, and conduct trading settlement for all users according to formula (14). In this example, it is assumed that the actual power generation capacity of seller B is 12.06kWh due to the fault of its own distributed generation equipment. The reputation value change and trading settlement of each prosumer are shown in Table 5, where $T_i^{\Delta t}$-$D_i^{\Delta t}$ represents the actual power transmission of each user, $D_i^{\Delta t}$ represents the deviation energy of each user, $H_i^{\Delta t}$ represents the reputation value of each user, $G_{\mathrm{LD}.i}^{\Delta t}$ represents the liquidated damages for users. (β_{pc}=0.8 ; α_{rvc}=1 ; μ_{pdtv}=0.2 .)

Table5. Trading settlement of participants in the market

Users	$T_i^{\Delta t}$-$D_i^{\Delta t}$ /kWh	$D_i^{\Delta t}$ /kWh	$H_i^{\Delta t}$	$G_{\mathrm{LD}.i}^{\Delta t}$ /yuan	Account balance /yuan
A	28.068	0	100	0	+21.2130
B	12.060	4.545	73	3.5686	+6.4920
C	-11.468	2.813	100	0	-9.8625
D	15.461	0	100	0	+13.2965
E	-7.906	1.732	100	0	-8.2949
F	-30.463	0	100	0	-24.1580
G	-11.066	0	100	0	-8.6315
W	5.314	0	100	0	+6.3768

8.6 Benefits analysis of prosumers

Compared with the direct grid connection, the relative return rate of each prosumer participating in the distributed trading is shown in Fig 9.

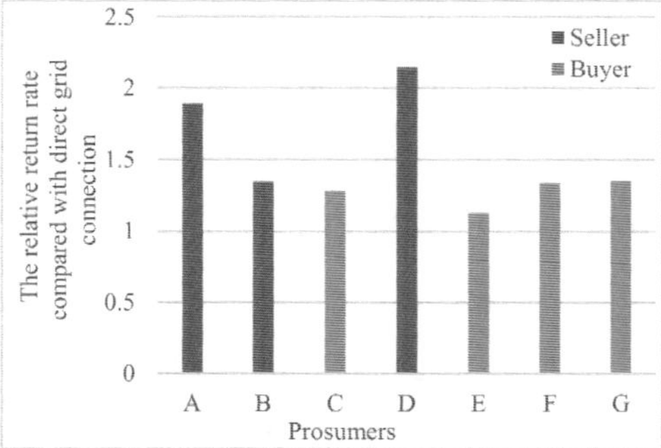

Fig9. The relative return rate of prosumers compared with direct grid connection

The specific analysis of the relative return rate of each prosumer in Fig 9 are as follows:

(1) Users B and D choose the same electricity selling price, while the relative return rate of user B is lower than that of user D. The main reasons are as follows: firstly, the trading reporting time of user D is shorter, so it is in front of user B in the trading matching queue, and can match with the buyer user with higher quotation; secondly, user B has certain liquidated damages due to equipment failure, so its return rate is relatively lower.

(2) User A chooses the lowest electricity selling price among all the seller users, but its relative return rate is very high. The main reasons are as follows: although user A chooses the lowest electricity selling price, according to the matching principle of "trading price priority", it will be at the top of the matching queue, so as to match with the user with the highest electricity purchase price, and obtain a trading price not too low. At the same time, a lower electricity selling willingness price has a higher trading willingness degree. When the central organization carries out congestion management by reducing the trading volume, the trading order of user A has a higher trading weight, which makes the reduction less, thus indirectly improving the electricity sales income of user A.

From the above analysis, we can see that under the distributed trading mechanism of two-way competition, the low willingness price does not mean the low final settlement price and the relative return rate of users, but also is related to the willingness price of the trading matching object, the reputation value of the prosumer and the trading weight factor. This requires that prosumers should enhance their enthusiasm to participate in distributed trading, and at the same time, they should maintain good electricity generation and consumption behavior, avoid the occurrence of liquidated damages, and strive for maximum revenue.

9 Conclusion

In this paper, a weakly-centralized electricity trading mode is proposed. In this mode, the central organization only manages the trading orders that fail to pass the security verification, and does not directly participate in the matching and settlement of electricity trading orders, which improves the transparency and fairness of the trading. In the part of security verification, the traditional SOR iterative method is improved to make the convergence rate of distributed algorithm faster. Finally, an example is given to analyze the return rate of multilateral electricity trading compared with direct grid connection, and the feasibility and economy of this power trading mode are verified.

This paper focuses on the power trading mode of prosumers based on blockchain, including the bidding strategy, security verification, congestion management model and so on. However, some specific details of the trading mode have not been studied, such as the preparation of smart contract, the construction of power trading platform based on blockchain, the data maintenance mechanism of energy blockchain, etc., which should be further studied in the future.

Acknowledgments

The authors would like to highly appreciate the financial support provided by the science and technology project 52199619000J of State Grid Sichuan Electric Power Company.

References

1. H.Sun, Q.Guo, Z.Pan. EnergyInternet Concept Architecture and Frontier Outlook[J]. Automation of Electric Power Systems, 2015,39(19):1-8.
2. Y.Yan, J.Zhao, F.Wen, X.Chen. Blockchain in Energy Systems : Concept, Application and Prospect[J]. Electric Power Construction, 2017,38(02):12-20.
3. W.Chen, Z.Zheng. Blockchain Data Analysis : A Review of Status, Trends and Challenges[J]. Journal of Computer Research and Development, 2018,55(09):1853-1870.
4. D.Zhang. Design and implementation of power trading platform based on blockchain[D]. University of Electronic Science and Technology of China, 2018.
5. S. Kar, G. Hug, J. Mohammadi and J. M. F. Moura, "Distributed State Estimation and Energy Management in Smart Grids: A Consensus${+}$ Innovations Approach," in IEEE Journal of Selected Topics in Signal Processing, vol. 8, no. 6, pp. 1022-1038, Dec. 2014.
6. X.Tai, H.Sun, Q.Guo. Electricity Transactions and Congestion Management Based on Blockchain in Energy Internet[J]. Power System Technology, 2016,40(12):3630-3638.
7. B.Wang, Y.Yan, F.Wen, Z.Zhou, S.Lin, X.Chen. A Blockchain Based Distributed Power Trading Mechanism[J]. Electric Power Construction, 2019,40(12):3-10.
8. B.Ji. Research on Retail Trading Mechanism and Model of Prosumers Based on Blockchain Technology[D]. Guangxi University, 2019.

E3S Web of Conferences **237**, 02009 (2021) https://doi.org/10.1051/e3sconf/202123702009
ARFEE 2020

Pricing of Special-Shaped Pearl Pendant Based on GABOR-GRANGER

Xuanrun Wu[1,a], Jianda Cao[2,b], Yuan Chen[1], Jing Ye[1], Duan Li[1]

[1] Design College, Jiaxing University, Zhe Jiang, China
[2] College of Materials and Textile Engineering, Jiaxing University, Zhe Jiang, China

Abstract:In order to gain high innovation value profit of special-shaped pearl pendant on the market, the optimum price of such new product was confirmed in the work using GABOR-GRANGER. Firstly, 200 target consumers were selected for the cluster analysis of their attitudes toward baroque pearls. Then, GABOR-GRANGER was applied to calculate the fake income and price threshold of set price point. Last, it was determined the lower limit of 95% confidence bounds of the maximum value of fake income as the conservative optimal price. Results show that the main target consumer groups were not concerned about the price because of the singular and beautiful shape of pearl. Through calculation by Granger Gabor, the optimal price of special-shaped pearl pendant was higher 60.8% than the price obtained by cost plus calculation. Therefore, it is possible to achieve high innovation value profit of new products on the market.

1. Introduction

For its unique personality, special-shaped pearl has become an important element and basic material in recent years. As an innovative variety developed in Chinese waters, it is an important issue to determine a price accepted by consumers while achieving a high value profit [1]. In the work, GABOR-GRANGER [2] was used for the price survey of special-shaped pearl pendant, with the final price of special-shaped pearl pendant determined.

2. Attitude Investigation and Cluster Analysis of Target Consumer Groups to Special-Shaped Pearl

To understand the attitude of target consumer groups to special-shaped pearl pendant, the seven-segment rating scale was used in the work.

A. Expenditure structure analysis of research object and consumers for pendant

Investigation was conducted on old female consumers from 20 to 40 years old in Yangtze River Delta region. While these consumers were selecting dresses in shop, it is possible to ask them questions.

Totally, 235 questionnaires have been sent out. Besides some incomplete questionnaires, there were 200 valid questionnaires.

As special-shaped pearl pendant belongs to non-necessities of life, the expenditure of consumers on pendant was measured in this study. Figure 1 shows the expenditure of consumers on pendant in one year.In Figure 1, the expenditure of the first consumer group on pendant is below 300 yuan, accounting for about 50%. The expenditure of the second consumer group is between 300 to 1,000 yuan, accounting for about 30%. About 12% of the consumers spend 1,000-2,000 yuan on pendant, and 8% of consumers prefer to spend 2,000-5,000 yuan on pendant. Such expenditure structure of consumers can be used as a reference to draft the price of special-shaped pearl pendant. Therefore, we choose a classic pendant with a low threshold of 300 yuan as research product.

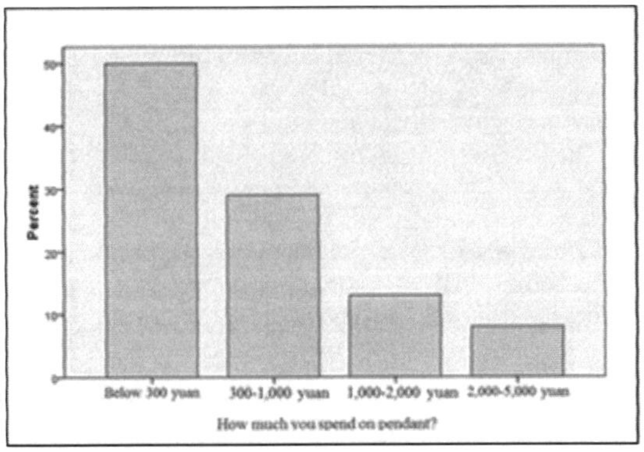

Figure1. Expenditure of consumers on pendant in one year.

B. Investigation and cluster analysis of target consumer groups' attitudes to special-shaped pearl

The investigation of target consumers' attitudes toward special-shaped pearl was conducted using a seven-segment rating scale (See Table I).With above

* Corresponding author: [a] wxr1988@zjxu.edu.cn;
[b] dacao@zjxu.edu.cn

questions, cluster analysis was conducted using SPSS17.0 statistics software. Table II shows the analysis results of 200 valid questionnaires.

Table.1. Investigation of target consumer groups' attitudes to special-shaped pearl

Problem statement	Strongly disagree	Disagree	Partially Disagree	Neutral	Somewhat agree	Agree	Strongly Agree
I like shaped pearl born singular form	-3	-2	-1	0	1	2	3
Wonderful design with a variety of special-shaped pearl inlay craft is interesting.	-3	-2	-1	0	1	2	3
I do not care about the price if the special-shaped pearl is strange and beautiful.	-3	-2	-1	0	1	2	3
I would pay more to special-shaped pearl coupled with attractive packaging.	-3	-2	-1	0	1	2	3
I like special-shaped pearl, and pay no attention to the brands.	-3	-2	-1	0	1	2	3
Although the pearls are shaped by nature, I care about the brand.	-3	-2	-1	0	1	2	3

Obviously, the investigation groups can be divided into two categories according to the concentration degree of rating. The first feature of investigation objects is that they are attracted by the singular form of special-shaped pearl and wonderful design of various inlays. Thus, they care little about the price and brand because of the singular and beautiful shape and exquisite design and packaging. Such consumers account for 52% of the investigation population.

The second feature is that consumers pay no attention to the shape of pearl. Although these consumers believe these special-shaped pearl is natural, they also concerned about the brand. Such consumers account for 48% of the investigation objects.

Above cluster analysis shows that most investigation objects little care about the price and brand because of the singular and beautiful shape and exquisite design and packaging.

For this type of new products, products the valuation of consumers is always uncertainty [4] when pricing need to consider the price strategy, consumer perception, price sensitivity [5], consumer prices quality balance [6], and many other factors, GABOR - GRANGER method is to determine the optimal new product price and the analysis of the influence of price on demand a effectively method, so this paper USES GABOR - GRANGER, formulate the heterosexual beads listed price of the product.

Table.2. Results of cluster analysis of Final Cluster Centers

	Cluster	
	1	2
I like shaped pearl born singular form	2	0
Wonderful design with a variety of special-shaped pearl inlay craft is interesting.	2	0
I do not care about the price if the special-shaped pearl is strange and beautiful.	1	0
I would pay more to special-shaped pearl coupled with attractive packaging.	1	0
I like special-shaped pearl, and pay no attention to the brands.	1	0
Although the pearls are shaped by nature, I care about the brand.	1	1

3. Summary of Simple GABOR-GRANGER

A. Research objective of simple GABOR-GRANGER

Generally, study on the price by Gabor Granger includes two aims: one is to test the influence of price changes of company's existing product on demand; the other is to determine the optimal price of new product and analyze the influence of such price on demand. In the work, Gabor Granger was used to determine the price of special-shaped pearl pendant.

B. Form of questions in questionnaires

While using Gabor Granger, the interviewees are firstly invited to watch the actual product, product samples or product concept. Then, they will be asked with following questions.

For example, "If the price of this product on the market is P1, P2, P3, P4 or P5, will you purchase it?"

This approach requires manufacturers to provide the prices of test. Of all the prices, one will be widely accepted by consumers.

C. Selection of price level

Generally, there should be 6 to 12 price levels in the investigation. If the price levels are too many, much time will be cost, resulting in high expenditure. Furthermore, interviewees may be reluctant to cooperate, resulting in incomplete information. If the price levels are too little, it is difficult to reflect the influence of price changes on demand.

	P1( )	P2( )	P3( )	P4( )	P5( )
Not to buy	1	1	1	1	1
Might not buy	2	2	2	2	2
Cannot decide	3	3	3	3	3
Likely to buy	4	4	4	4	4
Certainly buy	5	5	5	5	5

Note: P1<P2<P3<P4<P5, there are generally five or six prices.

A. Data analysis

Firstly, the percentage of interviewees who answer definitely or probably purchase should be tallied for each price level. Besides, the interviewees who answer definitely or probably purchase at high price level should be noted because they would give the same answer at low price levels. Then, fake income can be figured out according to such percentage. The definition is as follows.

Fake income=Price level×Percentage of interviewees answered definitely or probably purchase in this price level (1)

Based on Equation (1), the fake income of each price is figured out. The price with the largest fake income is regarded as the proposed price of new product.

4. Pricing of Special-Shaped Pearl Pendant Based on Simple GABOR-GRANGER

The investigation of the work is to price the new product — special-shaped pearl pendant. Before investigation, the price level of such new product should be determined. Based on the cost accounting of special-shaped pearl pendant, the price of pearl pendant was calculated as 29 yuan according to cost additive method.

This is also the threshold price of a non-essential fashion product [8]. Consumers of fashion products in this price section have little choice pressure. Then, eight price levels, including 29, 34, 39, 44, 49, 54, 59 and 64, were selected in this work to deeply observe the influence of price changes on demand. With these price levels, the investigation questions of special-shaped pearl pendant were designed as follows.

	29( )	39( )	44( )	49( )	56( )	64 ( )
Not to buy	1	1	1	1	1	1
Might not buy	2	2	2	2	2	2
Cannot decide	3	3	3	3	3	3
Likely to buy	4	4	4	4	4	4
Certainly buy	5	5	5	5	5	5

Q: Here is the special-shaped pearl pendant with silver chain (showing the product). If the price of such product on the market was 29, 34, 39, 44, 49, 54, 59 or 64, would you purchase it?

A. Fake income of special-shaped pearl pendant

Through analysis of investigation data of above questions, it is possible to obtain the percentages of consumers who certainly or possibly purchase at different price levels. Then, the fake income is obtained through calculation by Equation (1). Table III shows the fake income of special-shaped pearl pendant.

Table.3. Fake income of special-shaped pearl pendant

Price level (yuan)	Percentage of consumers who certainly or possible purchase (%)	Fake income (yuan)
29	42	12.18
34	45	15.3
39	50	19.5
44	47	20.68
49	52	25.68
54	42	22.68
59	34	20.06
64	27	17.28

B. Determination of the price threshold of special-shaped pearl pendant

Price threshold is a special price point. At this point, there is a significant reduction in the number of demand. Generally, the price threshold is determined through fake demand curve. Such curve is plotted with investigation prices as abscissa and the percentage of

https://doi.org/10.1051/e3sconf/202123702009

consumers who certainly or possibly purchase it as ordinate. Figure 2 shows the fake demand curve of special-shaped pearl pendant.

Figure 2 shows that the price threshold of special-shaped pearl pendant is 49 yuan. The percentage of consumers, who certainly or possibly purchase, decreases with the increase of price when the price surpasses 49 yuan.

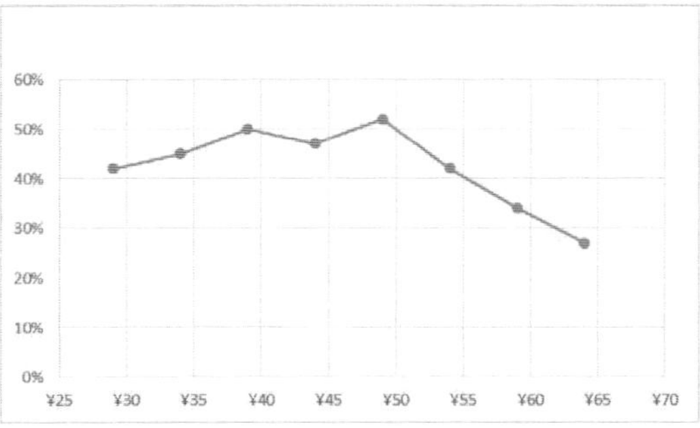

Figure2. Fake demand curve of special-shaped pearl pendant

According to consumers' psychological needs, the price of special-shaped pearl pendant should be determined as 49 yuan, thus achieving the highest fake income.

C. Determination of the final price of special-shaped pearl pendant

In Gabor Granger method, the optimal price is calculated using statistical method, which is difficult to avoid errors. Therefore, the lower limit of 95% confidence bounds of optimal price is generally adopted as the conservative price. However, the optimal price is also temporary due to the various inevitably factors on the market. To be conservative, there is a callback of 5%-10% based on the price threshold. Therefore, the final price of special-shaped pearl pendant is determined as 43 yuan.

5. Conclusions

According to Gabor Granger, the optimal price of special-shaped pearl pendant is higher than the price calculated using cost additive method. Thus, it is possible to obtain high value profit of new product on the market. Pricing the product by Gabor Granger can obtain the optimal price higher than that calculated by cost additive method. Besides, it can also avoid the price threshold which will lead to the reduction of demand. Therefore, it is an effective method to draft the price of new product.

The deficiencies of this study mainly include the following three aspects:

1.Gabor Granger method assumes that other marketing mix factors, such as channels, advertising, sales promotion and other aspects remain unchanged, and the price of competitive brands remains the same. When testing the impact of product price changes on sales, if other factors change, the conclusions obtained cannot be applied. Further studies will use the extended Gabor Granger method or BPTO (considering competition) for price testing.

2.Gabor Granger method assumes that other factors remain unchanged, but it is difficult to guarantee other aspects remain unchanged for a long time. Therefore, the optimal price obtained is temporary and cannot be guaranteed to remain unchanged for the whole period.

3.This time, the products of the special-shaped beads are the magnet products with low price, and the premium obtained is not too high in the pearl products. Further testing will select a type of high price products for empirical evidence.

References

1. Desmet Pierre , " Effectiveness of measures assessing response to price information" , Journal of Product & Brand Management , 2016, Vol. 25(7): 676-686.
2. Andre Gabor; Clive Granger, "The Pricing of New Products," Management Decision, 1979, Vol. 17(8):576-589.
3. Zhang li, Liu guolian" Clothing Market Research Analysis — Application of SPSS [M] Beijing: China Textile Press, 2012,7:151-158.
4. Stan Lipovetsky, Shon Magnan, Andrea Zanetti Polzi, "Pricing_Models in Marketing Research", Custom Research North America, Minneapolis, USA, September 2011, 167-174.
5. Zheng Zongcheng, Chen Jin, Zhang Wenshuang, "Market Research Practices & Methods," Guangdong Economic Publishing House Co. Ltd, 2011, 08, 141-144.
6. Peng Du , Qiushuang Chen , "Skimming or penetration: optimal pricing of new fashion products in the presence of strategic consumers", Annals of Operations Research, 2017, Volume 257 (1-2) : 275–295.
7. Michel Wedel, Peter S.H. Lee-ang, "A Model for the Effects of Psychological Pricing in Gabor–Granger Price Studies," Journal of Economic Psychology, 1998, Vol. 19(2):237-260.
8. Gabor, André; Granger, C. W. J., "Ownership and Acquisition of Consumer Durables: Report on the Nottingham Consumer Durables Project" European Journal of Marketing, 1972, Vol. 6(4):234.
9. Thomas Kneib, Bernhard Baumgartner, Winfried J. Steiner, "Semiparametric Multinomial Logit Models for Analysing Consumer Choice Behaviour," Asta Advances in Statistical Analysis, 2007, Vol. 17(3):225-244.
10. ChungKi Han , "The Effect of Consumer Innovativeness, Perceived Risk and Price Sensitivity on New Product Adoption: The Moderating Effect of Gender." Annals of Operations Research, 2016, Vol. 16(5): 99-114.

E3S Web of Conferences **237**, 02010 (2021) https://doi.org/10.1051/e3sconf/202123702010
ARFEE 2020

Space electromagnetic interference analysis of secondary equipment signal cable

Fan Kuang[1*], Yunfeng Xia[1], Tianyu Cheng[1], Rongfu Zhong[1], Haikun Wei[1]

[1]Dongguan Power Supply Bureau of Guangdong Power Grid Corporation, Dongguan 523000, China

Abstract. In the field of high-voltage transmission and distribution, the secondary equipments are influenced by the electromagnetic coupling effect generated by the primary circuit, which causes strong interference to the signal cable. In this paper, an equivalent circuit model is established for the secondary cables in different forms, namely conventional cable, coaxial cable and twisted pair cable, based on transmission line theory. Based on the model, the responses of these three kinds of signal cables under the impact of spatial plane electromagnetic wave are analysed in frequency domain. Different impedance characteristics of the cables at the near-end and far-end, and different cable layouts, and different injection directions of the electromagnetic wave are considered in the analysis. The response characteristics of different kinds of signal cables subjected to electromagnetic wave are obtained in frequency domain. By comparing the response characteristics of different kinds of signal cables in different conditions, measures to supress the interference from the coupling of electromagnetic wave are obtained. It can be used to guide the selection and cabling of the signal cables for the secondary equipment in power system.

1 Introduction

In recent years, with the large-scale construction of smart grid, more and more secondary equipments such as electronic devices, smart meters and smart devices are applied in the field of power transmission and distribution. Due to the complex and harsh electromagnetic environment in high-voltage transmission and distribution or substation, the signal cables on the secondary circuit will suffer severe interference problems through electromagnetic coupling. With the increasing of the capacity and voltage level of the power system, the electromagnetic interference generated by primary circuit is more and more strong. What's more, as the secondary equipments are becoming more miniaturized, digitized and electronic, they are more susceptible to the influence of electromagnetic interference. Especially in the high voltage switch environment, the signal cables between control room and secondary equipments are more vulnerable to serious electromagnetic interference.

Electromagnetic wave can transmit energy to the connected equipment through the coupling of signal cable. If the power of the radiated electromagnetic wave is very large, it will generate a significant impact on the secondary system. Therefore, it is of great value to analyse the characteristics of the interference from the radiated electromagnetic wave and provide theoretical guarantee for the normal operation of electronic and electrical equipment.

Generally, there are two main methods to analyse the problem of interference caused by spatial electromagnetic interference, namely field method and the equivalent circuit method. The field method is based on Maxwell equation. It is used to solve the boundry value of the inner and outer field of the cable system under given boundary conditions. The equivalent circuit method is on the basis of the transimission line theory. It is used to solve the response of external electromagnetic field inside the cable by equivalent the field-circuit coupling problem of spatial electromagnetic field to signal cable to the transmission line model[1].

Theoretically, the field method is very strict and has accurate calculation results. However, due to the special complexity of the calculation process and high requirements on the computer, there are some difficulties in practical application. In contrast, the equivalent circuit method is simple and practical in engineering applications because of its fast computing speed. Therefore, the equivalent circuit method is used to analyze the spatial electromagnetic interference of signal cables of secondary equipments in this paper.

When solving the field-circuit coupling problem, the transmission line theory mainly includes three models, the Taylor model[2] containing distributed voltage source and distributed current source, the Agrawal model[3] containing only distributed voltage source and the Rachidi model[4] containing only distributed current source. Different transmission line models are solved mainly through time-domain analysis and frequency-domain analysis[5]. In this paper, the equivalent circuit method is adopted to quantitatively analyze and calculate the conventional signal cable, coaxial cable and twisted-pair cable of the secondary equipment.

* Corresponding author: kuangfancsg@163.com

 https://doi.org/10.1051/e3sconf/202123702010

2 Analysis of space electromagnetic field radiation interference

2.1 Analysis model

In the analysis, it is assumed that the spatial electromagnetic radiation is in the form of a plane wave. Due to the linearity, the electric field intensity is assumed E=1V/m. As the free space characteristic impedance is Z_0=377Ω, the corresponding magnetic field intensity is $H=E/Z_0$ =2.65mA/m (corresponding to the magnetic flux density B=3.3nT). Considering the linear medium and calculating the interference in a unit length (i.e., a length of 1m), the real radiated interference can be estimated based on the calculation results of the above reference conditions and the actual stray field intensity.

The analysis model of the interference from radiated electromagnetic wave is shown in Fig. 1, N_1 and N_2 are two nodes at both ends of the cable, where N_1 connects to the far-end of the circuit (e.g., secondary equipment control system), and N_2 connects to the the near-end of the circuit (such as the secondary equipment sensor or transducer). The length of the cable is assumed 1m. Then we take the direction of electric field E parallel to the direction of the cable, and the direction of magnetic field H perpendicular to the cable but parallel to the ground plane, and the direction of plane wave S propagation perpendicular to the ground plane.

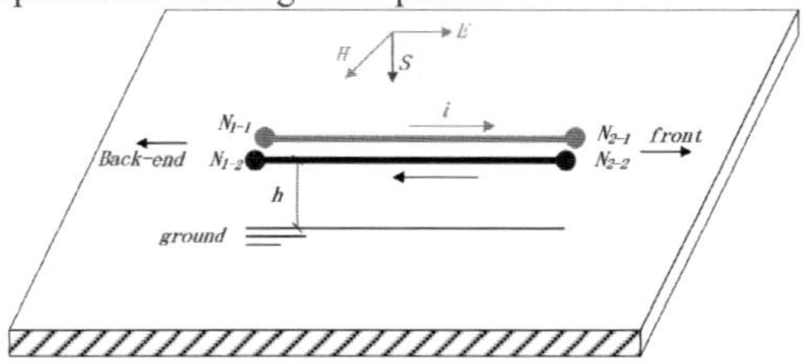

Fig1. Electromagnetic interference analysis model of conventional signal lines

2.2 Interference of conventional signal cable

In the analysis, the diameter and insulation thickness of the signal cable are 1.5mm and 1mm, and three layouts of the cables as shown in Fig. 2 are taken into account. The distance of the cables from the ground is h =1cm.

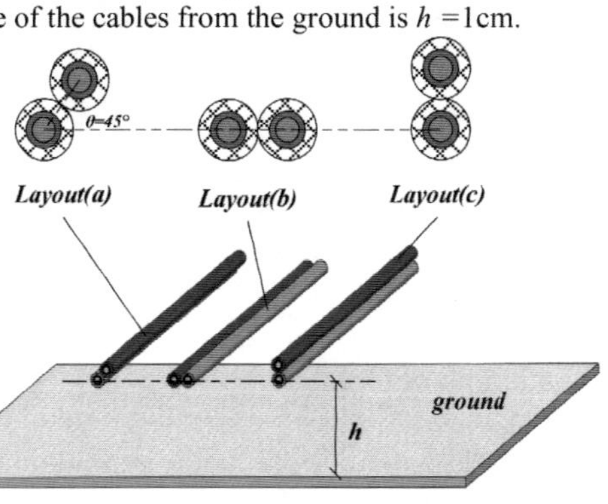

Fig2. Three layouts of conventional signal lines

Fig. 3 shows the equivalent circuit model of conventional signal cable. The cable contains two roots, the forward and the backward. Assuming that the cable is connected to the ground on both ends through 1pF stray capacitance, and both ends of the cable are connected with standard resistance R= 50Ω. The differential-mode interference voltage signals received at both ends of the cable (P_1 and P_2) are calculated, respectively.

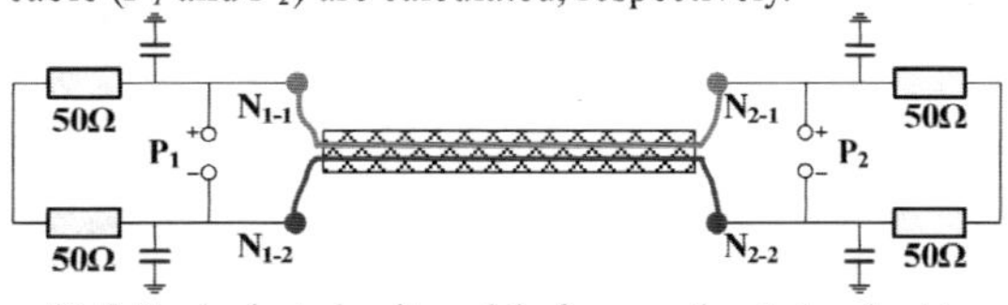

Fig3. Equivalent circuit model of conventional signal cable (standard impedance at both near-end and far-end)

The results of differential-mode voltage coupled by the conventional signal cable with standard impedance at both ends are shown in Fig. 4. It can be seen from the calculation results that the coupled differential-mode interference voltage basically increases linearly with the frequency within 100MHz in logarithmic coordinate. There are two resonance points around 120MHz and 370MHz. The differential-mode interference of the three kinds of cable layouts shown in Fig. 2 are basically the same. The cable layout (b) has the lowest differential-mode interference, about 3 orders of magnitude less than other two cases, and the result of cable layout (c) is slightly higher than (a). It is because the electric fields in all three layouts are parallel to the cable, in other words, the interfering voltage introduced by the electric are the same. In contrast, the effect of the magnetic field on various configurations is different. In layout (c), it is the strongest, followed by layout (a), and layout (b) is nearly zero. That is because that the loop area affected by the magnetic field is the largest in layout (c). As the magnetic field is parallel to the cable plane in layout (b), the magnetic field has little impact on it..

At the highest resonance point, the differential-mode voltage under the condition of unit electric field and unit length reaches to 20mV. But these resonance frequencies are beyond the frequency of concern. The interference voltage generated by 1kHz plane wave per unit strength on the cable with a unit length is about 1.6×10^{-13} V. Assuming the interference magnetic field is 1mT, the differential-mode interference voltage corresponding to a unit length is about 48nV. Then, assuming the transmission distance is 100m, the corresponding differential-mode interference voltage is about 4.8 μV, much lower than the general standard signal.

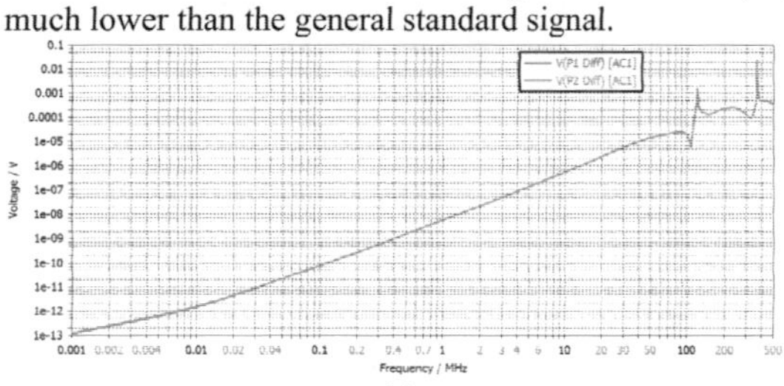

(a)

https://doi.org/10.1051/e3sconf/202123702010

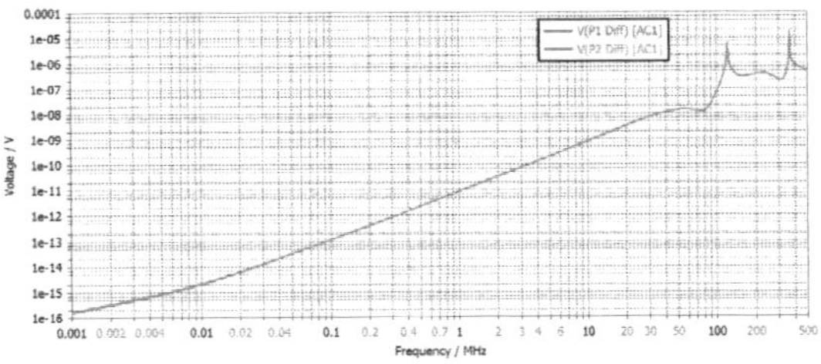

(b)

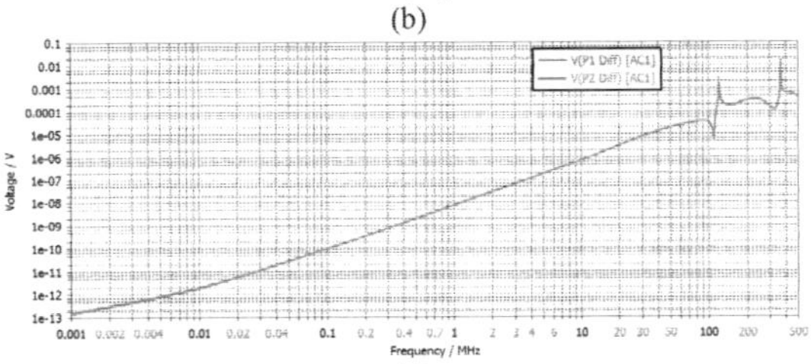

(c)

Fig4. Radiation-coupled differential-mode voltage of standard impedance of conventional signal cable (standard impedance at both near-end and far-end)

In the case of open at far-end, two resistances R=1MΩ are connected to the front-end to simulate this situation as shown in Fig. 5. The results of differential-mode voltage between two ports of conventional signal cables are shown in Fig. 6. The differential-mode interference voltage spectrum characteristics are similar to the ones in Fig. 4. However, due to the open-circuit state at the far-end, there is a certain difference between the voltage at near-end and the far-end. The voltage at the far-end is increased to some extent. The interference voltage generated by 1kHz plane wave per unit strength on a unit length cable rises to 3.3×10^{-13}V. At the same time, the peak differential-mode voltage at the resonance point reaches 40mV, which is double of the previous ones.

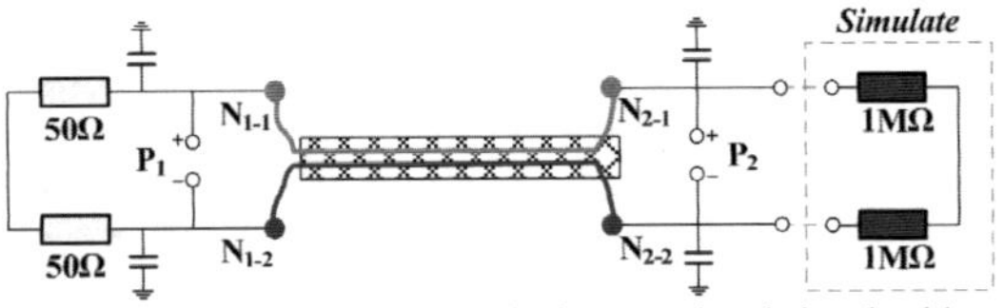

Fig5. Equivalent circuit Model of conventional signal cable (standard impedance at near-end and open at far-end)

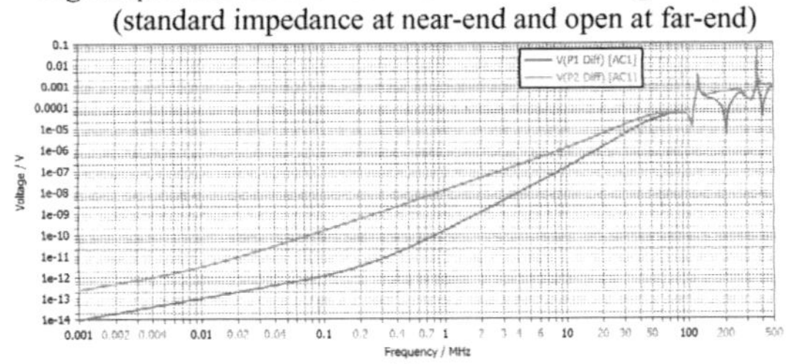

(a)

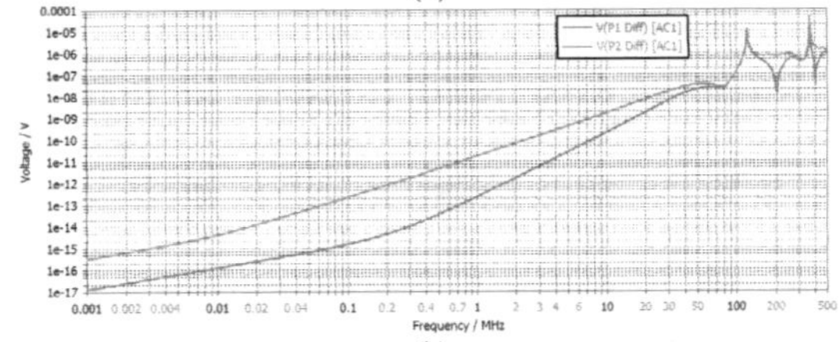

(b)

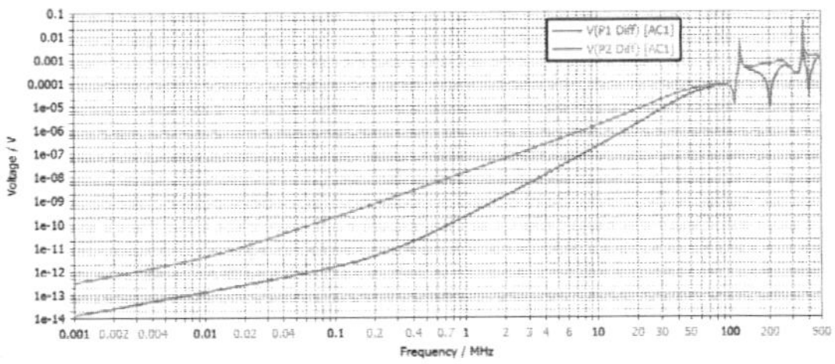

(c)

Fig6. Standard impedance radiation-coupled differential-mode voltage for conventional signal cables (standard impedance at near-end andopen at far-end

2.3 Interference of coaxial cable

For coaxial cables, the physical model of cable layout is similar to the one in Fig.2. But the equivalent circuit model is slightly different, as shown in Fig.7. The far-end is connected to standard resistors, and the near-end of the circuit is grounded.

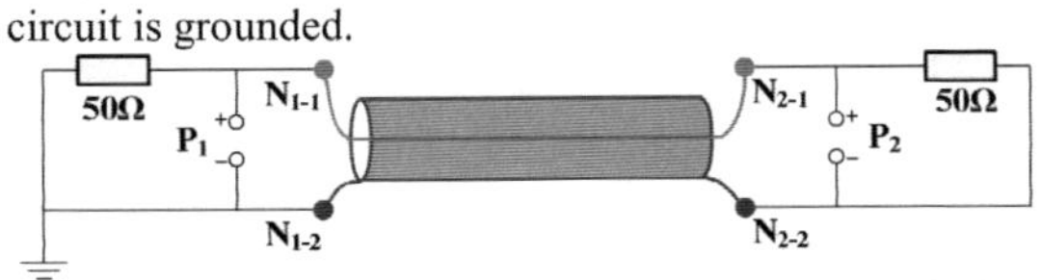

Fig7. Equivalent circuit model of coaxial cable (standard impedance at both near-end and far-end)

The differential-mode voltage analysis results of coaxial cable are shown in Fig. 8. According to the results, the differential-mode interference voltage of coaxial cable at low frequency is higher than that of conventional cable. The interference voltage generated by 1kHz plane wave per unit strength on unit length cable is about 1.5×10^{-12}V, which is about 10 times that of conventional cable. However, at high frequencies, the differential-mode interference voltage of coaxial cable is lower, and there are more resonance frequency points. The peak differential-mode interference voltage among the resonance points is about 4mV, which is about 1/5 of that of conventional cable (Fig. 4). At 1MHz, the differential-mode interference voltage of conventional cables is about 5.7nV, while that of coaxial cables is about 2.1nV.

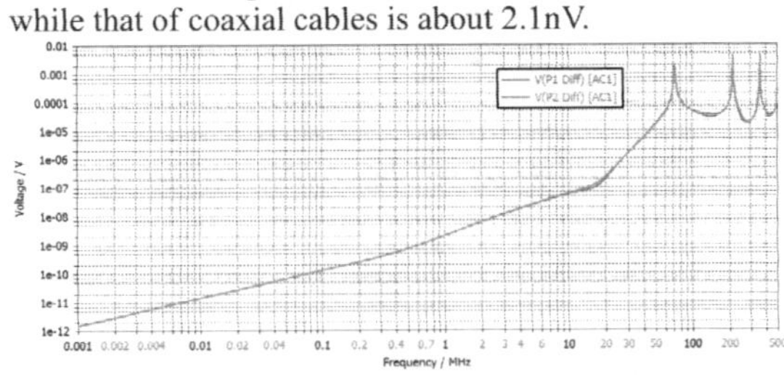

Fig8. Radiation-coupled differential-mode voltage of coaxial cable (standard impedance at both near-end and far-end)

The equivalent circuit of coaxial cable open at far-end is shown in Fig. 9. The result of differential-mode interfenrence voltage at two ports is shown in Fig. 10. According to the result, differential-mode interference voltage spectrum characteristics are similar to Fig. 8. As the circuit is open at the far-end, the voltage at the near-end is different from that at the far-end, and the voltage at the far-end is increased. At 1kHz, the corresponding front-end interference differential-mode voltage rises to $3\times10^{-}$

https://doi.org/10.1051/e3sconf/202123702010

[12]V. At the same time, resonance peak differential-mode voltage reachs to 8mV, enhanced to about double of that in the coaxial cable with standard resistance.

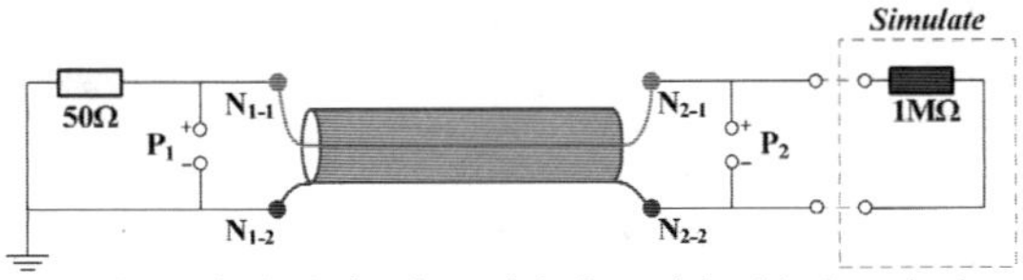

Fig9. Equivalent circuit model of coaxial cable (standard impedance at near-end and open at far-end)

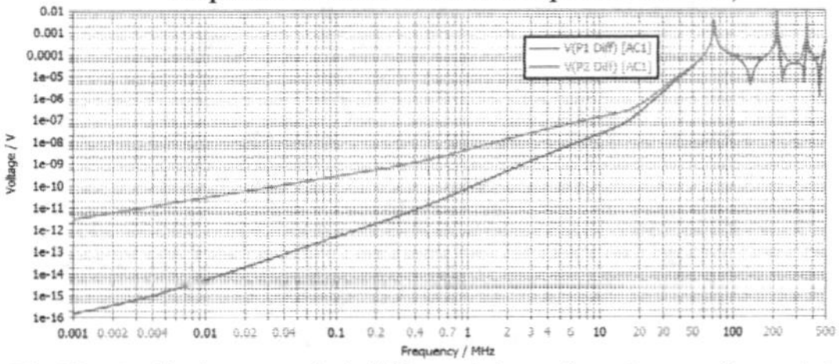

Fig10. Radiation-coupled differential-mode voltage of coaxial cable (standard impedance at near-end and open at far-end)

2.4 Interference of twisted-pair cable

For twisted-pair cables, the equivalent circuit is shown in Fig. 11. Both ends are connected to standard impedance, and connected to the ground through stray capacitor.

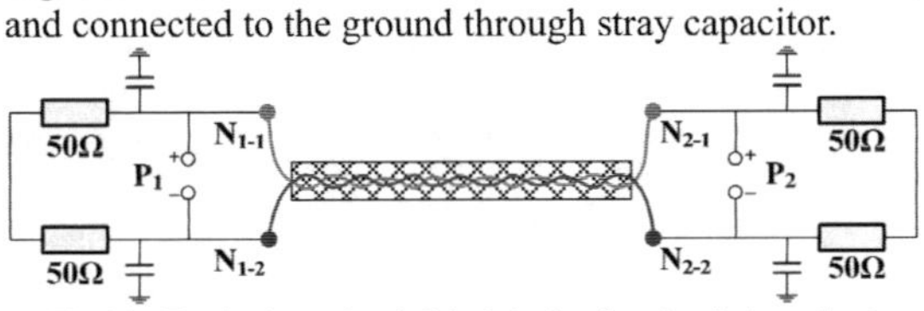

Fig11. Equivalent circuit Model of twisted pair (standard impedance at both ends)

The results of differential-mode voltage at two ports of twisted pair are shown in Fig. 12. According to the calculated results, the differential-mode interference voltage of twisted pair at low frequency is higher than that of conventional cable and coaxial cable. At 1kHz, the differential-mode interference voltage per unit length is about 2.3×10^{-11}V, which is about 140 times that of conventional cable and 15 times that of coaxial cable. However, in high frequency range, the differential-mode interference voltage of twisted-pair cable is lower, and there are more resonance frequency points than other cables. The peak differential-mode interference voltage of the resonance points is about 0.5mV, which is about 1/40 of that of conventional cable and 1/8 of that of coaxial cable. At 1MHz, the differential-mode interference voltage of the twisted pair is 23nV, that of the conventional cable is about 5.7nV, and that of the coaxial cable is about 2.1nV. The twisted pair has the highest differential-mode voltage, which indicates that the twisted-pair cablehas certain advantages only when it is above 1MHz.

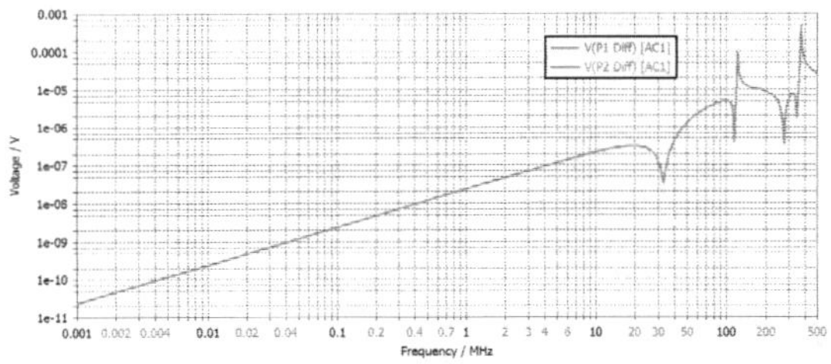

Fig12. Radiation-coupled differential-mode voltage of twisted pair (standard impedance at both ends)

The equivalent circuit of the twisted-pair cable open at far end, which is equivalent to 1MΩ resistor, is shown in Fig. 13. The results of differential-mode voltage between two ports of the signal cables are shown in Fig. 14. According to the result, differential-mode interference voltage spectrum characteristics are similar to Fig. 12. At 1kHz, the corresponding far-end differential-mode voltage rises to 4.7×10^{-11}V, and the resonance peak differential-mode voltage reaches to 1mV, which is about doubled.

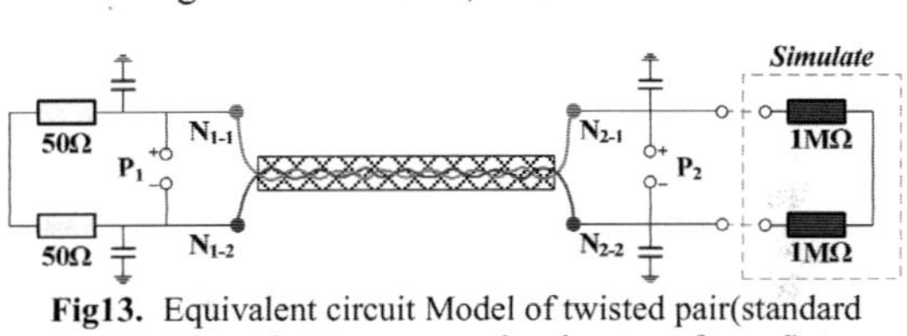

Fig13. Equivalent circuit Model of twisted pair(standard impedance at near-end and open at far-end)

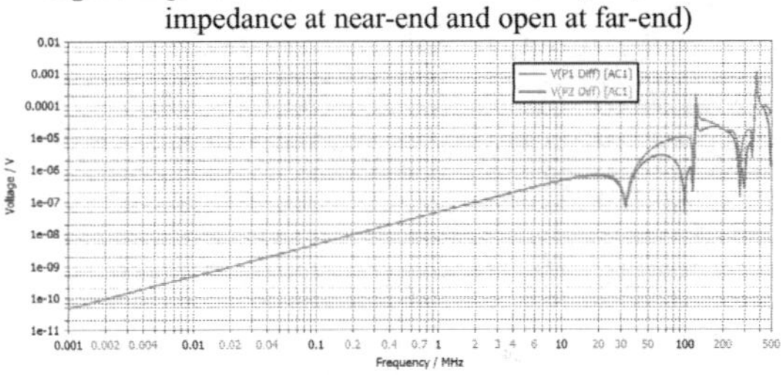

Fig14. Radiation coupled differential-mode voltage of twisted pair (standard impedance at near-end and open at far-end)

3 Analysis of influence of incident direction of electromagnetic wave

The direction of the plane wave has a great influence on the interference voltage. Taking the conventional cable layout as shown in Fig. 2 as an example, the influence of three different electromagnetic wave incident directions on the interference voltage is analyzed as shown in Fig.15.

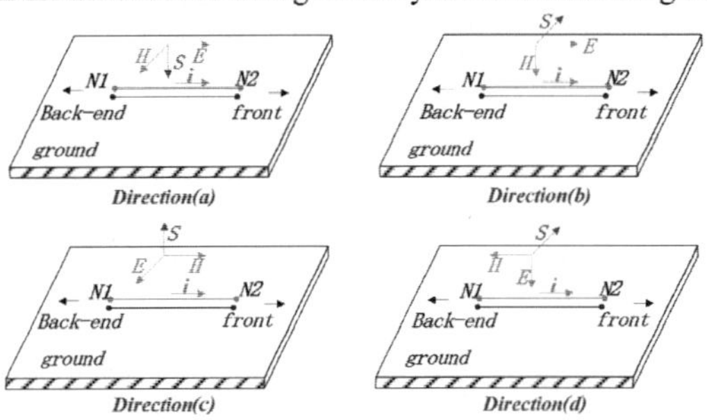

Fig15. Different incidence directions of electromagnetic waves

As shown in Fig. 16, the differential-mode voltage characteristics of (a) and (b) are very similar, but in high

 https://doi.org/10.1051/e3sconf/202123702010

frequency band, the differential-mode voltage of (b) is lower, because the electric field effect of both is exactly the same. However, in (b), the direction of magnetic field H is perpendicular to the ground plane, which will present a certain shielding performance. While in (a), there is no such effect. The differential-mode induced voltage of (c) is the smallest, which is six orders of magnitude lower than that of (a). This is because the electric field direction is perpendicular to the cable, and the magnetic field is parallel to the cable, so the differential-mode interference voltage introduced by both is small. The spectrum characteristics of the interferential voltage of (d) are similar to those of (b), which may be because the direction of the electric field E is perpendicular to the ground plane, and the interference voltage is generated by the capacitance effect of the cable to the ground.

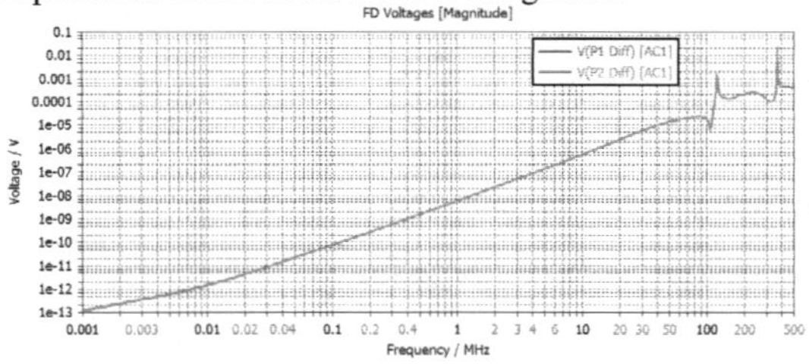

(a)

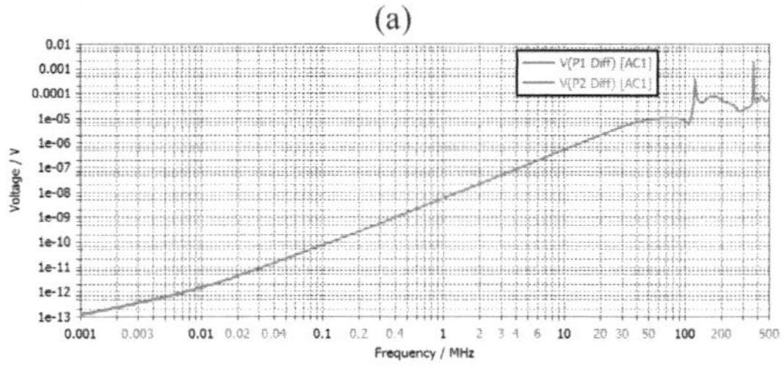

(b)

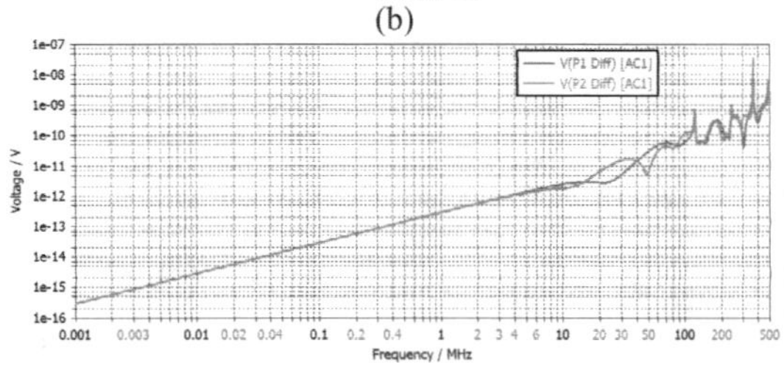

(c)

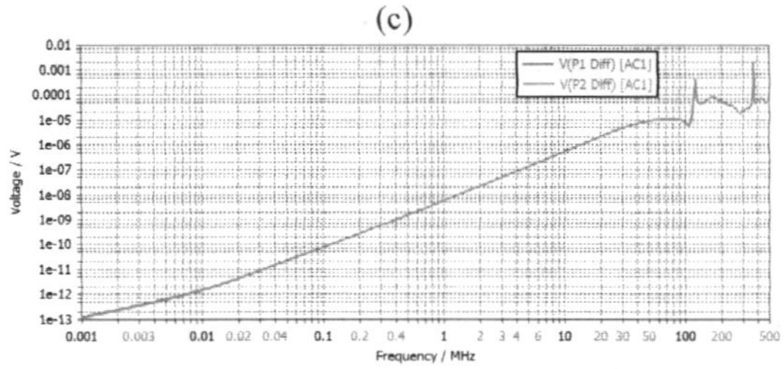

(d)

Fig16. Radiation-coupled differential-mode voltage of conventional cable under different plane wave incident directions(standard impedance at both ends)

4 Conclusion

This paper presents the analysis of the response characteristics of typical signal cables under the impact of electromagnetic wave. By comparing the response characteristics of different signal cables under different conditions, the following suggestions can be given to improve the immunity of secondary equipment in power system against the interference from electromagnetic wave.

(1) Coaxial cable and twisted-pair cable have a much better immunity than conventional cable. For sensitive or some important devices, coaxial cable or twisted-pair cable should be used.

(2) At high frequency (like higher than 1 MHz), twisted-pair cable has a better behaviour than coaxial cable, while at low frequency (like lower than 1 MHz), the coaxial cable will behave better.

(3) If the direction of the electromagnetic wave can be predicted in advance, the cable is better arranged parallel with the magnetic field, but perpendicular to the electric field.

(4) At extremely severe conditions, shielding technique can be used to reduce the interference.

References

1. K. C. Chen, K. M. Daumrau. Accuracy of approximation transmission line formulas for overhead wires[J]. IEEE Trans. On EMC. 1989, 31(4):396~397.
2. Taylor C, Satterwhite R, Harrison Jr C. The response of a terminated two-wire transmission line excited by a nonuniform electromagnetic field[J]. IEEE Trans. On EMC. 1965,13(6):987~989.
3. Agrawal A K, Price H J, Gurbaxani S H. Transient response of multiconductor transmission lines excited by a nonuniform electromagnetic field[J]. IEEE Trans. On EMC. 1980(2):119~129.
4. Rachidi F. Formulation of the field-to-transmission line coupling equations in terms of magnetic excitation field[J]. IEEE Trans. On EMC. 1993,35(3):404~407.
5. Tesche F M, Karlsson T. EMC analysis methods and computation models[M]. JOHN Wiley and Sons, 1997.

Breakage Mechanism Study of Overhead Ground Wire Under Lighting Stroke Based On Finite Element Analysis

Yunfeng Xia[1], Fan Kuang[1*], Ting Yang[1], Guanhao Chen[1], Rui Yang[1]

[1]Dongguan Power Supply Bureau of Guangdong Power Grid Corporation, Dongguan 523000, China

Abstract. Breakage of overhead transmission lines and distribution lines will lead to severe power outage problems, and clarifying the breakage mechanism of the overhead lines is of great importance and value for the reliability and stability of the power system. Taking overhead ground wire for example, this paper presents its breakage mechanism study under lighting stoke based on finite element analysis. A multi-physics analysis model is built first, which includes the model of contact points. The current density distribution and the temperature rise under the impact of short circuit current are analysed, and contact resistance effect is proved the primary cause for the breakage. Then some major factors that influence the contact resistance which in turn influence the temperature rise are discussed. Based on the discussions, some valuable suggestions are given for the protection of overhead lines.

1 Introduction

In the operation of overhead transmission lines and distribution lines, breakage of the lines usually occurs and it will lead to power outage. Statistics shows that, most of the breakage occurs at the connection fittings of the overhead lines [1]. Overhead ground wire and its suspension clamp is the most common case. However, the breakage mechanism is still unclear. Clarifying the breakage mechanism of the overhead ground wire is of great importance and value for the reliability and stability of the power system.

When a lighting occurs, flashover may occur on the overhead transmission lines to the ground (the tower or the overhead ground line) [2]. A large short circuit current to earth is possible to occur and flows in the overhead ground wire and its suspension clamp. Under the impact of the short circuit current, the temperature of the ground wire and its suspension clamp will be greatly increased, and this is the possible reason for the breakage.

The rapid temperature rise of the ground wire and its suspension clamp has two main explanations, namely skin effect [3] and contact resistance effect [4]. As the short circuit current is alternating at power frequency, it tends to flow on the surface of the conductor due to the skin effect. It will increase the equivalent resistance of the wire and increase the power loss. The contact resistance arises from imperfect contact surfaces, which is not smooth, but has many spots. When the short circuit current flows through two contact surfaces, it will flow through several spots, not the entire contact surface [5]. This will also lead to an increase of equivalent resistance and power loss.

Many researches have been performed to measure the contact resistance [6-9]. However, measuring the contact resistance can't directly reflect the temperature rise under the impact of the short circuit current, and can't give the direct evidence why the breakage occurs. So, a reasonable method is to directly measure the temperature rise at the contact points. However, as the short circuit time is very short, it is not easy to accurately measure the temperature rise. Even a measuring equipment with high precise is available, the measurement can't be realized as the contact points are in the deep channel and not easy to be accessed.

Luckily, finite element analysis provides an efficient and powerful tool to analyse the current density distribution and temperature rise of the overhead ground wire and its suspension clamp during the short circuit stage. This paper mainly presents the breakage mechanism study of overhead ground wire based on finite element analysis. This paper is organized as follows. Section 2 will briefly introduce the analysis model and corresponding assumptions. The analysis results, including current density distribution and temperature rise of a typical case from the analysis will be presented in Section 3. The contributions to the breakage from skin effect and contact resistance effect are compared, and the contact resistance effect is proved the primary cause why the breakage occurs. Detailed discussion of the contact resistance effect on the breakage will be presented in Section 4, and some valuable suggestions will be also provided. At last, a summary of this paper will be given.

2 Analysis model

2.1 Geometry Model

Fig. 1 shows the analysis model of overhead ground wire and its suspension clamp. The ground wire is usually made

* Corresponding author: kuangfan@dg.gd.csg.cn

 https://doi.org/10.1051/e3sconf/202123702011

of stainless steel and in stranded type. However, it is modelled as several parallel conductors to simplify the analysis, and each has a diameter of 4 mm. To avoid fatigue wear of the wire, a layer of aluminium armour tape is inserted between the wire and the clamp. In the analysis, the aluminium armour tape is simplified as a cylinder with a thickness of 1 mm.

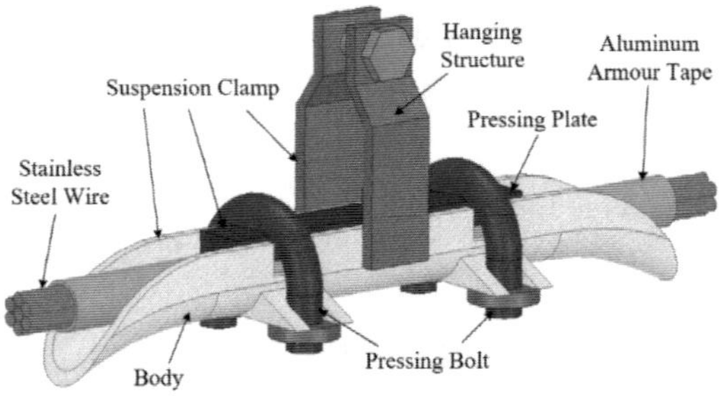

Fig1. Analysis model of overhead ground wire and its suspension clamp

In practice, the aluminum armour tape is used for the protection of the ground wire, not for the electrical connection. So no compression force is required on the aluminum armour tape except for the part in the suspension clamp. In the analysis, the aluminum armour tape is treated as a separation part from the ground wire except for the part compressed by the suspension clamp.

2.2 Model of Contact Points

Contact points act a significant role in the breakage of the overhead ground wire under the impact of short circuit current caused by lighting. So, it is necessary to include the effect of the contact points in the analysis. Fig. 2 shows the contact point distribution between the stranded ground wire and the aluminum armour tape. As the ground wire and the aluminum armour tape are compressed by the pressing plate and the body, the contact points mainly distribute on the upper and lower sides.

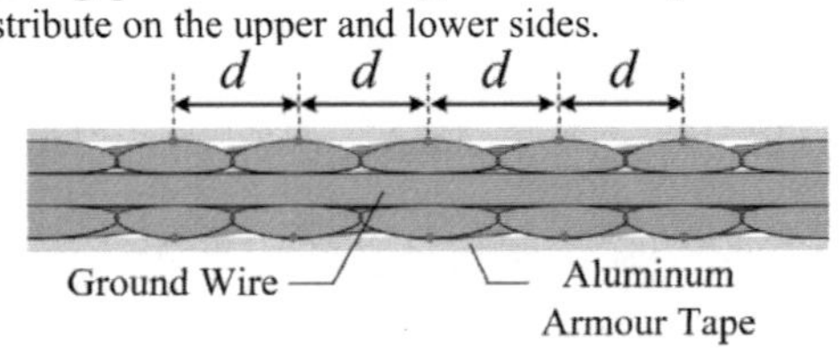

Fig2. The contact point distribution between the ground wire and the aluminum armour tape

If there are n conductors on the outer layer of the wire, there will be n contact points in a lay pitch L_p, and the distance between 2 adjacent contact points is $d=L_p/n$ as shown in Fig. 2. In the analysis model shown in Fig. 1, as the ground wire is simplified as several parallel conductors, the contact points are assumed uniformly distributed on the conductors as shown in Fig. 3. For each conductor, there will be one contact point in a lay pitch. In the analysis, the total number of contact points N_c is 60.

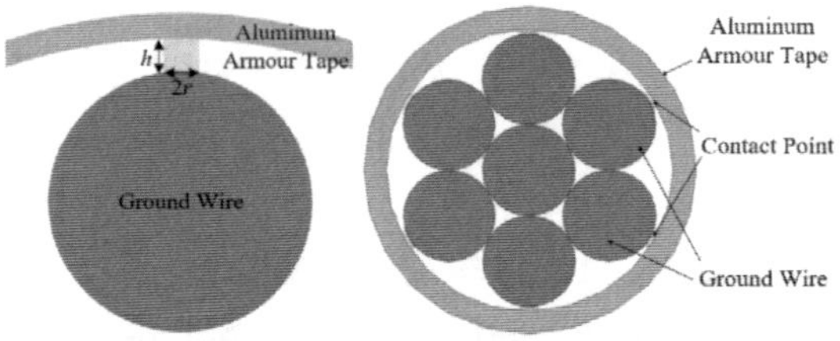

Fig3. Contact point distribution in the analysis model

The effect of contact point is modelled as an electrical bridge with a height of h and a radius of r. The material of the electrical bridge is the same as the ground wire. Its height h is set 0.2mm, and the radius r can be calculated by Holm equation as

$$r = \sqrt{F_c / \pi \xi H N_c} \tag{1}$$

where, F_c is the total compression force on the contact surface, ξ represents the contact condition at the contact surface, H is the Brinell hardness of the material at the contact surface and N_c is the total number of the contact points. The value of ξ usually ranges from 0.3 to 0.6, and it is selected at 0.5 in the analysis. For the Brinell hardness H, it is selected at 180 HB.

For the suspension clamp, the compression force can be calculated as

$$F_c = \frac{nT_c}{Kd_b}, \tag{2}$$

where, n is the number of the bolts and it is selected as 4, K is the friction coefficient of thread and it is selected as 0.25, d_b is the diameter of the bolt and it is selected as 10 mm, and T_c is the torque of the bolt and it is selected as 1 N·m.

2.3 Loads and Boundary Conditions

When a short circuit of a transmission line occurs to the ground, the short circuit current will flow through the tower and the overhead ground line. Previous study shows that, over 80% percentage of the short circuit current will flow into the ground line, and this is the main reason why the breakage occurs.

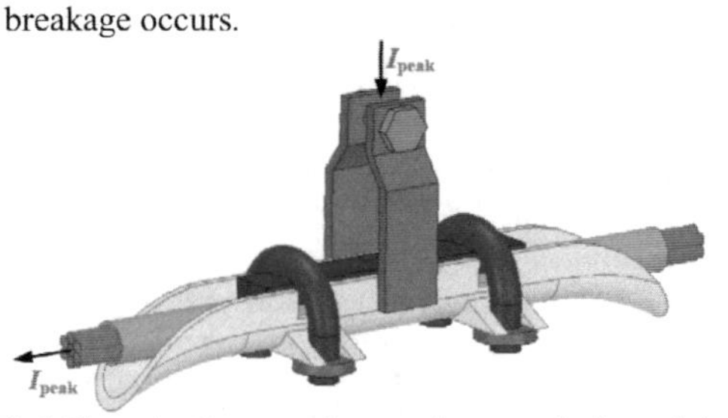

Fig4. Short circuit current loop on the ground wire and the suspension clamp

To evaluate the impact of the short circuit current, current load is applied and its amplitude is assumed as Ipeak. The current is assumed flowing from the hanging structure to one side of the ground wire as shown in Fig. 4. This situation represents the worst case as the current is concentrated on only half of the ground wire and the suspension clamp. In the analysis, the peak value of the short circuit current is assumed 20 kA and its lasting time td is assumed 100 ms.

When the short circuit current flows through the ground wire and its suspension clamp, especially through the contact points, a large amount of ohmic loss will be generated and give rise to the rapid temperature rise to the conductors. Thus, thermal analysis is also obligatory to evaluate the impact of the short circuit current. Due to the positive temperature coefficient effect, the resistance of the conductors will increase as the temperature rises. So in the analysis, the relation between the resistivity and the temperature is also considered as

$$\rho_T = \rho_0\left[1+\alpha\left(T-T_0\right)\right] \tag{3}$$

where, ρ_T is the resistivity at the temperature T, ρ_0 is the resistivity at the temperature T_0, and α is the temperature coefficient of resistivity.

As the duration time of the short circuit is very short, the convection and radiation is ignored in the analysis, only the thermal conduction in the conductors is considered. In addition, at the boundary of the analysis model, adiabatic boundary condition is applied.

3 Analysis results

3.1 Current Distribution

In the electromagnetic analysis, the total short circuit current flowing in the ground wire I_{peak} is 20 kA, the torque of the bolt is 1 N·m which corresponds to a radius of 0.3 mm for the electrical bridges. The conductivity of the ground wire and the suspension clamp is 6×10^6 S/m, and the conductivity for the aluminium armour tape is 3.8×10^7 S/m.

Fig. 5 shows the current density distribution on the ground wire and the suspension clamp. The short circuit current is injected from the hanging structure, and then flows to the body and then transmits to the ground wire through the pressing bolts, pressing plate and aluminium armour tape. As the short circuit current is assumed flowing to only one side of the ground wire, the current density on one side of the ground wire is very high, but very low on the other side. The total power loss of the ground wire and suspension clamp is about 59.2 kW, which indicates a total resistance of 296 μΩ.

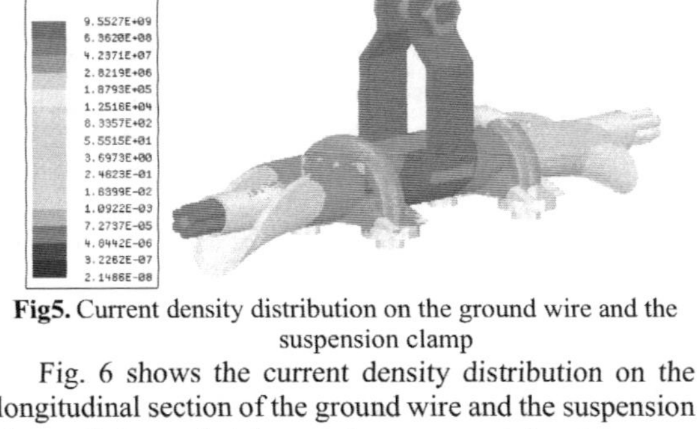

Fig5. Current density distribution on the ground wire and the suspension clamp

Fig. 6 shows the current density distribution on the longitudinal section of the ground wire and the suspension clamp. It shows that the maximum current density occurs at the electrical bridge, which represents the contact point, and the peak current density is as high as 9.6×10^9 A/m^2. As the current will flow to the end of the ground wire, the conductors on the outer layer of the ground wire has a very high current density. The centre conductor has a low current density due to the bad electrical connection with the outer ones.

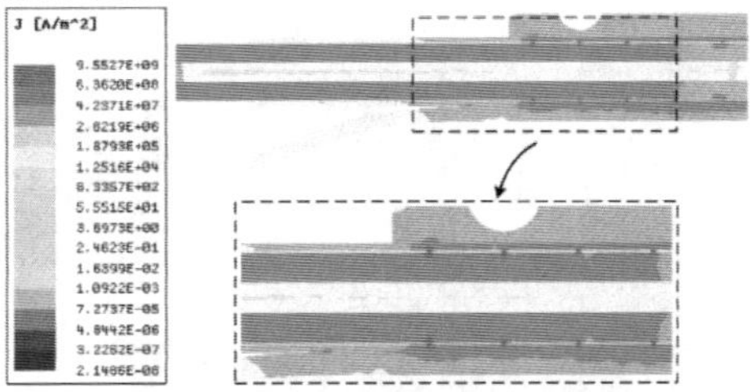

Fig6. Current density distribution on the longitudinal section of the ground wire and the suspension clamp

From Fig. 6, it can be also observed that, the closer to the end of the wire, the higher the current density of the electrical bridge will be. It is also because that the short circuit is assumed flowing to only one end of the round wire. In the conductors of the wire, no skin effect is observed as the current is almost uniformly distributed on the longitudinal section. It can be easily explained by calculating the skin depth of the conductor and compared with the size of the conductor. At the power frequency, the skin depth of the conductor is $d_s = (2/\mu\omega\sigma)^{1/2}$=29mm, which is much larger than the diameter of the conductor. It indicates that, the skin effect is not the primary cause for the breakage of the ground wire.

3.2 Temperature Rise and Distribution

In the thermal analysis, the mass density, thermal conductivity and specific heat of the ground wire and suspension clamp is 7750 kg/m^3, 50 W/m·K and 480 J/kg·K, respectively. For the aluminium armour tape, these values are set 2770 kg/m^3, 165 W/m·K and 875 J/kg·K, respectively. In addition, the lasting time of the shirt circuit stage is assumed 100 ms, and the initial temperature is assumed 30°C.

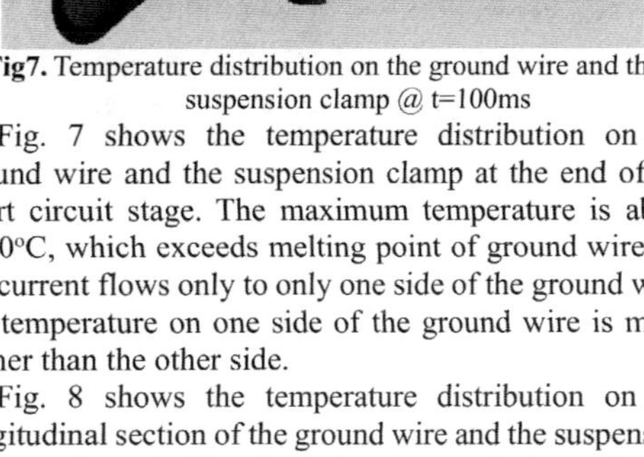

Fig7. Temperature distribution on the ground wire and the suspension clamp @ t=100ms

Fig. 7 shows the temperature distribution on the ground wire and the suspension clamp at the end of the short circuit stage. The maximum temperature is about 1590°C, which exceeds melting point of ground wire. As the current flows only to only one side of the ground wire, the temperature on one side of the ground wire is much higher than the other side.

Fig. 8 shows the temperature distribution on the longitudinal section of the ground wire and the suspension clamp at the end of the short circuit stage. It shows that the maximum temperature occurs at the electrical bridge, which represents the contact point, and the highest temperature exceeds the melting point. As the ground wire should bear the gravity force of itself, breakage will occur under the impact from the gravity force and the rapid temperature rise.

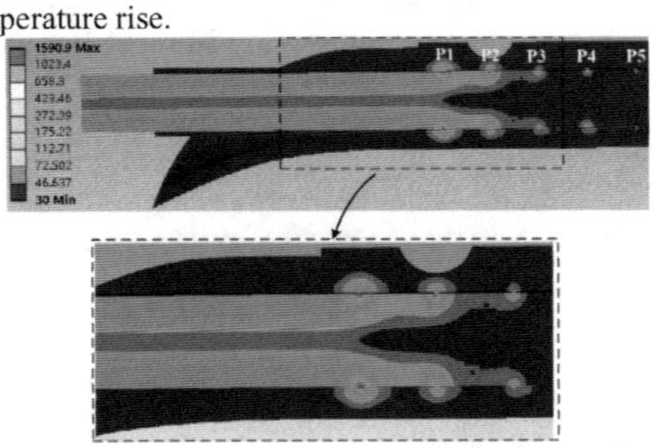

Fig8. Temperature distribution on the longitudinal section of the ground wire and the suspension clamp @ t=100ms

Fig. 9 shows the curve of the maximum temperature of the electrical bridges verse time. It shows that, the temperature goes up very rapidly within 20 ms, and then the rising rate slows down. It is due to the thermal conduction from the electrical bridges to the adjacent conductors where the temperature is relatively low. As the temperature difference increases, the thermal conduction will increase, and the temperature rising rate will be reduced.

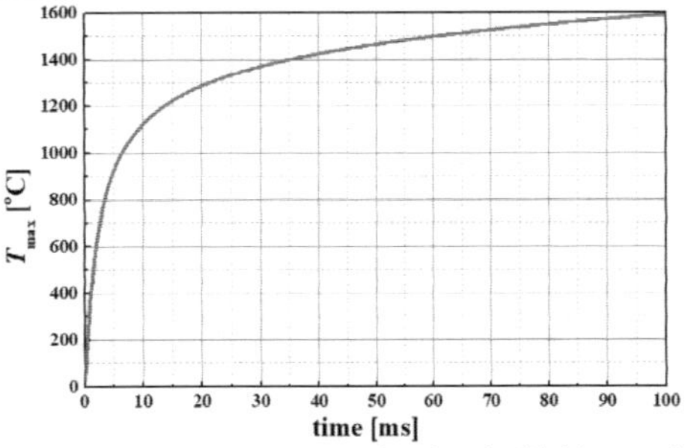

Fig9. Maximum temperature on the electrical brides vs. time

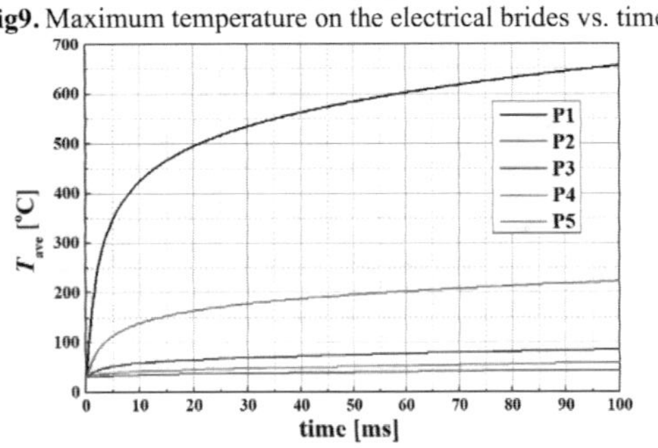

Fig10. Average temperature on the electrical bridges vs. time

Fig. 10 shows the curves of the average temperature of the selected electrical bridges as shown in Fig. 8 verse time. The average temperature is much lower than the maximum value as shown in Fig. 9. The electrical bridge closest to the end of the ground wire has the highest average temperature, which coincides with the analysis results shown in Fig. 6.

4 Discussions on major factors

Based on the analysis presented in Section 3, it can be concluded that the temperature rise of the ground wire depends on the contact resistance, the amplitude of the short circuit current and the lasting time of short circuit stage. This section will discuss the influence of these major factors on the temperature rise of the ground wire.

4.1 Contact Resistance

In the analysis, the contact points are modelled as electrical bridges with a cylindrical shape. Then the resistance of a contact point is

$$R_0 = \frac{\rho h}{\pi r^2} \tag{4}$$

As all the contact points are in parallel, then the total contact resistance of N_c contacts can be expressed as

$$R = \frac{\eta R_0}{N_c} = \frac{\eta \rho h}{N_c \pi r^2} \tag{5}$$

where η is the correction factor as the short circuit current flows through only half of the contact points, and the current density is not uniformly distributed among the contact points. Combining Eq. (1) (2) and (5), the total contact resistance can be expressed as

$$R = \frac{\eta \rho h \xi H K d_b}{n T_c} \tag{6}$$

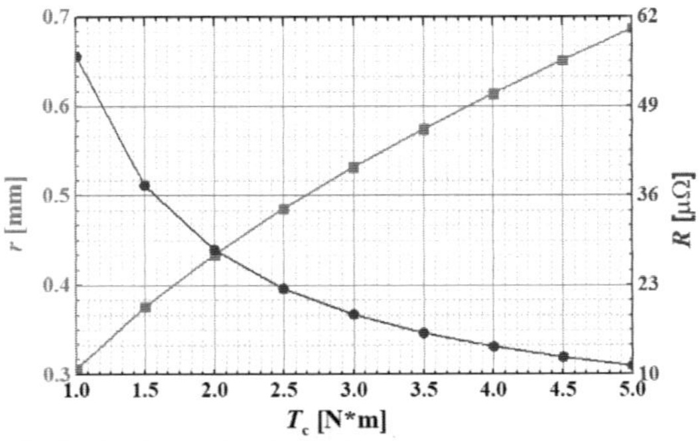

Fig11. Radius of electrical bridge and total contact resistance under different preloaded torques

In Eq. (6), only the preloaded torque on the bolt is easy to be adjusted, so the following will mainly focus on the influence of this factor. Fig. 11 shows the radius of electrical bridge and the total contact resistance under different preloaded torques, where the correction factor η is assumed 3. The total contact resistance has an inverse relation with the preloaded torque. When the preloaded torque varies from 1 N·m to 5 N·m, the total contact resistance will decrease from 56.3 μΩ to 11.2μΩ. Compared with total resistance of the whole system (296 μΩ @ T_c = 5 N·m), the contact resistance is much small and contributes little to the total power loss.

Fig. 12 shows the maximum temperature and the total power loss of the whole system under different preloaded torques. It shows that, with the increase of the preloaded torque, the total power loss of the whole system decrease only a little. It is because that the contact resistance contributes only a little to the total resistance. However, the increase of the preloaded torque will greatly reduce the maximum temperature on the contact points, and thus protect the ground wire from breakage. If the preloaded torque is increased to 5 N·m, the maximum temperature will decrease to only 378 °C, which is far below the melting point and ensure the reliability of the ground wire.

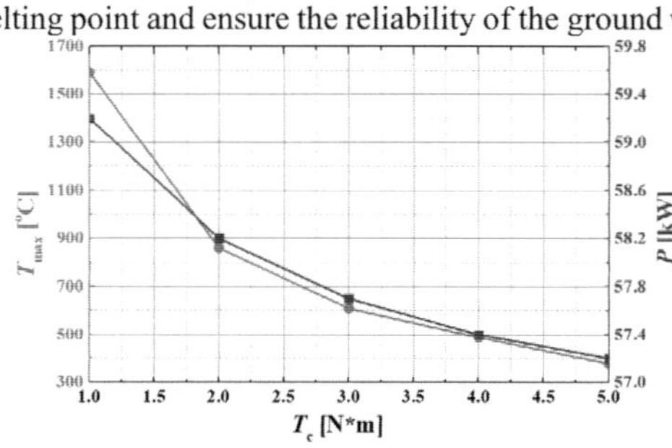

Fig12. Maximum temperature and total power loss of the whole system under different preloaded torques

4.2 Amplitude of Short Circuit Current

As the ohmic power loss is in direct proportion to the square of the short circuit current, the amplitude of short circuit current will certainly influence the temperature rise, and thus affect the breakage of the ground wire.

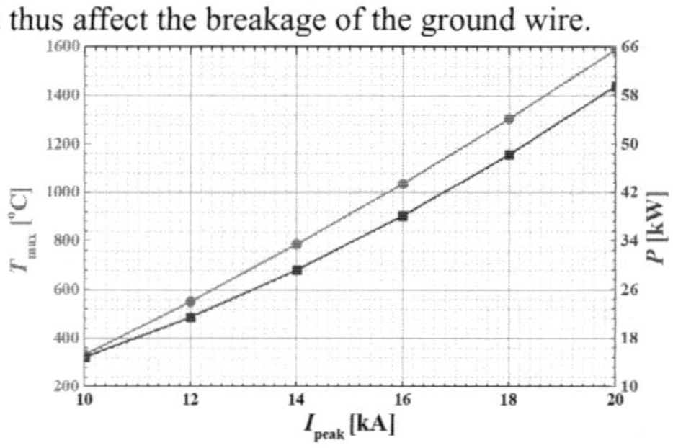

Fig13. Maximum temperature and power loss of the whole system under different amplitudes of short circuit current

Assuming the preloaded torque on the compressing bolts is 1 N·m, the power loss of the whole model and the corresponding maximum temperature on the contact point are calculated as shown in Fig. 13. As the amplitude of the short circuit current increases, the total power loss will increase in a quadratic manner, and the maximum temperature increases in a similar manner. It indicates that, limiting the amplitude of the short circuit current is very efficient to avoid the breakage of the ground wire.

4.3 Short Circuit Duration

When a short circuit occurs, the monitoring system will recognize it and relaying protection system will make an action to cut off the short circuit. However, it takes time to complete the protection procedure. As shown in Fig. 9 and 10, the temperature rises very quickly during the first 20ms of the short circuit duration, and the rising rate will be reduced due to the thermal conduction. This interval during which the temperature rises very quickly is determined by the thermal time-constant of the contact point and adjacent conductors. Reducing the time of the protection procedure will certainly reduce the maximum temperature. However, it is not easy to ensure an action time less than 20 ms. It is also unnecessary as the temperature rises relatively slow during the latter duration.

5 Conclusions

This paper presents the breakage mechanism study of overhead ground wire under lighting stoke based on finite element analysis. Analysis results shows that contact resistance effect contribute much more to the breakage than skin effect, and it is the primary cause for the breakage. Further analysis shows that, increasing the preloaded torque on the compressing bolts to reduce the contact resistance is a very efficient measure to reduce the maximum temperature on the contact point during short circuit duration. Besides, limiting the short circuit current is also efficient, but it is not easy to be realized.

References

1. Z. M. He, W. J. Chen, H. B. Shen. Experimental study of protection measures for preventing 10kV insulated conductors from breakage caused by lightning stroke. Electricity Distribution - Part 1, 2009. CIRED 2009.
2. Z. Mei, S. M. Chen, et al., Statistic of lighting accidents during 1998-2004 in China, High Voltage Engineering, 33(12), 2007.
3. Yucel, Mehmet K, et al. An ultrasonic guided wave approach for the inspection of overhead transmission line cables. Applied Acoustics 122: 23-34, 2017.
4. Yovanovich, M., and K. Nho. Experimental investigation of heat flow rate and direction on contact resistance of ground/lapped stainless steel interfaces. Thermophysics Conference 2013.
5. Kogut L. Electrical performance of contaminated rough surfaces in contact[J]. Journal of Applied Physics, 97(10): 103723.1-103723.5, 2005.
6. Wang W.G., Liang M.L., et al., Characteristics of static contact resistance between pantograph and catenary, High Voltage Apparatus, 48(1), 2012.
7. Chen Y.M., Song W., Assessment of the uncertainty of the measurement of conductor DC resistance, Electric Wire and Cable, 12(1), 2009.
8. Liu M., A new method for measuring contact resistance, Journal of astronautic metrology and measurement, 25(3), 2005.
9. Laor, A., Herrell, PJ., Mayer, M. A Study on Measuring Contact Resistance of Ball Bonds on Thin Metallization. IEEE Trans. Compon. Packag. Manuf. Technol. 2015, 5, 704-708.

Research on environmental Accounting Information Disclosure of energy Pollution Industry -- Case of the listed companies in coal industry

Hong Li[1]

[1]Shandong Women's University, China

Abstract. The coal industry plays a pivotal role in China's energy industry, which not only brings about rapid economic growth but also causes serious ecological environmental problems. This paper selects 25 listed companies in the coal mining and washing industry as the research objects to analyse the current situation and problems of environmental accounting information disclosure. This paper puts forward some suggestions for improving the environmental disclosure, so as to provide reference for promoting the transformation and upgrading of the coal industry.

1 The introduction

The report to the 19th National Congress of the CPC stated that green development is an important demand for building a high-quality modern economic system. The development of the energy industry should actively implement the spirit of the report of the 19th National Congress and promote the construction of a clean and low-carbon energy system. In 2019, raw coal production accounted for 69.6 percent of China's total one-time energy production, while coal consumption accounted for 57.7 percent of total energy consumption. It can be seen that the current coal industry occupies a major position in the entire energy industry. The coal industry has made outstanding contributions to the economic growth but also brought huge environmental pressure to the society. For example, the coal mining process will lead to the destruction of land resources, plant resources, air pollution and water pollution and other problems. Therefore, the disclosure of environmental accounting information about the coal industry has become the focus of scholars and the community.

2 Current situation of environmental accounting information disclosure of listed companies in the coal industry

2.1 Sample selection of environmental accounting information disclosure of listed companies in coal industry

According to the industry classification results of listed companies by China Securities Regulatory Commission in the first quarter of 2019, the coal mining and washing industry contains a total of 27 listed companies. After excluding the B-share listed Yintai B shares and *ST companies, this paper uses the remaining 25 listed companies as the research samples.This paper mainly analyzes the basic situation of environmental accounting information disclosure based on the sample companies' 2019 annual report, social responsibility report and the environmental accounting information disclosed in the prospectus. The basic information of sample companies is shown in Table 1.

Table1. Sample company basic information table.

Stock code	The company referred to as"	Stock code	The company referred to as"
000552	Jing yuan coal	600714	Jinrui mining
000780	PingZhuang energy	600758	Hongyang energy
000937	Hebei central energy	600971	Turbine coal
000983	Xishan coal and electricity	600985	Huaibei mining
002128	Open-air coal	601001	Datong coal
600121	Zhengzhou coal	601088	China shenhua energy
600123	Orchid kechuang	601101	HaoHua energy
600157	Yongtai energy	601225	Shanxi coal
600188	Yanzhou Coal	601666	Ping Coal

600348	Yangquan coal	601699	LuAn ring can
600395	Panjiang shares	601898	China coal energy
600403	Great energy	601918	Xinji energy
600508	Shanghai energy		

2.2 The disclosure of environmental accounting Information of listed companies in the coal industry

2.2.1 The form distribution of environmental accounting information disclosure of listed companies in coal industry

Environmental accounting information of listed companies in China's coal industry is mainly disclosed in the form of prospectus, annual report and social responsibility report. The detailed information disclosure form and distribution of environmental accounting information of listed companies in the coal industry in 2019 are shown in Table 2.

Table2. The distribution of environmental accounting information disclosure of listed companies in the coal industry in 2019.

Disclosure forms	Number of Disclosures (pieces)	The proportion
Prospectus	15	60%
The annual report	25	100%
Social responsibility report	13	52%

Through the investigation of the 25 sample companies mentioned above, it is found that the most common form of environmental accounting information disclosure of listed companies in China's coal industry is the annual report, and the 25 sample companies all made relevant disclosures in the 2019 annual report. The disclosure content is mainly distributed in the environmental information part of the notes. The specific content mainly includes information on the company's pollution discharge fee, environmental protection investment, solid waste, and safe disposal rate of hazardous waste. In addition, prospectus and social responsibility report are also the environmental accounting information disclosure forms that sample companies often choose. 15 companies disclosed environmental accounting information through the prospectus, and the sample number accounts for 60% of the total sample companies. The environmental accounting information in the prospectus is mainly distributed in the part that restricts environmental factors. The 13 sample companies released their social responsibility reports for 2019 and explained environmental accounting information.

2.2.2 The contents of environmental accounting information disclosure of listed companies in coal industry

The emphasis of environmental accounting information disclosed by different disclosure forms is different. This part summarizes and analyzes the characteristics of environmental accounting information disclosed in the three forms.

2.2.2.1 Environmental accounting information disclosed in the annual report

The content of environmental accounting information disclosed in the annual report is mainly distributed in chapters such as business discussion and analysis, important matters and notes to financial reports. The discussion and analysis part of the business situation mainly explains the impact of environmental governance on the amount of long-term receivables and estimated liabilities for the current period, and briefly explains the environmental protection tax payment, environmental protection facility investment and environmental governance. The important matters section mainly explains the pollution discharge information, the degree of environmental impact of construction projects, and the emergency plan for emergencies. The environmental accounting information disclosed in the notes to the financial statements mainly includes reclamation, disposal and environmental cleaning expenses, sewage discharge fees, environmental governance security deposit, investment in the acquisition of special equipment for environmental protection, environmental protection subsidies, environmental protection taxes, resource taxes and other contents. In general, the sample companies' 2019 annual report disclosed less about environmental assets, and mainly focused on environmental deposits. The disclosure level of environmental expenses and environmental liabilities is the highest. The disclosure of environmental benefits is mainly manifested in subsidies.

2.2.2.2 Environmental accounting information disclosed in the prospectus

The prospectus is an important basis for shareholders to purchase enterprise stocks. For listed companies engaged in coal mining and washing industry, it is necessary to properly explain the environmental information in the prospectus. The contents disclosed in the prospectus mainly include the environmental pollution problems that the company will bring, pollution treatment methods, and environmental governance cost estimates. In general, the disclosure of environmental accounting information in the prospectus is mainly written description and lacks certain quantitative indicators.

2.2.2.3 Environmental accounting information disclosed in social responsibility reports

The disclosure of environmental accounting information in the social responsibility report mainly involves pollution reduction and emission reduction, ecological protection, resource conservation and other aspects. But generally speaking, the content of environmental accounting information disclosure in social responsibility reports is mostly qualitative description, but less quantitative description.

3 Problems existing in environmental accounting information disclosure of listed companies in coal industry

3.1 Disclosure comes in many forms but lacks uniformity

China has not unified regulations on the form of environmental accounting information disclosure of listed companies in the coal industry. Listed companies in the coal industry can choose to disclose in various forms such as annual reports, prospectuses and social responsibility reports. The lack of uniformity in the form of information disclosure, on the one hand, reduces the efficiency of information users in collecting information, on the other hand, the overlapping of multiple disclosure methods also leads to an increase in corporate disclosure costs. In addition, the lack of uniform forms and standards also leads to the lack of comparability in environmental accounting information disclosure of the same enterprise in different periods, and the lack of comparability in environmental accounting information disclosure between different enterprises.

3.2 The content of information disclosure lacks comprehensiveness

China does not have strict regulations on the content of environmental accounting information disclosure, which is the enterprise's independent choice, subjective and arbitrary, lack of comprehensiveness and consistency. Listed companies in the coal industry tend to disclose projects and indicators that are conducive to the development of enterprises, such as the amount of government subsidies, environmental governance margin, etc., while less disclosure of adverse projects, such as the payment of fines. Moreover, the environmental accounting information disclosed by enterprises is mostly qualitative information description, while the monetary measurement of environmental accounting indicators is less.

3.3 The quality of information disclosure needs to be improved

Environmental accounting information should not only reflect the actual situation of enterprises in a real and objective way, but also possess the characteristics of information quality such as comprehensibility, substance over form and comparability. However, China's current accounting standards have not established a complete and unified disclosure standard for environmental accounting information disclosure. There are certain differences in the recognition and measurement of environmental accounting information of listed companies in the coal industry, and the comparability of enterprises in the same industry is poor. The environmental accounting information provided by the company cannot meet the decision-making needs of information users.

4 Suggestions on improving the environmental accounting information disclosure of listed companies in the coal industry

4.1 Improve the environmental accounting information disclosure system of listed companies in the coal industry

With the promotion of the concept of green development, listed companies in the coal industry have become more aware of environmental accounting information disclosure. However, due to the lack of environmental accounting information disclosure standards and systems, coal companies ' environmental accounting information disclosure methods are not standardized. Therefore, it is recommended that the national financial department and other institutions gradually standardize the standards for environmental accounting information disclosure in the coal industry, and clarify the relevant requirements for the confirmation and measurement of environmental accounting information, so as to provide a legal basis for environmental accounting information disclosure by listed companies in the coal industry.

4.2 Regulate the disclosure of environmental accounting information of listed companies in the coal industry

Although listed companies in the coal industry in my country have disclosed some environmental accounting information, they have a large selection of information disclosure. They disclose more positive environmental accounting information and less negative environmental accounting information. In order to ensure that information users can obtain comprehensive information, relevant state departments should clearly state the content of environmental accounting information that needs to be compulsorily disclosed by coal companies, and regulate the completeness and detail of environmental accounting information disclosure in the coal industry.

4.3 Improve the environmental accounting information disclosure and supervision mechanism of listed companies in the coal industry

The improvement of the environmental accounting information disclosure quality of listed companies in the coal industry requires the efforts of the enterprises themselves on the one hand, and the strengthening of supervision by the regulatory authorities on the other. Listed companies in the coal industry should strengthen their sense of social responsibility, take the initiative to assume responsibility, strengthen the construction of internal supervision mechanisms, and improve the quality of environmental accounting information disclosure. The government and other relevant departments must improve the environmental accounting information disclosure and supervision mechanism, while the supervisory department

https://doi.org/10.1051/e3sconf/202123702012

must increase its enforcement, strictly implement the rules and regulations to ensure the quality of environmental accounting information disclosure.

References

1. Chang jiao Liu.Finance and Accounting Research 1(2018)
2. Shufeng Wang, Lei Shang, Lijia Xi.Friends of accounting(2018)
3. Yamei Guan, Rui Zhao. Accounting communication(2017)
4. Xuguang Zhang, Yanna Xuan, Nan Zhang.Coal economy Research(2019)
5. Yutong Zhou.Shopping Mall modernization(2020)

Research on charging safety and early warning of intelligent networked electric vehicles

Lei Li*, WeiDong Liu, Dan Li, Xiaohui Li, and Xiaochen Liu

Marketing Service Center, State Grid Tianjin Electric Power Company, 300000 Tianjin, China

Abstract. With the continuous reduction of renewable resources and the increasing awareness of environmental protection, the phenomenon of replacing oil with electricity has become more and more common. Electric vehicles are an obvious example. In the past few years, electric vehicles have developed rapidly and become more and more popular in people's lives. However, imperfect technology has caused frequent safety accidents. The main reason for this is the charging safety risk. In this paper, there are many complex factors affecting the charging safety of electric vehicles in terms of the safety of electric vehicle charging and the energy and data exchange direction between charging piles and electric vehicles. Based on the analysis of the factors that affect the charging safety of electric vehicles, this paper combines big data technology to study the charging safety of electric vehicles.

1 Introduction

Environmental pollution is becoming more and more serious. In order to reduce carbon dioxide emissions and adhere to sustainable development, vigorously developing electric vehicles is an effective way. According to data from the China Charging Alliance: as of the end of 2019, China's electric vehicle ownership reached 3.368 million. It accounts for 1.46% of the total number of vehicles, and the supporting charging facilities reach 1.291 million vehicles per year, a year-on-year increase of 50.8%. With the increasing number of electric vehicles, there are more and more safety accidents, and the causes of accidents are also diverse. For example, if the battery management system (BMS) main board fails and the charger knows the battery cell voltage and total voltage protection limit, but the actual value exceeds the limit and no measures are taken, overcharging will occur. Failure to cut off the power in time during charging affects the service life of the battery. The charging equipment and battery of electric vehicles are the two main aspects that cause frequent safety accidents. However, there is still no effective method to evaluate its safety level. For this reason, the reference [1] puts forward the "safety protection qualification rate of electric vehicle charging equipment". The evaluation index is applied to the comprehensive evaluation system of electric vehicles to improve the safety protection capabilities of electric vehicles. Reference [2, 3] proposed a comprehensive evaluation method that considers both the safety of electric vehicles and charging stations. On this basis, Fabio Freschi et al [4] further discussed the safety of electric vehicle charging, and proposed to equip electric vehicle charging equipment with an insulating transformer to isolate the current from the grid for monitoring the safety of the electric vehicle charging process sex. In the reference [5], Claus-Christian Carbon et al improve the safety of charging infrastructure by providing data and materials within a safe range, thereby ensuring the safety of electric vehicle charging. Reference [6] pointed out the importance of battery safety for electric vehicle charging safety, and proposed a high-dimensional data flow outlier detection algorithm (DSOD) based on angular distribution for the safety assessment of battery systems. With the rapid development of big data technology, its simple and fast features have also been applied in the field of electric vehicle safety. In [7], Tomasz Wierzbicki proposed a data-driven security envelope algorithm to ensure its safety in order to solve the problem of easy damage to electric vehicle battery packs. Domestic research on charging safety is still in its infancy. It mainly studies charging equipment and environmental factors based on charging safety. Qian Lijun et al [8] analysed the influence factors of electric vehicle charging safety and used genetic wavelet neural network training principles and with the characteristics of multi-scale and multi-resolution, a charging safety warning model is designed to improve the safety warning capability of the charging system. Reference [9] summarizes the safety factors in the charging process of electric vehicles from four links, including equipment, technology, monitoring, and management. And a dynamic early warning method of battery short or micro short circuit based on monitoring data is proposed to control and eliminate potential faults. Reference [10] defines the safety level of charging facilities, and combined with the 5 new national standards released in 2016, analyses the factors affecting the charging safety of electric vehicles, and gives a solution to the charging safety. Generally speaking, most foreign studies focus on the evaluation

* Lei Li: leili@tj.sgcc.com.cn

methods of electric vehicle charging safety, while domestic research mainly focuses on factors such as the charging process of electric vehicles, charging equipment and batteries.

2 Functional requirements, logical architecture, communication architecture

Based on the ubiquitous Internet of Things, the functional requirements of electric vehicles and charging facilities networks should meet the five major characteristics of the ubiquitous power Internet of Things: terminal ubiquitous access, platform open sharing, computing cloud and fog collaboration, data-driven business, application free need to be customized. The logical architecture should follow the four-layer architecture of the power Internet of Things. Its architecture follows the three-layer architecture of the Internet of Things perception extension layer, network transmission layer, and platform application layer, adding an edge computing layer. The communication architecture should consider the network communication layer composed of two parts: the local communication network and the remote communication network.

3 Electric vehicle charging safety classification

According to the charging process of electric vehicles, charging safety can be divided into grid side safety, charging equipment side safety, vehicle side safety and monitoring system safety (Figure 1).

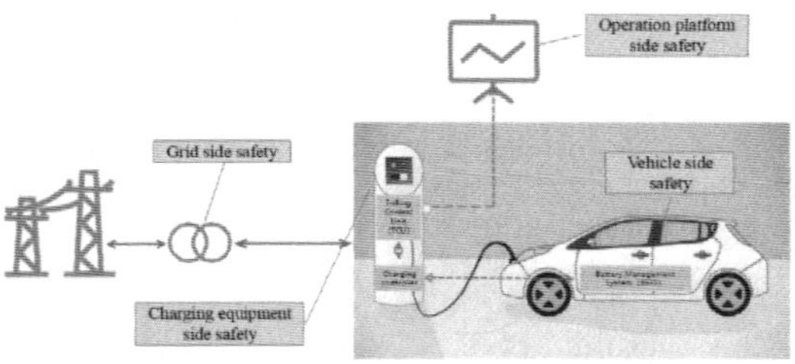

Fig1. Electric vehicle charging safety classification

3.1 Charging safety accidents

The main problem of charging safety on the grid side lies in the adverse effects of the charging behaviour of electric vehicles on the stable operation of the grid. Including: iolation of load limit, power quality limit, resonance risk, relay protection action and lack of anti-islanding protection function.

The charging safety issues on the charging equipment side are mainly the adverse effects of electric vehicle charging behaviour on personnel, electric vehicles, and charging facilities. Including: equipment leakage, charging equipment failure, charging incompatibility, charging start failure, abnormal protection measures are not in place.

The charging safety problem on the charging vehicle side is mainly the defective battery or battery management system. Including: excessive impurities in the electrolyte, unqualified process, defects of the diaphragm itself, and unreasonable structural design, the safety protection measures of the BMS management system are not in place.

The charging security problem of the platform monitoring system is mainly due to the lack of completeness of the system itself, which mainly includes the risk of electricity theft. Among them, time-sharing rental cars have the largest proportion of electricity theft in the early industry, abnormal charging transactions, too much or too little billing, and the charging equipment is offline without local charging countermeasures, lack of experience, hacker attacks on the operating platform, and the risk of service network paralysis.

3.2 Safety measures

Safety measures on the grid side. Promote coordinated charging technology in stations. Take the station area as the unit, hierarchically partition the charging load management, collect the charging load information and the electric load situation of the station area through the coordinated charging controller (energy controller, energy router, etc.), and according to the user charging and car time demand, arrange the charging plan to meet the cluster charging demand in the case of the distribution transformer of the station area without expansion, and achieve peak shaving and valley filling. (Figure 2)

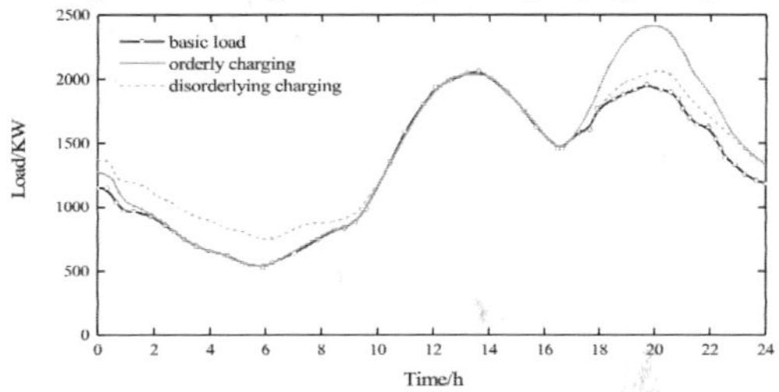

Fig2. Coordinated charging measures on the grid side

Safety measures for charging equipment. Before charging, recheck the size of interface, size and tolerance of mechanical locking mechanism. During charging, check the locking time and state of electronic lock (check whether the charging plug is pulled out by pressing the mechanical lock button) to avoid potential safety hazards. In the charging process, the content, format, cycle and time sequence of communication messages will be judged in real time. If there are abnormal messages or logical errors, the charging will be stopped or the power will be reduced according to the specific situation; in addition, the abnormal communication and error messages will be processed in real time, fault messages or charging suspension messages will be sent, and charging will be stopped directly if necessary. Research the classification of charging faults, fault characteristic values, safety thresholds, collection and judgment methods, quickly analyses the specific causes and handling methods of faults, improve the accuracy of charging fault location, and provide convenient charging fault solutions for operation and maintenance personnel.

Safety measures on the platform side. Add inter-regional access control, improve network boundary protection measures, improve system security configuration, and strengthen information security monitoring mechanisms.

Safety measures on the vehicle side. Battery status detection: it can effectively measure the cell voltage, total voltage, current, temperature, insulation resistance and other parameters of the battery system. The measurement accuracy and frequency should meet the requirements of national standards under normal working conditions and severe extreme working conditions.

4 Build database

Collect electric vehicle charging safety cases, collect daily driving rules, charging times, charging duration, and alarm operation information in a certain area through sampling, and combine specific cases and expert analysis to establish a characteristic database of electric vehicle charging safety accidents. From the perspective of charging safety, establish a network database of regional charging business information and equipment safety information; propose a quality evaluation method for the full life cycle of charging facilities; establish a database of regional charging service capability information and equipment operation safety information. However, with the rapid development of the ubiquitous Internet, if all the data generated by the Internet is transmitted to the cloud computing center, this will greatly increase the network load, cause information congestion, and thus lead to delayed information transmission and difficult to ensure the safety of charging. Edge computing can effectively solve this problem. Edge computing can meet the computing capabilities of edge devices by deploying small devices with computing, storage and network access capabilities at the edge of the network, and being closely connected with smart sensors, wired meters and users. Expanding demand can effectively reduce network load and data delay.

5 Security detection and early warning technology based on charging real-time data stream

Insulation resistance test: before the insulation resistance test, the external power supply connection of the charging equipment should be disconnected, and the input and output circuit switches should be closed. Between the independent live circuits of the electrical connection of the charging equipment, and between the independent live circuits and the protective grounding conductor (metal shell) Apply 500V direct current (DC) voltage, and check that the insulation resistance of the charging equipment is not less than 10MΩ.

Input overvoltage protection test: connect the charging equipment to the analog load, set the analog load at the rated power value, load and run and reach a stable state, adjust the output voltage value of the charging equipment power supply to be higher than 115% of the rated voltage value, and check whether the charging equipment starts input The overvoltage protection function cuts off the output and issues a fault alarm.

Input under voltage protection test: connect the charging equipment to the load, set the rated power to load and run and reach a stable state, adjust the output voltage of the power supply of the charging equipment to be lower than 85% of the rated voltage, and check whether the charging equipment has activated the input under voltage protection function and whether it has issued a fault alarm.

Output short circuit protection test: short-circuit the output of the charging equipment, load and start, check that the charging equipment should not allow output charging, and issue an alarm. Short-circuit test during operation: the charging equipment is connected to the load, the load is running and reaches a stable state, the output short-circuit failure of the charging equipment is simulated, and the charging equipment should be checked immediately to limit the current output or cut off the output, and issue an alarm.

Power cut off and restart test: connect the charging device to the load, set the rated power to load and run and reach a stable state, disconnect the alternate current (AC) side power supply, and check the output of the charging device. Keep the charging connection device fully connected, reconnect the AC side power supply, check that the charging device should not be able to continue to output charging, it needs to be manually replugged, and the connection is established before the next charging is allowed.

Output adhesion test: simulate the contact adhesion of the contactor, check that the charging equipment should not allow charging and issue an alarm.

Emergency stop test: connect the charging equipment to the load, load and run and reach a stable state, press the emergency stop button, and check that the charging equipment should be able to immediately cut off the output.

Connection and disconnection contact sequence test: in the process of checking the charging connection and disconnection, the ground terminal and contact of the charging connection device should be connected first and disconnected last.

Abnormal charging connection test: when an abnormal control pilot signal occurs during the charging process (such as signal short circuit, signal open circuit, signal grounding, etc.), the charging equipment should be able to disconnect the power supply circuit within 100ms.

6 Principles of information monitoring and safety alarms during charging

The electric vehicle charging process includes 6 stages: physical connection completion, low-voltage auxiliary power-on, charging handshake stage, charging parameter configuration stage, charging stage, and charging configuration stage. No matter in which phase the charger and BMS do not receive the other party's message or the correct message within the specified time, it will be judged as timeout and enter the error processing state. Using this principle, we can use efficient judgment software and calculation software to analyse the charging process. The CAN communication message between the charging equipment and the battery management system can realize the information monitoring and safety alarm of the charging process. Build a system local area network between on-site charging safety detection terminals, charging station monitoring systems, and fire safety early warning systems. The on-site charging safety detection terminal uses edge computing to quickly collect and analyse the life cycle operating information and fault information of the charging pile, and transmit the collected data and analysis results to the local area network in real time. The charging station monitoring system and the fire safety early warning system can obtain effective charging process information and accident early warning information in the local area network, and realize information sharing.

7 Conclusion

Electric vehicle charging safety involves four levels of electrical, communication, physical, and information safety factors such as vehicles, piles, networks, and operating platforms. Therefore, it is a system engineering. The design of protection functions, product design and material selection in accordance with standard requirements, passing safety test verification, regular maintenance and other factors will affect the safety of the charging system. Electric vehicles should be charged less quickly, and overcharged, and charging equipment should be output according to demand to improve the security risk resistance of all links. With the development of 5G technology, information transmission is more accurate and timely, and there are fewer and fewer factors restricting the development of electric vehicles. The research in this paper can effectively realize the charging safety control and eliminate potential dangers of electric vehicles, and is beneficial to the evaluation of electric vehicle charging safety indicators and fault location, and provide theoretical guidance for establishing a complete charging safety standard system.

Acknowledgments

This work was funded by the State Grid Tianjin Electric Power Company Science and Technology Project KJ20-1-10 Electric Vehicle Intelligent Charging Network for Electric Power Internet of Things Key Technology Research and Application Project

References

1. Kai Zhang, Zhongdong Yin, Xiaotian Yang, Zhuowu Yan, Yongzhang Huang. Quantitative Assessment of Electric Safety Protection for Electric Vehicle Charging Equipment[C]//International Conference on Circuits, Devices and Systems,2017. IEEE,2017:89-94.
2. Bo Wang, Payman Dehghanian, Shiyuan Wang, Massimo Mitolo. Electrical Safety Considerations in Large-Scale Electric Vehicle Charging Stations[J]. IEEE Transactions on Industry Applications ,2019:6603
3. Mohsen Ghavami, Somasundaram Essakiappan, Chanan Singh. A framework for reliability evaluation of electric vehicle charging stations. IEEE Power and Energy Society General Meeting,2016:1-5
4. Fabio Freschi, Massimo Mitolo, Riccardo Tommasini. Electrical Safety of Electric Vehicles[C]//2017 IEEE/IAS 53rd Industrial and Commercial Power Systems Technical Conference. IEEE,2017(24):58-63.
5. Claus-Christian Carbon n, Fabian Gebauer. Data and material of the Safe-Range-Inventory: An assistance tool helping to improve the charging infrastructure for electric vehicles[J]. Data in Brief,2017(14)
6. Changhao Piao, Zhi Huang,Ling Su,Sheng Lu. Research on Outlier Detection Algorithm for Evaluation of Battery System Safety[R]. Advances in Mechanical Engineering,2015.
7. Wei Li Juner Zhu Yong Xia ,Maysam B. Gorji,Tomasz Wierzbicki. Data-Driven Safety Envelope of Lithium-Ion Batteries for Electric Vehicles[J]. Joule,2019(3):2703-2715.
8. Qian Li Jun, Zhao Ming Yu, Zhang Wei guo. A design method of electric vehicle charging safety early warning model[J].Power System and Clean Energy,2016,32(12):114-119
9. Zhang Jing ,Shi Wei ,Zhang Yuan Xing, Zhang Jing ,Li Tao Yong. Safety factors and dynamic early warning during the charging process of electric vehicles[J].Power Technology,2019,43(05):863-868
10. He Chun Chen Zhuo Feng Jin Tao, Yin Qing wei ,Li Xiang .Analysis and Solution of Electric Vehicle Charging Safety[J].Power Supply,2017,34(01):12-18

Research on Safety Evaluation Method of Integrated Optical Storage and Charging Station

Jing Zhang[1,*], Junguo Jia[2], Hui Huang[3], Yi Long[3], Taoyong Li[1], Linlin Sun[4], Hao Sun[2]

[1]Beijing Electric Vehicle Charging/Battery Swap Engineering and Technology Research Center, China Electric Power Research Institute, 100192, Beijing, China
[2]SGCC Smart Energy and Electric Transportation Technology Innovation Center (SuZhou) Co., Ltd, 215000, Suzhou, China
[3]Marketing Service Center, State Grid Chongqing Electric Power Company, 400014, Chongqing, China
[4]EV Charging and Travel Department, Corporation of State Grid Electric Car service, 100053, Beijing, China

Abstract. This paper studies the correlation between charging process performance indicators and charging safety of Solar-Energy storage-Charge station, analyses the influence of environmental factors, technical factors, design factors, management factors and user factors on charging process safety of energy stations. The projection pursuit algorithm is used to evaluate the influence degree of each parameter on the safety of charging process; through the establishment of charging safety evaluation system, the safety risks of battery damage and even fire caused by excessive charging current and high battery temperature are identified. The monitoring parameters corresponding to the charging safety state are determined. According to the different battery types adopted by different vehicle models, the corresponding charging mode is determined. According to the monitoring results of key data of the core equipment in the charging process, such as charging current, charging voltage, battery temperature, etc., the charging strategies of different vehicle models and different battery types are analysed, which provides reference for the safe operation of charging process.

1 Introduction

In the aspect of charging safety protection, the main body of the research focuses on the charging equipment itself and the power battery. Safety accidents such as spontaneous combustion and explosion of power battery caused by battery failure; safety accidents caused by battery overcharge caused by communication fault between battery management system and charger; safety accidents caused by abnormal conditions such as over-current, short circuit and leakage of charger or failure to effectively deal with these conditions. In order to improve the electrical safety of charging equipment and the protection ability of charging battery, it is necessary to put forward a quantitative safety evaluation method of charging process, which aims to contribute to the electrical safety of electric vehicle charging equipment and guarantee the charging effect [1-3].

Recently, some scholars study and select the rating index system according to different types of charging infrastructure. Firstly, analytic hierarchy process (AHP) [4-5] is used to determine the comprehensive weight of each index, and then fuzzy comprehensive evaluation is applied to score the performance of charging infrastructure, and the evaluation value of charging infrastructure health status is obtained. [6] Different from the above, in this paper, the projection pursuit method [7-8] is applied to establish the relationship between the operating parameters of the main equipment in the Solar-Energy storage-Charge station [9-10] (referred to as the "energy station" in the following) and the charging safety, and a projection pursuit classification model based on real coded accelerating genetic algorithm is established to evaluate and classify the charging process safety.

2 System architecture

Before studying the charging safety evaluation method, it is necessary to clarify which factors will lead to safety problems in the charging process. In this paper, such problems are divided into five categories: environmental factors, technical factors, design factors, management factors and user factors, which are shown in Figure 1.

* Jing Zhang: zhangjing1@epri.sgcc.com.cn

https://doi.org/10.1051/e3sconf/202123702014

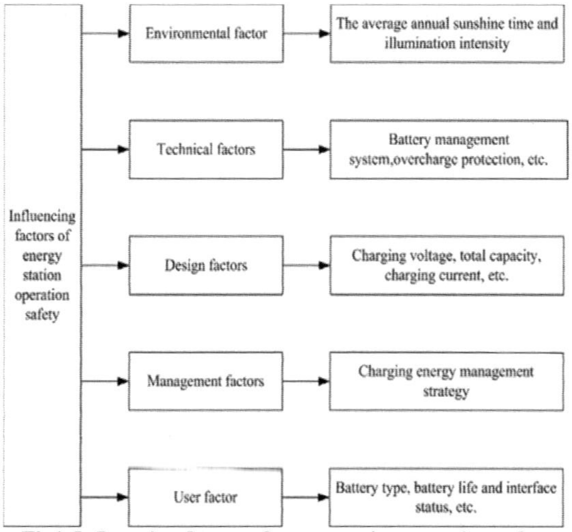

Fig1. Influencing factors of energy station operation safety

This paper introduces a new type of intelligent charging station, which can realize the optimal configuration of energy, classify and identify the charging vehicles. The station can realize the optimal configuration of energy, equipped with a variety of intelligent charging strategies for different batteries, and can achieve efficient utilization of energy. Firstly, the structure of the intelligent energy station is introduced, as shown in Figure 2.

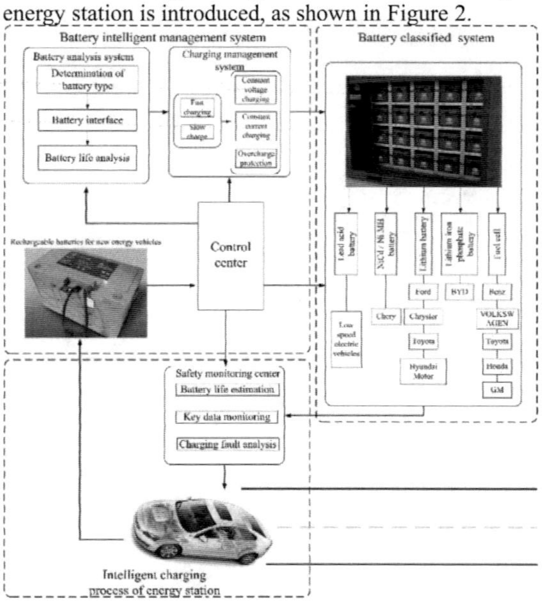

Fig2. Structure diagram of intelligent charging station

As is shown in Figure 2, the safety monitoring center is the core part of the whole power station. The center is responsible for monitoring the key operation data of a variety of core equipment and early warning of possible safety accidents. According to Figure 1, the battery overcharge accident caused by the communication failure between the battery management system and the charger can be classified as technical factors; the abnormal conditions such as over-current, short circuit and leakage of the charger itself can be classified as technical factors; the decline of charging capacity caused by insufficient load capacity of the charger can be attributed to design factors; the fault of battery itself can be classified into user factors, etc.

Due to the existence of the above security risks, the energy station needs to monitor the charging safety during the charging process. The specific safety monitoring system framework is shown in Figure 3.

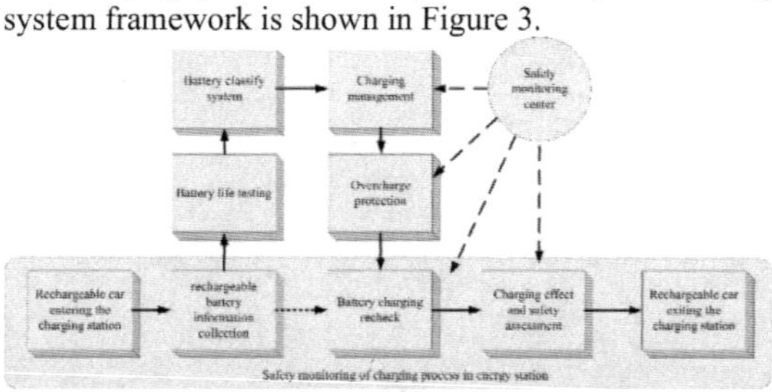

Fig3. Safety monitoring of charging process in energy station

As can be seen from Figure 3, the system will carry out safety monitoring on the whole process of charging vehicles from entering to leaving the station. Such as identifying battery types, detecting battery life, to charging strategy selection, charging voltage monitoring, current monitoring, overcharge protection, et al. Meanwhile, a large number of key data of core equipment will be obtained. Especially important, these key data will be used in the research of safety evaluation methods.

3 Algorithm selection

As mentioned in the previous article, there are many factors that affect the charging safety. However, which factors are the main ones and whether these factors can be evaluated by a mathematical model, so as to identify the potential charging safety hazards as soon as possible is a crucial issue. The appropriate algorithm selection can simplify the problem and find the core parameter anomalies quickly and effectively. In this paper, the projection pursuit algorithm is used to evaluate the influence degree of each parameter on the safety of charging process.

Project pursuit is an exploratory data analysis method directly driven by sample data. In this method, the high-dimensional data is projected into the low-dimensional subspace through some combination. For the projected configuration, the projection index function is used to describe the possibility of the projection exposing some sort structure of the original system, and the projection value that makes the projection index function reach the optimal (that is, to reflect the high-dimensional data structure or characteristics as accurately as possible) Then, according to the projection value to analyse the classification structure characteristics of high-dimensional data (such as projection pursuit clustering evaluation model), or according to the scatter diagram between the projection value and the actual output value of the research system, an appropriate mathematical model is constructed to simulate the output of the system (such as projection pursuit grade evaluation model). The projection pursuit grade evaluation model will be applied in this paper.

The application process of project pursuit method is shown in the Figure 4.

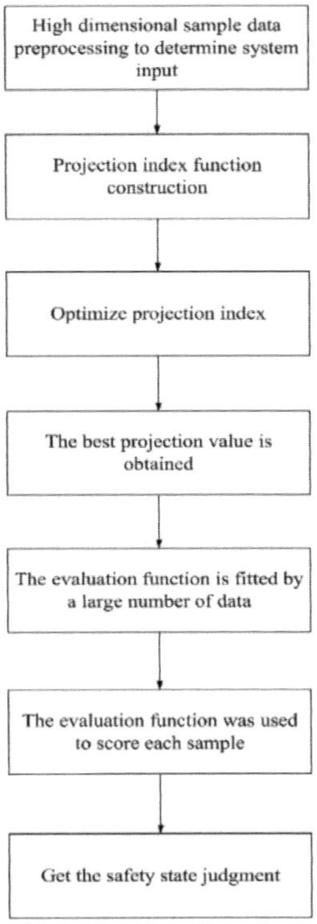

Fig4. Algorithm flow chart

4 Mathematical model

Before establishing the mathematical model, in order to find the relationship between the key parameters and charging safety, Firstly, the parameters should be normalized to integrate the data into (0, 1) through a large number of basic data:

$$a_i = \frac{a_t}{\sum a_{ti}} \tag{1}$$

Where the a_t represents monitoring data, a_{0i} represents optimal data, for example, a_t represents actual charging current, a_{0i} represents theoretical safe charging current. Similarly, the monitoring information such as charging voltage and battery temperature can also be processed according to this method. According to this, every car which is charged in this station will have a vector:

$$\vec{A_i} = \left\{a_{i,1}, a_{i,2}, \ldots a_{i,j}\right\} \tag{2}$$

Where *Ai* represents all the charging monitoring data of the ith vehicle.

Aiming at the charging safety detection problem of this project, various detection parameters such as battery temperature, charging voltage, charging current, battery life and other parameters can be constructed into multi-dimensional space. Different variable parameter values constitute different high-dimensional vectors. If these high-dimensional vectors are projected into low-dimensional space, different projection rules will get different low-dimensional point sets. Through optimization algorithm, the shadow index function is maximized to obtain the projection direction which can reflect the high-dimensional set. The process of finding the maximum projection function is the process of global optimization. After finding the best projection value by using genetic algorithm. The specific solution flow is shown in the Figure 5.

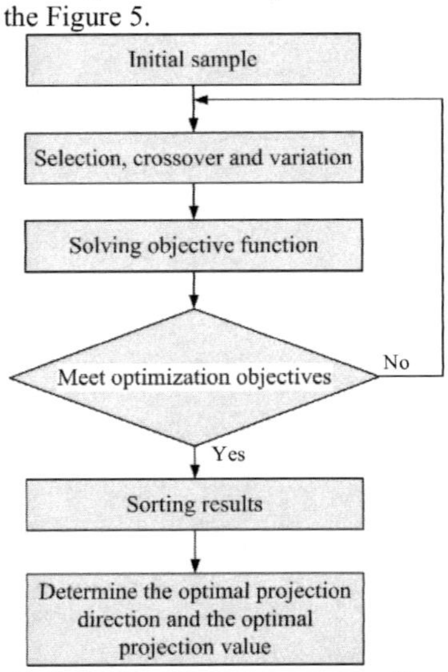

Fig5. Solution flow of genetic algorithm

As mentioned above, the contribution rate of the above parameters to the optimal projection can be calculated, which can be written as follows:

$$\eta_{s(m),i} = \frac{a_i}{\sum_{i=1}^{n} a_i} \tag{3}$$

Where, $s(m)$ represents the best projection value and represents each parameter. It can be found:

$$\sum_{i=1}^{n} a_i = s(m) \tag{4}$$

Through formula (3), the contribution rate of each parameter in the optimal projection can be obtained, that is, the weight vector of each parameter in the evaluation of charging safety can be obtained, which can be written as formula (5):

$$\left\{\eta_{s(m)}\right\} = \left\{\eta_1, \eta_2, \ldots \eta_n\right\} \tag{5}$$

When the best projection value is obtained under various working conditions, the scatter point set of charging state will be obtained, At the same time, we can get the safety evaluation grade according to the value of the best projection direction, and the best fitting function of scatter point can be found to establish the evaluation model of charging safety state.

$$D = k(a_1, a_2, \cdots, a_n) \tag{6}$$

Where "$a_1, a_2, \cdots, a_n$" represents the key parameters, such as battery temperature, charging voltage, charging current, etc., which parameters should be determined according to the data provided by the safety monitoring center. The charging process can be evaluated by publicity

(6), and finally the evaluation matrix can be obtained as follow:

$$\boldsymbol{D}=\begin{bmatrix} k(a_{11},a_{12},\cdots,a_{1n}) \\ k(a_{21},a_{22},\cdots,a_{2n}) \\ \vdots \\ k(a_{i1},a_{i2},\cdots,a_{in}) \end{bmatrix}=\begin{bmatrix} D_1 \\ D_2 \\ \vdots \\ D_i \end{bmatrix} \quad (7)$$

After comparing the data in the matrix with the best projection value, the safety evaluation grade can be obtained. For example, we can define three levels of safety, insecurity and danger according to the deviation between each data and the best projection value. Which is shown in Table 1.

Table1. Setting Word's margins.

Sample	Data range		
value	0-0.6	0.6-0.8	0.8-1
Security status	danger	insecurity	safety

When the dangerous state is found, the specific analysis is caused by the abnormal parameter, so as to identify all kinds of safety hazards early.

When the score is within the insecurity range, the system will find the abnormal parameters according to the fitting function. If it belongs to the category of charging station (such as charging current and charging voltage), the controller will inform the technician to adjust. If it belongs to user factors (such as battery failure), the controller will also inform the user. And when the charging process is in a dangerous state, the charging process will stop immediately and give an alarm. According to the fitting function, the abnormal parameters will be quickly located and the maintenance will be prompted. The safety hazard identification and implementation work-flow of safety assessment system is shown in Figure 6.

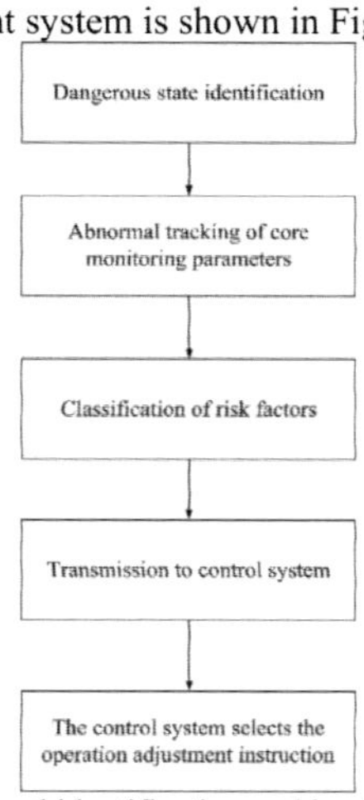

Fig6. Safety hazard identification and implementation flow chart

Randomly select 7 charging stations in a province, and obtain the corresponding projection values according to the above method. As shown in Figure 7. Corresponding to Table 1, it can be seen that charging station 2 is safe, charging stations 3 and 5 are unsafe, and the remaining charging stations are dangerous.

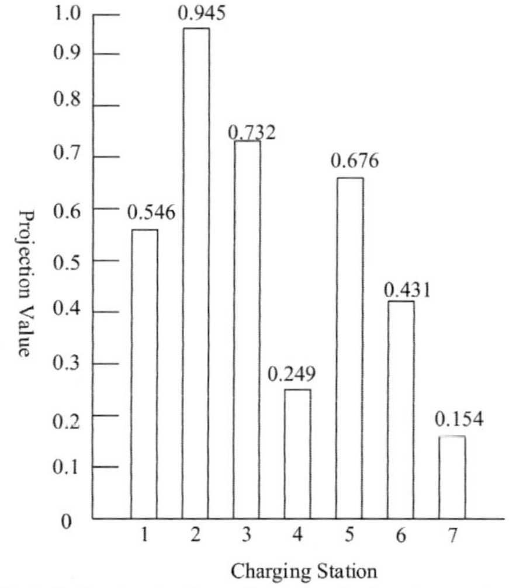

Fig7. Safety projection value of each charging station

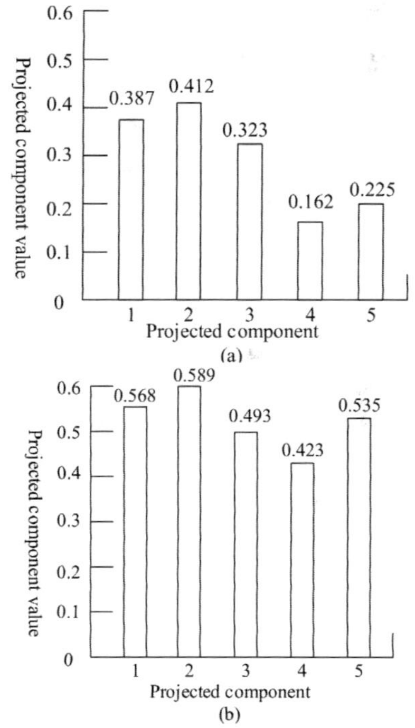

Fig8. Projected component of charging station

The projection component values of charging station 1 and 2 are shown in Fig. 8. The abscissa is the factors that affect the safe operation of charging station in Figure 1. The value of each component represents the degree of influence on safe operation. Through comparison, we can see that the safety evaluation model established in this paper is effective. After comparison, it can be seen that the effectiveness of the safety evaluation model established in this paper is.

https://doi.org/10.1051/e3sconf/202123702014

5 Conclusion

In this paper, in order to study the correlation between charging process performance index and charging safety of energy station, environmental factors, technical factors, design factors, management factors and user factors are set as the factors that may affect the charging process safety of energy stations. At the same time, the security monitoring system framework of charging station is introduced, and the important monitoring parameters of each core equipment are used to establish the mathematical model of charging process safety. Through the establishment of charging safety evaluation system, the safety risks of battery damage and even fire caused by excessive charging current and high battery temperature are identified. It provides a reference for safe operation of charging process.

Acknowledgments

Supported State Grid Corporation of China Science and Technology Project Funding (Research and application of key technologies for flexible control and active safety of electric vehicle energy Station based on high-power IGBT: SGCQKH00JSJS2000007).

References

1. X. M. Long, J. Yang, F. Z. Wu, X. P. Zhan, Y. J. Lin, J. Xu. "Prediction of electric vehicle charging load considering interaction between road network and power grid and user's psychology", Automation of Electric Power Systems, 2020, vol 44(14), pp 86-93.
2. S. C. Yan, Reaserch on operation of Solar-Battery-Charge DC micro grid system, Beijing: North University of Technology, 2018.
3. H. Li, Z. B. Du, L. D. Chen, L. Guan, B. R. Zhou. "Trip simulation based chargeing load forecasting model and vehicle-to-grid evaluation of electric vehicles", Automation of Electric Power Systems, 2019, vol 43(21), pp. 88-96.
4. D. H. Zhang, Y. W. Shi, D. H. Feng, Y. Zhou, H. J. Li. "A cooperative planning method of urban electric vehicle charging station group with multi-type charging facilities", Electrical Measurement & Instrumentation.2020,https://kns.cnki.net/kcms/detail/23.1202.TH.20200722.1141.022.html.
5. W. X. Wang, Z. Sun, M. Y. Pan, B. Q. Zhang, Z. W. Li, L. Ye. "Information security risk assceement method of electric vehicle charging pile based on fuzzy analytic hierarchy process", Electric Power,2020,http://kns.cnki.net/kcms/detail/11.3265.TM.20200401.1919.006.html
6. W. Chine, A. Mellit, V. Lughi, et al. "A novel fault diagnosis technique for photovoltaic systems based on artificial neural networks", Renewable Energy, 2016, vol 90, pp. 501-512.
7. H. Yang, P. Liu, Y. X. Li, W. C. Wang, W. H. Huang. "Disaster assessment and analysis of natural gas pipeline based on projection pursuit model", Oil-Gas Field Surface Engineering, 2019, vol 38 (s1): 111-115.
8. Z. Jin, B. B. Risk, D. S. Matteson. "The Power Transformer Operating Condition Evaluation Based on the Genetic Projection Pursuit Model", Statistical Analysis and Data Mining: The ASA Data Science Journal, 2019, vol 12(3), pp. 141-156.
9. W. L. Wu, A. J. Wu, W. Pan, Y. T. Yan, J. Xu. "Power supply capability evaluation of distribution network with PV-Energy-Storage charging station", Power & Energy, 2019, vol 40(2): 232-237.
10. R. X. Li, Z. C. Hu. "Stochastic optimization strategy for daily operation of electric bus charging station with pv and energy storage", Power System Technology, 2017, vol 41(12): 3772-3780.

The operating mode of electric vehicles and its impact on renewable energy consumption

Li Xiaoyu[1], Shang Shanshan [1,*], Liu Dunnan[1], Zou Mengjiao[1],Zhang Tingting[1],Wang Lingxiang[1],Zhang Yue [1], Liu Mingguang[1]

[1] North China Electric Power University, School of economics and management, Beijing, 102206, China

Abstract. Nowadays, under the dual pressure of fossil energy shortage and ecological environmental pollution, new energy vehicles have flourished in our country by virtue of their clean and low noise advantages. But correspondingly, the grid connection of electric vehicles will also bring many problems. Therefore, this article attempts to explore the operating mode of electric vehicles and its impact on the consumption of renewable energy, so as to provide reasonable planning of charging facilities and improve the top-level design of the grid.

1 Introduction

In recent years, with the increasing shortage of fossil energy and the pollution of ecological environment, people's awareness of environmental protection is gradually enhancing. Because new energy vehicles do not rely on traditional oil and low pollution, noise-free, market share generally climbed. It is estimated that the number of electric vehicles will exceed 25 million by 2025, 80 million by 2030 and 160 million by 2035,which means that electric vehicle will be one of the largest loads in the future power grid, and its scale effect makes the grid face many challenges.

On the one hand, the charging infrastructure of electric vehicles is connected to the power system on a large scale, which affects the distribution network transmission and distribution, power supply, power quality and load aggregation scheduling. The randomness of electric vehicle charging behavior will not only lead to the increase of power grid load, but also make the grid load gather in a certain period of time for charging, which leads to a series of problems, such as the sharp increase of power demand, the mismatch of power supply and demand, the decline of power quality, and the shortage of power supply[1].

On the other hand, the new round of energy reform focuses on the transformation of traditional fossil energy to clean energy. At present, with the rapid development of new energy power generation technology, wind and solar power generation is moving towards a new situation of large-scale development and utilization. However, the randomness and volatility of renewable energy integration will lead to the increase of peak valley difference of grid load. With the increase of renewable energy penetration in distribution network, it will cause great pressure on the safe and stable operation of grid voltage, and the difficulty of voltage control of distribution network will also increase sharply.

Although the wide access of electric vehicle charging and renewable energy has randomness, at the same time, electric vehicles are also energy storage units located at the end of the grid, which, as distributed energy storage, can provide power systems with considerable flexibility resources, thus effectively increasing the grid's ability to eliminate renewable energy sources.

To sum up, this paper mainly studies the operation mode of electric bus and its impact on renewable energy. Through the analysis of charging characteristics and actual data of electric bus, it guides the reasonable planning of charging facilities, solves the impact on the power grid when it is connected to the power grid, enhances the flexibility of the power grid, and improves the absorption capacity of renewable energy.

2 Operation mode of electric bus charging station

According to the charging types, electric buses at home and abroad can be divided into conductive (wired) charging and wireless charging, and conductive charging can be divided into vehicle charging and battery changing mode. Wireless charging mode refers to charging the vehicle power battery with radio energy transmission technology. At present, cable charging is the main charging mode of electric buses in China, which mainly includes four modes: online charging mode, centralized charging mode, slow charging mode and fast charging mode. The operation scale of online charging mode is small, and it mainly uses the urban trolley bus to charge for a certain distance, and uses the mode of on-board battery power supply for a certain distance. For the centralized power exchange mode, the charging station provides a special power exchange channel, so that each vehicle can change power in a short period of time, which

* Corresponding author: sss1771415@163.com

can ensure the high efficiency of the electric bus. However, the disadvantages are relatively obvious. The station covers a large area and requires a large number of backup battery. Frequent replacement will cause great loss to the bus frame, and the electrode plug has problems such as spark and wear. For slow charging mode, the power supply of electric buses is mainly at night, because of the large capacity of electric bus on-board power battery, the investment cost is higher.

At present, fast charging is a common charging mode adopted by most electric bus charging stations in China. In this charging mode, the electric bus can make use of it to realize fast power replenishment during the parking period.

Departure scheduling mainly includes train operation plan and real-time scheduling. Among them, train operation plan refers to arranging departure schedule according to line and vehicle conditions, and arranging various driving plans based on passenger flow parameters and speed parameters in different periods. The real-time dispatching requires that multiple lines can be switched, the departure mode can be edited (planned departure, manual departure, flow departure), and the plan adjustment (personnel, vehicle increase number, etc.) can be realized quickly.

3 Charging load characteristics of electric buses

Take Xiaoying charging station as an example. There are 25 sets of 360kW high-power DC chargers and 5 sets of 90kW DC chargers in the charging station, which can charge 30 buses at the same time. The charging station serves 192 electric buses. The power batteries for electric vehicles come from three manufacturers: Mengculi, Weihong and Yinlong. According to the statistics for the whole year of 2017, the total annual mileage of the three routes of buses is about 3 million kilometers, the total annual charging capacity of the charging station reaches 3.2 million kWh, the total number of charging is about 86,000 times, and the average daily charging is 12,000 kWh. , The average number of recharges per day is about 320.

The total power consumption of a bus is shown in the figure below. The average monthly electricity consumption is about 36100 kWh, and the annual total power consumption is 434000 kWh. Among them, the electricity consumption in January, February, November and December is generally higher than that in other months, which is mainly due to the large energy consumption of air conditioning heating in winter.

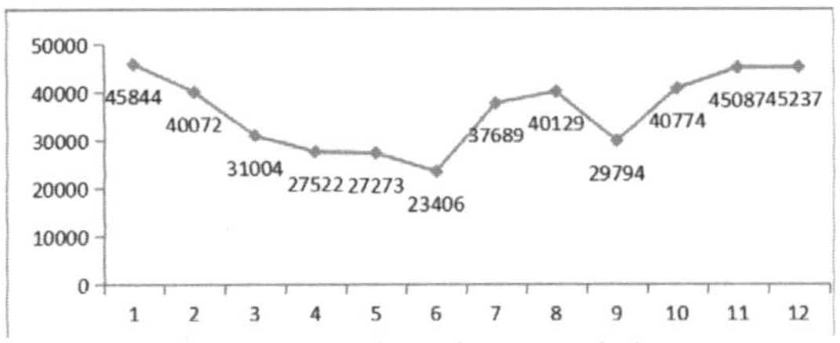

Fig1. Monthly electricity curve of a bus

According to actual statistics, the total annual frequency of a certain bus is 21,287. Due to the short mileage of the whole journey, it is generally charged once for two laps, and occasionally for three laps. The average value of the initial SOC when the bus is charging is 0.68, the actual distribution interval is [0.34,0.82], the average actual charging time is 10-11 minutes, and the distribution interval is [6,36] minutes.

The figure below shows the average energy consumption per unit mileage of a certain electric bus in each month. It can be seen that the value of energy consumption per unit mileage fluctuates around 1 kilowatt-hour per kilometer. Overall, the energy consumption in winter is slightly higher.

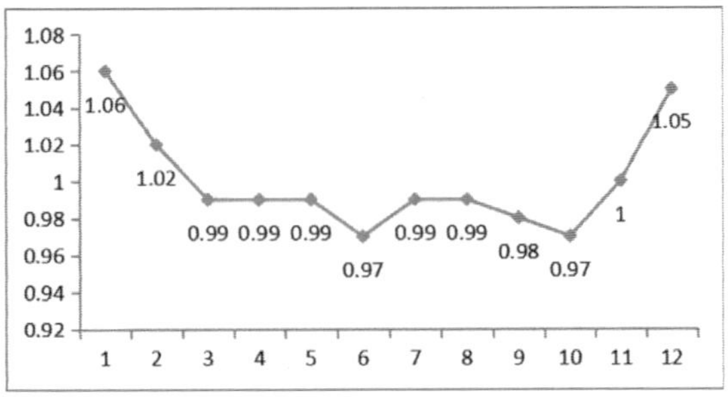

Fig2. Energy consumption curve per unit mileage of a bus

Then analyze the charging load law of the bus charging station.

(1) Due to the high-power charging method, the charging power of each vehicle is large and the charging duration is short. The charging power fluctuates with the number of vehicles being charged at the same time, and the fluctuations are relatively severe in general.

(2) Even on the same working day or weekend in the same month, the charging load varies greatly, indicating that the randomness of the charging load is also strong due to the randomness of the charging time of the bus arrival.

(3) There are short-term spikes in the charging load. If orderly charging control is adopted, the peak load can be reduced, thus reducing the distribution capacity required and the capacity cost that may be paid.

4 Electric vehicle charging mode

According to the purpose, electric vehicles are divided into private cars, business cars and taxis. According to the charging mode, electric vehicles are divided into fast charging mode and slow charging mode. For private cars and most business vehicles that do not have strict requirements on charging time, their driving routes and charging locations are relatively fixed, and they are suitable for slow charging in scattered charging piles.

For taxis and commercial vehicles that require fast charging due to business needs, when the electricity quantity is lower than the safety threshold, they should immediately go to the nearest charging station for quick charging.

The basis of OD analysis includes road traffic information, vehicle distribution information and OD matrix information. For a specific traffic area, an OD matrix can be generated to describe the travel rules of large-scale electric vehicles. The OD matrix B consists of

24 sub-matrices $B_{m\times m}^{(t,t+1)}$, where $0 \leq t \leq 23$, and m represents the number of nodes in the area. The matrix element b_{ij} ($1 \leq i \leq m$, $1 \leq j \leq m$) of each sub-matrix represents the number of cars from node i to node j during the period from t to t+1.To facilitate the simulation, $B_{m\times m}^{(t,t+1)}$ will be transformed into $C_{m\times m}^{(t,t+1)}$, where the element c_{ij} of $C_{m\times m}^{(t,t+1)}$ represents the probability of the electric vehicle from node i to node j during the period t to t+1.For the electric vehicles in the "vehicle network road" system, the OD matrix can be used to describe the travel probability of the vehicle in different periods, so as to simulate the driving trajectory of the electric vehicle.

$$c_{ij} = b_{ij} / \sum_{j=1}^{m} b_{ij} (1 \leq i \leq m) \qquad (1)$$

The OD matrix provides the driving probability from node i to j, but usually there are multiple roads to choose from between the two nodes. In the simulation, it is assumed that the drivers choose the shortest route during the iterative process of OD analysis. The shortest path l_{ij} between nodes i and j can be obtained by Floyd algorithm. Therefore, the travel time Δt from node i to node j can be derived from the formula (2), where $V_{ij}(t)$ is derived from the speed-flow formula according to the road grade.

$$\Delta t = l_{ij} / V_{ij}(t) \qquad (2)$$

For electric vehicles using fast charging, the time and space characteristics of fast charging load can be obtained by the following steps:

(1) Determine the basic parameters of electric vehicles. Four kind of electric vehicles are selected from the EV MERGE database as taxi simulation samples. The basic parameters are shown in Table 1.

Table1. Basic parameters of four kinds of electric vehicles

Manufacturer	Type	Battery	Battery Capacity	Electricity Consumption
Audi	R8	Lithium battery	45	0.1500
BMW	Mini-E	Lithium battery	35	0.2096
BYD	E6	Lithium battery	48	0.1846
BYD	E8	Lithium battery	72	0.1800

(2) The initial driving time t_s , the initial power SOC_0 , and the initial position L_t are generated.

(3) OD analysis is used to simulate the driving path of electric vehicles. In order to maintain the battery performance, the battery capacity should not be less than 20% of the battery capacity. When the termination condition (3) is satisfied, OD analysis is terminated and the position of electric vehicle at the termination time is recorded.

$$SOC_t \leq 0.2 \qquad (3)$$

(4) Determine the space-time information of fast charging. Assuming that the distance from the electric vehicle to the nearest distribution network node k is d_{nr} the remaining power during charging and the charging start time t_{sc} can be obtained by (4) and (5) respectively.

$$SOC_c = 0.2 - d_{nr} \cdot \Delta SOC \qquad (4)$$

$$t_{sc} = t - d_{nr} / v(t) \qquad (5)$$

(5) Due to differences in the parameters of electric vehicles, the number of fast charging may be more than once in a day. Repeat step (3) to step (4) to get the fast charging time and space information of the same electric car in one day.

For n electric vehicles, the above steps are repeated n times, and the charging time t_{sc} of each vehicle and the node k of the distribution network are recorded, then the fast charging load $P_{kf}(t)$ of node k at time t can be obtained by formula (6) , where m (m<n) is the number of electric vehicles that are quickly charged at node k at time t.

$$P_{kf}(t) = \sum_{i=1}^{m} P_{kf}^{i}(t), t = 1,2,\cdots,24 \qquad (6)$$

For electric vehicles using slow charging, the time and space characteristics of slow charging load can be obtained by the following steps:

(1) Generate initial power SOC_0

(2) Get the travel time t_s and return time t_f .

(3) Simulate the travel routes of electric vehicles during t_s and t_f

(4) Determine the remaining power SOC_t and remaining range D (SOC_t)

(5) Determine the starting charging time t_{sc}

Such as (7) formula, when the remaining power is less than the threshold or not enough to support the next driving section, it will be charged immediately. For private cars, D(n) is the shortest path from residence to work; for commercial vehicles, D(n) is the shortest path from the current location to the next destination.

$$SOC_t \leq 0.2 \text{ or } D(SOC_t) < D(n) \qquad (7)$$

When the formula (7) is not satisfied, the electric vehicle adopts the intelligent charging method after returning to the initial point O, and the charging time t_{sc} meets the probability distribution of formula (8). Among them, μ is the average charging start time of residents, and σ is the distribution coefficient.

$$f(t_{sc}, \mu, \sigma) = \frac{1}{\sqrt{2\pi\sigma^2}} e^{-(t_{sc}-\mu)^2/2\sigma^2} \qquad (8)$$

(6) Calculate the slow charge load. From equation (9), the distribution of slow charging load $P_{ks}(t)$ at each node can be obtained, where r(r<n) is the number of electric vehicles that are quickly charged at node k at time t.

$$P_{ks}(t) = \sum_{i=1}^{r} P_{ks}^{i}(t), t = 1,2,\cdots,24 \qquad (9)$$

For node k, the total charging load in any period of time can be expressed by equation (10).

$$P_k(t) = P_{kf}(t) + P_{ks}(t), t = 1,2,\cdots,24 \qquad (10)$$

 https://doi.org/10.1051/e3sconf/202123702015

5 The impact of electric vehicles on renewable energy

Because renewable energy has the characteristics of randomness, intermittent, when it is integrated into a large power grid, it is easy to cause problems such as grid voltage deviation, voltage fluctuation and flicker, frequency deviation, harmonic, and DC injection. Firstly, the orderly access of electric vehicles will improve the ability of the power system to absorb fluctuant renewable energy, reduce the amount of wind and photovoltaic abandonment, and reduce the cost of power generation and the level of pollutants and carbon emissions; Secondly, the orderly guidance of the charging load of electric vehicles will reduce the load peak and off-peak difference of the power system, improve the operating efficiency of generator , and save power generation capacity investment and fuel costs; Thirdly, the flexible charging and discharging of electric vehicles can be used as ancillary service resources in power system, reducing the cost of auxiliary services. Fourthly, effective management of electric vehicle charging will avoid or delay transmission and distribution investment costs to a certain extent.

If electric vehicle charging is not managed or guided, the charging load of electric vehicles will not help the grid-connected consumption of renewable energy, but will increase the peak regulating pressure of the power system, which will intensify the problem of renewable energy curtailment. If grid-connected methods such as orderly charging, two-way interaction between vehicles and the grid (V2G), and decommissioned battery energy storage are adopted, the energy storage potential of electric vehicles can be fully released, and the power system's ability to absorb renewable energy will be significantly improved.

At present, the trading modes of electric vehicles, distributed generation, energy storage, etc. participating in the power market at home and abroad can be summarized as the following several figures.

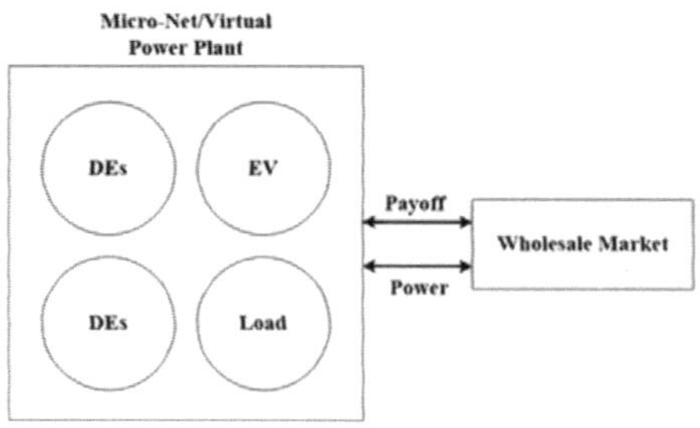

Fig3. Form a micro-grid or virtual power plant

Fig4. Trading with electricity distribution utilities

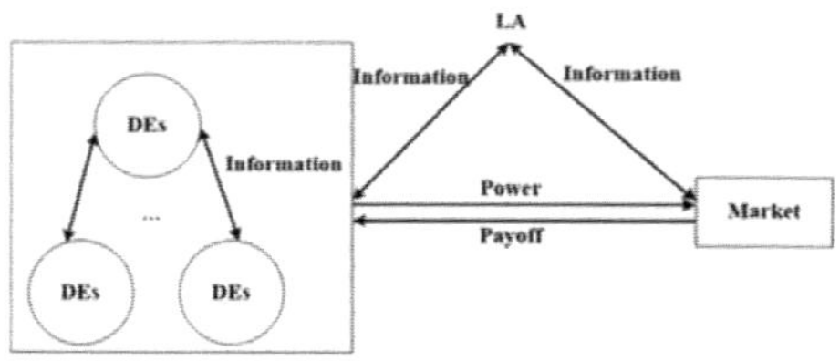

Fig5. Transaction based on information interaction

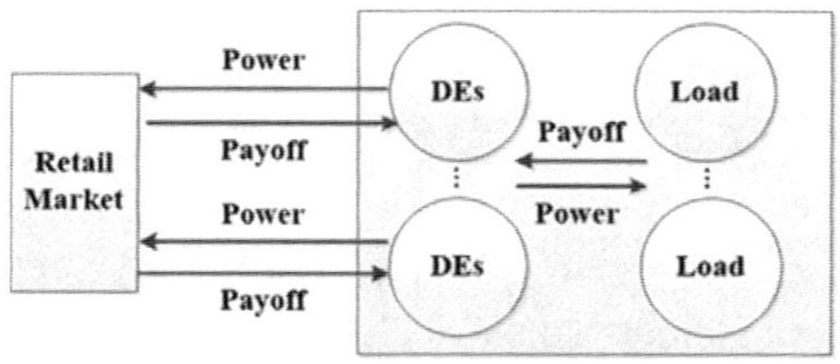

Fig6. Direct transactions with users

Fig.3.shows that electric vehicles, distributed energy, and energy storage constitute small entities such as virtual power plant agents or micro-grids to participate in day-ahead trading energy market through demand-side responses. Considering the impact of price fluctuations when electric vehicles participate in the day-ahead market and their own charging load fluctuations on the power purchase strategy, there are currently studies using conditional value-at-risk (CVaR) method to participate in day-ahead bidding in the energy market[2], and a two-stage stochastic optimization method is considered Day-ahead and real-time market participation strategies to minimize their own electricity purchase costs[3]. Besides, robust optimization method is used to deal with the uncertainty of market price to optimize the charging strategy of electric vehicles and minimize their own costs[4]. Foreign literature has carried out research on bidding strategy of electric vehicles participating in frequency modulation market. Electric vehicles are optimized by virtual power plants or agents in two time scales of day ahead and real-time[5].

In addition to participating in the wholesale market through virtual power plants, electric vehicles can also participate in the ancillary service market. The power balance adjustment of the power grid generally adopts the adjustment of unit output on the generating side, or the use of large-scale energy storage systems to provide peak and frequency modulation services. As a mobile energy storage device, electric vehicles can also participate in the ancillary service market to achieve the same effect. Taking into account the electric vehicles with load flexibility, the more suitable model for electric vehicles to participate in the market under the existing market mechanism is the medium and long-term market where agents participate in direct transactions with power generation companies. Or in the virtual power plant/micro-grid, participate in the spot market through demand-side response to promote the consumption of distributed new energy.

6 Conclusion

With the increasing number of electric vehicles in China,

it will be a challenging task to realize the orderly charging of electric vehicles. Fully exploiting the energy storage advantages of electric vehicles is of great strategic significance for power system transformation. Therefore, first of all, according to the charging characteristics and operation mode of electric vehicles, it is necessary to upgrade the distribution network, realize the technical adjustable and controllable, and gradually improve the top-level design of intelligent electric grid. Secondly, from the perspective of policy, we should build a competitive platform in the electricity market and give full play to the real-time charging and discharging pricing mechanism. Finally, the status of electric vehicles in the load resources should be clarified, its energy storage characteristics should be mobilized, and the natural advantages of V2G or the load aggregator agents should be involved through demand side response to improve the power grid's ability to absorb renewable energy.

References

1. Chao-Tsung Ma. Energies, 12(2019)
2. Z.c. Zhou. Journal of electric power science and technology,35,54-59(2020).
3. H.x.Guo, R.Gao, P.Yang. Power System Technology,43,2665-2674(2019).
4. B.b.Wang,Z.Qiu,X.h.Cong,Y.x.Zheng,S.m.Feng.Proceedings of the Chinese Society of Electrical Engineering,39,1-14(2020).
5. X.d.Zhang,H.j.Wang,G.Wu,M.Zeng,X.Zhou,Electric Power Automation Equipment,37,131-138(2017).

Optimal Configuration of Multi-type Energy Storage for Integrated Energy System Considering Multi-energy Trading With Load Substitution

Jun Wang[1,*], Wei Du[1], Dongmei Yang[1], Guoxin He[1], and Xiaochen Zhang[1]

[1]NARI Technology Co., Ltd., Nanjing 210032, Jiangsu Province, China

Abstract. How to consider the impact of load substitution on user-side participating in multi-energy trading on system operation when configuring multi-type energy storage (MES) is an urgent problem that needs to be solved to improve the economics and energy efficiency of the integrated energy systems (IES). This paper presents an optimal configuration method of MES considering multi-energy trading with load substitution based on the characteristics of different energy supply seasons. Firstly, the multi-energy trading framework and MES configuration principle are proposed based on the structure of the IES. Secondly, a MES optimal configuration model for IES considering multi-energy trading with load substitution is established, then the IES trading strategies and MES planning schemes are solved. Finally, the proposed method is verified by an example. The results show that the proposed method reduces the MES configuration cost by 8.9%. It can be seen that the MES configuration method proposed in this paper is helpful to repair the limitations of energy storage configuration research in IES and improve the economic benefits of the system.

1 Introduction

With the energy crisis and environmental pollution becoming more and more serious, the traditional fossil energy is increasingly depleted, and improving the efficiency of energy use, developing new energy sources and strengthening the comprehensive utilization of renewable energy are the inevitable choices for energy development. Integrated energy systems (IESs) can achieve energy interconnection, which is an important way to mitigate the energy crisis [1]. Energy storage systems as a key component of IES [2], with the ability to suppress load fluctuation to enhance renewable energy capacity, improve utilization of equipment, reduce energy costs [3]. However, with the diversification of user demand [4], the user-side integrated demand response resources are becoming more diverse. Load substitution can reflect the strong coupling relationship between electricity, gas and heat on the user side. Users can choose to use different energy sources to meet the same load demand, and when users change the demand for one or more energy sources, the demand for other energy sources will change accordingly. In addition, the differentiated production mode, supply system and consumption mode among different types of energy in the IES make the trading strategy and operation mode of multi-type energy storage (MES) including electric energy storage (EES), gas energy storage (GES) and heat energy storage (HES) more complex [5]. Therefore, it is an urgent problem that how IESs consider the multi-energy trading with load-substituting to configure the MES to improve the system economy and energy efficiency.

At present, a lot of research has been carried out on the optimal configuration of MES and the trading and operation of user side resources. In terms of configuration of MES, most of the existing researches are based on different benefit objectives to determine the equipment selection, equipment capacity and operation strategy of the system [6]. Reference [7] studied the optimal allocation of MES in regional IES considering multi-energy complementarity, but only discussed the economy and feasibility of HES configuration. Reference [8] established an optimal operation model of microgrid IES including MES, but only the configuration of MES capacity is considered, the selection of MES is not involved. An optimal allocation model of electricity-heat integrated energy storage in cogeneration microgrid including P2G was constructed in [9] aiming at the multi-objective such as economy, reliability and new energy consumption rate. It can be seen that most of the existing configuration objects focus on the energy system with electric and heat load. The optimal configuration of energy storage is carried out considering the uncertainty, economy and feasibility. The consideration of the diversity of MES selection and the trading strategy of system operators is lacking. In terms of the trading and operation with user side resources, reference [10] put forward the optimal trading strategy for the integrated energy suppliers including renewable energy to stabilize the fluctuation through demand side means. Reference [11-12] proposed targeted trading strategies for non-cooperative trading and joint trading among multi-agent markets. Reference [13]

* Corresponding author: wangjun18@sgepri.sgcc.com.cn

proposed an optimal market trading strategy of load aggregators based on comprehensive demand response considering market information and demand side resource scheduling potential. It can be seen that most of the existing studies on the trading and operation based on the market mechanism and the characteristics of load demand response considering the diversity of operation modes of energy conversion equipment in energy hubs. Few studies have considered the impact of demand side multi energy load substitution on the IES, especially the mutual substitution relationship between gas load and electric load in the system.

In view of the existing research problems above, this paper studies the optimal configuration method of MES for IES considering multi-energy trading with load substitution. The main innovations of this paper are as follows:

1) For the configuration of the MES, we consider load substitution to participate in multi-energy trading of IES operators, and put forward the optimization selection method of the MES considering the load substitution ability of user-side resources and the multi-energy trading strategy of system operators. It is beneficial to enhance the economy of the IES.

2) Based on the MES optimal configuration method of IES considering multi-energy trading with load substitution, the influence of user load and energy price on the MES configuration results is analysed.

The main structure of this paper is as follows, Section 2 constructs a multi-energy trading framework for IES, and expounds the principle of MES configuration. Section 3 establishes a MES optimal configuration model for IES considering load substitution to participate in multi-energy trading. Section 4 gives the solution flow of the model. Section Ⅴ analyses the effectiveness of the proposed method by example, and Section 5 summarizes the main findings of this paper.

2 Multi-energy trading framework and principle of MES configuration for IES

2.1 Basic structure of IES

The structure of IES has been elaborated in many literatures. Based on [14-15], the structure of IES in this paper is shown in Fig. 1. The IES include wind turbines (WT), combined heat and power (CHP), gas turbine (GT), power to gas (P2G), gas boiler (GB), electric boiler (EB), EES, GES and HES. All kinds of equipment supply, convert and store energy through electric supply bus (ESB), gas supply bus (GSB) and heat supply bus (HSB). In addition to supplying electrical, gas and heat load, it can also participate in energy trading through external energy supply center.

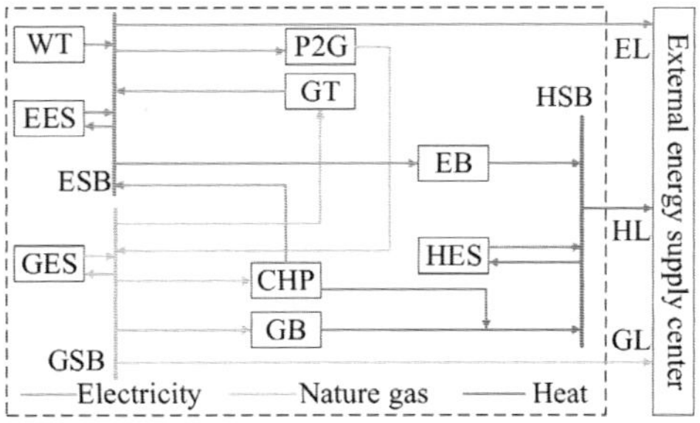

Fig1. IES structure.

2.2 Load substitution

Load substitution is reflected in the strong coupling between electricity, gas and heat. Users can choose to use different energy sources to meet the same load demand. When users change their demand for one or more energy sources, the demand for other energy sources may change [16]. This section considers the coupling between multi-energy sources and establishes the gas-electric and heat-electrical load substitution models, as shown in formulas (1) and (2):

$$\Delta P_{\mathrm{eg},t} = \varepsilon_{\mathrm{eg},t} P_{\mathrm{eg},t} \quad (1)$$

$$\Delta P_{\mathrm{eh},t} = \varepsilon_{\mathrm{eh},t} P_{\mathrm{eh},t} \quad (2)$$

Where, $\Delta P_{\mathrm{eg},t}$ and $\Delta P_{\mathrm{eh},t}$ are the changes of gas load and heat load produced by energy substitution in t period. $\varepsilon_{\mathrm{eg},t}$ and $\varepsilon_{\mathrm{eh},t}$ are electricity-gas and electricity-heat substitution coefficients in t period. $P_{\mathrm{eg},t}$ and $P_{\mathrm{eh},t}$ are the electricity load for substitution to gas and heat in t period.

2.3 Multi-energy trading framework

Based on the existing research, the IES trading framework is constructed as shown in Fig. 2. The system operators of the IES participate in the trading of the multi-energy market, and flexibly transfer the load substitute resources on the user side through the price signals of electric, natural gas and heat energy.

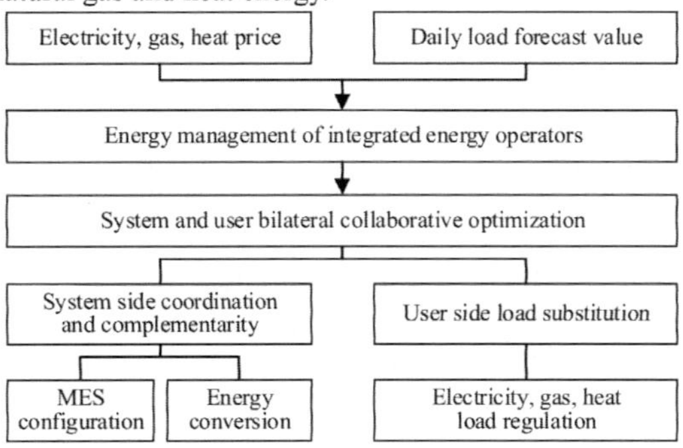

Fig2. IES trading framework.

According to the price signals of different energy markets, the following trading strategies are designed:

1) IES operators participate in the market for optimal decision-making considering load substitution of user-side resources.

2) Assuming that electricity and natural gas in IES can be sold to the outside, heat energy can only be consumed

internally.

3) Electricity trading mode: the electricity load is supplied by the WT first, if insufficient, the natural gas purchased can be converted into electricity energy through the CHP, or the electric energy purchased from the market or discharged through the EES. If there is electricity surplus, sell electricity, or to recharge the EES, or to supply P2G to convert to natural gas, or to supply EB to convert to heat.

4) Natural gas trading: to meet the demand of gas load, it may be purchased from the market, supplied by P2G or GES. The gas can be sold in excess of time, either to inflate the GES or to convert electricity and heat by CHP, or to convert heat through GB.

5) Heat trading: purchased from the market according to demand. Either use CHP to convert natural gas into heat energy, or through EB to convert electricity into heat energy to meet the heat load.

2.4 Principle of MES configuration

The principle of MES configuration in the IES in this paper is shown in Fig. 3. Because the system has different load characteristics in different seasons, the load curve and trading price of multi-energy is different in different seasons. Therefore, it is necessary to divide the scenarios to consider the characteristics of different seasons to optimize the configuration of MES. In this paper, four typical load scenarios are divided according to the seasons to form four typical daily load curves for the electrical, gas and heat loads in the system. The configuration period is the sum of four typical days multiplied by the number of days of the corresponding season. The configuration objects are EES, GES and HES. The multi-energy trading with load substitution of the system operator in section 2.3 is considered in the configuration.

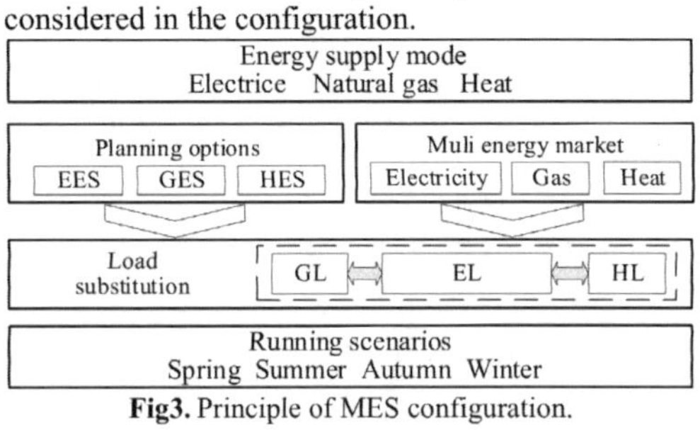

Fig3. Principle of MES configuration.

3 Optimal configuration model of MES for IES considering multi-energy trading with load substitution

3.1 Objective function

The optimal configuration model of MES aims at the minimum total investment operating cost of the IES. The objective functions include equipment investment cost, operation and maintenance cost, and multi-energy trading cost, as shown in formula (3).

$$\min C = C_{\text{inf}} + C_{\text{om}} + C_{\text{on}} \quad (3)$$

Where, C is the total investment and operation cost of the system. C_{inf} is the equipment investment cost. C_{om} is the operation and maintenance cost. C_{on} is the multi-energy trading cost.

The cost of equipment investment is shown in formula (4) and (5).

$$C_{\text{inf}} = \sum_{m}\sum_{s\in S}\frac{C_{s,m}I_{s,m}}{365T}D_m R_s \quad (4)$$

$$R_s = \frac{r(1+r)^{N_s}}{(1+r)^{N_s}-1} \quad (5)$$

Where, $C_{s,m}$ is the construction cost of sth equipment to be selected. s is the type number, indicating different capacity levels of the same equipment. S is the alternative set of the construction capacity of the equipment to be selected. $I_{s,m}$ is the 0-1 decision variable, indicating whether the equipment is invested or not. D_m is the number of days in the mth typical day. T is the number of typical day intervals. R_s is the investment recovery coefficient of sth equipment. N_s is the service life of equipment. r is the discount rate of 6.7%.

The operating and maintenance costs is shown in formula (6).

$$\begin{aligned} C_{\text{om}} = \sum_{m}\sum_{t=1}^{T}[\sum_{x\in X} c_x P_x(m,t) \\ + \sum_{s\in S} c_s \left|P_{\text{sc},s,t}(m,t) - P_{\text{sd},s,t}(m,t)\right|] \end{aligned} \quad (6)$$

Where, T is the interval number in a typical day. c_x is the unit power maintenance cost of built equipment. P_x is the output of existing equipment x in m scenario in t period. c_s is the unit power maintenance cost of the energy storage equipment s to be built. P_x is the output of the existing equipment x in the m scenario in t period. $P_{\text{sc},s,t}(m,t)$ and $P_{\text{sd},s,t}(m,t)$ represent the charging and discharging power of class s energy storage devices in t period under scenario m.

Multi-energy trading costs include the power cost of IES participating in the multi-energy market purchased from outside the system, as well as the heat and gas costs purchased, as shown in formula (7).

$$C_{\text{on}} = \sum_{m}[D_m \sum_{t=1}^{24}\sum_{y} c_{y,t} P_{m,y,t}\varphi_{m,y}] \quad (7)$$

Where, y is the type of energy. $c_{y,t}$ is the purchase price of energy y in period t. $P_{m,y,t}$ is the purchase amount of energy y in period t. The positive value is the purchase and the negative value is the sale. $\varphi_{m,y}$ is the energy price adjustment coefficient of energy y in m scenario.

3.2 Constraints

The constraints include power balance constraints of electrical, natural gas and heat, multi-energy trading constraints, MES operating constraints, CHP operating constraints, GB operating constraints, GT operating constraints, P2G operating constraints, and EB operating constraints. The specific model expression can be found in [17-18], which will not be repeated here.

3.3 Solving process

According to the MES optimal configuration model established in section 3.1 and 3.2, it can be seen that the optimization problem in this paper is a mixed integer nonlinear programming problem. Therefore, this paper takes the GAMS optimization software as the computing platform and calls the DICOPT solver to solve the problem. The specific solution process is as follows: first step, input wind power, load data and equipment parameters, energy price and other basic data. Considering the different energy supply season load and energy price characteristics, the specific application scenarios of IES planning are classified. The second step, for the energy trading problem of each scenario, consider the potential of user-side resources to participate in the trading and operation, and determine the optimal trading strategy. The third step, the multi-energy trading with load substitution participation is taken into account in the MES optimal configuration model of IES. The DICOPT solver is used to solve the problem on the GAMS platform, so as to determine the configuration scheme of MES and the optimal economic scheduling scheme. The solution flow is shown in Fig. 4.

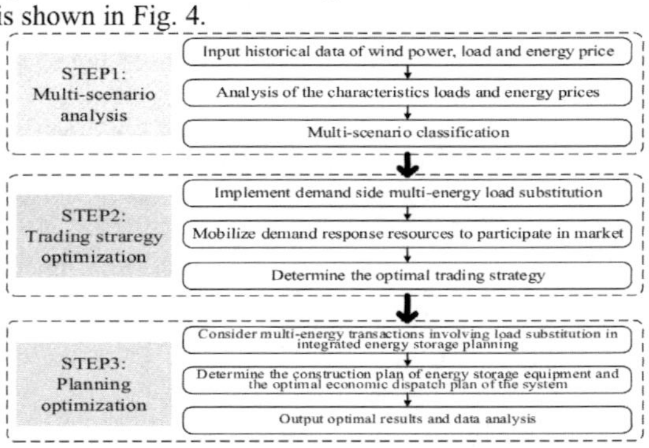

Fig4. Solution flow.

4 Simulation results and analysis

4.1 Basic data

This paper is based on the GAMS platforms on a Win10 operating system, i7CPU, 2.20GHz processor environment for the simulation and optimization analysis. To reduce the computational complexity of the model, the planning interval is reduced to four typical days. The investment cost takes into account the energy storage cost to be built, and it is amortized in order to balance its impact on simulation. The operating cost not only considers the running cost of energy storage, but also considers the operation and maintenance cost of other equipment, and the unit operating cost of the same equipment is the same in each scenario. The architecture of the example is shown in Fig. 1. Suppose the user side contains three types of energy load: electricity, gas and heat. The system has been equipped with WT, GB, EB, CHP and P2G. The equipment to be invested is three kinds of energy storage devices: EES, GES and HES. The load adjustment coefficient and the corresponding price adjustment coefficient under different scenarios are taken from [19]. The corresponding load curves, WT output, electricity and gas energy prices are referenced from [20]. The equipment types and parameters to be selected for MES are shown in Table 1.

Table1. Parameters of candidate MES.

Equipment		EES	GES	HES
Type1	rated capacity (MW)	1	0.8	0.8
	Investment costs (yuan/KW·h)	600	130	50
	Unit maintenance costs (yuan/KW·h)	0.018	0.017	0.016
Type2	rated capacity (MW)	3	1	1
	Investment costs (yuan/KW·h)	600	130	50
	Unit maintenance costs (yuan/KW·h)	0.018	0.017	0.016

4.2 Effectiveness analysis of results

4.2.1 Analysis of investment construction costs

In order to verify the effectiveness of the MES allocation method proposed in this paper in terms of investment construction cost, the following two situations are set up to compare:

S1: MES configuration for IES considering multi-energy trading without load substitution

S2: MES configuration for IES considering multi-energy trading with load substitution.

The MES configuration in two situations is shown in Table 2.

Table2. Scheme of MES configuration.

MES		Capacity in S1	Capacity in S2
EES	Type1	1	1
	Type2	-	-
GES	Type1	0.8	0.8
	Type2	-	-
HES	Type1	0.8	0.8
	Type2	-	1
Investment construction costs (10^4 yuan)		10.25	10.72
Total system cost (10^4 yuan)		1436.5	1307.3

From Table 2, we can see that S2 compared to S1, the EES and GES configuration is unchanged, but the HES configuration is increased. At this time, the overall energy storage configuration cost is increased, but the increase is small, only 4.38% is increased, and the total system cost is reduced by 8.9%. mainly because the multi-energy trading in S2 considers load substitution, and the demand for electricity, gas and heat loads changes, which in turn affects the configuration and operation scheme of MES. It can be seen that considering the multi-energy trading with load substitution has an impact on the construction cost of MES, which can change the investment and construction of MES equipment.

4.2.2 Analysis of operating costs

For verify the effectiveness of the MES configuration method proposed in this paper in terms of system operating cost, S3 is added on the basis of S1 and S2, specifically, neither load substitution nor investment in energy storage equipment is implemented. Cost of operation in three situations are 1426.23×10^4 yuan, 1296.55×10^4 yuan and 1447.01×10^4 yuan.

Compared with non-investment energy storage equipment, the total operating cost of the system that invest MES considering multi-energy trading without load substitution is reduced by 1.4%, the total operating cost of the system that invest MES considering multi-energy trading with load substitution is reduced by 9.09%. Therefore, the MES configuration and load substitution can play the advantage of multi-energy complementary and reduce the total operating cost of the system. However, the MES configuration considering multi-energy trading with load substitution is more able to improve the economic efficiency of the IES.

4.3 Sensitivity analysis of results

4.3.1 Effect of load substitution on MES operation mode

The operation of MES under different conditions in typical winter days is shown in Fig. 5.

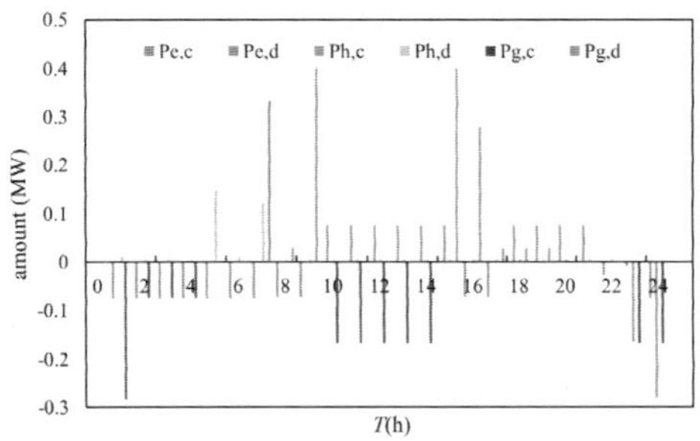

(a) MES operation under S1.

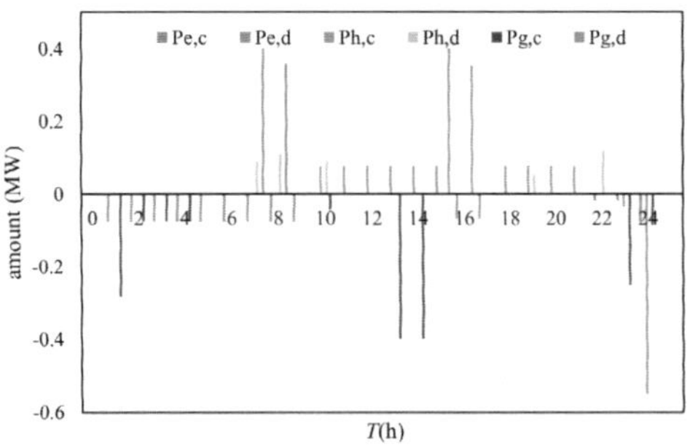

(b) MES operation under S2.

Fig5. MES operation under different situations.

It can be seen from Fig. 6 that after the load substitution participating the multi-energy trading, to meet the increased demand of heat load and gas load to implement electric-heat load substitution and electric-gas load substitution, the output force of EES changes in the flat period of electricity price, but the overall storage capacity of EES does not change. However, the total demand for the HES unit increased obviously, especially at the load valley period of 23:00~24:00, and the output of the HES unit increased significantly during the period, which increased by about 24%. Compared with EES and HES devices, GES devices are more frequently charged and discharged, and the energy storage period is more frequent. It can be seen that after considering the multi-energy trading with load substitution, the storage quantity of MES equipment is mainly optimized according to the change of the corresponding load under the premise of satisfying the overall demand of the multi-energy load.

4.3.2 Effects of different periods of typical day on multi-energy trading strategies

Because of the different load structure, the equipment output force of the typical day of different scenarios is different, and the corresponding trading strategy will be different. The specific analysis is carried out on the typical day of winter under S2. The trading strategies of electricity, natural gas and heat energy are shown in Fig. 6, Fig. 7 and Fig. 8.

As can be seen from Fig. 6, between 0:00 and 07:00, The main sources of electricity are market purchases and WT power. At this time the electricity load and prices are at a low point. The EES is charging, and P2G and EB are running at full power. Thereafter, as the electricity load and price increase, between 10:00-15:00 and 18:00-21:00, because of the low WT output, far from meeting the electricity load, EES and CHP equipment began to release electricity. At this point, as the electricity price is at its peak, the electricity load is partially replaced by gas load and heat load, the actual peak electricity load decreases after substitution, approximately 1.24 MW, decreased by about 34.3%. After 24:00, it is into the low period of electricity prices, electricity replacement is more economical, electricity load has increased. Therefore, through load substitution, it can be clearly seen that the demand for electricity load increases during the low period of electricity price, and the demand curve of electricity load decreases during the peak period of electricity price, which has the effect of cutting peak and filling valley.

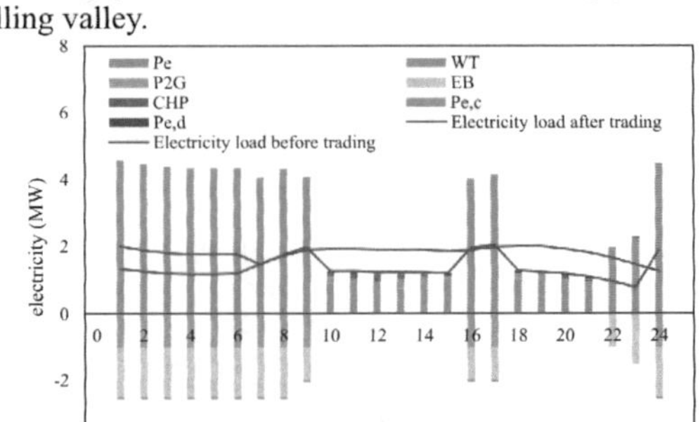

Fig6. Electricity trading.

It can be seen from Fig. 7 and Fig. 8 that the price of natural gas and heat energy is generally lower than the price of electricity, so we choose to buy more natural gas and heat energy. As an example, it can be seen that when

https://doi.org/10.1051/e3sconf/202123702016

the price difference is large, it is more economical to buy part of the heat energy from the market, while the other insufficient parts use the purchased natural gas to convert to heat energy through GB and CHP units. The EB heating is the main source of heat energy together. Through load substitution, the overall demand for heat energy and natural gas has increased to a certain extent, and the total demand has increased by 4.8% and 10.8% respectively. Of course, the corresponding heat energy and natural gas costs have also increased. However, despite the rising cost of natural gas and heat energy purchases due to load substitution, the scheme is still economically superior to the MES configuration of IES that do not take load substitution into account for multi-energy trading. The overall trading costs saved about 9.5%. Therefore, the multi-energy trading strategy should be combined with the price of multi-energy markets to optimize the choice. Energy trading is not only affected by the energy price, the price of other energy will cross affect the supply of this energy.

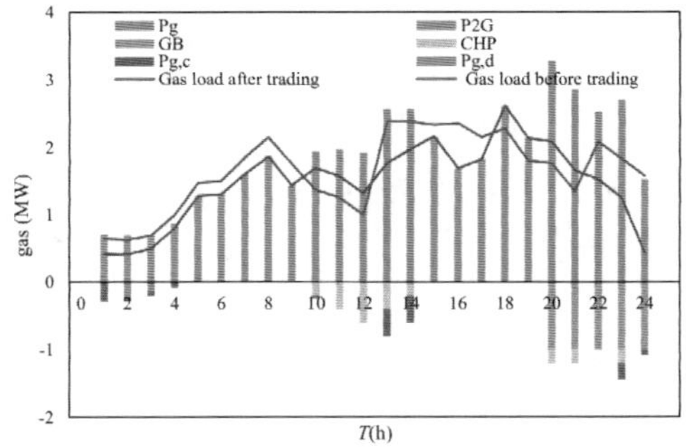

Fig7. Natural gas trading.

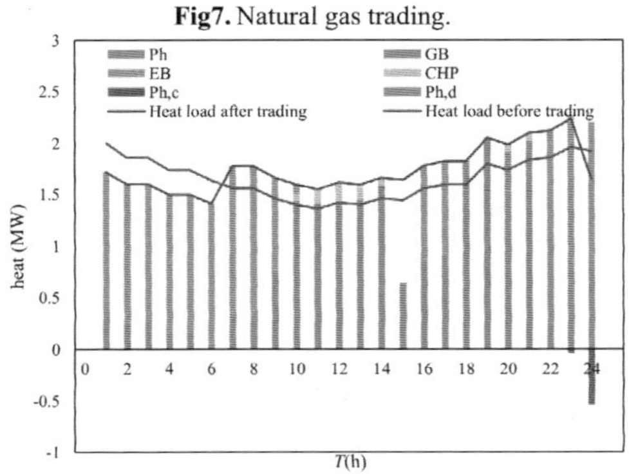

Fig8. Heat trading.

5 Conclusions

Aiming at the selection and allocation of the MES in IES, this paper proposes a MES optimal configuration method considering multi-energy trading with load substitution. The MES is configured to meet the energy supply and demand balance of different periods from different seasons. The following conclusions are obtained by examples:

(1) Load substitution participation in multi-energy trading can reduce the configuration cost of MES, which is 8.9% lower than that without load substitution.

(2) The characteristics of energy prices and loads affect the multi-energy load structure, the operation mode, and response capacity of EES, GES and HES.

(3) The IES network constraints will be taken into account to distinguish electricity, gas and heat trading to analyse the impact of network operating parameters on multi-energy trading at various times in the future work.

Acknowledgments

This work was supported in part by the Cooperative Research on Regional Integrated Energy Modeling Simulation and Operation Control Technology (No.BZ2019057).

References

1. H. J. Jia, Y. F. Mu, Y. Qi, INT J ELEC POWER, 54, 516 (2014)
2. C. Unishuay-Vila, J. W. Marangon-Lima, A. C. De Souza et al., IEEE Trans. Power Syst., 25, 1154 (2018)
3. Y. L. Xu, W. C. Wu, J. G. Zhou, IEEE Trans Smart Grid, 11, 686 (2020)
4. E. S. Parizy, H. R. Bahrami, S. S. Choi, IEEE Trans Smart Grid, 10, 3259 (2019)
5. G. Q. Li, R. F. Zhang, T. Jiang et al., Appl. Energy, 192, 408 (2017)
6. J. Qiu, Z. Y. Dong, J. H. Zhao et al. IEEE Trans. Power Syst., 30, 2119 (2015)
7. Y. H. Dai, Q. Chen, J. H. Hao et al. IEEE Trans Sustain Energy, 9, 1234 (2018)
8. C. Q. Ju, P. Wang, L. Goel et al. IEEE Trans Smart Grid, 9, 6047 (2018)
9. P. Rodrigo, B. Carlos, L. Fernando et al. IEEE Trans Smart Grid, 4, 996 (2013)
10. B. P. Hayes, A. Wilson, R. Webster et al. IET GENER TRANSM DIS, 10, 832 (2016)
11. C. Wang, Z. Y. Wang, Y. H. Hou et al. IEEE Trans. Power Syst., 33, 4958 (2018)
12. J. T. Chen, Q. Y. Zhu, IEEE Trans Smart Grid, 9, 6554 (2018)
13. D. Wang, Q. E. Hu, H. J. Jia et al. Appl. Energy, 248, 656 (2019)
14. C. C. Shao, M. Shahidehpour, X. F. Wang et al. IEEE Trans. Power Syst., 32, 4418 (2017)
15. C. B. Li, X. B. Liu, W. Zhang et al. IEEE Trans Smart Grid, 7, 481 (2016)
16. P. Y. Liu, T. Ding, Z. X. Zou et al. Appl. energy, 250, 512 (2019)
17. X. Dou, J. Wang, Z. Wang et al., J. Mod. Power Syst. Clean Energy, 8, 841 (2020)
18. X. Dou, J. Wang, Q. R. Hu et al., CSEE J POWER ENERGY (to be published)
19. A. Shahmohammadi, D. M. Moradi, H. Ghasemih et al. IEEE Trans. Power Deliv., 30, 878 (2015)
20. X. Dou, J. Wang, Z. Wang et al., INT J ELEC POWER, 124, 106356 (2021)

Classification of young males' shoulder shapes based on cross-sectional morphological characteristics

FangYuan Hu[1], JuanFeng Jin[1,2,3], ShuQing Wang[1], WenJing Dang[1], ZhenZhen Guo[1] and ZiChao Liu[1]

[1] Faculty of Clothing and Design, Minjiang University, Fuzhou, Fujian, 350108, China
[2] Fujian Province University Engineering Research Center of Textile and Clothing, Minjiang University, Fuzhou, Fujian, 350108, China
[3] Fujian Clothing Industry Technology Development Base, Fuzhou, Fujian, 350108, China

Abstract. This paper focused on the different characteristics of the shoulder cross-section curves closely related to the shape to subdivide the shoulder shapes. In this paper, 213 young college male students aged 18-26 were selected to measure the shoulder data with three-dimensional body scanner. With the help of imageware12.0 and matlabr2012b software, the cross-section curves which could be used to classify the shoulder shapes were extracted, and the method of subdividing the shoulder shapes with the curvature radius of the characteristic points of the cross-section curve and the ratio of sagittal to frontal diameter was established. K-means clustering method was used through dynamic clustering, the optimal classification number of shoulder shapes was determined to be 4 categories by variance analysis, and the shape differences of each shoulder shape were quantified; by comparing the curve shape of shoulder section, the curve change characteristics of 4 categories of shoulder section were further qualitatively described.

1.Introduction

In recent years, three-dimensional anthropometry has gradually become an important tool to obtain human body data [1], providing necessary conditions for the study of human body shape. The high-speed processing ability and accurate algorithm of computer also provide technical support for human body image and shape analysis [2]. There are differences in body shapes, although people with the same size wear the same clothes, because of different body shapes, the wearing comfort is often different. Therefore, somatotype analysis has always been a research hotspot [1]. In order to explore the body shape differences of various human bodies, Yong L C focused on the side contour of human body, put forward to use the side angle to represent the body shape characteristics of human body, and used the mathematical statistics method to divide the side shape into four categories [3]; Yang Y C introduced the body volume index (TVI), and divided the human body into four types: X, H, A and V [4].

The body shape analysis of shoulder, neck, chest, hip[5][6] and other local features has always been the focus of research. Zhang J H used three-dimensional measurement and mathematical statistics method to extract two typical shoulder indicators, subdivided the shoulder shape of young women into four categories [7]; Huang C Y extracted four typical neck indicators, subdivided the neck shape of male college students in school by K-means clustering method [8]. However, the body shape change can be divided into two directions: length (shoulder width, chest, waist, hip circumference) and girth (shape change of cross-section curve). Gu L proposed the method of body shape classification by parts (chest, waist, hip) based on the difference of body shape of cross-section curve, using the ratio index of characteristic point data [9].

The radius of curvature can be used to describe the degree of curvature change of the curve, and it is applied as a body type classification index. Lee H Y used reverse engineering software to extract the curvature radius of the feature points of the breast bottom contour and subdivide the breast shape [10].

This paper focused on the differences of the shape characteristics of the shoulder cross-section curves to subdivide the young males' shoulder shapes. With the help of three-dimensional anthropometry, reverse engineering software and MATLAB software, the ratio of sagittal to frontal diameter of shoulder section and curvature radius of curve feature points were extracted, and a method of subdivision of young males' shoulder shapes based on the shape indices of section curve is established.

2. Experiment

2.1 Subjects and instruments

In order to make the subdivision of shoulder shapes more targeted, 213 male college students were selected, aged 18-26 years old, height 165cm-185cm, weight 50-90kg, healthy. Three dimensional anthropometer [TC] [2], calibrated anthropometer and weight meter were used.

https://doi.org/10.1051/e3sconf/202123702017

The ambient temperature was (27 ± 3) °C and relative humidity was (60 ± 10) %, which accorded with the naked environment.

2.2 Extraction of section curve

The human body is composed of multiple parallel cross-section layered structure models, which can not only describe the overall shape of the human body through the characteristic curve of each cross-section, but also reflect the size and shape information of the local human body, as shown in Figure 1. Therefore, the cross-section curve of the end point of the shoulder is selected to analyze the shape characteristics of the male shoulder.

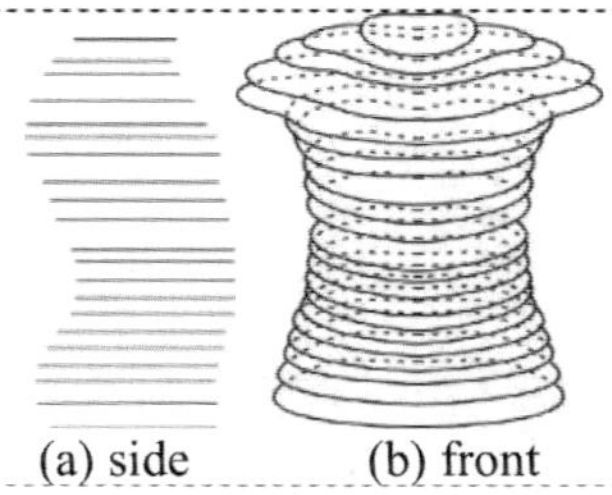

Figure 1. Cross-sections of human body.

Figure 2. Point cloud data of shoulder section curve.

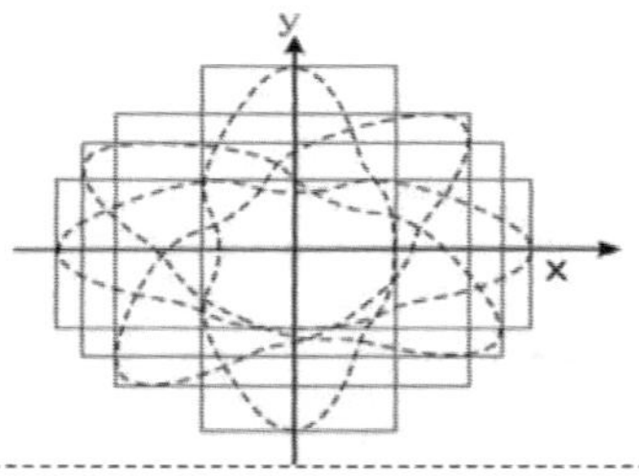

Figure 3. MER algorithm diagram.

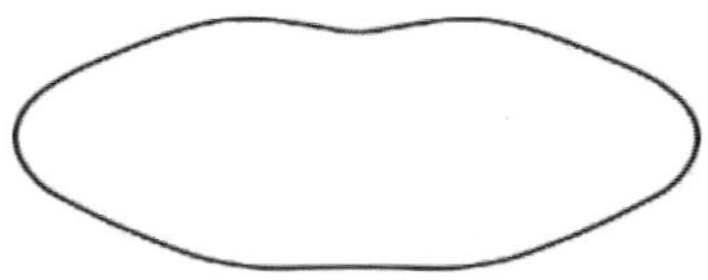

Figure 4. Fitting curve of shoulder section.

With the help of three-dimensional anthropometer, obtained the point cloud data of human body; used the reverse engineering software imageware12.0 to process the original point cloud data and extracted the section curve data that could represent the shape characteristics of shoulder, as shown in Figure 2; used MATLAB R2012b software to transform polar coordinates and used the minimum closing rectangle (MER) [11] to adjust the section axis of shoulder, as shown in Figure3, with symmetrical treatment, curve fitting of shoulder section was carried out by using the least square method [12], and a curve mathematical model with high goodness of fit was established to effectively characterize the shape characteristics of human shoulder, as shown in Figure 4.

2.3 Experimental data parameters

2.3.1. Shoulder sagittal frontal diameter ratio

In human orientation, the ratio of sagittal diameter to frontal diameter refers to the ratio of the short axis to the long axis. The ratio of sagittal to frontal diameter R of the cross section is the ratio of thickness a to width b of the cross section , that is, $R = a / b$, as shown in Figure 5. This characteristic value represents the degree of oblateness of the cross section. For the same section circumference, if the ratio of vector to frontal diameter is larger, it means that the ratio of vector to frontal diameter is larger, and the ratio of frontal diameter to frontal diameter is smaller, which means that the shoulder section is thicker.

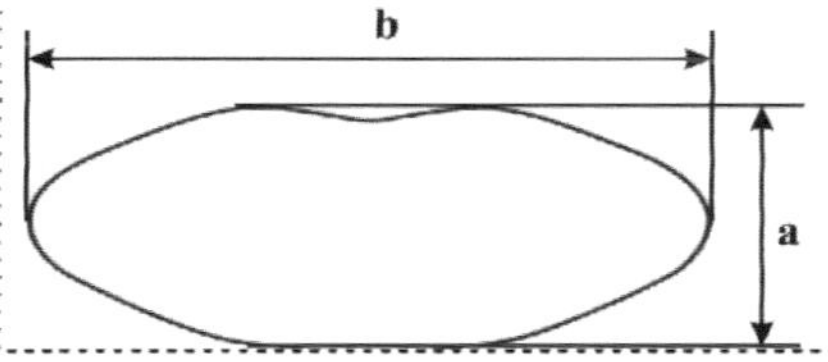

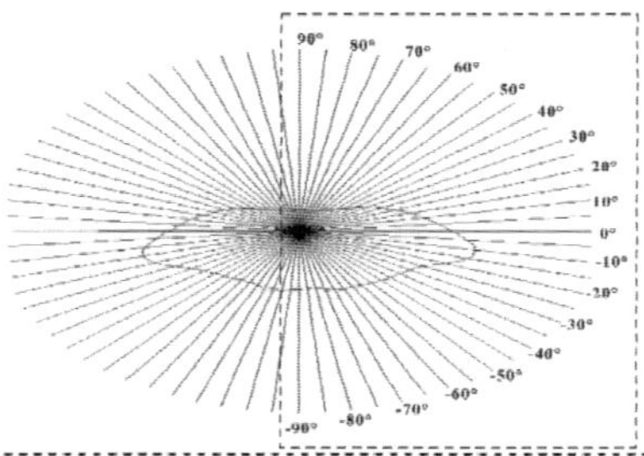

Figure 5. Schematic diagram of shoulder sagittal frontal diameter ratio.

Figure 6. Extraction of radius curvature of each point of section curve.

2.3.2. Curvature radius of shoulder section curve

Curvature refers to the degree of curvature of a curve. Curvature of a plane curve refers to the rate of rotation of the tangent direction angle of a point on the curve to the arc length. The greater the curvature is, the greater the degree of curvature of the curve is, and the greater the degree of deviation of the curve from the straight line is. The curvature formula of the curve [13] is as follows:

$$K=\frac{|y''|}{(1+y'^2)^{3/2}}$$

where y' is the first derivative of the fitting curve and y'' is the second derivative of the fitting curve.

The radius of curvature is equal to the reciprocal of curvature, that is, the radius of the circle formed when this arc is part of a circle. Because the curvature value of fitting curve of shoulder section was relatively small and there were many decimal places after the decimal point, it was not convenient for subsequent analysis, so curvature radius was selected to evaluate the curve bending degree. The larger the radius of curvature, the smoother the arc, the smaller the radius of curvature and the steeper the arc, so the radius of curvature could accurately reflect the shape characteristics of shoulder section.

The human body is approximately symmetrical to the left and right, so the left half of the cross-section curve of the male shoulder was analyzed, and the curvature radius was extracted with 5 ° as the interval angle [14][15], a total of 36 curvature radii were extracted to reflect the shape characteristics of the cross-section curve, as shown in Figure 6.

3. Data processing and analysis

3.1 Feature points extraction and analysis of shoulder section curve

In order to ensure the accuracy and reliability of the curvature radius of each feature point, it was necessary to preprocess it before data was analyzed, so as to eliminate the influence of singular value and wrong data on the experimental results. Spss19.0 was used to preprocess the data obtained by box chart, extracted useful information by eliminating the existing "abnormal data", and analyzed the data of 200 reserved experimental samples to get the change of mean curvature radius, as shown in Figure 7.

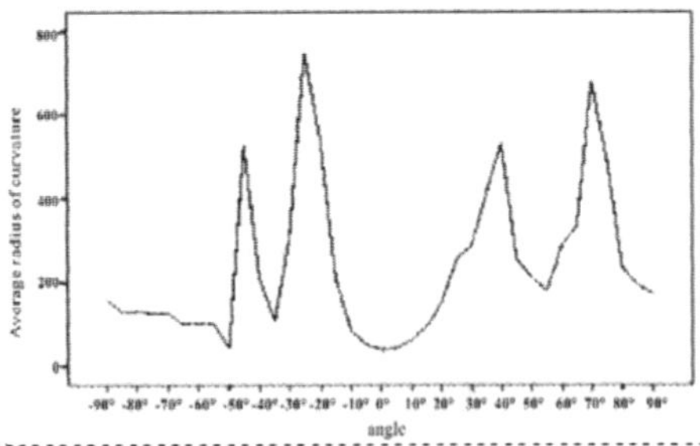

Figure 7. Mean distribution of curvature radius of shoulder.

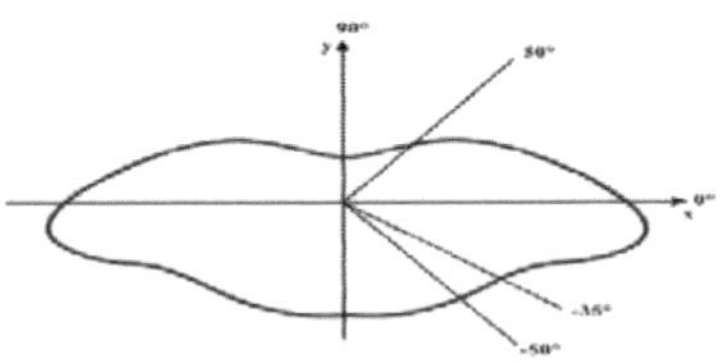

Figure 8. Characteristic points of section curve.

It could be seen from Figure 7 that the radius of curvature fluctuates continuously from - 90 ° to 90 °, among which the radius of curvature at - 50 °, - 35 °, 0 °, 50 ° and 90 ° were the local minimum values, and the curve in the local range had the largest bending degree, which could accurately reflect the local shape of the section curve and the overall change of the section shape. Therefore, the radius of curvature at these five angles, namely $RC_{-50°}$, $RC_{-35°}$, $RC_{0°}$, $RC_{50°}$ and $RC_{90°}$, were selected as the follow-up according to the classification of shoulder shape features, the corresponding position of shoulder section curve is shown in Figure 8.

3.2 Shoulder shape subdivision

In order to find the characteristics of the shoulder data, the k-means method was used to cluster the acquired shoulder data dynamically according to the six shape indexes of the cross-section curve, which were the curvature radius of the five characteristic points of the cross-section curve and the ratio of the section vector to the frontal diameter.

Table 1 showed the variance analysis results of curvature radius value and vector frontal diameter ratio of characteristic points obtained when the shoulder data were clustered into 3, 4 and 5 categories according to 6 section curve shape indices. It could be seen from table 1 that when the shoulder shape was divided into four categories, the probability of F test was less than 0.05, so four categories were the best choice for shoulder shape classification. When the shoulder shape was divided into four categories, the first category contained 49 samples, the second category contained 25 samples, the third category contained 106 samples, and the fourth category contained 20 samples.

Table 1. Analysis of variance about shape types of shoulder

Clusters	Indices	Mean square		F	Sig.
		Between groups	Within group		
3	R	0.093	0.103	3.124	0.001
	RC_{50°	135827.660	3525.145	49.852	0.000
	RC_{-50°	1105.480	421.218	8.526	0.000
	RC_{-35°	110376.109	258.187	601.938	0.410
	RC_{0°	175.311	119.275	2.133	0.008
	RC_{90°	201831.257	5088.021	112.038	0.000
4	R	0.198	0.015	3.395	0.037
	RC_{50°	143156.257	2375.692	53.818	0.000
	RC_{-50°	2651.428	380.726	6.284	0.000
	RC_{-35°	125347.166	221.462	535.372	0.000
	RC_{0°	287.163	74.420	3.426	0.005
	RC_{90°	263963.845	5007.271	52.893	0.000
5	R	0.248	0.031	7.407	0.025
	RC_{50°	131050.837	2964.435	51.342	0.000
	RC_{-50°	3274.242	371.161	5.598	0.000
	RC_{-35°	163591.510	200.344	580.225	0.081
	RC_{0°	386.528	54.484	2.753	0.000
	RC_{90°	419124.721	5011.225	150.148	0.000

Figure 9 was the cross-sectional curve of each type of shoulder shape, which clearly and intuitively showed that there were significant differences among the four types of shoulder shape.

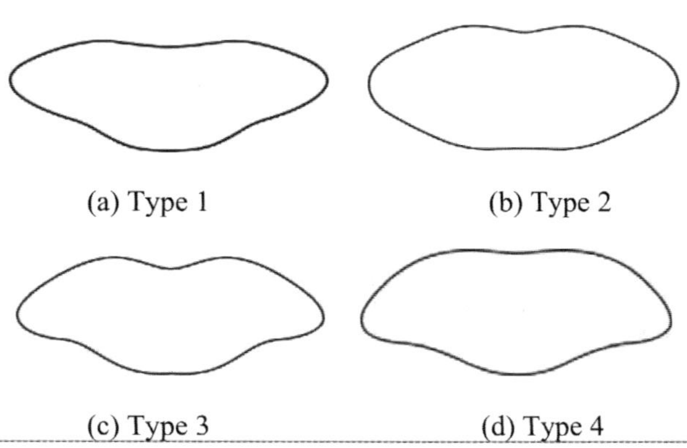

Figure 9. Sectional drawing of various shoulder shapes.

Type 1: moderate thickness of shoulder, obvious curve at both ends and middle part of shoulder, sharp angle at both ends of shoulder, smooth curve at back shoulder, obvious concave convex curve at front shoulder.

Type 2: thick thickness, the shoulder curve is relatively smooth, generally symmetrical, the two ends of the shoulder are round, the arc of the front and rear shoulders is smooth, the back is concave and convex, and the front shoulder is round.

Type 3: the shoulder curve is clear, the arc of the front and rear shoulders is concave and convex, and the two sides are prominent.

Type 4: the back is thick, the arc of the front and rear shoulders is gentle, and the two ends of the shoulders are slightly forward.

4. Conclusions

By analyzing the change of the mean value of curvature radius, the five characteristic point angles of shoulder section curve, namely - 50 °, - 35 °, 0 °, 50 ° and 90 °, were determined.

In this paper, a method was established to subdivide the body shapes of the male shoulder by using the curvature radius of the characteristic points of the section curve and the ratio of the sagittal diameter to the frontal diameter. By using k-means clustering method and variance analysis, the body shapes of the shoulder were divided into four categories, and the shape differences of the section curve were quantitatively distinguished, and the amount of samples was obtained. By comparing and analyzing the shape of the section curve of the shoulder, the curve change characteristics of the four categories of the shoulder section curve are qualitatively described.

Acknowledgments

This research was financially supported by the Program of Fuzhou Science and Technology Bureau (No.2018-G-97), the Open Project Program of Fujian Clothing Industry Technology Development Base, Minjiang University, China (No. MJXY-KF-201907), the Open Project Program of Fujian Province University Engineering Research Center of Textile and Clothing, Minjiang University, China (No. MJXY-KF-201902), the Rector Fund (No.103952019141), the Science and Technology Pre-research Program of Minjiang University (No. 31404305), and was supported by the corresponding author Doctor Juan Feng Jin, who can be contacted at jjf6859@163.com.

References

1. Vuuskan A and Bulgun E 2011 *Int. J. Cloth. Sci. Tech.* **23** 46-60
2. Costa F D L and Cesar R M 2001 *Shape Analysis and Classification Theory and Practice* (New York: CRC Press) pp 46-60
3. Young L C and Yun J 2010 *Hum. Factor. Ergon. Man.* **20** 378-90
4. Yang Y C, Chen M Z and Zou F Y 2009 *Journal of Textile Research* **30** 117-22
5. Jin J F, Yang Y C and Zou F Y 2016 *Fibres Text. East. Eur.* **24** 110-8
6. Jin J F 2019 *the 12th textile bioengineering and informatics symposium* (Soochow University, Suzhou) pp 442-9
7. Zhang J H and Wang H F 2011 *Journal of Textile Research* **32** 95-7
8. Huang C Y and Cui L N 2008 *China Textile Leader* **2**: 94-95.
9. Gu L and Zhang X 2008 *Journal of Beijing Institute of Clothing Technology* **28** 22-5
10. Lee H and Kyunghi H 2004 *Appl. Ergon.* **35** 353-9
11. Wu X G, Wang D Q and Sheng H 2004 *Computer Engineering* **30** 123-5
12. SHEN J H 2006 *Numerical Calculation Basis* (Shanghai: Tongji University Press) pp 37-45
13. Department of Applied Mathematics, Tongji University 2002 *Advanced Mathematics* (Beijing: Higher Education Press) pp 163-71
14. Deng W Y 2008 *Research on method of 3D personalized human body modeling based on photos and neural network* (Zhejiang: Zhejiang University) pp 61-2
15. Liu H, Zhang M and Chen D S 2011 *Journal of Textile Research* **32** 117-20

Research on the Capacity of Li-ion Battery Packer Based on Capacity Incremental Curve

Yu Tian[1], Zhengyuan Zhu[1], Shuangyu Liu[1], Dongpei Qian[1], Xiao Yan[2], Jonggeng Wang[1], Dong Dong[1], Zhongcai Liu[3]

[1]ZheJiang Huiayun Technology Information Co., Ltd,310000,China
[2]CUSPEA Institute of Technology, Wenzhou, Zhejiang, 325024, China.
[3]Shanghai Makesens Energy Storage Technology Co., Ltd.,201600,China

Abstract Lithium ion battery is the most widely used and reliable power source for electric vehicles. With the development of electric vehicles, the safety, energy density, life and reliability of lithium ion batteries have been continuously improved. However, in the field of vehicle power battery technology, battery monomers are combined in series and parallel to provide enough energy, but one of the major problems faced by group batteries is the consistency between battery monomers. Taking the capacity increment curve (IC curve) of lithium iron phosphate battery as the analysis tool, it is found that the characteristic peak of IC curve of different monomers in battery pack can reflect the relationship of monomer capacity. On this basis, the mathematical model is established, and the IC curve II peak characteristic point of a single cell are used as the reference to characterize the capacity of the single cell one by one. The results show that the method can be used in the normal charging process of the battery pack, and the capacity of the single cell in the battery pack can be characterized in real time during the whole life of the battery pack. It has certain research value for the ladder utilization and accurate management of battery pack.

1 Introduction

With the popularity of electric vehicles, power batteries have also been widely used in electric vehicle power systems, but also gradually exposed a series of problems, such as durability, reliability, safety and so on. the inconsistency between monomers after battery formation is one of the main causes of this series of problems.

lithium-ion battery capacity is one of the important performance indicators to measure battery performance. it represents the amount of electricity released by the battery under certain conditions (discharge rate, temperature, termination voltage, etc.), i.e., the capacity of the battery, usually in amperes • hours (abbreviated as A•H, 1A•H=3600 C). During the actual use of lithium ion batteries, there will be no full discharge of the battery, it is difficult to know the total capacity of the battery, thus affecting the estimation of the state (SOC) of the battery and nuclear power, and the accurate and real-time calculation of the battery capacity. It is of great significance for battery safety management.

There are many factors affecting the consistency of the battery. From the time order, the inconsistency of the single cell in the battery pack is mainly reflected in the inconsistency in the manufacturing process and the use process. Due to the immaturity of the preparation process, the state control of raw materials and the state control of production process equipment will lead to the inconsistency of single cell[1]. On the other hand, the inconsistency will become more significant due to the cycle aging of the battery during use. The inconsistency of lithium battery parameters mainly refers to the inconsistency of capacity, internal resistance and open circuit voltage. Inconsistent core string and use together, there will be the following problems.

(1) capacity decreases, the core unit forms the battery pack, the capacity conforms to the "barrel principle ", the worst core capacity determines the ability of the whole battery pack[2]. In order to prevent battery overcharge and over discharge, the logic of the battery management system is so set: when discharging, when the lowest monomer voltage reaches the discharge cutoff voltage, the whole battery pack stops discharging; when charging, when the highest monomer voltage touches the charge cutoff voltage, Stop charging. When charging, the small capacity of the battery must be filled first, to meet the charging deadline, the system no longer continues to charge[3]. When discharging, the battery with small capacity must release all available energy first, and the system immediately stops discharging.

In this way, the small capacity of the core is always full of discharge, but the large capacity of the core has been using part of the capacity. The entire battery pack is always partially idle.

(2) life loss, similar, battery life, is determined by the core with the shortest life. Very likely, the shortest life of the core, is the small capacity of the core. Small capacity core, each time is full of release, too strong, probably the first to reach the focus of life. Has been the end of the core life, a group of welded cores, also followed by the end of

life.

(3) internal resistance increases negative feedback, so that the high internal resistance core accelerates deterioration.

Therefore, how to diagnose the monomer inconsistency of battery pack has been an urgent problem. By analyzing the characteristic peak of capacity increment curve (IC curve) of lithium iron phosphate battery, it is found that the characteristic peak of IC curve of different monomers in battery pack can reflect the consistency between monomers[4].

2 Capacity increment curves

Capacity increment analysis is an important method to analyze the mechanism of battery decline, and it is also an important method to study the consistency of each cell in battery pack. Its core is the battery capacity increment curve (IC curve) [5]. The normal charging curve of lithium iron phosphate is shown in figure 1: the horizontal axis is the charging capacity and the longitudinal axis is the battery terminal voltage. It can be seen from the diagram that the voltage rises rapidly in the early and end of the charging curve, while in the middle of the charging curve, the voltage rises slowly. In fact, most of the charging capacity is filled in the platform part[6].

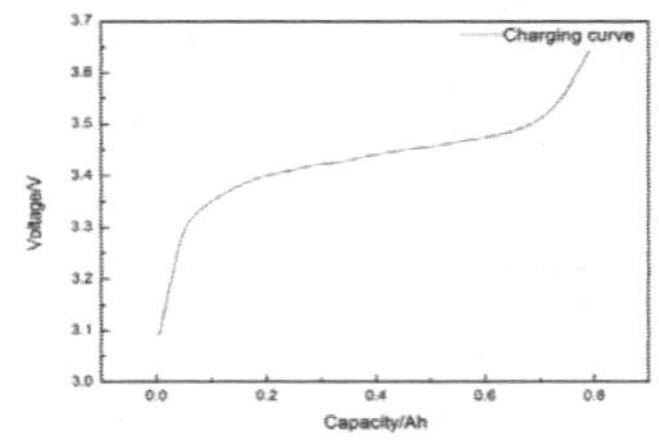

Fig1. Charging curve

The battery capacity increment curve is to process the original charging data, obtain the dQ/dV data, and then obtain the (dQ/dV)—V curve. As shown in figure 2, the advantage is that the voltage rise on the charging curve is slow, and the voltage platform region is easily observed and analyzed on the capacity increment curve[7]. In this paper, the IC curve of different monomer charging IC is studied, the capacity of each single cell in the series battery pack is obtained.

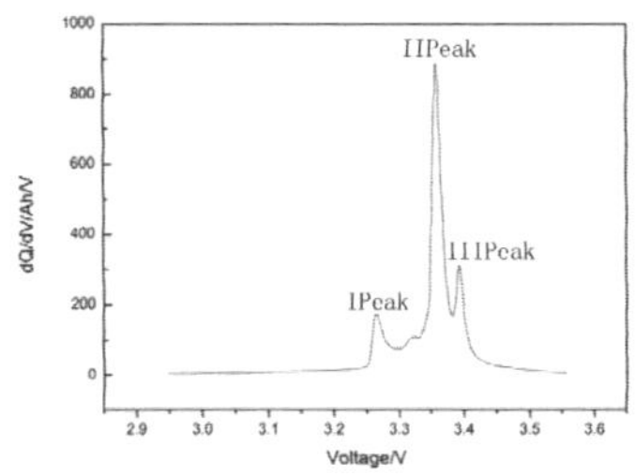

Fig2. Capacity increment curve

3 Research on the Consistency of Li-ion Battery Packages Based on Capacity Incremental Curve

In this paper, the voltage and current of each monomer IC curve and the eigenvalue position of the reference curve are taken as the input parameters of the algorithm model, so as to complete the real-time calculation of the capacity of all the monomers in the battery pack.

First, the charging data of the target lithium ion battery are processed, and the five-point and three-time smoothing filtering method is adopted (2 data before and after the smoothing filtering position are selected, a total of 5 data are selected, and the third order polynomial is used to fit the charging data of the target lithium ion battery. The capacity increment value of the position is obtained by taking out the value of the smoothing filter. The formula is

$$\frac{dQ}{dV}=\frac{\Delta Q_k}{\Delta V_k}=\frac{Q_k-Q_{k-1}}{V_k-V_{k-1}} \quad (1)$$

The characteristic peak position is the maximum position in the capacity increment curve, and the charging capacity Q1 and Q2, corresponding to the characteristic value position and the total battery capacity Q. are recorded

The capacity coefficient between the characteristic peaks of the battery is calculated. The formula is as follows:

$$\gamma=\frac{Q_1-Q_2}{Q} \quad (2)$$

When calculating the capacity of the j monomer, the charging capacity of the first eigenvalue of the IC curve is recorded Q_{1j}, the charging capacity of the second eigenvalue of the IC curve is recorded. Thus, the total capacity corresponding to the charging curve data of all single cells is calculated. The capacity calculation formula of the j monomer is as follows:

$$Q=\frac{Q_{1j}-Q_{2j}}{\gamma} \quad (3)$$

The Q_j is the total capacity value corresponding to the data of the charging curve line of the j single cell, and the charging capacity of the j single cell at the first characteristic peak position. The Q_{2j} is the charging capacity of the second characteristic peak position No. j single cell.

Calculate the total capacity of cell pack No .1 cell:

$$Cap=\frac{\Delta Q_1}{g}=\frac{|Q_{21}-Q_{11}|}{g}$$

$$=\frac{21.56-11.576}{0.32}=31.2Ah$$

 https://doi.org/10.1051/e3sconf/202123702018

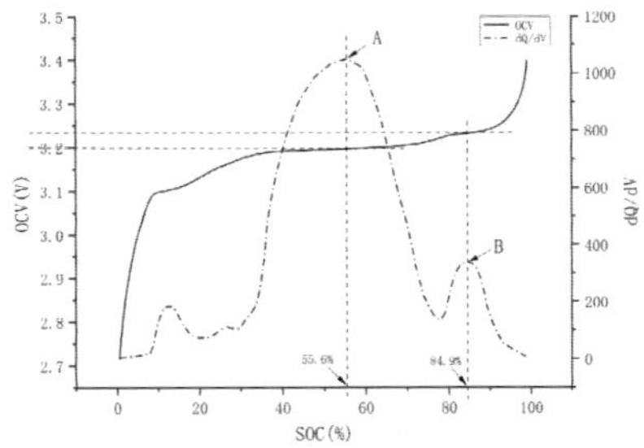

Fig3. Capacity increment curve

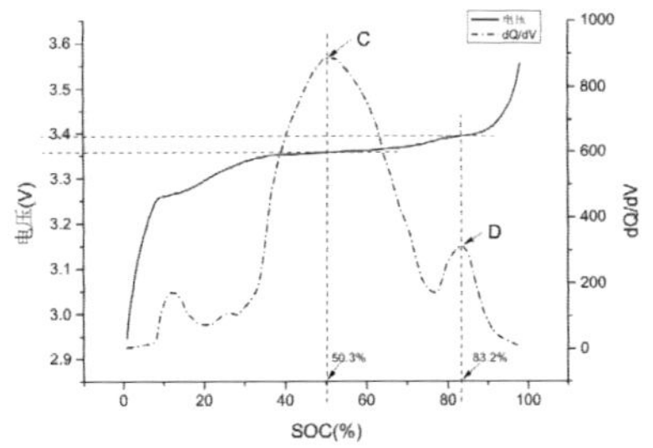

Fig4. Capacity Incremental Curve of Single Cell No .1

This paper uses the target lithium ion battery is the CATL lithium iron phosphate battery, the nominal capacity is Cap(initial)=180 Ah, the battery reference curve data uses the SOC-OCV relation curve data, the data interval is 2. The capacity value corresponding to each OCV point is calculated by formula: Q=SOC*180.

Adopt the five-point and three-time smoothing filtering method, Five data, Using the 3rd order polynomial, Find the value after smoothing filter) find the capacity increment curve of SOC-OCV relation data, As shown in the dQ/dV points in Figure 3, Figure 3 is based on SOC, And the left ordinate is OCV, battery open circuit voltage dQ/dV, is the right ordinate And the first eigenvalue position is the A point position in figure 3, Corresponding SOC is 55.6%, Volume Q_1= Ah,100.08 And the second eigenvalue is the B point in figure 3, Corresponding SOC is 84.9 per cent, Volume Q_2= Ah;152.82

The relationship coefficient between the battery characteristic value and the total capacity is calculated by the following formula.

$$g = \frac{\Delta Q}{Cap_{initial}} = \frac{|Q_1 - Q_2|}{Cap_{initial}}$$

$$= \frac{|152.82 - 100.08|}{180} = 0.293$$

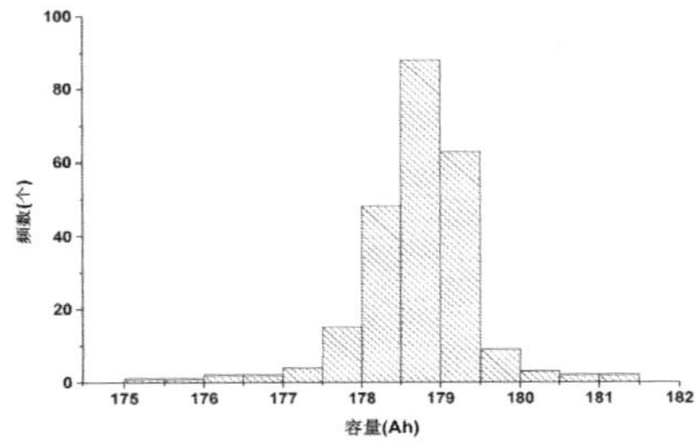

Fig5. Histogram of Single Cell Capacity

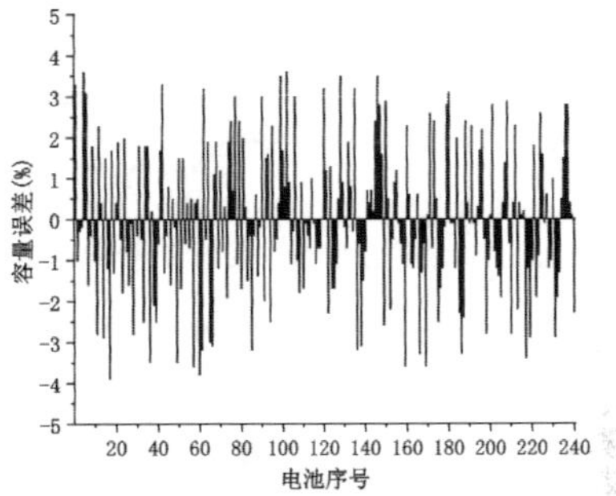

Fig6. Capacity error rate

The battery pack assembled by the target lithium ion battery is a commercial energy storage system composed of 240 single cells in series. The charging data condition of figure 3 is that the battery pack stops charging at 0.35 C,63A current to any single cell voltage of 3.6 V, the battery pack SOC is 20. The extraction condition of charging data is $\Delta V \geq 2$ mV, Extract the voltage and charge values of each cell, Using a five-point and three-time smoothing filter for the voltage and charge values of the battery, The capacity increment curve of single cell No .1 in the battery pack is shown in figure 4. Figure 4 is based on SOC, Left ordinate is battery voltage, dQ/dV, is the right ordinate And the first eigenvalue is the C point in figure 4, corresponding charging capacity is Q_11=67.66 Ah.The second eigenvalue is the D point in figure 4, corresponding charging capacity value is Q_21=119.2 Ah;

The same calculation method calculates the total capacity of No .2 single cell in the battery pack to the total capacity of No .240 single cell, and draws the total capacity of all single cells into histogram. Figure 5 shows the histogram of the capacity of all single cells in the battery pack.

The capacity value calculated by the invention method is compared with the real capacity value of the battery. The capacity error rate is shown in figure 6, and the maximum capacity error is 3.86%.

4 Conclusions

The capacity of lithium ion battery is related to the charge and discharge capacity of the battery. The capacity difference of the single cell in the series battery pack seriously reduces the performance of the battery pack. In order to better calculate the capacity of single cell in battery pack, this paper takes the capacity increment curve (IC curve) of lithium iron phosphate battery as the analysis tool.The capacity difference between the peak values of IC of different monomers in battery pack and the capacity value of each single cell are calculated by parameter analysis. Through experiments, the difference between the method and the actual capacity is calculated, and it is found that the maximum error between the method and the actual capacity is less than 5%. It shows that this method can quickly calculate the capacity of single cell in battery pack, and can be applied in the normal charging process of battery pack. The capacity of single cell in battery pack can be described and analyzed in real time. It has certain research value for the echelon utilization and accurate management of battery pack.

References

1. D Qian, Z Liu, D Liu, et al. Research on consistency of Grouped lithium batteries Based on Capacity Increment Curve [C]the 2019International Conference. (2019)
2. Z.C Liu, X Yan, W Yu, et al. New Method for Determination of Health Status of Li-ion Battery [J] PST 043(001):74-76,157(2019)
3. M Xu, Z.C Liu, Yan Xiao, et al. [J]. Method for On-line Detection of Internal Resistance Consistency in Capacity Increment Energy Storage Science and Technology ,008(006):1197-1203(2019).
4. W.H Lin, X Yan, Y Wang, et al. [J]. on the Characteristic Internal Resistance of Lithium Iron Phosphate Battery Power Technology 043(007):1116-1181,124(2019)
5. X.B Han, M.G OUYANG, L.G Lu ,et al .A comparative study of commercial lithium ion battery cycle life in electrical vehicle： Aging mechanism identification [J]. JPS, 251： 38-54 (2014)
6. M.G OUYANG, X.N FENG, X.B HAN, et al. A dynamic capacity degradation model and its applications considering varying load for a large format Li-ion battery [J]. AE,165： 48-59(2016)
7. C.R BIRKL, M.R ROBERTS, E MCTURK, et al. Degradation diagnostics for lithium ion cells [J]. J PS,341： 373-386(2017)

 https://doi.org/10.1051/e3sconf/202123702019

Research and practice of kinetic energy system risk control technology in tobacco enterprises

Wang Husong, Ni Zhenyu, Jin Wei, Huang Qizhong, Xiao Guihui*, Le Yuannan, Liu Jianmin, Hu Xiaohu, Chen Lifeng, Fang Biao

Ningbo Cigarette Factory, Zhejiang China Tobacco Industry , Jiapu West Road No.2001, Fenghua Economic Development Zone, Ningbo city, Zhejiang Province, China,postcode:315504

Abstract. The purpose of this paper is to provide a data-oriented risk prevention and control system for seven major kinetic energy supply systems (power distribution, dust removal, moisture removal and odor removal, air pressure, vacuum, process soft water and sewage station water manufacturing, boiler steam system, air conditioning). The traditional risk control means should be improved through systematic process and information technology, and the risk management of operation control should be gradually transformed from event-driven to data-driven, so as to strengthen the comprehensive warning and prevention and control of operational risks. The management mode of integrated management and control is "clear hierarchy, effective air separation, effective control in place, quick early warning and simple handling". Water, electricity, gas and steam of the power system are realized as part of the internal quality and process control of the cigarette brand, and all the control is pushed to the power supply terminal.

1 INTRODUCTION

According to the characteristics of the cigarette factories supply long-distance conveying, the stability of the power supply of energy is particularly important, long-distance conveying magnify kinetic energy supply instability, and unstable energy supply will directly affect the tobacco moisture in the manufacturing process, flow, temperature, or even indirectly affect the quality of cigarette products, especially the flavour and taste.Therefore, according to the requirement of high quality development put forward the new idea of power control: power seven related to production system (dust moisture removal, steam boiler, air compressor, high vacuum, the craft is cold water, and air conditioning) is an integral part of "cigarettes factory process control", dynamic services cannot be limited on the timely supply of energy, is to practise the service in the silk, the roll bag production line kinetic energy interface terminal.We need to study and practice the innovation of kinetic energy system risk control process in tobacco enterprises.

2 RESEARCH AND PRACTICE ON THE ESTABLISHMENT OF RISK CONTROL SYSTEM

2.1 Creatively establish the technical framework for risk point management and control of the seven major kinetic energy supply systems

The risk point management and control function adopts micro-service architecture, relies on advanced technologies such as streaming computing and big data, and integrates MES, energy management system, centralized control system, data center and other information resources on the basis of existing business processes and information implementation experience, so as to realize the whole-process management of "collection, early warning, analysis and response". In the production of perception, energy prediction, equipment control, balance between supply and demand, production security, etc, to realize information sharing, be based on production driven "precise kinetic energy supply, power control integration, scientific decision analysis", fully coordinated kinetic energy supply and demand of resources, gradually establish a power horizontal collaboration, vertical well versed in management, information connectivity, high share of resources integration service mechanism.

Through system integration, the risk point management and control system realizes the integration with MES system's production plan, work order information, process related information, products and other basic data information. Through the low-level development, the integration of centralized control system, OPC data, real-time database data, energy management system data and external meteorological information is realized, so as to provide comprehensive and complete data for the realization of unified monitoring, prediction, early warning and optimization of the platform.

As shown in Figure 1, the architecture of the system can be divided into data source, data processing layer, data storage layer, analysis and optimization layer, data service

* Corresponding author: 2535879510@qq.com

layer and application presentation layer from the perspective of data flow processing

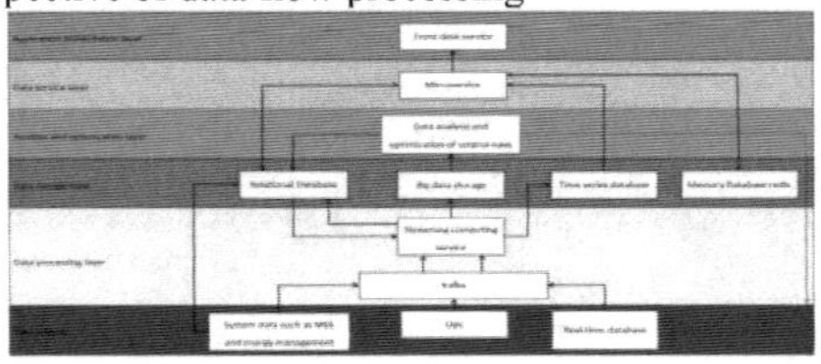

Figure. 1. System architecture diagram

The data sources for risk point control of power system come from many aspects, such as information of MES system and energy management system, external data such as weather service, as well as data acquisition information obtained from OPC server and sampling information extracted from real-time database.

The data of different information sources, according to a different approach, some data such as weather information, such as the repair order status information save directly to the relational database, the data need to processing, especially the flow calculation of data is sent to Kafka, after data processing, part of the data such as the alarm data to oracle, in history of monitoring information query data to influxdb temporal database, used in real-time monitoring of data sent to Kafka.

The data storage layer mainly includes relational database Oracle, big data storage Hadoop, time series database influxDB, and memory database RedIS. Oracle database mainly stores system related basic information, configuration information, model parameters, etc. Hadoop is used to store a large amount of processed data for analysis modeling and influxDB for historical query data; Redis to cache data that needs to be read and written repeatedly, such as warning rules.

The analysis and optimization layer mainly matches the specifications of equipment monitoring data and continuously optimizes them, and stores the optimized parameters in the risk point control standards.

The data service layer mainly provides data for micro-services deployed in this layer. Micro-services obtain relevant data from the data storage layer and provide data services for application presentation and information push.

The application layer provides display services for front-end PC and mobile terminals, etc.

As shown in Figure 2, the system includes 6 servers, among which 2 serve as application servers to deploy microservice framework and application services. 2 as data processing and optimization modeling servers for data flow calculation and data analysis modeling; Two servers serve as data storage service groups, including one deployed Oracle relational database, one deployed XDB timing data, and so on.

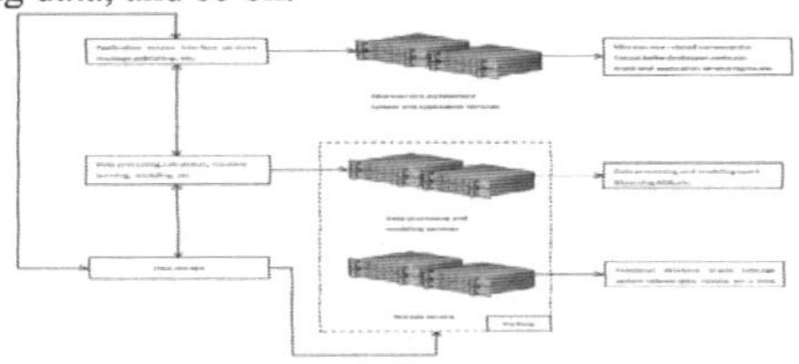

Figure. 2. System deployment diagram

Microservice architecture and application services mainly require response and processing performance. Therefore, clustering is recommended to improve the concurrency of the system and effectively balance the load.

Data processing and analysis service requires real-time calculation and monitoring of data from OPC, timing database and related system interface, and there are a large number of data processing and analysis jobs, so it is recommended to deploy the cluster.

Because it involves big data computing and machine learning, you need to make sure there is a lot of space and can be expanded.

Data storage service that stores data analysis and processing results, modeling information, etc., both from relational database Oracle and from real-time computation and processing results (influxDB timing database); In addition, hadoop is adopted to store massive information to ensure sufficient storage space, and the space can be continuously expanded with the continuous accumulation of system operation and data.

Risk point control covers the whole kinetic energy system, including real-time flow design of boiler steam system, vacuum, air pressure system, air conditioning and refrigeration system, dust removal and moisture removal system, process cold water, power distribution system, sewage treatment system, etc. Through real-time monitoring of key indicators of kinetic energy operation, the system can predict and warn risk points, and push the results to relevant personnel through mobile phones in real time, so as to achieve comprehensive control of kinetic energy management.

The risk point processing process is shown in Figure 3 below.

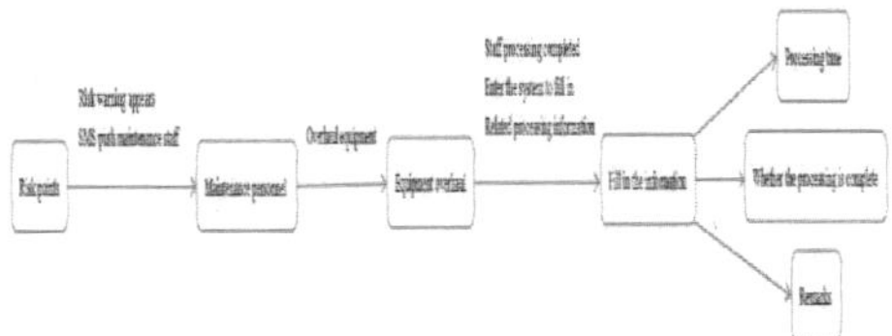

Figure. 3. Flow chart of risk points

According to the risk point detection standard, the system automatically detects the key risk point operation parameters of each kinetic energy system. When the operation index exceeds the control standard, the system will immediately push it to the operating team, relevant maintenance team and management personnel. After the operation team and maintenance personnel take on-site inspection, maintenance and other treatment measures, they will feed back the treatment process and measures to the system, so as to continuously provide data accumulation for the subsequent optimization of control standards and knowledge base.

2.2 Establishment of risk point control data service platform

In order to monitor, analyze and control the operation status of key nodes in the whole process of kinetic energy supply, this project designs a set of unified data aggregation and fusion architecture, establishes the core data support platform, and supports the operation of various business functions. The data processing process of the data service platform includes data acquisition, data management, data service and data application.

1) Data acquisition—Dynamic acquisition technology is used to gather real-time data of MES, energy management, power centralized control, equipment management and other systems and integrate historical data.

2) Data processing — To clean, convert, collect, organize, standardize and monitor the collected data to provide basic data meeting the requirements for data analysis;

3) Data modeling — The prediction, correlation analysis, anomaly detection and equipment control strategy of monitoring data under the background of kinetic energy big data were studied, and the data analysis model was established.

4) Data services—Design and implement data services based on the kinetic Internet of Things, and provide users and business systems with services or interfaces such as query, analysis, report, statistics and display; The upper data application makes authorized call to the data service and interface to support the operation of various business systems.

Based on the control requirements of power equipment, two modes, batch processing and flow calculation, are adopted for data on the platform to process data in different scenes respectively, so as to improve the processing capacity and layer expansion capacity of the system in the face of massive data.

2.3 Key Technologies

The integrated platform of intelligent power risk management and control is the implementation of the CPS system in the field of kinetic energy management and control in the tobacco industry. The platform adopts the framework system of micro-service and realizes efficient analysis, in-depth mining and application of massive data through flow computing.

1) Microservice framework

In recent years, with the rise and development of cloud computing, container virtualization, and DevOps integrating development, testing, deployment, and operation, microservice architecture has gradually become the mainstream technology. The basic idea is to divide the traditional single application into a series of software service units that can be independently designed, developed, deployed, operated and maintained according to the business functions, and the services cooperate with each other to realize the ultimate value. Microservices can be seen as a concrete implementation of a Service-Oriented Architecture (SOA), but they have advantages over Web services, such as decentralized management and lightweight container deployment, monitoring and early warning and multiple highly available fault tolerance strategies, and independent and free development of Service data.

The microservice architecture model has many advantages:

Componentization: Microservices can be thought of as components. The biggest difference from the traditional component approach is that the traditional component will apply modularity and build relatively independent units for it. Separate or extract common parts to build Shared libraries and achieve the effect of decoupling and reuse. Microservice architecture, on the other hand, decomposed the system into multiple services in a componentized way, with relatively independent and loosely coupled services. The change of a single function only requires the reconstruction and deployment of corresponding services.

Standalone deployment: Since microservices have separate running processes, each microservice can also be deployed independently. When a microservice changes, there is no need to compile and deploy the entire application, greatly reducing the application delivery cycle. The microservice architecture pattern makes continuous deployment possible.

Reducing complexity: try development status.Microservice architecture makes the complexity controllable and avoids unnecessary silos of data by decommissioning monolithic applications into multiple service methods. Under the condition of unchanged functions, the application is decomposed into multiple manageable services. Through the micro-service architecture pattern, the complex functions are presented in a modular way, providing modular solutions to the functions that are difficult to be realized by the single application coding method, and making the single service easier to develop and maintain

Technology diversification: The traditional development model is to use the same technology to build the entire application. Microservice architecture emphasizes a decentralized software organization architecture, enabling each service to independently choose the appropriate technology type and provide interface services according to its own service requirements and indus

Reduce the maintenance risk: Maintaining a large and complex single architecture system requires a lot of work, which is prone to lead to the whole system, so the risk is very high. In contrast, in the microservice architecture mode, when a component fails, there will be no defects such as in-process diffusion in the system of single block architecture, and the faults will be isolated in a single service, thus reducing the risk of maintenance and redevelopment.

2) Flow calculation

Streaming computing is a concept relative to batch computing, which refers to real-time calculation of incoming data stream in memory, while batch computing refers to centralized calculation of stored static data. Therefore, streaming computing has the characteristics of low delay, high throughput and continuous operation. In recent years, due to the high efficiency of distributed computing and real-time characteristics of data processing, streaming computing technology has been widely used in

 https://doi.org/10.1051/e3sconf/202123702019

many fields such as finance, Internet and Internet of Things, such as real-time video analysis and real-time warning of traffic flow. Common streaming computing frameworks include Storm and SparkStreaming, etc. In this project, after the data is acquired through the data acquisition framework, the flow calculation can calculate the values of all kinds of monitoring indicators in a certain window in real time (for example, the steam flow and energy consumption within 5 minutes, etc.). If the deviation between the monitored index value and the historical or preset value exceeds the threshold value, automatic warning will be given.

3 Case analysis and research on risk early-warning process control

With the continuous advancement of intelligent factory, higher requirements have been put forward for equipment maintenance, and the original manual detection method has been unable to meet the requirements of intelligent factories at the present stage. Only the real-time monitoring of the equipment and the predictive maintenance and preventive maintenance according to the monitoring data of the equipment can improve the risk control ability of the system to meet the control requirements of modern factories.

The existing dust removal, moisture drainage, odor removal equipment related maintenance work are manual equipment of the indicators of regular testing and regular maintenance of the equipment. Daily maintenance mainly took the form of point inspection, the dust, moisture removal equipment of motor shaft temperature monitoring and equipment overall PaoMaoDiLou phenomenon does exist, in addition to the daily maintenance, and there are periodic equipment maintenance, equipment maintenance time is according to the time period specified in the manual system, artificial search system manual to arrange the maintenance plan. This kind of maintenance has certain risks. First, there are errors in manual detection. Secondly, manual detection cannot guarantee that the real-time equipment state meets the standard. Third, part of the data is not suitable for manual inspection, resulting in incomplete equipment detection; Fourth, due to the difference in the time of regular maintenance of each part of each equipment, the manual inquiry and rearrangement plan is too complicated and prone to omission.

1) Real-time push of risk warning information

According to the collected data and equipment maintenance experience, set the risk warning value. When the corresponding data of the equipment exceeds the risk warning value, contact the relevant person in charge through SMS.

2) Real-time push of equipment maintenance information

According to the periodic maintenance benchmark and real-time data of the equipment of dust removal, moisture removal and odor removal system, when the equipment arrives at the maintenance cycle or the real-time data indicates that the equipment needs maintenance, remind the relevant person in charge to arrange equipment maintenance through SMS.

3) Establishment of equipment risk point database

For the discovered risk points, a risk point database is specially established, in which the occurrence time, occurrence point location, the specific situation of the risk, the solution and the effect after processing are recorded.

The following Figure 4 is the system flow chart, which briefly explains the specific implementation steps of the data-oriented risk prevention and control method for dust removal, moisture removal and odor removal equipment.

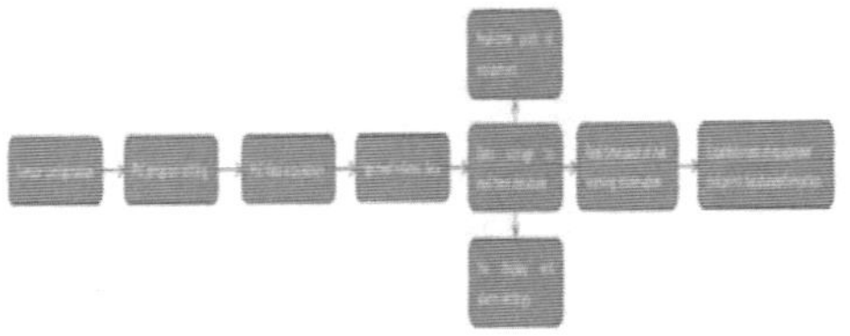

Figure. 4. System flow chart

The following is a brief introduction to the specific implementation.

1) Sensor configuration

As shown in Figure 5, install the additional sensors, such as SMT platinum thermal resistance MIK-PT100 (US Control) to monitor the temperature of the fan bearing, and complete the wiring connection with the PLC.

Figure. 5. Effect of sensor installation

2) PLC program editing

Program the PLC corresponding to the added sensor, such as the wafer platinum thermal resistance MIK-PT100 (US Control) for fan bearing temperature monitoring.

3) IGS reads relevant data

Through IGS, read the temperature data, negative pressure value, cloth bag pressure difference, dust concentration, material level data and converter current value, voltage value and frequency value in PLC.

4) Data will be stored in real-time database

Create points in a real-time database through Proficy client, and store all relevant data in a real-time database for subsequent data processing.

5) IFIX display and alarm Settings

Create a corresponding point on the iFIX interface to read the data received by IGS and display it, and then configure the alarm Settings for that data.

6) Real-time push of risk warning information

When the data exceeds the set warning value during the operation of the device, the relevant person in charge will be notified via the company's 0A platform SMS, which includes the time, point and specific situation of the risk.

7) Establishment of equipment risk point database

After the risk point is processed, the processing situation of the risk point shall be explained in detail, and relevant contents such as processing method and effect after processing shall be entered into Oracle database. The next time a similar situation occurs, it will be automatically called so that the risk can be quickly identified and dealt with.

The establishment of the system completely breaks the shackles of traditional manual risk identification based on experience and the inability to quickly and effectively judge potential risks. Through the identification, control and early warning of operational system risks of power energy supply, the following objectives are realized in terms of functions: 1.Real-time data monitoring (second level) and historical data modeling analysis; 2.Multi-platform collaborative control; 3.Rapid and accurate prediction and early warning; 4. Intelligent reference to the current situation of equipment and equipment maintenance plan prompt equipment maintenance; 5. Monitor the instability factors in the production and process control. Finally, it can reduce the equipment failure rate, reduce the potential safety hazards, and keep the kinetic energy supply stable and efficient for a long time.

4 Conclusion

This paper first introduces in detail the research content of risk control system establishment: system architecture, system deployment risk point system, control data platform, data processing mode; Then the practical application and beneficial analysis of the system in the equipment of dust removal, moisture removal and odor removal in our factory are introduced. Since the equipment of tobacco industry has a high degree of similarity, the system has a high reference value in the practical application of our factory.

References

1. Guo Chaohao.Research on smart Grid Security Control based on intermittent renewable energy interconnection [D]. North China Electric Power University, North China Electric Power University(Beijing),2017.DOI: 10.7666/d.Y3264420.
2. Zheng Mingyu, Dong Zheng, Hu Yongsheng.Discussion on quality assurance method of Equipment Supervision based on risk Management [J] China Equipment Engineering,2020(10):53-54.
3. Huang Lijie, Brief Discussion on Environmental Safety Risk Management of Power Supply Equipment operation [J]. Science and Technology Innovation and Application,2017,(35):133-134.
4. wang huifang, Yang hejuan, he pentium, et al. Improvement analysis of state failure rate model of power transmission and transformation equipment [J] power system automation 2011,3(16:27-31,43).
5. Zhang Yige, Risk control business process analysis and optimization Strategy research of Central and Southern Air Traffic Control System [D]. South China University of Technology,2018.
6. Zhang Guangbiao, Discussion on the Establishment of silk Control alarm System in Tobacco Factory [J]. China Equipment Engineering,2015(09):56-57.
7. liao huan's application of IFIX in centralized monitoring system of tobacco factory [J]. Journal of hunan university of science and technology,2012,33(11):109-11.
8. Jie Daoliang, Research and Practice of Automatic Data Acquisition System for Tobacco Enterprises [J]. Automation Instrument,2006(03):52-55.
9. Chaoyang Fu'Ji ashen Zheng, Jingmao Zhao, et. al. Application of grey relational analysis for corrosion failure of oil tubes [J]. Corrosion Science,2001,43(5):81-89.
10. Miguel Aria-Albomoz, Heman.Sanhueza Hardy. Distribution Networl Con69uration for Minimum Enemy Supply Cost[J].IEEE Transactions on Power Systems. 2004,19(1): 38-542.
11. Liu LAN, Research and Application of port production operation Risk Control Training and Examination System [D]. Tianjin: Tianjin University of Technology,2014, DOI: 10.7666/ d.Y2519605.

Fault characteristics of Inverter-Interfaced Distributed Generation

Xiaoqin Ma[1,*], Hui Lv[1], and Wenjuan Xiao[1]

[1] Northwest Minzu University, 730030, No.1 northwest new village, Lanzhou City, Gansu Province, China

Abstract. Distributed generations can be divided into traditional synchronous generators and inverter interfaced distributed generations (IIDG) according to their different operation mode. While the fault characteristics of IIDG is different from traditional Distributed Generators, the in-depth analysis on output characteristics and fault characteristics of IIDG are the foundation of micro grid protection. The mathematical models under P/Q and V/F control strategy are discussed. After simplifying its model, the fault characteristics of IIDG under P/Q and V/F control strategy are studied using the simulation of PSCAD/EMTDC. The analysis on the fault characteristics of IIDG that lays the theoretical foundation for the micro-grid relay protection.

1 Introduction

As a key technology in smart grid, distributed power supply technology has attracted more attention [1].With the rapid development of power electronics technology, inverter interfaced distributed generation (IIDG) has been widely used [2]. The research on traditional rotating distributed generation (DG) is relatively mature for the time being, while limited research has been conducted on IIDG. Therefore, it is necessary to analyze IIDG and its fault output characteristics for the distribution network protection [3, 4].

2 Modelling of IIDG

In order to study the fault characteristics of IIDG, it is necessary to simplify its model reasonably. Firstly, there is always a capacitor on the bus in front of the DC side of the IIDG, the existence of this energy storage device makes the output voltage of the IIDG DC side basically constant; Secondly, the capacitor between the IIDG inverter and its DC side can maintain the output voltage of IIDG and also provide energy for the IIDG in the transient state; Finally, environmental factors such as illumination and wind speed can be considered as invariable in tens of milliseconds of power system relay protection action, so the impact of environmental factors on IIDG can be ignored. Therefore, in the process of IIDG transient analysis, DC source can be approximately used to replace the output of IIDG DC side without loss of generality [5].

The simplified IIDG model is shown in Fig.1. It can be seen from the figure that the transient characteristics of IIDG mainly depend on the control strategy of the inverter, other factors can be ignored.

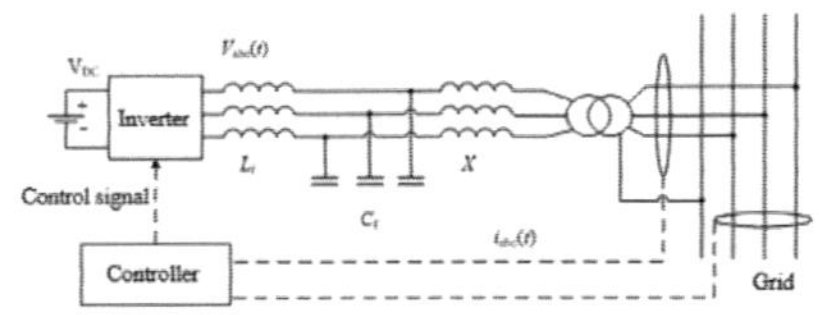

Fig1. The schematic diagram of IIDG

This shows that the influence of the control strategy of IIDG inverter on its fault characteristics is more important. Therefore, the mathematical model of IIDG is established respectively under P/Q and V/F control strategy.

2.1 Interface model of IIDG under P/Q control

When the inverter adopts P/Q control strategy, that is, the controller in Figure 1 adopts P/Q control strategy, its control links mainly include three parts, namely power control link, PI control link and limiting control link. The specific control process is shown in Fig.2.

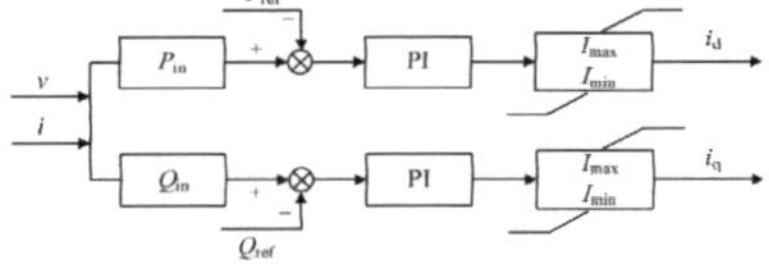

Fig2. P/Q control strategy

The calculation formula of power is:

$$P_{\mathrm{in}} = v_d i_d + v_q i_q \quad (1)$$

$$Q_{\mathrm{in}} = v_d i_q - v_q i_d \quad (2)$$

The PI control formula is:

$$i_{\mathrm{dr}} = (k_{\mathrm{p}} + \frac{1}{sT_{\mathrm{p}}})\Delta p \quad (3)$$

$$i_{\mathrm{qr}} = (k_{\mathrm{q}} + \frac{1}{sT_{\mathrm{q}}})\Delta q \quad (4)$$

*Xiaoqin Ma: 657995062@qq.com

2.2 Interface model of IIDG under V/F control

When the inverter adopts V/F control strategy, its control links mainly include four parts, power control link, drop control link, PI control link and integral control link, the specific control process is shown in Fig.3.

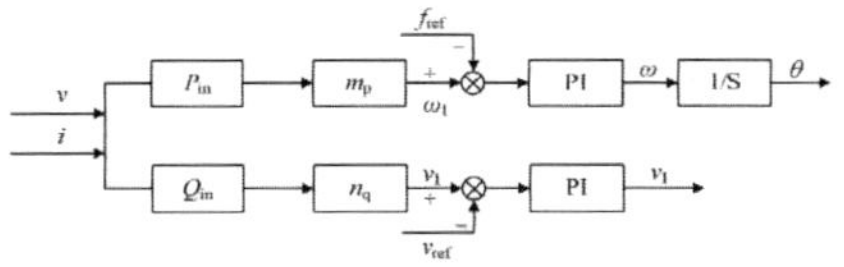

Fig3. V/F control strategy

The power calculation process is the same as that of P/Q control.

The drop control formula is:

$$\omega_1 = \omega_n - m_p p \qquad (5)$$

$$v_1 - v_n - n_q q \qquad (6)$$

Where m_p is the slope of active power droop curve, n_q is the slope of reactive power droop curve.

3 Characteristic analysis for IIDG

Since the control strategy of the inverter is the main factor affecting the transient characteristics of IIDG, it is necessary to analyze the fault characteristics of IIDG under different control strategies [6]. In addition, since DG has two operation modes, independent DG generally adopts V/F control strategy to maintain the stability of microgrid frequency and voltage, while grid connected DG usually adopts P/Q control strategy [7, 8]. Therefore, this section mainly analyzes the fault characteristics of IIDG under P/Q control and V/F control strategy in the power system transient simulation software PSCAD /EMTDC without considering the protection of inverter and IIDG itself.

3.1 Simulation analysis on fault characteristics of IIDG under P/Q control strategy

The simulation model is established under PSCAD, as shown in Fig. 4. At this time, IIDG adopts P/Q control strategy for grid connected operation.

The P / Q control parameters are k_p=1, k_q=1, T_p=0.4, T_q=0.5, I_{max}=1, I_{min}=-1. The voltage of the line is 380V and the length of the line is 0.2km, the parameters of the line are R=0.797Ω/km, X=0.105Ω/km. The fault characteristics of IIDG under P / Q control strategy are obtained by simulation analysis of various short-circuit faults in transmission lines under P / Q control strategy.

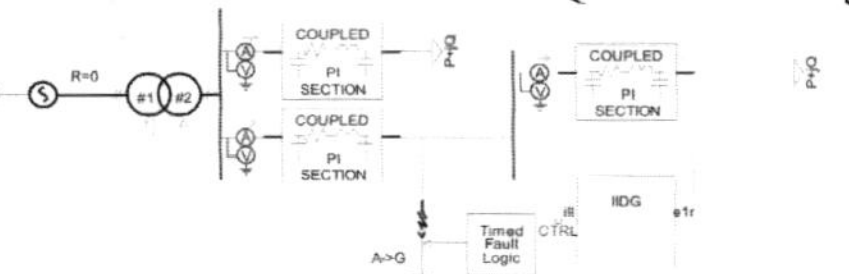

Fig4. The simulation model under P/Q control strategy

3.1.1 Three-phase short circuit fault

When the three-phase short circuit fault occurs at 50% of the line, the output current, active power and terminal voltage waveform of the fault phase of the IIDG inverter under P/Q control strategy are shown in Fig. 5.

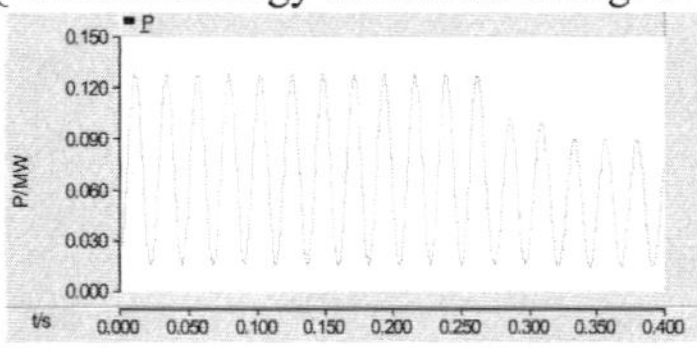

(a) active power waveform

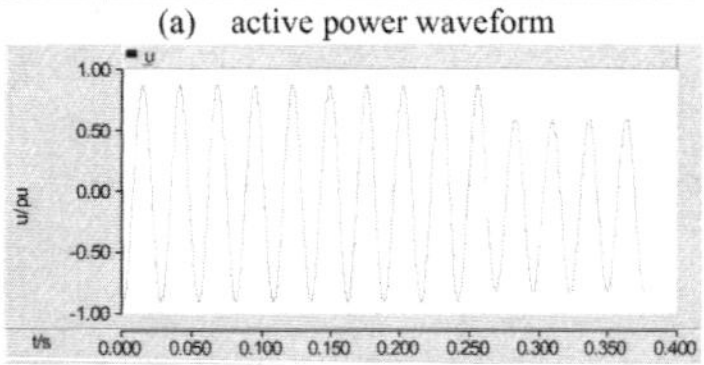

(b) terminal voltage waveform

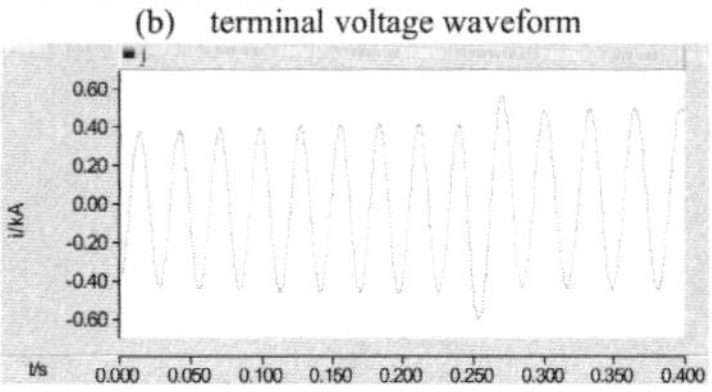

(c) output current waveform

Fig5. Waveforms of P/Q controlled inverter when three-phase short circuit fault occurred

This shows the voltage and output active power of IIDG after fault are significantly reduced compared with normal operation. It can be seen from the figure that the instantaneous maximum fault current is about 1.3 times of that under normal operation.

3.1.2 Two-phase short circuit fault

When the two-phase short circuit fault occurs at 50% of the line, the output current, active power and terminal voltage waveform of the fault phase of the IIDG inverter under P/Q control strategy are shown in Fig. 6.

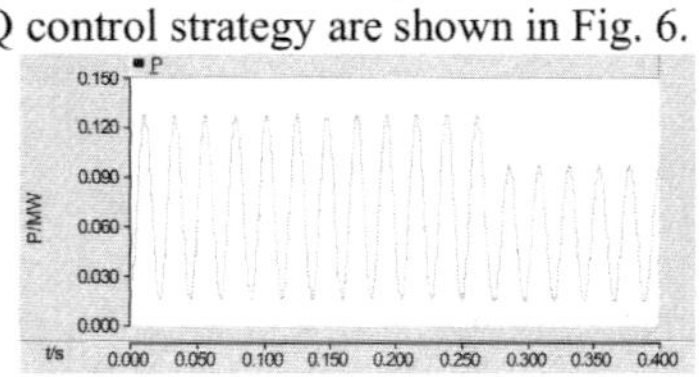

(a) active power waveform

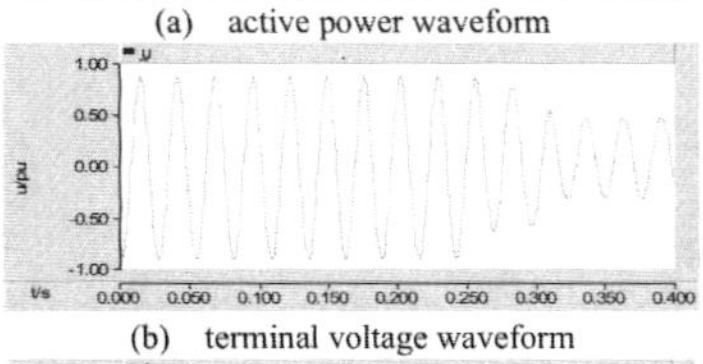

(b) terminal voltage waveform

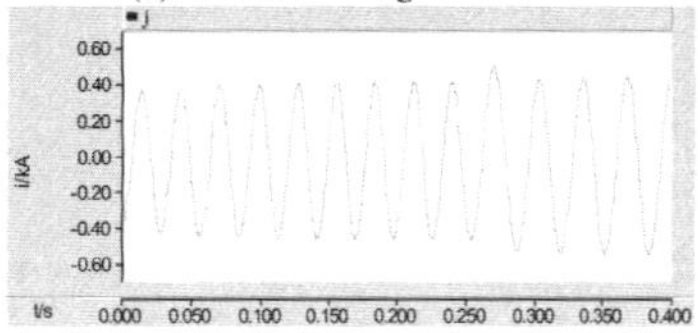

(c) output current waveform

Fig6. Waveforms of P/Q controlled inverter when two-phase short circuit fault occurred

At this time, the output power of the fault phase of the IIDG inverter decreases slightly, and the terminal voltage

 https://doi.org/10.1051/e3sconf/202123702020

drops greatly. The fault current is about 1.3 times of the normal operation current.

3.1.3 Two-phase grounding short circuit fault

When two-phase grounding short circuit fault occurs at 50% of the line, the output current, active power and terminal voltage waveform of the fault phase of the IIDG inverter under P/Q control strategy are shown in Fig. 7.

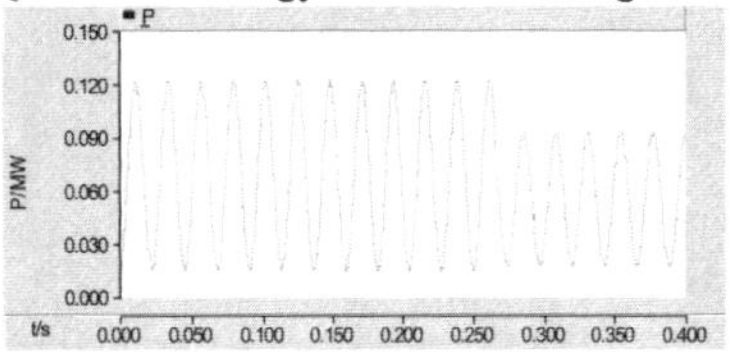

(a) active power waveform

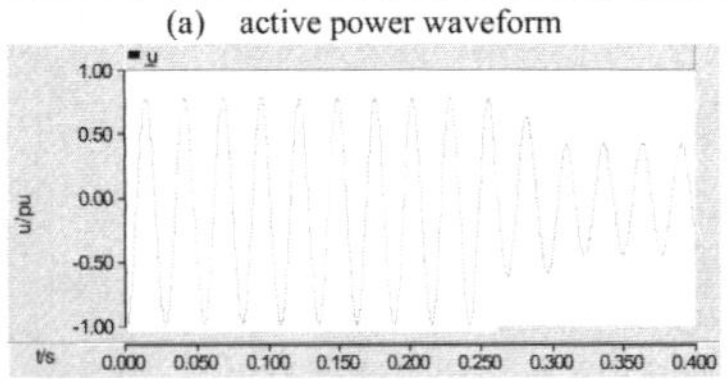

(b) terminal voltage waveform

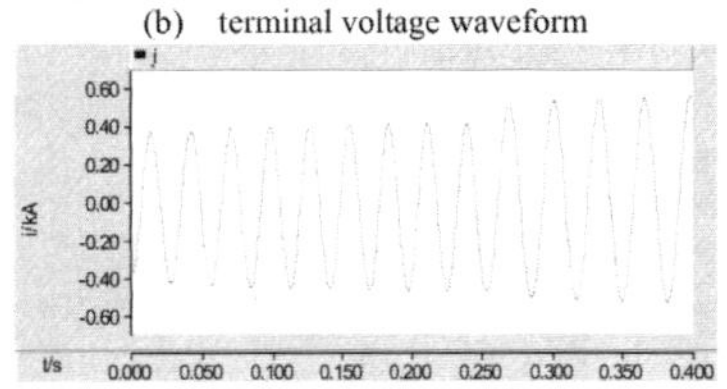

(c) output current waveform

Fig7. Waveforms of P/Q controlled inverter when two-phase grounding short circuit fault occurred

At this time, the output power and terminal voltage of the fault phase of the IIDG inverter decreases greatly. The fault current is about 1.5 times of the normal operation current.

3.1.4 Single-phase grounding short circuit fault

When single-phase grounding short circuit fault occurs at 50% of the line, the output current, active power and terminal voltage waveform of the fault phase of the IIDG inverter under P/Q control strategy are shown in Fig. 8.

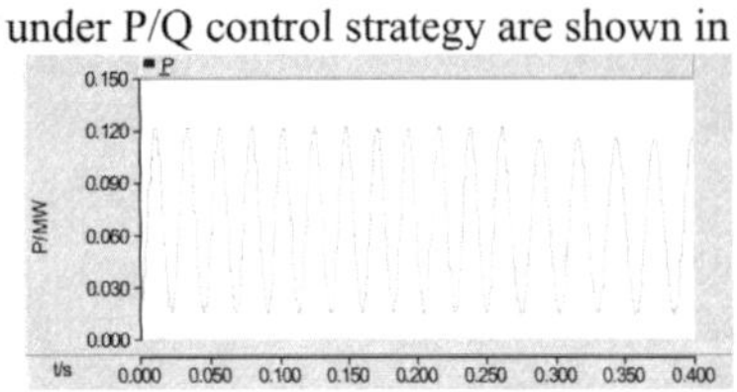

(a) active power waveform

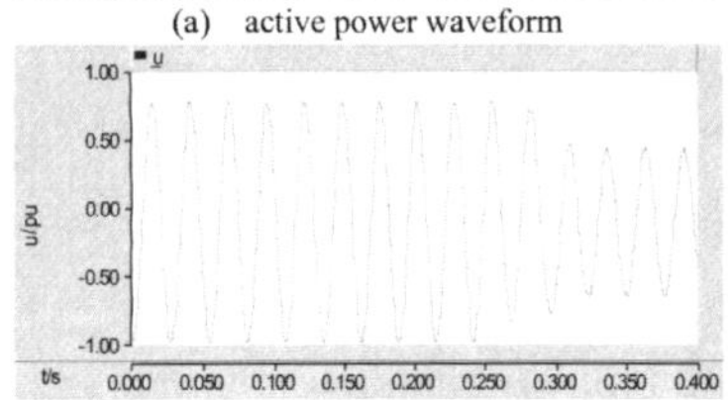

(b) terminal voltage waveform

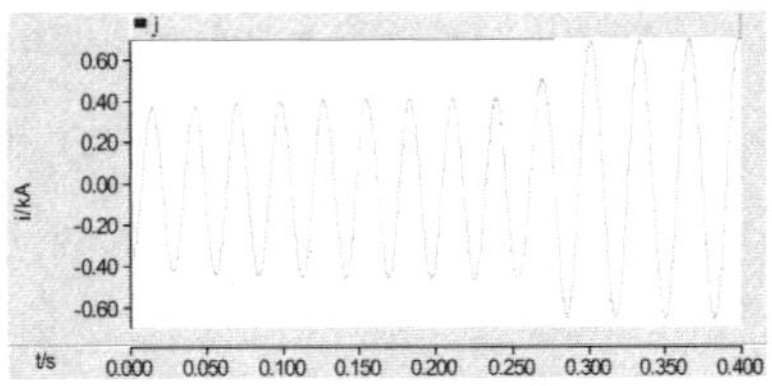

(c) output current waveform

Fig8. Waveforms of P/Q controlled inverter when single phase grounding short circuit fault occurred

At this time, the output power of the fault phase of the IIDG inverter is not changed basically, and the terminal voltage decreases slightly, the fault current is about 1.5 times of the normal operation current.

According to the simulation analysis, the fault characteristics data of IIDG under P/Q control strategy are shown in Table 1.

Table1. Normal and fault data under P/Q controlled inverter

	P (MW)	U (pu)	I (kA)
Normal operation	0.13	1.0	0.40
Three-phase short circuit	0.08	0.42	0.62
Two-phase short circuit	0.09	0.51	0.51
Two-phase grounding short circuit	0.09	0.49	0.62
Single-phase grounding short circuit	0.11	0.52	0.72

It can be seen from Table 1 that the maximum fault current generated by IIDG is about 1.74 times of rated load current in case of single-phase grounding short-circuit fault, and the minimum fault current generated by IIDG under two-phase short-circuit is about 1.24 times of rated load current, indicating that the zero sequence impedance of IIDG is less than its positive sequence impedance.

The terminal voltage of IIDG decreases the least when single-phase grounding short circuit occurs, which is 0.52 times of the rated voltage; the terminal voltage of IIDG decreases the most when the three-phase short circuit occurs, which is 0.42 times of the rated voltage. This shows that the closer the fault point is to the IIDG, the greater the influence on the terminal voltage is, and the output power is reduced compared with the normal value. At this time, the output power of IIDG cannot maintain the setting value.

To sum up, the terminal voltage of IIDG under P/Q control strategy will decrease when its adjacent feeders are short circuited, so the output active power of IIDG will be affected, which is lower than the rated value. The output fault current of IIDG is not only affected by the maximum allowable current of internal power electronic devices, but also closely related to the output power and voltage of IIDG.

 https://doi.org/10.1051/e3sconf/202123702020

3.2 Simulation analysis on fault characteristics of IIDG under V/F control strategy

The simulation model is established under PSCAD, as shown in Fig. 9. IIDG with V/F control strategy and synchronous generator with active power output of 0.01MW are the two power sources of the system.

The V / F control parameters are k_p=1, kq=1, Tp=10, Tq=10, mp=0.001, nq=0.001.

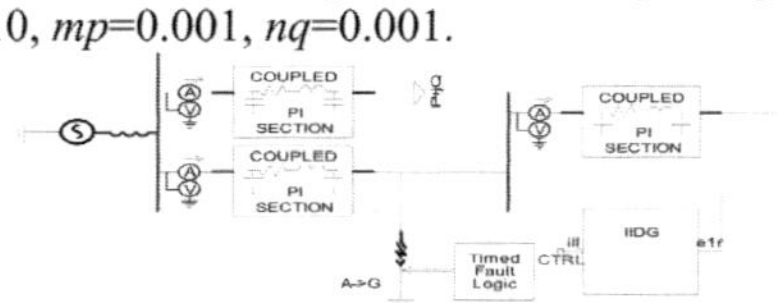

Fig9. The simulation model under V/F control strategy

3.2.1 Three-phase short circuit fault

When the three-phase short circuit fault occurs at 50% of the line, the output current, active power and terminal voltage waveform of the fault phase of the IIDG inverter under V/F control strategy are shown in Fig. 10.

The voltage and output active power of IIDG after fault are significantly reduced compared with normal operation. It can be seen from the figure that the instantaneous maximum fault current is about 1.3 times of that under normal operation.

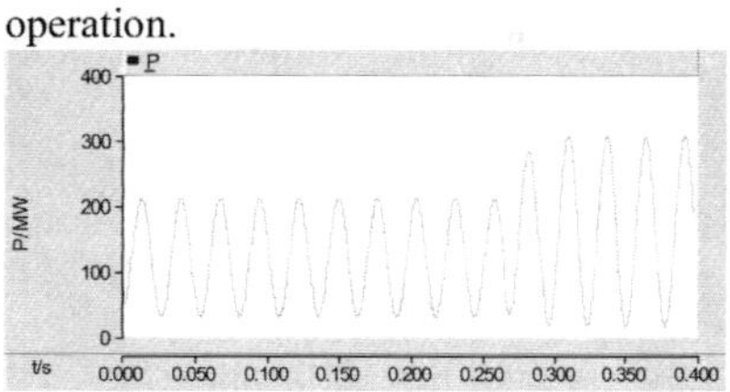

(a) active power waveform

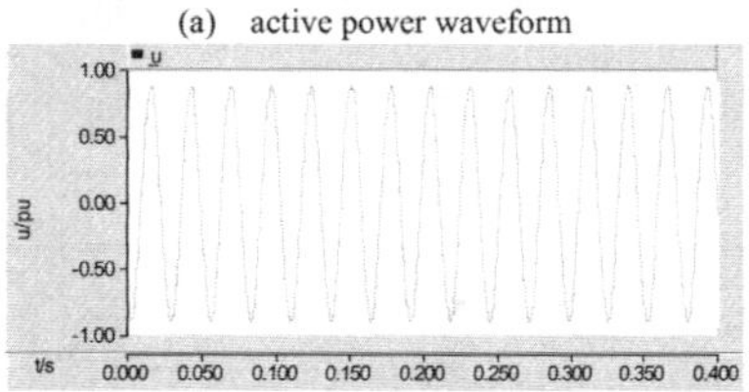

(b) terminal voltage waveform

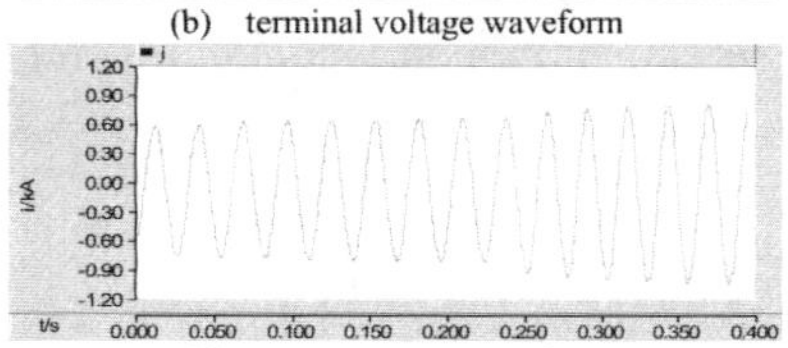

(c) output current waveform

Fig10. Waveforms of V/F controlled inverter when three-phase short circuit fault occurred

It can be seen from Fig.10 that the terminal voltage of IIDG is basically unchanged at this time.

3.2.2 Two-phase short circuit fault

When the two-phase short circuit fault occurs at 50% of the line, the output current, active power and terminal voltage waveform of the fault phase of the IIDG inverter under V/F control strategy are shown in Fig.11.

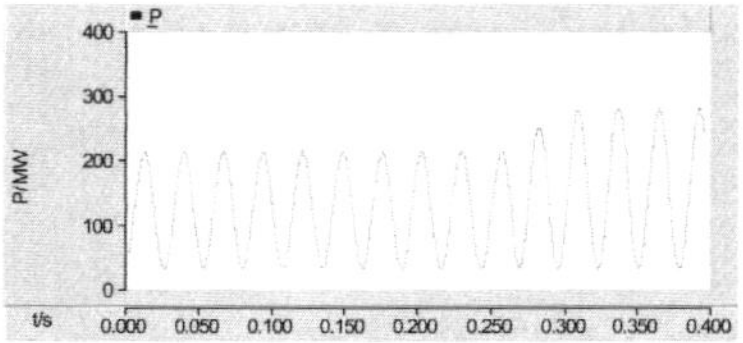

(a) active power waveform

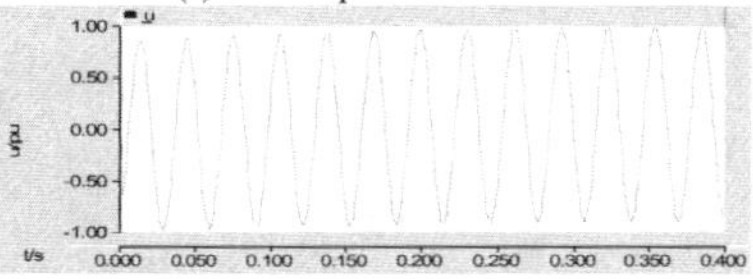

(b) terminal voltage waveform

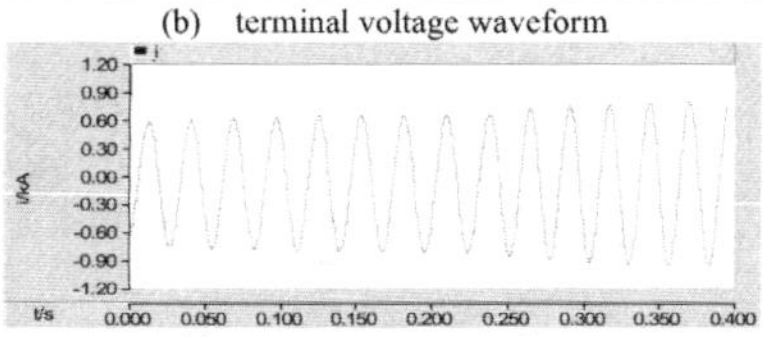

(c) output current waveform

Fig11. Waveforms of V/F controlled inverter when two-phase short circuit fault occurred

It can be seen from Fig.11 that the terminal voltage of IIDG is basically unchanged at this time.

3.2.3 Two-phase grounding short circuit fault

When two-phase grounding short circuit fault occurs at 50% of the line, the output current, active power and terminal voltage waveform of the fault phase of the IIDG inverter under V/F control strategy are shown in Fig. 12.

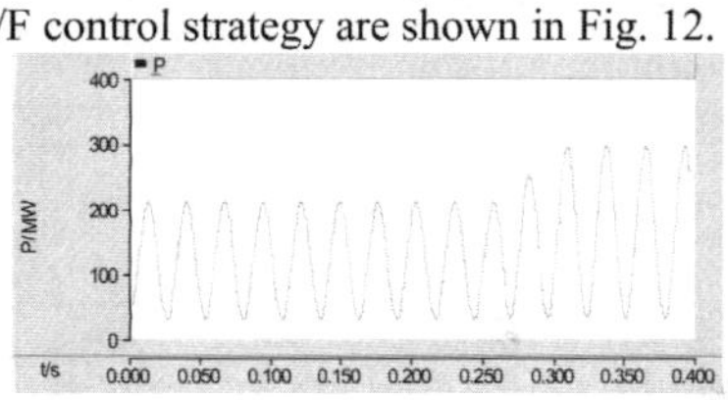

(a) active power waveform

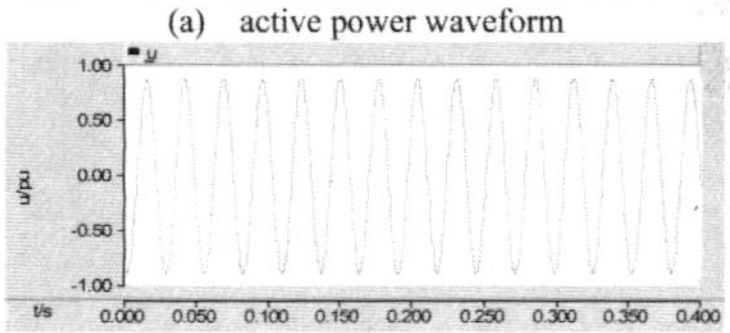

(b) terminal voltage waveform

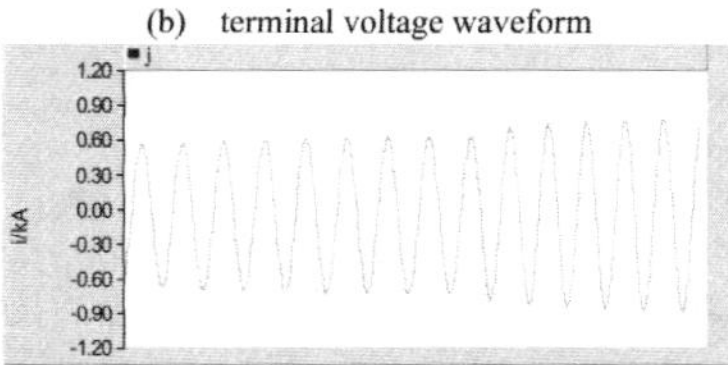

(c) output current waveform

Fig12. Waveforms of V/F controlled inverter when two-phase grounding short circuit fault occurred

It can be seen from Fig.12 that the strategy of V/F control has excellent performance.

3.2.4 Single-phase grounding short circuit fault

When single-phase grounding short circuit fault occurs at 50% of the line, the output current, active power and terminal voltage waveform of the fault phase of the IIDG inverter under V/F control strategy are shown in Fig. 13. It can be seen from Fig.13 that the strategy of V/F control play a good role in maintaining terminal voltage of IIDG stability.

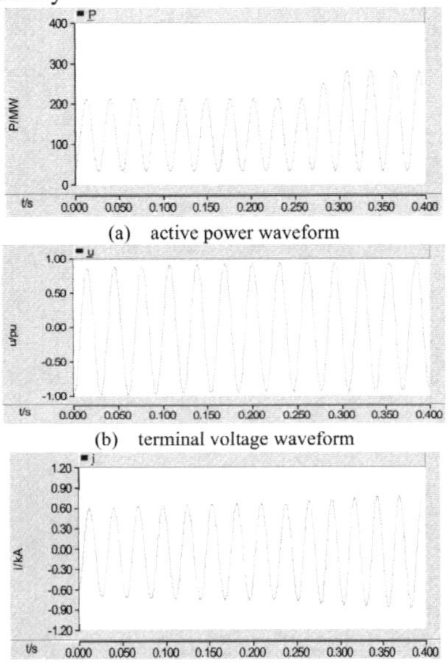

(a) active power waveform

(b) terminal voltage waveform

(c) output current waveform

Fig13. Waveforms of V/F controlled inverter when single phase grounding short circuit fault occurred

According to the simulation analysis, the fault characteristic data under V/F control strategy are shown in Table 2.

Table2. Normal and fault data underV/F controlled inverter

	P (MW)	U (pu)	I (kA)
Normal operation	0.21	1.0	0.60
Three-phase short circuit	0.36	1.01	1.04
Two-phase short circuit	0.30	1.01	1.01
Two-phase grounding short circuit	0.31	1.01	1.02
Single-phase grounding short circuit	0.30	1.01	1.01

The simulation experiment shows that that IIDG under V/F control strategy can play a better role, so that the terminal voltage of IIDG can be basically kept constant when the system has a short-circuit fault. In addition, due to the upper limit of active power in this control strategy, and its terminal voltage is stable around the rated voltage, the fault current of IIDG has little difference.

4 Conclusions

Firstly, the model of IIDG is simplified, and the mathematical models of IIDG are established under P/Q control strategy and V/F control strategy. Then, the simulation model of IIDG is established by power system transient simulation software PSCAD. The fault characteristics of IIDG under P/Q control strategy and V/F control strategy are analyzed respectively. Through the above analysis, the following conclusions are obtained:

(1) IIDG under P/Q control strategy has the lowest terminal voltage when three-phase short circuit occurs, and the highest terminal voltage when single-phase ground short-circuit occurs; the minimum fault current occurs when two-phase short-circuit occurs, and the maximum fault current occurs when single-phase ground short-circuit occurs. The positive sequence impedance of IIDG is greater than its zero sequence impedance. The controlled voltage source after reactance can be equivalent to IIDG. The controlled condition is that the output power of IIDG is constant, and the constraint condition is that the outlet current value of IIDG is less than 2 times of its rated current value.

(2) IIDG under V/F control strategy, the fault voltage of IIDG is basically maintained near the rated voltage when the fault point is far away from the outlet of IIDG. Since the upper limit of active power is set in V / F control strategy, the difference of fault current at IIDG outlet is not obvious when various short circuit faults occur. In this case, the controlled current source after reactance can be used to equivalent to IIDG. Obviously, the controlled condition is that the terminal voltage remains constant, while the constraint condition is that the outlet current value of IIDG is less than 2 times of its rated current value.

References

1. B. Zhao, C.S. Wang, J.H. Zhou, AEPS, **38**, 125-135(2014)
2. G. Pepermans, J. Driesen, D. Haeseldonckx, Energy Policy, **33**, 787-798(2005)
3. X. Yang, X. Guo, H.Y. Zang, Journal of Electrical Engineer, **27**, 63-68(2012)
4. C.H. Liang, X.Z. Duan, AEPS, **25**, 53-56(2001)
5. C.H. Cheng, JGSCASS, 113-118(2009)
6. L.C. Wang, H.L. Gao, PCMPS, **2**, (2017)
7. V. Sridhar, S. Umashankar, RSER,78(2017)
8. E. Gnani, M. Visciarelli, A. Gnudi. Solid State Electronics (2019)

Economic Evaluation of CCUS Retrofitting of Coal-fired Power Plants Based on Net Cash Flow

Jingqi Jin[1], *Feng* Xue[1], *Bin* Cai[1,*], *Xinxin* Yang[2], *Yening* Lai[1], *Dalin* Jiang[3], *Yalin* Mao[3], and *Yi* Tao[3]

[1]NARI Group Corporation/State Grid Electric Power Research Institute, Nanjing 211106, PR China
[2]School of Electrical Engineering, Southeast University, 2 Sipailou, Nanjing 210096, PR China
[3]Shenhua Science and Technology Research Institute Co.,Ltd.,201 Shenhua S&T Innovation Base, Beijing 102211, PR China

Abstract. Carbon Capture, Utilization and Storage (CCUS) is one of the key technologies for realizing large-scale low-carbon utilization of coal-fired power plants in service. How to evaluate its economics is crucial to the decision-making of traditional coal-fired power enterprises. This paper analyzes the changes in the physical, emission and economic parameters of in-service coal-fired power plants without and with the CCUS retrofit. A method for evaluating the economic feasibility of coal-fired power plants retrofitting based on net cash flow is proposed, which compares the impact of CCUS retrofit on the net present value of the remaining life cycle of the power plant. The impact of uncertain parameters such as carbon dioxide sales unit price, carbon capture device operating cost, free carbon quota, and carbon emission right price on the evaluation results are analyzed.

1 Introduction

The clean and low-carbon transition has become a global consensus, but large-scale coal-fired power will still occupy an important position in the power production structure for a period of time in the future[1-2]. After coal-fired power has achieved ultra-low emission of pollutants, carbon emissions have become the primary issue affecting the sustainable development of coal-fired power enterprises[3].

Carbon Capture, Utilization and Storage (CCUS) refers to an industrial process that separates carbon dioxide (CO_2) from industrial emission sources or directly uses or stores it to achieve CO_2 emission reduction[4]. The International Energy Agency points out that CCUS is the only way to significantly reduce CO_2 emissions from fossil fuels[5]. The Intergovernmental Panel on Climate Change also emphasizes that CCUS technology is an important lifeline to combat climate change, and it will be difficult to achieve emission reduction targets without CCUS[6-7]. According to the energy revolution roadmap of The National Development and Reform Commission/ National Energy Administration, China is expected to realize the commercial application of CCUS in 2030[8]. The literature [4] has formulated China's CCUS technology development roadmap. It is believed that the cost of CCUS is expected to drop rapidly, and the application scale of CCUS is expected to reach 1-3 billion tons/year by 2050.

Allowing investors to obtain positive economic returns is the key to the large-scale commercial application of CCUS technology. Existing research generally uses real option method and net present value method to evaluate CCUS investment. The literature [9] evaluates the cost of electricity loss caused by CCS retrofit through the net profit per unit power generation of coal-fired power plants given exogenously. The literature [10] regards the investment cost of CCS project as the only cost of CCS retrofit, but none of the above studies has conducted a detailed analysis of the cash inflow and outflow of the entire coal-fired power plant after CCS retrofit. The literature [11] carries out an economic evaluation on the complete process of CCS after combustion, and analyzes the cost and benefit of CCS technology in detail, but does not analyze coal-fired power plants and CCS technology as a whole. In view of the shortcomings of the above research, this paper analyzes the coal-fired power plant and CCUS retrofit as a whole, and calculates its net cash flow year by year during the remaining life cycle.

The remainder of the paper is structured as follows: section 2 introduces the evaluation method. Section 3 is the case study, which introduces the baseline scenario and uncertain parameters. Section 4 is the result analysis, which analyzes the research results under the baseline scenario and the sensitivity of uncertain parameters. The last section is the conclusion.

2 Methodology

2.1. Evaluation method

The basic idea of this assessment method is: first, calculate the net cash flow of the retrofit without CCUS and retrofit with CCUS during the remaining life cycle

* Corresponding author: caibin@sgepri.sgcc.com.cn

from the current moment to decommissioning for the in-service coal-fired power plant to be assessed; then, compare the net present value difference after discounting to the current year at a certain discount rate; finally, the economic feasibility of CCUS retrofit for coal-fired power plants is evaluated through the change of the net present value difference.

2.2. Calculation method

Net cash flow refers to the difference between cash inflow and cash outflow in a certain period. In this study, the year is taken as the calculation step. The current year is recorded as t_0, the year when the coal-fired power plant is decommissioned is recorded as t_f, and the year when CCUS retrofit is implemented is recorded as t_r, where $t_0 \leqslant t_r \leqslant t_f$. The calculation is divided into the following three parts.

(1) The net cash flow of the coal-fired power plant in each period without the retrofit can be expressed as:

$$NCF_i = C_i^{in} - C_i^{out} \quad (1)$$

$$C_i^{in} = R_i^{pgs} + R_i^{VATs} + R_i^{rvfa} \quad (2)$$

$$C_i^{out} = C_i^{ce} + C_i^{rlp} + C_i^{rli} + C_i^{o} + C_i^{pcVAT} + C_i^{st} + C_i^{pit} + C_i^{moi} \quad (3)$$

Where NCF_i represents net cash flow, C_i^{in} represents cash inflow, C_i^{out} represents cash outflow, R_i^{pgs} represents power generation sales revenue, R_i^{VATs} represents VAT subsidy income, R_i^{rvfa} represents recovery of residual value of fixed assets, C_i^{ce} represents capital expenditure, C_i^{rlp} represents repayment of loan principal, C_i^{rli} represents repayment of loan interest, C_i^{o} represents operating costs, C_i^{pcVAT} represents pay current VAT, C_i^{st} represents sales tax and surcharges, C_i^{pit} represents pay income tax, C_i^{moi} represents maintain operating investment.

(2) The net cash flow of the coal-fired power plant in each period with the retrofit can be expressed as:

In the time range t_r to t_f, the net cash flow NCF_i^{ccus} of the target coal-fired power plant in service should be calculated after the CCUS retrofit for each time period. CCUS and coal-fired power plants are decommissioned together, and CCUS has completed repayment and depreciation before decommissioning. Update related technical and economic parameters for the coal-fired power plant after CCUS retrofit.

$$NCF_i^{ccus} = C_i^{in,ccus} - C_i^{out,ccus} \quad (4)$$

$$C_i^{in,ccus} = R_i^{pgs,ccus} + R_i^{si} + R_i^{rvfa,ccus} + R_i^{cds} \quad (5)$$

$$C_i^{out,ccus} = C_i^{ce,ccus} + C_i^{rlp,ccus} + C_i^{rli,ccus} + C_i^{o,ccus} + C_i^{pcVAT,ccus} + C_i^{st,ccus} + C_i^{pit,ccus} + C_i^{moi,ccus} + C_i^{cdt} + C_i^{r} \quad (6)$$

Where NCF_i^{ccus} represents net cash flow with CCUS retrofit, $C_i^{in,ccus}$ represents cash inflow with CCUS retrofit, $C_i^{out,ccus}$ represents cash outflow with CCUS retrofit, $R_i^{pgs,ccus}$ represents power generation sales revenue with CCUS retrofit, R_i^{si} represents the total subsidy income with CCUS retrofit, $R_i^{rvfa,ccus}$ represents recovery of residual value of fixed assets with CCUS retrofit, R_i^{cds} represents carbon dioxide sales revenue with CCUS retrofit, $C_i^{ce,ccus}$ represents capital expenditure with CCUS retrofit, $C_i^{rlp,ccus}$ represents repayment of loan principal with CCUS retrofit, $C_i^{rli,ccus}$ represents repayment of loan interest with CCUS retrofit, $C_i^{o,ccus}$ represents operating costs with CCUS retrofit, $C_i^{pcVAT,ccus}$ represents pay current VAT with CCUS retrofit, $C_i^{st,ccus}$ represents sales tax and surcharges with CCUS retrofit, $C_i^{pit,ccus}$ represents pay income tax with CCUS retrofit, $C_i^{moi,ccus}$ represents maintain operating investment with CCUS retrofit, C_i^{cdt} represents carbon dioxide transportation cost with CCUS retrofit, C_i^{r} represents running cost with CCUS retrofit.

(3) The net present value of the coal-fired power plant without and with retrofit can be expressed as:

$$NPV_{without\ CCUS} = \sum_{i=t_r}^{t_f} \frac{NCF_i}{(1+r)^{t_r-i}} \quad (7)$$

$$NPV_{with\ CCUS} = \sum_{i=t_r}^{t_f} \frac{NCF_i^{ccus}}{(1+r)^{t_r-i}} \quad (8)$$

Where $NPV_{without\ CCUS}$ represents net present value without CCUS retrofit, $NPV_{with\ CCUS}$ represents net present value with CCUS retrofit, r represents discount rate.

Calculate and discount the net cash flow without and with the CCUS retrofit to the current year's net present value at a certain discount rate, and compare the changes in the net present value without and with the CCUS technology retrofit in the remaining life cycle. If $NPV_{with\ CCUS} - NPV_{without\ CCUS} > 0$ and $NPV_{with\ CCUS} > 0$, then it is judged that CCUS retrofit is worth investing in; if $NPV_{with\ CCUS} - NPV_{without\ CCUS} = 0$, then it is expressed as a critical state; if $NPV_{with\ CCUS} - NPV_{without\ CCUS} < 0$, then it is judged that CCUS retrofit is not worth investing in.

3 Case study

3.1. Baseline scenario parameters setting

This paper takes an in-service coal-fired power plant as an example for analysis. The parameters of the baseline scenario are set as follows:

(1) Time parameters

The coal-fired power plant has put a thermal power unit into operation in 2010 and it will be decommissioned at the end of 2044. The current year is 2019. The design depreciation period is 20 years and the repayment period is 15 years.

(2) Physical parameters

The installed capacity is 600MW, the annual power utilization hours are 4000 hours, and the auxiliary power rate is 5%.

(3) Emission parameters

Carbon emission right price is 40 yuan/ton, and it will increase at a rate of 1 yuan/ton per year; the proportion of free carbon quota in 2019 is 100%, which will be reduced by 1% per year.

(4) Economic parameters

The coal price is 500 yuan/ton of standard coal, and the benchmark loan interest rate is 4.9%.

The specific data are shown in Table 1.

The parameters for the CCUS retrofit of in-service coal-fired power plant are as follows:

(1) Time parameters

The coal-fired power plant will complete the CCUS retrofit in 2030 and it will be decommissioned together with the thermal power unit at the end of 2044. The design depreciation period of CCUS equipment is 10 years, and the repayment period is 10 years.

(2) Physical parameters

The annual carbon dioxide capture capacity of the CCUS device is 300,000 tons. The additional coal consumption due to the operation of the CCUS device is 0.1 ton of standard coal/ton carbon dioxide.

(3) Economic parameters

The cost of CCUS is 500 yuan per ton of carbon dioxide. The annual operating cost is 188 yuan/ton carbon dioxide. All the captured CO_2 is transported and sold at a unit price of 450 yuan/ton carbon dioxide.

The specific data are shown in Table 2.

Table 1. Parameters table of coal-fired power plant under the baseline scenario.

Parameters	**Value**	**Unit**
Installed capacity	600	Megawatt
Annual power utilization hours	4000	Hours
Auxiliary power rate	5	%
Coal consumption per unit of power supply (standard coal)	330	g/kwh
Coal price (tax included)	500	Yuan/tce
Value-added tax rate	13	%
Carbon emission right price	40	Yuan/ton carbon dioxide
Proportion of free carbon quota	100	%
On-grid electricity price (tax included)	0.391	Yuan/kwh
Discount rate	5	%
Comprehensive tax rate	12	%
Operation and maintenance cost	48	Yuan/MWh
Manufacturing cost	3390	Yuan/kw
Total investment	2034	Million yuan
Capital	406.8	Million yuan
Capital ratio	20	%
Residual value	91.53	Million yuan

Table 2. CCUS parameters table under the baseline scenario.

Parameters	**Value**	**Unit**
Annual carbon dioxide capture	30	Ten thousand tons
Coal consumption due to wastage	0.1	Tce/ton carbon dioxide
CO_2 sales unit price (tax included)	450	Yuan/ton carbon dioxide
CO_2 transportation unit price (tax included)	50	Yuan/ton carbon dioxide
Carbon capture device operating cost	188	Yuan/ton carbon dioxide
Manufacturing cost	500	Yuan/ton carbon dioxide·year
Total investment	150	Million yuan
Capital	30	Million yuan
Loan	120	Million yuan
Residual value	6.75	Million yuan

3.2 Uncertain parameters

A large number of uncertain parameters are involved in the evaluation of CCUS retrofit of coal-fired power plants. The uncertain parameters that affect the economic feasibility of the retrofit can be divided into: the parameters that only affect the net present value with the CCUS retrofit and the parameters that both affect the net present value without and with the CCUS retrofit. The former mainly includes the parameters related to CCUS, such as CO_2 sales unit price, CO_2 transportation unit price, carbon capture device cost and carbon capture device operating cost, etc.; the latter mainly includes carbon emission right price, free carbon quota, coal price, and annual power utilization hours.

The key uncertain parameters analyzed in this study are shown in Table 3.

Table 3. Uncertain parameters table.

Parameters	**Value**	**Unit**	**Variation range**
CO_2 sales unit price (tax included)	450	Yuan/ton carbon dioxide	50-700
CO_2 transportation unit price (tax included)	50	Yuan/ton carbon dioxide	-
Carbon capture device cost	500	Yuan/ton carbon dioxide·year	-
Carbon capture device operating cost	188	Yuan/ton carbon dioxide	50-400
Annual increase in carbon emission right price	1	Yuan/ton carbon dioxide	1-10
Annual reduction rate of free carbon quota	1	%	1-5
Coal price (tax included)	500	Yuan/tce	-

 https://doi.org/10.1051/e3sconf/202123702021

4 Results

4.1. Results analysis of the baseline scenario

Figure 1 shows the estimated annual net cash flow of coal-fired power plants from 2019 to 2044 under the baseline scenario. The orange and blue curves represent the situation without and with CCUS retrofit respectively. The net present value of the coal-fired power plant has increased from 1.55 billion yuan when the CCUS retrofit is not carried out to 1.72 billion yuan when the retrofit is carried out. Therefore, it is economically feasible to carry out CCUS retrofit in 2030.

Compared with no CCUS retrofit, both the cash inflow and cash outflow of coal-fired power plant have changed with CCUS retrofit. Among them, the sales income of the captured carbon dioxide increased the cash inflow, and the investment and operation of the carbon capture device (including the increase of coal consumption) and carbon dioxide transportation cost increased the cash outflow (see Figure 2). During the remaining life cycle of the coal-fired power plant, the discounted value of cash inflows increased by 866.24 million yuan, and the discounted value of cash outflows increased by 690.29 million yuan. Among them, the largest proportion of the increase in cash outflow is operating costs, which is 4.5406 million yuan (66%), and the total value-added tax and income tax paid increased by 134.78 million yuan (20%), and the principal and interest of CCUS loans is 73.3 million yuan (11%), which added a total of 436.8 million yuan in net cash flow.

It should be pointed out that the impact of CCUS retrofit on cash inflows and outflows in different years is different. For example, in 2029, the net cash flow decreased by 30 million yuan due to a one-time expenditure of capital for the construction of carbon capture device but no new income; in 2030, the cash inflow from the sale of carbon dioxide was increased by 135 million yuan, but additional cash outflows such as operating costs (73 million yuan), principal and interest repayment (18 million yuan), and taxes (20 million yuan) were added at the same time, and net cash flow increased by 23.70 million yuan.

The two curves in Figure 1 fluctuate to varying degrees in some years. Coal-fired power plants have completed repayment, coal-fired power plants have completed depreciation of fixed assets, CCUS investment deductible value-added input tax has been deducted, and CCUS installations have completed depreciation and repayment, payments and the residual value of fixed assets recovered when power plants and carbon capture devices are decommissioned are the reasons for the changes in net cash flow in 2025, 2030, 2034, 2040 and 2044, respectively.

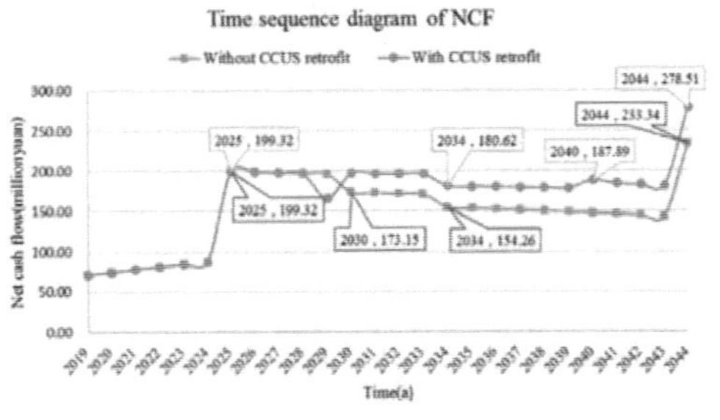

Fig. 1. Time sequence diagram of NCF with or without CCUS retrofit.

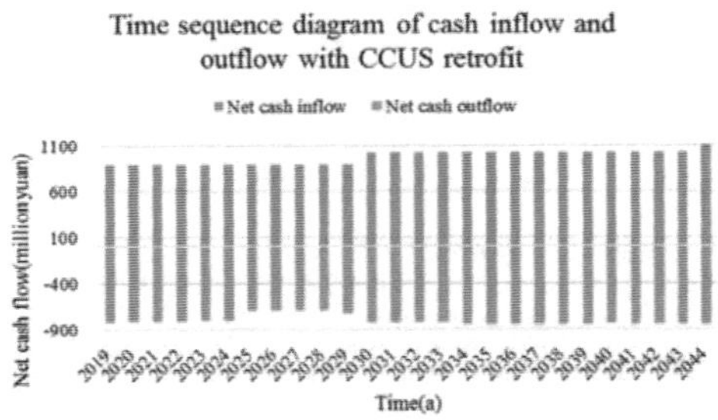

Fig. 2. Time sequence diagram of cash inflow and outflow with CCUS retrofit.

4.2. Sensitivity analysis of uncertain parameters

(1) Parameters that only affect CCUS retrofit

In the case of CCUS retrofit, the sales revenue of carbon dioxide is the new revenue of coal-fired power plants. The higher the sales unit price, the more conducive to the economics of CCUS retrofit (see Figure 3). In the process of increasing carbon dioxide unit price from 50 yuan/ton carbon dioxide to 700 yuan/ton carbon dioxide, the change in net present value brought about by the capture and sale of carbon dioxide in coal-fired power plants increased from -0.33 billion yuan to 0.48 billion yuan. For each increase of 10 yuan/ton, the difference of net present value increases by 0.012 billion yuan, and unit price of 300 yuan/ton of carbon dioxide is the critical point for CCUS retrofit to be economical.

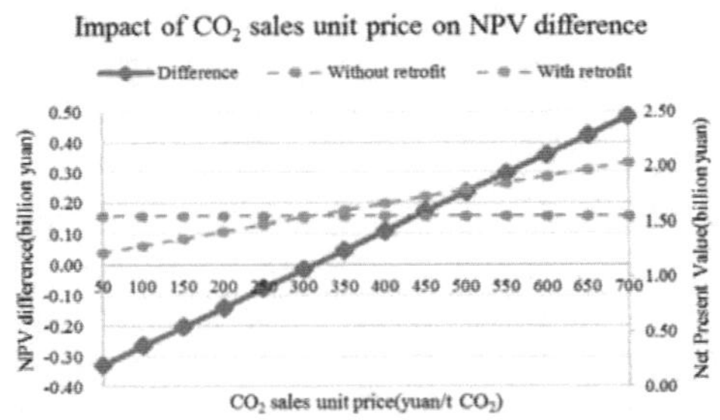

Fig. 3. Impact of CO_2 sales unit price on NPV difference.

The higher the operating cost of the carbon capture device, the more unfavorable the economics of CCUS retrofit (see Figure 4). Carbon capture device operating cost decreased from 400 yuan/ton carbon dioxide to 50 yuan/ton carbon dioxide, and the net present value of coal-fired power plant increased from -0.13 billion yuan to 0.37 billion yuan. The difference of the net present value increased by 0.07 billion yuan for each decrease of 50 yuan/ton. The operating cost of the carbon capture device is 307 yuan per ton of carbon dioxide, which is the critical point for CCUS retrofit to be economical.

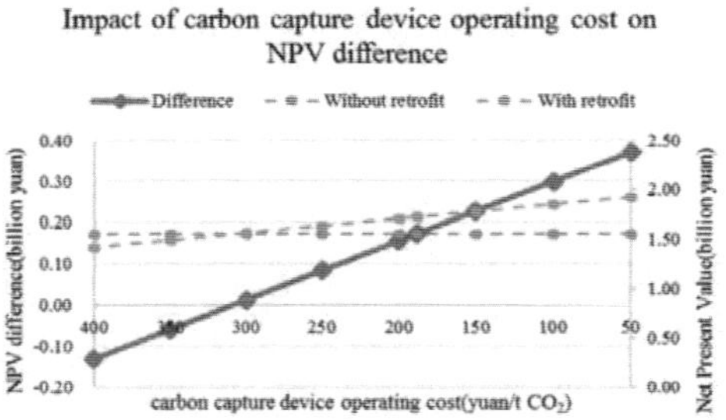

Fig. 4. Impact of carbon capture device operating cost on NPV difference.

(2) Parameters that affect situation with and without CCUS retrofit

In the process of the annual increase in carbon emission right price from 1 yuan per ton of carbon dioxide to 10 yuan per ton of carbon dioxide, the change in net present value caused by the capture and sale of carbon dioxide by coal-fired power plants increased from 0.17 billion yuan to 0.39 billion yuan (See Figure 5). The increase of carbon emission right price will increase the carbon emission cost of the coal-fired power plant. In other words, more carbon emission costs can be saved by capturing carbon dioxide. Therefore, the higher carbon emission right price, the more conducive to the economy of CCUS retrofit.

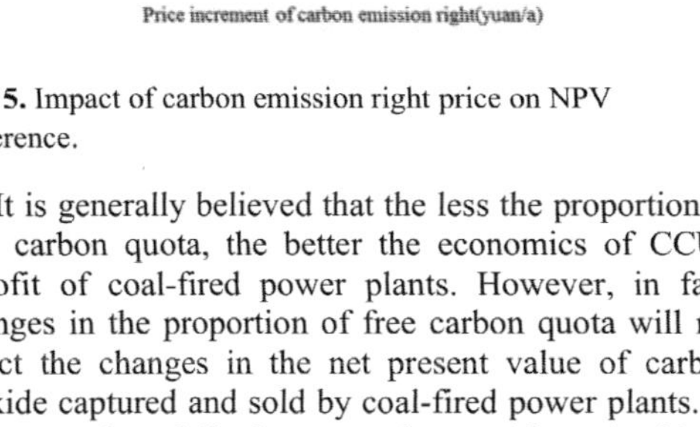

Fig. 5. Impact of carbon emission right price on NPV difference.

It is generally believed that the less the proportion of free carbon quota, the better the economics of CCUS retrofit of coal-fired power plants. However, in fact, changes in the proportion of free carbon quota will not affect the changes in the net present value of carbon dioxide captured and sold by coal-fired power plants. In the case of coal-fired power plants with and without CCUS, the reduction in the proportion of free carbon quota will reduce the net cash flow of coal-fired power plants. However, the net present value difference remains unchanged since the reduction is the same. In this example, the annual reduction rate of free carbon quota increased from 1% to 5%, the net present value of coal-fired power plants dropped from 1.55 billion yuan to 1.11 billion yuan without CCUS. And the net present value dropped from 1.72 billion yuan to 1.28 billion yuan with CCUS, and the difference between the two remained at 0.17 billion yuan (see Figure 6).

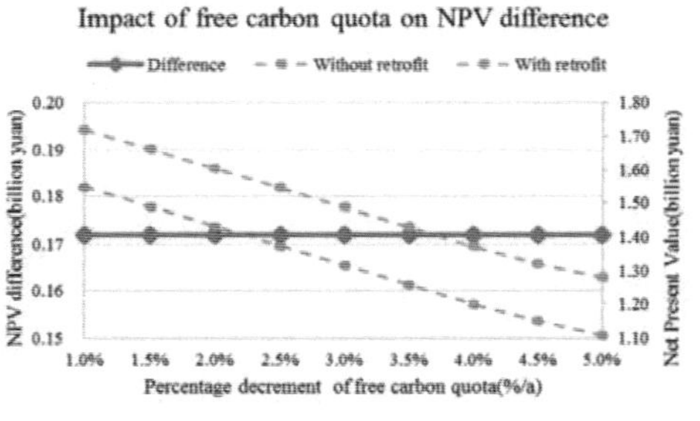

Fig. 6. Impact of free carbon quota on NPV difference.

5 Conclusion

Coal-fired power plants are the most important potential application targets of CCUS. Whether the income of coal-fired power plants can be improved is the key to the commercial application of CCUS. This paper proposes a quantitative economic evaluation method based on the net cash flow of the CCUS retrofit by coal-fired power plants. The feasibility of the retrofitting is evaluated by comparing the difference in the net present value of the remaining life cycle of the coal-fired power plant with or without CCUS retrofit. For a typical coal-fired power plant, the impact of uncertain parameters such as the carbon dioxide sales unit price, the carbon capture device operating cost, the amount of free carbon quota, and the carbon emission right price are analyzed. Among them, the proportion of free carbon quota under the carbon market mechanism does not affect the economy and enthusiasm of CCUS retrofit of coal-fired power plants is an important discovery of this paper. This method can provide decision support for the CCUS retrofit of coal-fired power plants, and it can also lay a foundation for how to coordinate and optimize the CCUS retrofit of coal-fired power plants and new energy investment during the low-carbon transition process of power generation enterprises.

Acknowledgments

This work is supported by the pilot programs for major science, technology and innovation projects toward 2030 of China Energy Investment Corporation- Clean and efficient utilization of coal: Research on medium and long-term carbon emission reduction paths and energy structure optimization strategies of China(Grant No. GJNY2030XDXM-19-20.1), and the Science and Technology Project of NARI Technology Co., Ltd "Research on Key Technologies for Simulation and Evaluation of Comprehensive Energy Systems".

References

1. J Yuan, K Zhang, *Research and future forecast of coal power transformation in the middle and late period of the 13th five-year plan*, China Coal, J. E **08**, 13-19 (2019)(in Chinese)
2. L Liang, J Sun, M Yue, H Geng, *Comparative analysis of global energy consumption mix in recent ten years*. Cnpc. J. E **03**, 41-47 (2020)(in Chinese)
3. J Kang, Y Wang, J Li, W Zhang, *Current Situation, Trend and Orientation Analysis of Coal Power Development in China,* Engry of China, J. E **08**, 9-13 (2019)(in Chinese)
4. *China's carbon capture, utilization and storage technology development roadmap 2019* (2019)(in Chinese)
5. International Energy Agency. *The role of CCUS in low-carbon power systems*, EB/OL https://webstore.iea.org/download/direct/4028?fileName=The_role_of_CCUS_in_low-carbon_power_systems.pdf.(2020)
6. Intergovernmental Panel on Climate Change. *Global warming of 1.5°C*, EB/OL https://www.ipcc.ch/site/assets/uploads/sites/2/2019/06/SR15_Full_Report_High_Res.pdf.(2019)
7. Intergovernmental Panel on Climate Change. *Carbon Dioxide Capture and Storage*, EB/OL https://www.ipcc.ch/site/assets/uploads/2018/03/srccs_wholereport-1.pdf.(2018)
8. *Energy production and consumption revolution strategy (2016-2030)*(2016) (in Chinese)
9. X Zhang, M Xu, J Fan, *Evaluation of incentive measures for carbon capture and storage technology transformation investment in coal-fired power plants*, China Coal, J. E **12**, 22-26 (2017)(in Chinese)
10. Z Lin, S Wen, B Song, *Research on the Effect of Carbon Tax Policy on CCS Projects Investment Decision Based on Real Option*, China Population,Resources and Environment, J. E **09**, 13-20 (2015) (in Chinese)
11. Z Xu, *Study on Economic Assessment of a Full-Chain CCS-EOR Project based on CO_2 Post-Combustion Capture of Coal-Fired Power Generation* , Journal of Cleaner Production, (2016) (in Chinese)

E3S Web of Conferences **237**, 02022 (2021) https://doi.org/10.1051/e3sconf/202123702022
ARFEE 2020

Neural network-based thermal comfort prediction for the elderly

JinJin Zhang[1], *Hong Liu*[2,*], *YuXin Wu*[1],*Shan Zhou*[1],*MengJia Liu*[1]

[1] Joint International Research Laboratory of Green Buildings and Built Environments (Ministry of Education), Chongqing University, Chongqing, China 40004
[2] National Centre for International Research of Low-carbon and Green Buildings (Ministry of Science and Technology), Chongqing University, Chongqing, China 400045

Abstract. Machine learning technology has become a hot topic and is being applied in many fields. However, in the prediction of thermal sensation in the elderly, there is not enough research on the neural network to predict the effect of human thermal comfort. In this paper, two neural network algorithms were used to predict the thermal expectation of the elderly, and the accuracy of the two algorithms was compared to find a suitable neural network algorithm to predict human thermal comfort. The dataset was collected from the laboratory study and included 10 local skin temperatures of the subjects, thermal perception voted at three temperatures (28/30/32℃), different wind speeds, and two forms of wind. Thirteen subjects with an average age of 63.5 years old were recruited for the subjective survey. These subjects sat for long periods of summer working conditions, wore uniform thermal resistance clothing, and collected votes on thermal sensation, as well as skin temperature. The results showed that the prediction accuracy of the two algorithms was related to the added influence factors, and the RBF neural network algorithm was the most accurate in predicting thermal sensation of the elderly. The main influencing factors were average skin temperature, wind speed and body fat rate.

1 Introduction

With the development of computer and artificial intelligence, data acquisition is becoming more and more easy due to the large number of data monitoring systems and acquisition platforms, and the application of machine learning algorithms in the field of thermal comfort is also attracting more and more attention. Due to the complex correlation between thermal comfort, behavior habits, thermal environment and energy consumption, it is difficult to find the inherent law, so the traditional thermal comfort evaluation methods or models are increasingly challenged in application and prediction accuracy. In contrast, machine learning algorithms have attracted more and more attention because they can extract the implicit unknown, valuable or regular knowledge from a large amount of data.

At the same time, some scholars have compared the subjective and physiological responses of young and old people to different thermal environments and found that different groups had different responses to the same thermal environment. For example, some scholars found that older people had higher perceptual thresholds and lower sensitivity to cold and warm environment than younger people[1]. Some people found that in warm and cold environments, older people typically had a thermal sensation 0.5 units lower than younger people[2]. This suggested that older people respond differently to thermal stimuli. Based on the above literature, there may be differences between the elderly and the young in thermal comfort, but so far, we have not found an appropriate thermoregulation model that can reflect the physiological and physiological characteristics of the elderly, so we cannot accurately predict human thermal comfort through the existing model [3]. In order to fill this gap, the goal of this paper is to use neural network to build a thermal sensation model for the elderly to predict their thermal responses under different thermal conditions.

2 Methodology

2.1 The experiment

This experiment was conducted in an artificial climate chamber in summer. Experimental data collection included thermal environment, measurement of human physiological parameters and subjective questionnaire. The thermal environment data includes indoor dry-wet bulb temperature, wind speed, black-bulb temperature, relative humidity, etc. Human physiological parameters includes skin temperature of ten parts (forehead, upper arm, lower arm, back of hand, chest, back, thigh, calf, abdomen and instep). Recording instrument of temperature and humidity test using contact temperature (temperature range of 0 to 50℃, the error of plus or minus 0.5℃, the appropriate measurement range of 10-90% RH, error of plus or minus 2.5% RH, black ball

* Corresponding author:liuhong1865@163.com

temperature range: 100-400℃, the error of plus or minus 0.3℃), the wind speed test using a handheld hot wire anemometer (measuring range: 0~20 m/s, precision: plus or minus 5% (0.03 m/s + measurements)), and the skin temperature test using HOBO tester (temperature range: -40-70 ℃ , precision: plus or minus 0.21 ℃).Test instruments and methods shall be used in accordance with relevant ASHRAE55-2013 regulations.

Under different indoor dry bulb temperatures (28/30/32℃), different wind speeds (face wind speeds of 0.6, 1.4, 2.2 m/s) and different wind forms (direct and sinusoidal winds), thermal perception polls were conducted under different indoor dry bulb temperatures (28/30/32℃). Subjects were recruited for a subjective survey in which they sat for a long time, wearing a uniform thermal resistance of 0.5 clo to collect thermal sensation votes, and skin temperatures in ten body parts.

Data collection was conducted at the same time as the subject questionnaire survey, including the thermal sensation vote, thermal expectation, etc. Subjects' background information includes whether they suffer from diseases, living habits, etc. The statistical table of physiological information of the subjects is shown in Table 1. After screening, the effective sample size meeting the requirements was: 443 for men, 522 for women, and 964 for the total sample size.

Table 1. Subject'anthropometric information.

Age(years)	Height(cm)	Weight(kg)
Mean±S.D.	Mean±S.D.	Mean±S.D.
63.5±7.2	160.2±8.7	63.2±10.2

3 Model

3.1. Selection of neural network types

Artificial neural network (ANN), commonly referred to as neural network, is an information processing system based on the structure and function of the human brain. Common neural network models include perceptron, feedforward network, radial basis (RBF), regression neuron network (Hopfield), etc. BP and RBF neural networks belong to feedforward neural network. BP and RBF neural networks are the most widely used neural networks at present, and their algorithms and models are relatively mature. Moreover, existing research results have proved that BP and RBF neural networks not only have good prediction accuracy for large samples, but also have good prediction effect for small samples. Therefore, BP and RBF neural networks are selected to establish human thermal comfort prediction model.

3.1.1 BP neural network

BP network is a kind of multilayer forward of one-way transmission network, layered neurons of the input layer, hidden layer and output layer, each layer of neurons only accept input from a layer of neurons before, input model is based on the progressive transformation of each layer by the output layer after final output, there is no feedback between neurons, each neuron in single internal connection and order delivery.

The learning process of BP network: is divided into two process forward and reverse transmission, when the spread of the positive input of the sample data from the first starting from the input layer, then the implicit layer through layer upon layer processing, finally comes out, so that each layer of neurons state affects only the adjacent layer neurons, when in the output layer can't get output when it changes direction into a back propagation, using gradient search technique to make the neural network's actual and desired output error of the mean square value to a minimum. The network structure is shown in Figure 1. In the case of forward propagation, the input value is from left to right, and the output value of the network is calculated layer by layer by combining the weight and excitation function. In the case of back propagation, parameters such as network weights are adjusted reversely according to the defined error function, so that the error decreases gradually.

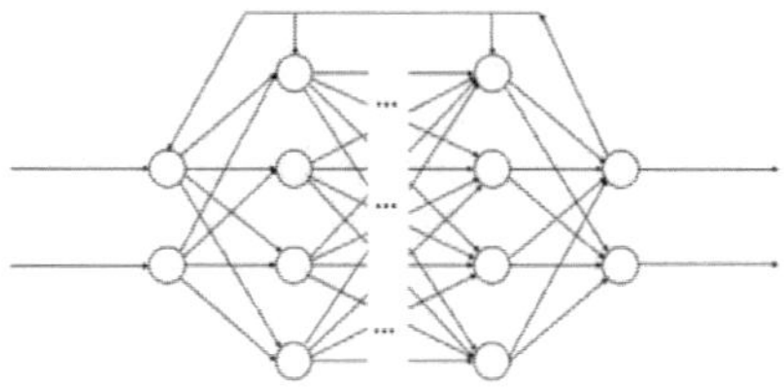

Fig. 1. BP neural network structure

When designing BP network, consideration should be given to the number of layers, number of neurons at each layer, initial value, activation function and learning rate.

Firstly, the number of layers of the network is analyzed. It is proved theoretically that the network with bias and at least one hidden layer plus one linear output layer can approximate any function. Increasing the number of layers can reduce the error and improve the accuracy, but also complicate the network. Therefore, a three-layer multi-input single-output network prediction model with a hidden layer is adopted. At present, there is no clear formula for the determination of the number of neurons in the hidden layer, only the empirical formula, and the final determination of the number of neurons should be determined according to experience and experiment. In this paper, the number of hidden layer neurons is selected with reference to empirical formula 1:

$$l = \sqrt{m+n} + a \tag{1}$$

n: Number of input layer neurons

m: Number of output layer neurons

a: Constants greater than 1 and less than 10

According to the above formula, the number of neurons is between 2 and 12. The number of hidden layer neurons was selected as 6. Secondly, the initial weight of BP neural network is analyzed and selected. Generally, the initial weight is selected between -1 and 1.

When the selection of learning rate of BP network is considered, it is generally selected between 0.01 and 0.8.

3.1.2 RBF neural network

The radial basis neural network (RBF neural network) is a feedforward back-propagation neural network. The structure of radial basis neural network is similar to that of forward network, as shown in Figure 2. There are three network layers. From left to right, there are input layer, hidden layer and output layer. The function of the hidden layer is radially symmetric with its origin. RBF network is a local approximation network, and only a small part of the values are modified, so the training time is short. The radial basis neural network has a nonlinear relation from the input layer to the hidden layer, and a linear relation from the hidden layer to the output layer. The hidden layer node of RBF neural network uses the distance between the input mode and the origin vector as the independent variable and the Gaussian function as the activation function. The further away the input layer neuron is from the origin of the function of the radial basis, the lower its activation degree will be.

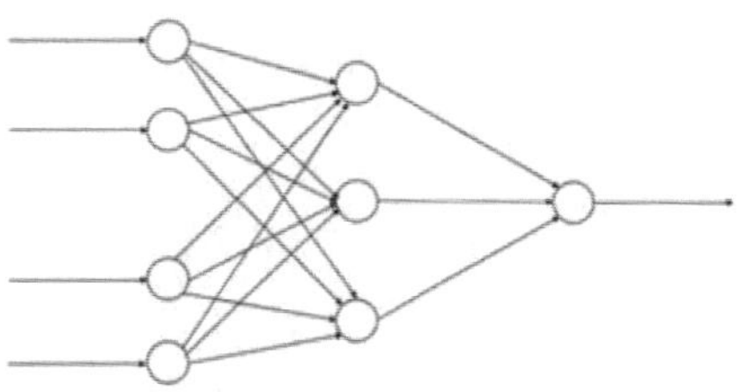

Fig. 2. RBF neural network structure

There are three unknown parameters in the training method of RBF network: the center point of the radial basis function, the variance and the weight between the hidden layer and the output layer. RBF is a process of repeated experiments by the computer. In the establishment process, there is no need to set the number of hidden layer neurons, and the computer automatically adds the number of hidden layer neurons until the output error of the radial basis neural network reaches the preset requirement.

3.2. Thermal comfort prediction model based on neural network

Although through preliminary data screening, selected the 13 factors affecting thermal sensation, including ten parts of the skin temperature with the average arithmetic average skin temperature, said but in the actual model, introducing the more variables, the interpretation of the results, the more difficult, model application is also more difficult, in principle, should as far as possible to introduce less variable to maximize explain the results. Therefore, the 13 selected variables need to be further analyzed to identify the most important variables used to predict the results and introduce the model. Through feature selection, 13 predictive variables were selected, including gender, age, body fat rate, wind speed, blowing state, duration of blowing, duration of no wind, temperature condition, indoor dry bulb temperature, indoor relative humidity, indoor black bulb temperature, filling time of questionnaire, and average skin temperature. Due to the mutual influence between some factors, or some variables with strong correlation, such as the correlation between indoor temperature and humidity, indoor black-bulb temperature, etc., it needs to be further simplified. Table 1 shows the correlation analysis of some factors. It can be seen that indoor relative humidity, dry bulb temperature and black bulb temperature are strongly correlated with each other, and the correlation values are relatively large. Moreover, their influence on thermal sensation can be largely reflected by T_{air} parameters. In addition, the subjects' body fat rate is highly correlated with gender and age, so the factor of body fat rate can also explain the factors of age and gender to some extent. The duration of wind blowing was also correlated with the duration of no wind. Comprehensive analysis, the model to cut out the gender, age, duration, there is no wind temperature, relative humidity, black ball temperature variables, model choice variables for body fat rate, wind speed, wind speed, wind length, dry bulb temperature and average temperature and so on six skin as the input variables of neural network model, the subjects at this time of thermal sensation as output variables of the model.

4 Results analysis

4.1. Distribution of human thermal sensation

By sorting the experimental data, the average skin temperature of the subjects and the corresponding thermal sensation were plotted as Figure 3.Figure in white dots represent the median, the subjects in thermal sensation vote in feel cold, neutral, hot, median average skin temperature, respectively is 32.89, 33.25, 33.78℃, the skin temperature can be obtained, the higher the participants feel more warm, and each group of data is normally distributed, can also prove the effectiveness of the experimental data.

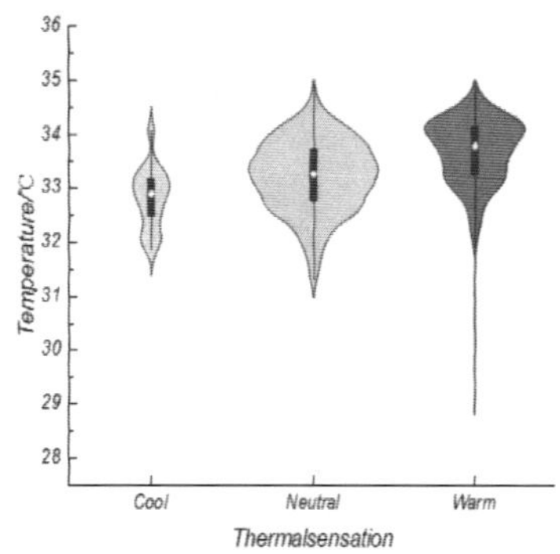

Fig. 3. Thermal sensation distribution of subject

4.2. Prediction of thermal sensation and analysis of influencing factors

4.2.1 Analysis of influencing factors

In order to obtain the influence degree of each influencing factor on the thermal sensation of the experimenter, and rank the influencing factors according to the importance degree. In this paper, the variables included in the prediction model in the experimental data were analyzed by MATLAB, and the result was drawn as Figure 4. The order of influence degree was: Body fat rate = Skin temperature = Wind regime > Wind speed > Blowing duration > Indoor dry bulb temperature.

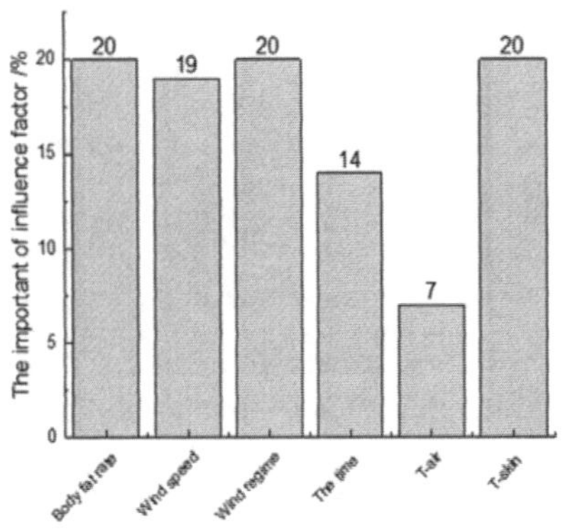

Fig. 4. Distribution of factors affecting thermal sensation

4.2.2 Prediction accuracy comparison

In order to evaluate the predictive performance of the model, the samples were partitioned before modeling, 80% of which were used for training the model, and 20% for prediction and verification. In this paper, BP and RBF artificial neural network prediction programs were first written to train and predict human thermal sensation votes respectively, and the prediction results were obtained. Finally, 200 samples were selected to verify the prediction results, as shown in Figure 5, 6 and Table 2. It can be seen from Table 2 that the prediction result of RBF neural network is slightly more accurate than that of BP neural network. The average accuracy of the RBF neural network is about 10 percent higher than that of the BP network.

Table 2. Subject'anthropometric information.

Network model	Prediction accuracy (%)
BP neural network	77.33
RBF neural network	87.82

As can be seen from Figure 5, the model established by BP neural network is more accurate in the comfort zone of human body, while in the extreme environment, the prediction accuracy of BP network decreases significantly. However, the prediction accuracy of the RBF neural network model decreased even under extreme circumstances.

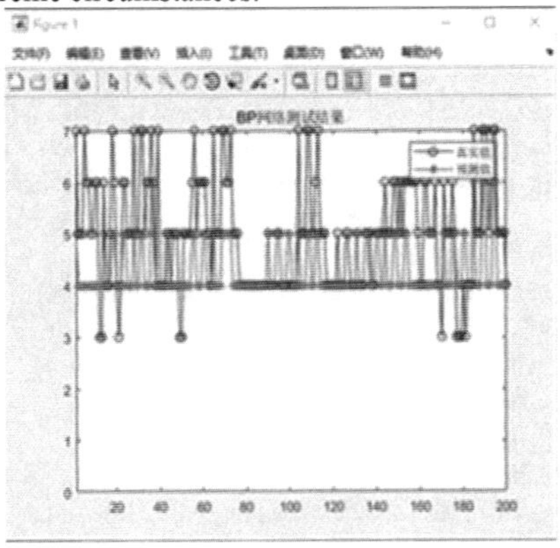

Fig. 5. Comparison of BP network prediction results

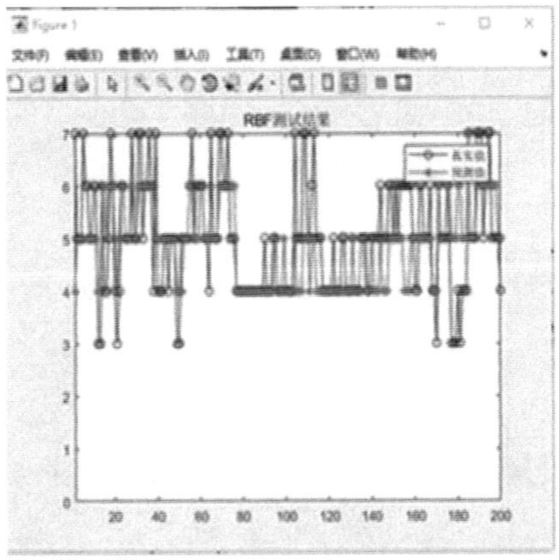

Fig. 6. Comparison of RBF network prediction results

5 Conclusions

Through the evaluation of the accuracy of the established thermal sensation models, we found that the accuracy of RBF model is higher than BP model. It also can be seen that through the analysis of the predicted value of the model and the value of thermal sensation, it is found that both models have high error rate when the predicted thermal sensation outside the range of [-1, 1]. Combined with the thermal sensation distribution in Figure 5and 6, the majority of the thermal sensation votes in the experimental data are within the range of [-1, 1], but the sample outside the range is small, which may also make the model fail to fully train and learn and master the data features and results in a large prediction error rate.

https://doi.org/10.1051/e3sconf/202123702022

Funding

Graduate Scientific Research and Innovation Foundation of Chongqing, China, Grant/Award Number: CYS20024.

Acknowledgments

This work was supported by the Graduate Scientific Research and Innovation Foundation of Chongqing, China (Grant No. CYS20024).

Appendix 1. Correlation analysis of different variables

	v2	v3	v4	v5	v6	v7
v1 Age	0.242**	-0.802**	0.013	-0.005	0.008	-0.002
v2 Gender		-0.416**	0.016	-0.014	0.004	-0.004
v3 Body fate rate			-0.016	0.014	-0.006	0.006
v4 Experimental condition				0.009	0.343**	0.006
v5 Wind speed					0.490**	0.342**
v6 Wind regime						0.552**
v7 Blowing duration						
v8 No blowing duration						
v9 Temperature condition						
v10 Tair						
v11 Tglob						
v12 RH%						
v13 Mean skin temperature						

	v8	v9	v10	v11	v12	v13
v1 Age	0.003	-0.023	-0.026	-0.001	0.031	0.087**
v2 Gender	0.002	0.002	-0.005	0.079*	-0.017	0.053
v3 Body fate rate	-0.006	0.02	0.013	-0.049	-.087**	-0.091**
v4 Experimental condition	-0.001	0.491**	0.503**	0.083**	0.569**	0.417**
v5 Wind speed	-0.525**	0.001	0.014	0.086**	-0.023	-0.228**
v6 Wind regime	-0.829**	0.009	0.013	0.058	0.183**	-0.099**
v7 Blowing duration	-0.597**	0.003	-0.004	-0.033	-0.018	-0.218**
v8 No blowing duration		-0.003	0.008	-0.031	0.044	0.140**
v9 Temperature condition			0.994**	0.086**	0.170**	0.716**
v10 Tair				0.082*	0.178**	0.717**
v11 Tglob					0.037	0.074*
v12 RH%						0.074*
v13 Mean skin temperature						

Annotation:** means the significance level is greater than 0.01;* means the significance level is greater than 0.05.

References

1. Stevens J C, Choo K K, Temperature sensitivity of the body surface over the life span. Somatosensory & motor research **15(1)**, pp 13-28(1998).
2. Schellen L, van Marken Lichtenbelt W D, Loomans M G L C, et al, Differences between young adults and elderly in thermal comfort, productivity, and thermal physiology in response to a moderate temperature drift and a steady - state condition. Indoor air **20(4)**, pp 273-283(2010).
3. Wang Z, Yu H, Luo M, et al, Predicting older people's thermal sensation in building environment through a machine learning approach: Modelling, interpretation, and application. Building and Environment, pp 106231(2019)

Analysis of base characteristics of trench gate field termination IGBT

Wang Bo[1,*]

[1]Wuhan Donghu University, CHINA

Abstract. Trench gate structure represents the latest structure of Insulated Gate Bipolar Transistor(IGBT). Because there are great differences in model analysis coordinate system and carrier transport between trench gate structure and planar gate structure, the modeling method using planar gate structure will inevitably have great deviation. Based on the characteristics of trench gate structure and model analysis coordinate system, the base region is divided into PNP and PIN by considering the two-dimensional effect of carriers. According to whether the trench of PIN part can be covered by depletion layer of PNP part, the specific base region current is analyzed. Finally, simulation and experimental verification are carried out.

1 Introduction

Insulated Gate Bipolar Transistor(IGBT) is a kind of composite device which combines the structure of field effect transistor and bipolar transistor, and absorbs their advantages at the same time. It is widely used in various medium and high power electronic devices. With the continuous improvement of the design level and manufacturing process of power semiconductor devices, the internal gate structure of IGBT has changed from the early planar gate to the later trench gate[1]. This trench gate structure absorbs the advantages of previous generations, and adopts some new structures and technologies. Its performance indexes, such as withstand voltage capability, conduction current capability, on-state voltage drop and switch damage, have been significantly improved, and it has excellent comprehensive performance[2-3].

Since the advent of IGBT in 1980s, many literatures have carried out relevant research on IGBT mathematical model, and the early theory of planar gate structure has been relatively mature. The most representative Hefner model studies the steady-state and transient mathematical models of planar gate structure by using the theories of bipolar carrier transport in base region, quasi-static approximation and transient carrier linear distribution, and the existing IGBT mathematical models are basically developed on the basis of this model[4-5]. However, this model only makes a one-dimensional approximation to the distribution of excess carriers in the base region, and seldom studies the influence of the change of trench gate structure on the two-dimensional distribution. Other literatures have adopted the method of two-dimensional distribution of carriers, which promotes the research of trench gate structure[6].

In this paper, based on the characteristics of trench gate structure and model analysis coordinate system, by dividing the base region into PNP and PIN and considering the two-dimensional effect of carriers in transient process, the transient specificity of the base region current is analyzed according to whether the trench of PIN can be covered by depletion layer of PNP. Finally, simulation and experimental verification are carried out.

2 Theoretical analysis

2.1 Trench gate IGBT model coordinate system

The planar schematic diagram of trench gate field termination IGBT structure is shown in Figure 1, which is usually regarded as a BJT structure driven by MOSFET. The vertical width L_x of the trench reflects the depth of the trench into the base region, and the horizontal width L_y of the trench reflects the proportion of the base region occupied by the trench. Among them, the P body region on the upper part of the base region is very small relative to the width of the base region, which is similar to the planar gate type. While the trench gate on the upper part of the base region goes deep into the base region, and the depth can't be ignored compared with the width of the base region. Therefore, different from planar gate structure, trench gate structure needs to consider the influence of trench.

* Corresponding author: author@e-mail.org

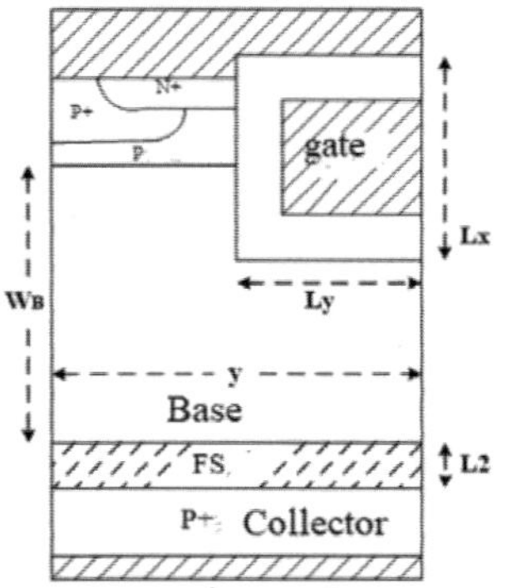

Fig1. Schematic diagram of trench gate IGBT structure

The mathematical model analysis coordinate system of trench gate field termination IGBT is shown in Figure 2. The horizontal axis of FS layer coordinate system is x^*, the horizontal axis of base area coordinate system is x, and the vertical axis of coordinate system is y. The width of base region is W_B, and the width of FS layer is W_H. According to the carrier movement characteristics of p body region and trench gate in the base region, the base region is divided into PNP at the lower part of p body region and PIN at the lower part of gate. The depletion layer width of PNP part is $W_{dep,PNP}(t)$, and the quasi-neutral base region width is W_L. The electron current and hole current of PNP part and PIN part are calculated respectively, and then the electron current and hole current in the base region are summarized.

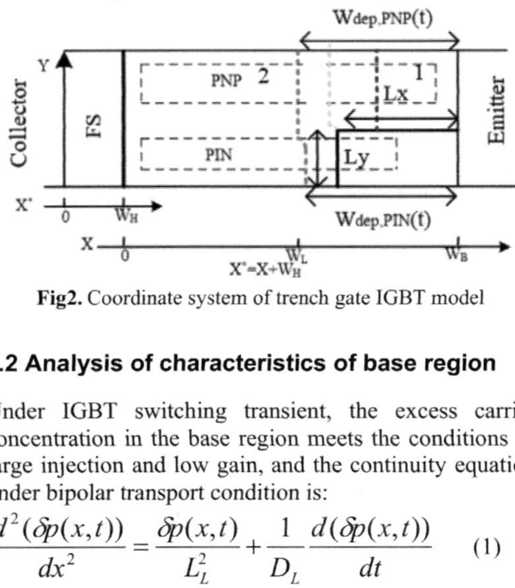

Fig2. Coordinate system of trench gate IGBT model

2.2 Analysis of characteristics of base region

Under IGBT switching transient, the excess carrier concentration in the base region meets the conditions of large injection and low gain, and the continuity equation under bipolar transport condition is:

$$\frac{d^2(\delta p(x,t))}{dx^2}=\frac{\delta p(x,t)}{L_L^2}+\frac{1}{D_L}\frac{d(\delta p(x,t))}{dt} \qquad (1)$$

Where: $\delta p(x,t)$ refers to excess carrier concentration; L_L is the base carrier bipolar diffusion length; D_L is the bipolar diffusion coefficient of base carriers.

When the depletion layer in the PNP part of the base region widens to the PIN part, it bends due to the influence of the gate voltage. The width of the depletion layer formed by the collector-emitter voltage and the gate voltage in the PIN part is:

$$W_{dep,PIN}(t)=\sqrt{\frac{2\varepsilon_{si}(V_{bc}(t)-V_{GS}(t)+V_{Td})}{qN_L}} \qquad (2)$$

In which: ε_{si} is the dielectric constant of silicon; $V_{bc}(t)$ is BJT base-emitter voltage drop; $V_{GS}(t)$ is the gate voltage; V_{Td} is that threshold voltage of the gate-drain overlap depletion layer; q is the electronic quantity; N_L is the base doping concentration.

According to whether the depletion layer of PNP part can be widened to cover the trench gate of PIN part, that is, the relationship between the width of depletion layer of PIN part and the width of trench gate deep into the base region, two cases are analyzed.

The first case: the depletion layer of PNP part failed to widen the trench gate area covering PIN part, that is $W_{dep,PIN}(t)\le L_x$.

The depletion layer of PNP part expands along the Y-axis direction and fails to surround the trench area of PIN part, as shown in Figure 2, the depletion layer expands to the 1 position. Then the PNP part and PIN part are described separately, and then the electron current flowing into the trench and the hole current flowing into the cathode are synthesized. The movement process of carriers is understood as follows: electrons are injected from the trench gate and are continuously converted into holes in the process of moving to the FS layer; When holes are injected from the anode and move to the cathode after passing through the FS layer, the hole concentration in the PNP part decreases gradually, and is considered to be approximately zero at the depletion layer. The hole concentration in the PIN part decreases gradually similar to that in the PNP part, and turns to flow into the depletion layer along the trench wall at the trench.

For PNP, the depletion layer width and quasi-neutral width are:

$$\begin{cases} W_{dep,PNP}(t)=\sqrt{\dfrac{2\varepsilon_{si}V_{ce}(t)}{qN_L}} \\ W_L(t)=W_B-W_{dep,PNP}(t) \end{cases} \qquad (3)$$

The two boundary conditions of excess carrier concentration are:

$$\begin{cases} \delta p(x=0,t)=P_0(t) \\ \delta p(x=W_L(t),t)=0 \end{cases} \qquad (4)$$

The carrier concentration distribution in the transient process is approximated by a moving linear distribution, and the expression of space-time distribution of excess carrier concentration in the base region is obtained. The distribution process can be represented as Figure 3.

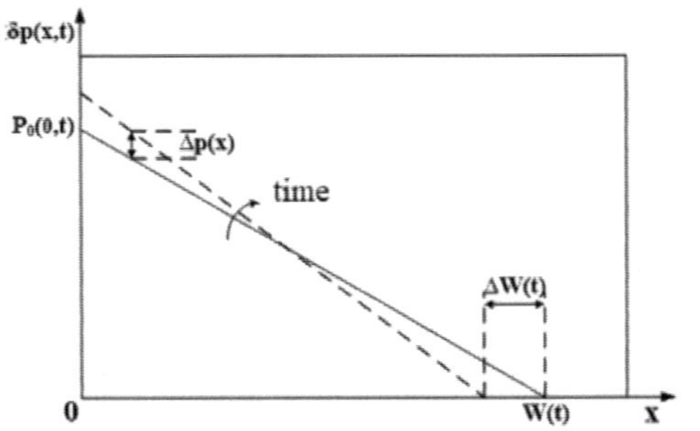

Fig3. Distribution of excess carrier concentration in base region during switching transient process

For the PIN part, the excess carrier concentration has the same distribution as that of PNP part in most areas of X axis. When near the trench, the excess carrier bends to PNP part and is swept into the depletion layer together with the excess carrier of PNP part. Therefore, the excess carrier concentration distribution of the PIN part is approximately considered to be the same as that of the PNP part at the same distance ($0 \sim (W_B - L_x)$).

The quasi-neutral width of the PIN portion is determined by the width of the base region and trench metallurgy junction, and the excess carrier concentration distribution of the PIN portion has the same form as that of the PNP portion. The expressions of electron current and hole current in PNP part and PIN part are obtained by using bipolar transport equations of electron current and hole current respectively.

Among them, the currents of PNP part and PIN part satisfy:

$$\frac{I_{PNP}}{I_{PIN}} = \frac{y - L_y}{L_y} \tag{5}$$

The second case: the depletion layer of PNP part can widen the trench gate area covering PIN part, that is $W_{dep,PIN}(t) > L_x$.

The depletion layer formed by PNP expands in the Y-axis direction to cover the trench area of PIN. As shown in Figure 2, when the depletion layer expands to the 2 region, the expressions of carrier concentration distribution, electron current and hole current in the base region are obtained by a processing method similar to that of planar gate. The difference from planar gate is the calculation of junction capacitance of depletion layer under the gate.

The transient description of trench gate structure is similar to that of planar gate structure. The bipolar transport equation of electron current and hole current in base region is:

$$\begin{cases} I_{nL}(x,t) = \dfrac{b_L}{1+b_L} I + qD_L \dfrac{d(\delta n(x,t))}{dx} \\ I_{pL}(x,t) = \dfrac{1}{1+b_L} I - qD_L \dfrac{d(\delta p(x,t))}{dx} \end{cases} \tag{6}$$

The expression of hole current is obtained, and the hole current flowing into and out of the base region is obtained.

Therefore, for the above two cases, the hole current flowing into the base region and the hole current flowing out of the base region are obtained respectively.

2.3 Simultaneous analysis of base area and FS layer

The hole current and electron current flowing through the boundary between FS layer and base region are equal, that is, the current at the boundary is continuous, but the excess carrier concentration is discontinuous, as shown in Figure 4.

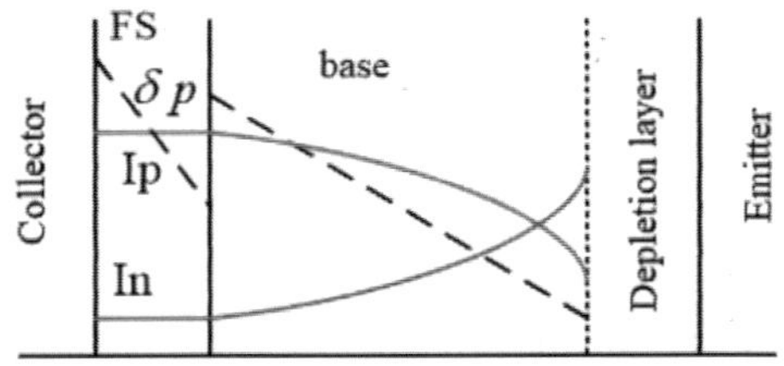

Fig4. Simultaneous analysis of base zone and FS layer

The carrier concentration at the boundary satisfies the relation:

$$I_{pL}(0,t) = I_{pH}(x^*,t) \tag{7}$$

According to whether the depletion layer of the base region can be widened to cover the trench gate region and the relationship between the total charge, the expression of the charge of the base region from the total charge is obtained.

Then, according to the quasi-thermal balance relationship between FS layer and base region and the current at the boundary, the relationship between the hole current and the total charge of internal BJT collector and base is obtained continuously, thus eliminating the intermediate quantity of the assumed boundary concentration.

3 Experimental verification

A trench gate field termination IGBT module with KW60N65S model is selected, and the rated voltage and current levels are 650V and 60A respectively. The experimental test conditions are as follows: the bus voltage is set to 400V, the driving voltage is 15V when switching on, the load inductance is 500uH, and the room temperature. Double pulse test method is used to test IGBT collector-generator voltage and collector current, and the IGBT turn-off waveform is shown in the Figure 5.

With the trench gate model established in this paper, the parameters required for simulation are provided by the device manufacturer, and the stray inductance of the line is set to 1uH according to the experimental test circuit. After the initial value is input, it is solved, and the turn-off transient simulation waveforms of IGBT are obtained as shown in the Figure 5.

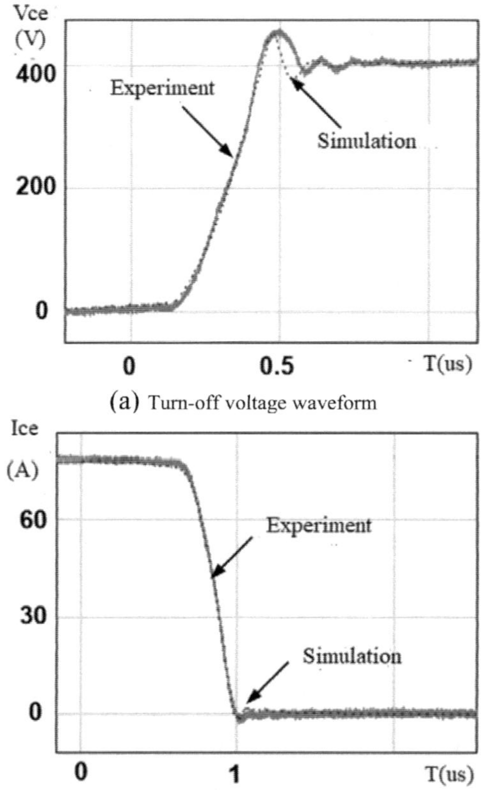

Fig5. Comparison of shutdown waveforms

It can be seen from Figure 5 that the IGBT model considering trench gate structure is basically consistent with the experimental waveform.

The error causes are analyzed as follows:

① The gate drive circuit, load circuit and stray inductance of the device itself are difficult to be accurate;

② It is difficult to unify the average effect of carrier lifetime in the model with the dispersion in practical work;

③ It is difficult to give consideration to the idealization and uniqueness of structural parameters and the dispersion of device technology in model simulation;

④Reasonable assumptions and simplification in the modeling process;

⑤Errors introduced in the process of experimental testing.

4 Summary

Compared with the traditional planar gate structure, this paper divides the base region into PNP and PIN according to the characteristics of carrier movement in the trench gate structure, and then analyzes the carrier transport and current in the base region according to whether the depletion layer of PNP can cover the trench of PIN, which reflects the characteristics of trench gate structure.

Acknowledgments

This work was supported by the 2018 year Youth Foundation of Wuhan Donghu University(2018dhzk005), and the Scientific Research Project of Hubei Education Department in 2019(B2019267).

References

1. Chen Xudong, Cheng Jianbing, Teng Guobing, et al. Novel trench gate field stop IGBT with trench shorted anode[J]. Journal of Semiconductors, 2016, 37(5): 61-64.
2. Qin Mengliang, Li Zehong, Zhang Bo, et al. Trench gate IGBT structure with floating P region[J]. Journal of Semiconductors, 2010,31(2):11-13.
3. Fu Qiang, Zhang Bo, Luo Xiaorong, et al. A dual-gate and dielectric-inserted lateral trench insulated gate bipolar transistor on a silicon-on-insulator substrate[J]. Chinese Physics B, 2013,22(7):1-5.
4. Bryant A T, Liu Liqing, Santi E, et al. Modeling of IGBT resistive and inductive turn-on behavior[J]. IEEE Transactions on Industry Application, 2008,44(3):904-914.
5. Iannuzzo F, Busatto G. Physical CAD model for high voltage IGBTs based on lumped-charge approach[J]. IEEE Transactions on Power Electron, 2004,19(4):885-893.
6. Udrea F, Chan S, Thomson J. Development of the next generation of insulated gate bipolar transistors based on trench technology[C]. The 27th European Solid-State Device Research Conference,1997:504-507.

https://doi.org/10.1051/e3sconf/202123702024

Analysis of junction capacitance characteristics of trench gate IGBT

Wang Bo[1,*]

[1]Wuhan Donghu University, CHINA

Abstract. Trench gate field termination IGBT represents the latest structure of insulated gate bipolar transistor (IGBT). Because the internal current of IGBT includes the charging and discharging current of gate capacitance and internal junction capacitance during switching transient, the influence of junction capacitance should be considered. The conductive channel of trench gate structure is different from that of planar gate structure, and the analysis method of junction capacitance using planar gate structure will inevitably bring some deviation. Based on the characteristics of trench gate structure, this paper analyzes the different expressions of internal gate-drain junction capacitance in two cases according to whether the base depletion layer can be widened to cover the trench gate, and finally carries out simulation and experimental verification.

1 Introduction

Insulated gate bipolar transistor (IGBT) is a composite device which combines the structure of field effect transistor and bipolar transistor, and absorbs the advantages of both. It has high input impedance, fast switching speed, low driving power, low saturation voltage, simple control circuit, large current bearing and so on. Widely used in new energy technology, energy-saving technology and other fields, it has gradually become the leading device of medium and high-power power electronic devices[1-3].

With the continuous improvement of design level and manufacturing level of power semiconductor devices, IGBT has also experienced the development process from planar gate type to trench gate type, and punch-through type to field termination type. The new generation of trench gate field termination IGBT has achieved a leap in power grade and comprehensive performance, and gradually occupied the dominant position of IGBT application. Due to the structural characteristics of trench gate field termination type, which is different from the traditional planar gate type, the two-dimensional distribution of carriers in the base region and the influence of injection and depletion layer broadening on junction capacitance are different. This method of directly adopting the existing model will inevitably bring great deviation[4-5]. In IGBT switching transient process, the change of external collector-emitter voltage will significantly affect the internal junction capacitance, and the charging and discharging current of junction capacitance constitutes a part of IGBT transient current. Therefore, accurate modeling of internal junction capacitance will significantly improve the accuracy of IGBT transient current, which has certain theoretical and engineering practical significance.

Based on the equivalent circuit analysis of internal junction capacitance of trench gate field-terminated IGBT, this paper discusses the calculation method of internal gate junction capacitance in two cases according to whether the base depletion layer can widen and cover the trench gate. The calculation method considers the influence of different collector-emitter voltages on the width of base depletion layer in IGBT transient process, which causes the change of internal gate junction capacitance of IGBT and improves the accuracy of IGBT transient model.

2 Theoretical analysis

2.1 GBT junction capacitance characteristics

IGBT, as a fully controlled power electronic device, can switch between on-state and off-state conveniently by charging and discharging the gate capacitance through an external driving circuit, so the gate capacitance directly affects the switching transient process of IGBT. The gate capacitance consists of gate-source capacitance, gate overlapping oxide capacitance and gate overlapping depletion capacitance, which varies with external voltage.

Schematic diagram of planar structure of trench gate and planar gate IGBT is shown in Figure 1, which can be regarded as BJT structure driven by MOSFET. The conductive channel of planar gate structure is horizontal, while that of trench gate structure is vertical. As the direction of MOS conductive channel changes, the corresponding internal junction capacitance changes. The comparative analysis of junction capacitance of two gate structures is also shown in Figure 1.

* Corresponding author: author@e-mail.org

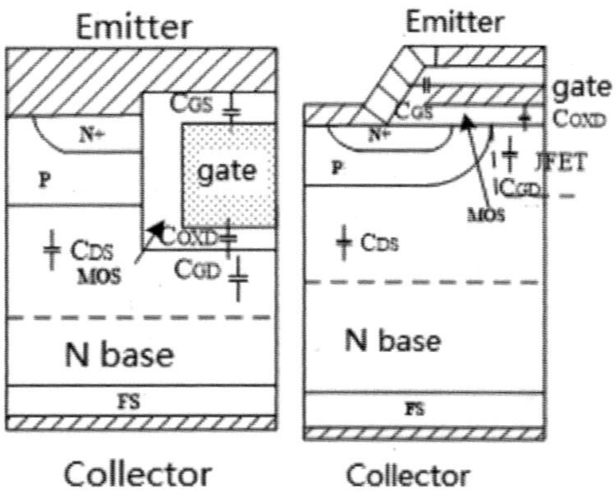

Fig1. Comparative analysis of junction capacitance of trench gate and planar gate

For planar gate structure, because the width of P-type body region is relatively small, it is considered that the width of depletion region covers the horizontal conductive channel, and the junction capacitance of gate-drain depletion layer of internal MOSFET can be expressed by a unified expression:

$$C_{GDJ}(t)=\frac{A_{GD}\varepsilon_{si}}{\sqrt{2\varepsilon_{si}(V_{bc}(t)-V_{GS}(t)+V_{Td})/qN_L}} \quad (1)$$

In which: A_{GD} is the lateral lateral area of trench gate deep into the base region; ε_{si} is the dielectric constant of silicon; $V_{bc}(t)$ is BJT base-emitter voltage drop; $V_{GS}(t)$ is the gate voltage; V_{Td} is that threshold voltage of the gate-drain overlap depletion layer; q is the electronic quantity; N_L is the base doping concentration.

The corresponding gate-drain junction capacitance is expressed as:

$$C_{GD}(t)=\begin{cases}C_{OXD}\\ \dfrac{C_{OXD}C_{GDJ}(t)}{C_{OXD}+C_{GDJ}(t)}\end{cases} \quad (2)$$

However, for the trench gate structure, due to the large gate thickness, the width of the depletion layer may not cover the trench in IGBT switching transient, so it is necessary to analyze it according to different conditions.

2.2 Analysis of trench gate junction capacitance

In switching transient state, the internal current of IGBT includes the charge and discharge currents of gate capacitance and other internal junction capacitance besides hole current and electron current, which together constitute the total IGBT current. Therefore, the influence of capacitance must be taken into account in transient model analysis. The equivalent circuit diagram of IGBT including gate capacitance and internal junction capacitance is shown in Figure 2.

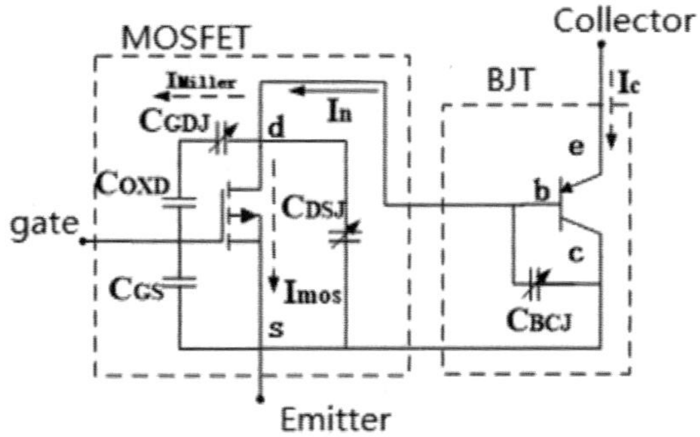

Fig2. IGBT equivalent circuit diagram

In Figure 2, C_{GS} is the capacitance between gate and source; C_{OXD} is the capacitance of grid overlapping oxide layer; C_{GDJ} is the grid overlapping depletion layer capacitance; C_{OXD} and C_{GDJ} are connected in series to form the capacitance between gate and drain, that is C_{GD}, which is also feedback capacitance; C_{DSJ} is the capacitance between drain and source; C_{BCJ} is the junction capacitance between the P+ emitter and the N-base region, which can be approximately regarded as the sum of C_{GDJ} and C_{DSJ}. Among them, the capacitors C_{GS} and C_{OXD} are all fixed values, while the capacitors C_{GDJ}, C_{DSJ} and C_{BCJ} all change with the change of the width of the depletion layer.

According to the relationship between depletion layer width and trench gate width without considering trench gate structure, the junction capacitance of trench gate structure in Figure 2 is also analyzed in two cases.

(1) The depletion layer is not widened to cover the trench gate, that is $W_{dep,PIN}\le L_x$.

The internal junction capacitance of the trench gate IGBT is shown in Figure 3.

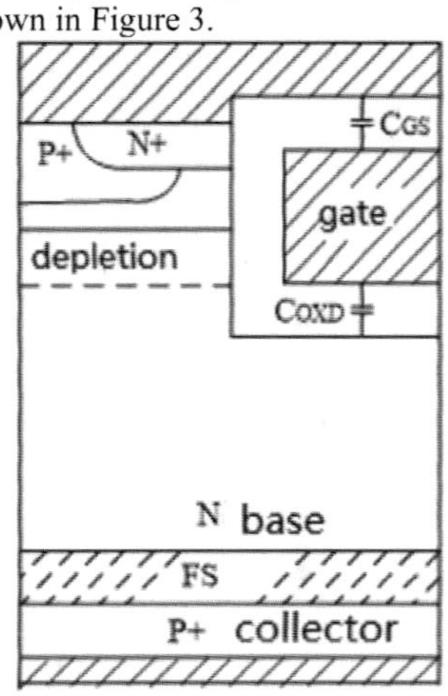

Fig3. Internal junction capacitance under condition 1

Because the depletion layer fails to widen and cover the trench gate, there is no depletion layer at the lower part of the trench gate, and the grid-drain junction capacitance only needs to consider the oxide layer capacitance at the lower part of the trench gate. At this time, the junction capacitance is expressed as:

$$C_{GD} = C_{OXD} \tag{3}$$

Where: C_{OXD} represents the capacitance of oxide layer at the lower part of trench gate.

(2) The depletion layer can be widened to cover the trench gate, that is $W_{dep,PIN} \le L_x$.

The internal junction capacitance is as shown in Figure 4.

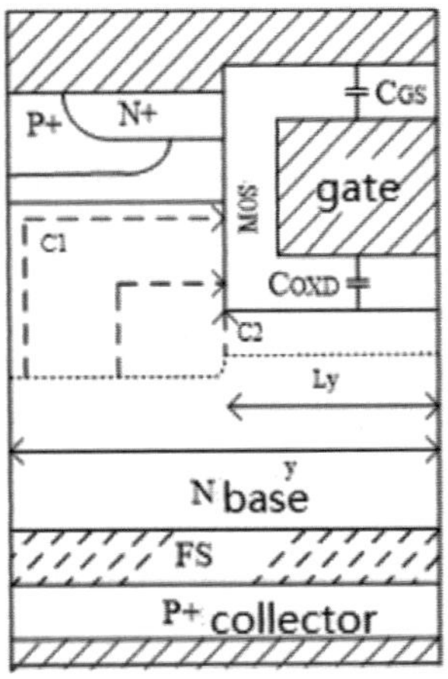

Fig4. Internal junction capacitance under condition 2

Because of carrier movement and current flow, it is necessary to consider the longitudinal depletion layer capacitance at the lower part of trench gate, the lateral depletion layer junction capacitance at the side, and the trench gate oxide layer capacitance.

The vertical depletion layer junction capacitance of the gate-drain can be expressed as:

$$C_{GDJ,Z}(t) = \frac{A_{GD}\varepsilon_{si}}{\sqrt{2\varepsilon_{si}(V_{bc}(t) - V_{GS}(t) + V_{Td}) / qN_L} - L_x} \tag{4}$$

Where: L_x is the longitudinal depth of trench gate.

The lateral depletion layer junction capacitance of gate-drain is the average of C1 and C2 depletion layer capacitances.

$$C_2(t) = \frac{A_{GDH}\varepsilon_{si}}{\sqrt{2\varepsilon_{si}V_{bc}(t) / qN_L} - L_x} \tag{5}$$

$$C_1(t) = \frac{A_{GDH}\varepsilon_{si}}{\sqrt{2\varepsilon_{si}V_{bc}(t) / qN_L} + y - L_y} \tag{6}$$

$$C_{GDJ,H}(t) = \frac{C_1(t) + C_2(t)}{2} \tag{7}$$

The grid-drain depletion layer junction capacitance is obtained by connecting the horizontal and vertical depletion layer capacitances in parallel

$$C_{GDJ}(t) = C_{GDJ,Z}(t) + C_{GDJ,H}(t) \tag{8}$$

The gate-drain Miller capacitor is formed by connecting the gate-drain depletion layer capacitor and the oxide layer capacitor in series.

Combining the above two cases, the expression of gate-drain capacitance is obtained as follows:

$$C_{GD}(t) = \begin{cases} C_{OXD} & W_{dep,PIN} \le L_x \\ \dfrac{C_{OXD}C_{GDJ}(t)}{C_{OXD} + C_{GDJ}(t)} & W_{dep,PIN} > L_x \end{cases} \tag{9}$$

Because the junction capacitance of depletion layer is affected by external voltage, when the voltage rises, the junction capacitance of depletion layer will drop rapidly. Therefore, in the linear region, the feedback capacitance is basically equal to the oxide layer capacitance, while in the cut-off region and saturation region, the feedback capacitance is basically equal to the depletion layer capacitance.

Gate-drain junction capacitance $C_{GD}(t)$, drain-source junction capacitance $C_{DS}(t)$ and base-collector junction capacitance $C_{BC}(t)$ are all variable junction capacitances, which have high capacitance in the linear region of low voltage and decrease with the increase of depletion layer width, i.e., applied voltage.

Three capacitors are usually given in IGBT manual, namely input capacitor C_{iss}, output capacitor C_{oss} and feedback capacitor C_{rss}. The relationship between these capacitors and the internal junction capacitance in Figure 2 is as follows:

$$\begin{cases} C_{iss} = C_{GS} + C_{GD} \\ C_{rss} = C_{GD} \\ C_{oss} = C_{DS} + C_{GD} \end{cases} \tag{10}$$

The MOS channel current meets the following requirements according to the conditions that MOSFET works in cut-off region, linear region and saturation region:

$$I_{mos}(t) = \begin{cases} 0 \\ K_P(V_{GS}(t) - V_{th})V_{ds}(t) - K_P V_{ds}^2(t)/2 \\ K_P(V_{GS}(t) - V_{th})^2/2 \end{cases} \tag{11}$$

The expression of base current obtained from the equivalent circuit of fig. 2 is:

$$I_{nL}(x = W) = I_{mos}(t) + (C_{DS} + C_{GD})\frac{dV_{ds}(t)}{dt} - C_{GD}\frac{dV_{GS}(t)}{dt} \tag{12}$$

The expression of gate voltage change rate is:

$$\frac{dV_{GS}(t)}{dt} = \frac{I_G(t)}{C_{GS} + C_{GD}(t)} + \frac{C_{GD}(t)}{C_{GS} + C_{GD}(t)}\frac{dV_{bc}(t)}{dt} \tag{13}$$

The expression of voltage change rate under different conditions can be obtained by introducing the expressions of gate-drain junction capacitance under two different conditions.

3 Experimental verification

A trench gate field termination IGBT module with KW60N65S model is selected, and the rated voltage and current levels are 650V and 60A respectively. The experimental test conditions are as follows: the bus voltage is set to 400V, the driving voltage is 15V when switching on, and the load inductance is 500uH. The collector-generator voltage and collector current of IGBT are measured by double pulse test method. As the accurate calculation of junction capacitance mainly affects the turn-off current in IGBT transient process, the turn-off current is compared. The test circuit is shown in Figure 5.

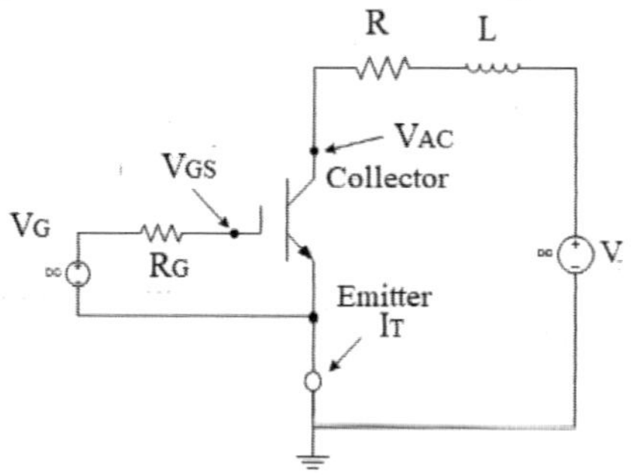

Fig5. Experimental test circuit

Using the junction capacitance calculation method of planar gate structure, the simulation waveform of IGBT turn-off transient current and the comparison of experimental waveforms are shown in Figure6.

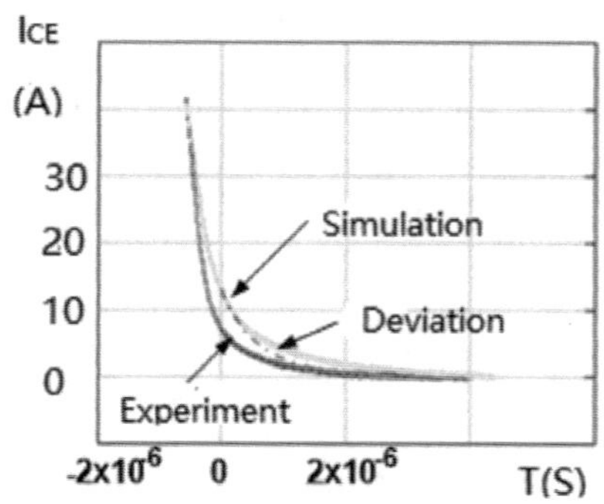

Fig6. Comparison of shutdown currents

Using the calculation method of junction capacitance in this paper, the simulation waveform of IGBT turn-off transient current and the comparison of experimental waveform are shown in Figure 7.

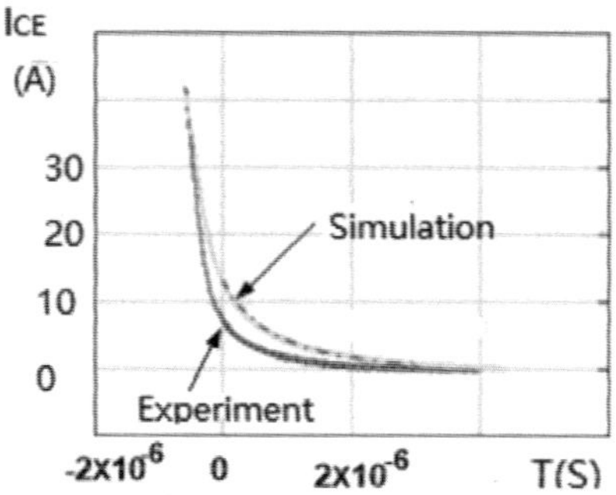

Fig7. Comparison of shutdown currents

It can be seen from Figure 6 and Figure 7 that the simulated turn-off transient current obtained by the calculation method of junction capacitance in this paper is closer to the experimental waveform.

4 Summary

According to the difference between the traditional planar gate structure and the new trench gate structure, this paper discusses the calculation method of junction capacitance in IGBT in two cases according to whether the depletion layer in the base region can widen and completely cover the trench gate, which overcomes the single junction capacitance treatment method in planar gate structure.

Acknowledgments

This work was supported by the 2018 year Youth Foundation of Wuhan Donghu University(2018dhzk005), and the Scientific Research Project of Hubei Education Department in 2019(B2019267).

References

1. Chen Xudong, Cheng Jianbing, Teng Guobing, et al. Novel trench gate field stop IGBT with trench shorted anode[J]. Iournal of Semiconductors, 2016, 37(5): 61-64.
2. Bazzi A, Philip T, Jonathan W. IGBT and diode loss estimation under hysteresis switching[J]. IEEE Transactions on Power Eletronics, 2012, 27(3): 1044-1048.
3. Chamund D J, Coulbeck L, Newcombe D R, et al . High PowerDensity IGBT Module for High Reliability Application[C] . Proceeding of IPEMC, 2009: 274-280.
4. Hu Shao-wei, Zhu Yang-jun, Duan Yao-yu. An Impact Analysis of Gate Resistance on Static and Dynamic Dissipation of IGBT Modules[C]. Proceeding of ICECC, 2011:715-718.
5. Vinod Kumar Khanna, The Insulated Gate Bipolar Transistor-IGBT Theory and Dedign[J].IEEE Society Press. 2003:105-107.

Public Space Sharing Strategies for Youth Education Building

Misheng Lin[1,*], and *Yang* Wang[1,2]

[1]School of Architecture, South China University of Technology, 510640 Guangzhou, Guangdong, China
[2]South China University of Technology Architectural Design and Research Institute Co., Ltd, 510640 Guangzhou, Guangdong, China

Abstract. Education building can be activated by space sharing strategies to satisfy the large demand of outside school education needs. The public space of youth education building has a more flexible social adaptability than the full-time school. Through case collection, data research, statistical analysis methods, design elements of public space of youth education buildings are studied. Based on the consideration of design elements, the spatial layout modes and design strategies are summarized. These results are helpful for the similar space design in outside school education buildings.

1 Introduction

1.1 Public space sharing of educational buildings under social needs

Due to the insufficiency of Chinese per capita distribution of education resources [1], conventional public schools are hard to satisfy modern education. Parents pay lots of attention to the growth of teenagers through outside school education. Sharing economy promotes the sharing of education building's idle spaces, better serving the society. Space sharing usually refers a transaction to have the equal right to use or acquaintance a same place with others [2-3]. A public building must be the core platform as an intermediary for the connection and distribution of education resources. The most basic task of space sharing should be revitalizing idle resources. Considering the safety of students under the unified management of schools, most public full-time education schools are closed management and not open to the public [4-5]. With the emergence of education economy, various education building spaces have expanded their original possibilities [6]. Among various public buildings that can be shared for public education, the youth education building is more flexible and adaptable than the traditional closed-managed school. The youth education building holder provides space as a functional platform. Through the platform, the usage of the space can be negotiated, thus multiple activities can be reasonably arranged.

1.2 Review of youth education building in China

In China, a youth refers to a statutory minor who are among 6 to 18 years old. Youth education building, such as the Youth Activity Centre, Youth Science Museum, and Youth Education Base, refers to a type of independent outside school education institute [7]. It provides a systematic, standardized supplementary education activity. Nowadays, the education industry is highly competitive, youth education buildings are constantly increasing their range of openness. In 1949, the completion of the Dalian Youth's Cultural Centre, which was reformed by a private school, marked the birth of China's first professional youth education building prototype [8]. After the education revolution from 1969 to 1979, outside school education has entered a stage of comprehensive reform [9]. Outside school education buildings for youths are gradually regularized. Since 2000, with the improvement of domestic architectural technology and modern design methodology, the architectural design of the youth education building has shown a diversification in its public and opened characteristics [10]. Up to 2020, youth education buildings have experienced more than 70 years of development. Gradually, youth education building attaches great importance to the role of space sharing design in improving teaching experience. This trend greatly arouses the building resources, benefit the society and increase the extra income of the youth education building.

2 Design Elements of space sharing

2.1. Openness of public space

The youth education building is a public building that almost all spaces are possible for sharing [11]. However, an excessive openness of space sharing will interfere with the normal operation of the original function of the building. Generally, the public space of a youth education building is suitable for sharing. As shown in

* Corresponding author: 568685279@qq.com

the Table 1 below, fully open space refers to the space open to the public during its' whole operation period. This type includes spaces like the plaza, open courtyard, open overhead floor, porch and open viewing deck. Usually it is a transitional space between the building and the urban public space. It is a space that citizen can enter for legal activities most of the time. Usually, the sharing of this space is free of charge under official applications. Fully open space breaks through the limitations of time management, thus further enhances the physical accessibility of the public space. Semi-open space can be entered and used reasonably within some certain restrictions. This type is usually half indoor and outdoor, including auxiliary spaces such as public foyer, passage hall, traffic space and commercial space. It provides the public with a place to rest, chat or cross. Specific open space is formed by the building's special interface, streamline or segmentation, which has a character of inner openness. This type includes public laboratories, temporary offices, multimedia conference rooms, etc. These spaces are of control that require maintenance and access management. Such space is only shared for specific usage with applications, or within a specific period of official working time. And not all youth education buildings share this type of space.

Table 1. Openness of public space in youth education building (Resource: by authors)

Openness	Diagram	Case photo
Fully open	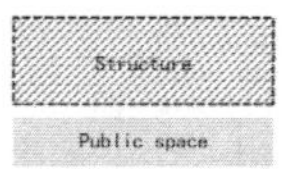	
Semi-open	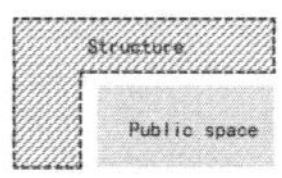	
Specific open	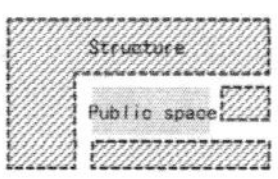	

2.2 Physical and mental needs of youth education

Youth students are in the transition period from young to mature. They have a strong sense of independence and try to protect their personal privacy. At the same time, they have social curious on the surrounding people. During education, youth students want to be respected and get satisfaction in communication. The youth education building provides a basic study platform for youth students. The building space needs to be accessible and convenient for students. It is necessary to create a comfortable environment which is suitable for the body scale of youth students. Design should be considered from the aspects of space scale, colour, texture to the history, culture, and region of the building. It is necessary to meet the diversity of education functions, provide corresponding places for youth education activities.

2.3 Specific facility for educational activity

In order to have an easy sharing access of the public space, safe, practical and durable basic facilities should be provided. Basic infrastructures are necessary for space sharing, such as shading plate and fire protection facilities. Usually, fully open space includes the basic hydropower interface, area delimitation, landmark indication, etc. Semi-open space includes water and electricity pipes, air conditioning, HVAC (Heating Ventilation and Air Conditioning), lighting, intelligent security, self-service machines, etc. Besides, public facilities should create an active atmosphere. A safe and interactive atmosphere can be enhanced according to specific facilities design for youths, teachers and caretakers. Results show that the sharing rate of well-equipped youth education buildings are relatively high.

2.3.1 Youth body scale furniture

Many spontaneously education activities need furniture combinations for youth. Furniture inside public space shall be fit for youths' body scale and can be safely assembled for different range of use. For example, combination of body scale seats in the youth education building should meet the requirements of the height and behaviour habit of youths. The weight of seats should not be to light to be thrown, either too heavy to be moved by teenagers.

Fig. 1. Combination of body scale seats for youth study (Sources: archgo.com)

2.3.2 Safe guard

Safe guard is an essential complement especial for outside school education security. It is common for the aged to look after the youth in outside school education. Pre-protection and barrier-free design should be considered for public spaces that may prevent users from impacted, dropped or slipped, etc. Due to foreseeability of public space sharing, a reasonable safety reminder for the caretaker is also a kind of strengthening protection design. For example, anti-slip treatment shall be performed on the floor, reflective warning on height difference floor shall be added, and all power sockets shall be covered with a safety cover.

Fig. 2. Safe guard considering youth behaviours (Sources: architecturalrecord.com)

2.3.3 Education game facility

Education games are very suitable for youth teaching. Education game facilities shall be made of non-angular and anti-drop materials. Youths between the ages of 3 to 6 can learn to communicate in the game within protections of their caretakers. Youths among 6 to 10 years old can choose sport activities under monitoring. And youths over 10 years old are allow to choose adventure activities independently according to their own interests.

Fig. 3. Educational game facilities in youth education buildings (Sources: by authors)

3 Design strategies of public space sharing

3.1 Layout mode

The layout mode refers to the coordination of the overall spatial relationship of the public space sharing and building function management. Public space sharing is usually designed clearly so as not to interfere with the original building function. As shown in the following Table 2, the space sharing layout strategy of the youth education building can be divided into 3 modes. It is presumed that the Central has an advantage of centrally management. The obvious division of the public space sharing and the private space management is easier to implement with less interference, which can solve the problem of functional partition. Youth education buildings with complex heights, environments, or rooms tend to use the Multi-Cores. In fact, Multi-Core is more flexible for modern public space sharing. This mode reorganizes the scattered idle space independently and extend them to the society. The Net Format is a kind of spatial pattern with interlacing and fluidity. Youth education building with flat and spacious field is more likely to apply this mode. The public space sharing in this mode is not completely fixed and can be adjusted according to the scale or form of activities.

Table 2. Layout modes of public space sharing (Resource: by authors).

Mode	Diagram	Legend	Character
Central	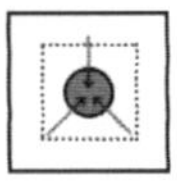	Stable space sharing	The space area is concentrated to form one public sharing core.
		Inner function	
		Building range	
Semi-Cores	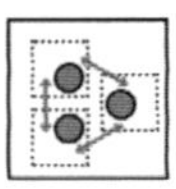	Half stable space sharing	The function partition is flexible, which is convenient for multiple activities.
		Inner function	
		Building range	
Net Format	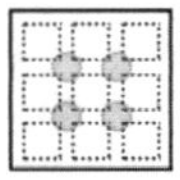	Potential space sharing	The free and versatile network organization provides a variety of options.
		Inner function	
		Building range	

3.2 Space connection

The connection strategy is to set link of the public space with social needs. For the sake of teaching quality, youth education building generally needs to be regularly managed and maintained. Not all public space of youth education building can be shared to the public. Considering educational requirements, the integrated connection of space can increase the plasticity of the building public space into the urban space system. The space sharing limitation can be compensated through this strategy. Youth education building shall provide a high convenience of participation. This strategy can be achieved by developing an isolated space from the building management system. The isolated space can be enclosed by buildings structures, tall vegetation or guiding structure. As shown in the Table 3, 3 types of connection strategy are listed. It should be noticed that these strategies are side-by-side relationships, thus multiple strategies can be adopted according to actual needs. For example, the Qingpu Youth Activity Center in Shanghai adapts the Virtual space in the entrance. An interior wall void of public opening space has been customized. The wall void encloses an internal virtual space, enhancing the visual enjoyment. The observation from the outside to the inside is relatively weak, which protects the privacy of youth activities. This design allows the public space to be integrated into the multi-directional visual cross network, which ensures the rationality of the internal sharing activities by social environment.

Table 3. Types of connection strategy (Resource: by authors)

Type	Diagram	Enclosure character
Central corridor	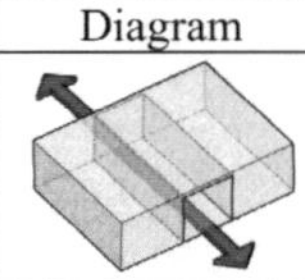 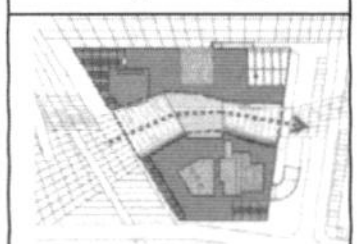 Nanjing Youth Centre	Inside the building, a space corridor for people to walk through is opened, combined with the surrounding urban transportation network and urban residents' habits.
Internal court		An open-air courtyard is enclosed by building blocks, so that the public can reach

https://doi.org/10.1051/e3sconf/202123703001

Type	Diagram	Enclosure character
	 Baotou Youth Base	directly without passing through the internal space of the building.
Virtual space	 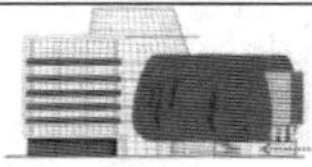 Guangzhou Second Youth Institute	Space with a covered top is provided at the edge of the building. The projection area of the upper cover is enclosed to form a transition space between the interior and exterior.

3.3 Humanity optimization

The humanity optimization of public space is the promotion content of space sharing. It is an ergonomics of space through certain design principles and methods. The humanization role is to provide users with psychological buffer space in space sharing. As shown in Table 4, there are respective humanity optimization methods for the 5 spatial problems. Combining with these methods, there are 2 strategies that can be used according to the vertical and horizontal space system.

Table 4. Humanity optimization methods of public space sharing problems (Resource: by authors)

Problem	Humanity diagram	Case analysis	Description
Too excessive or too empty. difficult for social activities.	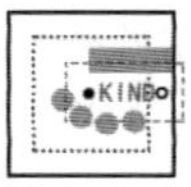	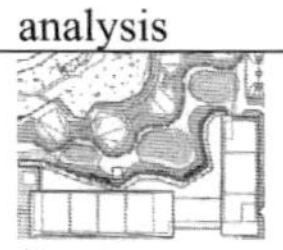 Dongguan Changan Town Youth Base	Properly divide and enclose suitable space for activities.
Too narrow, crowded, or even losing the sense of distance.		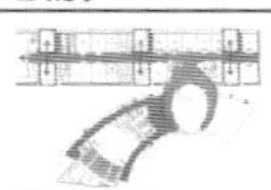 Guangzhou Second Youth Institute	Increase space sense by opening space joint, borrowing outdoor scenes.
The channel-like space is too short to provide communicating space.		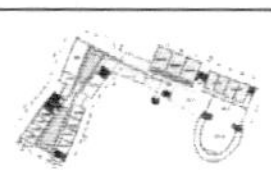 Jiangmen Youth Institute	Increase the feasibility of the channel by adding interior landscapes or joints.
Distance between passages is too long, not conducive.	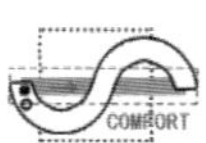	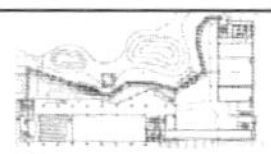 Dongguan Changan Town Youth Base	Partial design convenient access for direct path.
The division of intersections, ramps, and trails are too fragmented		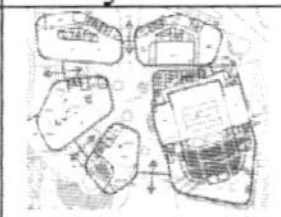 Beijing Youth and Teenager Educational Base	Enhancing the guiding and direct system through key nodes.

3.3.1 Communication joint

When the range of public space provided by the youth education building is not reasonably to be shared, a communication joint with proper dimensions can be appropriately implanted to refine this situation. If the joint is placed in the middle of the space, the whole space shall be relatively separated into several smaller zone. Activities nearby the zone have less interference with the other functions of the original space. Else if the joint is placed at the edge or corner, it helps to create a visible spatial zone. When a user passes, it is possible to recognize this special zone. And this joint can be designed to activate the atmosphere in combination with specific facilities. For example, Guangzhou Second Youth Institute combines some colourful corners to enlarge the corner space for the youths. There are rest seats with some game facilities placed in the corner of the rest zone. Natural skylight adapted to subtropical climate is adopted at the top of this space, which can introduce air and natural light. Joints create a sense of comfort in the visual experience for young people.

Fig. 4. Corner communication joint in Guangzhou Second Youth Institute (Sources: by authors)

3.3.2 Vertical segmentation

Vertical segmentation is an intuitive design strategy for human visual perception. The ascending or sinking of the floor interface can divide the spatial content by enriching the level of space through the micro-height difference. The vertical space is effectively separated to meet the temporary private scope during sharing. And under the premise of satisfying the use of the layer height, a stage, platform or a mezzanine can be added to divide the part of the space into several levels. When adding the sharing area of the original space in the same area, the space can be more useful. For example, in the Qinghai youth education centre in Guangzhou, an performance stage is set to define different heights in the vertical direction of the atrium. Spaces with different heights are distinguished by contrasting the colour and material of the paving. The slightly higher space highlights the stage as a designated activity area for youths, creating a strong sense of exclusive communication space for them.

Fig. 5. Vertical segmentation of Qinghai youth education centre in Guangzhou (Sources: by authors)

4 Conclusion

As a public welfare institution, the youth education building has gradually gained attention to the social sharing needs of its public space. Results show that at present, most of the space sharing of youth education buildings in China only allows public space to be shared to the public. The internal function space such as classroom, office and multimedia room are still not ready for the sharing mode. However, with the advent of the internet information and the development of mass media, education socialization has become an inevitable trend of education reform. The socialization of educational resources also requires the youth education building to increase its openness to the society. This paper summarizes the elements and design strategies of public space sharing by analysing engineering cases of youth education buildings. Three layout modes are proposed based on the condition of spatial agglomeration and morphological characteristics. The strategies of space connections and humanity optimization are summarized. These strategies may enhance the social education and community service functions of the youth education buildings. The open sharing characteristics of the public space will be further highlighted. Space sharing will not only meet the needs of the original use, but also meet the needs of the public and become a friendly neighbourhood of citizens.

References

1. C. Heyin, H. Yu, L. Jingwei, and S. Xiaojie, *The impact of environmental regulation, shadow economy, and corruption on environmental quality: Theory and empirical evidence from China*, Journal of Cleaner Production, **Vol. 195 No. 1**, pp. 200-214. (2018)
2. J. Hamari, M. Sjöklint, and A. Ukkonen, "*The sharing economy: Why people participate in collaborative consumption*", Journal of the association for information science and technology, **Vol. 67 No. 9**, pp. 2047-2059. (2016)
3. H. Heinrichs, "*Sharing economy: a potential new pathway to sustainability*", GAIA-Ecological Perspectives for Science and Society, **Vol. 22 No. 4**, pp. 228-231. (2013)
4. W. Dan, W. Jingying, L. Hui, and L. Ling, "*School context and instructional capacity: A comparative study of professional learning communities in rural and urban schools in China*", International Journal of Educational Development, **Vol. 52**, pp. 1-9. (2017)
5. R. Morton, "*Home education: Constructions of choice*", International electronic journal of elementary education, **Vol. 3 No. 1**, pp. 45-56. (2017)
6. A. Malhotra, and M. Van Alstyne, "*The dark side of the sharing economy and how to lighten it*", Communications of the ACM, **Vol. 57 No. 11**, pp. 24-27. (2014)
7. J. Sheldon, A. Arbreton, L. Hopkins, and J. B. Grossman, "*Investing in success: Key strategies for building quality in after-school programs*", American Journal of Community Psychology, **Vol. 45 No. 3-4**, pp. 394-404. (2010)
8. W. Ying, and K.H. C. Raymond, "*Autonomy and control: The struggle of minban schools in China*", International Journal of Educational Development, **Vol. 45**, pp. 89-97. (2015)
9. L. Yang, Z. Yajun, C. Sitong, Z. Jialin, G. Zongchen and C. Peijie, "*Associations between parental support for physical activity and moderate-to-vigorous physical activity among Chinese school youths: A cross-sectional study*", Journal of Sport and Health Science, **Vol. 6 No. 4**, pp. 410–415. (2017)
10. W. Xiaogang, and Z. Zhuoni, "*Population migration and youths's school enrollments in China, 1990–2005*", Social Science Research, **Vol. 53**, pp. 177-190. (2015)
11. M. A. Holleman, M. J. Sundius, and E. J. Bruns, "*Building opportunity: Developing city systems to expand and improve after school programs*", American journal of community psychology, **Vol. 45 No. 3-4**, pp. 405-416. (2010)

Study on Characters of Rural Residential Buildings Based on the Investigation in Shaanxi, Guangxi and Jiangsu in China

Li Zhihong[1,*], Huang Xianming[1], Yuan Duo[1], and Ma Yingying[1]

[1]Architectural Design and Research Institute of Tsinghua University Co., LTD, Chaoyang District, China

Abstract. This paper is supported by the national key R&D program（2018YFD1100205） of China during the 13th Five-Year Plan "Research on the Construction Technology System of Green Livable Rural Residential Buildings". This paper is focused on the characteristics of rural residential buildings form, functional space and villagers' demands of buildings based on different regions, climate and economic performance. The research is processed in the field with spot plotting and questionnaire survey including interviews about the rural residential buildings in Luonan County of Shaanxi, Longan County of Guangxi and Wujiang District of Jiangsu. According to the statistical results, there is a contrastive analysis of rural residential buildings in time and space, and then the characteristics, influencing factors, modular spatial type of residential buildings have been teased out. Meanwhile it refines the development trends of rural residential buildings eventually combined with the demands of villagers.

1 Introduction

To build the beautiful village, improve human settlement environment in village and comprehensively build a well-off society is one of the mission goals in China at this stage. With China's economy stepping into a new normal, a new look of urban and rural construction has been gradually opened in the vast rural areas and it is accelerating the transformation from the urban-centered to the coordinative development of urban and rural areas. The 2014 National New Urbanization Plan (2014-2020) clearly puts forward that the basic principles of urbanization construction, which are people-oriented, overall planning of urban and rural areas, optimizing the layout, an intensive and efficient development [1]. The government work report of the State Council in 2019 states that rural construction should be compared with the goal of building a moderately prosperous society in all respects, and effectively promoting the poverty alleviation and rural revitalization [2].The green and livable rural residential buildings have become the focus of attention along with the deepening and acceleration of new countryside's development. However, according to the research about the current state in the wide rural areas of China, the development has many negative influences such as lacking unified planning and design, chaotic layout of buildings, a serious waste of land resources and so on[3]. Also there are many of the problems in construction of residential buildings, such as unplanned and crude construction, low quality, less secure and lower quality of indoor environment caused by the traditional construction model in rural which is normally grouped by the house owner, relatives and friends or temporary team composed by villagers without professional training. But from another angle, it fits actual demands of villagers for production and living greatly as the villagers are taking part in the whole process of construction. Therefore, it has realistic needs for the research of technology on green and livable rural residential building construction based on the studies of current situation, spatial characteristics and demands for buildings in rural areas.

2 Research path

This paper conducts literature studies, field observation and interviews to understand the external environment factors such as the status quo of rural planning, economic development, main source of income, rural policy of the government and the cultural customs. Meanwhile, by questionnaire survey and mapping, the characters of functional space, construction materials, technology and demands of villagers for the houses are summarized. On the basis of sorting out the relationship between external environment factors and characters of rural residential construction, this paper carries out initial exploration in internal mechanism of evolution of rural residential buildings and summarizes characters of green and livable rural residential buildings in the future combining with the target of green, livable and demands of villagers.

3 Research overview

In order to comprehensively understand the current situation and demands of rural residential buildings in different regions and periods, the research team carried out field research in Luonan County of Shaanxi Province, Longan County of Guangxi Province and Wujiang District of Jiangsu Province. The layout of rural areas investigated is shown in Fig. 1. The principle and method of survery is

* Corresponding author: lizhihong0217@foxmail.com

 https://doi.org/10.1051/e3sconf/202123703002

to select the typical buildings of village settlement to be representative. Eventually there are 196 questionnaires and 78 drawings completed.

The social and economic background research of three regions mentioned above is summarized in Table 1. In terms of regional industries, the local industries in Luonan, Longan are mainly traditional cultivation and breeding while Wujiang is dominated by manufacturing and new industries. In terms of infrastructure, Wujiang district sets up a relative perfect infrastructure and Luonan and Longan need to improve water supply and drainage system.

- Luonan Shaanxi is located in the hill area of loess plateau of northwest China and has an arid and semi-arid continental monsoon climate. This region is poor in vegetation cover and water resources.
- Wujiang Jiangsu locates at the east coast of China, mild climate, four seasons, with plenty of rainfall. The landscape is mostly flat.
- Longan Guangxi is located in the southern mountainous hilly region of China. The climate belongs to sub-tropical humid monsoon climate with hot and rainy, short winter.

Fig. 1. Distribution map of survey villages

Table 1. Social and economic background survey

	Luonan Shaanxi	Longan Guangxi	Wujiang Jiangsu
Major industries in the region	Faiming and Breeding	Faiming and Breeding	Manufacturing and new industries
Main types of family economy	Farmer , Migrant work, Half labour and half pessant	Farmer, Migrant work, Half labour and half pessant	Local job, self-employed, Half labour and half pessant,Farmer
Infrastructure	Roads, electricity, communications and public services are relatively complete, inadequate water supply and drainage facilities	Roads, electricity, communications and public services are relatively complete, inadequate water supply and drainage facilities	Roads, water supply and sewage, electricity, communications, public services and other aspects are very perfect

Fig.2 shows the gross domestic product and disposable income per capitat of local rural residents in Luonan, Longan and Wujiang in 2017. It is apparent form the chart that the GDP and PDI of Wujiang are much higher than that of Luonan and Longan, and Wujiang is at a higher level of rural economic development in China. The types

https://doi.org/10.1051/e3sconf/202123703002

of rural family income are shown in Fig. 3. The ratios of rural labor migration family are respectively 33% and 21% of the households surveyed in Luonan and Longan, and the hollowing and aging phenomenons are observed in these areas during the survey. In Wujiang, only 4% of the households surveyed are rural labor migration family, and most of them are locality workers or self-employed. It is obviously that the urbanization level of Wujiang is much higher than Luonan and Longan.

According to the analysis above, Luonan and Longan have the typical characteristics of traditional rural, while Wujiang is the new-type developed rural at the present stage in China.

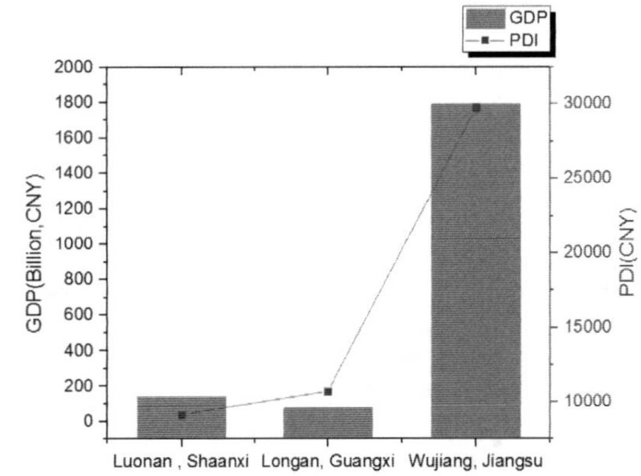

Fig. 2. The personal disposable income and gross domestic product in 2017

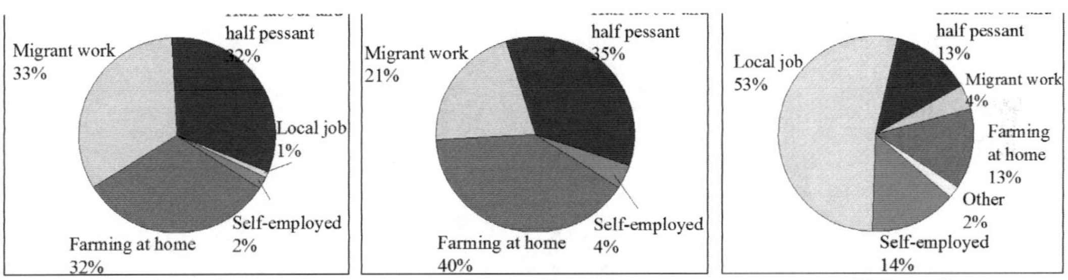

Luonan, Shaanxi Longan, Guangxi Wujiang, Jiangsu

Fig. 3. Types of rural family income

4 Analysis on the evolution of rural residential construction

Since the reform and opening up 40 years, the rural development consists five stages in China [4]. At each stage, the national rural policies have greatly promoted the rural economic level. The economic development brings about the changes in technology, awareness, people living styles and other aspects. And meanwhile, times features are responsed in the form of buildings accordingly. The characteristics of residential buildings in this survey are summarized and shown in Table 2 and Fig. 4.

Table 2. The investigation of building form in rural areas

	Period		The 1970s and before	The 1980s	The 1990s	From 2000 to 2009	Since 2010
Luonan, Shaanxi	Sample photos						
	Residential units (94 in total)		7 units	12 units	11 units	20 units	44 units
	Layer	Single	100%	100%	55%	35%	23%
		Multi	0%	0%	45%	65%	77%
	Shapes of Roof	Flat	29%	25%	45%	50%	50%
		Pitched	71%	75%	55%	50%	50%
Longan, Guangxi	Sample photos						
	Residential units		7 units	5 units	12 units	22 units	11 units

https://doi.org/10.1051/e3sconf/202123703002

	(57 in total)						
	Layer	Single	100%	40%	8%	18%	27%
		Multi	0%	60%	92%	82%	73%
	Shapes of Roof	Flat	100%	100%	100%	86%	82%
		Pitched	0%	0%	0%	14%	18%
Wujiang, Jiangsu	Sample photos						
	Residential units (45 in total)		2 units	2 units	24 units	7 units	10 units
	Layer	Single	50%	0%	0%	0%	0%
		Multi	50%	100%	100%	100%	100%
	Shapes of Roof	Flat	50%	0%	29%	14%	9%
		Pitched	50%	100%	71%	86%	91%

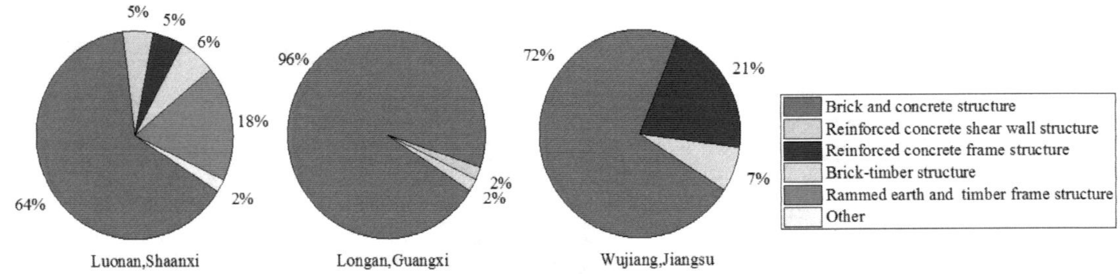

(a) Structure type

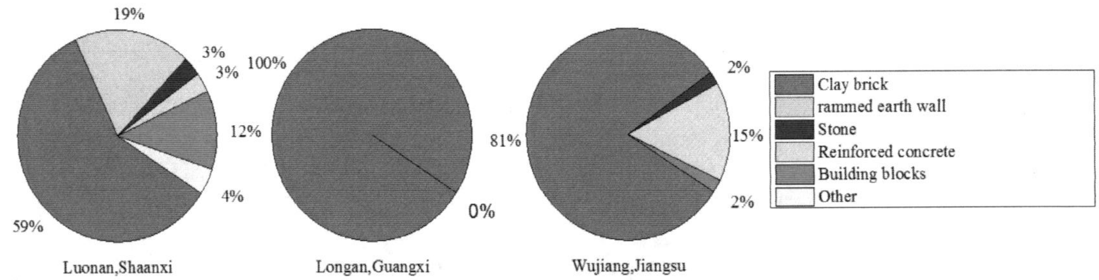

(b) Exterior wall type

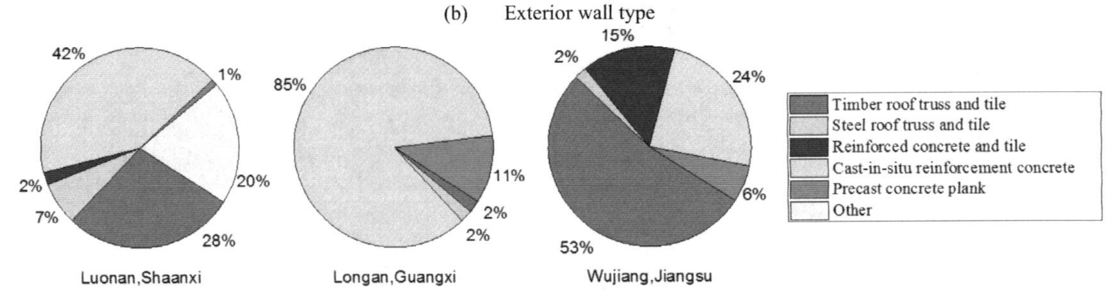

(c) Roof type

Fig. 4. Form characteristics of residential buildings surveyed

4.1 Building form

The 1970s and before which is just before the reform and opening up of china, the overall social environment is relatively closed and the level of economy, technology is generally low in rural. The construction mainly continues the traditional mode and technology. In this stage, the differences of buildings in the surveyed areas are generally small, Single-storey, sloping roof with timber roof truss and tile. During the investigation, it is obvious that at this stage, the rural residential buildings are generally with a lower wall, smaller window and door holes and a raw shell appearance without the cladding. However, there are still some obvious characters of buildings which are restricted by the climate and natural resource in different regions. The features of vernacular buildings are presented such as rammed earth and timer frame structure in Shaanxi, stone and wood structure in Guangxi, brick and wood structure in Jiangsu. Regional climate and traditional construction techniques are fully reflected in the residential construction at this stage.

In the 1980s and 1990s, with the reform and opening up, the productivity and thoughts have been greatly liberated. Living standards of villagers are improved and the development of buildings construction begins a new stage. Reinforced concrete and some multilayer buildings in rural have appeared in developed and less developed regions at this stage. In the regions surveyed, most of the multi-storey houses are two-storey with brick and concrete structure. In term of roof construction, the flat roof is widely used in Longan due to waterproofing technology, convenient construction and small construction difficulty. Though there are some buildings with flat roof in Luonan and Wujiang, the pitched roof is still dominant which is affected by the construction experience, local culture customs and climates. In this stage, clay bricks are commonly employed as the exterior walls in rural. With the strength of exterior walls improved, the height of wall is higher about 3 meters or more and the window and door holes are larger than before. At the same time, influenced by urban social, economic and cultural, the materials used on facades become various, such as tiles, exposed concrete, paint, clay brick and so on. But in Guangxi, the facades are greatly influenced by humid climate and the surface of the facade is generally exposed without any decorations to avoid falling out and stained due to the humid weather.

Since 2000, Along with the rapidly development of reform and opening up, the implementation of new rural construction and revitalization strategy, the rural human settlements have been greatly improved. In recent years, the rural policy has become one of the main factors affecting the form of building in rural areas. For example, new houses are required to be exploited on former homestead in some villages of Luonan, and in some areas of Longan, only one-storey houses are allowed for new buildings. In Wujiang, with the implementation of policies such as "beautiful village" "characteristic and idyllic village" and "livable villages", the styles and features of new buildings have been controlled, and the differences have been shrunk gradually in a certain region. Due to the further tightening of rural homestead policy, the vertical expansion of rural residential buildings has been intensified and embodies in this survey. Rural residential buildings surveyed are generally 2-3 floor in this period. Brick and concrete is the major structural system, and there has a certain proportion of buildings with reinforced concrete frame structure. Most of them are mainly distributed in economically and technologically developed areas as the construction cost of concrete frame structure is relatively high. In Wujiang, the proportion of the residential buildings with frame concrete structure has reached 21% of the samples. With the concrete construction technology promoted and applied in rural areas, some of buildings constructed in recent years also have adopted concrete frame structure in Luonan. The roof form of residential buildings in Luonan shows that the ratios of flat roof and slop roof are close. Flat roof is still the major style of rural buildings in Longan, while slope roof is the main form in Wujiang affected by the policy, construction experience, and local culture customs. The proportion of wall with exterior finish is increased in the survey regions and the exterior finish is mainly cement mortar, coating and tiles.

Throughout the evolution of rural residential buildings, the form of buildings is mainly affected by economic, resources and construction technology level in the early stage, while at the present stage, it is mainly affected by policies and consciousness of builders. From the past to the present, most of rural residential buildings in china are still mainly self-built, the builders are grouped by blood relationship, neighbourhood or local craftsmen, so it is greatly influenced by the construction technology and the consciousness of builders. Information and technology are developed slowly in the past, therefore, the residential form has been fully continued in traditional rural areas. However, in developed villages, the development of Information and technology are relatively fast and building construction has begun to trend to the urbanization mode with professional design and construction team. The buildings are usually with higher construction level and variety of forms but at the same time, the regional characteristics and traditional culture are lacking seriously. With the process of urban-rural integration and rural construction in China, a great deal of information and technology are imported into vast rural areas. It will be further shrinking the distance of residential buildings form between traditional and newly developed rural based on the development of infrastructure and logistics in rural areas.This phenomenon has been reflected in this survey, as shown in Fig. 5.

 https://doi.org/10.1051/e3sconf/202123703002

Case in Luonan, Shaanxi　　Case in Longan, Guangxi　　Case in Wujiang, Jiangsu

Fig. 5. Appearance of rural residential surveyed

4.2 Functional space

78 buildings were mapped randomly in the surveyed residential buildings during this survey. The plane of these residential buildings can be generalized into 4 types- L type, U type, flat plate and others and the statistical is shown in Fig. 6. It shows the main type of survey plane in Luonan is L type. Normally, the L type is composed by principal room and subsidiary space and it forms a courtyard as a semi enclosed space. While the principal forms of plane in Longan is flat plate which is largely affected by topography and generally without courtyard. Flat plate and L type are the main types in Wujiang and there is usually an enclosed courtyard formed by wall and/or fences.

The traditional residential buildings in China is three-bay house with a hall and two bedroom or accessory spaces. Functional space of residential buildings surveyed mainly continues the traditional model with local evolution according to the analysis of residential functional space surveyed. The transformation regulation of the functional space is summarized and present in Table 3.

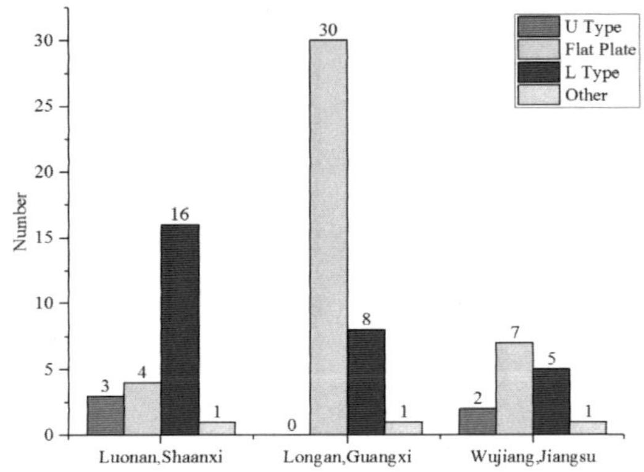

Fig. 6. The statistical of residential plane type

Table 3. Types of functional space layout

Type	Diagrammatic layout	Spatial composition
three-bay house (Traditional)	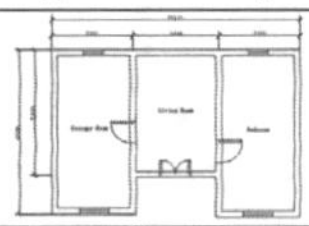	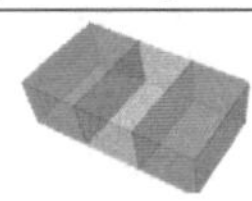
Shear-type	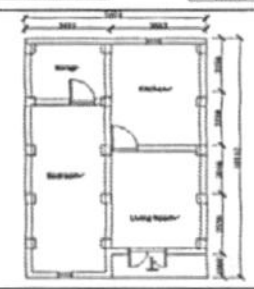	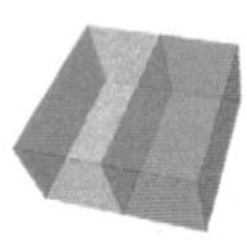
Extension-type	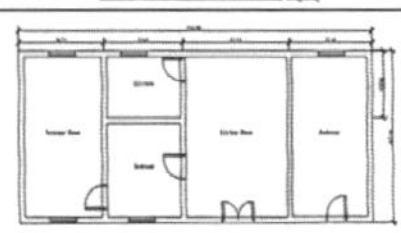	
Internal change-type	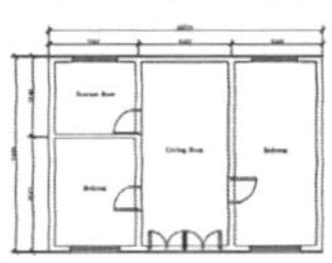	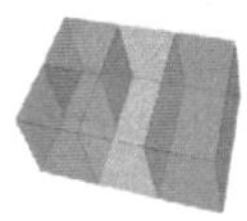

Splice-type	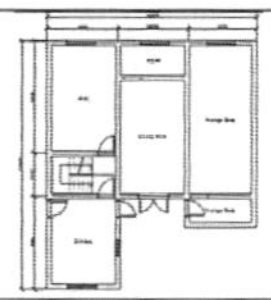	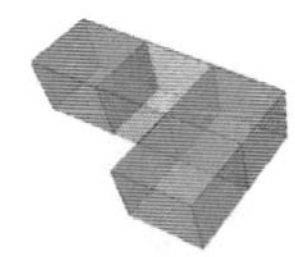
Comprehensive-type		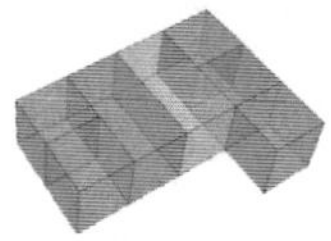

- Shear-type. This type is simplified to two bay or a small bay with large depth by the traditional type. It is normally affected by topography. Generally the large depth is divided vertically, and it is more common in Guangxi region.
- Extension-type. Generally this type extends horizontally on a relatively flat and open site, and the extension can be unilateral or bilateral based on traditional type. Usually, the three-bay house is mainly as the core area of daily life and the extension zone is set as the secondary bedroom or ancillary space such as storage ,kitchen, staircase and so on.
- Internal change-type. Interior space is divided and changed on the basis of traditional type. This type is a common spatial type in samples and it is compatible with the rural residential design approaches which is from big to small [5]. The overall space is divided into several small and suitable spaces to enable the functional spaces to be variety.
- Splice-type. L or U-shaped plane is formed by connecting ancillary space to traditional space type and such semi enclosed or enclosed space form is advantage in resisting adversary impacts of the external environment. This space type is common in Luonan region during the survey and the attached ancillary space is usually as kitchen, staircase or bedroom with kitchen.
- Comprehensive-type. It is combined freely with the space types above. This space type shows that the stronger adaptability of complex environment and diversified demand. A large number of rural residential buildings adopt this space type in the survey.

The functional space of rural residential buildings features modularization obviously according to the summarized and classified space layout. Bedroom and living room are the fundamental units of residential functional spaces, while other Spaces are set up based on the basic units. So the spatial scale of fundamental unit investigated has been studied and shown in Fig. 7 and Fig. 8.

It is shown in Fig. 7 that there is about 80% of the surveying and mapping bedroom samples which area is centralized to 10-25 ㎡, and the bedroom samples within 10-15 ㎡ account for a relatively large proportion about 43%. The spatial scale characteristics of bedroom samples have been analysed further by the temporal and spatial as shown in Fig.9.

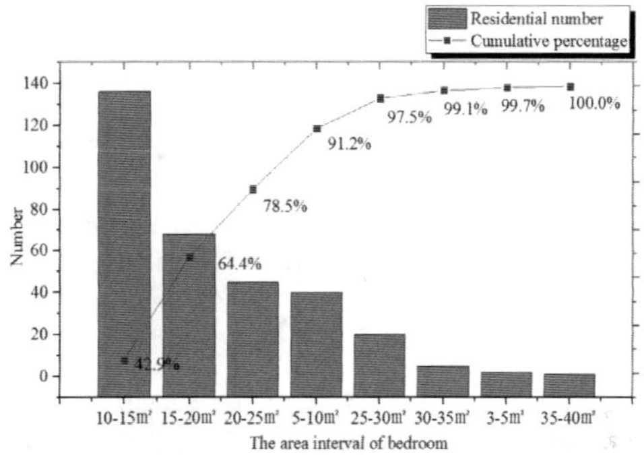

Fig. 7. Distribution of the bedroom area statistics

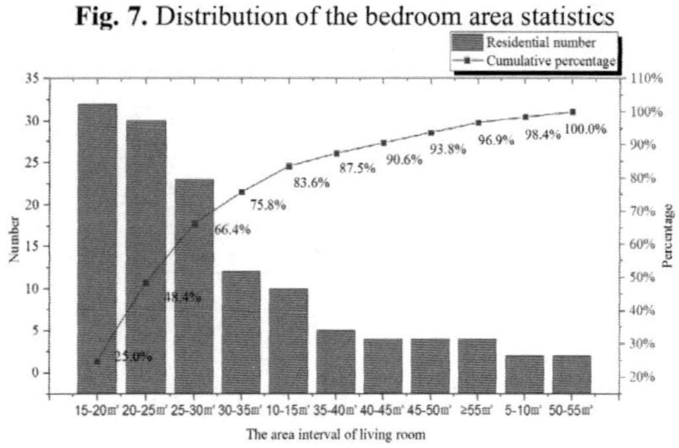

Fig. 8. Distribution of the living room area statistics

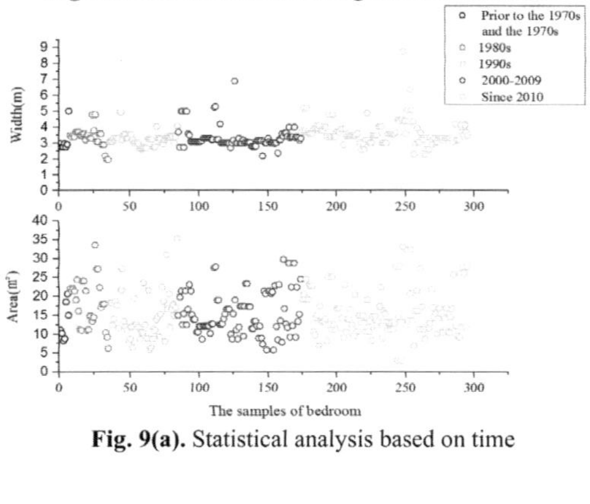

Fig. 9(a). Statistical analysis based on time

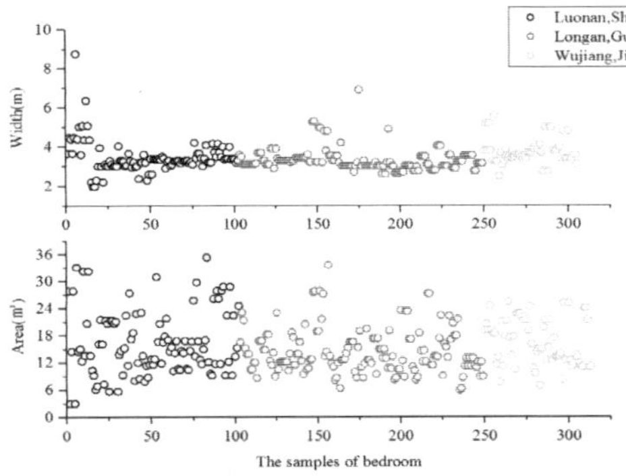

Fig. 9(b). Statistical analysis based on regions

The area of bedrooms is centralized to 10-25 m^2 and the width focus on 3-4 m in various periods without obvious variation trend analysed from the time angle. From the perspective of regions, the area of bedroom samples in different regions is evenly distributed in 5-30 m^2, and the distribution of width is basically 3-4 m consistent with each other. Although the width is very concentrated in Luonan, the area has a higher dispersed degree. It reflects that the depth is discrete and to a large extent is due to some of the bedrooms are mixed with kitchen in Luonan.

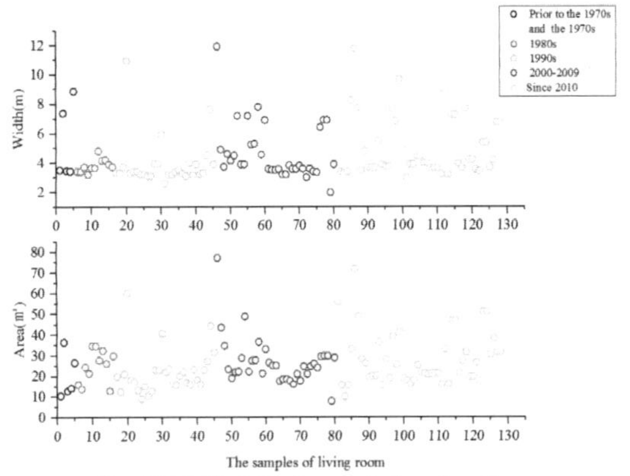

Fig. 10(a). Statistical analysis based on time

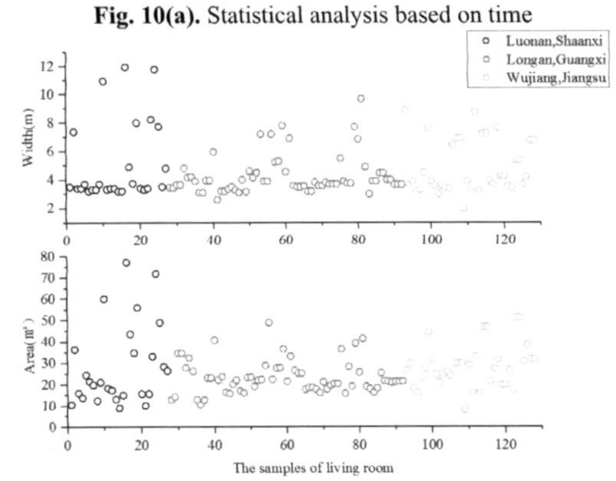

Fig. 10(b). Statistical analysis based on regions

80% of the living room samples are concentrated in 15-35 m^2 and the samples within 15-20 m^2 account for a large proportion about 25% as shown in Fig. 8. According to Fig. 10, the area distribution of living room samples has no significant changes over time too. The width distribute of samples is generally centralized to 3-4 m in each period while the proportion of samples with a width larger than 6 m has been increased since 2000 due to the development of construction technique and building materials. From the regional perspective, the area of samples is mainly concentrated in15-20 m^2 in Luonan, 20-25 m^2 in Longan and 25-30 m^2 in Wujiang. The width distribution of samples is generally consistent in all regions.

The spatial scale of residential core units surveyed in different regions basically have a constant modulus according to the analysis results from the different temporal and spatial. And most of it has due to the continuation of the traditional residential modulus and the people are already used to this spatial scale in their daily.

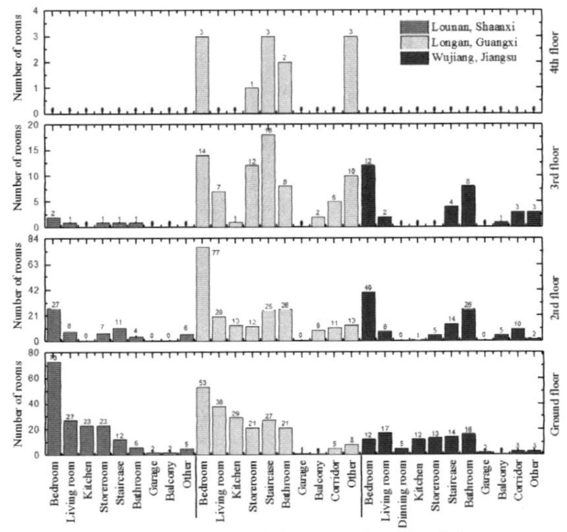

Fig. 11. The spatial layout of residential

The functional layout of residential surveyed is shown in Fig. 11. Most of the main space for activities is located on the ground floor in Luonan region and most of the villagers have a courtyard outside for leisure and communication, and generally the toilet is set independently in the courtyard. So it is more convenient for daily life and communication with neighbors to set the main space for activities on the ground floor. The main activities space of buildings in Longan and Wujiang is located on 2nd or above. The main reason is that there is normally no courtyard in rural residential in Guangxi or the size of courtyard limited and personal privacy in Jiangsu. Another reason is the climatic which major characteristics of Guangxi and Jiangsu is torridity and raininess and living above the ground floor can avoid moisture intrusion. Therefore, the daily life for villagers of those regions are mainly indoors and most of the residential buildings surveyed are set two or more living room interior. For most of villagers, the daily living activities such as accommodation and communication are on the 2nd floor and the main function of space on the ground floor is served as reception for guests, dining room,

https://doi.org/10.1051/e3sconf/202123703002

and storage for bicycle or electrical vehicle in Longan and Wujiang. The function layout of residential buildings surveyed in Wujiang is further improved and refined for the division of space is getting closer to urban house such as independent dinning space.

5 Demands analysis

The demands of the residential form and performance are investigated and the results are shown in Fig. 12 and Fig. 13. Single-family residential with 2-3 layers about 100-200 ㎡ is the most popular and it accounts for a large proportion in three different regions with various economic performance, geographical location and policies according to Fig.12. This desired form of residential buildings is in conformity with the fact about land policy tightening and vertical development trend. Single-family with single-storey, which is traditional buildings in rural, is also popular and second only to 2-3 layers. Some of the villagers preferred to remain the same because they have already used to the way of life during the interviewer.

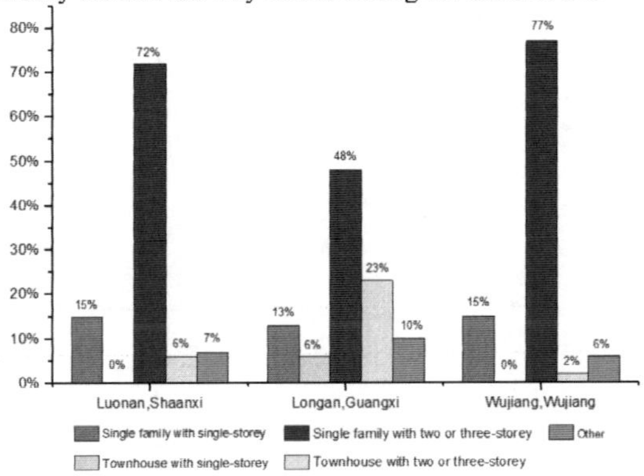

Fig. 12. The statistical demand for residential forms

The residential performance demands focused on are not identical in three region shown in Fig. 13. Due to the climate and social circumstances. It is mainly focused on area, orientation, sectorization, security and function requirement in Luonan. Ventilation and lighting, water and moisture resistance, orientation, comfortable environment and sectorization are more concerned in Longan. In Wujiang, ventilation and lighting, green, water and moisture resistance, function requirement and comfortable environment are mainly focused on. The concern to indoor environmental quality is obviously in this investigation such as orientation, ventilation and lighting, sectorization, water and moisture resistance and so on. Accordingly, continually improving indoor environment quality of rural residential buildings is one of the most important task at present and in the future.

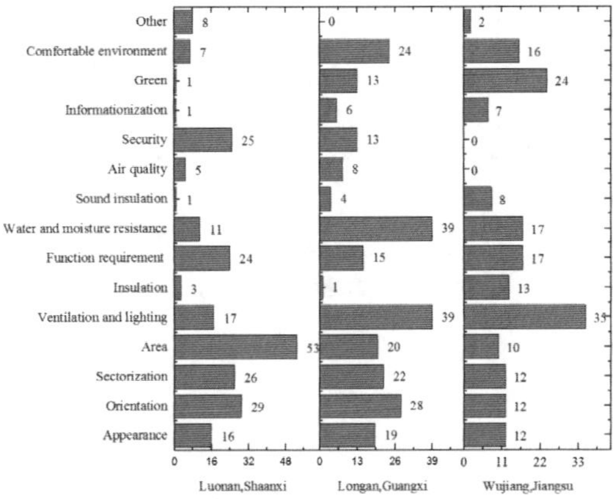

Fig. 13. The statistical demand for residential performance

6 Conclusion

The characters of rural residential buildings can be summarized as following based on analysis of the preceding context.

- The differences of residential construction in different phases are due to a variety of factors constraints at that time. Currently, the unbalanced development of construction techniques and engineer's working experience are major contributors to the differences in vast rural areas of China. With the development of urbanization and the spread of technology, the differences between city and countryside, traditional and new form of rural tend to be slightly while the impact of region factor, culture and policies are still not ignorable.
- Rural residential buildings surveyed in China have a modular structure. The functional space has been concluded to 5 types; the traditional type, shear-type, extension-type, internal change-type, splice-type and comprehensive-type. The spatial scale of residential basic unit is generally consistent in different phases and regions, and most of it has to do with the continuation of the traditional residential modulus and the people are already used to this spatial scale in their daily.
- Based on the present situation, land policies and villagers' demands of rural residential, single-family residential with 2-3 layers is one of the most popular types at this stage as well as the future development trends of rural.
- The functional division of rural residential tends to urbanization and be refined. The environment quality indoor of rural residential has become one of the focuses in the rural residential construction at present and in the future.

References

1. The national plan for new urbanization (2014-2020). *People Publishing House*. (2014)
2. The Government Work Report 2019. http://www.gov.cn/premier/2019-03/16/ content _5374314.htm. (2019)

3. L.Y. Liu, Y.H. Zhu, X.C. W. The problems and the strategies in the self-establish buildings in rural. Chin. Home, 02 (2014)
4. S.A. Huang. The gradual evolution and theoretical summary of China's rural development strategy in the past 40 years of reform and opening up. Econ. Res., volume **53**, 12 (2018)
5. B.S. Jia. "From big to small" and "From small to big": two approach about the systematization of residential. Arch. J., (1992)

Comprehensive Exploration and three-dimensional construction technology of water plug based on directional horizontal drilling caved column

Gan Tian[1*]

[1]Xi'an Research Institute of China Coal Technology & Engineering Group, Xi'an City, Shaanxi Provence, China

Abstract. The ordovician limestone water inrush accident occurred during the excavation of 8228 working face in Tashan Coal Mine, Tongmei Group. On the basis of comprehensive geophysical exploration above and below the well, in order to find out the development boundary of the water flowing subsided column and the distribution characteristics of the broken body filled in it, and to control the subsided column by Grouting, the development boundary, the influence zone, the cementation of the fractured body and the distribution characteristics of the water flowing subsided column are found out by means of multi-layer and multi-branch surface directional horizontal drilling speed logging, drilling fluid leakage, cuttings logging and logging while drilling, and combined with data statistical analysis. The length and short axis of the collapse column is 410m ~ 200m, and the broken area is mainly distributed in the right front of the driving head of the belt roadway near the water inrush roadway. The 3d geological modeling of the spatial structure and development characteristics of the collapse column is carried out by using SURPAC software, and the main passageway area, fracture area and secondary fracture area are divided according to the crushing degree of the collapse column filling material. In view of the different zones of the collapse column broken body, the directional horizontal drilling control technology and the three-dimensional block grouting technology are used to construct the separate zones of the grouting technology. The safe mining of the working face is finally realized by the comprehensive treatment of cutting the source and reinforcing the water-flowing collapse column. The comprehensive exploration and treatment technology of directional horizontal drilling multi-branch collapse column provides a good reference for other similar projects.

1 Introduction

Karst water hazard is one of the main water hazards affecting the safety production of coal mines in China. With the gradual depletion of China's shallow coal resources, the depth of coal mining continues to increase, the impact of karst water hazard on coal production is more prominent. According to the statistic data of Karst water-inrush accidents in China, 80% of the water-inrush accidents are structural water inrush, and most of them are water inrush from collapse column, for example, water inrush from the world's largest subsided column occurred in Fangezhuang Coal Mine of Kailuan Group in June 1984, water inrush from the subsided column occurred in Dongpang Mine of Xingtai in April 2003, water inrush from the subsided column occurred in Luoshan subsided column of Shenhua Group in March 2010, etc. .[1-6] The water inrush volume in the above water inrush accidents exceeded 1000m^3/min, which caused great loss to people's lives and property.

The location and water conductivity detection of collapse column are the foundation of treatment, and the method of combination of downhole geophysical exploration and drilling is usually used to detect the location and scope of collapse column.[7] At present, the treatment technology of collapse column mainly includes the closure of roadway, the construction of "water stop plug" , the "three-stage" water plugging of collapse column, direct plugging, backflow grouting and drainage grouting, etc.[6] The development process of control technology can be divided into three stages. The first stage (1984-1996) is represented by 2171 fully mechanized mining face in Fangezhuang Coal Mine of Kailuan Group, which is the first case of water inrush disaster at home and abroad. The treatment technology is to block up the water inrush collapse column by subsection downward grouting. This method has the disadvantages of low drilling precision, large engineering quantity and high cost. In the second stage (1996-2006), the technology of "water plugging" in the collapse column is widely used, that is, building a certain length of "water plugging" in the pre-set layer of the water-inrush column, and cutting off the water inrush channel of the collapse column, such as the 7277 working face collapse column Treatment Project of Wanbei Renlou coal mine. The disadvantage of this method is that it can not control drilling accurately, only "Serpentine" drilling can be used, and the engineering quantity is relatively large. In the third stage (since 2006) ,

* Corresponding author: ccritiangan@126.com

the Directional branch hole-making technology is widely used, which makes the "water blocking" construction technology in the collapse column more mature and perfect, such as the control project of the water bursting collapse column in 2903 working face of Dongpang coal mine. In this stage, directional drilling technology was used to drill accurately and directionally in the loose body of the collapse column, and a "water plugging" construction technology was formed, in which directional branch drilling was the main method. The exploration precision is greatly improved, and the engineering quantity and the management cost are greatly reduced in this stage.[6,8]

At present, the Directional Branch inclined hole "water plugging" construction technology is mostly used to detect and control the water hazard of many collapse columns. In general, this method has the advantages of small engineering amount, easy to put aggregate, and can be widely used in the stereoscopic detection of subsided columns. This technique is suitable for the application under the condition that the position and boundary of the collapse column are relatively clear. However, due to the limitation of the terrain construction environment, the Directional branch inclined hole "water plugging" construction technology can't be used in this collapse column exploration and treatment project. Combined with the characteristics of site construction conditions, this project adopts a special multi-level multi-branch directional horizontal construction "water blocking" technical method. This method is generally suitable for the collapse column location and boundary conditions are not clear. The prospecting and harnessing technology has a good reference to the structural prospecting and harnessing engineering under similar conditions.

2 Project design of collapse column exploration and treatment

2.1 Survey of the study area

The TASHAN coal mine of Datong Coal Mine is a coal mine of Datong Coal Field of carboniferous-jurassic-series. The length of 8228 coal face is 3101m, the dip length is 230.5 m. The average thickness of coal seam is 19m. According to the analysis of the geological and hydrogeological conditions of the working face, the No. 3-5 coal seam is 80m away from the top of ordovician limestone Aquifer, and the ordovician limestone water pressure in the floor is 1.84 MPa. The water inrush coefficient is 0.023 MPa/m, and the working face is in the relative safe zone. At the end of November 2016, water inrush from floor occurred when 8228 working face belt roadway was driven to 2060 m in two-plate area of Tashan mine. The maximum water volume was about 272 m^3/h at the heading face, which caused part of the roadway to be flooded, seriously affecting and threatening the safety of mine production. The water level of G1 observation hole in the ordovician limestone aquifer dropped about 60m after water inrush occurred in the belt roadway of the working face. Based on the comprehensive analysis of underground electrical method, Rayleigh wave exploration, surface 3D seismic data and controlled source exploration data, the water inrush in the belt roadway of the working face is the water inrush accident in the ordovician limestone aquifer of the coal seam floor caused by the water flowing collapse column in front of driving.[9-10]

In order to find out the development boundary and filling characteristics of the water flowing collapse column in the working face, and to treat the collapse column by Grouting, the comprehensive exploration and treatment of the collapse column based on the technology of directional horizontal drilling on the ground were carried out in the mine[11-12].

2.2 Exploration and treatment plan of caved column

The design of the Collapse Column Control Scheme is based on the principle of "side exploration, side treatment and side design", and is guided by the principle of "taking treatment as the main task, combining exploration with treatment, stopping off water and cutting off water sources, and curing water disaster at the root". The goal is to achieve safe mining under pressure in No. 3-5 coal seam and No. 8 coal seam of lower group after treatment of collapse column. According to the geological analysis of the working face, 3-5 coal seam is 80m away from the top of the ordovician Limestone Aquifer, and 8 coal seam is 38m away from the ordovician limestone aquifer under the floor.No. 8 coal seam is about 6m thick, and 8 coal seam is basically connected with the ordovician limestone aquifer. In order to ensure the follow-up safe mining of 3-5 and 8 coal seams after grouting treatment of subsided column, the level of grouting treatment of water-flowing subsided column exploration and water-blocking plug construction should be selected from 10m to 30m above the top of ordovician limestone, and three layers (interval 20m) with 15 branch holes are arranged, of which 10 branch holes are arranged 30m below top surface of ordovician limestone and the spacing of branch holes is 40m. Three branch holes are arranged 10m below top surface of ordovician limestone and the spacing of branch holes is 120m. Two branch holes are arranged on top surface of ordovician limestone and the distance between the branch holes is 160m. A total of 2 main holes are designed in the ground. Each main hole is constructed with 7-8 branch holes (as shown in fig. 1). At the same time, the water-resisting layer of coal floor in the scope of collapse column is increased and strengthened, and finally the safe mining of coal face is realized.

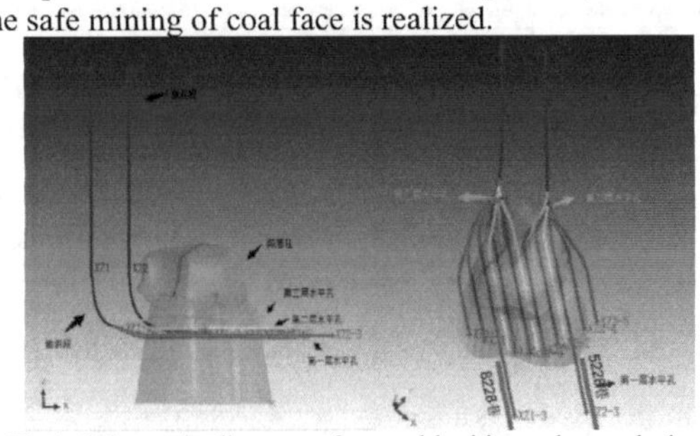

Fig1. Schematic diagram of water blocking scheme design.

 https://doi.org/10.1051/e3sconf/202123703003

3 Comprehensive exploration technology of caved-in column based on directional drilling

3.1 Microanalysis of borehole cuttings logging

Cuttings logging is an important technical means to judge formation in directional drilling. In the course of directional drilling, cuttings are taken out of the ground by drilling fluid after the rock is broken, filtered by vibrating screen, and salvaged at regular intervals (generally 2~4m) .

In the process of exploration, the dense cuttings are retrieved (once for 1~2m) when the branch hole is drilled to the target zone and the ordovician limestone is close to 50m outside the geophysical exploration boundary of the collapse column. The cuttings are washed with clear water and dried after they are salvaged. And then the cuttings are initially judged by the naked eye and the method of dilute hydrochloric acid titration dissolution. On this basis, the cuttings are magnified 500 times by electron microscope to identify the lithology of cuttings.

Fig. 2 is a 500x magnification micrograph of a partial branching cuttings microscope. The lithology of drilling formation is distinguished by analyzing the purity and crystal characteristics of cuttings of different lithology in the picture. The normal formation cuttings in the borehole are the same bright white crystal-like lithic cuttings containing dolomite and calcite. The cuttings are mixed cuttings such as gray sandstone, coal and mudstone after the branch hole enters the collapse column. The lithology and color of cuttings are disordered. According to the position distribution of cuttings of each branch hole, the development range of collapse column can be determined (Table 1).

Fig2. Microcrystalline analysis of cuttings from partial branched boreholes.

Table1. A list of the scope of collapse column identified by cuttings of each branch borehole.

Bore number	Collapse column range /m	Bore number	Collapse column range /m
XZ1-1	985~1195	XZ2-1	1010~1145
XZ1-2	970~1185	XZ2-2	1090~1195
XZ1-3	988~1198	XZ2-3	994.72~1140
XZ1-4	975~1175	XZ2-4	1070~1155
XZ1-5	975~1140	XZ2-5	1062~1155
XZ1-6	960~1190	XZ2-6	1110~1215
XZ1-7	975~1200	XZ2-7	975~1180

According to the microcrystalline analysis of different lithology cuttings, the length and short axis of the collapse column is 410m ~ 180m.

3.2 Analysis of logging data of branch hole while drilling

During the construction of each branch hole, Gamma logging while drilling was carried out in the exposed formation, and the gamma value suddenly increased when the hole entered the formation such as mudstone from limestone. In this way, the development range of the collapse column can be analyzed and judged by the variation of the gamma curve of the borehole.

Fig. 3 is the logging curve of XZ2-3 branch borehole with drilling. From the XZ2-3 logging-while-drilling curve, it can be seen that the gamma value is less than 100 API when the branch hole is drilled in limestone. But the gamma value suddenly increases to about 250 API when the branch hole is drilled to about 1000 m depth and enters the collapse column. When the branch hole enters the limestone stratum from the collapse column Broken Body (the hole depth is about 1150m), the gamma value suddenly drops below 100API, the branch hole exits the collapse column boundary and enters the normal limestone stratum. The well logging curves show that the boundary of the collapse column is obvious.

In the same way, the boundary of collapse column can be clearly determined by the curves of other branch boreholes changing with well logging. The length axis of the collapse column is about 335m ~ 220m.

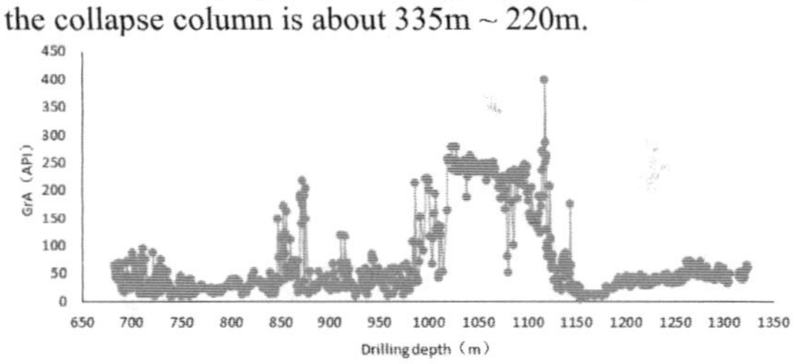

Fig3. Gamma change curve of XZ2-3 branch borehole while drilling.

3.3 Analysis on variation of drilling speed of branch hole

To some extent, the speed of drilling reflects the formation integrity and cementation, and the development of voids, pores and fractures. The variation of penetration rate of each branch hole can be used to analyze the broken distribution of the filling material in the collapse column, and to provide a reference for the selection of appropriate grouting technology for mine water disaster control.

Fig. 4 is the time curve of XZ2-2 branch hole drilling. It can be seen from the variation curve of drilling speed of XZ2-2 that the drilling speed reaches 0.5 m/min at about 930 m and 0.3 m/min at about 1100 m depth, which indicates that the formation is relatively broken and the degree of cementation is poor in these sections. It is the impact area of collapse column and the developed filling area. In the same way, we can analyze and judge the broken section of formation in the course of drilling by other drilling time curves.

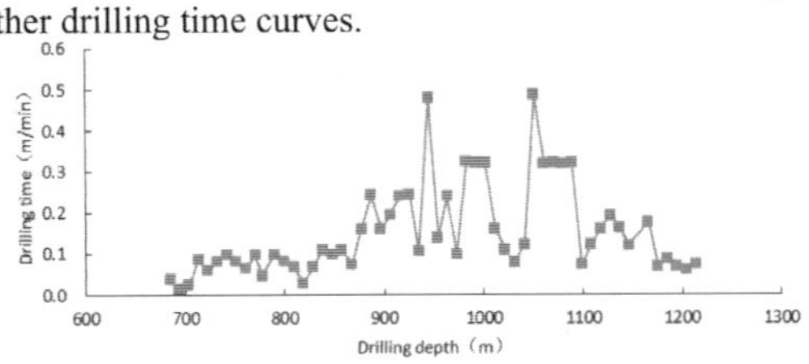

Fig4. Variation curve of drilling time for XZ2-2 branch hole.

3.4 Analysis of variation of drilling fluid consumption

In the process of drilling, the amount of drilling fluid consumption to a certain extent reflects the development of formation fractures. The larger the consumption of drilling fluid, the more developed the fractures and fractured the formation. Therefore, the development characteristics of collapse column are analyzed by using the distribution characteristics of drilling fluid consumption of each branch hole.

Fig. 5 shows the curve of drilling fluid consumption during XZ2-1 branch drilling. From the change curve of drilling fluid consumption of XZ2-1 branch hole, it can be seen that the water leakage from the hole suddenly increases and even completely loses when the branch hole is drilled to the sections of 750m, 883m and 1010m depth, which indicates that the formation is broken and the pores and fractures are developed in these sections, filling cementation difference. In the same way, the section with broken formation and poor cementation can be determined by analyzing the change curve of drilling fluid consumption in other branch holes.

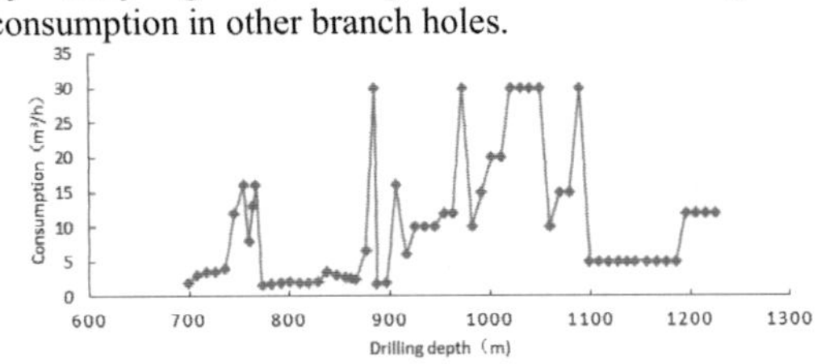

Fig5. Variation curve of drilling fluid consumption in branch of XZ2-1.

3.5 Comprehensive analysis of development characteristics of collapse column

Based on the results of directional drilling, it is found that the size of collapse column is 410m ~ 200m, and the broken area is mainly distributed in the direction of water inrush roadway. Combined with the comprehensive analysis of the ground inclined hole exploration and logging results, some sections of No. 3-5 coal seam moved down 25 m, and some sections of No. 8 coal seam collapsed 55 m. This indicates that the collapse column was formed relatively young, the collapse of the lower strata was carried away by water flow, the collapse was larger than that of the upper strata, and the cavity on the top of the collapse column was not completely formed.

The above exploration results provide a basis for the three-dimensional characterization of the fractured body of the collapse column and the construction of water blocking blocks.

4 Three-dimensional description and zoning technology of collapse column broken body

Based on the data of drilling rate and drilling fluid consumption during the exploration of each branch of directional horizontal drilling, a 3D geological model of the Development Section of collapse column is established by using SURPAC software. The fractured body of the collapse column is divided into three-dimensional sections (Fig. 6, Fig. 7) according to the distribution characteristics of voids, pores and fractures. The red areas are the areas with high penetration rate and drilling fluid consumption, indicating that the filling body of the collapse column is very broken and the cavities and pores are developed. They are classified as the main passage area. The yellow areas are the areas with high penetration rate and drilling fluid consumption, and the filling body of the collapse column is relatively broken. The green area is the area with small amount of fracture and the fracture is the area with secondary fracture. The main passageway area is cavity and pore-type area, which is the main passageway of water conduction of collapse column. The fracture area is poor cementation area of broken body, and the fracture area is relatively developed. The secondary fracture area is the influence zone of collapse column, and the fracture is not developed.

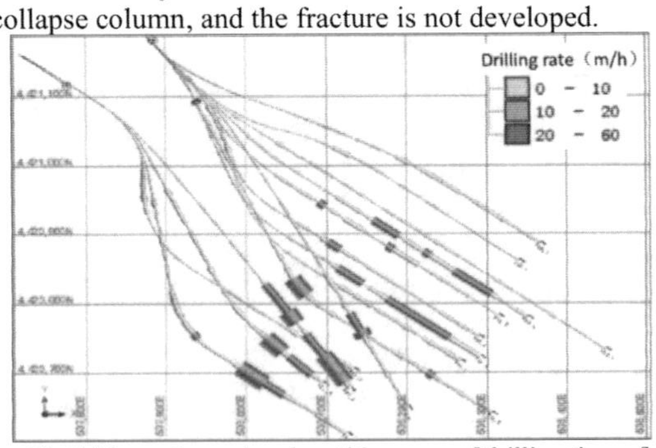

Fig6. Three-dimensional partition map of drilling time of collapse column broken body.

 https://doi.org/10.1051/e3sconf/202123703003

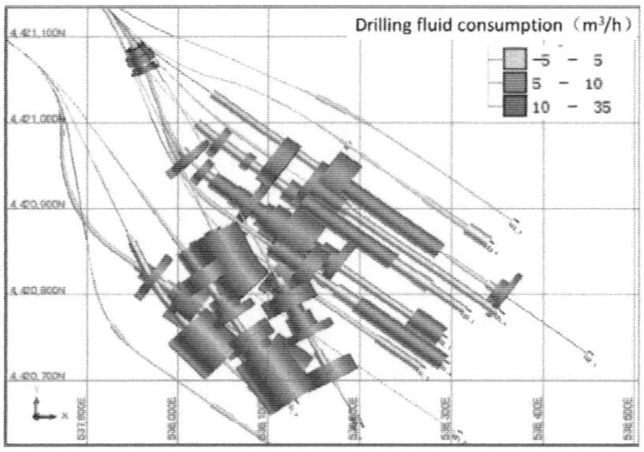

Fig7. Three-dimensional partition map of drilling fluid consumption of collapse column broken body.

5 Three-dimensional construction technology of water stoppage in collapse column

5.1 Selection of grouting technology for water blocking construction

In the process of directional horizontal drilling comprehensive exploration of collapse column, the area, where each branch hole drilling is located, is analyzed in real time according to the three-dimensional description of the broken body of collapse column. according to the different degree of cementation of filling body in different areas, different grouting techniques such as filling, squeezing and cracking are adopted (such as Table 2) , to block the construction of water, and block the water source.

Table2. A list of grouting technology for construction of water blocking plug.

Grouting area	Grout type	Grouting parameter	Grouting mode	Target
Main passage area	Cement fly ash slurry	No pressure	Quantitative grouting, Intermittent grouting	Blocking the main passage and build the water blocking frame
Fracture area	Single liquid slurry	Slurry concentration 1.3-1.5	Step-up pressure, continuous grouting	Sealing large fissures and dissolve them
Secondary fracture area	Single liquid slurry	Slurry concentration 1.2-1.3	Step-up pressure, continuous grouting	Sealing the cracks at the edges

5.2 Construction Technology of blocking water plug grouting

1) Directional precision control drilling technology. The ground directional screw precision drilling technology is adopted in the construction of water-blocking plug, and the directional precision control drilling is carried out on the subarea of collapse column broken body.

2) Three-dimensional construction technology of layered multi-branch blocking water plug. The construction method of blocking water plug is bottom-up, hierarchical, directional and horizontal branch hole construction. Firstly, 30m first-order horizontal branch holes under the top surface of ordovician limestone were constructed, then the second-order horizontal branch holes were constructed, and then 10m directional horizontal branch holes under and above the top surface of ordovician limestone were constructed.

3) A variety of grouting process control technology. On the basis of Directional Control of multi-branch hole stereo construction, various grouting materials, slurry ratio and grouting pressure are used to depict different zones according to the fractured body of collapse column in three dimensions, and grouting techniques such as filling, squeezing and cracking are adopted. Three-dimensional construction of water blocking is realized.

5.3 Analysis on construction effect of water blocking plug

1) The amount of grouting at the end of each branch hole is less than 50L/min, and the pressure at the end of the hole is 3MPa and lasts 30min. It has reached the end of grouting standard for blocking water plug construction.

2) According to the analysis of the distribution of the order grouting amount of the branch hole, the grouting amount is mainly concentrated in the first order branch holes. The grouting amount of one sequence branch holes accounts for 60% of the total grouting amount, while the leakage amount of the second sequence holes is small. The results show that the pre-grouting treatment has a good effect on filling and compaction of the fractures and fissures in the collapse column, and plays a key role in the water-blocking construction of the water-flowing collapse column.

3) The water absorption of branch holes injection test is less than 0.01l/(min·m·m). It met the requirements of treatment, and blocking water plug construction effect is good.

4) The water inrush point has been effectively controlled. The ordovician limestone water level in G1 observation hole has been basically restored, and finally the safe mining of 8228 working face has been realized.

6 Main conclusions

The following main conclusions are drawn:

1) The collapse column was carefully explored base on the comprehensive exploration techniques such as surface directional horizontal drilling cuttings microcrystal analysis, penetration rate analysis, drilling fluid loss and logging while drilling data analysis. The size of the collapse column was basically found to be 410m ~ 200m. The broken zone is mainly distributed in the direction of Lane 2228 in water inrush roadway.

2) The distribution characteristics of voids, pores and fractures in the fractured body of caved column are divided into three-dimensional geological modeling sections by using SURPAC software, which provides a basis for water-blocking grouting construction.

3) According to the different pore development

characteristics of the fractured body of the collapse column, the precise three-dimensional construction of the water blocking plug of the collapse column is realized by using different grouting techniques in different subareas.

4) The safe and smooth mining of 8228 working face in Tashan mine is realized by blocking water plug solid construction. The comprehensive exploration technology based on the directional horizontal drilling collapse column and the three-dimensional construction technology of blocking water plug have a good reference to similar projects.

References

1. Yin Shangxian, Wu Qiang, Wang Shangxu. Water-bearing characteristics and hydro-geological models of Karstic collapse columns in north China [J]. Chinese Journal of Rock Mechanics and Engineering, 2005, 24 (1): 77-82.
2. Wu Qiang. Progress problems and prospects of prevention and control technology of mine water and reutilization in China [J]. Journal of China Coal Society, 2014(05): 795-805.
3. Du Bingjian. Fathering practice of hidden water inrush subsided column in Jiulong Coalmine, Fengfeng mining area [J]. Coal Geology of China, 2012(3): 130-131.
4. Shi Xianzhi,Yuan Dezhu,Ge Jungang. On the prevention and control of mine water induced by collapse column in # 21301 coal face in Chensilou Coal Mine[J].China Coal, 2009(4): 104-105.
5. Wang Zecai,Wu Qiang. On Karst collapse column and tactics of mine water prevention [J]. Geotechnical Investigation & Surveying, 2002(01):29-30,21.
6. Zhao Suqi, Wu Qiang. Integnted management technology of water guide collapse column looded wells [J] . China Coal, 2004, 30(7):27- 29,6.
7. MENG Zhaoping, YI Wu . Water inrush characteristics of Fangezhuang coalmine field in Kailuan and its geological condition analysis of water inrush from coal seam floor[J] . Chinese Journal of Rock Mechanics and Engineering, 2009, 28(2) : 228 – 237 .
8. Yin Shangxian, Lian Huiqing, Liu Demin, Yin Huichao. 70 years of investigation on Karst collapse column in North China Coalfield:cause of origin, mechanism and prevention[J]. Coal Science and Technology, 2019, 47 (11): 24-29.
9. Gou Yongquan. Research on grou nd detection and treatment technology of Ordovician limestone water damage in extra-thick coal seam of Carboniferous[J]. Datong Coal Science &Technology, 2018,161(3):8-10.
10. Xu Xiaohong. Study on Technology of Coal Mine Synergistic Detection of Ordovician limestone Water[J].Coal Technology, 2019,38(10):115-117.
11. Yao Ningping. Development trend of nearly horizontal directional drilling technology in coal mines of China [J]. Coal Geology & Exploration, 2008 ,36(4) :78-80.
12. Zheng Shitian. Application of ground directional borehole technology to control prevention karst collapsed column water inrush in coal mines[J]. Coal Science and Technology, 2018,46 (7): 229-233.

https://doi.org/10.1051/e3sconf/202123703004

Component optimization of porous permeable brick in "sponge city" based on rainfall area division

Xiaofu Wang[1,*], Xiong Zhang[1]

[1] Institute of Building Materials, School of Materials Science and Engineering, Tongji University, Shanghai , China

Abstract. The present study aims to prepare resin-based permeable bricks with micron-sized pores using fine aggregate with a particle diameter of 0.08mm-0.6mm and bisphenol-A epoxy resin, a polymer binder. The rainfall area of China was divided into 5 levels by using ArcGIS and Geoda geographic analysis software; The performance indexes of key components of resin-based permeable brick were selected by correlation with relevant standards, and the performance of materials is adjusted and designed according to the corresponding performance indexes, so as to meet the design requirements of "Sponge City". The main method of material design is to adjust the key parameters of raw materials, such as aggregates, binder materials and admixtures, by using Image Pro Plus software and SEM and other test methods, so as to optimize the selection of "graded" aggregates and admixtures for different performance requirements scenarios to achieve the material selection goal. The highest permeable rate of resin-based permeable brick is 6.22× $10^{-2}cm/s$ s, and the maximum compressive strength is more than 60MPa.

1 Introduction

In the process of global urbanization, more than 50% of the world's population live in cities ([1]). The rapid development of economy and the acceleration of urbanization have brought convenience to human beings, but also brought serious environmental pollution, resulting in many urban diseases. The typical urban diseases are "Heat island effect" and "Urban waterlogging". The heat island effect is caused by human factors, which changes the local temperature, humidity, air convection and other factors on the urban surface, and then causes the urban microclimate change; Urban waterlogging refers to the phenomenon of waterlogging disaster in the city due to the heavy precipitation or continuous precipitation exceeding the urban drainage capacity. The causes include the objective concentration of precipitation, imperfect drainage facilities and lagging drainage system construction. Therefore, urban climate and hydrological problems restrict the further development of the city, so it is urgent to find a reasonable solution in the process of urban construction and governance.

Based on the above reasons, China proposes to build a "Sponge City". "Sponge City" is a special appellation in China, and there are many different appellations for such urban pollution control programs in the world ([2-8]) . BMP for Stormwater Management in the United States is also called Low Impact Development (LID); Sustainable Drainage Systerm (SuDS) in the United Kingdom; Water Sensitive Urban Design(WSUD) in Australia and Low Impact Urban Design＆Development （LIUDD） in New Zealand, etc.

Considering that urban environmental governance is a systematic project, it is necessary to find a fit point in a specific field, and the problem of "precipitation-pervious-drainage" in urban hydrology is just the key. When the rainwater enters the river through the urban drainage system, it will increase the pressure of the urban drainage system; while the rainwater that cannot be drained will generate a large amount of water on the road surface, which will lead to the frequent occurrence of urban flood. In addition, a large number of pollutants on the road enter the river with the runoff through the urban drainage pipe or overflow under the rain wash, forming a typical urban rainfall runoff pollution, which is the main non-point source pollution source in the city, and has a serious impact on the urban ecological environment. These are mostly environmental problems caused by hardened pavement ([9, 10]). These environmental problems are mainly caused by urban waterlogging, traffic paralysis, water pollution, etc. caused by the lack of water seepage when facing the precipitation ([11-15]) . A large number of studies show that ([16]), compared with impermeable pavement, permeable pavement can alleviate surface runoff, supply underground water, purify pavement rainwater, improve urban thermal environment, absorb urban noise, improve urban surface soil ecological environment, etc ([17-19]).

In this context, permeable brick is widely used as an efficient permeable pavement material in China. In the construction of pavement system, permeable brick is a part of the system, which has the following advantages: 1. Good water permeable performance, which can make rainwater quickly penetrate into the ground, supplement soil water and groundwater, maintain soil moisture, and

* Corresponing author: 1710048@tongji.edu.cn

 https://doi.org/10.1051/e3sconf/202123703004

improve the living conditions of urban ground plants and soil microorganisms; 2. It can absorb water and heat, adjust the temperature of local space on the ground humidity, large-scale application can adjust the urban microclimate and alleviate the effect of urban heat island; 3. It can reduce the pressure of urban drainage and flood control, prevent the pollution of public waters and make the road free of water, etc. Permeable brick is mainly divided into three types: cement-based permeable brick, sintered permeable brick and resin-based permeable brick. Their main materials are aggregate and binder. However, the first two kinds of permeable bricks have some defects, such as: the production and maintenance cycle of cement-based permeable bricks is long, the quality is low; the energy consumption of sintered permeable bricks is very high, and the secondary pollution is serious. Therefore, the use of resin based permeable brick has become the mainstream.

However, since the sponge city was built in 2012, there are still many deficiencies in the national overall urban planning, such as the imbalance of construction caused by the climate difference between the north and the south, the irrationality of construction caused by the selection of the same construction system, and the exclusiveness of manufacturing and selection caused by the differences in the origin of materials. However, the "Six Words Policy" of "Sponge City" of "infiltration, stagnation, storage, purification, utilization and drainage" is often mixed together, and there is no very obvious boundary. Based on this, it is necessary to implement the sponge city plan of "adjusting measures to local conditions" for different regions according to different rainfall and material properties. On the basis of the above logic, this study conducted an in-depth study on the selection scheme of "sponge city" in accordance with the classification of "macro; meso; micro", and formulated a specific guidance scheme according to the reasonable combination of these three levels. The main research contents are as follows:

Macro research: collect the rainfall data of each city in recent 10 years in China, and divide the rainfall area in China; Meso research: study the permeable brick based on resin and obtain the regulation scheme for the permeable and mechanical properties; Micro research: study the admixtures to obtain the further regulation method.

2 Experimental

2.1 Raw material

Binders and aggregates are the main materials for making resin-based permeable pavement. Admixtures are chemical substances that can significantly improve the performance of permeable bricks.

2.1.1 Aggregate

The aggregate is 10 kinds of natural sand with similar particle size but different distribution obtained from different regions in China. Their particle size is controlled within 0.08mm-0.6mm by vibrating screen.

2.1.2 Binder materials

Binders: bisphenol-A epoxy resin, white transparent viscous liquid, epoxy value is 0.44, industrial brand is E44; Amine curing agent (adduct of diethylenetriamine and butyl glycidyl ether), industrial grade is 593, relative density is 0.985, viscosity (25℃) is 90 ~ 150mPa·s, total amine value is 500 ~ 700mgKOH/g. In a dry environment at room temperature, the 1-day strength can reach 90% of the maximum value.

2.1.3 Admixture

The additives were selected from four aspects: alcohol additives, filler additives, reaction additives and water-resistant additives, and the macro performance of permeable brick was tested to explore the influence of different additives on the epoxy resin binder material system. For the convenience of research, the above additives are classified according to the molecular weight and functional differences as shown in Fig. 1:

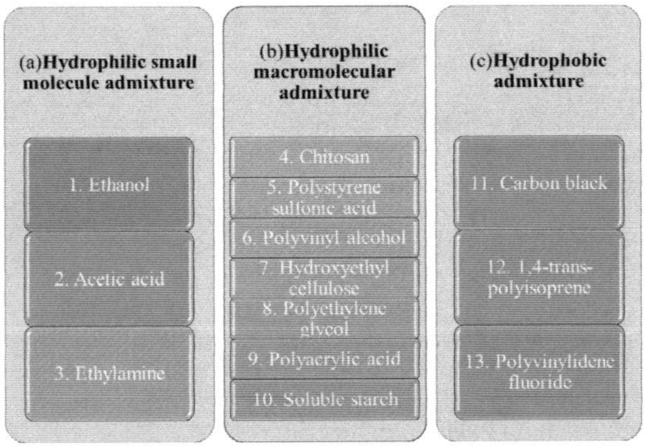

Fig1. Admixture selection scheme (a) according to different functional groups (b) according to molecular weight (c) according to functional group characteristics

2.2 Sample preparation

For the preparation of resin-based permeable bricks (RBPB for short), firstly, the aggregates shall be weighed according to the formula and poured into the mixer, epoxy resin and curing agent shall be added, and then the mixture shall be placed in the mixer for 240s until the materials are well mixed. In order to make the pores well-distributed, samples need to be vibrated. The mix proportion is shown in **Table 1**.

Table1. The mix proportion of resin based permeable brick

Composition	Sand	Resin	Curing agent
Weight fraction (wt%)	94.3	4.4	1.3

When the total volume of porous materials is constant, the mechanical properties increase with the increase of the content of binder materials, but the water permeability will decrease. Through some exploration experiments, it was found that the proportion of the total mass of resin and curing agent accounting for 6% of the aggregate can meet the requirement of the mechanical properties, water permeability and cost better. So the proportion of 6% was used for subsequent research and the proportion of curing agent and resin is as follows: m (curing agent) : m (resin) = 3:10 .

2.3 Methods

2.3.1 Test methods for water permeability and mechanical properties of RBPB

The test methods and indicators of water-permeable properties of permeable materials are based on GB/T 25993-2010 ([20]) for ordinary permeable brick. The permeability coefficient K(cm·s-1) is calculated by the formula as shown in Eq(1). The compressive strength of the samples was measured by using a universal testing machine at a crosshead speed of 2 mm/min([21]) . The flexural strength was measured according to the loading rate of 0.04MPa/s ~ 0.06MPa/s ([22]) .

$$K = \frac{QL}{AHt} \tag{1}$$

2.3.2 Test method for porosity

The porosity measurement method refers to the experimental method of Jiang ([23]) , and the calculation is as follows:

$$\varepsilon = \frac{m_{12} - m_{22}}{V \cdot \rho_w} \times 100\% \tag{2}$$

Where ε represents porosity, m_{12} represents the wet weight of permeable brick, m_{22} represents the dry weight of permeable brick, V represents the apparent volume of permeable brick, ρ_w is the density of water.

2.3.3 Particle group parameter test

The optical stereo microscope and Sony ILCE-6000 24.3 megapixel camera were used to collect the image information of the sand particle group, and then the image information of the particle group is analyzed and processed with Image Pro Plus software to obtain the particle group parameters. The image collection and analysis process is shown in **Fig. 2** .

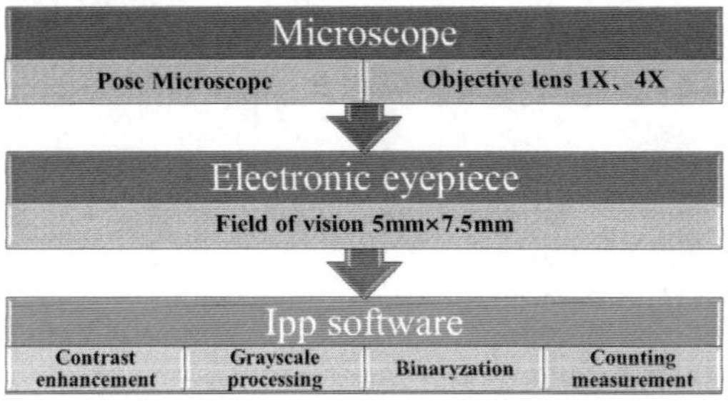

Fig2. Image acquisition and analysis process

A modified Gaussian distribution function is used to describe the particle size distribution of the aggregates. At the same time, data such as the most probable particle size and the standard deviation of the distribution can be obtained. The equation is shown in Eq(3) below. It can be seen that in the ideal state, it is still a standard normal distribution.

$$y = y_0 + \frac{A}{w\sqrt{\pi/2}} e^{-2\frac{(x-x_c)^2}{w^2}} \tag{3}$$

Where:

x_c (mm) is the most probable particle size of the aggregate and w is the distribution standard deviation of the aggregate.

2.3.4 Division method of rainfall area

Climate types in China are complex and diverse, with significant regional precipitation differences. The sum of the average monthly precipitation in a region is the "Annual Precipitation". The average annual precipitation in a region for many years is the "Average Annual Precipitation" in this region. Usually, the "Annual Precipitation" reflects the amount of precipitation in a region in a year. Only by investigating the rainfall situation of the whole country and dividing the corresponding rainfall area can we reasonably equip permeable materials according to local conditions. According to the above ideas, the precipitation data of 34 provinces, municipalities, autonomous regions and special administrative regions of China in 2009-2018 are found in the research platform of China's economic and social big data, and the average annual precipitation of each province in China in recent 10 years is calculated. Then, the spatial pattern analysis and regional division of the national rainfall areas are carried out by using ArcGIS 10.2 and Geoda spatial data analysis software.

3 Results and discussion

3.1 Micro research

3.1.1 Pore structure analysis

SEM was used to study permeable brick, and the results are shown in **Fig. 3**.

(a)

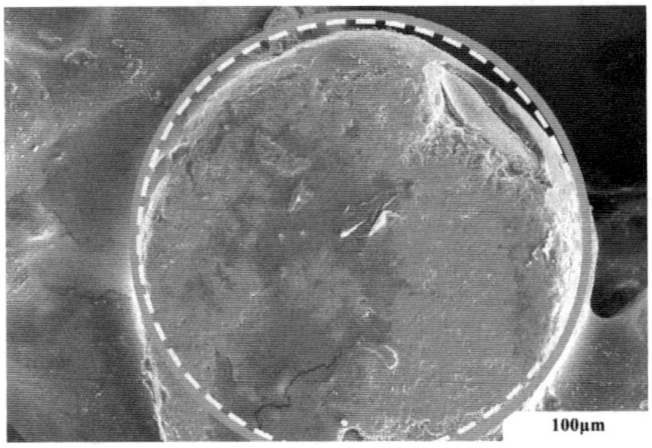

(b)

Fig3. Micrograph of permeable brick specimen. (a) Two different types of pores, (b) Roundness of particle

It can be seen from **Fig. 3** that the roundness of aeolian sand particles is high, so it is reasonable to consider it as a sphere. It can be seen in the SEM image that the pores and large gaps caused by the overlapping particles are very obvious. Image Pro Plus software was used to calculate more than 100,000 particles, the average roundness of this desert aeolian sand is 0.92. In addition, the pore size of the lapped pores is 30-100 microns, of which the three-particle lapped pore diameter is 30-50 microns, and the four-particle lapped pore diameter is 100 microns. Kayhanian's ([24]) study shows that when the diameter of harmful particles is larger than 38 microns, permeable concrete will be blocked, while Siriwardene's ([25]) study shows that when the diameter of particles is smaller than 6 microns, particles in porous media will migrate with water, and a deposit layer will be formed at the bottom of porous materials to prevent water from penetrating underground. The permeable bricks prepared in this study have two micron-level pores, which have smaller pore sizes than millimeter-scale permeable concrete/bricks, which can effectively block the clogging and migration of harmful particles, making the permeable pavement service life longer.

3.1.2 Admixture research

The experimental results of 13 different admixtures are shown in **Fig. 4**.

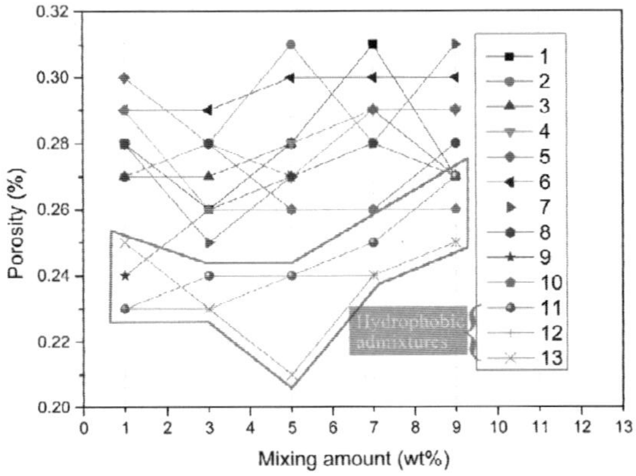

Fig4. Relationship between the amount of admixture material and the porosity of permeable bricks

Adding small molecule admixtures within the range of 3%-7% will increase the porosity, and the increase of porosity is significant when carboxyl group is added in the low dosage of admixtures. Macromolecular additives have no special effect on porosity; Adding hydrophobic material will increase the porosity, indicating that the addition of hydrophobic material is conducive to the dispersion of binder materials, so that the effective porosity is improved. With the increase of additives, the porosity shows an upward trend, and then drops rapidly after reaching the highest point, which shows that additives can improve the fluidity of binder materials and disperse more evenly in the process of use, so the binder materials are not easy to block the pores, which leads to the increase of the effective porosity of the permeable materials and the increase of the connecting pores.. However, if the amount of additives continues to increase after achieving the best coupling of fluidity and distribution uniformity, it will lead to the plugging phenomenon of large amount of binder material, so the porosity will further decrease.

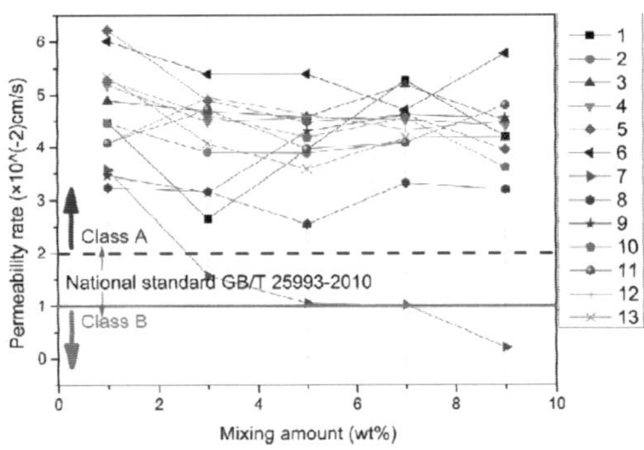

Fig5. Relationship between the amount of admixture material and the permeability of permeable bricks

https://doi.org/10.1051/e3sconf/202123703004

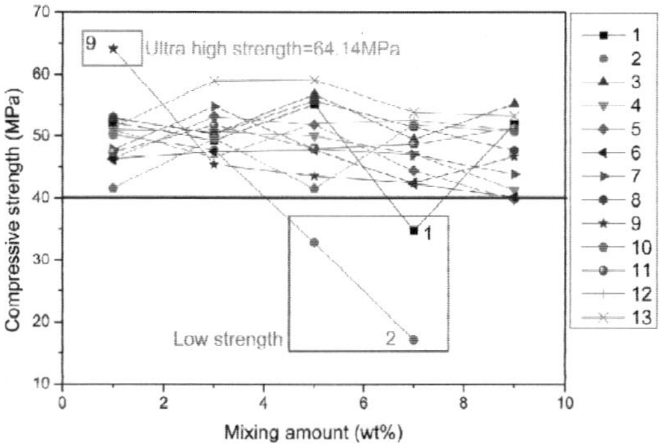

Fig6. Relationship between resin content and mechanical properties

The use of ethylamine as an additive does not affect the water permeability and mechanical properties of the permeable material, which is mainly because amine as a reactive additive only increases the concentration of the curing agent in the chemical reaction of cross-linking curing, only accelerates the reaction speed, but does not affect its functionality. There is a negative correlation between the porosity of ethanol and acetic acid and the compressive strength, which indicates that the structure of the binder material will be destroyed when the content of acetic acid is very large.

One of the highest points in **Fig. 6** is that in the group with a small amount of polyacrylic acid added, its compressive strength is much higher than that of other groups, and after further adding polyacrylic acid, its strength drops rapidly, which is very special. Due to the strong polarity of the carboxyl group of polyacrylic acid, it can react with the epoxy group on both sides of the epoxy resin, and can also activate the cross-linking curing reaction. At the same time, the molecular weight of polyacrylic acid is moderate, adding it to the body-shaped macromolecules increases the structural stability, but adding more polyacrylic acid containing the carboxyl group will reduce the strength of the material.

It can be seen from the above figures that with the increase of the dosage, the water permeability will be improved, and the compressive strength will also be increased accordingly. In the study of hydrophobic admixtures, the measurement results of water retention rate are very interesting. Through the water retention rate, an important conclusion can be obtained as shown in **Fig. 7**.

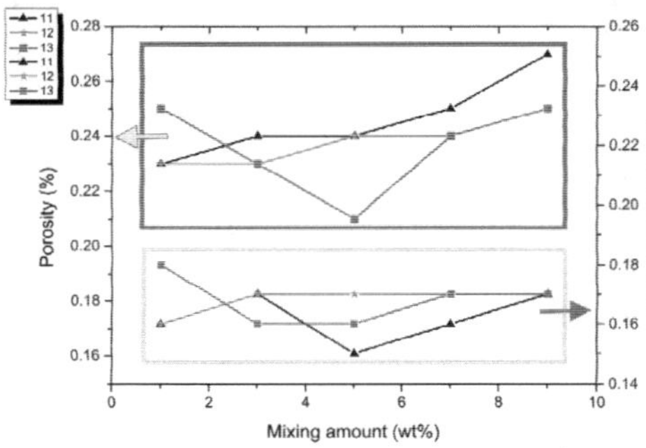

Fig7. The relationship between the change of porosity and the change of water retention rate of the three hydrophobic materials

It can be seen from **Fig. 7** that the water retention rate does not change with the change of porosity after adding hydrophobic materials, and it always fluctuates within the range of 0.12-0.14, which is different from other additives. Because the hydrophobic material can't be well wetted by water, the pores in the water permeable material added with the hydrophobic material will form a hydrophobic layer, so that the water permeable material does not have good water retention, and the water has a quick drainage effect after entering the water permeable material.

Through the selection of functional groups, hydrophilic and hydrophobic properties, it is found that hydrophilic small molecular substances are used as additives to modify the permeable pavement brick, and the test specimens mixed with methanol and ethylamine shows relatively good water permeability, which is named as "Class A" high water permeability and high water retention additive; The water permeable brick was modified with hydrophilic macromolecule as additive, and the soluble starch as additive. The test specimens had good water permeability and compressive strength over 40MPa, so it was named as "Class B" high strength additive; The test specimens modified by hydrophobic substances as additives did not show relatively excellent water permeability, but this kind of test block generally has high compressive strength. In general, porosity = water retention rate, but after adding hydrophobic substances, the water retention rate is less than porosity, which can promote drainage. They are named as "Class C" drainage promoting admixture.

3.2 Meso research

3.2.1 Aggregate division method

The distribution standard deviation w in the formula reflects the dispersion degree of particle size to x_c .Define α as the aggregate distribution coefficient, which is a dimensionless quantity, the calculation formula of α is shown in Eq(4).

$$\alpha = \frac{w}{x_c} \tag{4}$$

Two parameters x_c and w in the modified Gaussian distribution function determine the particle size distribution of the particle group. From the calculation formula of α, it can be seen that the smaller α is, the smaller w is, and the larger x_c is, which means that the particle size distribution tends to be narrow, and the most probable particle size tends to be large. Such a single aggregate grading and larger particle size will lead to the larger effective pore diameter, which is conducive to the improvement of the material's water permeability; On the contrary, the smaller α means that the particle size distribution tends to be wider and the most probable particle size tends to be smaller. Aggregates with such characteristics can play a good role in pore filling, with more continuous grading and fewer pores, and better strength. As the mean value of flexural strength is 10.75MPa, which is far greater than the 3.0MPa specified in the national standard mentioned in **Table 1**, so it will not be discussed here.

According to the compressive strength value and permeability rate, the characteristic parameter range of 10 kinds of sands is divided as shown in **Fig. 8 (a)** and **Fig. 8 (b)**. Taking the strength boundary (35MPa and 40MPa) and permeability rate boundary (0.02 cm·s-1 and 0.04 cm·s-1) as the standard, the compressive strength of RBPB is divided into " Ⅰ、 Ⅱ、 Ⅲ " three areas, and the permeability rate is divided into "A、 B、 C" three areas.

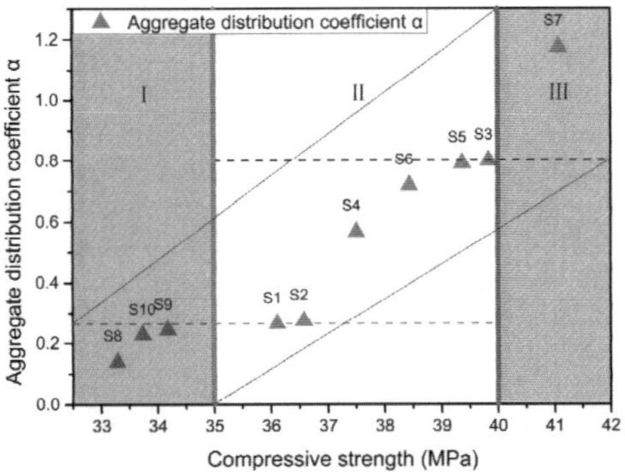

(a)

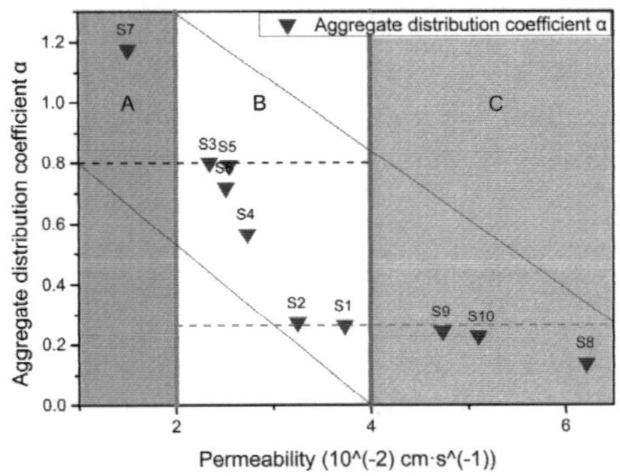

(b)

Fig8. RBPB's macro performance area division.(a) Relationship between compressive strength and aggregate distribution coefficient α;(b) Relationship between permeability and aggregate distribution coefficient α

The characteristic parameter values of S1 and S3 are used as the classification nodes. The critical values in the two figures are compared and the intersection of them is taken. The aggregate distribution coefficient α value of aggregate is below 0.264, and the properties of RBPB prepared with it tend to be high permeable rate and low compressive strength (the compressive strength can still reach the national standard), which can be divided into high permeable aggregate(HPA for short); When the aggregate distribution coefficient α value of aggregate is above 0.8001, it is high compressive strength and low permeable rate (permeable rate can meet the national standard), which can be divided into high strength aggregate(HSA for short) ; When the aggregate distribution coefficient α is in the range of 0.264-0.8001, its strength and permeable rate are in a relatively balanced level, which can be divided into comprehensive aggregate (CA for short). S7 belongs to HSA, S8-S10 belongs to HPA and S1-S6 belongs to CA. The range of characteristic parameters of particle groups of various aggregates is shown in **Table 2**.

Table2. Performance index and particle group characteristic parameter range of three types of aggregates

Aggregate category	Permeability/cm·s^{-1}	Compressive strength/MPa	α
HPA	≥0.040	≥30	<0.264
CA	≥0.020	≥35	0.264-0.8001
HSA	≥0.010	≥40	> 0.8001

There are many studies to explore the mechanical properties and water permeability of porous materials from the porosity([26-28]), but few studies from the perspective of particle group parameters. There are some defects in the design of permeable brick only by the parameters of pore. Many properties of pore have influence on the strength and permeability of porous permeable material, such as pore size, pore distribution, pore shape, etc. Moreover, the key parameters of pore structure such as effective porosity, tortuosity and connectivity are difficult to obtain and the accuracy is low. This method of judging the performance of bricks by measuring porosity is a lagging method, that is, a material must be prepared to measure its performance, so a predictive method of predictability is required. This paper proposes a method to design the various functions of porous materials in advance by obtaining the particle group parameters of aggregates, which is beneficial to selecting materials and controlling materials in advance according to the design parameters.

3.2.2 Selection principle of binder materials

It can be seen from the above that the strength of permeable bricks increases with the increase of epoxy resin content. When the content of resin is less than or equal to 3%, the surface of aggregate can not be fully covered due to the small amount of binder material, and the high-strength cementation layer can not be formed between aggregate particles after solidification, resulting in the strength of resin based permeable pavement material can not reach the minimum strength standard of pavement material; When the amount of epoxy resin is more than or equal to 7%, excessive binder materials, in addition to fully covering the aggregate surface, will have excessive epoxy resin mixture seepage at the bottom of the mixture after mixing, filling the pores between the aggregates, so that the water permeability of the material can not meet the standard of water permeable materials.

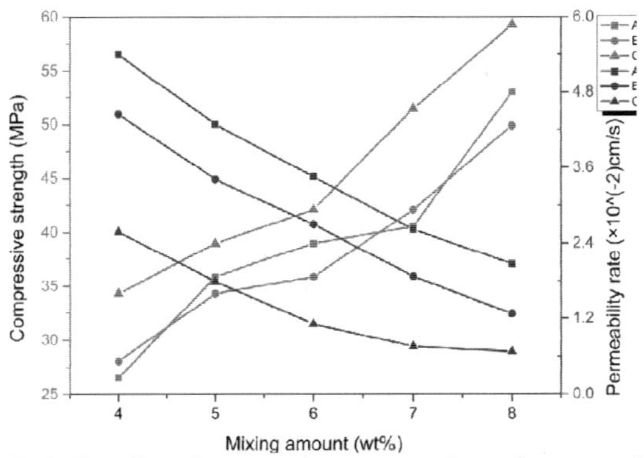

Fig9. The effect of epoxy resin content on the performance of resin-based permeable pavement materials

Fig. 9 shows the experimental study on the representative aggregate selected from three types of aggregate, including HPA-S9 (A), CA-S4 (B) and HAS-S7 (C). It can be seen that the curves in the image like "X shape", such X-shape is often seen in the study of porosity, that is, with the increase of porosity, the compressive strength and permeability rate of permeable materials show the opposite trend. Therefore, in the preparation of resin-based permeable brick, epoxy resin content in the range of 4% - 8% is the most appropriate.

3.3 Macro research

China is located in the east of Asia and the west of the Pacific Ocean, with tropical monsoon climate area, subtropical monsoon climate area, temperate monsoon climate area, plateau mountain climate area, temperate continental climate area and tropical rainforest climate area. In such a country that spans many climate regions, there is a great gap in the atmospheric weather conditions faced by different regions, and there are many kinds of technical measures for sponge city construction. Therefore, the principle of material selection is to "adjust measures to local conditions", which can not be controlled and designed in one stroke. Some areas need rapid drainage, some need water storage, and some need water filtration. The most reasonable way is to arrange the selection of materials as a whole. In this context, we need to divide the national rainfall area.

3.3.1 Rainfall area data and spatial pattern

According to the precipitation data of 34 provinces, municipalities, autonomous regions and special administrative regions in 2009-2018 found by China's economic and social big data research platform, the average annual precipitation of each province in China in the past 10 years is calculated. The values are shown in **Table 3**.

Table3. Average annual precipitation of 34 provinces, municipalities, autonomous regions and special administrative regions in China in recent 10 years

No	The 34 provinces, autonomous regions, municipalities directly under the central government and special administrative regions of China	Average annual precipitation in recent 10 years (unit: mm)
1	Beijing	574.77
2	Tianjin	566.70
3	Shanghai	1371.29
4	Chongqing	1193.91
5	Jiangsu	1140.64
6	Guangdong	1708.16
7	Sichuan	1039.28
8	Zhejiang	1581.13
9	Hubei	1200.00
10	Shandong	723.72
11	Xinjiang	176.04
12	Anhui	1231.00
13	Yunnan	660.43
14	Henan	678.97
15	Guangxi	1709.35
16	Hebei	533.03
17	Hunan	1412.77
18	Shaanxi	658.152
19	Jiangxi	1852.98
20	Fujian	1600
21	Shanxi	838.425
22	Jilin	640.84
23	Liaoning	800
24	Heilongjiang	577.80
25	Gansu	410.79
26	Guizhou	1162.26
27	Neimenggu	319.96
28	Ningxia	587.19
29	Hainan	1639.00
30	Qinghai	412.67
31	Xizang	400.00
32	Taiwan	1800.00
33	Xianggang	1708.16
34	Aomen	1790.00

Using ArcGIS 10.2 geographic information system, spatial interpolation analysis and three-dimensional visualization of the average annual precipitation data of China's provinces and cities in the past 10 years are shown in **Fig. 10** and **Fig. 11**. It can be seen that the precipitation in different regions of China varies greatly, the coastal rainfall is more than that in the inland, and the rainfall in the south is more than that in the north. Decreasing from southeast coast to northwest inland is the general trend of annual precipitation spatial distribution in China.

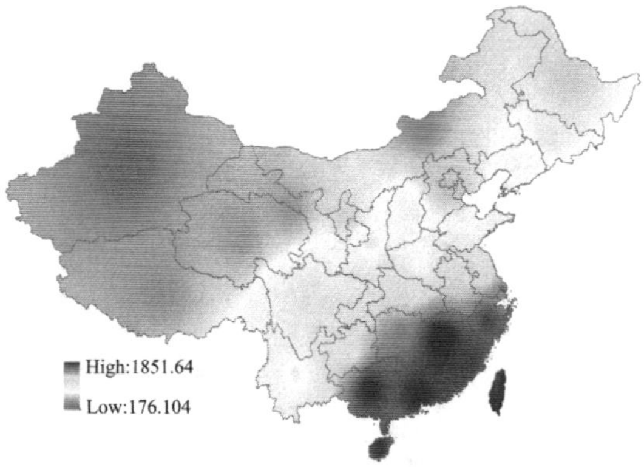

Fig10. Spatial pattern analysis of average annual precipitation in recent 10 years in China using spatial interpolation method

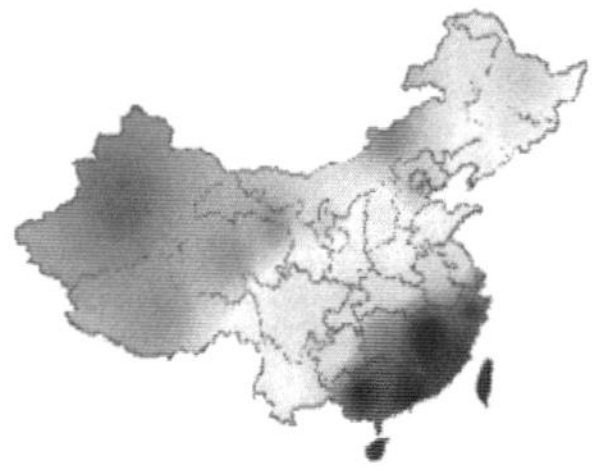

(a)

(b)

(c)

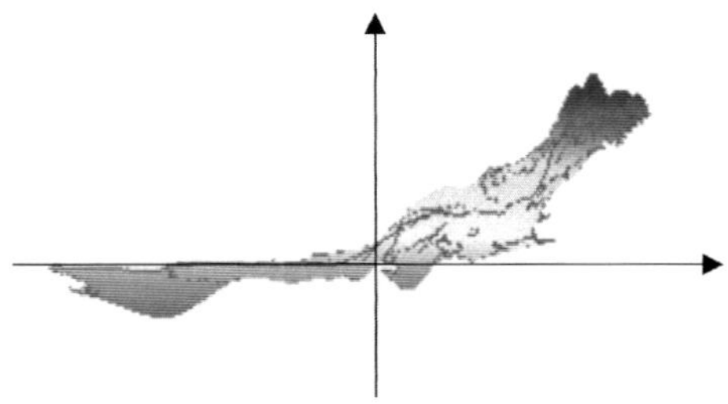

(d)

Fig11. Spatial pattern analysis of China's average annual precipitation in recent 10 years based on 3D visualization

Fig. 11 is a three-dimensional visual analysis of the spatial pattern of China's average annual precipitation in the past 10 years. The graphic changes from (a) to (d) can directly see the height differences of precipitation in different regions of China. **Fig. 11 (d)** shows that although the area of southeast coastal area is much smaller than that of northwest, the absolute height of southeast coastal area in quadrant I in the positive direction of Y axis is much higher than that of northwest inland area in quadrant III in terms of rainfall. In the southeast and northwest directions, there are several areas with high and low peaks. Therefore, in the process of sponge city construction, not only the material properties need to be divided, but also the rainfall areas need to be divided throughout the country. and different permeable material system deployment schemes need to be used in different regions.

3.3.2 Division of rainfall areas in China

Using Geoda spatial data analysis software, the equal interval map method in its mapping tool is used to make the average annual precipitation distribution map of China's provinces and cities in recent 10 years, as shown in **Fig. 12**. According to **Fig. 12**, China's average annual precipitation can be divided into five grades, as shown in **Table 4**.

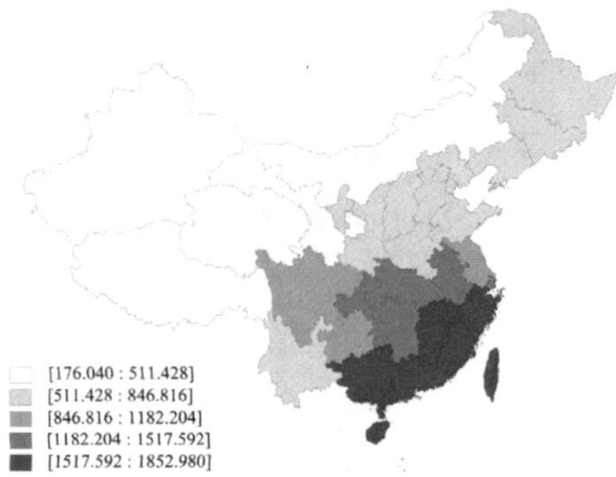

Fig12. Distribution of average annual precipitation in China in the past 10 years

Table4. Classification of average annual precipitation in recent 10 years in China

Grade	Average annual precipitation range	provinces
Level 1	176.040mm-511.428mm	Neimenggu、Gansu、Qinghai、Xinjiang、Xizang
Level 2	511.428mm-846.816mm	Heilongjiang、Jilin、Niaoning、Beijing、Tianjin、Hebei、Shandong、Henan、Shanxi、Shaanxi、Ningxia、Yunnan
Level 3	846.816mm-1182.204mm	Sichuan、Jiangsu and Guizhou
Level 4	1182.204mm-1517.592mm	Shanghai、Anhui、Hubei、Hunan and Ghongqing
Level 5	1517.592mm-1852.980mm	Zhejiang、Fujian、Jiangxi、Guangdong、Guangxi、Hainan、Xianggang、Aomen、Taiwan

From **Fig. 12**, we can visually observe the division of China's 5-level rainfall area, and from **Table 4**, we can see the average annual precipitation range of 5-level rainfall area. On this basis, the following two principles should be clear:

(1)Increase the permeation rate in rainy areas, so that the maximum precipitation can meet the precipitation-permeability balance while taking into account the mechanical properties.

(2)Economically applicable aggregates and binder materials are used in areas with little rain to fully reduce costs while satisfying basic performance.

According to the above principles, it is reasonable to build permeable material storehouse in different rainfall areas, but the requirements of specific pavement quantity are not within the scope of this paper.

3.4 "Sponge city" material selection plan

In order to promote the concept and material system of "Sponge City" in China, it is necessary to design the pavement system and material of "Sponge City" in a targeted way. Through geographical division and sponge system preset, each division can choose the road material use plan of "Sponge City" according to local conditions, so as to make the collection and utilization of rainwater and more efficient drainage. According to the research in 3.3.2, the whole country is divided into five specific rainfall areas according to the rainfall gradient, and a material storehouse needs to be established to correspond to these five rainfall areas, so that the construction of "Sponge City" can match the materials according to the area or the materials. The idea of establishing material storehouse is shown in **Table 5**.

Table5. Design scheme of permeable materials for "sponge city"

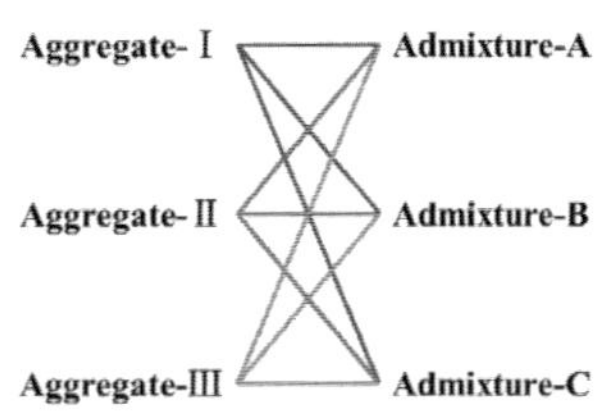

Type	Performance
Aggregate -I	HPA
Aggregate -II	HSA
Aggregate -III	CA
Admixture-A	HPA
Admixture-B	HSA
Admixture-C	DPA

According to the previous study on aggregate and admixture, aggregate can be divided into different categories according to the particle group parameter α value of aggregate, including HPA , HSA and CA； At the same time, three kinds of admixtures with different functions are obtained in the study of admixtures, which are high permeable admixtures (HPA-A), high strength admixtures (HSA-B) and drainage promoting admixtures (DPA-C). They can modify the binder materials to some extent, so that the performance of the water permeable materials can also be improved, Therefore, according to the collocation method in **Table 5**, 9 different material selection schemes can be obtained as shown in **Table 6**:

Table6. Material selection plans

Aggregates	Admixtures	Targets	Performance	Plans
Aggregate-I	Admixture-A	A I	Ultra HP	Plan 1
	Admixture-B	B I	HPHS	Plan 2
	Admixture-C	C I	HPDP	Plan 3
Aggregate-II	Admixture-A	A II	HSHP	Plan 4
	Admixture-B	B II	Ultra HS	Plan 5
	Admixture-C	C II	HSDP	Plan 6
Aggregate-III	Admixture-A	A III	CHP	Plan 7
	Admixture-B	B III	CHS	Plan 8
	Admixture-C	C III	CDP	Plan 9

Note: the abbreviation H means High, P means Permeability, S means Strength, DP means Drainage Promoting.

From the above 9 schemes, raw material combinations with different performance characteristics can be obtained. In 5 rainfall areas or different demand scenarios, detailed screening can be carried out according to the above schemes. For example, the fifth level rainfall area is located in the southeast coastal area. Taking Hainan as an example, its annual average rainfall is more than

1639.00mm. Therefore, the selection of permeable materials in this area obviously tends to be highly permeable, while the demand for water storage performance is not large, so the drainage promotion scheme can be selected. According to **Table 6**, scheme 1 and scheme 3 can be selected in the pedestrian road system, while scheme 4, 5 and 6 can be selected in the light load vehicle road system. The rainfall area in Hainan is shown in **Fig. 13** .

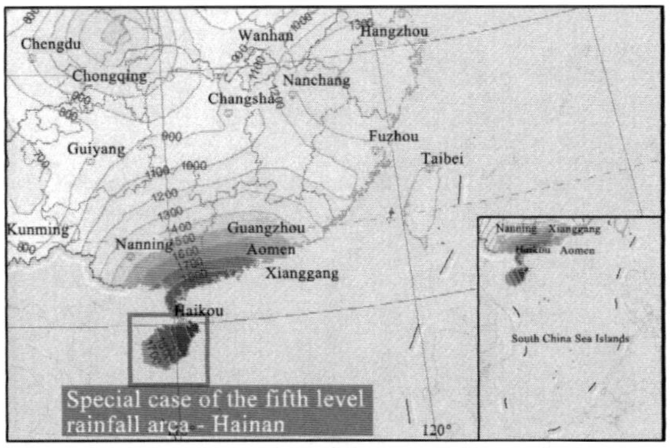

Fig13. The fifth level rainfall area -- Taking Hainan as an example

4 Conclusion

In this study, the permeable brick with permeability rate of $6.22 \times 10^{-2} cm/s$ and the compressive strength is more than 60MPa was prepared by using aeolian sand and epoxy resin adhesive materials with particle size of 0.08mm-0.6mm. After in-depth study on the aggregate and admixture of permeable brick, the performance division methods were found. By using the geographic analysis software to divide the rainfall area in China, and combining the different rainfall demand with the material bank plan reasonably, a "Sponge City" water permeable brick component material compatibility optimization plan is formed according to the local conditions. The specific conclusions are as follows:

(1) The pore shape of permeable brick are three side pore and four side pore. The pore size is 30 μm-100 μm. The micron level pore can effectively prevent large particles from blocking and small particles from migrating in the brick body, and extend the service life of permeable brick;

(2) Aggregate can be divided into three types according to particle group parameters and properties, namely, high permeable HPA, comprehensive CA and high-strength HSA; Permeable brick admixture can be divided into three types, namely, high permeable type, high-strength type and drainage promoting type.

(3) China's "Sponge City" rainfall area is divided into five levels, each of which corresponds to nine different material compatibility optimization schemes.

References

1. Cohen J E. Human population: the next half century[J]. Science, 2003,302(5648):1172-1175.
2. Strecker E W, Quigley M M, Urbonas B R, et al. Determining Urban Storm Water BMP Effectiveness[J]. Journal of Water Resources Planning and Management, 2001,127(3):144-149.
3. Dietz M E. Low Impact Development Practices: A Review of Current Research and Recommendations for Future Directions[J]. Water, Air, and Soil Pollution, 2007,186(1-4):351-363.
4. Norton B A, Coutts A M, Livesley S J, et al. Planning for cooler cities: A framework to prioritise green infrastructure to mitigate high temperatures in urban landscapes[J]. Landscape and Urban Planning, 2015,134:127-138.
5. Abbott C L, Comino Mateos L. IN - SITU HYDRAULIC PERFORMANCE OF A PERMEABLE PAVEMENT SUSTAINABLE URBAN DRAINAGE SYSTEM[J]. Water and Environment Journal, 2003,17(3):187-190.
6. Coutts A M, Tapper N J, Beringer J, et al. Watering our cities: Thecapacity for Water Sensitive Urban Design to support urban cooling and improve human thermal comfort in the Australian context[J]. Progress in Physical Geography: Earth and Environment, 2013,37(1):2-28.
7. van Roon M. Water localisation and reclamation: Steps towards low impact urban design and development[J]. Journal of Environmental Management, 2007,83(4):437-447.
8. Lim H S, Lu X X. Sustainable urban stormwater management in the tropics: An evaluation of Singapore's ABC Waters Program[J]. Journal of Hydrology, 2016,538:842-862.
9. Liu W, Chen W, Feng Q, et al. Cost-Benefit Analysis of Green Infrastructures on Community Stormwater Reduction and Utilization: A Case of Beijing, China[J]. Environmental Management, 2016,58(6):1015-1026.
10. Wang M, Zhang D, Adhityan A, et al. Assessing cost-effectiveness of bioretention on stormwater in response to climate change and urbanization for future scenarios[J]. Journal of Hydrology, 2016,543.
11. Walsh C J, Fletcher T D, Burns M J. Urban stormwater runoff: a new class of environmental flow problem[J]. PLoS One, 2012,7(9):e45814.
12. Weng Q. Modeling urban growth effects on surface runoff with the integration of remote sensing and GIS[J]. Environ Manage, 2001,28(6):737-748.
13. Jiang Y, Zevenbergen C, Ma Y. Urban pluvial flooding and stormwater management: A contemporary review of China's challenges and "sponge cities" strategy[J]. ENVIRONMENTAL SCIENCE & POLICY, 2018,80:132-143.
14. Jamali B, Löwe R, Bach P M, et al. A rapid urban flood inundation and damage assessment model[J]. Journal of Hydrology, 2018,564:1085-1098.
15. O Driscoll M, Clinton S, Jefferson A, et al. Urbanization Effects on Watershed Hydrology and

https://doi.org/10.1051/e3sconf/202123703004

In-Stream Processes in the Southern United States[J]. Water, 2010,2(3):605-648.

16. Zhang Xiong, Wang Xiaofu. Research progress on the influence mechanism of some factors on the performance of permeable brick [J]. Materials Reports, 2019,33 (23): 3949-3954.
17. Liu Yanan. Research on application and promotion of permeable pavement in sponge City [D]. Beijing University of architecture, 2017.
18. Ao Jing. Study on the regulation of urban permeable pavement system on local geothermal wet climate [D]. Harbin University of technology, 2014.
19. Qin Yuchao. Study on the effect of runoff reduction and pollutant reduction of typical permeable pavement in cities [D]. Xi'an University of technology, 2017.
20. State Administration for Quality Supervision and Inspection and Quarantine Permeable Paving Bricks and Permeable Paving Flags[J]. GB/T 25993-2010, Standards Press of China, Beijing,2011(in Chinese).
21. Zhu M, Wang H, Liu L, et al. Preparation and characterization of permeable bricks from gangue and tailings[J]. Construction and Building Materials, 2017,148:484-491.
22. Sand-based water permeable brick[S]. JG/T 376-2012, Standards Press of China, Beijing, 2012.
23. Jiang Z, Sun Z, Wang P. Effects of Some Factors on Properties of Porous Pervious Concrete[J]. Journal of Building Materials, 2005,8(5):513-519.
24. Kayhanian M, Anderson D, Harvey J T, et al. Permeability measurement and scan imaging to assess clogging of pervious concrete pavements in parking lots[J]. Journal of Environmental Management, 2012,95(1):114-123.
25. SIRIWARDENE N, DELETIC A, FLETCHER T. Clogging of stormwater gravel infiltration systems and filters: Insights from a laboratory study[J]. Water Research, 2007,41(7):1433-1440.
26. Lian C, Zhuge Y, Beecham S. The relationship between porosity and strength for porous concrete[J]. Construction and Building Materials, 2011,25(11):4294-4298.
27. Chindaprasirt P, Hatanaka S, Chareerat T, et al. Cement paste characteristics and porous concrete properties[J]. Construction and Building Materials, 2008,22(5):894-901.
28. Ba M, Qian C, Guo X, et al. Effects of steam curing on strength and porous structure of concrete with low water/binder ratio[J]. Construction and Building Materials, 2011,25(1):123-128.

Study on the influencing factors and rules of the phased evacuation of building

Li Yang[1], Daoliang Zhao[1*], Meiting Zeng[1],Yun Lu[1], Lei Shi[1], Hu Zhou[1]

[1]College of Urban Construction and Safety Engineering, Shanghai Institute of Technology, Shanghai, China

Abstract. The whole evacuation of building fire personnel is always a research hotspot. However, in special occasions and conditions, evacuation to the first safety zone as soon as possible is a powerful way to ensure the safety of people's lives, that is, the study of the phased evacuation of building is also of great practical significance. In this paper, the BuildingEXODUS software was used to establish the evacuation model, which mainly analyzed the influence and law of two physical factors on the evacuation of people from the room to the corridor: personnel density, personnel distribution in the corridor. In addition, MassMotion software was used to simulate and compare some conclusions under the same evacuation scene, aiming to study and compare the similarities and differences of the influence of the same physical factors on the phased evacuation of building under different simulation mechanisms.

1 Introduction

When an emergency occurs in a specific building, organizing personnel to flee from the building in a short time can minimize or avoid casualties, that is, orderly evacuation is an important guarantee for public safety[1]. Xue et al.[2]mainly discussed the impact of the width and position of the exit of the room and the width of the corridor on the whole evacuation efficiency. Liao et al.[3]first conducted a questionnaire survey, and then simulated the evacuation of a teaching building through Anylogic, it is obtained that the evacuation efficiency of the whole teaching building has been improved under the additional sign.

The overall evacuation of construction personnel is important in the event of an accident, but in some special circumstances, for those who are on the scene of the first accident, it is more important to evacuate from dangerous places to relatively safe areas in the shortest possible time.

There are many kinds of fire hazard sources inside the hospital, so there are certain fire safety hazards, however, apart from medical staff and family members, there are also patients with mobility difficulties, which leads to great difficulty in evacuation[4,5,6]. A common scene in a hospital is a doctor in an outpatient department in a room to diagnose a patient, while people outside the room wait in the hallway for a call. If a fire breaks out in a room, evacuating the entire building can cripple the health care system, the most unfavorable situation is to cause a certain number of people panic, and patients already with mobility difficulties may be trampled or pushed by the crowded crowd, which will aggravate their condition, obviously with disastrous consequences. At this time, it is crucial for people in the room to evacuate quickly from the room to the corridor. Li et al.[7]believed that the temporary safety area obtained from the transformation of the evacuation walkway could solve the problem of people escaping due to insufficient evacuation width of the building when meeting certain functional requirements. Tian et al.[8]took the evacuation walkway as a quasi-safety zone to solve the problem that the fire protection zone in the underground cultural heritage storehouse of a museum exceeds the standard, on the premise that the fire parameters of the walkway do not pose a danger to the safety of evacuators, finally, based on the simulation, it was concluded that the performance-based design of the quasi-safety zone was feasible. Moreover, it is pointed out in 《the Performance-based Design of Building Fire Prevention》[9] that the evacuation walkway is called the first safety zone in the evacuation design. The quasi-safe area is a relatively safe area that meets certain performance indicators[10], the first safety zone is the area with certain safety performance that evacuees first arrive at, therefore, under reasonable structural conditions, the evacuation walkway can become a safer area compared with the danger area during evacuation.

The first stage of the evacuation route in the building is from the burning room to the door of the room[9], then arrive at the evacuation walkway (corridor), evacuation efficiency and evacuation time in this stage may be related to personnel density, personnel distribution and other factors. In this paper, BuildingEXODUS software was used to simulate and analyze the influence of these physical factors on the first-stage evacuation process, and MassMotion software was used to simulate and compare, in order to obtain the behavior rule of the first-stage evacuation of building personnel.

*Corresponding author's e-mail: zhaodl@sit.edu.cn

2 Software Comparison and Parame- ter Introduction

In BuildingEXODUS, the attribute of an individual searching for the destination during evacuation is set as "Nearest door", there is no specific direction bias during the movement.

In MassMotion, the attribute of an individual searching for the destination during evacuation is set as "lowest cost". In addition, evacuees are set to have "Direction Bias", that is, the preferred direction to avoid oncoming traffic, and set to "Right strong".

In this paper, BuildingEXODUS software was used to simulate and study the impact of physical factors on the efficiency of room staff evacuation to the corridor. It is mainly analyzed according to the simulated scenario and the last entry data of each internal exit.

Last entry means the last usage time of the internal exit. In this paper, the time of the last person in the room arriving at the internal exit is considered according to the last entry and the simulated scenario, that is, how long it takes the most dangerous people to escape from the room. The following are the arrival time of the final personnel at internal exits.

3 Evacuation Simulation Model

The evacuation model in this article is created based on the geometric plane shown in Fig.1, the effective size is as follows: the corridor is 13m×2.5m, the room is 4m×2m, the external exit on the left is 1.5m, the external exit on the right is 2.5m and the internal exit is 1m.The maximum capacity of the corridor is 130 people, the maximum capacity of each room is 32 people, the evacuation speed was set as 1.2m/s~ 1.5m /s, and the response time was 0~30s. In order to reduce the difference of the data generated by the random distribution of personnel, each set of simulations for the full text was run five times.

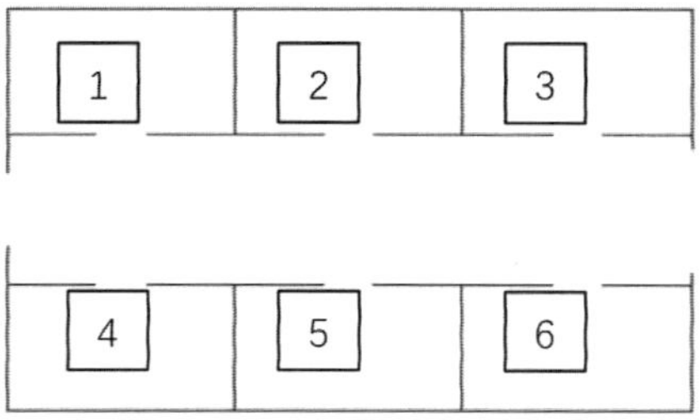

Fig1. Geometric plane of evacuation model

4 Simulation Results Analysis

4.1 The Influence of Personnel Density on the Evacuation of Room Personnel

The simulation in this part mainly considers the impact of personnel density on the evacuation of room personnel. Setting the number of people in each room at five initial moments in BuildingEXODUS, that is, 6, 12, 18, 24 and 30. According to the specific number of people in each room, increase the number of people in the corridor successively from 0-120, 20 people were increased each time, and finally five groups of simulations were carried out. The origin software was used to fit the quadratic polynomial of the arrival time of the final personnel at internal exits, as shown in Fig.2.

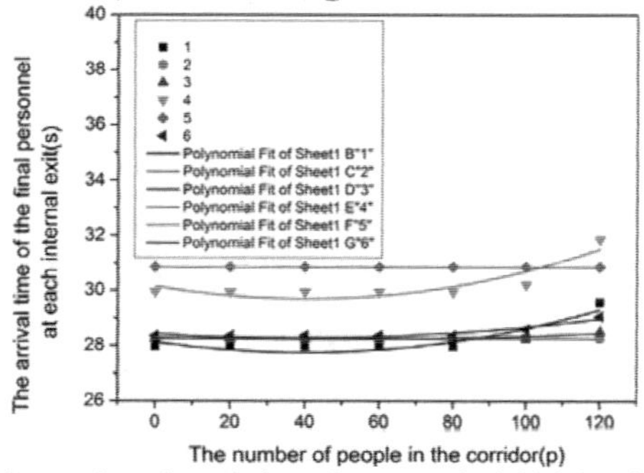

a. the number of people in each room at the initial time is 6

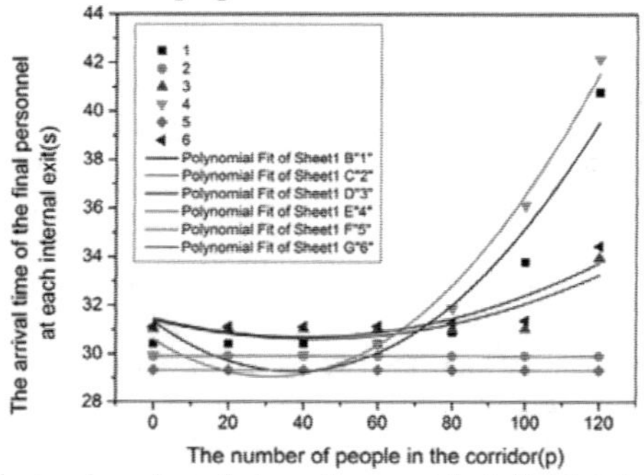

b. the number of people in each room at the initial time is 12

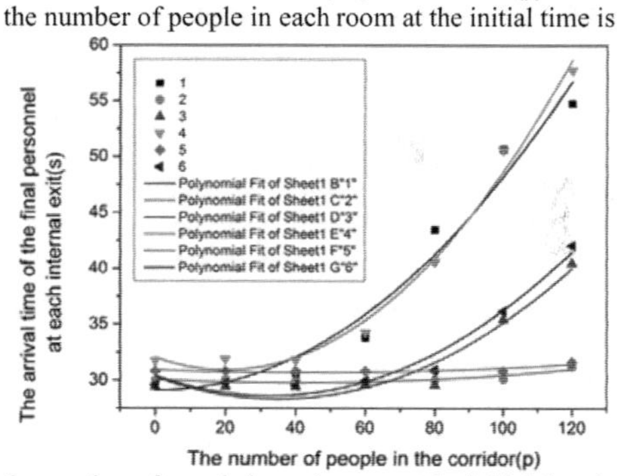

c. the number of people in each room at the initial time is 18

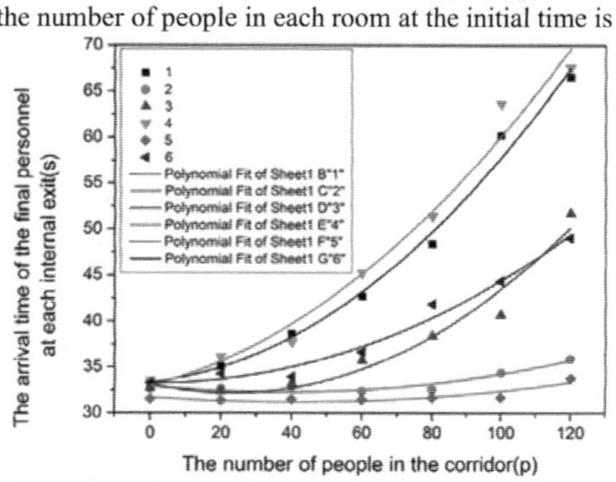

d. the number of people in each room at the initial time is 24

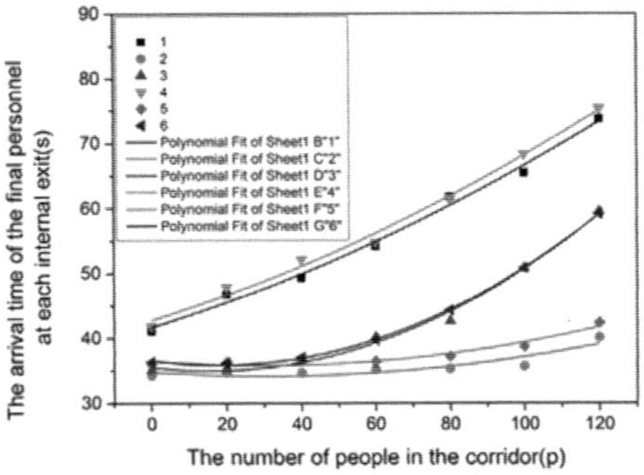

e. the number of people in each room at the initial time is 30

Fig2. The effect of change in personnel density on the arrival time of the final personnel at internal exits

It is easy to see from Fig.2 that all internal exits fall into three categories. The first type is No. 1 and 4, the second type is No. 2 and No. 5, and the third type is No. 3 and No. 6.

With the increase of the personnel density of rooms, the influence of changing the personnel density of the corridor on the arrival time of the final personnel at internal exits of the first type and the third type are more and more obvious, b、c、d and e are shown in Fig.2. In addition, the arrival time of the final personnel between the first, second and third types of internal exits gradually starts to produce a certain gap under a smaller personnel density of the corridor, because the number of people in rooms increases and the width of the external exits on the left and right is different. When the personnel density of rooms is high, as shown in Fig.2e, the arrival time of the final personnel at internal exits of the first type increases almost linearly.

On the premise that the personnel density of rooms is low, if the personnel density of the corridor is not high, the arrival time of the final personnel at internal exits of these three types is not different, if the personnel density of the corridor is high, the general rule of the arrival time of the final personnel at internal exits is 1、4 > 3、6 > 2、5, as shown in Fig.2b. On the contrary, when the personnel density of rooms is relatively high, change the personnel density of the corridor, and the arrival time of the final personnel generally presents such a rule. For individuals looking for destination in BuildingEXODUS is " Nearest door ", finally the difference in the number of people who pass through the two external exits is small, but because the width of the external exit on the left is smaller than that of the external exit on the right, when the evacuees gradually approach external exits, the left side of the corridor is more likely to form congestion, resulting in the delay of the arrival time of the final personnel at internal exits of the first type. In addition, with the evacuation to both sides of the corridor, the number of people in the middle gradually decreases, so that the change of personnel density in the corridor has little impact on the arrival time of the final personnel at the second type of internal exits.

According to the simulation of BuildingEXODUS software, when the personnel density of rooms is relatively high, change the personnel density of the corridor, and the arrival time of the final personnel at each internal exit is generally 1、4 > 3、6 > 2、5. At the initial moment, 30 people in the room and 100 people in the corridor were set, and the simulation results were compared under the same scene with MassMotion software, as shown in Fig.3.

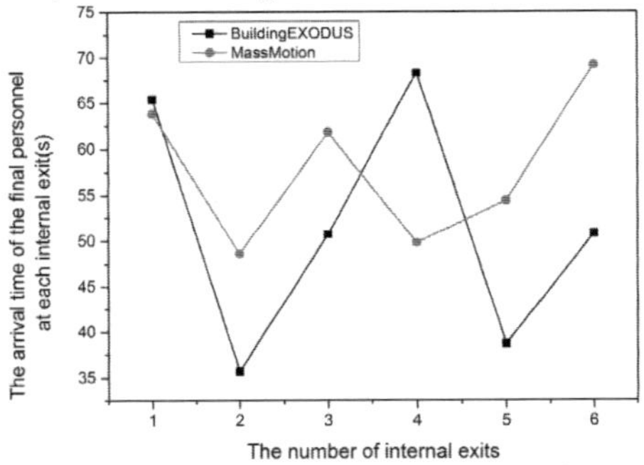

Fig3. The final arrival time of each internal exit with 30 people in the room at the initial time and 100 people in the corridor

In this part of the simulation, the "lowest cost" attribute of individual searching for the destination in the evacuation process in MassMotion improves the overall evacuation efficiency of the building. Facing the external exits with different widths on the left and right, the number of people passing through the external exits varies greatly, while the difference in the final passing time is small.

As can be seen from Fig.3, when the personnel density of both the room and corridor is high, what is different from the simulated result of BuildingEXODUS is that the arrival time of the final personnel at the internal exit 1 and 4 near the external exit on the left is greatly different and does not belong to the same category, so are those at the internal exit 3 and 6 near the external exit on the right, while the arrival time of the final personnel at the internal exits of 2 and 4 is similar, and the difference of 1 and 3 is also small. In MassMotion simulation, there is a big difference in the number of people passing through external exits on the left and right, making the total number of people crowded near the internal exit 1 and 4 less, and the number near the internal exit 3 and 6 more. However, because the direction bias of individuals is right strong, when people are evacuated in the corridor and facing the stream of people at two internal exits at the same time, they tend to be on the right side of the evacuation direction of the corridor, and the density near the internal exit 1 is also high. At the same time, the direction bias of individuals is also the reason why two internal exits located in the same position of the corridor do not belong to the same kind.

4.2 The Impact of the Personnel Distribution in the Corridor on the Evacuation of Room Personnel

The simulation in this part mainly analyzes the impact of the personnel distribution in the corridor on the evacuation of the room. When using BuildingEXODUS to simulate, set the number of people in the corridor at 80 and set three initial personnel distribution states of the corridor, that is, random, middle and on both sides , as shown in Fig.4.

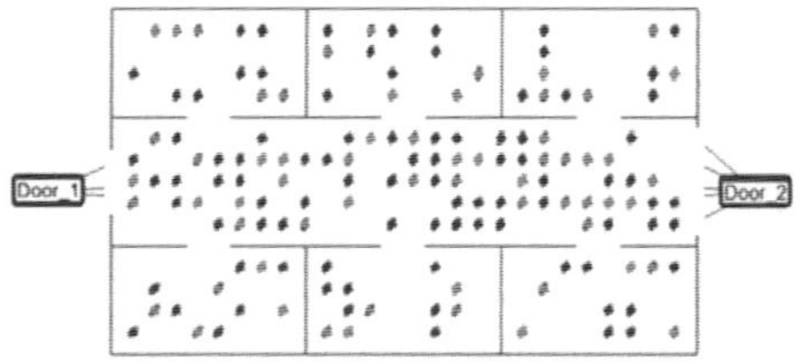

a. The corridor personnel were randomly distribute

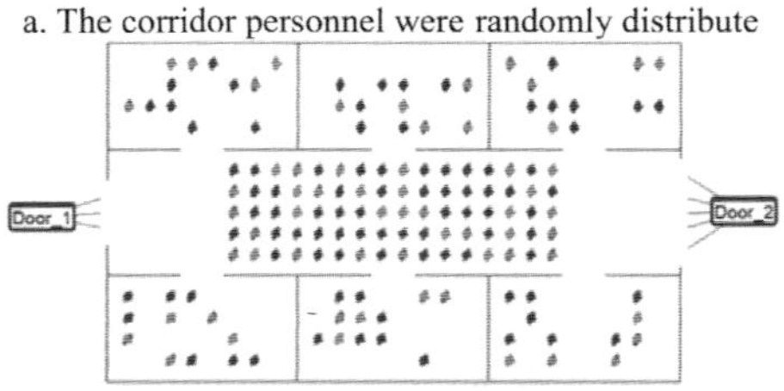

b. The corridor personnel were distributed in the middle

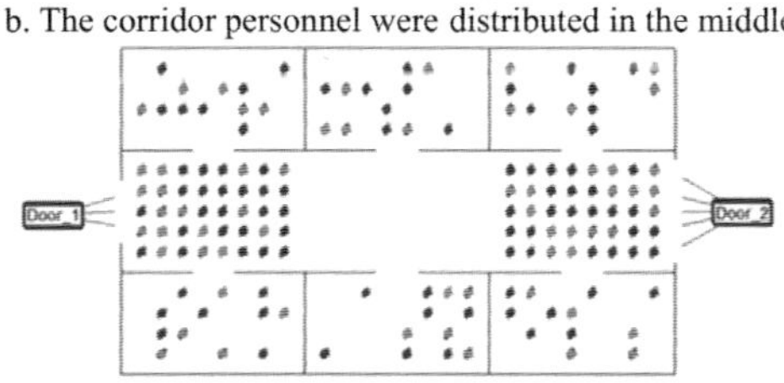

c. The corridor personnel were distributed on both sides

Fig4. Three types of corridor personnel distribution of 12 people in the room at the initial moment

The origin software was used to fit the quadratic polynomial of the arrival time of the final personnel at internal exits, as shown in Fig.5.

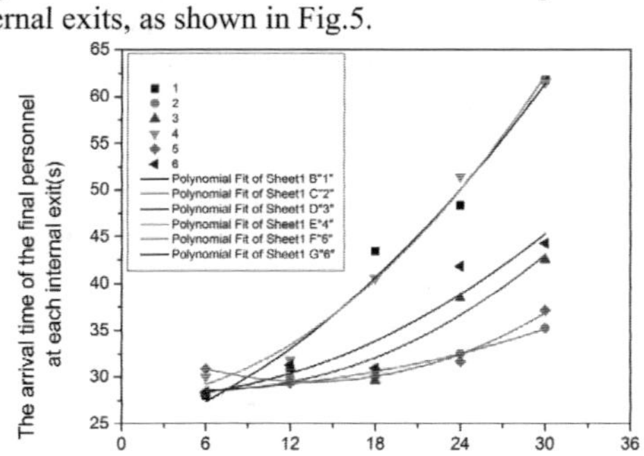

a The corridor personnel were randomly distribute

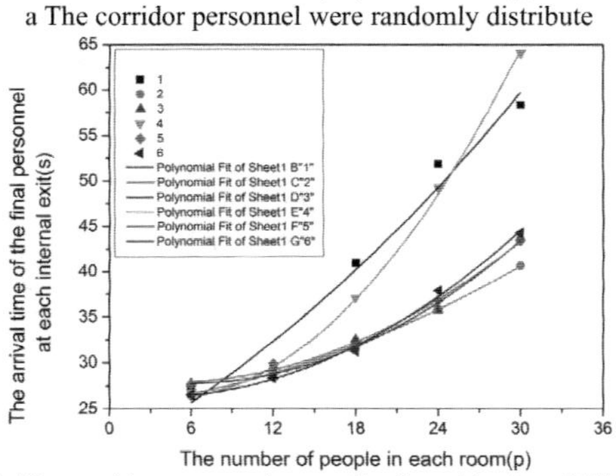

b The corridor personnel were distributed in the middle

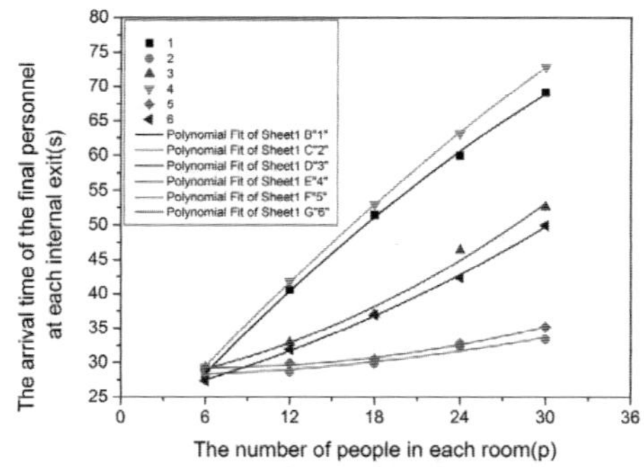

c The corridor personnel were distributed on both sides

Fig5. Influence of different corridor personnel distribution on the final arrival time of internal exits

As in the BuildingEXODUS simulation in Section 4.1, all internal exits are divided into three categories: the first type is No.1 and No. 4, the second type is No. 2 and No. 5 and the third type is No. 3 and No. 6.

It can be seen from Fig.5 that, with the increase of the personnel density of rooms, the difference between the arrival time of the final personnel at internal exits of the first type in the random distribution of corridor personnel and the middle distribution is relatively small, which is about 3.5s at most. When people in the corridor are distributed on both sides, on the premise of low density of people in the room, there is a small difference between the arrival time of the final personnel at internal exits of the first type and the two situations above. For example, when the number of people in each room is 6, the difference is small because the number of people is low. The personnel density of the room gradually increases, and the arrival time of the final personnel at internal exits of the first type is higher when the corridor personnel are distributed on both sides. In this case, the people in the vicinity are in a dense state at the initial moment, and the number of people in each room is relatively large, so it is not easy for the people in the room to evacuate quickly, thus increasing the arrival time of the final personnel. In addition, the average evacuation efficiency of rooms near the external exit on the left under the previous two distributions increased by about 27.6% at the highest level compared with that of the distribution on both sides. The situation of the third type of internal exits is similar. The difference between the arrival time of the final personnel in the random distribution of corridor personnel and the middle distribution is small, the most is about 4s. Moreover, the average evacuation efficiency of rooms near the external exit on the right under these two distributions increases by about 20.08% at the highest level compared with that of the distribution on both sides.

It is also easy to conclude that, as the personnel density of the room increases, the arrival time of the final personnel at internal exits of the second type is similar in the case of the random distribution of corridor personnel and the distribution on both sides, with a maximum difference of about 2s. When the corridor personnel are distributed in the middle, if the personnel density of the room is low, the difference between the arrival time of the final personnel and that of the previous two distributions is small, such as the number of people in each room is 18. As can be seen from Fig.6, the middle distribution of corridor personnel makes the initial time near the internal exits of the second type more dense, but people around the 20s are evacuated and separated, which reduces the density. If the number of people in the room is not particularly large, they will evacuate to the corridor more quickly. If the corridor personnel are distributed in the middle, the arrival time of the final personnel is greater in the case of a larger number of people in the room than in the case of the other two distributions, for example, the number of people in each room is 30. Although people distributed in the middle of the corridor have not been evacuated to both sides for a long time, there are enough people in the room. In addition, the impact of dense personnel in the corridor before evacuation and separation leads to the growth of the arrival time of the final personnel at internal exits of the second type. The average evacuation efficiency of the rooms in the middle under the random distribution of corridor personnel and the distribution on both sides increased by about 16.76% at the highest level compared with the middle distribution.

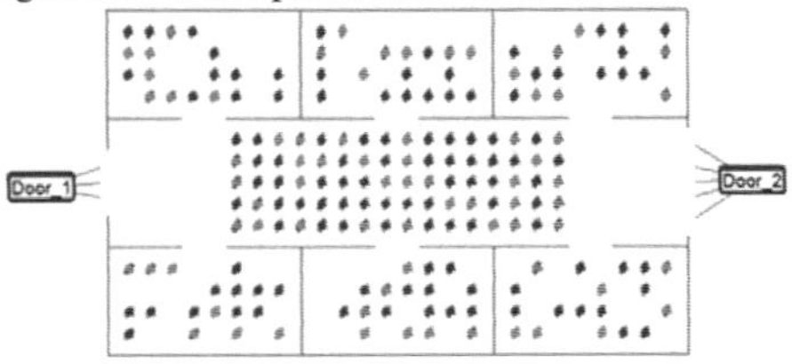

a. When the simulation time was 0s

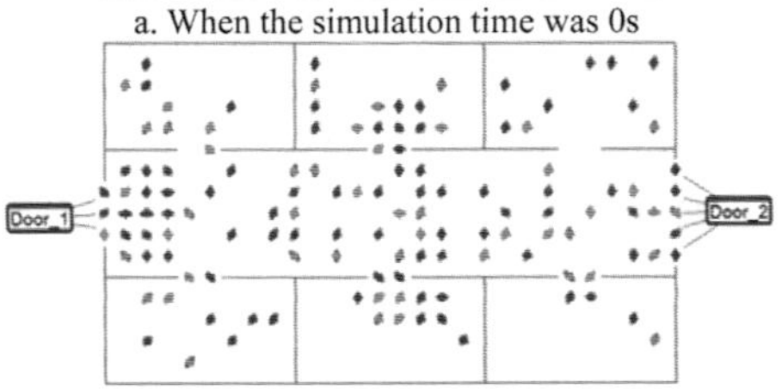

b. When the simulation time was 20s

Fig6. A simulation scenario with 18 people in the room at the initial time and the people in the corridor distributed in the middle

According to the results of BuildingEXODUS, the difference of the arrival time of the final personnel at internal exits of the first type(1,4) is small in the case that the corridor personnel are randomly distributed and in the middle. In addition, as the personnel density of the room increases gradually, the arrival time of final personnel is higher when the corridor personnel are distributed on both sides than in the previous two distribution situations. The third type of internal exits (3, 6) have similar situation. Examples are shown in Fig.7. MassMotion software was used to simulate in the same scene, and the results are shown in Fig.8.

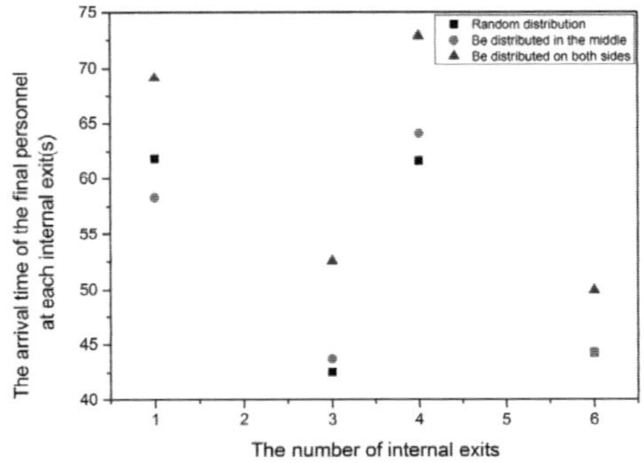

Fig7. Simulation results of BuildingEXODUS for 30 people in the room at the initial time

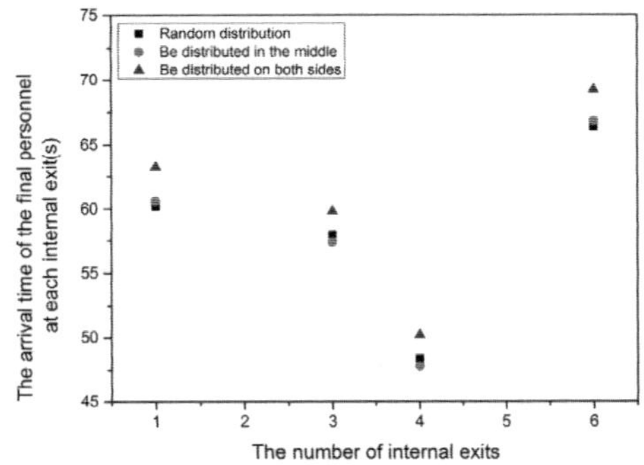

Fig8. Simulation results of MassMotion for 30 people in the room at the initial time

It is the same as that obtained in MassMotion simulation in Section 4.1, because of the problem of direction bias, the No.1 and No.4 internal exit close to the external exit on the left in MassMotion do not belong to the same category, and the same is true for No.3 and No.6 internal exit.

It can be seen from Fig.8 that when the density of people in the room is high, the difference in the arrival time of the final personnel at the No.1 internal exit is small in the three distribution forms of corridor personnel, and the same is true for No.3、No.4、No.6 internal exit. However, it is also easy to find that the arrival time of the final personnel at four internal exits is slightly higher than that under the random and middle distribution when the corridor personnel are distributed on both sides. For example, the reason of No.1 internal exit is the same as that in BuildingEXODUS, because the corridor personnel are distributed on both sides at the initial moment, the people near the internal exit will have a long time of congestion, but in MassMotion, there are fewer people in the final moment, which has a small impact, so the difference between the three distributions is not big.

5 Conclusion

1) BuildingEXODUS was used to simulate the initial personnel density of the room and corridor, which has a great influence on the evacuation efficiency of rooms near the two external exits, but a small influence on the evacuation efficiency of the middle rooms. And the change in the arrival time of the final personnel at two internal exits near the external exit on the left belongs to

the same category, that of two middle internal exits also belongs to the same kind, the same is true for two internal exits near the external exit on the right. In MassMotion, when the personnel density of both the room and corridor is high, the difference of the arrival time of the final personnel at two internal exits near the external exit on the left is larger, that of two internal exits close to the external exit on the right is also larger, and that of two middle internal exits is not similar.

2) In the simulation of BuildingEXODUS, the initial personnel distribution in the corridor will affect the phased evacuation efficiency of room personnel. The difference of evacuation time of people in rooms near the external exits is smaller in the random distribution of corridor personnel and the middle distribution, while the difference of evacuation time of people in middle rooms is smaller in the random distribution of corridor personnel and the distribution on both sides. In MassMotion, the evacuation time of each room near external exits varies little among the three distribution forms of corridor personnel.

Acknowledgments

Project of Shanghai Natural Science Foundation (15ZR1440900);

Science and Technology talent Development Fund for young and middle-aged teachers of Shanghai Institute of Technology (ZQ2018-21)

References

1. W.H. Wang, S.X. Wu, J. Cheng. Mathematical Model of Evacuation Program in a Building[J]. Journal of Wuhan University of Technology, 32(11):155-158+162(2010).
2. P. Xue, J.W. Zhou, K.Z. Bai, L.J. Kong, M.R. Liu. Cellular Automaton Model of Occupant Evacuation from Multi-room in Corridor[J]. Journal of Guangxi Normal University(Natural Science Edition), 27(03):1-4(2009).
3. H.M. Liao, X.J. Luo, H. Su. Research on cognitive coping regularity of evacuation signs in fire of teaching building[J]. Journal of Safety Science and Technology, 15(08):131-136(2019).
4. F. Yang, J. Chen. On the Risk Assessment of Fire Hazards in Hospitals[J]. The Science Education Article Collects, (16):190-192(2015).
5. S.B. Li. Discussion on hospital fire safety management and emergency[J].Jiangsu Health System Management, (1):133-134(2014).
6. H.B. Liang. Fire safety management and emergency of large hospital buildings[J].Modern Preventive Medicine, 34(7):1327-1327(2007).
7. F. Li, Y. Zhang. Solutions on the lack of evacuation width in existing building[J]. Fire Science and Technology, 33(08):894-896(2014).
8. Y.H. Tian, G. Li. Fire performance-based design of a museum[J]. Fire Science and Technology, 32(4):391-393(2013).
9. F. Liu, S.J. Liao. Safe evacuation of personnel. In: H.M. Chen, H.G. Zhou(Eds.), the Performance-based Design of Building Fire Prevention[M].CHONGQING UNIVERSITY PRESS, Chongqing, 147-151(2007).
10. G.W. Zhang, G.Q. Zhu, W.H. Wu, L.L. Huang. Judging method of "quasi safe area" of building based on performance-based fire protection design[J].China Safety Science Journal, 21(04):60-65(2011).

Influence of the Passive Design of a Building Facade on the Indoor Thermal Comfort of Residential Buildings

Zhou Pengfei[1], Zhang Chi[1], and Jiang Wang[2]

[1]Faculty of Built Environment, University of New South Wales, Sydney, Australia
[2]Faculty of Built Environment,Yancheng Institute of Technology, Yancheng , China

Abstract. Building facades have evident effects on indoor thermal comfort. Hence, on the basis of a multifunctional residential building in Sydney, Australia, this research uses DesignBuilder software to optimize passive system design on building facades. This research also analyses the influences of changing window glazing type, adding additional shading devices and changing the material of the exterior wall on indoor thermal comfort. Results show that the number of uncomfortable hours can be reduced by 446, 186 and 874 hours by using a double-layer Low-E glass, adding extra shading device and adopting an external wall material with low thermal conductivity, respectively. When the three aforementioned passive design strategies are combined, indoor thermal environment discomfort time can be reduced by 24%. Therefore, the indoor thermal comfort of a building can be considerably improved through effective passive designs of the building facade.

1 Introduction

The operation stage of a building accounts for a higher proportion of its entire life cycle compared with the architecture design and construction stages. Therefore, energy consumption during a building's operation stage comprises a major part of its energy consumption in its entire life cycle [1]. With the global energy crisis becoming increasingly serious, ensuring and improving indoor thermal comfort whilst reducing energy consumption during a building's operation stage are important goals [2]. At present, energy conservation in most buildings is focused on reducing energy consumption, whereas indoor thermal comfort is less frequently considered [3].

Building facades affect indoor thermal comfort. The dual effects of reducing building energy consumption and improving indoor thermal comfort can be achieved by adopting a passive design in a reasonable manner. In this study, a passive design is applied to a multifunctional residential building in Sydney, Australia to optimize its facade. The computer simulation software DesignBuilder is used to simulate indoor thermal comfort during the operation stage[4]. Corresponding passive construction optimization strategies are proposed based on the analyses of factors that influence the original indoor thermal comfort of a building. Subsequently, indoor thermal comfort is simulated and analysed after adopting passive optimization strategies. Without increasing energy consumption, passive design strategies on the facade are optimized and the indoor thermal comfort of the building is improved. This study provides references for the design and construction of local buildings of the same type.

2 Project overview

A multifunctional residential building with eight stories located in Bondi Beach (33°89′S, 151°27′E) is selected as the research case. The project location exhibits the characteristics of a subtropical humid monsoon climate. The annual mean temperature in the region is 20.7 °C. July is the coldest month, with a mean temperature of 8 °C. January is the warmest month, with a mean temperature of 19 °C[5]. As shown in Figure 1, the building faces northwest, with an area of 2961 m^2. Natural ventilation is adopted in the entire building, with no heating or refrigeration equipment. The building design meets the requirements of the Building Codes of Australia (BCA). This study develops optimization
design based on three factors that are closely related to building facades: facade glazing, shading devices and exterior wall materials. This study also uses DesignBuilder software for quantitative simulation analysis and the comparison of indoor thermal comfort before and after implementing the optimization design. Meanwhile, the new building energy efficiency design ASHRAE standard 55-2013[6], which stipulates that the total time should be within the acceptable temperature range, is used to evaluate indoor thermal comfort. Thermal comfort is the mental state that shows satisfaction with the thermal environment. It is evaluated based on the ASHRAE Standard, which points out that in a neutral environment (i.e. wind speed lower than 0.2 m/s, indoor humidity of 50%), the comfortable temperature range during winter is 20 °C–24 °C and that during summer is 24 °C–27 °C. However, this standard is influenced by

* Corresponding author: yczpf1412956361@qq.com

individual activities, clothing and outdoor temperature and humidity. Therefore, inhabitants living under natural ventilation are more likely to accept a wider temperature range, combine indoor comfortable temperature with outdoor temperature, and define 80% and 90% acceptability limits as satisfactory temperature range, as shown in Figure 2. On the basis of an experimental analysis, the aforementioned range is the range of thermal comfort, and any point outside this range will cause thermal discomfort. In the current experiment, the total time of indoor thermal comfort of a building with natural ventilation is used as the test datum, and the satisfactory range for 90% acceptability of the inhabitants in ASHRAE Standard 55-2013 is adopted as the acceptable temperature range.

A software analysis is performed based on the satisfactory temperature range for 90% of the inhabitants in ASHRAE Standard 55-2013. A conclusion is drawn that in the satisfactory temperature range, the uncomfortable time that does not meet this criterion is the uncomfortable time of the indoor thermal environment.

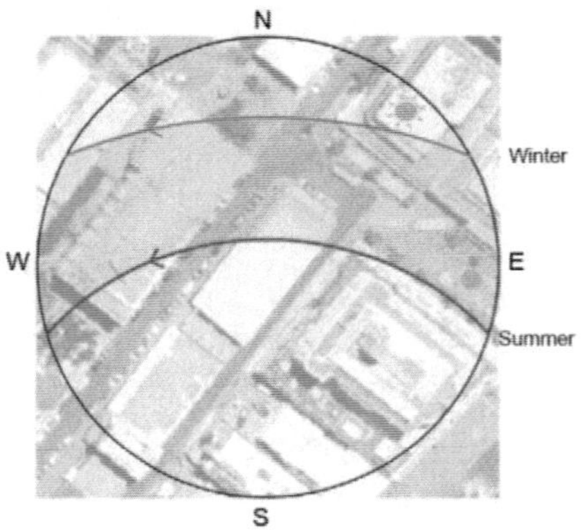

Fig1. Site orientation and year-round sunshine tracking range.

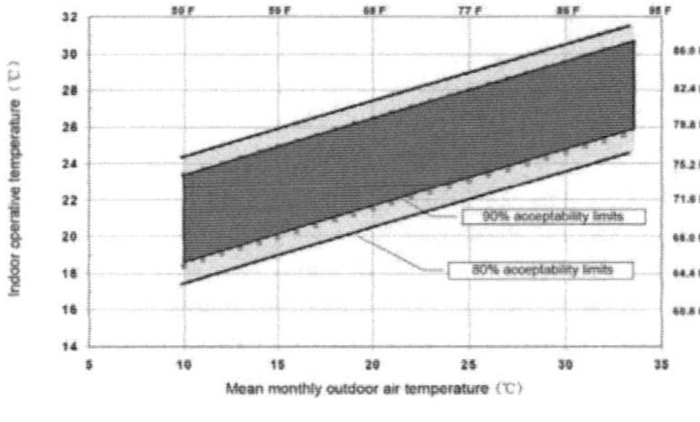

Fig2. Definition of the temperature range that corresponds to thermal comfort time in ASHRAE Standard.

3 Influence of building facade on indoor thermal comfort

Building facades are directly exposed to the external environment. Different building facades are related to varying indoor comfort and energy consumption levels. A reasonable facade design helps reduce energy consumption and improve indoor thermal comfort [7]. Therefore, the passive design of building facades is an important aspect in managing building energy consumption [8].

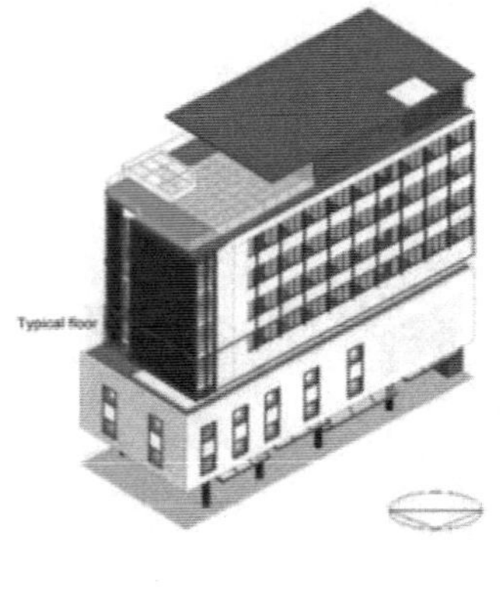

(a) Building model

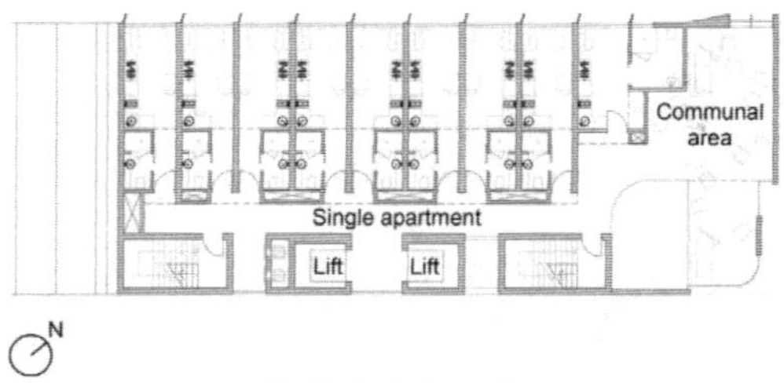

(b) Typical floor plan

Fig3. Building model and typical floor plan.

As shown in Figure 3(a), the building has a concrete frame structure with sand-coloured facing bricks. The ground floor is a whole open space for public services.The first and second floor contain an auditorium for 150 people and offices for community parties and meetings. The third and higher floors are typical floors for residences, where each floor is composed of eight single apartments and a public activity area. The building faces northwest; hence, an analysis of the typical floor plan of the original building shows that a considerable amount of solar radiation is absorbed in the afternoon. To reduce the accumulation of solar radiation, vertical shading devices are included in the original design, as shown in Figure 4.

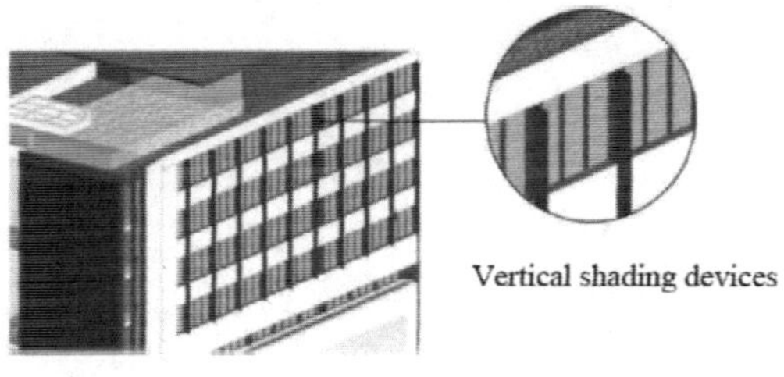

Fig4. Vertical shading devices on the original building.

In DesignBuilder, division is implemented based on the actual building design parameters and functions, and modelling is performed for the selected typical floor plan.

https://doi.org/10.1051/e3sconf/202123703006

The properties used in each area are defined, as shown in Figure 5(a). Simulation analysis on the indoor thermal comfort of the original building indicates that the major factor that influences indoor thermal comfort is the excessive accumulation of solar radiation through the facade, particularly through the external window, as shown in Figure 5.

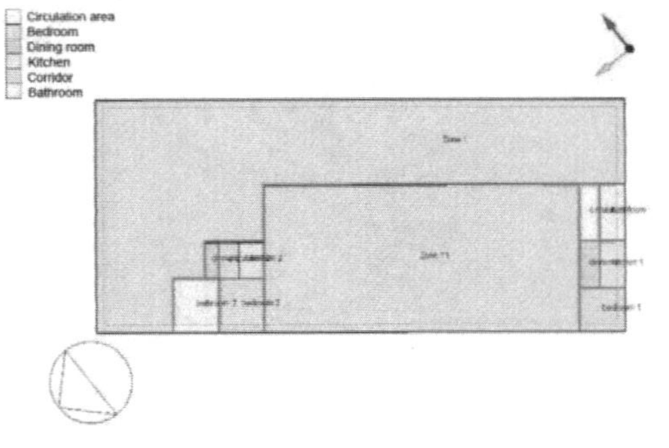

(a) Division of different areas based on a typical floor model

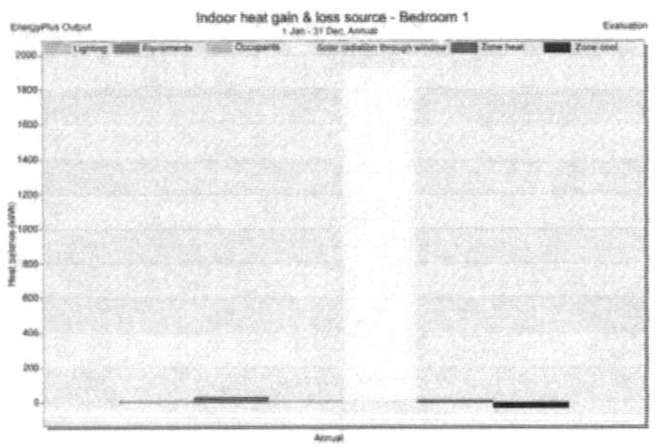

(b) Bedroom 1 Sources of heat gain and loss analysis

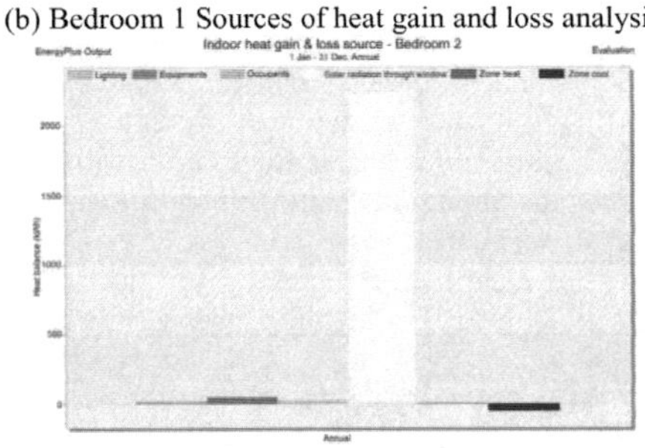

(c) Bedroom 2 Sources of heat gain and loss analysis

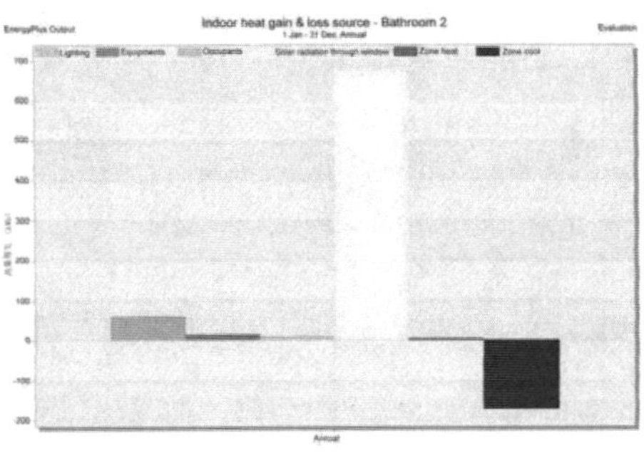

(d) Bathroom 2 Sources of heat gain and loss

Fig5. Simulation model of the original building and heat gain and loss.

On the basis of the building orientation and using unchanged functions, optimization design is implemented on elements of the typical floor facade, as shown in Figure 6. Optimization design includes replacing the common single glass with Low-E double glass for the external window, adding horizontal shading devices to the vertical shading devices, forming integrated shading devices and replacing facade materials to decrease the heat conductivity coefficient.

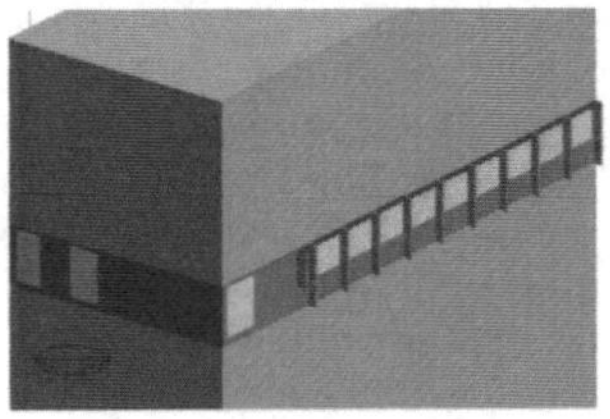

(a) Facade before optimization

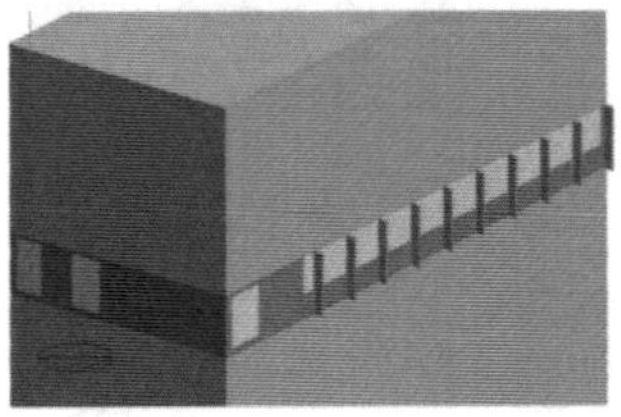

(b) Facade after optimization

Fig6. Facade before and after optimization.

4 Passive design strategy on the facade

The aforementioned simulation analysis of the original building indoor thermal comfort shows that heat that causes indoor thermal discomfort is mostly due to direct solar radiation through the facade. The optimization of passive design can effectively improve the heat-insulating performance of the external part of the building. The influence of indoor overheating due to direct solar radiation on indoor thermal comfort can be enclosed and reduced.

4.1 Changing window glass type

The original building uses a common single glass, which is similar to those used in most local residential buildings. Common single glass exhibits an evident shortcoming of having a high U value (5.88 W/m^2•K) and cannot effectively block heat brought by long-wave light[9]. To reduce heat gain due to solar radiation, the optimization design adopts a Low-E glass coating, which demonstrates excellent thermal barrier effect and outstanding light-admitting quality with a high reflection of medium-distance and far infrared rays. After optimization (Figure 7), the air layer of Low-E double-layer glass is 13 mm thick, and the glass is 6 mm thick. The heat conductivity coefficient of the external window is reduced from 5.88 W/m^2•K to 1.67 W/m^2•K when the common single glass is replaced with Low-E double glass. As shown in Table 1, the heat gained through the external window after optimization is reduced, and the total time of indoor thermal discomfort for the entire year decreases from 6286.5 h to 5840.5 h, which accounts for 7.1%.

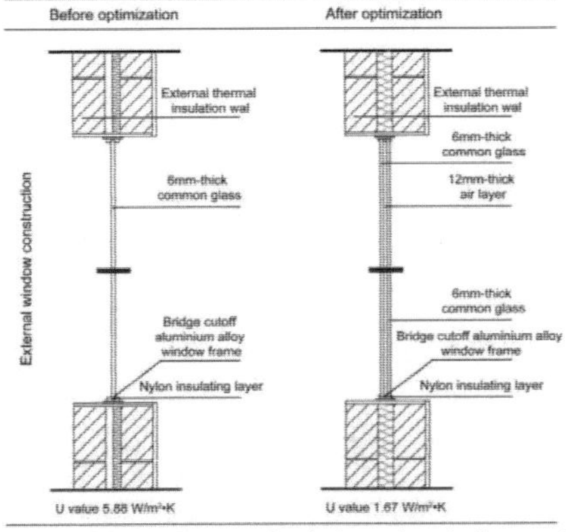

Fig7. Comparison of external window glass before and after optimization.

Table1. Comparison of time of indoor thermal discomfort.

under different external glass conditions (unit: h)

Space	Before optimization	After optimization
Bedroom1	945.5	810
Bedroom2	335	220
Dining room1	677.5	592.5
Dining room2	639.5	554.5
Kitchen1	531.5	498.5
Kitchen2	563	564
Corridor1	664	665
Corridor2	936	896.5
Bathroom1	619	639
Bathroom2	375.5	400.5
Total	6286.5	5840.5

4.2 Addition of shading devices

In addition to changing the external window glass type, installing reasonable shading devices on the external wall can also reduce the negative influence of solar radiation on indoor thermal environment. The simulation analysis indicates that using only the vertical shading method cannot provide a satisfactory shading effect; therefore, horizontal shading devices are added to the optimization design. Consequently, integrated shading is adopted. As shown in Figure 8, the horizontal and vertical shading baffles have a ledge size of 500 mm. With the adoption of integrated shading based on the software simulation statistics, the total time of indoor thermal discomfort for the entire year is reduced by 186 h, which accounts for 3%.

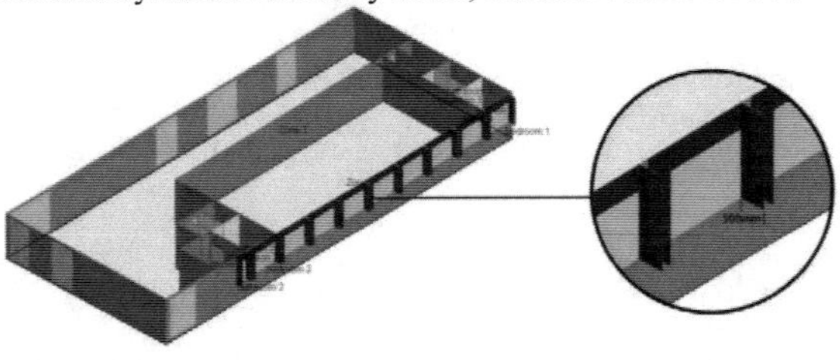

Fig8. Application of integrated shading devices on the facade after optimization.

4.3 Optimization of building facade wall materials

In addition to the window glazing and shading devices, another important element of the facade is the external wall. The entire building adopts a reinforced concrete structure, with double brick external walls and external thermal insulation. In the original building design, a 25 mm asbestos board insulating layer and a 25 mm air layer are adopted. With an increase in thermal insulating layer thickness, the thermal performance of the external wall can be improved to a certain degree. The thickness of the asbestos board insulating layer is increased from 25 mm to 50 mm, thereby canceling the air layer, as shown in Figure 9. Hence, the thermal performance of the external wall is considerably improved. The simulation analysis on indoor thermal comfort after optimization indicates that the total time of indoor thermal discomfort for the entire year is decreased by 874 h, which accounts for 14%.

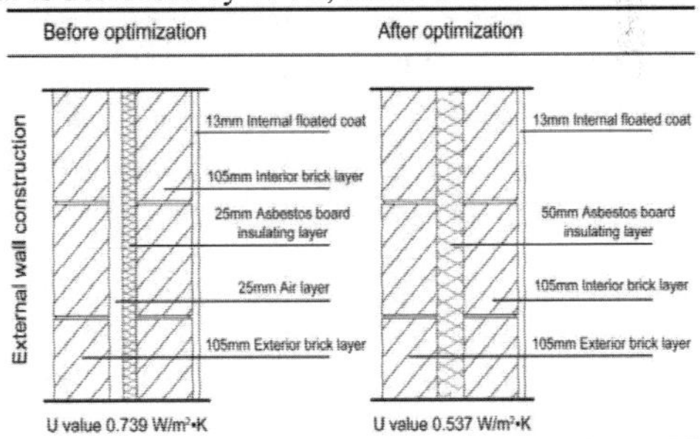

Fig9. Comparison of the exterior wall structure before and after optimization.

5 Conclusion

With the three passive designs on building facades, namely, double-layer Low-E glass, shading devices and exterior wall materials, the total time of indoor thermal discomfort for the entire year is reduced by 1,506 h, which accounts for 24%. When double-layer Low-E glass is used, the total discomfort time is reduced by 446 h (accounting for 29.6% in total reduction time). When shading devices are added,

 https://doi.org/10.1051/e3sconf/202123703006

the total discomfort time is reduced by 186 h (accounting for 12.4% in total reduction time). When exterior wall materials with a low heat conductivity coefficient are used, the total discomfort time is reduced by 874 h (accounting for 58% in total reduction time). Therefore, the choice of external wall materials exerts a considerable influence on indoor thermal comfort. Furthermore, the indoor thermal comfort of existing buildings can also be improved by changing the external window glass type and adding shading devices. The results of this research provide a technical reference for local residential building design and construction.

Acknowledgments

This work has been supported by Jiangsu province university outstanding science and technology innovation team project.

References

1. .A.R. Lewis, E. David, Int. J. of Facility Management. (2010) , 1(2):145-156.
2. J Xiao, J Zhou , F Xu et al, J. of Central South University. (2012) , 43(9):3693-3697.
3. ZF Wu, LD Chen, J. of Ecology. (2016) , 35(5): 1364-1371.
4. HH Yin, Passive Design Strategy of Public Building Transition Space of climate in Areas Warm in Summer and Cold in Winter.D. Chongqing: Chongqing University, (2010)
5. Climate Data of Sydney. http://gb.weather.gov.hk/wxinfo/climat/world/chi/australia/au/sydney_c.htm[EB/OL].
6. ASHRAE Standard 55- (2013) . Bondi Beach Climate, Averages And Extreme Weather Record[S].
7. A.Aksamija, ABER.J. (2015) , 10 (2): 240-262.
8. A.Aksamija, High-Performance Building Envelopes: Design Methods for Energy Efficient Facades.M. (2015)
9. HR Xie, Theoretical Analysis and Experimental Study on Energy Consumption of Filmed Window Glass .D. Changsha: Hunan University, (2007)

Study on compatibility optimization of resin-based permeable pavement bricks

Menghao Luo [1], Xiong Zhang [1,*], Heng Zhang [2], Xiaofu Wang [1], and Rongrong Du [1]
[1]Key Laboratory of Advanced Civil Engineering Materials of Ministry of Education, Tongji University, Shanghai, China
[2]CCCC (Shanghai) Harbor Engineering Design and Research Institute Co., Ltd, Shanghai, China

Abstract: The permeable pavement plays an important role in mitigating urban flooding. In order to explore the relationship between materials and properties to better guide the practical production of resin-based permeable bricks, 10 kinds of representative aggregate samples with obvious different characteristics were selected for preparation. In this study, Image Pro Plus was used to binarize the acquired image pictures of the aggregate so as to obtain particle group characteristic parameters. The properties and porosity of the brick were measured in order to describe the influence of the material. The results are as follows. The relative standard deviation of aggregate and the amount of cementing material are negatively related to the compressive strength of permeable bricks, but positively related to water permeability and filtration performance. The roundness and roughness of the aggregate are the opposite. Furthermore, the porosity of the permeable brick is the essential reason for this phenomenon, that is, as the porosity increase, the compressive strength decrease, but the water permeability and the filtration performance become better. In the end, an optimization method for the compatibility of resin-based pavement permeable bricks was proposed through reflecting all factors in a two-bit flat grayscale image which can be applied in performance prediction and guidance of material selection.

1 Introduction

The Sponge City Project plays an important role in alleviating "urban heat island effect", saving water and improving the ecological environment[1]. In the process of sponge city construction, permeable pavement as an important construction facility has become an important part of the sponge city construction process[1,2]. Due to the high strength, good permeability and long-term effectiveness of resin-based permeable bricks, it has caused extensive research by scholars[3].

Wang Pingping[4] et al. used epoxy resin GCC135 and curing agent GCC137 to bond aeolian sand to produce a permeable brick with a compressive strength of 41.1MPa and a water permeability coefficient of 2.1×10^{-2}cm/s; Zhang Weimin[5] used the cemented material with aeolian sand as aggregate to develop a permeable brick with a compressive strength of up to 82.3MPa and a water permeability coefficient of more than 2.0×10^{-2}cm/s. However, neither of them has conducted in-depth research on the relationship between aggregate particle group parameters and permeable brick performance. KayWille[6] et al. used the recommended equations involving matrix strength, total porosity, aggregate size and average pore size to predict the compressive strength of permeable materials. However, the study did not comprehensively consider the relationship between raw material parameters and various properties of permeable materials. The performance of resin-based permeable pavement bricks is excellent, but the long-term effectiveness of resin-based permeable bricks and the compatibility optimization of raw materials need to be studied, which restricts its application in actual production.

Based on this, on the basis of exploring the relationship between aggregate particle parameters and the amount of cementing material and the performance of resin-based permeable bricks, this paper discusses its internal influence mechanism, and considers the optimization relationship between the aggregate and cementing material, which can guiding the preparation of resin-based permeable pavement bricks with excellent performance.

2 Materials and Methods

2.1 Materials

2.1.1 Aggregate

In this paper, 10 kinds of representative aggregate samples with obvious different characteristics (particle size, RD, RG, etc.) were selected. The particle gradation of them is shown in Fig.3.

2.1.2 Resin

AB-1 type epoxy resin (two-component), its component A is bisphenol A epoxy resin, component B is amine curing agent 593. The epoxy is cured in a dry environment at room temperature, and the 1d strength can reach 90% of

* Corresponding author: xiong.zhang@vip.sina.com

 https://doi.org/10.1051/e3sconf/202123703007

the maximum value. The performance is shown in Table 1.

Table 1 Properties of AB-1 epoxy resin

Viscosity (20°C)/(mPa·s)	Compressive strength/MPa	Tensile strength/MPa	Bond strength/MPa	Elongation/%
60—100	≥35	>15	>3.0	8

2.1.3 Kaolin powder

The kaolin powder was used to prepare simulated flow path materials for water filtration performance test, the gradation requirements are shown in Table 2.

Table 2 Gradation of Kaolin powder

Particle size/mesh	Mass ratio/%
200~400	25
140~200	20
100~140	25
70~100	25
40~70	5

2.2 Sample preparation

According to the formula, weigh the aggregate and resin in the mortar mixer, mix it evenly, pour it into the mold and place it in the press at 2.5MPa for 90s. The molding size is 160.0mm×132.0mm×17.3mm for permeable test and 40.0mm×40.0mm×160.0mm for strength test, it can be removed after 24 hours of natural curing to obtain a sample of permeable pavement brick.

2.3 Test method

2.3.1 Particle group characteristic parameters

This article uses an optical stereo microscope and a Sony ILCE-60002430 million pixel camera to acquire images of 10 kinds of aggregates (respectively marked as S1 to S10). The results are shown in Fig.1.

Image Pro Plus is used to binarize the acquired image pictures to extract the sand grain contour and convert the microscope image into a black and white binary image[7,8], as shown in Fig.2. After the image is measured by the software, the required parameter value is obtained.

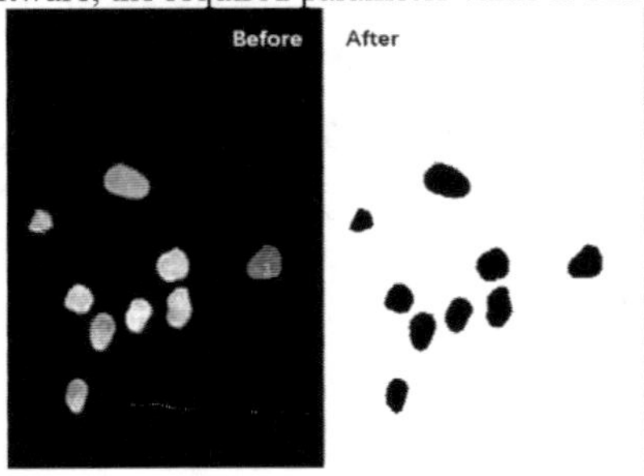

Fig.2 Microscope image & black and white binary image

2.3.2 Test method for water permeability

According to the standard of "Permeable Pavement Bricks and Permeable Road Panels" (GB/T25993-2010) [9], a water permeability coefficient tester is made to test the water permeability of test blocks with a size of 160.0mm×132.0mm×17.3mm.

2.3.3 Test method for water filtration performance

According to the standard of "Sand-based Permeable Brick" (JG/T376-2012) [10], Kaolin powder and deionized water are used to prepare a standard simulated road path flow sample for water filtration performance test.

2.3.4 Test method for compressive strength

According to the standard of "Sand-based Permeable Brick" (JG/T376-2012) [10], an electro-hydraulic pressure testing machine is used to conduct strength test on the test block with a size of 40.0mm×40.0mm×160.0mm.

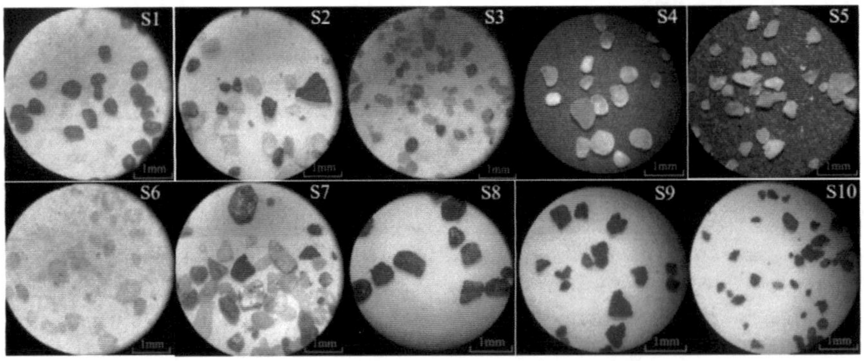

Fig.1 Photograph of 10 aggregate samples taken with a microscope lens

2.3.5 Porosity test method

Putting the test block soaked in water for 48h and then take it out. Wiping off the water attached to the surface and weigh it immediately to obtain the mass of the test block after absorbing water for 48h (m_{12}). Then put the test block in an oven at 60±5℃ to dry it, and weigh it once every 24h until the difference between the continuous weighing values is less than 0.1%, which is regarded as the quality of the dry test block (m_{22}). Measure the size of the test block, calculate the volume of the test block, and calculate the porosity of the test block according to formula (1)

$$K = \frac{m_{12}-m_{22}}{V \cdot \rho_w} \cdot 100\% \qquad (1)$$

3 Results and discussion

3.1. Effect of aggregate particle group parameters on the properties of resin-based permeable bricks

The test used 10 kinds of aggregates (S1-S10), and the resin-based permeable brick was prepared under the dosage of 4.5% epoxy resin, and their compressive strength (R), water permeability coefficient (V) and water filtration frequency (n) was measured. Three preferred aggregate particle characteristic parameters: relative standard deviation (α), roundness (R_D) and roughness (R_G) were used to discuss the macrophysics of aggregate gradation, shape and surface structure.

3.1.1 Influence of Aggregate Gradation on Properties of Resin-based Permeable Brick

The sieve residue of 10 kinds of aggregates obtained by the vibration sieve test is shown in Fig. 3.

It can be seen from Fig.3, the gradation of 10 kinds of aggregates is mainly concentrated above 0.4mm, and the distribution is different, but the frequency distribution of the number of particles is basically close to the modified Gaussian distribution. The modified normal distribution density function is expressed as:

$$y = y_0 + \frac{A}{w\sqrt{\pi/2}} e^{-2\frac{(x-x_c)^2}{w^2}} \qquad (2)$$

x_c—the most likely particle size,

w—the standard deviation of the distribution.

In addition, the relative standard deviation of the distribution is calculated by the formula as follows:

$$\alpha \equiv \frac{w}{x_c} \qquad (3)$$

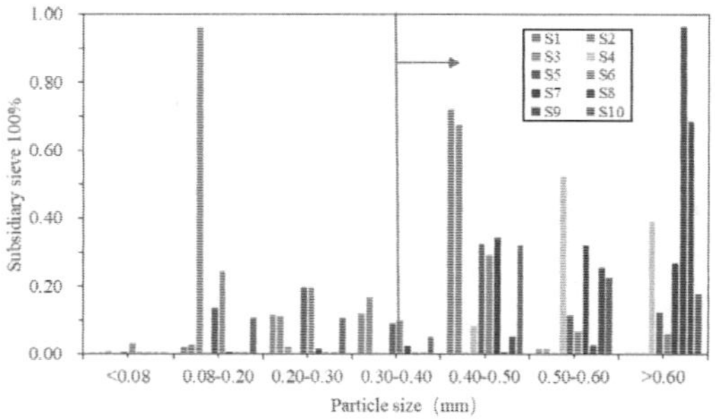

Fig.3 The sieving surplus of 10 kinds of aggregates

Using the modified Gaussian distribution to fit the particle size of 10 kinds of aggregate samples to indicate their particle size distribution, and then calculate the value of the relative standard deviation through formulas (2) and (3) to characterize the distribution of the particle group.

Table 3 shows the fitting parameters of the gradation normal distribution of the 10 aggregates obtained by the fitting calculation.

Figure 4 shows the relationship between the relative standard deviation of aggregate and the performance of permeable bricks.

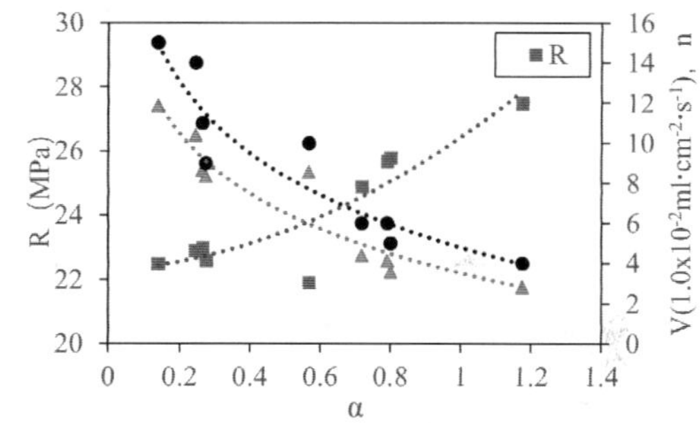

Fig.4 Relationship between aggregate relative standard deviation and permeable brick performance

The characteristic parameters of the aggregate can affect the thickness of the resin layer and the number of cured joints, thereby affecting its properties[11]. Therefore, the greater the relative standard deviation of the aggregate, the lower the porosity, the greater the number of curing contact, resulting in the higher strength of the permeable brick, but the permeability and water filtration performance is lower under the same type and amount of epoxy resin, which are shown in Fig.4.

Table 3 Fitting parameters of normal distribution of gradation of 10 kinds of aggregates

Number	S_1	S_2	S_3	S_4	S_5	S_6	S_7	S_8	S_9	S_{10}
w	0.066	0.0735	0.1329	0.2746	0.2419	0.131	0.2098	0.0697	0.1042	0.0823
x_c	0.25	0.2686	0.1661	0.4844	0.3059	0.1826	0.1785	0.5026	0.4273	0.3616
α	0.264	0.2736	0.8001	0.5668	0.7907	0.7174	1.1753	0.1386	0.2438	0.2276

3.1.2 Effect of aggregate shape and surface structure on the properties of resin-based permeable brick

With reference to the research on the morphological characteristics of coarse aggregates by Li Xiaoyan[12] et al, the R_D and R_G were selected to characterize the particle shape and surface structure of the resin-based permeable brick aggregate. The data of R_D and R_G obtained by image analysis is shown in Table 4 and Table 5.

This paper calculates the average R_D and R_G according to the particle weight of each particle size in each aggregate sample, and uses the gray correlation analysis method to analyze the correlation between the parent sequence of ten aggregates (macroscopic performance index of resin-based permeable pavement bricks) and the child sequence (R_D and R_G of aggregates) is obtained. Table 6 shows the gray correlation calculation results.

Table 6 Gray correlation calculation of 10 kinds of aggregates

Number		S1	S2	S3	S4	S5
R_D	R	-0.473	-0.505	-0.512	-0.572	-0.542
	v	+0.645	+0.663	+0.672	+0.593	+0.677
	n	+0.732	+0.742	+0.706	+0.634	+0.731
Number		S6	S7	S8	S9	S10
R_D	R	-0.443	-0.562	-0.508	-0.522	-0.487
	v	+0.682	+0.634	+0.652	+0.663	+0.652
	n	+0.752	+0.714	+0.682	+0.742	+0.670
Number		S1	S2	S3	S4	S5
R_G	R	-0.531	-0.587	-0.575	-0.524	-0.543
	v	+0.563	+0.592	+0.631	+0.631	+0.621
	n	+0.602	+0.578	+0.654	+0.628	+0.583
Number		S6	S7	S8	S9	S10
R_G	R	-0.507	-0.473	-0.505	-0.512	-0.542
	v	+0.531	+0.645	+0.663	+0.572	+0.630
	n	+0.540	+0.732	+0.742	+0.554	+0.617

From the results in Table 6, it can be seen that the effects of R_D and R_G on resin-based permeable pavement materials are isotropic, negatively correlated with compressive strength, and positively correlated with water permeability coefficient and filtration times. That is, the rounder the aggregate shape and the rougher the surface, the more the permeable brick prepared tends to have a larger water permeability coefficient and higher water filtration performance, but the compressive strength is lower. It is speculated that the R_D and R_G will affect the porosity of the permeable brick, which in turn will affect the performance of the material.

3.2 The effect of cement material on the performance of resin-based permeable brick

The three macroscopic properties of resin-based permeable bricks with the amount of epoxy resin (EP) ranging from 4.0% to 6.0% with a gradient of 0.5% are discussed. The selected resin is modified epoxy E-44, curing agent 593, and the aggregates are S9 (A) with better permeability, S4 (B) with better comprehensive performance and S6 (C) with better strength. Fig. 5 shows the relationship between the compressive strength (R) water permeability coefficient (V) of resin-based permeable bricks and the amount of epoxy resin (EP).

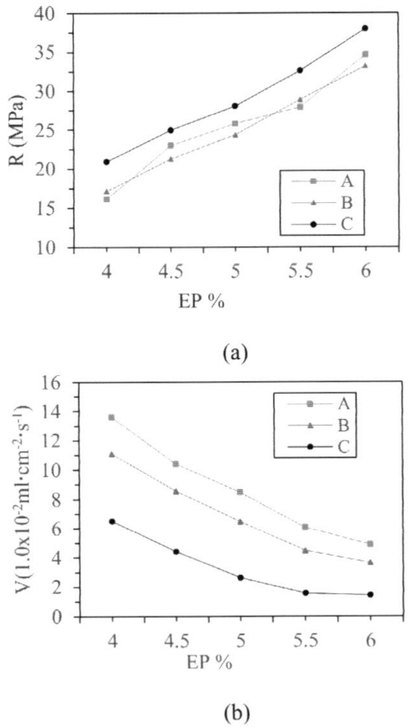

Fig.5 Relationship between EP, R and V

Table 4 Raw data of R_D analysis of 10 kinds of aggregate sand samples

Particle size/mm	R_D									
	S1	S2	S3	S4	S5	S6	S7	S8	S9	S10
<0.1	0.902	0.925	0.912	0.931	0.924	0.925	0.955	0.886	0.922	0.925
0.1-0.2	0.883	0.950	0.922	0.911	0.895	0.926	0.940	0.872	0.915	0.930
0.2-0.3	0.878	0.959	0.905	0.902	0.932	0.914	0.925	0.865	0.913	0.892
0.3-0.4	0.896	0.955	0.901	0.889	0.866	0.895	0.905	0.890	0.899	0.897
0.4-0.5	0.882	0.946	0.882	0.916	0.824	0.872	0.904	0.866	0.903	0.899
0.5-0.6	0.881	0.920	0.887	0.921	0.901	0.830	0.914	0.879	0.912	0.912
Average	0.887	0.945	0.907	0.928	0.898	0.889	0.895	0.878	0.910	0.902

Table 5 R_G of 10 kinds of aggregate sand samples

Particle size/mm	R_G									
	S1	S2	S3	S4	S5	S6	S7	S8	S9	S10
<0.1	1.034	1.012	1.022	1.013	1.010	1.016	1.018	1.066	1.021	1.034
0.1-0.2	1.032	1.018	1.027	1.022	1.018	1.023	1.016	1.069	1.016	1.038
0.2-0.3	1.040	1.023	1.033	1.021	1.024	1.027	1.015	1.073	1.013	1.043
0.3-0.4	1.046	1.026	1.028	1.012	1.031	1.028	1.010	1.083	1.022	1.042
0.4-0.5	1.033	1.017	1.032	1.018	1.023	1.026	1.011	1.075	1.024	1.033
0.5-0.6	1.032	1.013	1.021	1.013	1.026	1.024	1.014	1.080	1.018	1.037
Average	1.044	1.009	1.035	1.011	1.026	1.020	1.014	1.080	1.021	1.042

It can be seen from Fig.5(a), as the EP increases, the compressive strength of the three types of aggregate resin-based permeable bricks gradually increases. As shown in Fig.5(b), as the EP increases, the water permeability coefficient gradually decreases, and its trend is exactly opposite to the change trend of strength. From the mechanism analysis, with the increase of the epoxy resin content, the surface of the aggregate can be fully wetted by the epoxy resin, the resin layer between the aggregates becomes thicker and some pores are filled, the bonding area and the number of contact points Increase, the overall porosity of the resin-based permeable brick decreases, the original connected pores will gradually decrease and become unconnected, the compressive strength increases but the permeability coefficient decreases accordingly[13].

The influence of EP on the water filtration performance of resin-based permeable bricks is shown in Fig.6.

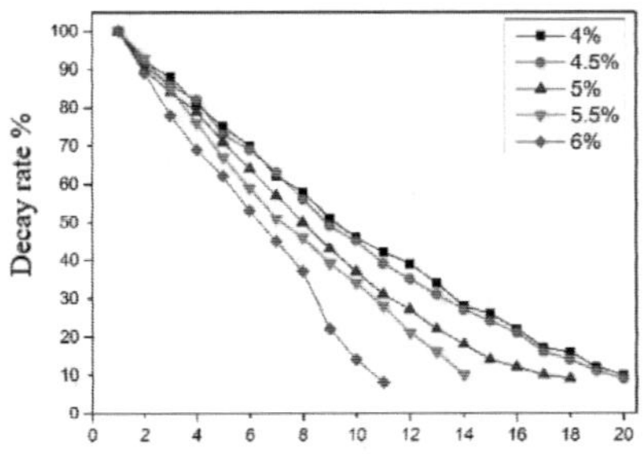

Sample (A)

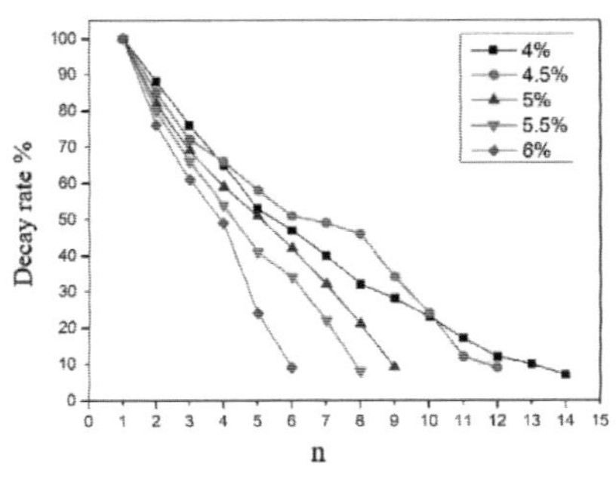

Sample (B)

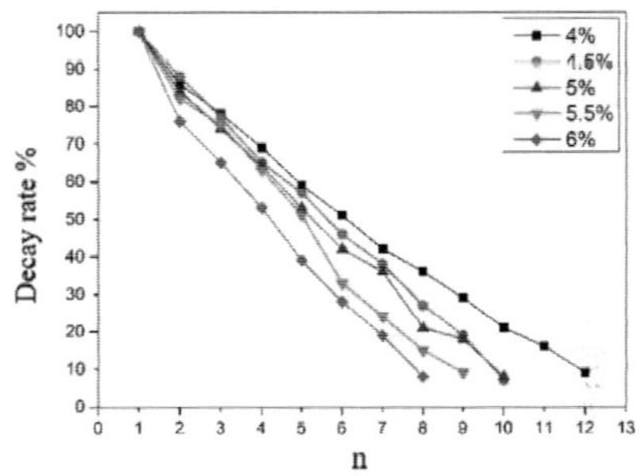

Sample (C)

Fig.6 Effect of EP on the water filtration performance of resin-based permeable bricks

It can be seen from the Fig.6 that the change trend of the water permeability coefficient is similar. The number of water filtration decreases with the increase of the epoxy resin content, and the higher the resin content is, the faster the water permeability declines. This is because when the epoxy resin is added in a large amount, the size and number of the permeable pores are reduced, and it is easily blocked by smaller impurities, the attenuation of the permeable performance is accelerated, and the number of times of water filtration is reduced.

3.3 Relationship between porosity, aggregate particle group parameters, resin dosage and properties of resin-based permeable bricks

3.3.1 Effect of different aggregates on porosity of resin-based permeable brick

Using ten kinds of aggregates (S1-S10), a set of resin-based permeable bricks with an epoxy resin content of 5.0% were prepared, and the porosity (K) of each set of test blocks was calculated according to formula (1) And take the average for analysis.

The characteristic parameters of the three particle groups are now integrated as a single parameter for discussion. According to the aforementioned research, the value of α has a positive correlation with R, and has a negative correlation with V and n; while the R_D and R_G

have a negative correlation with R, which has a positive correlation with V and n. Thus, a parameter Q describing the aggregate morphology level is defined, which is used to indicate the degree of influence of any aggregate particle group parameter on the performance of the resin-based permeable brick.

$$Q = \frac{\alpha}{R_D \cdot R_G} \tag{4}$$

According to the influence trend of the three aggregate parameters on the performance of resin-based permeable bricks, it can be seen that the larger the Q of the aggregate, the more the resin-based permeable bricks tend to have a higher strength; the worse the permeable performance and filtering performance of resin-based bricks. Table 7 illustrates the original data of the particle group characteristic parameters, Q and K of the 10 kinds of aggregates. The relationship between the Q and K of the test block is shown in Fig.7.

Table 7 Characteristic parameters and Q value of 10 types of aggregates, and K value of test block

Characteristic Parameters				Q	K/%
Number	α	R_D	R_G		
S_1	0.2640	0.887	1.044	0.272	18.72
S_2	0.2736	0.945	1.009	0.416	19.34
S_3	0.8001	0.907	1.035	0.693	16.66
S_4	0.5668	0.928	1.011	0.595	15.27
S_5	0.7907	0.898	1.026	0.885	5.52
S_6	0.7174	0.889	1.02	0.791	7.81
S_7	1.1753	0.895	1.014	1.087	3.92
S_8	0.1386	0.878	1.08	0.134	28.35
S_9	0.2438	0.91	1.021	0.225	29.48
S_{10}	0.2276	0.902	1.042	0.246	24.91

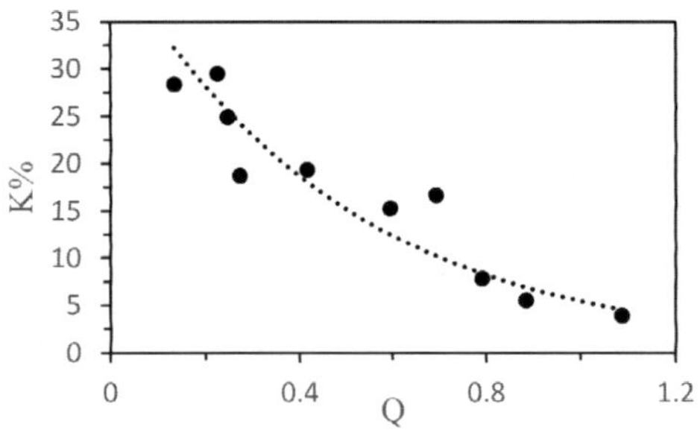

Fig. 7 Relationship between Q and K

It can be seen from Fig.7, as the Q increases, the porosity of the test block gradually decreases. From the perspective of microscopic mechanism, for permeable bricks, at a certain amount of resin, the larger the α is, the greater the packing density and the smaller the pores formed, so the smaller the porosity shows. The lower the R_D of the aggregate particles is , the finer the pores generated after curing and molding shows, that is, the greater the R_D is , the greater the porosity shows; and the greater the R_G of the aggregate is, the greater the degree of unevenness of the surface shows. When molding with the same amount of resin, the larger the amount of resin consumed on the surface of the aggregate and the smaller the amount of resin between the voids, the greater the porosity, that is, the greater the R_G, the greater the porosity. In summary, the larger the Q is, the smaller the porosity of the resin-based permeable brick shows.

3.3.2 Effect of different epoxy resin dosage on porosity of resin-based permeable brick

The value of EP discussed in this article ranges from 4.0% to 6.0%. Aggregate S4 with a moderate Q value and epoxy resin with dosage of 4.0%, 4.5%, 5.0%, 5.5% and 6.0% were used to prepare resin-based permeable bricks, and the porosity was measured. The relationship between the porosity and the epoxy resin EP is shown in Fig.8.

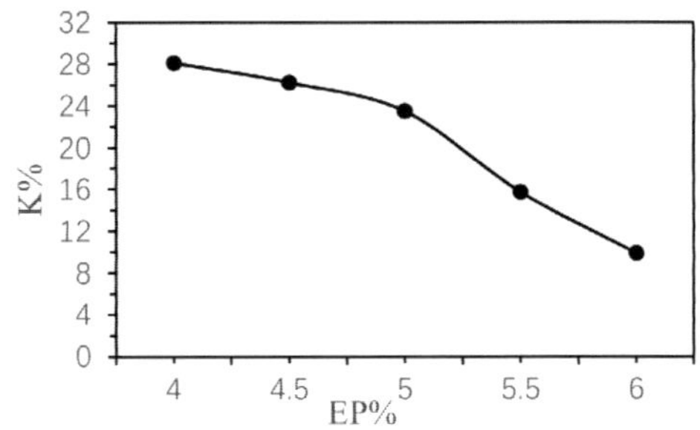

Fig.8 Relationship between EP and K

It can be seen from Fig. 8, as the EP increases, the porosity of the test block decreases. This is because, with the increasing amount of epoxy resin, the surface of the aggregate particles is gradually coated or even partially filled with resin, resulting in a decrease in the porosity. Throughout the process, the trend is to gradually reduce the porosity of resin-based permeable bricks.

3.3.3 Analysis of the relationship between the porosity of resin-based permeable brick and its macroscopic properties

By measuring K, R, V and n, the relationship between the porosity of the resin-based permeable brick and three macroscopic properties was discussed. Ten kinds of aggregates (S1-S10) were used to prepare resin-based permeable brick test blocks under the formulations of epoxy resin with EP values of 4.0%, 4.5%, 5.0%, 5.5% and 6.0% for porosity testing and its macroscopic performance. For some samples of resin-based permeable bricks whose macro performance does not meet the standard, remain the samples whose R≥20MPa, V≥0.025cm/s, and n≥8.

Fig.9 shows the results that as the porosity of the sample increases, its R shows a general trend of decline, while V and the n show an upward trend. The greater the porosity of the material is, the smaller its packing density shows, the smaller the R is, the larger the diameter of the permeable channel pores shows, the more favorable for the

https://doi.org/10.1051/e3sconf/202123703007

permeable performance, and the same for the curing of epoxy resin, the greater the porosity, the less of solidified contact points is ,the better of the water penetration performance shows and less of the strength exists[14].\.

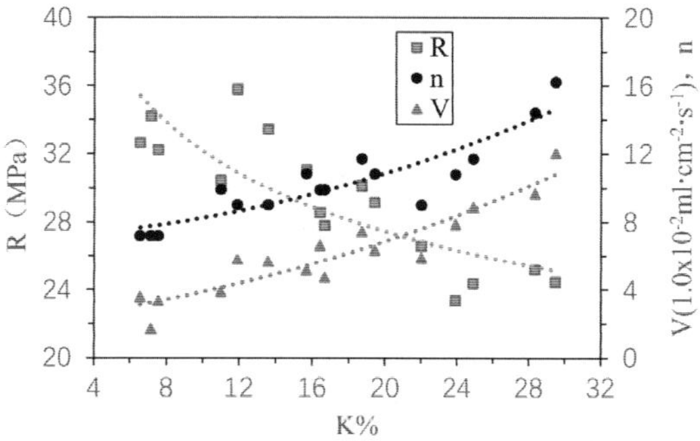

Fig.9 Relationship between K and macro performance

For resin-based permeable bricks, the porosity of material samples whose performance meets industry standards ranges from 6% to 30%. Whose R⩾30Mpa is divided into high-strength resin-based permeable bricks, and its porosity is in the range of 6.00%-15.74%. Whose V ⩾0.08ml·cm^{-2}·s^{-1} is divided into high permeability resin-based permeable bricks[9,10], and its porosity ranges from 23.96% to 30.00%. When the porosity of the brick is in the range of 15.74%-23.96%, it belongs to the comprehensive performance bricks.

3.4 Compatibility optimization of resin-based permeable pavement bricks

Based on the above conclusions, taking the Q as the x-axis, the EP as the y-axis, and the compressive strength of the resin-based permeable brick as the z-axis, the relationship of strength with Q and EP is shown in Fig.10.

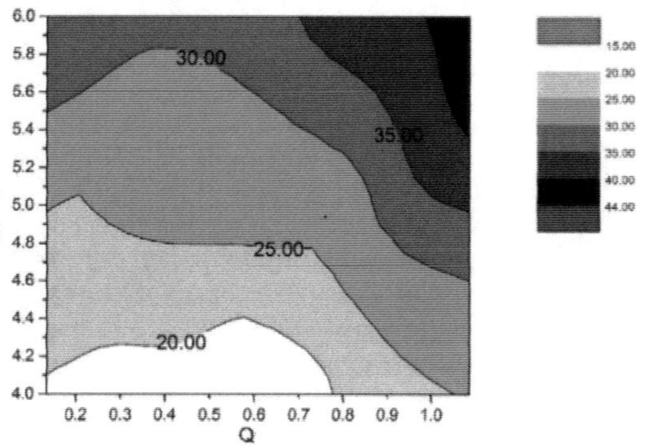

Fig.10 Relationship between Q, EP and R

From Fig.10, it can be seen that as the color of each area changes from white to gray to black, the R is getting larger. Among them, the white area represents the range of R <20MPa, which cannot meet the minimum standards of the construction industry and is unusable[9,10].

Similarly, Fig.11 is a two-dimensional grayscale graph of V (a) and n (b) of the resin-based permeable brick.

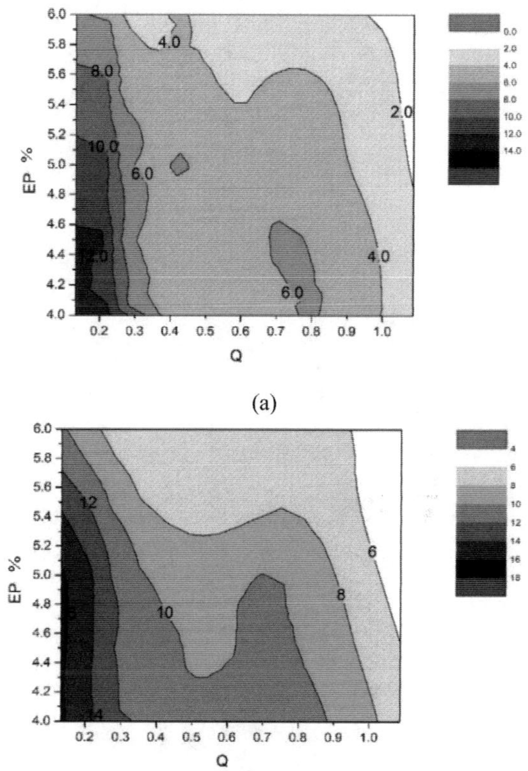

(a)

(b)

Fig.11 Relationship between Q,EP and V (a) or n (b)

The comparison shows that the change trend of Q and n are basically the same. As the color of each area changes from white to gray to black, the permeability and water filtration performance are getting higher and higher. Among them, the white area in Fig.11 (a) and the range below the contour line 8 in Fig.11 (b) are areas that do not meet the requirements and are unusable.

In summary, for resin-based permeable bricks, there is a certain degree of contradictory relationship between the compressive strength and the performance of water permeation and water filtration. Therefore, in the selection process of resin-based permeable bricks, its aggregate Q value and epoxy resin content EP value are not desirable to one end, and its usable range is where the two overlap.

Based on the above analysis, the relationship between the three macroscopic properties and the two parameters is obtained in Fig.12. The gray-scale distribution of the green, blue and red colors in the figure represents the relationship between Q, EP, R, V and n. Using green, red, and blue lines to indicate the limits of R, V, and n. So the range of available resin-based permeable bricks can be visually observed from the figure.

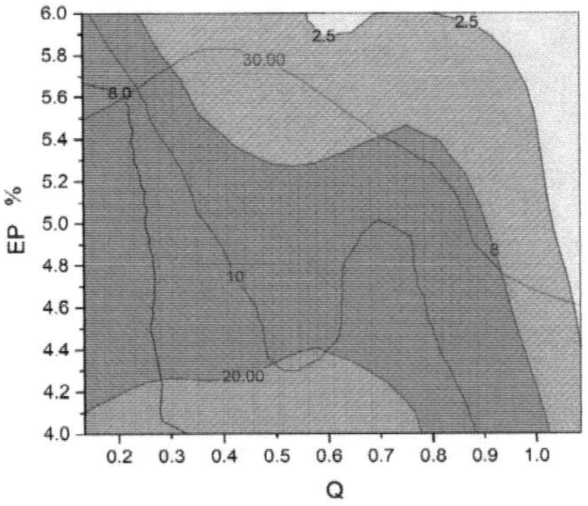

Fig.12 Model of coordination relationship of resin-based permeable brick system

The area can be divided into zone I of high-strength, zone II with excellent performance in both aspects, zone III of high permeability and zone IV with general performance but meeting industry standards. The compatibility optimization of resin-based permeable pavement bricks is shown in Fig.13.

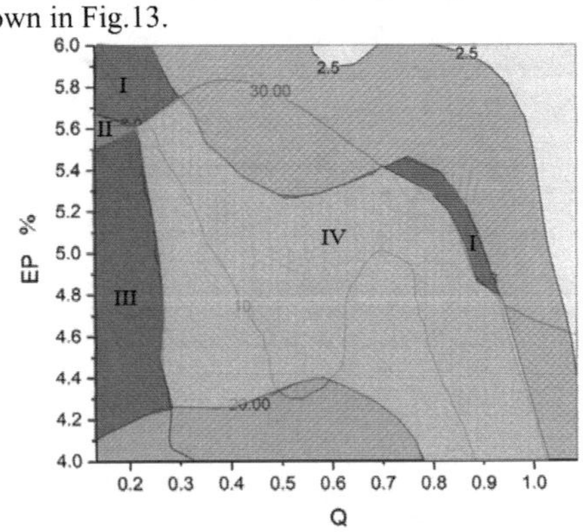

Fig.13 Compatibility optimization of resin-based permeable pavement bricks

Actually, the model can be used to predict the compressive strength, water permeability coefficient and water filtration performance of resin-based permeable pavement bricks according to the characteristic parameters and the epoxy resin dosage. In addition, raw materials can be optimal selected according to the application requirements.

4 Conclusion

(1) Under the same type and amount of epoxy resin, the smaller the relative standard deviation is, the greater the R_D exists, the greater the R_G shows, the better the water permeability coefficient and water filtration shows, but the compressive strength is lower.

(2) With the increase of epoxy resin content, the permeable brick's compressive strength increases, while the water permeability and water filtration performance decreases.

(3) Within a certain range, the smaller the α is, the greater the R_D shows, the greater the R_G exists, the smaller the amount of epoxy resin, the greater the porosity of the permeable brick, resulting in a decrease in the compressive strength of the permeable brick, and an increase in the permeability and water filtration performance; Thereby, an effective compatibility optimization of resin-based permeable pavement bricks is obtained.

References

1. Yong L, Tian L, Peng H. A new structure of permeable pavement for mitigating urban heat island[J]. Science of the Total Environment, 2018, 634:1119-1125.
2. Chaoqun Zhou. Production of eco-friendly permeable brick from debris[J]. Construction and Building Materials,2018,188.
3. Wang Ping, Li Guochang. Study on the preparation and properties of sand permeable bricks in desert [J]. Nonmetallic Mine, 2010, 33 (5): 61-63.
4. Wang Pingping. Preparation and properties of resin sand-based permeable brick [D]. Dalian University of Technology, 2013.
5. Zhang Weimin, Li Lihong, Zhu Yonggang, Wang Fang, Qin Chuan, Chen Jing. Research on sand-based permeable materials [J]. Thermosetting resin, 2012, 27(06): 57-60.
6. R. Zhong,K. Wille. Linking pore system characteristics to the compressive behavior of pervious concrete[J]. Cement and Concrete Composites,2016, 130-138.
7. ALTUHAFI F,O SULLIVAN C,CAVARRETTA I. Analysis of an image-based method to quantify the size and shape of sand particles[J]. Journal of Geotechnical and Geoenvironmental Engineering, 2013, 139 (8): 1290–1307.
8. MORA C F,KWAN A K H. Sphericity,shape factor and convexity measurement of coarse aggregate for concrete using digital image processing[J]. Cement and Concrete Research,2000,30(3):351–358.
9. GB/T 25993-2010, Permeable pavement tiles and permeable pavement panels[S].
10. JG/T 376-2012, Sand-based permeable brick [S]
11. Jiang Zhengwu, Sun Zhenping, Wang Peiming. Influence of several factors on the properties of porous permeable concrete [J]. Journal of Building Materials, 2005,8 (5): 513-519.
12. Li Xiaoyan, Bu Yin, Wang Hainian et al. Research on quantitative evaluation index of morphological characteristics of coarse aggregates [J]. Journal of Building Materials, 2015,18 (3): 524–529.
13. Zhang Xiong,Wang Xiaofu.Research progress on the influence mechanism of several factors on the performance of permeable brick[J].Materials Review:1-20,2019.
14. Wang Zhongjia, Yang Ce, Weng Daqing, Zheng Zhijie, Meng Tao, Hong Yongpeng. Study on the effect of target porosity on the performance of recycled permeable bricks[J].New Building Materials,2019,46(04):51-54.

Effect of polycarboxylate superplasticizer on fluidity and rheology of cement slurry containing silica fume

Qingen Meng[1], Juan He[2,*], Congmi Cheng [2], and Xiaofen Zhu[2]

[1] Guangzhou City Construction & Development Co. Ltd. 510085, P. R. China
[2] School of Civil Engineering, Guangzhou University, Guangzhou 510006, P. R. China

Abstract. The effect of polycarboxylate superplasticizer on the fluidity and rheology of cement - silica fume - water paste was investigated. The changes of dispersion degree, yield stress and plastic viscosity of paste with different superplasticizer content were analyzed. The results show that the rheological properties of paste with different superplasticizer content conform to Herschel-Bulkley model. The shear thinning of the slurry is manifested as a typical yielding pseudoplastic fluid characteristic. When the content of superplasticizer is less than 1.0%, the plastic viscosity and yield stress decrease and the fluidity increase with the increase of plasticizer content. When the content of superplasticizer is more than 1.0%, the yield stress decreases slightly and the plastic viscosity increases with the increase of plasticizer content. The fluidity decreases with the increase of yield stress, and there is a good correlation between them.

1 Introduction

High cement content, low water binder ratio and the use of silica fume(SF) give RPC a very dense microstructure. The inherent defects of RPC are greatly reduced, so it has ultra-high strength[1]. However, with the decrease of water binder ratio, the viscosity of cement paste gradually increases, causing a series of construction problems such as difficult mixing and pumping, which greatly limits its application[2]. The ball effect, activity effect and micro-aggregate filling effect of silica fume are used to optimize the rheological properties of cement paste, and the macro defect free structure of cement matrix is formed. However, the existence of ultrafine silica fume aggravates the formation of flocculated structure in the slurry. Polycarboxylate superplasticizer has strong electrostatic repulsion force and steric repulsion force[3], which can not only disperse flocculation structure, but also form hydration film on the surface of silica fume and cement to reduce the friction between particles.

The dispersion of polycarboxylate superplasticizer in cement- silica fume- water system was investigated. The influence of polycarboxylate superplasticizer on the dispersion degree, yield stress and plastic viscosity of cement-silica fume-water slurry was determined, and the influence law of its content on the fluidity and rheology of cement-silica fume-water system was grasped.

2 Experimental

2.1 Materials

Polycarboxylate superplasticizer, solid content of 40%, water reduction rate of 35%, Guangdong Hualixin Building Materials Technology Co., Ltd. Portland cement, P.I 42.5, Fushun Cement Co., Ltd. Silica fume, silica dioxide content of 93%, particle size of 0.1-1.0μm, density of 2.21g/cm^3, specific surface area 2.0 × 104m^2/ kg, Guangdong Qinshi new energy Co., Ltd. (Fig. 1).

Fig1. Morphology of silica fume

2.2 Mix design

In order to grasp the influence of polycarboxylate superplasticizer on the fluidity and rheology of cement paste containing silica fume, six mix proportions with different superplasticizer agent contents were prepared by keeping the silica fume content and water binder ratio unchanged. The content of silica fume is 25% and the water binder ratio is 0.19. The mix proportions are given

* Corresponding author: congmi2008@sina.com

 https://doi.org/10.1051/e3sconf/202123703008

in Table 1.

Table1. Mix proportion of composite binder paste

Sample	Content of superplasticizer/%	Mix proportion / g			
		Cement	Silica fume	Water	Superplasticizer
M3-1	0.4	750	250	190	4
M3-2	0.6	750	250	190	6
M3-3	0.8	750	250	190	8
M3-4	1.0	750	250	190	10
M3-5	1.2	750	250	190	12
M3-6	1.6	750	250	190	16

2.3 Test methods

The fluidity test is carried out according to GB/T 8077-2012. The paste was poured into a slump cone. The cone was lifted up and the spread diameter of paste was recorded as the average of two perpendicularly crossing diameters at the set time. The rheological properties of paste were measured by a model RST-CC rheometer (Brookfield Co. Ltd., USA). In order to obtain the same shear state of each paste before rheological test, the test is divided into two stages: pre-shearing and data collection. The rheological mechanism is given in Fig. 2.

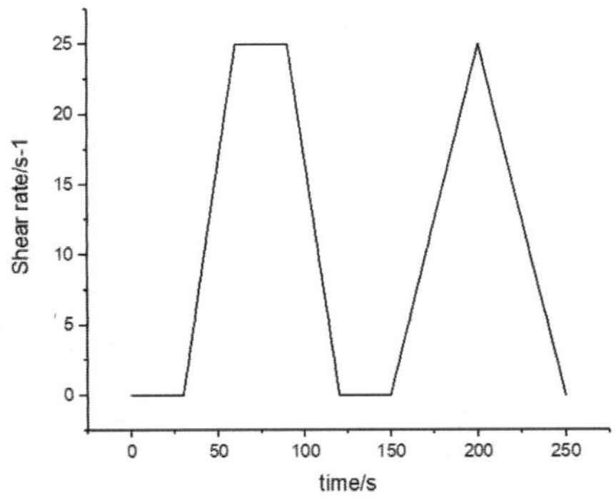

Fig2. The rheologica mechamism

3 Results and discussion

3.1 Fluidity

When the water-to-binder ratio is very low, the tiny particles will wrap water to form a flocculation structure, an there is little free water in the paste. The superplasticizer adsorbed on the surface of the particles produced electrostatic repulsion to disperse the particles and release the water wrapped in the flocculation structure. At the same time, the adsorption layer has a strong steric hindrance, which reduces the friction between particles. The combined effect improves the fluidity of the mixture[4]. Figure 3 exhibits the dispersion-time curve of cement-silica fume-water paste with different super-plasticizer content. When the content of superplasticizer is small, the fluidity increases with the increase of superplasticizer content. When the content of superplasticizer increased from 0.4% to 1.0%, the dispersion degree of paste in 150s increased from 84 mm to 159 mm, and the growth rate reached 89.3%. This is due to with the increase of the superplasticizer content, the flocculation structure was destroyed, the water was released, and the fluidity of paste increased. When the content of superplasticizer is more than 1.0%, the fluidity would not be improved, but slightly decreased. This is due to excessive additives increase the viscosity of the paste, resulting in the decline of fluidity.

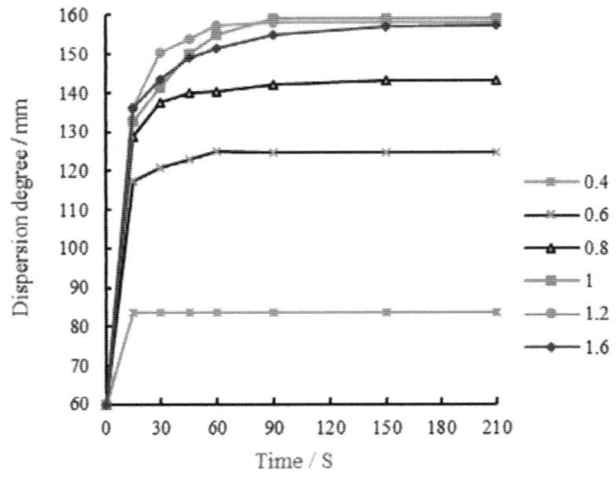

Fig3. Fluidity of cement-silica fume-water paste with different superplasticizer content

3.2 Rheology

3.2.1 Rheological curve

The rheological curve can better reflect the influence of superplasticizer on the cement-silica fume- water paste. The paste exhibited shear thinning in the rheological experiment. The downward section of the rheological curve is located below the upward section to form a hysteresis loop (Figure 4). This is due to the failure of internal microstructure in the shear process, which leads to the decrease of shear stress when the shear rate is constant [5].

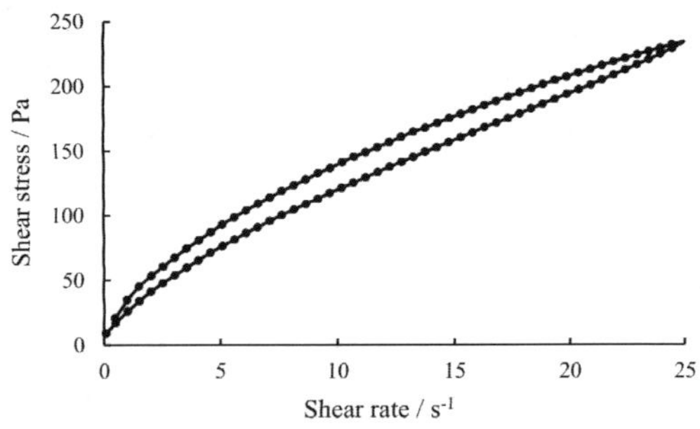

Fig4. Hysteresis loop

The data measured in the downward section of the shear curve are more reliable and repeatable. Therefore, the data of downward section is used for rheological analysis[6-8]. Figure 5 exhibits the downward section of the rheological curve of cement-silica fume-water paste with a superplasticizer content of 0.4% to 1.6%. The rheological curve of paste with 0.4% superplasticizer is different from other samples. This is because when the content of superplasticizer is small, there are a lot of flocculation structure in the paste, free water is less, the distance between particles is close, and the shear resistance is correspondingly very large.

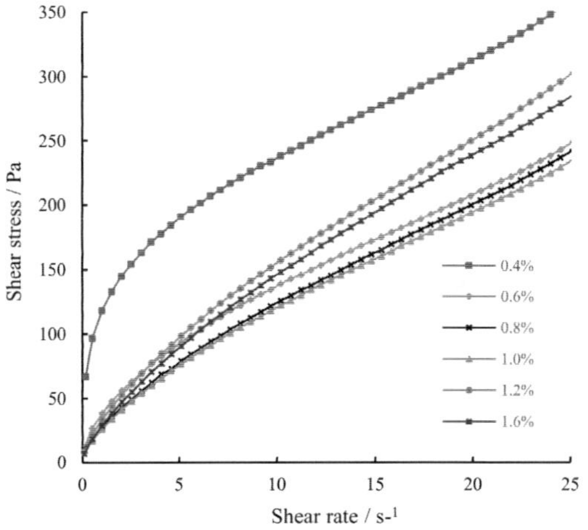

Fig5. Downward section of rheological curve of slurry with different superplasticizer content

3.2.2 Rheological model

Table2. Fitting results of rheological parameters

Sample	Content of superplasticizer/%	τ_0/Pa	m/Pa·s^n	n	R^2
M3-1	0.4	59.46	55.28	0.51	0.99447
M3-2	0.6	24.59	20.44	0.74	0.99896
M3-3	0.8	7.65	20.69	0.75	0.99959
M3-4	1.0	5.99	20.84	0.74	0.99961
M3-5	1.2	5.80	27.76	0.73	0.99951
M3-6	1.6	2.06	26.90	0.73	0.99992

Many scholars believe that the fresh cement-based composite can be regarded as a plastic fluid, which can be analyzed by Bingham fluid model [9-11]. There are also views that it is not feasible to analyze fresh mortar by Bingham model [12]. Fig. 4 presents that the rheological curves of cement-silica fume-water paste with different superplasticizer contents show typical yield pseudoplastic fluid characteristics, which are consistent with Herschel-Bulkley model. The equation $\tau = \tau_0 + m(\dot{\gamma})^n$ based on Herschel-Bulkley model was used to fit the paste with each superplasticizer content. The fitting results of rheological parameters are shown in Table 2.

It exhibits that the correlation coefficients of each rheological curve fitted by Herschel-Bulkley model are greater than 0.99. It shows that Herschel-Bulkley model can be applied to cement-silica fume-water paste containing polycarboxylate superplasticizer very well.

3.2.3 Rheological parameters

Table 2 presents the rheological index (n) of pastes with different superplasticizer contents is less than 1.00. It presents that the paste is a pseudoplastic fluid with shear thinning. This is due to the flocculation structure in the paste is broken and more free water is released. The plastic viscosity cannot be obtained directly from Herschel-Bulkley model. Ferris and de Larrard gave the empirical formula of plastic viscosity $\mu = \frac{3m}{n+2}{\gamma_{max}}^{n-1}$ through a large number of experimental studies, where γ_{max} is the maximum shear rate in the test. Figure 6 exhibits the effect of superplasticizer content on the plastic viscosity and yield stress.

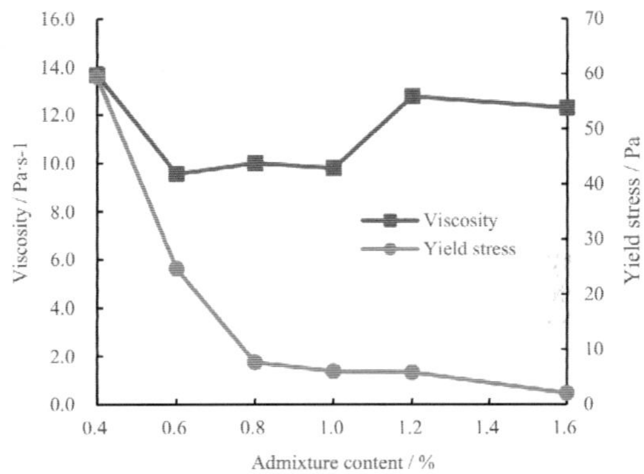

Fig6. The effect of superplasticizer content on the plastic viscosity and yield stress

When the content of superplasticizer is 0.4%, the cohesion in the paste is large, and the slip resistance between particles under shear is large, so the yield stress is very high [13]. When the content of superplasticizer is less than 1.0%, the yield stress and plastic viscosity decrease with the increase of superplasticizer content. This is due to the flocculation structure of paste was destroyed by superplasticizer. However, when the content of superplasticizer is more than 1.0%, the yield stress decreases slightly, but the plastic viscosity increases slightly with the increase of superplasticizer content. This is due to the fact that the excess superplasticizer no longer disperses the flocculation structure, but increases the viscosity of the slurry.

3.3 Correlation analysis of polycarboxylate superplasticizer on fluidity and rheology of cement paste containing silica fume

The fluidity indicates the ability of particles to flow under external force. It can be directly applied to engineering. However, it is only one index, which cannot reflect the interaction between particles in the paste. The rheological can indicate the interaction between particles under shear stress. Many rheological parameters can be obtained to describe the movement and deformation of paste under the external force from different aspects. This is a good method to characterize the workability of the mixture. However, the rheological analysis is more complex and difficult to be directly applied in engineering.

The change of rheological parameters determines the change of fluidity. Figure 7 shows that the fluidity decreases with the increase of yield stress. When the content of superplasticizer is low, the yield stress and plastic viscosity are large, and the dispersion of paste stops quickly. With the increase of superplasticizer content, the flocculation structure in the paste decreases, the hydration film on the surface of particles becomes thicker, and the yield stress decreases greatly. Hence, the dispersion resistance of the paste decreases, the dispersion time becomes longer, and the dispersion degree increases greatly. When the content of superplasticizer exceeds 1.0%, the excess superplasticizer only slightly reduces the yield stress and increases the plastic viscosity, so the dispersion degree of the paste is no longer increased. Both yield stress and fluidity indicate the interaction between particles in the paste without high-speed shear failure. There is a good correlation between them. The correlation coefficient (R2) is 0.9686 (Fig. 7).

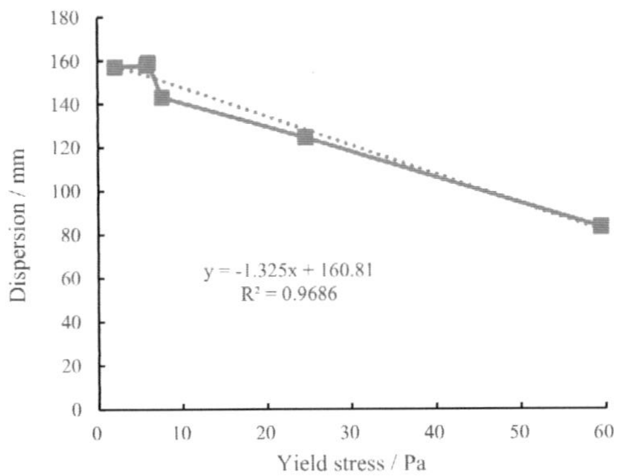

Fig7. Effect of superplasticizer on plastic viscosity and fluidity of cement paste containing silica fume

4 Conclusion

The effect of polycarboxylate superplasticizer on fluidity and rheology of cement slurry containing silica fume was investigated, and the following conclusions were drawn.

(1) The rheological curves of cement-silica fume-water paste with different polycarboxylate superplasticizer content exhibit typical yield pseudoplastic fluid characteristics. The correlation coefficients fitted by Herschel-Bulkley model were all greater than 0.99. The pastes with different superplasticizer content all show shear thinning, and the rheological index (n) is all less than 1.00.

(2) When the superplasticizer content is 0.4%, a large number of flocculation structures are formed in the cement paste. The fresh composite lacks free water and the slip resistance between particles is high. Accordingly, the yield stress and plastic viscosity are large, which makes the paste difficult to dispersed.

(3) When the content of superplasticizer increased from 0.4% to 1.0%, more flocculation structures were destroyed. At the same time, the hydration film on the surface of particles becomes thicker, the steric hindrance increases, and the slip resistance between particles decreases. The plastic viscosity and yield stress decreased and the fluidity increased. On the one hand, the friction force between the flowing particles is reduced, on the other hand, the flocculation structure is constantly destroyed when the paste moves at high speed.

(4) When the content of superplasticizer is more than 1.0%, too much superplasticizer increases the viscosity of the paste. The yield stress decreases slightly, but the plastic viscosity increases slightly with the increase of superplasticizer content.

(5) Both yield stress and fluidity indicate the interaction between particles in the paste without high-speed shear failure. There is a good correlation between them. The fluidity decreases with the increase of yield stress, and there is a good correlation between them.

References

1. C.G. Long, Y.J. Xie, P.M. Wang, Z.W. Jiang, J. Ceram Soc, 33, 4 (2005), (In Chinese)
2. V. Živica, Constr. Build. Mater, 23 (2009)
3. Y.Q. Jiang, Application basis of concrete admixtures (Chemical Industry Press, 2011), (In Chinese)
4. R.J. Flatt, Y.F. Houst, Cem Concr Res, 31 (2001)
5. M.L. Cao, L. Xu, C. Zhang, J. Ceram Soc, 44, 2 (2016), (In Chinese)
6. H.H.C. Wong, A. Kwan. W.W.S. Fung, Adv Cemt Res, 22, 1 (2010)
7. A.K.H. Kwan, W.W.S. Fung, Cem Concr Compos, 34, (2012)
8. H. Vikan, H. Justnes, Cem Concr Res, 37, (2007)
9. C. Hu, F. Larrard, Cem Concr Res, 26, 2 (1996)
10. T. Yen, C.W. Tang, C.S. Chang, K.H. Chen, Cem Concr Compos, 21, (1999)
11. J. Golaszewski, J. Szwabowski, Cem Concr Res, 34, (2004)
12. J.Z. Liu, W. Sun, Q.Q. Zhang, J.P. Liu, Concr Cem Prod, 1, (2014), (In Chinese)
13. D.P. Bentz, C.F. Ferraris, M.A. Galler, A.S. Hansen, J.M. Guynn, Cem Concr Res, 42, (2012)

Influence of Loading Peak and Loading Rate on Twin Fibers Pullout Test of Concrete Matrix

Zhang Yafang [1,*],Luo Feihua [1], Lu Juan[1], He Juan[1], Zhuang Yuhai[2]

[1]School of Civil Engineering, Guangzhou University, Guangzhou 510006, China;
[2]Guangzhou City Construction & Development Co. Ltd. 510085, P. R. China

Abstract. Numerical analysis software based on linear finnite element method and statistical damage theory concrete material has been adopted in this study. Dynamic numerical model of twin fibers pullout test of concrete matrix have been created by considering the meso-heterogeneity of concrete material. The whole process from micro-cracks initiation, propagation to crack penetration has been simulated. The influence of stress wave loading peak and loading rate on twin fibers pull-out test of concrete matrix under dynamic load has, therefore, been scrutinized. Results show that loading rate has effect on failure mode and crack propagation along interface of the twin fibers pull-out specimens. With increasing loading rates, the rate of interface crack propagation and damage area increase either.

1 Introduction

Fiber reinforced concrete is widely used in civil and structural engineering. Due to the physical and chemical interactions between the fibers and the matrix, a nanoscale thin layer is formed during the processing of the composites, which is called the interface [1].The interface between fiber and matrix in fiber reinforced concrete is an important microstructure. The efficiency of stress transfer between interfaces plays an important role in the overall properties of composites under different loading conditions[2-3]. There is a significant difference between the mechanical properties of concrete materials under dynamic load and quasi-static action[4]. At present, some scholars[5-8] consider the drawing performance experiments of deformed steel fibers with loading rate.in addition, because the bond mechanism between steel fiber and concrete is equivalent to that between steel bar and concrete[9], some scholars have analyzed the mechanism of steel fiber and concrete by studying the bond properties of steel bar and concrete at different loading rates [10-11].Aiming at the complexity of the double-wire dynamic pull-out problem, this study, considering the influence of the fiber interface properties, establishes the dynamic numerical model of the double-wire pull-out of concrete by using RFPA2D-Dynamic（Realistic Failure Process Analysis）numerical analysis software, and probes into it the influence of loading peak and loading rate on the pull-out performance of concrete double wire is discussed.

2 Numerical model

Most of the existing studies use macroscopic elastoplastic theory to describe the nonlinear behavior of concrete load damage and cracking process based on the uniformity of materials, ignoring the mesoscopic non-uniformity of concrete phase materials, it is difficult to accurately express the entire failure process of concrete complexity [12]. The calculation method of the RFPA^{2D}-Dynamic system is mainly based on linear finite element theory and statistical damage theory. The Weibull distribution function is adopted to consider the non-uniformity of the mechanical properties of the mesoscopic elements of concrete materials. The Weibull statistical distribution function expression is:

$$f(x) = \frac{m}{x_0}\left(\frac{x}{x_0}\right)^{m-1} exp\left(-\frac{x}{x_0}\right)^m \qquad (1)$$

In the formula: m is the homogeneity of the material, the greater the homogeneity, the more uniform the material; x_0 represents the average value of the mechanical properties of mesoscopic units of each phase material.The dynamic numerical model of twin fibers shown in Fig 1. The steel fibers in the model were defined as No. 1 and No. 2 respectively. The geometric parameters of the numerical model were as follows: the matrix size was 40mm × 40mm, which was divided into 102400 units（320 × 320 ） , the diameter of the steel fiber was 0.5 mm, the distance between the twin fibers was 10 mm, the embedding depth was 20 mm, and the interface size was 0.125 mm × 20 mm. The mechanical parameters and homogeneity of the materials in each phase in the numerical model shown in table 1 , and the values were referenced [12-13].

* Corresponding author: zhangyafang2004@163.com

https://doi.org/10.1051/e3sconf/202123703009

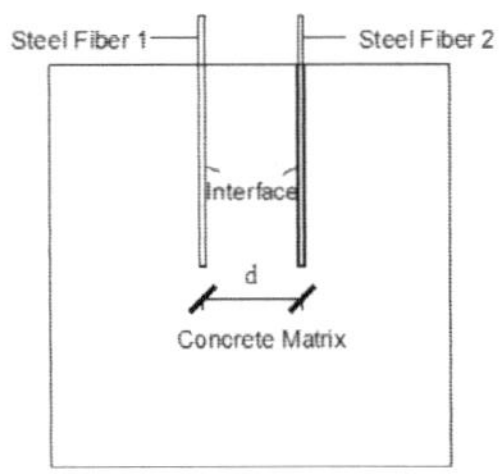

Fig1. Numerical simulation model

Table1. Mechanical parameters of numerical model

Component	Elastic Modulus/MPa	Ultimate tensile strength/MPa	Poisson ratio
Cement mortar	20400 (3)	3.2 (3)	0.25 (100)
Interface	3500 (1.5)	0.5 (1.5)	0.25 (100)
Steel fiber	210000 (15)	1000 (15)	0.30 (100)

Note: The values in parentheses represented the homogeneity of the material.

In order to explore the influence of the peak load and the change of the load rate on the performance of the twin fibers pullout of concrete matrix , a triangular dynamic stress wave as shown in Fig 2 was applied to the steel fiber in the model, with a total loading time of 20 μs. In the numerical calculation, the propagation distance of the internal wave in each time step should be smaller than the unit size. In this study, the time step was 0.1 μs, and the total number of loading steps was 200 steps.

As shown in Fig 2, in order to study the effect of loading peaks, the stress wave rise time was fixed at 5 μs, and the loading peaks were taken to be 30 MPa, 35 MPa, 40 MPa, 45 MPa, 50 MPa, 55 Mpa, for future discussion , followed by the numbers Peak 1, Peak 2, Peak 3, Peak 4, Peak 5, Peak 6. In order to study the effect of loading rate, the peak load was fixed at 50 MPa, and the rise time was taken to be 2 μs, 6 μs, 10 μs, 14 μs, and 18 μs. The effect of loading rate was considered by changing the lifting time. The larger, the number was Rate 1, Rate 2, Rate 3, Rate 4, Rate 5.

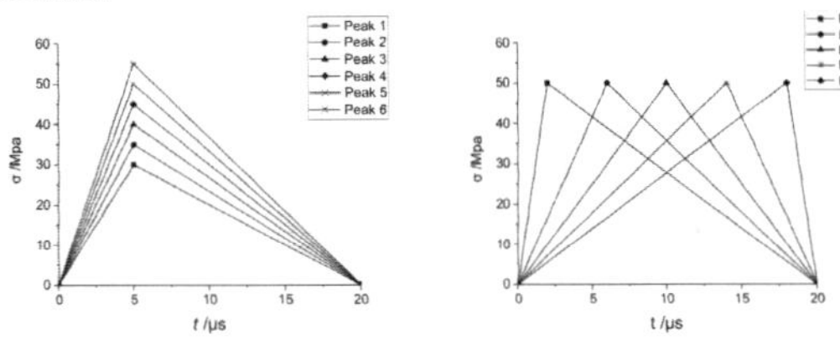

(a) Loading peak curve (b) Loading rate curve

Fig2. Dynamic loading curve

3 Results discussion

3.1 Dynamic failure mode

Peak 2, Peak 4, Peak 6, and Rate 1, Rate 2, and Rate 5 specimens are shown in Fig 3 at the time of loading time t = 10 μs and 20 μs, respectively. The stress wave in the specimen continuously propagates in the matrix in the shape of a hemisphere. The brighter the graph is, the greater the stress value is. As can be seen from Fig. 3, with the continuous propagation of stress waves, the interface crack gradually expands along the direction of fiber depth from the twin fiber embedding end (the end of the steel fiber exposed to the matrix). In general, the loading peak and loading rate have significant effects on the crack growth and failure mode of the specimen interface. When the loading peak and the loading rate are small (Peak 2, Rate 5), the load value generated by the stress wave is small, which is not enough to achieve the strength required when the interface unit is completely destroyed, the interface crack propagation is slow, the interface debonding area and the specimen amount of damage is small, the failure mode of the specimen is that the fiber is not pulled out, and the specimen still has a certain bearing capacity. With the increase of loading peak and loading rate (Peak 4, Rate 2), the amount of element damage in the matrix increases, and the interface of the twin fiber is basically debonded. The macroscopic transverse main crack of the matrix between the fiber and fiber is pulled out and destroyed.When the loading peak and the loading rate are higher (Peak 6, Rate 1), the energy released during the failure of the specimen unit is higher, the interface crack propagation speed is accelerated, the interface debonding area and the damage amount of the base unit are also significantly increased, and the interface is completely debonded ,the twin fiber is pulled out.

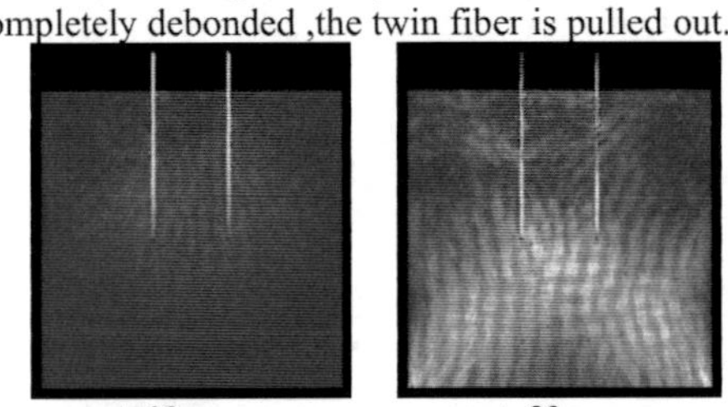

t=10 μs t=20 μs

(a) Peak 2

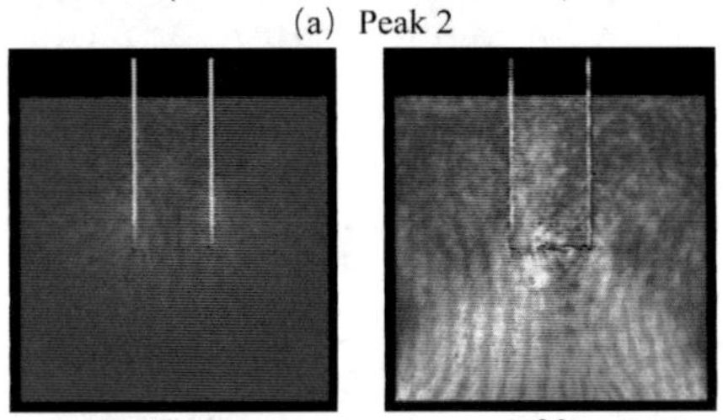

t=10 μs t=20 μs

(b) Peak 4

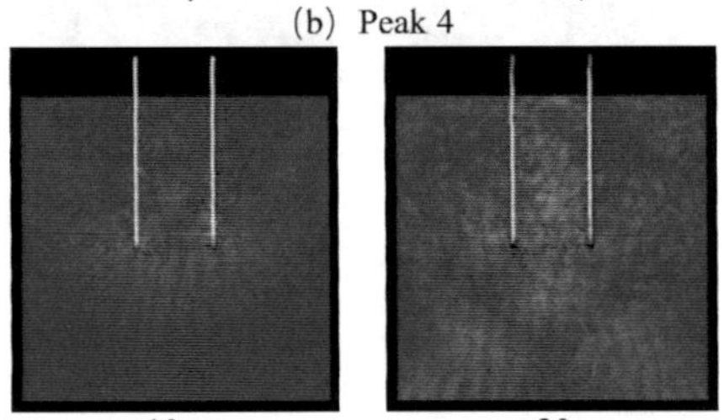

t=10 μs t=20 μs

(c) Peak 6

 https://doi.org/10.1051/e3sconf/202123703009

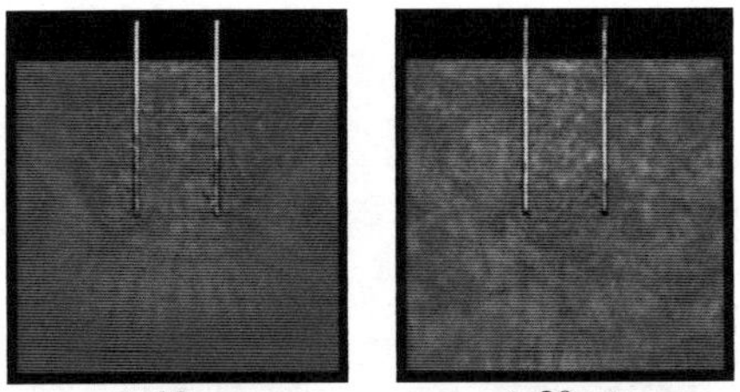

t=10 μs t=20 μs

(d) Rate 1

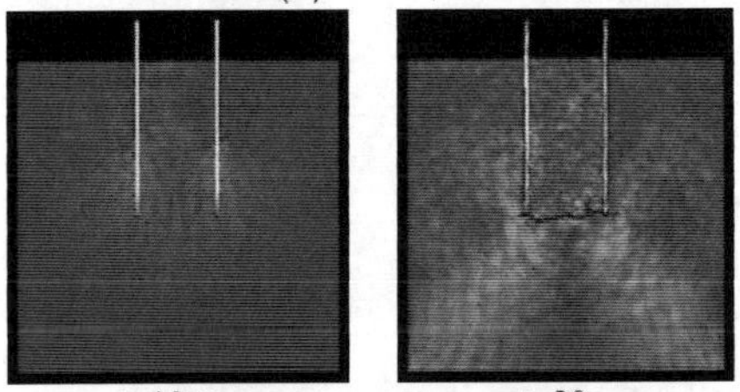

t=10 μs t=20 μs

(e) Rate 2

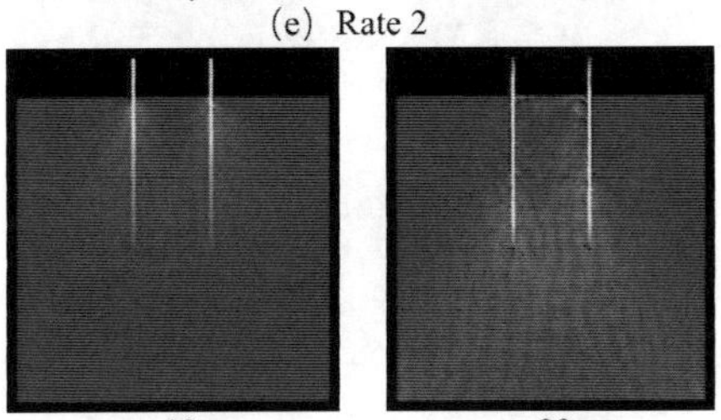

t=10 μs t=20 μs

(f) Rate 5

Fig3. Failure process of twin fibers pullout specime

3.2 Acoustic emission

Acoustic emission is the phenomenon that brittle materials such as concrete and rock are micro-ruptured under stress, strain energy released in the form of elastic waves[14-15](Acoustic Emission,AE). In this study, based on the two basic assumptions that the acoustic emission number is proportional to the element damage and the acoustic emission energy is proportional to the element damage strain energy release [16], the relationship between the acoustic emission number and acoustic emission energy and the evolution of concrete damage is established.The acoustic emission counts of the specimens numbered Peak 2, Peak 6, and Rate 1, Rate 5 shown in Fig 4. It can be seen that when the loading peak value is low , the interface of the drawn specimen is not completely debonded, the damage amount of the base unit is small, and the overall acoustic emission count generated by the specimen is low from Fig. 4 . With the increase of stress wave loading peak, the debonding failure of the interface element is complete, the damage of the matrix element caused by the load increases, the number of acoustic emissions produced by the specimen is relatively dense and the value is generally large.

It can be seen that the peak of the acoustic emission of the drawing specimen is delayed as the loading rate decreases. Due to the high loading rate, the stress wave reaches the peak load in a short time, which shows that the number of damaged elements rises sharply when the specimen is damaged. When the loading rate is slow, the peaks of the acoustic emission of the specimens all appear at the end of loading, and in the early period of loading, the number of acoustic emissions is small and relatively flat.

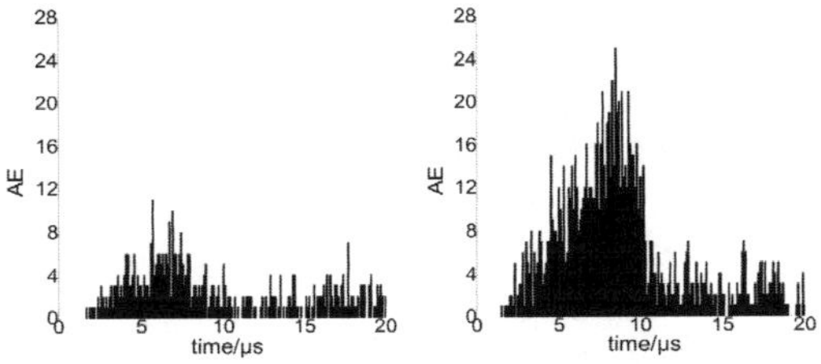

(a) Peak 2 (b) Peak 6

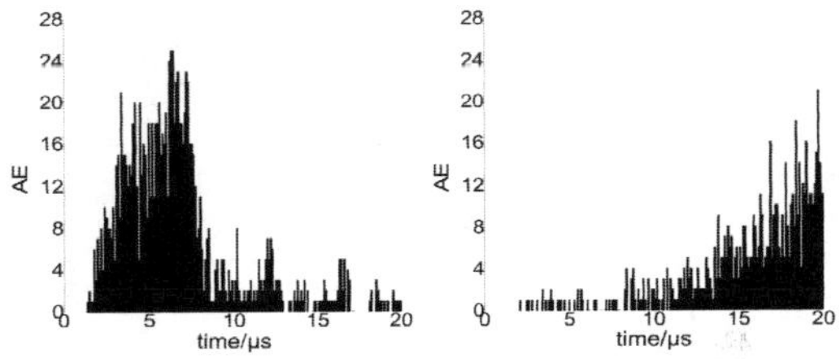

(c) Rate 1 (d) Rate 5

Fig4. AE of twin fibers pullout specimens

for the acoustic emission counts of the drawn test pieces, as shown in Table 2 and Table 3, the unit crack time, peak acoustic emission time and peak acoustic emission peak of the test piece under different loading peaks and loading rates. Table 2 shows that when the loading peak is large, the cracking time of the unit is slightly earlier, but it can be basically ignored. In addition, with the increase of the loading peak, the peak time of the acoustic emission is delayed, but the number of peaks of the acoustic emission continues to increase, that is, the peak number of the acoustic emission is the maximum when the peak value of the loading test is Peak 6.Table 3, the faster the stress wave rise time, the smaller the loading rate, and the significantly higher crack initiation time and acoustic emission peak time, indicating that the reduction in loading rate can effectively delay the initial failure of the unit and the time when the acoustic emission peak appears. In addition, the number of peaks of acoustic emission basically does not change much at different loading rates, indicating that the loading rate has little effect on the peaks of acoustic emission in this study.

Table2. Elements damage initiation time and AE peak of varied loading peak

Numbering	loading peaks /MPa	Unitinitial cracking time/μs	Time of AE peak /μs	AE peak
Peak 1	30	1.7	3.7	7
Peak 2	35	1.7	5.7	11
Peak 3	40	1.6	7.3	17
Peak 4	45	1.6	6.8	21
Peak 5	50	1.6	9.4	23
Peak 6	55	1.5	8.5	25

Acoustic emission results are shown in Fig 5、6. With the increase of loading peaks and loading rates, the cumulative energy of acoustic emission and the cumulative number of acoustic emissions of the specimen increase, indicating peak stress waves and the greater the loading rate, the more the interface debonding area and the base unit damage, the more energy is released when the specimen is damaged by loading, and the specimen damage unit increases. As shown in Fig. 5 (b) and Fig. 6 (b), when the loading peak and the loading rate are maximum, the curve of the cumulative number of acoustic emissions rises more slowly in the later stage of loading. Because the loading peak and the loading rate are relatively large, the interface of the drawn specimen has been completely debonded and destroyed, and subsequent stress waves cannot continue to be transmitted into the matrix.Gokoz[17], Banthia et al.[18,19] carried out pull-out tests for various types of fibers at different loading rates. The results of the study show that the increase in loading rate increases the energy dissipation capability of the material specimen. It is consistent with the results of this study that the cumulative energy of acoustic emission increases with increasing loading rate.

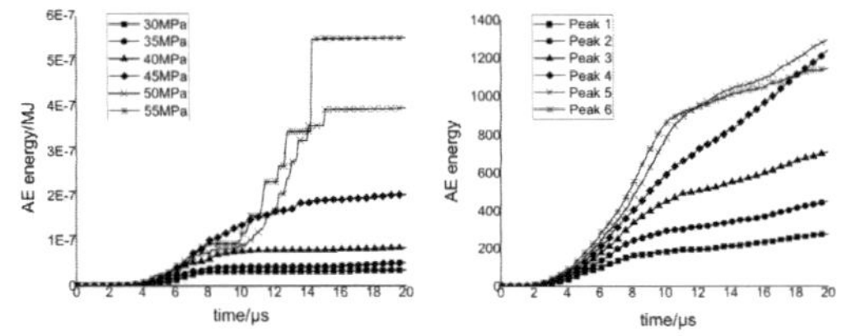

(a) Acoustic emission cumulative energy curve (b) Acoustic emission cumulative number curve

Fig5. AE of specimens of varied loading peak

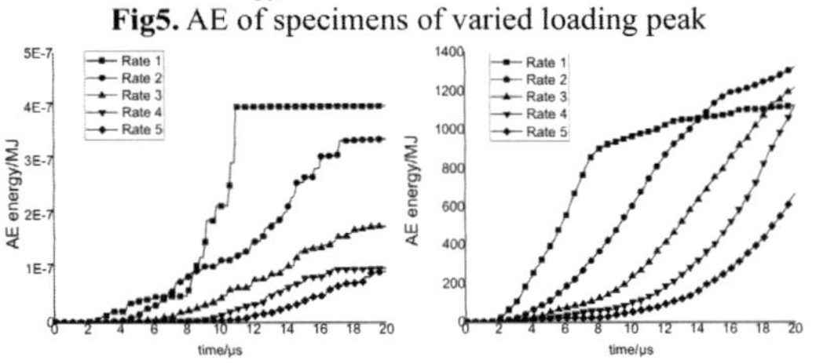

(a) Acoustic emission cumulative energy curve (b) Acoustic emission cumulative number curve

Fig6. AE of specimens of varied loading rate

4 Conclusion

(1) The dynamic failure mode of twin fibers pullout of concrete matrix has obvious correlation between loading peak and loading rate. When the loading peak and loading rate are large, the twin fibers is debonded and pulled out. The loading peak and loading rate are reduced, the twin fibers are debonded, and a macroscopic main crack that penetrates the matrix between the twin wires is formed at the buried end. When the loading peak and the loading rate are further reduced, the fiber is not completely debonded, and the test piece still have a certain bearing capacity.

(2) With the increase of the loading peak and the loading rate, the energy released when the specimen is damaged by the load increases, the interface crack growth rate of the specimen increases, the number of unit failures increases, and the specimen is seriously damaged.

(3) The loading rate affects the unit cracking time of the drawn specimen. The peak of the acoustic emission number of the test piece increases with the increase of the loading peak.

Acknowledgment

This research work is financially supported by the National Natural Science Foundation of China, Grant No. 51678170 and 51878190.

References

1. Q Wu,M Li,Y Gu. Composites Part A: Compos Part A: Appl Sci Manufac.56,143-149.(2014)
2. Kim, J K., Mai, Y W. Elsevier, (1998)
3. Jiang Y P, Yue ZF, Wan JS.J Comput Mech.20,743-748.(2003)
4. Ning JG, Shang L, Sun YX. Adva Mech. 36, 389-405.(2006)
5. Kim D J, El-Tawil S, Naaman A E. AC MATERJ.105,576-584.(2008)
6. Abu-Lebdeh T, Hamoush S, Zorning B. AM. J. Eng. Appl. Sci.3,(2010)
7. Zhu DJ, Li XY, Shi CJ. J. Hunan. Univ (Natural Science Edition).(2018)
8. Xu M, Hallinan B, Wille K. Cement Concrete Composites.70,98-109(2016)
9. Li LJ, Zhang HZ, Liu F. New Buil.Mater10,1-5 (2004)
10. Zheng XY,Wu SX.J.Civ.Eng.39, (2006)42-46.
11. Zhang WP, Luo DY, Chen H. China J.Highw.Trans.27,58-64(2014)
12. Tang CN, Zhu WC.Beijing: Sci. Press. (2003).
13. Chen PR, Zhang YF, Li G.J.Sun Yat-Sen Univ.(Natural Science Edition).52,68-80 (2013)
14. Wang LL. Beijing: Nat.Def. Ind. Press.(2005).
15. Wan GX, Wang QS, Li XB.Exper.Mech. 27,727-733(2012)
16. Tang CA, Xu XH.J.Seismo.Res.13, 203- 213(1990)
17. Gokoz UN, Naanman AE. Intern.J.Cement Composites.Lightweight Concrete.3,187-202.(1981)
18. Banthia N, Trottier JF.Cement Concrete Research.21,158-168(1991)
19. Bindiganavile V, Banthia N. Canadian J.Civ. Eng.32,924-933(2005)

https://doi.org/10.1051/e3sconf/202123703010

Comparison and analysis of pile-soil separate calculation and composite foundation calculation method for gravel pile reinforcement

Li Hong-yuan[1*]

[1]School of Engineering Science, University of Chinese Academy of Sciences, Beijing 100049, China
Institute of Mechanics, Chinese Academy of Sciences, Beijing 100190, China

Abstract: This paper uses Indraratna seepage theory, Han pile-soil seepage theory and composite foundation theory to establish three plane strain models. Through finite element calculations, the changes in pore pressure and settlement are compared, and the following laws are found: (1) The model established by the composite foundation theory cannot reflect the objective laws of seepage and deformation; (2) In the case of multiple piles, there is no difference in settlement between Han pile-soil seepage theory considering the bearing characteristics of gravel piles and Indraratna seepage theory without considering the bearing characteristics of gravel piles. (3) The settlement value obtained by the principle of composite foundation is much smaller than the settlement value obtained by the separate calculation of pile and soil.

0. Preface

Nowadays, in engineering design, the foundation reinforced by gravel piles is generally analyzed based on the theory of composite foundation. In addition, some researchers have adopted the method of separate calculation of piles and soils, using the plane strain model of gravel piles to strengthen the foundation for calculation and analysis. Based on Indraratna [1] sand well seepage theory, Han [2] pile-soil seepage consolidation theory solution and gravel pile reinforced composite foundation principle, this paper establishes plane strain model under the condition of multiple piles, and compares the settlement and pore pressure changes of different models when they are consolidated.

1. Theory

Assuming that the gravel piles are distributed like plum blossoms, the spacing of gravel piles is 4m, and the effective pile diameter is 1m, the influence radius of a single gravel pile is 2.1m. The conversion formula given by Tan [3] is converted into an equivalent plane strain model, the 1/2 pile spacing B is 1.86m, the gravel pile radius B_c is 0.10556m, and the soil thickness is 10m. Based on this, the establishment of its equivalent two-dimensional generalized model of gravel piles to strengthen the foundation is shown in Figure 1. Due to the axis symmetry, 2.5 piles are selected in the figure.

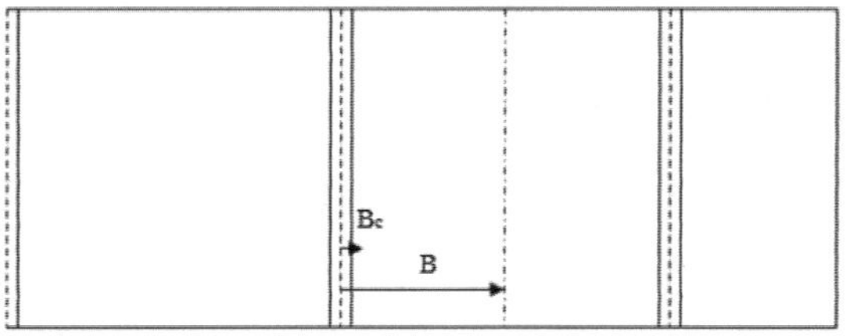

Figure 1 Two-dimensional generalized model of the foundation reinforced by gravel piles

1.1 Indraratna equivalent plane strain model

Indraratna uses Hansbo's sand well seepage consolidation theoretical solution, only considering the drainage effect of gravel piles, and obtains the permeability coefficient under plane strain conditions.

$$k_{x,\,pl} = \frac{2\alpha}{\ln(n) - 0.75}\frac{B^2}{R^2}k_{x,\,ax} \qquad (1)$$

In the formula: n is the aperture ratio, $\alpha = \frac{1}{3} + \frac{b_c}{2B} - \frac{b_c}{6B} - \frac{b_c^2}{6B^2} + \frac{b_c^2}{2B^2} - \frac{b_c}{B}$, $k_{x,\,pl}$ and $k_{x,\,ax}$ respectively represent the permeability coefficient under axisymmetric conditions.

1.2 Han pile-soil seepage consolidation theory solution

In Han's theoretical solution of pile-soil seepage consolidation, he considered the two roles of gravel piles to share additional load and drainage. In the conversion, the genetic algorithm is used to solve the equation (2), so

Li Hong-yuan: lihongyuan231@foxmail.com

that the average degree of consolidation under the plane strain condition matches the average degree of consolidation under Han's pile-soil axisymmetric condition.

$$\sum_{m=1}^{\infty}\frac{2}{M^2}\exp(-M^2T_v)=\exp(\frac{-8T_r}{f(n)}) \quad (2)$$

Consolidation matching is shown in Figure 2

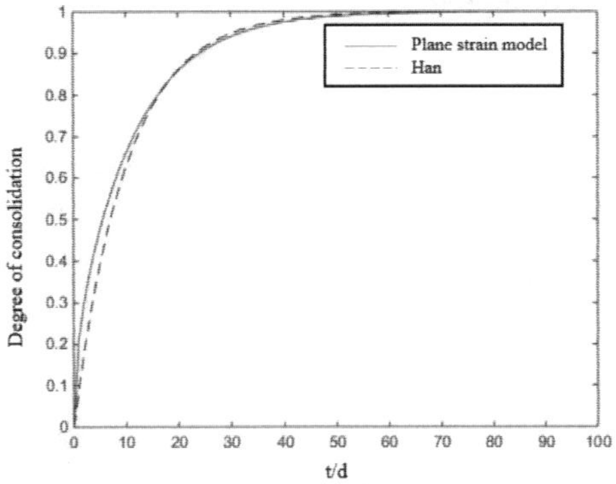

Figure 2 Schematic diagram of consolidation degree matching

1.3 Composite foundation

In engineering, the elastic modulus and deformation modulus of composite foundation are generally calculated according to the weighted average method according to the area proposed by Schweiger and Pande [4], as shown in equation (3).

$$E_{cs}=(1-\alpha)E_s+\alpha E_c \quad (3)$$

In the formula: α is the area replacement rate.

Regarding the permeability coefficient, the permeability coefficient of the composite foundation is generally adjusted so that the degree of consolidation obtained by the sand well consolidation theory matches the degree of consolidation obtained by the one-dimensional seepage consolidation theory of Terzaghi in the composite foundation.

Barron [5] obtained that the consolidation degree of seepage in sand wells at equal strain is

$$U_r=1-\exp[\frac{-8T_r}{f(n)}] \quad (4)$$

2. Modeling

According to the above method, determine the corresponding model size and material parameters. The corresponding model size is shown in Figure 1, and the determined parameters are shown in Table 1. The first line in the table is the original parameter, and the second, third, and fourth lines are the calculated parameters converted according to the theory described above. Among them, the permeability coefficient of the composite foundation is adjusted to match the solution of the Terzaghi seepage and consolidation theory, as shown in Figure 3.

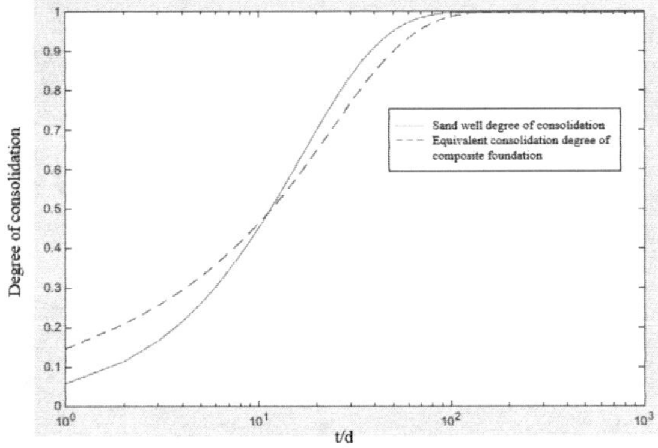

Figure 3 Comparison between consolidation degree of sand well seepage and theoretical consolidation degree of composite foundation

Table 1 Material parameter table

Model	$\gamma_s=\gamma_c$ kN·m^{-3}	ν'	E_s MPa	E_c MPa	k_h m·s^{-1}	k_v m·s^{-1}	c_s' kPa	φ_s' deg	c_c' kPa	φ_c' deg
Original parameter	15	0.3	3	30	3.00·10^{-9}	3·10^{-9}	0.1	22	1	40
Separate calculation of pile and soil 1	15	0.3	3	30	2.04·10^{-9}	3·10^{-9}	0.1	22	1	40
Separate calculation of pile and soil 2	15	0.3	3	30	1.55·10^{-9}	3·10^{-9}	0.1	22	1	40
composite foundation	15	0.3	9.48	/	4.86·10^{-8}	4.86·10^{-8}	0.1	22	1	40

The finite element model is shown in Figure 4, with 100kPa load applied, and the consolidation changes of three different plane strain models are obtained.

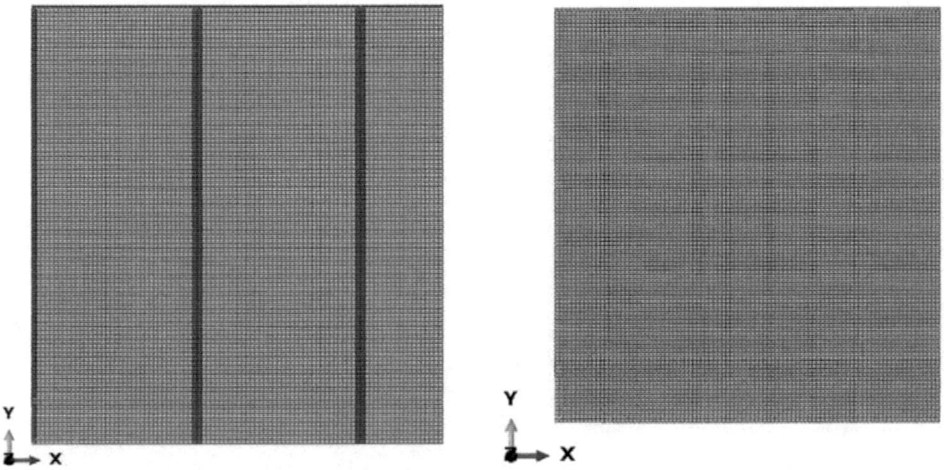

(a) Separate calculation of pile and soil (b) Composite foundation

Figure 4 Finite element model

3. Analysis

The pore pressure and settlement conditions of the three plane models after 100 days of consolidation are shown in Figs. 4-6. It can be seen from the figure that under the condition of separate calculation of pile and soil, there is a clear gap between the pore pressure of gravel pile and the soil between piles. It shows that there is a tendency for water in the soil between the piles to seep into the gravel piles. Under the composite foundation, this tendency is obviously not reflected. It can be seen from the settlement diagram that near the surface of the soil layer, there is a significant gap between the settlement of the gravel pile and the soil between the piles, which is also unreactive in the composite foundation.

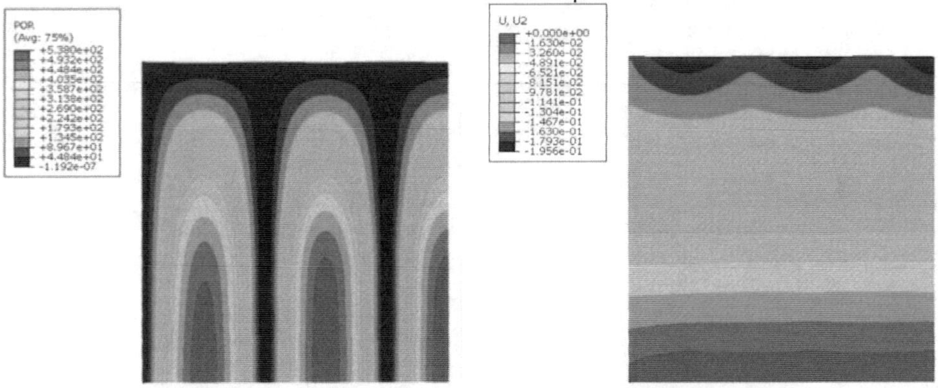

(a) Pore pressure /Pa (b) Settlement /m

Figure 5 Pile-soil split calculation 1 (Indraratna equivalent plane strain)

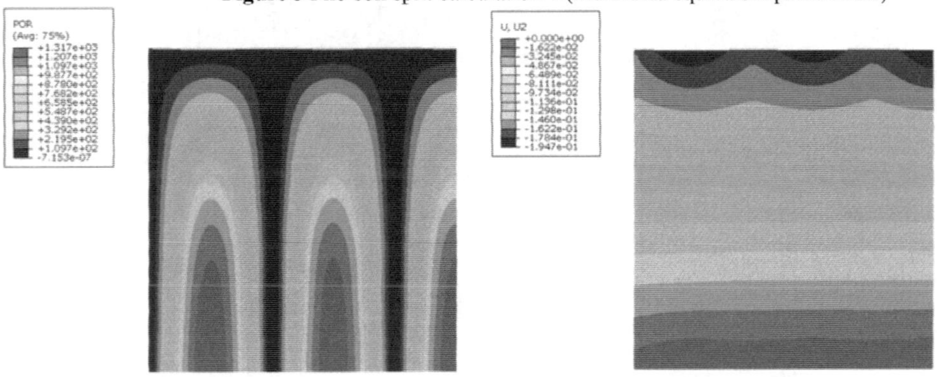

(a) Pore pressure /Pa (b) Settlement /m

Figure 6 Pile-soil separation 2 (Han pile-soil seepage consolidation)

https://doi.org/10.1051/e3sconf/202123703010

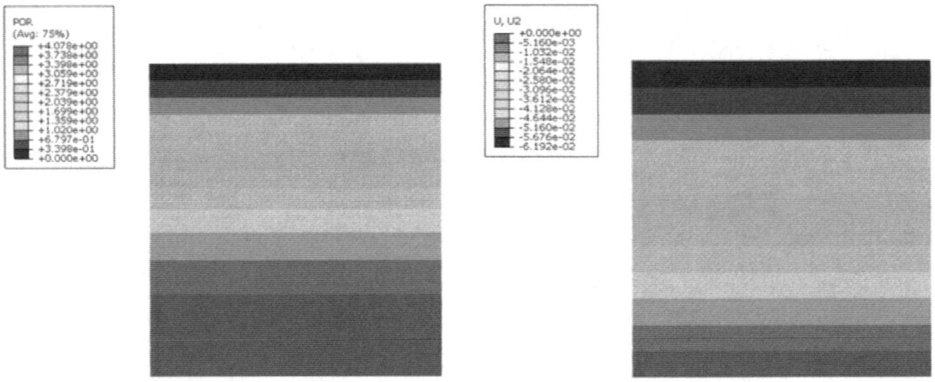

(a) Pore pressure /Pa (b) Settlement /m

Figure 7 Composite foundation

According to the distribution law of displacement and pore pressure, choose a point near the surface of the soil layer as the settlement curve with time and a point near the bottom surface of the soil layer as the curve of pore pressure with time as shown in Figure 7 to further understand its changes. It can be seen from the figure that the changes of excess pore pressure and displacement of the two methods of pile-soil separate calculation are close. It can be analyzed that the Indraratna's pile-soil separation method, which does not consider the gravel piles to share the additional load, has a 31.6% larger permeability coefficient than Han's pile-soil separation method, which considers the gravel piles to share the additional load. But under the condition of multiple piles, it has little effect on the settlement result. In addition, considering that the theoretical settlement value of the composite foundation is much smaller than the settlement value calculated separately for the pile and soil, the settlement value of the composite foundation is about 56mm at the moment when the settlement of the consolidation 100d stabilizes. The settlement value of the two pile-soil separate calculation models is 166mm, and the settlement value of the pile-soil separate calculation is about 3 times the settlement value of the composite foundation. Analyze the specific reasons. First, it can be seen from the excess pore pressure curve that the composite foundation is consolidated faster and the excess pore pressure dissipates faster; second, it can be seen from the deformation modulus of the material parameters that the composite foundation theory greatly increases the deformation modulus of the soil.

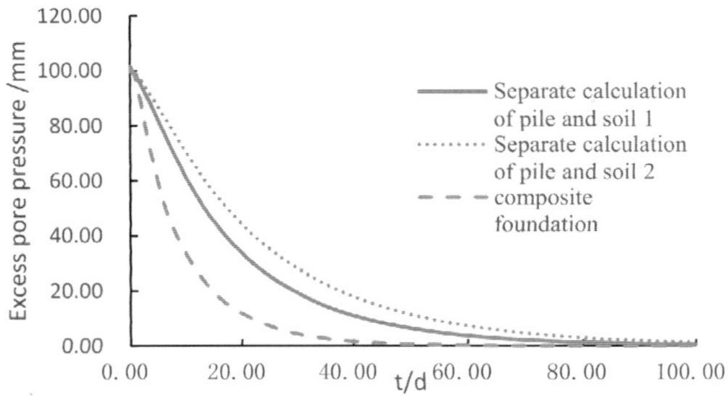

（a）Excess pore pressure /kPa

 https://doi.org/10.1051/e3sconf/202123703010

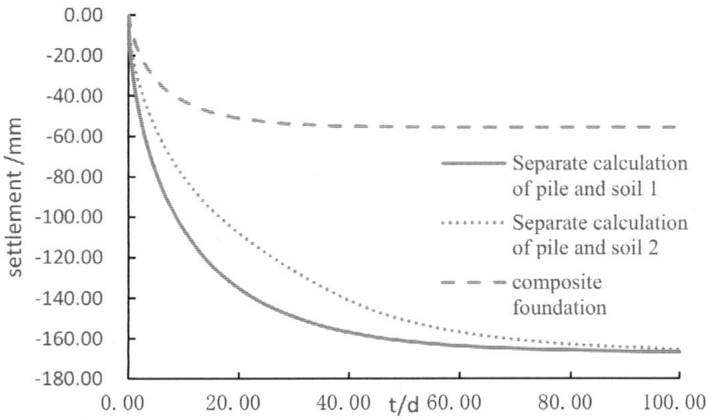

(b) settlement /m

Figure 8 Changes in excess pore pressure and settlement with time

4. Conclusion

In this paper, three plane strain models are established through Indraratna seepage theory, Han pile-soil seepage theory and composite foundation theory. By comparing the changes in pore pressure and settlement, the following conclusions are made:

(1) The model established by the composite foundation theory cannot reflect the objective law of seepage, and cannot reflect the seepage of soil between piles into gravel piles. In addition, the settlement diagram cannot reflect the settlement difference between the gravel pile and the soil between the piles.

(2) Under multiple pile conditions, the Han pile-soil seepage theory considering the pressure-bearing properties of gravel piles and the Indraratna seepage theory without considering the pressure-bearing properties of gravel piles have basically no difference in settlement analysis.

(3) The settlement value obtained by the principle of composite foundation is much smaller than the settlement value obtained by the separate calculation of pile and soil. There are two reasons for this. First, the equivalent permeability coefficient is large and the excess pore pressure dissipates faster; second, the composite foundation theory greatly increases the deformation modulus of the soil.

References

1. Indraratna B, Redana I W. Plane-Strain Modeling of Smear Effects Associated with Vertical Drains [J]. Journal of Geotechnical & Geoenvironmental Engineering, 1997, 123(5):474-478.
2. Han J, Ye S L. Simplified Method for Consolidation Rate of Stone Column Reinforced Foundations [J]. Journal of Geotechnical & Geoenvironmental Engineering, 2001, 127(7):597-603.
3. Tan S A, Tjahyono S, Oo K K. Simplified Plane-Strain Modeling of Stone-Column Reinforced Ground [J]. Journal of geotechnical and geoenvironmental engineering, 2008, 134(2):185-194.
4. Schweiger H F , Pande G N . Numerical analysis of stone column supported foundations[J]. Computers & Geotechnics, 1986, 2(6):347-372.
5. Pyrah I C, Hird C C, Russel D. Finite element modelling of vertical drains beneath embankments on soft ground [J]. Géotechnique, 1992, 42(3):499-511.

Settlement simulation of soft clay in the subway under dynamic load based on Midas GTS NX

Quan Cao[1], Yu Hang[1,*]

[1] School of Railway Transportation, Shanghai Institute of Technology, Shanghai 201418, China

Abstract. The subway has become the main way for people to travel nowadays. The saturated soft clay area has a large population and subway construction is the most extensively distributed. The saturated soft clay foundation will settle under the dynamic load of the subway train, which will affect the service life and structural safety. To study the settlement characteristics of soft clay under the dynamic load of the subway, a three-dimensional dynamic finite element model was established based on the finite element software Midas GTS NX, and the soft clay under the normal design speed (80 km/h) and the high-speed (120 km/h) were compared and analyzed. The research results show that the higher the train speed, the larger the lateral influence range of the surface settlement trough, but the settlement decreases with the increase of the subway running speed. The settlement of the subway line after one year of operation is about 45mm, and the settlement after 20 years is about 58mm. The growth rate of the settlement prediction curve decreases gradually, and the settlement increment is very small for a long time in the later period of operation.

1 Introduction

The construction of the subway has a significant effect on improving the utilization rate of urban land and alleviating the increasingly crowded urban traffic. Cities in the saturated soft clay area (take Shanghai as an example) have a large population, and the subway has gradually become the main way for people to travel. With the popularization of subway trains, a series of engineering problems have also caused. When the soft soil foundation is subjected to continuous dynamic load, even the soft soil foundation after a long-term consolidation process will produce different degrees of settlement. Once the saturated soft clay layer has a large settlement, the tunnel structure will suffer from water leakage, mud leakage, and pipe segment fracture, which will seriously affect the normal operation safety and efficiency of the subway. Under the action of long-term cyclic load, the soft soil foundation will produce a post-construction settlement, which will affect the performance and service life of the subgrade. Shanghai Metro Line 1 put into operation in 1995, the settlement problem is more serious, the settlement reached 6cm after 8 months of opening to traffic, and the settlement of several sections is still showing a developing trend, with the maximum annual settlement value reaching 40mm [1].

At present, the commonly used research methods for soil settlement under subway dynamic loads include theoretical calculation methods, empirical methods combined with field measurements, indoor model test methods, and numerical simulation methods. Shirlaw [2] studied the long-term settlement data of a lot of tunnels and found that the long-term settlement of soft soil tunnels can usually account for more than 30% of the total settlement. Yan Chunling and Tang Yiqun [3] reviewed the dynamic characteristics of silty soil under the subway load and the long-term subsidence of subway tunnels. They found that the subway tunnels will be subjected to intermittent cyclical loads after they are completed and put into operation. Fujiwara et al. [4-5] comprehensively discussed the effects of total load, load cycle, load increment ratio, loading method, cementation degree, and cycle times on clay deformation, and found that the consolidation deformation under cyclic load was greater than that under static load. Ding Zhi et al. [6], Liu Ming et al. [7], Yildirim A et al. [8], Yilmaz et al. [9], and He Shaoheng et al. [10] also conducted related studies on the settlement of soft soils under subway dynamic load.

Midas GTS NX is a finite element software developed for geotechnical engineering. The software has powerful pre-processing and post-processing modes that can meet most engineering needs. In this article, Midas GTS NX is used to establish a three-dimensional finite element calculation model for the interaction between train lining, roadbed, and soil to simulate subway vibration, and analyze the settlement variation law of soft soil foundation under the normal design speed (80 km/h) and high speed (120 km/h) of subway trains. Based on the simulation study of the influence of subway vibration on the strata along the line and the subsidence of the strata, the article predicts the settlement of the ground during operation by

* Corresponding author: sakuragihy@163.com

combining the theoretical and empirical formula, which provides a reference for the design and long-term service safety evaluation of subway engineering in soft soil area.

2 Geological conditions

The soil of this model was selected in the section from Fanghua Road Station to Longyang Road Station on Shanghai Metro Line 7, and the typical geological section was selected as the soft clay settlement simulation object. The soil in the 30m depth range of the survey borehole is mostly soft clay, and there are 7 soil layers in the exploration depth. The typical interval geological cross-section is shown in Fig. 1. The buried depth of the tunnel is 15m.

The survey samples were subjected to indoor geotechnical tests and statistical analysis of the results. The physical and mechanical indexes of each soil layer were shown in Table 1.

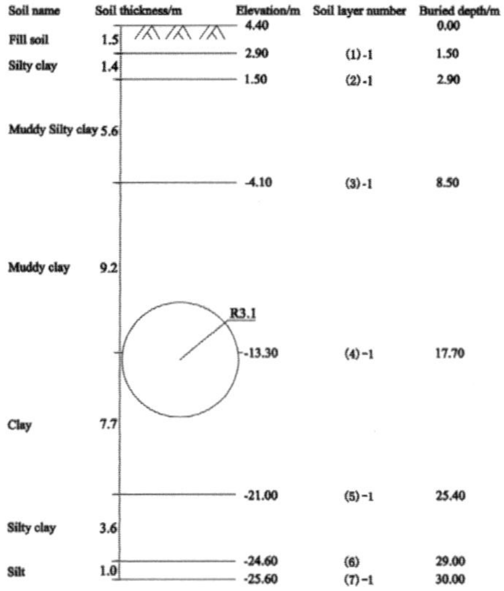

Fig1. Typical interval geological section

Table1. Physical and mechanical parameters of soil layers

Soil name	Bulk density γ/ kN/m³	Poisson's ratio v	$E_{s0.1\text{-}0.2}$ /MPa	CU	
				C_{CU} /kPa	φ_{CU} /(°)
Silty clay	18.8	0.35	5.64	21.0	23.0
Muddy Silty clay	17.4	0.25	3.22	12.5	17.3
Muddy clay	16.7	0.25	2.32	14.0	11.8
Clay	17.4	0.35	3.02	16.0	13.8
Silty clay	19.2	0.35	6.55	48.0	17.0
Silt	18.6	0.25	12.49	2.0	32.5
Silt	18.6	0.25	13.03	0.5	33.5

3 Numerical simulation

The outer diameter of the shield tunnel of the subway line is 6.2m, the inner diameter is 5.5m, the segment thickness is 350mm, the trackbed thickness is 0.450m, the standard gauge is 1.435m, and the strength grade of the precast concrete lining segment is C55. Midas GTS NX was used to establish a soil model, according to the characteristics of finite element simulation and considering the actual force situation and boundary effect of the subway station, the model can be simplified appropriately. The upper layer including the fill layer is merged into a silty clay layer, the middle layer is merged into a clay layer, which contains the subway tunnel, and the lower part is a silty layer. When establishing soil parameters, the soil layer with weaker elastic modulus is taken as the most unfavorable situation, and the established model is safer and more reliable. The model size is 62m×50m×120m, as shown in Fig. 2. The tunnel structure model of the lining and trackbed is shown in Fig. 3.

Fig2. Soil model

Soil parameters of typical sections completed by laboratory tests were extracted, physical and mechanical parameters of soil layers were shown in Table 1. The model range of numerical simulation is as follows: take 10 times the outside diameter of the tunnel in the horizontal X direction, that is 62m, take 120m in the vertical Y direction, and take 50m in the longitudinal Z direction. The five loads of train dynamic load were calculated, and each load was recorded and analyzed respectively.

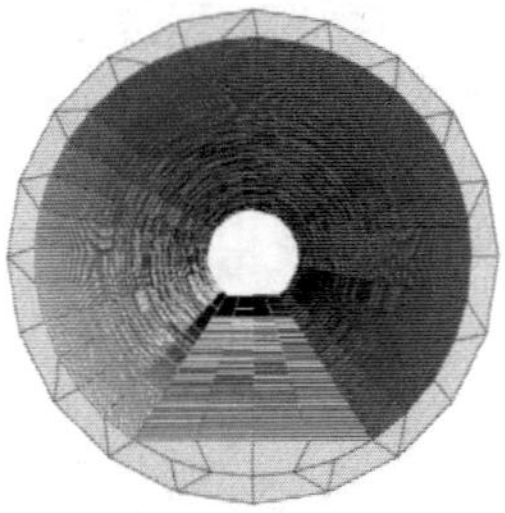

Fig3. Tunnel structural model

Eigenvalue analysis is used to analyze the dynamic characteristics of the structure itself, and is also called "free vibration analysis". The periods of the first and second main modes of the eigenvalue analysis will be used in the time history analysis to calculate the damping matrix. We select all meshes to establish an elastic boundary with ground springs, select "Eigenvalue" as the solution type, activate all meshes and elastic boundary conditions, and check the eigenvalue analysis result table. The periods of the first and second main modes should be focused on.

In static analysis, it is necessary to ensure that the boundary is sufficiently far away from the main area for models containing infinite materials such as rock and soil. But in the dynamic analysis, the wave will be reflected. If the boundary setting is the same as the static analysis, it is likely to be wrong. To avoid errors, we use approximate methods, such as artificial boundary, transitive boundary or hyperfinite element, infinite element, and boundary element. This article adopting a "sticky boundary".

The train dynamic load is applied to the model. The moving train dynamic load is used to simulate the operation of the subway train. Select the node where the subway load is applied (the wheel passing point). Select the starting and ending nodes according to the direction of the subway and enter the proposed train speed. Select "linear time history (direct integration method)" to solve. Enter the results of the previous eigenvalue analysis, and drag all meshes, viscous conditions, and dynamic loads to the activation group to start the calculation.

4 Analysis of calculation results

4.1 Vertical settlement of the soil at the bottom of the tunnel

In this settlement simulation, we calculated the case where the vehicle dynamic load was cyclically loaded five times at two different speeds of 80km/h and 120km/h. The soil settlement is shown in Fig. 4-Fig. 6. It can be seen from the comparison of the figures that as the depth increases, the vertical cumulative settlement of the soil gradually decreases. For the average settlement value of the soil at the same depth, the settlement at the normal speed is greater than that at the faster speed. The trainload frequency is relatively low. The lower the frequency, the slower the load change, and the longer the load action time, which is more conducive to the compaction of the soil. The more energy acts on the soil, the lower the frequency, the greater the accumulated strain of the soil will be generated, thus resulting in greater deformation of the soil.

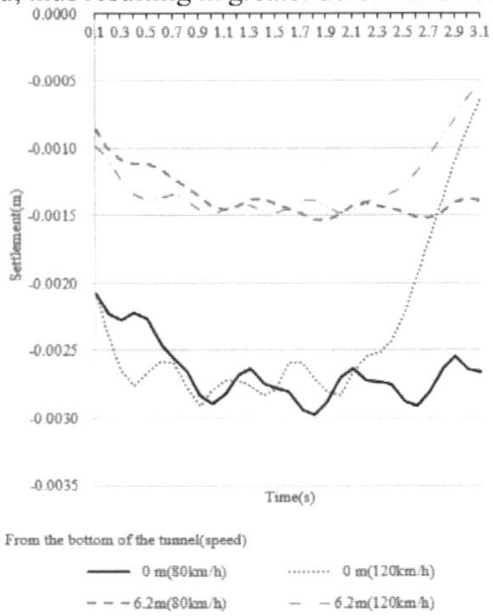

Fig4. Soil settlement at different speeds (First load)

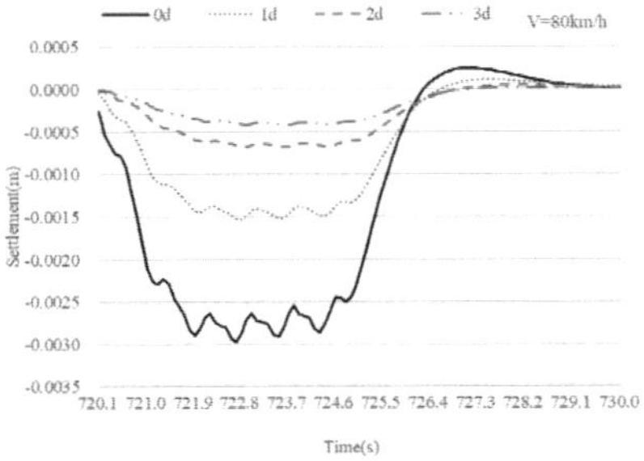

Fig5. Settlements at bottom of tunnel (5th time)

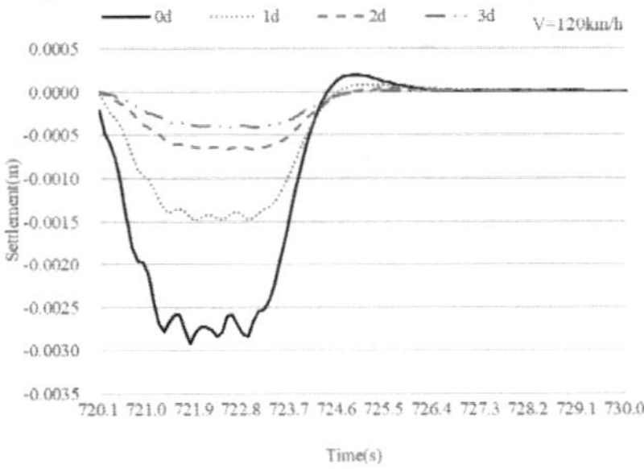

Fig6. High-speed settlements at bottom of tunnel (5th time)

4.2 Vertical settlement of lateral soil

Fig. 7 shows the vertical settlement curve of the soil in the surface horizontal settlement trough at different speeds. It can be seen from the figure that the surface settlement trough under normal speed is more obvious, the center depth of the settlement trough is greater than the faster speed, and the settlement decreases as the distance from the tunnel axis increases. When the train increases to a faster speed, the lateral influence range of the settlement trough gradually develops from "deep and narrow" to "shallow and wide".

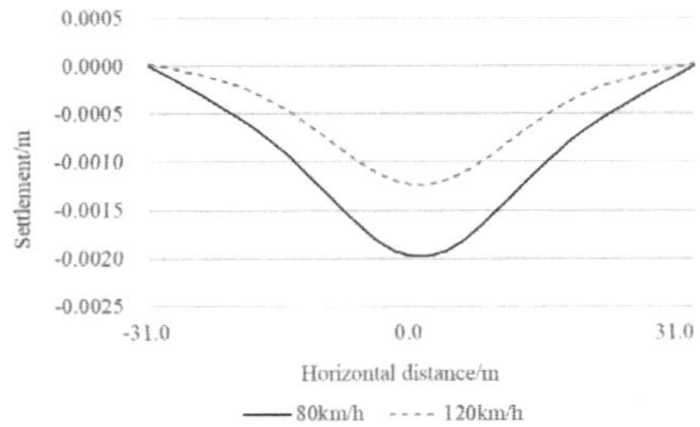

Fig7. Horizontal settlements of the ground surface

5 Settlement prediction

Based on the numerical simulation results and empirical prediction formulas, in this article we propose an empirical prediction model of construction settlement during operation time:

$$s = \gamma \sum_{i}^{n} \varepsilon h_i \quad (1)$$

$$\varepsilon = 0.96N^{0.085} \quad (2)$$

Where ε is the cumulative plastic strain of the calculated soil layer, N is the number of vibrations, S is the predicted settlement value (mm), γ is the correction coefficient considering experience, generally can be 0.85. h_i is the thickness of the calculated soil layer, generally can be considered 3 to 4 times the tunnel diameter lying depth, or maximum 20m. According to the proposed settlement prediction empirical model, the settlement prediction during the post-construction operation period is carried out. The number of vibrations per year is calculated as 400 thousand times. Fig. 8 shows the settlement evolution curve of this line from the beginning of the operation to 20 years of operation. It can be seen from the figure that the settlement prediction value in the first year is 45 mm, and the settlement prediction value in the 20-year operation period has reached 58mm. The growth rate of the settlement prediction curve gradually decreases, and finally, tends to zero. In the later period of operation, the settlement increment is very small in a long time increment.

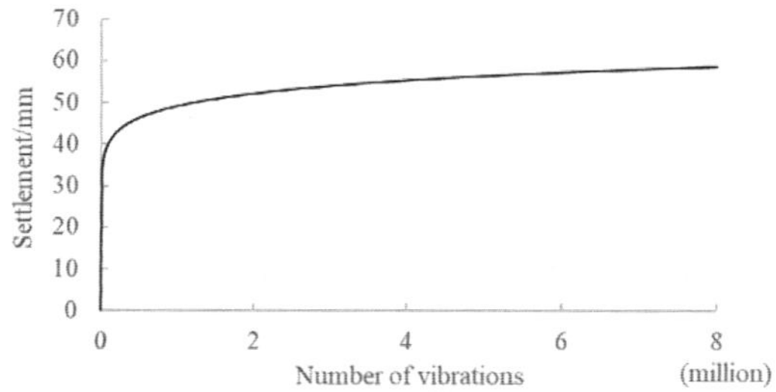

Fig8. 20 years development curve of predicted tunnel settlement

6 Conclusion

This article discusses the settlement characteristics of soft clay under the dynamic load of the subway at a different speed, analyzes the cumulative plastic deformation of typical soil layer under the dynamic load. The settlement prediction model of the shield tunnel was established, and the post-construction settlement of the soft ground under the tunnel was predicted and analyzed. The main conclusions are as follows:

(1) For the vertical settlement value of soil at the same depth, the settlement at the normal speed is larger than the faster speed. Under the normal speed, the strain response of the soil under the tunnel is more obvious, and the settlement curve is relatively "high and steep". Compared with the normal speed, the post-construction operation settlement caused by the faster subway train is smaller.

(2) When the train increases from normal speed to the faster speed, the lateral influence range of the settlement trough gradually develops from "deep and narrow" to "shallow and wide". The lateral influence range of the surface settlement trough is larger, but its maximum settlement value is relatively reduced.

(3) Under the action of the subway dynamic load, the long-term settlement development of the tunnel conforms to the exponential growth law. As the operating time increases, the influence of the trainload on the settlement of the soft soil around the tunnel gradually tends to be stable. The settlement in one year after the operation is about 45 mm and is predicted to be up to 58mm in 20 years of operation, which provides a reference for engineering decision-making.

Acknowledgements

The author thanks the financial support from School of Railway Transportation, Shanghai Institute of Technology.

References

1. Y.D. Ye, H.H. Zhu, R.L. Wang. The present situation and cause analysis of the diseases of the railway operating tunnel in soft soil. Journal of Underground Space and Engineering, 3(1): 157-160 (2007)
2. J.N. Shirlaw. Observed and calculated pore pressures and deformations induced by an earth balance shield. Canadian Geotechnical Journal, 32(30): 476-490 (2011)
3. C.L. Yan, Y.Q. Tang. Research progress of dynamic characteristics of silt soil under the subway load. Journal of Earthquake Engineering, 33(2): 200-205 (2011)
4. H. Fujiwara, T. Yamanuochi, K. Yasuhara. Consolidation of alluvial clay under repeated loading. Soils and Foundations, 25(3): 19-30 (1985)
5. H. Fujiwara, U. Shunji, Y. Kazuya. Secondary compression of clay under repeated loading. Soils and Foundations, 27(2): 21-30 (1987)
6. Z. Ding, J.L. Fan, M.Y. Zhang. Model test study on undisturbed soil pore pressure and strain under the subway trainload. Journal of the China Railway Society, 39(3): 96-103 (2017)
7. M. Liu, M.S. Huang, Y.H. Liu. Analysis of long-term settlement of cross-river tunnel in soft soil caused by vehicle vibration. Chinese Journal of Geotechnical Engineering, 31(11): 1703-1709 (2009)
8. A. Yildirim, H. Ersan. Settlements under consecutive series of cyclic loading. Soil Dynamics and Earthquake Engineering, 27(6): 577-585 (2007)
9. M.T. Yilmaz, O. Pekcan, B.S. Bakir. Undrained cyclic shear and deformation behavior of silt clay mixtures of Adapazari in Turkey. Soil Dynamics and Earthquake Engineering, 24(7): 497-507 (2004)
10. S.H. He, Q.Q. Zheng, T.D. Xia. Experimental study on long-term dynamic characteristics of marine soft soil under subway train load considering time intermittent effect. Chinese Journal of Rock Mechanics and Engineering, 1-11 (2018)

Evaluation of green building incremental cost and benefit based on SD model

Guoqiong Long[1,*], Tianxiang Xu[2], and Chong Li[1]

[1]School of Economics, Yunnan University of Finance and Economics,650221 Kunming, China
[2]Economics Development Research Center of Yunnan University of Finance and Economics,650221 Kunming, China

Abstract. In order to eliminate the characteristics of long payback period and external economy of green building investment, and realize the accurate evaluation of green building cost and benefit. Based on the new estimation method, this paper constructs the system dynamics(SD) model of incremental cost-effectiveness of green building, and the rationality of the model is further tested and judged. This paper studies the change law of incremental cost and benefit of green building energy saving from the perspective of SD model. The incremental benefit of green building is greater than the incremental cost, and the time balance between them is about 8.0 years. The conclusion can provide enlightenment for the further development of green building.

1 Introduction

Under the pressure of the decline of environmental quality and the sharp increase of energy consumption, in order to change the current situation of the construction industry, it has become a consensus to establish sustainable green building. Therefore, green building has been rapidly promoted. However, due to the use of energy-saving technology, intelligent system, recycling and other technical facilities, the construction cost of green buildings is much higher than that of ordinary buildings, and the direct and indirect benefits generated in the use process are also higher than ordinary buildings. Therefore, it is particularly important to study the incremental cost-effectiveness of green buildings. At the same time, the current theory and understanding of green building is not mature enough, and the knowledge of incremental cost and benefit of green building is still vague. Most of the evaluation of green buildings is from the perspective of technology and economy. There is no systematic evaluation system of comprehensive benefits of green buildings, and the evaluation of comprehensive benefits also has different degrees of deviation[1]. In view of this, it is necessary to scientifically evaluate the cost-benefit level of green building and study the comprehensive benefit of green building on the basis of current domestic and international research on the cost-effectiveness of green building.

In addition, the government, property developers and consumers are the participants of green building construction in China. Among them, the government plays a leading and promoting role. As the executive body, developers will be restricted by the problem of high incremental cost. Although consumers can obtain incremental benefits by purchasing green buildings, they should also share the incremental costs transferred from developers[2]. Therefore, in the construction design of green building, it is necessary to construct the cost control model of green building. Combining with cost control and the estimation method of system dynamics parameters, the cost prediction and engineering cost control of green building are carried out. It has important practical significance to improve the quality of green building construction and reduce the cost of green building[3]. This paper studies the incremental costs and benefits of green buildings and ordinary buildings by using the method of "with or without comparison". Based on the theory of system dynamics, the constrained parameter model of building cost-benefit evaluation is constructed, which can accurately estimate the cost-benefit results and draw effective conclusions.

2 Incremental cost and benefit analysis of green building

Cost benefit evaluation of green building is a comprehensive evaluation including incremental benefits of economy, environment and society, and incremental costs of management and maintenance. This requires a systematic induction of different types of cost-benefit elements. See Figure 1 for details.

2.1 Evaluation index of incremental cost

The incremental cost of green building refers to the incremental investment brought by adopting green building technology different from the benchmark building in order to realize the Green function of buildings[4]. It mainly includes two aspects: The first is the purchase cost, which mainly includes the design cost and material cost of the developer. The developers transfer

* Corresponding author: lichong@ynufe.cn

the increased green building construction cost to consumers, mainly by increasing the sales price to make consumers bear the purchase cost. The second is the operation and maintenance cost, mainly including the incremental cost of the installation, management, repair, replacement and so on of the energy-saving facilities in the use process. Operation and maintenance is the main stage to generate incremental costs and benefits. Due to the adoption of some green technologies and facilities, the operating costs increase. Through the comparative analysis of energy-saving technologies in more than 20 green building projects, and combining with the relevant evaluation indexes of "energy conservation and energy utilization" in GB/T50378-2014, the following indicators are finally screened out. See Table 1 for details.

Table1. Incremental cost index system

Primary indicators	Secondary indicators
Energy saving and energy utilization	Energy saving maintenance structure (Energy saving roof, energy saving doors and windows and building floor cost)
	Energy saving lighting system (Cost difference between natural lighting and green lighting)
	Renewable resource systems (Cost of solar energy, geothermal energy and wind energy)
Land saving and outdoor environment	Site utilization (Increased cost of abandoned sites and underground space)
	Outdoor environment (Cost of noise control, ground landscape design, three-dimensional greening)
	Pervious ground (Cost increase caused by permeable paving method and permeable floor tile)
Water saving and utilization of water resources	Water saving system (Water source collection, reasonable discharge, etc.)
	Water saving equipment (Cost difference of water-saving faucet and water-saving sanitary ware compared with ordinary equipment)
	Water recycling (Increased cost of clear roads, toilet cleaning, air conditioning cooling water)
Material saving and material utilization	Green building materials (Increased cost of construction materials with little environmental impact and no pollution)
	Material recycling (Cost increase to improve the utilization of building materials)
	Waste recycling (Increased cost of waste treatment)

2.2 Evaluation index of incremental income

The incremental benefits of green buildings are mainly reflected in three aspects: economic benefits, social benefits and environmental benefits. As the direct benefit of green building, economic benefit is reflected in water saving, energy saving, government subsidies and so on. The cost-effectiveness of using process is usually due to the use of various green building equipment during the construction period, such as energy saving, water saving, material saving and other cost saving, which are manifested in the improvement of economic benefits. The benefits of government subsidies can be realized in the form of deed tax return or other forms. Environmental benefits mainly refer to the benefits from the use of green building materials and the implementation of environmental protection measures in the whole construction process, which can reduce pollution, prolong the life of buildings and reduce the occurrence of diseases. The social incremental benefits of green buildings are due to the use of green building materials and the implementation of environmental protection measures in the whole process of construction. Consumers can reduce diseases caused by living environment, reduce medical expenses, ensure physical and mental health, improve work efficiency, and increase welfare and income.

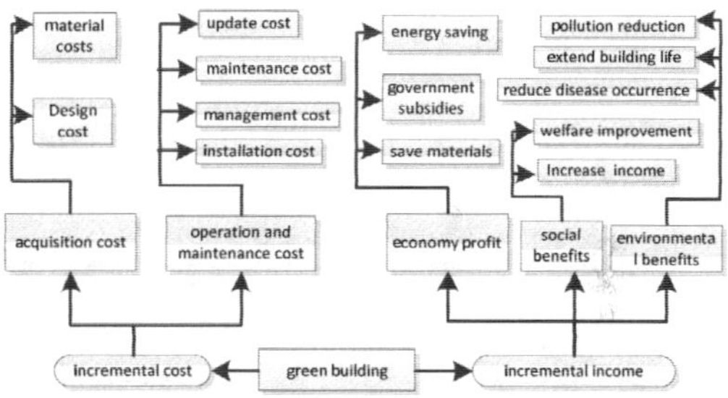

Fig1. Components of incremental costs and benefits

3 Construction of SD model for incremental cost and benefit of green building

System dynamics method is an important part of system engineering theory, which can reflect the whole process and simulate[5]. Therefore, the system dynamics method can be used to simulate the incremental cost-benefit analysis of green building in this paper, and has good applicability.

3.1 System objectives

Based on the dynamic theory, the incremental cost-benefit evaluation model of green building is constructed, which considers the influence of consumers, government and developers on the research object. Green building provides a more reasonable and healthy green working environment, so consumers should pay extra for it. At the same time, in order to promote the popularization of energy-saving technology in green building industry, the government not only introduced relevant subsidy policies, but also formulated mandatory laws and regulations to promote the orderly development of green building market. Whether green building developers use energy-saving technology depends on the cost-benefit ratio of incremental cost and benefit and the payback period of investment. Therefore, the model should not only consider the relationship between green building operation cost reduction and energy-saving technology utilization, but also pay attention to the environmental and social benefits contribution.

3.2 System boundary

According to the theory of system dynamics, the interaction between the internal elements of the system is the fundamental driving force to promote the dynamic behavior of the whole system. Under the influence of external environment, the whole system is easily affected, so it shows certain complexity and dynamics. In order to ensure the accuracy of the model calculation results, the boundary of the incremental cost-benefit index system of green building should be defined to reduce the complexity of the system as much as possible. Therefore, the incremental cost-benefit system is divided into incremental cost subsystem and incremental benefit subsystem. By determining whether the incremental costs and benefits have a greater impact on the objectives under study, the larger and more significant incremental costs and benefits are included within the system boundary, while some less significant incremental costs and benefits are ignored and divided outside the system boundary. When determining the boundary of incremental income index system of green building, the most basic unit of incremental benefit subsystem of green building is formed by referring to the construction form of incremental cost index in Table 1.

Table2. Incremental income index system

Primary indicators	Secondary indicators
Energy saving benefit and lighting	Heating air conditioning (Energy consumption difference of heating and air conditioning in green buildings compared with traditional buildings)
	Energy saving lighting system (Energy saving difference between energy saving lighting system and ordinary lighting system)
	Renewable resources (Electricity saved by using renewable resources in green buildings)
Land saving and outdoor environment	Land purchase (Saving land resources and purchase cost by using waste sites)
	Welfare benefits (Benefits from noise control, ground landscape design and three-dimensional greening)
	Underground space(Relieve the pressure of land use and improve the land utilization rate)
Water saving and utilization of water resources	Comprehensive utilization of rainwater (Water saving)
	Comprehensive utilization of reclaimed water (Reclaimed water recycling capacity)
	Water saving facilities (Amount of rainwater recycling)
Material saving and material utilization	Green building materials (Building material savings)
	Material recycling (Structural material savings)
	Waste recycling (Material cost savings)

3.3 Model construction and verification

3.3.1 Feedback loop.

This paper qualitatively studies the feedback process and relationship among the components within the system boundary by drawing the causality diagram, which is the basis of constructing the incremental cost-benefit evaluation model of green building. The specific drawing process is as follows: firstly, combining with the relationship between incremental cost and incremental benefit defined by the previous system, the causal relationship of incremental benefit system is drawn. Secondly, the causality analysis of incremental benefit system shows that the energy consumption level of green building is reduced due to the adoption of energy-saving technology, which reduces the emission of polluting gas, and correspondingly reduces the medical cost of residents, which can strengthen the willingness of residents to purchase the green building and improve the income of developers. Thirdly, the causal relationship between incremental benefits and incremental costs is analyzed. Developers can not only obtain direct benefits such as government subsidies and energy-saving economic benefits, but also obtain indirect benefits such as the increase of house price income and the extension of building life. By drawing the causal relationship and combing the relationship between the elements in the incremental cost-benefit system of green building, three feedback loops are obtained, all of which are positive feedback loops. The circuits are as follows:

Loop 1: Incremental cost →+ Energy saving technology level →+ Reducing emissions of polluting gases →+ Medical cost reduction →+ Willingness intensity →+ Incremental income →+ Incremental cost.

Loop 2: Incremental cost →+ Energy saving technology level →+ Building life extension →+

Efficiency improvement →+ Income increase →+ Willingness intensity →+ Housing purchase rate →+ House price income →+ Incremental income →+ Incremental cost.

Loop 3: Incremental cost →+ Investment rate of energy saving technology →+ Government subsidy →+ Willingness intensity →+ House price income →+ Incremental income →+ Incremental cost.

3.3.2 Drawing of stocks and flows change chart.

Compared with the incremental cost-benefit causality of green building, the stocks and flows chart is adjusted according to the system dynamics equation. The specific drawing process is as follows: Firstly, considering the availability of data and the operability of the model, the constant of usage, maintenance, operation, management and update incremental cost is added to the inventory flows chart. Secondly, four state variables of economy, society, environment and incremental benefit are introduced into the stocks and flows chart.

3.3.3 Construction of system dynamics equation.

The system dynamic equation between variables is established by dynamo language in software Vensim combined with stocks and flows chart. After the variables and parameters of the model are determined, the variable equation of the model should be edited. There are many variables and equations involved in this model. The details are as follows:

Final time = 25, units: year, initial time = 0, units: year.

(1) Incremental cost (L_1) = (maintenance incremental cost L_3 + operation incremental cost L_4 + management incremental cost L_5 + update incremental cost L_6) × pulse (0, 4), units: RMB / m^2; (2) Incremental maintenance cost (L_3) = (maintenance charge standard × actual charge area) × pulse (0, 2), units: yuan / m^2; (3) Operating incremental cost (L_4) = (energy saving product consumption reduction rate × operation charging standard) × pulse (0, 2), units: yuan / m^2;(4) Management incremental cost (L_5) = (energy saving product consumption reduction rate × management cost standard) × pulse (0, 2), units: yuan / m^2; (5) Update incremental cost (L_6) = (equipment update rate × equipment update standard) × pulse (0, 2), units: RMB / m^2; (6) Incremental cost (B_1) = (acquisition cost (L_2) + usage cost (L_1))× pulse (0, 2), units: RMB / m^2; (7) Incremental benefit (B_2) = IF THENELSE (time<=3,0, (social benefit (B_3) + economic benefit (B_4) + environmental benefit (B_5)) × pulse (0,3), units: RMB / m^2; (8) Social incremental benefit (B_3) = (efficiency improvement benefit + disease reduction benefit) × pulse (0,3), units: yuan / m^2; (9) Economic increment benefit (B_4) = (government subsidy benefit + energy saving benefit + water saving benefit + material saving benefit + other saving benefit) × pulse (0,5), units: yuan / m^2; (10) Environmental incremental benefit (B_5) = (domestic waste treatment benefit + carbon emission reduction benefit) × pulse (0,3), units: yuan / m^2.

3.3.4 Model test.

After the construction of the system model, it is necessary to further test and judge the rationality of the model. By adding or deleting the set variables in the model, we can observe whether the system can still form a closed loop to test the model. First of all, make sure that the variables related to the research problem have been set in the system model. Secondly, check each variable in the estimation model. When one of the variables in the model is eliminated, whether the closed loop can be formed is monitored. When a closed loop cannot be formed due to the elimination of a variable, it means that the variable is endogenous and cannot be deleted. Similarly, this method is used to test other variables in the evaluation model, and the final variables in the model have passed the test, and the model construction is reasonable.

3.3.5 Empirical analysis.

This paper selects a residential district which has obtained green building certification in Guangzhou City as an empirical analysis case. The project is 10 residential buildings with a height of 77 meters, and has been awarded two-star green building design logo. A variety of green energy-saving technologies and equipment are used in the internal facilities of the whole residence. Renewable materials and green environmental protection materials are used in the building. The air conditioning system is equipped with automatic air conditioning system, and the fresh air system which can filter and purify the air is installed. In terms of light energy utilization, the external building is installed with point hanging curtain wall solar panels, which can be used to collect solar energy and reduce energy consumption. Each building is equipped with intelligent monitoring sensors, which can detect the working state of air conditioning, water, electric energy and elevator, and timely understand the working state of air quality, water consumption, lighting, etc., and can automatically adjust in case of abnormal conditions.

3.3.6 Empirical results.

Through the simulation of its incremental cost and incremental benefits, and the estimated results are studied. Through the analysis of incremental cost and benefit in green building simulation period, the annual incremental cost will be stable by 2025, with an annual incremental cost of 45 yuan / m^2. The incremental cost in this stage mainly includes the maintenance of new energy-saving equipment and replacement of parts. The incremental benefits of green buildings gradually increase every year, exceed the incremental costs in 2021, and tend to be stable in 2028. The annual incremental benefits of green buildings are 100 yuan / m^2. According to the calculation method of payback period, the incremental cost-benefit balance time point of the residential building project is 8.0 years.

4 Conclusions

Based on the analysis of incremental cost and benefit of green building, the system dynamics model of green building is constructed and tested. Incremental cost includes two stages of incremental cost. Incremental benefit includes economic benefit, social benefit and incremental benefit. The conclusions are as follows:

Firstly, based on the system dynamics model, this paper studies the change law of incremental cost and incremental benefit of green building energy conservation from different perspectives of government, open business and consumer, and confirms that the total incremental benefit of green building is greater than the incremental cost.

Secondly, the two main sources of incremental benefits of green buildings are house appreciation and energy consumption reduction. By analyzing the causal feedback relationship between incremental cost and incremental benefit, and quantitatively studying the external economic benefits of green building projects, the accuracy of the model estimation results is improved.

Thirdly, the benefit of green building to residents is greater than the cost of initial investment. Consumers should have an objective and correct understanding of this. The government should strengthen the publicity and guidance of green buildings, increase subsidies and investment in green buildings, and support and encourage the development of green buildings.

References

1. Xiaoling Z., A. Platten , and L. Shen, Green property development practice in China: Costs and barriers. Building and Environment, **46**(11):p.2153-2160 (2011)
2. Ai D., Z. Hong, and Y. Wan, Research on the incremental cost-benefit of green buildings from the perspective of consumers. Value Engineering, **038**(022):11-15(2019)
3. Eichholtz P., N. Kok , and J. M. Quigley, The Economics of Green Building. Review of Economics and Statistics, **95**(1):50-63(2013)
4. Coma J., G.Perez, and A. De Gracia, Vertical greenery systems for energy savings in buildings: A comparative study between green walls and green facades. Building and Environment, **111**(JAN.):228-237(2017)
5. Yongguang Z., J. Xiaoqing, and Q. ying,*System Dynamics (Second Edition)*. Science Press(2013)

https://doi.org/10.1051/e3sconf/202123703013

Study on the influence of radius of curvature on flow resistance in spiral pipe

Wenqiang Li[1], Hui Qi[1], Yongfeng Yang[1], Guojun Zhao[1], Rong Liu[2,*], and Haonan Wang[2]
[1]Power China Road Bridge Group Co, Haidian District, Beijing, 10089, China
[2]State Key Laboratory of Coal Mine Disaster Dynamics and Controls, Chongqing University, Chongqing, 400030, China

Abstract. With the continuous development of west of China, a new kind of tunnel, namely spiral tunnel, has appeared in the expressway. Due to the special linear pattern, the resistance of airflow in spiral tunnel has been found to change, but there are few relevant researches at present. Therefore, numerical calculation method is used to study the variation of flow resistance in spiral pipe with different curvature. The results show that when the fluid flows in the spiral pipe, the wind speed is not uniformly distributed. The highest speed is not in the center of the pipe, but on the outside of the pipe, and the offset distance decreases with the increase of the radius of curvature. In addition, the change of flow resistance in spiral pipe is studied, and it is found that the change rate of flow resistance decreases with the increase of curvature radius. It shows that the radius of curvature is negatively correlated with the flow resistance.

1 Introduction

When the highway is built in two places with a large difference in height, the spiral tunnel will be built to take the driving safety into consideration. At present, spiral tunnels have appeared in western China. Table 1 lists the spiral tunnels in China. A lot of pollutants will be generated during the construction of the tunnel. However, the spiral tunnel is different from the existing straight tunnel. The special line shape may change the flow resistance of the wind flow in the tunnel. Before the construction of the tunnel, the fan must be selected. The fan selection needs to calculate the maximum air demand and maximum ventilation resistance of the tunnel. The reliability of ventilation system directly affects the safety of tunnel construction and the physical and mental health of workers. The calculated ventilation resistance is lower than the actual ventilation resistance, resulting in the inability of fresh air to enter the tunnel construction area. The calculated ventilation resistance is higher than the actual ventilation resistance, and the ventilation system energy consumption is too high. However, there are few studies on the influence of spiral pipes on resistance.

Ventilation system is the basic guarantee of coal mine safety in the production process, so it is paid much attention in the production process. Ventilation resistance test and distribution analysis will be carried out regularly.[1-4] Shu Li and Xuefeng Liu[5] studied the factors affecting stroke obstruction of urban pipe corridor by means of physical model. According to the method of computational fluid dynamics, Isidro Diego[6] had calculated 138 models with different section dimensions, wind speed and wall surface characteristics, studied the airflow resistance in various situations. Jin Wenliang et al. conducted resistance system tests on zhongnanshan tunnel, Jinping tunnel, Beibei tunnel, and Zhongliangshan tunnel in Qinling Mountains, China.[7-11] As mentioned above, although a lot of literatures have studied the along resistance, few have studied the relationship between the along resistance and the

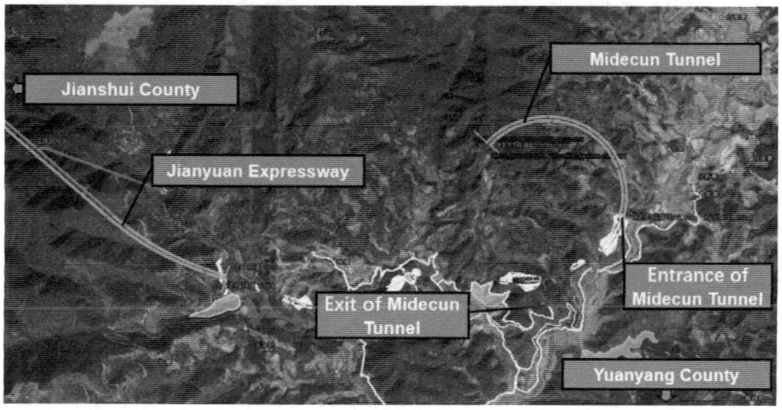

Fig.1. Tunnel location map

* Corresponding author: cqu_liurong@cqu.edu.cn

radius of the spiral tunnel.

The purpose of this paper is to obtain the variation of ventilation flow resistance in tunnels with different radius of curvature by numerical calculation. The wind flow in the tunnel is the pipe flow in the final analysis. Therefore, this study is not limited to the study of the flow resistance changes in the spiral tunnel, but also to the study of the resistance of the wind flow in the spiral pipe under different curvature radii.

Table 1. The spiral tunnel in China.

Tunnel name	length/m	radius of curvature/m	expressway
Changhe lu tunnel	left：415	100	Changhe road
	right：475	110	
Ganhaizi tunnel	left：1713	600	Yaxi expressway
	right：1798	618	
Tiezhaizi tunnel	left：2792	600	Yaxi expressway
	right：2940	618	
Wolong gou tunnel	left：2626	700	Daxun expressway
	right：2554	690	
Jinjiazhung tunnel	left：4228	890	Yancong expressway
	right：4104	860	

2 Numerical model and calculation

2.1. Calculation method of pipe resistance.

In the process of viscous fluid movement, energy loss is inevitable. In the Bernoulli equation of total flow, the mechanical energy loss of unit weight fluid is expressed as the head loss. Viscosity is the root cause of head loss, ideal fluid flow, because of no viscosity, so no head loss. According to the different flow boundary conditions and mechanism of the head loss, the head loss can be divided into the on-way resistance and the local resistance. The experiment shows that the water head loss h_f along the pipe and non-pipe flow is directly proportional to the flow length l and inversely proportional to the pipe diameter d. The calculation formula is as follows:

$$h_f = \lambda \frac{l}{d} \frac{v^2}{2g} \quad (1)$$

And local head loss calculation formula：

$$h_f = \zeta \frac{v^2}{2g} \quad (2)$$

2.2. Introduction of Midecun Tunnel

Spiral tunnel site ventilation resistance test is selected in Yunnan Province (a) yuan expressway mi village tunnel entrance right hole. The left line (Z3K35+076 ~ Z3K38+906) long 3830 m, the right line (K35+110 ~ K39+080) long 3970 m, the tunnel is a spiral tunnel, the height difference between the entrance and exit is about 80 m, the longitudinal slope of the tunnel is -2.08%, -2.05%(one-way slope) tunnel entrance axis direction 147°, the maximum buried depth of 366 m, is a deep buried long tunnel. The location area of Midecun tunnel is in Hani-Yi Autonomous Prefecture of Honghe, Yunnan Province and the curve characteristics are shown in Table 1. Figure 1 is the map of Midecun Tunnel.

Table 2. Curve elements of the Midecun Spiral Tunnel.

		Left tunnel	Right tunnel
Boundary mileage		ZK34+752.611~ ZK37+926.346	K34+781.034~ K36+424.870
Curve elements	R	710	730
	L	3173.735	1643.836
Boundary mileage		ZK37+926.346~ ZK38+569.533	K36+424.870~ K38+068.706
Curve elements	R	710	730
	L	643.187	1643.836
Boundary mileage		ZK38+569.533~ ZK39+770.859	K38+706.635~ K39+898.578
Curve elements	R	750	750
	L	1201.326	1191.943

2.3. Physical model and boundary conditions

Pipelines with different radius of curvature are constructed according to the calculation requirements.

Table 3. Detailed parameters of pipeline.

serial number	Length L/m	Pipe radius r /m	Radius of pipe curvature R /m

 https://doi.org/10.1051/e3sconf/202123703013

S-1			
R-1			50
R-2			100
R-3			200
R-4	298.3m	1	400
R-5			800
R-6			1200
R-7			1600

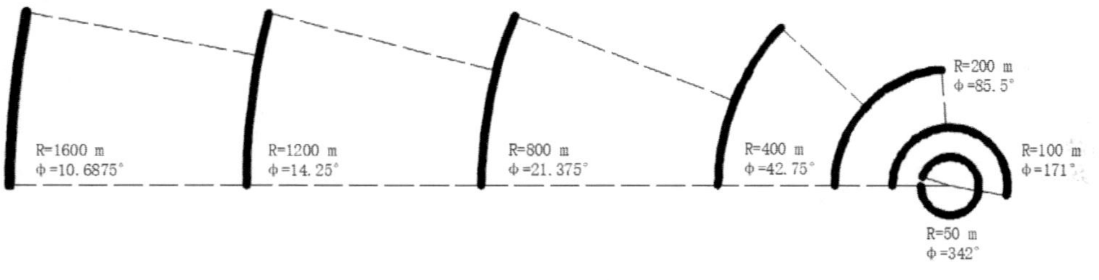

Fig.2. Different pipe corresponds to the radius and Angle

Detailed parameters of pipeline are shown in Table 2. The pipes with different radius of curvature R have the same length L and the same radius r. The variation trend of flow resistance in spiral pipe under different radius of curvature is studied. The pipes with different radius of curvature are shown in Figure 2. It can be seen from the diagram, along with the increase of radius of curvature, the same as the length of the pipe is more and more straight. The boundary condition of the pipe inlet is set as the pressure inlet, and the pressure is 0 Pa; the pipe outlet is set as the velocity inlet, and the velocity is set as 1, 3 and 10 m/s respectively.

3 Numerical Simulation Results

3.1 Flow field distribution in pipeline

After the calculation, the wind speed distribution in the pipeline under different curvature radii was observed, as shown in Figure 3. Through observation, the maximum wind speed in the pipe was found and marked, as shown in Figure 3. It is obvious that as the radius of curvature increases, the position of the maximum velocity gradually moves towards the center of the pipe. When the radius of curvature is small, the maximum velocity is obviously close to the outside, while when the radius of curvature is larger, the maximum velocity is closer to the center.

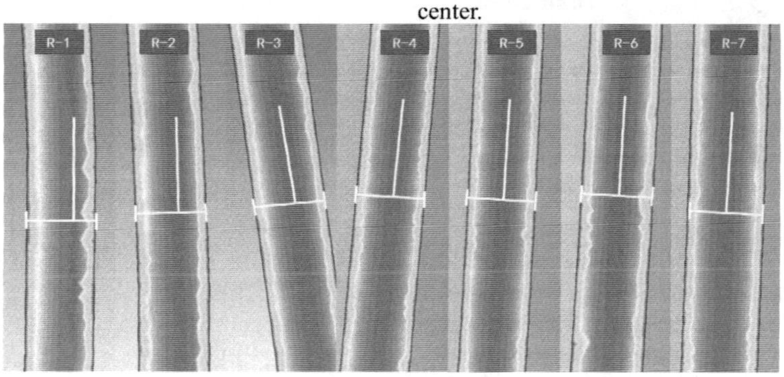

Fig. 3. Flow velocity distribution in pipe with different radius of curvature

3.2 Flow resistance study

Two conditions are set in the calculation. The pipe inlet is the pressure inlet, and the outlet is the speed outlet, respectively setting 3 and 10 m/s. After the calculation is completed, the pressure value on the center line of the pipeline is derived and the pressure change curve is drawn in Figure 4.

The slope in the figure represents the change of resistance in the flow process. The same resistance coefficient is set in the pipeline during calculation, and there will be no local resistance in the pipeline with regular shape . Therefore, it can be considered that the slope in the figure represents the influence of the helix on the flow resistance. Therefore, it can be considered that the slope in the figure represents the influence of the helix on the flow resistance. It can be seen that the flow resistance changes at two different flow rates have similar trends. With the increase of the radius of curvature, the flow resistance of the same length spiral pipe is smaller.

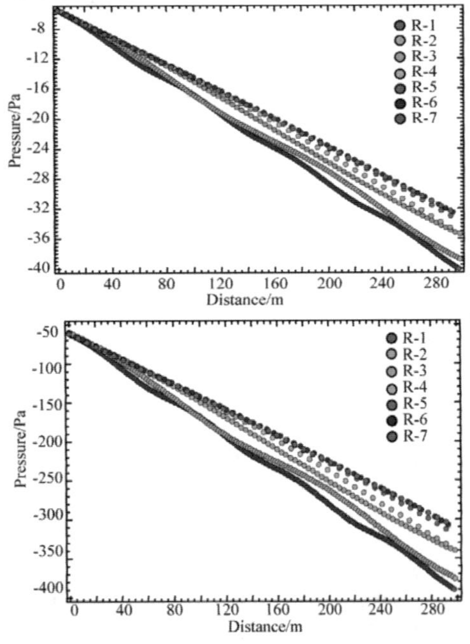

Fig.4. Curve of pressure in pipe with distance

4 Conclusions

In this paper, the flow resistance in spiral pipe is studied based on the village tunnel of Jian-ge-yuan expressway. It can be found from the calculation results that when the fluid flows in the spiral pipe, the flow velocity in the pipe is not distributed symmetrically like the linear pipe, but has an offset, and this offset gradually decreases with the increase of the spiral radius. It is found that no matter what the flow velocity is in the pipe, the change rate of flow resistance in the spiral pipe decreases with the increase of the spiral radius. This indicates that the spiral shape brings extra resistance to the flow, and the resistance decreases with the increase of the radius of curvature.

References

1. X.D. Huang, Z.M. Liu, S.L. Chen, J.K. Cui. Testing Methods and Application for Mine Ventilation Friction. Coal Mine Safety, **35**, 13-15 (2004)
2. S.R. Cheng, J.J. Cheng. Measurement of mine ventilation resistance and analysis of several problems. Coal Mining, 72-74 (2006)
3. Y.H. Zeng, X.X. Ma. Study on the accuracy of ventilation resistance measured by the total differential pressure method. Coal Engineer, 5-6 (1998)
4. Shenyang Coal Mine Design and Research Institute. Friction resistance coefficient of mine ventilation roadway (A standard) table. (1985)
5. S. Li, X.F. Liu, J.X. Wang, Y.L. Zheng, S.M. Deng. Experimental reduced-scale study on the resistance characteristics of the ventilation system of a utility tunnel under different pipeline layouts. Tunnelling and Underground Space Technology, **90**, 131-143 (2019)
6. I. Diego, S. Torno, J.Torano, M.Menendez, M. Gent. A practical use of CFD for ventilation of underground works. Tunnelling and Underground Space Technology, **26**, 189-200 (2011)
7. W.L. Jin. Study on ventilation environment Parameters test of zhongnanshan Extra-Long Highway Tunnel in Qinling Mountains; Chang 'an university, (2006)
8. Y.H. Yu. Study on flow resistance and jet fan lift pressure in Jinping Tunnel; Southwest Jiaotong University, (2006)
9. Y.H. Yu, B. Lei, H.Q. Bi. Numerical simulation of flow field and local resistance in emergency stop zone of tunnel. Modern tunneling technology, **43**, 49-52 (2006)
10. Chongqing Expressway Group, Southwest Jiaotong University. Longitudinal ventilation study of highway long tunnel; (1998)
11. Chongqing Expressway Group, Southwest Jiaotong University. Evaluation report on the effect of the Intelligent control system of Changchang Road Tunnel; (2006)

https://doi.org/10.1051/e3sconf/202123703014

Research on Flocculation of Cement Particles in Water

Zuo Yanfeng[1], Zhang Yijie[2], and Wang Dongmin[3]

[1] Department of Civil Engineering, Institute of Disaster Prevention, Langfang, China
[2] School of Civil Engineering & Transportation, Beijing University of Civil Engineering and Architecture, Beijing, China
[3] School of Chemical & Environmental Engineering, China Mining & Technology University(Beijing), Beijing, China

Abstract. Nowadays, workability is absolutely necessary for piping concrete, which is influenced by characteristics of structure of fresh cement paste. In this paper, the movement of particles in electric field and flocculent units at different W/C were observed, and rheologic characteristics of fresh cement paste at different W/C and the pastes with naphthalene sulphonate (NS) and poly carboxylate (PC) superplasticizers were investigated using rotational viscometer (NXS-11A). It is found that at initial hydration age cement particles carry different electric quantity and electrical property; increment of W/C induces flocculent units more separated, less hysteresis loop area and lower crush point; application of superplasticizers brings much less loop area and indistinct crush point. It is concluded that 1) fresh cement paste is filled with flocculation units of different strength; 2) different units are connected with one another loosely; 3) tunnels for bleeding form in the weakest connecting area. Water can disperse flocculent units to some extent, especially for superplasticizers. However, the latter can't separate flocculation of cement particles completely.

1 Introduction

Workability of fresh concrete is required in the process of placing concrete, especially for high-rises[1]. Some of the construction requires self-compacting placement[2]. The factors influencing workability are characters of cement, water content, sand percentage, chemical admixtures[3][4] and so on. Fresh cement paste is cohesive material between fine and coarse aggregates, which affects workability to large extent and is also one of the important sources making fresh concrete flow. Water reducing agent, particularly superplasticizer, can change the structure of cement paste in fresh concrete, and enables concrete to be plasticized.[5][6][7]

Before addition of water, cement particles are separated. Tattersal and Banfill [8] reported that there is weak surface force between the particles and they are connected with each other loosely. After water addition, surface of cement particles is changed, and a type of paste with yields stress, named weak solid by Heimuth [9], establishes. Rheologic characteristics of fresh cement paste are relevant to its inner structure, which is associated with the nature of particles in it.

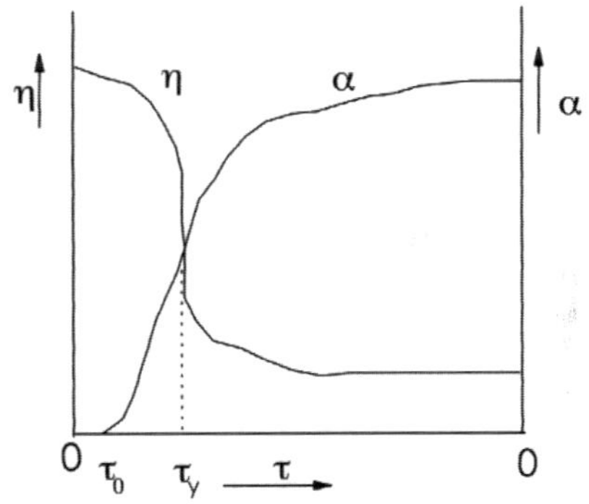

Fig1. Variation of viscosity, shear stress and structure of fresh cement paste.

T.C. Powers's sedimentation experiments [10] showed that in the watery suspension cement particles usually were flocculated and sedimentated sharply. Helmuth [9] considered that water molecules were adsorbed on cement particle surface, and every particle was packed within a continuous water film, and once the paste is stirred or water reducing agent exits, this film would tend to be uniform. Tattersall and Banfill's [8] rheology experiments and SEM observation indicated that hydrate film enwrapped the flocculation structure after cement and water mixed.

Zhang [11] draw a conclusion from the viewpoint of correlation between viscosity and fresh paste structure. His results indicated that viscosity varies with the structure of fresh paste, which is shown in Figure 1. αrepresents structure crushing process, and η viscosity changing process. When shear stress is less thanτ0, structure is not crushed, and while reaching τy, namely crush point, the structure will be destroyed sharply. If τ is large enough, viscosity becomes constant, which means the structure is destroyed completely.

This paper would illustrate the formulation of fresh cement paste through rheological and optical method, which is important for design of piping or self-compacting concrete.

2 Experimental Details

2.1 Materials

Portland cement is applied. And its chemical, mineral content are shown in Table 1 and Table 2.

UNF-5 is β-naphthalene sulphonate (NS) superplasticizer in powder state. COPOCA202 is poly carboxylate (PC) superplasticizer, in liquid state, and its solid content is 30%. Water reducing effect of the former is 17% at dosage of 0.70% (by mass), the latter 26% at 0.60% (by solid mass) [12].

Table1. Mineral and chemical components of Portland cement (%)

SiO_2	Al_2O_3	Fe_2O_3	CaO	MgO
21.24	4.73	2.63	63.94	1.78

Table2. Mineral and chemical components of Portland cement (%)

SO_3	Na_2Oeq	f-CaO	C_3S	C_3A	Ignite loss
2.73	0.54	1.00	54.01	7.36	2.06

2.2 Observation of Flocculent Unit

Cement powder was added in deionized water, stirred for 20s to a homogenous suspension, a drop of which were left on a glass slide to be investigated. XJZ-6 metallosope was applied to survey the specimens.

As for the observation of interface among flocculent units, extra stewing for 30s is needed. And then filter paper was employed to suck the above water slightly, and the remainder is the sample required.

The experimental temperature was 26°C.

2.3 Electric Characters of Cement Particles at Very Early Age

Cement was added in deionized water, stirring for 20s. BDL-B zeta potential apparatus was applied. The experiment temperature was 20°C. W/C is 1000.

2.4 Rheologic Characteristics of Fresh Cement

The fresh cement pastes were agitated for 3min by a stirring machine at same speed. Rotational viscometer (NXS-11A) was employed to measure rheologic performance. The experimental temperature was 17°C. The dosages of UNF-5 and COPOCA202 are 0.70% and 0.60% (by solid mass) respectively.

3 Experimental Results and Discussions

3.1 Electric Characteristics of Cement Particles at Very Early Age

Figure 2 shows that moving directions of particles are different in the electrophoresis tank, and so is velocity, which means some particles present positive, others negative, and the values are also different. And that confirms the different electric characteristics of every cement particles during their initial hydration period.

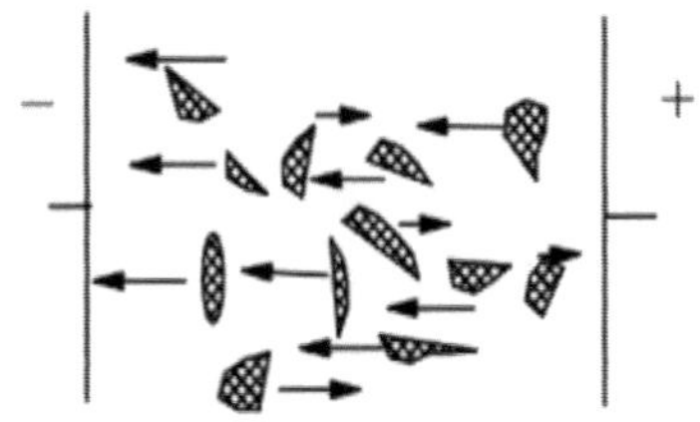

Fig2. Illustration of cement particles movement in electric field.

3.2 Flocculation of Cement Particles

From the Figure 3, it is indicated that at W/C=5.0, structures can't be shown definitely; as for (b), large flocculation units are presented, and small units are filled among large ones; for (c), the units become smaller, and the smallest in (d). Therefore, the strength of flocculation units in fresh cement paste is different. And they can be destroyed progressively by water. Firstly, some of the weak units are crushed and at the same time the rest of harder ones alters to be weak. Then with more water applied, the weaker ones of the former residual are destroyed, and it brings much weakness to the remainder. And the process will cycle until all of them being crushed.

(a) W/C=5.0

 https://doi.org/10.1051/e3sconf/202123703014

(b) W/C=10.0

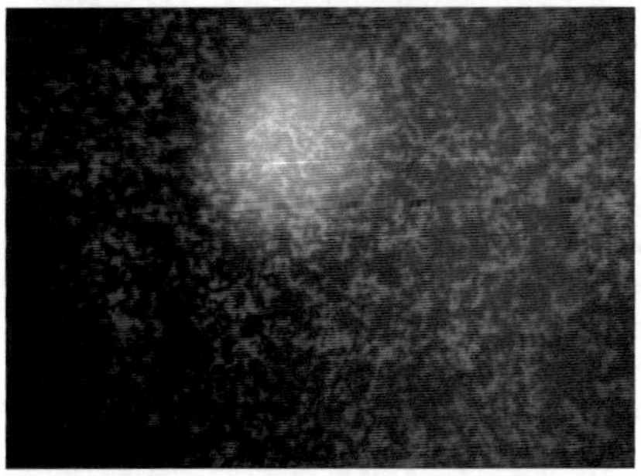

(c) W/C=30.0

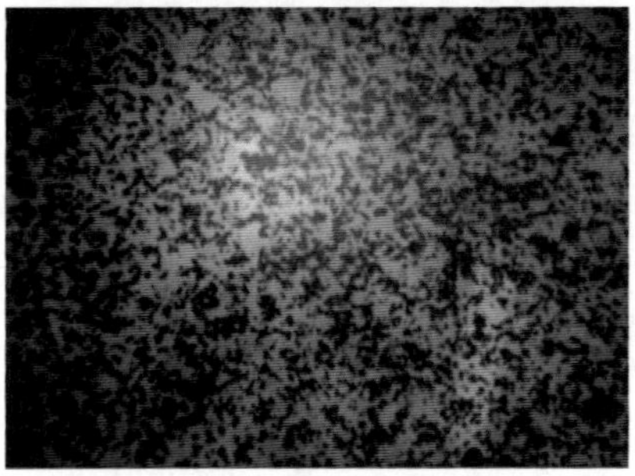

(d) W/C=100.0

Fig3. Flocculent units at different W/C, 50×.

It is shown in Figure 4 that flocculent units are connected incompactly, which conduces the formation of bleeding tunnels between them.

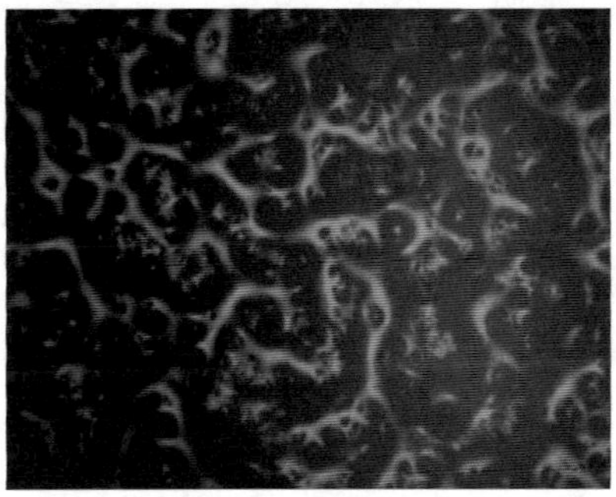

Fig4. Interface between flocculent units, W/C=20.0, 50×

3.3 Rheologic Properties of Fresh Cement Paste

3.3.1 Rheologic Characters of Pure Cement Paste

Typically, fresh cement paste has hysteresis loop, shown in Figure 5, which means that shear rate (DS) increasing process (upward) curve and decreasing (downward) process curve are not in superposition. And it is also indicated that the shear stress of upward process is higher than downward, which is called shear thinning or thixotropy phenomenon. And loop area is considered to represent flocculation degree, and the larger area, the harder flocculation.

With different W/C, cement paste shows different flocculation. Figure 6 indicates that loop area decreases with W/C increasing and so does shear stress at the same shear rate. And in Figure 7, it is shown that with W/C increasing, viscosity decreases at the same shear stress, and crush point also decrease.

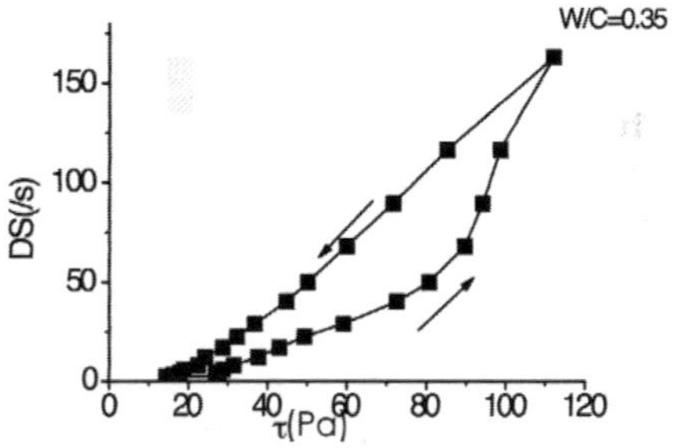

Fig5. Hysteresis loop of fresh cement paste.

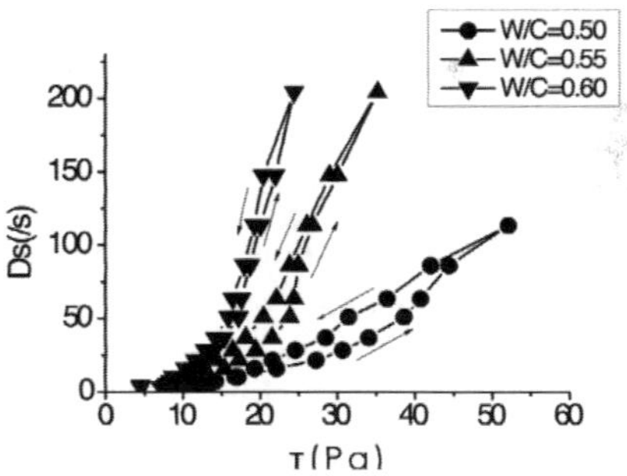

Fig6. Hysteresis loop of fresh cement paste at different W/C.

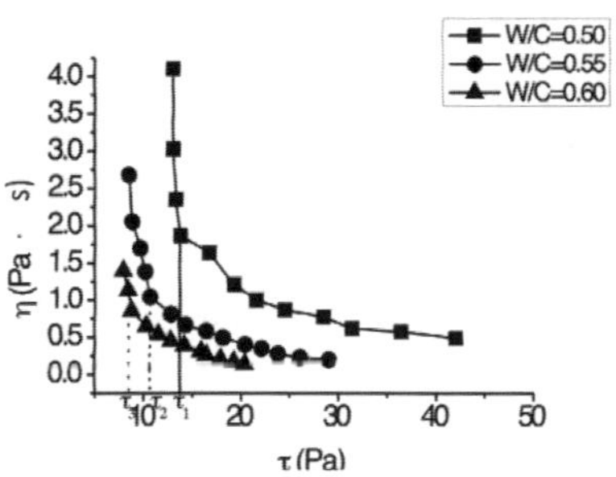

Fig7. Crush point of fresh cement paste at different W/C, andτ_1, τ_2, τ_3 are the crush points of paste of W/C =0.50, 0.55, 0.60 respectively.

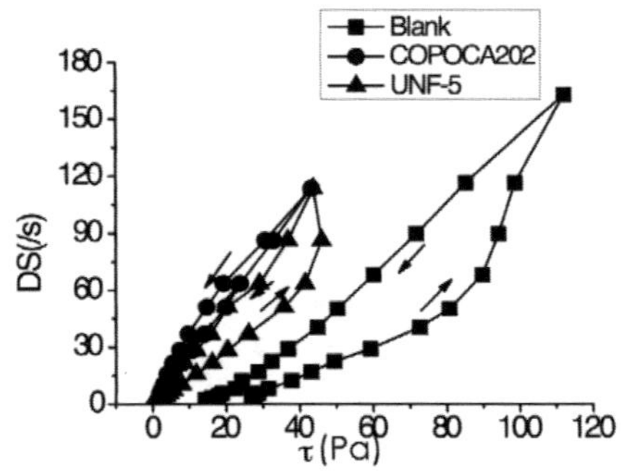

Fig8. Hysteresis loop of fresh cement paste with superplasticizers, W/C=0.35.

3.3.2 Rheological Characteristic of Cement Paste with Superplasticizers

Figure 8 shows that addition of superplasticizer leads to a decrease in the hysteresis loop, and PC superplasticizer with higher water reducing effect shows the least area.

As to Figure 9, viscosity of the samples with superplasticizers is almost unchanged with shear stress increasing compared with blank sample, and the crush points are indistinct either, which means that superplasticizer can destroy flocculation structures to large extent.

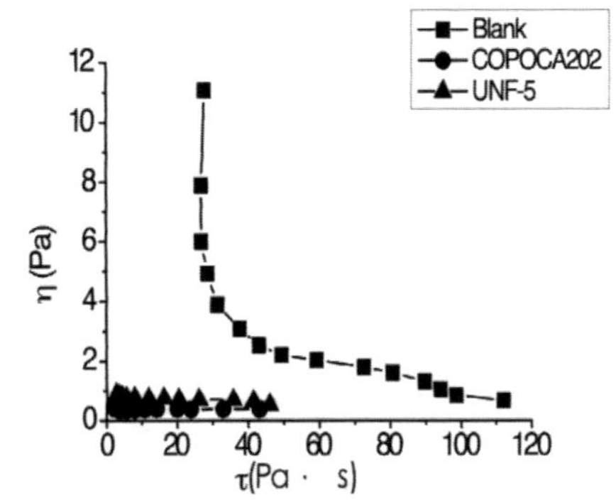

Fig9. Crush point of fresh cement paste with and without superplasticizers.

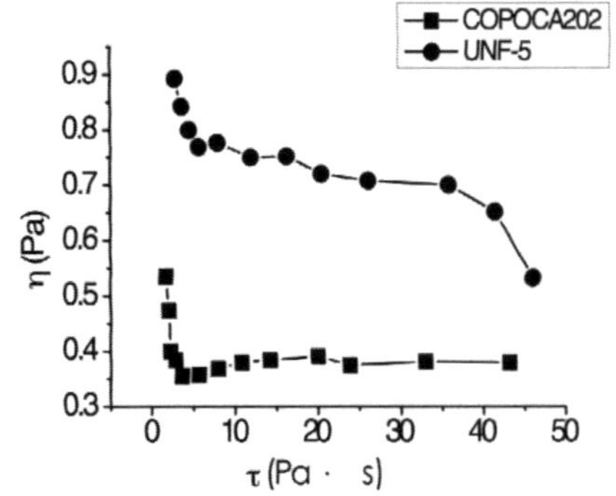

Fig10. Crush point of fresh cement paste with COPOCA202 & UNF-5.

Figure 10 shows that the specimens with superplasticizers have a similar curve of Figure 1. Namely, superplasticizers just destroy a part of flocculation structures, while the structures with high strength do not diminish and just alters to be weak ones. Compared with Figure 9, Figure 10 indicates that the crush points of the sample with superplasticizers are much smaller than the former, because of the residual infirm structure that can be destroyed by weak shear stress.

At high shear stress, UNF-5 paste exhibits decline of viscosity, which COPOCA202 sample doesn't have. The dispersion of the former is less than the latter, which induces more and stronger flocculent units exiting in UNF-5 specimen. High shear stress destroys them, which brings forth decreasing of viscosity.

3.4 Discussions

The size of cement particles is various and the shape is irregular. Moreover, mineral distribution in every particle is nonuniform. Thus, every particle has its own characteristics of hydration process, which makes surface characters (such as electric properties, solvation film, etc) of every particle different. Reports [13-15] show that hydration of aluminate phase contributes to positive electricity, however, silicate phase negative, and at very early stage of hydration cement particles show positive.

After water addition, the aluminate phase in cement particles makes them show positive, and however, some of fine particles containing more silicate phase represent negative. Particles with different sign connect with each other through static electric attractive force, which bring forth flocculent units, and the higher static electric attractive force, the higher strength of the unit. With time elapsing and hydration process going on, gradual formation of hydrates at particle surface will engender chemical osculation once they develop large enough in size. Actually, during the initial hydration period, electric quality and sign of cement particles also vary with time [16], which may influence strength of flocculent unites to some extent.

Therefore, there are two types of important forces during formation of cement paste structure, i.e. static electric attractive force and chemical bond. At the first stage, the former dominates, which generates flocculation; then, the latter plays more important role, which engenders hardened cement paste eventually. Another one for attraction is Van der Waals force, which is much less than electric force at very early age of hydration.

The whole fresh cement paste is filled with different strength flocculation units, which are connected with one another loosely. And tunnels for bleeding form in the weaker connecting area.

During stirring process, weak flocculation structures are crushed fractionally. At higher shear stress, more of them would be crushed. When shear rate decreasing, many of them can't be reconnected, and the whole paste is filled with many separated flocculation units. During downward process, the decreasing shear stress can't crush flocculent units so greatly as that of the upward process, and therefore, the downward curve is more linear with DS than the upward. And because of destruction of large flocculation unites, under the condition of same DS, shear stress during downward is lower than upward.

The paste at high W/C has more incompact flocculent units whose solidity is lower, and maybe the amount of the units is less than the former. When shear stress acts on the paste with high W/C , flocculent units are apt to be broken up. And it induces smaller counterforce to balance the outside force, that shows lower shear stress. And also at low shear stress the flocculent units can be destroyed to large extent, which engenders low crush point. If W/C is high enough, flocculent units may be destroyed to great extent in stirring period. And the upward line can be near to superposition with the downward one.

Naphthalene sulfate and poly carboxylate superplasticizers are anionic surfactants, which tends to adsorb on positive cement particles, and enlarge repulsive force between them through increasing static-electric or steric repulsive force, which brings forth the destruction of some flocculent units. The adsorption induces identical electric sign of cement particles, which brings difficulty to flocculation for the electric repulsive force between particles. However, Fig. 10 shows that they cannot disperse the flocculent units entirely, and a few very weak ones remain. And PC superplasticizer, which has higher water reducing effect than NS, induce much weaker units than NS. This type of flocculation is maybe formed by collision of particles at their irregular tips or edges. And therefore, they are very weak.

4 Conclusions

During the initial hydration period, cement particles show different electrical performance, some carrying positive charge, and some negative, and the electric quantity of every particle is also different, which induces formation of flocculation by static electric attractive force. And collision at the irregular tips or edge of particles is maybe another cause of the formation.

The whole fresh cement paste is filled with flocculation units of different strength. Different units are connected with one another loosely. And tunnels for bleeding form in the weakest connecting area.

During stirring process, a part of weak flocculation structures are been destroyed. With higher shear stress, more of them would be crushed. That is reason why fresh cement paste has hysteresis loop. And the paste with high W/C has weak flocculation, which brings a small value of hysteresis loop area and crush point.

The crush point becomes indistinct when superplasticizers are applied, which means they can destroy flocculation structure greatly. Particularly, PC is able to destroy the structures more compared with UNF-5. However, they can't destroy flocculation completely for the existence of crush points.

References

1. Liu.Rentai,Zhang.Chunyu,Liu.Haojie,Pei.Yan,Li.Xiu hao, Wang.Zhenjun, Effects of flocculation of cement slurry on groutability of porous media Construction and Building Materials, **v 233** （10 February 2020）
2. Rao.S.Venkateswara.,Rao.M.V.Seshagiri, Ramaseshu.D,Kumar, P.Rathish, Durability performance of selfcompacting concrete Magazine of Concrete Research, **v 64**, n 11, p 1005-1013 (November 1, 2012) Construction and Building Materials, **v 237** (20 March 2020)
3. Han.Kaidong,Xiao.Jia,Zhang.Zedi,Wang.Jialei,Hu.X iaoyan, Effect of particle size distribution on flocculation and its growth in cement-ground limestone suspensions Construction and Building Materials, **v 262** (30 November 2020)
4. Lucia,Influence of superplasticizers on the flocculation degree of cement suspensions Ferrari, Boustingorry, Pascal American Concrete Institute, ACI Special Publication, **v 2015** (January,2015)
5. Zeyad, Abdullah M, Almalki.Ali Influence of mixing time and superplasticizer dosage on self-consolidating concrete properties (Open Access) Journal of Materials Research and Technology (2020)
6. Ma.Baoguo,Qi.Huahui,Tan.Hongbo,Su.Ying,Li.Xian gguo, Liu.Xiaohai, Li.Chunbao, Zhang.Ting, Effect of aliphatic-based superplasticizer on rheological performance of cement paste
7. Ke.Guoju, Zhang.Jun, Journal of Advanced Concrete Technology, **v 18**, n 1, p 17-26 (January 2020) Effects of retarding admixture, superplasticizer and supplementary cementitious material on the rheology and mechanical properties of high strength calcium sulfoaluminate cement paste
8. G. H. Tattersall and P. E. G. Banfill, Rheology of Fresh Concrete ISBN:0-273-08558-1, Pitman Advanced Publishing Program, Pitman Press (1983).
9. R. A. Heimuth. Flocculation and rheology of fresh cement paste. Chinese version of 7th international cement chemistry conference collective papers. Beijing: China Architecture & Building Press (1983)
10. T. C Powers. The bleeding of Portland cement paste, mortar and concrete, ACI Journal, **Vol 35**, p465-480, (1939).
11. G. L. Zhang, Y. L. Zhang. Principles of Chemical Admixtures for Concrete and Their Application, ISBN: 7-112-00682-1, Shanghai Science & Technology Press, （1985）
12. Y. F. Zuo, D. M. Wang, S. Z. Wu. Effects of Cross-linking Poly Carboxylate Superplasticizer (COPOCA202) on Fresh Concrete, Annual Conference on Cement Based Materials in 2003. p 418-424, Beijing, China, （Sept 19-21, 2003）
13. Y. F. Zuo. Effects of Poly-carboxylic Superplasticizer on Fresh Concrete and Its Mechanism. Master thesis. Beijing: China Building Materials Academy, （2004）
14. X. C. Hu, C. Y. Liu, B. P. Yao. Effect Mechanism of Water Reducing Agent for Concrete, Proccedings of 1985 Beijing international symposium on cement and concrete, **Vol 2**, p179-190, Beijing, China, （May 14-17, 1985）.

15. J. Plank, C. Hirsh. Impact of zeta potential of early cement hydration phases on superplasticizer adsorption Cement and Concrete Research, **Vol 37**, No. 4, p 537-542.
16. J. K. Chen. Principles and Application of Chemical Admixtures for Concrete (2nd edition) ISBN: 7-80058-500-X, Chemical Industry Press, （ April 2004）.

https://doi.org/10.1051/e3sconf/202123703015

Ancient Chinese architecture in the future –An investigation with parametric modeling tools

Yuan Li[1], Zhi Li[2], Chang Xu[3]

[1]No.2 Department of Architectural Engineering, Planning and Design institute of Forest Products Industry, 100010 Beijing, China
[2]College of Animation and Digital Arts, Communication University of China, 100024 Beijing, China
[3]No.1 Department of Urban Planning, Planning and Design institute of Forest Products Industry, 100010 Beijing, China

Abstract. With the flexibility and modularity of the construction system, the ancient Chinese architecture was a configurative system until the modern time when it became figurative with nationalistic symbolism, and lost its flexibility to adapt in different functions. Through this study, the researchers try to exploit and release the powerful adaptation of this configurative system by using the method of parametric design. First of all, according to the modern parametric theory, the logic of the traditional structural system is found out by abandoning the feudal social hierarchical factors. And then, a 3D model of parametric structure with wide adaptability is established in Rhino by using its parametric design plug-in Grasshopper. Finally, on the basis of the modern parametric skin theory, two skin systems are attached to this new structure separately, in order to prove the feasibility of the new skin idea to this structure. Through the introduction of parametric modeling steps, this paper focuses on mining the modern adaptability of Chinese traditional architecture.

1 Introduction

1.1 The ancient Chinese architecture and the configurative architecture system

1.1.1 The flexibility of construction system in Ancient Chinese architecture

Ancient Chinese architecture has a wonderful construction system.

The structure of ancient buildings has flexibility to adapt to nearly all kinds of building types and in different natural environment of China. For example, the structural flexibility can also be found in Buddhist or Taoist temples, since they are usually located in mountainous areas. "Xuankong Si", a Buddhist and Taoist temple in Shanxi province was built on a precipitous cliff with the wood structure and mortise and tenon joints. The temple was suspended in the air without support from ground.

However, the wonderful flexibility has been limited by some factors.

Firstly, the application of the building system has been limited to rectangular forms as a result of Chinese cosmological schema of a square shaped earth in the ancient times. This culturally defined conceptual representation of the universe overshadowed the potentials of the flexibility nature of the structural system. Nevertheless, the flexibility and potential of the structural system show brilliantly in the Temple of Heaven where the circular structures were required to reflect the cosmological schema of heaven that is circular.

Secondly, it was limited by the traditional single use of material and the feudal social hierarchy, although the truss structure can meet the spatial needs of people. Yet the feudal social hierarchy is related to the modular system in the traditional Chinese structural system.

To sum up, the flexibility of construction system is the natural quality in ancient Chinese architecture, although there were some limiting factors.

1.1.2 The modular structural system

The traditional Chinese structural system is modular. The basic module determines all the sizes structural components so as to control the scale of the entire building. When the dimensions of the basic module change, so do other parts of a building. Because of these features, the structure system is a configurative system. For example, structural components in the Temple of Heaven are similar to those used in the rectangular palaces of the Forbidden City are used to form the circular structure with similar modular rules of construction.

1.1.3 The traditional Chinese architecture should be a configurative architecture system.

The delicate wood bracket system and graceful hyperbolic roof draw so much attention that the logic behind the appearance is ignored. Most of people, even architects, think of old building's essential characteristics in terms of these external features. Therefore, in modern time, ancient Chinese architecture is treated as a figurative system with nationalistic symbolism.

However, because of the flexibility and modularity of the structural system, the traditional Chinese buildings have a lot of striking formal features as Aldo van Eyck points out that configurative architecture is figurative[1].

Therefore, the traditional Chinese architecture should be treated to be a configurative architecture system.

1.2 The construction system and parametric design

Since the traditional Chinese structural system is a rule based modular system, it can be seen as a parametric system with the basic modular unit as its main parameter. When the value of the parameter changes, all the parts change according to the rule. In the theory of parametric design, swarm intelligence is the property of a system whereby the collective behaviors of agents interacting locally with their environment cause coherent functional global patterns to emerge. In the case of a traditional Chinese building, the individual components can be seen as the agents whose behavior is determined by the construction rules, and the whole building can thus be seen as the global pattern of emergence.

1.3 The aim of this study

The intention of this study is to investigate the adaptability of a parametric model of a structure based on the traditional Chinese structural system in order to demonstrate that when the traditional Chinese architecture is treated as a configurative building system, it can better serve contemporary functions in more irregular sites, and result in new figurative images. In such a way, the configurative nature of the traditional Chinese architecture can be rediscovered, and the ancient tradition can be reconnected with the contemporary life to give the traditional architecture a new life. Then the parametric design tools can be used to explore the future Chinese architecture based on the logic of the traditional constructional system.

1.4 The research status

In recent years, many Chinese architects have used parametric methods to study traditional Chinese architecture. However, some of them were concerned with on the form of the bracket sets of the traditional Chinese architecture. One direction is to explore the automation of the generative process of the bracket sets as a figurative system representing the Chinese architectural tradition[23]. And other studies were interested in how to use parametric tools to recover the ancient buildings[3]. These studies may still miss the more fundamental aspect of the Chinese architectural tradition. Therefore, this study will concentrate on the configuration of the post and beam structure in its adaptive use.

2 Process

2.1 Logic and rules

According to studying the module and structure system, the logic of Chinese antique was found out. The unit of the structure system is a piece of truss, composed with two pillars supporting several lays of beams and roof curves by linking the points of beam heads. Two pieces of units, hyperboloid roof and the clearance between them compose a space. This is the logic of the ancient structure. Because of the logic, the structure can create a series of space which china people live with. The royal palaces were built by this structure, so were the dwellings. The temples of Buddhism, whatever large or small and imperial or folk, were built by the structure. The buildings, the corridors and the pavilions were built by the same structure. However, there is hierarchy in China whatever past or present. However, this is impossible that there is no difference between royal palaces and folk dwellings, temples and residential buildings. Strict hierarchy system controlled all of architecture. Ancient Chinese made a special module rules to control the scale of a building and the luxuriant degrees. According to it, the difference can exist.

To sum up, the dimension of every components and clearance between two units are controlled by a modular system. And how did the special module rules work? It was ruled that there are two kind of architecture – big type building system and small type building system. The big type system applied to the official and royal architecture, while the small type system applied to the folk and some subsidiary architecture, even of royal. And the most distinguish is whether there is Dougong, a system of brackets inserted between the top of a pillars and a crossbeam, which can let the eaves more profound. In the official building system, the basic module is the dimension of Douko, the smallest dimension of the mortise and tenon joint between beam and pillars. There are 10 basic module graded the architecture to 10 levels. The dimension of basic module is bigger, the class is higher, the ornament is more luxuriant. In the small type system, the basic one is the diameter of pillar. Every component of a building has the ratio of diameter. As a result, the dimension of pillar diameter and these radios control a building. Because of these module systems, the logic of Chinese ancient architecture can be studied by parameter methods.

2.2 modeling

A model has been build in Rhino by using plug-in Grasshopper to finish this study.

In the studying, I dislodge these factors what exist due to strict hierarchy system, yet reserving those original elements. And it should be gotten rid of the limit of materials. Consequently, these factors are reserved: the unit of the structure system, which is a piece of truss, composed with two pillars supporting several layers of beams and roof curves by linking the points of beam heads, the module system without those 10 levels. The radio will be reserved, too. It displays the relation of each component

https://doi.org/10.1051/e3sconf/202123703015

clearly.

2.2.1 To build a structure

The structure was built, by constructing the pillars, beams and roof keels. The first step is drawing a curve. It simulates the wavy terrain and different spatial needs.

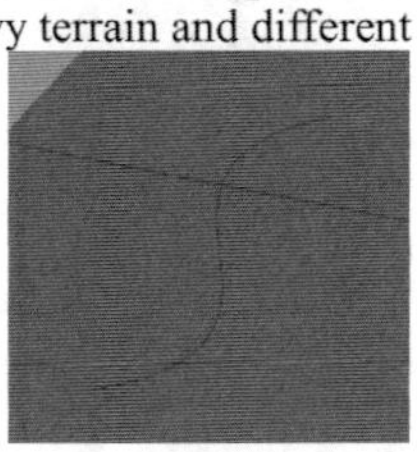

Fig. 1. The original curve (be made by author).

And then the curve was divided (Figure 2). One group of pillars would be built on these dividing points. There were another curve and two new points which were drew. The function of the controlling points and curve is to simulate some special influence factors to the building. Therefore, some relations would be set up between controlling factors and the building. It is known that the height and width can control the scale of a space, so that they can control a building. Therefore, it was set that the height of the pillars were controlled by the dimension between one point and theses dividing points, and the width would be controlled by the relations between another point and these points in the curve between old group of pillars and new group. It means that the height were not the same with every pillars and there would be a curve in Z direction, and so did between two group of pillars (Figure 3).

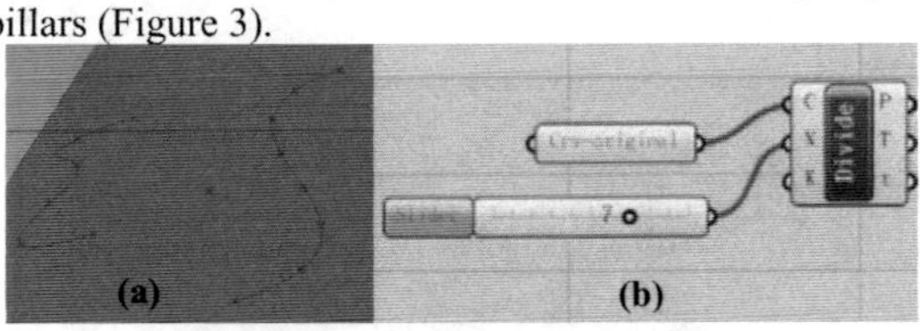

Fig. 2. To divide the curve (be made by author).

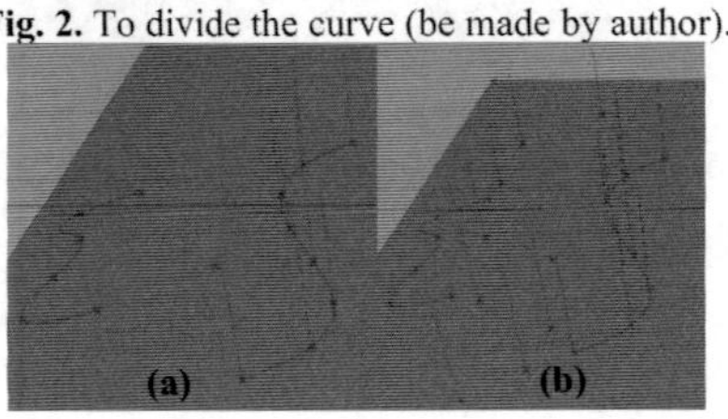

Fig. 3. (a). To build a group of pillars. **(b).** To build another group of pillars. (Be made by author)

The third step is to finish the last layer of the truss structure, the multilayer triangle beam frames. Like the Figure 4, a group of pieces of post-and-beam structure would be set up.

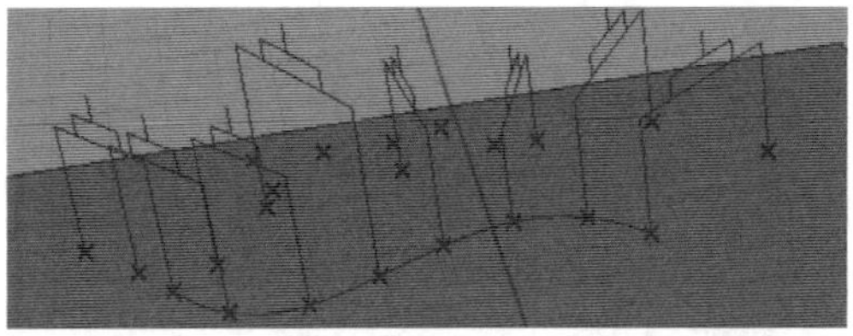

Fig. 4. The multilayer triangle beam frames (be made by author).

In addition, it is the forth step to connect these post-and-beam structure pieces by setting the roof purlins in sequence (like Figure 5 and Figure 6).

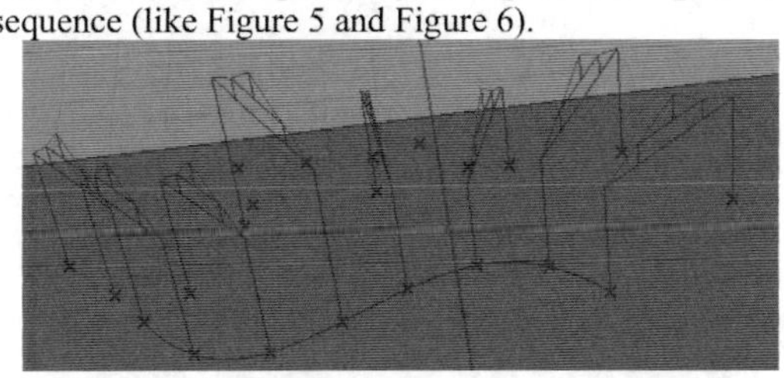

Fig. 5. To connect these post-and-beam structure pieces (be made by author).

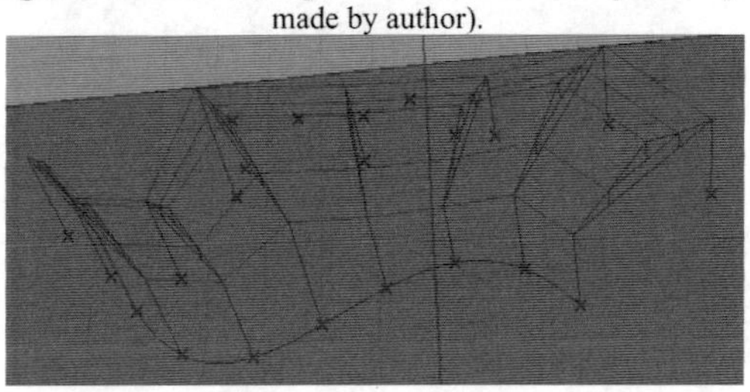

Fig. 6. The primary structure (be made by author).

By the end of this step, the primary structure was built (like Figure 6).

The fifth step is connecting all the purlins to make the roof.

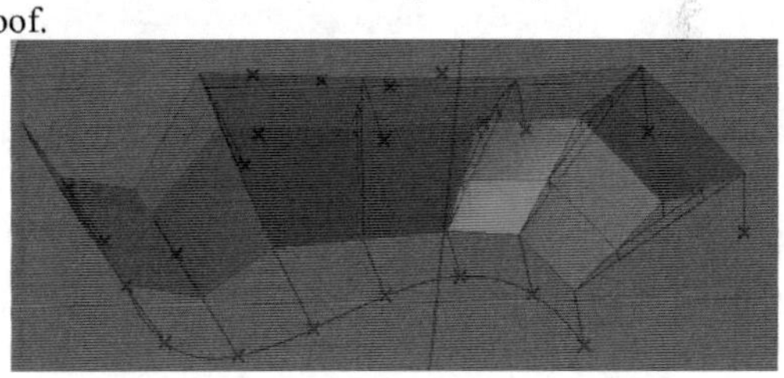

Fig. 7. To loft a roof (be made by author).

For protecting the wall, the cornices would be set up (Figure 8). It is the last step. The angle between cornices and the eave curves would be adjusted in order to adjust to some surrounding requirements. It was carried out by offsetting the eave curves, lofting the two curves and setting a slide bar to control the angle(like Figure 9 and Figure 10).

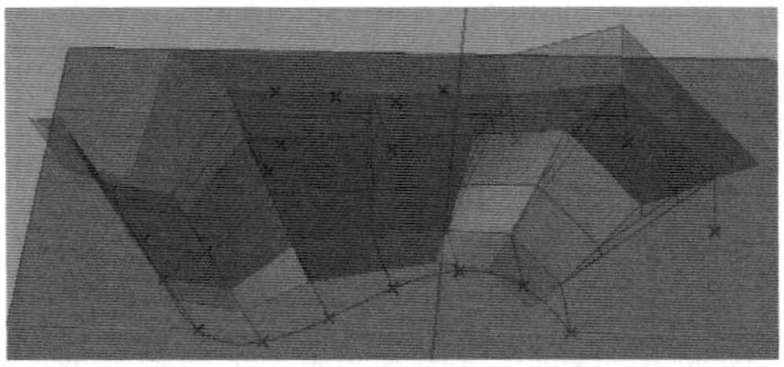

Fig. 8. To overhang eaves (be made by author).

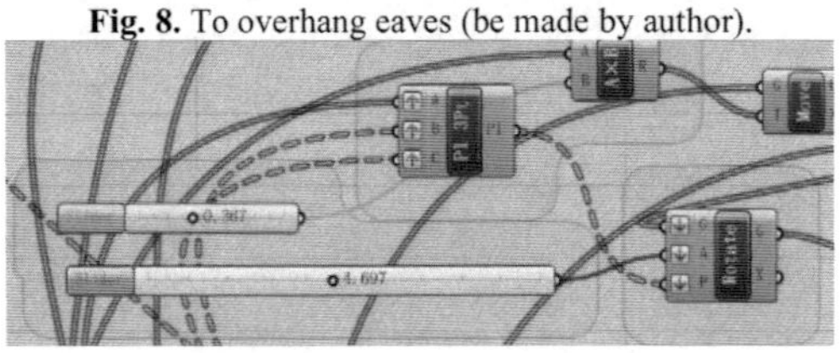

Fig. 9. The slide bars to control the angle (down bar) and the length (up bar). (Be made by author)

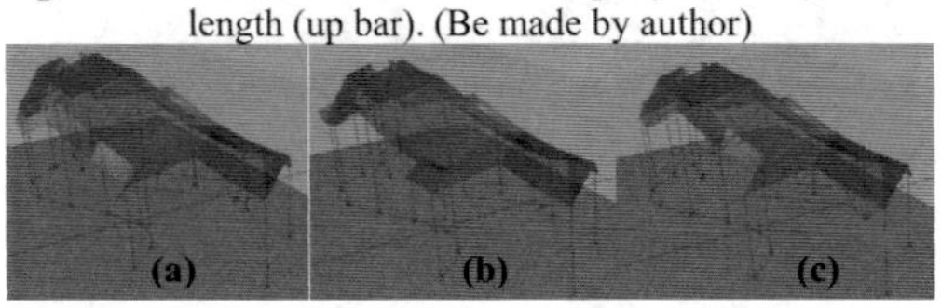

Fig. 10. The process of the eave's angle changing (be made by author).

2.2.2 The vertical space was divided and created.

To divide and create the vertical space. In other words, the floors were built. In one group of pillars, select some pillars, set one slideable point in one pillar and link these points to a floor curve (figure 11-a). In the other pillar group, by doing the same thing, another curve will be drawn (figure 11-b). Then lofting the two curves set up a floor (Figure 11-c). By controlling these slidable points, the height of the floors could be changed (figure 12-a). By using this method, multiple floors can be built basic on the different requirements. And the important thing is that the vertical space would be changeful and interesting. In ancient Chinese architecture, these parts of hangover outside floors were considered. In order to protect these walls and the floors, large cornices are the solution. The making method is similar with making the roof cornice with the slidable angle (Figure 13 and Figure 14).

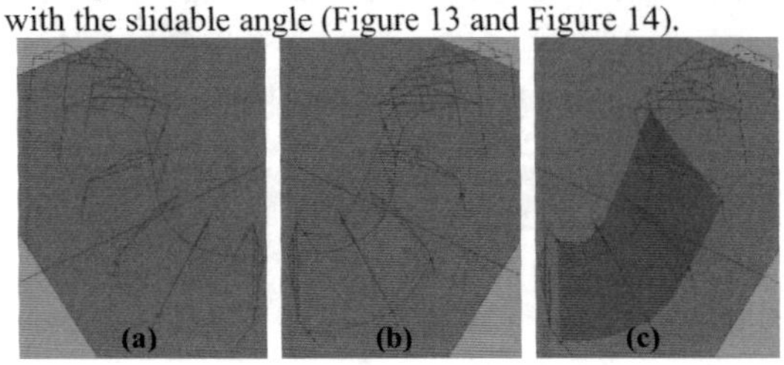

Fig. 11. The process of the first floor was built (be made by author).

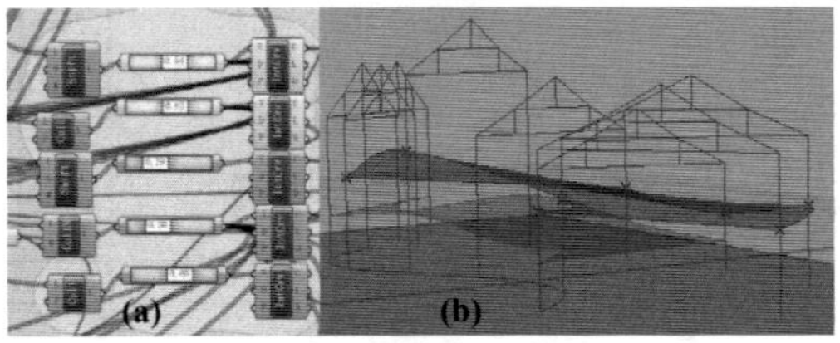

Fig. 12 (a). slide bars to build two floors in Grasshopper; **(b).** the second floor was built in Rhino. (Be made by author)

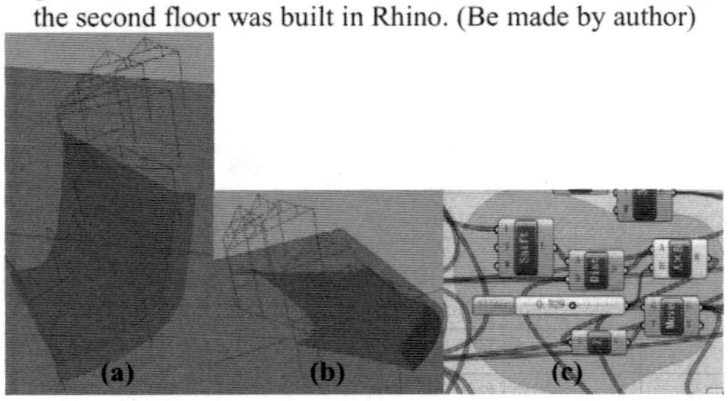

Fig. 13. (a)&(b). the overhanging floor; **(c).** The slid bars in Grasshopper to control the eave's angle of overhanging floor (be made by author).

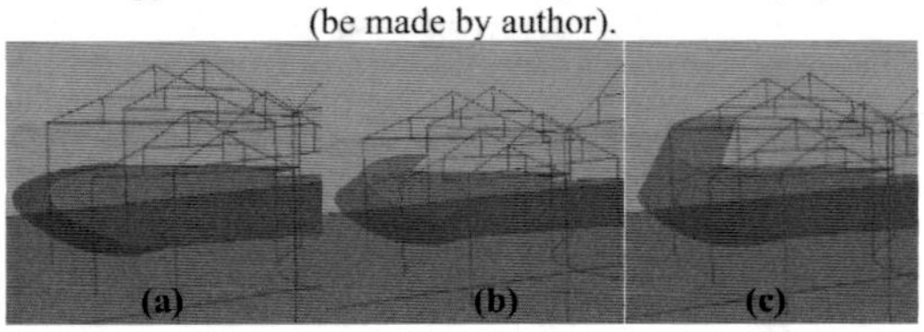

Fig. 14. The process of the eave's angle changing about the overhanging floor primary structure (be made by author).

At last, the structure modeling which is simplified is created .

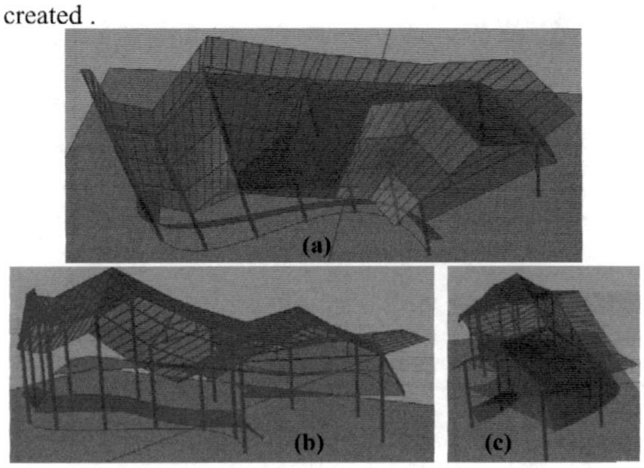

Fig. 15. The simplified structure modeling (be made by author).

2.2.3 Building envelopes would be built.

Building envelopes would be built. Because the truss structure system is the supporting system, the building skin is unconstrained, whatever in the form or in material. The current methods and skin theories are adaptable/ applicative.

According to the current theories of skin system, these steps would be followed:

1. To build a unit of skin.

https://doi.org/10.1051/e3sconf/202123703015

2. To made or choose a wall as a target wall.
3. To build the skin system by letting the unit of skin be attached to the target wall following some logic rules.

2.2.3.1 To build the first skin system with "亞"

Considering the cultural factors,"亞"(Ya), as one of the ancient Chinese words, is chosen to build the first skin system.

Firstly, the reason why "亞" was chosen is that the pattern of "亞", as one of the traditional patterns in ancient China, was often used on the grid/window/door of ancient Chinese buildings. In this study, a attempt to apply the pattern of "亞" to the surface of architecture in a new way will show that Chinese tradition and traditional architecture can be revived in the contemporary era.

Secondly, a skin system of "亞" is built in Rhino by following the modeling steps:

1. Under the guidance of multi-dimensional mindset, the "亞" pattern is made into a three-dimensional unit, like Figure 16(a). In this attempt, the middle part of the "亞" three-dimensional unit is set to be raised, so that the unit will have multiple faces . By giving different attributes to different faces, the adaptability of skin system can be improved.
2. To make a target wall by connecting two curves, like Figure 16(b). The one curve is the original curve in the ground, and the other curve is made by linking apexes of columns. In this way, the target wall is simplified a curved surface, so that the wall thickness is 0mm.
3. The "亞" three-dimensional unit is attached to the target wall in order to build the skin system, like Figure 16(c).

Thirdly, through giving different attributes to different skin units, the external windows can be set in the architectural skin which is building envelopes. And there are two ways.

The one way is that: some "亞" units are defined as entities, while others are defined as void bodies which can be regarded as windows of buildings. Therefore, the light can enter into the building through the parts of skin attached by void units. Following different light requirements of architectural space and other factors, the suitable materials would be chosen or created to fit the requirements of light transmittance.

According to this logic, for a unit, we can choose some surfaces in a "亞" unit to be solid, while others to be virtual. Because the angle and area of each surface of "亞" unit can be set and changed by parameters, amount of light and incidence angle can be controlled according to the actual situation .

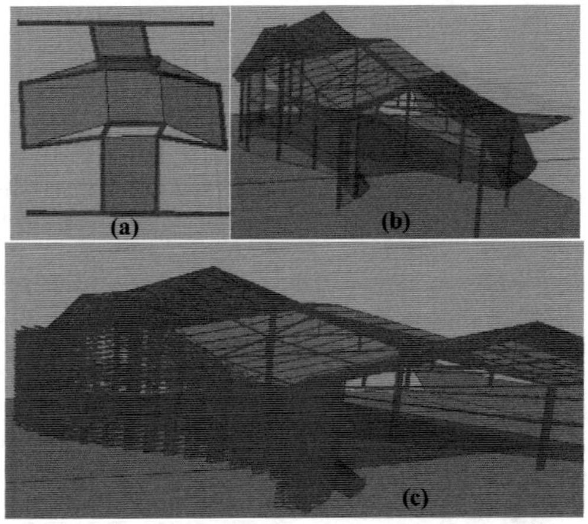

Fig. 16. To build one skin system (be made by author).

2.2.3.2 To build the second skin system with light as the main factor.

Another attempt is made to adapt the skin of this structural system. Without cultural factors , light is chosen to be the unique/main influencing factor. At first, to make a unit with a hole in the middle which the hole's area can be controlled according to the illumination of light, like Figure 17(a). And then , the unit is attached to the exterior wall, like Figure 17(b) and (c). This is the most common method of making skin method in contemporary times.

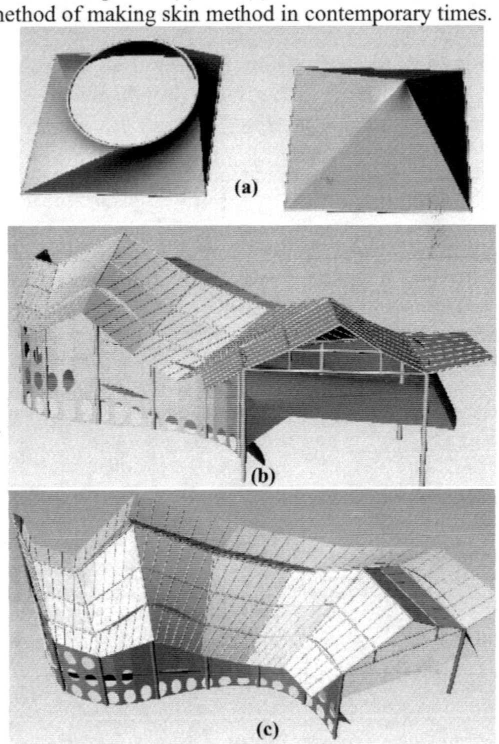

Fig. 17. To build the other skin system (be made by author).

Above that, it is viable that new methods of architectural skin design adapted to the truss structure system from ancient Chinese Architecture.

2.3 Result

In this study, the model was controlled by 3 parameters.

The first one is basic curve and the dividing numbers of this curve. The curve comes from the simulation of the site conditions. In a real project, the simulate cure could come from abstracting the site conditions, such some geography condition and so on. By changing the numbers in the curve, the number of truss pieces could be changed and the dimension between two pieces could be altered; by changing the trend of the curve, the trend of the building would be changed. And the rule which control the number is basic on the requirement of the materials and the load condition.

The second factor is the controlling curves and points. By moving the points and the curves, the height of each pillars and the depth of each room could be altered. Those two alterable factors were set to simulate the varied terrain and other effects which influence the building volume and spatial arrangement.

The third one is the slide bar to control the angle of the pitch to the eave. In old buildings, for protecting the roof and the wall from the rain and snow, the hangover eaves occupied. And the raked roof is good for dewatering naturally. As a result, with the structure supporting, the eaves were raked and hangover. The deeply hangover eaves are necessary. By offsetting the eave curve and lofting the two curves, the hangover eaves were built. And set two slide bars, one is to control the pitch to the eaves and the other is to control the hangover length. It is good to find the adaptable angel and length of the eaves, according to the local natural surrounding factors. For example, if the sunlight is one of the main influences to a building, the length and angle can alter following the changes of light.

However the model abandons the old ratio between each dimension of components. The old ratio rules related to the vertical and horizontal loads of building, because the ratio rules control the dimension of the pillars, beams and other structure components which made of wood to adopt to the vertical and horizontal loads of a building. Moreover the ratio rules came from the experiences. Consequently, when the material science and mechanical system are developed, the ratios will be no more adoptable, so the old ratios are given up. And the dimension of each component could be controlled by a new ratio system basic on new material and load situation. In another word, the method is continued that a ratio system control the dimensions. The only improvement is that a new ratio system will instead of the old one.

3 Conclusion

The ancient Chinese architecture is a configurative system. In the modern time, it is important to exploit and release the powerful adaptation of this configurative system.

The first key factor which must be reserved is the unit of the structure system, which is a piece of truss, composed with two pillars supporting several layers of beams and roof curves by linking the points of beam heads. By abstracting the specific condition of a program to a curve and the key influence to points, the controlling parameter could be found out. And these parameters will manage the permutation and combination of the unit structure pieces. Because of this, the form and spatial combination will be controlled by parameters. At the same time, the advantages can emerge.

The second key point is to continue the method that the dimensions are controlled by a ratio system. The improvement is that the new ratio system would instead of the old ratio system. These rates are the other group parameters.

I am greatly indebted to my supervisor, professor Feng Jin who is teaching in Lawrence Technological University, for his valuable instructions, suggestions and criticism on my paper in the academic studies as well as his careful guidance of the translation. In addition, I also owe a special debt of gratitude to professor Zhang Bo who once offered me valuable courses and advice during my study. Last but not least, my gratitude also extends to my family members, especially my mother and husband. Their encouragements and unwavering supports have sustained me through frustration and depression. Without their pushing me ahead, the completion of this paper would be impossible.

REFERENCES

1. A.van Eyck, Forum, **3**, 81-93 (1962)
2. P. Ma, W. Zhao, J. Chongqing Univ. Sci. Technol.(Natural Sci. Edition), **11(6)**, 94-97(2009)
3. Z. Wang, J. Lu, B. Zheng, Sichuan Building Sci., **45(2)**, 78-82 (2019)
4. Y. Chen, J. Liangshan Univ., **4(1)**, 19-21(2002)
5. R. Wang, W. Sun, X. Zhang, J. Xi'an Univ. of Arch. & Tech.(Natural Sci. Edition), **45(4)**, 479-486(2013)

 https://doi.org/10.1051/e3sconf/202123703016

Research on Safety Measures of Shallow-buried and Undercut Tunnel

LIU jian-jun [1], LIU Xue-ying[1*], FAN Zhong-jing[2], and Yang Bao-feng [3]

[1]Zhejiang University of Water Resources and Electric Power, 310018 Hangzhou, Zhejiang China
[2]China Railway Tunnel Bureau Group Construction Co., Ltd, 530000 Quzhou, Zhejiang China
[3] Ningbo Longyuan Shenghong Ecological Construction Engineering Co., Ltd, 315327 Ningbo, Zhejiang, China

Abstract. The application of mine method shallow burying and undercutting method in subway and tunnel engineering is becoming more and more significant, but shallow burying and undercutting method also has its own limitations: the shallow burying and undercutting method can easily cause the subsidence of nearby surface, there is a security threat to surrounding buildings, the instability of the tunnel inside, easy to cause the collapse of construction tunnels, endangering construction safety, so ensuring construction safety becomes the key to the project. By combining the examples of the Heyi River Tunnel in Quzhou, the project focuses on the safety measures of tunnel surrounding rock, and simulates tunnel excavation with BIM technology.

1 Introduction

With the development of economy, people are more and more perfect in the construction of infrastructure, and the construction of infrastructure is becoming more and more complicated. The shallow burying and undercutting method of mining method is increasingly used in subway and river-crossing tunnel projects. The shallow burying and undercutting method is convenient for the operation of the used machines and has low cost requirements, the cost of construction is greatly reduced, but the shallow burial method also has its own limitations: the shallow burying and undercutting method is likely to cause the settlement of the nearby ground surface to cause the instability of the tunnel and the collapse of the construction tunnel, which is dangerous to the construction safety.

Many scholars have conducted researches on the limitations of the shallow burying method in design and construction. *Yi Xiaoming*[1]etc. conducted a mechanical analysis on the overall subsidence of the overlying stratum of the shallow buried tunnel through the analysis of the on-site monitoring data of urban tunnel construction, and believed that the overall subsidence and collapse of the overlying stratum of the tunnel was caused by the surrounding area. The result of the combined effect of the mechanical properties and geometric properties of the rock. *Shi Yaxin* [2] etc. used the theory of random medium method to study the prediction of surface movement and deformation caused by the construction of large-section shallow buried tunnels, and obtained relevant prediction formulas for surface movement and deformation. *Zhou Shunhua*[3] took the tunnel Based on in-situ observations and indoor centrifugal model tests, the scaffolding principle of the pipe roof method in the construction of shallow burial excavation in soft ground is studied. It is believed that the pipe roof mainly strengthens the surrounding rock and spreads the pressure of the surrounding rock, while reducing the excavation Relieve stress. *Li Shuchen*[4] and others used fracture damage finite element and numerical software to study the minimum rock cover thickness of submarine tunnels, and established a displacement convergence method to determine the minimum rock cover thickness of submarine tunnels. In the 1970s, *Lebuschwitz* based on actual Observation of the tunnel failure phenomenon, proposed the tunnel rupture wedge theory[5], which means that the deformation pressure is the main pressure of the deep tunnel.

The research on shallow burying and undercutting at home and abroad has become mature, but there are relatively few researches on shallow burying and undercutting under relatively special geological conditions. Because the shallow burying and undercutting tunnels are located under the Qujiang River, the The safety of tunnels digging across the river needs further study.

2 Project overview

2.1 Project location

The construction section of the underground excavation section of the Heyi River Crossing Tunnel is located below the Qujiang River. During the construction process, the tunnel excavation will cause uneven ground settlement, and the probability of tunnel collapse will be greatly deepened. In view of the safety problems in this project, the During the construction of the project, the surrounding rock and BIM simulation are carried out to ensure the safety of the construction.

* LIU *Xue-ying* : liuxy@zjweu.edu.cn

2.2 Engineering scheme

During the construction of soft soil layer tunnels, scientific researchers have conducted a lot of research on the soft soil layer tunnel construction method, structural stress system, settlement and deformation control, etc[6,7].Heyi River Crossing Tunnel Crossing Qu Jiang Duan adopts the mining method of underground excavation. In order to reduce the impact of surface settlement during construction, the tunnel adopts a unidirectional single-hole section. Long pipe shed support method is used for grouting to pre-reinforce the surrounding rock to ensure the stability of the tunnel during construction. BIM simulates the construction law of shallow buried and undercut excavation to avoid instability risks during tunnel construction.

3 Safety measures

3.1 Construction safety measures

3.1.1 Initial support

The cross-river tunnel adopts one-way single-hole section. In order to strengthen construction quality control, sufficient support preparations must be made before construction. The initial support (Figure 1 initial support distribution map) adopts modified polyester The fiber C25 concrete spraying ensures that the support rigidity is in line with the actual situation and controls the settlement of the stratum and the surface[8]. To prevent the problem of ground subsidence, it is necessary to improve the soil to ensure the positive movement of the soil, use the long pipe roof support method for grouting, and pre-reinforce the surrounding rock to achieve the stability of the soil structure and avoid the large movement of the soil.

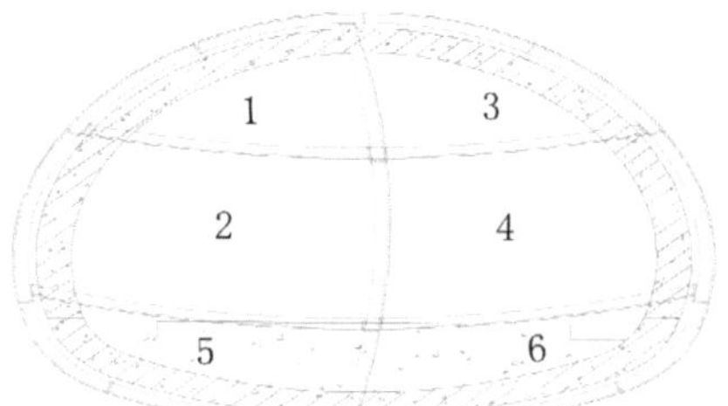

Fig.1. Initial support distribution map

3.1.2 Secondary lining

The secondary lining is made of waterproof concrete, sprayed with C35 reinforced waterproof concrete, and the arch, side wall, and invert are tested and found to be C35 reinforced waterproof concrete. After the surrounding rock and the initial support deformation become stable, the secondary lining is carried out to strengthen the stability of the lining structure and avoid collapse in construction. The initial support link focuses on the load bearing capacity of all structures, and the value of safety reserve is used in the secondary support. These two links need to carry out joint bearing on the structural load capacity. Support occupies a major position in the support to support the stability of the tunnel and avoid collapse.

3.2 Application of BIM in engineering safety

The comprehensive application of BIM auxiliary construction technology ensures that the foundation reinforcement achieves the desired effect, guarantees the tunnel construction quality, and effectively controls the project cost[9].

3.2.1 Deepening the design

The process of engineering modeling for the passage across the river is a comprehensive "three-dimensional proofreading" process. During this process, a large number of problems hidden in the design can be found. These problems often do not involve specifications, but are closely related to professional cooperation, as shown in Figure 2. As shown, through the process of "three-dimensional verification", safety threats in construction are avoided.

Fig. 2. BIM deepening design before and after comparison

3.2.2 Simulation roaming and construction plan simulation

With the help of 3D roaming, the tunnel can be displayed in three dimensions from the whole to the part, key nodes, and complex procedures, so as to improve the sensory perception of the various parts of the tunnel. Secondly, provide technical guidance to the construction party to avoid blind construction safety hazards. According to the design drawings and construction plan, establish a detailed model of the special technical plan and the special safety plan, simulate its construction method, identify unreasonable and problematic construction methods in the design, verify the interference and conflicts in the construction, and optimize the construction resources Configure, shorten the construction period, save costs, and ensure construction safety (Figure 3 simulation of the construction plan of bored piles).

https://doi.org/10.1051/e3sconf/202123703016

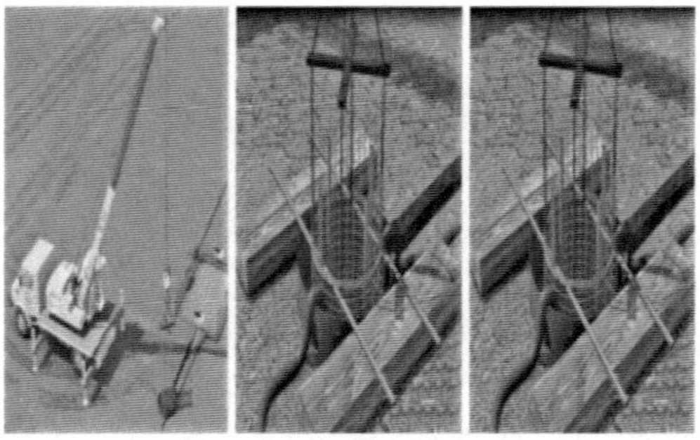

Fig.3. Simulation of the construction plan of bored piles

3.2.3 Three-dimensional technical clarification

The dynamic construction process is demonstrated with the aid of a three-dimensional visualization image, and the safety construction precautions and technical points are visually expressed, so that the construction personnel can more accurately understand the design intent, so as to better control the on-site construction quality. The video construction technology of the steel support of the buried section is explained as shown in Figure 4. After the civil structure is completed, through the application of three-dimensional scanning technology, the required structural section point data can be quickly collected, and the theoretical and actual comparative analysis results can be quickly obtained through information technology (Figure 5). Through BIM technology, each stage and each key index , the current supply information of each organization, each specialty, and each project is no longer limited to adjacent members, ensuring the timeliness and interactivity of information. With the deepening and progress of engineering projects, the BIM information model is gradually deepened and refined, and continuously updated, revised and supplemented to ensure the integrity of the safety, quality, and inspection process results in the project construction process.

Fig.4. Video construction of steel support in open cut and buried section

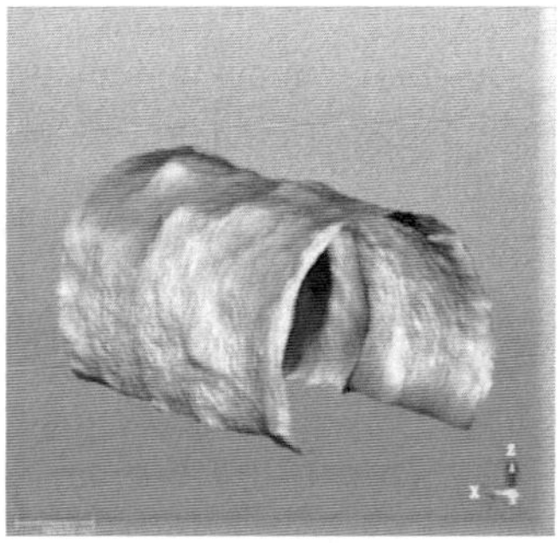

Fig.5. Tunnel status scan

4 Conclusion

The engineering geological environment of the river-crossing tunnel is complex and the risk of underwater digging is relatively high. The safety guarantee is particularly important. During the construction, the tunnel is prone to collapse due to surface settlement, which affects the safety of tunnel construction. In this paper, combined with the construction of Heyi Project, the surrounding rock is supported by long pipe sheds and secondary lining, which can effectively achieve the stability of the soil structure and avoid the large-scale movement and collapse of the soil. The application of BIM technology in the project can effectively avoid construction risks and ensure the safe progress of the project.

References

1. YI Xiaoming, Zhang Dingli, Pang Tiezheng & Luo Jianjun. (2009). Mechanical mechanism analysis of the overall subsidence of the overlying stratum of an urban tunnel. *Chinese Journal of Rock Mechanics and Engineering* (S1), 2860-2867.
2. SHI Yaxin, Tao Dejing, Wa Mingnian . Surface movement and deformation prediction caused by the construction of shallow buried tunnels [J] . *Rock and Soil Mechanics*,2008,**29**(2) : 465-474 .
3. ZHOU Shunhua. (2005). Scaffolding principle of pipe roof method in construction of shallow buried and undercut in soft ground. *Chinese Journal of Rock Mechanics and Engineering* (14), 2565-2570.
4. LI Shucai, Li Shuchen, Xu Bangshu, Wang Hanpeng & Ding Wantao. (2007). Research on the method for determining the minimum rock cover thickness of subsea tunnels. *Chinese Journal of Rock Mechanics and Engineering* (11), 2289-2295.
5. YU Xuefu, ZHENG Yingren, LIU Huaiheng, et al. *Stability analysis on surrounding rocks of underground works*[M]. Beijing: Coal Industry Press, 1983: 245–259.
6. DONG Yunpeng & Sun Xiaojing. (2018). Research on the impact of small clear distance subway tunnels

through buildings at short distances. *Journal of Railway Engineering Society* (11), 87-91.

7. Xi Shujuan & Huang Lin. (2015). Research on the stability of the tunnel underneath the existing building with small clear distance. *Railway Standard Design* (06), 117-122.
8. LUO Ganxiao. (2019). Shallow buried tunnel construction technology and its ground subsidence control strategy. *Housing and Real Estate* (03), 157
9. WANG Zhijun. (2018). Application of Shallow Buried Excavation Method in Tunnel Construction Technology. *Transportation World* (21), 136-137.

Study on the Leakage Damage Mechanism and Repair Technology of Shangmaji Weir

GAO Jing[1], LIU Xue-ying[1*], and LIN Jiang-yuan[2]

[1]Zhejiang University of Water Resources and Electric Power, 310018 Hangzhou, Zhejiang, China
[2]Ningbo Longyuan Shenghong Ecological Construction Engineering Co., Ltd, 315327 Ningbo, Zhejiang, China

Abstract: As a type of dam with a long history and a wide range in the construction of water conservancy projects in my country, weir dams have the functions of flood control, irrigation and diversion. Due to the limitation of construction conditions and environment at that time, the treatment of weir and dam is not standardized enough, which leads to different degrees of problems in weir dam. Based on the example of Shangmaji weir, the article focuses on the analysis of the leakage damage mechanism of the weir, and proposes to choose the corresponding technology to repair it from two angles of the water-facing slope.

1 Introduction

As an important water conservancy project facility, the weirs in Zhejiang area can not only solve the problems of drinking water for humans and animals and irrigation of farmland, but also have the function of blocking sea water and gather fresh water and eliminating flooding. According to investigation, there are thousands of large and small weirs in Zhejiang. However, in the actual construction process, the quality of the weirs cannot be guaranteed due to the influence of on-site construction environment and construction technology, and various problems are prone to occur. Such as leakage, cracking, damage, etc, which seriously threaten the life and property safety of the people[1]. For the anti-seepage treatment of the weir, many scholars at home and abroad have proposed many methods to deal with it: There is an option to use circular grouting, that is, insert two pipes in the borehole, and the slurry that is pressed into the borehole will fill the gap, and the other part will return to the mixing drum and other boreholes from the gap, which makes the slurry always flow State, no precipitation, so the grouting effect is good[2]; The filling grouting method is mainly used to inject the grout into the hidden troubles of the dam body by using the grout's own weight[3]; The relatively complete high-pressure jet grouting technology can easily form an impermeable wall in the soil to achieve the ultimate goal of imperviousness. This type of technology is mainly suitable for silty sand with large sand content and gravel with a particle size of less than 15cm Fine sand[1]; The anti-seepage technology for backfilling the impervious wall and the clay inclined wall of the punching sleeve well mainly uses punching-type well drilling tools to make holes in the impermeable body of the earth-rock dam, and then use the cohesive soil for layered backfilling to form a continuous clay anti-seepage wall and the soil layer of the shaft wall are compressed to make the soil around the wellhead compact, so as to achieve the purpose of anti-seepage and reinforcement[4]; The "top-down and bottom-up integrated grouting method" used in the dam of the South Russia 3 Hydropower Station in Laos can not only build curtain walls that meet the design requirements, but also grouting method reduces the risk of the plinth being lifted and also reduces the project cost[5], compare to the traditional "orifice closure".

At present, the grouting technology and research in the dam body has gradually been perfected and matured, and more scholars are conducting research in this filed, but these are all for the repair and reinforcement of large dams, and for small stone masonry, the research on repairing technology of dams and weirs is particularly lacking. The application of grouting technology to small weir dams, on the one hand, is due to the limited construction site of the dam, and the transportation of materials and equipment is not convenient enough, which is not conducive to construction and has a long construction period, another aspect is to consider economic applicability. The case studied in this paper -- Shangmaji weir, the elevation of which is 130.7m, the main material of the dam body is filled with C20 concrete and masonry, and the surface layer is cast-in-place concrete of 1m thick C25; The dam adopts an open, broken-line practical weir design, with five levels of water drop, the drop height is 0.64m, the drop platform is 1.6m, and the top of the weir is cobbled; the cross-sectional length of the dam body is 9.4m, and 7m long and 1m thick C25 concrete bottom protection is installed downstream of the dam. In response to the above problems, based on the results of field investigations, carefully analyze their causes, and design a reinforcement plan suitable for the weir based on the desired treatment effect to extend its service life.

* LIU *Xue-ying*: liuxy@zjweu.edu.cn

2 Analysis of leakage failure mechanism

2.1 Influence of seepage diameter

Shangmaji Weir is located in Wanyao Irrigation District. The irrigation area is located in valley basins and hilly areas on both sides of the middle and lower reaches of Jiangshan Port. It has a low geographical latitude and belongs to a subtropical monsoon climate zone. The main climate characteristics are warm and humid, four distinct seasons, and abundant rainfall. The main disasters are low temperature, cold current, rainy season and typhoon[6]. Therefore, the water flow of the weir is the largest in April to June, and the cross-section of the entire weir is only 9.4m. The designed seepage diameter is too short to meet the requirements of the seepage diameter length in the high water period, which results in the weir a certain degree of leakage occurs. (Figure 2)

Fig 2. Cross-sectional view

2.2 Foundation deformation

The Shangmaji Barrage is located in the lower part of the Nanxiong Formation of the Upper Cretaceous. The lithology is mainly brick red and blue-gray siltstone, argillaceous siltstone, glutenite, etc[7]. This kind of rock belongs to geological soft rock with low strength. For example, siltstone will degenerate into schist or gneiss under high temperature and high pressure environment, which will cause the internal stress of the dam body to change due to different pressures, which causes due to the difference in load, the foundation will gradually deform, and the small cracks will gradually expand to large cracks, so the phenomenon of leakage will also occur, and the porosity of this type of rock is large, and the degree of cementation is low, which will also affect the weir to a certain extent. The overall anti-seepage function leads to the destruction of the weir.

2.3 Lack of anti-seepage treatment at intake

There is a water intake on the left side of the weir. The survey data shows that the water intake is 1.8m wide and 1.4m deep. It is only lined with concrete and has no anti-seepage measures (Figure 1). Therefore, when the water flows through the water intake, it is easy to seep to both sides. Under the long-term water infiltration, the weir will also leak and damage.

Fig 1. Leakage at the intake

2.4 Construction impact

During the construction of the weir, there may also be many problems leading to leakage. For example, the grouting stones are not dense enough during the construction process to make them staggered; the water stop on the water surface of the weir is broken or damaged; the concrete is not mixed uniformly during the transportation process, causing the poured concrete to produce pitted surfaces and air bubbles; Vibration is not dense enough to cause cracks and collapse in the concrete between the grouting stones. These construction influences will eventually lead to the emergence of cracks. If the cracks run through the upstream and downstream, water leakage and seepage will be caused, and eventually the whole weir will be gradually damaged, unable to play its due function, endangering the safety of people's life and property.

3 Reinforcement and repair technology research

In the prevention and reinforcement of weirs and dams, in view of the problems that arise, we must first conduct a serious investigation of damage and start from the root cause; rely on the results of field investigations to analyze the causes and take targeted measures to ensure that the weirs reliable construction for removing danger and strengthening.

3.1 Principles of leakage treatment

(1) According to the leakage investigation, cause analysis and the expected effect achieved through the

requirements of leakage treatment, evaluate the shape, specification and direction of the crack, and select a reasonable reinforcement plan based on the structural characteristics of the weir, the actual situation, and the climate and environmental conditions, repair materials and construction time.

(2) The leakage treatment should be as close as possible to the source of the leakage. If conditions permit, anti-seepage treatment can be carried out on the front surface.

(3) Leakage treatment should first be constructed during the dry season of the river, first stop the leakage and drain, and then install the waterproof layer.

(4) When selecting repair materials, compare and analyze five aspects: treatment effect, durability, easeof construction, pollution-free, and economy[8].

3.2 Leakage treatment method

(1) For the treatment of cracks on the waterfront surface, on the one hand, surface covering materials can be applied. This method is the simplest and easy to operate method in adding an impermeable layer on the waterfront surface, which can greatly shorten the construction period and is economical commonly used surface covering materials include asphalt mortar, rubber sheet and acrylic emulsion mortar[9]. In the construction, firstly, the original dam surface must be hemped, washed, etc; and the polymer mortar application process must be carefully mastered to achieve a good leakage treatment effect; on the other hand, you can choose to use sprayed concrete panels. This method does not need to erect a template, the construction is relatively simple, and the construction period is short.

(2) For the anti-seepage treatment of the water intake, because it is necessary to ensure the amount of water diversion, it is not suitable for large-scale thickening repair, so a relatively simple anti-seepage plan can be adopted, that is, laying geomembrane. Geomembrane is a coiled material, which is convenient for transportation, convenient for construction, and short construction period. At the same time, the cost of using in an anti-seepage system is 30% or more lower than that of a traditional reinforced concrete anti-seepage structure. And the geomembrane has strong acid and alkali resistance, anti-aging, anti-ultraviolet abilities, and can be used quickly in various complex environments and is durable. Geomembrane can be used together with non-woven fabrics. Non-woven fabrics have a rough surface and can generate greater friction with the contact surface to ensure that the geomembrane and the protective layer are in a good and stable state.

(3) For cracks and leakage in the dam body, drilling and grouting can be used. In the past, high-pressure jet grouting or filling grouting was generally used, but because the width of the cracks changes with temperature and load conditions, cement mortar and grout are rigid materials, and it is difficult to adapt to the changes in the width of the cracks with changes in the external environment. So the repair effect is not good. Therefore, it is now beginning to promote the application of chemical materials for grouting, and have achieved better results. The materials that can be selected are lignin grouting materials, water glass grouting materials, epoxy grouting materials, acrylamide grouting materials, etc[10].

(4) During the implementation of the reinforcement construction, the mixing and pouring of the concrete should be strictly inspected to avoid the occurrence of pitted surfaces and air bubbles due to insufficient mixing, which may cause secondary damage to the weir.

3.3 Benefit Analysis

Because there are few domestic research literatures on the analysis of small-scale dams and their related reinforcement, and there is no unified design specification and technical standard to provide reference for the prevention and remediation of dams, this article is a comprehensive study of large-scale dams. Infiltration analysis and other relevant water conservancy facilities research experience to discuss and analyze. Through the scheme discussed in this article, the leakage problem of the weir can be effectively solved, and the construction is simple, fast, and the treatment effect is good, and it has sufficient durability, safety and economy.

4 Conclusion

As a small stone masonry dam in Zhejiang area, Shangmaji Barrage is representative of its leakage damage. Combining the scheme given in this article, the anti-seepage treatment from the water-incoming surface and the water-intake can effectively solve the local leakage and dispersion problems of the dam body, and the construction is convenient and easy to operate; chemical grouting from the inside of the dam body, it can greatly avoid the problems caused by load changes and deformation, and enhance the anti-seepage ability of the weir.

References

1. TANG Jiyong, LIU Rongfang.(2020).Research on anti-seepage and reinforcement treatment technology of reservoir earth-rock dam. *Pearl River Water Transport* (03), 73-74.
2. TAN Dejun. (2019). Discuss the application of grouting technology in seepage prevention and reinforcement of grout stone dams. *Low Carbon World* (10), 113-114.
3. XU Zhigang. (2017). Management, reinforcement and anti-seepage management measures for earth-rock dams of small and medium-sized reservoirs. *Technological Innovation and Application* (36), 97-98.
4. SHEN Hui. (2015). On the reinforcement of earth-rock dams in small reservoirs. *Heilongjiang Science and Technology Information* (08), 120.

https://doi.org/10.1051/e3sconf/202123703017

5. LIU Wei, MA Xu & LUO Ying. (2020). Production test and result analysis of curtain grouting for dam of South Russia 3 Hydropower Station in Laos. *Sichuan Hydropower* (04), 58-60+73.
6. ZHANG Gang, QIU Yaolin. (1994). Current situation and construction measures of water supply in villages and towns in our province. *Zhejiang Water Conservancy Science and Technology* (03).
7. WANG Yin. (2012). Late Cretaceous-Early Paleocene paleoclimatic changes in Nanxiong Basin (Master's thesis, Nanjing University).
8. ZHOU Xunjian. (2012). Application of stone-masonry dam leakage treatment technology in dam reinforcement. *Inner Mongolia Water Resources* (02), 139-140.
9. CUI Rujing. (2007). Talking about the prevention and repair methods of concrete cracks in hydraulic structures. *Water Conservancy Construction and Management* (04), 53-54+38.
10. MA Zhe, PANG Hao, YANG Yuanlong & XU Yuliang. (2014). A review of the research progress of chemical grouting materials. *Guangzhou Chemistry* (01), 9-13.

 https://doi.org/10.1051/e3sconf/202123703018

Study on the Influence of fly ash, mineral powder and silicon powder on the performance of high-performance concrete

YANG Hua-quan[1], LIU Xue-ying[1*], and CHEN Xiao-dong[2]

[1]Zhejiang University of Water Resources and Electric Power, 310018 Hangzhou, Zhejiang, China
[2]Zhejiang Institute of Water Resources and Ocean Engineering, 310018 Hangzhou, Zhejiang,China

Abstract. To study the effect of fly ash, mineral powder, and silica fume on the working performance and mechanical properties of C70 high-performance concrete by adding the same amount of fly ash, granulated blast furnace slag powder, and silica fume as a composite admixture to replace the amount of cement. Influencing the law, at the same time, the optimal dosage ratio of various admixtures is determined through the orthogonal experiment. The results show that: when adding 6% silica fume, it can improve the performance of high-performance concrete. When the amount is increased, the viscosity of the concrete increases and the fluidity decreases. Incorporating an appropriate amount of silica fume can greatly increase the compressive strength of concrete. When blended with fly ash in the proportion of 20%, the performance of high-performance concrete is better. When the same amount of fly ash replaces cement, fly ash reduces hydration and improves the cohesion of concrete, 7d, 28d the compressive strength of the cube increases significantly. Adding 10% mineral powder, mineral powder can affect the early compressive strength of high-performance concrete, extend the setting time of concrete, and improve the pumping capacity of concrete.

1 Introduction

High-Performance Concrete (abbreviated: HPC) is a new concept of concrete proposed by some developed countries based on the durability design of concrete structures in the late 1980s and early 1990s. It is also a new type of high-tech concrete. The concrete made with modern concrete technology is used on the basis of improving the performance of ordinary concrete. Since it was put forward for more than 20 years, countries around the world have carried out a lot of research, development, and popularization and application of it. High-performance concrete has good working performance, high strength and high durability. Adding fly ash, granulated blast furnace slag powder, and silica fume can replace cement. The combined use of admixtures can refine the pore structure of concrete, improve compactness, and improve the performance of high-performance concrete. At the same time, it can reduce the amount of cement and concrete, reduce the energy consumption and material consumption of production and transportation, and reduce the damage to the environment. It not only uses resources rationally, optimizes economic benefits, but also well for environment [1] [2].

In order to study the influence of different admixtures and admixtures in different proportions on the performance and compressive strength of high-performance concrete, this paper uses low water-binder ratio, high-efficiency polycarboxylic acid water-reducing agent and mineral admixtures take C70 high-performance concrete as an example[8]. By orthogonal test method, different dosages of fly ash, slag powder and silica fume are mixed respectively, and the dosage combination is tested to determine the optimal dosage of various admixture at the same time, analyze its influence on the working performance of high-performance concrete and the compressive strength of the cube.

2 Test method and mix ratio

According to the *"Technical Regulations for High Strength Concrete Application" (JGJ/T281-2012)* and the full calculation method of concrete mix proposed by Professor Chen Jiankui and Professor Wang Dongmin, we can calculate C70 high-performance concrete with a water-binder ratio of 0.3 and a sand ratio of 40%. [11] The design value of the mix ratio [3], and conduct trial mixing to determine the amount of various materials. On the basis of trial mixing, keeping the total amount of cement material-unchanged, determine the mixing amount of the three admixtures by orthogonal experiment. Choose fly ash content of 10%, 20%, mineral powder content of 10%, 20%, and silicon powder content of 6%, 8%. The material consumption of the test concrete is shown in Table 1.

Table 1. Concrete material consumption

Num	*C* /kg	*F-A* /kg	*M-P* /kg	*S-F* /kg	*W* /kg	*S* /kg	*G* / kg	*WR* /kg

* LIU *Xue-ying* : liuxy@zjweu.edu.cn

1	389.98	52.7	52.7	31.6	158	706	1059	7.9
2	379.44	52.7	52.7	42.	158	706	1059	7.9
3	337.28	52.7	105.4	31.6	158	706	1059	7.9
4	326.74	52.7	105.4	42.2	158	706	1059	7.9
5	337.28	105.4	52.7	31.6	158	706	1059	7.9
6	326.74	105.4	52.7	42.2	158	706	1059	7.9
7	284.58	105.4	105.4	31.6	158	706	1059	7.9
8	274.04	105.4	105.4	42.2	158	706	1059	7.9

3 Concrete working performance test method

When using concrete slump and slump expansion to evaluate the workability of concrete, slump is the distance between the height of the slump cylinder and the highest point of concrete collapse, and the expansion is the maximum diameter of the concrete collapsed and expanded [3]. At the same time, it is necessary to observe the water retention and cohesion of concrete. The slump and expansion of concrete reflect the fluidity and consistency properties of the concrete mixture.

According to the *"Standard for Test Methods for the Performance of Ordinary Concrete Mixtures" (GB/T50080-2016)*, the concrete mixture should be evenly loaded into the slump cylinder in three layers. Each layer of the mixture should be evenly inserted 25 times in a spiral shape from the edge to the central. After tamping, the height of each layer of concrete mixture sample is about one-third of the cylinder height. When the concrete no longer slump or the slump time reaches 30s, measure the height difference between the cylinder height and the highest point of the concrete sample after the slump with a steel ruler as the slump value of the concrete mixture. When the concrete mixture no longer diffuses or the diffusion time reaches 50s, a steel ruler should be used to measure the maximum diameter of the expanded surface of the concrete mixture and the diameter in the vertical direction as the expansion degree of the concrete mixture. The effect of different admixtures on the performance of high-performance concrete is shown in Table 2.

Table 2.Test results of performance of high-performance concrete

Num	*Slump* /mm	*Expansion* /mm	*Slump* / *Expansion*	*Cohesion*	*Water retention*
S1	190	480	0.39	good	good
S2	180	480	0.38	good	good
S3	185	530	0.35	good	good
S4	210	500	0.42	good	good
S5	230	580	0.40	good	good
S6	210	500	0.42	good	good
S7	205	520	0.39	good	good
S8	200	500	0.40	good	good

Through the slump and expansion test of high-performance concrete mixture, it is known that concrete has good fluidity. The sample numbers from Table 2 are: S1 and S2, S5 and S6, S7 and S8, when the content of fly ash and mineral powder remains unchanged, the content of silicon powder increases from 6% to 8%, and the slump and the degree of expansion decreases with the increase of silicon powder content. The particle size of silica fume is very small, and the specific surface area is large. The addition of silica fume in high-performance concrete can better fill the pores between cement, which is beneficial to prevent concrete bleeding and improve concrete cohesion and water retention [4]. When 6% silica fume is added, the fluidity of concrete is better. When excessive silica fume is added, the viscosity of concrete will increase, the fluidity of concrete will decrease rapidly, and the overall workability will decrease. Therefore, more water reduction is needed. Concrete achieves the required fluidity.

From the sample number in Table 2: S1 and S5, S2 and S6, it is known that when the content of mineral powder and silica fume is unchanged, the fly ash increases from 10% to 20%, and the concrete slump and spread the increase in fly ash increased significantly. When 10% fly ash is mixed, the ratio of slump to expansion is ≤0.4, and when 20% fly ash is mixed, the ratio of slump to expansion is ≥0.4, which shows that when the amount of fly ash increases at the same time, the fluidity of the concrete mixture also increases. When the same amount of fly ash replaces cement, it can reduce hydration, play a role of lubrication, and improve the viscosity of concrete.

The sample numbers in Table 2 are: S1 and S3, S2 and S4, S6 and S8, when the content of fly ash and silica fume is unchanged, the slag powder increases from 10% to 20%, and the slump and expansion the degree increases with the increase of mineral powder content, but the increase is not obvious. Incorporating an appropriate amount of mineral powder can improve the fluidity of concrete, but the main function of mineral powder is to prolong the solidification time of the cement material, improve the internal structure of the concrete, and improve the impermeable of the concrete.

4 Test methods for concrete mechanical properties

According to *"Specifications for Design of Concrete Structures" (GB50010-2019)* and *"Standards for Test Methods of Physical and Mechanical Properties of Concrete" (GB/T50081-2019)*, the standard value of compressive strength of cubes refers to the production and maintenance according to standard methods. In this test, a cube specimen with a side length of 100mm is used. According to the conversion formula $f_{cu,150}=0.95f_{cu,100}$, the concrete specimen is cured to the specified age in the standard curing room, and the compressive strength of 3d, 7d, and 28d is measured according to the standard method. The compressive strength results are shown in Table 3:

Table3. Compressive strength of high-performance concrete

Number	*Compressive strength of concrete cube* (MPa)		
	3d	7d	28d
S1	40.3	62.9	71.3
S2	44.9	64.7	76.3
S3	44.1	66.1	78.4
S4	50.2	71.2	86.8
S5	51.3	80.2	89.7
S6	50.6	77.8	78.7
S7	48.3	69.8	76.6
S8	54.5	75.3	80.2

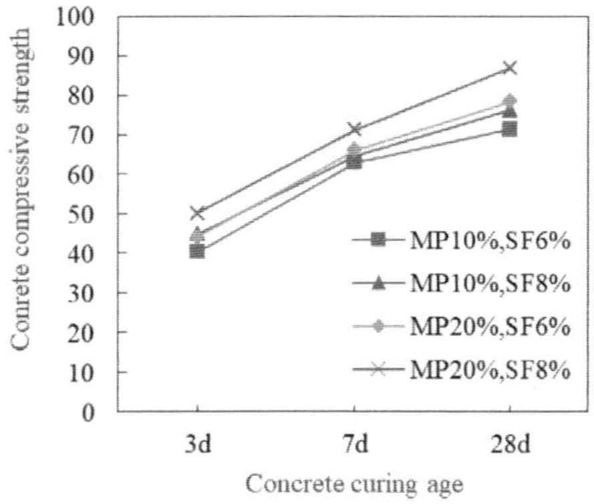

Fig1. The effect of mineral powder and silica fume at 10% fly ash

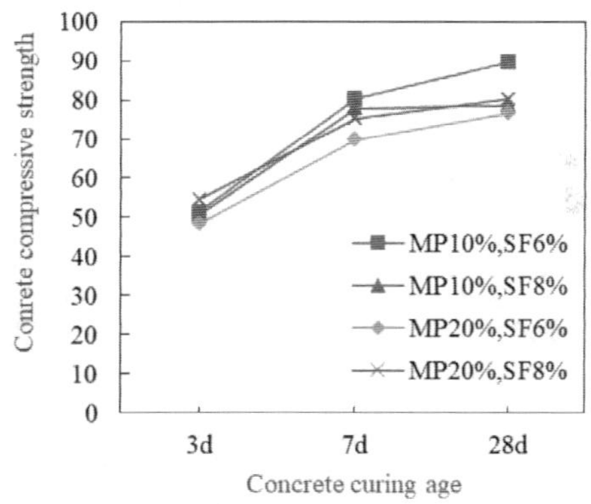

Fig2. The effect of mineral powder and silica fume at 20% fly ash

Result analysis: It can be seen from Figure 1 that when both fly ash and mineral powder are added at 10%, with the increase of silica fume, the 3d, 7d, and 28d concrete compressive strengths all show an increasing trend. This is due to the large specific surface area of silica fume. After adding silica fume, the capillary water inside the concrete is constrained by the silica fume, which reduces the bleeding inside the concrete and improves the bonding force between the cement slurry and the aggregate, thereby making the concrete the compressive strength is increased. With the increase of mineral powder, the compressive strength growth rate of concrete in the middle and late stages is relatively fast, while the strength growth in the early stage is relatively slow. This is because the mineral powder contains more SiO_2, Al_2O_3 and $Ca(OH)_2$, which chemically react to form stable calcium silicate hydrate and calcium aluminum plastic hydrate, which helps concrete harden and increase strength [6].

It can be seen from Figure 2 that when 20% fly ash and 10% mineral powder are mixed, with the increase of silica fume, the compressive strength of 3d, 7d, and 28d concrete will decrease slightly. The superimposition effect of the mixture increases the hydration effect, which reduces the fluidity of the concrete and reduces the overall work performance. With the increase of silica fume, the compressive strength of high-performance concrete first increases and then decreases. The early strength development is faster [5]. It can be seen from Figure 1 and Figure 2 that with the increase of fly ash, the late strength of high-performance concrete (after 7 days) increases faster, and the early compressive strength contrast is lower. Due to the increase of fly ash, the amount of cement decreases. The hydration rate of cement decreases, so the initial strength growth is slower, but as the age increases, fly ash reduces the heat generated by the internal hydration of the concrete, improves the density of the concrete, and reduces the porosity of the cement paste. The pore structure of the hardened body is improved, so the later strength of the high-performance concrete is greatly increased. [5] Since each component of different admixtures such as fly ash, mineral powder and silica fume has active molecules, under the action of the hydration reaction of cement, it promotes the hydration of different admixtures such as fly ash, mineral powder, silica fume, etc. So that different admixtures play their reactions at each age, thereby enhancing the strength of high-performance concrete at each age.

5 Conclusions

The compound mixing of fly ash, mineral powder and silica fume can make the high-performance concrete have better workability. Superimposed reaction of various admixtures, fly ash can improve the cohesion and plasticity of concrete. When 20% fly ash is added, the ratio of slump to expansion of high-performance concrete is greater than 0.4, and the fluidity is the highest. Adding 6% silica fume, silica fume improves the workability of performance concrete. As the silica fume increases, the fluidity of the concrete decreases. The effect of mineral powder on the fluidity of concrete is not obvious. The main function of mineral powder is to extend the setting time of concrete and improve the pumping capacity. The three mixing of fly ash, mineral powder and silica fume can significantly improve the overall fluidity of high-performance concrete.

The content of fly ash, mineral powder and silica fume on the compressive strength of high-performance concrete varies with the content of the mixture. From the above test, it can be seen that when 20% fly ash, 10% slag powder, 6% silica fume are added to replace cement, the 28d compressive strength of C70 high-performance concrete is the largest. The compressive strength in the early stage (before 7d) and the compressive strength of the late stage (after 7d) are significantly higher than other proportions. The amount of fly ash mixed with 20% instead of cement can improve the growth rate of the compressive strength of the high-performance concrete bigger.

The three-blending use of fly ash, mineral powder and silica fume is beneficial to improve the pores of the molecular particles of the cement material. The three mineral admixtures and cement form a good particle gradation, making the cement interior more dense and solid. The admixture has a volcanic effect, which is beneficial to improve the mechanical properties of high-performance concrete.

References

1. Wu Zhongwei. (1998). The development trend and problems of high-performance concrete (HPC). Building technology (01), 8-13.
2. Chen Lei, Xiao Jia, Tang Xianyan, Chen Feng , Liu Jing. (2007). Research on the effect of fly ash and slag on the performance of concrete. Comprehensive utilization of fly ash (02), 22-25.
3. Chen Jiankui, Wang Dongmin. (2000). A new method of high-performance concrete (HPC) mix design-full calculation method. Journal of The Chinese Ceramic Society (02), 194-198.
4. Li Qingfu, Xue Yanxin , Zhang Haiyang. (2012). Experimental study on the influence of different admixtures on the performance of high-performance concrete. Concrete (01), 58-61.
5. Xu Ke. (2013). Experimental study on the influence of different admixtures on the compressive strength of high-performance concrete. Highway Transportation Science and Technology (Applied Technology Edition) (08), 226-228.
6. Li Zhigang, Li Jiahe, Zhang Honggui. (2009). The influence of fly ash and slag composite admixture on the strength of concrete. Low temperature construction technology (04), 17-19.
7. Xiao Jia, Chen Lei , Zhao Jinhui. (2011). The influence of admixtures on the mechanical properties of C60 high-performance concrete. Low temperature building technology (04), 10-13.
8. Fu Baohua, Gai Dongfang & Li Jun. (2016). The influence of the superposition effect of fly ash and mineral powder on the mechanical properties of high-performance concrete. Juye (07), 32-34.
9. Li Siqi, Meng Zhenya & Zhang Shouzhi. (2013). The influence of fly ash and mineral powder on the permeability of high-performance concrete. Concrete (11), 119-121.
10. Zhang Haiyang, Guo Jun, Zhang Xuhui & Li Weidong. (2014). The influence of fly ash and silica fume on the compressive strength of high-performance concrete. China and Foreign Highway (03), 312-316.
11. Tan Feng. (2017). C90 high-strength and high-performance concrete mix design and experimental research on its mechanical properties (Master's thesis, Chongqing Jiaotong University).

Typology in Vernacular Architecture—Qianmo Tower Post Station in Mingyue Village

Li Yunzhang[1], Zhu Yuhong[1], Yu Lirong[2], Bi Zhongsong[1,3], and Huang Guanjun[3]*

[1]College of Architecture and Environment, Sichuan University,610000 Chengdu,China
[2]College of Physical Education, Sichuan University,610065 Chengdu, China
[3] *College of Architectural and Engineering & Research Center of Huizhou Architecture, Huangshan University,245041 Huangshan, China

Abstract. The acceleration of the modernization process has promoted the improvement of rural residential construction technology and efficiency, but the rural buildings have not inherited and developed the regional traditional characteristics on this basis, instead, there has been a blind trend of urban architectural styles in different regions, resulting in a phenomenon of the homogenization of development. The design of qianmota post station uses the typological analysis method to excavate and inherit the characteristics of the local traditional buildings in Mingyue village, then combines the characteristics with new technologies and materials to restore the local architecture in the hearts of the villagers in Mingyue village. It is restored in two aspects: the architectural appearance of the traditional houses in Western Sichuan and the spiritual symbol of the church.

1 Field crossroads, tall bamboo forest—the ancient Mingyue village.

Mingyue village, an ancient village located in Ganxi Town, Pujiang County, Chengdu, 90 kilometers from downtown Chengdu, has been the resting post station on the Southern Silk Road and the Tea-horse Ancient Road since the Sui and Tang dynasty. The history of pottery production in the village is as long as that of tea.（Fig. 1）The best-preserved "Mingyue Kiln" in the village, whose firing process still retains the skills of the Tang Dynasty, is currently considered as the only "existing Qiong Kiln" in Sichuan province[1]. （Fig. 2）Here, the villagers work at sunrise and rest at sunset, and their pottery-making, tea-growing, and land-farming are passed from generation to generation; There is an ancient charm in this village: the old pine trees are beautiful, and the old kiln is rest in peace,and the elderly and children are living a happy and satisfying life.

However, like many other villages in China, Mingyue village, with the acceleration of urbanization, has also faced with a series of problems in terms of the destruction of the rural natural environment, the gradually disappearing traditional culture, the massive outflow of local population, and the constant loss of vitality in the village. Therefore, started with Mingyue Kiln, Mingyue village officially launched the Mingyue International Pottery village project in the year 2014, relying on its unique rural environment and cultural resources. While great concern has been put into developing ecological agriculture of phyllostachys praecox and tea, cultural and creative tourism projects and art facilities were also introduced to form its brand of Mingyue village "Cultural and Creative village". Relying on its unique development model, Mingyue village has now successfully got rid of the title of poor village in the past and became an icon for rural revitalization and rural development [2] .

Fig. 1 Mingyue village

Fig. 2 The best-preserved "Mingyue Kiln" in the village

Although the change of business format has made Mingyue village a well-known celebrity online and this ancient village is once again vibrant. Local villagers are also losing the memories of the traditional farming life and their attachment to neighboring relationships.

* Corresponding author: 1445033373@qq.com

Therefore, in the design of Qianmo Tower post station, the team hopes to extract and utilize the elements of different local vernacular architecture from the perspective of architectural typology—to restore the "archetype" of vernacular architecture in local residents' memory, thus creating a place that can carry homesickness and a place where the soul belongs.

Fig. 3 General layout of Qianmo Tower

Fig. 4 Perspective of Qianmo Tower

2 Architectural typology: extraction of archetypes of rural architecture in western Sichuan

Architectural typology is an important and general way to analyze architecture. Typological theory would find certain types from a wide range of materials, then summarize and categorize the existing types as to graphically convert them into simple geometric figures, namely archetypes, and discovers its variants. Finally the design would be proceed based on common elements of archetypes and its variants. Typological theory tends to extract and find architectural images from collective memory and human archetypes in its design. By establishing architectural intentions, typological methods allow the reproduction of the conventional forms in people's memory, and provide a followable way fro the combination of this conventional form with modern design activities[3].

Mingyue village has a long history and unique regional characteristics. This village is located in the depths of the Pujiang River. It is quiet and full of humanistic feelings. In such a comfortable environment, the collective memory of the villagers must carry a unique understanding of the local vernacular architecture archetype. Vernacular buildings, constructed by humans, have the universal value of being local characteristics and shaping the memories of villagers, thus are vividly expressed as a unique impression of a certain period. Furthermore, this image has become the most paramount concept in the field of vernacular architecture, given its interconnections with the local history and culture and the behaviour of local villagers[4]. Meanwhile, it carries specific spiritual values. Although the vernacular architecture in Mingyue village' s historical period is gone forever, as architects, we seek to discover the "archetype" form of vernacular architecture that exists in the memory of local residents and find eternal value from it, so as to generate new ideas

For this ancient village, what is the archetype of the local vernacular architecture?

Before designing the Qianmo Tower post station in Mingyue village, the design team went to Pujiang county for on-site investigations. Text materials and drawings were sorted out through visits and exchanges, on-site surveying and mapping, document collection and other methods. （Fig. 5,6）After typological analysis of the data, we summarized the current features of the local buildings in Mingyue village, and compared these features with the same types of buildings in the historical period, so as to sum up the universal archetype of Mingyue village's vernacular architecture. These preliminary surveying and drawing papers and the archetype research of Mingyue village's vernacular architecture have played a guiding role in the design process of Qianmo Tower post station.

The typological approach emphasizes the establishment of architectural intentions to restore conventional forms in people's collective memory[5]. The residential buildings in western Sichuan, represented by Mingyue village, are mostly through type timber frame, with sloped roofs and thin eaves, creating an environment that is open and transparent, light and delicate, elegant and simple. However, the external shape of the building alone is insufficient to establish the archetype of Mingyue village's vernacular architecture. Western Sichuan architectural culture is a highly compatible fusion culture, which has not lost its tradition while absorbing foreign cultures. Modern church buildings in Chengdu are a typical example.

Fig. 5 The local buildings in Mingyue village

Fig. 6 The local buildings in Mingyue village

Therefore, in the design process, the team decided to maintain the traditional houses of Mingyue village in Western Sichuan, while drawing inspiration from church architecture to introduce this centralized and symbolic architectural type. And eventually, restore the vernacular architecture in the minds of the villagers in Mingyue village from the two aspects: the architectural appearance of the houses and the spiritual symbolization of churches.

 https://doi.org/10.1051/e3sconf/202123703019

3 Combination of Chinese and western architectural styles—The blending of Basilica and traditional western Sichuan residence

"Sichuan Chronicles. Architecture Chronicles"[6] states that "the earliest modern building in Sichuan was the Catholic Church". The church architecture reflects the level of Chinese construction at that time, the spread of Western architectural art in Sichuan, and the development of local architectural art. During the early state when Catholicism was introduced into Sichuan, it could only be spread in the countryside far from the city in a low profile due to the influence of etiquette disputes, retreat policies, and Sichuan religious cases. Rural churches, therefore, shaped its unique form of combining Chinese and Western architectural styles under this background[7].

The design of Qianmo Tower post station follows the combination of Chinese and Western characteristics of rural churches in Western Sichuan. It adopts a Western Basilica layout, while the house structure adopts the traditional Western Sichuan architectural form, showing an ingenious integration of these two cultures. (Fig. 7,8)

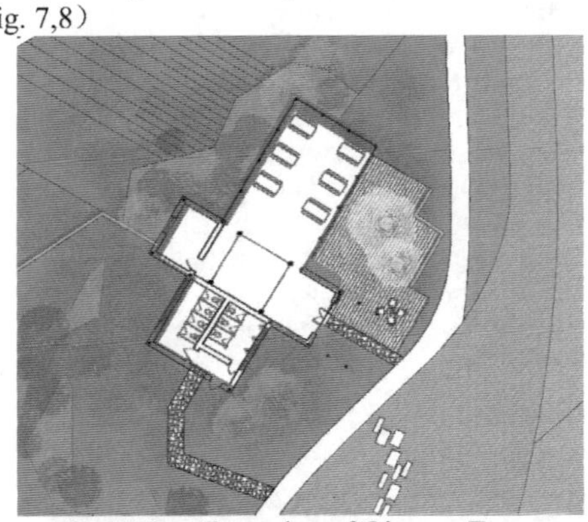

Fig. 7 First floor plan of Qianmo Tower

Fig. 8 Elevation of Qianmo Tower

The plane composition of the building is inspired by Western churches, using a typical Basilica layout. The longitudinal hall of the post station is rectangular, and the lateral halls are added to the ear rooms on both sides, which are cross-shaped as a whole. There are four columns of pillars inside the longitudinal hall, which divide the hall into three long spaces. The central one is wider, which is the central hall, and the sides are slightly narrower, which are side corridors. Although the design simplifies the two columns in the middle, the Basilica style representing the form of the central hall higher than the side corridors is still retained—the tower structure in the center. However, in terms of the facade structure, Qianmo Tower post station retained as much as possible the characteristics of western Sichuan residential houses, and redefined the component materials of the facade according to the modern functional requirements and the element in through type timber frame of western Sichuan residential buildings. The facade of the steel frame structure not only follows the graphic elements of the through type timber frame in Western Sichuan, but also incorporates the stained glass window elements of Western churches. Between light and shadow, the colors are colorful and mottled, as if a world away. (Fig. 9,10)

Fig. 9 The stained glass window of Qianmo tower

Fig. 10 The stained glass window of Qianmo tower

The bell tower in the centre of the post station takes the towering structure of the Basilica-style central hall as the starting point and combines the intention of "tower" to shape the composition centre and spiritual fortress of the post station. The exterior of the bell tower follows the through type timber frame of the western Sichuan residential houses. (Fig. 11) The roof of the green tile slope, the eaves are thin, forming a light architecture style; the wooden pillars are cylindrical, surrounding the building, slender and stable; the walls are built with soil, natural and simple; the walls are decorated with wooden square windows, enriching the structure of the facade.

The external sense of form could affect psychological activities of human. The tall, straight, and pointed structure of the bell tower shows a strong upward momentum. Such a structure is often easy to become the focus of people's vision. The utilization of the through type timber frame in western Sichuan and local materials in the post station make this place full of historical vicissitudes and human warmth. （Fig. 12） As time goes by, the bell tower will be repeatedly used by local residents as an identifying structure of the station; And as residents become more familiar with Qianmo Tower Station, their reliance on the bell tower would increase accordingly. Finally, the bell tower is likely to become a landmark in the site, and then gradually merge into the memory of local residents.

Fig. 11 The bell tower

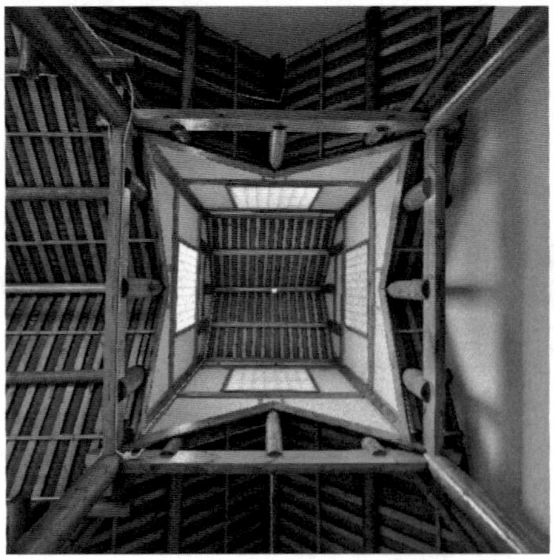

Fig. 12 Interior structure of the bell tower of Qianmo Tower

4 Attach importance to nature and return to the native land—attention and response to the local natural environment

Paying attention to and responding to specific natural environments is one of the key characteristics of vernacular architecture[8]. Architecture does not exist in isolation, it always be in a certain environment. Therefore, issues such as " how the building combines the original environment? " , " How to use the intangible factors such as local climate conditions? " , And " what posture the building itself appears in the tangible environment? " are aspects that must be considered in architectural creation. In the design process, Qianmo Tower Station especially focus on combining the external appearance of the building with the local environment of Mingyue village, giving people an overall image, and at the meantime, enabling people to feel the external nature while inside the building. This would also deepen people's experience and impression of the local characteristics and natural environment of Mingyue village and create a sense of belonging.

Pine forests, tea valleys, and rice fields slowly stretch out along the texture of the hills of western Sichuan, and the unique residential houses of western Sichuan are scattered among them — the simple and comfortable natural environment of Mingyue village. The design of Qianmo Tower Station is based on the concept of local houses and rural churches in western Sichuan. It absorbs the Basilica-style plane composition of the Catholic Church and uses the structure and materials of the houses in western Sichuan. Meanwhile, it is dominated by natural landscapes and is coordinated with the surrounding environment, including the use of overall colors, the use of topography, the combination with the terrain and the reflection of the stream, etc. These allow the station as compatible as possible with the surrounding environment without sacrificing the centrality and symbolism of the building. （Fig. 13）

Fig. 13 External environment of Qianmo Tower

5 Vernacular architecture and its future development

Vernacular architecture is a materialized product of the development of rural culture under a specific natural and social context. In the process of evolution, the rural buildings have condensed unique regional characteristics, some of their characteristics have been preserved, while others have gradually disappeared in the evolution. " How to generalize and abstract the archetype of the rural architecture? ", " How to give the beauty displayed in it an eternal charm? ", And " how to create a building that enhances historical significance and meets the needs of modern development? ". Those are issues those run through the design process of Qianmo Tower Station.

With the acceleration of the modernization process and the implementation of the country's policy of socialism new rural reconstruction, the construction technology, construction materials and construction efficiency of rural dwellings have been noticeably promoted. However, the buildings in the countryside have not inherited and developed the traditional characteristics of the region on this basis. Instead, there has been a phenomenon of homogenization development caused by the blind tendency of different regions to urban architectural styles. Local architecture is also accelerating its loss of the unique regionality in this process. As a rational analysis method, typology simultaneously constructs buildings and cities, connects tradition with the present, and integrates architecture and nature. It helps to excavate and inherit the characteristics of local traditional architecture, and to use new technologies to combine it with materials.

In Qianmo Tower, the original architectural form of the western Sichuan residential houses retains the architectural archetype of the western Sichuan residential houses, and retains the original outlook of the hometown architecture existed in the memories of local residents; the Latin cross Basilica-like plane layout of the western church also attaches the construction with religious iconicity and centrality. The QianMo Tower combines the types of traditional houses in Western Sichuan and the Basilica style of Latin cross, adding a cohesive force to the local architecture in the hearts of residents.

Strolling in the QianMo Tower to feel the combination of Chinese and western culture, the overlap of tradition and modernity, and the blend of homesickness and foreign places. It is like a double-sided mirror, illuminating the past and the future of this land.

Acknowledgment

This paper has been supported by the following fund projects. MOE (Ministry of Education of the People's Republic of China) Youth Fund Project of Humanities and Social Sciences (Project No.18YJC850002). Bidding Project of Key Research Base of Humanities and Social Sciences in Colleges and Universities of Anhui Province (Project No. SK2015A170).

References

1. Gao Xin. Analysis on the role of cultural and creative industries in rural community construction -- Taking Mingyue village in Pujiang County as an example [J]. China market, 2019 (17): 25-26
2. Guan Xiuyuan. Feasibility analysis on the development of rural tourism driven by cultural and creative industries: a case study of Mingyue village, Ganxi Town, Pujiang County [J]. Anhui agronomy bulletin, 2019, 25 (01): 22-23
3. Urban architecture [M]. China Architecture press, (Italy) Rossi, 2006
4. Zhao Xing. Contemporary "construction" road of traditional vernacular architecture [D]. Tianjin University, 2005
5. Tian Rongrong. Research on the practice of new vernacular architecture in Southwest China from the perspective of construction [D]. Guizhou University, 2018
6. Sichuan local chronicle Compilation Committee. Sichuan Provincial annals · architectural records [M]. Chengdu: Sichuan Science and Technology Press, 1996
7. Cao Lun. Modern Catholic Church Architecture in Western Sichuan [D]. Southwest Jiaotong University, 2003
8. Zhu Jinliang. Research on the creative practice of contemporary Chinese Vernacular Architecture [D]. Tongji University, 2006
9. Zhang Jizhen. Applied research on Typology in the renewal and development of rural settlements in western Henan [D]. Hunan University, 2010. Li Bin. Analysis of inheritance and revival of local architectural culture [D]. Tianjin University, 2018
10. Wu Liangyong. Modernization of vernacular architecture and regionalization of modern architecture: on the way to explore new architecture in China [J]. Central China architecture, 1998 (01): 3-5
11. Wu Zhihong. Context, problems and prospects of Chinese vernacular architecture research [J]. Journal of Kunming University of science and Technology (SOCIAL SCIENCE EDITION), 2014,14 (01): 103-108
12. Han Dongqing. Types and vernacular architectural environment -- on the environmental understanding of villages in southern Anhui [J]. Acta architecturae, 1993 (08): 52-55

Analysis on the Influence Factors of Construction Linear Control of Continuous Rigid Structure Bridge

Yang Liu[1], Chunmo Zheng[1], Huaiqiang Ba[2],Guangli Xu[2],Chao Li[2],Qinyun Xie[3]

[1]School of Architecture and Planning, Yunnan University, Kunming 650091, China
[2] Yunnan Kunchu expressway investment and Development Co., Ltd, Kunming 650030, China
[3] Yunnan Dazhu Technology Co., Ltd, Kunming 650030, China

Abstract. In order to study the influence of prestress on cantilever deflection and construction linear control of continuous rigid frame bridge in construction stage, this paper introduces the significance of continuous rigid frame bridge's linear control, the calculation principle and deflection influence analysis of vertical formwork elevation in cantilever construction. According to a specific continuous rigid frame bridge, this paper use the finite element software to simulate and calculate the deflection of prestress to the cantilever construction of continuous rigid frame bridge. The influence of friction coefficient between prestressed steel bundle and bellows and prestress loss on cantilever deflection and construction line control of continuous rigid frame bridge is also analyzed, furtherly brings out the solution to deal with the problems due to the change of prestress.

1 Introduction

The prestressed tendons are tensioned before the combination of concrete, so as to change the working state of concrete, put pressure on concrete in advance, making the high strength steel bar play a full role, also make the concrete's compressive performance better. Meanwhile, restressing the concrete members in advance can offset part of the used load or even all because of the direction of action is exactly opposite to the direction of the used load. When the torque generated by the external action on the beam section is offset by the torque generated by the applied prestress, the flexural member can be designed as a compression member, thus improving the mechanical performance of the concrete member. The prestressed continuous rigid frame bridge has the advantages of large stiffness, small deformation, less expansion joint, good driving comfort, high integrity and high seismic performance.[2]In the construction of prestressed rigid frame bridge, the cantilever construction is represented by the hanging basket construction. The two adjacent piers move the hanging basket to both sides at the same time, advancing horizontally and synchronously. The suspension is connected with the long symmetrical construction section by section along the span direction until the span is closed, and the beam sections are tightly connected to the whole by tensioning prestressed steel bundles. Under the action of prestress, the lower deflection brought by each construction stage can be offset. By controlling the prestressing tendons, the reverse deflection is produced, and the concrete is pre-loaded, thus offsetting the lower deflection brought by the later use of the load, Meet the design requirements. Therefore, the linear control of continuous rigid frame by prestress is the main factor.

2 Continuous rigid frame construction control

In order to ensure the construction quality of the bridge and make the structure internal force and displacement meet the design requirements, it is necessary to control the construction of prestressed concrete continuous rigid frame bridge, Includes deformation control and internal force control. Deformation control is mainly to strictly control the vertical deflection and transverse deviation of each section of box girder. When there is a deviation and the deviation is large, the error analysis is carried out immediately and the adjustment scheme is made to prepare for the more accurate construction of the next section. Because the structural state of continuous rigid frame can not be adjusted after construction, prediction control is mainly used in construction control, including construction control structure simulation analysis, construction monitoring, construction error analysis and subsequent state prediction, etc[1].

3 Analysis of deflection and calculation principle of vertical die elevation

3.1 Deflection analysis

In the actual construction, there is often a situation that does not accord with the actual design, and the actual situation deviates from the theoretical calculation. For

 https://doi.org/10.1051/e3sconf/202123703020

continuous rigid frame, deflection control is the basis of elevation control. In order to judge whether the beam force and deformation are normal and whether the theoretical calculation model is in line with the reality. The deflection of continuous rigid frame bridge during construction includes beam weight, pre-stress, concrete shrinkage and creep, and hanging basket deformation. Deflection calculation mainly includes: a. deflection and cumulative deflection of each node section in each construction stage. b. the cumulative deflection of each node section in the bridge state. c. maximum (minimum) live load deflection of each node section[3].

3.1.1 Calculation of deflection of dead load, live load and prestress

The deflection of cantilever beam with variable cross section is calculated by conjugate beam method, and the bending moment M generated by dead load, live load and prestress on virtual beam can be obtained X arbitrary sectionjDeflection at:

$$f_j = \sum_{i=1}^{j} \frac{M_i}{E_i I_i}(x_i - \xi_i)d_i \quad (i < j) \qquad (1)$$

M_i: Average bending moment of section i beam

I_i: i moment of inertia of section of beam section

3.1.2 Calculation of shrinkage creep deflection

At present, the code "Design Code for Highway Reinforced Concrete and Prestressed Concrete Bridges and culverts "(JTG3362-2018) is used to calculate.

3.2 Pre-arch setting

In order to make continuous rigid frame under the influence of comprehensive deflection such as dead load, live load, prestress load and concrete creep, it is necessary to set pre-arch for continuous rigid frame during construction, so as to offset the deformation caused by dead weight of bridge structure, bridge deck pavement and railing. The general value is equal to the vertical deflection of all dead loads and half of dead loads.

$$f_{camber} = \left(-f_y + f_{g1} + f_{g2}\right)\left(1 + \phi(\infty,\tau)\right) + \frac{1}{2}f_p \qquad (2)$$

f_y: deflection resulting from prestress

f_{g1}: Self-weight, temporary load and deflection of hanging basket in first stage dead load beam section

f_{g2}: Deflection of secondary dead loads, guardrails and paving

Deflection resulting from fp: live loads

$\phi(\infty,\tau)$Final value of shrinkage creep of concrete

3.3 Calculation Principle of Vertical Model Elevation

Elevation control is the basic guarantee to ensure the smooth closure of continuous rigid frame, and it is also the decisive factor to ensure that the bridge alignment meets the design requirements. The deflection of each stage in cantilever pouring construction is inevitable, so the elevation of vertical formwork is based on the deflection of each stage, and the pre-arch degree is set in the construction process. The deflection deformation caused by self-weight and other loads is offset by the elevation of vertical die in each stage.

4 Prestressed deflection analysis

4.1 Upper deflection due to prestress

Because of the characteristics of prestress, the tension of longitudinal prestressed tendons in the stage of construction will produce upward negative deflection, which can effectively offset the deflection caused by other loads, such as self-weight, and is the key to linear control[5].

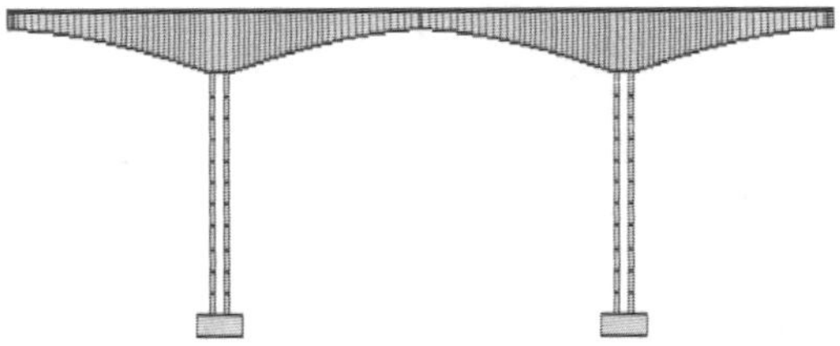

Fig.3.1 Midas civil full bridge model.

Taking prestressed concrete continuous rigid frame bridge (120 m+230 m+120 m) as an example, the highway-class I, double two-way four-lane, single bridge width 13.85 m, and double thin-walled pier height 55 m, are modeled by finite element Midas Civil software. The longitudinal prestressed steel bar of the main beam of the bridge is constructed by high strength and low relaxation rate steel strand and fpk=1860Mpa, E=1.95*105Mpa, using hanging basket cantilever pouring construction, the whole bridge main beam construction has 33 construction stages, according to the actual construction steps to establish the whole bridge model analysis.

According to the construction model of the whole bridge, two working conditions are established, namely, normal prestressed concrete reinforced structure and ordinary prestressed concrete reinforced structure. The deflection deformation curve before the closure is obtained by calculation. Through two kinds of cumulative deflection deformation, it can be clearly found that the cumulative deflection of ordinary concrete reinforced structure in the construction stage is positive deflection, and the vertical deformation is large. Negative deflection can be offset by the deflection of the structure itself, more in line with the design line. The maximum difference in the cumulative deflection between the two is near the maximum cantilever section. The deflection difference at the maximum cantilever is shown in Table 3-1 as below.

Table.3.1 Cumulative deflection difference at maximum cantilever.

Node Number	12	13	70	71
Common concrete structures (cm)	-25.43	-25.77	-19.04	-18.06
prestressed concrete structures (cm)	-5.8	-5.8	1.94	4.3

Difference (cm)	19.63	19.97	20.98	22.36

Fig .3.4 Comparison of cumulative deflections before closure.

Through the deflection contrast curve, it is clearly found that the deflection difference between the two is small at block 0. With the increase of segment construction, the cumulative deflection becomes larger and larger. At this time, the deformation of concrete stiffness becomes smaller and more stable, which provides favorable conditions for the closure stage. By comparing the slope of the curve, it can be concluded that for continuous rigid frame, prestress has the main influence on the construction line control. Under the action of prestress, the relative deflection difference of each stage is small, which is convenient for the actual construction line control. More in line with the design line.

4.2 Analysis of pipe friction parameters on deflection

At present, the prestressing continuous rigid frame generally adopts the construction technology of tensioning at both ends of the post-tensioning method. Because of the characteristics of the construction, corrugated pipes are reserved in the box girder in advance, and the pipeline positioning is not accurate during the construction process. The negative friction caused by steel bundle tension and hole wall slip will produce greater friction loss for bending prestressing tensioning[6]According to the new code, the loss of prestress friction caused by pipeline parameters is based on the formula:

$$\sigma = \sigma_{con}\left(1 - e^{-(\mu\theta + kx)}\right) \quad (4)$$

σ_{con}tensile stress at the end of tension

μFriction coefficient between prestressed steel bundle and hole

k: local deviation coefficient per meter length

θ: The sum of the bending angles from the end of the tension to the channel of the calculated section

x: the length of the channel from the tension end to the calculated section .

μBy analyzing the influence of the friction coefficient and k on the deflection of the continuous rigid frame, the friction coefficient of the channel is analyzed by using five cases.

Working conditions 1: k=0.0015=0.17μ

Working conditions 2: k=0.002=0.17μ

Working conditions 3: k=0.002=0.35μ

Working conditions 4: k=0.003=0.35μ

Working conditions 5: k=0.0015=0.15μ

The deflection curve of the main beam can be seen that the deflection of the main beam becomes larger when the μis increased, and the deflection of the main beam is relatively larger when the k is increased. Therefore ,μ friction coefficient and deviation coefficient of prestressed pipe will cause prestress loss and increase vertical deflection of main beam. The reverse deflection changes caused by prestressing forces under different friction parameters of the main sections are worthy of Table 3-2 below.

Table.3.2 Reverse Deflection Value of Prestressed under Section Corresponding to Friction Parameters.

Deflection value (mm)	k=0.002 μ =0.35	k=0.002 μ =0.17	k=0.0015 μ =0.17	k=0.003 μ =0.35
Section 6	0.805	0.815	0.817	0.802
Section 10	1.481	1.538	1.543	1.473
Section 14	2.604	2.613	2.619	2.594
Section 18	5.788	5.844	5.857	5.763
Section 22	11.828	11.927	11.962	11.766
Section 26	19.437	19.533	19.553	19.408
Section 30	35.715	35.838	35.914	35.587

Through the analysis of the reverse deflection value caused by prestress under different friction parameters of the hole, it can be concluded that the deflection value of prestress under different friction coefficients of the same section is affected by friction coefficient. When the μ、k increases, the deflection value of prestress becomes smaller; when the k remains unchanged, when the μ increases, the deflection value of prestress becomes smaller; when the μ remains unchanged, when the k increases, the deflection value of prestress becomes smaller; when the friction coefficient increases, the loss of prestress increases, the reverse deflection value of prestress becomes smaller, and the vertical deflection of main beam becomes larger.

4.3 Change in deflection due to loss of prestress

Long-span continuous rigid frame bridge deflection is a prominent disease, among which the loss of prestress is one of the main reasons. Because of the linear control and deflection control of continuous rigid frame, the reverse deflection of prestress is mainly used to offset the deflection caused by the structure itself and long-term use. The loss of prestress is affected by construction method, material and temperature [7]. To study the deflection changes resulting from the loss of prestress, Take the state river bridge, Drawing 1395 MPa, at both ends of post-tensioning method used in actual construction The finite element model, Under the conditions of 10%,20%,30% and all prestress loss, respectively, the side span closure (section number 1), the side span 3/4(section number 2), the side span 1/2(section number 3), the side span block 0(section number 4), the middle span 1/2(section number 5), the middle span 3/4(section number 6), the middle span closure (section number 7) are selected, Vertical deflection of main beam. By comparison, it can be clearly concluded that with the increase of prestress loss, the lower deflection of the main beam is increasing, and the loss of prestress has a significant effect on the alignment of continuous rigid frame bridges. In the same section, with the increase of prestress loss, the vertical deflection of the main beam is 4.71 cm. It can be seen from the chart that the greater the loss of prestress, the greater the deflection difference, which is consistent with the lower deflection of the main beam.

5 Response measures under prestress

Based on the theoretical analysis of the influence of the deflection of continuous rigid frame on the linear control in the construction stage, combined with the actual project list of the Zhouhe Bridge, the measures to deal with the change of prestress on the linear control are summarized:

(1) Because the reverse deflection caused by prestress is the main factor in the linear control of continuous rigid frame, it is necessary to configure the reasonable number of prestressed tendons according to the bridge type at the beginning of the continuous rigid frame design. Ensure that the line shape after tensioning prestressing bundle is more in line with the design line.

(2) The post-tensioning method is usually used in the construction of cantilever casting. The friction coefficient of the hole k、 μ need to be tested, the friction coefficient between the prestressed steel bundle and the hole is measured μ and whether the local deviation coefficient per meter length of the hole k, meet the design requirements. If the experimental results are abnormal, the reasons should be found in time to solve the problem[9]The value of k、 μ is 0.001~0.003 and 0.1~0.35 according to the Design Code for Highway Reinforced Concrete and Prestressed Concrete Bridges and culverts.

(3) The loss of prestress is very important to the linear control of continuous rigid frame bridge, which affects the linear change in the construction stage and the long-term use of the bridge. In the process of tensioning prestressed steel bundles, it is necessary to stretch accurately, to overtension in the design code, to use double control of tension and extension, to predict the loss of prestress, and to monitor the line shape according to the actual construction stage. Adjust the elevation of the model accordingly.

6 Concluding remarks

Through the above calculation and analysis, prestress is the main influencing factor of linear control of long span prestressed continuous rigid frame.

(1) Because the reverse deflection brought by prestress significantly changes the deflection value of the main beam and makes the whole bridge more in line with the design line, the monitoring of prestress is the key to the control of the whole line.

(2) Under the action of prestress, the pre-arch should be set according to the actual situation. Because of the deviation between the finite element simulation and the actual situation, the elevation of the vertical die should be corrected in real time.

(3) Because the control of alignment is affected by many factors, it is necessary to combine the prestress comprehensive consideration, use the mathematical statistical method to combine the influencing factors of alignment, select the most unfavorable factors, and carry out linear prediction to ensure safety.

(4) In the field line monitoring, to real-time deflection measurement, timely feedback.

References

1. YongXia Dong. Shijiazhuang Tiedao University (2013)
2. Yugen Chen. Southeast University (2016)
3. Jiangang Tong. Southwest Jiaotong University (2007)
4. Jiyao Zhang. Cantilever Prestressed Concrete Continuous Beam Bridge (2004)
5. Jianli Pang. Fujian Building Materials 04, 4 (2013)
6. Huabin Fang, Qixian Tian. Bridge Construction 02, 81 (2006)
7. Jianguo Shao. Lanzhou Jiaotong University (2012)
8. Hongchun Han. Southwest Jiaotong University (2007)
9. Jian Huo. Northeast Forestry University (2013)

A Preliminary Study on Green Building 3D Modelling Energy-saving Design Combined with Traditional Ecological Concepts

Runqi Qin*[1], Xinjian Li[1]

[1]School of Architecture, Southeast University, Nanjing, Jiangsu Province, China

Abstract. By combining the green ecological wisdom on traditional Chinese residential buildings, this paper summarized the ideas that can be used for references and applications in modern green energy-saving design. In addition, in view of the inherent shortcomings of traditional residential buildings, this paper proposed a green energy-saving design method for traditional residential buildings based on 3D modelling technology, which is complementary to the traditional ecological construction concept. Based on the establishment of a 3D building information model of traditional residential buildings, the optimal solution for the energy-saving design of green buildings is found through energy-saving adjustments to the building space structure and building spacing parameters.

1 Introduction

With the rapid development of China's economy, buildings are facing major energy supply challenges, and the concept of sustainable development continues to deepen. Green ecological buildings have become an important part of sustainable development.

Many efforts have been made to achieve the goal of reducing energy consumption. Among them, green roofs have been widely used since ancient times in both hot and cold climates [1]. Castleton et al. [2] extensively reviewed the existing literature to assess the potential building energy benefits of green roofs, and they reported that green roofs can significantly reduce the energy consumption from both winter heating and summer cooling in buildings with poor insulation. In contrast, other investigations [3] show that the feasibility of green roofs is very scarce when natural gas, electricity and water tariffs are considered in warm climates; and green roofs are very inconvenient when water costs are considered in cold climates. Furthermore, if the investment costs are considered, green roofs are not convenient energy refurbishments.

Although green building technology has made some progress, it still needs improvement [4]. Among buildings of different functional types, residential buildings can best reflect the close relationship between people and buildings. Traditional residential buildings also contain considerable green and energy-saving ecological wisdom, which is worthy of research and reference by contemporary designers.

Furthermore, with the development of computer technology, there is a new green energy saving methods. Recently, building information modelling (BIM) technology has been widely used as a digital design platform in green building design [5-8].

Traditional residential buildings with ultralow energy consumption need to combine local living habits and climatic conditions, learn from the inherent ecological ideas in traditional construction, and form targeted green and energy-saving designs. Moreover, since traditional houses use preindustrial construction technology and methods, there will be many shortcomings in the modern living environment that need to be improved. A previous study that analysed the occupants' responses to thermal discomfort in green and conventional buildings in New Zealand was reported that occupants in the green buildings had fewer environmental adjustments and more personal and psychological coping mechanisms than those in conventional buildings, which means that occupants in the green buildings had more access to building controls while the occupants in conventional buildings are not as dependent on the building design to moderate temperature and lighting [9].

Three-dimensional modelling technology can establish two-dimensional and three-dimensional information of a building at the same time and interface, can quickly and accurately obtain various parameters of the building and can adjust related parameters according to later needs with high accuracy and reliability as advantages.

Based on this concept, the author takes traditional houses in the area south of the Yangtze River as an example, analyses the spontaneously formed and perfected traditional construction wisdom in traditional houses, summarizes green ecological ideas that can be used for reference and discusses a green building energy-saving design based on the 3D modelling technology method.

* Corresponding author: 220170185@seu.edu.cn

2 Green ecological ideas in traditional houses

2.1 Ecological wisdom in traditional residential buildings

Since ancient times, China has had the traditional concept of "the unity of human and nature". This concept was used by ancient people in the construction of human settlements, forming a highly harmonious architectural form with nature.

At the macro level, when selecting sites for traditional residential buildings in China, people pay attention to local conditions, back mountains and face waters, coordinate natural factors such as the wind direction and solar radiation, and choose the best orientation of the local environment to ensure that the buildings get good sunlight and have comfortable indoor temperatures. In terms of spatial form, courtyards and patios are important spatial organization factors for traditional houses in southern China.

The climate in the area south of the Yangtze River is hot and humid, and a patio can be ventilated and dissipated, which can also help the layout of residential houses with narrow spacing and close lighting, as shown in Fig. 1. The "water in four directions converges into the courtyard" reflects the organization and drainage of the courtyard.

Fig1. Patio of traditional architecture in Nanxun

Traditional residential buildings also pay attention to "retaining wind and gathering gas". For example, a screen wall opposite a gate or inside a gate slows down the wind speed and avoids draughts [10].

In terms of building materials, we pay attention to local materials and have a strong sense of green ecology. For example, the area south of the Yangtze River is rich in water resources, so the commonly used building materials have low water absorption, strong moisture resistance and good durability. In the treatment of external walls, hollow wall masonry structures are commonly used as thermal insulation measures.

2.2 Reference of traditional ecological thought to the design of green energy-saving buildings

Under the guidance of traditional ecological wisdom, traditional Chinese houses have formed architectural forms that are compatible with local climatic conditions, geographical conditions, and humanistic conditions [11] and spontaneously formed a healthy balance between humans and nature. These traditional ecological wisdoms are also worthy of our reference and research today, and they have influenced and inspired the green ecological concept of modern buildings.

The first point is the layout between the buildings. The layouts of buildings in traditional settlements seem to be free and flexible, but in fact, they are reasonable. The layout of a building is generally related to the local wind direction, forming an air duct between the buildings [12]. This affects the natural ventilation inside and outside the building by removing heat, bringing in fresh air and increasing comfort in the hot summer [12].

The second point is the orientation of traditional houses. Buildings generally face south. That is, in accordance with the sunlight, more solar radiation can be obtained in winter. Combined with the locations and sizes of the windows, this can avoid western sunlight and receiving excessively high intensity and long sunlight during the hottest summer.

The third point is the layout of the space under eaves and patios. The space under the eaves of traditional houses in the area south of the Yangtze River is low. This forms a height difference with the high and deep patio, generates air pressure, and forms convection and good heat dissipation and ventilation effects[12].

The fourth point is the choice of building materials. Traditional residential buildings usually take local materials, account for the local climate and utilize material properties. Traditional houses in the area south of the Yangtze River mostly use brick and wood structures. Wood is relatively easy to obtain. Bricks are fired from clay. Bricks are usually used on the periphery to form an empty brick wall, which has a certain effect on heat preservation and heat storage.

3 Shortcomings of traditional houses

Traditional houses were spontaneously built people's ancestors, and they were constantly improved and transformed according to their living conditions. These houses were the result of simple design and planning. However, because there is no systematic coordination and planning, there are many shortcomings and defects in primitive traditional settlements and residential buildings.

First, the spacing of the buildings does not meet the relevant regulations. As time passes, the number of households gradually increases, and additional houses will appear. The traditional settlement streets that originally had a "land and material saving strategy [13]" became narrower and do not meet the requirements of fire protection regulations.

Second, the spatial structure is unreasonable. Through an investigation of the traditional residential settlements in the Nanjing area, it was found that the residents in the buildings have considerable debris piled up, there is no reasonable spatial division, and the evacuation channel is long and not clear. There is no proper placement of the kitchen, bathroom or bath space.

Although under the guidance of traditional wisdom the ecological awareness of residential buildings is worth learning, there are still shortcomings under modern norms and conditions. Researchers are required to combine traditional ecological thinking with modern technology, continuously improve energy-saving design methods, and create more standardized, liveable and energy-saving

 https://doi.org/10.1051/e3sconf/202123703021

building forms.

4 Energy-saving design of traditional residential green buildings based on 3D modelling technology

The ecological wisdom of traditional residential buildings will inspire the green and energy-saving design of modern buildings, such as the layout and orientation of the buildings mentioned above, the choice of building materials, and the design of building shapes, including the ratio of eaves to patios and windows to walls. There is also the design of the building exterior wall structure.

Cho et al. [14] used the Energy Plus and IES software to conduct modelling and energy consumption simulations, respectively, and explored the best building orientation and window area. As shown in Table 1 and Table 2, the final results indicate that when the building model is facing the 0° and 180° directions, the total energy consumption including cooling and heating is the lowest. Therefore, the best building orientation is true south or true north. In addition, as the window area increases, energy consumption also increases.

Table 1. Energy Simulation Results Using Energy Plus. [14]

Parameters		Heating (GJ)	Cooling (GJ)
Orientation (window size 48"x36")	0°	417,764.1	275,803.1
	45°	417,765.1	275,805.2
	90°	417,765.8	275,805.2
	135°	417,766.1	275,804.6
	180°	417,765.2	275,802.3
Window (Building orientation 0°)	24x36	417,755.5	275,803.1
	48x36	417,764.1	275,803.1
	72x36	417,772.9	275,803.1

Table 2. Energy Simulation Results Using IES <VE>. [14]

Parameters		Heating (GJ)	Cooling (GJ)
Orientation (window size 48"x36")	0°	77.8	6.4
	45°	78.0	6.5
	90°	78.3	6.5
	135°	78.3	6.5
	180°	77.9	6.3
Window (Building orientation 0°)	24"x36"	77.8	5.9
	48"x36"	77.8	6.4
	72"x36"	77.9	6.9

Combining the factors in traditional energy-saving technology with 3D modelling technology can effectively assist building energy-saving design.

First, we must establish a three-dimensional building information model corresponding to traditional houses. Second, based on the model, the space layout, building shape, material and structure of traditional residential buildings are designed; and the corresponding parameter values are adjusted and calculated to achieve the best energy-saving effect. The design steps are shown in Fig. 2.

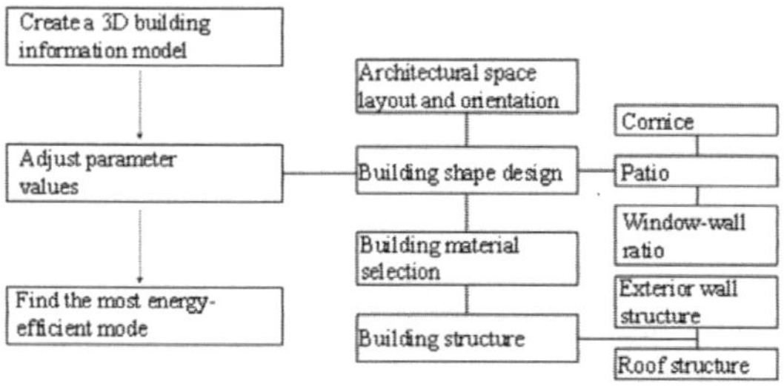

Fig2. Steps of building energy-saving design

4.1 Establishment of 3d building information model

Taking the traditional residential buildings in Nanjing as an example, the Revit-Architecture software is used to build a three-dimensional building information model. In addition, the Ecotect software is used to analyze the environment in the area where a building is located; and on this basis, the parameters including building spacing, roof shape, building stories, window type, etc. are adjusted. Through the adjustment of the parameters, the best energy saving effect is achieved. Table 3 shows the green building energy-saving design parameters of some traditional residential buildings.

Table 3. Energy-saving design parameters of traditional residential green buildings in the Nanjing area.

Item	Parameter name	Parameter value
The outer wall structure of Nanjing's traditional residential buildings	Cement mortar protective layer	thickness 5.0 mm
	Autoclaved fly ash brick	thickness 240.0 mm
	Cement mortar	thickness 20.0 mm
	Lime cement mortar	thickness 20.0 mm
The roof structure of Nanjing's traditional residential houses	Clay blue tile	thickness 20.0 mm
	Cement mortar (inner layer)	thickness 40.0 mm
	Extruded polystyrene board	thickness 50.0 mm
	Cement mortar (outer layer)	thickness 20.0 mm
The exterior window structure of Nanjing traditional residence	Ordinary glass (outer layer)	thickness 6.0 mm
	Air layer	thickness 12.0 mm
	Ordinary glass (inner layer)	thickness 6.0 mm
The windows of Nanjing's traditional houses face love condition	South-facing window-to-wall ratio	0.4
	North window-to-wall ratio	0.1

As shown in Table 3, the 3D building information model of traditional residential buildings in Nanjing includes information on the exterior wall structure, roof structure, exterior window structure and window orientation. Combined with actual local conditions, through the continuous adjustment of these model parameters, the green and energy-saving design of traditional residential buildings can be realized.

4.2 Spatial layout and body shape design among residential buildings

In macro planning and design, the layout of a building should reasonably use solar radiation and wind direction to reduce energy consumption and create a comfortable environment.

Furthermore, the size of a building has an impact on energy savings. A building should choose a volume with a small figure coefficient, which will provide more heat in winter and less heat in summer [15].

Combining the local topography and landforms of traditional residential buildings, the building spacing is set to the best value. In addition to meeting the requirements of fire protection regulations, each building must receive sufficient sunlight to ensure that the winter temperature is not too low and save considerable air conditioning energy consumption.

The average environmental parameters of the Nanjing area in recent years were input into Ecotect software for environmental analysis. Then, according to the analysis results, the distance and angle between two adjacent traditional residential buildings were adjusted to achieve the best results.

4.3 Energy-saving design of single residential buildings

Traditional houses in the area south of the Yangtze River are generally 2-3 storey buildings, mainly brick-wood structures. Based on the abovementioned 3D modelling analysis, combined with the humid and rainy climate in the area south of the Yangtze River, which includes humid and hot summers, and cold winters, the following methods are briefly described:

First, the orientation of traditional houses determines how much solar radiation they receive. By adjusting the angle of the building, as mentioned above, the light intensity can also be increased by adjusting the window orientation angle to save costs. Double-layer hollow glass can be used to appropriately increase the window area of traditional houses. A bay window design can be adopted to select the best convex distance to achieve the best lighting efficiency.

Second, considering the thermal insulation requirements of residential buildings in the area south of the Yangtze River, the body shape coefficient of a building and the heat transfer coefficient of the outer envelope structure can be adjusted to achieve a benign balance, which can meet the thermal insulation demand.

Third, the design of overhanging eaves can be optimized. There are many overhanging eaves in the area south of the Yangtze River dwellings. Overhanging eaves affect indoor brightness and have the effect of allowing rain to roll off. Combined with the 3D model, the width of the eaves can be adjusted to ensure that a room is not affected by the wind and rain and the indoor brightness is appropriate.

In summary, a 3D building information model corresponding to traditional residential buildings is established, its own spatial structure and body shape are designed for energy savings, and the building spacing is adjusted to create an energy-saving and comfortable building form.

5 Conclusion

By analyzing the traditional ecological construction methods in the area south of the Yangtze River, summarizing traditional ecological thoughts worthy of reference and research, and addressing the shortcomings caused by the inconsistency of traditional methods with the modern environment, a traditional residential green building energy-saving design method based on 3D modelling technology is proposed. This method can systematically optimize the building environment, materials, building spacing, window-to-wall ratio and other related parameters so that modern technology and traditional ecological construction concepts complement each other.

Acknowledgments

The authors wish to acknowledge the financial support of the National Natural Science Foundation of China (NSFC, No. 51678129) and the Jiangsu Provincial Department of Culture and Tourism (No. 2019SK03).

References

1. R.M. Lazzarin, F. Castellotti, F. Busato, Energ Buildings. **37**, 1260-1267 (2005).
2. H.F. Castleton, V. Stovin, S.B.M. Beck, J.B. Davison, Energ Buildings. **42**, 1582-1591 (2010).
3. F. Ascione, N. Bianco, F. de Rossi, G. Turni, G.P. Vanoli, Appl Energ. **104**, 845-859 (2013).
4. N. Zhang, X. Liu, Architecture Technology. **48**, 158-160 (2017).
5. H. Xia, K. Yi, Study on Collaborative Design of Green Building Based on BIM Technology. 2013, Springer Berlin Heidelberg: Berlin, Heidelberg. p. 921-928.
6. T. Wei, Y. Chen, J Amb Intel Hum Comp. **11**, 3699-3706 (2020).
7. K.S. Abhinaya, V.R.P. Kumar, L. Krishnaraj, International Journal of Renewable Energy Research. **7**, 1675-1681 (2017).
8. Z. Yuan, J. Zhou, Y. Qiao, Y. Zhang, D. Liu, H. Zhu, Sustainability (Basel, Switzerland). **12**, 7862 (2020).
9. N.S.M. Azizi, S. Wilkinson, E. Fassman, Energ Buildings. **104**, 191-198 (2015).
10. J. Wang, The ecological concepts of Chinese traditional residential buildings in the application of energy-saving design of the green building, East China Jiaotong University, (2013).
11. M. Wang, Research on influencing factors of green

https://doi.org/10.1051/e3sconf/202123703021

building promotion, Capital University of Economics and Business, (2014).

12. L. Bao, Architect. **132**, 5-12 (2008).
13. J. Xu, Z. Wu, Y. Liu, Z. Shu, W. Yang, Y. Wei, Industrial Construction, **49**, 1-6 (2019).
14. C. Cho, D. Chen, S. Woo, Journal of KIBIM. **2**, (2012).
15. T. Liu, China Forest Products Industry. **44**, (2017).

 https://doi.org/10.1051/e3sconf/202123703022

Theoretical and experimental study on vibration modal for simply supported bridge

Ying Wang[1], Li Peng[1], Xuewei Zhang[1], Lingfeng Chen[1], Tao Chen[1], and Congqi Fang *[1]

[1]College of civil engineering, Shanghai Normal University, Shanghai 201418, China

Abstract. For the highway bridges with small or medium span, a vehicle will stay on the bridge after the excitation. So the measured values of the vibration characteristics of the bridge are related to the vehicle-bridge system. In order to analyse the natural vibration characteristics of simply supported bridge and the loaded frequency of vehicle-bridge system, theoretical analysis was carried out firstly, then the simply supported beam calculation model was built based on finite element method, and the results were compared with the measured values. The finite element model of a mass block-simply supported beam system was built by regarding the vehicles as a concentrated mass block attached. Both the vibration modals of the bridge system and the loaded frequencies of the vehicle-bridge system under diffident vehicle positions were obtained by software analysing. As a result, the variation characteristics of the loaded frequency for this simply supported bridge were extracted and discussed through the comparison of the results from both finite element method and the measured.

1 Introduction

With the sustainable development of national economy and engineering technology, achievements of bridge construction change with each passing day. The amount of existing bridges is increasing rapidly since a large number of newly built bridges were putting into operation. In recent years, the on-site testing technology is used widely in bridge test and evaluation. As an on-site test for highway bridges with short or medium span, their vibrations were excited usually by a single vehicle, and the vehicle will stay on the bridge after excitation, so the measured values of the bridge under the action of a single vehicle are the vibration characteristics of vehicle-bridge system. In other words, the measured frequencies are the loaded frequencies of the vibration system. The literatures show that the frequency differences between the vehicle-bridge system and bridge structure cannot be neglected [1], of even larger [2] than the changes of structural damage.

In this paper, the authors firstly analysed a simply supported steel beam theoretically, then a finite element model was built based on the software of Ansys, and the analysis results were compared with the measured values. Next, the finite element model of a mass block-simply supported beam system was built by regarding the vehicles as a concentrated mass block attached. At last, the analysis results from software were compared with the tested values, and the authors gave some useful conclusions which are appreciated for reference.

2 Profile of the simply supported beam

The simply supported steel beam with a span of 690mm between two points, and with a rectangular cross section of 50mm (width) ×8.5mm (height) is shown in Fig.1. The elastic modulus and Poisson ratio of the steel beam are 2.06×10^5MPa and 0.3 respectively. Furthermore, the mass density of steel is 7850kg/m^3.

3 Modal frequency analyses

3.1. Theoretical analysis

As a simply supported beam with uniform section, its natural frequencies can be described as following [3]:

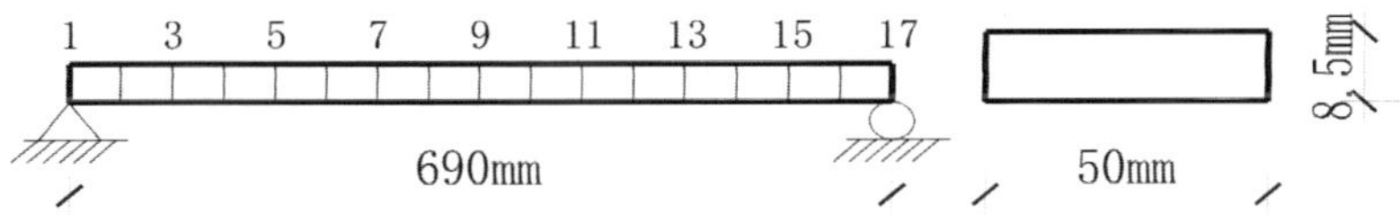

Fig.1. Sketch of a simply supported steel beam

*Corresponding author: congqifang@msn.com

$$f_n = n^2 \pi^2 \sqrt{\frac{EI}{mL^4}} \quad n=1,2,3,\cdots (1)$$

Where f_n is the frequency of each order, EI is the flexural stiffness of cross section, m is mass for per unit length, and L is the span between two support points.

According to the physical property parameters and size parameters of the steel beam described as above, the first 5 order natural vibration frequencies can be calculated as in Table 1.

Table 1 The first 5 order natural frequency from theoretical analysis（Hz）

order	frequency	order	frequency
1	41.449	2	165.797
3	373.044	4	663.189
5	1036.225		

3.2 Finite element analyses

The finite element model using commercial software Ansys [4-5] was built as shown in Fig.2. The conventional beam element was used to simulate the steel beam, and block Lanczos algorithm was adopted for modal extraction.

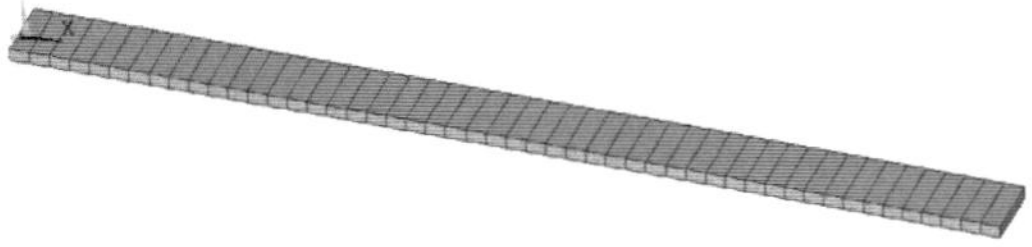

Fig.2 Finite element model for steel beam

The vibration modal parameters can be obtained by running Ansys modal analysis module. The first 5 order natural vibration frequencies are showing in Table 2. And the related vibration deformation diagrams are showing in Fig.3.

Table 2 The first 5 order natural frequency from Ansys（Hz）

order	frequency	order	frequency
1	41.469	2	165.840
3	373.030	4	663.880
5	1035.200		

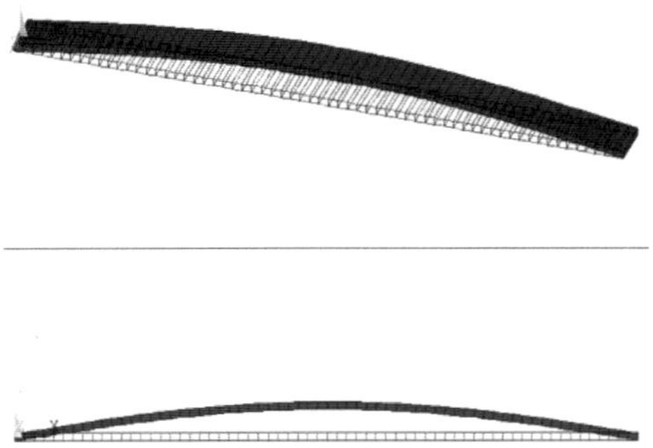

（a）The 1st order

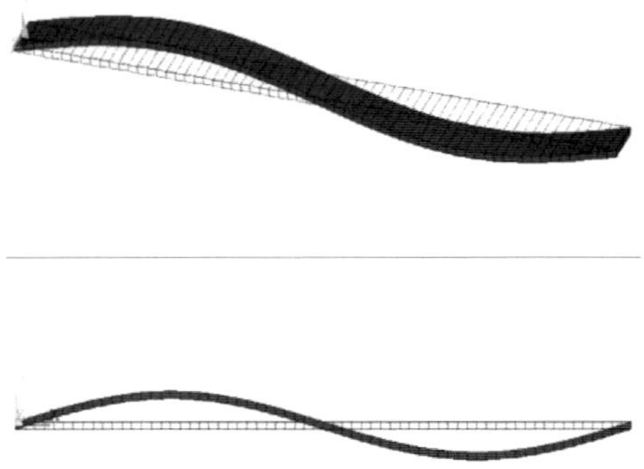

（b）The 2nd order

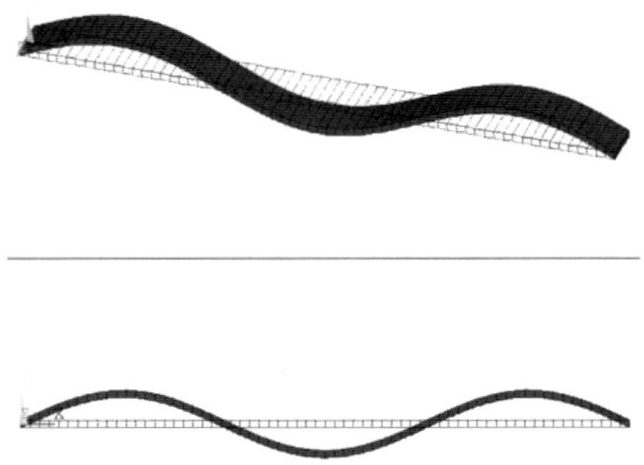

（c）The 3rd order

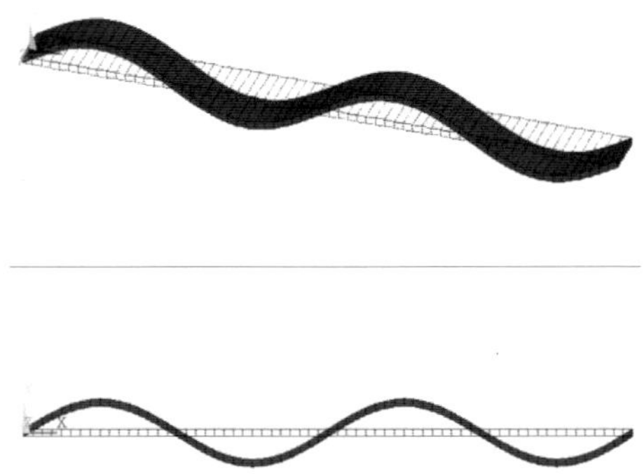

（d）The 4th order

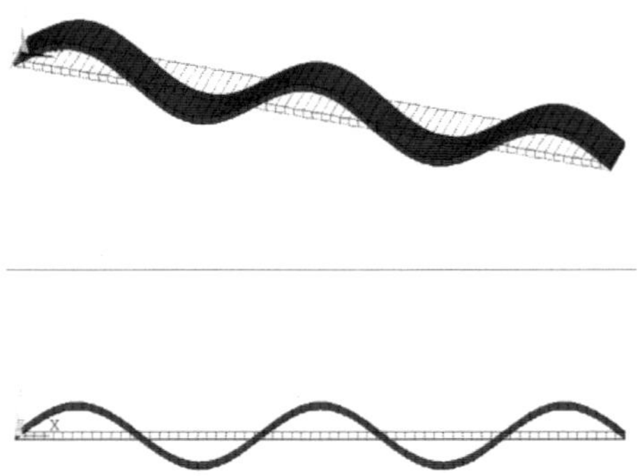

（e）The 5th order

Fig.3. The first 5 order vibration deformation diagram

Fig.4. The forced vibration test for steel beam

3.3 Experimental test

The ZHT-1 integrated experiment teaching system was adopted for vibration test. The size and the material parameters are the same with the Ansys model mentioned above. In this paper, the natural frequencies of the simply supported steel beam were obtained by free vibration and forced vibration methods. The test results were the mean values of the both methods, showing in Table 3. And Fig.4 is the picture of forced vibration test.

Table 3 The first 5 order natural frequency from test (Hz)

order	frequency	order	frequency
1	41.653	2	167.009
3	371.523	4	657.912
5	1028.969		

3.4 Results analysis

Three approaches are involved in carrying out the natural frequencies of the simply supported steel beam. The results are showing in Table 1 to 3. Supposing the experimental test results are the real values, the relative errors were calculated shown in Table4.

Table 4 Relative error for natural frequency (%)

order	Finite element method	Theoretical analysis method
1	0.441744892	0.489760641
2	0.699962277	0.725709393
3	0.405627646	0.409395919
4	0.755116186	0.802082953
5	0.605557602	0.705171876

From Table 4, it shows that the relative errors are all less than 1% for both methods. Furthermore, the errors from finite element method are less than theoretical analysis, it means, the natural frequencies from finite element method coincide well with test results. So the finite element method using beam element from Ansys can be adopted to simulate the steel beam vibration.

4 Loaded frequency analyses

4.1 Finite element method

The finite element model of a mass block-simply supported beam system was built by regarding the vehicles as a concentrated mass block attached. For convenience of comparing the results from experimental tests, the simply supported steel beam was divided in 16 equal parts, including 17 notes, as shown in Fig.1. It's 40.295 gram for each mass block. For simulating the passing vehicles on the bridge, the position of the mass block changed along the length of the beam. The loaded frequencies of the mass block-beam system were carried out by running the Ansys software. The first 5 order of loaded frequencies are showing as Fig. 5.

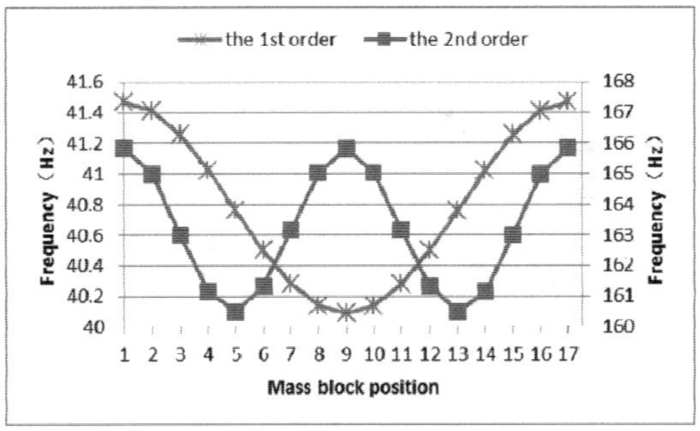

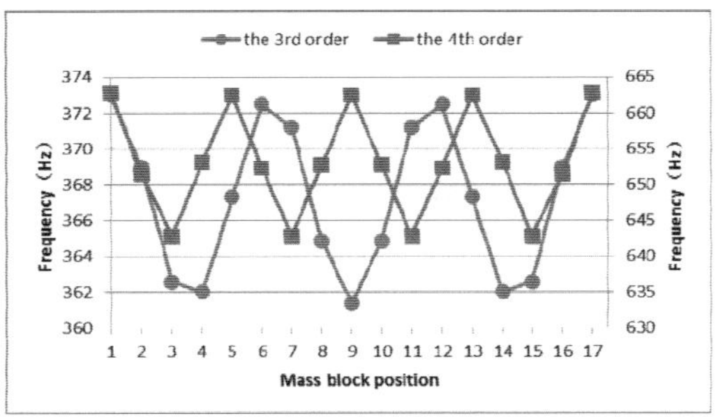

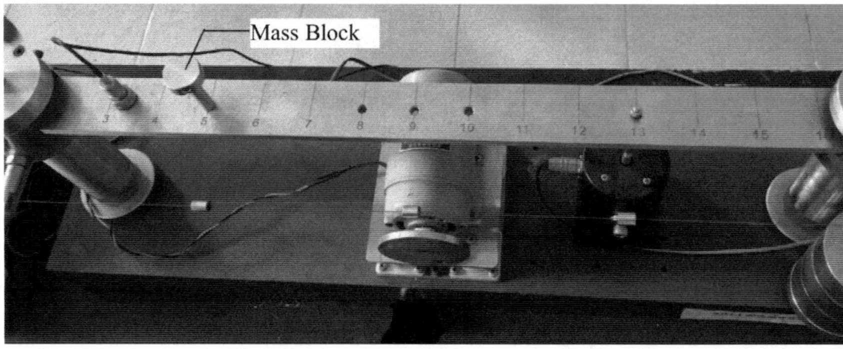

Fig.6. Mass block-simply supported beam system

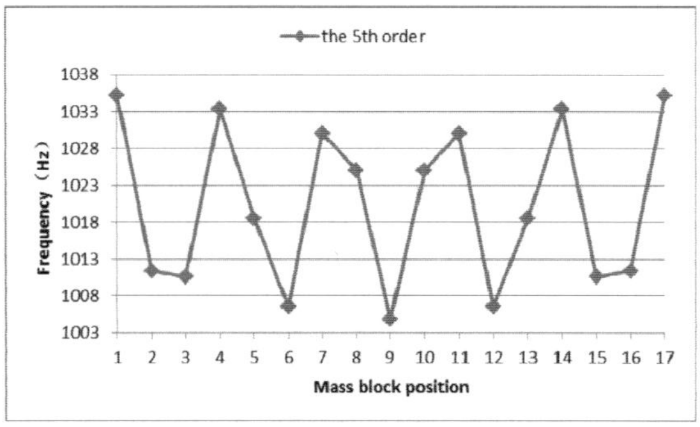

Fig.5. The loaded frequency of mass block-beam system

4.2 Experimental test

The ZHT-1 integrated experiment teaching system was adopted for vibration test again. The 80.59 gram mass block was fixed at note No.5 firstly, then at No. 8, No. 9 and No. 10 at last in order to simulate the passing vehicle on the bridge. Fig.6 is the picture of vibration testing when the mass block is at note No.5.

The natural frequencies of the mass block-simply supported steel beam system were obtained by free vibration and forced vibration methods respectively. The test results were the mean values of both methods, showing in Table 5.

Table 5. The first 5 order natural frequency for mass block-beam system from test（Hz）

Note / order	5	8	9	10
1	41.079	40.760	40.150	40.680
2	162.323	165.250	165.810	165.240
3	369.380	365.200	364.180	365.210
4	662.790	656.260	662.390	656.240
5	1023.540	1027.92	1007.460	1027.910

4.3 Loaded frequency results analysis

4.3.1 Finite element results analysis

Both the loaded frequencies of the mass block-beam system and the natural frequencies of the steel beam are obtained by Ansys software analysis. The relative errors of the two kinds of frequencies can be calculated and shown in Fig. 7.

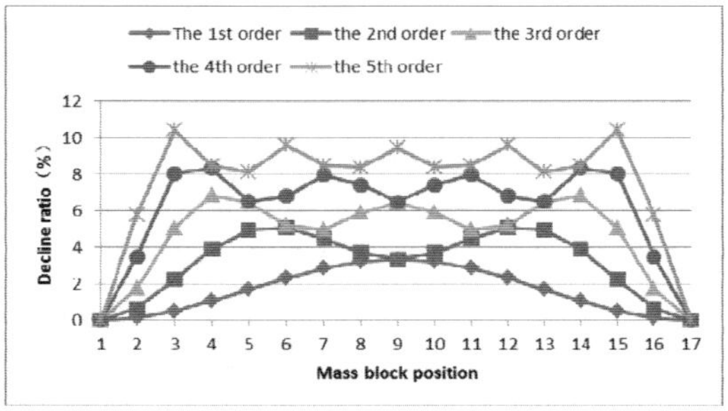

Fig.7. Relative error for loaded frequency by Ansys

4.3.2 Test results analysis

Regarding the limited positions can fix the mass block on the steel beam, four different positions carried out for loaded frequency testing. Comparison the loaded frequencies of the mass block-beam system with the natural frequencies of the steel beam, the relative errors can obtain, showing in Table 6.

Table 6. The relative error of loaded frequency for test (%)

Vibration order	Note NO.5		Note NO.8		Note NO.9		Note NO.10	
	Ansys	Test	Ansys	Test	Ansys	Test	Ansys	Test
1	1.71	0.94	3.21	1.71	3.33	3.18	3.21	1.90
2	3.23	2.12	0.49	0.36	0.01	0.02	0.49	0.36
3	1.54	0.98	2.20	2.10	3.13	2.37	2.20	2.10
4	0.05	0.01	1.54	1.00	0.05	0.07	1.54	1.00
5	1.61	1.13	0.99	0.70	2.94	2.68	0.99	0.70

5 Discuss and conclusions

In this paper, the authors selected the simply supported beam as research object. Firstly, the theoretically analysis was carried out for natural frequencies, and then the finite element model was built using Ansys. By regarding the vehicle on the bridge as mass block, the finite element model was built for mass block-beam system as well. And the comparison of the loaded frequencies of the mass block-beam system with the natural frequencies of the beam was carried out.

The following conclusions can be drawn from the analysis above:

(1)The loaded frequencies are slightly less than the natural frequencies of the simply supported beam. Fig 7 shows that the relative error drops of each mass block position except two ends. The same conclusion can be drawn from the theoretical analysis equation (1).

(2)Curve shape of the relative error for loaded frequency of each vibration order is similar with the vibration deformation diagram. For example, the double-wave curve shape of the 2nd order in Fig 7 is the similar with the 2nd order of Fig5.

(3)The position of the vehicle on the bridge has influence on the loaded frequencies. From Fig7, it shows that the relative error is larger than other position when the mass block is on the position of maximum amplitude. In other words, the position of maximum amplitude is the most sensitive position for loaded frequency.

(4)The beam element of Ansys can be adopted to simulate the vibration characteristics of simply supported bridge. Both Table 4 and 6 show that the frequency results from Ansys are very close to the results from experimental test. So the beam element from Ansys can be used to simulate the loaded frequency of vehicle-bridge system.

Acknowledgment

The authors are grateful to Mr. Cui Qi who has offered selfless support on experimental tests.

This paper was supported partially by the project from Shanghai Science and Technology Committee (NO. 18070502900).

References

1. *Ying* Wang, *Li* Peng. Research on vertical loaded frequency for haif-through arch brisge. Structural Engineers. Vol.**35**, No.4: 37-41 (2019)
2. *Yongchun* Chen, *Guojin* Tan, *Hanbing* Liu, et al. Test frequencies freely supported beam of highway bridge under effect of vehicles. Journal of Jilin University (Engineering and Technology Edition). Vol. **39**, No.6: 1492-1496 (2009)
3. Ray W. Clough， Joseph Penzien. Dynamics of structures (Second edition). Computers and Structures, Inc. California (2003).
4. *Xinmin* Wang. Structural Dynamic Analysis and Application with ANSYS. China Communication Press Co., Ltd. (2014)
5. *Ying* Wang, *Youting* Li, *Li* Peng, etc. An analysis of irregular rigid-frame bridge's responses to the lateral vibration caused by moving vehicular loads. Journal of Shanghai Normal University (Natural Sciences). Vol.**45**, No.1: 103-107 (2016).

Evaluation Index for Preventive Maintenance of Highway Asphalt Pavement

Weiling Teng[1]*, Xiaofang Li[1]

[1]College of Civil Engineering, Xijing University, Xi 'an, Shaanxi

Abstract. In order to make the decision of preventive maintenance of asphalt pavement, this paper analyses the related research, and finds the problems in the evaluation index system of preventive maintenance. Based on the characteristics of asphalt pavement and the actual maintenance demand, the evaluation index system of preventive maintenance of asphalt pavement is constructed.

1 Introduction

Pavement prevention maintenance is an active protection project taken in advance. Preventive maintenance can effectively prevent the diseases, slow down the deterioration, and extend service life. Preventive maintenance can alleviate the financial pressure of maintenance departments, which has important economic and social significance[1]. Preventive maintenance should follow the principle of scientific decision-making and prevention first. The focus of preventive maintenance is the pavement with good overall performance but slight diseases. Therefore, it is particularly important to make scientific and reasonable timing selection and evaluation criteria for preventive maintenance.

The Technical Specification for Maintenance of Asphalt Pavement (JTG 5142-2019) stipulates that the section of asphalt pavement technical Condition index (PQI) and its sub-indexes are evaluated as "excellent and good", and daily maintenance, prevention maintenance or repair maintenance can be carried out.[2], the evaluation standard could not distinguish between daily maintenance and preventive maintenance and lacked practical guiding significance; According to the Highway Asphalt Pavement Maintenance Design Code (JTG 5421-2018), each evaluation unit can be divided into maintenance types such as prevention maintenance and repair maintenance by using pavement technical condition data. However, the classification method is too complicated and the operation is relatively complex. Many researchers are adapted to the preventive maintenance of the evaluation index system and standard for more detailed analysis, but it has not yet formed a mature domestic pavement preventive maintenance evaluation standard, generally determined according to experience in maintenance, preventive maintenance and key points in the "prevention", should be based on technology of grade, use fixed number of year, road conditions, traffic volume size and composition, factors such as climate condition, to determine asphalt pavement preventive maintenance time[2].

Only the single or comprehensive performance index of asphalt pavement is used as the evaluation index, but the concept of "average" exists in the comprehensive index of road performance evaluation. When a certain disease is more serious and other aspects are in good condition, the overall evaluation is better after average treatment, but actually preventive maintenance is needed. Therefore, further research is needed on the evaluation index suitable for preventive maintenance. For scientific decision-making and the priority of the asphalt pavement preventive maintenance, the best time to reasonable judgment of asphalt pavement preventive maintenance and preventive maintenance quantitative guidance specific measures, this paper studies the evaluation index of asphalt pavement preventive maintenance system, based on the analysis of the current evaluation index of asphalt pavement, according to the characteristics of the asphalt pavement preventive maintenance, build the evaluation index system.

2 Evaluation for preventive maintenance in specifications

In the relevant specifications, pavement condition index (PQI) and each sub-index are generally taken as the evaluation criteria of maintenance scheme, which are summarized as follows in three aspects.

2.1 Evaluation indexes and standards for pavement technical conditions

Maintenance project plan and maintenance countermeasures should be formulated according to the investigation and evaluation results of each sub-index of asphalt pavement technical condition index (PQI).In the Technical Specification for Highway Asphalt Pavement Maintenance, sections with the technical Condition index (PQI) of asphalt pavement and each sub-index are evaluated as "excellent and good" can be maintained daily,

* Corresponding author: weiling_happy@126.com

prevented or repaired. Sections with any sub-index evaluated as "medium" or below should be repaired and maintained. According to the Highway Technical Condition Evaluation Standard, taking the expressway as an example, the grade standard of "excellent and good" pavement is shown in Table 1.

Table 1. Classification of highway Technical Conditions.

Evaluation index	Optimal	Good
PQI	≥90	≥80, < 90
RQI, RDI, PBI, PWI, SRI, PSSI	≥ 90	≥80, < 90
PCI	≥92	≥80, < 92

According to the Technical Specification for Maintenance of Asphalt Pavement, the technical condition index of asphalt pavement should meet the requirements, as shown in Table 2. If the requirements are not met, a reasonable maintenance plan should be arranged and comprehensive maintenance measures should be taken.

Table 2. Technical requirements for highway asphalt pavement.

Technical Performance Index	Network level	Basic unit				
	PQI	PQI	PCI	RQI	RDI	SRI
Highway	≥90	≥80	≥80	≥80	≥75	≥75

2.2 Standard of preventive maintenance measures for asphalt pavement

According to the Technical Specification for Maintenance of Highway Asphalt Pavement, the application of pre-maintenance measures such as sealing layer, micrometer and cover is shown in Table 3.

2.3 Pavement technical condition data index

According to the Highway Asphalt Pavement Maintenance Design Code, after the evaluation and analysis of the evaluation units based on the pavement technical status data, the evaluation units can be divided into preventive maintenance level repair maintenance and other maintenance types, as shown in Table 4.

Table 3. Pre-maintenance Technology and Road Performance Level (Highway).

Mesures	Road conditions index	Highway
Fog Sealing Layer	PCI, RQI, RDI	≥90
	SRI	≥75
Micro-Surfacing	PCI, RQI	≥85
Composite sealing layer	PCI, RQI, RDI	≥80
Ultra-thin surface	PCI, RQI	≥85
	RDI	≥80
Thin overlay	PCI, RQI	≥80
	RDI	≥75
Overlay	PCI, RQI	≥80

Table 4. Evaluation method of unit maintenance type division.

Evaluation Index				Maintenance type
PCI	RQI	RDI	SRI	
≥ 90	≥ 90	≥ 80	< 75	Preventive
		< 80		Repair
	85-90			Preventive
	< 85			Repair
85-90	≥85			Preventive
	< 85			Repair
< 85				Repair

In conclusion, in the relevant rules and regulations, generally based on pavement condition index (PQI) and each component index as an evaluation standard maintenance plan, but the PQI scope is not unified, comprehensive, PQI and each component index is larger than 80 or 85 above, can carry out preventive maintenance, obviously, the current evaluation standards is too general, and did not form a unified index system and practical operation is difficult.

3 Pre-maintenance evaluation index of asphalt pavement in relevant studies

Many domestic researchers have analyzed and studied the evaluation index system and standard of the preventive maintenance of asphalt pavement on the basis of the current specifications.

Guangzhou ShiYan expressway co., LTD, with good strength of pavement structure and above (SSI acuity 0. 85) as the basic premise, on the basis of comprehensive index of pavement performance, increase the crack rate of CR, rut RD, anti-sliding performance of SFC, flatness IRI four single index, specific criteria for PQI > 92, CR < 0. 5%, RD < 5 mm, SFC > 50, IRI 2 or less 0, preventive maintenance should be considered[3]; Based on the existing maintenance specifications, the evaluation index was expanded and refined, and the evaluation standard improved the sensitivity of disease recognition.

Xuying Fan Hebei construction engineering college, etc., in the traditional performance and asphalt pavement material performance as a two level of evaluation indexes, and puts forward the three secondary evaluation index of asphalt material performance under, the thermal properties, tar, asphalt performance at low temperature, high temperature performance and using the improved entropy weight of the analytic hierarchy process to determine each index weight[4]; However, the evaluation criteria were not studied. For different regions or new road sections, they had to be re-assigned according to their environmental requirements for calculation.

China highway engineering consulting group co., LTD., marshal dong, etc, put forward the road conditions trigger method is used to determine preventive maintenance time, on the basis of traditional effect - cost than, put forward the effect ratio and energy consumption effects - carbon emissions than two new evaluation indexes, using life cycle 3 indexes of maximum value to determine preventive maintenance time, it is concluded that the best opportunity to implement the micro table place maintenance for PCI = 90[5].

Chunan Zhang second highway engineering co., LTD., etc., put forward from the aspects of technology, environment and economy three indicators establish preventive maintenance of macroscopic and microscopic evaluation system, the cost - benefit cost evaluation criterion, building the appraisal model of asphalt pavement preventive maintenance decision-making, according to the plan layer selected 10 kinds of typical curing technology index matrix model is established and the forecast of cost - benefit of index weight calculation, the preventive maintenance program[6].

4 Construction of preventive maintenance evaluation index system

In current specification and related research, damaged pavement condition index PCI was adopted as a judge maintenance evaluation index, but PCI is comprehensive, includes the deformation, cracks, loose kind of diseases, it cannot accurately reflect the road, under the condition of single disease and preventive maintenance technology focuses on a single disease, such as rut, crack, and the comprehensive index PCI technology seriously weakened the preventive maintenance of pertinence and accuracy.

4.1 Initial construction of evaluation index system

In view of the complexity of asphalt pavement, the selection of preventive maintenance evaluation index should not only integrate the selected index with the existing highway performance evaluation system, but also fully consider the particularity of asphalt pavement. If only pavement damage status index (PCI) is used to evaluate the timing of preventive maintenance, the results will be too rough, and the correlation between road performance index and disease condition and maintenance measures is not strong, which is not conducive to the refined decision of preventive maintenance.

Preventive maintenance on pavement maintenance measures before the damage was not damaged, should be considered: (1) the road asphalt pavement preventive maintenance should meet the basic precondition for preventive maintenance, pavement condition index PQI and the sub-index achieve good, (2) the particularity of asphalt pavement early mild disease, loose, rut, crack, etc.

According to the specification and relevant scholars' research, the "possible complete set of indicators" for the evaluation of preventive maintenance of asphalt pavement is obtained, as shown in Table 5.

Table 5. Possible complete set of evaluation indicators.

Macro technical	Pavement technical condition index PQI
	Pavement structure strength index PSSI
Micro technical	cracks
	loose
	rut
	flushing
Economic	Unit maintenance cost
	The service life
	Times of major and medium repairs
Environnemental	Maintenance and energy consumption
	Carbon emissions

4.2 Analysis and select of evaluation indicators

The analysis and selection of preventive maintenance evaluation index should include the following characteristics: the index is easy to detect, it can represent pavement performance, and the index value changes with the adoption of preventive maintenance measures.

Combined with domestic and foreign research results showed that the comprehensive evaluation index is the base of asphalt pavement performance, under its guarantee compliance, and the selection of the single index to carry on the reasonable and effective evaluation, which can realize real and effective grasp of actual road conditions, and accordingly makes the corresponding road surface quality evaluation, so as to determine whether or how to make scientific and effective maintenance measures. Therefore, based on the comprehensive evaluation index, this paper puts forward a scientific and reasonable preventive maintenance system combining the comprehensive evaluation index as the objective standard and the sub-index as the micro standard.

(1) Pavement damage index: refers to a can truly reflect the physical characteristics of pavement damage due to the influence of objective factors, such as: vehicle load, geographical environment, construction factors, such as pavement damage showed a variety of forms, mainly ruts and cracks both damage is the most common,

 https://doi.org/10.1051/e3sconf/202123703023

so in this study, sets the pavement damage single evaluation index to crack, rut.

(2) Pavement functional index: It mainly reflects a comfortable, safe and economic experience provided by the vehicle in the driving process through service level and driving quality. The pavement function mainly reflects the real condition of the pavement by means of flatness, rutting and anti-skid property. The corresponding indexes can be selected according to the specifications. The integral index IRI is taken as a single evaluation index of the driving quality of the pavement. The transverse force coefficient (SFC) is used as a single evaluation index for the safety performance of asphalt pavement.

(3) Pavement structure bearing capacity and pavement materials evaluation indexes: at present our country specification on pavement deflection as validation indexes of the bearing capacity of pavement structure in this paper, we explore the pavement performance index can reflect the real road conditions, real and effective for judgment basis to formulate reasonable, scientific and effective preventive maintenance measures, the pavement structure bearing capacity and pavement material attribute evaluation index in evaluating this but much of the.

5 Conclusion

In the paper, two simple and easy indicators were added, , on the basis of studying the current domestic maintenance specifications and relevant studies, and a relatively scientific and complete evaluation index system was established. The index considered the actual situation of preventive maintenance of asphalt pavement, providing reference for the decision making of preventive maintenance of asphalt pavement.

References

1. JTG 5421-2018.Specifications for Maintenance Design of Highway Asphalt Pavement[S]
2. JTG 5142-2019.Technical Specification for maintenance of Highway Asphalt Pavement[S]
3. Shi Yan, Ling Tian-qing, Cui Li-long, Ge Hao

   Chen Qiao-qiao. Study on Evaluation Standard for Asphalt Pavement Preventive Maintenance and

   Decision-making Optimization [J].Journal of Highway and Transportation Research and Development
   .2020.(10):25-34
4. Fan Xu-ying, Gao Feng-chun, Wang Hai-long, Li Yu-zhong. A Model for Asphalt Pavement Preventive Maintenance Evaluation Based on Improved EW-AHP [J]. Journal of Highway and Transportation Research and Development.2017(9):8-13
5. Marshal Dong et al. Time Decision Optimization of asphalt pavement pre-maintenance based on life Cycle [J]. Highway.2020(4):325-330
6. Zhang Chu nan, Tian Zhi peng. Based on preventive maintenance decision research of asphalt pavement of high grade highway. Road Engineering.2019(12):77-80
7. Gao Feng chun. Research on preventive maintenance Evaluation and Maintenance Timing of Asphalt pavement in Zhangjiakou region [D].Hebei Institute of Architecture and Engineering.2019
8. Chen Sidi. Research on pavement preventive maintenance Decision Model based on Multi-objective particle swarm optimization algorithm [D].Beijing jiaotong university. 2019

 https://doi.org/10.1051/e3sconf/202123703024

Numerical Simulation on the Heat Dissipation of the Outdoor Unit in the Groove of High-Rise Building

Yuan Wang [1,2*], Xiaoqing Zhou [2], Ju Tian [1], Chuansheng Liu [1] and Jian Wang [1]

[1] School of Aircraft Maintenance Engineering, Guangzhou Civil Aviation College, Guangzhou, 510403, China
[2] School of Civil Engineering, Guangzhou University, Guangzhou, 510006, China

Abstract. In order to study the influence of the heat dissipation of the outdoor unit in the groove of the high-rise building. Numerical simulation of the thermal environment of a high-rise building is carried out. The results show that: When the outdoor mean wind speed is 2.7m/s, the maximum inlet air temperature of the outdoor unit under full load condition is about 43.3 °C, which is 2 °C lower than that under no wind condition. The monsoon is beneficial to the heat dissipation of the outdoor unit. When the hot air in the building groove rises, the corner flow is induced at the lower side of the refuge layer. Due to the reverse pressure gradient, the flow separation occurs, resulting in the blockage in the flow channel. When the air flows through the refuge layer, the flow expands suddenly due to the increase of the cross-section of the channel. The turbulent boundary layer is separated. The vortex region is induced on the upper side of the refuge layer. Since the heat on the upper and lower sides of the refuge floor cannot be effectively dissipated, the inlet air temperature of the outdoor unit is higher than that of other floors, which reduces the working efficiency of the outdoor unit.

1 Introduction

Multi-split air conditioning system is highly automated and flexible, which is widely used in high-rise buildings [1-2]. Generally, the wind environment directly affects the microclimate of the urban environment and people's comfort, so the wind environment should be considered in the planning and design of super high-rise buildings. However, the structure and distribution of high-rise buildings are relatively complicated. At present, there are few studies on the air distribution of air conditioning systems in super high-rise residential buildings [3-4]. Moreover, under the effect of regional monsoon and the interaction between buildings, the study of airflow distribution in the special position of the external structure of the building becomes very complex. Due to the influence of the building groove and the location of the outdoor unit, the ventilation of the outdoor unit is poor. The heat cannot be dissipated in time, which increases the return air temperature of the outdoor unit. In serious cases, the air conditioner shuts down [5-6]. Besides, the lateral exhaust or upper exhaust of the outdoor units on low floors may affect the return air temperature environment of adjacent upper units, resulting in an airflow short circuit. Therefore, it is necessary to study the heat dissipation of the outdoor unit in high-rise buildings.

In this paper, PHOENICS numerical simulation software is used to study the distribution of heat dissipation and the working conditions of the outdoor units in the groove of high-rise buildings.

2 Numerical simulation model

2.1 Geometric model

The building height in the numerical simulation of the thermal environment is 183.15m. The geometric model of the building is established according to the architectural design plan, and the size and location are consistent with the design plan. The geometric model of the building is shown in Figure 1. The first floor of the building is overhead. The 16th, 32nd and 48th floors are refuge floors, and there are no outdoor units on these three floors; the other floors are equipped with outdoor units. The location of the outdoor unit is shown in Figure 2.

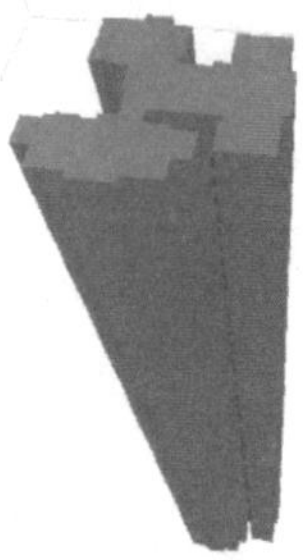

Fig. 1 The building model

* Corresponding author: wangyuan_nuaa@163.com

 https://doi.org/10.1051/e3sconf/202123703024

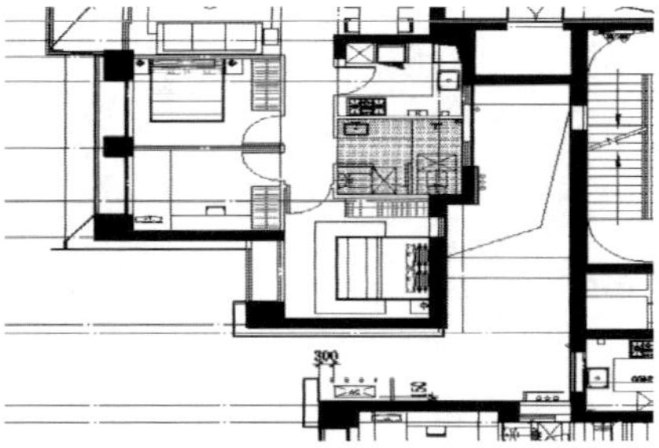

Fig. 2 Location of the outdoor unit

2.2 Numerical method

2.2.1 Turbulence model

The standard $k-\varepsilon$ turbulence model is selected to simulate the outdoor wind environment of buildings. The second-order upwind scheme is used to discretize the equations. The standard wall function method is used near the wall. The simple algorithm is used for velocity and pressure coupling. The standard $k-\varepsilon$ governing equation is as follows:

$$\frac{\partial}{\partial t}(\rho k)+\frac{\partial}{\partial x_i}(\rho k u_i)=\frac{\partial}{\partial x_j}\left[\left(\mu+\frac{\mu_i}{\sigma_k}\right)\frac{\partial k}{\partial x_j}\right]+ G_k+G_b-\rho\varepsilon-Y_M+S_k$$

$$\frac{\partial}{\partial t}(\rho \varepsilon)+\frac{\partial}{\partial x_i}(\rho \varepsilon u_i)=\frac{\partial}{\partial x_j}\left[\left(\mu+\frac{\mu_i}{\sigma_k}\right)\frac{\partial \varepsilon}{\partial x_j}\right]+ C_{1\varepsilon}\frac{\varepsilon}{k}\left(G_k+G_{3\varepsilon}G_b\right)-C_{2\varepsilon}\rho\frac{\varepsilon^2}{k}+S_\varepsilon \quad (1)$$

2.2.2 Distribution of inlet wind speed

In order to make the simulation close to the actual situation, the outdoor wind speed varies with the height. The wind speed at the entrance of the near surface layer obeys the exponential distribution.

$$u_h=u_0\left(\frac{h}{h_0}\right)^a \quad (2)$$

Where:

u_h The wind speed at h height;

u_0 The wind speed at h_0 height;

h_0 The reference height of meteorological typhoon observation point. Generally, the value is 10m;

a A constant related to the type of ground roughness. The value of a in this paper is 0.2.

According to the meteorological parameters of Shenzhen, the dominant wind direction in summer is a southeast wind, and the outdoor mean wind speed is 2.7m/s. According to formula (2), the gradient wind boundary conditions are set.

In this paper, two cases are simulated: 1. there is no wind in the environment; 2. the outdoor mean wind speed is 2.7m/s.

2.2.3 Boundary conditions

According to the design code for heating ventilation and air conditioning of civil buildings, the outdoor temperature in summer in Shenzhen can be used as the simulation temperature, which is 33.7°C. In order to simplify the model, the following assumption is adopted: the building envelope can be regarded as an adiabatic boundary. Solar radiation, ground reflection and heat transfer between the walls are ignored. The outdoor unit of the air conditioner runs at full load.

The relevant parameters of the outdoor unit are shown in Table 1.

Table 1. The parameters of the outdoor unit

Heat dissipation (KW)	Exhaust velocity (m/s)	Exhaust air rate (m^3/min)	Size (mm)	Outlet diameter (mm)
22.063	4.1	2.04	1380×950×370	560

3 Results and discussion

3.1 Heat dissipation distribution in the building groove

Numerical simulation analysis is carried out for the high-rise building under wind (2.7 m / s) and no wind conditions. The temperature distribution at different building heights is shown in the following figure:

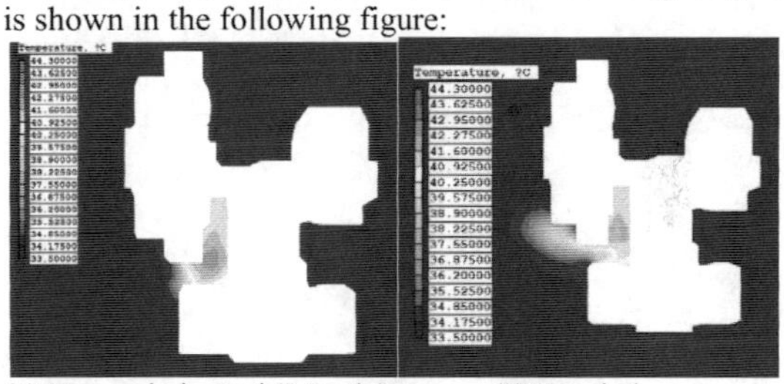

(1) Mean wind speed (2.7m / s) (2) No wind

Fig. 3 Temperature distribution at a height of 170m

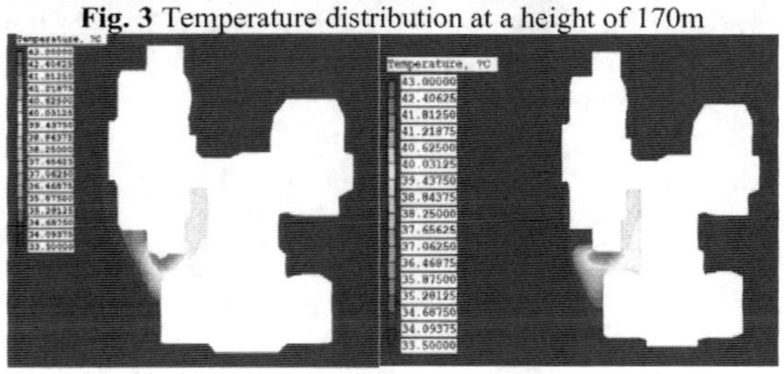

(1) Mean wind speed (2.7m / s) (2) No wind

Fig. 4 Temperature distribution at a height of 100m

https://doi.org/10.1051/e3sconf/202123703024

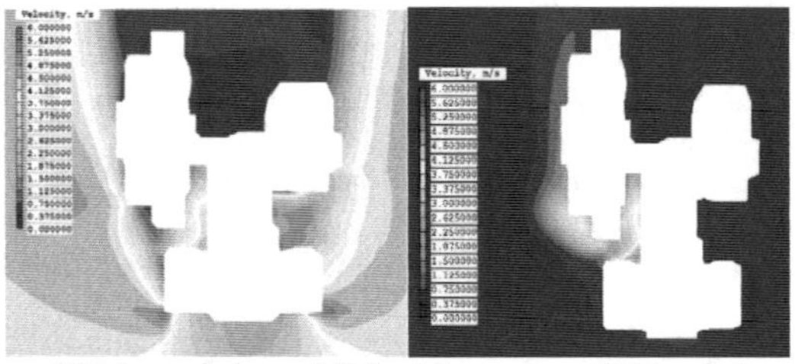

(1) Mean wind speed (2.7m / s) (2) No wind

Fig. 5 Velocity distribution at a height of 170m

It can be seen from the above three figures that the outdoor unit in the groove is placed on one side, and the temperature distribution is uneven. In general, the temperature in the area near the outdoor unit is higher.

In windy conditions, the maximum temperature can reach 44.2°C, and the maximum temperature appears on the wall facing the direction of the outdoor unit exhaust. This is because the wall is only 1.8m away from the exhaust louvers of the outdoor unit, and the exhaust air of the outdoor unit blows directly on the wall, causing heat to accumulate near the wall, making the temperature higher than the ambient air temperature. The temperature at other locations in the groove is slightly lower than that of the exhaust area of the outdoor unit, but higher than the outdoor environment temperature. This is because the fluidity in the groove is relatively poor compared to the outside of the building. Due to poor heat dissipation, heat accumulates near the installation location of the outdoor unit, and the intake and exhaust temperatures are relatively high. The airflow outside the building groove is influenced by the monsoon and other building structures, so the heat can't spread out smoothly. There is a low velocity region. Under the condition of no wind, the maximum temperature in the groove can reach 46 °C. Due to the high exhaust speed of the outdoor unit, the heat diffuses to both sides, and the influence range of high temperature area is larger than that under the windy condition.

3.2 Analysis of the working condition of the outdoor unit in the building groove

As shown in Fig. 6, the temperature distribution in the groove is as follows: the temperature at the bottom of the groove is relatively low, and the temperature distribution increases with the increase of the height, regardless of the working conditions with or without wind. When the air-conditioning equipment is running, the heat of the outdoor unit is directly discharged into the air. Due to the effect of thermal buoyancy, the ambient temperature of the upper layer of the building rises, which affects the operational performance of the air conditioner on the upper floor. The maximum temperature in the groove under no wind condition is slightly 2°C higher than that under wind condition, and the influence range of high temperature area is larger. The reason is that under the influence of monsoon, the velocity of airflow in the groove is accelerated. The heat accumulated in the groove is dispersed to the external environment, which makes the temperature in the groove drop.

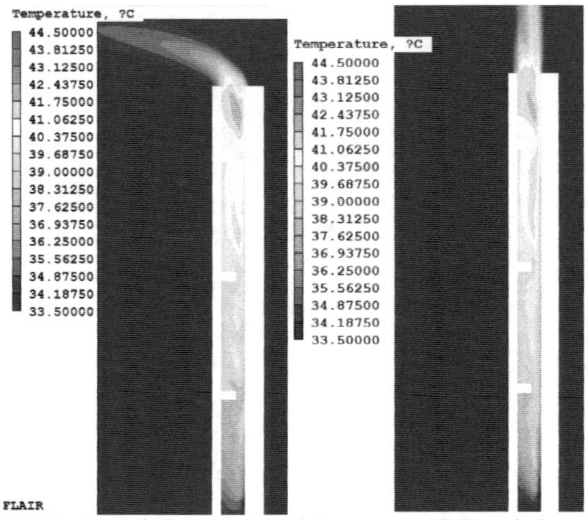

(1) Mean wind speed (2.7m / s) (2) No wind

Fig.6 Temperature distribution at the middle section of building groove

Figure7 - (1) shows the temperature distribution at the air inlet on the right side of the outdoor unit under windy condition. It can be observed in the figure that the upper and lower sides of the 48th floor of the refuge layer are affected by high temperature airflow, and the high temperature area covers 44-52 floors. Among them, the temperature at the side air inlet of the outdoor unit on the 50th and 51st floors is higher, and the maximum temperature is about 43.3 °C. The inlet air temperature of the outdoor units on the other floors is close to 43 °C. The influence range of the high temperature area near the 32nd floor of the refuge floor is 28-35 floors; Among them, the temperature at the side air inlet of the outdoor unit on the 33rd and 34th floors is relatively high, with the highest temperature being about 43.2°C , and the inlet air temperature of the outdoor unit on the other floors is close to 42°C. In general, the inlet air temperature of each outdoor unit increases with the increase of height. For the units near the upper and lower sides of the refuge floor, the inlet air temperature will be slightly higher than that of the nearby outdoor unit due to the influence of airflow around the refuge layer. It can be seen from Fig. 7 - (2) that the temperature distribution at the inlet of the right side of the outdoor unit under no wind condition is similar to that under wind condition. However, the inlet temperature of some outdoor units has reached 45.3°C. Due to the high condensation temperature of the air conditioner, shutdown protection may occur, which seriously affects the normal operation of the outdoor unit.

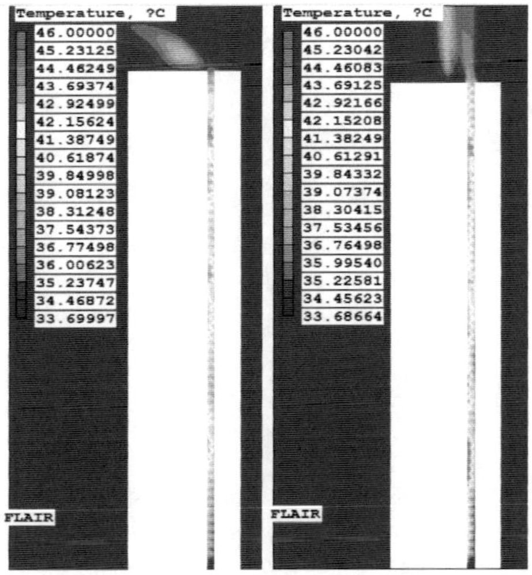

(1) Mean wind speed (2.7m / s) (2) No wind

Fig.7 Temperature distribution at the side air inlet of the outdoor unit in the groove

As shown in Fig. 8 and Fig. 9, when the hot air in the building groove rises, the corner flow is induced at the lower side of the refuge layer. Due to the reverse pressure gradient, the flow separation occurs, resulting in the blockage of the flow channel. When the air flows through the refuge layer, the flow expands suddenly due to the increase of the cross-section of the channel. The turbulent boundary layer is separated. The vortex region is induced on the upper side of the refuge layer. Since the heat on the upper and lower sides of the refuge floor cannot be effectively dissipated, the inlet air temperature of the outdoor unit is higher than that of other floors, which reduces the working efficiency of the outdoor unit.

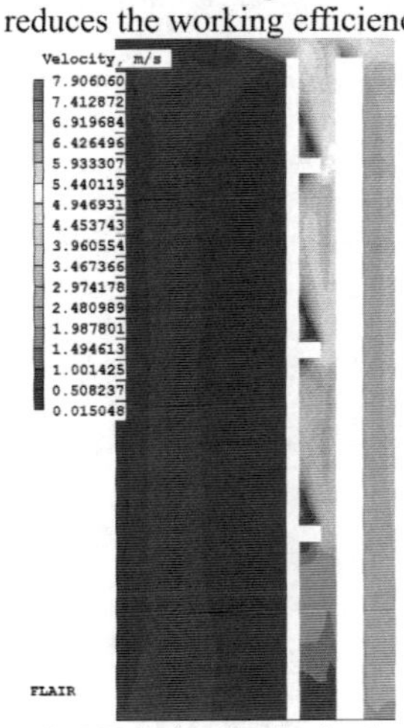

Fig.8 Velocity distribution in the groove

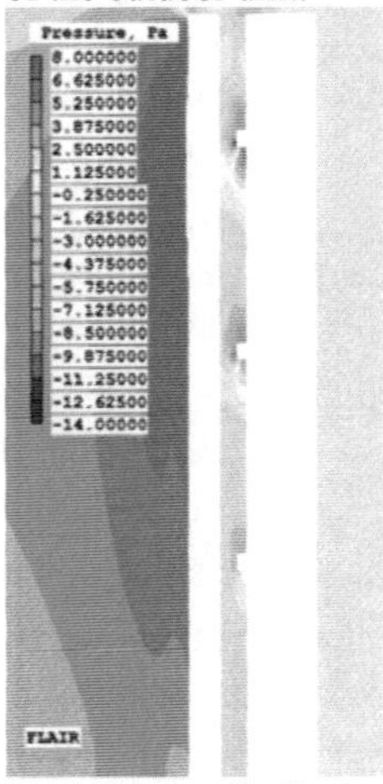

Fig.9 Pressure distribution in the groove

4 Conclusion

In this paper, the numerical investigation of the thermal environment of the outdoor units in the groove of high-rise building is carried out. It is found that it is affected by the outdoor wind speed and the shape of the building structure.

The conclusions are as following:

1. When the outdoor mean wind speed is 2.7m/s, the maximum inlet air temperature of the outdoor unit under full load condition is about 43.3 °C, which is 2 °C lower than that under no wind condition. The monsoon is beneficial to the heat dissipation of the outdoor unit.
2. When the hot air in the building groove rises, the corner flow is induced at the lower side of the refuge layer. Due to the reverse pressure gradient, the flow separation occurs, resulting in the blockage of the flow channel. When the air flows through the refuge layer, the flow expands suddenly due to the increase of the cross-section of the channel. The turbulent boundary layer is separated. The vortex region is induced on the upper side of the refuge layer. Since the heat on the upper and lower sides of the refuge floor cannot be effectively dissipated, the inlet air temperature of the outdoor unit is higher than that of other floors, which reduces the working efficiency of the outdoor unit.

References

1. Zhou Dehai, Shao Xiaoliang, Zhang Xiaoling, et al. Numerical simulation of thermal environment for outdoor unints at building re-entrant [J]. Journal of Guangzhou University(Natural Science Edition), 2010, **9**(6) :17-22.
2. Jiang Yuebo, Jin Wufeng, Piao Wankui, et al . CFD analysis and optimization of thermal environment for outdoor units at re-entrant of residential building [J]. Building Energy Conservation, 2012, **40**(9) :4-8.
3. Jin Wufeng, Zheng Yafei, Jiang Yuebo, Bai Xiansan. Experimental study on the effect of the outdoor unit installation conditions on the heat dissipation in the building groove [J]. Journal of Xi'an University of Architecture & Technology(Natural Science Edition), 2015, **47**(2) :282-286.
4. Hang Yin, Liu Dong, Huang Yan, Liu Jun. Numerical Simulation of Airflow and Optimization of Outdoor Unit of the Air Conditioner in a Certain High Building [J]. Building Energy & Environment, 2006, **25**(3) :12-16.
5. Shen Lin, Wan Xiongfeng. Simulation of the Operating Environment around Outdoor Air Conditioner Unit in a High Rise Building [J]. Building Energy & Environment, 2010, **29**(4) :82-88.
6. Jia Xuefeng, Liu Dong. Numerical Simulation of Ventilation Effect of Outdoor Units in Air Conditioning System for High-rise Building [J]. Building Energy & Environment, 2010, **29**(1) :69-76.

 https://doi.org/10.1051/e3sconf/202123703025

Seismic Damage Model of High-Strength Reinforced Concrete Frame Columns

Guojun Zhang[1, *]

[1]College of Civil Engineering, Shanghai Normal University, Fengxian Shanghai 201418, China,zhgjxj@163.com

Abstract. In this paper, several existing seismic damage models as well as their characteristics were discussed first of all. And then it was figured out that variational history of incremental dissipated hysteretic energy of tentative frame columns under different loading cycle levels; the influence of axial compression ratio, type of stirrups, stirrup ratio, reinforcement ratio, strength grade of concrete and shear span ration on the incremental dissipated hysteretic energy was discussed and analysed. The damage index of tentative frame columns was compared and analysed according to the existing seismic damage models, and the seismic damage models adapt to high-strength concrete columns and ordinary-strength concrete columns were offered. The influence of shear span ration, axial compression ratio and stirrup ratio on the damage was discussed and analysed according to the variational history of the damage index under different loading cycle levels. Lastly through analysing all sorts of seismic damage models comparatively, the seismic damage model of high strength concrete frame columns is put forward, which offers good basement for further damage analysis.

1 Seismic damage model

The most intuitive manifestation of building structure under earthquake action is damage. It is not economical and realistic to expect the structure to have no damage or no damage under earthquake action. On the premise of allowing structural seismic damage, according to the historical seismic data, the seismic fortification level and importance of the structure, and on the basis of fully estimating the future seismic characteristics, it is a natural and reasonable pursuit for people to carry out seismic design of structures according to certain damage degree limits. Generally, the properties of seismic structures include elastic properties and inelastic properties [1], and inelastic properties are also called damage properties. For the elastic properties, the elastic deformation reaction of the structure (such as the maximum elastic relative displacement between layers, the maximum elastic displacement of the vertex) is generally used to describe, which is caused by the seismic design codes of all countries in the world. For the evaluation of inelastic properties, the most widely used parameter is the maximum elastic-plastic displacement or displacement ductility coefficient of the structure (defined as the ratio of the maximum elastic-plastic displacement to the yield displacement of the structure). The investigation and experimental study of earthquake damage examples show that the structural earthquake failure forms are mainly divided into two categories :1) the first time to exceed the damage and 2) the cumulative damage. Because structural seismic damage is a broad term, the definition and calculation of damage are very complicated. choose what kind of parameters can reflect the change of structural mechanical properties to describe the damage degree of the structure, which is not only related to the concept and definition of damage, but also makes the selected parameters easy to calculate, can be measured directly or indirectly, and can describe the development of structural seismic damage, reflecting the of damage mechanism [2].

From the point of view of the number of damage parameters, the seismic damage model of the structure can be divided into single parameter damage model and double parameter damage model.

(1) Single-parameter earthquake damage model

The single parameter seismic damage model holds that the seismic damage of a structure or component can be described by a single damage parameter Δ , and the calculated value of the member is destroyed Δ_c it is greater than its limit allowable value Δ_u . The damage index model described by a single damage parameter can generally be expressed as:

$$DM = f(\Delta_c, \Delta_u) \qquad (1)$$

(2) Two-parameter earthquake damage model

Domestic and foreign scholars have put forward a lot of double-parameter seismic damage models, more mature and widely used or Park models. To consider the accumulation of hysteretic energy dissipation based on deformation analysis, in 1985, American scholars Park Y.J. Hong Watson (AngA.H.S.) and Wen Yigui (WenY.K.) first

*Corresponding author: zhgjxj@163.com

proposed a linear combination of maximum deformation and cumulative hysteretic energy dissipation [3]. They defined the damage index DM as:

$$DM = \frac{X_m}{X_u} + \beta \frac{E_h}{F_y X_u} \quad (2)$$

The X_u is the ultimate deformation of the member under the action of single withering load, the F_y is the yield shear force of the member, the X_m and E_h are the maximum elastic-plastic deformation and cumulative hysteretic energy consumption of the member under the action of earthquake; and β is the energy dissipation factor of the member. For shear structure, the following formula is calculated:

$$\beta = \begin{pmatrix} -0.447 + 0.073\lambda \\ +0.24\lambda_N + 0.314\rho_s \end{pmatrix} \times 0.7^{100\rho_w} \quad (3)$$

The λ is the shear span ratio of the member, the ρ_s is the reinforcement ratio of the longitudinal reinforcement, and the ρ_w is the volume hoop ratio.

For curved structures β can be calculated by following formula:

$$\beta = \begin{bmatrix} 0.37\lambda_N + \\ 0.36\left(k_\rho - 0.2\right)^2 \end{bmatrix} 0.9^{100\rho_W} \quad (4)$$

Formula k_ρ normalized reinforcement ratio, $k_\rho = \frac{\rho_t f_y}{0.85 f_c'}$.

2 Damage to frame columns

2.1 Comparison of cumulative damage of frame columns

2.1.1 Comparison of cumulative hysteretic energy consumption

Cumulative hysteretic energy consumption of the specimen is of the area surrounded by the hysteretic curve [4], calculated from the following formula:

$$E = \int F dx \quad (5)$$

Materials [6] according to the characteristics and practical engineering experience of the frame column in the frame arrangement system of the main workshop, 9 high strength concrete frame columns and 7 general concrete cantilever test frame columns are tested repeatedly at low cycle. The concrete strength grades are C60 and C30, respectively different shear span ratio, axial compression ratio, stirrups form and hoop ratio are adopted. According to these test results, the damage model of frame column is analysed and studied.

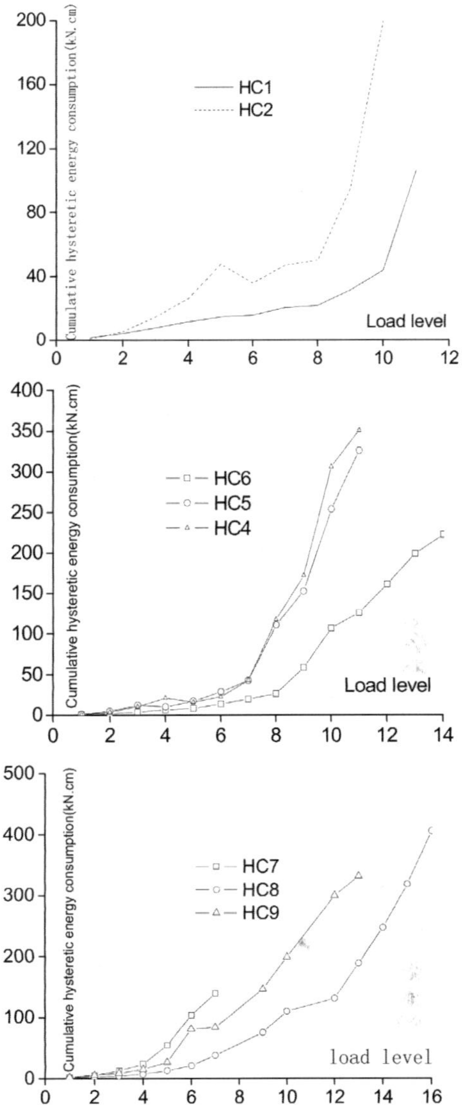

Fig.1. Variation of Accumulated Hysteresis Energy Consumption with Loading Level of High Strength Concrete Frame Column

Figures 1 and 2 show the variation of cumulative hysteretic energy consumption with the loading cycle level. The loading cycle level of the transverse coordinates in the diagram corresponds to the single cycle before yielding, and the yield is corresponding to the two to three weeks cycle of each amplitude.

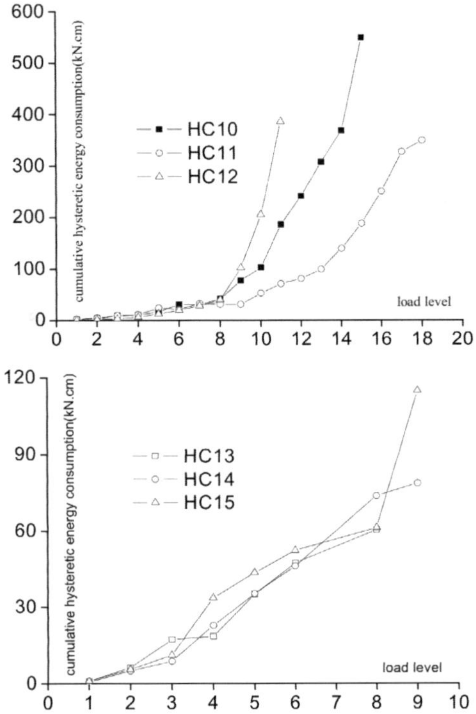

Fig .2. Variation of cumulative hysteretic energy dissipation with loading level of ordinary concrete frame columns

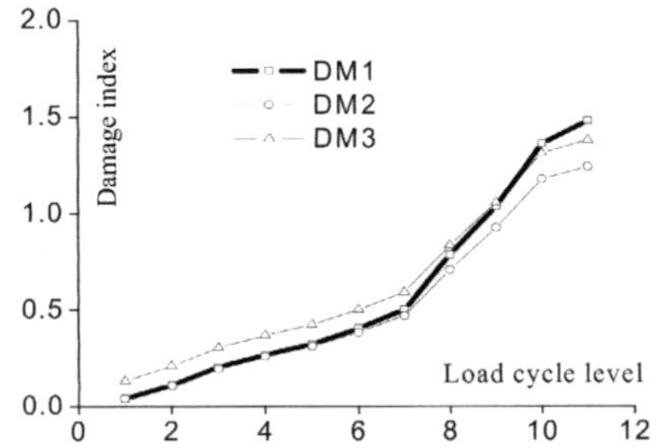

Damage Model of High Strength Concrete Frame Column

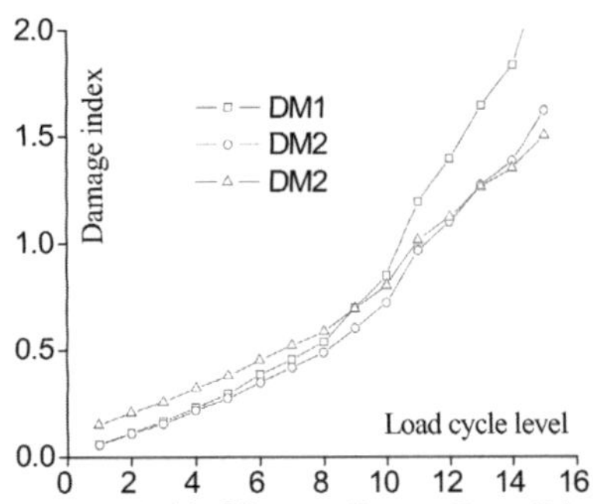

Damage Model of Common Concrete Frame Column

Fig. 3. Comparison of three damage indices

We can see from Figures 1 and 2 that with the increase of axial compression ratio, the cumulative hysteretic energy consumption becomes smaller and smaller. The HC4, HC5 and HC6 of high strength concrete members and the HC10, HC11 and HC12 of ordinary concrete members have the same conclusion. However, high strength concrete members have less hysteretic energy consumption than ordinary concrete.

Under the same axial compression ratio, the cumulative hysteretic energy consumption is related to the form of stirrups, hoop ratio, longitudinal reinforcement ratio, concrete strength and shear span ratio. Hysteretic energy consumption of rectangular hoop, cross-pull hoop and well-shaped hoop members increases in turn, and increases with the increase of hoop ratio and reinforcement ratio [6].

2.1.2 Selection of earthquake damage model

For high strength concrete frame columns, damage index calculated by formula (2) and β value calculated by formula (3) are in good agreement with experimental phenomena; For ordinary concrete frame columns, damage index calculated by formula (2) are in good agreement with experimental phenomena. The damage index calculated by formula (5) is too large, if 0.1% after a week of loading. The concept of damage is quite different. In figure 3, damage index β is calculated by formula (2); damage index $DM2$ is calculated by formula (2) and formula (3).

2.1.3 Comparison of damage indices

The damage model of formula (2) is used in figure 4. For ordinary concrete frame column. It can be seen from figure 4 that the damage of frame column is mainly affected by shear span ratio, axial compression ratio, stirrups form and hoop ratio [4-5].

(1) As the shear-span ratio increases, energy dissipation factor β also increased accordingly. Such as high strength concrete specimen HC1 and HC2 are smaller than other specimens, the HC10, HC11 and HC12 of ordinary concrete specimens are smaller than that of HC13, HC14 and HC15.

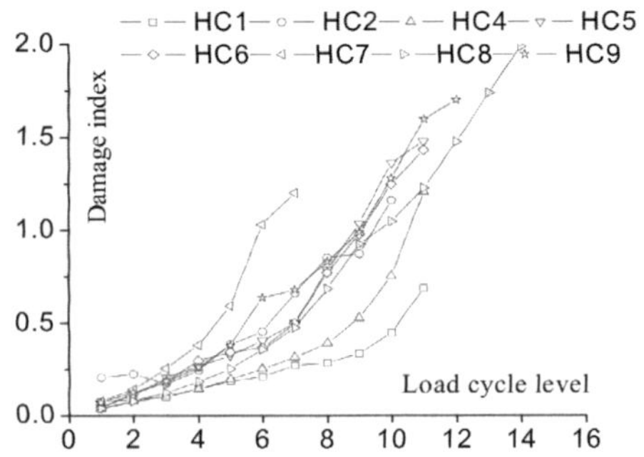

Change of damage index of high strength concrete frame column with loading cycle level

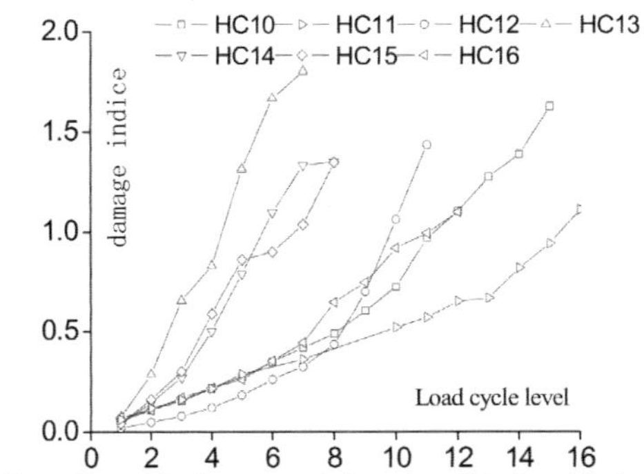

Change of damage index of common concrete frame column with loading cycle level

Fig. 4. Damage index with loading level

(2) As the axial compression ratio increases, the damage also increased accordingly. Such as high strength concrete specimen HC4, HC5 and HC6 and ordinary concrete specimen HC10, HC11 and HC12 damage increased in turn.

(3) As the hoop ratio decreases, the damage of the specimen increases; the damage of well type hoop, cross tension hoop and rectangular hoop increases in turn. Such as high strength concrete specimen HC9, HC8 and HC7 and ordinary concrete specimen HC15, HC14 and HC13, the damage value increases in turn.

2.2 Damage Model of High Strength Concrete Frame Column

From the above test results, the damage model of formula (2) can be adopted for both high strength concrete frame columns and ordinary concrete frame columns. Through the study of the damage of high strength concrete frame column, it is found that the energy dissipation factor β calculated by formula (5) is slightly larger, and the energy dissipation factor β is obtained:

$$\beta = 0.66^{\rho_w}\left(-0.266 + 0.18\rho_s + 0.06\lambda + 0.45\frac{N}{f_{c0}A_c}\right) \quad (6)$$

Type: ρ_w — transverse hoop ratio, ρ_s — longitudinal reinforcement ratio, λ — shear span ratio, $\frac{N}{f_{c0}A_c}$ — axial compression ratio.

The β value of energy dissipation factor calculated by formula (6) is related to transverse hoop ratio (ρ_w)、longitudinal reinforcement ratio (ρ_s), shear span ratio (λ) and axial compression ratio $(N/f_{c0}A_c)$, but not to concrete compressive strength f_{c0}. It can correctly reflect the damage performance of high strength concrete frame column.

3 Conclusions

Based on the calculated cumulative hysteretic energy consumption of high strength concrete test frame columns with the change of loading cycle level, it can be found that with the decrease of axial compression ratio and the increase of hoop ratio and reinforcement ratio, the cumulative hysteretic energy consumption increases accordingly. Under the same axial compression ratio, the cumulative hysteretic energy consumption is related to the form of stirrups, hoop ratio, longitudinal reinforcement ratio, concrete strength and shear span ratio. The hysteresis energy consumption of rectangular hoop, cross-pull hoop and well-shaped hoop members increases in turn, and the hysteresis energy consumption increases with the increase of hoop ratio and reinforcement ratio. The main damage of frame column is shear span ratio, axial compression ratio, stirrups form and hoop ratio. As the shear-span ratio increases, the energy dissipation factor β increases accordingly; with the increase of axial compression ratio, the damage increases accordingly; with the decrease of hoop ratio, the damage of specimen increases; the damage of well hoop, cross hoop and rectangular hoop increases in turn. Finally, through the comparison and analysis of each earthquake damage model, the earthquake damage model of high strength concrete frame column is put forward.

References

1. B.Q. Liu. Failure Criteria and Reliability Analysis of Seismic Structures. Beijing: China Building Materials Industry Press (1996).
2. G.J. Zhang, B.Q. Liu, Y.B. Li, T. Wu. Advances in the Study of Damage Criteria for Seismic Structures. Architectural Science Research, Sichuan, No.**3**:68~71(2002).
3. Park Y. J. and Ang A. H-S, A Mechanistic Seismic Damage Model for Reinforced Concrete, J. Structure Engineering, ASCE, **111**(4): 740 ~ 756(1985).
4. G C Beolchini, D Galeota, M M Giammatteo and M Zulli. Seismic Behavior of high Strength RC Columns. 12th World Conference on Earthquake Engineering. Auckland, New Zealand: 2000 New Zealand Society for Earthquake Engineering, **0593**: 1 ~ 8(2000).
5. H.Q. Li, L.H. Lei, X.L. Lu. Study on Resistance to Low-cycle Fatigue Damage of Reinforced Concrete Bending Members, Journal of Nanjing Institute of Architectural Engineering, No. **3**:10~16(1998).
6. G.J. Zhang. A Study on Seismic Performance of High-strength Concrete Frame Columns in Large-scale Thermal Power Plant. Doctoral thesis of Xi'an University of Architectural Science and Technology: Xi'an University of Architectural Science and Technology (2003).

Study on Design of Concrete Box Girder of A Railway Swivel Cable-Stayed Bridge

Yang Liu[1],Tenghui Ding[1], Qinyun Xie[2],Guangli Xu[3],Chao Li[3]

[1]School of Architecture and Planning, Yunnan University, Kunming 650091, China
[2] Yunnan Dazhu Technology Co., Ltd, Kunming 650030, China
[3] Yunnan Kunchu expressway investment and Development Co., Ltd, Kunming 650030, China

Abstract. A swivel cable-stayed bridge over the existing railway is a span across the existing railway. The recommended scheme for the main bridge is (128 + 388 + 128) m steel mixed composite beam swivel diagonal pull bridge with span. The cables of the diagonal pull bridge are arranged according to the fan-shaped central double cable plane, taking into account the mechanical performance and aesthetics. The bridge structure adopts semi floating system. The concrete swivel diagonal pull bridge is adopted in the comparison scheme. The design of the bridge is three spans and (138 + 268 + 138) m prestressed concrete box girder is adopted. The cables are arranged according to the central double cable plane, and the bridge composition adopts the consolidation system. Considering the needs of bridge operation and maintenance in the later stage of the bridge, when the dead weight of concrete diagonal pull bridge is within the ideal range, the concrete swivel diagonal pull bridge can be preferred. In order to calculate the dead weight of the selected bridge, the author uses the finite element software to model the whole bridge and calculate the weight of the bridge. The results show that the dead weight of the concrete swivel diagonal pull bridge is too large, which has exceeded the maximum bearing capacity of the existing spherical hinge. In order to continue to use the concrete swivel diagonal pull bridge scheme, it is necessary to optimize the design of the concrete swivel diagonal pull bridge scheme.

1 Introduction

With the rapid development of China's transportation industry, a large number of new passengers dedicated lines, intercity railways and urban expressways have been built. At the same time, along with the high-speed expansion of the main urban areas of some large and medium-sized cities, it is bound to lead to the emergence of more overpass railway Urban Bridges. Overpass railway bridge is an important project often encountered in highway construction. The selection of bridge construction technology has a great impact on the quality of bridge construction project. Reasonable structural design can ensure the safety performance of the whole project. And the railway that the bridge crosses generally carries a great traffic capacity, which can't be interrupted casually. Therefore, how to carry out the normal construction of the bridge without affecting the railway traffic is particularly important. Usually, the solution is to start with the design scheme of the bridge. The completion of urban bridge over railway will improve the coverage of urban road network, and the operation efficiency of the whole urban road network will also be improved, which will make an important contribution to the development of the city. At present, for the design of urban bridges over railway, the single tower concrete girder cable-stayed bridge constructed by horizontal rotation method is mostly considered from the perspective of reducing the impact on the existing railway operation under the bridge. This type of bridge has the characteristics of beautiful shape and flexible structure[1 ~ 12].

2 Methods

Step1: On the premise of ensuring the feasibility, the project should also meet other requirements put forward by the owner as far as possible, strive to be beautiful in appearance, and arrange the bridge scheme on the premise of minimizing the impact on railway operation.

Step2: As the bridge is located at the side of small pile number, there is planned urban road, and there is planning station East Road on the side of large pile number. Therefore, it is considered to reserve a passage under the bridge side of the bridge.

Step3: Collecting and consulting the design scheme and construction technology of the existing bridge, combining with the project overview of the bridge, the process cases with reference value are selected. And collect the construction technology and experience of the cable-stayed bridge swivel, combined with the current development of the bridge swivel, and then determine the structure of the bridge to be used.

Step4: The bridge design scheme shall minimize the

railway operation loss caused by the later operation and maintenance of the bridge, and shall not deteriorate the existing railway operation conditions.

Step5:According to the requirements of the railway department, the long-term development of the railway should be taken into account.

3 Scheme comparison and selection

3.1 Recommended scheme

The span of the bridge is (128 + 388 + 128) meters steel mixed composite beam swivel cable-stayed bridge. In order to reduce the weight of swivel, the middle span of bridge is designed as steel box girder. In order to make the weight difference between the two ends of the spherical hinge as small as possible, the side span is prestressed concrete box girder. As a new type of structural bridge, hybrid girder cable-stayed bridge has significant advantages in stress and structural design. It has excellent performance in self weight and span ability. However, in the later operation of the bridge, a large amount of resources are needed for maintenance, and the maintenance requirements are high. Because the project spans several existing railways, it will bring certain safety risks to the railway operation when it is put into operation in the later stage and regular maintenance and coating are carried out.

3.2 Trial design of bridge scheme 1

Considering the needs of bridge operation and maintenance in the later stage of the bridge, the (138 + 268 + 138) meters prestressed concrete box girder swivel cable-stayed bridge is proposed for the trial design of the bridge. Compared with the steel-concrete hybrid girder cable-stayed bridge, the self-weight of the concrete beam cable-stayed bridge will increase significantly. Therefore, the requirements of spherical joint will be greatly improved. However, the concrete box girder has obvious advantages in the later maintenance, and the driving experience is relatively good, the maintenance cost is low, and the overall cost is lower than the economic maintenance requirements.

3.3 Trial design of bridge scheme 2

Considering the factors of landscape design, the shape of bridge tower adopts H-type bridge tower. In scheme 2, (128 + 388 + 128) meters steel-concrete composite girder is used to push the cable-stayed bridge. The main beam is a double side box structure. The side box is connected with the side box by a large cross beam. At the same time, a small beam is set between the large beams to increase its stability. The deck of main girder is made of precast concrete. Although the cost scheme of this scheme is lower, temporary piers need to be set between the lines in operation during the construction of the scheme, which will have a greater impact on the line operation. At the same time, the economy is general and the construction period is long.

After the comparison and selection of schemes, the author selects the first scheme of bridge for finite element modeling analysis.

4 Trial design of bridge parameters

4.1 Grid Selection

The main girder of the selected bridge is prestressed concrete box girder. The main girder section is W-shaped web section, the top plate width is 45m, the floor width is 20m, and the box girder height at the center is 3.8m. In the standard section, the thickness of top plate is 30cm, the thickness of bottom plate is 35cm, the thickness of outer inclined web is 40cm, and the thickness of inner inclined web is 30cm. The main beam is provided with a diaphragm at the connection between the bridge tower and the beam end. The thickness of diaphragms at different joints is different, which is 12m at the parent tower, 11m at the connection between the sub tower and the main beam, and 2 meters at the beam end. The concrete grade of bridge is C55.

4.2 Modeling and foundation design of main tower

Single column pylon is recommended for this bridge. The shape of the upper tower adopts the traditional national musical instrument Sheng as the design idea. Through the evolution of its shape, combined with the characteristics of its own bridge tower structure. However, in order to simplify the modeling process, the inverted Y-shaped main tower is proposed. The inverted Y-shaped main tower can not only meet the design requirements of the main tower in the comparison scheme, but also meet the design requirements in terms of stress and stiffness, and can resist large bending moment. The tower height above the bridge deck is 68m, in which the vertical height of the left and right sides of the tower body is 50m. The anchorage zone of stay cable is set in the middle tower column section, and the central tower column adopts hollow section. The 3d model of bridge tower is shown in Figure 1

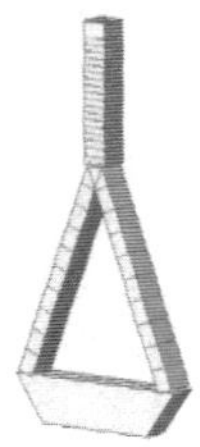

Fig.3.1 Finite element model of main tower

The left and right sides of the main tower body extend upward and meet the middle tower column. The vertical height of the tower body is 50m and the height of the middle tower is 18m, all of which are hollow sections.

5 Self weight calculation of whole bridge model

Midas civil is used to establish the model of main beam and bridge tower, and the self weight of the whole bridge model is roughly estimated. If the dead weight can be controlled within 60000 tons, it can be proved that the scheme is feasible in terms of swivel weight. At present, the domestic swivel tonnage has reached 46000 tons, and the swivel balance control of cable-stayed bridge structure is relatively mature. After a certain period of design and manufacturing, it can reach 60000 tons of bridge swivel.It can be seen from Figure 4 that the reaction force at the bottom of the bridge tower is 708712.3kn, which can be converted into a mass of about 71000 tons, that is, when the bridge rotates, the weight of the spherical hinge is 71000 tons

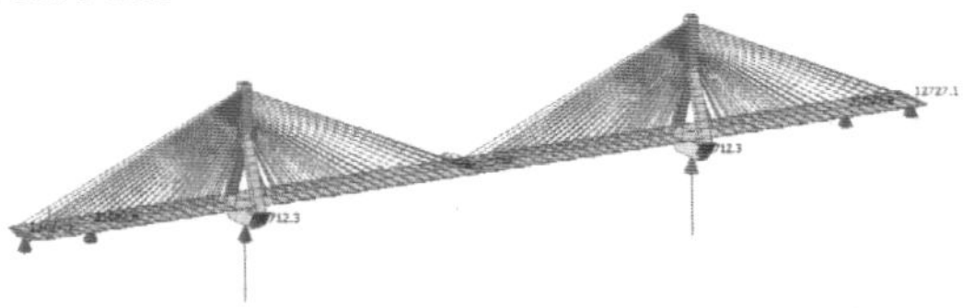

Fig.5.1 Calculation results of finite element model

6 Development status of swivel ball joint

The swivel hinge is located at the bottom of the pier and can be used to rotate. Swivel hinge plays an irreplaceable role in the swivel of the bridge. The self weight of the bridge basically acts on the swivel hinge. Therefore, the manufacture and bearing capacity of the swivel hinge should reach a high standard. There are also different structural forms of swivel hinge. We mainly classify it into two types: spherical hinge with shape similar to spherical surface and plane hinge whose action surface is plane. Because the materials used for making swivel hinges are different, we can roughly divide them into two categories according to different raw materials. The raw material is the swivel hinge of concrete, so we call it concrete hinge. If the raw material is steel hinge, we call it steel hinge. When we choose the swivel hinge, we should design the swivel hinge according to the needs of the bridge. Each swivel hinge should be made according to the swivel bridge, so as to make the selected swivel hinge suitable for the actual needs of the project as far as possible, so as to ensure the stable and safe rotation of the bridge during the swivel process. Because of the large span and self weight of the swivel bridge, the spherical hinge has the best bearing effect. Therefore, this paper focuses on the development status of the spherical hinge.

6.1 Concrete Swivel Ball Joint

In the swivel ball joint, the ball joint can be classified according to different manufacturing materials. The spherical hinge made of concrete is called concrete hinge. When the raw material belongs to steel, it is steel ball hinge.Among them, in order to ensure the bearing capacity of the spherical hinge, the concrete spherical hinge is generally made of high-grade concrete, and the strength grade of concrete is generally above C50. There are some differences between concrete spherical hinge and steel spherical hinge in structural design. The concrete spherical hinge is generally divided into two parts: the upper and the lower. The upper ball gap is called "upper grinding plate". The upper part of the ball gap and the lower part of the ball need to fit enough to meet the force and rotation requirements. In order to make them meet the rotation requirements, we usually scrape off the surface of the ball gap to make a perfect fit. Before pouring and setting with concrete, the grinding disc and grinding center should be checked manually to ensure that the contact surface can run in well. When rotating, the friction force during rotation should be minimized to make the rotation process smoother. First of all, it is not necessary to obtain the advantages of concrete ball in terms of material selection and technology. Especially in some places where the working condition is not good and some large-scale instruments cannot reach, the concrete spherical hinge can reach the construction site by ordinary methods. In the existing swivel bridge, it is very common to use the concrete ball joint in the swivel bridge. Therefore, according to these projects, the technical parameters of the concrete spherical hinge also have a certain standard. The static friction coefficient of concrete spherical hinge is less than 0.05, and the dynamic friction coefficient is less than 0.03. Among these existing swivel bridges, the self-weight of swivel bridges has reached 9000 tons. The weight of the swivel is very different from that of this project, so the use of concrete spherical hinge is not considered.

6.2 Steel ball joint

The steel ball hinge is also composed of two parts, which is different from the concrete ball hinge. The upper and lower parts of the steel ball hinge are not the shape of solid ball gap, but are made of thick steel plate. After the steel plate is pressed into a spherical surface on the special equipment, the upper and lower spherical hinge is connected through the steel shaft, and the circular c2f4 (tetrafluoroethylene) mushroom head and c2f4 sliding blade are arranged at the joint of the spherical joint. Steel itself has great strength, so the bearing capacity is obviously better than the concrete hinge. Therefore, in the long-span swivel bridge, the steel ball hinge has been widely used. At present, when steel ball hinge is used to rotate the bridge, the bearing capacity has reached about 50000 tons. In theory, it is possible to design the swivel of the bridge through the design of the spherical joint and the overall optimization design of the bridge, so the construction risk is relatively small.

7 conclusion

The calculation results show that the weight of the bridge in scheme I has greatly exceeded the weight of existing swivel bridges in China. Because of the simple weight estimation of the model, the boundary conditions are also applied to the end of the bridge, that is, the fulcrum of the bridge beam end also shares part of the reaction force. Therefore, in the actual project, the weight of the spherical hinge is greater than 70000 tons. If we want to continue to consider the use of prestressed concrete box girder, it is necessary to optimize the design of the girder section and the size of the bridge tower to reduce the weight of the structure. At the same time, the ball joint, the core part of the swivel structure, also needs to have high quality requirements.

Acknowledgements

Thanks for the support of the fund: Research project of Yunnan Provincial Department of Transportation Yun Jiao Science and Education [2018]No.43

References

1. Qin shaozong. Research on monitoring technology for Swivel Construction of cable stayed bridge across railway lines [D]. Shijiazhuang Railway University, 2019.
2. Zhang Li. Study on mechanical properties of composite tower and steel-concrete composite section of main beam of hybrid girder cable-stayed bridge [D]. Hunan University, 2019.
3. Han Feifei. Structural system research and parameter analysis of partial cable-stayed bridges [D]. Hefei University of technology, 2019.
4. Wang Xuxu. Research on key design issues of Large Span Composite Girder single tower swivel cable stayed bridge [D]. Tianjin University, 2018.
5. Yao Youlin. Study on technical and economic characteristics of girder types for cable stayed bridges with different spans [D]. Hunan University, 2017.
6. Li Haijun. Design and construction control analysis of prestressed concrete single cable plane cable stayed bridge [D]. Shijiazhuang Railway University, 2016.
7. Tang zhunzun, Tan Ke. Discussion on comparison and selection of preliminary design schemes for a super large bridge [J]. China high tech enterprises, 2014 (19): 31-32.
8. Dong Zhuangzhuang. Study on Optimization of construction scheme of stay cable and main girder steel bundle of horizontal rotation cable-stayed bridge [D]. Chongqing Jiaotong University, 2013.
9. Lv yaoxiu. Comparison and selection of cable stayed bridge design schemes [J]. Shanxi architecture, 2010,36 (02): 322-323. [2] Chen Yifei. Analysis, evaluation and identification of cable bridge structure based on long-term monitoring [D]. Dalian Maritime University, 2018.
10. Cheng W U Study of Cable and Pylon Structure Types of Cable-stayed Bridge using Swivel Construction with the Characteristics of Broad Bridge Deck and Steel Girder[J]. Transportation Science & Technology, 2017.
11. Quan-Sheng S , Shi-Jie W , Yong-Cun S , et al. Static Load Test and Research on Cable-Stayed Bridge with Horizontal Swivel,Single Tower and Single Cable Plane[J]. Highway, 2008.
12. Luo R D , Wang H L , Hu J H . Deformation and Stress Analysis of an Asymmetric Single Tower Cable-Stayed Bridge on High Speed Railway in Swiveling Process[J]. Advanced Materials Research, 2015, 1079-1080:359-363.

A Study on Meteorological Architecture

Yuqian Wang[1,2], Yi Yang[1,2*]

[1]College of Civil Engineering and Architecture of Zhejiang University
[2]The Architectural Design&Research Institute of Zhejiang University

Abstract. Under the background of global warming and energy crisis, exploring the new relationship between architecture and climate has always been a hot spot in the construction industry. Thus , a new design pattern "meteorological architecture" guided by a localized meteorological condition is developed .The paper starts with the distinction of the concepts of "meteorology" and "climate" to explains the connotation of "meteorological architecture". Through the analysis of current theories and practices related to meteorological architecture, a new principle of architectural space division and architectural form generation logic have been summarized. Finally, the enlightening effect of meteorological architecture on other fields is proposed.

1 Introduction

Throughout the history of architecture, there is a long and complicated relationship between architecture and the natural environment. Introducing the concept of meteorology into architecture allows us to understand contemporary architecture from a new perspective.

On the other hand,responding to climate change, saving energy and reducing carbon dioxide emissions has become a global consensus. The high energy consumption of the construction industry has become an important culprit of greenhouse gas emissions. Since the Industrial Revolution, the invention and widespread use of air conditioners and other mechanical equipment has broken the basic balance between people's original construction methods and the natural environment(1). Humans can freely adjust the indoor environment of buildings through these devices. Climate considerations are no longer a necessary part of architectural design. Architecture and nature are in an isolated and opposed relationship. Although air-conditioning equipment does provide people with a "pleasant" living environment to a certain extent, it brings an energy crisis at the expense of consuming a lot of energy.

In the 1940s and 1950s, the rise of the global ecological movement caused us to re-examine the relationship between architecture and climate.Architects kept trying to break the state of isolation between architecture and environment. Concepts such as green buildings, low-carbon buildings, sustainable buildings, passive buildings, and near-zero energy buildings have been put forward one after another.However Some scholars believe that today's energy-saving buildings represented by "energy rating certification" are still a continuation and strengthening of the tradition of isolation from the environment(2). They tried to establish a new relationship between architecture and nature. Thus , a new design pattern "meteorological architecture" guided by a localized meteorological condition is developed .

2 The Connotation of "Meteorological Architecture"

What is "Meteorological Architecture"? What is the difference with "climate architecture"? In the past, scholars did not make a strict distinction between "climate" and "meteorology". In fact, there are subtle connections and differences between meteorology and climate. Climate is a natural environmental factor formed during the long-term energy exchange and mass exchange between the earth and the atmosphere. It can also be said to be the long-term average state of meteorological elements. For example, temperate maritime climate, tropical monsoon climate. In the "confrontation" between architecture and climate, people are often subjective responses to the natural environment based on their accumulated life experience. Therefore, in different climate zones, buildings have the same expression, such as thermal insulation culture in cold climates and tropical sun-shading architectural aesthetics. These are all terroir aesthetics formed based on the experience of natural conditions such as climate and geography(3).

Now scholars are more willing to use "meteorological architecture" to explain their architectural practice. Meteorology refers to the general term for various physical phenomena and physical processes in the atmosphere, such as cold , heat, dryness ,humidity, wind, clouds, rain, snow, frost, fog, and lightning. Compared with the study of climate,

* Yi, Yang: yyilion@163.com

meteorology pays more attention to the study of the logic of energy operation in nature. Incorporating it into the architectural design, the logic of building space division and building form generation comes from the logic of energy operation in nature, which are also two important directions of current meteorological architecture exploration. The current climate-related global warming and resource consumption are not an issue worthy of continuous concern. It will become the driving force for more innovative initiatives. Compared with the previous use of it as a mandatory additional technical restriction, meteorological architecture is to take the physical elements of pressure, temperature, and relative humidity as a real part of architecture and turn it into a new architectural design tool. Thinking about meteorological architecture means mapping architecture to a brand-new field, establishing a perceptible relationship with space, and treating indoors as an atmospheric environment with abundant climatic conditions, climate changes and climate gradients (4) .It is no longer the original climate-oriented empirical construction. The meteorological building performs more precise control of the built environment. The research object is reduced from the original broad climatic zone to the small area where the building is located, allowing the building in each different small area more personalized. Among the architect firms currently engaged in meteorological architectural practice, the most representative ones are Philippe Rahm and ROEWU (ROEWU is an design-focused office as a collaboration between Stephen Roe and Chiafang Wu .) Their exploration of the space and form of meteorological architecture has produced a wide range of international influences.

3 Probe into the Practice of Meteorological Architecture

3.1 Exploring the potential of architectural space division

According to the findings of Rahm's related theories and practices, he is more inclined to explore the potential of using meteorology to divide architectural space. Rahm believes that architecture is not a pile of bricks and cement, but a partitioned reinforced concrete. The purpose of the building is to provide a comfortable living environment. However, most of the problems of modern greenhouse effect and environmental pollution are generated during the construction and use of buildings (5) . Rahm's design introduces modern technology to make the atmospheric environment around us participate in the design through physical methods such as sensing, convection, heat conduction, pressure, and radiation. On the basis of following the weather conditions, firstly design a suitable atmospheric environment, then put the building shape in the weather frame, and finally give it different functions according to the needs. This has changed the previous method of mechanically dividing space by function, forming the principle of space organization of " meteorology first, function last".

The House of Convection is one of Rahm's masterpieces of meteorological architecture. The physical term "convection" is often used to describe the combined effect of conduction and fluid flow. In free convection, an increase in temperature causes a decrease in air density, so hot air rises and cold air decreases. The design of the House of Convection is based on the above Archimedes' natural law of warm air rising and cold air falling. In actual situations, different spaces in the room have a certain temperature difference, and the maximum temperature difference between the floor and the ceiling can even reach 10 ° C. In addition, the temperature required for different spaces is also different according to different human activities and clothing levels. For example, when we have a quilt on the bed to keep warm, the temperature in the bedroom only needs to be 16℃; when we sit quietly on the sofa and the human activity is low, the living room usually needs 20℃; in the kitchen where cooking, the temperature only needs to be maintained At 18℃; and the bathroom without clothes should be the warmest space in the building. Based on the above indoor weather and human activity data research, arranging functional spaces with different temperature requirements according to the actual temperature difference in the house can effectively save energy. For example, changing the height difference of the horizontal floor, different depths and heights produce different indoor temperatures, and placing the bathroom that needs the most heat at a higher place, and placing the kitchen that does not need heat at a lower place. In this way, the house becomes an energy-saving landscape with different temperatures and scattered patterns. The single and homogeneous indoor space state of the air-conditioned building is broken. The logic of energy operation becomes the principle of space division. It can be seen that meteorological architecture requires architects to have an accurate grasp of local meteorological conditions and human physiological characteristics. Meteorological architecture presents a state of interdisciplinary integration.

Fig.1. Schematic diagram of cold and warm air exchange

 https://doi.org/10.1051/e3sconf/202123703027

Fig.2. The effect picture of The House of Convection

3.2 Exploring the innovation of architectural form

According to the theoretical and practical research of the ROEWU office, although they also involve the exploration of meteorological space, and even coincide with Rahm, their exploration in architectural form innovation is more in-depth. They believe that due to topographical conditions, vegetation conditions, and surrounding buildings each area has a special microclimate. Buildings have the function of regulating wind pressure, wind speed, and solar radiation. These energies are visualized through computer technology, and the logic of energy operation becomes the logic of building form generation.

The book *Weather Formations -An Architectural Meteorology* which they wrote introduced their theory and practices of meteorological architecture to readers. The author grew up on the northwest coast of Ireland.The stone walls on the coast can stand upright in strong winds for many years. It can be seen that the stones are full of pores, and small air volume can pass through these pores. Therefore, he was inspired to carry out a new structural system for this area. The author uses fluid mechanics simulation to study the relationship between the ratio of holes , wind speed and pressure to explore the stability of the open space, while allowing people to perceive the dynamics of "wind blowing". The author explored a design prototype and their topological relationship. The basic shape are a hexagon and a rectangle. Three vertices are selected as "anchor points". The rule is that the rectangle should be connected with the "anchor points" (the trapezoid is protruding). Then it starts to change. The changed parameters include the angle of rotation and the width of the rectangle. Several dynamic units are generated[6]. The wall combines the weight of the material with the lightness of the wind, giving people a sense of sheltering while blowing. Faced with different wind environments, different effects will be obtained by adjusting the parameters and combining different units.The state of isolation between the building and the environment was broken, showing an open state. This new type of structure can be called meteorological architectural structure, showing the power of an architectural "prototype".

Fig.3. The stone walls on the northwest coast of Ireland

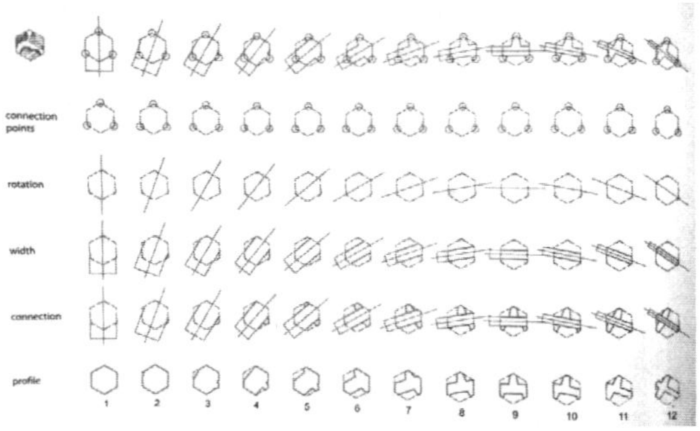

Fig.4. The change process of basic shape

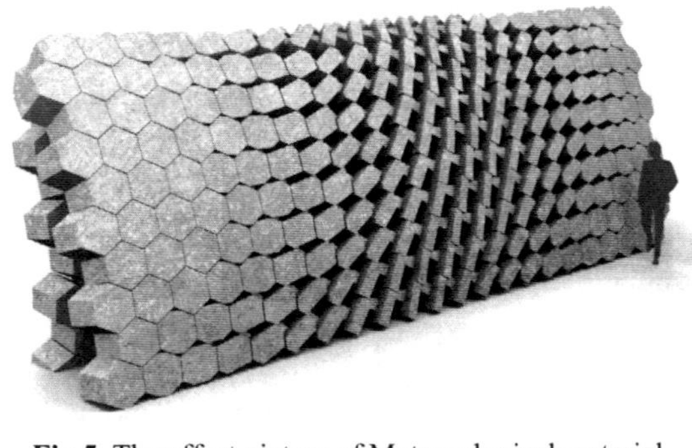

Fig.5. The effect picture of Meteorological material

4 Enlightenment from Meteorological Architecture

Although meteorological architecture was developed in the context of the energy crisis, it also has enlightening effects in other fields. In the past 30 years, "healthy buildings" have been a hot spot in the world. In particular, since the new crown virus epidemic, people have been exposed to buildings and their indoor environment for a long time.There are higher requirements for the improvement of building indoor health performance. How to comprehensively promote the physical and mental health of users through the air, water, comfort, fitness, humanities, services and other

 https://doi.org/10.1051/e3sconf/202123703027

aspects in the building is an important proposition for achieving healthy buildings(7). According to the relevant theories of meteorological architecture, we should start with "building a comfortable climate" as a part of the design. "building a comfortable climate" is a method not a target .Compared with a series of measures to improve the indoor environment after the building is completed, the construction according to this method will greatly improve the health performance of the indoor environment of the building. Meteorological architecture provides new ideas for the development of healthy buildings,which we should change cogitation and solve problems from the source.

5 Conclusion

Meteorological architecture makes meteorology the most important part of architectural design and the logical source of architectural space division and architectural form generation. The exploration process of meteorological architecture is a process of thought transition from passive response to active creation in the face of energy crisis. This mode of thinking also provides a lot of inspiration for related fields. Under the background of the new era, when weather has become a new architectural vocabulary to participate in the design of architectural environment, how to establish a stable system model to guide the construction of meteorological buildings is what we need to constantly explore.

Image Source

Fig.1, Fig.2:http://www.philipperahm.com/data/projects/convectiveapartments/index-c.html.
Fig.3, Fig.4, Fig.5: https://www.roewu.com/weather-formations

References

1. Q.Song,L.Yang,B.Yin,HuaZhong Arch, **1003-739X04（2014）-0011-05**(2014)
2. L.Li,Time+Arch,**1005-684X(2018)03-0036-006**(2018)
3. X.Lin, *Green Building* (China Construction Industry Press, 2011)
4. P.Rahm, Z.Yu, Time+Arch,**1005-684X(2015)02-0032-06** (2015)
5. Y.Qian,J.Zhai,The Architect,**107**(2019)
6. ROEWU Arch, *Weather Formations -An Architectural Meteorology* (Garden City Co., Ltd, 2017)
7. Q.Wang,Y.Deng,G.Li,C.Meng,L.Xie,M.Liu,L.Zeng, Chinese Science Bulletin,**246**(2020)

https://doi.org/10.1051/e3sconf/202123703028

Experimental Study on Mechanical Properties of Small Size Viscoelastic Damper

Miao Han[1], Richard Twizeyimana[1], and Hongkai Du[1,*]

[1]Beijing Advanced Innovation Center for Future Urban Design, Beijing University of Civil Engineering and Architecture, Beijing 100044, China

Abstract: Long span roofs are very likely to oscillate when subjected to wind load that can lead to structure fatigue and endanger structures safety. Dampers have been used for long time to dissipate wind and earthquake induced energy in structures. This research work aims to present experimental study of small size viscoelastic damper that can be installed in truss of long span roof. Small size viscoelastic dampers that can be used to dissipate wind induced energy in large span roof structure need to be tested to know their performance behavior and mechanical properties at different loading amplitudes and frequencies. A kind of viscoelastic dampers were manufactured and tested under horizontal cyclic loads. Resistance and deformation of the damper were measured to study the viscoelastic damper properties dependence on frequency and amplitude. Mechanical properties including shear storage modulus, shear loss modulus, loss factor and energy dissipation are studied. Experimental results show that the small size damper's mechanical properties are significantly related to its loading frequency and amplitude. The energy dissipation capacity of the damper was stable under different loading frequency and amplitude.

1 Introduction

Earthquakes and heavy winds have occurred in different parts of the world, resulting in serious damages and destruction of buildings and infrastructures, which inflicted serious injuries, human lives loss, and properties loss [1]. Researchers have discovered that seismic and wind-induced energy in structures can be dissipated by the addition of dampers. Dampers are usually used to retrofit and strengthen new structures by dissipating induced energy hence reduce structural damages [2,16].

Dampers can be classified as passive, active, and semi-active damper. where viscoelastic dampers are classified as a passive energy dissipation device as they do not need an external source of energy [3,15]. Viscoelastic damper has a viscoelastic material which is used as the medium transaction where mechanical energy transfers to heat energy [4]. Viscoelastic dampers have a special characteristic of functioning at a very small displacement, hence are very effective to be used against winds-induced oscillation.

Viscoelastic dampers started to be used in civil engineering since 1960. Viscoelastic dampers are widely used in the wind-resistant design of high-rise structures due to its effectively energy dissipation at any displacement, simple structure and low cost [5]. Reduction of wind vibration was achieved by installing more than 10,000 viscoelastic dampers in each tower of World Trade Center in New York [2]

The viscoelastic damper has received much popularity for its special characteristic to dissipate energy from lower to high displacement or amplitude [6]. Hence, resistance against winds, earthquakes, and assurance of structure habitability during wind and earthquake can be achieved by using the devices. Researchers have discovered that the mechanical properties of viscoelastic material are heavily dependent on loading frequency, amplitude, and temperature [7,14]. To ensure the successful application of viscoelastic damper, appropriate evaluation of their mechanical properties is needed. However, the mechanical model used for property analysis are relatively complex [8].

Long-span roof structures like stadium, airport lobby and exhibition hall are widely used worldwide. Lightweight, flexibility, low damping ratio and close natural frequencies are the main characteristic of long-span roofs, hence make them more likely to oscillate when they are under wind load, which may lead to structure fatigue or endanger structure safety [9]. Control of wind-induced vibration in long span roof is of urgent need. Although, many researchers have been studying mechanical properties of viscoelastic damper which are mainly used in framing structures as braces, aim of this research is experimental investigation of mechanical properties of small size damper which can used in long roof truss structure.

To investigate the mechanical properties of small size damper, a kind of viscoelastic dampers with same parameters were manufactured and tested under

* Corresponding author: duhongkai@bucea.edu.cn

 https://doi.org/10.1051/e3sconf/202123703028

horizontal cyclic loads. Mechanical properties, dependency on frequency and amplitude, including shear storage modulus, shear loss modulus, loss factor and energy dissipation are studied in this paper.

2 Properties of Viscoelastic Materials

Viscoelastic materials, as their name suggests, have to main properties which are viscous and elastic properties [10]. The term "viscous" means that they deform slowly when they are subjected to an external force. The term "elastic" means that once the force has been removed, the material will return to its original configuration (For purely elastic materials, loading and unloading "stress versus strain" curves (lines) are superimposed). For viscoelastic material, they form a "hysteresis" loop [11]. Viscoelastic material total resistance to deformation is studied in term of complex modulus. The contribution of elastic component to deformation is studies as shear storage modulus (S_S) while contribution from Viscous component is studies as shear loss modulus(S_L). Simply, shear storage and shear loss modulus significantly affect the damping performance of viscoelastic dampers [12]. For clinical application, it is very important to know the response behavior of viscoelastic material (specifically shear storage, shear loss and loss factor) at a corresponding loading rate and temperature [13]

The strain of a viscoelastic body is out of phase with the stress applied, by the phase angle, δ [18]. This phase lag is caused by the excess time necessary for molecular motions and relaxations to occur.

Dynamic stress, σ, and strain, γ, given as:

$$\sigma=\sigma_0\sin(\omega t+\delta), \quad (1)$$

$$\gamma= \gamma_0 \sin(\omega t), \quad (2)$$

where ω is the angular frequency. Using this notation, stress can be divided into an "in phase" component ($\sigma_0\cos\delta$) and an "out-of-phase" component ($\sigma_0\sin\delta$) and rewritten as,

$$\sigma = \sigma_0 \sin(\omega t) \cos \delta + \sigma_0 \cos(\omega t) \sin \delta \quad (3)$$

Dividing stress by strain to yield a modulus and using the symbols S_S and S_L for the in phase (real) and out-of-phase (imaginary) moduli respectively:

$$\sigma = \gamma 0\ S_S \sin(\omega t) + \gamma 0\ S_L \cos(\omega t), \quad (4)$$

$$S_S=\frac{\sigma 0}{\gamma 0}\cos\delta. \text{ and } \quad S_L=\frac{\sigma 0}{\gamma 0}\sin\delta, \quad (5)$$

$$\gamma = \gamma 0 \exp(\omega ti). \text{ and } \sigma= \sigma 0 \exp(\omega t+\delta)i, \quad (6)$$

$$S=\frac{\sigma}{\gamma}=\frac{\sigma 0}{\gamma 0}ei\delta=\frac{\sigma 0}{\gamma 0}(\cos\delta+i\sin\delta)= S_S+iS_I, \quad (7)$$

Equation (7) shows that the complex shear modulus(**S**) consists of "real" and "imaginary" parts. The real "shear storage modulus (S_S)" part describes the ability of the material to store energy and representing the elastic portion. The imaginary "shear loss (S_L)" part is associated with energy dissipation in the form of heat upon deformation and representing the viscous portion.

The phase angle δ is given by

$$\tan\delta=\frac{SL}{Ss} \quad (8)$$

The storage modulus is related with the stiffness of a material and is related to the Young's modulus, the dynamic loss modulus is related to internal friction and is sensitive to molecular motions, relaxation processes, transitions, morphology and other structural heterogeneities. Understanding the dynamic properties gives information that help to understand polymer mechanical behavior [19].

3 Damper Configuration

Viscoelastic dampers are non-load-carrying structure elements, whereas they reduce the mechanical energy of structure motion and transfer them into heat energy [20]. This energy transfer achieves only through an energy transfer medium or a damping medium, i.e., viscoelastic material. Standard viscoelastic dampers are made of three-layer of rigid parts, and two-layer of viscoelastic materials (Figure 1~2). The damper should be installed so that the viscoelastic material undergoing pure shear deformation. In this research, the damper consists two viscoelastic materials of 100 mm × 50 mm× 10mm. The steel plate is 362 mm × 50 mm and 10 mm thick. The viscoelastic material is firmly bonded to the steel plate.

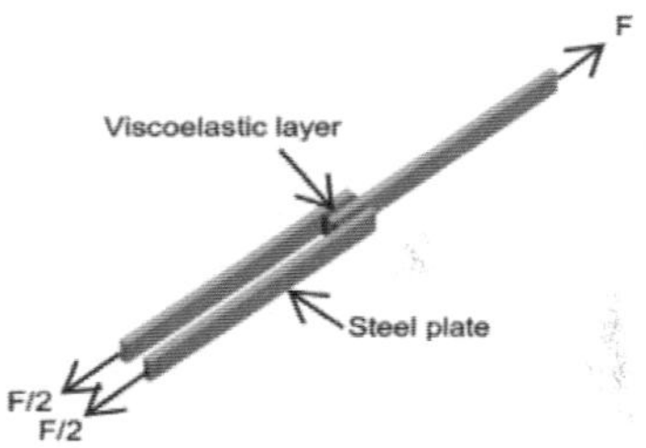

Figure 1. Sketch of standard viscoelastic damper

Figure 2. Photo of standard viscoelastic damper

4 Test Setup and Protocol

The two outer steel plates were fixed on the truss of the base, and the inner steel plate was connected to the MTS Hydraulic actuator by holes at the end of the specimen. The inner steel plate was relatively displaced by the MTS Hydraulic actuator, which causes the viscoelastic material

 https://doi.org/10.1051/e3sconf/202123703028

to shear and deform (Figure 3). The displacement gauge was placed on the damper between the steel plates connected to the base and the actuator to record accurately deformation of the specimen.

Figure 3. Test setup.

Loading amplitude and frequency are control variables in the tests, and test protocol is shown in Table 1. The test specimen was subjected to 4 cycles of different amplitudes at a frequency of 1.0 Hz to study the correlation of the deformation amplitude (strain) and the damper performance. The loading amplitudes were: 3mm, 5mm, 7mm, 9mm, 11mm, 13mm, and 15mm. The test specimen was loaded at 3mm strain amplitude at different loading frequencies to study damper performance dependency on the loading frequency. The test specimen was tested at 0.5 Hz, 1.0 Hz, 1.5 Hz, and 2.0 Hz.

Table 1. Test Protocol

Condition	Amplitude(mm)	frequency(Hz)	Temperature	cycle
Amplitude	3,5,7,9,11,13,15	1.0	20^{0}C	4
Frequency	3	0.5, 1, 1.5, 2	20^{0}C	4

5 Test Results and Analysis

There was no obvious change in the test specimen, and the viscoelastic body and the steel plate were intact after loaded in various test conditions.

5.1 Damper property related to amplitude.

The shear storage modulus **(S_S)** is defined as the ratio of stress at the peak strain to peak strain of the damper, which reflects the elastic performance of the damper in a shear deformation, and used to measure the energy stored during each cycle loading. The shear loss modulus **(S_L)** is defined as the ratio of the stress corresponding to zero displacement to the peak strain in the hysteresis loop. Shear loss modulus **(S_L)** reflects the viscous characteristics of the damper and is used to characterize the energy dissipation during loading cycle. The loss factor is defined as the ration of shear loss Modulus to shear storage modulus. The loss factor is the main indicator to measure the energy dissipation capacity of viscoelastic dampers.

The viscoelastic damper was subjected to the amplitude of 3, 5, 7, 9, and 11 mm at 20 degrees centigrade to analyze the correlation between damper properties and loading amplitudes. The relationship curves of shear storage modulus, shear loss modus, loss factor, energy dissipation and loading amplitudes are shown in Figure 4a~4e.

Test result as shown in Figure 4e, the load-deformation hysteresis curve of the damper is stable and full, and the curve area increased as amplitude increases. The increase in the area of the curve show the increase in energy dissipated. Throughout the amplitude loading condition, the damper showed a stable performance. Figure 4a-4d, is obvious that damper properties are related to loading amplitude. As loading amplitude increases, there is a decrease of shear storage modulus, shear loss modulus and loss factor.

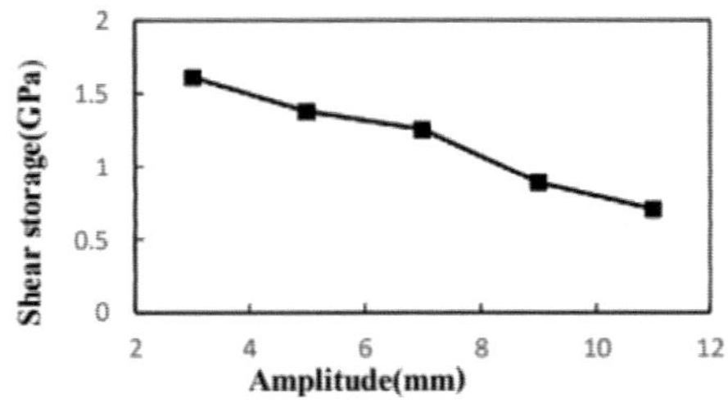

4.a. Shear storage modulus

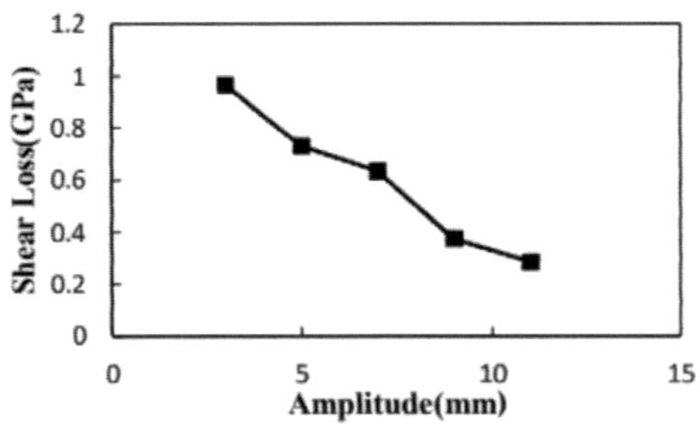

4.b. shear loss modulus

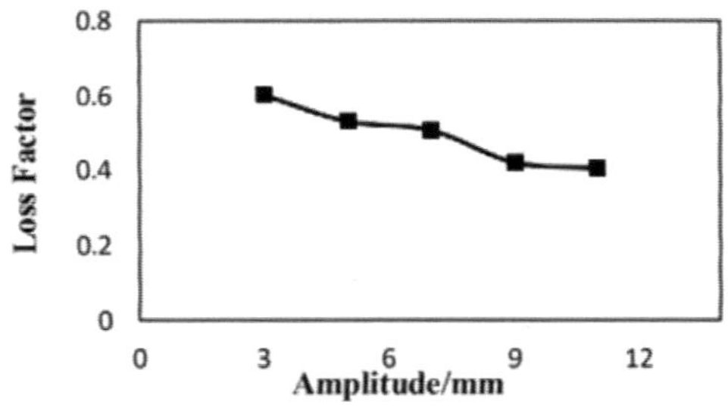

4.c. Loss factor

https://doi.org/10.1051/e3sconf/202123703028

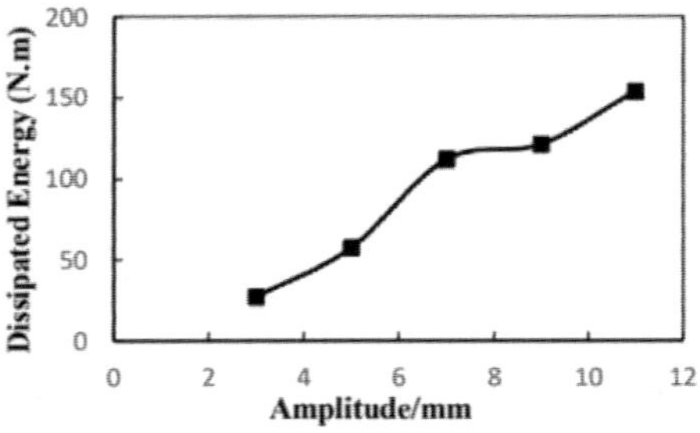

4.d. Dissipated energy

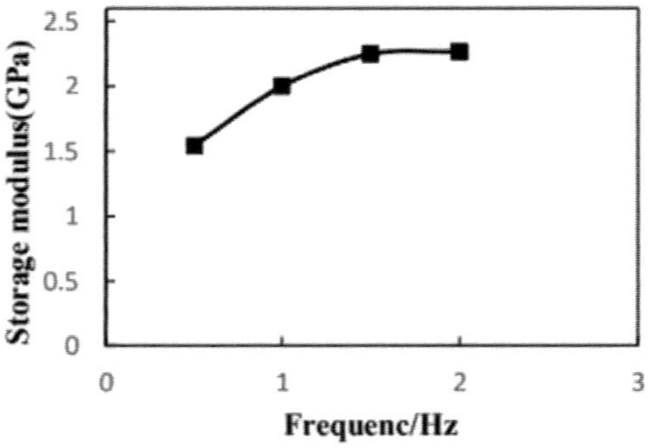

5.a. Shear storage modulus

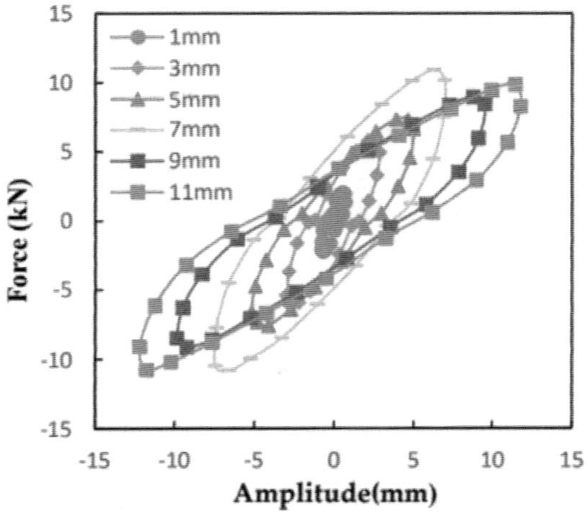

4.e. hysteresis curve

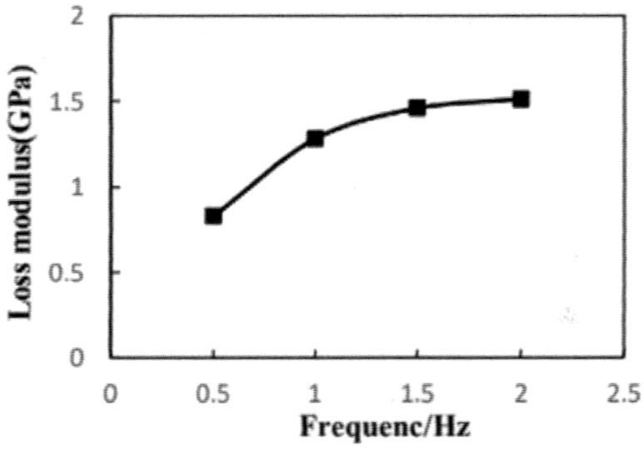

5.b. shear loss modulus

5.2 Damper property related to frequency

The damper was loaded with a sine wave of 0.5Hz, 1.0Hz, 1.5Hz, and 2.0Hz at a loading amplitude of 3mm at 20 degrees centigrade. Figure 5.e shows the load-deformation hysteresis curve at the different loading frequencies. Figures 5a～c respectively presents the shear storage modulus (**S_S**), the shear loss modulus (**S_L**) and the loss factor D. Figure 5.d shows energy dissipated. Test results show that shear storage, shear loss modulus, loss factor and dissipated energy increases as frequency increases. However, Shear storage and loss moduli increase very fast at frequency less than 1Hz after that start to increase slowly.

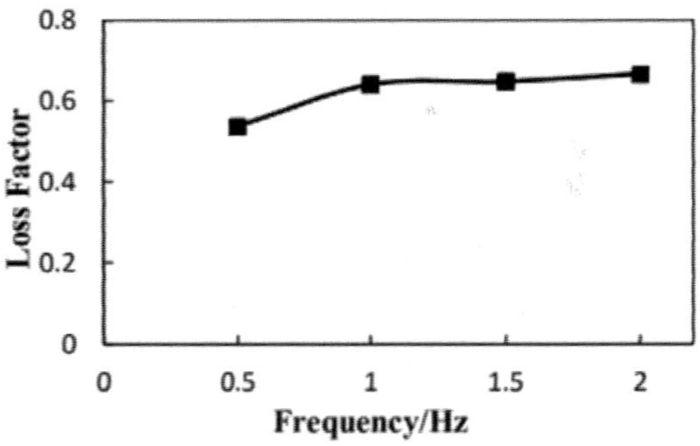

5.c. Loss factor

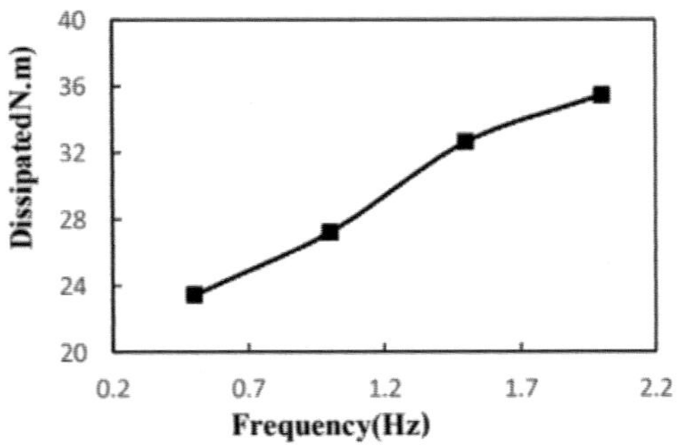

5.d. dissipated energy

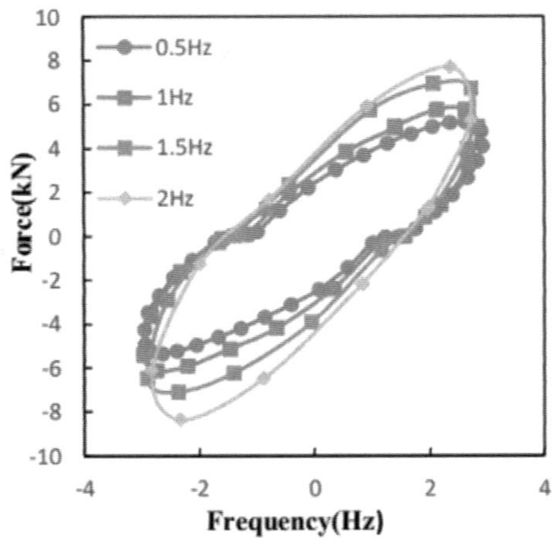

5.e. hysteresis curve

6 CONCLUSION

The experimental research of small size viscoelastic damper demonstrates that hysteresis curve under different loading amplitude and frequency are stable, good energy dissipation capability. Test result showed that small size damper subjected to an increasing amplitude will experience relatively a constant shear storage and shear loss modulus and loss factor decreases as amplitude decrease.

The experimental research demonstrates that small size viscoelastic damper subject to a changing frequency have a stable performance. Mechanical property such as Shear storage, Shear loss moduli and loss factor increase as frequency increases. The increase in shear and loss modulus increase from 1.54GPa at 0.5Hz to 2.24GPa at 1.5 Hz and 0.82GPa at 0.5 Hz to 1.45GPa at 1.5Hz respectively but when reaches 1.5Hz, the increase is too small, where there is an increase up to 2.26 at 2Hz for shear storage and increase up to 1.51GPa at 2.0Hz for shear loss modulus, Loss factor shows a fast increase between 0.5Hz to 1Hz and then show a very slow increase up to 2Hz. Generally, the test results showed a stable hysteresis curve which justify good energy dissipation capacity.

Authors Contribution: Han Miao conceived the presented idea and give the methodology used in this research, Richard Twizeyimana carried out the experimental and write the manuscript by the help of HongKai Du, HongKai Du supervised the findings of this work.

Funding: The National Key Research and Development Program of China, No. 2016YFC0700706-06.

Acknowledge: This experiment test was successful by the help of Beijing University of Civil Engineering and Architecture, Structure Lab technician.

Conflict of Interest: The authors declare no conflict of interest.

References

1. Adnan, A. R. *Disaster Management and Mitigation for Earthquakes*. **9th** Asia Pacific Structural Engineering and Construction Conference (APSEC), Kuala Lumpu, Malesia, (2015)
2. Samali, B. *Use of viscoelastic dampers in reducing wind and earthquake induced motion of building structures. Engineering Structure* (1995) **1**.
3. Faravelli, L. *Passive Energy Dissipation Systems in Structural Engineering*. Journal of structure Control (1999), 2-3.
4. Jones, D. *Viscoelastic Vibration Damping*, John Wiley & Sons Ltd: New York, USA, **1st Ed.**; 25-28, (2001).
5. Zhao Gang, Pan. Peng. *Experimental study on large deformation performance of viscoelastic dampers*. Journal of Building Structures, 2-4, 2012.
6. J.A Inuadi, M. Heat Generation on Viscoelastic damper in Structure. *World Conference on Earth Quake Engineering*,**11**,11, 1996.
7. Ramorino, G. *Developments in dynamic testing of rubber compounds: assessment of non-linear effects*. Polymer Testing, 5, (2003).
8. Mitsuo ASANO, H. M. *The experimental study on viscoelastic material dampers and the formulation of analytical model, World Conference on Earthquake Engineering*, Auckland, New Zealand, (2000).
9. Yaoyao Clare Duan, L. H. *Vibration control for large-span roof structure using pounding tuned mass damper*, Semanticscholar ,1 (2018).
10. Subic, A. *Materials in Sports Equipment,* Woodhead Publishing: Sawston, UK,124-125(2019).
11. Gauthier, M. M. *Deformation and Viscoelasticity of Polymers* . Asm Internation (1995)
12. Tom Ehrig, K. H. *Dynamic Behaviour Adaptation of Lightweight Structures by Compressible Constrained Layer Damping with Embedded Polymeric Foams and Nonwovens*. Applied Science (2019).
13. Sarwar, w. *viscoelastic material as energy dissipater viscoelastic damper for building structures to mitigate the seismic vibration* . Sciendo, 2-4 (2019).
14. Francisco Paulo Lépore Neto, M. B. A *Procedure for the Parametric Identification of Viscoelastic Dampers Accounting for Preload*. Journal of the Brazilian Society of Mechanical Sciences and Engineering 4-6,(2011).
15. M.C. Constantinou, M. S. *Fluid Viscous Dampers In Applications Of Seismic Energy Dissipation And Seismic Isolation*. Journal of Structure control 1-3 (1993).
16. SARWAR, W. *Viscoelastic Material As Energy Dissipater Viscoelastic Damper For Building Structuresto Mitigate The Seismic Vibration* . Sciendo 2-3, (2019).

https://doi.org/10.1051/e3sconf/202123703028

17. Shedbale, N. *Review on Viscoelastic Materials used in Viscoelastic Dampers*. International Research Journal of Engineering and Technology 1, (2017).
18. Menard, K. P. *Dynamic Mechanical Analysis: A Practical Introduction*, CRC Press, NewYork, USA, (1999).
19. Shrivastava, A. *Introduction to Plastics Engineering*. NewYork: William Andrew, (2018).
20. Y. Zhou, F. Shi, X. Deng. *Experimental Study on Properties of High-damping Rubber Damper*. WCEE, Santiago Chile, **16,** (2017).

https://doi.org/10.1051/e3sconf/202123703029

DIFFERENTIAL QUADRATURE METHOD FOR VIBRATION ANALYSIS OF PRESTRESSED BEAMS

Li Peng[1], Ying Wang[1]

[1]College of Civil Engineering, Shanghai Normal University, Shanghai 201418, China

Abstract. This paper investigates natural frequencies of free transverse vibrations of prestressed beams, and the governing equations and natural frequencies of the free vibration in related literatures are discussed and corrected. The differential quadrature methods (DQ) are applied directly to the corrected governing equations to get the the values of natural frequency numerically. Under the simple supported boundary conditions, the natural frequencies of model beam are numerically studied, and the physical parameters of the beam are analyzed respectively. The numerical results show that the natural frequency values increase with the growth of concrete strength and eccentricity of prestressed steels. But with the increase of the span length of beam and values of original prestressing force, the natural frequency values decrease.

1 Introduction

Prestressed concrete technique can effectively improve the cracking of concrete structure, improve the stiffness and durability of the structure, and give full play to the high strength performance of materials, which has been widely used in civil engineering[1-3]. In recent years, many researchers have studied the static performance of prestressed concrete structure, but the investigations on the complex dynamic performance are relatively insufficient.

The differential quadrature (DQ) method is a powerful tool for dealing with dynamical problems. Firstly it was introduced for structural analysis by Bert et al.[4]. and since then it has been successfully employed for the analysis of vibration of beams[5–6].

For the vibration characteristics of prestressed beams, the formulas of different linear vibration frequencies are derived by applying the theory of material mechanics and structural dynamics[3,7], However, these formulas are based on simplified vibration control equations, and their accuracy is worth discussing. In this paper, the existing vibration control equations of prestressed beams are analyzed and corrected, and the differential quadrature method (DQ) is directly applied to the corrected equations. The differential equations are transformed to a generalize d eigenvalue problem, and the numerical solutions of natural frequency are obtained based on the method of DQ. Under the simply supported boundary conditions, the first four order numerical solutions of natural frequency of model beam are given, and the different effects of physical parameters of prestressed beam and prestressed steel on natural frequency are analyzed.

The present paper is organized as follows. Section 2 establishes the governing equation for the transverse free vibration of a finite elastic prestressed beam. Section 3 develops the DQ schemes to descretize the governing equation. Section 4 presents numerical examples for investigations on the effects of the different parameters of the relevant beam and prestressed steel on the natural frequencies. Section 5 ends the paper with the concluding remarks.

2 Equation of motion

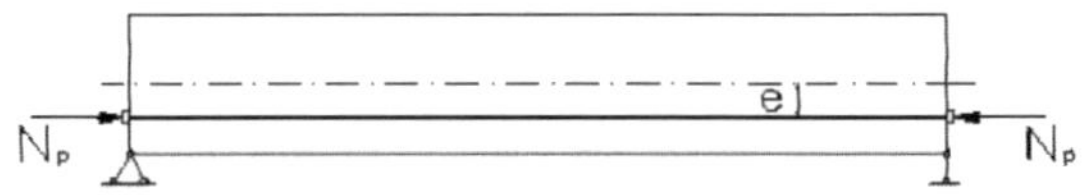

Fig. 1 The model of prestressed beam

Free transverse vibration of a prestressed beam can be described by transverse deflection of the cross section due to bending moment, depending on axial coordinate x and time t.

Fig.1 shows a finite elastic prestressed concrete beam with a length l, the bending moment and the axial force can be expressed as

$$\begin{cases} N_p = N_{p0} + \Delta N_p \\ M = N_p e = (N_{p0} + \Delta N_p)e \end{cases} \quad (1)$$

The free vibration equation of a prestressed beam is governed by the equation[1]

$$\frac{\partial^2}{\partial x^2}\left(EI\frac{\partial^2 y}{\partial x^2}\right) + \frac{\partial^2}{\partial x^2}(N_p y) - \frac{\partial^2}{\partial x^2}(M) + m\frac{\partial^2 y}{\partial t^2} = 0 \quad (2)$$

Substituting Eq. (1) into Eq. (2) yields

$$\frac{\partial^2}{\partial x^2}\left(EI\frac{\partial^2 y}{\partial x^2}\right) + \frac{\partial^2}{\partial x^2}[(N_{p0} + \Delta N_p)y] - \frac{\partial^2}{\partial x^2}\left[(N_{p0} + \Delta N_p)e\right] + m\frac{\partial^2 y}{\partial t^2} = 0$$

(3)

where y, m and EI are vibration displacement, unit mass, and bending flexural rigidity of the beam. e, N_{p0} and ΔN_p are eccentricity, initial prestressing force and increased prestressing force due to flexural vibration.

It is generally accepted that y is much less than e, so the value of $\Delta N_p y$ is negligibly small, and Eq.(3) can be rewritten as

$$EI\frac{\partial^4 y}{\partial x^4}+\frac{\partial^2}{\partial x^2}(N_{p0}y)-e\frac{\partial^2}{\partial x^2}(\Delta N_P)+m\frac{\partial^2 y}{\partial t^2}=0 \quad (4)$$

Based on the assumption that increased prestressing forces ΔN_P is proportional to midspan deflection y, the horizontal displacement due to the applied midspan load F can be calculated from the equation

$$\delta=\int\frac{m_1 M}{EI}dx=\frac{Fl^2 e}{8EI} \quad (5)$$

The horizontal displacement caused by the unit force acting on the anchorage point can be obtained as

$$\delta_1=\int\frac{m_1^2}{EI}dx+\int\frac{N_1^2}{EA}dx=\frac{le^2}{EI}+\frac{l}{EA}+\frac{l_t}{E_t A_t} \quad (6)$$

The change of anchoring force ΔN_P caused by load F is

$$\Delta N_P=\frac{\delta}{\delta_1}=\frac{\int\frac{m_1 M}{EI}dx}{\int\frac{m_1^2}{EI}dx+\int\frac{N_1^2}{EA}dx}=\frac{\frac{Fl^2 e}{8EI}}{\frac{le^2}{EI}+\frac{l}{EA}+\frac{l_t}{E_t A_t}}=\frac{Fle}{8(e^2+\lambda)} \quad (7)$$

where $\lambda=\frac{I}{A}+\frac{EI}{E_t A_t}\frac{l_t}{l}$, A, I, l and EI are cross-sectional area, cross-sectional moment of inertia, span length and bending flexural rigidity of the beam. l_t and $E_t A_t$ are length and flexural rigidity of prestressed reinforcements.

From Eq. (7), the load F can be given in the form

$$F=\frac{8(e^2+\lambda)}{le}\Delta N_p \quad (8)$$

The deflection of the midpoint of a beam under the action of force F is

$$y_F=\frac{l^3}{48EI}F \quad (9)$$

Substituting Eq. (8) into Eq. (9) yields

$$y_F=\frac{l^2(e^2+\lambda)}{6EIe}\Delta N_p \quad (10)$$

By the displacement reciprocal theorem, the upward displacement at midpoint due to ΔN_P is

$$y_{\Delta N_p}=\frac{l^2 e}{8EI}\Delta N_p \quad (11)$$

Deflection y can be calculated as

$$y=y_F-y_{\Delta N_p}=\frac{l^2(e^2+\lambda)\Delta N_P}{6EIe}-\frac{l^2 e\Delta N_P}{8EI}=\frac{l^2(e^2+4\lambda)}{24EIe}\Delta N_p \quad (12)$$

Therefore increased prestressing forces ΔN_P can be expressed as

$$\Delta N_p=\frac{24EIe}{l^2(e^2+4\lambda)}y \quad (13)$$

Substitution of Eq. (13) into Eq. (4) yields

$$EI\frac{\partial^4 y}{\partial x^4}+\left[N_{p0}-\frac{24EIe^2}{l^2(e^2+4\lambda)}\right]\frac{\partial^2 y}{\partial x^2}+m\frac{\partial^2 y}{\partial t^2}=0 \quad (14)$$

Eq. (14) is the free vibration equation of prestressed beam with straight prestressed steel. It is derived by applying the theory of material mechanics and structural dynamics.

With method of separation of variables, the vibration solution to Eq. (14) can be assumed as

$$y=X_{(x)}\cdot e^{i\omega_n t} \quad (15)$$

Substituting Eq. (15) into Eq. (14) yields

$$EI\frac{d^4 X}{dx^4}-\left[N_{p0}-\frac{24EIe^2}{l^2(e^2+4\lambda)}\right]\frac{d^2 X}{dx^2}=m\omega^2 X \quad (16)$$

For a simply supported beam, $X(x)$ can be expressed as

$$X(x)=A\sin\frac{n\pi}{L}x,\ (n=1,2,3,4\mathrm{L}) \quad (17)$$

Substitution of Eq. (17) into Eq. (16) yields

$$EI\left(\frac{n\pi}{L}\right)^4+\left[N_{p0}-\frac{24EIe^2}{L^2(e^2+4\lambda)}\right]\left(\frac{n\pi}{L}\right)^2=m\omega^2 \quad (18)$$

Solving Eq. (18), the explicit formulae for natural circular frequencies can be obtained

$$\omega_n=\sqrt{\frac{EI}{m}}\cdot(\frac{n\pi}{l})^2\xi \quad (19)$$

where

$$\zeta=\sqrt{1-\left(\frac{l}{n\pi}\right)^2\frac{N_{p0}}{EI}+\frac{24}{(n\pi)^2}\frac{e^2}{(e^2+4\lambda)}} \quad (20)$$

With Eq. (19), the natural circular frequencies of free vibration of a prestressed beam can be easily calculated. But it must be noticed that these formulas are based on simplified vibration control equations, and their accuracy is worth discussing.

In paper [2], Eq. (14) is rewritten as[2]

$$EI\frac{\partial^4 y}{\partial x^4}+m\frac{\partial^2 y}{\partial t^2}+\left[N_{p0}-\frac{6EIe^2}{(e^2+4\lambda)(lx-x^2)}\right]\frac{\partial^2 y}{\partial x^2}$$
$$+\frac{12EIe^2(l-2x)}{(e^2+4\lambda)(lx-x^2)^2}\frac{\partial y}{\partial x}-\frac{12EIe^2(3x^2-3lx+l^2)}{(e^2+4\lambda)(lx-x^2)^3}y=0 \quad (21)$$

It is a complex partial differential equation with variable coefficients, which cannot be directly solved to obtain the analytical expression of natural frequency.

In paper [3], considering ΔN_P is proportional to the midpoint ($x=l/2$) deflection of the beam, the simplified equation is given by

$$EI\frac{\partial^4 y}{\partial x^4}+\left[N_{p0}-\frac{24EIe^2}{l^2(e^2+4\lambda)}\right]\frac{\partial^2 y}{\partial x^2}+m\frac{\partial^2 y}{\partial t^2}=0 \quad (22)$$

Solving the equation with method of separation of variables, the explicit formulae for natural circular frequencies can be obtained[3]. Comparing two equations, it can be found that the last two items of Eq. (21) disappear in Eq. (22). However, when $x=l/2$ is substituted into Eq. (21), the coefficient before y term is not zero, and this term mainly reflects the influence of eccentricity on vibration performance. So the accuracy of the analytical expression of natural frequency obtained by Eq. (22) is also worth discussing.

Considering the coefficient before y term of Eq. (21) is not zero, this paper corrected the Eq. (22) as

$$EI\frac{\partial^4 y}{\partial x^4}+\left[N_{p^0}-\frac{24EIe^2}{l^2(e^2+4\lambda)}\right]\frac{\partial^2 y}{\partial x^2}-\frac{192EIe^2}{l^4(e^2+4\lambda)}y+m\frac{\partial^2 y}{\partial t^2}=0 \quad (23)$$

To simplify the expression, the Eq. (23) can be rewritten as

$$\rho A y,_{tt}+EIy,_{xxxx}+k_P y,_{xx}-k_e y=0 \quad (24)$$

where

$$k_P=N_{p0}-\frac{24EIe^2}{l^2(e^2+4\lambda)},\ k_e=\frac{192EIe^2}{l^4(e^2+4\lambda)} \quad (25)$$

Introducing the non-dimensional quantities into Eq. (24) results in the non-dimensional form

$$y,_{tt}+k_f^{\,2}y,_{xxxx}+k_1 y,_{xx}-k_2 y=0 \quad (26)$$

For simply supported beams, the dimensionless boundary conditions can be expressed by

$$y(0,t)=y(1,t)=0,\quad y''(0,t)=y''(1,t)=0 \quad (27)$$

3 Differential quadrature method

Eq.(26) is a complex partial differential equation, and the analytical expression of natural frequency cannot be obtained directly. In here, the method of DQ is employed to numerically solve the problem. This method requires to descretize the domain of the problem into N points. Then the derivatives at any point are approximated by a weighted linear summation of all the functional values along the descretized domain.

Consider the computational domain $0\leq x\leq 1$ of the beam. The number of sampling point is N in the x direction. Introduce N sampling point as [5]

$$\begin{aligned}&x_1=0,\ x_N=1,\\&x_i=\frac{1}{2}\left[1-\cos\left(\frac{i-1}{N-1}\pi\right)\right]\ (i=2,3,\cdots,N-1)\end{aligned} \quad (28)$$

The quadrature rules for the derivatives of a function at the sampling points yield

$$\begin{aligned}&y,_{xx}(x_i,t)=\sum_{j=1}^{N}A_{ij}^{(2)}y(x_j,t),\\&y,_{xxxx}(x_i,t)=\sum_{j=1}^{N}A_{ij}^{(4)}y(x_j,t)\end{aligned} \quad (29)$$

Substitution of Eq. (29) into Eq. (26) yields

$$\ddot{y}_i+\sum_{k=1}^{N}\left(k_f^2 A_{ik}^{(4)}+k_1 A_{ik}^{(2)}\right)y_k-k_2 y_i=0 \quad (30)$$

where

$$y_i(t)=y(x_i,t) \quad (31)$$

under simply supported boundary conditions

$$y_1=y_N=0,\ \sum_{k=1}^{N}A_{1k}^{(2)}y_k=\sum_{k=1}^{N}A_{Nk}^{(2)}y_k=0 \quad (32)$$

where the weight coefficient is defined by

$$A_{ik}^{(1)}=\begin{cases}\prod_{\mu=1,\mu\neq i}^{N}(x_i-x_\mu)\Big/\left[(x_i-x_k)\prod_{\mu=1,\mu\neq k}^{N}(x_k-x_\mu)\right]\\(i,k=1,2,\cdots,N,\ i\neq k)\\\sum_{\mu=1,\mu\neq i}^{N}\frac{1}{x_i-x_\mu}\qquad(i=1,2,\cdots,N,\ i=k)\end{cases} \quad (33)$$

and in case of r =2, 3, ..., N-1

$$A_{ik}^{(r)}=\begin{cases}r\left(A_{ii}^{(r-1)}A_{ik}^{(1)}-\frac{A_{ik}^{(r-1)}}{x_i-x_k}\right)\ (i,k=1,2,\cdots,N,\ i\neq k)\\-\sum_{\mu=1,\mu\neq i}^{N}A_{i\mu}^{(r)}\qquad(i=1,2,\cdots,N,\ i=k)\end{cases} \quad (34)$$

To overcome difficulties in the implementation of the boundary conditions, the modified weighting coefficient matrix are introduced as

$$[\tilde{A}^{(2)}]=\begin{bmatrix}0&0&\cdots&0\\A_{21}^{(2)}&A_{22}^{(2)}&\cdots&A_{2N}^{(2)}\\\vdots&\vdots&\ddots&\vdots\\A_{(N-1)1}^{(2)}&A_{(N-1)2}^{(2)}&\cdots&A_{(N-1)N}^{(2)}\\0&0&\cdots&0\end{bmatrix} \quad (35)$$

$$[\tilde{A}^{(r)}]=[A^{(1)}][\tilde{A}^{(r-1)}]\qquad(r=3,4,\cdots) \quad (36)$$

Substituting matrix (35) and (36) into Eq. (30), one obtain

$$\ddot{y}_i+\sum_{k=1}^{N}\left(k_f^2\tilde{A}_{ik}^{(4)}+k_1\tilde{A}_{ik}^{(2)}\right)y_k-k_2 y_i=0 \quad (37)$$

Eq. (37) leads to a generalized eigenvalue problem that can be numerically solved.

4 Numerical results

Consider a finite elastic prestressed concrete model beam, the parameters of the beam are listed in Table 1. Under pinned boundary conditions, from Eq. (37), the values of natural frequency can be found by numerical method.

Table 1 Properties of the beam

Item	Notation	Value
Young's modulus of concrete	E_c	31.5 GPa
Young's modulus of prestressed steel	E_t	195 GPa
Mass density	ρ	2.6×10^3 kg/m^3
Cross sectional area	A	8×10^4 mm^2
Second moment of area	I	2.67×10^8 mm^4
Eccentricity	e_0	30mm
Length	L	7 m

For the free motion of the model beam, the effects of parameters on the natural frequencies are respectively studied. In Table 2, the first four natural frequencies of the model beam with various concrete strength are given. For same concrete strength, with the growth of the orders, the natural frequencies increase greatly, but for the beam with higher concrete strength, the natural frequencies increase slightly. In Table 3, it shows that for same span length, with the growth of the orders, the natural frequencies increase significantly, but with the increase of span length, the natural frequencies decrease a little, and the natural frequency values of higher orders decrease more quickly.

Table 2 Natural frequencies with various concrete strength

	f_1	f_2	f_3	f_4
C20	19.381	74.476	166.271	294.780
C30	20.700	79.988	178.778	317.080
C35	21.085	81.602	182.441	323.612
C40	21.308	82.542	184.577	327.423

Table 3 Natural frequencies with various beam length

L	f_1	f_2	f_3	f_4
6.5	24.335	94.517	211.465	375.190
7.0	21.085	81.602	182.441	323.612
7.5	18.461	71.182	159.024	282.000
8.0	16.312	62.652	139.858	247.942

Table 4 Natural frequencies with various original prestressing force

N_{P0}	f_1	f_2	f_3	f_4
600	21.394	81.923	182.764	323.936
800	21.240	81.763	182.603	323.774
1000	21.085	81.602	182.441	323.612
1200	20.929	81.441	182.279	323.450

Table 5 Natural frequencies with various eccentricity

e_0	f_1	f_2	f_3	f_4
20	20.976	81.490	182.328	323.499
30	21.085	81.602	182.441	323.612
50	21.326	81.889	182.793	324.051
80	21.684	82.265	183.171	324.432

In Table 4 and Table 5, with the increase of the values of the original prestressing force, the natural frequencies decrease slightly, but with the growth of eccentricity, the natural frequencies increase a little.

5 Conclusions

This paper investigates natural frequencies of free transverse vibrations of a finite prestressed beams under simply supported boundary conditions, and the free vibration governing equations in related literatures are discussed and recorrected. As for the complex partial differential equation, the analytical expression of natural frequency cannot be obtained directly, the differential quadrature methods are applied to get the numerical results, and the effects of the physical parameters of the beam on the natural frequencies are respectively analyzed. The numerical results show that the natural frequency values increase with the growth of concrete strength and eccentricity of prestressed steels. But with the increase of the beam span length and values of original prestressing force, the natural frequency values decrease.

References

1. Ayaho M., Katsuji Tei. *Behavior of Prestressed Beam Strengthened with External Tendons*. J. Struct Eng **9**, ASCE(2000).
2. Menglin L., Tingting H. *Mode perturbation method for lateral vibration analysis of prestressed beams*. J. Prestress Technology **1**, (2010).
3. Xueyu X., Shousheng W. *Analysis and research of externally prestressed concrete beam vibration behavior*. J. Earthq Eng Eng Vib **2**, (2005).
4. Bert, C.W., Malik, M. *Differential quadrature method in computational mechanics: A review*. Appl Mech Rev., **49**, (1996).
5. Malekzadeh P., Karami G. *DQEM for free vibration analysis of Timoshenko beams on elastic foundations*. Comput Mech **31**, (2003).
6. Magrab E.B. *Natural Frequencies and Mode Shapes of Timoshenko beams with attachmens*. J. *Vib Control* **13**, 7, (2007).
7. Li P. *Analysis of natural frequency of externally prestressed concrete beams with asymmetric configuration*. J. Shanghai Normal University (Natural Sciences) **40**, 5, (2011).

https://doi.org/10.1051/e3sconf/202123703030

Experimental study on interface friction between geosynthetic clay liner (GCL) and geomembrane under different hydration conditions

Xuan Li [1,*] ,Yue Zhang [2]

[1]Hubei Business College, 430000 Wuhan, Hubei Province, China
[2]Hubei Business College, 430000 Wuhan, Hubei Province, China

Abstract. The interfacial friction properties of GCL and geomembrane under different hydration conditions are discussed. In addition, the ionic solubility of hydration solution, the size of hydration normal stress and test conditions will affect the shear strength of GCL. In this paper, the friction properties of GCL upper geotextile and geomembrane interface are discussed from the following aspects: ① hydration solutions with different ionic concentrations (deionized water, 0.05mol/l $CaCl_2$ solution and 1:1 ethanol solution by volume); ② different types of hydration solutions (0.05mol/l $CaCl_2$ solution and 1:1 ethanol solution by volume); ③ different hydration normal stress conditions (pressureless hydration condition, 5kPa hydration condition and 10kPa hydration condition).

1 Research background

The shear failure of GCL generally occurs in the following three parts: inside the GCL; between the top surface and the overlying material; and between the bottom surface and the underlying material.[1]Among them, the latter two belong to the interface shear strength failure, and the former is more damaged in practical engineering. The research structure shows that the internal failure of GCL caused by reinforcement will not affect the slope environment. However, when the bentonite in GCL interlayer is hydrated, the shear strength of the system will decrease rapidly. This will lead to the relative sliding between the surface layer and the bottom layer of GCL. [2] If GCL is laid on the slope, not only the relative sliding between materials is easy to occur, but also the hidden sliding surface can easily appear due to the poor shear strength at this time. Considering that the bentonite in GCL may be extruded, the friction coefficient of other synthetic materials will also decrease. Therefore, the study on the interfacial shear strength between GCL and other geosynthetics has practical significance and engineering value for the application of GCL in landfill. [3]

2 Preparation of test specimens

The test materials include GCL, geomembrane and hydration solution.

In the process of the experiment, GCL and geomembrane are the experimental objects of interface friction. In terms of hydration solution, the composition of landfill leachate is complex, including a variety of organic matter, inorganic matter and toxic and harmful components. With the exchange of ions, the concentration of pollutants in the filtrate has a strong fluctuation, whether in a short time or in a long term, and the process is extremely complex. In addition, the composition of leachate is also affected by the location and distribution of the area, different components of garbage and different landfill methods, and its composition also shows different changes.[4]Based on the laboratory test and the limitation of GCL material as sodium bentonite waterproof blanket, in order to analyse and compare the test results, the simulated leachate is deionized water, 0.05mol/l $CaCl_2$ solution and ethanol solution (the volume ratio of ethanol to water is 1:1). The reasons are as follows: ① Calcium ion and chloride ion are the main ions in landfill leachate. ② When landfill leachate interacts with sodium bentonite waterproof blanket, the ion exchange mainly shows that calcium ion replaces sodium ion. ③ Deionized water is selected for the test to provide a comparative basis for the shear strength analysis of GCL under the action of solution. [5]

3 Analysis of test results of interface friction between GCL and geomembrane under different hydration conditions

3.1 Analysis of test results of interface friction between hydrated GCL and geomembrane

It can be seen from Fig. 1 that under the same normal stress, the interfacial shear stress between GCL and geomembrane under pressureless hydration condition is larger than that between GCL and geomembrane under

* Corresponding author: lixuanfengguoer@126.com

other conditions, the interfacial shear stress between GCL and geomembrane under 5kpa hydration condition is the second, and that between GCL and geomembrane under 10kPa hydration condition is the smallest. The interfacial shear stress of GCL is very close to that of geomembrane at 5kPa and 10kPa. The interface peak strength of GCL and geomembrane under pressureless hydration state is the largest, followed by 5kPa normal stress hydration state, and the minimum interface peak strength of GCL and geomembrane under 10 kPa normal stress hydration state. [6]

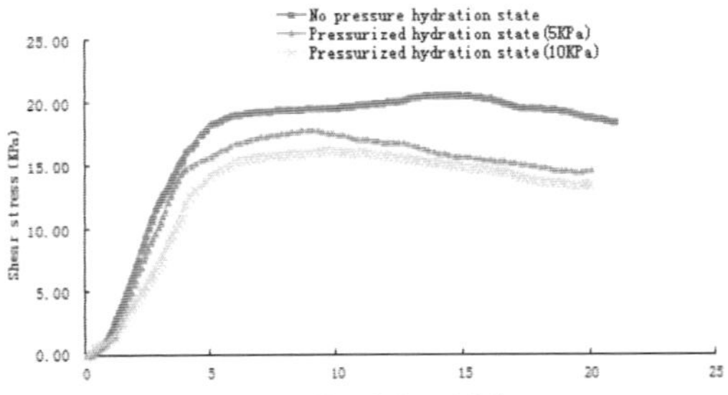

Fig. 1. Different hydration normal stress conditions of (pressureless hydration condition, 5kPa hydration condition and 10kPa hydration condition) GCL and geomembrane graph

Table 1 shows that under the same normal stress, the peak value of interfacial shear stress of GCL under pressureless hydration condition is higher than that of GCL under 5kPa hydration condition and 10kPa hydration condition. The peak value of interfacial shear stress between GCL and geomembrane under pressureless hydration condition is 18.37% higher than that under 5kPa normal stress hydration condition and 8.05% higher than that under 10kPa normal stress hydration condition. It can be seen that the interfacial shear stress between GCL and geomembrane decreases with the increase of hydration normal stress.

Table 1. The improvement rate of GCL in different hydration conditions

Normal stress	50kPa	100kPa	150kPa	200kPa
5kPa Hydration conditions	25.57%	25.40%	18.37%	25.44%
10kPa Hydration conditions	11.29%	13.19%	8.05%	16.64%

Under the condition of pressureless hydration, the cohesion between GCL and geomembrane is the largest, followed by 5kPa normal stress hydration, and the lowest under 10 kPa normal stress hydration condition; while under the three conditions, the number of internal friction angles of GCL and geomembrane interface friction under pressureless hydration is the largest, and the internal friction angle under 5kPa normal stress hydration is several times, The number of internal friction angle is the smallest under the condition of 10kPa normal stress hydration. In comparison, the interface shear strength between GCL and geomembrane under pressureless hydration is the largest, followed by 5 kPa normal stress hydration, and the minimum under 10 kPa normal stress hydration.

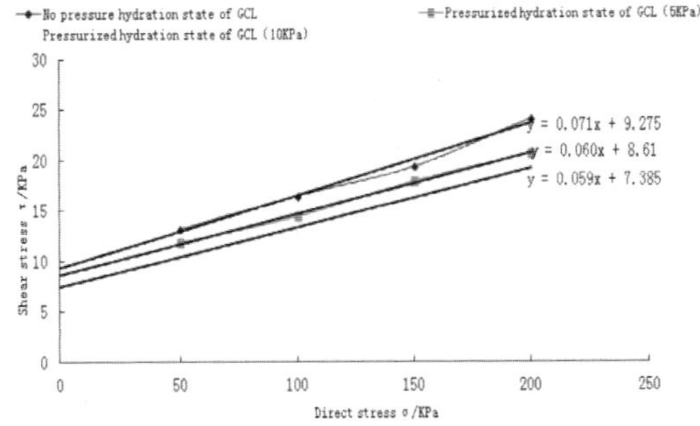

Fig. 2. GCL and geomembrane maximum shear stress and normal stress ($\tau_f - p$) relationship graph

Table 2. Test results of interface friction between different hydration solutions GCL and geomembrane

Geotechnical materials	Hydration solution	Cohesion	Friction angle
		c/kPa	φ (°)
GCL and geomembrane	No pressure Hydration conditions	9.275	4.11°
	5kPa Hydration conditions	8.61	3.43°
	10kPa Hydration conditions	7.385	3.38°

3.2 Analysis of GCL test results of 0.05mol/l $CaCl_2$ solution hydration

It can be seen from Figure 3 that under the same normal stress, the interface stress curve between GCL and geomembrane under pressureless hydration state is higher than that under other conditions, the interfacial shear stress value between GCL and geomembrane under 5 kPa hydration condition is the second, and that between GCL and geomembrane under 10 kPa hydration condition is the smallest. The interfacial shear stress of GCL is very close to that of geomembrane at 5 kPa and 10KPA. The interface peak strength between GCL and geomembrane is > 5kPa normal stress hydration state > 10 kPa normal stress hydration state under pressureless hydration state.

It is shown in Table 3 that under the same normal stress, the peak value of interfacial shear stress under pressureless hydration condition of GCL is higher than that of GCL under 5kPa hydration condition and 10kPa hydration condition. The peak value of interfacial shear stress between GCL and geomembrane under pressureless hydration is 0.84% higher than that under 5kPa normal stress hydration condition, and 14.96% higher than that under 10kPa normal stress hydration condition. It can be seen that the interfacial shear stress between GCL and geomembrane after hydration with 0.05mol/l $CaCl_2$ solution decreases with the increase of hydration normal stress.

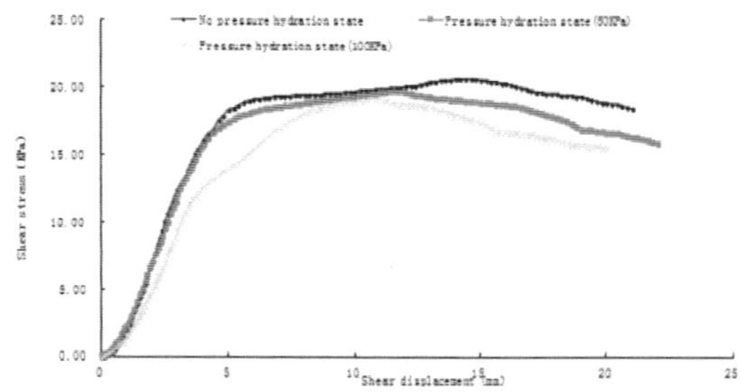

Fig. 3. Different hydration normal stress conditions of (pressureless hydration condition, 5kPa hydration condition and 10kPa hydration condition) GCL and geomembrane graph

Table 3. The improvement rate of GCL in different hydration conditions

Normal stress	50kPa	100kPa	150kPa	200kPa
5kPa Hydration conditions	0.84%	6.21%	2.93%	4.22%
10kPa Hydration conditions	20.18%	26.18%	14.96%	19.35%

Under the condition of pressureless hydration, the cohesion between GCL and geomembrane is the largest, followed by 5kPa normal stress hydration, and the lowest under 10kPa normal stress hydration; and under the three conditions, the number of internal friction angles of GCL and geomembrane under the condition of pressureless hydration is the largest, the internal friction angle of 5kPa normal stress hydration is several times, and that of 10kPa normal stress hydration is the highest The number of internal friction angles is the smallest. In comparison, the interface shear strength between GCL and geomembrane under pressureless hydration is the largest, followed by 5kPa normal stress hydration, and the minimum under 10kPa normal stress hydration.

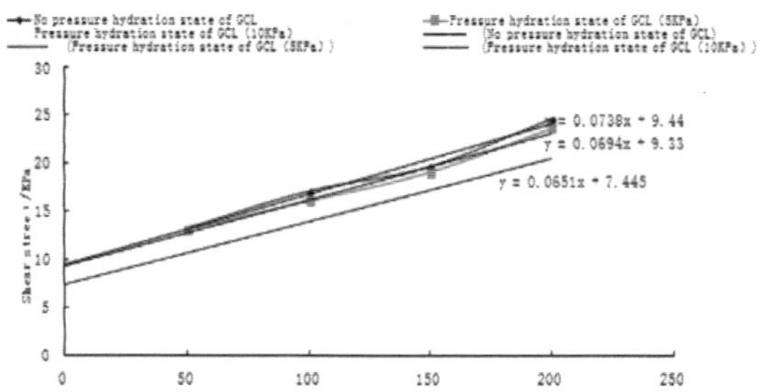

Fig. 4. GCL and geomembrane maximum shear stress and normal stress ($\tau_f - p$) relationship graph

Table 4. Test results of interface friction between different hydration solutions GCL and geomembrane

Geotechnical materials	Hydration solution	Cohesion	Friction angle
		c/kPa	φ (°)
GCL and geomembrane	No pressure Hydration conditions	9.44	4.18°
	5kPa Hydration conditions	9.33	3.94°
	10kPa Hydration conditions	7.445	3.72°

3.3 Analysis of hydration test results of ethanol (volume ratio 1:1)

It can be seen from Fig. 5 that under the same normal stress, the interface shear stress curve between GCL and geomembrane under pressureless hydration state and between GCL and geomembrane under 5kPa hydration condition are higher than that between GCL and geomembrane under 10kPa hydration condition. The interfacial shear stress curves of GCL and geomembrane under pressureless hydration state and GCL with geomembrane under 5kPa hydration condition are intertwined together, without obvious regularity. However, it can be seen from the figure that the interface peak strength between GCL and geomembrane is the largest under 5kPa hydration condition, followed by GCL and geomembrane under pressureless hydration condition, and the minimum interface peak strength between GCL and geomembrane under 10kPa normal stress.

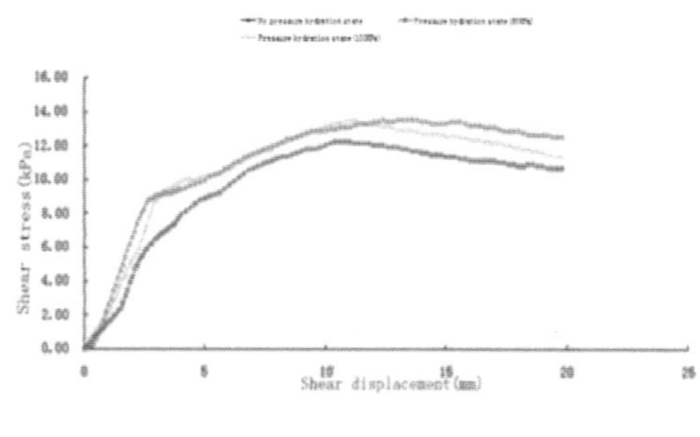

Fig. 5. Different hydration normal stress conditions of (pressureless hydration condition, 5kPa hydration condition and 10kPa hydration condition) GCL and geomembrane graph

It is shown in Table 5 that under the same normal stress, the peak value of interfacial shear stress under pressureless hydration condition of GCL is higher than that of GCL under 5 kPa hydration condition and 10 kPa hydration condition. The peak shear stress ratio of GCL to clay under pressureless hydration condition is 3.24% less than that under 5kPa normal stress hydration condition, and is at least 8.04% higher than that under 10 kPa normal stress hydration condition. It can be seen that the interfacial shear stress between GCL and clay decreases with the increase of hydration normal stress.

Table 5. The improvement rate of GCL in different hydration conditions

Normal stress	50kPa	100kPa	150kPa	200kPa
5kPa Hydration conditions	-3.24%	10.90%	10.22%	0.45%
10kPa Hydration conditions	8.04%	8.04%	19.78%	10.04%

Under the condition of pressureless hydration, the cohesion between GCL and geomembrane is the largest followed by 5kPa normal stress hydration and 10kPa normal stress hydration. Under the three conditions, the number of internal friction angles of GCL and geomembrane interface under pressureless hydration is the largest, the internal friction angle under 5kPa normal stress hydration is several times, and the internal friction angle number under 10kPa normal stress hydration condition is the minimum. In comparison, the interface shear strength between GCL and geomembrane under pressureless hydration is the largest, followed by 5 kPa normal stress hydration, and the minimum under 10 kPa normal stress hydration.

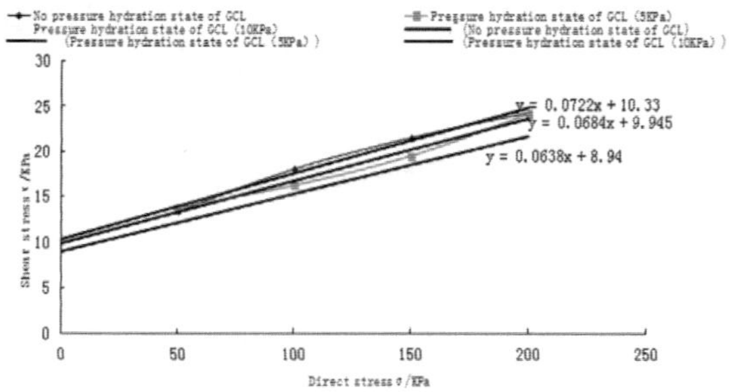

Fig. 6. GCL and geomembrane maximum shear stress and normal stress ($\tau_f - p$) relationship graph

Table 6. Test results of interface friction between different hydration solutions GCL and geomembrane

Geotechnical materials	Hydration solution	Cohesion	Friction angle
		c/kPa	φ (°)
GCL and geomembrane	No pressure Hydration conditions	10.33	4.12°
	5kPa Hydration conditions	9.954	3.89°
	10kPa Hydration conditions	8.94	3.60°

4 Summary

The conclusions obtained through the above-mentioned experimental analysis are as follows:

1. the shear strength of GCL and clay is the largest under pressureless hydration condition; the interfacial shear stress between GCL and geomembrane decreases with the increase of hydration normal stress after deionized hydration, and the maximum shear strength of GCL and geomembrane under pressureless hydration is 5kPa the shear strength under normal stress hydration is the second, and the shear strength under 10kPa normal stress hydration is the lowest.

2. The interface shear strength between GCL and geomembrane is the largest under pressureless hydration, followed by 5kPa normal stress hydration, and the lowest under 10kPa normal stress hydration.

3. the shear strength of GCL and geomembrane under pressureless hydration is the largest, followed by 5kPa normal stress hydration, and the lowest under 10kPa normal stress hydration.

4. Under the condition of pressureless hydration, 5kpa normal stress hydration and 10kPa normal stress hydration, the shear resistance of interface friction between GCL and geomembrane is ethanol > 0.05mol/l $CaCl_2$ solution > deionized water.

References

1. X. Li, Laboratory investigation on the interface of GCL and Geomembrane. (2013)
2. C. Xu, H. Xing. Geosynthetics. (2010)
3. L. Yu. Analysis of interface friction characteristics between geosynthetics and sand. (2012)
4. X. Kong, Experimental study on the anti-seepage performance of GCL in liner of waste landfill. (2011)
5. J.Zhou, Environmental Geotechnical Engineering. (2009)
6. X. Li, Y. Zhang. Study on the effect of hydration on the shear strength of the interface between geosynthetic clay liner (GCL) and geomembrane. (2020)

https://doi.org/10.1051/e3sconf/202123703031

Analysis on the unit space design of primary and secondary school campus building teaching area

Jitao Tang[1,*],Yuan Zhang[2]

[1]Colledge of Civil Engineering,Nanjing Technology,Nanjing 210009,China
[2]Faculy of Architecture and Civil Engineer,Huaiyin Institute of Technology,Huaian 223001,China

Abstract. This paper takes the teaching area unit space of primary and secondary schools as the main research object, analyzes, discusses and studies the reasonable use size of teaching unit space, commonly used space combination mode and internal detail design of education space, analyzes the actual use of various space units, and discusses the advantages and disadvantages of different design methods. Finally, from the perspective of teachers and students, combined with the analysis of specific project cases, this paper puts forward the optimization design strategy of school teaching unit space in line with the use requirements and architectural aesthetics.

1 Main idea of the article

Campus building is an indispensable architectural form in the city. With the continuous increase of population in recent years and the opening of the two child policy, the construction scale of primary and secondary schools is also increasing. At the same time, the development of society and the renewal of education concept require the all-round development of students, which makes the architectural space of primary and secondary schools more and more abundant. As the main functional space of teaching area and the main place for students' study and teachers' office, more attention should be paid to the design of unit space in teaching area. Therefore, it is right and wrong to optimize the design of unit space in teaching area for the space design of the whole campus Often important. This paper mainly discusses the use area, combination mode and detail design of unit space in teaching area.

2 Use area of unit space

The teaching area can be divided into several types of unit space, including ordinary classroom, professional classroom, teacher's office and toilet. Among them, ordinary classroom and professional classroom are students' concentrated activity space, and teacher's office is teacher's activity space. The toilet is a service space.

2.1 Ordinary classroom

According to 5.2 of "code for design of primary and secondary schools (GB 50099-2011)" 5.2, combined with the number of students in each class, we can get the minimum classroom area: 67.5 m2 for primary school classrooms, 75 m2 for middle school classrooms, i.e. 1.5 m2 per student. According to the requirements of the specification, it can be seen that the column network of ordinary classroom is at least 7900 * 8400 (primary school) and 7900 * 9300 (middle school). The continuous enrichment and optimization of educational resources has brought about changes in teaching space, which makes the general classroom area and the per capita use area of students increase, and the comfort of teaching space is improved.

2.2 Professional classroom

Nowadays, the requirements for students' professional skills are constantly improving, and the types of professional classrooms are also increasing. Such as calligraphy classroom, piano room, woodworking, metalworking classroom, 3D printing classroom, vr virtual room, etc. Traditional professional classrooms mainly include labor skills, art, music, dance classroom, computer network classroom, multimedia classroom, science laboratory, etc. The new professional classroom can be designed according to the size of the traditional classroom or combined with the specific needs of the school.

According to the provisions of "code for design of primary and secondary schools (GB 50099-2011)" and "standards of educational technology and equipment for primary and secondary schools in Jiangsu Province", the usable area of traditional professional classrooms is required, while the area of different professional classrooms is also different. According to the number of students and the average requirement of students, the area of professional teaching rooms is mostly 96120110, In order to modularize the classroom units, we usually unify

* Corresponding author: 75525400@qq.com

the classroom area of 96 m2 and 120 m2 in design. When the area requirement is met, the column network is usually set at 9600 * 10000 and 9600 * 8400.

2.3 Teachers' Office

The teacher's office here refers to the teacher's office and the school administrative office. Teachers' offices are mostly scattered around the teaching area, which is convenient to serve all kinds of classes, and the use area varies with different schools. The teacher's office can be divided into grade group office and teaching and research group office. The teacher's office of grade group is located near the ordinary classroom of each grade, and the office of each course teaching and research group is generally located near its professional classroom, which is combined with the special classroom. The teaching and research groups of other courses can be set up in the vicinity of administrative office or library. According to the regulations, the per capita use area of teachers should not be less than 5 square meters. The administrative office of a school generally refers to the office that undertakes the daily school affairs and the use of educational administration personnel, such as educational administration and administrative office, general affairs office, conference room, general warehouse, duty room, etc. Among them, the school administration office should be set up in the convenient transportation place because it undertakes the daily contact between teachers and students; the educational administration office undertakes the daily teaching affairs, which can be combined with the teacher's office or set in a relatively close position; the general affairs office is responsible for logistics tasks, and its setting can be combined with warehouse, maintenance room, canteen, etc., and warehouse and other rooms are the users It is often set at the school gate, so the general affairs office and other rooms can also be set up at the entrance and exit of the school; the on duty room is generally used for school teachers and administrative staff, which can be used as supporting facilities for teaching teachers' office and administrative office.

2.4 Toilet

As an essential living and service room, toilet should be set in each floor of the teaching building. If conditions permit, toilet for male and female students and toilet for male and female teachers should be set. Men's and women's toilets can also be set up on the partition according to the actual conditions. The convenience of the toilet should be ensured while the sanitary environment quality of the whole teaching area should not be affected. The toilet should be provided with ventilation conditions or installed with ventilation exhaust and sewage pipes.

3 The combination mode of unit space

The above four types of units are distributed in the teaching area of the campus, providing teaching conditions for the students of the whole school. Students and teachers can carry out corresponding teaching and learning activities. Due to the requirements and regulations of our country's specifications for the design of classroom windows, as well as the limited requirements for the coordination of various functions, various unit spaces are combined in different modes.

3.1 The combination of ordinary classroom and professional classroom

According to the previous chapter, we can see that different classrooms have different use areas and functions. There are several ways to arrange classrooms with different areas and functions:

3.1.1 Block layout

It is a common mode that ordinary classrooms and professional classrooms are set up separately. In this way, it is not only convenient for the general classroom teaching building and the professional classroom teaching building to be separated from each other, but also to have a certain connection with each other. At the same time, the column network can also be set up in groups. But the first requirement of such a layout is that the land area is abundant, which is not suitable for some campus with limited land use. Moreover, this kind of layout increases the number of teachers' offices, which leads to unnecessary waste of space. At the same time, it may lead to long traffic flow between ordinary classrooms and professional classrooms, which is not convenient in actual use.

3.1.2 Middle corridor layout

In some campuses with limited land use, teaching buildings are often unable to design too many buildings because of the sunshine requirements and 25m spacing requirements of teaching buildings. In the design, we often combine the ordinary classroom and the professional classroom in a teaching building. The common design method is the middle corridor layout, which is the ordinary classroom in the South and the professional classroom in the north. While ensuring the daylighting of ordinary classrooms, professional classrooms can also be arranged accordingly. The traffic flow line between the two is relatively short, and the two types of classrooms are more closely connected, and the use is more convenient and convenient. Ease the land shortage and other issues

The biggest advantage of this mode is to save land to the greatest extent, but there are also some disadvantages. The layout of middle corridor leads to poor lighting in the inner corridor, which will bring depression to the teaching space. The middle corridor layout also makes the plane layout too compact and provides less public space for users. In addition, some professional classrooms need to have good lighting and sunshine requirements, which can not be met in such layout mode, and need to be considered separately. Due to the different size requirements of professional classrooms and ordinary teachers, this kind of layout may bring difficulties in the layout of plane column network.

3.1.3 Zigzag layout

In the current campus design, there are many teaching buildings will adopt the back shape layout. Generally speaking, the number of ordinary classrooms is far more than that of professional classrooms. Therefore, in the design, the general classrooms will be set in the south of the Huixing teaching building, and the professional classrooms will be arranged in the east-west direction. The corridor will be widened, which can not only increase the public activity space of the teaching building, but also optimize the layout of the classroom space and reduce the cost of land use. However, with the increasing attention of the whole society to education, the comprehensive development of students is required, more professional courses are opened, and the quality and quantity requirements of professional classrooms are also higher and higher. This layout may not meet the increasing requirements of the number of professional classrooms.

3.1.4 Vertical layout

Ordinary classroom and professional classroom are set in the same teaching building, hierarchical and partition setting is also a common method. It should be noted that in this mode, due to the different area required by different classrooms, the occupied column span is not uniform, which will lead to some difficulties in facade design.

In order to avoid such problems, we can put the professional classroom on the low floor, and the ordinary classroom on the second floor and above. The professional classroom still adopts a column span as a class, but the depth can be appropriately increased to meet the use needs. In addition, many professional classrooms for landing projects do not meet the requirements for classrooms in the specification, so they are often considered as multi-functional rooms to avoid many restrictions on facade and window opening in the design.

3.1.5 Focus on the choice of combination form

Each of the above-mentioned space combination forms has its advantages and disadvantages, which requires the designer to select the appropriate space combination form according to the actual land use scale and use requirements, and in some specific cases, these space layout forms can also be combined. When choosing the space combination form, it is necessary to meet the teaching requirements to the greatest extent, with students as the first consideration object.

3.2 Combination of teachers' offices

There are not too many restrictions on the combination of classroom and teacher's office. The teacher's office is set up as a supporting room for teachers. In the existing design concept, the priority of the teacher's office is lower than that of the classroom, so the teacher's office will be set up in some locations with relatively weak sunshine conditions in the teaching building, or the teacher's office will be set in the location with weak East-West lighting.

Of course, the setting of the teacher's office should also be designed according to the school's teaching mode. In the actual case, part of the school's teaching mode serves two classes for one teacher's office, so the teacher's office is located between the ordinary classrooms and is arranged at intervals; or the teacher's office is located at the end of the corridor, which is convenient for teachers to directly observe the students' going on and off class.

4 Detail design of unit space

4.1 Interior design

The interior of the classroom unit includes the platform, seats, student cabinets, lockers, etc., among which the podium and seats are common furniture in the classroom design. When setting these two kinds of furniture, attention should be paid to the relationship between the platform and the seats in the classroom, so as to ensure that each student can see the platform. Student cabinets are often designed together with classroom units, as shown in figure 1. In the design, attention should be paid to the relationship between the height of cabinets and the height of students. Generally speaking, the scale of three-tier cabinets is more appropriate.

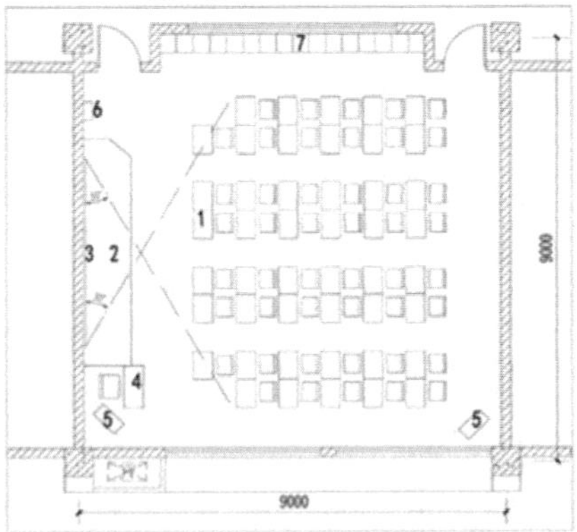

Fig. 1. enlarged picture of classroom
1. Desk (800mm in height) 2. Podium (height: 200mm) 3 finished blackboard (900mm from the bottom edge to the platform, 2200mm in height) 4. Seats for teachers; 5. Cabinet type air conditioner; 6. Broadcasting stereo; 7. Storage cabinet (400x350mm)

In addition, the wall skirt of the classroom unit should also be consistent with the interior design. The main function of wainscot is to protect the wall, and the height is usually 900-1100mm. In the design, we should consider the color and height, not only to meet the functional requirements, but also to make the teaching building corridor have a good visual effect.

4.2 Window design of facade

There are many technical regulations restricting the opening of windows on the exterior facade of classroom units. Among them, article 5.1.8 in the specification has an impact on the facade: the length of the end wall of the front side window of each classroom should not be less than

1.00m. The width of the wall between windows shall not be greater than 1.20m. In the design, this article mainly affects the distribution of ordinary classrooms and professional classrooms in different floors of the same teaching building. In the design, it is necessary to pay attention to this kind of classroom in advance and deal with it on the facade, such as controlling the window wall between windows, avoiding the end wall of 1m classroom window in staggered floor design.

4.3 Air conditioning design

Air conditioning design is a problem that needs to be considered in every classroom. Generally, there are two treatment modes: one is to use the window end wall and the window wall to design and hide the external air conditioner. However, due to the size of the external air conditioner often needs to be larger than 1300mm, it needs to be considered uniformly; the other is to design the position of the windowsill under the window, which has the advantage of having enough length to hide the external unit, but because it is in the classroom To prevent the appearance of bay window, it is necessary to retract the window. It should be noted that different regions have different calculation methods for sunshine. Therefore, when using this method, whether there is self shielding should be considered and adjusted in combination with local regulations.

5 Summary

Education has always been an important social issue. In the construction industry, the discussion and thinking of educational buildings have never stopped. In recent years, China's attention to education has been increasing. As the main facilities for educational activities, schools are an important part of the education system. With the improvement of social concern, the responsibility of the school is increasing, and the requirements for school space are also higher and higher. Therefore, in the campus design, we should constantly adjust the thinking, explore and optimize the design of teaching unit space under the premise of meeting the norms, and create a relatively rich teaching space from the perspective of the educatee, so as to adapt to the social development Exhibition needs.

This paper analyzes and discusses the use area, layout mode and detail design of unit space in the teaching area. Based on the existing school space design, this paper analyzes its advantages and disadvantages, and puts forward a set of educational facilities space optimization design concept, which has certain design features and is easy to achieve from the perspective of the educated. Most of the optimal design schemes are combined with the engineering application.

References

1. Code for design of primary and secondary schools (GB 50099-2011). Beijing: China Construction Industry Press,(2010-12)
2. Standard of educational technology and equipment for primary and secondary schools in Jiangsu Province (12th Five Year Plan). Department of education of Jiangsu Province.(July 2011)
3. Qian Xiaofei. Designing educational space: the formation of American modern school architectural concept . Comparative education research, (2020,42 (11): 45-52)
4. Shao Lu. Reflection on teaching reform of "office space design" course in Colleges and universities. Road to success,(2020 (28): 20-21)
5. Che Zhiyuan, Osama. Space design of teaching buildings in Xiamen primary and secondary schools under the new concept of Education [J]. Urban architecture, （2020,17 (22): 151-153）
6. Xia Bing. Review and Prospect of innovative campus space planning and design [J]. Urban architecture, （2020,17 (16): 70-72）
7. Tian Yuan, Jia Xiaohu. Thinking on the design strategy of high plot ratio primary and secondary schools [J]. Architecture and culture, 2020 (03): 181-182

Research on Comprehensive Evaluation Methods of Existing Public Buildings Based on Diamond Evaluation Method

Gao Yubo [1,*], Mao Mingwei [1,a]

[1] Department of Architecture, Taiyuan University of Technology, Wan Berlin, Taiyuan, Shanxi, China

Abstract. Based on the analysis of characteristics of existing public buildings, a comprehensive method based on the diamond evaluation is established, which is integrated by the physical performance and the user's physical experiences. The method combines objective and subjective evaluations. It can divide the comprehensive value of existing public buildings in four aspects: function, economy, satisfaction, and value. Then, it explores the process and technical route of the existing public buildings using the analytic hierarchy of comprehensive evaluation. The article intends to use this method to conduct an comprehensive evaluation of Taiyuan Workers' Cultural Palace.

1 Introduction

There is a huge stock of existing public buildings in our country. Many existing public buildings cannot fully meet the needs of production and life at this stage, due to the constraints of their construction years and conditions. To transform and reuse the existing public buildings, scientific comprehensive value evaluation is an important link. Although a large number of evaluation studies are used in existing public buildings, none of methods gains a universal recognition. In the process of practical operation, the evaluation of existing public buildings often apply some evaluation standards of historical preservation buildings. Facing the existing public buildings with abundant types, large numbers, and different preservation states, how to construct a scientific comprehensive evaluation method for comprehensive evaluation of existing public buildings is a problem. There is a great significance to establish an index system and evaluation model for existing public buildings. Using a method to provide some basis, it can help the evaluation and reconstruction of existing public buildings.

2 Connotation of comprehensive evaluation of existing public buildings

Existing public buildings refer to public buildings that have been completed for a certain period of time. Compared to the concept of new buildings, it may exist generally 20 years or more. The forms of reuse the existing public buildings are mainly divided into three types: restoration, protection and renew. Restoration is to restore the inherent appearance with a specific historical period. Protection is mainly to protect the material foundation of the existing public building body while continuing its original basic style. Renew refers to the reinforcement and transformation of the buildings. The basis of building reuse is to evaluate existing public buildings and obtain their comprehensive status information. This research is based on our country's actual national conditions and practical requirements, which cannot be divorced from actual empty talk theory. We strive to achieve scientific and reasonable evaluation conclusions by objective and subjective evaluation perspectives. The index system should be comprehensive and well-defined to ensure the applicability of the evaluation system.

3 Construction of a comprehensive evaluation system for existing public buildings

3.1 Evaluation mechanism of the diamond law

The diamond method[1] was proposed by William Pena[2], who created a practical and feasible self-evaluation method in the planning stage and applied it to practice. Pena believed that evaluation should be measured from four aspects: function, form, economy and time. In these four aspects, score the quality of each aspect and set it from 1 point of "complete failure" to 10 points of "perfect". As shown in Figure 1, the quadrilateral is composed of function, form, economy, and time. The area of the quadrilateral is determined by the following formula: area = 0.5 (function + time) (form + economy). The area of the quadrilateral represents the evaluation result (Fig.1). In this way, the indicators were quantified and the evaluation results were obtained[3]. The diamond evaluation method is simple and clear, easy to operate, and avoids the overall problem of being too trivial and paying attention to details. On the other hand, the person who conducts building evaluation is often an architect. Too much complexity may cause the architect to be out of control.There is no

* Corresponding author: 491372453@qq.com
a maomingwei@126.com

https://doi.org/10.1051/e3sconf/202123703032

feasibility in actual project practice.

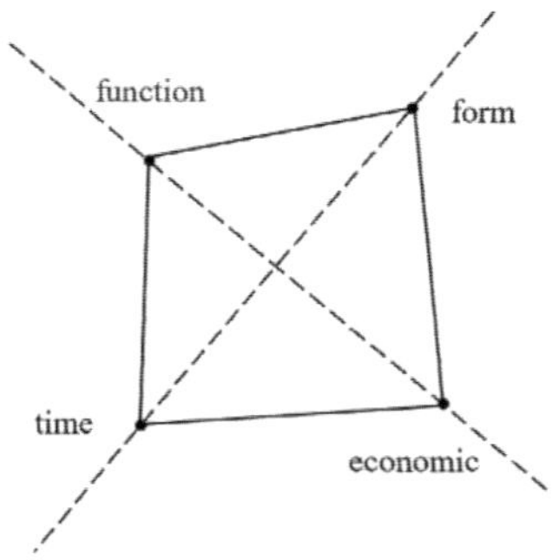

Fig. 1. Diamond Evaluation Method

3.2 Evaluation index system for existing public buildings

The four aspects(form, function, time, and economy) are included in the diamond evaluation method, which cannot cover the characteristics of existing public buildings. The diamond evaluation method is an evaluation method used in the construction planning stage. Based on the content of the Diamond Evaluation Method, and combined with the characteristics of existing public buildings, the author has constructed a comprehensive evaluation index system for existing public buildings. The index system is divided into three levels of indicators. The overall target level is the evaluation target "the comprehensive evaluation index system for existing public buildings. The first level indicators include 4 items, namely function, economy, satisfaction, and value, among them, function and economy are more important. The objective evaluation of building physical performance category, value and satisfaction are mainly the subjective evaluation of the user's physical experience, there are 12 secondary indicators. The specific indicator system is shown in Figure 2.

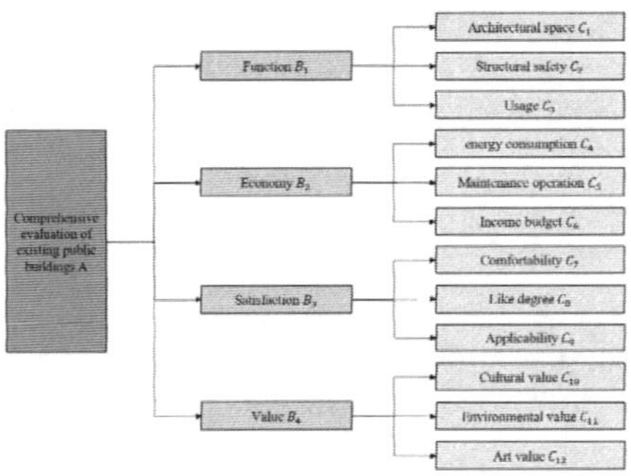

Fig. 2. Index system for comprehensive evaluation of existing public buildings

After multiplying the score of the second-level index multiply its corresponding weight, the score of the four first-level indexes can be obtained respectively. According to the calculation method of the diamond evaluation method, the quadrilateral area are enclosed by the corresponding score values of the 4 first-level indicators. Because the four first-level indicators take the full score of 10, the highest value of the quadrilateral area is 200, and converting it to a percentage system can make the result more intuitive and easier to understand. It can be obtained that the final score of comprehensive evaluation of existing public buildings is:

$$A=\frac{1}{4}(B_1+B_2)(B_3+B_4) \quad (1)$$

4 Validated research on comprehensive evaluation of existing public buildings

4.1 Overview of evaluation objects

Taiyuan Workers' Cultural Palace (Figure 3), which is also called Nangong, covers an area of 60,000 square meters, and with a building area of 30,000 square meters. It is the largest cultural activity center for employees in Shanxi Province. The Nangong was completed on February 15, 1958. It was designed by the Shanxi Provincial Architectural Engineering Design Institute. It is a cultural and leisure place for the city's employees. It was also a larger public civil building in Taiyuan at that time. The whole building has one auditorium (theater), two lecture halls, two production technology research rooms, two east and west exhibition halls, and two east and west corridors, which can accommodate more than 6000 employees for cultural and entertainment activities at the same time. The building is a Soviet-like building with three stories high in the middle. The corridors on both sides are connected to two-story small buildings. The whole building is like an airplane, which is eagered to fly. Nangong is the center of Taiyuan's cultural, sports, and entertainment activities. The colorful square cultural activities and large-scale theatrical performances has greatly activated the leisure life of the majority of employees and citizens.

Front elevation

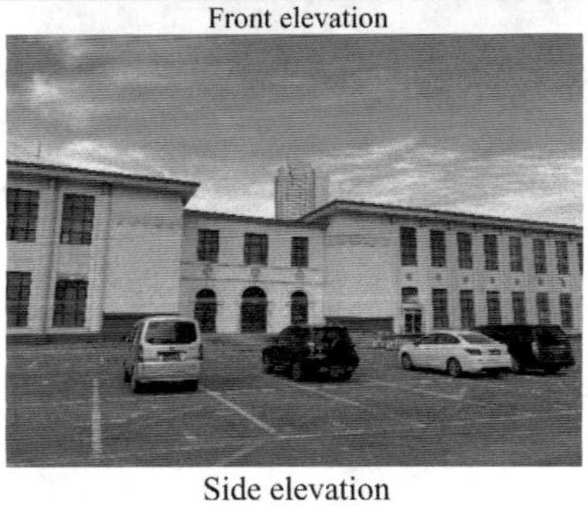

Side elevation

https://doi.org/10.1051/e3sconf/202123703032

Lobby
Fig. 3 Taiyuan Workers' Cultural Palace

4.2 Index score

According to the characteristics and circumstances of the case, the Delphi method[4] was used to organize experts to score each building index, and the quantitative assignment of each index was between 0 and 10 (Table 1).

Table 1. Setting Word's margins.

Overall target	First-level index	Second-level index	Sco re	Weight
Comprehen sive evaluation of existing public buildings A	Function B_1	Architectural space C_1	6	0.2500
		Structural safety C_2	5	0.5000
		Usage C_3	5	0.2500
	Economy B_2	energy consumption C_4	6	0.4000
		Maintenance operation C_5	7	0.2000
		Income budget C_6	8	0.4000
	Satisfaction B_3	Comfortability C_7	7	0.3333
		Like degree C_8	7	0.3333
		Applicability C_9	8	0.3333
	Value B_4	Cultural value C_{10}	9	0.3333
		Environmental value C_{11}	8	0.3333
		Art value C_{12}	10	0.3333

It can be seen from Table 1 that the index score vectors R corresponding to the first-level indexes are:

Function index score vector: $R_1=\begin{bmatrix}6\\5\\5\end{bmatrix}$

Economic indicator scoring vector: $R_2=\begin{bmatrix}6\\5\\5\end{bmatrix}$

Satisfaction index score vector: $R_3=\begin{bmatrix}6\\5\\5\end{bmatrix}$

Value index score vector: $R_4=\begin{bmatrix}6\\5\\5\end{bmatrix}$

4.3 Weight setting

According to the Delphi method and analytic hierarchy process, experts are organized to judge the importance of each building index. The corresponding weights of the secondary indexes are shown in Table 1. The weight vector W corresponding to the first-level index is:

Function index weight vector:
$W_1=[0.4000\quad 0.2000\quad 0.4000]$
Economic indicator weight vector:
$W_2=[0.4000\quad 0.2000\quad 0.4000]$
Satisfaction index weight vector:
$W_3=[0.3333\quad 0.3333\quad 0.3333]$
Value index weight vector:
$W_4=[0.3333\quad 0.3333\quad 0.3333]$

4.4 Weight setting

Multiply the weight vector W correspond to the first-level index by the score vector R to obtain the evaluation result of the first-level index, as follows:

Function index score:
$$B_1=W_1\times R_1=5.25 \tag{2}$$
Economic indicator score:
$$B_2=W_2\times R_2=7 \tag{3}$$
Satisfaction index score:
$$B_3=W_3\times R_3=7.3 \tag{4}$$
Value index weight vector:
$$B_4=W_4\times R_4=9 \tag{5}$$

According to the rules of the index system constructed above, the comprehensive evaluation score of Taiyuan Workers' Cultural Palace is equal to the quadrilateral area enclosed by the four values of function B_1, economy B_2, satisfaction B_3, and value B_4:
$$A=\frac{1}{4}(B_1+B_2)(B_3+B_4)=50 \tag{6}$$

According to the full score of 100, the comprehensive evaluation of Taiyuan Workers' Cultural Palace is good. Specifically, the function of the building is general, because the original space inside the house is seriously damaged.There are major safety hazards due to leaks in multiple roofs.The economy of the building is good, and its scale is large, and the storey height is high. The space is flexibly arranged, and its reconstruction and utilization have considerable economic benefits.The satisfaction of the building is good, and its social identity in the hearts of the city residents is high. It is a typical cultural landmark building.The value of the building is also excellent.It not only has distinctive features in appearance and style, but also contributes a lot to its rich historical background information and cultural symbolic alignment.

5 Summary

The comprehensive evaluation index of existing public buildings is based on the summary of domestic and foreign related evaluation systems.The diamond evaluation method is the core concept, which combined the actual characteristics of existing public buildings in my country. The evaluation indexes are divided into four aspects : function, economy, satisfaction, and value. This division can take into account the basic logic and content of the diamond evaluation method, and it also integrate the objective evaluation of the physical properties of the building and the subjective evaluation of the user's experience.The existing comprehensive evaluation index system for public buildings has a comprehensive characteristics.

In terms of research methods, the Delphi method and analytic hierarchy process, which is widely used in the industry,established a quantitative mechanism for evaluation indicators[5].A judgment matrix is constructed to determine the corresponding weights of the indicators ,and to increase the credibility and scientific of the evaluation. Through the evaluation of existing public buildings, we can have a more comprehensive grasp of the current advantages and disadvantages of the buildings, so as to assist in determining the strategy and focus of the protection and reuse of the building.

Acknowledgment

This article is one of the phased achievements of the key research and development project in Shanxi Province, China, "Shanxi Province Existing Building Performance Evaluation and Renovation Technology Research" (201903D32035).

References

1. William Pena, Steven Parshall. Problem Seeking: An Architectural Programming Primer[M]. John Wiley & Sons Inc. 2001:209
2. Mallows Edward teaching a technology[M]. Johannesburg: Witwatersrand University press,1965
3. Zhuang Weimin, Zhang Wei, Liang Sisi. Architectural planning and post-evaluation[M]. China Construction Industry Press. 2018
4. Jiang Nan. Technology and method of industrial heritage value evaluation based on adaptive reuse[J]. New Building.2016(3):4-9
5. Zhang Chengxing, Wang Xiaofeng, Li Yitong. Research on the comprehensive performance evaluation index system of existing public building renovation[J]. Building Energy Conservation. 2019 (10): 138-141

Research on prediction of concrete frost resistance based on random forest

Lei Wang[1,*], Tiemei Zeng[2], Yanyan Tao[3], Xianguo Wu[1], and Hongyu Chen[4]

[1] School of Civil Engineering & Mechanics, Huazhong University of Science and Technology, Wuhan, Hubei, 430074, China
[2] Wuhan Metro Group Co Ltd, Wuhan, Hubei, 430030, China
[3] School of Urban Construction, Wuchang Shouyi University, Wuhan, Hubei, 430074, China
[4] School of Civil and Environmental Engineering, Nangyang Technological University, Singapore, 6397898, Singapore

Abstract. Poor frost resistance of concrete will accelerate the deterioration of concrete structure and shorten the service life of concrete. This paper predicts the frost resistance of concrete and selects the initial index of factors that affect the frost resistance of concrete. The random forest algorithm is introduced to remove the unimportant indicators in the initial indicator system of the concrete mix ratio and determine the optimal indicator combination. The relative dynamic elastic modulus is used as the output index of random forest to predict the frost resistance of concrete. The proposed random forest model provides a reasonable and effective method for concrete frost resistance prediction.

1 Introduction

Compared with other building materials, concrete is economical, easy to manufacture on site, and has good durability. It is one of the most widely used materials in construction[1]. Under the conditions of freezing and thawing, the frost resistance of concrete is a consideration that cannot be ignored in engineering practice. Improving the frost resistance of concrete has very important engineering value. In order to ensure the safe service life of buildings, accurate and rapid prediction of concrete frost resistance is of great significance to actual projects.

At present, many domestic and foreign scholars are conducting research on concrete prediction. Amini K proposed statistical univariate and multivariate regression models to predict concrete durability[2]. Lin et al. used the relative dynamic elastic modulus to calculate the performance degradation index and showed that the relative dynamic elastic modulus has a linear relationship with the number of freeze-thaw cycles[3]. Using formula fitting to perform regression prediction, the prediction process is relatively simple, but the biggest flaw is the poor fitting accuracy. Ji established a concrete strength and slump prediction model based on artificial neural networks (ANNs)[4]. Yan et al. used regression technology to introduce a precision-insensitive loss function, and studied the application of SVM in the relative dynamic elastic modulus of concrete[5]. Behforouz B et al. used ANN prediction model to predict the durability of concrete such as frost resistance[6]. The above traditional machine learning algorithms are easy to deviate from the global optimal situation, making the prediction results inaccurate.

Therefore, aiming at the deficiencies of the above methods, this paper proposes a random forest frost resistance prediction model based on the ratio of raw materials. This paper establishes initial indicators of factors affecting concrete frost resistance, introduce random forest algorithm (RF) to remove unimportant indicators in the initial indicator system of concrete mix ratio, and determine the optimal factor indicators. The relative dynamic elastic modulus is used as the output index of the random forest to predict the frost resistance.

2 Preliminaries

Random Forest (Random Forest) model is based on the theoretical basis of Classification Tree[7]. Random forest adopts Bootstrap re-sampling method to establish a decision tree model for each sample set. When modeling, each decision tree randomly selects features to split the attributes of internal nodes, thereby forming a random forest. The final output result is the result of the comprehensive decision tree. For the regression algorithm, the final prediction result is the mean value of the output results of all decision trees, and the regression results are expressed as[8]:

$$\bar{H}(x)=\frac{1}{k}\sum_{i=1}^{k}\left(h_i\left(x=Y\right)\right) \quad (1)$$

Among them, $\bar{H}(x)$ represents the combined random forest model, $h_i(x)$ represents the single decision tree model, Y represents the output variable (or called the target variable), and k is the number of decision trees in the random forest model.

One of the biggest advantages of the random forest prediction model is its good generalization ability, which can effectively prevent overfitting when predicting data.

* Corresponding author: 1063609358@qq.com

The generalization error of the model will not cause over-fitting as the number of decision trees increases, but will approach a finite upper bound[9]. This finite upper bound is:

$$E^* = \frac{\bar{\rho}\left(1-s^2\right)}{s^2} \quad (2)$$

Among them, ρ is the average correlation coefficient of the decision tree, and s is the average strength of the decision tree.

Constructing a random forest model can be used to rank the characteristics of indicators. Assuming that the random forest has N decision trees, use the corresponding out-of-bag data to calculate its out-of-bag error r1, and randomly calculate the out-of-bag error r2 after randomly changing the order of a feature in the out-of-bag data, then the importance of a certain feature I is:

$$I = \frac{\sum_{i=1}^{N}\left(r_1 - r_2\right)}{N} \quad (3)$$

After obtaining the importance degree of each feature, the features are sorted reasonably, and the feature of the least importance is sequentially removed from the feature set using the sequence backward method until the optimal number of features is reached, and feature selection is realized.

3 Equations and mathematics

3.1 Random forest parameter settings

About 4/5 of the data in the original data set is used as the training sample set of the RF model, and the rest contains 1/5 of the data in the original data set as the test sample set of the RF model. In the modeling process, two parameters are mainly involved: the number of regression trees ntree and the number of random features mtry. The value of ntree will affect the training degree and accuracy of the random forest model. Generally, the larger the value of ntree, the higher the accuracy of the model; mtry controls the degree of disturbance of the random forest model attributes. mtry is generally determined according to empirical formulas. The number of variables is P, by default mtry=P (classification model) or mtry= P/3 (regression model). After the two parameters are determined, the RF algorithm can be used to establish a random forest regression model for the training sample set and the test sample set.

3.2 Importance evaluation

Variable importance measure can evaluate the degree of influence of characteristic variables on dependent variables, so as to determine the key variables that need to be controlled. This paper calculates the importance of feature variables according to formula (3), that is, uses OBB error to evaluate the importance of feature variables. It evaluates the importance of feature variables by measuring the reduction in mean square residual error (%Inc MSE) and the reduction in model accuracy (Inc Node Purity) after random replacement[10].

3.3 Evaluation of prediction results

The prediction result of the model is obtained by taking the average of all decision trees according to formula (2). In order to evaluate the prediction accuracy of random forests, two evaluation parameters, root mean square error (RMSE)[11] and goodness of fit (R^2)[12] can be used. It can be used to describe the deviation between the predicted value and the measured value, and it can be used to describe the degree of correlation between the predicted value and the measured value. The expressions of mean square error and goodness of fit are as equations (4) and (5).

$$MSE = \frac{1}{n}\sum_{i=1}^{n}\left(y_i - \hat{y}_i\right)^2 \quad (4)$$

$$R^2 = 1 - \frac{\sum_{i=1}^{n}\left(y_i - \hat{y}_i\right)^2}{\sum_{i=1}^{n}\left(y_i - \bar{y}_i\right)^2} \quad (5)$$

Where n is the number of samples in the test set; $y_i\left(i = 1, 2, \ldots n\right)$ is the true value of the i-th sample; $\hat{y}_i\left(i = 1, 2, \ldots n\right)$ is the predicted value of the i-th sample.

4 Case study

4.1 Background

Taking a civil engineering project as an example, this paper selects relative dynamic elastic modulus as the output index of random forest prediction. With reference to a large number of related documents[13] and the actual situation of construction engineering[14], 11 indicators are selected as the initial input indicators of the random forest, as shown in Table 1:

Table 1. Monitored sample data.

Variables	Data		
	Min	Max	Ave
Cement dosage (kg/m^3)	358	457	407.5
Fly ash dosage (kg/m^3)	27.6	127.9	73.6
Fine aggregate (kg/m^3)	675	701	687.2
Coarse aggregate (kg/m^3)	1102	1143	1121
Water reducing agent (%)	0.8	1	0.9
Air entraining agent (%)	0	1.79	0.56
Water-binder ratio	0.31	0.35	0.33
Cement strength (Mpa)	55.5	61.5	58.5
Steel fiber(kg/m^3)	0	48	16
Expansion agent(kg/m^3)	0	80	5.3
Silica fume (kg/m^3)	0	25	9.4
Relative dynamic elastic modulus %	91.0	96.1	92.5

https://doi.org/10.1051/e3sconf/202123703033

4.2 Random forest for index selection

In this case, there are 100 sets of sample data, the training sample set is 80 sets, and the test sample set is the remaining 20 sets. The generalization error of the random forest regression model will gradually converge as ntree increases. Therefore, a large enough ntree value should be selected to build the model to ensure that the error of the training set can stabilize. Use OOB (out of bag data) error rate estimation method to determine the value of ntree, as shown in Fig. 1. In this study, the optimal ntree is determined to be 600, which can not only ensure the sufficient amount of decision trees, but also control the calculation amount of the algorithm within a relatively reasonable range.

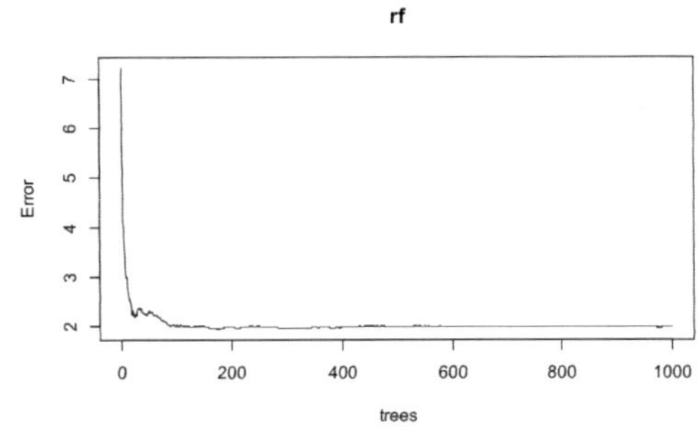

Fig. 1. Selection of the number of decision trees

Through the importance evaluation, the importance of the influencing factors to the relative dynamic elastic modulus can be determined, as shown in Fig. 2. Variables such as water-binder ratio, cement strength, cement dosage, fly ash dosage, coarse aggregate and fine aggregate are of high importance, which should be given priority in actual engineering projects.

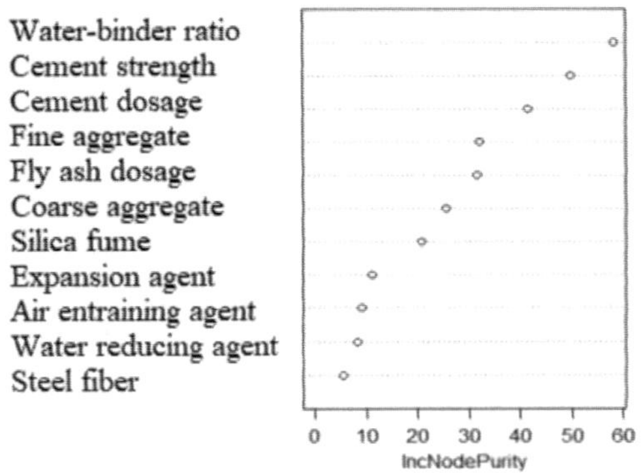

Fig. 2. Importance ranking

In order to screen out the number of variable combinations of concrete raw material mix ratios based on random forest, recursive feature elimination (RFE) can be used to select the characteristics of the training data set, and finally obtain the optimal number of variable combinations. Fig. 3 shows the trend diagram of the relative dynamic elastic modulus variable combination. When the number of deformation combinations is 9, the root mean square error RMSE reaches the lowest value and the model has the highest accuracy. The top nine concrete raw material mix ratio parameters are selected as the key factors selected by the random forest and will be used in the construction of the random forest prediction model later.

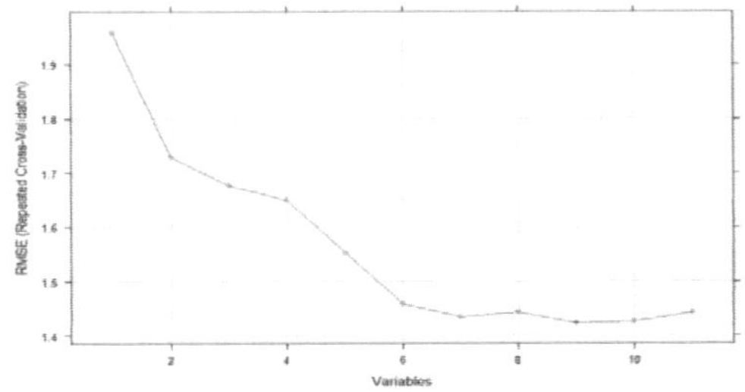

Fig. 3. Relative dynamic elastic modulus variable combination trend chart

4.3 Frost resistance prediction based on random forest

The input characteristic indexes are water-binder ratio, cement strength, cement dosage, fine aggregate, fly ash dosage, coarse aggregate, silica fume, expansion agent and air-entraining agent. The Randomforest package of R software was used to establish a random forest model of concrete frost resistance prediction, and the two parameters of the model were set to mtry=3 and ntree=600. The regression fitting curve of the training sample set is shown in Fig. 4. The trained random forest model is used to perform regression fitting on the test sample set, and the prediction result of the test sample set regression fitting is shown in Fig. 5.

It can be seen intuitively from Fig. 4 and 5 that the established random forest prediction model for concrete frost resistance is very consistent with the actual value of the test set samples. After calculation, the root mean square error (RMSE) between the actual value and the predicted value of the random forest model on the test sample set is 0.001, and the goodness of fit is 0.907. The closer the RMSE is to 0, the closer the goodness of fit is to 1, indicating that the prediction effect is better, which can indicate that the model has a good effect on the prediction of concrete frost durability and has good generalization ability.

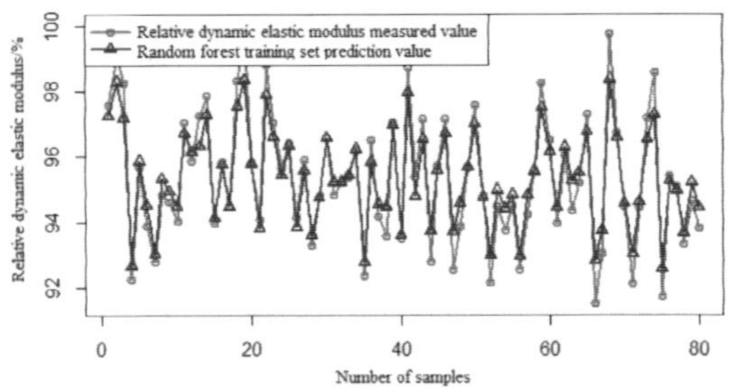

Fig. 4. Training set sample fitting results

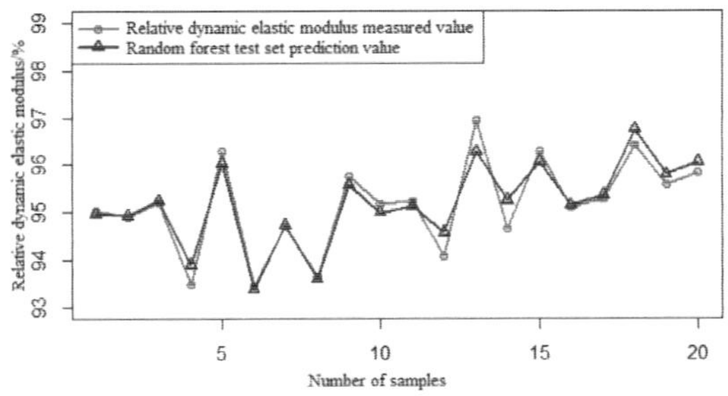

Fig. 5. Test set sample fitting results

In order to further verify the reliability of the random forest model, support vector machines and BP neural network models are selected to predict the frost resistance of concrete. The prediction results are compared with the results of the RF model, and the root mean square error RMSE and the goodness of fit R^2 are selected to measure the prediction effect of the model. The error comparison of the prediction results of different models is shown in Table 2.

Table 2. Comparison of prediction results of different models.

Model	Performance	
	RMSE	R^2
RF	0.0010	0.907
SVM	0.3216	0.842
BP neural network	0.4567	0.806

It can be seen from Table 2 that the root mean square error and goodness of fit of the prediction model based on the random forest relative dynamic elastic modulus are 0.0010 and 0.907, respectively. The root mean square error and goodness of fit of the support vector machine prediction model are 0.3216 and 0.842, respectively. The root mean square error and goodness of fit of the BP neural network prediction model are 0.4567 and 0.806, respectively. Compared with the support vector machine prediction model, the random forest prediction model reduced the root mean square error by 0.3206 and the goodness of fit increased by 0.065. Compared with the BP neural network prediction model, the root mean square error is reduced by 0.4557, and the goodness of fit is increased by 0.101. Therefore, in the random forest prediction model, support vector machine prediction model, and BP neural network prediction model, the random forest prediction model predicts the early frost resistance of concrete and can obtain prediction results with higher accuracy and reliability.

5 Conclusion

This paper proposes a random forest algorithm, which fully demonstrates the complex nonlinear relationship between the frost resistance and mix ratio of concrete, and makes high-precision prediction of concrete. Compared with other forecasting models, it reflects the reliability of the random forest forecasting model.

(1) The random forest method is introduced to predict the frost resistance of concrete, and the corresponding modeling process and steps are proposed. Use the variable importance measure of the random forest algorithm to rank the feature variables. After removing the variables of small importance, the dimensionality of the training model is reduced and the prediction accuracy is increased.

(2) In this paper, taking a concrete project as an example, 9 factors are determined as important factors after selection of characteristic variables. Taking into account that the data comes from the project, it indicates that in the actual project, attention should be paid to strengthen the control of these factors. Taking these 9 factors as the optimal set of input variables, a random forest-based concrete frost resistance prediction model was constructed. The final output predicted value and actual value verify the accuracy and reliability of the model.

(3) The prediction results of random forest model, support vector machine and artificial neural network model are compared and analyzed. The results show that the random forest feature selection has successfully improved the prediction and fitting accuracy of the model. The random forest model can obtain more accurate and stable prediction results, which further illustrates the good application prospects of the model.

References

1. Sun, Y., et al., A new mixture design methodology based on the Packing Density Theory for high performance concrete in bridge engineering. Construction and Building Materials, **182**(2018)
2. Amini, K., et al., Development of Prediction Models for Mechanical Properties and Durability of Concrete Using Combined Nondestructive Tests. Journal of Materials in Civil Engineering, **31**(2019)
3. Lin, S., H. Li, and H. Zhang, Experimental Study on Frost Resistance Durability and Service Life Prediction of Normal Cement Concrete. Advanced Materials Research, **368-373**: p. 2425-2429(2011)
4. Ji, T., T. Lin, and X. Lin, A mortar mix proportion design algorithm based on artificial neural networks. Cement and Concrete Research, **36**: p. 1399-1408(2006)
5. Yan, K. and C. Shi, Prediction of elastic modulus of normal and high strength concrete by support vector machine. Construction and Building Materials - CONSTR BUILD MATER, **24**: p. 1479-1485(2010)
6. Behforouz, B., P. Memarzadeh, and M. Eftekhar, Regression and ANN models for durability and mechanical characteristics of waste ceramic powder high performance sustainable concrete. Computers and Concrete, **25**(2020)
7. Zhang, P., A Novel Feature Selection Method based on Global Sensitivity Analysis with Application in Machine Learning-based Prediction Model. Applied Soft Computing, (2019)

8. Breiman, L., Random Forests. Machine Learning, **45**(1): p. 5-32(2001)
9. García-Pedrajas, N., C. Martínez, and J. Muñoz-Pérez, Multi-objective cooperative coevolution of artificial neural networks (multi-objective cooperative networks). Neural networks : the official journal of the International Neural Network Society, **15**: p. 1259-78(2003)
10. Genuer, R., J.M. Poggi, and C. Tuleau-Malot, Variable selection using random forests: Elsevier Science Inc(2010)
11. Hyndman, R. and A. Koehler, Another look at measures of forecast accuracy. International Journal of Forecasting, **22**: p. 679-688(2006)
12. Boddy, R. and G. Smith, Statistical Methods in Practice: for Scientists and Technologists(2009)
13. Kwan, A., Concrete mix design based on water film thickness and paste film thickness. Cement and Concrete Composites, **39**: p. 33–42(2013)
14. Kalla, P., et al., Mechanical and durability studies on concrete containing wollastonite–fly ash combination. Construction and Building Materials, **40**: p. 1142-1150(2013)

Research on displacement prediction of shield tunneling through existing tunnel based on LSSVM

Xianguo Wu[1,*], Tiemei Zeng[2], Yanyan Tao[3], and Hongyu Chen[4]

[1] School of Civil Engineering & Mechanics, Huazhong University of Science and Technology, Wuhan, Hubei, 430074, China
[2] Wuhan Metro Group Co Ltd, Wuhan, Hubei, 430030, China
[3] School of Urban Construction, Wuchang Shouyi University, Wuhan, Hubei, 430074, China
[4] School of Civil and Environmental Engineering, Nangyang Technological University, Singapore, 6397898, Singapore

Abstract. The construction scale of the shield tunnel underpass is expanding day by day. In order to study the safety influence and deformation control of the shield tunnel underpass on the existing tunnel, the LSSVM model is established. Based on the collected soil storage pressure, foam volume, simultaneous grouting volume and other six shield construction parameters and corresponding sample data of the tunnel bottom displacement, the horizontal displacement and settlement displacement of the existing tunnel bottom caused by the approach construction are predicted. Taking a subway project as an example, the research results show that the prediction model of shield tunneling under the existing tunnel bottom level and settlement displacement has strong generalization ability and rapid and accurate prediction effect. This method can provide reference for similar projects.

1 Introduction

With the widespread rise of my country's underground rail transit system, there are more and more shield tunnel construction projects. The greatest risk that may occur during the shield construction of the tunnel is excessive ground level and settlement displacement. Therefore, in order to ensure the safety of subway tunnel construction and operation, it is of great significance to adopt reliable measures to predict horizontal displacement and settlement displacement[1].

At present, a large number of domestic and foreign documents study the influence of tunnel construction from the perspective of surface settlement. There are three main research methods: theoretical formula[2], numerical simulation[3], and model test[4]. Theoretical formula method such as Wang et al.[5]derived the differential equation of adjacent pipeline deformation caused by tunnel excavation based on the Winkler foundation model to analyze the uplift or settlement deformation. Numerical simulations such as Li et al.[6] used FLAC3D to establish a three-dimensional numerical simulation model, and analyzed the impact of the staged construction of the foundation pit on the station and the existing subway tunnel. Model tests such as Ma et al.[7] conducted a series of three-dimensional centrifuge model tests to study the effect of parallel double tunnels of different depths on existing buried pipelines in dry sand. The above methods all have a certain value for the study of the settlement of the adjacent shield tunnel construction. However, the theoretical formula method is usually only suitable for specific situations and has poor accuracy. The numerical simulation accuracy is general but time-consuming and labor-intensive, and the model test accuracy is high but the instruments are expensive and time-consuming.

Chen et al.[8] predicted the maximum surface settlement based on artificial neural network method. However, the generalization performance of ANN in prediction is poor[9]. Therefore, in view of the shortcomings of the above methods, in order to control the impact of the shield construction on the existing tunnels, this paper proposes an intelligent algorithm framework based on the LSSVM algorithm. A shield construction parameter index system is established to make high-precision nonlinear predictions of the horizontal displacement and settlement displacement of the existing tunnel bottom.

2 LSSVM

The least squares support vector machine (LSSVM) improves the traditional SVM, and uses the least squares linear system as the loss function to transform the inequality constraints of the optimization problem in the SVM into equation constraints[10]. Suppose the training sample set (x_i, y_i) , $x_i \in R^n$, $y_i \in \{-1, +1\}$, $i = 1, 2, \cdots, N$ Where N represents the total number of training samples, n is the dimension of the sample space, and y is the class label of the sample. According to the principle of structural risk minimization, the optimal classification surface of the classification problem of LSSVM is obtained by the following optimization problem[11]:

* Corresponding author: wxg0220@126.com

$$\begin{cases} \min R(\omega,\xi) = \frac{1}{2}\omega^T\omega + \frac{1}{2}c\sum_{i=1}^{N}\xi_i \\ y_i\left[\omega^T\varphi(x)+b\right] = 1-\xi_i, i = 1,2,\cdots,N \end{cases} \quad (1)$$

Among them, $\varphi(\cdot)$ is the nonlinear mapping, ω is the weight, ξ_i is the error variable, and $c > 0$ is the penalty coefficient. Therefore, the Lagrange function corresponding to the optimization problem of LSSVM is:

$$L(\omega,b,\xi,a) = \frac{1}{2}\omega^T\omega + \frac{1}{2}c\sum_{k=1}^{N}\xi_i^2 - \sum_{k=1}^{N}a_i\left\{y_i\left[\omega^T\varphi(x_i)+b\right]-1+\xi_i\right\} \quad (2)$$

Among them, the *Lagrange* multiplier $a_i > 0(i = 1,2,\cdots,N)$. Optimize the above formula to make the partial derivatives of ω, b, a_i, and ξ_i equal to 0, and then use the Mercer condition to get: $\varphi(x_j) = y_i y_j K(x_i,x_j)$, where $K(x_i,x_j)$ is the core function. The optimal regression function can be obtained by the least square method as[12]:

$$f(x) = \sum_{i=1}^{l} a_i y_i K(x_i,x_j) + b \quad (3)$$

3 Prediction of bottom displacement of existing tunnel based on LSSVM

3.1 Data acquisition and preprocessing

By consulting a large number of related documents and according to the actual engineering situation, the factors that are more sensitive to the deformation of the existing tunnel are selected as the input parameters of the tunnel bottom displacement prediction. The horizontal displacement of the arch bottom and the settlement displacement of the arch bottom are used as prediction output parameters.

In order to prevent the sample from being overwhelmed or not converging because the data is too large or too small, the sample data needs to be preprocessed. In this paper, the input variables and output are normalized to the interval [-1,1], so that each parameter can be effective in the prediction process.

3.2 Kernel function parameter optimization

The kernel function has a great influence on the prediction accuracy of GA-LSSVM. The research should select the appropriate kernel function according to the characteristics of the experimental object. The Gaussian kernel function has the advantages of the radial basis kernel function while maintaining good anti-interference ability. Therefore, this paper will use the Gaussian kernel function as the kernel function of the prediction model for research, and the expression is as in formula (4).

$$K(x_i,x) = \exp(-\frac{\|x_i - x\|^2}{2\sigma^2}) \quad (4)$$

Among them, x_i is the input variable, and x is the output variable.

After determining the kernel function, in order to ensure the generalization level of LSSVM, this paper will use K-fold cross-validation, which can avoid the under-learning or over-learning state of the LSSVM model.

3.3 Evaluation of prediction results

In order to verify the prediction accuracy of the LSSVM model, the goodness of fit R^2 is introduced to test. The goodness of fit R^2 represents the fitting effect between the predicted value and the true value. The closer the goodness of fit is to 1, the higher the prediction effect.

$$R^2 = 1 - \frac{\sum_{i=1}^{n}\left(y_i - \hat{y}_i\right)^2}{\sum_{i=1}^{n}\left(y_i - \bar{y}_i\right)^2} \quad (5)$$

Where n is the number of samples in the test set; $y_i\,(i = 1,2,...n)$ is the true value of the i-th sample; $\hat{y}_i\,(i = 1,2,...n)$ is the predicted value of the i-th sample.

4 Case study

4.1 Background

This research is based on a subway line 4 under the subway line 2. Line 2 and Line 4 are upper and lower crossing tunnels, and the angle of the plane projection is approximately 90°. The diameter of the shield machine is D=6.2m, the length is 7.5m, and the outer diameter of the existing tunnel is also 6.2m. The soil layer of the existing tunnel is soft clay, and the Poisson's ratio is 0.28. The elastic modulus of the soil is 24.5Mpa, the thickness of the lining structure is 0.3m, the buried depth of the tunnel is 18m, and the concrete of the lining structure is C50.

The spatial position relationship between the shield and the existing tunnel is: the buried depth of the upper tunnel vault is 8.2m, and the buried depth of the lower tunnel vault is 14.4m. The clear distance between the upper and lower tunnels is 3.1m, and the length of the two tunnels is 6.2m in diameter. The problem to be solved is to study the influence of the optimization of shield construction parameters on the deformation of the adjacent existing tunnel for the construction of the shield tunnel at a close distance.

4.2 Data acquisition and preprocessing

A large number of documents have shown that the factors that are more sensitive to the deformation of existing tunnels include cutter head torque, jacking force, foam volume, grouting volume, driving speed and earth pressure of chamber[13].

Increasing the torque of the cutter head will increase the disturbance range of the surrounding soil.

As the jacking force increases, the force of the shield machine on the soil will increase.

Foam is usually used to improve the properties of the soil in order to reduce the torque of the cutter head of the shield machine and reduce the disturbance to the ground.

With the continuous increase of the grouting volume, the maximum ground settlement will continue to decrease.

The change of the soil stress field is related to the change of the driving speed. The large change of the driving speed can easily cause the change of the soil stress field.

After the earth pressure of chamber changes, the deformation of the soil around the shield tunnel will change due to the interaction, which will cause the ground settlement.

Therefore, this paper will take the above six parameters as input parameters, and use the horizontal displacement of the existing tunnel bottom and the settlement displacement of the arch bottom as the predicted output parameters. The selected LSSVM training sample data is shown in Table 1, and the value range of each input parameter in the training sample is shown in Table 1.

Table 1. Monitored sample data.

Variables	Data		
	Min	Max	Ave
Cutter head torque (kPa)	1927	3264	2291
Jacking force (kN)	15049	26667	21711
Foam volume (m^3)	10.1	14.2	12.1
Grouting volume (m^3)	14.0	18.9	17.1
Driving speed (mm/min)	17	34	24
Earth pressure of chamber (kPa)	243	332	291
Horizontal displacement of arch bottom(mm)	0.4	1.1	0.78
Settlement displacement of arch bottom(mm)	1.8	2.9	2.26

4.3 Kernel function parameter optimization

StandardScale is used to standardize the training samples, and GA algorithm is used to optimize the model parameters c, g, and p of LSSVM. The search range of the penalty coefficient c in the LSSVM model parameters is set to [0,100], the search range of the kernel function width parameter g is [0,1000], and the search range of p is [0.01,1]. The 5-fold cross-validation method was selected to model the kernel function width parameter g and the penalty coefficient c obtained by GA optimization, and the optimal parameters of the LSSVM prediction model of the arch bottom horizontal displacement and settlement displacement were obtained. The optimization results of the prediction parameters of the arch bottom horizontal displacement and the arch bottom settlement displacement are shown in Fig. 1 and Fig. 2 respectively.

It can be seen from Fig. 1 that the penalty coefficient best c=17.4949, the kernel function parameter best g=915.08, p=0.43452, and the minimum root mean square error is mse=0.083303.

Similarly, it can be seen from Fig. 2 that the penalty coefficient best c=31.1065, the kernel function parameters best g=0.13542, p=0.4438, and the minimum mean square error value is mse=0.07707.

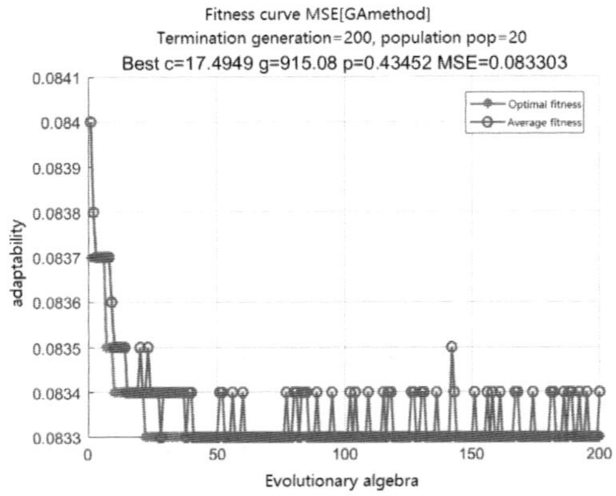

Fig. 1. Horizontal displacement of arch bottom model parameters

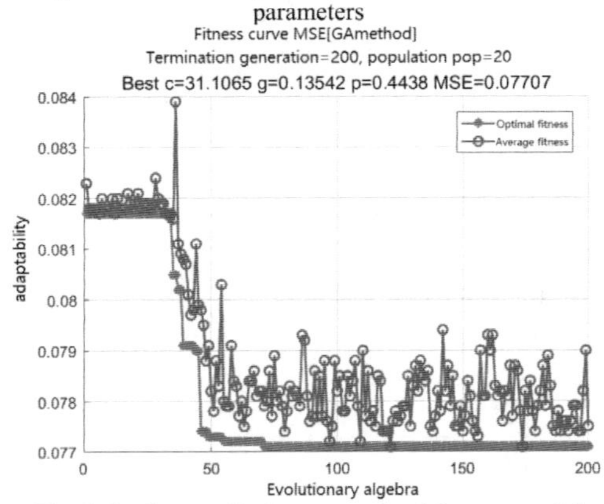

Fig. 2. Settlement displacement of arch bottom model parameters

4.4 Forecast result analysis

Based on the optimization results of the LSSVM kernel function parameters, the training set is used for learning simulation. The LSSVM arch bottom horizontal displacement and arch bottom settlement displacement prediction models are established respectively. According to the above steps, the prediction result of the arch bottom horizontal displacement is shown in Fig. 3, and the prediction result of the arch bottom settlement displacement is shown in Fig. 4.

It can be found from Fig. 3 that the LSSVM model can predict the changes in the horizontal displacement of the arch base well. Based on the LSSVM model, the predicted goodness of fit for the arch bottom horizontal displacement is 0.9987. It can be seen that the model fits well, and the error between the predicted value and the actual value is very small.

It can be found from Fig. 4 that the LSSVM model can make a good prediction of the change of the arch bottom settlement displacement. Based on the LSSVM model, the predicted goodness of fit for the arch bottom settlement displacement is 0.9964. It can also be seen that the model fits well, and the predicted value of the sample is very close to the experimental value..

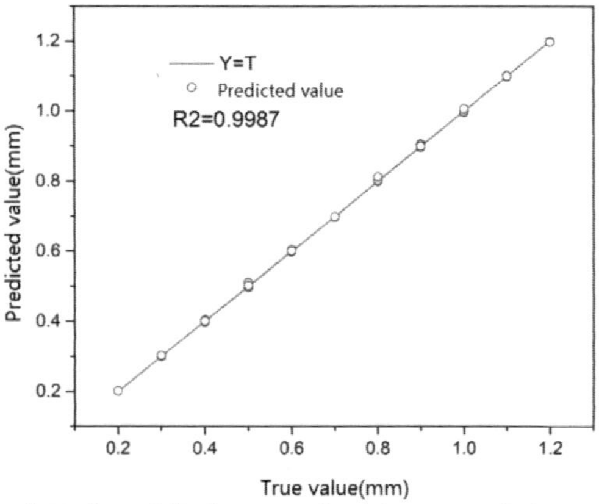

Fig. 3. Horizontal displacement of arch bottom prediction result

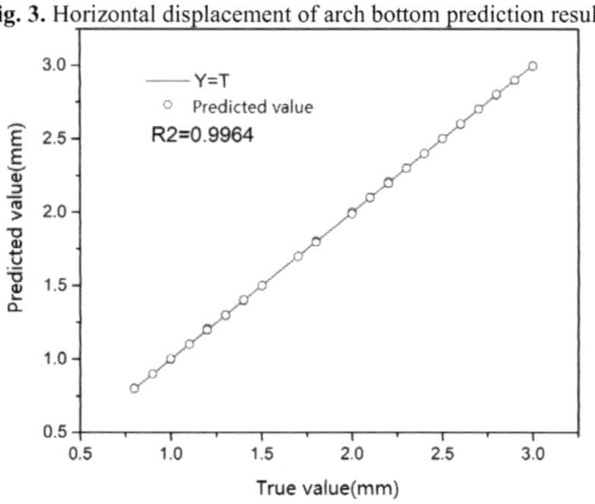

Fig. 4. Settlement displacement of arch bottom prediction result

5 Conclusion

In this paper, an intelligent prediction model of LSSVM is developed, which realizes the high-precision prediction of the horizontal displacement and settlement displacement of the shield under the existing tunnel bottom, which has important engineering value.

Taking the Wuhan subway project as an example, in view of the shortcomings of the traditional grid search method for parameter optimization, the soil tank pressure, foam volume, simultaneous grouting volume, tunneling speed, cutter head torque and jacking force are selected as input parameters. A prediction model based on LSSVM for the displacement of the existing tunnel bottom is proposed. The goodness of fit for the prediction of the horizontal displacement at the bottom of the tunnel is 0.9987. The goodness of fit for the prediction of the horizontal displacement at the bottom of the tunnel is 0.9964. The high accuracy of the model and good prediction effect reflect the feasibility of this method..

References

1. Zhang, D.M., et al., *Predicting the grouting effect on leakage-induced tunnels and ground response in saturated soils.* Tunnelling and Underground Space Technology, **65**: p. 76-90(2017)
2. Liang, R., et al., *Simplified method for evaluating shield tunnel deformation due to adjacent excavation.* Tunnelling and Underground Space Technology, **71**: p. 94-105(2018)
3. Jin, Y.-F., et al., *Three-dimensional numerical analysis of the interaction of two crossing tunnels in soft clay.* Underground Space, **4**(4): p. 310-327(2019)
4. Huang, H., X. Huang, and D. Zhang, *Centrifuge modelling of deep excavation over existing tunnels.* Proceedings of the ICE - Geotechnical Engineering, **167**: p. 3-18(2014)
5. Wang, Y., Q. Wang, and K. Zhang, *An Analytical Model for Pipe-Soil-Tunneling Interaction.* Procedia Engineering, **14**: p. 3127-3135(2011)
6. Li, M., et al., *Numerical study on responses of an existing metro line to staged deep excavations.* Tunnelling and Underground Space Technology, **85**: p. 268-281(2019)
7. Shaokun, M., et al., *Responses of pipeline to side-by-side twin tunnelling at different depths: 3D centrifuge tests and numerical modelling.* Tunnelling and Underground Space Technology, **66**: p. 157-173 (2017)
8. Chen, R.-P., et al., *Prediction of maximum surface settlement caused by earth pressure balance (EPB) shield tunneling with ANN methods.* Soils and Foundations, **59**(2): p. 284-295(2019)
9. Zhou, J., et al., *Feasibility of Random-Forest Approach for Prediction of Ground Settlements Induced by the Construction of a Shield-Driven Tunnel.* International Journal of Geomechanics, **17**: p. 04016129(2016)
10. Zhai, Y., et al., *Adaptive LSSVM based iterative prediction method for NOx concentration prediction in coal-fired power plant considering system delay.* Applied Soft Computing, **89**: p. 106070(2020)
11. Shao, M., et al., *Prediction of energy consumption in hotel buildings via support vector machines.* Sustainable Cities and Society, **57**: p. 102128(2020)
12. Prayogo, D., et al., *Combining machine learning models via adaptive ensemble weighting for prediction of shear capacity of reinforced-concrete deep beams.* Engineering with Computers(2019)
13. Zhou, C., et al., *Visibility graph analysis on time series of shield tunneling parameters based on complex network theory.* Tunnelling and Underground Space Technology, **89**: p. 10-24(2019)

Impact and Assessment of Urban Wind Environment Change on Architectural Heritage Protection

Yi Xiaolie[1,*], Zheng Lipeng[2]

[12]School of Architecture, South China University of Technology; State Key Laboratory of Subtropical Building Science ,Guangzhou 510640,Guangdong,China

Abstract. In view of the contradiction between urban construction and protection of architectural heritage, this paper analyzes the hidden dangers of wind and fire protection of architectural heritage from the perspective of wind environment impact of high-rise buildings. The wind environment simulation and fire simulation based on computational fluid dynamics are used to study the typical building heritage cases in Guangzhou. The wind environment indicators that affect the comfort, safety and fire development of the wind environment of the building heritage are explained, so as to form a strategy for evaluating the impact of wind environment of high-rise buildings on the protection of building heritage.

1 Introduction

The change of urban wind environment has always been a hot research and concern at home and abroad. On the one hand, the rapid urban expansion and high-intensity construction cause the wind speed to slow down in the urban area, resulting in a large area of quiet wind area aggravating the phenomenon of air pollution and heat island effect. Many cities have put forward improvement measures such as wind corridor planning [1]. On the other hand, urban high-rise buildings have become the main factors affecting the wind environment. At present, urban renewal and construction are in full swing in China. The sharp increase of urban building height and planning adjustment will significantly change the regional wind environment conditions, producing adverse wind environment effects, disturbing the surrounding air flow conditions, damaging the comfort of the wind environment and the safety of pedestrians and buildings [2], making the regional architectural heritage protection face new challenges.

In view of the interference phenomenon of urban high-rise buildings on the wind environment of surrounding buildings, scholars at home and abroad mainly focus on the comfort and safety of pedestrian wind environment around high-rise buildings [3], the wind load interference effect of group high-rise buildings, and the structural wind load interference effect of adjacent low-rise buildings. The research methods are also extended from the traditional field measurement to wind tunnel experiments and CFD simulation method. Xie Zhenyu summed up the adverse effects of high-rise buildings on the outdoor wind environment which including increasing wind speed and causing airflow chaos, and result in "narrow pipe effect" ,"Wind funnel effect", "shelter effect" and "street canyon effect"[4]. Zhang Min and Guan Jiping studied the wind load interference effect of group high-rise buildings, and compared the wind tunnel test results with the numerical simulation method. They explained the applicable conditions and conditions of the numerical simulation method, but the research focus is on the wind resistance design of high-rise buildings [5-6]. Gu Ming and shangluxi pay attention to the wind load interference effect of high-rise buildings on the adjacent low flat roof buildings. They systematically study the lifting coefficient and wind pressure coefficient of low-rise buildings under different wind direction angles, building height and spacing, but do not involve the situation of sloping roof [7-8]. It can be seen that there is little research on the impact of wind environment of high-rise buildings on the surrounding architectural heritage. Due to the particularity of building heritage, only some scholars study from the perspective of architectural heritage disaster prevention [9-10], and fail to systematically analyze and evaluate the impact of wind environment changes on pedestrian wind environment and building wind load under the interference of high-rise buildings.

In this paper, the typical building heritage cases in Guangzhou are selected, and the impact of wind environment effect of regional high-rise buildings on architectural heritage is enumerated. It is shown that the wind environment index affecting the protection and utilization of building heritage can be obtained by using the wind environment simulation and fire simulation method based on computational fluid dynamics (CFD), so as to form the assessment methods and Strategies of the effect of architectural heritage in the wind environment of urban high-rise buildings.

* Corresponding author: hillyyi@163.com

2 Change of urban wind environment and hidden danger of architectural heritage

When the urban wind flow meets the obstruction of high-rise buildings, it will produce strong vortex wind at the corner of the building and the windward front, and produce a long distance wind shadow area at the back of the building [11].The wind speed in the wind shadow area will be significantly reduced if the building heritage is located in the windward downstream of the high-rise building, but the negative wind pressure on the roof of the building heritage will be amplified. When the building heritage is located in the windward upstream of high-rise buildings, the wind speed will increase within a certain distance, and the roof of architectural heritage will even show obvious positive pressure. In addition, there is a complex interference phenomenon of wind environment between high-rise buildings, especially when the wind channel between high-rise buildings becomes narrow. It will produce a slit effect resulting in the rapid increase of wind speed and the decrease of wind pressure [12].

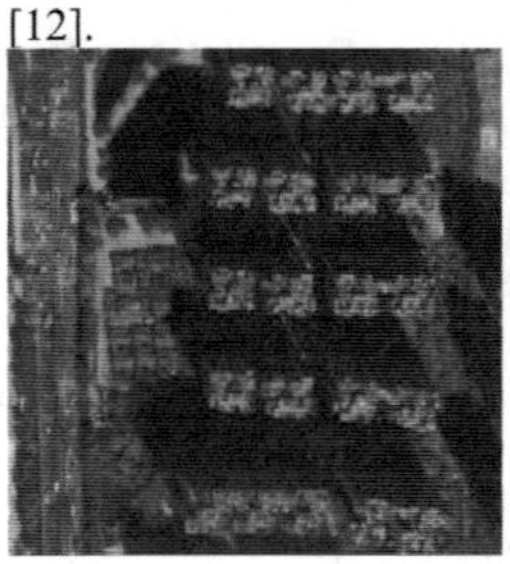

Fig1. Liede village in Guangzhou.
Fig2. Yangji village in Guangzhou.

According to the layout characteristics of high-rise buildings and architectural heritages in Guangzhou, the common hidden dangers of wind environment are as follows: first, the layout of buildings adjacent to high-rise buildings or super high-rise buildings. For example, after the overall reconstruction of Liede village in Guangzhou, the newly-built villagers' houses are close to the ancestral hall area (Fig.1); the second is that the architectural heritage is affected by the slit effect between high-rise buildings, such as the removal of the ancestral hall in Yangji village, Guangzhou. It's located between the gaps of high-rise residential buildings (Fig. 2).

Therefore, due to the layout of high-rise buildings in the city, adverse wind environment effects will appear, resulting in abnormal wind speed and wind pressure, which will affect the protection and utilization of regional architectural heritage. First of all, the building heritage is located in the wind shadow area of high-rise buildings. Low ambient wind speed and the increase of quiet wind days are not conducive to the loss of pollutants and reduce people's thermal environment comfort; secondly, the higher wind speed of the environment is not conducive to crowd activities, and strong wind may hurt people. At the same time, excessive wind speed will promote the spread of building fire, because wind speed is closely related to fire risk; Finally, the strong environmental wind leads to the increase of wind pressure and wind load on architectural heritage, which is easy to lead to the wind disaster damage. Especially in the coastal typhoon prone area of Guangzhou, we should pay more attention to the amplification effect of wind environment on the destructive force of high-rise buildings.

3 Evaluation methods and standards

3.1 Simulation of outdoor wind environment of Architectural Heritage

Computational fluid dynamics (CFD) simulation is a research method of building wind environment which is often used in the stage of urban planning and design. It can quickly and accurately calculate the urban wind environment. The mature CFD simulation software includes ANSYS fluent, PHOENICS and so on. The research on the calculation model shows that the prediction results of SST κ-ε turbulence model based on Reynolds average method are close to the experimental results, and the simulation results of RNG κ-ε model are also in good agreement with the measured data of wind environment.

Computational domain, boundary conditions and mesh generation are the main factors affecting the simulation results:

- The wind field calculation domain is divided into inflow, outflow and building area. The building area is based on the height H of high-rise building, centered on the assessed building heritage, and the radius of no less than 2H is the simulated building area. According to the building wind environment test and evaluation standard of Guangdong Province, the inflow distance is greater than 5H, the outflow distance is greater than 10H, the calculation domain height is greater than 4H, and the maximum blocking ratio of wind direction incoming flow is not more than 5%.
- The exponential rate model is used to simulate the gradient wind at the entrance boundary of the calculation domain. The roughness index α=0.22 is taken as the geomorphic type. The outlet boundary is the pressure outlet, and the top and both sides are symmetrical walls.
- The minimum accuracy of the calculation grid in the key assessment area should be 1 /10 of the building scale, at least 3 or 4 layers of grid within the vertical 1.5 m height range. The grid should be densified in the area where the flow field is expected to change greatly, and the growth ratio of the grid size should not exceed 1.2.

3.2 Architectural heritage fire simulation based on environmental wind speed

Large eddy simulation (LES) is used to describe the movement of fire plume in architectural heritage fire dynamics simulation (FDS). The three-dimensional unsteady Navier Stokes equation driven by heat is solved by numerical method. The process of smoke flow and heat transfer in fire can be calculated accurately, and the physical quantities such as temperature, velocity and gas concentration in each part of the fire can be obtained. At the same time, FDS simulation process can consider the influence factors of environmental wind speed and wind direction.

• The simulation domain of FDS is mainly the interior space of the building, but additional calculation domain should be set at the top of the building outdoor to meet the fire spread through the roof. At the same time, the calculation domain at the external opening of the building is extended outward by more than 1m, so as not to affect the smoke flow and air inflow at the opening [13].

• Considering the influence of ambient wind speed on fire development, the boundary is set as periodic to simulate infinite flow field.

• The density of calculation grid directly affects the accuracy of simulation solution. The appropriate mesh size is related to the characteristic diameter of fire source [14]. When the minimum grid size is 1 / 4 to 1 / 16 of the characteristic diameter of fire source, the balance between simulation accuracy and calculation efficiency can be achieved. The characteristic diameter d * of the fire source is calculated from the heat release rate Q of the fire source (formula 1).

$$D^{*}=\left(\frac{\dot{Q}}{\rho_{\infty}c_{p}T_{\infty}\sqrt{g}}\right)^{\frac{2}{5}} \quad (1)$$

Where: q is the heat release rate of the fire source; ρ ∞ is the air density of $1.204kg/m^3$, cp is the air specific heat capacity of 1.005kJ/kg.k, t ∞ is the ambient air temperature of 293k (20 ℃), and g is the gravity acceleration of $9.81m/s^2$.

3.3 Evaluation standard of wind environment

3.3.1 Comfort and safety assessment of pedestrian wind environment

The wind environment comfort assessment is based on the average wind speed V at pedestrian height. Firstly, the average wind speed ratio R = Vr / V0 is calculated (The ratio of the average wind speed Vr to the average wind speed V0 at the height of 10 m corresponding to the local standard landform), while the average wind speed at pedestrian height is $V = RV_{10}$ (V_{10} is the average wind speed at 10m height of local standard landform).

The average wind speed ratio r under the dominant wind direction in summer should not be less than 0.2, otherwise the wind environment would feel hot. In winter, the average wind speed ratio R should not be greater than 2.0, otherwise the wind environment would feel cold. When the maximum average wind speed is greater than 5m / s, the pedestrian wind environment is not comfortable.

Gust wind speed is adopted for wind environment safety assessment. Gust wind speed at pedestrian height V '= β_{gh} V, β_{gh} is the fluctuation coefficient of wind speed. In urban areas with dense buildings, when the basic wind pressure is greater than or equal to $0.5KN/m^2$, the wind speed fluctuation coefficient at 1.5m height is 1.69; when the basic wind pressure is less than $0.5KN/m^2$, the wind speed fluctuation coefficient at 1.5m height is 1.58. The safety degree of pedestrian wind environment is judged according to the annual frequency of gust wind speed v 'greater than 15m / S (Table 1).

Table 1. Safety assessment of pedestrian wind environment

P（V'> 15m/s）	Safe	<0.05%（< 2 times/year）
	Risky	0.05-0.3% (≥2times/year ,≤12times/year）
	Danger	>0.3%（>12times/year）

3.3.2 Wind protection valuation of architectural heritage wind environment

The influence of regional wind environment change on the wind resistance performance of architectural heritage mainly depends on the effect of wind pressure and wind load. The height and slope of roof, whether there are gables and vertical ridges will lead to different wind load distribution of building roof.The results show that the height of the ridge can increase the average wind pressure and fluctuating wind pressure of the windward roof, and reduce the overall upward lifting force of the roof. The gable out of the roof is beneficial to reduce the average negative pressure and fluctuating wind pressure at the vertical ridge and the roof corner area, and greatly reduce the most unfavorable wind pressure of the roof.

The wind environment simulation can calculate the average wind pressure coefficient distribution of building surface. If the wind pressure distribution varies greatly, and abnormal positive pressure occurs, it is necessary to check whether the wind load intensity exceeds the maximum limit value of regional wind load. When there is obvious negative pressure area, it is necessary to judge whether the position is located at the roof corner, roof structure and other weak parts.

3.3.3 Fire protection valuation of architectural heritage wind environment

The spread of building heritage fire is mainly affected by fire load, heat release rate and fire space condition. Different fire environment leads to different fire development. By comparing the influence of fire plume on buildings in downwind direction under different wind speeds, whether the high temperature smoke flow reaches the critical radiation intensity of ignition is analyzed. When the radiant heat is low, the temperature of the outer surface of the building cannot reach the ignition point, so it is impossible to ignite.

4 Case analysis and evaluation

4.1 Influence of wind environment of high-rise buildings on architectural heritage

Taking the planning and construction of Guangzhou baietan international financial center as an example, there are architectural heritages of Guangzhou Tongsheng machinery factory in the plot, and several super high-rise building towers will be built nearby. According to the meteorological data of Guangzhou, the typhoon weather, rainy and humidity season and the dominant wind direction in summer and winter are selected for analysis. CFD simulation analysis can obtain the wind speed ratio, average wind speed and gust wind speed of each measuring point in the wind environment of building model, as well as the cloud chart of wind speed ratio, wind speed vector diagram and streamline diagram under different wind directions. In addition, the calculation results of wind pressure, wind speed and wind pressure coefficient on the external surface of buildings can be displayed intuitively (Fig.3), which is helpful to judge the change of wind environment and the disturbance degree of architectural heritage. The results show that:

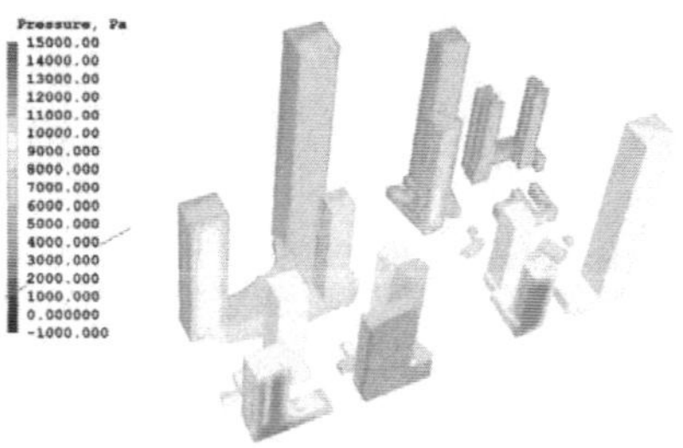

Fig3. Wind pressure distribution on facade of building complex

- Typhoon weather belongs to extreme strong wind weather, and the amplification effect of wind environment will enhance the damage of typhoon. Through the wind environment simulation to verify the safety of structure wind load, the positive and negative pressure parts of wind load are obtained. For sloping roof buildings, the negative pressure is the main reason for the roof overturned by strong wind.
- Under the weather of rainy and humidity, if the architectural heritage is located in the wind shadow area of high-rise buildings, the ventilation performance of the area will become worse, the wind speed will slow down and the air humidity will be more obvious, which is not conducive to the maintenance of the architectural heritage.
- Under the dominant wind direction in summer and winter, affected by the surrounding building layout and high-rise buildings, the outdoor wind speed reaches good thermal comfort in summer, but there is a strong wind downward along the high-rise buildings in winter.

4.2 Influence of environmental wind speed on fire spread of architectural heritage

In order to evaluate the impact of environmental wind speed on the fire protection performance of architectural heritage, taking Chen's Academy, a typical ancestral temple building in Guangzhou, as an example, the main hall and east hall with the largest fire load were selected for fire simulation. The main hall and the east hall is made of dense wooden frame, and the roof height accounts for about 1/3 of the elevation.The fire scenario is that the wooden frame is ignited after an electrical fire in the main hall, and the roof will burn through after the fire flashover. The roof fire is affected by the wind speed of the environment, which will affect the East Hall building in the downwind direction.When the ambient wind speed is 2,3,4,6,8m/s, the influence of roof flame on the roof temperature of east hall is calculated to judge whether the gable ridge and building spacing can prevent the fire spreading.

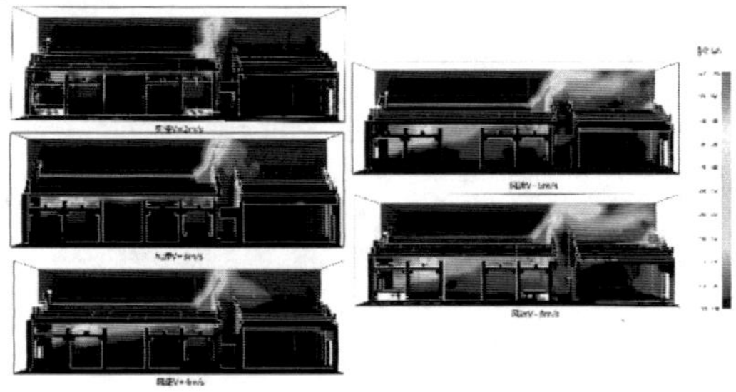

Fig4. Wind speed and roof fire spread

The results show（ Fig.4） that when the wind speed is 2m/s, the fire smoke rises vertically, the gable blocks the flow of high temperature smoke, and the roof temperature changes little. When the wind speed is 3m/s, the roof flame exceeds the gable, and the smoke enters the lane space, causing the roadway space and east hall gable to heat up. When the wind speed is 4m/s, the height of fire flame increases, and the fire plume flows downward under the influence of wind. The tunnel temperature reaches 140℃ and the roof temperature of the east hall reaches 120℃. When the wind speed is 6m/s, the influence distance of the fire plume increases, and the high-temperature smoke is blocked by the vertical ridge of the east hall roof, forming a high-temperature part of 400 ℃. When the wind speed is 8m/s, the fire plume will accumulate seriously on the roof of the east hall, and the roof temperature can reach 800 ℃, which will cause roof damage and lead to long-distance ignition.

It can be seen that after the flashover of the main hall fire in the ancestral hall building group, the roof burning through flying fire is obviously affected by the environmental wind speed. When the wind speed is greater than 4m/s, attention should be paid to the fire safety of the building group. At this time, the fire of single building is easy to spread by the wind.

5 Conclusion

From the perspective of architectural heritage protection, this paper analyzes the hidden dangers of wind environment changes caused by urban high-rise buildings, and evaluates two architectural heritages in Guangzhou by using wind environment simulation and fire simulation, which illustrate the application of specific evaluation methods and evaluation standards.

•The layout of urban high-rise buildings is the cause of adverse wind environment effect, which not only affects the comfort and safety of pedestrian wind environment, but also causes interference effect on the wind load of adjacent architectural heritage. Severe typhoon weather will enlarge the damage caused by wind disaster.

•The wind environment simulation and fire simulation based on computational fluid dynamics (CFD) can effectively reflect the interference effect of high-rise buildings on regional wind environment and the impact of environmental wind speed on the spread of fire, which is conducive to evaluate the impact of wind environment on the protection and utilization of architectural heritage in the early stage of urban renewal planning and design.

•The simulation analysis can obtain the wind environment indicators that affect the comfort of pedestrian wind environment, the safety of building structure and the development of fire. The wind environment comfort evaluation index adopts the average wind speed of pedestrian height, and the wind environment safety evaluation index adopts the gust wind speed of pedestrian height. The safety assessment of architectural heritage structure is based on the wind pressure distribution on the external surface of the building and the standard wind load. The influence of wind speed on the spread of architectural heritage fire can be judged by temperature distribution and thermal radiation.

References

1. A Ren Chao, Yuan Chao, he Zhengjun, et al. Study on Urban Ventilation corridor and its planning application [J]. Urban Planning Journal, 2014 (03): 52-60.
2. Zhuang Zhi, Yu Yuanbo, ye Hai, et al. Research status of CFD simulation technology for building outdoor wind environment [J]. Architectural science, 2014,30 (02): 108-114.
3. Xu Xiaoda. Distribution characteristics of average wind speed at pedestrian height around super high-rise buildings and wind environment assessment [D]. 2019.
4. Xie Zhenyu, Yang ne. Shape optimization design strategy for improving outdoor wind environment of high-rise buildings [J]. Acta architecture, 2013, 000 (002): 76-81.
5. Zhang Min, Lou Wenjuan, he gejun, et al. Numerical study on wind load disturbance effect of high-rise buildings [J]. Engineering mechanics, 2008 (01): 179-185.
6. Guan Jiping, Chen Wei, Zhu Ledong. Study on group interference effect of wind environment in high-rise buildings [C] National Conference on structural wind engineering. 2009
7. Shang Luxi. Wind load interference effect of high-rise buildings on low rise flat roof buildings [D]. 2017
8. Gu Ming, Huang Peng. Research status and Prospect of wind load disturbance on high-rise buildings [J]. Journal of Tongji University (NATURAL SCIENCE EDITION), 2003, 31 (7): 762-766
9. Liang Haoyan, Meng Qinglin, Li Xiaohui, et al. Study on wind environment assessment method of Lingnan old city renewal and Reconstruction Planning -- a case study of Yuzhu old city renewal planning in Huangpu District of Guangzhou City [J]. Southern architecture, 2018 (04): 34-39
10. Liu Pengfei, Zou Fang, Ouyang Huiting, et al. Research on the local renewal mode of old urban area based on wind environment simulation: a case study of Taiping old street in Changsha City [J]. Southern architecture, 2020 (01): 47-52
11. Yang Li, Liu Xiaodong, sun Biman. Research progress of building wind environment [J]. Building science, 2018, 034 (012): 147-156
12. Fu Xiaojian, Yan Xiaoping, Xu Minghui. Numerical study on slit effect of wind load on twin tower buildings [J]. Shanxi architecture, 2007 (15): 82-84
13. Zhu Shi, Huo Ran, Hu Longhua, et al. Influence of grid generation and calculation area extension at opening on FDS simulation results [J]. Journal of safety and environment, 2008 (04): 133-137
14. Deng Ling. Grid analysis in FDS field simulation [J]. Fire science and technology, 2006, 25 (2): 207-210

Design of the three-forked-shape pedestrian landscape cable-stayed bridge with egg-shape arch tower

Biqing Li[1*], Jinbao Li[1] and Jinfa Xu[2]

[1]Jiangsu Southeast Special Technical Engineering Co., Ltd, Nanjing, China,210008
[2]Jiangsu traffic engineering consulting and Supervision Co., Ltd, Nanjing, China,211800

Abstract. According to the landscape requirements of the pedestrian cable-stayed bridge of the three-forked-shape canyon, this paper puts forward a kind of pedestrian landscape cable-stayed bridge with egg-shape arch bridge tower, and two kinds of spatial curve design elements are introduced, which are egg-shape arch bridge and three-forked-shape curved bridge deck. The cultural connotation of three-forked-shape canyon has been greatly improved. Combined with the actual project, the engineering parameters are designed and the Midas finite element analysis model is established, static analysis and dynamic modal analysis were carried out, it verified that the three-forked-shape pedestrian landscape cable-stayed bridge with egg-shape arch bridge tower has good mechanical performance.

1 Introduction

Built in 1981, the Three-forked-shape Midland Bridge in Michigan is located at the confluence of the Chippewa river and Tittabawassee river.

Fig.1. Midland Bridge

A special three-forked-shape bridge connects the three banks. The bridge has three branches, which lead to three different directions.Three-forked-shape bridge naturally fits in the environment and vividly illustrates the concept of unity of man and nature[1-3].

Canyon refers to the valley with steep slope and deeper than width. At the confluence of the mainstream and tributaries of the canyon River, a fork shaped Canyon will be formed.The construction of the three-forked-shape bridge is conducive to the tourism traffic of the fork shaped canyon scenic spot.

With the development of cable-stayed bridge structure design and construction technology, people have higher and higher requirements for cable-stayed bridge landscape[4-6]. Arch steel bridge tower has been widely used because of its advantages of beautiful shape, energy saving, high efficiency and green environment protection[7-10].

Based on three-forked-shape canyon terrain, this paper proposes a three-forked-shape pedestrian glass bridge landscape cable-stayed bridge, the modified for eggs-shape arched bridge tower cable-stayed bridge tower and bridge deck form uses annular three-forked-shape canyon, cross the bridge, the bridge the central set up a huge circular tour patio, deck stiffening girder cross-section of the stomach beam structure, bridge panel for transparent tempered laminated glass, railing for the silver metal rail. Space cable-stayed cables crisscrossed suspended glass bridge deck, forming a unique shape of three bifurcated pedestrian landscape glass cable-stayed bridge.

Fig.2. Three-forked-shape Canyon

In this paper, combined with the design of pedestrian glass cable-stayed bridge in three-forked-shape canyon, the paper carries out the configuration research of three fork type pedestrian landscape glass cable-stayed bridge and the engineering parameter design. The Midas finite element model was established, and carries out the dynamic modal research, so as to verify the structural superiority of the three fork pedestrian landscape glass cable-stayed bridge.

*Corresponding author:Li Biqing: umabank@163.com

2 Configuration study

The bifurcated bridge deck of three-forked-shape canyon pedestrian glass cable-stayed bridge is composed of straight-line bridge deck and curved transition section bridge deck. The straight-line bridge deck forms 120° each other. The three straight-line bridge decks are connected by arc-shaped curved transition section. A huge circular sightseeing patio is set in the center of the bridge deck, and the center point of the bridge deck is 100 meters away from the center line of the egg-shape arch bridge tower root.

Egg shape curve is an important geometric curve in mathematics. Egg curve is a symmetrical and smooth curve used in mathematics, drawing and art. Egg-shape symmetrical and smooth curve is its charm. The mathematical equation of egg shape curve is as follows:

$$\frac{x^2}{a^2}+\frac{y^2}{(ky+b)^2}=1 \quad |k|\le 1 \tag{1}$$

In the equation: x、y are coordinate axis dimensions
a is the egg width parameter;
b is the egg height parameter
k is the egg shape parameter

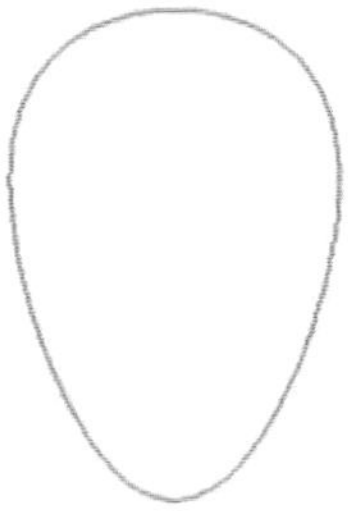

Fig.3. Geometry of Egg Curve

The traditional pedestrian landscape cable-stayed bridge structure is reformed, and the egg-shape arch bridge tower is used instead of the portal bridge tower structure, which is beautiful in shape and can improve the landscape grade of the pedestrian suspension bridge.

The arched pylon of egg-shape bridge is different from the common parabolic arch bridge tower. The lower part of the parabolic arch bridge tower is divergent and peeled outward. The lower part of the egg-shape arch bridge tower is convergent. The egg-shape arch bridge tower covers a small area, so the egg-shape arch bridge tower is very suitable for the narrow mountain landscape platform environment.

Egg-shape arched structure is beautiful, the structure is novel and light, smooth and smooth, the shape is smooth, the form is changeable, it is convenient to modify according to the actual needs, and can be widely applied to the bridge towers of various landscape bridges. three-forked-shape canyon pedestrian glass cable-stayed bridge adopts the unique egg-shape arch bridge tower, which can coordinate the scenic environment.

Fig.4. Cable Stayed Bridge in Three-forked-shape Canyon

With the change of bridge deck curve, the span of stiffening beam near the center circle of bridge deck is larger. Fish bellied steel beam with variable cross-section is adopted to increase the height of cross-section, but not more weight and cost. At the same time, it also enriches the bridge deck landscape viewed from the canyon.

The bridge deck is paved with toughened glass on the stiffening beam system of the bridge deck. The bridge deck is transparent and crystal clear. It is extremely dangerous and exciting to see the rapid flow of the canyon at the foot of the bridge.

The curved bridge deck with three bifurcations and the arched pylons with curved egg-shape can coordinate with the environment of three-forked-shape canyon. The architectural collocation is reasonable and the advantages are complementary. They are the ideal combination of structure and architecture.

3 Parameter design

The pedestrian glass cable-stayed bridge of three-forked-shape canyon is a pedestrian glass landscape cable-stayed bridge with three bifurcated bridge deck and egg-shape arch bridge tower. The horizontal distance from the center of the bridge deck to the bottom of the egg-shape arch bridge tower is 100 meters, and the span of approach bridges is 24 meters.

Egg-shape arched bridge tower adopts concrete-filled steel tube tower structure. The diameter of steel pipe is 2.5 meters, the wall thickness of steel pipe is 18 mm, and C40 concrete is filled inside. The inclination angle of egg-shape arch bridge tower is 15 degrees, the highest part is 55 meters, and the widest part of bridge tower is 40 meters. The width of the tower top in upper part section is 30 meters, the rise height is 8 meters, and the height of tower column in lower part is 47 meters. A horizontal cross bar is set everywhere Support the bridge deck.

The stay cables are scattered and anchored on the arched pylons. Each arched pylon has two groups of 11 pairs of stay cables, with a total of 66 cables. Each arched bridge tower has two groups of 5 pairs of cable-stayed cables, a total of 30 cables. The diameter of the cables varies linearly from 0.15 to 0.2m. The cable-stayed cables are made of 1670mpa galvanized high-strength steel wire.

https://doi.org/10.1051/e3sconf/202123703036

Fig.5. Stiffening Beam

The width of the bridge deck in the straight section is 15 meters, and the radius of the opening of the central patio is 20 meters. The curve sealing beam of the bridge deck changes according to the arc curve of the bridge deck, and its arc radius is 175 meters.

Steel grillage structure is adopted for the bridge deck, and the beam spacing is 8m. Due to the curve change of the bridge deck, the span of the stiffening beam gradually increases from the approach bridge to the center. The bridge deck stiffening beam in straight section is 15m long. The stiffening beam adopts steel beam structure of rectangular steel tube. The beam width is 1m and the beam section height is 2-3.5m. The stiffening beam of curved bridge deck is fish bellied variable cross-section steel beam. The section height of steel beam is increased to ensure the bearing capacity requirements of glass deck at midspan.

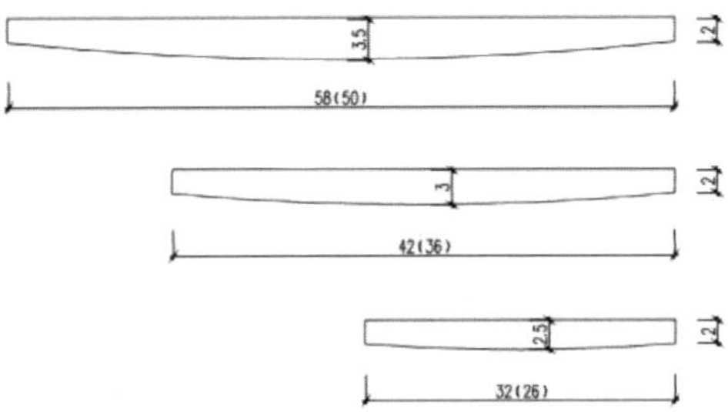

Fig.6. Fish Bellied Steel Beam

Midas model of pedestrian glass cable-stayed bridge in three-forked-shape canyon is shown in the figure below.

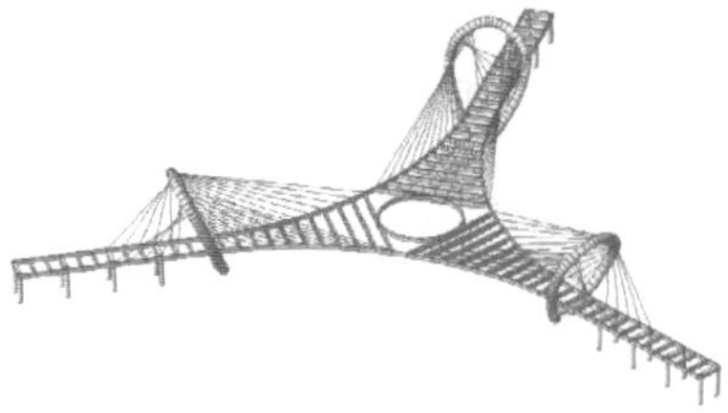

Fig.7. Midas Finite Element Model

4 Calculation results under vertical load

Load the main span bridge deck with full load. The standard value of uniform constant load is 5kn / m^2. The standard value of uniform load 4kn / m2 is adopted for the live load of bridge deck.

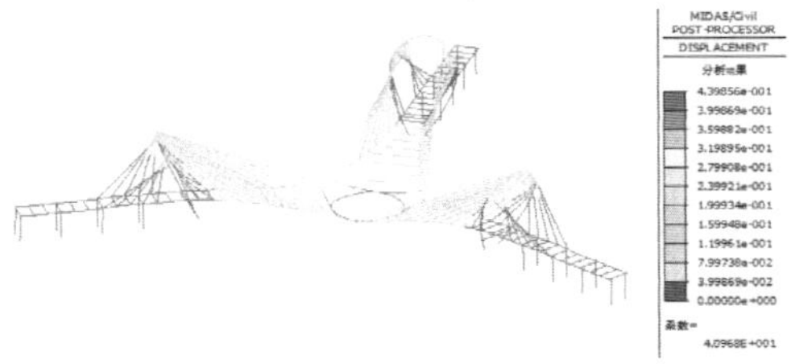

(a) Deflection

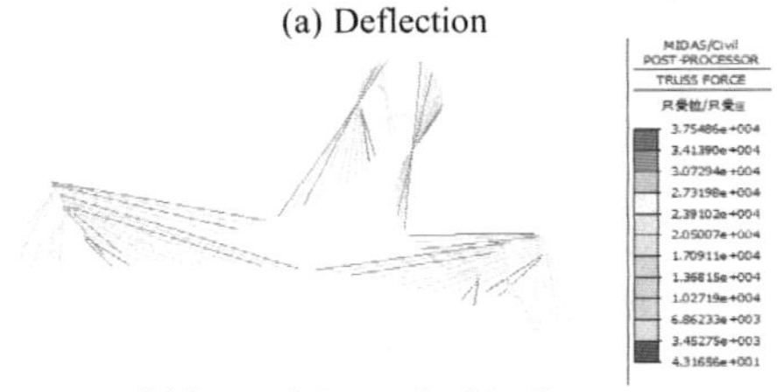

(b) Internal force of cable element

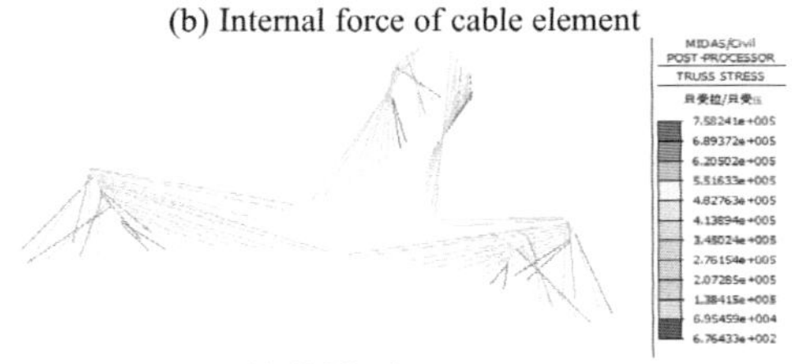

(c) Cable element stress

(c) Internal force of bridge tower

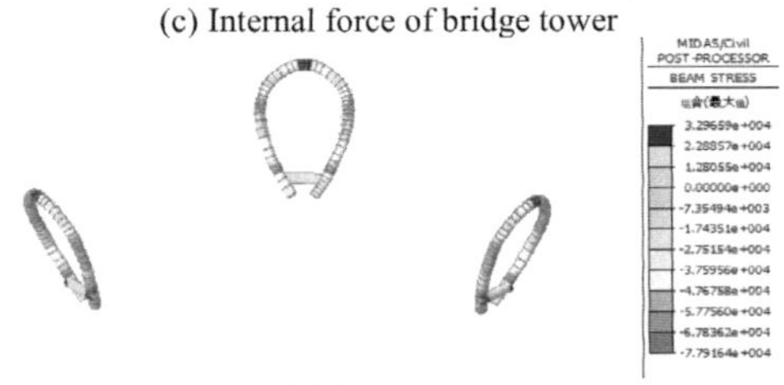

(e)Tower stress

Fig.8. Calculation Results under Vertical Load

The maximum vertical displacement occurs in the middle of the span. After deducting the form finding error of the completed bridge state, the vertical deflection under the vertical dead load is 0.224 m, which meets the requirement of 1 / 500 limit specified in the code.

The maximum internal force of the cable is 37548.6 KN, and the maximum stress of the cable is 758.2 MPa, which meets the strength requirements.

The maximum internal force of bridge tower is 537663 KN, and the maximum stress of bridge tower is 77.9 MPa, which meets the strength requirements.

5 Modal calculation results

The main cable is the main load-bearing structure of cable-stayed bridge, and the internal force of main cable and

 https://doi.org/10.1051/e3sconf/202123703036

sling is included in the form of initial tension in dynamic characteristic analysis and modeling. Based on Midas' functions of nonlinear static analysis, stress stiffening effect and modal analysis, the natural vibration characteristics are analyzed.

In order to avoid omitting any vibration mode, the sub block method is used to solve the characteristic equation.

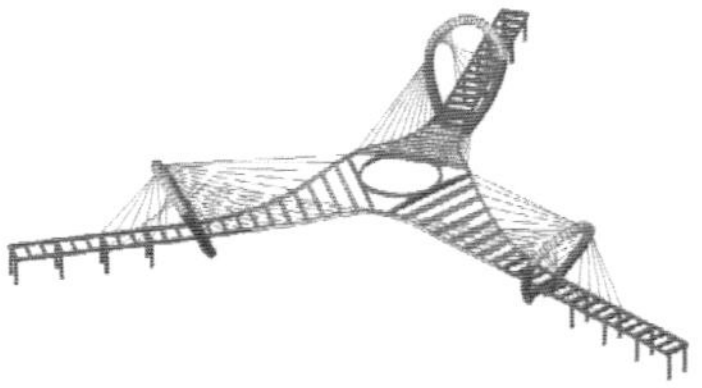

(a)first mode shape（Symmetrical vertical bending）（0.549Hz）

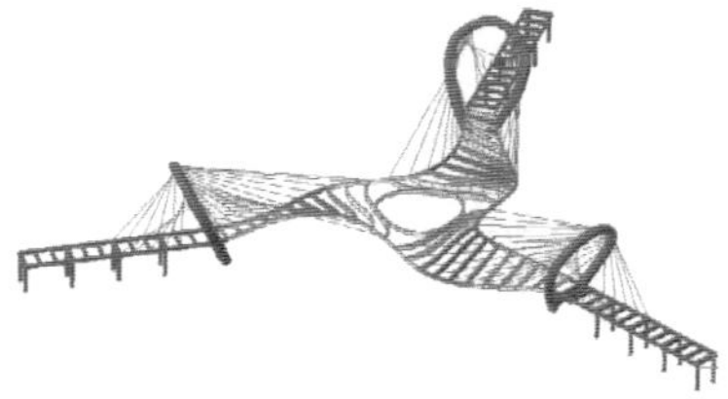

(b)The second mode shape（0.684Hz）

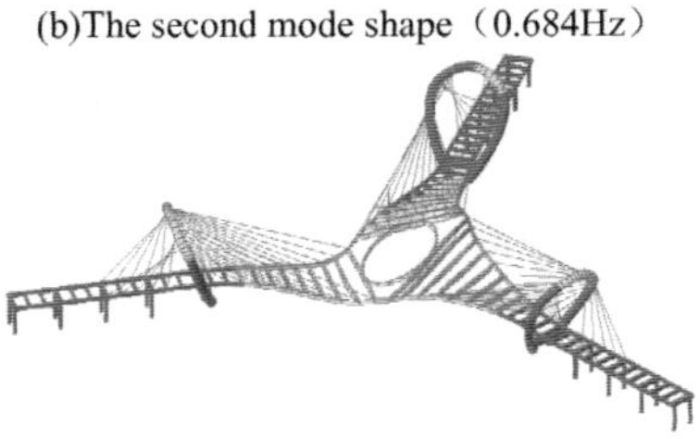

(c)The third mode shape（0.747Hz）

(d)The fourth mode shape（0.747Hz）

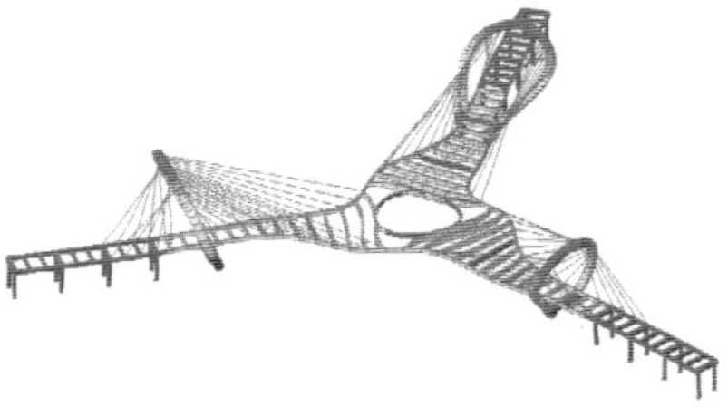

(e)The sixth mode shape（1.112Hz）

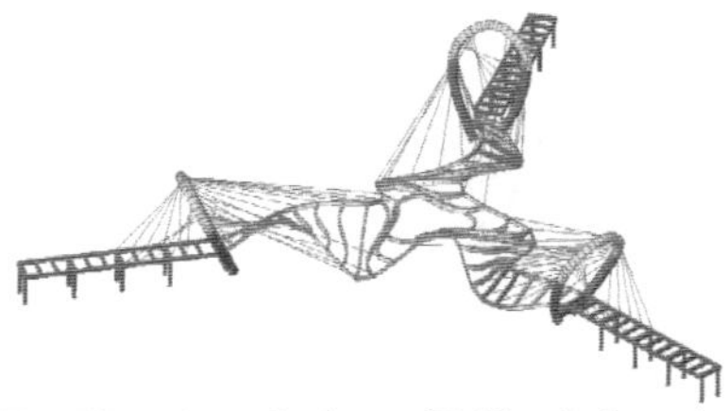

(f)The fifteenth mode shape（Bridge deck torsion）（1.568Hz）

Fig.9. Typical Dynamic Modal

According to the calculation results in the figure above, the first mode shape is symmetrical vertical bending, whose frequency is 0.549Hz; the fifteenth mode shape is bridge deck torsion,whose frequency is 1.568Hz.

On the whole, the vibration modes are dense, and obvious mode grouping phenomenon appears. The vibration modes are mainly vertical bending and side bending vibration, and the positive symmetric torsional vibration mode does not appear until the 15th order. The torsion bending frequency ratio is 2.86, which indicates that it has good wind resistance stability.

6 Conclusion

In this paper, taking the pedestrian landscape glass cable-stayed bridge of three-forked-shape canyon as the engineering background, the paper studies the configuration of the cable-stayed bridge with three fork deck of egg-shape arch tower, establishes the Midas finite element model, carries out the static analysis under the vertical load, and carries out the dynamic modal analysis:

(1)The traditional portal cable-stayed bridge tower of cable-stayed bridge is reformed, and a new type of egg-shape arch cable-stayed bridge tower is proposed. The egg-shape arch has the advantages of beautiful shape, small occupied area, convenient anchoring and dispersing cables, excellent mechanical properties and fast construction. The egg-shape arch bridge tower is very suitable for the narrow mountain landscape platform environment.

(2)The three forked pedestrian landscape glass bridge deck and egg-shape arch bridge tower match each other, and the spatial curved architectural structural elements coordinate with the environment of three-forked-shape curved Canyon, which greatly improves the cultural connotation of the scenic spot, and is the ideal combination of structure and architecture.

(3)The maximum vertical displacement occurs in the middle of the span, and the vertical deflection under the vertical live load is 0.224 m, which meets the requirements of the code. The maximum internal force of the cable is 37548.6 KN, and the maximum stress of the cable is 758.2 MPa, which meets the requirements of the specification.

(4)The results show that the main vibration modes are vertical bending and side bending vibration, and the positive symmetric torsional vibration mode does not appear until the 15th order. The ratio of torsional frequency to bending frequency is 2.86. The structure of the three bifurcated bridge is firm and the three legs stand at the same time, which has good wind stability.

References

1. Chen Ning, Zhu Jiaqi, Li Yongle. Conceptual design and implementation of long span pedestrian cable stayed bridge in Pescara, Italy [J]. *World Bridges*(in China),2011(05):4-6.
2. Vladimír, Š., Michal, P., & Tomáš, P. A Dynamic Analysis of the Cable-Stayed Footbridge in Čelákovice Town[J]. *Procedia Engineering*, 2017,199, 2877–2882.
3. Luo Xiaotian. Design of pedestrian landscape bridge in Nanxun Jinxiang Lake Park [J]. *World bridge*, 2013,41 (3): 33-37
4. Lai Yaping, Long Hao. Landscape and view: space construction of urban river crossing bridges [J]. *Journal of Chongqing Jiaotong University (NATURAL SCIENCE EDITION)*, 2019, 38 (03): 14-20 + 26
5. Yi, J., Li, J . Experimental and numerical study on seismic response of inclined tower legs of cable-stayed bridges during earthquakes [J]. *Engineering Structures*,2019, 183, 180-194.
6. Wan Yufan, Xu Zhiqin, Zheng Lijun. Design and Construction of Wide Deck Ovate Single-pylon Cable-stayed bridge[J]. *Urban Roads Bridges & Flood Control(in China)*,2012(08):69-74+380.
7. Yang Shanhong, Xu Hongguang, Mei Yinghua, Wang Shengbin. Design originality of arched tower cable-stayed bridge of main branch of maanshan Yangtze river bridge [J]. *Highway traffic technology (applied technology version)*,2009,5(05):192-193.(In Chinese)
8. Wang Fan, LAI Yaping, Qiao Yunqiang, Yan Fucheng. Research on the Design of Triangular Arch Bridge tower of Hechuan Qujiang Landscape Bridge [J]. *Chinese and Foreign Highways*, 2008,38(02):109-112.(in Chinese)
9. Wang Ruirui, Han Da-zhang, Zhou Yanfeng.Stress analysis of steel and concrete mixing tower of Nanjing Youth Olympic Landscape Bridge [J]. *Shanghai Highway*, 2017,(4):21-24.(in Chinese)
10. Xu Ruifeng, LI Zheng, ZHOU Yan-feng, HAN Da-zhang.Design of steel and Mixed Section of main Tower of Nanjing Youth Olympic Landscape Bridge across the River [J]. *Modern Traffic Technology*, 2015,(4):38-41.(in Chinese)

Research on Failure mechanism and Reinforcement Technology of Shangmaji Weir

MA Jia-bao[1], LIU Xue-ying[1*], and YAO Jian-feng[1]

[1]Zhejiang University of Water Resources and Electric Power, 310018 Hangzhou, Zhejiang, China

Abstract. After the reinforcement and repair of the Shangmaji weir in Wanyao Irrigation District of Jiangshan City in 2014, the results of the repair did not improve much. The surface layer of the weir fell off, these not only affected the drainage function of the weir, but also the appearance was different from what is now advocated "Green ecology" is incompatible. This paper analyzes the causes of damage to this problem, and obtains two solutions: drilling, grouting, vertical paving, and to extend the width of receiving water step. After adopting the research programme, the degree of shedding of pebbles is greatly reduced, and the strength of the weir is significantly increased.

1 Introduction

What people care most about whether the weir is reliable, durable, and whether it can fully play its role. For this phenomenon, many scholars at home and abroad have conducted in-depth studies on the repair and reinforcement of the weir. The rubber dam has a simple structure. The use of rubber dams to reinforce the stone dam is a common method, but it has some shortcomings. As time goes by, the rubber dam has serious leakage. For northern China, the water in the rubber dam freezes into ice in winter and needs to be broken., and maintenance is time-consuming and laborious[1]; the construction of permanent barrages to reinforce the dam body is also common, but this method has a long cycle and high input cost, and its effect is not very obvious[2]; for valley areas, the mud in the river bed Sand flow can cause damage to the dam body. Lowering the height of the silt dam is also a measure to protect the dam body. However, this method is only suitable for specific areas. For the barrage in plain areas, it is not practical[3]. A relatively primitive way to restore the dam body is to use the deep- rooted characteristics of plants to grow aquatic organisms to strengthen the dam body. This method is novel, but its cycle is relatively large. Plants need a cycle from planting to growth and rooting, and different water areas Plants that adapt to growth are also different, and they have very large uncertainties[4]; domestically, the method of strengthening the dam body is commonly used the method of punching and grabbing wells to backfill the clay anti-seepage technology. This technology has simple construction technology and easy quality control. The soil layer has high requirements for soil materials, and it is not suitable for strengthening earth-rock dams with larger rocks[5].

The above analysis shows that the damage situation of the dam body is different, and the solution measures are also different. Shangmaji Weir is located in Hecun Town, Jiangshan City, Zhejiang Province. The main material of the dam is made of concrete-filled masonry. The surface is made of cast-in-place concrete with a thickness of 1m. The surface is covered with pebbles. The water height is 0.46m, the drop platform is 1.6m, and the cross-sectional length of the dam is 9.4m. There are five steps in the overflow area of the Shangmaji Weir. Except for the top platform, the remaining four platforms all have damaged pits of different sizes. The pebbles on the surface are washed away, resulting in the continuous expansion of the gap area, exposing the concrete layer below; The same problem exists in the sand sluice area. The surface pebbles fall off, causing the entire step to be washed away, exposing the sand and gravel layer below, which not only causes soil erosion in the sand and gravel layer, but also endangers the block stone slope protection on the right bank. The situation will endanger the safety of residents on the shore. In response to the above problems, this article investigates and analyzes the damage mechanism of the weir, and gives a reasonable plan for reinforcement and repair to ensure the safety of the weir.

2 Damage Cause Analysis

2.1 Excessive water flow

The Shangmaji weir belongs to the Qiantang River system in Jiangshan City. The area of the Qiantang River system accounts for 91.3% of the total area of the city. Jiangshan Port is a first-level tributary of the upper reaches of the Qiantang River, with a total length of 134km, of which 105km in Jiangshan. It has the characteristics of short source and rapid flow, uneven distribution of precipitation from year to year, and large changes in flood and drought.

Jiangshan City belongs to the subtropical monsoon area, with four distinct seasons, sufficient sunshine, warm

* LIU *Xue-ying*: liuxy@zjweu.edu.cn

and humid, abundant rainfall, the multi-year average temperature is 17.1°C, the frost-free period is 253 days, the multi-year average rainfall is 1842.6mm, and the rainfall distribution is uneven. The rainfall from March to June is about It accounts for more than 57% of the total annual rainfall. It is prone to city-wide continuous rainstorms and heavy rainstorms, which can cause floods. After the rainy season, the precipitation decreases significantly, the temperature is high, the evaporation is large, and droughts are prone to occur in midsummer and early autumn, and are affected by typhoons during the period. External influences also often cause floods. The main stream of Jiangshan Port is a mountain river with large flood peaks. Floods often occur from May to July during the flood season. The floods rise and fall rapidly, and the time for water collection is short. The time from the maximum period of heavy rain to the peak of the flood is usually about 6 hours, it took 3-4 days. The maximum peak flow rate appeared in 1942 and was 4900 cubic meters per second; the maximum flow rate in five years was 1930 cubic meters per second.

It can be seen that the Shangmaji weir encounters a large water flow during the flood season every year. After many years of such erosion, the pebbles on the surface of the weir are easily damaged. As shown in Figure 1, although measures are taken every year to repair, they can only maintain for a short time. Around the year, the surface damage is more serious.

Fig1. Barrage damage map

2.2 The pebble laying method affects its stability

Since the surface of the pebbles is very smooth, the binding force of the concrete to the pebbles depends on the way the pebbles are laid. There are two ways to inlay the pebbles into the concrete surface, horizontal laid and vertical laid (Horizontal laying: big side up, vertical laying: small side up). Due to the large water flow of the Shangmaji weir, in this situation, the surface of the weir should be pavement and concrete, and the proportion of cement should be large, so that the pebbles insert the cement deeper and ensure that the combination is stronger. The pebbles are not washed away by the flood[6].

The overflow platform of the Shangmaji Barrage adopts the laying method of "leveling one layer, standing three layers, and leveling one layer"[7], so that the laying method will make the pebble on the outermost edge easily washed away by the large flow, and the inner side Laying pebbles is relatively safe. When a large flow of river water passes through the top platform of the embankment, the arc will fall on the middle or the outermost side of the lower platform, so the laying method of the inner pebbles is not restricted. As shown in Figure 2, the pebbles of the barrage basically fall off the outermost part first, causing the damage gap to become larger and larger, extending to the inside of the platform, exposing the concrete cushion.

Fig2. Shedding of outermost pebble

2.3 The pebble concrete layer and the cast-in-situ concrete have low adhesion

From the site damage, it can be seen that the pebble concrete layer and the surface cast-in-situ concrete with a thickness of 1m are a staged construction process. After the surface cast-in-situ concrete layer is completed, the curing and solidification of the surface layer are completed, and then the pebble concrete layer is laid for the second time. The process will cause the pebble layer and the cast-in-place layer to be unable to form a whole, and the bonding force is not strong enough. When a pebble is washed away, the surrounding pebble layer will suffer joint damage.

After the maintenance operations in recent years, partial repairs were also carried out, and the pebble layer and the lower cast-in-place concrete layer could not be formed as a whole, and the maintenance effect was not obvious.

2.4 The impact of concrete shrinkage

Concrete shrinkage is divided into three situations: drying shrinkage, spontaneous shrinkage and carbonization shrinkage. Among them, the drying shrinkage has the greatest impact, and is affected by the external environment, and the adsorbed water inside will evaporate over time[8]. Concrete shrinkage is directly related to internal factors such as water-cement ratio, cement dosage, aggregate, etc. External factors such as age, loading stress, and duration also directly affect concrete shrinkage.

The surface of the Shangmaji weir is inlaid with pebbles, and the shrinkage of the pebbles is negligible. The dam is directly exposed to the outside for a long time each year. In high temperature weather, the concrete thickness of the pebble concrete on the surface is about 5 cm, which is relatively thin. The drying shrinkage will reduce the bonding force between the concrete and the pebbles. When the overflow area of the weir is large, the scouring force of the pebbles on the surface of the weir will be greater, causing local damage.

3 3 Reinforcement and repair measures

3.1 Drilling, grouting and vertical paving cobble reinforcement method

3.1.1 Construction preparation

Construction tools include hand air drill, mixer, grout pump, grout pipe[9]. Before construction, clarify the spacing of the holes, the number of holes and the position of each hole. The width of receiving water step is 1.6m and the length of the weir is 180m. Therefore, two rows of vertical holes are used. The outer row is 0.45m from the outer edge, and the inner row is The distance between the holes is 0.45m from the inner edge, the distance between the two holes is 0.61m, the number of holes in each row is 90, the spacing is 2m, and there are 180 holes in total[10]. The borehole diameter is 45mm and the depth is 1.1m. Before construction, the pebble concrete within the radius of 0.2m with the center of the hole as the center is knocked out. When grouting, the concrete is poured in this area together, and then the pebble is erected to Make the pebble concrete and the lower layer become a whole, strengthen the bonding force between the pebble concrete and the concrete layer, and strengthen the stability of the weir.

3.1.2 Construction process

The drilling and grouting vertical paving pebble reinforcement method uses a 40mm PVC pipe inserted into the grouting hole, and the penetration depth is not less than 30cm[11]. When grouting, seal the contact point between the PVC pipe and the borehole, and keep the grouting pressure between 0.2 and 0.5Mpa. When grouting, the principle of first outside and then inside is used[12]. Now the outermost hole is grouted, and then the inner hole is filled. After the grouting overflows, pour a circular area with a surface diameter of 0.4m to a thickness of 6cm, and then pave the pebbles tightly into the concrete so that the upper surface of the pebbles exposes the concrete, and the exposed height shall not exceed 1cm.

3.1.3 Benefit Analysis

It can be seen from other similar domestic cases that this reinforcement measure is convenient to construct, has a short period, and has obvious effects. The pebble concrete on the surface and the concrete layer are tightly bonded and reliable, so that it can play its role even in the large flow of water, without damage.

3.2 Method of increasing the width of receiving water step

The height of receiving water step of the Shangmaji weir is 0.46m and the width of receiving water step is 1.6m. Therefore, the area is a flood-prone area. Combined with the damage of the weir, it can be seen that when the height of receiving water step remains the same, the width of receiving water step needs to be lengthened to relieve the scouring force of water on the step.

3.2.1 Construction preparation

Understand the structure of the dam before construction, and redesign the receiving water step of the dam. Since the pebble concrete and the concrete layer are not well bonded, the width of receiving water step needs to be lengthened to alleviate the shear damage caused by the impact of the water flow on the pebble layer. The width of the designed receiving water step is 2.2m, and the length of each receiving water step must be widened.

3.2.2 Construction process

The method of increasing the width of the receiving water step is to knock off the pebbles on the top surface of each layer on the weir. Construction according to the height of receiving water step is 0.6m and the width of receiving water step is2.2m, a total of five layers of platforms are used to Pouring concrete. After the maintenance is completed. Apply 6cm thick concrete on the surface, re-use the knocked-out pebbles and lay them on the concrete layer, so that the concrete fully wraps the pebbles, and the exposed pebbles at the top do not exceed 1cm. The side pebbles are laid with the same process, and concrete is used between each layer of pebbles. Pack tightly so that it fully envelops the pebbles.

According to the construction of the weir dam according to this operation, the concrete bottom plate downstream of the weir will not reach 7m in length, so the bottom plate needs to be lengthened. Excavate the outer edge of the bottom plate until the total length of the bottom plate is 7m, and the thickness is controlled to 1m thick. The support formwork is poured with concrete. After solidification, the formwork is removed and maintained to the design strength.

3.2.3 Benefit Analysis

It can be seen from similar domestic cases that extend the the width of the receiving water step has a significant buffering effect on water flow. Many engineering cases failed to buffer the impact of water flow, resulting in damage to engineering buildings. Increasing the length of the drop platform has the same effect as reducing the height of the drop platform, but the former is simple to construct, has a short cycle, saves a lot of human resources, and brings considerable economic benefits.

4 Conclusion

The damage of Shangmaji barrage is very common in our country, but the solutions are only partial repairs, which can't solve the actual problem. This thesis gives two solutions for this. The drilling, grouting, vertical paving fundamentally solves the pebble the problem of shedding is applicable to a variety of infrastructure construction cases. It not only preserves the ecological pebble layer on the surface, but also strengthens the overall structural stability of the weir and increases the service life of the weir; extend the width of receiving water step can effectively alleviate the flow of water the impact force not only meets the mechanical performance requirements, but also fundamentally solves the problem of pebble shedding.

References

1. YANG Xiaochen.(2016).Repair and reconstruction technology of rubber dam in urban river. Northeast Water Resources and Hydropower(02),14-16.
2. LIN Feng.(2014).Research on repair and repair technology of flooded sluice dam. Private Technology(05),166.
3. Marta Anna Lapuszek.(2019).Analysis and Modeling the Streambed Evolution After the Check Dam Restoration: the Case of Krzczonówka Stream. 20(1), 188-196.
4. & P.S. Abhilasha.(2020). Issues in using geosynthetics in hydraulic structures under restoration. 9(1), 22-27.
5. LI Long.(2018).Comparison and selection of reinforcement schemes for earth-rock dams in Wannian County. Sichuan Cement(12),358.
6. HUANG Yamei,WANG Lihua,CHEN Xirong. (2017). Coarse aggregate surface structure and its influence on concrete performance. People's Yellow River(09),129-134.
7. YE Lin.(2018).Analysis and Design Practice of Damage Forms of Sandy Gravel Weirs on Mountain Streams. Zhejiang Water Conservancy Technology(06), 42-44.
8. DING Xiaobin,WU Haotian & HU Chengjie. (2020). Summary of shrinkage and creep mechanism and influence factors of concrete. Sichuan Cement(08),22-26.
9. LI Wei.(2020).Application of Grouting in the Reinforcement of Stone Masonry Dam Body of Shuikou Reservoir. Journal of Anhui Technical College of Water Resources and Hydropower(03), 23-25.
10. LI Wenyi,WANG Jiangli.(2009).Application of Grouting Method for Reinforcement of Stone Masonry Dam in Reinforcement of Houhe Reservoir. China Water Resources(02),35-36.
11. LI Zongjie,JIANG Honghua.(2000).Grouting for Leakage Control and Reinforcement Technology of Stone Masonry Dam. Progress in Water Conservancy and Hydropower Technology(06), 50-53.
12. ZHAO Xiaoli,YANG Jinlu.(2007).Grouting reinforcement treatment of dry stone dam of Tapangou Reservoir. Journal of Water Resources and Civil Engineering(02),72-75.

Study on the influence of fly ash and silica fume with different dosage on concrete strength

GUO Yin-le[1], LIU Xue-ying[1*], and HU Yue-ping[2]

[1]Zhejiang University of Water Resources and Electric Power, 310018 Hangzhou, Zhejiang, China
[2]Zhejiang Huanlong Environmental Technology Co., Ltd, 324400 Quzhou, Zhejiang, China

Abstract. In order to improve the compressive strength of concrete, mineral admixtures can be added to concrete to replace part of the cement(C). Commonly used mineral admixtures include fly ash(FA) and silica fume(SF). Through a series of experiments, the effect of adding different amounts of fly ash and silica fume and double mixing on the strength of concrete was studied by using the same amount of substitute for cement under the same water-binder ratio. The results show that fly ash has a certain influence on improving the fluidity of concrete, and the influence on the strength of concrete is mainly in the later period; the influence of silica fume on the fluidity and strength of concrete is better than that of fly ash, and the main influence of silica fume is on the increase in strength. The early stage and the later stage are slower; the strength of concrete mixed with fly ash and silica fume at the same time increases first and then decreases with the increase of mineral admixtures.

1 Introduction

With the development of economy and the progress of engineering technology, roads and bridges and other projects continue to extend to areas with complex geology, which requires concrete to have higher performance of concrete[1]. Modern building construction is convenient and fast, with high structural strength, diverse structural systems, and complex and changeable use environments. The buildings show the characteristics of high-rise, large-span, lightweight, underground and harsh use environments. The strength of concrete used in the application is increasing day by day[2]. Mineral admixture is an important component of modern concrete, which plays an important role in improving the strength of concrete. Reasonable selection of the type and mixing ratio of mineral admixtures is of great significance to improve the construction quality of concrete[3].

Fly ash and silica fume are wastes of industrial development. They are discharged randomly to pollute the environment and endanger health. However, as admixtures of concrete, they have good benefits for improving the strength of concrete[4]. Many scientific researchers have carried out in-depth research on this and found that mixing industrial waste such as fly ash and silica fume in concrete has a certain impact on various performance indicators such as concrete strength. This research will study the influence and regularity of different dosages of fly ash and silica fume on the compressive strength of concrete through experiments, and provide reference conditions for the configuration of high-performance concrete.

2 Materials

The cement adopts Haolong brand P·O42.5 grade cement, with a specific surface area of 330m^2/kg, 3days flexural strength of 4.2MPa, compressive strength of 23.7MPa; 28days flexural strength of 7.4MPa, and compressive strength of 47.5MPa.

Fly ash adopts grade III fly ash, with a fineness of 17%, a moisture content of 0.2%, a loss on ignition of 5%, and a water demand ratio of 105%. All indicators have reached national standards.

The silica fume is produced by a processing plant in Shanxi.

The gravel(G) from a stone factory in Quzhou, Zhejiang Province, where the coarse aggregate is produced, uses a continuous gradation of 4.75-19mm, of which 9.5-19mm large stones account for 70%, 4.75-9.5mm small stones account for 30%, and the crush value is 8.5%.

The fine aggregate is made of river sand(S), which is produced in a sand field in Quzhou, Zhejiang, with a fineness modulus of 3.0. The gradation zone of this sand belongs to the second zone sand with a mud content of 1%.

The admixture adopts polycarboxylic acid high-performance water reducing agent (WR), solid content is 10.4%, pH=6.0, density is 1.024, and water reduction rate is 27%.

The water(W) is local tap water.

*LIU *Xue-ying*: liuxy@zjweu.edu.cn

 https://doi.org/10.1051/e3sconf/202123703038

3 Experimental Proportioning Scheme

In order to study the influence of fly ash and silica fume on the strength of concrete, concrete with a water-binder ratio of 0.3, a sand ratio of 40%, and a total amount of cementitious materials of 527Kg is the research object. Choose 10%, 20%, 30% of fly ash content, 6%, 8%, 10% of silica fume content, and study its influence on concrete strength by changing the content of fly ash and silica fume. The concrete mix ratio is shown in Table 1. The mineral admixture adopts the same method to replace cement. The mineral admixture of group A is fly ash, the mineral admixture of group B is silica fume, and the mineral admixture of group C is powder both coal ash and silica fume are mixed.

Table1. Concrete mix ratio.

Num	C/Kg	FA /%	SF /%	S/Kg	G/Kg	W/Kg	WR/ %
A0	527.00	0	0	686	1029	158	2
A1	474.30	10	0	686	1029	158	2
A2	421.60	20	0	686	1029	158	2
A3	368.90	30	0	686	1029	158	2
B4	495.38	0	6	686	1029	158	2
B5	484.84	0	8	686	1029	158	2
B6	474.30	0	10	686	1029	158	2
C7	442.68	10	6	686	1029	158	2
C8	421.60	10	10	686	1029	158	2
C9	379.44	20	8	686	1029	158	2
C10	337.28	30	6	686	1029	158	2
C11	316.20	30	10	686	1029	158	2

Each group of mix ratios were tested for compressive strength. The molding size of the test piece was 100mm×100mm×100mm. After the concrete aggregates were evenly mixed by a concrete mixer, they were manually vibrated on a vibrating table to form them. Demolding after 24hours. Indoor maintenance to the prescribed age. The pouring, curing, and compressive strength tests of the cube specimens are tested in accordance with GB/T50081-2019 "*Standard for Test Methods for Physical and Mechanical Properties of Concrete*".

Table2. Test workload.

Performance	Test age /d	Number per age	Specimen specifications
Compressive strength	3d、7d、28d	3	100mm × 100mm × 100mm

4 Analysis of results

4.1 Influence of different content of fly ash and silica fume on the performance of concrete

The two performance indicators of slump and expansion of each group of concrete have been tested. The experimental results are as follows.

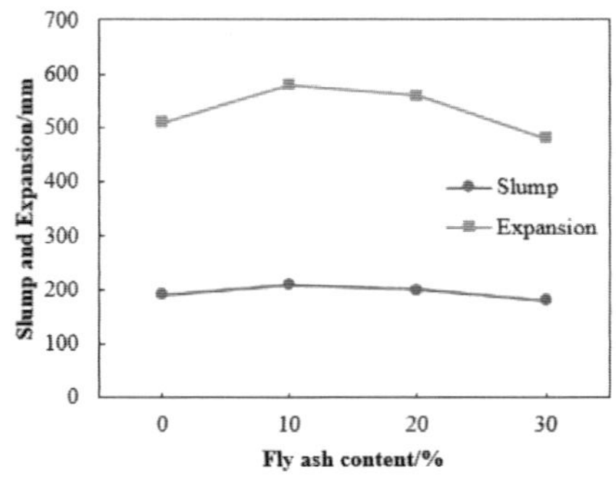

Fig1. Comparison of working performance of concrete with different content of fly ash.

It can be seen from Figure 1 that after replacing part of the cement in the concrete with the same amount of fly ash, the slump and expansion of the concrete mixture have certain changes. With the increase of fly ash content, the slump and expansion degree are also increasing. But when the fly ash content reaches about 15%, with the increase of fly ash, the slump and expansion are continuously reduced, and the concrete is too viscous, resulting in lower fluidity.

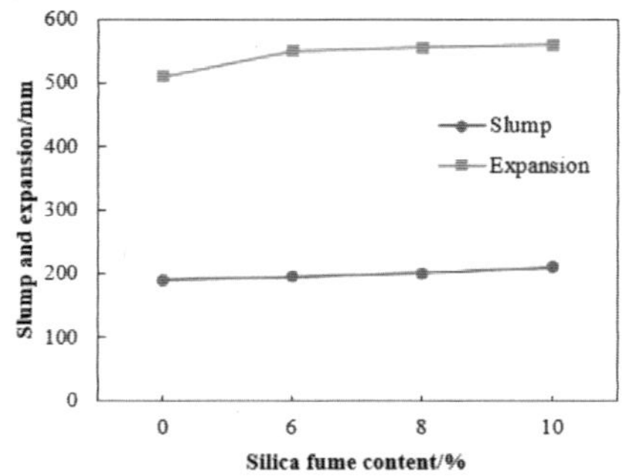

Fig2. Comparison of working performance of concrete with different content of fly ash.

It can be seen from Figure 2 that after replacing part of the cement in the concrete with the same amount of silica fume, the slump and expansion of the concrete mixture are significantly improved. With the continuous increase of silica fume, the degree of slump and expansion are also increasing. This is because silica fume can improve fluidity. On the one hand, the surface of silica fume is covered by a layer of active material, and the silica fume produces a ball between the cement particles, which increases the fluidity of the mixture; on the other hand, silica fume can replace the filling water in the voids of cement particles, turning it into free water, thereby improving the working performance of the mixture. The activity of silica fume is high. When the amount is too high, the viscosity of the slurry will increase and the fluidity will decrease. When the amount of silica fume is further increased, due to the large specific surface area of the silica fume, the excess silica fume will absorb a large amount of water, resulting in an increase in the water demand, and the excess water of the hydration reaction exists in the closed pores of the concrete, which will cause the decrease of its relative density and affect the strength of the concrete[5].

https://doi.org/10.1051/e3sconf/202123703038

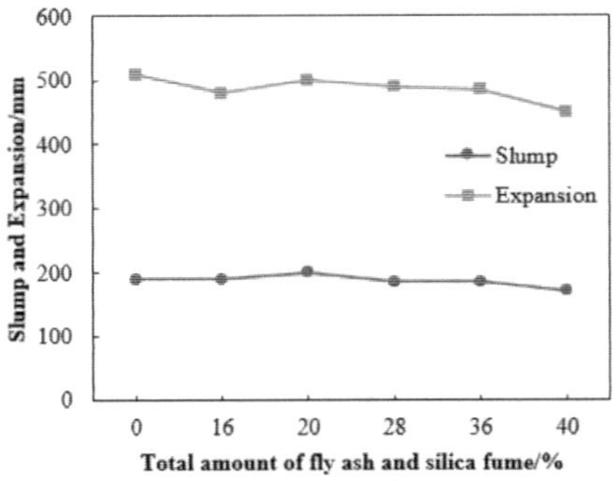

Fig3. Comparison of the performance of concrete with different admixtures.

From Figure 3, it can be found that when fly ash and silica fume are mixed together, the slump and expansion of concrete decrease with the increase of the admixture.

To sum up, the single addition of fly ash can improve the fluidity of concrete, and it will be reduced after excess. The single addition of silica fume can improve the workability of concrete, and the dual addition of fly ash and silica fume can also improve the fluidity of concrete.

4.2 Influence of different content of fly ash and silica fume on the strength of concrete

According to the test operation procedure, the test blocks of different ages were measured, and the results are shown in the following table:

Table3. Concrete compressive strength/MPa

Serial number	3d	7d	28d
A0	51.4	55.8	62.1
A1	50.5	62.9	77.5
A2	48.6	57.4	70.8
A3	31.5	41.3	55.2
B4	58.2	77.7	88.5
B5	61.4	80.1	89.8
B6	63.7	82.8	90.2
C7	40.2	57.2	63.1
C8	44.5	59.6	75.4
C9	43.7	58.7	72.1
C10	39.7	48.7	61.5
C11	31.5	42.7	58.5

4.2.1 Influence of Single Mixing Fly Ash on Concrete Strength

Figure 4 shows the compressive strength of concrete when only fly ash is added. It is observed that as the amount of fly ash increases, the compressive strength of concrete first increases and then decreases. Because fly ash has a significant active effect, the active ingredients can react with cement hydration products to form a pozzolanic reaction to form CSH gel, which improves the interfacial bonding strength correspondingly. There are less cement hydration products in the early stage, and fly ash participates in the secondary There is less chemical reaction, which leads to the decrease of the compressive strength of the concrete in the early stage. In the later stage, with the continuous increase of cement hydration products, it promotes and accelerates the pozzolanic reaction, which greatly promotes the growth of the later strength of the single-mixed fly ash concrete[6][7][8]. However, the amount of fly ash should not be too much. Obviously, the amount of 10% has a greater impact on the strength of concrete.

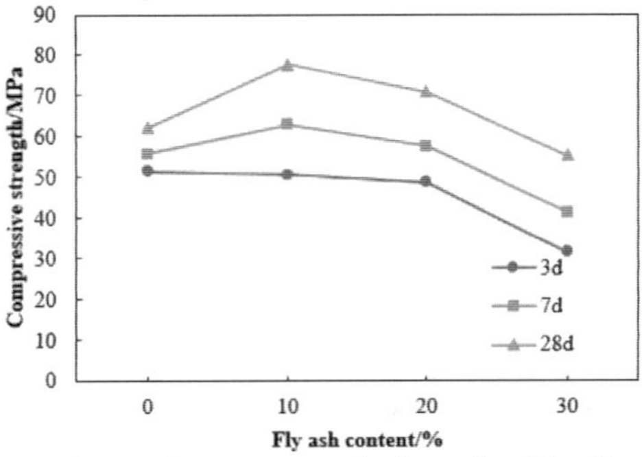

Fig4. Change of concrete strength when only adding fly ash.

4.2.2 Influence of Single Silica Fume on the Strength of Concrete

Figure 5 shows the strength of concrete when only silica fume is added. Observed from Figure 5, the strength of concrete is also increasing with the continuous increase of silica fume. After analyzing the strength of 3d, 7d and 28d, it is found that silica fume has a greater influence on the strength of concrete in the early stage, and the later influence is smaller than that of fly ash. Among them, 10% has the greatest influence on the strength of concrete. Compared with the influence of fly ash alone on the strength of concrete, the influence of silicon fume alone on the strength of concrete is greater in both the early and late stages.

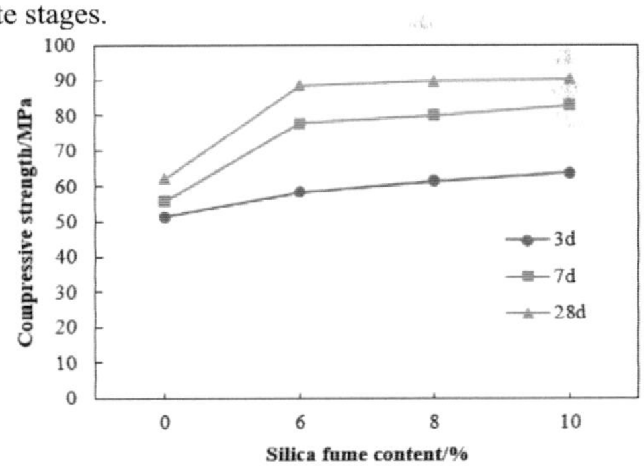

Fig5. Change of concrete strength when only adding silica fume.

4.2.3 Influence of double mixing fly ash and silica fume on the strength of concrete

Figure 6 shows the double mixing of 10% fly ash and 6% silica fume, 10% fly ash and 10% silica fume, 20% fly ash and 8% silica fume, 30% fly ash and 6% silica Powder, 30% fly ash and 10% silica fume concrete compressive strength. Analysis of this shows that with the increase of mineral admixture, the strength of concrete first increases and then decreases. It is lower than the strength of fly ash or silica fume alone. Although the strength has increased in the later period, the strength value is still low with the increase of the dosage, indicating that when the dosage is too large, the amount of cement is small, and relatively few high-strength gel products are generated, resulting in lower strength.

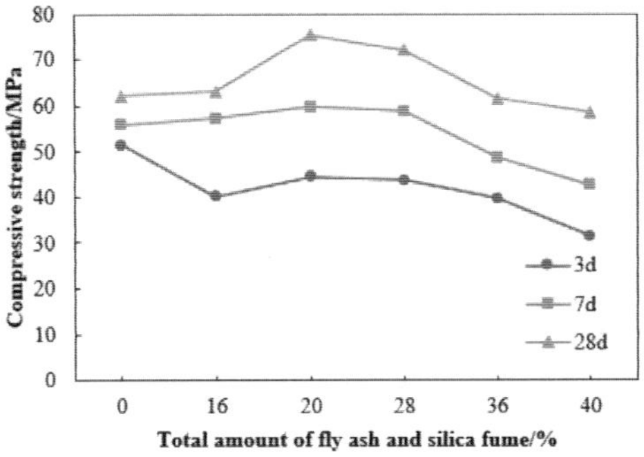

Fig6. Concrete strength change when double mixing.

5 Conclusion

In the case of a fixed water-binder ratio and total amount of cementitious materials, adding a certain amount of fly ash and silica fume can increase the strength of concrete and other physical and mechanical properties by replacing cement in the same amount.

When the same amount of fly ash is added separately to replace cement, the strength change trend is to increase first and then decrease. The strength in the early stage increases slowly, and the increase in the later strength increases. When the fly ash content is 10%, the concrete strength reaches the maximum and the fluidity is the best.

Single-doped silica fume replaces part of the cement, and the same curing time in the tested dosage range increases with the increase in the dosage, and the strength increases in the early stage and the later stage is relatively slow. When the content of silica fume reaches 10%, the performance of concrete reaches the best.

Double-doped silica fume and fly ash replace part of the cement. The strength of concrete increases first and then decreases. With the increase of mineral admixtures, the amount of cement decreases, resulting in relatively few high-strength cemented products, resulting in higher strength of concrete. low. When the content of fly ash is 10% and the content of silica fume is 10%, the compressive strength of concrete reaches the best at each age.

References

1. Tian Shuai, Lu Jianfeng, Li Shihua, Wang Junjie & Wang Mobi. (2020). The influence of mineral admixtures on the durability of high-performance concrete. *Concrete and cement products* (10), 12-16.
2. Li Qingfu, Sun Zhenhua & Zhang Haiyang. (2011). Experimental study on the influence of fly ash and silica fume on the strength of concrete. *Concrete* (05), 77-79.
3. Wang Junwei, Li Qiuyi, Qi Xiushan & Yue Gongbing. (2015). Research on the influence factors of high-performance concrete. *Concrete* (02), 94-97.
4. Li Shiyang. (2018). *Research on the influence of mineral admixtures on concrete strength and permeability* (Master's thesis, Harbin Engineering University).
5. Gu Zifeng, Lu Guanzhong, Huang Pai, Zhang Hao & Yu Guanglong. (2018). Experimental study on the influence of fly ash and silica fume on the strength of concrete. *Shanxi Construction* (23), 104-106.
6. Zhang Haiyang, Guo Jun, Zhang Xuhui & Li Weidong. (2014). The influence of fly ash and silica fume on the compressive strength of high-performance concrete. *China and Foreign Highway* (03), 312-316.
7. Ma Shaojun & Zhang Huili. (2010). Preparation technology of double-doped silica fume fly ash high-performance concrete. (eds.) *The second national hydraulic discharge building safety and anti-scouring new material development new technology application collection* (pp.107-110).
8. Gao Wei & Tai Surong. (2011). On the influence of silica fume on high-performance concrete. *Brand and standardization* (16), 49.

 https://doi.org/10.1051/e3sconf/202123703039

Influence of Silica Fume Content on Performance of High - Performance Concrete

ZHANG Heng-yuan[1], LIU Xue-ying[1*], and WEI Dan-feng[2]

[1]Zhejiang University of Water Resources and Electric Power, 310018 Hangzhou, Zhejiang, China
[2]Zhejiang Huanlong Environmental Technology Co., Ltd, 324400 Quzhou, Zhejiang, China

Abstract. The influence of different silica fume content on the slump, expansion, compressive strength and flexural strength of high-performance concrete was studied. The results show that with the increase of silica fume content, the slump, expansion, compressive strength, and flexural strength of high-performance concrete show a trend of increasing first and then decreasing. The optimal silica fume content of HPC(High-performance concrete) is 3% to 6%.

1 Introduction

With the rapid development of the global economy, the modernization construction is getting faster and bigger, and the requirements for the quality of the project are getting higher. For the safety of the project and the life extension, the government and experts pay more attention to the high-performance concrete. Due to its high workability, high durability, high volume stability and high strength, high performance concrete has been regarded as the best concrete in the world[1]. High-performance concrete reconfiguration requires low water-binder ratio, high-efficiency carboxylic acid water reducer, sufficient quantity of high-quality mineral admixtures and raw materials, and well-graded gravel[2]. In order to obtain high-performance concrete with good workability and rapid growth in the early stage of the project, a certain amount of silica fume is usually added. Silica powder is collected by special treatment from the fumes escaping from the exhaust gas during the process of smelting industrial silicon at high temperature in an industrial electric furnace. The chemical composition of silicon powder is mainly SiO_2, which is composed of extremely fine amorphous spherical particles. The particle size is usually about 0.1 ~ 0.15 μm, which is two orders of magnitude smaller than cement[3]. Small spherical silica fume particles are filled between the cement paste, so that the cementitious material has a better assembly, reducing the harmful pores in the cement paste[4], making the internal structure of the concrete denser and increasing its strength. The specific surface area of silica fume is large, and the free water molecules in the fresh concrete are restricted by silica fume particles, which greatly reduces the amount of bleeding and the accumulation of moisture at the concrete interface, thereby improving the bonding performance of the interface[5]. In this paper, the influence of silica fume content on HPC slump, expansion, compressive strength, and flexural strength is studied by mixing different amounts of silica fume.

2 Experiment

2.1 Selection of raw materials

2.1.1 cement

The cement(C) adopts Haolong brand P·O42.5 grade cement, and the measured physical properties of cement are shown in Table 1.

Table1. Cement physical performance index

Detect project	Ratio tablear ea (m^2/kg)	Pure pulp (mm)	Liquid ity (mm)	Flexural strength (Mpa)		compressivestr ength (Mpa)	
				3d	28d	3d	28d
Test results	325	310	280	6.4	9.2	27.3	47

2.1.2 Fly ash and mineral powder

Grade II fly ash(FA) and mineral powder(MP) produced by Guo Long are used. The physical properties of fly ash and mineral powder are shown in Table 2 and Table 3 respectively. After testing, the indicators meet the national standards of "*Fly Ash Used in Cement and Concrete*" (GB/T596-2017) and "*Granulated Blast Furnace Slag Powder Used in Cement, Mortar and Concrete*" (GB/T18046-2017) Indicator requirements.

*LIU *Xue-ying*: liuxy@zjweu.edu.cn

Table2. Fly ash physical performance index.

Test items	Fineness (%)	Ignition loss (%)	Moisture content (%)	Water demand ratio (%)
Detection value	17	5	0.3	103
National standard	≤30	≤8	≤1	≤105

Table3. Physical performance index of slag powder.

Test items	Liquidity ratio (%)	Moisture content (%)	Ignition loss (%)	7d activity index (%)	28d activity index (%)
Detection value	96	0.1	0.2	90	105
National standard	≥95	≤1	≤1	≥70	≥95

2.1.3 Silica fume

Using silica fume(SF) produced by a company in Shanghai, its main chemical composition is shown in Table 4.

Table4. Chemical composition of silicon powder.

Project	SiO_2	FeO_3	Al_2O_3	CaO	Ignition loss
Content /%	90.5	3.0	3.1	<1.7	<3.8

2.1.4 River sand and coarse aggregate

The river sand(RS) used has a mud content of 0.8%,awatercontent of 4%, an apparent density of 2580kg/m^3, a fineness modulus of 3.0, the gradation zone to which this sand belongs is the second zone sand.

The coarse aggregate(CA) particle size required for the configuration of high-performance concrete should not be greater than 20mm. The crushed stone adopts 5-20mm continuous gradation produced by Yangchen, with a mud content of 0.5%, a needle-flaky content of 3%, a crushing index of 6.5%, and an apparent density of 2610kg/m^3.

2.1.5 Water reducing agent

The water reducing agent（WR） adopts the high-efficiency polycarboxylic acid water reducing agent produced by Subote Company, the water reducing rate reaches 27%, the PH value is 6, 10.2% of the solid content.

2.2 Experimental program

HPC with a water-to-binder ratio of 0.28 is used as the basic mixing ratio, and 2.0% carboxylic acid water reducing agent is added to ensure its fluidity. Adding 11.4% of mineral powder and 8.6% of fly ash to replace cement to reduce the early hydration heat of concrete. Fixed water-binder ratio and total amount of cementitious materials (535kg), using internal mixing method[6], respectively 0%, 3%, 6%, 9%, 13%, 18%, 23% silica fume, etc. Equivalent replacement cement, a total of 7 sets of mixing ratios are shown in Table 5.

Table5. HPC mix design.

Num	*SF*/ %	*C* /kg	*MP*/kg	*FA*/ kg	*CA* /kg	*RS* /kg	*W* /kg	*WR* /kg
A0	0	428	61	46	1063	652	150	10.7
A1	3	415	61	46	1063	652	150	10.7
A2	6	402	61	46	1063	652	150	10.7
A3	9	389	61	46	1063	652	150	10.7
A4	13	372	61	46	1063	652	150	10.7
A5	18	351	61	46	1063	652	150	10.7
A6	23	330	61	46	1063	652	150	10.7

2.3 Experiment method

2.3.1 Specimen preparation

The laboratory mixer can make 20 liters of cement each time. According to the experimental design, the various raw materials are weighed according to 1:0.02, and the big stones and small stones in the coarse aggregate are weighed according to 8:2. First, pour the coarse aggregate and fine aggregate into the mixer, then add 50% water and stir for about 25 seconds, pour the admixture into it, mix the remaining 50% water and the water reducing agent, and then slowly pour Into the blender. After testing the degree of slump and expansion, install the mold, vibrate evenly with a vibrator, smooth with a shovel, and then classify and label it. After 24 hours, remove the mold and put it in a standard curing room for curing to the age and take out the test strength. The experiment is carried out according to relevant specifications or standards.

2.3.2 Testing method

The fluidity experiment is measured in accordance with the "*Standard for Test Methods for the Performance of Ordinary Concrete Mixtures*" (GB/T50080-2016). The strength test is determined in accordance with the "*Standard for Experimental Methods of Concrete Physical and Mechanical Properties*" (GB/T50081-2019). When testing the strength, in order to reduce errors, 3 test blocks are prepared for each age, and calculate its representative value. During the test, if the strength difference between the three test blocks is too large, it may be the result of uneven mixing of the concrete mortar placed in the installation mode or uneven vibration of the mold on the vibrator. The sample needs to be remade for the test.

3 Result analysis

3.1 The influence of silicon powder content on HPC fluidity

Figures 1 and 2 show the influence of silicon powder content on HPC expansion and slump.

From the figure, we can see that as the content of silica fume increases, the fluidity and slump of HPC show a trend of first increasing and then decreasing. When the content of silica fume is 0-6%, fluidity and expansion are better than without There is a significant increase in silicon powder, and the fluidity and expansion degree increase the most when the content of silicon powder is 3%. When the content of silica fume is more than 9%, the fluidity is lower than when the silica fume is not added. On the whole, adding a small amount of silica fume HPC is the reason for the increase in fluidity. The small spherical particles of silica fume can fill part of the space between the coarse particles of cement and replace the filling water in the gap. This part of the water can improve the fluidity of the concrete[7] . However, because the specific surface area of silicon powder is too large, with the increase of silicon powder content, the water demand increases sharply. While keeping the water consumption constant, the concrete becomes viscous, resulting in a rapid decrease in slump and expansion.

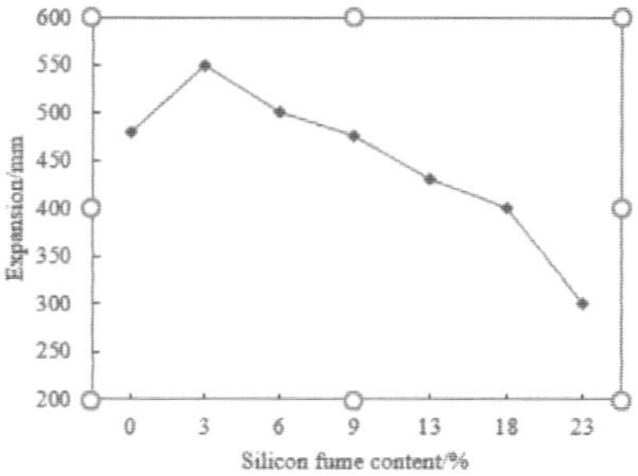

Fig1. . The influence of silica fume content on HPC expansion.

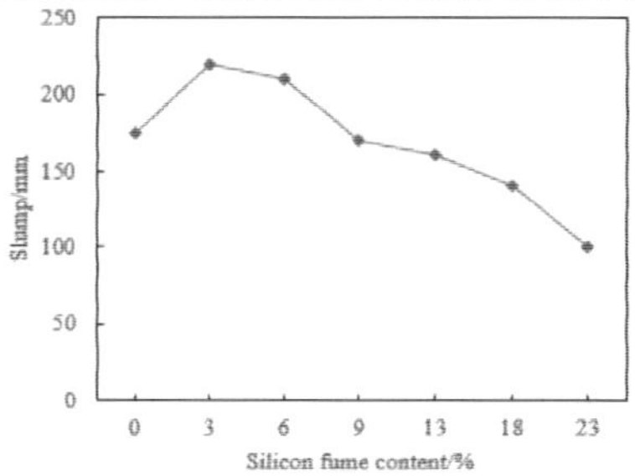

Fig2. The influence of silica fume content on HPC slump.

3.2 The influence of silicon powder content on the mechanical properties of HPC

The high-performance concrete test block with W/B=0.28 is placed in the standard curing room for 3d, 7d, 28d compressive strength and 28d flexural strength as shown in Figure 3 and Figure 4.

It can be seen from the figure that with the increase of silicon powder content, the compressive strength and flexural strength of HPC show a trend of first increasing and then decreasing. Comparing A0 and A1, it can be seen that the 3d compressive strength of A1 specimen is 18.46% higher than that of A0, the 7d compressive strength is increased by 6.5%, and the 28-day compressive strength is increased by 3.2%. 28 flexural strength increased by 17.93%. It can be seen that after adding a certain amount of silicon powder, the compressive strength and flexural strength of HPC have been significantly improved. On the one hand, because the particle size of the silica fume particles is small, the filling effect is achieved. The incorporation of silica fume reduces the porosity, improves the pore gradation, and reduces the pores larger than 0.1μm, making the HPC more compact, reducing internal defects, and improving strength. On the other hand, the key to increasing the strength of concrete with silica fume is to strengthen the bond between cement paste and aggregate. Silica fume has strong pozzolanic properties, and can react with cement hydration product $Ca(OH)_2$ to form a C-S-H gel with higher strength. C-S-H is type III in the early stage and type IV in the later stage. It acts as a skeleton in the cement paste [8], so the strength of HPC is very high. No silicon powder is added to A0, and fly ash is weak in activity, and sufficient $Ca(OH)_2$ is needed to cause secondary hydration. Therefore, the early fly ash does not participate in the cement hydration process. When the age reaches about 7 days, the concentration of $Ca(OH)_2$ in the test block increases, and the fly ash particles slowly hydrate with $Ca(OH)_2$. Generate CSH gel to increase the strength of concrete [9]. Therefore, by adding a certain amount of silicon powder, the early strength of HPC increases rapidly. Adding excessive silica fume, the compressive and flexural strength of HPC decreases because: the increase in silica fume content also increases the water demand. Under the condition that the water consumption remains unchanged, the excess silica fume and incomplete cement hydration lead to the internal HPC Unable to form a whole, internal defects increase and strength decreases. Moreover, adding excessive silica fume will greatly accelerate the early hydration process of cement, and the risk of cracks in the early concrete will greatly increase.

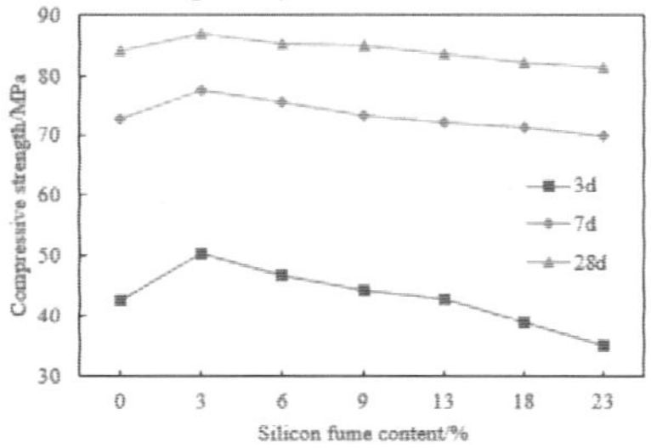

Fig3. Influence of Silicon Fume Content on HPC Compressive Strength.

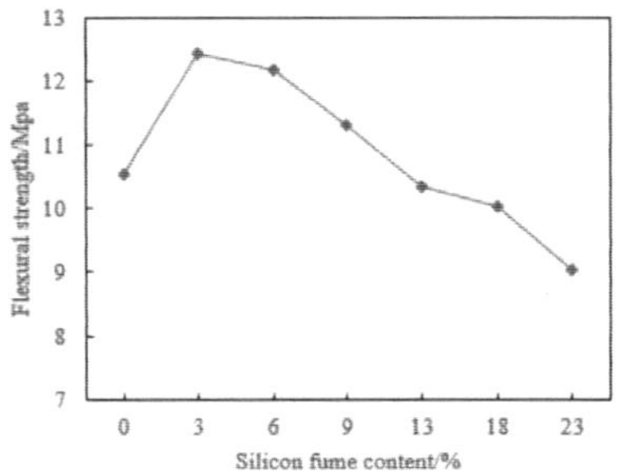

Fig4. Influence of Silicon Fume Content on the Flexural Strength of HPC28d.

4 Conclusion

With the increase in the amount of silicon powder incorporated, the slump and expansion of HPC show a trend of increasing first and then decreasing. When the silicon powder parameter is 3%, the HPC slump and expansion reach the maximum.

After HPC is mixed with a certain amount of silicon powder, the compressive strength and flexural strength of HPC will increase first and then decrease. When the content of silicon powder is 3% to 6%, the strength increase is most significant. When the content of silicon powder is greater than 9%, HPC will have an adverse effect and its strength will decrease accordingly.

Comprehensive workability, fluidity, and mechanical properties, the optimal content of HPC silica fume in this experiment is 3% to 6%.

References

1. WANG Man,LI Shiqi & DING Rui.(2017). Research progress of high performance concrete. Sichuan Cement(07),337.
2. DONG Fangyuan, ZHENG Shansuo & SONG Mingchen.(2018). Research progress of high performance concrete I: Design method of raw material and mix ratio. Material Guide (01),159-166.
3. MA Shaojun,ZHANG Huili & DONG Peng. (2005). Mixing technology of high performance concrete with double silica fume and fly ash. China Rural Water Conservancy and Hydropower (04), 64-67+69.
4. CUI Qiwen.(2008). Influence of silica fume on the performance of high performance concrete. Shanxi Architecture(13),193-195.
5. YANG Ping,PENG Zhenbin.(2002).Discussion on Application of Silica Fume in Concrete. Concrete (01),11-14.
6. GAO Wei,TAI Surong.(2011). Talking about the influence of silica fume on high performance concrete. Brand and standardization (16),49.
7. LI Qingfu,SUN Zhenhua & ZHANG Haiyang. (2018). Experimental study on the influence of fly ash and silica fume on the strength of concrete. Concrete (05),77-79.
8. LI Luhua,WANG Hailong & ZHAO Qingkuan. (2009). Influence of silica fume on high performance of concrete. Shanxi Water Conservancy Technology(03),28-30.
9. YAO Yan,WANG Lin & TIAN Pei.(2006). High performance concrete. Beijing: Chemical Industry Press, 22-40.

Overview of The Sustainable Technology System about Rural Residential Construction

Yuan Ling[1], Liu Xiaohui[1], Wang lisha[1], and Huang Xianming[1]

[1]Architectural Design and Research Institute of Tsinghua University Co., LTD, Chaoyang District, China

Abstract. This paper is based on the national key R&D project of China during the 13th Five-Year Plan—"Research On The Construction Technology System of Green Livable Rural Residential Buildings" (2018YFD1100205). Based on the quantitative analysis and systematic sorting of the research journals on sustainable rural residential construction, this paper summarizes the research progress and development trend of sustainable rural residential construction, analyzes the basic characteristics and common problems of rural residential construction in China at present, and proposes a sustainable construction strategy and research system according to local conditions.

1. Relevant research and practice

The main content of this paper is the sustainable technology of rural residential construction. We summarized the relevant researches on China and established the research framework, so as to provide a foundation for further research.

1.1 Literature analysis

The Research before 2012 focused on the overall rural construction in China, such as earthquake relief and new rural construction, natural disasters and policies have strong guidance. After 2013, the research focus has gradually been put into specific technologies, especially energy saving technologies, including thermal performance of building envelope, renewable energy utilization, indoor thermal environment and comfort,etc. After 2016, research on rural energy conservation is becoming mature, the latest research hotpots include building construction mode, main structure of building (including prefabricated building, industrialized building), function space,etc([1]).

In China, the research of building science on rural residence is relatively extensive, among which indoor thermal environment, energy saving, building envelope structure, flat planning design, solar energy and prefabricated building are the main concerns. The research on the construction technology of rural residential buildings mainly includes two aspects: the overall strategy and specific technologies, the research of energy saving and thermal environment-related technology takes the largest proportion in the specific technology research, and they are mainly concentrated in the super-cold zone and cold zone where the envelope is the key factor of energy saving.

Compared with the research in China, the international studies mainly in western developed countries has little research in the main structure and prefabricated building of rural residence, they pay more attention to the comfort of the built environment, sustainability, landscape ecology, the protection and renewal of rural cultural heritage. And the thermal performance of building envelope, indoor physical environment, pollution emissions and renewable energy utilization are also concerned internationally.

1.2 Other studies

By extracting the important scientific research and demonstration projects of rural residence in China in the past 20 years, and sorting them out according to the climate zone and time sequence, we found that the research focuses mainly concentrated in the plane optimization design, thermal performance of building envelope, improvement of local building materials, solar energy, heating system optimization,etc. And light steel structure system has been widely used in demonstration projects([2,3]). In south of China, the improvement of local building materials and structures, the inheritance of traditional architectural features and techniques, natural ventilation, reconstruction of existing buildings, and spontaneous construction are the main features of the research([4,5]).

2. Summary of problems

Based on the literature research and the demonstration projects, we summarizes the current construction characteristics of rural residence as follows:

- The living mode of rural residence is based on agricultural activities, and the function space is more

https://doi.org/10.1051/e3sconf/202123704001

diverse and complex than the living space in the city. Agricultural production, storage and even transportation all takes up part of the space of the farmhouse.

- The activities of village residents are open, most activities of villagers are outdoors, especially in the courtyards. Courtyards play an important role in rural life. Compared with indoor space, the use frequency of courtyard is higher and the function is more complex.
- As the homestead is owned by villagers, it will be built and expanded independently according to the changes of their family structure, economic level and lifestyle. In the long-term life, the use mode of the house is constantly changing along with the major life events such as "son marriage" and " have children ", people and environment influence each other to achieve a dynamic equilibrium([6]).
- Compared with the city, the regionality of rural residence is more obvious. The rural self-built houses in different regions have formed unique regional features under restrictions of population density, climate, topography, economic development level and other conditions. For example, through the investigation of self-built houses, Zhou Xiaohong([7]) and others found that the houses are mainly single-storey in north China, while those in the southern of Qinling Mountains-huaihe River are mostly multi-storey buildings.
- At present, China's rural aging problem is widespread. Most of the young and middle-aged workers choose to go out to work, while most of the old and children stay in the village. According to the analysis of the sixth census data([8]), the proportion of the elderly over 65 years old in villages and towns was 60.06 percent of the total population. Yao Yao([9]) made a survey of self-built houses in Anhua and found that 69 percent of the households have elderly people, and the elderly's intention is to provide for the aged at home. Therefore, the research on suitable aging of rural houses is also one of the main problems in sustainable village construction.

3. The technology system of rural residential construction

At present, most of the research on rural residential construction in China is aimed at specific content. Such as the safety of structural system, the energy-saving effect of building envelope, or the introduction of demonstration projects in specific areas and their applied technologies. There are few studies on the whole building system and the whole country. Which can be referred to is the research that studied by Liu Fang et al([10]), they proposed a technology system of well-off residential housing in villages and towns. However, since this system was put forward in 2015, which were introduced emphatically were the hotspot at that time, such as the energy-saving, new energy utilization and disaster prevention. After nearly 5 years of development, energy-saving, new energy and disaster prevention technology have become the normal design, which do not need separate chapter. Therefore, on the basis of referring to the above system, this paper proposes a concise framework for sustainable housing combining with the current construction characteristics, as shown in the following figure:

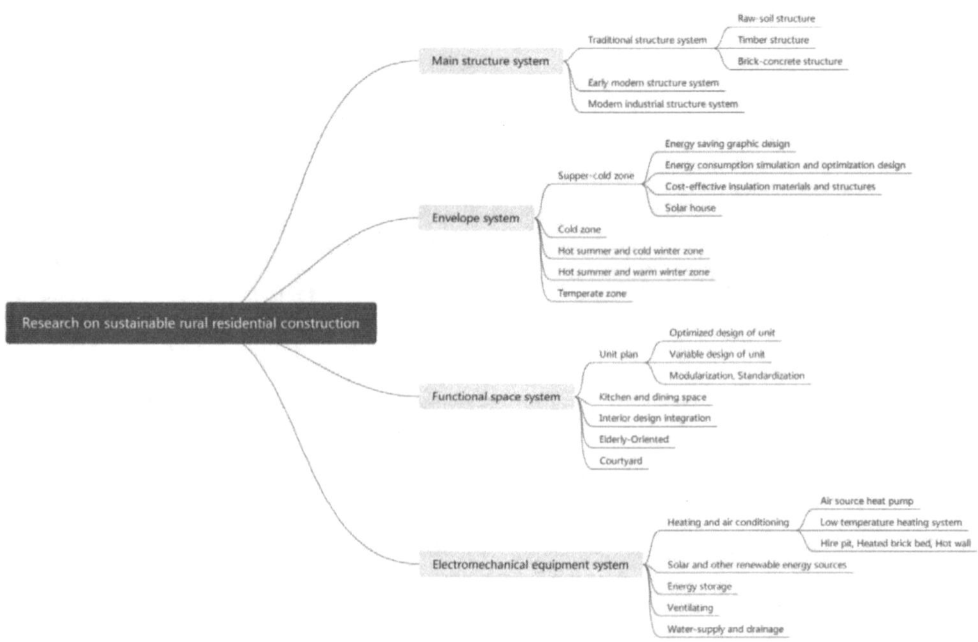

Fig. 1. Research framework for sustainable rural residence

The construction research system consists of three levels. The first level is classified according to the basic components of architecture, the second part is the key content of each basic component of architecture, and the last level is specific sustainable technologies.

3.1 The main structure of buildings

In recent years, more and more studies have been conducted on the residential structural system of villages and towns in China. The research contents most concentrated on the hot spots such as prefabricated building and industrialized building. In addition, Due to low price and low technical requirements, the traditional raw soil structure and brick-concrete structure still occupy a large proportion of use in villages and towns. Therefore, many scholars and designers are carrying out optimization research on the traditional system([11,12]). Based on the current research results, the author believes that the co-development of multiple structural forms is still the main trend in China in the near future. As long as the principle of adapting to local conditions is satisfied, various structural forms can be used as sustainable construction techniques.

Chinese traditional structure system includes raw soil structure, stone structure, timber structure and so on. The raw soil structure adopts adobe, lime or rammed earth to form load-bearing walls. It is characterized by the convenience of materials and construction, low cost, good heat preservation and fire resistance. The disadvantage is that the strength of raw soil structure is low and the structural integrity is poor, which is suitable for the areas with abundant raw soil resources. The stone structure system adopts stone block as load-bearing wall, which has similar advantages to raw soil structure and better weather resistance than raw soil structure. The disadvantages are no seismic structural measures, poor tensile strength and structural integrity of materials. Stone structure system is suitable for areas where is rich in stone resources. The timber structure is mainly loaded by timber, characterized by light weight, good earthquake resistance, rapid construction and ecological. But the fire resistance, insect control and anticorrosion ability of timber is poor. The use of timber structure is restricted by timber resources, which is more suitable for areas rich in timber resources.

Fig. 2. Raw soil structure([13])

Picture source:Wang Y, research on the sustainability development of traditional village planning and raw soil construction in yimeng mountain area.

Fig. 3. Stone structure
Picture source: Baidu pictures

The early modern structure system includes brick-wood structure, brick-concrete structure and cast-in-situ concrete structure. The vertical load-bearing wall and column of brick wood structure are brick masonry, and the floor slab and roof truss are made of wood. It is characterized by simple construction technology, low cost, flexible space separation, poor earthquake resistance and poor durability. It is suitable for areas with relatively rich wood resources and low seismic requirements. It should be noted that clay bricks are banned in China and should be replaced by recycled bricks. In brick-concrete structures, reinforced concrete is used on vertical load-bearing walls, constructional columns, floor slabs and roof panels. It is characterized by simple construction technology, low cost, good durability and high structure weight. The disadvantages are poor earthquake resistance, waste of resources and pollution of environmental. It is suitable for areas with low seismic requirements, and it is also necessary to pay attention to the problem that clay bricks need to be replaced by recycled bricks. Cast-in-situ concrete structures bear loads though cast-in-situ concrete. The advantages are simple construction technology, low cost and good comprehensive performance. The disadvantages are high labor cost, waste of resources and pollution of the environment. Cast-in-situ concrete structure is widely used in China at present.

Fig. 4. Brick-wood structure
Picture source: Baidu pictures

 https://doi.org/10.1051/e3sconf/202123704001

Fig. 5. Cast-in-situ structure
Picture source: Baidu encyclopedia

Modern industrial structure system includes precast concrete structure, steel structure, light steel structure, light timber structure and so on. Precast concrete structure adopts precast components to form the frame or shear wall to bear the load, which has the advantages of good seismic performance, heat preservation, sound insulation and other comprehensive performance. In addition, precast concrete structures perform well in resource intensive utilization and environmental pollution reduction. However, it is expensive to build, requires a higher level of industrial chain and workers' skill, and requires closer system integration. The steel structure system is a frame structure composed of shaped steel components. It adopts plate-type building envelope. The advantages of steel structure are similar to precast concrete, and the structural performance is superior. However, it also has the problem of high cost, high skill requirements for industrial workers, additional requirements for fire prevention and corrosion prevention. In addition, it is difficult to deal with the joints of the structure. Light steel structure is a composite load-bearing system composed of thin-walled light steel material. The advantages and disadvantages are similar to steel structure. These two structural forms are both not widely recognized and accepted. The light timber structure is a composite load-bearing system composed of light timber, with good comprehensive performance and more ecological and environmental protection than steel. But it is expensive and the timber resources need to be imported. Generally speaking, the above-mentioned modern industrial structure system is suitable for regions with relatively developed economy, relevant industrialization foundation, policy and financial support, and good transportation conditions.

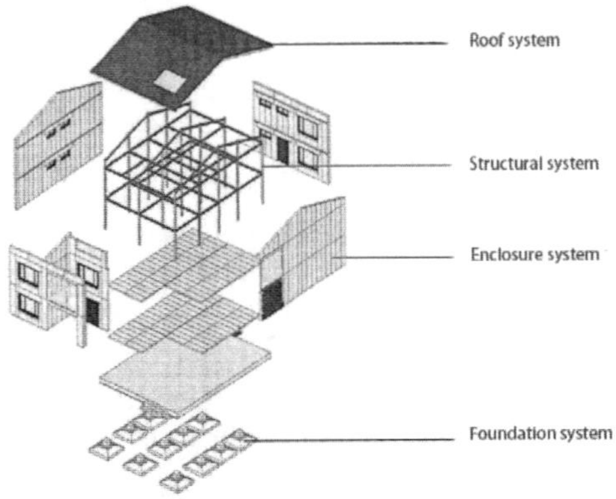

Fig. 6. Steel structure([14])
Picture source: Zhang L, Application discussion of the assembled steel structure in rural housing of severe cold areas.

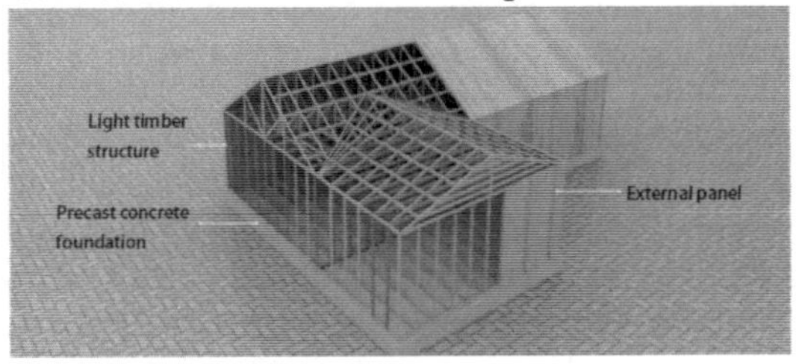

Fig. 7. Light timber structure([15])
Picture source: Wang L X, Study on the design of new wooden structure housing in jilin province.

3.2 Building envelope

The study of building envelope is mainly based on the climatic and regional characteristics, it can be summarized according to the climatic zones in China. Supper-cold zone and cold zone have similar research emphasis and suitability technology, but it should be noted that there are some differences between the two regions: most of the supper-cold zone are sparsely populated, with agriculture and animal husbandry as the main industry and rich natural resources, which are suitable for the application of renewable and natural materials such as raw soil, wood and straw. Most of the cold regions have relatively dense population, high degree of industrialization, relatively poor natural resources and great pressure on environmental protection, so it is more suitable to promote the utilization of industrial recycled materials and the improvement of traditional local materials and structures. In temperate zone, the performance requirements of envelope structures are not high, and there are few relevant studies, so there is no in-depth discussion in this paper.

https://doi.org/10.1051/e3sconf/202123704001

Fig. 8. Compressed straw building slabs([16])
Picture source: Jin H, A study on the construction technics of strawbale walls in severe cold rural areas of northeast China (in Chinese).

Fig. 9. Sandwich straw composite wall([16])
Picture source: Jin H,A study on the construction technics of strawbale walls in severe cold rural areas of northeast China (in Chinese).

In super-cold zone, the research emphases of building envelope mainly includes: energy saving design, energy saving renovation, thermal performance of materials and structures, indoor thermal comfort, passive technology, (super) low energy consumption technology, and renewable materials. The sustainability technologies of envelope in supper-cold zone mainly include:

- Energy-saving graphic design (system coefficient control, distinguishing heating and cooling areas,etc.);
- Energy consumption simulation and optimization design;
- Cost-effective thermal insulation materials and structures, such as composite insulation blocks, straw brick, thermal insulation curtain, roof composite thermal insulation system,etc;
- Solar house, passive house;
- Application of natural renewable materials.

On the basis of supper-cold zone, the research of heat insulation performance is added in cold zone. The sustainability technologies of envelope in cold zone mainly include:

- Energy consumption simulation and optimization design;
- Cost-effective thermal insulation materials and structures, such as composite insulation blocks, improved raw soil materials, roof composite insulation system, energy-saving doors and windows,etc;
- Solar house, passive house, and passive design combined with greenhouse;
- Industrial recycled materials, integrated insulation structure;
- Improvement of local materials and structures.

Research on building envelope in hot summer and cold winter zone mainly includes: energy-saving design and renovation, heat preservation and insulation performance of envelope structure, moisture-proof design and construction, combination with regional nature, culture and local skills. The sustainability technologies of envelope in hot summer and cold winter zone mainly include:

- Energy consumption, ventilation simulation and optimization design;
- Thermal insulation materials and structures, such as external thermal insulation materials, composite thermal insulation blocks, energy-saving doors and windows, elevated overhead roof;
- The ground moisture-proof and heat insulation composite technology;
- Industrial recycled materials;
- Improvement of local materials and structures.

Research on building envelope structure in hot summer and warm winter zone mainly includes: energy saving design, thermal insulation performance of envelope structure, ventilation, thermal environment and shading,etc. The sustainability technologies of envelope in hot summer and warm winter zone mainly include:

- Energy consumption, ventilation simulation and optimization design;
- Shading of windows;
- Roof and external wall structure;
- Patio ventilation, the pulling air shaft.

3.3 Function space

Functional space is the most important part of the whole residential construction research, because the function of the building plays an integral role in the study of other parts. It mainly includes architecture, interior design, landscape and parts design,etc. Based on the quantitative analysis of relevant literature, we divided it into five main fields, namely: unit plans, kitchen and dining room, interior design integration, elderly-oriented and courtyard.

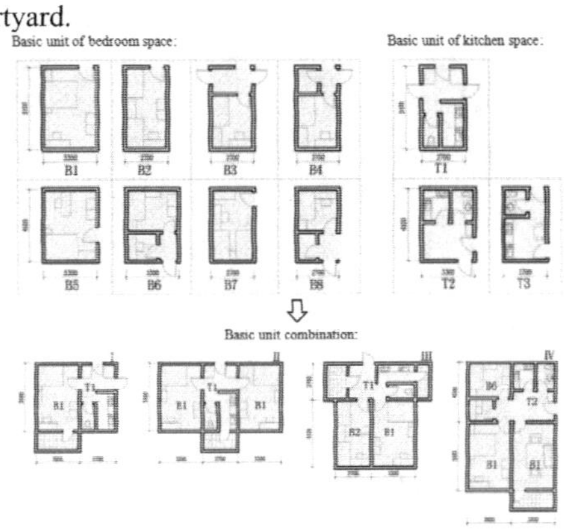

Fig. 10. "Basic room" setting and combination([17])

Picture source: Chen T, Layout standardized design and diversity application research of light steel prefabricated rural house.

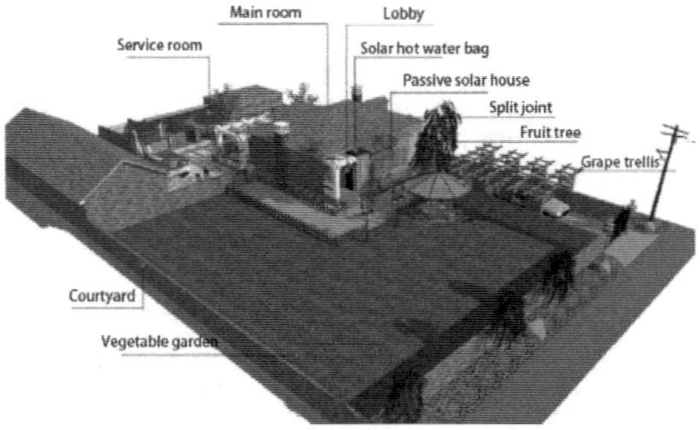

Fig. 11. Example of rural residential graphic design([18]) Picture source: Cong Z C, The research on green mode modeling of northeast rural dwelling.

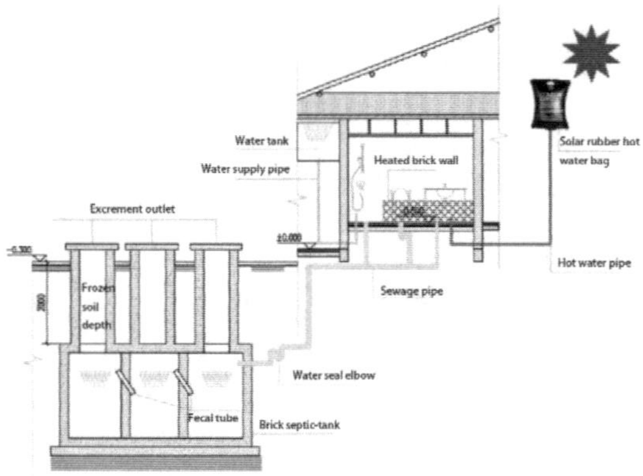

Fig. 12. Example of rural residential toilet equipment([18]) Picture source: Cong Z C, The research on green mode modeling of northeast rural dwelling.

The research emphases of unit plans are: historical evolution and current situation, demands of modern living, adaptability and variability of whole life cycle, connection between unit plans and industrialization (prefabricated construction). Sustainability research technologies include optimized design of unit plans, variable unit plans design, modular and standardized design. The research focus of kitchen and dining room is traditional customs and living habits, public and family communication space, dynamic line design,etc. Sustainable technologies include integrated design, improved kitchen design, kitchen storage system,etc. The research of the integration of internal assembly is mainly related to the industrialization of internal assembly and system integration. Sustainable suitability technologies include: system integration, integral storage system, overhead floor and wiring system, integral kitchen and bathroom system, furniture system,etc. The research on elderly-oriented focuses on the characteristics and needs of home-based care, the psychological and physiological characteristics of the elderly in rural areas, and the sharing of the elderly and children. Sustainable adaptive technologies include retrofit and design for aging, aging parts,etc. The research about courtyard focuses on the physical environment, use demand and efficiency, land conservation and ecology, courtyard economy,etc. Sustainable adaptive technologies include integrated design, variable design, physical environment comfort, courtyard ecological environment and so on.

3.4 Equipment and pipeline

Equipment and pipeline is mainly aimed at HVAC, water supply and drainage, electrical engineering research. Due to the particularity of rural residence, there are few studies on electromechanical engineering. The main contents of equipment and pipeline research include: heating and air conditioning, solar energy and other renewable energy utilization, energy storage, ventilation, water supply and drainage pipeline system,etc.

The research emphases of heating and air conditioning mainly include: heating load calculation and simulation, system optimization design, combination and selection of different heat sources and air conditioning terminal, improvement of traditional heating mode. Sustainable suitability technologies include air source heat pump system, low temperature heating system, kang (fire wall) and so on. Solar energy and other renewable energy research focus on solar heating and building integration design. Sustainable technologies include solar heating and hot water utilization, integrated design of buildings, solar energy combined with other heat sources for heating, photo-voltaic power generation and biomass energy. The main research content in the field of energy storage are energy storage materials and components, the combination of energy storage and solar energy. Sustainability technologies include phase change materials and modules, thermal storage floors and walls and other components. Research on water supply and drainage focus on improvement of aqua privy and sewage treatment,etc. Specific technologies include water-saving appliances, pipes for water supply and drainage, rainwater collection and utilization, and utilization of small-scale sewage treatment.

4. Summary

The research on "sustainable rural residential construction technology" has a rich foundation. Based on these studies, we propose a technology system and draw up a detailed list of construction technologies. Finally, we summarize the following measures and suggestions:

- According to the general problems, it is necessary to further study the people's different demands for rural residence under different economic development levels, locations and industrial conditions in the next 5-10 years;
- Based on these requirements, to improve and perfect the corresponding construction mode, and form a systematic construction strategy;
- Based on the above summarized technology system, to put forward relevant supporting technical

standards and key index parameters for construction, and develop the key database for the design and construction of green livable towns and villages;

- It is necessary to pay attention to the protection of traditional villages, rural buildings, historic and cultural villages and towns, so as to maintain regional characteristics;
- The promotion of green technology needs the government's support, and the guidance of national policy plays an important role in the construction of green livable villages and towns.

References

1. Yuan L, et al. Literature Review of Construction Technologies for Green and Livable Rural Housing. Community,(2019).
2. Jin H, et al. Designing rural houses of low energy consumption and high comfort in response to the extremely cold climate: a case study of an immigrant village in woniuhe town, Zhalantun. Architectural journal, (2015):74-77.
3. An S, The construction and application of the light assembly plate housing system:a case study of the rural house in zoucheng,shandong, Master's thesis of Southeast University,(2017).
4. Zhang S H, Construct new countryside: experience from "New Zhenshan environmental housing project"in Guizhou province, Huazhoong architecture,(2007):63-65.
5. Zhang M Z, Ren W Z, Practice and thinking of sustainable rural construction:a case study on Anji eco-house. Eco-city and green building,(2012):80-85.
6. Li B, Chen C, Form and use of rural house in natural village:a case study of village J in Shanghai Suburb. Architectural journal,(2011):116-120.
7. Zhou X H, Chu B, Research on the usage of living rooms and the configuration of furniture in self-built rural housing. Architectural journal,(2011):12-17.
8. Cai Q, Xiang HL, Analysis on the current situation and development trend of China's rural ageing: Based on the data of the sixth national population census. Journal of hubei polytechnic institute,(2013):99-104.
9. Yao Y, The research on the current situation and the adaptability for the elderly of self-built houses in rural areas of Anhua, Master's thesis of Hunan University,(2018).
10. Liu F, Yi B Y, Technology System Research Approach and Establishment,Applicable in Building Well-off Residential Housing in Villages and Towns. Housing science.(2015):29-32.
11. Li W J, et al. Research and ponderation on the precast building structure system based on the new rural construction. Structural engineers,(2018):173-178.
12. Zhang P P, Exploration on prefabricated residential buildings in rural development. Master's thesis of Zhe Jiang University,(2018).
13. Wang Y, Research on the sustainability development of traditional village planning and raw soil construction in yimeng mountain area. Chang'an University,(2015).
14. Zhang L, Application discussion of the assembled steel structure in rural housing of severe cold areas. Dalian University of technology,(2017).
15. Wang LX, Study on the design of new wooden structure housing in jilin province. Jilin Jianzhu University,(2018).
16. Jin H, Kang J, Jin Y M, et al. A study on the construction technics of strawbale walls in severe cold rural areas of northeast China (in Chinese). Sci Sin Tech,(2016),46:1079-1085.
17. Chen T, Layout standardized design and diversity application research of light steel prefabricated rural house, Beijing University of technology.
18. Cong Z C, The research on green mode modeling of northeast rural dwelling:Dwellings in heilongjiang province as an example. Xi'an University of architecture and technology,(2014).

https://doi.org/10.1051/e3sconf/202123704002

Analysis of urban spatial cognition in Luzhou city based on city image theory

Lin Lu, Peijia Jiang, Xu Zhang, Yu Zhao

College of Landscape Architecture, Sichuan Agricultural University, No.211, Huimin Road, Wenjiang District, Chengdu, China

Abstract. This paper uses the methods of picture recognition, sketch map and questionnaire to understand the citizens' evaluation of Luzhou's urban environment, obtain the overall city image and influencing factors to explore the spatial composition and city characteristics of Luzhou.Based on the analysis of the deficiencies in the process of urban development, this paper puts forward some appropriate suggestions to construct a good city image of luzhou city space.

1 Introduction

City imagery[1] is a thought of urban planning and design proposed by Kevin Lynch after thorough research and analysis of cities such as Boston, Jersey City and Los Angeles.The "city image"is a pioneer in modern city space research by introducing the psychological factors of citizens instead of the previous seemingly objective academic composition rules.

While luzhou is located in the south of sichuan province, is the "Chinese famous historical and cultural city", "National Double Support Model City", "National Health City", "Sichuan Province Social Security Comprehensive Management Model Area", with outstanding location advantages and geographical conditions,is the commercial center and important material distribution center in the junction of Sichuan, Yunnan,Guizhou and Chongqing.This paper investigates the image of Luzhou,aiming to provide the basis for urban design, establish a good city image space to bring enough attraction for citizens and create a "safe, convenient, comfortable and distinctive" city pattern.

2 The research methods

This study uses picture recognition,sketch map[2] and questionnaire[3] to analyze the city image of Luzhou.The picture recognition is the first step of the survey.We selected 30 points from the map of Luzhou and took 80 photos of landscapes and buildings to these points through field investigation.Select 40 of them and conduct a small sample survey.The small sample was 10 people, all luzhou residents who had lived in Luzhou for more than 10 years.Then selected 26 photos that can be recognized by more than 10 people from 40 photos and numbered as the photos used in the survey.Then the investigators presented the photos to the respondents and asked them to say the name or the general location of the scene in Luzhou. One of the two answers was correct, and put the picture aside. Finally, numbered the photos recognized by the respondents and mark them on the questionnaire (Figures 1).The second method is the sketch map. Respondents need to outline a simple map of their city(Figure 2).Such a simple map of the cognitive city is called the "Sketch Map", which used to obtain data on how people perceive the environment.For the convenience of research, this paper uses the term cognitive map and limits the concept of cognitive map to "People's perception of imagery".

Fig. 1. Picture recognition

The basic data of the survey consists of photo identification data, cognitive map data and questionnaire data.A total of 122 questionnaires were issued in the survey that obtained photo data, and the effective rate of the questionnaires was 82%.Cognitive map data was obtained by asking respondents to draw a map of Luzhou. Questionnaire interview data were obtained by filling in a questionnaire(survey form).

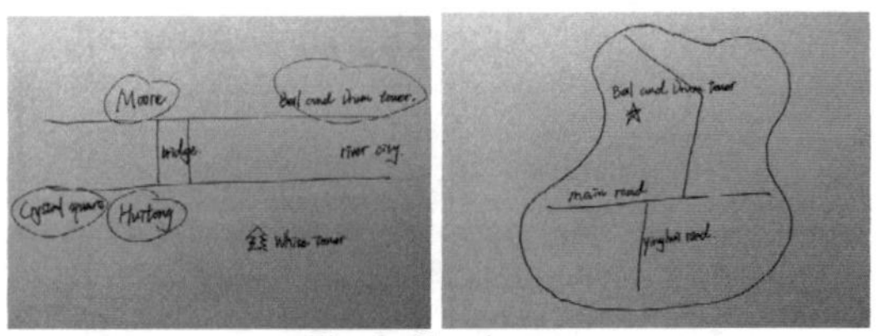

Fig. 2. Sketch map

 https://doi.org/10.1051/e3sconf/202123704002

3 The results of survey

3.1 The results of picture re-read

The 26 pictures were divided into 7 categories according to their functions,namely landmark buildings,leisure and entertainment places,commercial places,polpulation center,cultural and sports places,scenic spots and political landmarks.The overall identification of photos by the social group is shown in Table 1.

3.2 Sketch map

We received 33 sketch map.According to the scope of the cognitive map drawn by the respondents, it can be divided into four categories(Figure 3).According to the analysis of the sketch map,the city image is based on the pattern of common interaction between city and people. Because people have the closest interaction with the region where they live and work,the sketch map is mostly based on the starting point or scope of residence and work place,and is more detailed than other regions.Commercial centers are closely related to the daily life of citizens that have more opportunities to contact with them,so their intentionality is also very high.

The most common element in the sketch map is the road,which shows that the road is the main frame of Luzhou's spatial image. It plays a role in connecting other elements and helping people to conduct spatial location.Commercial buildings, such as Huitong - Moore and other large shopping malls,appear more frequently in the logo elements, followed by buildings with certain external features (such as bell and drum towers).The boundaries of Luzhou Administrative district are very obvious on the map, but administrative districts appear very rarely in sketch map,which are significantly lower than commercial districts and residential districts.Due to the special geographical location of Luzhou, the appearance of boundaries is mostly combined with rivers and paths.Thus,it can be seen that the macro spatial structure image of luzhou is supported by the main roads,with the commercial district and residential area as the background,and the three-dimensional frame structure as the characteristic bulidings.

Table.1. Photo identification number statistics

Classification of city functions	spot	Number of people recognized (person)	College Students (%)	Residents (%)	Floating population (%)
landmark buildings	1.Yangtze river bridge	72	55.88	95.00	57.69
	2.Tuojiang Second Bridge	86	79.41	90.00	88.46
	3.Luzhou Old Cellar Hotel	100	100.00	100.00	100.00
	4.Bell and Drum Tower	100	100.00	100.00	100.00
	5.White Tower	100	100.00	100.00	100.00
leisure and entertainment places	6.Culture square	90	88.24	95.00	84.62
	7.National Pits square	75	82.35	80.00	57.69
	8.Hundreds of children square	61	52.94	75.00	50.00
	9.Riverside road	88	85.29	92.50	84.62
commercial and shopping places	10.Huitong - Moore Business District	92	94.12	92.50	88.46
	11.Small wholesale market	59	44.12	80.00	46.15
	12.Wang's Mall	23	14.71	30.00	23.08
	13.Pedestrian street	36	41.18	45.00	15.38
polpulation center	14.Luzhou Bus Station	40	35.29	32.50	57.69
	15.Luzhou Lantian Airport	76	79.41	82.50	61.54
	16.Luzhou Extra-Long Passenger station	32	32.35	22.50	46.15
cultural and sports places	17.Luzhou Library	33	64.71	17.50	15.38
	18.Wenxuan bookstore	65	82.35	62.50	46.15
	19.Luzhou Gymnasium	45	52.94	52.50	23.08
	20.Youth Activity Center	22	14.71	37.50	7.69
scenic spots	21.Zhangba Longgan Woods	38	32.35	52.50	23.08
	22.Sweet park	21	5.89	42.50	7.69
	23.Zhong Shan Park	56	52.94	72.50	34.62
	24.Dragon through customs	24	20.59	32.50	15.38
	25.Longmatan District	22	14.71	30.00	19.23
political landmarks	26.martyr cemetery	41	38.24	55.00	23.08

 https://doi.org/10.1051/e3sconf/202123704002

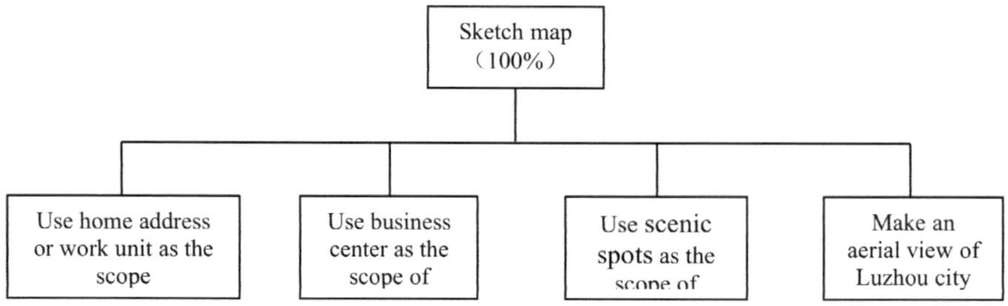

Fig.3. Sketch map range classification map

4 Image analysis of Luzhou

4.1 Analysis of Image elements

(1).**Path and edge.**The Yangtze River and Tuojiang River are not only the waterway, but also the natural boundary of the town, which has become the basic reference system for residents to remember the urban environment.All main roads are parallel or perpendicular to the Yangtze River or tuojiang River, which is conducive to the judge the general direction.At the same time, they also play the role of traffic passage and boundary,together with the Yangtze River and Tuojiang River, constitute the urban cognitive structure of the whole Luzhou City.

(2).**District.**Luzhou is located at the confluence of the Yangtze River and the Tuojiang River. The two rivers clearly divide the city into three regions -- Shizhong District (part of Jiangyang District), Jiangyang District and Longmatan District.Therefore, respondents are familiar with the environment of their own region, while the other two regions are relatively vague. In the public image of the whole city, the occurrence rate of all city elements is not very high.

(3).**Node and landmark.**As the direction of the Yangtze River and The Tuojiang River determines the direction of city roads,the respondent does not have a clear sense of direction like northerners,so the nodes and markers play a more prominent role in the identification of the environment.

Riverside road constitutes the largest node in Shizhong District.It is built along the river, where the Tuojiang River merges into the Yangtze River,the scene is spectacular.Although the respondents' memory of the city were more focused on the area where they were located , their memory of the environment in this area was particularly clear.Almost all of the environmental elements that appear most frequently in public imagery are in this area.Among them,Riverside road is 88%, Luzhou Old Cellar Hotel is 100%, Yangtze River Bridge is 72%, Tuojiang Bridge is 86%, culture Square is 90%.

In addition,the city's landmarks include the White Tower,the Bell and Drum Tower, the National Pits square, and the Huitong-Moore Business Center.Although Luzhou is a medium-sized city, the number of nodes is relatively small,but from the image statistics, most of the nodes have clear meaning.The most representative ones are the National Pits square,Riverside road and the Bell and drum Tower.

4.2 Analysis of evaluative image of Luzhou

The first question of the interview is to obtain people's overall evaluation of Luzhou(Note: The first question is "When you see the word'Luzhou', what do you think of first?")and the main factors influencing people's evaluation image of Luzhou.The answer to this question can be divided into three types:①"Wine city", 56 people answered; ②one or several symbolic places (buildings) in Luzhou,among which Riverside road is the most, 22 people answered, others are bell and drum tower,Luzhou Old Cellar Hotel;③food categories,such as "longan", "tofu fish","The piglets dumplings"and so on, are relatively few (1-3 of each kind) and are not representative. Some people also answered both the first and second types of answers.

The second, third, fourth, and fifth questions in the interview include which places in Luzhou are most liked and dissatisfied with. The investigation shows that the factors affecting the overall evaluation image of Luzhou are mainly economic, cultural, traffic and environmental,and also influenced by security and belonging.

4.3 Investigation on city characteristics of Luzhou

City characteristics is an image that people acquire through the experience and perception of space and social activities.

In order to obtain information about Luzhou's characteristics from the public,we have designed two questions: "What would you introduce to someone who has never been to Luzhou"and "Which location do you like most in Luzhou". These two questions are significantly different (the coincidence rate is about 25%).The reason is that residents are more inclined to answer "The symbol of Luzhou" for the former while "personal preference" for the latter.On the whole,the former more reflects the overall image,while the latter more reflects the diversity of images,that is, individual's different feelings towards the city.Combining two questions, the answers can be divided into three categories: ① It clearly points out some places and buildings, such as Riverside road, the National Pits square, Bell and Drum Tower, white Tower, etc.It is worth pointing out that most of the locations, places, and buildings mentioned by the respondents are landmarks, nodes or areas that have an impact on the structural images,and new commercial districts are the main ones(Table 2).②General term for a type of place,such as: historical sites, scenic spots, etc. ③ Non-spatial characteristics,such as economic development, local customs,cultural atmosphere, etc.Among the non-spatial characteristics,"wine culture" was mentioned most,while some of them also mentioned history and food culture.

Table.2. Luzhou influences the place and architecture of critical images

Place name	number
Riverside road	35
Zhangba	28
National Pits square	22
White Tower	15
Fang mountain	10
Bell and Drum Tower	8
Huitong - Moore	8
Yang Qiao reservoir	5

5 Analysis of formative factors affecting urban intention evaluation of Luzhou

Through this investigation and analysis of Luzhou image space, we can get the conclusions:Economy, history, cultural atmosphere, transportation and environment are important factors that affect the residents' judgmental images[4].There are great differences in the image space of different social groups.

In general,the public in Luzhou has similar structural image of the city, that is, there is a "public image".The main structure of "public image" in Luzhou is clear,and the main factors are roads, landmarks, nodes and cultural atmosphere.The Yangtze River and Tuojiang River are not only the waterway, but also the natural boundary of the town, which has become the basic reference system for residents to remember the city environment.All the main roads are parallel or perpendicular to the Yangtze River or The Tuojiang River, and they also play the role of traffic passage and boundary, which together with the Yangtze River and the Tuojiang River constitute the city cognitive structure of Luzhou.In the immaterial image, the century-old wine culture supports the whole charm of this city. In this way, both inside and outside image have created a delicate, characteristic riverside wine city.

6 Suggestions on constructing good city image

As a considerable carrier of a city,the author believes that there are still some deficiencies in the city image of Luzhou.

First of all, the roads in Luzhou are very efficient,but lack individuality.The importance of the road is not only to provide the main visual observation point, but also become the direct perception of city image. The overall perception of roads in Luzhou has a strong sense of direction, but there are still many areas to be improved. For example, there is almost no street furniture on both sides of the main road, no small landscape belt and space for pedestrians to rest; no cultural sculptures and cultural relics, also no road landscape for appreciation. The single function and low recognizability lead to street lacks the cultural perception of city life, which turn into purely vehicular traffic roads.

Second, the city nodes lack the overall appearance.Node is the spatial focus of the city with the meaning of connection and concentration. A node can be the core of area, a central square, a central garden, or a building in a central location.In Luzhou,such as Hundreds of children square, Bell and Drum Tower, White Tower, Luzhou Old Cellar Hotel are the most image nodes, however,due to their single architectural form and function,they can't be the center of city image ,nor can they be composed together,so the city lacks an overall style that can be imaged by people.Therefore, from the perspective of nodes,Luzhou is a continuous city complex that is difficult to remember.

Based on the above analysis, the author puts forward the following Suggestions for planning,and city management of Luzhou:

First, the road pattern of Luzhou not only inherits the traditional road structure, but also adapts to the characteristics of modern city, playing a decisive role in the formation of city image.Therefore, no matter whether it is expansion or redevelopment of the city, this pattern should not be broken.Instead, focus on the cultural construction of the road itself.

Second, according to the obvious characteristics of marginal elements in Luzhou image,the marginal elements should be more prominent in areas with favorable natural conditions. As for the "inner" and "outer" city pattern divided by river, the difference is conducive to the formation of the "outer" city image and the shaping of the "inner" city image, the author suggests to preserve rather than bridge this boundary.

Third, combine nodes and highlight the characteristics of the city.In the process of characteristic investigation,the image that wine culture as a symbol of Luzhou has been rooted in the hearts of Luzhou citizens.However,high imagery and integrity of nodes become the biggest contradiction in city construction,so how to reflect the characteristics and continuously increase the image elements in the nodes has become the focus of node planning.

References

1. K.Lynch.*The image of the city*[M].Beijing: Huaxia Press,(2001).
2. X.M.Li,J.H.Li.Analysis of Urban Space in Dalian[J].Acta Geographica Sinica.**61**,8(2006)
3. J.Feng.Spatial Cognition and the Image Space of Beijing's Residents[J].Scientia Geographica Sinica.**25**,12(2005)
4. Y.R.Shen.City characteristics and city image[J].Urban Problems,03(2004)

Strategies for Activating Historic District: Micro-reform Implement in Qiulin District, Harbin

Binyu Lei[1,*], *Qi* Gao[1], and *Yanyao* Cui[1]

[1]Urban Planning and Design Institute of Shenzhen, Shenzhen, China

Abstract. This paper explores urban strategies that address existing problems in Qiulin District, which is famous as a historic district with plenty of cultural memories in Harbin, China. In the paper, it firstly introduces the definition of historic district and made a brief review of academic literatures about space activation. It then takes Qiulin District in Harbin as a case to further discuss the current dilemma. In terms of land use, Qiulin District is primarily commercial and residential that increases the difficulty of activating this area. In accordance with demonstrated issues: 1) heritage buildings with limited value capture, 2) diverse business types challenged with urgent upgrading, 3) multiple high-quality resources utilized in weak synergy, it proposes three main strategies to solve problems. With introducing the concept of micro-reform, the strategies: 1) activate heritage buildings to recapture historic value, 2) extend outdoor space and upgrade commercial structure to enhance characteristic streets, 3) design for fun life to build inclusive network, compressively explore a number of possibilities to reflect the current situation and achieve visionary planning. Finally, this paper draws discussion about the significance of space activation for historic districts towards environmental sustainability and human beings.

1 Introduction

Historic district, as part of cultural heritage, significantly contributes to urban context. Basically, historic district is considered as an important carrier of urban cultural environment [1]. According to the State Ministry of Housing and Urban-Rural Development in China (2015), it lists a total of 30 historic and cultural streets and districts regarding cultural value and economic benefits [2]. However, during the process of urban development, it has become growing concern about activation of historic districts. On the one hand, traditional spatial patterns have increasingly decayed, which are hard to adapt to modern built environment. On the other hand, the shortage of land resource and the mass construction of real estate inevitably pose a threat against urban heritage preservation [3].

Researches on activating historic districts are mainly concluded in two perspectives: renewal and protection [4]. Mumford (1961) makes a critique of urban construction in macro-scale ignoring city's vitality [5]. Jacobs (1961) also demonstrates the significance of historic areas during the process of urban regeneration movement [6]. It literally emphasizes preservation of urban heritage in terms of its integrity. However, since the twenty-first century, historic district conservation is further challenged by more complicated conditions [7]. The concept of organic renewal is raised to promote micro-scale guide planning towards historic areas [8]. It addresses every historic district as a whole with integrating numerous cultural elements and modern parts as well. Also, the activation of historic districts not only remains in the discussion of government or other stakeholders, but also associates with public participation [9]. In 2015, United Nations Educational, Scientific and Cultural Organization (UNESCO) proposes "Operational Guidelines for the Implementation of the World Heritage Convention", adapting Historic Urban Landscape (HUL) as an innovative approach to value urban heritage [10]. HUL approach is dynamic in response to changes over time including built environment, human behaviors and historic heritage itself. The present of HUL approach extends the scope of historic conservation which aims to include it as the media reconnecting culture and society [11]. It largely relieves contradiction between urban renewal and heritage preservation, and also suggests new perspective to activate historic streets and districts. In recent decade, studies on activation of urban historic districts more focus on social cohesion and activation tactics in micro-scale, rather than static perspective on protection [12]. Space activation nowadays pays more attention to revitalize historic district without destroying its original features. Moreover, it attaches importance to public engagement with consideration of numerous demands to achieve an inclusive and sustainable neighborhood.

2 Case study: Qiulin District

Qiulin District is a typical example of heritage conservation in Harbin, located in the heart of the city boasting a history spanning over more than a hundred

* Corresponding author: leiby@upr.cn

years (Figure 1). The development of Qiulin District entirely reflects the history of culture and society. In 1900, the Churin Foreign Trading Company established by the Russians and became the most popular shopping mall. With the rapid growth of Churin Company, it conducted diverse business across countries and attracts a large number of tourists from Europe and Asia. Subsequently, more types of buildings were built by foreigners for residence and commerce, that forming a cluster of foreign-culture buildings. The company and other constructions have sustained a history of more than 120 years while nowadays they have no functions as before only reserving the heritage building. A hundred years flashing by, Qiulin District, as the origin of the city, inherits the initial layout and preserves a number of delicate historical buildings, delivering its unique urban texture. At present, Qiulin District has become an essential identify of Harbin through one hundred year of construction and preservation.

Fig. 1. The location of Qiulin District.

Qiulin District occupies an area of 1.43 in the center of Harbin. The land use in this area is mainly commercial and residential, as well as a small proportion of public space for services like education and administration (Table 1).

Table 1. The provision of land use in Qiulin District.

Categories	Area/m^2	Proportion/%
Residential	1181202	43.68
Commercial	1261066	46.64
Public space	253059	9.36
Residential	1181202	43.68

In general, today's Qiulin District is highly mixed with diverse urban functions. Historic buildings are gradually transformed into museums, schools, malls and other functions while a number of new dwellings are built for residence (Figure 2). The Museum of Heilongjiang Province, the Youth Palace of Harbin, and the Activity Center, together with other public infrastructures in Qiulin District, manifest the potential to rebuild a cultural circle in the contemporary era. In terms of heritage buildings, Qiulin District gathers a cluster of Modern European Style buildings, represented by the consulates and consulate generals of several countries, including Russia, Britain, Italy, France, and Japan. They concentrically distribute in the west and east of Qiulin District, mainly along the Dazhi Street and Guogeli Street. Such distinct architectural heritage reflects the cultural communication between China and the Western countries. Although the original purposes have passed, the cultural buildings themselves still have a profound impact on the development of Harbin and play a significant role in the new era.

Fig. 2. The architectural nature of Qiulin District.

However, with the rapid development, the heritage buildings have lost their original charm. The historic value has not been fully captured. A part of historical buildings is faced with total vacancy or insufficient utilization, some of which are poorly maintained due to illegal extensions or reconstructions. The building clusters scatter across this area with poor connectivity, disrupting the intrinsic ambience. In terms of industrial structure, a large proportion of the business types in Qiulin District stagnate in the traditional formats, in want of transition and progression to the latest forms, that fail the synchronous development of new era (Figure 3). As consumption upgrades and innovative economy grows, the young generation has shifted their predilections for various exhibitions, brand shops, fashion events. The old business model is hard to continue with a lack of attractiveness to satisfy people's demand at the present. The functions of multiple heritage buildings in Qiulin District, such as commercial shopping centers, and cultural facilities, lack inner linkage that poorly maintain the integrality as a whole. Some cultural facilities and historical buildings are relatively enclosed, sealing the built-in yards and other open space off from interaction with the public. Pedestrian access to public space in this area is blocked by vehicle pathways, resulting in fragmentary spaces and diminishing dynamics. In consideration of numerous problems existed, Qiulin District urgently needs adaptation to the up-to-date demands, deeply mining its own features and value, so as to augment its regional competitiveness and appeal to more diverse groups.

https://doi.org/10.1051/e3sconf/202123704003

Fig. 3. The status quo of Qiulin District.

3 Strategies for space activation

In 2016, "Measures for Urban Renewal of Guangzhou" is introduced by the Guangzhou Municipal Government to guide and manage the activation of historic districts, which nowadays widely applied to practice in China. It firstly introduces "micro-reform", which is implemented through partial demolition, preservation or renovation on the premise of maintaining the basic construction patterns [13]. Literally, micro-reform is a flexible way to capture historic value and adapt heritage for new era development. In spatial pattern, it addresses the architectural features with continuing traditional culture. In addition, micro-reform implementation of historic districts considerably emphasizes environmental sustainability, as primarily keeping heritage as an essential part of natural resources. [14]. It considers the regeneration of heritage that provides physical inputs into social services and experience. Meanwhile, the strategy of micro-reform activation takes indigenous peoples into serious consideration. It largely manages the cost of construction without threatening the security of original property right [15]. Micro-reform implement provides a comprehensive method to activate historic districts regarding historic value and multiple subjects in the process of urban renewal. According to status quo of Qiulin District, it develops three main strategies to draw micro-reform, including cultural core regeneration, historic streets preservation and life enhancement (Figure 4).

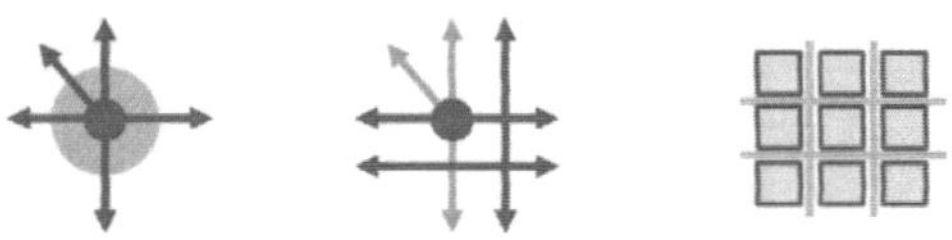

Fig. 4. Three main strategies (cultural core, historic streets and life network).

3.1 Strategy 1: Build Sky Walk to recapture value in the culture core

The culture core in Qiulin District is surrounded by Youzheng Street, Jianshe Street, Huayuan Street and Haiguan Street. There are 16 heritage buildings in this area, which include 2 Embassies and Consulates buildings, a history museum, the children's palace and 7 public squares. Since the lag of construction from the last century, the culture core nowadays presents a chaotic spatial pattern losing attraction in the new era. The primary thing for the existing area is to recapture its value, in order to awaken the vitality and promote attractive atmosphere. Through building the vertical floor-connection system (Sky Walk), it connects historical buildings and integrate cultural gardens (Figure 5). In general, it bridges scattered elements to increase connectivity and provide more access to explore historic value in this area.

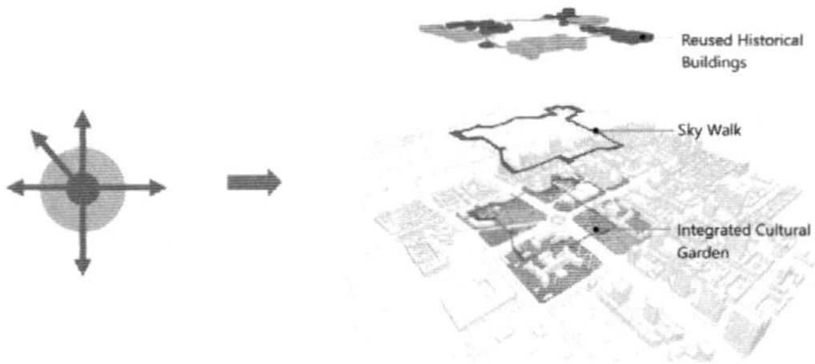

Fig. 5. Build the culture core.

In terms of representative buildings in Qiulin District, there are two tactics to reuse them and stimulate their vitality. Firstly, the replacement of original functions conforms to current urban development. It is widely received as a contemporary way towards environmental sustainability, which largely maintains the historic elements and introduces modern functions, like innovation and creative design, to promote cultural integration. For example, it encourages the conservation of historical features and transforms the inner function to exhibition or experience, like the Creative Museum and the VR Experience Center (Figure 6).

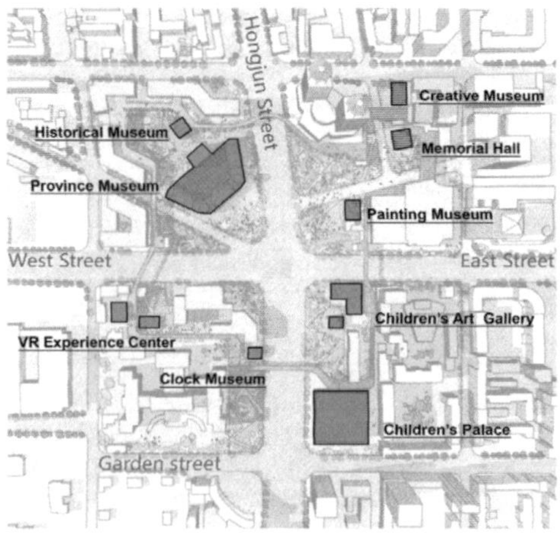

Fig. 6. Reused heritage buildings.

The other tactic is to extend existing heritage buildings to satisfy more demands in the new era and support

 https://doi.org/10.1051/e3sconf/202123704003

tourism development as well, like the Historical Museum and the Province Museum. It attaches public services to heritage buildings where people have more chances to learn about the precious history [16]. In order to improve access to reused buildings, the Sky Walk system with corridors ensures the comfortable wandering of pedestrians in the historic district to freely enter heritage buildings [17]. It also links scattered small squares and plenty of green space in this area, to basically provide sufficient activity space regarding walking consistency (Figure 7).

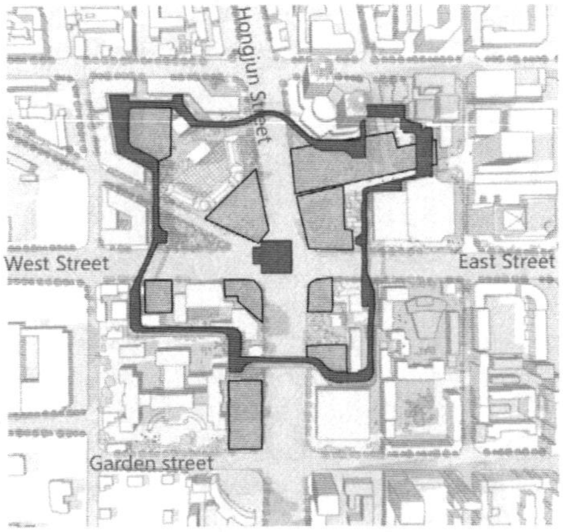

Fig. 7. The Sky Walk.

3.2 Strategy 2: Enhance characteristic streets to satisfy contemporary development

Dazhi Street, Huayuan Street and Guogeli Street are arterial streets in Qiulin District, presenting unique landscape and cultural environment in this area (Figure 8). Basically, the three streets combine numerous historic elements including heritage buildings and public infrastructures, which entirely reflect the flux of social history. However, the existing problems limit the sustainable development. The traffic-oriented development largely weakens the cultural ambience of commercial pedestrian streets. The heritage buildings scattered throughout the streets fail to demonstrate their historical value. Thus, the enhancement of historic characteristics is a considerable method to address the competitiveness on account of current situation.

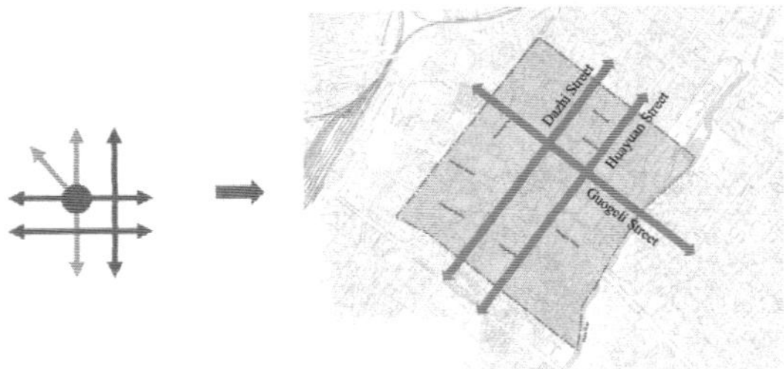

Fig. 8. Enhance characteristic streets.

It takes the Guogeli Street as an example. Without changing the red line, road separation will fully release more public space through separating the current streets. It alleviates traffic jams against the heavy use of private cars, and ensures the safety of pedestrians as well [18]. The released commercial space is open to the public designed as a catalyst to drive the revival of heritage buildings, potentially engaging people in walking along the street and pursuing leisure activities [19]. Also, the outdoor space provides a panoramic view to completely enjoy historic landscape. For instance, depending on diverse historic features, cultural events, like Russian Food Festival, will be regularly held in the extension commercial space to improve the vitality of Guogeli Street. In accordance with emerging consumer demands, the industrial structure needs to introduce new industries, such as fashion shopping, snacks and bars, in order to complete upgrading. The vertical business model is regarded as an effective way to upgrade the industrial formats for the sake of full mix-use in order to well maintain the historical façade along the Guogeli Street. It insets office functions in dwellings and adds first-floor business to meet daily demands, like bookshop or café (Figure 9). It sufficiently builds diverse industrial forms to improve functions to enhance multi-purpose development, and preserves the historic atmosphere along the arterial street as well.

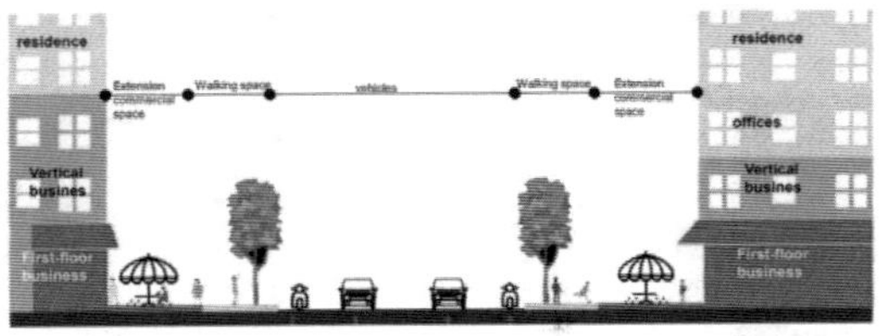

Fig. 9. Conduct road separation.

3.3 Strategy 3: Design for fun life

On the basis of existing open small blocks and courtyard-style space in Qiulin District, a more interesting and high-quality urban living neighborhood will be built (Figure 10). It sets up small facilities in the block to satisfy various activities and makes full use of inefficient corner space to build pocket parks. The entire network is systematically designed to improve the quality of life with activation of the historic district. People are attracted to engage in this area and enjoy city life with traditional ambience. The future growth of population in next decades will stimulate various demands for public facilities. Outdoor space is essential for people where they have opportunities to interact with others and enjoy physical activities [20]. It requires to provide adequate facilities along the street, such as benches, where people feel free to seat and have recreation. Scattered corner space is designed for pocket parks, that intend to provide more public space for people to resume social activities and explore historic value. Pocket parks make use of roadside green belts and other negative space in line with local conditions to refine landscape. For example, around the culture core, there will be built numerous pocket parks providing sufficient space for leisure activities, where people are able to experience the essence of history as well as natural resources with aesthetic design. Green space along the street and within small blocks are addressed as an important component in the city where people have open access to contact with nature through an ecological environment [21].

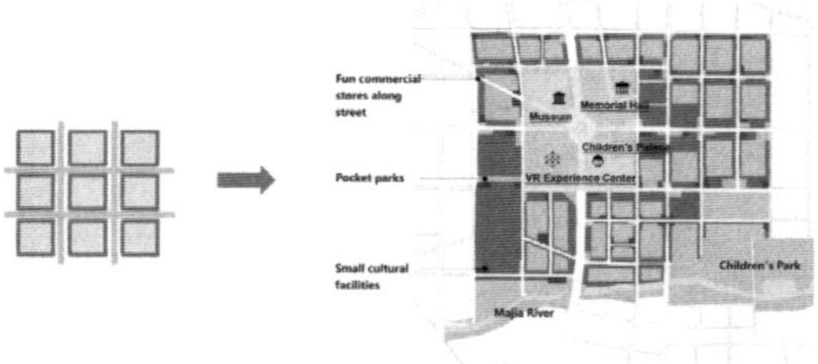

Fig. 10. Design for fun life.

4 Discussion

Micro-reform implement is widely considered as an effective method to address the activation of historic district that preserves the cultural value and adapts multiple functions for new development. With taking Qiulin District as an example, it should be pointed out that the origin of Qiulin District is prestigious regarding heritage buildings in collaboration with cultural elements and aesthetic sense. The activation of this special area is not a simple matter of regeneration to keep abreast of times. More importantly, it has to explore a number of possibilities to awaken the vitality towards the historic value and future demands. In consideration of land use in Qiulin District, it highly mixes residential and commercial space as well as some public services. There are three main strategies proposed to revitalize this area in order to achieve a historic and inclusive neighborhood. Some strategies emphasize the combination of historic elements and new functions, like shopping and other relaxation, to stimulate further tourism development. Other proposed actions aim to explore more possibilities for current and future demands of people living in this area. The micro-reform to activate Qiulin District comprehensively involves historic value into every aspect of physical buildings and casual experience. It seeks not only to solve the existing problems led by the lag of development, but also to enhance its characteristics and eventually build better life network.

References

1. B. S. Yeoh, & S. Huang, Cities **13(6)**, 411-422 (1996)
2. T. Liu, R. J. Butler, C. Zhang, J. Archit. Conserv. **25(3)**, 228-251 (2019)
3. R. M. Leichenko, N. E. Coulson, D. Listokin, Urban Stud. **38(11)**, 1973-1987 (2001)
4. L. Dai, B. Xu, B.Wu, Area **49(1)**, 94-105 (2017)
5. L. Mumford, *The city in history: Its origins, its transformations, and its prospect* (1961)
6. J. Jacobs, *The death and life of great American cities* (1961)
7. B. Zhai, M. K. Ng, Cities **35**, 14-25 (2013)
8. H. Kou, J. Zhou, J. Chen, S. Zhang, Sustainability **10(12)**, 4645 (2018)
9. K. Winson-Geideman, D. Jourdan, S. Gao, J. Real Estate Res. **33(1)**, 25-47 (2011)
10. K. Taylor, Landsc. Res. **41(4)**, 471-480. (2016)
11. F. Bandarin, R. Van Oers (Eds.), *Reconnecting the city: the historic urban landscape approach and the future of urban heritage* (2014)
12. F. Zeng, Z. Shen, Int. J. Environ. Res. Public Health **17(12)**, 4367 (2020).
13. Z. Xiaoyang, W. Shifu, F. Yan, South Architecture, **5**, 15 (2017)
14. S. Wang, Y. Hu, H. He, G. Wang Sustainability **9(10)**, 1847 (2017)
15. A. Mozaffari, T. Jones (Eds.), *Heritage Movements in Asia: Cultural Heritage Activism, Politics, and Identity* (2019)
16. M. K. Smith, *Issues in cultural tourism studies* (2009).
17. M. Johansson, C. Sternudd, M. Kärrholm, J Urban Des (Abingdon) **21(2)**, 256-275 (2016)
18. Y. Liao, A. Shibata, K. Ishii, M. J. Koohsari, S. Inoue, K. Oka, J. Transp. Health **11**, 73-79 (2018)
19. G. Ferilli, P. L. Sacco, G. Tavano Blessi, S. Forbici, Eur. Plan. Stud. **25(2)**, 241-258 (2017)
20. J. Wilks, Int. Res. Geogr. Environ. Educ. **19(1)**, 25-38 (2010)
21. K. Shriver, Transp. Res. Record **1578(1)**, 64-75 (1997)

Research on Public Spaces Reconstruction of Old Communities under the Concept of Micro- renewal

Duan Xiang[1,*] , Xu Tai-meng[1] ,Duan Chen-hao[2]

[1]School of Engineering and Architecture,Wuhan Ploytechnic University,Wuhan,China
[2] School of Architecture,Southeast University,Nanjing,China

Abstract. Community public space is an important place for daily communication between residents. In the context of the current historical change of "urban double repairs", community renewal has raised new heights, and public space renewal has shifted from the surface of the physical space transformation layer to the protection, inheritance and environmental quality of the community. This article summarizes the concept and principles of micro-renewal, analyzes the related influencing factors of the reconstruction of the old community public space, further proposes the strategy of reconstructing the old community public space and enumerate several typical public space renewal schemes in old communities under the micro-renewal concept.

1 Introduction

Public space narrowly refers to the outdoor place where people conduct social activities, and broadly refers to the ideology of social relations that combines people, environment, and society [1].Simple material transformation makes the shape of community public space the same, lacks features, and public space renewal needs to change its mind.Old communities generally take a long time, the overall space structure is basically complete, the space density is almost saturated, and public spaces need to meet the dual needs of people 's material and spiritual. The "big move" renewal model is obviously not suitable for old communities.The concept of micro-renewal is more in line with the development needs of communities and people, and is the theoretical basis of this study. The reconstruction of public space in old communities is a reconstruction of the relationship between people, society and space,which needs to pay attention to the community's ecological development and humanistic feelings.The purpose of this study is to reconstruct the public space of the old community according to the concept of micro-renewal to promote the sustainable development of the community.

2 The concept and principle of micro-renewal

2.1 The concept of micro-renewal

"Micro-renewal" is a strategy adopted by the West to abandon large-scale urban renewal and implement urban renewal through small-scale renewal. Micro-renewal is a small-scale, gradual, low-impact update method. The update behavior is closer to the user's space, and it is more able to find problems in community development and reflect residents' demands.Micro-renewal is a long-lasting update process. The project is constantly adjusted and updated according to the needs of the times and life to meet the requirements of use.

2.2 The principle of micro-renewal

In micro-renewal, the scale and cost of the project are reduced, which lower the risk of the project. Micro-renewal focus on areas that are more connotative, close to the crowd, and more refined in all aspects[2].The concept of micro-renewal emphasizes that "people-oriented" is always implemented during the renewal process, pays attention to the living needs of residents, and encourages the public to participate in the renewal activities to renew and reconstruct the harmonious relationship between people, environment and society. Micro-renewal respects ecology, inherits historical culture, develops local characteristics, and pays attention to the project's flexible transformation and self-adjustment during and after the renewal process.

3 Influencing factors of the reconstruction of public spaces in old communities

3.1 User's needs

The main subjects closely related to the community are government, non-governmental organizations and community residents. Based on multiple subjects, listening to the voices of all parties and promoting social equity, participation of multiple subjects is an important feature of micro-renewal and reconstruction. Relying only on government forces can often only form temporary actions, and the update results only stay on the surface. Government power should shift from control to guidance, improve government service awareness and level, and listen to the opinions of residents. Only by fully and truly understanding the needs of community residents can we truly understand the problems in community development and grasp the main contradictions in community development.

*Corresponding author:287264@qq.com

3.2 The current situation of public spaces in old communities

Due to the erosion of wind and rain and the disruption of urban development, the explosion of population and the change of people's lifestyles, most old communities have gradually become weak in the city, swallowed by dense large-scale urban development, plus overuse and lack of protection consciousness, the overall appearance of the community is broken. The following table lists the current status of public spaces in old communities (Table1)

Table1. The current situation of public spaces in old communities

Type	Explanation
Community lacks public space	The construction standards of the old community were mainly economic and practical. The internal density of the community was large, and there was no public space available for residents 'activities. Most of them were linear spaces for streets and lanes with transportation functions.
Low utilization of public space	The scale of public spaces is relatively small, and most of the public spaces in the community are used by community residents to dry clothes and pile up waste products.
Deteriorating public space environment	The mobile operation of merchant stalls in the old community not only occupies public space, but the fumes and garbage generated by barbecue stalls and fruit stalls pollute the environment.
Public facilities are imperfect or aging	The construction standards of supporting facilities in old communities often cannot meet the living and living needs of modern people. The quality of road surfaces in the community varies, the local roads are damaged, subsidence, and water accumulation. Lighting, maintenance railings, information and other facilities are seriously aging and damaged.
Low greening rate and poor quality	Most old communities neglected the construction of greenery and its role in the construction. In addition to the lack of public space, part of the greenery space was compressed, making the greening of old communities very poor. Some community residents independently build flower garden spaces, but the overall landscape quality is poor

3.3 The form and function of public space

The interface height and interface rhythm of the space form the space form. The height of the space interface is the height of the building surrounding the space. The interface rhythm refers to different forms of building interfaces. The high degree of interface control and rich interface rhythm can make the space full of vitality. Space function is its own vitality and usability. The functions undertaken by community functions are actually the nature and changing patterns of people 's activities. [3].Jane Jacobs clearly put forward the concept of "functional blending" from the perspective of "urban diversity": In a successful and beloved urban area, there are people's activities at different times of the day. In the reconstruction and update of public spaces in old communities, the repair and setting of mixed functions are essential. Repairing the basic functions of the community can make the old community adapt to different activities and provide multi-level comprehensive service functions for local residents and tourists.

3.4 Inheritance and continuation of public space culture

Old community buildings are declining, so the buildings bearing the historical and cultural marks may be rudely pushed down and rebuilt during the renewal process, so that the cultural marks attached to the architectural landscape are destroyed [4].On the issue of public space reconstruction, we should focus on the excavation and continuation of the community's history and culture, so that the old community, with its local cultural characteristics, can become a space with unique local characteristics and gain a foothold in regional development.

4 Strategies for reconstructing public spaces in old communities under the concept of micro-renewal

4.1 Reconstructing the public space order

The public space structure of the old community is not very perfect. Due to the lack of unified space planning and the random construction of private residents, the sequence of public space in the old community is chaotic. In the reconstruction of public spaces in old communities, we should first sort out a clear spatial structure which is Open Space-Semi-Open Space-Semi-Private Space-Private Space. A clear sense of spatial hierarchy can improve the utilization rate of community space and promote the formation of residents' sense of belonging. In addition, we should establish a continuous spatial sequence, open up unreasonable roads, improve the accessibility of street spaces, and connect spatial nodes to form a complete spatial sequence, which not only increases the interest of public spaces in old communities, but also provides more communication opportunities and interaction time for residents to go out.

4.2 Reconstructing the form of public space

4.2.1 Maintaining continuity of the interface

The continuous space interface makes the people coming and going to stay, most people like to stay on the edge of the space and observe the activities in the space. The space formed by the enclosed and continuous boundaries can better guide people's activities and make full use of the combined effect of the space. Therefore, when reconstructing the public space of the old community, we should pay attention to the continuity of the space interface, establish an effective and clear transition within the horizontal horizon of people, and form a virtual and real space through various forms such as bumps, retreats, overhangs to attract people.

4.2.2 Improving space accessibility

Space accessibility includes sight accessibility and path accessibility. Sight accessibility can effectively guide people's behavior and have a positive impact on spontaneous activities. When people can see the activities in public places, the combined effect will attract more people to participate. The barrier-free sight will make the activities more natural, and the use of open spaces can be improved to a certain extent. Path accessibility is represented by a walking system, a short walking length, an effective and convenient entry path, and a reasonably unobstructed road. Opening up blocked roadways and using more effective spatial connections can make it easier for people to reach interesting spaces

4.2.3 Moderately transforming the shape of the space

Different forms of space will bring different psychological influences and hints to people. Intermittent public space means that people will stop, linear public space means that people will have a tendency to move forward in the space, the enclosed space implies that residents can stay here. According to the community development and the needs of the residents, the appropriate transformation of the space form can enhance the subject's sense of use and satisfaction. Public spaces should try to avoid long and straight lines, so that the walking experience becomes an interesting thing. Creating rhythmic changes on the interface of the ground floor space and cooperating with the landscape greening, will enrich people's space experience.

4.2.4 Humanized spatial scale

The spatial form in the old community is more complicated and chaotic. Therefore, in the reconstruction of the public space in the old community, humanized spatial scale should be established and maintained."the hidden dimension"mentions, when the distance exceeds 25m, it is difficult for people to recognize the language and facial expressions of others, and it is difficult to produce communication.Jan Gehl proposed that 100m is the pole of the experience distance[5]. For a linear space, a width less than 25m can help people passing by to perceive each other better. However, a narrower spatial scale will cause discomfort to the two streams of people who meet in opposite directions. This space is only a temporary space and cannot produce activities such as staying and communicating.

4.3 Reconstruction of public space functions

In the long-term development process of old communities, many functional spaces have spontaneously formed, such as sports spaces, farmers 'markets, supermarkets, breakfast shops, etc. These spaces conform to the trajectory of the residents 'lives, and should be retained during the reconstruction of public spaces, not only the needs of the indigenous people 's social relations, but also the necessary conditions for the continuation of the indigenous people 's lifestyle. At the same time, integrating the original function configuration and dismantling the privately built space will make the disordered space function orderly. The function of the original public space in the old community is relatively simple. According to the resources of the community and the needs of the residents, the single function can be replaced with small and medium-sized inclusive functions, and the original functional space can be integrated and reconstructed to give the space new and diverse functions and increase the variety of choices for space users.

4.4 Reconstructing the continuation of public space

The spatial structure and scale of the old community have formed a unique space texture in the community for a long time. Streets, courtyards and nodes are the public space of the community. The organic extension of the space texture can be carried out in the following ways. We can precisely control the "scale" of public space, protect historical buildings, and give it new connotation and significance. Also, we can select good quality building samples from the existing public spaces, execute a reasonable transformation process, and replace effective functions with them. Under the low-cost and low-risk concept advocated by micro-renewal, the parts of old communities that need to be demolished can make full use of the original building materials to restore vitality in new buildings and establish an organic connection between old and new spaces.

5 Renewal plan for some typical public spaces in old communities under the concept of micro-renewal

5.1 Community entrance

5.1.1 Space characteristics

The entrance forms of old communities can be divided into street-style entrances and open entrances. The entrance form of the partial overhead of the first and second floors of the building is the entrance to the street, which residents are more likely to stay chat. Open entrances are more fluid than street entrances, and it is easier to shape the brand characteristics of the community. The disorderly parking of non-motorized vehicles, publicity boards, rest facilities cause a mess in the open entrance.

5.1.2 Renewal plan

The main point of updating the street-style entrance is to increase the convenience of residents passing here and the improvement of interpersonal communication. For old communities where is other spaces to park non-motorized vehicles, we should move and park non-motorized vehicles to expand travel space; For old communities where there is no other space to park non-motor vehicles, we should divide the parking places of non-motor vehicles and regulate the parking order.

The main points of the reneal of the open entrance are the community image and the separation of people and vehicles. The entrance image should be combined with the historical and cultural characteristics of the community and coordinated with the architectural style of the community; The pedestrian-vehicle intersection should regulate the diversion route; Publicity boards, rest seats, etc. should not be placed near the entrance to avoid affecting the smooth travel.

5.2 Street space

5.2.1 Space characteristics

Street space can be divided into street space and roadway space. The street is the widest road in the community, with a large spatial scale, which can be used by vehicles. It is generally a two-lane or single-lane road, with one or two sidewalks attached, which is the link between the community and the outside world. The laneway is the secondary road of the community. The scale of the space is small, usually a single lane or a narrow road only for non-motor vehicles and residents.

5.2.2 Renewal plan

The main point of the street space update is to improve the safety and comfort of the space. We should strictly divide the pedestrian and vehicle routes, promote the separation of pedestrians and vehicles, and separate pedestrians and vehicles with green belts to ensure the smooth vision of road intersections, and buildings, facilities, greening, etc. [6] Community motor vehicle lanes with a width of less than 10m can regulate the order of motor vehicles parked on the street. Community motor vehicle lanes with a width of more than 10 meters can increase the marking line to subdivide the vehicle space and add non-motor vehicle lane. Also, we should add emergency traffic routes to meet firefighting and first aid. The design of this route should avoid interference with residents' daily routes. Generally, the width of the walkway is about 2 ~ 3m, and the sidewalk with a width of more than 3m can be set on the side of the green belt for the leisure space of the pedestrians.

5.3 Other functional public spaces

5.3.1 Amusement park space transformation

The quality of the amusement park space in the old community is generally poor, mainly due to the relatively single space form, messy plant species, and outdated leisure facilities. Transformation of amusement park space should be designed according to local conditions, and combined with actual plot status and community needs.We should compact layout, improve land utilization. If the plot area is small, the park road should adopt a regular pattern; if the plot area is large, a natural tortuous path can be used. Through the rational addition of different living and leisure facilities to enrich the functions of the amusement park space; Through plant configuration and landscape transformation to improve the greening quality of the space.

5.3.2 Sports space transformation

The renovation and reconstruction of the sports space should focus on practicality and safety.

First of all, according to the main types of activities of community residents such as dance, fitness, etc., rationally divide the venues and update the corresponding sports equipment; Secondly, the venue should pay attention to the anti-slip property of the surface layer, and the sports plastic surface layer. The children's field can be laid with colored plastic surface layer to increase the fun of the venue. Finally, the plant landscape should be properly configured to reduce the impact of pollen and insects on the residents.

5.3.3 Micro space transformation

The renovation of the micro-space is based on the current situation, revitalizing the stock, and invigorating the vitality as the starting point, oriented by the problem, and detailed design according to local conditions. For example, the micro space next to the residential building should be cleaned of debris first, and the functional quality of the space should be improved through landscape reconstruction combined with leisure facilities. Street green space should focus on reducing the interference of dust, car exhaust, and noise to meet the functional needs of residents' rest and communication. Residents of old communities often spontaneously open up flower garden spaces next to their houses. Due to the difference in the size, planting type, and repair materials of each flower garden space, they appear complicated, complicated, and chaotic. In order to reduce the clutter of the flower garden space, flower pots of the same material and repair materials of the unified flower garden can be used to combine scattered flower and pot plants.

6 Conclusion

The ideal spatial form, on the one hand, must inherit the local history and culture, on the other hand, it needs to become a place to effectively gather social relations. Effective public space renewal can inject new blood into the community and promote the community's new vitality. Taking renewal and reconstruction of public space as an entry point is an important way to solve the disorderly development of public space, cultural mining, and reshaping of sustainable social relations in the development of old communities, and also provides a valuable model for the development of public spaces in old communities.

Funding

This work was supported by the Hubei Social Science Foundation(2018q124),Wuhan Committee of Urban and Rural Construction(201739) and Wuhan Social Science Foundation(19005).

References

1. Gao Y.B. Analysis of Urban Public Space [J]. Shanxi Architecture, 10 5-6. (2013)
2. Zhu W, Chiu R L H. The Planning and Design of Environmentally Sustainable High-Rises[M]// High-Rise Living in Asian Cities. Springer Netherlands,553-560. (2011)
3. Shi S.J. "Environmental micro-renewal" historical location protection model exploration [D]. Southeast University (2015)
4. Lu W. Research on the Humanistic Design of the Landscape Reconstruction of the Old City Community under the Sustainable DevelopmentD]. Northeast Normal University (2014)
5. Jan Gehl,Ouyang W. Cities for People [M]. China Construction Industry Press,(2010)
6. Ren Y, Zhang X.K. Research on Road Landscape Design Method [J]. Urban Construction Theory Research, 17 (2013)

Investigation and Research on Building Diseases in Yueliangdi Village, Yinggebao Township, Mulei Based on GIS and Infrared Thermal Imaging Technology

Zhang Baoqing[1,2,3,]

[1]Ph.D. Candidate, Huazhong University of Science and Technology, 430047 Wuhan, China
[2]lecturer, Shihezi University, 832000 Shihezi, China
[3]Ph.D. Candidate, Hubei urbanization Engineering Technology Research Centre, 430047 Wuhan, China

Abstract. In this paper, the representative buildings built in different periods of the village are observed and their types and conditions are recorded according to the phenomenon that the defects and diseases of the local historical settlements are increasingly serious due to blind transformation and excessive commerce in the protection work of moon village in Mulei County, Xinjiang. Then, the main building diseases of moon village are studied based on the theory of architectural pathology and GIS platform The coupling relationship with local microclimate, and the degree of disease of the village houses are classified into three categories: light, medium and severe diseases. Based on the analysis of the change of macro climate and local microclimate conditions in the South and north of the East Tianshan Mountains, it is concluded that the main reason for the defect is the damp disease. The infrared thermal imaging technology is introduced to analyze the damp disease in the settlement, and the preliminary judgment method of the damp disease is obtained, which should be paid attention to in the future protection work.

1 Historical evolution and geographical and climatic conditions of yueliangdi village, yinggebao Township

Fig1. Main village pattern

1.1 historical evolution of yueyuedi Village

Yueyuedi village, yinggebao Township, Mulei Kazak Autonomous County, Changji Hui Autonomous Prefecture, Xinjiang, is located in the northern foot of the East Tianshan Mountains and the west edge of the Turpan Hami basin. It was built in the late Qing Dynasty and the early Republic of China. It has a history of 100 years. It is named yueyuedi village because the land is cut into crescent shape by the river around the village. Due to its integration of traditional dwellings, folk customs, handicrafts and natural landscape culture, the village was listed in the list of Chinese traditional villages by seven ministries and commissions of the State Council in 2014. It is also one of the most complete traditional villages in Mulei County.

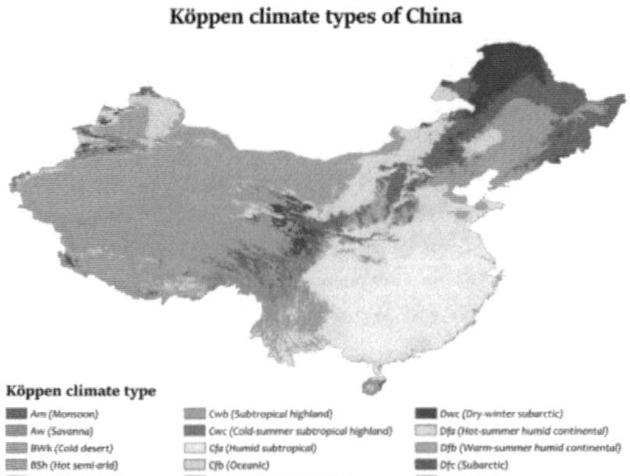

Fig2. China cobain climate partition

 https://doi.org/10.1051/e3sconf/202123704005

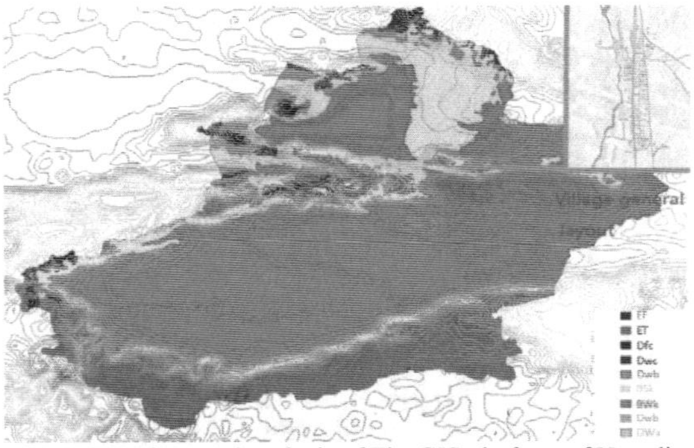

Fig3. Superimposed analysis of The GIS platform of Yuandi Village

Al though the settlement has experienced many years of change and repair, the settlement pattern and features still reflect the characteristics of the village during the initial construction period of the late Qing Dynasty and the early Republic of China[1], that is, most of the village houses are arranged along the road in order to facilitate contact and communication, and some houses are arranged freely along the river to facilitate water intake. Most of the dwellings in the village are dominated by balang buildingwith a strong Shaanxi Gansu style. Compared with other parts of Xinjiang, such as green tile, yellow clay wall, wooden fence courtyard walland other components, it also tells the common people its origin.

1.2 The geographical and climatic conditions of Yueliangdi Village

In the past, when mentioning the climate in Xinjiang, the general impression of scholars was traditional concepts such as dryness, less rain, and more wind and sand. However, recent studies by geographers have shown that the climate in Xinjiang, especially in the Tianshan Mountains, is changing from warm and dry to warm and wet[2] And in recent years, the precipitation in mountainous areas has shown an obvious upward trend[3].

In this study, the author used the Köppen climate zone theory and ArcGIS software to superimpose the Xinjiang Köppen climate zone with Xinjiang terrain contours and the geographic coordinates of Moon di Village and the author's general plane drawn by Google Maps. The analysis showed that Compared with other regions in northern and southern Xinjiang, the climate in Tianshan Mountains is more diverse and complex: the Tianshan Mountains have four types of semi-arid climate, monsoon climate, humid continental climate and tundra climate from low to high altitude. According to the geographic coordinates of Yueliangdi Village, the author judged from Fig. 2 and Fig. 3 that the climate type of Yueliangdi Village is a warm summer humid continental climate (Köppen classification Dfb climate). The main characteristics of this climate are: the coldest The average monthly temperature is below -3°C (27°F), the average temperature for all months is below 22°C (71.6°F), and the average temperature for at least four months is above 10°C (50°F). There is no significant difference in rainfall between seasons, and the data collected by the author verifies this judgment: Yueliangdi Village ' s monthly average temperature is above 10℃ in June, July, August, and September, and its precipitation varies in each season. There is no significant difference. From the collation of meteorological data, the author found that the annual precipitation in Yinggebao Township, where Moondi Village is located, can reach about 400mm, which can reach the standard of semi-humid climate. Because of the climatic conditions, Yinggebao Township and Mu Lei County became one of the main granaries in Xinjiang in the era of planned economy. Based on the above climatic materials, the author proposes that there is a crisis of damp disease at the northern foot of East Tianshan Mountain represented by Mu Lei.

Fig4. General layout of yueyuendi village and selected residential buildings

2 Analysis of the pathological characteristics of residential buildings based on infrared thermal imaging technology

According to this study building pathology classification and diagnostic measures about building disease (4, 5) and settlement survey the scene of the local-style dwelling houses building disease degree settlement houses pathological conditions can be divided into light, medium and heavy at three levels (Table 1), which houses wall paint layer empty drum and flake paint materials classified as "light", the soil cracking and peeling as "in", and structural components such as post stairs will luck receiving gate, luck receiving palace of corrosion cracks and classified as "heavy".

This article selects two representative residential buildings, and introduced the infrared thermal imaging study, from the village general layout (Fig. 4), the two houses in the village of the moon to middle, because of the distance from rivers are ruled out the influence of river foundation, structural system, the same are adobe brick wall structure, Moreover, the two residences were built in the 1980s for local residents to live in for a long time, rather than built in recent five years for tourists to stay in the family hotel. The only difference is that the wall structure of Building 1 is larger than that of Building 2. In terms of wall structure, the outermost side of Building 2 is painted with a new imitation Clay Facing due to the requirements of landscape protection in recent years. Visual observation, feature evaluation and moisture content measurement were used to investigate the degree of building diseases and pathological analysis on site.

Fig5. Investigation and analysis of Building 1 disease

According to the pathological investigation results of Building 1 (Fig. 5), the east side of the main facade of Building 1 (the part on the right side of Fig. 5 that was not painted with imitation clay finish) was affected by rain and freeze-thaw. Most of the painted wall layer fell off, and the cracking and falling off of the overlying soil layer appeared in some parts. And on the west facade because its new finishes painting and no peeling phenomenon occurred, but under the eaves stylobate up part is a large painting layer infiltration, the phenomenon of empty drum, at the same time the entire building in the stylobate part, the porch pillar and window mouldings cracks occurred in the part of the crack and the material corrosion phenomenon, a few parts in the mould. To choose a building 1 eaves column and girder junction in judgment, temperature cracks found relative to other parts of the surface temperature is lower, while the eaves columns, girders and wooden fang also presents the low temperature condition, and from the hot like part of the larger area relatively visible moisture, shown in the figure shows that in the stylobate internal to the naked eye can't • wet condition is still there.

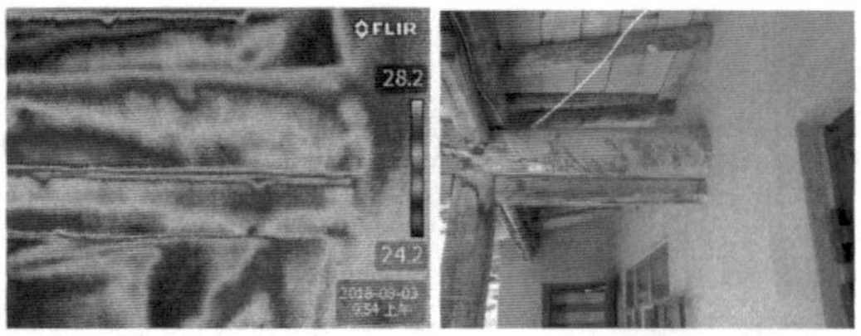

Fig6. Thermal imaging judgment of Building 1 disease

In Building 2, due to the failure to refresh the veneer outside the overlying soil layer, a large number of empty drums and peeling of the painted layer appeared. In addition, the cracking and shedding of the overlying soil layer in this building was significantly more than that in Building 1, and there were also many cracks on the cornice. As shown in Fig. 3, from the direction of the gable, take the north facade as an example. The overlying earth cracks appear on the gable wall subjected to freeze-thaw action, and some cracks are long, up to 1.1-1.2m, with a width of 10-15mm and a depth of about 15mm. East facade direction its disease situation is relatively serious, appeared in the stylobate turns the soil loss, caused a large number of adobe bricks, exposed the relate to s oil moisture stylobate is larger, and the influence of the freeze-thaw action on, at the same time, the border of stylobate and ground in a small amount of moss, the longitudinal wall part of the soil loss relative gable is more apparent, some parts of the fixed effect of oblique cross wood has also exposed long; In addition, the cornice part of Building 2 also suffered from deformation of cornice sealing plate and screw falling off due to rain water, freeze-thaw effect and old disrepair. Building 2, the author find out the typical stylobate and entrance door window frame based on infrared thermal image (Fig. 8), from the infrared thermal image shows: first, the temperature differences on the stylobate in conformity with the stylobate in vertical and horizontal cracks, and relatively intact surface temperature is higher, the more susceptible to cracks and rain, soil water infiltration and shade shadow; The shedding degree of the plastering layer at the base is also shown in the infrared thermal image because of the difference of heat transfer coefficient of different materials (exposed structural members).

Fig7. Investigation and analysis of Building 2 disease

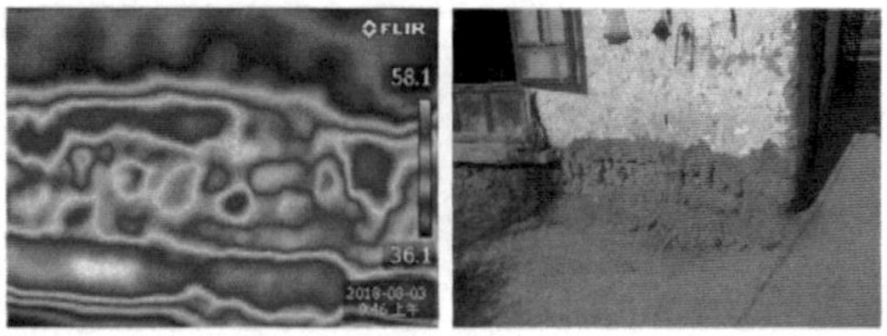

Fig8. Thermal imaging judgment of Building 2 disease

Table1. Disease classification

Severity	Light	Moderate	Heavy

Pathological name	The surface painting layer is empty and the painting material is peeling off	Overburden falling off and cracking	Cracks in foundation and column	Exposed corrosion of structural components
Pathological symptoms	The internal water seepage, vaporization and expansion of the plastering material cause the material to bulge and deform, and then peel off	Due to the larger variation range of the temperature in the day and night, cracks and shedding of the covering soil of residential buildings are caused by freezing and thawing	The cracks of structural components are caused by moisture and freeze-thaw, and then local looseness and collapse	
Location	Surface coating	General wall and window frame	Lime sand brick platform foundation and exposed wood column	Sand lime brick foundation, cornice and windowsill
legend				

3 Analysis of the causes of building defects and diseases in settlements

3.1 Pathological analysis of hollow drum and peeling of wall painting layer

Mentioned above, because the moon is a village at the northern foot of east Tianshan warm humid continental climate in the summer of climate classification (cobain Dfb) climate, the environment temperature difference and large temperature difference between day and night, and precipitation on September 4 - more, combined with the village layout along the river, the surrounding soil moisture content is bigger, makes the settlement building by late spring or early summer rain, soil water and the resulting freeze-thaw effect produced a lot of diseases.

As local precipitation more focused on September 4 -, make settlement in the period of focus is washed by the rain, because of their special civil structure, combined with the roof overburden soil water storage ability is limited, the rain is through the roof into the wall, at this time the moon river surrounding the village just thaw, water reached the peak of my year, makes the soil moisture content is affected by river also showed a trend of rise, soil water down the stylobate to rise due to capillary action, make its damp is in serious condition at the point; Moreover, the temperature difference between day and night in the valley is huge, especially in April, when the daytime temperature is relatively high, the seepage vaporizes and expands, causing the deformation of the painted layer and the empty drum. Because the night temperature can fall below 0℃, the seepage can freeze and expand, causing the peeling layer to break off. As the facade wall of building 1 is shielded by the front eave gallery, the evaporation rate of infiltration rainwater on its surface is slow, which makes the rainwater penetrate into the painted layer more than other parts, so the empty drum phenomenon is more obvious. The west facade is due to the perennial upward adsorption of soil water, which causes the large area of its painted layer to fall off, and the obvious infiltration marks of its base, and mildew spots also appear in some areas.

3.2 Analysis of cracking and shedding of overburden soil

The cracking of the overburden soil layer in Building 2 is obvious, and the reason of the cracking of the overburden soil layer is similar to that of the peeling of the overburden layer, both of which are caused by rain scouring, soil moisture and freeze-thaw action. However, the cracking and shedding of the overburden soil layer relative to the whitewash soil layer need to go through a slow and long process, which is often the result of years of accumulation.

3.3 Cracks, deformation and erosion of structural members

Cracks are mostly found in the base and columns of the two dwellings, and the pathogenic mechanism is the same as the former two. Deformation of the components from the west side of building 2 verge board found that verge board bending deformation due to more people ignore the maintenance set screw, together with concentrated rainfall make wooden strip infiltration, and then affected by temperature changes and the effect of freezing and thawing, the bending deformation, the verge of building 2 board shall promptly repair, to avoid falling and injuring residents; The exposed corrosion phenomenon of the component is more than that found at the base and windowsill thread foot. The reason for this phenomenon is that the aforementioned cracks have not been treated in

 https://doi.org/10.1051/e3sconf/202123704005

time, which leads to their continuous development and finally causes the external plastering layer mortar to break off.

To sum up, the causes of "light", "medium" and "severe" diseases in Moonland village are related to local climate characteristics and soil moisture caused by rain erosion.

4 Study on infrared thermal image determination method of colony moisture disease

Come to a conclusion from the above analysis, the moon to village settlement main disease with local climate and environment more damp conditions, according to the author's research results before[6], the introduction of infrared thermal imaging technology for wet disease diagnosis and analysis, as shown in Fig. 9, the author choose at building a stylobate is analyzed, through visual found its most serious stylobate damaged area (turns the soil loss, adobe bricks exposed), as shown in the right yellow area is further proof of the author studied before flooding bricks into a nonlinear relationship between moisture content and moisture distance, and then by the infrared thermal imager and determination of moisture meter, It is found that the seriously damaged area of the platform should be the hottest area in summer and the coldest area in winter, that is, the part of residential buildings with disguised thermal Bridges. At the same time, according to the determination of moisture meter and previous research results, the author can determine the minimum temperature area and the location of the maximum moisture content of the platform moisture disease[4].

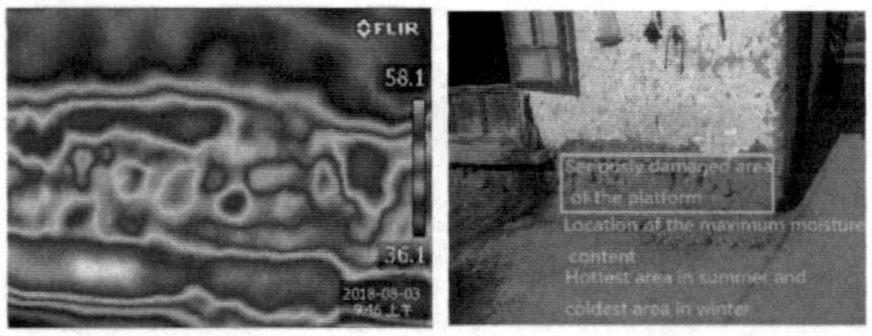

Fig9. Analysis of moist infrared technology of platform base

5 Conclusion

From what has been discussed above, the author based on GIS platform to the moon village, it is concluded after the investigation and analysis of the settlement site is located in the east Tianshan north, soil relative humidity, and the relative average annual precipitation, precipitation is bigger, in the September 4 - are greatly influenced by wet conditions, and the weather gets warmer in April because of the snow melt water run down along the eaves column presents the concentration characteristics, present various wet disease characteristics, according to these characteristics from the perspective of building pathology using infrared thermal imaging technology preliminary quantitative detection of the settlement damp disease research, provide the foundation for the further study.

References

1. C. Zhendong, Xinjiang Folk Houses (China Building Industry Press, Beijing, 2009)
2. P. Chao, Z. Shanqing, W. Shenglan, et al, Variation characteristics of potential evapotranspiration in Tianshan Mountains and its comparison with Southern and northern Xinjiang in recent 36 years. Studies on arid regions. 26, 3 (2009)
3. L. Youcun, J. Keqin, Z, Kui, L. Yan, H. Tian ding, Z. Yu, S. Yongping, H. Yonghong, Y. Baisheng. Response of precipitation in Tianshan region of China to global climate change. Glacial-permafrost. 39,4(2017)
4. Z. Baoqing, L. Zukang. Experimental Study on water-bearing Characteristics detection of Traditional Masonry based on infrared Thermal image. Huazhong Architecture. 36,3 (2012)

Connotation Analysis on Sustainable Development Strategy of Garment Enterprises — Case Study on the Famous Brand JNBY

Yuqian Dai[1], Yan Cui[1,2], Xiaogang Liu[1,*]

[1]Fashion and Art Design Institute, Donghua University, Shanghai 200051, China
[2]Department of Fiber Science & Apparel Design, College of Human Ecology, Cornell University, NY, United States

Abstract. Domestic original designer brand JYBY adheres to environmental protection and social responsibility actions, and integrates sustainable development into its strategic planning. This paper mainly USES case analysis method to analyse the composition and implementation measures of JNBY's sustainable development strategy, so as to provide experience for other garment enterprises in China to formulate sustainable development strategies.

1. Introduction

In recent years, with the increasing influence of various fashion climate innovation forums, sustainable fashion has become a new trend. With the upgrading of the consumer market and the rapid rise of China's fashion industry, new requirements have been placed on the reform of enterprises. Internal and external factors such as government and investor support for sustainable design and low-carbon industrial upgrades have driven the rapid development of domestic sustainable fashion brands.

1.1. Superior policies promote the expansion of the reach of sustainable apparel markets

In recent years, government departments have begun to support, guide and encourage of sustainable fashion development by setting up special support funds, regulating industrial structural development, and accelerating green industry talent development. Enterprises are the main body of innovation and market, then the government is the promoter of innovation[1]. China attaches great importance to the 2030 sustainable development agenda. In March 2016, the government integrated the sustainable development agenda with the national long-term development plan. The policy environment for the fashion industry is increasingly perfect. The government supports most of the sustainable fashion industry's operating projects, and the domestic sustainable fashion brand development has a stable policy foundation.

1.2. Textile and apparel corporate social responsibility promotes sustainable fashion development

With the increasing attention of consumers and the media on environmental issues, social responsibility and sustainable environmental protection measures have become the focus of the development strategy of apparel companies. China's textile and apparel industry is also gradually in line with international standards, and is produced in accordance with international industry standards during production to ensure the environmental protection and sustainability of products. At present, 2 brand companies, 14 group companies and their value chain partners have joined the "Climate Innovation 2030" industrial action organized by China Textile Association. Advocating sustainable fashion is conducive to promoting the stable and sustainable development of enterprises and improving the level of social development of enterprises [2].

1.3. Change of consumer concept to expand domestic demand for sustainable clothing market

The promotion of sustainable fashion has become the focus of consumers, while the awareness of environmental protection among Chinese consumers is also increasing. According to the report "New Trends in China's Consumption: Three Major Motives Shaping New Customer Groups for China's Consumption", consumers support green consumption and environmental protection. A survey by GFK, a market data analysis agency, showed that 80% of Chinese respondents insisted that brands and companies must be responsible for the environment. The eco-friendly design of clothing aims to develop environmentally friendly products and meet consumers' demand for fashion trends [3].

* Corresponding author: liuxg@dhu.edu.cn

 https://doi.org/10.1051/e3sconf/202123704006

2. Strategic analysis of sustainable development of JNBY

The company is an excellent designer brand, that mainly focus on designs, promotes and sells fashion clothing, shoes and accessories. As the basic unit of society, enterprises have the responsibility to reduce the negative impact on the environment in addition to producing products and creating value. The elements of sustainable development of clothing companies are composed of economic factors, social factors and environmental factors. Therefore, the three aspects of responsibilities that companies need to undertake are economic responsibility, social responsibility and environmental responsibility. The sustainable development strategy of JNBY (Figure 1) revolves around three elements. The company actively promotes sustainable initiatives, adopts sustainable corporate management strategies, and assumes environmental and social responsibilities.

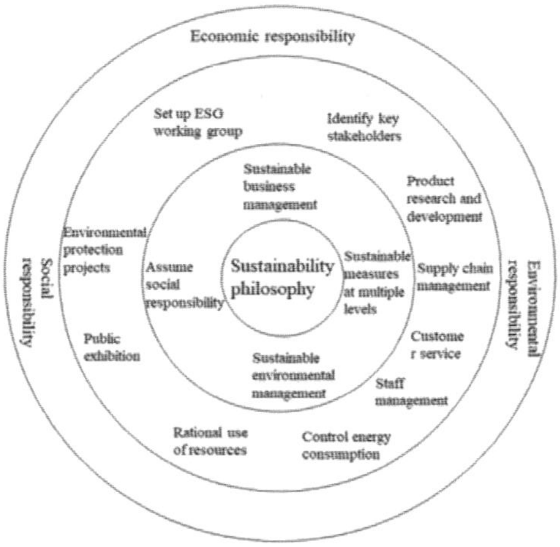

Figure 1. Sustainable development strategy of JNBY

2.1. Incorporate the concept of sustainable development into the concept of enterprise management

2.1.1. Establishment of ESG working group references

In order to carry out environmental, social and governance management in an all-round way, JNBY has set up an ESG working group led by internal control and internal audit, and formed an ESG working group including the group affairs department, group brand department, headquarters human resources department, direct human resources department, production and procurement center, research and development center and e-commerce operation center. The heads of departments participat and appoint special personnels to be responsible for ESG management and reporting. The ESG working group regularly reports to the company's management and puts forward appropriate improvement suggestions.

JNBY pays attention to the sustainable development of the environment, and regards the social responsibility of enterprises as an obligatory responsibility and an extension of enterprise values. The company is committed to running the green concept into the company's operations, advocating the reduction of energy consumption and resource conservation. The company is very concerned about the sustainable development of society. JNBY believes that the development of art is regarded as an essential part of the brand value of the company. It always adheres to the designer's artistic pursuit and social responsibility, which is also the governing concept of ESG Working Group.

2.1.2. Identifying key stakeholders

Stakeholder communication is way for enterprises to fulfill their social responsibilities and achieve sustainable development. Based on the characteristics of the industry and the group's operations, JNBY has identified eight major stakeholder groups. It has maintained close communication with key stakeholders through various channels and actively responded to the requirements and expectations of stakeholders.

2.2. Take measures to implement the sustainable development strategy of enterprises

2.2.1. Sustainable product development.

JNBY values product quality and service quality, and uses a design-led retail model to provide full freedom for the creation of the designer team, allowing designers to use their unique aesthetic vision to pursue artistic beauty. The company adheres to the design concept of being close to nature and insists on selecting high-quality raw materials, such as natural organic cotton certified by the Global Organic Textile Standards (GOTS) (Figure 2) and pure wool certified by the Woolmark Company. The company chooses environmentally friendly raw materials, such as recycled polyester, which are extracted from plastic bottles for recycling (Figure 3), recycled nylon extracted from discarded fishing nets for decomposition and recycling.

Figure 2. Organic cotton

Figure 3. Regenerated polyester

2.2.2. Sustainable supply chain management.

JNBY attaches importance to supply chain and supplier management. The company conducts hierarchical management of suppliers and actively promotes suppliers to fulfill their social and environmental responsibilities in order to achieve sustainable development values that are mutually beneficial to supply chain partners. Suppliers of JNBY include four major categories: surface accessories suppliers, finished product suppliers, processing suppliers and other suppliers. The company has established strict procedures and standards for supplier access, and regularly inspects and evaluates suppliers. JNBY is concerned about the environmental and social performance of suppliers and investigates the employment and environmental protection effects of suppliers.

2.2.3. Sustainable customer service.

Increasing consumer awareness of environmental protection consumption and resource reuse is an important part of achieving sustainable development. JNBY is committed to constructing a sustainable lifestyle ecosystem. It has established a full-channel interactive platform consisting of three parts, physical retailing stores, online platforms and WeChat-based social media interactive marketing service platform. It insists on taking "fan economy" as the core. The company provides attentive customer service and personalized member activities to strengthen contact with customers, thereby improving customer satisfaction and enhancing service competitiveness. In FY 2019, driven by the WeChat-based social media interactive marketing service platform, JNBY already has more than 3.6 million member accounts, including 3.1 million electronic member WeChat users. The retail sales contributed by members accounted for 70% of total retail sales. The brand's fans are constantly expanding and have a high fan loyalty. Driven by the "fan economy" strategy, offline retail stores grew by 3.4% compared with the same store in 2019.

2.2.4. Sustainable customer service.

The success of an enterprise depends on the high-quality and reliable service of its employees. Good employee relations are the foundation of its development and will have a strong impact on its development potential. JNBY has established good cooperative relations with employees and is committed to providing employees with a safe, healthy and comfortable working environment. The company protects the legitimate rights and interests of employees, and cares about the health and safety and sustainable development of employees.

2.3. Emphasis on environmental management extension connotation of enterprise sustainable development

2.3.1. Sustainable customer service.

In order to reduce energy consumption, JNBY encourages employees to reduce consumption and waste from the source, and advocates green office. For example, the company strictly requires employees to work without paper, so as to cultivate employees' awareness of resource conservation.

2.3.2. Sustainable customer service.

(1) Recovery of office resources

In order to increase the utilization of resources, JNBY encourages the secondary or even multiple use of resources to raise the awareness of environmental protection. In response to the call for garbage classification, the company set up a sorting box for plastic bottles and paper in the REVERB brand design office, and carried out office recyclable waste processing and reuse activities. By reducing solid waste emissions, the company improves waste recycling and utilization rates and contributes to the environment and circular economy.

(2) Recovery of production resources

JNBY takes measures to classify and recycle waste fabrics and inventory fabrics, thereby reducing waste of resources and increasing resource reuse. The company has set up a fabric recycling box at the entrance of the production purchasing center and the design department to collect the waste fabric for classification and recycling, and classify it according to the color for employees to use in the platform classroom. The JNBYHOME brand has produced the "Replay Series" designed and developed from stock fabrics, aiming to adhere to and promote the environmental protection concept of sustainable development and recycling utilization of resources.

JNBY implemented a series of energy-saving and consumption-reducing measures and paid attention to environmental management. In 2019, emissions of the company have reached the national emission standards. The company is a leader in the sustainability sector of the apparel industry. The environmental management performance indicators are shown in Tables 1 and 2.

Table 1. Emissions

Key performance indicators	Fiscal year 2019
Waste Water Quantity (Tons)	19364.00
Total greenhouse gas emissions (tons)	1751.19
Direct greenhouse gas emissions (tons)	139.93
Among them: gasoline (tons)	139.93
Indirect greenhouse gas emissions from energy sources (tons)	1611.26
Among them: purchased power	1611.26
Total greenhouse gas emissions per square meter of floor (tons/square meter)	0.09
Total amount of harmless wastes (tons)	40.50
Emissions of harmless wastes per square meter of floor (kg/square meter)	2.13

Source: JNBY 2019 Financial Report

Table 2. Resource usage

Key performance indicators	Fiscal year 2019
Total energy consumption (megawatt-hours)	2861.03
Of which: electricity consumption (megawatt-hours)	2288.71
Gasoline consumption (megawatt-hours)	572.32
Water consumption (tons)	24205.00
Energy consumption per square meter of floor (megawatt-hours/square meter)	0.15

Source: JNBY 2019 Financial Report

2.4. Undertaking social responsibility to promote sustainable value of enterprises

JNBY takes service and return to society as its own responsibility, and fulfills its social responsibility and mission. According to the business characteristics of the enterprise, a variety of public welfare activities have been carried out. In 2019, JNBY donated about 560,000 yuan in total. While fulfilling its social responsibilities, JNBY demonstrates its responsibilities and responsibilities as an enterprise, and enhances its sustainable value.

JNBY pays more attention to the concept of sustainable development, and actively promotes the cooperation between brands and environmental protection projects and brand environmental protection activities. REVERB and the well-known visual installation artist Olafur Eliasson launched the "Little Sun" project for cross-border cooperation. The company's sales of solar-powered lamps and lanterns stimulate the public's thinking about the fair distribution of energy and electricity, and respond to the UN's sustainable development goals. REVERB sponsors the sale of "Little Sun" in Burkina Faso. The small sun lamps will be provided at low prices to communities without electricity or lack of electricity, bringing people pollution-free and inexpensive light sources. The distribution plan eventually includes more than 200 small and micro enterprises selling small solar lights and mobile charging devices for mobile phones, which will have the following impact in the future:

- More than 13,000 solar products will be sold.
- The cost of 770,000 euros for flashlights will be saved.
- 355,000 batteries will be saved, thus reducing the possibility of environmental pollution caused by their disposal.
- 60 new dealers can be employed and trained.
- The local tax revenue of 17,000 euros was generated from the sales of small solar lamps.

3. Enlightenment for the sustainable development of other domestic clothing brands

3.1. Reconstruction of enterprise value system

The development of JNBY has always followed the concept of sustainability for society, economy, environment and employees. The concept of sustainability has been integrated into the enterprise's operation and management process. The company identifies stakeholders and proposing differentiated measures to deal with it, which has laid a foundation for the enterprise to realize the vision of sustainable development. JNBY achieves sustainability of enterprise growth by creating value and pursuing economic benefits. Since listing in Hong Kong in 2016, JNBY has handed over 3 annual reports. The economic returns from FY17-19 can be seen in Figure 4, which has maintained a healthy growth. After years of rapid development, China's brand clothing enterprises have entered the stage of industrial upgrading, structural adjustment, innovation drive and responsible production. First, the company should distinguish the situation and think about the problem from the overall pattern of the entire apparel industry chain. Secondly, the company needs to identify key stakeholders on time and collaborate with stakeholders at different levels to integrate as many key resources as possible. Finally, the company should use industry responsibility and social responsibility to build an enterprise value system to achieve coordinated and sustainable development of the branded apparel industry chain in the economic, environmental and social aspects[4].

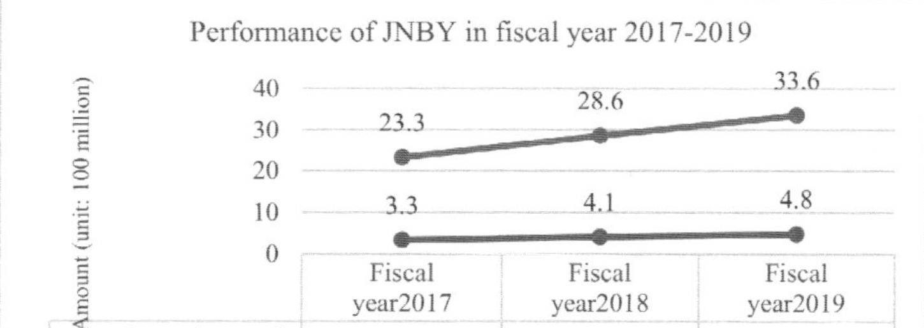

	Fiscal year2017	Fiscal year2018	Fiscal year2019
Operating profit	23.3	28.6	33.6
Net profit	3.3	4.1	4.8

Figure 4. Performance of JNBY

3.2. Strengthen contact with stakeholders

The further promotion of sustainable fashion for domestic clothing brands cannot be achieved without the support of relevant parties such as the government, non-governmental organizations, enterprise employees and consumers. Enterprises will provide clothing products to meet the social needs as their business philosophy, while also providing good working conditions for employees to realize the social responsibility of enterprises as a basic unit of society [5].Apparel brand enterprises should use their influence to strengthen cooperation with various social organizations and promote all aspects of sustainable measures to improve accessibility standards. Brand enterprises should work with the government to accelerate the formulation and implementation of environmental protection policies in the supply chain in low-carbon life [6]. Enterprises promote the concept of green consumption in marketing, and guide consumers to form environmentally friendly consumption habits to achieve the goal of promoting the coordinated development of economy and environment. Sustainable apparel companies should expand the scope of social responsibility, such as carrying out charity, and jointly promote the sustainable development of the economy, environment and society.

3.3. Strengthen contact with stakeholders

From the perspective of the international market and the domestic market, environmental management has become an environmental responsibility that apparel companies must perform. Apparel companies should change from passively complying with environmental standards to actively taking on and fulfilling environmental responsibilities. "Made in China 2025" put forward the goal of promoting green industrial development, through green product development, building green factories, building green supply cha ins, building my country's green manufacturing system. Apparel companies fulfill their environmental responsibilities and achieve coordinated development with the environment. The environmental responsibility of apparel companies is reflected in two aspects: resource conservation and environmental protection. From the perspective of material resource consumption, the enterprise can save resources and reduce waste in product design, production and use. Through the closed-loop development model of "resource-product-renewable resources", the resources can be used continuously. From the perspective of environmental protection, enterprises must reduce the damage of waste water, waste gas and waste to the environment in production and operation, reduce energy consumption, save water, and become an environmentally friendly enterprise. [7].

4. Summary

The case of JNBY indicates that the company's sustainable development strategy has expanded from sustainable product design and production, sustainable supply chain management, and sustainable consumer management to sustainable enterprise employee management and sustainable social responsibility. The company integrated the concept of sustainability into its management, forming a comprehensive strategy for sustainable development, and achieved obvious results. As a well-known and original brand in China, JNBY's sustainable development strategy has certain guiding significance to other brands in China. Textile and apparel enterprises should respond to climate change and assume social and environmental responsibilities.

References

https://doi.org/10.1051/e3sconf/202123704006

1. Liu Xiaogang , Brand Innovation Theory [M]. Donghua University Press, 2010.5-6.
2. Kim, Hanna, Eun-Jung Lee, and Won-Moo Hur. "The normative social influence on eco-friendly consumer behavior: The moderating effect of environmental marketing claims." Clothing and Textiles Research Journal 30.1 (2012): 4-18.
3. Joy, Annamma, et al. "Fast fashion, sustainability, and the ethical appeal of luxury brands." Fashion theory 16.3 (2012): 273-295.
4. Yang Guangyong, Ji Guojun. Constructing Green Supply Chain Based on Triple Bottom Line: Comparison of Environmental Regulations between EU and US [J]. China's Industrial Economy, 2011(02):122-132.
5. Yang Jie. Patagonia clothing brand sustainable development strategy analysis [J]. renewable resources and recycling economy, 2019, 12(05):41-44.
6. Xu Yangyang. Exploring Sustainable Fashion Design Based on Fast Fashion Brand [D]. Beijing Institute of Clothing Technology, 2019.
7. Shao Zhengyan, Zhang Jiayue. A case of sustainable supply chain management and its enlightenment to Chinese brand clothing enterprises [J]. Wool Textile Technology, 2017, 45(08):69-74.

 https://doi.org/10.1051/e3sconf/202123704007

Analysis of the experience of renovating industrial enterprises into hotels in China

Zehong Gong* and Larysa Bridnia

Architecture, Architecture Department, Kyiv National University of Construction and Architecture, Ukraine

Abstract. The adaptive reuse of abandoned industrial buildings is a very hot topic in the post-industrial modern society. The re-equipment of these buildings and territories can be carried out in different directions for different functions: exhibition complexes, public centers, museums, trade enterprises. This study focuses on the renovation of such industrial buildings into hotels in China. The experience of reconstruction and related literature has been studied. The study found that since most of the post-industrial buildings in China are small in size, they are suitable for repurposing as boutique hotels. Large industrial buildings and complexes in mega cities in China can be converted into luxury hotels. Renovation of old industrial buildings into hotels makes it possible to double the average lifespan of buildings - approximately up to 60-65 years.

1 Introduction

In today's high-tech society, information technology industries are constantly supplanting more traditional industrial manufacturing industries in China. Accordingly, a large number of industrial buildings remain abandoned and destroyed. At the same time, most of these buildings are of great interest in terms of preserving the historical and cultural heritage of the region, as well as important for preserving the urban planning integrity of the city. The renovation of such enterprises takes place in different directions. The experience of converting industrial enterprises into hotels remains poorly studied.

Adaptive building reuse is a special form of renovation that brings many special challenges and challenges for designers [1]. Very often, modern technologies are difficult to fit into the existing historical structure of the old building. Difficulties arise in two directions - compliance with modern rules and regulations and modern design requirements [2]. Some elements of buildings are reworked several times to find a worthy embodiment for them in a new building, which significantly affects the final cost. Designing reusable buildings takes more time and effort from the design team [2]. Designers must understand the importance of preserving historically, culturally and architecturally valuable parts and elements of buildings, as well as anticipate the consequences of using reclaimed elements. Aware of the significant increase in construction costs involving the re-use of old elements, they should strive to use new design and construction methods to reduce potential adaptation costs. Otherwise, this approach may not be widespread and successful.

Pollution of abandoned industrial enterprises is one of the main problems of re-adaptive use of such territories [3]. The cost of cleaning them significantly affects the total cost of the renovation. The experience of adapting industrial buildings for high-class housing has shown that often excessive cleaning, renovation and hiding old walls behind new ones leads to emasculation and loss of cultural value for society. [4].

Until 2006, little attention was paid to the protection of industrial heritage in China. As of 2006, 2,352 buildings were under protection (information from the Ministry of Culture of China), and only 140 of them were associated with industrial heritage. April 18, 2006 ' Wuxi Suggestions' were put forward at the Forum of China's Industrial Heritage Protection [5]. The concept of "industrial heritage" in China was defined and seven programmatic steps were identified for the protection and reuse of industrial heritage. This opened a new path for China in terms of preserving, protecting and using industrial cultural heritage.

Que studied the experience of protecting industrial heritage around the world and put forward proposals for its application in the Chinese context. He paid particular attention to the need to preserve Chinese agricultural industrial enterprises, as well as national industrial enterprises of the industrial construction period. [6]. Yu and Fang proposed to identify, study and document all of China's industrial heritage in terms of the prospects for its further use. The history of the formation of China's industrial heritage was studied, four periods before and three periods after the founding of the People's Republic of China were identified. In accordance with these periods, types, costs were identified and the potential for reuse of such establishments was considered. They emphasized the public interest and value of the protection and adaptive reuse of industrial heritage [7].

These systemic studies have created the basis for further exploration and reuse of China's industrial heritage.

This study defines the main boundaries of protection

* Corresponding author: zehong.g_a@knuba.edu.ua.

and preservation of cultural industrial heritage in combination with its reconstruction and adaptation to a new function. The stages of systematization and evaluation of such enterprises are highlighted, the need to change the attitude of society towards industrial heritage is revealed, as well as all the advantages of their reuse for a hotel function. The modern tendencies of re-profiling industrial enterprises for hotels have been studied, ways of preserving industrial relics, taking into account the economic effect of their reuse as hotels, primarily as boutique hotels, have been determined.

2 Methods

This study is based on the methods of statistical analysis, which includes a documentary study of cases of reconstruction of industrial buildings for hotels, their field examination and photographic recording. Researched project documentation and accompanying early-journalistic literature, as well as Internet sources. A survey of stakeholders (designers, developers, hoteliers and consumers) was conducted, and the websites of hotels rebuilt from industrial enterprises were studied. The research was carried out on 52 industrial buildings converted from 1990 to 2020 into hotels in China.

3 Results

3.1 General information on the reuse of industrial buildings as hotels in China.

The study included 52 cases of adaptive reuse of industrial buildings for hotels in China (identified as of 2019). The total area of such renovation exceeded 360,000 square meters. Such buildings are found in the first-, second-, third-tier cities and in the countryside. The capacity of such hotels ranges from 1 to 398 rooms. The comfort level of hotels varies from 2-5 stars, and the cost of living is from 118 CYN (for 2-star hotels) to 5000 CYN (for 5-star hotels).

The first case of reuse of an industrial building for a hotel in China was in 1995, at that time the life cycle of the building was 55.5 years. As of the beginning of 2020, the average lifespan of such buildings was 65.2 years. The longest lifespan was 214 years, and the shortest was 30 years.

The largest number of such examples is in Shanghai (18 cases), in Beijing and Hangzhou - 5 cases each, in Xi'an - 4 projects.

Table1. The statistics information on 52 repurposed industrial hotels.

NO	Name of project	lifespan	Location	price	Number of rooms
1	Jinjiang Inn (Shanghai Zhongshan Park)	37	Shanghai	288	190
2	Manshan hotel	62	Shanghai	358	2
3	One Hundred Holiday Hotel	91	Shanghai	188	40
4	Jinchang hotel	32	Shanghai	203	266
5	Hanting hotel	73	Shanghai	218	8
6	24K International Hotel	Na	Shanghai	170	29
7	Jinjiang Inn (Shanghai Huamu road)	Na	Shanghai	368	189
8	Yu Chuan Chateau hotel	41	lan tian	332	249
9	motel168 Hotel	27	Shanghai	137	
10	Yayue hotel (Aiben)	50	Shanghai	98	
11	China Community Art and Culture Hotel	50	Qindao	479	255
12	CityInn hotel	40	Chengdu	155	152
13	motel 268 Hotel	88	Shanghai	368	77
14	Nanqun hotel	Na	Shanghai	542	162
15	Jinjiang Inn	Na	Shanghai	188	39
16	3+1 bedroom	225	Beijing	218	518
17	Gracie Art Hotel 798	63	Beijing	580	24
18	7 Days InnGuangzhou Panyu City Bridge	Na	Guangzhou	198	97
19	Fortune Ancient Canal Hotel	73	Wuxi	389	49
20	Mingdi Boutique Hotel	170	Shanghai	485	2
21	Mutianyu tile factory Hotel	59	Beijing	359	216
22	WaterHouse	90	Shanghai	268	117
23	Pier One Hotel	87	Shanghai	418	25
24	Scholars Hotel Suzhou Pingjiangfu	35	Suzhou	688	31
25	Hansheng hotel	85	Shanghai	789	110
26	Phoenix Mountain Hotel	50	Hangzhou	272	130
27	Zuoyouke Theme Hotel (Xi'an Gaoxin)	40	Xianan	215	65
28	Sheng Gao Hotel	47	Hangzhou	590	30
29	Chunqiu hotel	59	Xianan	449	193
30	UHome	99	Tianjin	420	24
31	Dahua 1935 Theme Hotel	85	Xianan	199	19
32	Qingdao Tiger Hostel	110	Qindao	269	88
33	Hui hotel	40	Shenzhen	1050	96
34	Youth Hotel of iD Town	31	Shenzhen	399	1

35	*Ovolo Southside hotel*	44	Hongkong	888	84
36	1 Hugu Hotel	62	Hangzhou	713	88
37	Jiuhou Theme Hotel	89	Ningbo	293	16
38	Cheery Canal Hotel	69	Hangzhou	548	3
39	Huanyuan hotel	28	Shanghai	239	17
40	OneHouse	70	Shanghai	629	66
41	joyardhotel	38	Guangzhou	569	55
42	Zuoyouke Theme Hotel (Steel factory)	62	Xianan	589	4
43	Ouwei Design Hotel	35	Hangzhou	980	54
44	Taoxichuan International Trade Hotel	62	Jingdezhen	228	26
45	Beijing Backyard Inn	40	Beijing	523	13
46	Cangge Holiday Inn Beijing Shougang	60	Beijing	316	88
47	The Journey Hotel	30	Dongguan	399	108
48	Wuguan hotel	64	Wenzhou	1238	35
49	Alila Hotel	60	yangshuo	2565	159
50	Narada ResortSpa Nanxun Giant	69	Huzhou	668	131
51	Sure 24 rooms hotel	62	Chengdu	279	140
52	Holiday inn riverside	83	Wuhan	90	32

3.2 Location

Industrial buildings converted to hotels in one-tier cities accounted for 53.8%, which exceeds the total number of such examples in other cities. The number of repurposed projects in the cities of the first and second tier reached 86.5%, as shown in Figure 1.

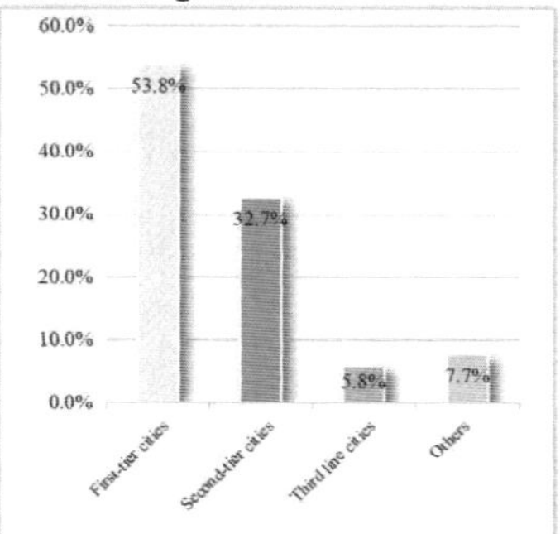

Fig1. The city where the industrial buildings is repurposed as hotels in China.

From this information, we can conclude that such renovation may be the most attractive for cities of the first- and second-tier cities.

Studies of the influence of climatic features have shown that the largest percentage of such reconstructions occur in areas with warm climates.

When it comes to refurbishing industrial buildings in a hotel, which includes the installation of additional doors, vestibules and heating systems - to prevent the ingress of cold air, it is economically much more profitable than the demolition of existing buildings and new construction. Nevertheless, the percentage of renovations of industrial buildings in the hotel in Southern China (40 facilities) is higher than in the North of China (12 facilities). It seems that the level of urbanization affects this factor more than environmental protection and energy conservation. At the same time, this indicates a high potential for the development of this area in the cold regions of northern China.

At the same time, an important indicator is the location of the object relative to the historic city center. The largest share of reusable objects falls on the areas close to the historic center. This attraction is due to both the prestige of the district and the higher percentage of historic industrial buildings that need to be preserved. More detailed information on this issue is given in Figure 2.

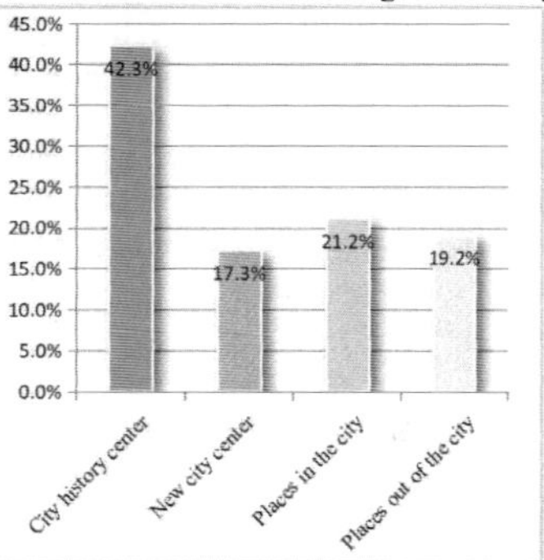

Fig2. Location of industrial buildings repurposed as hotels in the city.

Among the 52 examples of renovation of industrial buildings for hotels, the largest percentage falls on detached buildings (61.5% or 32 cases). At the same time, during the reconstruction of industrial parks, approximately 9.6% are hotel projects as part of other public buildings and 28.8% - individual hotel buildings.
A very small percentage, only about 10% (including 6 cases), are examples where an industrial building was converted into a hotel, and then still such a hotel closed and the building was demolished or converted again. Among these 6 cases - 2 examples - were erected under the project of reconstruction of a separate building, and 4 - as a part of renovation of industrial parks.

3.3 Lifespan

Industrial buildings of different construction periods in China differ greatly in terms of the use of building structures and materials. Therefore, it was important to explore the potential for conversion of industrial buildings of different construction periods into hotels.

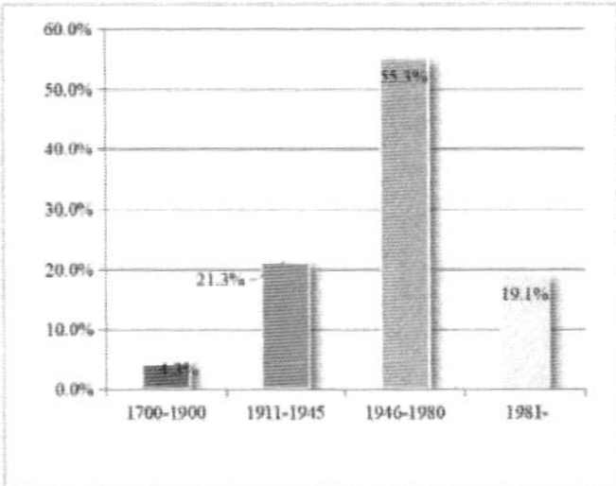

Fig3. The original completion time of industrial buildings.

Most examples of renovation of industrial buildings in the hotel belong to buildings built in 1945-1985. This is due to the fact that China has the largest number of industrial buildings during this period of construction, as well as such buildings are well preserved and suitable for further reconstruction.

Significantly fewer reconstructions of industrial buildings for hotels occurred in the building of 1700-1900. Although such industrial buildings are artistic and architectural expressive, contain many features of national culture and may seem attractive for reconstruction in a boutique hotel, but at the same time they are under enhanced protection of China's cultural heritage. In addition, the reconstruction of the hotel function requires significant changes in the original structure of buildings, and therefore it is very difficult in such a reconstruction to preserve the original characteristic structure, which is under protection.

Such buildings are often converted into museums, exhibition complexes or community centers.

Most industrial buildings built after 1985 continue to operate as industrial enterprises and do not require reconstruction.

Reconstruction of industrial buildings for hotels in China has a pronounced upward trend. The first such project was Jinjiang Inn (on Zhongshan road) completed in 1995, which is still used as a hotel. In total, from 1990 to 2008, 15 projects were implemented. From 2008 to 2014, 20 projects were completed, and from 2015 to 2019 - 17 projects.

The design period for the construction of new buildings is 50 years, at the same time, the average life cycle of buildings in China is 25-30 years. Studies have shown that the average lifespan of industrial buildings converted into hotels is 65.5 years, and therefore twice the average and much higher than the design cycle of new construction.

3.4 The dimensions of repurposed hotels

Among the 50 projects where the exact area could be measured, 57% had an area of less than 5,000 m2, 24% had an area of 50,000 to 10,000 m2, and 22% had an area of more than 10,000 m2.

As Figure 4 shows, among the 52 projects for the reconstruction of industrial buildings for hotels, more than 60% are 4-6-storey hotels.

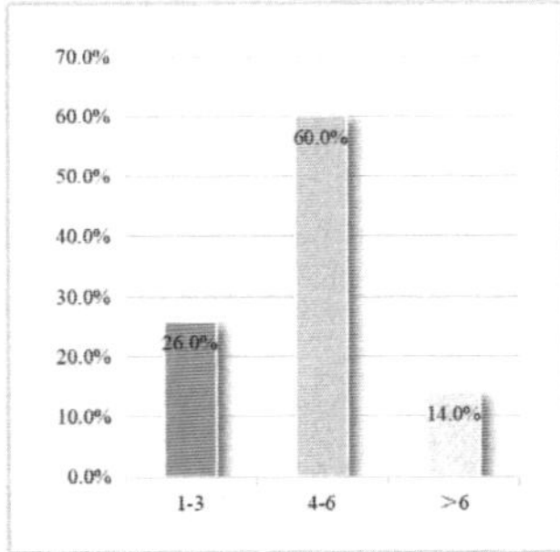

Fig4. The superficiality of repurposed industrial hotels.

Hotels that are reconstructed on the basis of industrial buildings mostly have a relatively small capacity. See Figure 5. More than 60% of such hotels have less than 100 rooms. This is due to both the structure of the industrial buildings and the fact that boutique hotels, which are usually converted to such buildings are generally small capacity, which allows such institutions to have an individual atmosphere of industrial culture.

3.5 Comfort level of the hotels

The star rating of 52 reconstructed hotels is shown in Table 2. 52 examples of post-industrial hotels were developed, of which: 28 - in first-tier cities, 24 in the cities of other tiers. In first-tier cities, 75% are 2-3-star hotels, 4-star hotels are 21%, and five-star hotels are only 3.4%, in non-first-tier the situation looks the opposite: 58.3% account for 4-5 star hotels and 41.7% - for 2-3 star hotels. And in such cities a large percentage of 5-star hotels - 25%.

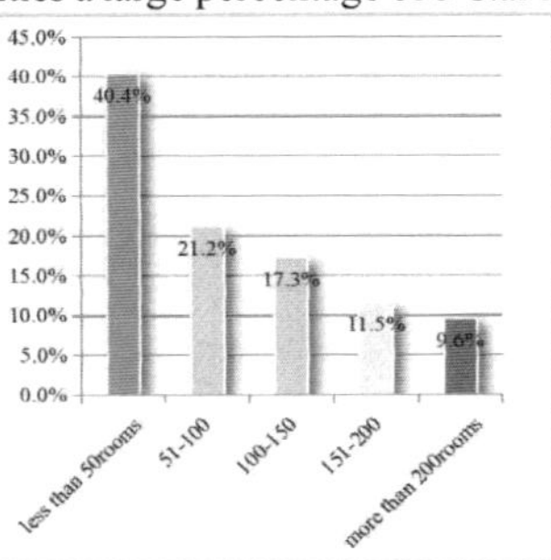

Fig5. The room number of repurposed industrial hotels.

Table2. The statistics information on 52 repurposed industrial hotels.

	2-star	3-star	4-star	5-star	Number of cases
52 cases	28.8%	30.8%	26.9%	13.5%	52
In first-tier cities	39.3%	35.7%	21.4%	3.6%	28
Not in first-tier cities	16.7%	25.0%	33.3%	25.0%	24

The analysis of pricing in such hotels is also carried out. Among the 52 projects, the price of a standard room per night in a 2–3-star hotel in a city with a population of more than 10 million is similar to that of a city with a population of less than 10 million. At the same time, the cost of living in standard rooms in a 4-5-star hotel is 1.52 times higher than the price of the corresponding hotels in cities with a population of more than 10 million (see

Figure 6).

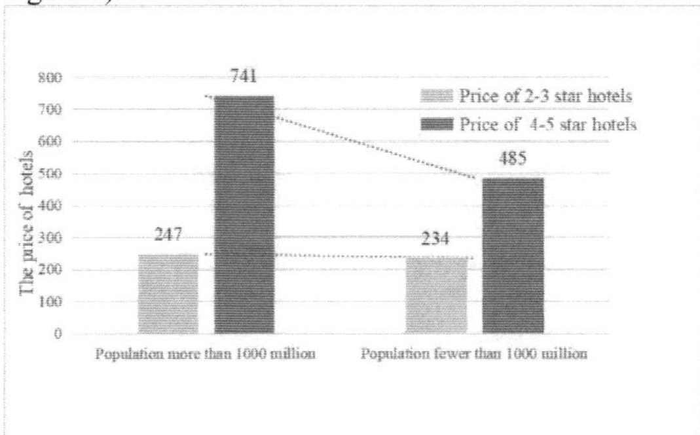

Fig6. The correlation between population and star ranking of hotels.

4 Conclusion

The successful repurposing of industrial buildings as hotels can extend the life of the building and avoid damage to the building caused by the second reconstruction. The following aspects are worth considering via statistical study. It is more suitable to be repurposed as a boutique hotel with low floor area ratio. In such a hotel, it creates the best conditions for retaining the industrial culture. To excavate the cultural value of the legacy buildings and use them to enhance the added value of hotel products, so as to achieve the unity of cultural appeals and business models for enterprises to win higher profits.

Based on statistical analysis, it has been found that in cities with a permanent population of more than 10 million, most industrial buildings are converted into 2-3-star hotels, while in other cities buildings are more often converted into 4-5-star hotels. This model seems to be the most economically determined. This is due to the higher flow of middle-class customers and the low consumer capacity of China as a developing country. The price per room per night of 4-5 star hotels in first and second level cities with a population of over 10 million is almost twice as high as in smaller cities, as shown in Figure 6. In such first-tier cities, the occupancy rate of 4–5-star hotels in first-tier cities is quite low, with a record low level of 8.3% in April 2020 [8].

The nightly price of 52 repurposed hotels is higher than non-repurposed hotels for the same star ranking and the same region. At the same time, their lifespans are 40.5–35.5 years longer than the average life span of all buildings in China, which is twice the average life span of all buildings in China. City managers guide investors to participate in the repurposing of industrial buildings as hotels in suitable locations, which can achieve the continuation of the industrial building heritage and economic benefits.

In terms of various urban, socioeconomic, structural, and styles aspects, a statistical study and detailed data is of vital importance for the analysis and decision-making of stakeholders in the repurposing and reusing of industrial buildings. At the same time, the collection of data and the establishment of databases pose challenges for future city managers.

References

1. Fisher-Gewirtzman, D. Adaptive reuse architecture documentation and analysis. Journal of Architectural Engineering Technology, 5(3) (2016). https://doi.org/10.4172/2168-9717.1000172
2. Conejos, S., Langston, C., Chan, E. H. W., & Chew, M. Y. L. Governance of heritage buildings: Australian regulatory barriers to adaptive reuse. Building Research and Information, 44(5–6), 507–519.(2016). https://doi.org/10.1080/09613218.2016.1156951
3. Bachman. L. R. New professionalism: The post-industrial context. Building Research and Information, 41(6), 752–760.(2013). https://doi.org/10.1080/09613218.2013.804778
4. Clark, J. Adaptive reuse of industrial heritage: Opportunities & challenges. Heritage Council of Victoria, 1–7.(2013).
5. Xu, Y. Reflections on industrial heritage protection in China. e-Phaïstos, I(1), 111–119. (2012).https://doi.org/10.4000/ephaistos.300
6. Que, W-M. On International Protection and Research of Industrial Property. Journal of Peking University (Natural Science), 04, 523-534.(2007). (In Chinese)
7. Yu K.-J., & Fang W.-L., Preliminary Exploration of Chinese Industrial Heritage. Chinese Journal of Architecture. 08,12-15. (2006). (In Chinese)
8. Beijing Municipal Commission of Tourism Development. China Beijing: Star-Rated Hotel: Room Occupancy Rate, 5 Star Economic Indicators. (2020).https://www.ceicdata.com/en/china/starrated-hotel-beijing/beijing-starrated-hotel-room-occupancy-rate-5-star

Research on SWMM Runoff Control Index Decomposition Based on Constraint Optimization Method

Shiyuan Huang[1,*], *Pengfei* Liu[2], *Hongqin* Zhang[2], and *Zhipeng* Ding[2]

[1]School of Civil Engineering,University of South China,Hengyang 421001,China
[2]School of Architecture,University of South China,Hengyang 421001,China

Abstract. The "Sponge City Construction Technical Guide" mentions the method for decomposing and implementing the annual total runoff control rate: volume method and model simulation method. Through research, it is found that the two are based on the control detailed planning level in decomposing the scale parameters of LID facilities. The indicators to each block are allocated through repeated tests through experience and trial calculations, resulting in a lack of scientific rationality in the process and results of the indicator decomposition, and because the Guide is still in the trial stage, there are few researches on the decomposition of runoff control indicators represented by SWMM. Therefore, with the help of MATLAB's constraint optimization module and SWMM hydro-hydraulic model, this study proposes a complete set of decomposition ideas and methods for the decomposition and implementation of the annual runoff total control rate in the special planning of sponge city, and the index decomposition process through relevant cases Elaborate.

1 Introduction

In December 2013, General Secretary Xi Jinping proposed to build sponge cities that can naturally accumulate, permeate and purify, so as to improve the quality of cities and improve the quality of people's lives, in view of the current water pollution, water shortage, flood and other problems encountered in urban development. In October 2014, the Ministry of Housing and Urban-Rural Development, together with other major design institutes, issued the Technical Guide for Sponge City Construction -- Construction of Low-Impact Rainwater System for Development (Trial) (hereinafter referred to as the Guide). The Guide has given four control indexes for different water control problems, among which the total runoff control will directly affect the effectiveness of the other three controls, so the total runoff control can be taken as the primary control object to promote sponge city planning and design. In the actual planning, the annual runoff control rate can be further refined and decomposed into the unit area control volume (comprehensive index) of each sub-plot, and then the comprehensive index can be further refined into single indicator such as sunken green land rate according to the actual situation of the plot[1]. All these indicate that the total runoff control rate plays an indispensable leading role in the planning and construction of sponge cities, which is the main line and basis throughout all planning stages. Correctly understand the specific meaning of control index in different planning stages; How to select the relevant control indexes and how to decompose and implement the control indexes has become the key task of further deepening the sponge research.

2 Decomposition and implementation of indicators at each planning level

According to the planning principles of protective development, minimization of hydrological interference and overall coordination, the Guidelines put forward corresponding index decomposition methods for different planning levels.

2.1 Overall planning level

Explicitly mentioned in the guide overall urban planning to consider the local actual situation, the organization carries out is given priority to with low impact development of related research, put forward the low impact development strategy and objectives, the comprehensive analysis of the rate, building density, space, function layout, etc, on the basis of the total annual runoff control into the overall urban planning level. According to the nature and functional layout of land parcels in the study area, combined with local self-heating geographical conditions, short-term and long-term development planning, location of runoff control zoning map, and comprehensive economic strength, the target of annual runoff control rate is determined. In addition, it is necessary to consider the development intensity, rain pattern and law, abundant water resources and other related factors of the research area.

* Shiyuan Huang : 550903597@qq.com

When the above conditions are considered comprehensively, if it is found that the study area does not have the space conditions for runoff control or the reconstruction cost is too high, the annual runoff total control target can be appropriately reduced. At the same time, it should also be noted that the target of runoff control is not as high as possible. Excessive emission reduction of rainwater will cause the water body in the basin to shrink, which is not conducive to the ecological cycle of the rainwater system. When the runoff control rate exceeds a certain level, the economic benefits of investment will decline, which is likely to cause problems such as oversized facilities and waste of resources.

2.2 The level of controlled detailed planning

The main task of regulatory detailed planning is to determine the runoff control indicators of each sub-plot, including comprehensive indicators and single indicators. All relevant majors should coordinate with each other at the level of regulatory detailed planning. By analyzing the functional layout, spatial structure, economic strength and other factors of each block in the planning area and combining with the actual runoff management and control ability of each sub-block, the preliminary evaluation of each block is made, and the comprehensive index of each block is determined by weighted accounting. In addition, factors such as building density and green land rate are combined to determine the low-impact development facilities to be selected, and the runoff control objectives proposed in the previous stage are further decomposed and implemented. Finally, single indicators such as lower concave green land rate are obtained to guide the development and construction of various blocks.

2.3 Site detail planning level

Constructional detailed planning shall be made in accordance with the superior plan control requirements, actively coordinate related major, formulate reasonable design scheme, in order to promote regional sustainable development as the goal, the control requirement of specific fulfil LID facilities type and composition, spatial layout, construction scale, construction sequence, funding arrangement, etc.In order to ensure that the runoff discharge after the block development and construction meet the expected runoff control objectives

2.4 Existing problems

With the continuous promotion of sponge city construction in the whole country and the deepening of sponge city construction in various regions, the planning and construction ideas of sponge city at various planning levels are also constantly improved. From urban master planning to regulatory detailed planning to constructive detailed planning, how to scientifically and rationally decompose and implement the control indexes of sponge city runoff has always been the focus and difficulty of planning authorities. However, according to the current research results, there are few researches on the decomposition system and methods of the sponge city planning and control index. In Table 3-1, the Guide explains in detail the different control objectives, indicators and corresponding assignment methods at different planning levels. Through practice, it is found that there are still some problems in the decomposition process of control indicators, no matter the trial algorithm or the model method.

2.4.1 The decomposition process of single index is too complicated

Index decomposition in regulatory detailed planning can be divided into comprehensive index decomposition and single index decomposition. Related concepts have been described above. In the Guide, two decomposition methods are proposed for single index: trial algorithm and model method. At present, most of the researches are based on trial algorithm, which can't get rid of the disadvantages of trial algorithm, such as tedious calculation process, strong subjectivity, inaccurate decomposition results and so on. The size of the index is based on different LID facilities around the regulatory level decomposition to block index, using the method of volume through experience and trial test repeatedly to allocate[2, 3], as long as around the conform to the overall total annual runoff control block shows that the index decomposition conforms to the regulations, in order to meet the assessment requirements, constantly repeated set index, the sponge in the urban planning control index decomposition of assignment process is too complicated.

2.4.2 The model simulation method is not mature to decompose the index process

Although it is mentioned in the Guide that qualified cities can decompose and implement the control indicators by means of model simulation, because the Guide is still in the trial stage, it does not give a detailed answer on what model to use and how to use the model to decompose the indicators.Due to the fact that the decomposition of control indicators should be implemented in different scales of LID, and there are few domestic hydrologic models with LID simulation function, SWMM, HSPF, SUSTAIN and other models are widely applied in this field. By reading the literature and practice of SWMM the direction of the research can be found, but the current of SWMM is used to study the urban flood control and drainage[4, 5], water environmental governance, rainwater pipe network optimization[6], LID facilities efficiency simulation assessment [7, 8].The research on the decomposition of runoff control index in sponge city is almost blank.

3 Technical route

In view of the above two problems, this research reference related knowledge, mathematical statistics with the help of MATLAB and constrained optimization module of SWMM hydrologic hydraulic model, the special planning of sponge city middle-aged runoff amount control decomposition and implement the system puts forward a set of complete decomposition and the method, hope to be able to deepen the planning of urban special sponge depth, reduce the workload planning personnel, improve the feasibility of runoff control index decomposition results. The decomposition work starts from the overall planning and regulatory detailed planning, and is gradually decomposed according to mandatory indicators, comprehensive indicators and individual indicators (Figure 1).

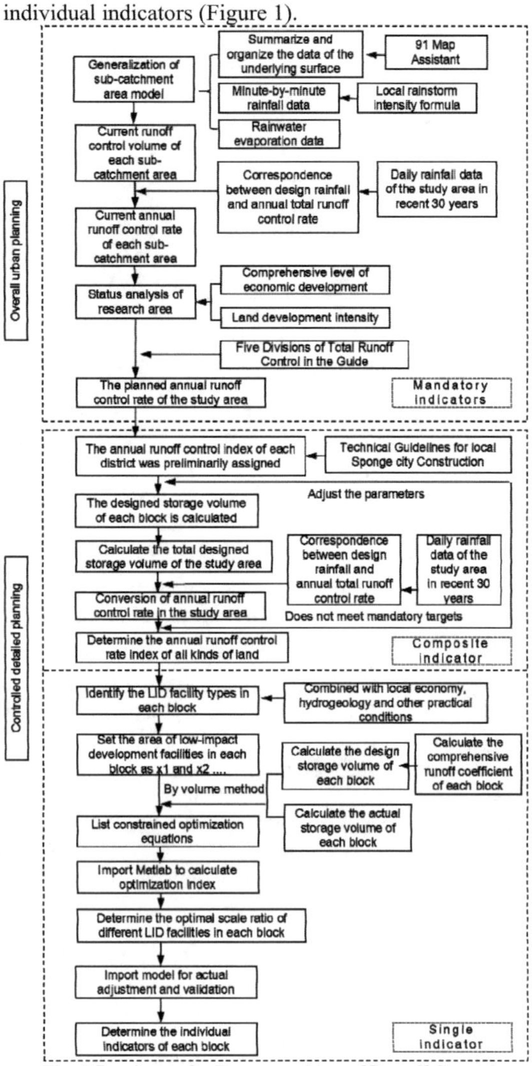

Fig. 1. Road map for Decomposition of Runoff Control Indicators

4 Case interpretation

This study takes Hengyang city center as an example. The study area starts from Xiangjiang North Road in the east, Hongxiang North Road in the west, Chuanshan Avenue in the north, and Yancheng Road in the south. The total planned area is 333.5HA (Figure 2), with residential land as the main land and park land in the south.

4.1 The determination of the general objective of runoff control at the general planning level

According to the study of regional soil type, topographic slope, land development nature and intensity, hydrogeology and other factors analysis of the current plot of actual runoff control ability, combined with hengyang city in the runoff control zoning map location, the level of economic development to comprehensively determine the total annual runoff control target (mandatory indicator).

Firstly, basic data should be collected and collated. DEM data of the research area should be obtained through 91 Health map assistant, and relevant parameters such as average slope and area of each sub-catchment area can be obtained through GIS analysis (Figure 3). The impermeable percentage parameters of each sub-catchment area required in the model are calculated according to the high-precision remote sensing image, including green space, road, building and water surface area (Figure 4), and obtained through the weighted calculation of the corresponding runoff coefficient. Model simulation required by minute rainfall data is according to hengyang city storm intensity formula fit out of Chicago rain pattern, rainfall design recurrence period for 5 years, peak ratio r=0.4, rainfall lasts 120 minutes, rainfall total of 67.4930 mm (Figure 5); The houghton model was selected for infiltration, and other relevant parameters were obtained by referring to relevant literature and the value of SWMM model user manual[9-11]. Finally, the stormwater model of hengyang central urban area was established (Figure 6). The actual rainfall situation of the model was simulated, the rainwater storage capacity of each sub-catchment area was analyzed, and the rainfall runoff control volume of each block was calculated.

Fig. 2. Image of planning area

 https://doi.org/10.1051/e3sconf/202123704008

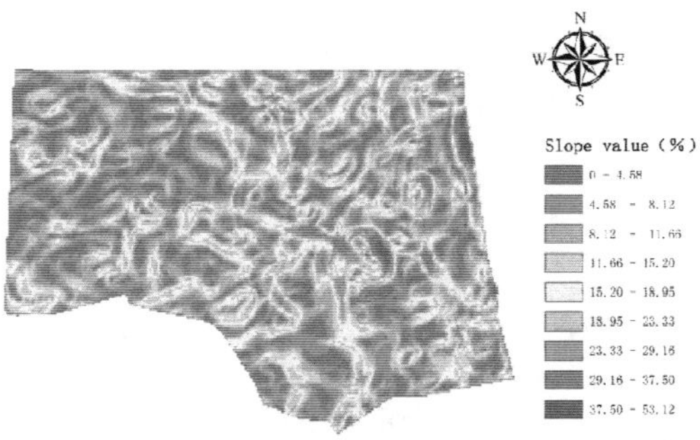

Fig. 3. Slope nephogram of study area

Fig. 4. Interpretation of Buildings in Planning Area

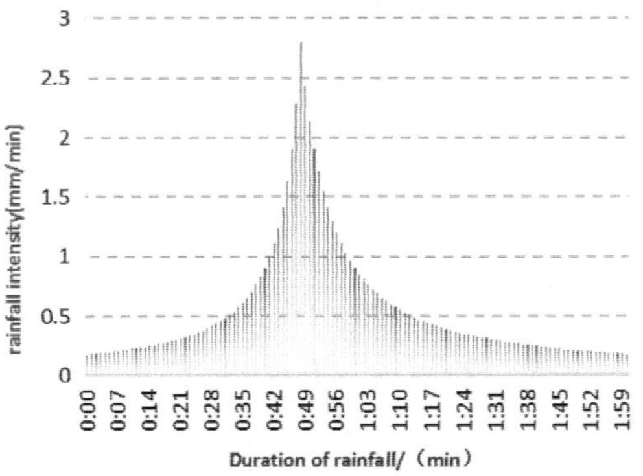

Fig. 5. Ainfall process line of Hengyang rainstorm recurrence period P=5a

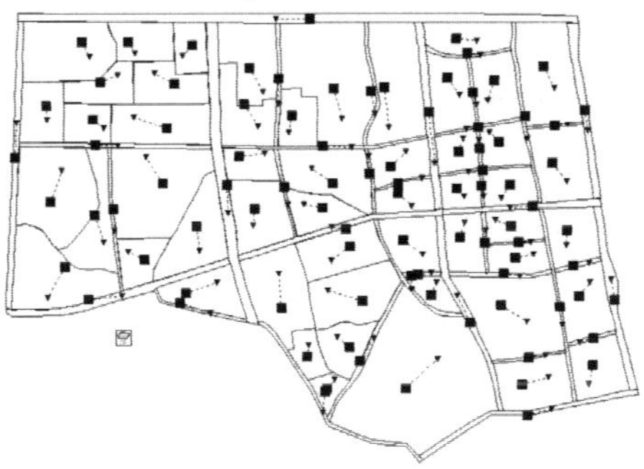

Fig. 6. Overview of SWMM Model in Planning Area

Due to the "Guide" appendix 2 does not give hengyang city annual runoff control rate and its corresponding design rainfall value, considering the different natural conditions of different cities, rainfall characteristics are also different, annual runoff control rate and the corresponding design rainfall should be calculated separately. Reference related literature, from the national meteorological information center web access to hengyang 1989.1.1-2020.4.30 11413 daily rainfall data, according to the guide mentioned[12, 13],statistical method combined with linear interpolation method, calculate the hengyang's latest step were 5%, 1% for corresponding design rainfall runoff amount control table values (Table 1 and Table 2), according to the rainfall runoff control volume of each block calculated above, the annual runoff control rate of each block's current status can be obtained by referring to the table (Figure 7).

Table 1. Annual Runoff Total Control Rate with 5% Step Size and Design Rainfall Value in Hengyang City (1989-2020)

Annual runoff control rate/%	Design rainfall/mm	Annual runoff control rate/%	Design rainfall/mm
15%	2.1	55%	10.8
20%	2.8	60%	12.6
25%	3.6	65%	14.7
30%	4.6	70%	17.1
35%	5.6	75%	20.1
40%	6.6	80%	23.7
45%	7.8	85%	28.6
50%	9.2	90%	35.7

Table 2. Table of Total Annual Runoff Control Rate and Design Rainfall Values for Some 1% Steps in Hengyang City (1989-2020)

Annual runoff control rate/%	Design rainfall/mm	Annual runoff control rate/%	Design rainfall/mm
16%	2.2	81%	24.7
17%	2.4	82%	25.7
18%	2.5	83%	26.6
19%	2.7	84%	27.6
51%	9.5	85%	28.6
52%	9.9	86%	30.0
53%	10.2	87%	31.4
54%	10.5	88%	32.8

Fig. 7. Distribution chart of current annual runoff total control rate in the planning area

The terrain in the study area is relatively flat, with an average slope of 11.1%. The soil is mainly red soil with little or no water permeability and poor permeability. The actual runoff control rate of each block is shown in Figure 7. The eastern part of the study area is dominated by old residential areas with high building density, with high degree of hardening and poor runoff control ability; the western part is dominated by newly built residential areas with low degree of hardening; the southern part is park green space, with strong runoff control ability. After conversion, the annual runoff control rate of the whole region is 23.7%. Considering that Hengyang City is in the third interval of annual runoff control, with the index limit of 75%-85%, and considering its comprehensive economic strength, the overall goal of planning control is determined to be 75%.

4.2 Decomposition of runoff management control indicators at the regulatory level

Combined with the general goal of planning control mentioned in the overall planning level, the control regulations gradually decompose and refine the hard indicators proposed by the superior planning through comprehensive indicators and single indicators.

4.2.1 Composite index decomposition

According to the runoff control requirements of all kinds of land construction, reconstruction and expansion projects in the local sponge city construction technical guidelines, the preliminary index decomposition of each block was carried out. Some cities have not issued relevant technical guidelines, so the values can be used for reference according to the relevant guidelines of neighboring cities and combined with local actual conditions. This research region refers to the various requirements of land control mentioned in The Technical Guidelines for The Construction of Sponge Cities in Changsha. The specific assignment situation is shown in Table 3. According to the corresponding relationship of Table 1 and Table 2 and the area of each block, the runoff control volume of each block was calculated. After weighted calculation, the annual runoff control rate of the study area was 75.9%, which met the mandatory index mentioned in the general plan level. The specific assignment of each block was shown in Figure 8.

Table 3. Preliminary Evaluation Table of Various Land Use Control Indicators

Nature of the land	New construction	reconstruction
Residential land	80%	65%
Public green space	90%	90%
Commercial service land	75%	65%
Educational and scientific research land	80%	60%
Land for municipal facilities	75%	70%
Administrative office space	80%	75%
Urban road land	60%	55%
Land for railway	—	45%
Square land	70%	65%
Water area	100%	100%
Industrial land	70%	60%
Logistics and warehousing land	70%	60%
Medical and sanitary land	80%	75%

Fig. 8. Decomposition and assignment diagram of comprehensive index in planning area

4.2.2 Single index decomposition

In order to ensure the feasibility of index decomposition, the comprehensive index should be further decomposed into a single index. According to the current runoff management and control situation of each block in Figure 7, the sunken green space and permeable pavement should be selected as the main LID facilities in consideration of reconstruction cost, too little green space that can be reconstructed in the planning area and high building density. In the SWMM model, LID facility parameters are set as follows: surface stagnant water depth of permeable pavement is 15mm, Manning N value is 0.013, surface slope is 1.5%; The thickness of the surface layer is 150mm and the pore ratio is 0.2. The thickness of the gravel layer is 300mm, the pore ratio is 0.3, the drainage coefficient of the bottom drainage layer is 3.3, and the drainage index is 0.5. The surface depth of sunken green space was 200mm, the vegetation cover

https://doi.org/10.1051/e3sconf/202123704008

coefficient was 0.8, and the roughness was 0.25. The sand and gravel thickness of the aquifer is 450mm, and the pore ratio is 0.75. The thickness of the soil layer is 250mm and the pore ratio is 0.3.

At the same time, in order to solve the problems of tedious decomposition process and inaccurate results caused by repeated trial calculation of LID facility scale in SWMM, this study, with the help of MATLAB constraint optimization module, listed the constraint equation according to the basic data of underlying surface of various blocks collected in the early stage and the correlation of volume method, and the specific steps are as follows:

First, set area parameters for LID facilities pre-selected for each block, such as:

$S_{\text{Sunken green space}}=X_1$; $S_{\text{Permeable pavement}}=X_2$

Secondly, the comprehensive rainfall runoff coefficient of each block should be calculated, as shown in Formula 1.1:

$$\varphi = \frac{\sum_{i=1}^{n} \varphi_i s_i}{\sum_{i=1}^{n} s_i} \quad \cdots\cdots\cdots \; 1.1$$

Where φ_i is the runoff coefficient of different catchment surface, which can be taken as the value in Table 4-3 of the Guide according to the actual situation of each block; S_i is the area of different catchment surface. According to the calculation relation mentioned in the volume method, the relevant constraints are listed.

Constraint Condition 1: Designed storage volume (Formula 1.2) :

$$V_1 = 10\varphi HF \quad \cdots\cdots \; 1.2$$

Actual storage volume (Formula 1.3) :

$$V_2 = \sum^{n}_{i=1} h_i p_i X_i \quad \cdots\cdots \; 1.3$$

We want V_2 minus V_1 to be greater than or equal to 0; Where H is the designed rainfall corresponding to the comprehensive control index of each block, h is the designed storage depth of LID facility, and p is the porosity of LID facility aquifer.

Constraint condition 2: Actual design rainfall (Formula 1.4) :

$$H = V_2 / 10\varphi_1 F \quad \cdots\cdots \; 1.4$$

Then H_1 is required to be less than or equal to H_2, and H_2 value can be selected according to the corresponding relationship between the total annual runoff control rate and the designed rainfall. The corresponding total annual runoff control rate is slightly higher than that of H_1, in order to limit the search interval that meets the requirements.

Constraint 3: The maximum size of LID facilities in each block can be arranged (Formula 1.5) :

$$0 \le X_1 \le S_1 ; \; 0 \le X_2 \le S_2 \quad \cdots \; 1.5$$

Where S_1 is the area of ordinary green space or other land that can be transformed into sunken green space; S_2 is the area of concrete or other hard pavement that can be transformed into permeable pavement. If other LID facilities are to be installed on the lot, the same can be done. Finally, the constraint optimization model was obtained (Figure 9). The constraint inequality was converted into a matrix and substituted into the MATLAB linear optimization module, and a single index was preliminarily obtained.

$$\min F = \sum_{i=1}^{n} h_i p_i x_i \quad \begin{cases} V_2 - V_1 \ge 0 \\ H \le H_2 \\ 0 \le X_1 \le S_1 \\ 0 \le X_2 \le S_2 \end{cases}$$

Fig. 9. Constrained optimization model

According to the size of LID facility obtained, it is put into the model to simulate and adjust, and finally individual control indexes of each block are obtained. A total of 91 sub-catchment areas were divided in this research area, with a large number of them, and only some indicators were presented. The specific results are shown in Table 4.

Introduction of constrained optimization module can be met quickly recessed the rate under this sub catchment area or permeable pavement rate (single index), If the calculated recessed green space rate or permeable pavement rate of special plots is too high, can limit the covering rate in constraint conditions, to substitute other LID facilities for its high part, and to get the LID size to meet the low scale of control, to prevent the scale facilities is too big, because the trial by the investment waste phenomenon. Thus, a single index satisfying the runoff control of each sub-catchment area can be preliminarily obtained, and on this basis, indexes can be adjusted according to the actual situation. According to the calculation, the actual volume of runoff control in the study area reached 45673.6m3 after the decomposition index, and the corresponding annual total runoff control rate was 77.4%, the total sunken green land rate was 37%, and the total permeable pavement rate was 35.6%, which met the requirements of runoff control in the planned area.

Table 4. Decomposition Results of Runoff Control Indicators for Some Plots

Plot Numbers	Nature of the land	Composite indicator		Single parameter	
		Annual runoff control rate/%	Control volume per unit area ($m^3 \cdot m^{-2}$)	Sunken green rate/ %	Permeable pavement rate/ %
1	Residential land	80	0.01313	43	60
9	Land for education	85	0.01398	19	35
14	Medical use of land	80	0.01451	40	30
26	The park land	95	0.02311	21	35
34	Residential land	70	0.01006	25	60
46	Road land	60	0.00941	62	10

68	Road land	60	0.0095	68	10
90	Square land	70	0.00751	6	30

5 Conclusion

With the help of constraint optimization module of MATLAB and SWMM hydrological and hydraulic model, this paper puts forward a complete set of decomposition ideas and methods for decomposition and implementation of total runoff control rate in sponge city special planning according to mandatory index, comprehensive index and single index. This set of methods simplifies the complicated process of decomposing indexes by trial algorithm and model method, and reduces investment waste. Moreover, the index obtained by this method can not only meet the control rate, but also not excessively limit the layout of LID facilities; It can not only solve the problem of operability of plot planning, design and construction in the later period, but also avoid making the index too specific, restricting the development and construction of plot, leaving enough space for planning. The constrained optimization module can greatly simplify the process of calculating indexes by trial algorithm without model conditions. Because the model simulation focuses more on the actual process, the trial algorithm focuses more on the whole, and the indexes obtained by optimization also need to be adjusted in the simulation, which is more suitable for the traditional trial algorithm.

At the moment, the runoff pollution control target can be realized through controlling the total runoff mostly, so most of the area to total annual runoff control as the main planning index decomposition, but actually there will still be satisfied runoff control doesn't meet the pollution control situation, other scholars can on this basis, the perfect model data, using MATLAB modules related constraints runoff control target goal and the runoff pollution in consideration, in order to continue to improve the sponge city special planning; In addition, since the LAYOUT of LID facilities in THE SWMM model is not simulated in accordance with specific plans, other software is needed to further solve the index decomposition at the repair level.

Reference

1. Beijing University of Civil Engineering and Architecture. Technical Guide for Sponge City Construction-Construction of Rainwater System for Low Impact Development (Trial) [M]. Beijing: China Building Industry Press, 2015
2. Yang Ye, Shen Min, He Junchao, Zhu Jiang, Cao Pan. Study on optimization of decomposition method of annual runoff total control index in sponge city [J]. Chinese Landscape Architecture, 2019,35(06):89-93.
3. Zhang Gaoyou, Gao Bin, Wang Xinliang. Deepening and implementation of annual runoff total control rate in sponge city in control regulations [J]. China Water Supply and Drainage, 2018,34(06):1-5.
4. Minmin Huang, Shuanggen Jin. A methodology for simple 2-D inundation analysis in urban area using SWMM and GIS. 2019, 97(1):15-43.
5. Sui Zhang, LI Zhe. Simulation of Urban Rainstorm Waterlogging and Pipeline Network Drainage Process Based on SWMM. 2019, 1213(5)
6. F. Javier Martínez-Solano, Pedro L. Iglesias-Rey, Juan G. Saldarriaga, et al. Creation of an SWMM Toolkit for Its Application in Urban Drainage Networks Optimization. 2016, 8(6)
7. Tielan Huang, Yunpeng Wang, Jinlan Zhang. Simulation and Evaluation of Low Impact Development of Urban Residential District Based on SWMM and GIS. 2017, 74(1)
8. Jing Zhang, Yu Zhang, Shi-Jun Sun, et al. Analysis of the Effect of Low Impact Development on Urban Runoff Control Based on the SWMM Model. 2019, 96(sp1):62-67.
9. Gaurav V. Jain, Ritesh Agrawal, R.J. Bhanderi, et al. Estimation of sub-catchment area parameters for Storm Water Management Model (SWMM) using geo-informatics. 2016, 31(4):462-476.
10. N.A Zaghloul, M.A Abu Kiefa. Neural network solution of inverse parameters used in the sensitivity-calibration analyses of the SWMM model simulations. 2001, 32(7):587-595.
11. Tohru KANDA, Kazuo KANKI, Masahiro WATANABE, et al. ON MODEL PARAMETERS OF SWMM FOR SMALL URBAN DRAINAGE BASIN. 2001, 45:79-84.
12. Liu Pengfei, Huang Shiyuan, Liu Hui, Li Yingjie, Huang Xiao. Error analysis of annual runoff control rate estimated by linear interpolation method [J]. Water Resources Planning and Design, 2020(04):55-58.
13. Liu Hui, Sun Bo, Zeng Fanyong, Xie Shuibo. Fast estimation of annual total runoff control rate and design rainfall based on linear interpolation method [J]. Water Resources Planning and Design, 2019(10):41-43+148.

Re-Exploration of the Sustainability of Traditional Village Spatial Development in The Epidemic Era – From the Perspective of Jiangnan Region, China

Xinqu Liu[1,*], Yongfa Wu[1], Junyang Liu[2], and Zaiyi Liao[3]

[1] Gold Mantis School of architecture, Soochow University, Jiangsu 215123, China
[2] School of Architecture and Planning, University of Auckland, 1010, New Zealand
[3]The Faculty of Engineering and Architectural Science (FEAS), Ryerson University, ON M5B 2K3

Abstract. The global spread of the COVID-19 has given severely challenge to the contemporary human settlement environment. Traditional villages, as an essential human settlement, which has been significant in contemporary life. This article takes the traditional village space in the Jiangnan region, China as an example to analyse the positive interaction between humans and nature, health, and space in its traditional design. This research points out that in the traditional village space, healthy human settlements have two levels: "material form" and "internal sprite". In addition to focusing on the transformation, innovation, and application of the technology, designers should exceed the material level to seek non-material dialogue and integration, to explore the important role of inner spiritual art for healthy living. Finally, this research combine with the urban epidemic prevention function and pointed out that the symbiosis mechanism between urban and rural areas should be actively explored. The penetration and integration of urban and rural areas can better cure, and promote the sustainable protection, activation, utilization and innovation of traditional villages in the future.

1 Introduction

At the beginning of 2020, the global outbreak of the COVID-19 broke the original stable social situation and the traditional thinking mode as a designer. This is a warning and an important opportunity for change. A healthy and harmonious human settlement has always been an important goal of human development, and it is also the unshakable foundation of life[1]. Traditional villages, as a typical living environment space for rural revitalization in China[2], are also the foundation and important practice model for contemporary healthy living activities. Although traditional villages lack advanced science and technology compared with urban spaces, the living wisdom that is harmonious with nature formed during long-term human settlement activities cannot be ignored. Therefore, analysing the advantages and disadvantages of the traditional village space under the emergent epidemic and its role in large-scale public health emergencies is of great significance to sustainable development. On the one hand, it is conducive to reflect on and improve the lack of contemporary human settlement space in public health security. Moreover, it can deeply reflect on the possibility of the transformation of traditional urban and rural pluralist values, so then to explore the contemporary urban-rural symbiosis mechanism to further promote contemporary rural revitalization. Therefore, this paper uses architectural typology[3,4], taking traditional villages in Jiangnan region as a typical case, analyses the important types and spiritual performances of healthy human settlements in their spaces. Finally, this research tries to explore the sustainable development of contemporary traditional villages through comparative research with contemporary urban space epidemic prevention.

2 Spatial types and characteristics of Jiangnan traditional villages

2.1 Ecosystem

2.1.1 Taking water as the origin

In terms of the developed water system and flat terrain in the Jiangnan region villages tend to grow towards the water, forming a village space dominated by water systems and river networks[5]. The village conforms to the distribution of water networks and roads and uses abundant water resources to form a powerful irrigation and transportation system. As the basic condition of traditional farming life, water act an important role in the traditional culture of Jiangnan. It is not only in ancient books such as "The Classic of Mountains and Seas"(called"山海经" in Chinese) [6] and " The Shi King"(called" 诗经 " in

* Corresponding author: liuqicrystal@foxmail.com

https://doi.org/10.1051/e3sconf/202123704009

Chinese) [7] that many traditional mythological images and gods are derived from water. Even in the square, temples, pagodas, and other sacrifice places also worship the water like the god. The diverse forms and flowing rhythm of the water system enrich the space of traditional villages and form corresponding design techniques, using water to form a dynamic and static artistic expression of water landscaping (see Figure1).

Fig1. Ge Xin village: Water system in village of Jiangnan region

2.1.2 Take the mountain as a protective barrier

This kind of traditional villages are mainly distributed in areas dominated by mountainous landforms in southern Anhui and northern Zhejiang[8,9]. These villages are built on the basis of the mountain terrain. Therefore, the villages are mainly distributed at the foot of the mountain, between the mountains or the ridge, and are closely integrated with the terrain (see Figure2).

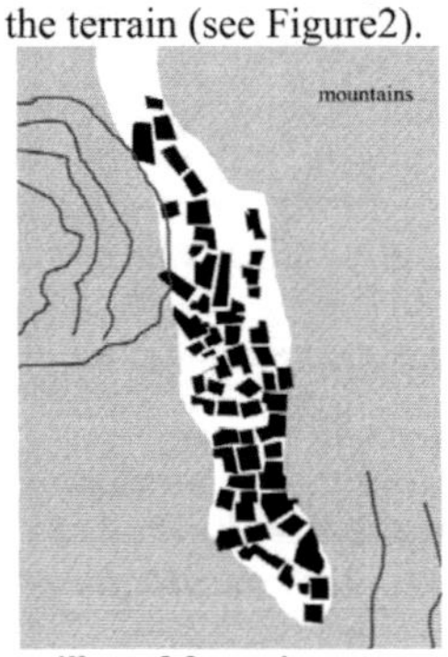

Fig2. Mu Ligong village: Mountain system around village of Jiangnan region

2.1.3 The combination of mountains and waters

The villages are mainly distributed in areas with both mountains and water. They are favourable geomorphic conditions for the formation of traditional villages in the Jiangnan region, surrounded by mountains and waters. The soil in the area is fertile, suitable for the growth of rice and other crops. In this drastic change, traditional craftsmen rely on Feng Shui principles to form a visual landscape axis, which shows the beauty of ecology by combining multiple natural elements[10]. The overall shape of the village is free and provides favourable conditions that are used to form a boundary traffic and water network system. (see Figure3).

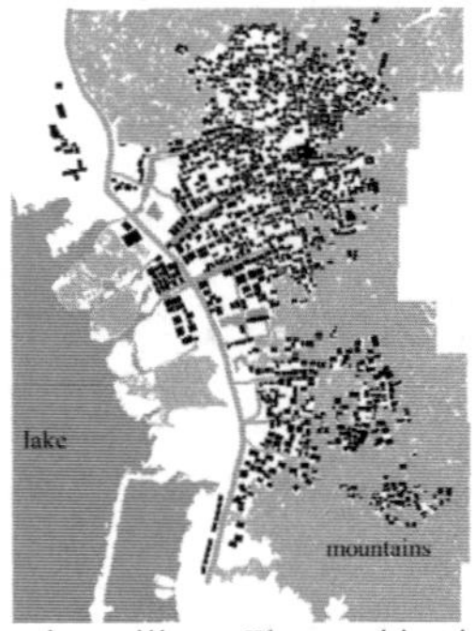

Fig3. Luxiang ancient village: The combination of mountains and waters in village of Jiangnan region

2.2 Spatial pattern

2.2.1 Centripetal aggregation style

This type of spatial pattern has formed one or more important spatial focal points in the natural development of the village, mainly composed of ancestral halls, temples, and partly composed of village squares or important landscape nodes (see Figure4). This kind of village spatial layout is tidier and more orderly, and the village core is prominent, which has a strong guiding effect on villager's behaviours. Inside the village, the village buildings and roads are spread out around the centre layer by layer, and the overall form is in a tightly gathered form.

Fig4. Shang Wufang village: The centripetal aggregation style in village of Jiangnan region

2.2.2 Reticular radiation style

This type begins to radiate around a point in the village or an important public building or landscape node in the village as the centre. The shape presents the characteristics of spreading and radiating shapes, just like a spread net attached to the natural landscape (see Figure5). Traditionally, this type of village is affected by the degree of population concentration and is generally large in scale. Moreover, the form of the spatial pattern presents a flexible and outward-looking visual aesthetics, which as a whole highlight a yearning tension and inner vitality.

Fig5. Weng Xiang village: The reticular radiation style of Jiangnan region

2.2.3 Free scatter style

This type of village is generally located in an area with extreme topographical conditions or severe climatic conditions, which prevents the villages from being able to gather in a certain area on a large scale or restricts traditional construction techniques. Affected by this typical external condition, the village can only use roads as the main connection, and gather in a small area under suitable terrain conditions, showing a relatively scattered form (see Figure6).

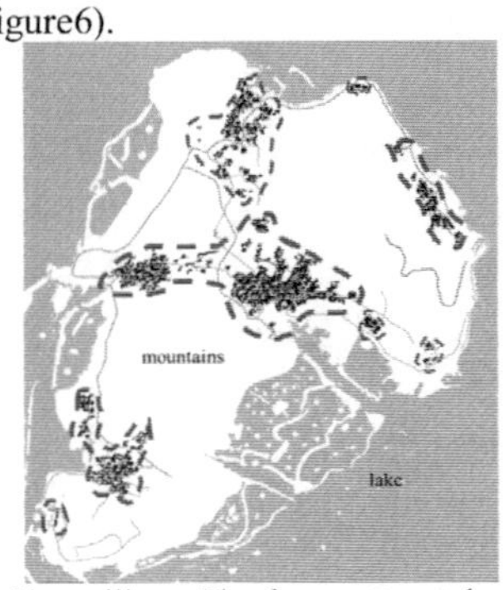

Fig6. San Shan village: The free scatter style of Jiangnan region

2.3 Typical traditional residence

2.3.1 Waterfront residences

Residences are built near the water, borrowing space to take space, usually by introducing water system or adjusting the microclimate. The scale of the housing is proportional to the size of the family and the economic situation. The architecture structure is mainly a combination of lifting beam and wearing bucket (see Figure7).

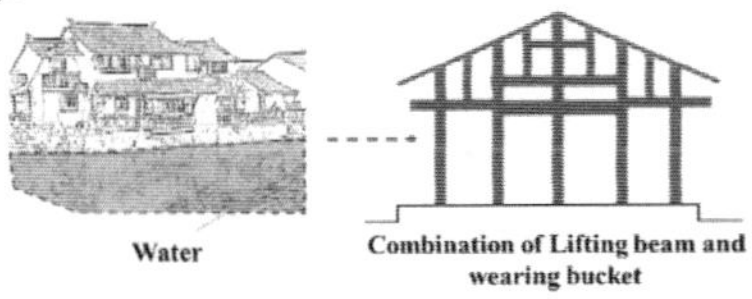

Fig7. The form and structure of waterfront residence

2.3.2 Street-facing residence

The residence takes the street orientation as the main space layout, free form, and realizes the purpose of saving land and optimizing space by using the terrain. The architecture structure is combination of lifting beam and wearing bucket or wearing bucket (see Figure8).

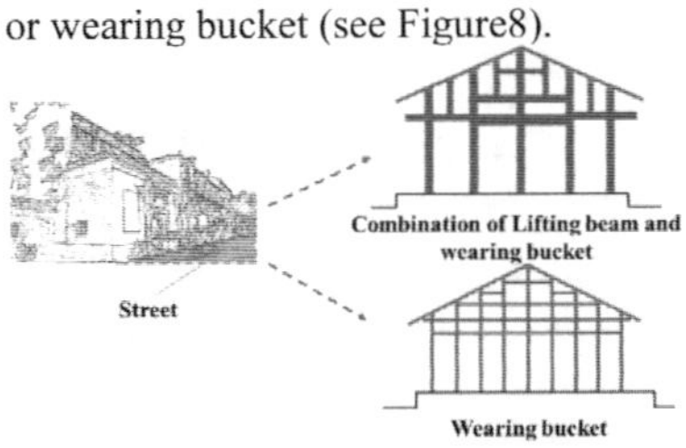

Fig8. The form and structure of street-facing residence

2.3.3 Courtyard residence

This type of residence is mostly built in a flat area, with courtyards that have unique themes as the main form of closed buildings. The structure of the building is mainly combination of lifting beam and wearing bucket or wearing bucket (see Figure9).

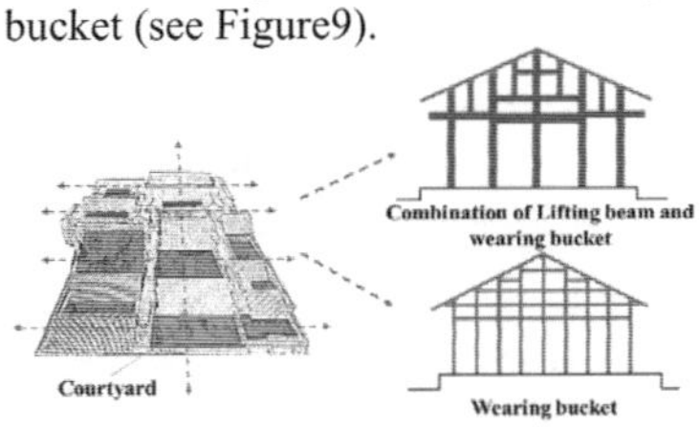

Fig9. The form and structure of Courtyard residence

2.3.4 Chinese and Western combined dwellings / New Shikumen

The residences are mainly distributed in economic development areas influenced by Western culture. The architectural form features prominent western elements, simplified facade, obvious functional distinction, clear traffic line. The building structure is brick-wood structure (see Figure10).

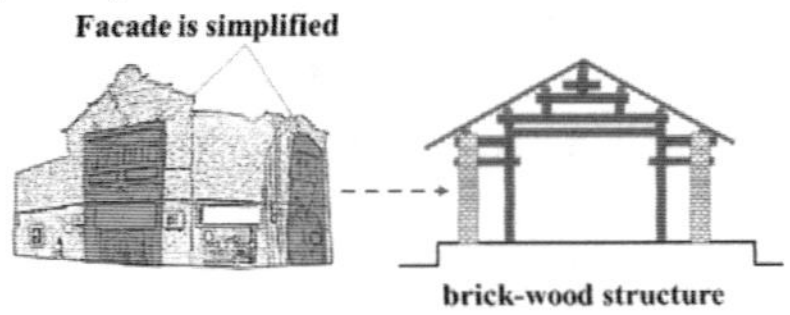

Fig10. The form and structure of Chinese and Western combined dwellings / New Shikumen

2.3.5 Garden residence

The dwellings were mainly built in the Ming and Qing dynasties, and were influenced by the historical scholars, and were built in the economically developed areas of Jiangnan. The structure of the building is mainly combination of lifting beam and wearing bucket (see Figure11).

https://doi.org/10.1051/e3sconf/202123704009

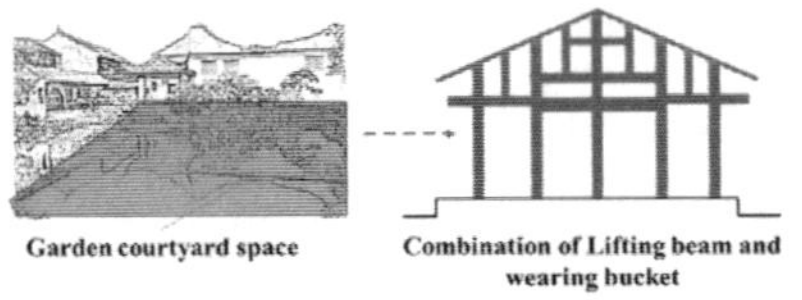

Fig11. The form and structure of Garden residence

3 Jiangnan Traditional Village Space and Contemporary Healthy Human Habitat

3.1 Advantages of space epidemic prevention in traditional villages

Jiangnan has had unique geographical advantages since ancient times. The humid climate and favourable landscape pattern have made the traditional villages ingeniously integrated with nature[11]. Compared with modern cities, traditional villages have low population density, are close to nature, and have open and flexible spaces. Therefore, the spatial structure of the countryside has less destructive power and influence on the natural environment, forming a circular and sustainable integration model of survival, production, and life. Under the influence of the epidemic environment, rural areas dominated by agriculture can ensure self-sufficiency in daily life without relying on external forces, and the cohesion of the rural population is more convenient for unified management under the impact of emergency security incidents.

3.2 Traditional village space is weak in epidemic prevention

In order to specifically study the thermal comfort of the natural environment of Jiangnan villages, this research used 186 questionnaires and 50 traditional residences as the survey samples to conduct a field survey on the thermal comfort of residences in the Jiangnan region. In terms of indoor comfort in four seasons (see Table1).

Table1. Thermal comfort analysis of Jiangnan traditional village residences

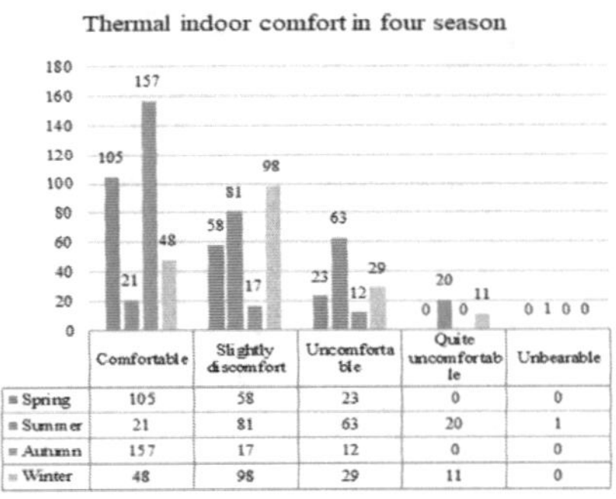

	Comfortable	Slightly discomfort	Uncomforta ble	Quite uncomfortab le	Unbearable
Spring	105	58	23	0	0
Summer	21	81	63	20	1
Autumn	157	17	12	0	0
Winter	48	98	29	11	0

56.4% of people feel comfortable indoors in spring, and 84.4% of people feel comfortable indoors in autumn. On the contrary, 43.5% of people feel slightly uncomfortable in summer, 33.8% feel uncomfortable, 10.8% feel very uncomfortable; in winter 52.6% feel slightly uncomfortable, 15.6% feel uncomfortable 5.9% of people feel very uncomfortable. Therefore, in Jiangnan traditional residences, most users think that the indoor comfort is comfortable in spring and autumn, but the indoor comfort is weak in summer and winter.

In order to accurately measure the thermal comfort in summer, the indoor and outdoor temperature, humidity and wind speed were measured in 50 residences without air-conditioning and other equipment in summer, and finally the comfort index of 50 households was calculated. The results show that the thermal comfort index of these residences in summer are all above 80 (see Table2). Comparing with the human body comfort index grading table, the range of comfort index felt by a normal human body is 50-70. Therefore, in summer, the comfort index of Jiangnan traditional residences far exceeds the normal comfort index of the human body. Consequently, the thermal comfort of traditional residences in Jiangnan still needs to be improved.

Table2. Analysis on the comfort of 50 residences in Jiangnan traditional village

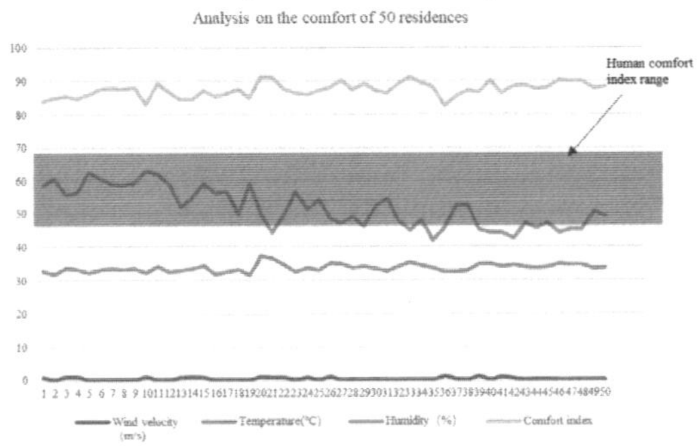

To sum up, although traditional villages provide a favourable ecological and healthy human settlement foundation and have strong natural restoration capabilities, they still expose many shortcomings. Due to the traffic congestion, the economic level of the village has been lagging behind that of the city for a long time, resulting in insufficient awareness of the residents of the public health environment, so they lack the ability to respond effectively to the impact of the epidemic.

4 Urban-rural interaction under the epidemic

4.1 The symbiosis mechanism of urban and rural areas

At the level of the energy field, there are multiple apparently interdependent energy elements between contemporary urban and rural areas, and the mutual influence and interaction between the two are realized through the flow or conversion of energy[12]. It mainly includes material field, information field, capital field, human field, technology field, etc. The ultimate goal is to achieve maximum resource sharing under limited material conditions. Taking Space field to effectively serve different types of contemporary communities the Jiangnan villages during the epidemic period as an example. Villages dominated by agriculture, farming, fisheries, etc. provide necessary living supplies for cities, while cities import new resources into the countryside. Science and technology and medical resources, while realizing the monitoring and resource exchange of urban and rural epidemics through platforms such as the internet and media.

At the spatial field level, urban and rural areas are important spaces for contemporary healthy human settlements. In the spatial framework, the transition from the natural environment to the physical environment of human settlements has been realized[13]. Different spatial elements reasonably constitute different spatial scales and have certain laws and logics. Therefore, the space field can serve different types of contemporary communities effectively. In this field, both urban and rural areas have their own spatial independence and form an organic whole. Because any changes in the environment and public space will affect the shock of the overall space system, it means that any reversal and disorder of spatial elements will cause the balance of urban and rural spaces to become unbalanced.

At the level of consciousness field, it is gradually formed under the combined action of urban and rural energy field and material field to guide urban and rural practitioners and users, and it also reacts to energy field and space field[14]. As an intangible product of thought, its power is deeper, and it is a comprehensive manifestation of the collective consciousness response. The transformation and expansion of the field of consciousness can directly and effectively feed back to the positive behaviours of practitioners and users, and it has a deeper guiding role in the contradiction of thinking in the two perspective of urban and rural areas. Finally, through the positive practice of behavioural activities, reasonable planning and improvement of the overall urban and rural system space are carried out, and the contradiction is gradually transformed into an effective driving force for development.

4.2 Reconsideration on the contradictions and symbiosis of contemporary urban and rural areas

For a long time, there have been conflicts between the poor and the rich, the slow and the fast, the new and the old that cannot be ignored between the urban and rural areas in Chinese society. Consequently, the social, consciousness and other multi-level conflicts have been produced, which restricts the possibility of mutual benefit and win-win in the development process of urban and rural areas, and has formed a limited understanding of their relationship for a long time. Through the analysis of the characteristics of urban and rural areas in the epidemic, it can be clearly seen that the shortcomings of cities are the advantages of the villages, while the disadvantages of the villages are the advantages of the city, which shows an obvious opposite complementary relationship (see Figure12). If contemporary designers can consider the pros and cons of urban and rural space from a holistic perspective. Complementing each other's strengths between urban and rural areas can just make up for the lack of development, achieving healthy and sustainable development of urban and rural areas. The purpose is to improve the contradictions caused by the confrontation in the past, and realize the transformation from "urban-rural contradictions to urban-rural integration". By transforming the contradiction between urban and rural areas, it can help improve the resilience of urban and rural spaces. The symbiotic between urban and rural spaces provides more convenient and more diverse possibilities for future development.

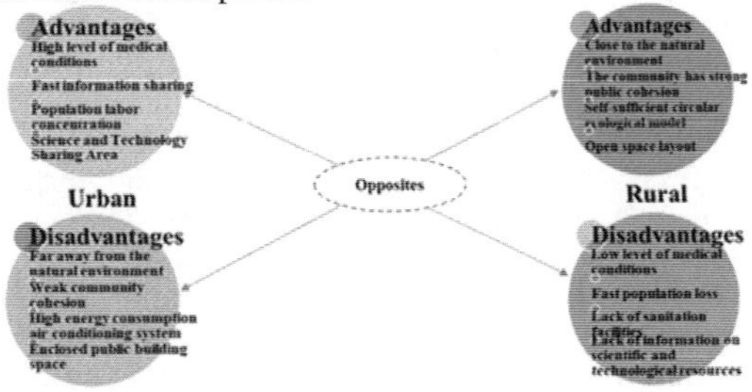

Fig12. Analysis of the opposite relationship between urban and rural areas under the epidemic

From the urban and rural system and antagonism to internalize into speculation, it is necessary to establish a "city-rural" integration concept. According to the characteristics and deficiencies of the urban-rural spatial relationship, a balance urban-rural relationship system can be constructed from four levels: co-construction, sharing, symbiosis, and co-creation (see Figure13).

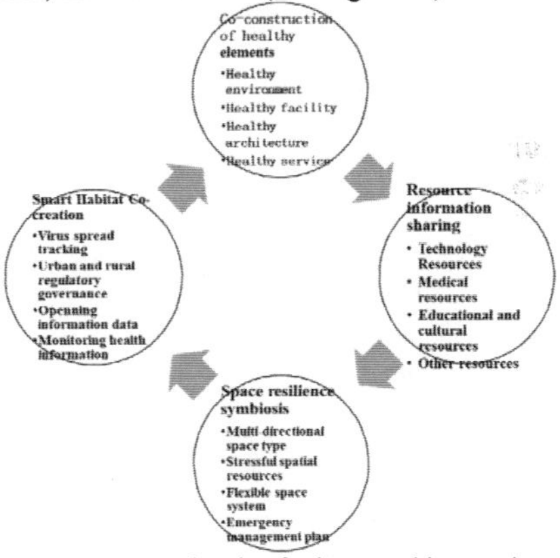

Fig13. Four levels of urban-rural integration

First, the joint construction of urban and rural health elements is to improve the weak links of the primary medical environment and facilities. Ensure the necessary hardware equipment in the entire disease prevention and control process, and build the medical space structure based on the principles of pathological transmission and the needs of healthy living. Gradually establish the foundation of a stable and healthy urban and rural area, and think deeply about the essential elements of a healthy and healthy circulation space according to the corresponding shortcomings, and realize the design requirements of healthy building space while paying attention to health psychology. Second, the sharing of urban and rural resource information, that is, the sharing of multiple resources such as technology, medical care, education and culture, maximizes the utilization of various resources. The use of various information platforms to serve different population structures enables the fair, open and free use of effective resource structures to achieve efficient production and construction while helping to balance social and economic imbalances. Third, the symbiosis of urban and rural resilience, which mainly cuts into the variability and flexibility of urban and rural spaces, and explores the possibility of transforming urban and rural static spaces into dynamics. It is based on the use of effective resources to actively construct a multi-functional and spatial multi-dimensional urban-rural integration model, which improves the stability of the society by ensuring the fluidity and variability of the human settlement environment under stress conditions. Fourth, co-creation of smart human settlements. That is, through the use of contemporary intelligent information technology to realize the monitoring of healthy human settlements and the development of a modern model of urban and rural integration. It is convenient for effective supervision and real-time monitoring of known diseases. And it also has a predictive effect on the emergence of unknown new viruses in the long-term development.

5 Conclusion

The large-scale spread of the COVID-19 has sounded a new round of alarm for us. Compared with this epidemic, contemporary chronic diseases such as obesity, diabetes, and depression have gradually eroded human health. The stressful and fast-paced living conditions, we have to reflect on whether rural space has played a positive role in human health in this process. According to the logic of historical development, with the advancement of science and technology, modern urban space should be able to solve many contradictions. However, it is a pity that the rapid urbanization construction has made the life of contemporary people a kind of "regression"[[15]]. Moreover, objective facts tell us that the "side effects" caused by the epidemic have obviously not faded away, and that problems will have a continuous impact on the human settlement environment for a period of time in the future. Therefore, as designers, we still need to improve vigilance and sensitivity.

Since China proposed rural revitalization in 2017, the enthusiasm for protecting and developing the countryside has not faded, and the interactive relationship between the countryside itself and urban and rural areas is still in the exploratory stage. Through the in-depth analysis of the different feedback mechanisms of two typical human settlements under the epidemic situation, it can be clearly distinguished that the opposition between urban and rural areas has the possibility of complementary symbiosis. In addition, contemporary villages are facing a self-awakening, and the contradictory urban-rural values of multiple contradictions also need to be re-examined and effectively converted. However, urban-rural practices cannot be achieved just one day. Since the construction of the urban-rural relationship is the same as architectural design, there is a time gap between theory and practice, which leads to disconnection. In this epidemic, the display of forces from all parties has reduced this gap as much as possible, so there are many possibilities for future urban-rural integration. It is necessary to comprehensively consider its overall operating mode in space, and also to deeply consider the effectiveness of independent individuals.

References

1. Zeng Wei , Gao Xinyi , Zhao Yingjie. A Brief Analysis of "Healthy Human Settlements" Research Framework. *J. Hum. Settlements West China* **2**, 33–42 (2020).
2. Weng, L. Spatial Distribution of Traditional Chinese Villages and Factors Affecting Their Distribution. *J. Landsc. Res.* **11**, 29–37 (2019).
3. Diane Ghirardo. *The Architecture of the City*. (The MIT Press, 1984).
4. Wang Lijun. *Architectural typology*. (Tianjin University Press, 2005).
5. Chinese, C., Town, W., Zhijian, G., Chang, G. & Qiao, G. Z. Jiangnan: Views of a Contemporary Chinese Water Town. *Cross-Currents East Asian Hist. Cult. Rev.* **20**, 127–135 (2016).
6. Birrell, A. *The Classic of Mountains and Seas*. (Penguin, 2000).
7. James Legge. *The Shi King*. (Oxford University Press, 1876).
8. Fenzhong, W. *Eastern Residence Legend Zhejiang Dongyang Residence*. (Tianjin: Tianjin University Press, 2008).
9. Bian, L. One thousand years of historical relics : the Huizhou Documents. *J. Mod. Chinese Hist.* **10**, 248–268 (2016).
10. Case, J. A., Liukeng, A., County, L., Huan, W. & Zhongrong, M. I. N. Study on the Inheritance of Landscape Culture in Traditional Ancient Province. *J. Landsc. Res.* **7**, 64–66 (2015).
11. Yi, C. Renovation of Traditional Water Villages in Jiangnan: a Casa study of Renovation Planning of Wenchang Village. *J. Landsc. Res.* **9**, 15–17 (2017).
12. Matthias Naumann1;David Rudolphb. Conceptualizing rural energy transitions: Energizing rural studies, ruralizing energy research. *J. Rural Stud.* **73**, 97–104 (2020).
13. Zhu, J. ;Zhu M. ;Xiao Y. Urbanization for rural development: Spatial paradigm shifts toward inclusive urban-rural integrated development in China. *J. Rural Stud.* **71**, 94–103 (2019).
14. Cody, S. 'A new consciousness of the countryside'? Elite ruralism in contemporary China. *Asian Anthropol.* **18**, 21–36 (2019).
15. Wang Dong. Countryside: As a Kind of Power of Critique and Thought. *Archit.* **6**, 100–108 (2017).

Design on smart water meter system based on Cloud Computing

Xiye Feng[1,*], Dawei Zang[2], Lelin Yan[1], Wenfeng Xie[3], Hailin Zhu[1], Xiaoyan Yin[1] and Meihui Xu[4]

[1]School of Information Science & Engineering, Qilu normal university, 250013, Jinan, China
[2]Presidents' Office, Qilu normal university, 250013, Jinan, China
[3]Office of Educational Administration, Qilu normal university, 250013, Jinan, China
[4]School of Information Science & Engineering, University of Jinan, 250022, Jinan, China

Abstract. The smart water meter based on cloud computing is a new instrument that uses modern sensor technology, communication technology and modern microelectronics technology to measure water volume, transfer water data and settle transactions. The traditional water meter only has the function of flow collection and pointer mechanical display of water consumption. Compared with traditional water meter, smart water meter has great progress. With the development of wireless transmission technology and mobile communication network, remote water meter technology begins to develop rapidly.

1 Introduction

For a long time, the water meters we use are generally low-grade. This kind of instrument has low price and single performance. With the progress of society, people's demand is higher and higher. The shortcomings of traditional water meter are more and more obvious. The traditional meter reading method requires staff to measure and verify the data in person, which consumes a lot of manpower and material resources.

With the rapid development of information technology, the collection and transmission of all kinds of information realize gradually intelligent. The era of using manual door-to-door meter reading and water meter rotary valve to count water consumption will become history. The traditional mechanical water meter could not meet the new business needs. It needs to report the water information timely through the smart water meter. It can sense the turnover state of the urban water supply system in time, and can load the monitoring functions of water quality and water pressure. The smart water meter will become the new development trend of water meter with the application of Internet of things technology, electronic technology, wireless transmission technology, network technology and computer technology. The smart water meter system is based on the Internet of things technology and integrates NB-IOT data transmission module. The system also integrates upper computer software, regional control cabinet, shunt controller and transmitter. It can realize intelligent control through organic combination with water supply system.

Based on the development of smart city, this research proposes a set of reasonable, low-cost and easy-to-use smart water meter system solutions. With the help of cloud computing, the water meter system can realize remote monitoring, data transmission and intelligent control. The smart water meter system is an automatic, digital network management system for water meter reading and water fee solution in water industry.

2 Overall system architecture

2.1 Whole framework

The platform can transmit accurately the information of multiple distributed units to the integrated system of the next base station. That is the data processing center. The system includes the medium of transmitting information, the high precision instrument of acquiring information and the efficient and fast server etc. The whole framework of the intelligent water meter system is shown in Figure 1.

2.2 Main functions of the system

Smart water meter system includes real-time monitoring, equipment support, statistical analysis, historical view, file management and other functional modules. The main function modules are as follows.

1. Data connection.

It can transfer the acquired data to the computer for corresponding data processing.

2. Meter reading management.

Meter reading times and time can be set according to certain rules.

3. File management.

It can measure water meter by category.

4. Real time monitoring.

Users can observe the use of meters at any time.

* Corresponding author: lkfxy@163.com

https://doi.org/10.1051/e3sconf/202123704010

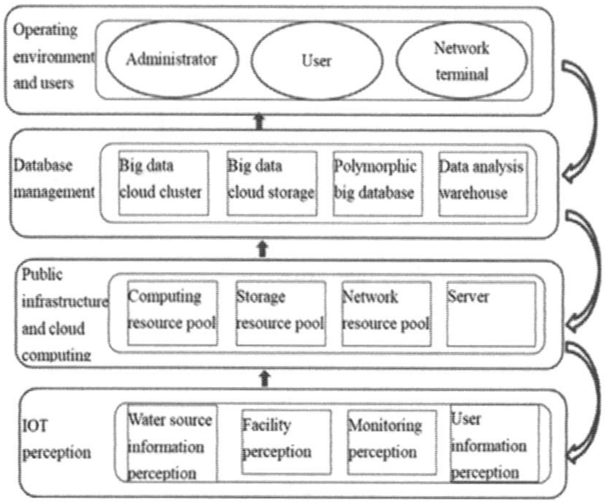

Fig1. The whole framework of the intelligent water meter system

5. User management

The user information can be input into the form, and then directly imported into the system. The user ID, name, geographical location and water consumption of the user are concentrated in a table, which is convenient for management and query.

6. System settings.

It can set the administrator's information and related system data.

7. Settlement and billing.

According to the usage, it can settle the user's expenses of each table and print the consumption list. It can also facilitate the user to pay.

8. Historical query.

Historical water consumption can be queried, and the historical water consumption can be exported to save the data for future data analysis.

9. Statistical analysis.

It can carry out statistical analysis of user water consumption according to the date and present it in the form of chart.

The application software structure of smart water meter platform is shown in Figure 2.

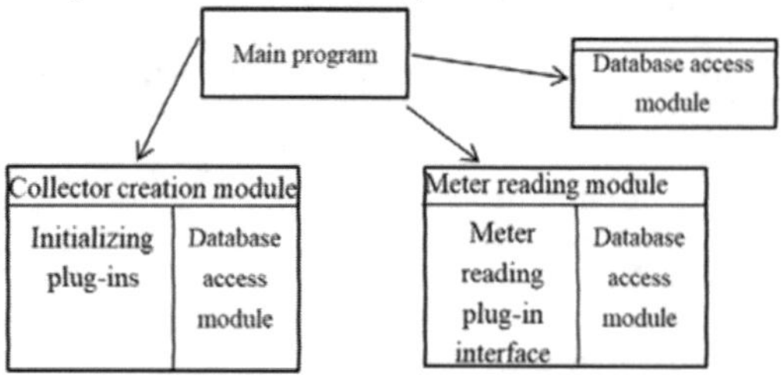

Fig2. The application software structure of smart water meter platform

3 Equations and mathematics

Remote meter reading module is the core of the whole smart water meter system. The main function of the remote centralized meter reading module is that the information of each scattered base meter is transmitted to the base station of the general integrated system without error. It includes high-precision instruments to collect information, accurate transmission media and efficient cloud processing computer. Therefore, the remote centralized meter reading system is mainly composed of remote meter reading control valve module, data acquisition module, concentrator and cloud processor.

3.1 Remote centralized meter reading

Remote meter reading can realize remote collection of equipment data without substantial contact. The collection process can be carried out automatically, so as to realize the high-frequency data collection, and then master the water used rules more precisely. In order to meet the special needs of different locations, there are usually many different configuration schemes in a system. The mechanism of remote meter reading system requires that all components must be compatible with other communication equipment, and can be extended and upgraded. The network topology of remote centralized meter reading system is shown in Figure 3. At present, Wireless water meter is widely used in youth apartment, white-collar apartment and hotel apartment, which directly connects to remote server. Of course, there are also GPRS water meters, which can be used directly through APP application. The real smart water meter in the future of Internet of things is NB-IOT water meter. It is a smart water meter directly connected to remote server through NB-IOT network.

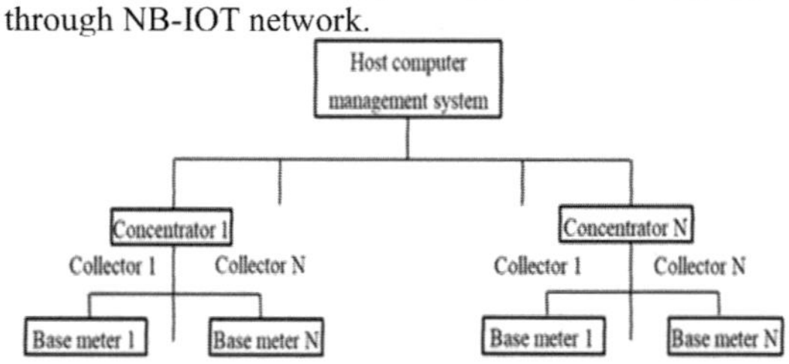

Fig3. Network topology of remote centralized meter reading system

3.2 System communication mode

To solve the problem of data communication is the key to solve the remote centralized meter reading. Therefore, the development of various communication methods has also affected the development of remote meter reading. According to the different communication modes, there are mainly wired transmission mode and wireless transmission mode. The wired transmission mode includes bus, power carrier, data network, modem and so on. The wireless transmission mode includes wireless transmission and GSM wireless transmission.

The platform adopts wireless transmission mode and simplifies NB-IOT protocol and signaling process based on LTE technology. In terms of network communication function, NB-IOT network is facing the needs of deep coverage, low power consumption, low cost and mass access of mobile Internet of things. After the adjustment of data transmission scheme, signaling process optimization and mobile management scheme, the network adapts to the data transmission requirements of

NB-IOT terminal at low speed and low frequency. And it assists the terminal to achieve energy saving and consumption reduction.

3.3 Key technology NB-IOT

The wireless communication of smart water meter adopts NB-IOT network. This is shown in Figure 4.The water meter is powered by battery. According to the industry standard, the service life of the battery is not less than six years. NB-IOT technology has great advantages. It has ultra-low power consumption, precise micro power operation module. And it focuses on dual sensing, software error correction, encryption and anti-interference technology. This ensures accurate data transmission. The key technical process of NB-IOT mainly includes: Non-IP, data transmission, mobility management, etc.

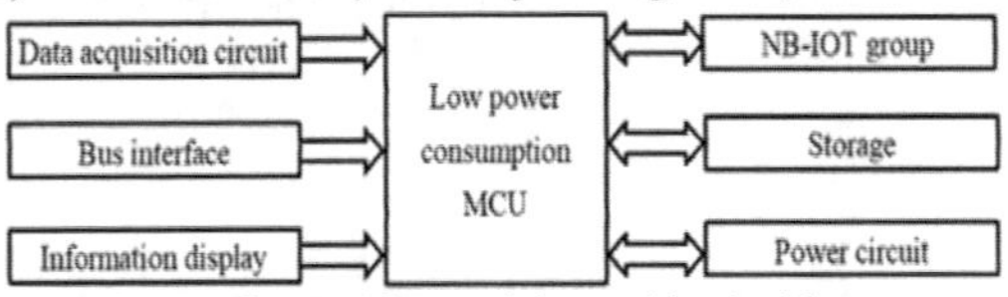

Fig4. Scheme of NB-IOT module combined with water meter

This kind of water meter is composed of the following parts: ultra-low power single chip microcomputer, NB IOT module, data acquisition circuit, information display, power supply circuit, bus interface, storage module.

1. Data acquisition circuit.

It uses analog-to-digital conversion technology to convert the collected information into digital information that can be recognized by the computer.

2. Bus interface.

A general interface is used to transmit the acquisition modules of multiple smart water meters to the upper computer through NB-IOT.

3. Information display.

The water meter information display module adopts LCD output. For low power consumption consideration, the water meter uses LCD display because the power consumption of LCD is far less than LED.

4 System implementation

4.1 System user data query

After the user fills in the relevant selection information, the database is connected after verification. If the connection is not successful, the relevant page will pop up. If the query is successful, the current information is saved and the program is ended.

4.2 The system module implemented

There are many functions in the system. The main functions listed below are water management, user management, recharge and payment records, message push and remote control valve.

1. Water query module.

After logging in, the manager or user will directly display the success rate of meter reading, water meter dosage and analysis.

2. Water management module.

In this module, the administrator can query the water consumption of each household, increase the information of users, edit the daily water consumption of users for control, and generate excel reports. Users can also check their water consumption balance and other information. The water display interface of water meter is shown in Figure 5.

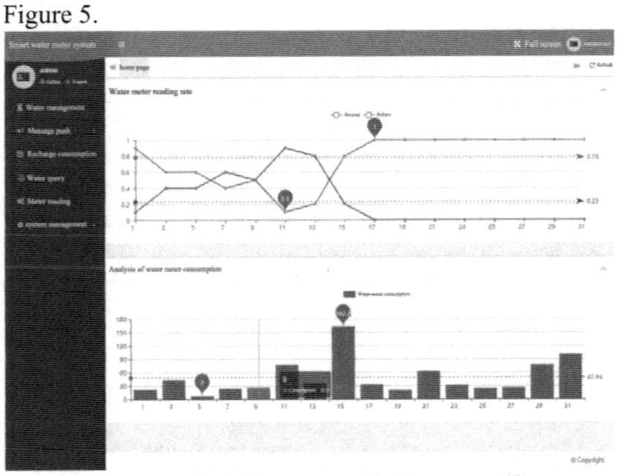

Fig5. Water meter display

3. File management.

This module can retrieve qualified users according to the status of water meter and water consumption time. And it can remotely close the valve of user's water meter.

4.3 Algorithm implementation

The smart water meter platform based on cloud computing can initialize data and complete remote meter reading. It can read and save and refresh data. Through algorithm design, it can initialize the tree structure control and its properties to display all user information.It can be copied by clicking on the user tree structure and selecting the corresponding region.It can prompt to save the monthly report and call the stored procedure to create the monthly report and save the call to the background stored procedure.After reading the data, it can refresh the read water meter information. In the aspect of algorithm implementation, remote meter reading module algorithm is more complex. Part of the algorithm implementation is as follows.

```
* Controller
@Controller
@RequestMapping("/water/status")
public class WaterStatusController extends BaseController
{
    private String prefix = "water/status";
    @Autowired
    private IWaterStatusService waterStatusService;
    @Autowired
    private IUserService userService;
    @RequiresPermissions("water:status:view")
    @GetMapping()
    public String status()
    {
        return prefix + "/status";
    }
}
* Add and save meter reading controller
@RequiresPermissions("water:status:add")
@Log(title = "Controller", businessType = BusinessType.INSERT)
@PostMapping("/add")
@ResponseBody
public AjaxResult addSave(WaterStatus waterStatus)
{
     user = userService.selectUserByLoginName(waterStatus.getLoginName());
     if (user == null)
     {
          return error("User does not exist.");
     }
     waterStatus.setUserId(user.getUserId());
     return toAjax(waterStatusService.insertWaterStatus(waterStatus));
}
```

5 Conclusion and prospect

This research designs the overall architecture of smart water meter system based on cloud computing. It realizes the system management, remote meter reading, data communication, cloud server processing and other related functions with the help of cloud computing. And it describes the corresponding implementation of algorithms. However, the smart water meter system is a large system. Many aspects need to be further improved, and it will develop towards a more intelligent direction. Using modern internet information technology, measurement sensor internet of things technology, big data and cloud computing application technology, the smart system has high practical value in business optimization and production management.

Acknowledgements

This work is supported by the Shandong Higher Education Undergraduate Education Reform Project (z2018x067) and the Ministry of Education Cooperative Education Project (201602028018). It is also supported by the Social Science Planning and Research Project of Shandong Province (17CGLJ10).

References

1. S. Uzayr, N. Cloud,T. Ambler, Vue.js, 1(2019)
2. R. Cobacho, Water meter and measurement (2015)
3. Huang Yuhong, Technical analysis and case analysis of Nb IOT Internet of things (2019)
4. Fan yuan, Smart city and information security in the era of big data (2019)
5. Sun Guoxiang, On the development and overall design of remote intelligent water meter reading system, Shijiazhuang water supply group (2017)
6. Huang Kai, Cloud computing system and application of artificial intelligence (2014)
7. Liao Zhengwei, Research and practice of intelligent water affairs. (2014)
8. Wen Ying, design and implementation of distributed data mining system based on cloud computing, Electronic technology and software engineering (2016)
9. Wei Feng, Tang Qiang, An urban intelligent water service architecture, computer knowledge and technology (2015)
10. Ji Zhu, software design of intelligent water integration platform, Hydropower plant automation (2016)

On the Innovative Application of China's Traditional Arts and Crafts in the Industry of Fashion Arts

Jin Lan

Shanghai Art & Design Academy

Abstract: The language concerning traditional arts and crafts is increasingly applied in modern fashion designing, which has become the important design ideas by focusing on the traditions. By elaborating the recognition of traditional arts and crafts, its diversified emerges in fashion arts including its classification s, materials, production methods, design methods and the diversity of design themes, the author discusses the creative application by combining the traditional arts and crafts and the modern fashion design and artistry. By explaining the traditional culture and analyzing its applications, the author explores the innovative application of traditional arts and crafts in the industry of modern fashion arts. The terminal aim is to link up the traditions and fashion, protect and inherit China's traditional arts and crafts, promote modern fashion brands and extend the cultural connotation of fashion brands.

Introduction

With the rapid development of China's economy, the industry of fashion arts has become the key loop in the China's industry chains. Here we have two hot issues. The first one is how to closely combine traditional arts and crafts with modern designing to fulfill their advantages with their development and promotion; The second one is how to rationally use China's traditional arts and crafts in various fashion designing to be "culture endorsement" of China's traditional culture in the fashion industry. Aiming at the above two problems, the author will elaborate in the following three aspects: One, China's traditional arts and crafts is the carrier of cultural innovation in the industry of modern arts; Two, the diversified emergence is the application of China's traditional arts and crafts in modern fashion arts; Three, China's traditional arts and crafts can and be able to actively promote the brand consciousness in modern fashion.

1. China's traditional arts and crafts is the carrier of cultural innovation in the industry of modern fashion

The industry of fashion arts has become the industry with potentials at home and abroad. The fashion arts must have its own cultural connotations, while cultural creativity is helpful in forming the unique designing styles, which can inspire the innovative thought and make it more competitive. If the modern fashion arts in one country wants to survive and develop in the global competition, it needs to focus on the design creativity and innovate them in the process of inheriting China's arts and crafts. Moreover, it needs to derive and dig the cultural connotation from the traditional culture in the country. The people's identity is shaped by sharing the history and forming its relative significance. Thus the designing philosophy with its national characteristics is formed1.

The cultural consumption begins to emerge in both the materials and spirits. The traditional arts and crafts in fashion arts is playing an important role in the development of one country. As the core strategy, the fashion art is always being popularized and promoted in economically developed countries or regions. The traditional arts and crafts in North Europe emphasizes the concept of simplicity. In early 19th century, the Nordics put forward that, no matter what the design was, it must meet the two requirements in the first place: practical and suitable. Therefore, the practicality and suitability is the unique design style in the fashion arts of North Europe. The designing concept of the fashion arts in Germany is its preciseness and rationality, which results in more high-class products by integrating its traditional arts and crafts and is the symbol of German Designing. From Victorian Style to Naturalistic Style advocated by the movement of arts and crafts, the arts and crafts of Great Britain influenced and is influencing the Europe Continent and even the United States of America. Up to now, these cultural and designing concepts are still influencing the fashion arts in these countries.

[1] the legacy café the holistic benefit of reviving lost arts, crafts and traditional skills through an early childhood intergenerational sustainable skills project[J].Diane Boyd.Education3-13.2020(2)

* Corresponding author: 1325@sada.edu.cn

 https://doi.org/10.1051/e3sconf/202123704011

In China's designing and creative industry, the modern fashion arts is one of the most potential fields. The strategy must be established in order to research and develop, deconstruct and reconstruct the traditional arts and crafts and coordinate the development of modern fashion arts. The product design of our neighboring country, Japan, can be quickly integrated into the development trend of fashion industry, which is based on the research and inheritance of traditional crafts. In recent time, the proposal of enhancing cultural confidence has made China's fashion arts focus more on the traditional culture and connotation. By embedding China's traditional cultural elements, or using some traditional crafts and techniques which are identified as intangible cultural heritages has become a new design trend and is more and more accepted by the public. As the direct design concept and means, traditional arts and crafts covers many intangible spiritual culture, which are embodied in people's living habits, customs, values and aesthetics. Intangible cultural heritages such as grass weaving art, lacquer art and even embroidery art in traditional arts and crafts can be introduced into modern fashion art. On the one hand, fashion has more creativity. On the other hand, it is helpful to protect and promote the traditional arts and crafts. The combination of the industry of modern fashion design and China's traditional arts and crafts should be the demand of the development of the times, which is also the demand of Chinese people in the dual activities of aesthetics and consumption. It profoundly reflects the aesthetic and social significance of the forms of fashion art. Moreover, it extends and expands the main connotation of the fashion art innovation. Therefore, the innovative modern fashion art with abundant cultural connotation is formed.

2. The diversified emergence is the application of China's traditional arts and crafts in modern fashion arts

The traditional arts and crafts is the accumulation of culture, while fashion is the people's temporary worship of something in the society, which represents the latest trend in a certain time. In the field of fashion art, traditional arts and crafts can be integrated with the history, the material, the technology, the manufacturing, etc. The craft variety, the materials, the manufacturing techniques, the design techniques are quite different from each other.

2.1The diversity of process classifications

In modern fashion art, the ceramic process has been more and more used in the home furnishing; the metalwork in the jewelry designing; the colored glaze in clothing decoration; the lacquer art in jewelry boxes; the cloisonné technique in vase decoration; the woolen embroidery in bag decoration, to name just a few. There are more process varieties are waiting for the designers to develop and utilize. The classifications of traditional arts and crafts are various, which can used in every aspect of fashion designing.

2.2 The diversity of arts and crafts materials

The materials used in the fashion designing includes the gold, the silver, the wood, the jade and the herbaceous plants in nature; the man-made or fired materials includes the glass, the ceramics, the cloths, the silk etc.; The applied materials includes both the soft materials and the hard materials with different textures. Sometimes, the use of unique materials represents the culture in a certain area because of its unique geology and climate, the mining and digging the materials are endowed with unique cultural characteristics. The application of materials should be based on the understanding the fundamental characteristics of the culture, which is in line with the modern design concept to reflect the instinct in selecting the materials in traditional culture in the long term. For example, in China's clothing industry, the traditional materials such as agate are used in designing necklaces and rings, which can endow the clothing with the new cultural connotation. In ancient time, the red agates symbolized love and hope. Even more, it is said that they can reduce stress and fatigue. Embroidery is increasingly used in the cultural and creative products. The embroidery is rich in diversities with different techniques, which is strongly expressed in modern fashion art.

2.3 The diversity in the manufacturing

Different crafts results in the different manufacturing methods in processing varieties, such as firing, engraving, weaving, freehand sketching, inlaying, molding etc, which can be applied in the fashion art. The fashion requires perfect presentation. Thus, the exquisite traditional arts and crafts needs to be integrated with the fashion designing. The exquisite techniques can give full play to the artistic appeal of design works, which is the unity of techniques and beauty in modern fashion designing. Techniques represented the advancement and productivity of the science and technology at that time. With the application of advanced design concepts and the high-tech design means, the fashion industry is booming while the traditional culture is being protected and inherited.

2.4 The diversity in designing methods

The fundamental elements were derived in China's traditional costume culture. In which, the various design elements, such as the rotundity and square are the common patterns. Auspicious patterns have rich connotations. There are different patterns in different categories. Fashion design is done with the features of abundance and completeness, which express the richness, happiness and completeness in aesthetics. The traditional fundamental elements and patterns derived from traditions in the modern clothing and cultural and creative products are the expressions of traditional languages. The spiritual meaning of the concepts such as less is better than more and the simplicity in fashion designing is expressed in new ways by the natural, simple and frugal artistic conceptions. The designs in China focus on the planarization while those in the West focus on the three-dimensional objects. In traditional culture, the designs are usually symbolic and meaningful, which is quite different from the realism in the West. Therefore, the fashion industry in China has its own features because of these differences.

2.5 The diversity in designing themes

The innovative themes give fashion more languages. For example, the series pens The Little Prince manufactured by Montblanc, are the high-quality products by digging the theme from the stories. The reason why the cultural and creative products in the Imperial Palace are best-seller is the creative themes. The fashion design products with China's traditional cultural characteristics will stand out and catch the eye of the public. With the support of powerful statistical technology of databases, the collected information has become the asset which is easy to extracted and used2,which makes it possible to introduce China's traditional arts and crafts into the new designing themes. Its values can be reflected in fashion designing and take a seat in world economy, which is still an important part in global cultural economy3.

3. China's traditional arts and crafts can and be able to actively promote the brand consciousness in modern fashion

China's traditional arts and crafts has its profound cultural background with a history of thousands of years. Fashion has a long history. However, for a long time, there is no brand consciousness in China's traditional arts and crafts. More people believe that China's industry of arts and crafts is small-sized, most of them are individual workshops. The mode that apprentice are trained by masters lasted and is existing now. The development of traditional arts and crafts markets is hindered. But if the brand consciousness of fashion industry can be integrated into traditional arts, in return, the connotation of fashion can be promoted by the traditional arts and crafts, both of them can possibly be developed well. In Japan, for instance, the lacquer wares by its traditional arts and crafts are researched and developed in molding, technology and pattern designing, by which its lacquer wares are sold globally. The reason is clear, its products are integrated into modern fashion life and can meet the requirements of modern society.

In France, in order to protect traditional arts and crafts, the organization of precious masters is established, the master list is compiled, the heritage history and works by the masters are exhibited. The most important is that their traditional arts and crafts are directly used in the field of modern designing and different design fields. From which, more modern products are actively developed, such as the handiworks using enamel craft have become one of most luxury brands. More examples are suitcases and bags manually weaved by straws and the costumes by using the technology of handmade laces.

Figure1. French brand Chanel's packaging and printing process

Figure 2. French hand-made decorative folding fans

[2] Lindy Ryan. The visual imperative: creating a visual culture of data discovery[M]. Morgan kaufmann.2016

[3] Paris and fashion: reflection on the role of the Parisian fashion industry in the cultural economy[J]. Solange montagne villette irene hardill international journal of sociology and social policy 2010(9/10)

https://doi.org/10.1051/e3sconf/202123704011

Germany is also the origin area of ancient culture in Europe, itself has a very unique source of artistic development. In Coburg, more products are developed by its rattan weaving process, such as the rattan chairs, the rattan tables, which have become the modern home fashion furniture (See **Figure 3.**).

Figure 3. Modern wicker chair design

Figure 4. Czech glass products on display in the Glass Museum

The designers in China begin to emphasize the cultural heritage in designing step by step. The traditional concepts are combined more with the modern elements. More fashion products with rich and profound cultural connotations are designed by Chinese designers. The design of modern fashion in China needs to protect and inherit China's unique national characteristics. Because traditions have their tenacious vitality and won't disappear in the impact of modern high-tech industry. The application of 3-D printing technology and Computer Numerical Control only benefit the traditional technology and process. It won't impact but start another design integrated with modern designing, which will endow China's fashion design with distinctive features and unique artistic temperament.

With the time flying, the lasting features have become one kind of recognizable cultural icons. The application of traditional arts and crafts in modern fashion art needs to establish the brand consciousness, more globalized Chinese brands enlightened by China's traditional arts and crafts needs to be founded. Together with fashion art, to found the sustainable growth of cultural economy is the correct thought at present. The humanity attainment is improved while developing the economy. Excellent traditional culture can be compared to a kind of nutrition of fashion industry. The values should be diverse and abundant. To carry forward the traditional arts and crafts, we should protect them and innovate them at the same time. Starting from the brand consciousness, the carrier of traditional arts and crafts can make the products more vital for a long time. Only by developing the products with cultural artistry, from traditional materials and by traditional handcrafts can the international and influential Chinese brands be established. It is impossible to reinforce the promotion of Chinese brands, to inherit the culture and innovate the designing in a short time. We need more time to incubate, breed, adjust and improve before prestigious Chinese brands are established.

Figure 5. Baoku Jiangxin Art Center

Figure 6. Tapestry craftsmanship exhibited in Baoku Jiangxin Art Center

Baoku Jiangxin Art Center, which is located in Shanghai Tower, devotes itself to protecting and inheriting China's traditional arts and crafts. Its exhibits of traditional arts and crafts range from decoration design, home design, weaving arts to jewelry design and so on. Almost every traditional process has its own brand. For example, the rosewood art brand of Yinzun Pavilion; the woodblock printing of Zhuxian Town; the enamel brand of Trends, are all the famous Chinese brands. This is a good phenomenon, in the past, some arts and crafts disappeared because of the lack of brands. Globally, we can find the exquisite enamel wristwatches brought by the Swiss Brand Patek Philippe and created by the enamel master Ms. Anita; And more, the neoteric enamel jewelry designed by the Italy Brand Bvlgari. The traditional enamel technique didn't disappear but became the mainstream craft in luxury brands. Of course, the common fashion brand products are updated frequently. Thus, how to rapidly integrate the traditional arts and crafts and apply them in the fashion products and quickly appear in the markets at home and abroad, which needs us to think about.4

The traditions are integrated with the fashion designing. After that, the concept of cyclic economy and sustainable design needs to be imbedded. The sustainable design includes not only the sustainability of resources but also the sustainability of social culture5, and more, the innovation is only way to develop. The ideas in China's traditional aesthetics are influencing the modern fashion design ideas, the sustainable, recyclable concepts in traditional arts and crafts are innovatively integrated into modern elements6, which urges the continuous renewal of innovation consciousness. Thus, the establishment of brands supplies the opportunity to build up China's national consciousness. The application of China's traditional cultural craft design is diversified in the ancient life. The traditional arts and crafts needs to be integrated into the works of modern fashion crafts. Inspired by this, the designed fashion arts will have their historical imprint and be a good inheritance of traditional arts and crafts.

4.Conclusion

The fashion design in China is comparatively weak at present, the designers need to know more about traditional culture and their humanity attainment needs to be improved in order to dredge the essence and spirit of China's traditional culture. The investigation, research, development and utilization of China's traditional handicrafts need to be strengthened. By understanding China's traditional culture and its features, the designers can understand the internal aesthetics of traditional arts and crafts and think about it in a creative way. By integrating designing ideas and manufacturing methods, the designers can construct the new modern fashion design concepts. The designers pay more attention to traditional arts and crafts and even love traditional arts. With the help of advanced process engineering and diversified material development, the designers can actively construct the fashion design arts with China's cultural characteristics and propel the updating of fashion art industry. It is impossible for the above-mentioned innovation application research to cover all ranges. However, it is quite important to build traditional design consciousness and create the unique fashion art products in the international markets. The follow-up research and development is the innovative application of traditional arts and crafts. All in all, the present traditional arts and crafts is playing an important role in fashion designing. If so, more design and design brands will created and have a profound impact on the world.

Acknowledgment

This paper is the research result of 2019 Research and Innovation Project in Shanghai Art & Design Academy.

Author brief: Jin Lan, female, graduated from Donghua University with Master of Engineering, the Associate Professor and the Senior Arts & Crafts Artist, working as the Assistant Dean of WPP School of Visual Arts in Shanghai Art & Design Academy, as the Deputy Secretary General of Special Committee of Graphic Vision in Teaching Guiding Committee of the Major of Art Design of Vocational Colleges in the Ministry of Education of the People's Republic of China.

References

1. The legacy café the holistic benefit of reviving lost arts, crafts and traditional skills through an early childhood intergenerational sustainable skills project[J].Diane Boyd.Education3-13.2020(2)
2. Lindy Ryan. The visual imperative: creating a visual culture of data discovery[M].Morgan kaufmann.2016
3. Paris and fashion: reflection on the role of the Parisian fashion industry in the cultural economy[J]. Solange montagne villette irene hardill international journal of sociology and social policy2010(9/10)
4. Fast fashion: response to changes in the fashion industry[J].Vertica Bhardwai Ann Fairhurst the international review of retail, distribution and consumer research 2010(1)
5. Sustainable Design Futures: An open design vision for the circular economy in fashion and textiles[J]. Paul Smith, Jen Baille, Lynn-Sayers McHattie. The design journal. 2017(supl)
6. AMFI's Reality School: A circular economy agenda for fashion education [J] Nicholas Hall Fiona Velez-Colby. Art design & communication in higher education 2018(1)

[4] Fast fashion: response to changes in the fashion industry[J].Vertica Bhardwai Ann Fairhurst the international review of retail, distribution and consumer research 2010(1)

[5] Sustainable Design Futures: An open design vision for the circular economy in fashion and textiles[J]. Paul Smith, Jen Baille, Lynn-Sayers McHattie. The design journal. 2017(supl)

[6] AMFI's Reality School: A circular economy agenda for fashion education[J]Nicholas Hall Fiona Velez-Colby. Art design & communication in higher education 2018(1)

https://doi.org/10.1051/e3sconf/202123704011

7. Ludden, G.D., P. Hekkert. and H. N. Schifferstein. Surprise & emotion. in proceedings of the 5th conference on design and emotion society. Gothenburg, Sweden: Chalmers University of technology. 2006
8. Condition analysis and forecasting in the fashion industry Mariusz Czekala; Agnieszka Bukietynska; Marek Gurak International Journal of Economics, Finance and Management Sciences. 2019
9. Green product development under competition: A study of the fashion apparel industry. Shu Guo; Tsan-Ming Choi; Bin Shen. European Journal of Operational Research. 2020
10. Agreement by design: The effect of visual harmony on responses to surveys Ulrich R. Orth. Journal of consumer behaviour. 2020

https://doi.org/10.1051/e3sconf/202123704012

Regional Human Settlement Environmental Ethical Features of Dayi Liu's Manor in the Western Sichuan Plain

Ziwei Luo[1], Wenfeng Xie[1*], and Guohua Wang[2]

[1]College of Landscape Architecture, Sichuan Agricultural University, No.211, Huimin Road, Wenjiang District, Chengdu, China
[2]Sichuan Guangyuan landscaping affairs center, Guangming Apartment, Dongba Binhe North Road, Guangyuan City, China

Abstract. Manors are the most representative human settlements in the history of Western Sichuan, with complete regional tradition and ethical order. As a protective heritage in the revitalization and development of rural areas, manors have been recognized as construction achievements for their artistic and practical references. This paper starts from the perspective of human settlement environmental ethics, focusing on the environmental ethical order between the microcosm of the residential environment of Liu's Manor and the macrocosm of the regional pattern of the Western Sichuan Plain. The result shows that the environment characteristics of Liu's manor, under the regional settlement pattern, actually are the images of human subjective ethical choice; gathering in an ideal environment, dispersing to the origin, and the integration of classical styles. On this basis, this study proposes a means through which to realize sustainable development in rural settlement construction in the process of modernization in China.

1 Introduction

In ancient China, early settlements built for feudal lords were called manors. Manors evolved from folk houses, it represented the highest achievement in the development of traditional residences. After the Qin and Han Dynasties, manors gradually developed into 'landlord manors' [1]. From the perspective of social influence, the landlord manor was an entity. It represents the landlord class' oppression and exploitation of the peasant. From the perspective of material environment construction, the manor often represents a superior form of local microcosmic living environment. Compared with other traditional residences, manors pay more attention to the creation of regional traditions and ethical orders. This paper takes Dayi Liu's Manor as the object, which is the largest landlord manor on the Western Sichuan Plain and the typical model of historiographical research, aiming to interpret the relationship between residential and regional environments from the perspective of ethics, and further providing a reference to the regional construction of residential settlements in the new era.

2 Human settlement environmental ethics

Ethics located in philosophy. It aims to understand orders and norms that people believe in for interpersonal communication. Since the 20th century, in order to solve increasingly environmental problems, people have begun to care about the stakes between human beings and the natural environment, resulting in the development of the field of environmental ethics [2]. On this basis, an ethical perspective of the human settlement environment has been further developed to study the relationship between human beings and the living environment. The purpose of applying ethical philosophy to the science of human settlements is to extend moral concern to living spaces and the natural environment and to establish a new moral and ethical order as well as a tacit understanding to guide construction planning in an environmentally conscious direction [3,4].

Human settlement environmental ethics offers a concrete description of the human settlement environment. People's subjective demands for residence and subject cognition of the environment form the subject-oriented ethical choices, so as to coordinate the placement order between the microcosm residence environment image and the macrocosm regional pattern, and ensure the harmonious coexistence between the human and the environment (Figure 1).

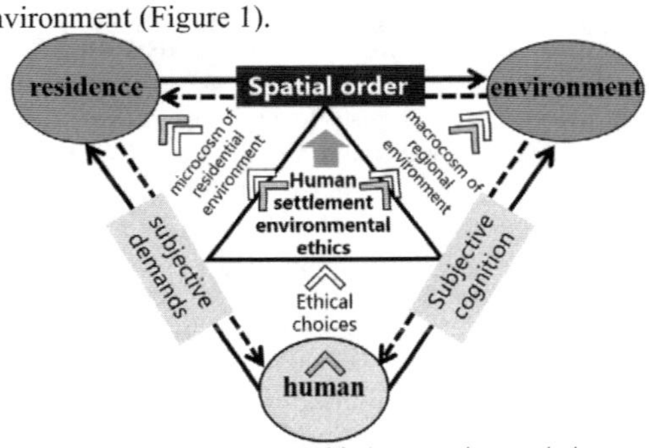

Fig1. The ethical relationship between human beings, residence, and the environment.

Manors as a kind of human settlements, developed

* Corresponding author: xwenfeng@sicau.edu.cn

with a certain economic basis and artistic integration, are usually a combination of regional characteristics and other human cultures. They are usually built spontaneously by local people when their economic conditions are relatively good. Such buildings not only depict the original environment of the local users, but also reflect their views on environmental selection in pursuit of culture and art. The environmental image presented by the architecture is the embodiment of its ethical characteristics, especially in the area where the culture is diversified and blended. In Western Sichuan, studying the construction of human settlements through the residences represented by Liu's Manor offers insight into the historical experience of those who built, owned, and lived in the manor. The spatial order between residence and environment, formed by the cognition and the integration of the regional environmental culture, is a portrayal of the ethics of human settlements in historical heritage [5,6].

3 Macro-settlement patterns of the Western Sichuan Plain

With Chengdu as its center and including the vast surrounding rural areas, the Western Sichuan Plain is an agricultural settlement area that integrates living and production (Figure 2). The excellent farming conditions have made it the main settlement area for immigrants from other provinces, and the settlement pattern has gradually recovered and developed steadily [7].

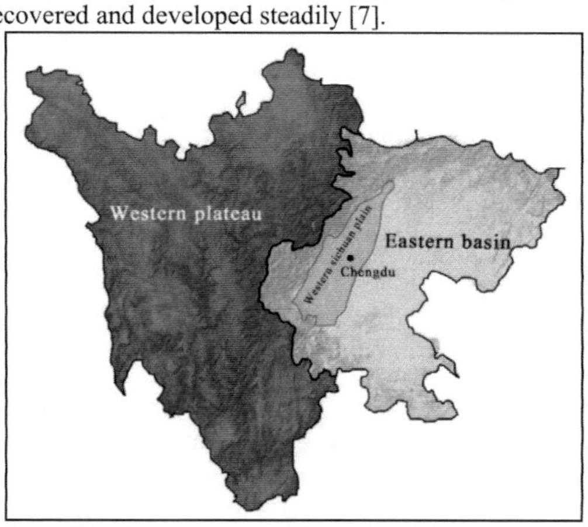

Fig2. The location of the Western Sichuan Plain in Sichuan Province.

Immigrants who entered Sichuan at the beginning of the Qing Dynasty took the lead in the land reclamation, preparing fields for crops and building houses, thereby forming a settlement pattern with clans as the core community. By the middle of the Qing Dynasty, the increase in population led to a gradual saturation of cultivated land. At this time, farmers had to rent land from landlords both for agricultural and residential purposes, which gradually lead a dispersion outwards from the center of the initial settlement. By the late era of the Qing Dynasty, the landlord economy had further developed, with land subleased and subdivided. The pressure of population growth against the availability of arable land eventually led to the development of smaller settlements, which increased the density and dispersion of settlements in the region (Figure 3). Against the background of migration, the dispersion and integration of rural areas became apparent, forming a stable high-density settlement pattern. This pattern tended to be consistent at the macro-level, and formed a regional environment characterized by gather, dispersion, and integration.

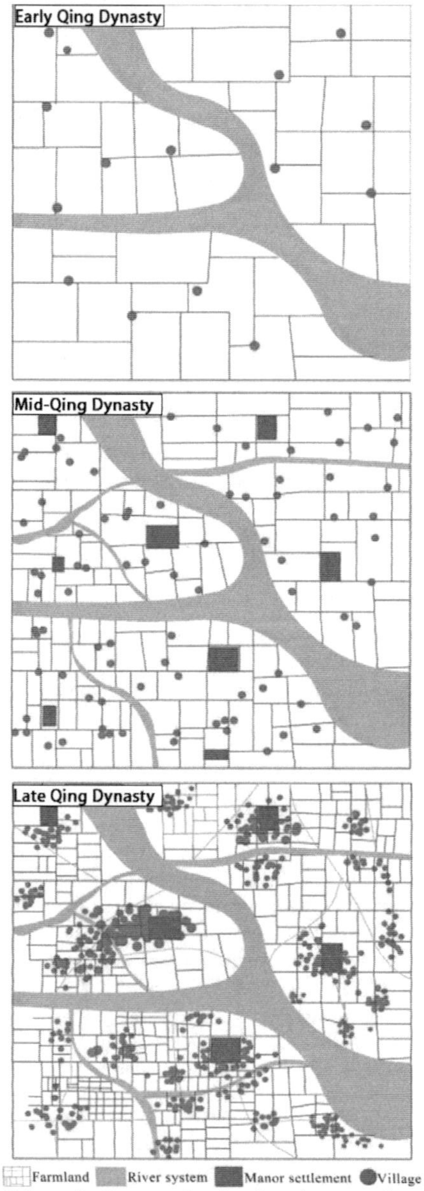

Fig3. Schematic diagram of regional environmental evolution in the Western Sichuan Plain.

4 The environmental ethical features of Dayi Liu's Manor

Dayi Liu's Manor is the ancestral home of Wencai Liu, a substantial landowner, whose manor is known as "the largest landlord's manor in western Sichuan". Over the

 https://doi.org/10.1051/e3sconf/202123704012

past 100 years, the environmental features continue to reflect the owner's choices that how to harmonize the living space and the regional environment; gathering in an ideal environment, dispersing to the origin, and the integration of classical styles.

4.1 Gathering in an ideal environment

Liu's Manor is an advanced form of living in the history of western Sichuan's residential development. The manor reflects an environmental optimization in terms of pattern, the spatial aggregation of architecture, and identifying an ideal environment in the composition of the whole settlement.

4.1.1 Settlement location

Liu's Manor is located in the famous historical and cultural town of West Sichuan-Anren Ancient Town in Dayi County. In the 1930s, the construction of the mansion began. It gradually formed into the old mansion (including Liu's ancestral yard as well as Wencheng Liu, Wencai Liu, and Wenzhao Liu's house yard) and the new mansion (Wenhui Liu's house yard) (Figure 4).

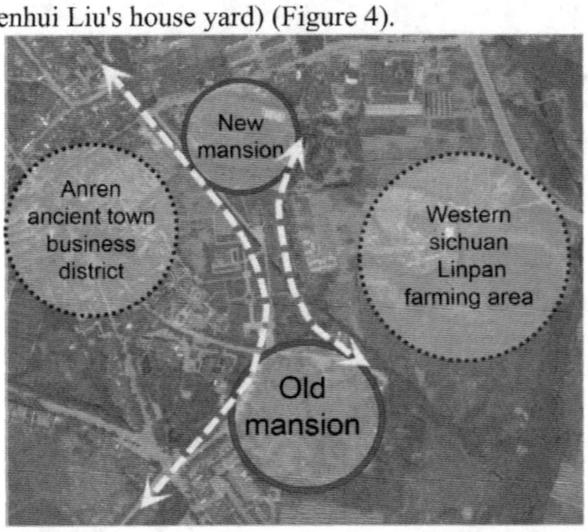

Fig4. The surrounding environment of Liu's manor.

Anren Ancient Town represents the mature model in the development of rural settlements in western Sichuan. It has distinct street spaces and architectural texture, primarily for public rallies and commercial activities, and its function is relatively holistic. In ancient times, the town was too far away from the imperial rule, and Liu's Manor was located at the intersection between the town of Anren and the vast Linpan, so that the social environment has not restricted its development. This prime location enjoyed the advantages of a perfect network system—on the edge of a natural environment, on one side, and a business and residential district, on the other—enabled the Liu family to expand the scale of their manor around the core of Liu's ancestral home using the resources of the land and relying on local transportation, markets, manpower, and other resources. From the perspective of the living environment, the manor could enjoy a vast agricultural landscape as well as the nearby commercial landscape that had gathered (Figure 5), forming a harmonious living atmosphere between gathering and dispersing. The flexible external environment supported its independent development and optimized the expression of the living space in line with the manor's status.

Fig5. The town and farming fields between which the manor is located.

4.1.2 Settlement space

Liu's Manor settlement has an undeniable spatial continuity and integrity. It has formed a very successful enclosure of woodland resources. The combination of plants and architecture is conducive to the formation of a microclimate environment, and the manor space is the greatest beneficiary of this environment.

In order to adapt to the humid, hot, and rainy climate, Liu's Manor adopted a very typical architectural system—the Southern Siheyuan (a quadrangular courtyard)—which is the highest in form of traditional dwellings. Traditional rural buildings have little space for a significant continuous courtyard space, but Liu's manor presents us with two very distinct types of continuous quadrangular courtyards: the differently sized and free-form layout of the old manor complex and the very obvious axis layout of the new manor complex (Figure 6).

a.The old manor complex.

b.The new manor complex.

Fig6. The layout of the two manors showing the courtyards.

In the flexible layout of the old mansion, the family rules of etiquette were hidden in the quadrangle arranged in multiple directions, this is the embodiment of Chinese family order and hierarchy. The courtyards are connected, forming a continuous, flexible path, joining the scattered garden spaces, incorporating green space as far as possible in the limited configuration of buildings, and constructing a habitable living space. While the new manor situates the enclosed buildings around the green space under the governance of clear rules of order, the linear path forms a symmetrical axis of the three courtyards, and this garden corridor extends residents' view and enhances the experiential sense of hierarchy in the architectural space. The two layouts of Liu's Manor clearly reflect the importance of the courtyard space in creating a cohesive spatial environment. The patios, both large and small, hold the key to the ventilation of the whole manor and offer bright spots within the innermost spaces. In the courtyards, one can feel the spatial communication between the pastoral environment and the courtyard, linking the living spaces and the nature that surrounds it (Figure 7).

Fig7. The manor's internal spaces.

4.2 Dispersing to the origin

4.2.1 Family origin

The family system has been integral to traditional Chinese culture since ancient times, especially in the immigrant society of western Sichuan, in which the inheritance of family culture is reflected. According to Liu's Clan Genealogy, Liu's ancestors, Yingliang Liu, emigrated to Sichuan from Huizhou, Anhui Province, in the early Qing Dynasty and were the first settlers in Dayi, laying the foundation for the development of the Liu clan. As recorded in the Great Bulletin in 1949, Wencai Liu is noted as a famous landowner in Sichuan Province, possessing about 800 hectares in farm fields, spread across Dayi and 11 counties in western Sichuan. Furthermore, the article notes numerous properties, including 13 mansions, scattered across Wenjiang, Chongqing, Chengdu, Dayi, and Ya'an. Nevertheless, the manor in Dayi was said to be Wencai Liu's favorite—clearly determined since it was the origin of the family establishment. Liu's manor is divided into two settlements, about 300 meters apart from the north and the south, and comprises five manors built by Wencai Liu and his brothers, which also form part of Liu's ancestral residence. The settlement expansion mode—ancestral residence, the old mansion, and the new residence—embodies the inheritance and development of clan culture. Liu's ancestral residence is the origin of the manor, and the subsequent construction of the old mansion on the periphery reflects the consolidation and support of the family strength, while the new mansion, as a spatially independent entity, has fully outgrown the power of its family. Thus, the whole history of the family is imprinted in the very construction of the manor complex. The diasporic community, formed by the manor as a whole, highlights the fundamental idea of returning to the origin of the clan, which is the most important conception for the construction of the small-family living environment.

4.2.2 Agricultural origin

In addition to the family's origin, the cultural heritage of farming in the Western Sichuan Plain unites the diasporic culture as a whole. The construction of a private estate should form a microcosm of the local social environment so that the sentiment of the family can be amplified into the whole living environment and establish a harmonious sense of Regional Environmental Ethics.

Although Liu's Manor is considered to be the most luxurious private manor in western Sichuan, it is not entirely out of the origin land sentiment of agriculture contained in the dwellings. The ancestral residence in Liu's Manor represents the primitive and plain residential style of western Sichuan, with typical sections for farm production and residential living, including a brewery, dam, vegetable field, and pig pen, which is an authentic portrayal of farming culture in the Western Sichuan Plain (Figure 8). Subsequently, the old mansion and the new mansion expanded one after the other, and are similar to the ancestral residence in terms of space, color, and material. The farming culture is hidden in the landscaping, which preserves the original conception of environmental construction and simultaneously improves the artistic quality of space. In terms of regional pattern, the manor's architecture shows that attention has been paid to a sense of harmony with the surrounding environment. The environmental choices and adaptability of the manor are consistent with the simplicity of a common farmhouse, thus eliminating the class difference in the living environment pattern and appearing as a somewhat symbiotic form attached to the land. The supremacy of the land is reflected in the atmosphere of the microcosmic environmental image to the view of the whole regional environment.

Fig8. The simplicity of Liu's ancestral residence.

4.3 The integration of classical styles

The evolution of the residential environment in the Western Sichuan Plain is a turbulent and extraordinary process, which has also determined its regional inclusiveness, not only of immigrant culture but also of Western culture. Combining architectural characteristics of China and the West, as well as those of Sichuan and Anhui, Liu's Manor captures the essence of the region and becomes the classical form of rural residential space construction in the Western Sichuan Plain.

4.3.1 The integration of the traditional building styles of Sichuan and Anhui

Liu's ancestral residence was the earliest of all the manor buildings and follows the typical style of column-and-tie construction in western Sichuan. Since the family of Liu is native to Anhui, where the traditional architecture style is distinctive from Sichuan, so the later mansions in the manor complex have very obvious features of the Hui-style, the original tone of hui-style architecture is a harmonious combination of white walls, black tiles, special Ma Tau walls, brick, wood and stone carvings, as well as laminated courtyards, high ridges and cornice, winding corridors and pavilions. While in the manor, the features mainly embodied in the shape of the wall and the stone and brick carving. The whole estate was surrounded by brick walls more than six meters high, and the blue-gray brick joints were decorated with white wire, giving the manor a sense of a rustic but dignified image in the outside environment. The inner courtyard of the manor is primarily enclosed by the smooth and round "Ma Tau Wall" (Figure 9), which brings a different style to the manor. In addition, the body of the manor buildings remain styled in a wooden column-and-tie structure, with a brick and wood outer corridor, which is a typical southern feature. In such a framework, the two regional styles may have a clear contrast, but share a simple and elegant tone, which is the result of a combination of adaptive choices. The two different cultures are preserved by the most distinct or core characteristics, reconciled at the subordinate and secondary aspects of culture, which eventually transforms the geographic differentiation into spatial dominance.

Fig9. The Ma Tau Wall.

4.3.2 The integration of building styles from China and the West

The expansive development and construction history of Liu's Manor shows the typical style of the Republic of China in the 1920s and 1930s. However, the combination of Chinese and western characteristics distinguishes the manor from other houses, showing its status, which extends to the family. The gate of the old mansion is a mixture of western Baroque stylistic decoration—columns, arch coupons, domes, and triangular windows—and Chinese stylistic elements, including a calligraphy plaque to enrich the image of the facade. The combination of the two cultures is harmonious. It is not too extravagant but embellished the status of the host (Figure 10).

Fig10. The front door and decoration of the old mansion.

Compared to the old mansion, the elegant western style of the new mansion is more obvious. The manor gate has a gothic style, evoking height, slenderness, sharpness and a cold quality. The gate appears perfectly integrated with the blue sky, in the way of a western castle or church. This stylistic choice also reflects the fusion of feudal customs and western religious thought. Furthermore, the formation of the architectural form and style is also a result of the fusion of technology from the cultural background, itself (Figure 11).

Fig11. Gate modeling showing the combination of Chinese and western stylistic elements.

Currently, the architectural style of the manor is not only a symbol of status, but its environmental aesthetic value also enhances the quality of life of the rural society that surrounds it and enriches the image of the rural environment. The combination of Chinese and western elements in Liu's manor was not exaggerated or deformed but designed to respect the difference in culture between China and the west. In some sense, the design evokes a new classical style by drawing from these original classical styles. By incorporating a limited number of western elements within the greater skeleton of Chinese architecture, the contrast of cultural symbols reflects the residents' ethics in the order of the development of regional space and cultural origin, that is all innovation and integration beyond the region should be carried out around the local original environment, so as to ensures the consistency of the macroscopic spirit of the rural human settlement environment in the West Sichuan Plain.

5 Conclusion

Residences can be understood as encompassing a microcosm of the residential environment within a macrocosm of regional pattern orientation. As the largest landlord manor in western Sichuan, Liu's Manor's special economic and cultural background endows it with a high level of environmental selection goal.

Liu's Manor originates from the living environment of rural people in the western Sichuan plain. It gathers in an ideal environment, dispersing to the origin, and the integration of classical styles. The overall settlement form, spatial atmosphere, and architectural style are closely linked to the macrocosmic pattern. This mode of construction preserves the collective memory of the western Sichuan, optimizes and extends the environmental

https://doi.org/10.1051/e3sconf/202123704012

texture, location, layout, and cultural elements. Furthermore, the construction forms a personalized integration with the similarity and continuity of the original environment, refining the core concept of the regional environment construction and contributing to the goal of healthy and sustainable development. The analysis of its microcosmic environment, the order formed between its environmental characteristics, and the macrocosmic natural pattern of the Western Sichuan Plain reveals that the owner's selection and integration of regional elements used in Liu's Manor reflect optimal ethical choices. The manor, in terms of settlement environment, family, society, and cultural integration, displays a sense of ideology and virtuosity superior to the general form of living environments in the Western Sichuan Plain.

With increasing attention paid to the construction of the rural residential environment, implementation of the rural revitalization plan needs not only to adapt to local conditions but also to draw on experience to the fullest possibility. In the upsurge of rural construction, Chengdu has listed the restoration of Linpan as a key project. As a typical demonstration of a rural living environment in western Sichuan, the Liu's Manor embodies the origin and integration of the region, setting a standard for the planning and construction of rural settlements. Therefore, the interpretation of the environmental ethical features conveyed by historical architectural classics, like Liu's Manor, explores the harmony and mutually beneficial integration of individuality and commonness, and microcosmic and macrocosmic, so as to provide a reference for the sustainable development of regional human settlements.

The continuously evolving human settlement environment has both a past and a present, the rural human settlement environment needs to establish a reasonable developmental order. As the goal of environmental construction, quantitative changes caused by man's innovation in the living environment should not be separated from the actual situation, leading to the absence of collective memory and disturbing the order of development. The purpose of the present research, on the environmental ethics of historic human settlements, aims to inject collective memory and spirit into the fast-paced trend of thought in construction. Rural construction is rapidly developing, so we should make appropriate use of the history of building in local conditions, develop village settlements that accord with the ethics of the regional human settlement environment, and maintain it. This is essential because this approach to understanding human settlements is helpful to the continuous and stable development of the overall regional environmental order in the time dimension, and can provide a reference for the construction of the rural human settlements environment system, holistically, in the future and contribute to the rejuvenation of the nation.

References

1. X. Wang, X. Hou, Evolution and Modality of Fortified Lord-Manor Settlements[J]. Journal of Qingdao University of Technology, 31(03), 6-11 (2010).
2. Y.F. Chen, Ethics of human settlements and greenness of building materials[J]. Jiangxi building materials, 14, 35+39(2010).
3. S.Y. Wu, Application of Environmental Ethics in Tourism Planning[J]. Northern literature (second half month), 02, 228 (2012).
4. W.M. Li, X.Y. Ye, W.F. Qi, A Review of the Development of Human Settlements Theory and practice at Home and abroad[J]. Journal of Zhejiang University (Science Edition), 02,205-211(2000).
5. Y.X. Fan, A review of the present situation of environmental and ecological ethics in China[J]. Environmental Science & Technology, 09,108-111 (2007).
6. P.C. Li, J.W. Zhou, The theoretical Origin of Environmental Ethics in the West[J]. Journal of Social Science of Hunan Normal University, 06,22-29+37 (2002).
7. Y. Feng, Research on modern rural construction model based on traditional environmental ethics[D]. Kunming University of Technology, (2012).

Research on Design of the Landscape inside Mountain Torrent Basin Based on Ecological Restoration – Take the Research Base in Wenling City, Zhejiang Province as an Example

Zhihao Zhang[1], Xidong Ma[1,*]

[1]Zhejiang Sci-Tech University, Xiasha Higher Education Zone, 310018 Hangzhou City, Zhejiang province, China

Abstract. In order to reduce the ecological damage and economic loss caused by mountain torrent disasters in the flooded basin, it's very important to realize the self-repair of the ecosystem in a short time. Based on the analysis of the causes of mountain torrents, this paper summarizes three main parts of ecological restoration in the river basin, including rainwater storage, Sedimentation and wetland conservation. Taking a research base in Wenling, Zhejiang Province as an example, this paper also discusses the design methods of the landscape inside mountain torrent basins based on ecological restoration from these parts above. Finally, these methods were obtained and applied in design practice, providing more solutions and theoretical references for the ecological restoration design in the future.

1 Introduction

In recent years, the sudden and heavy rainfall has caused many natural disasters in the flood season, such as mountain torrents, debris flows and landslides. And it has caused serious ecological damage as well as economic losses to the river basin, which arises widespread concern from all walks of life[1]. The formation and eruption of these mountain torrent disasters are not only affected by natural factors, like heavy rainfall and geological characteristics, but also caused by excessive intervention of human factors such as construction activities[2].

Through the analysis of the causes of mountain torrents, this paper summarizes the three main parts of ecological restoration in the basin, including rainwater storage, sedimentation and wetland conservation. And taking a research base in Wenling, Zhejiang Province as an example, this paper explores the design methods of the landscape inside mountain torrent basin based on ecological restoration. Finally, three theoretical design models, named Water Tree System, Zigzag Ditch and Habitat Archipelago, were obtained and applied in design practice, providing more solutions and theoretical references for the ecological restoration design in the future.

2 Causes of mountain torrents

Mountain torrents refer to explosion floods occurring in mountain creeks. The formation of mountain torrent disasters in various places mainly depends on the topographical and geological conditions of the mountain[3]. However, it is generally believed that natural factors and human factors are the two main causes of flash floods. The former include heavy rainfall and underlying surface conditions or so, and the latter mainly includes human unreasonable economic activities.

The heavy rainfall is the most active initial driving force for mountain torrents. What's more, the long-term rainfall will destroy the stability of rock formations and the soil on the surface of the mountain. The conditions of the underlying surface of the mountain include topography, geology, vegetation and other elements. Among them, the terrain conditions with steep slopes and large differences of elevation have the effect of increasing the potential energy of runoff and increasing the destructive power of mountain torrents. And for mountains with easily weathered bedrock, dissociation creates favorable conditions for collapse and debris flow. Besides, the sparse vegetation and the exposed rocks, which increase soil erosion, are also important factors leading to flash floods[4].

Long years ago, the disorderly and unreasonable development activities of the natural resources of the mountain by human beings caused the mountain to be devastated. A large number of naturally formed mountain ponds have undergone arbitrary development by humans. And their area has been drastically reduced, with the capacity of rainwater storage weakened severely. In the event of extreme rainfall, the runoff on the mountain surface will increase rapidly in a short time, which will pose a huge risk of flash floods to residential areas and farms around the mountain.

In 2017, the "Guiding Opinions on Strengthening Ecological Restoration Urban Rehabilitation" issued by the Ministry of Housing and Urban-Rural Development put forward the goal of ecological restoration: "Ecological restoration needs to implement the concept of sponge city construction, strengthen the protection of the natural form

* Corresponding author: mxd2813@163.com

https://doi.org/10.1051/e3sconf/202123704013

of the urban water system, avoid blind transformation and over development, control river pollution, restore and maintain the natural connectivity and fluidity of the river and lake water system." [5, 6]It can be seen that ecological restoration aims to transform canalized rivers according to local conditions, restore natural shorelines as well as waterfront vegetation communities, and enhance the self-purification capacity of water bodies.

For the ecological restoration of mountain torrents, it's vital to consider both natural and human factors. On the basis of satisfying human intervention, it's not only necessary to adopt various measures, like planting trees, to achieve rainwater conservation and mountain restoration, but also to carry out ecological conservation of the underlying surface to enhance the water purification capacity. Therefore, the ecological restoration design can be divided into three parts: rainwater storage on the mountain surface, sediment deposition and wetland conservation on the underlying surface, from the causes of mountain torrents. Then this paper takes a research base in Wenling to discuss the design methods of the landscape and restoration methods of mountain torrent basins based on ecological restoration.

3 Project background

Wenling City is located in East China, on the coast of southern Zhejiang Province. Affected by the regulation of the ocean water and the blocking of the cold currents by the northwest mountains, it has abundant rainfall and a mild climate. However, due to the East China Sea, a total of 43 typhoons have landed on Wenling City in the past 50 years, with an average annual rainfall of 1632 mm, of which the rainfall during the typhoon period is about 840 mm, accounting for 54.8% of the year[7]. And the heavy rainfall brought by typhoons has resulted in urban flooding and flash floods for many times, causing heavy economic losses and immeasurable damage to the ecosystem.

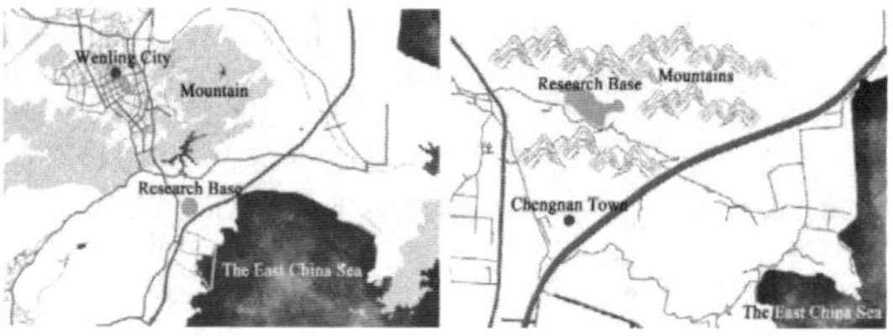

Fig1. The location of the research base.

As shown in Fig.1, the research base is located in a suburban scenic area of Chengnan Town, with only 4 km away from the East China Sea. And it's a typical mountain and hilly terrain, surrounded by mountains on three sides, and there is a creek passing through it. Because of the construction area built along, the mountain is the most important ecological element. The total land area is about 60750 m^2.

For the last few years, the rapid development and construction process in the southern part of the city has neglected the natural elements, such as lakes and ponds. Then the blind development has severely damaged the original mountain drainage system and weakened its ability to regulate and store rainwater. At present, the flood interception design of the mountain is not yet sound. Therefore, when faced with sudden heavy rainfall, a large amount of rainwater would collect in the valley, scouring the shallow soil, and turn into flash floods finally. And the valley floor, through which the mountain torrents flowed, degenerates into a wasteland covered with sand and gravel slowly.

To sum up, in order to protect the rural scenery and biodiversity within the research base, it's very important to carry out the restoration design of the pond, mountain and the underlying surface of the valley to alleviate the risk of mountain torrents in the basin, with the regional water problem as the goal-oriented.

4 Design method

4.1 "Water Tree" system: precipitation storage in layers

On the treatment method of mountain rainwater storage, firstly, the team identified and stated the depressions in the mountains surrounding the base. These depressions include natural ponds, lakes and creeks, which are potential floodwater points. In order to achieve effective remediation and ensure the restoration, it is necessary to analyze the rainwater runoff flow direction and the catchment of the surrounding mountains in the early stage before fill and excavate according to local conditions. As shown in Fig.2, it's the current analysis of the depressions in the surrounding mountains of the base.

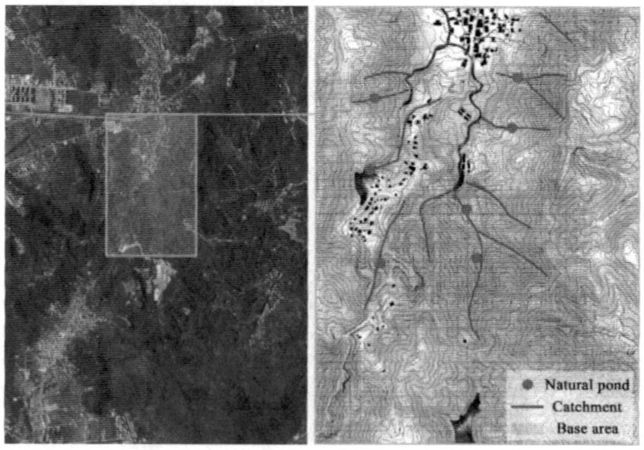

Fig2. The current analysis of the depressions in the surrounding mountains of the base.

For the protection and restoration of mountain ponds, the purpose is to improve the capacity of the mountain to regulate and store rainwater, so as to reduce the peak flow of mountain torrents and ensure the surrounding safety. And the specific protection and restoration strategy is shown in Fig.3. Firstly, the depressions on the water catchment line were repaired, combined with the previous analysis of rainwater runoff. And the main method is to expand existing mountain ponds appropriately and add small artificial ponds locally, which forms the "leaves" on the "water tree" system. Secondly, to enhance the "storage and detention" capacity of mountain ponds, walls are rammed on the periphery of mountain ponds to increase

 https://doi.org/10.1051/e3sconf/202123704013

the storage capacity, intercept floods, and delay flood peaks. At last, ecological flood interception ditches should be constructed along the Catchment, connecting mountain ponds with adjacent depressions on the same catchment, to increase the storage space for mountain torrents. The connecting ditches between mountain ponds are like the "branches" of the "water tree" system.

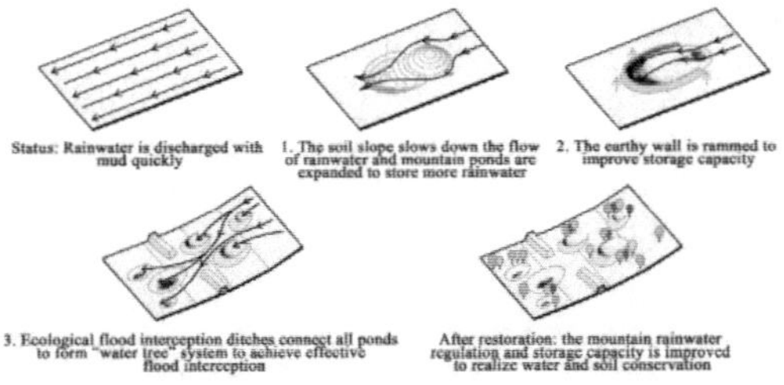

Fig3. The protection and restoration strategy of mountain ponds.

The function of the "Water Tree" system is to store rainwater in layers. When encountering extreme rainfall, rainwater will be stored in mountain ponds from top to bottom, which can intercept floods effectively and restrain soil erosion. Finally, only the excess rain will reach the valley and enter the next step of purification.

4.2 Zigzag ditch: sedimentation and utilization

Rainwater flows through the mountain with gravel and other impurities in it, which is the main cause of soil loss in the mountain. In this design, the valley floor is used as the construction land for the base. Therefore, how to combine the architectural design with the ecological restoration of the underlying surface is the key. The team used the rainwater filtration system in industrial production as a prototype and found that the precipitation effect of Sedimentation is determined by the flow rate of water and the residence time of the water in the sedimentation tank. And the zigzag streamline can provide a long streamline with a stable flow rate in a limited area.

As shown in Fig.4, the bottom floor of the building is used as a trench for rainwater precipitation, and unique architectural space is formed at the same time. Then the rainwater with impurities slowly flows into zigzag ditches at the bottom of the building for long-term sedimentation. Eventually, the sediment collects at the corners and becomes fertilizer for plant growth. Meanwhile, the clear rainwater flows out from the end of the building and spreads around the building. And plank roads are installed on the ditches to create an architectural space that blends with nature.

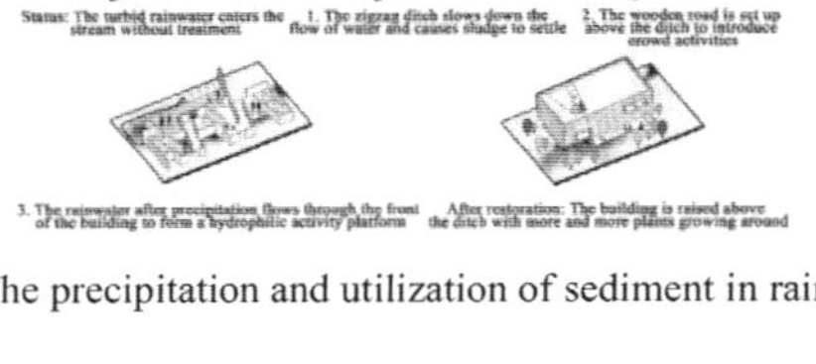

Fig4. The precipitation and utilization of sediment in rainwater.

4.3 Habitat archipelago: wasteland landscape restoration

The underlying surface of the valley floor is abandoned farmland. And due to a lot of gravel left by mountain torrents, there is almost no vegetation here. The team hopes to introduce the rainwater from the surrounding mountains into the base through manual control, which can replenish soil moisture, improve the local microclimate and enrich the habitat types of the research base. Exactly, the water in the base comes from rainwater from the mountains, which is a natural, ecological and reasonable way of using water resources.

On the basis of the original wild vegetation, the soil and stones in the wasteland were trimmed to form habitat islands, as shown in Fig.5. Having flowed through zigzag ditches of the building, the rainwater is introduced into the wasteland to form a seasonal wetland habitat. At the same time, on the basis of retaining better growing trees, native trees and shrubs with fruits will be planted to form a rich forest habitat, which is hoped to create habitats for birds as well as amphibians and also provide natural teaching places for research bases. The planting of native herbaceous plants, such as reeds and lotus, has a positive effect on the conservation of wetland water.

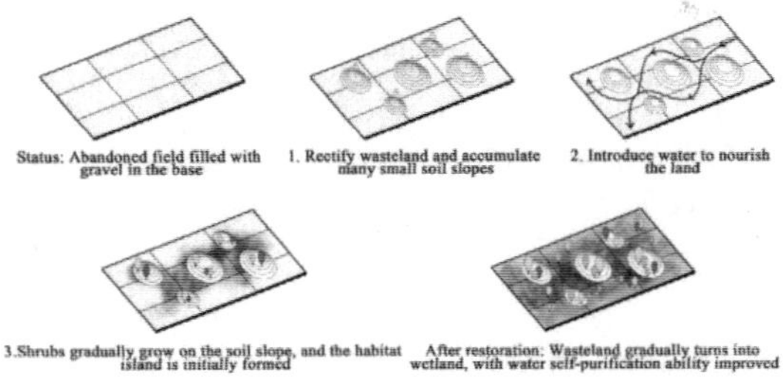

Fig5. The restoration of the wasteland at the bottom of the valley.

The establishment of the Habitat Archipelago not only reconstructs the self-renewal ability of nature, completes rainwater purification and soil conservation, but also shows the subtle and mysterious succession process of the natural ecosystem, becoming a place for tourists to experience the evolution of the natural and cultural landscape.

5 Balanced linkage between ecological landscape and human activities

The human activities that trigger mountain torrents are mainly the blind development of mountain resources, because people often sacrifice precious natural resources to obtain short-term economic benefits. As the economic artery of Chengnan town, the research base is located in the mountains, but intervening in nature in a reasonable and gentle way. The building and the mountain are fused, which means that the building becomes a container for the ecological restoration and the ecological settlements at the mountain become a part of the research base, to achieve the ideal state of harmonious coexistence.

After the base is completed, it will establish a cooperative relationship with surrounding schools and become a second classroom with "natural ecological science" as the main content. And as shown in Fig.6, combined with the three parts of ecological restoration, a variety of activities are implanted in the mountain. Around the mountain pond, multiple rest stations and sightseeing platforms can be set up to form nodes in the hiking and hiking route. What's more, there will be lots of workshops and exhibition halls on this base. People can make specimens by themselves and carry out simulation experiments about mountain torrents in the workshop, while the exhibition hall focuses on exhibiting specimens and introducing the formation of mountain torrents. In addition to indoor learning areas, the wetlands around the building community are also classrooms for people to study and get close to nature.

Fig6. The ecological landscape and human activities in the research base.

The research base will provide tourists and scholars with a platform to understand nature and complete courses on ecology and landscape, spreading and expanding the concept of sustainability among the younger generation. At the same time, the base steadily achieves profitability with the support of tourists, thereby achieving a healthy balance of the ecological environment, the social culture and economic benefits.

6 Conclusion

In recent years, the non-ecological development of mountains by humans has caused serious damage to the capacity of mountains to regulate the storage and intercept floods. And the triggered flash floods also caused immeasurable losses to residential areas around the basin. This paper analyzes the causes of mountain torrents at first and summarizes the three main parts of mountain ecological restoration. Furthermore, combined with a design case of a research base in Wenling City, Zhejiang Province, the methods of the landscape inside the mountain torrent basin were discussed based on the concept of ecological restoration.

In conclusion, before carrying out ecological restoration, it's necessary to analyze the restrictive construction elements of the mountain and grasp the natural conditions, such as the direction of rainwater and the current situation of mountain ponds. Secondly, stabilizing the water and soil through planting is a vital and effective measure, and it's also essential to transforming the existing mountain ponds to meet the requirements of rainwater storage. On the underlying surface of the valley bottom, the ecological flood interception channel and zigzag sedimentation ditch are connected in series to delay and weaken the flood peak during extreme rainfall, so as to reduce the impact of mountain torrents on the surrounding area. Finally, through improving the diversity of vegetation communities at the bottom of the valley, multiple habitat islands are set up to create a rich wetland habitat and achieve self-purification of water bodies.

Acknowledgement

This study was supported by the Student Science and Technology Foundations of Zhejiang Province (Code:2020R406029).

References

1. J. Chen, Current situation, problems and countermeasures of mountain torrent disaster prevention and control (China Water Resources Press, Beijing, 2016)
2. Z.T. Zhang, Mountain torrent disaster prevention measures and effectiveness (Water Resources and Hydropower Press, Beijing, 2016)
3. Z.L. Chen, Research on the causes and prevention technology of mountain torrent disasters in small watersheds (South China University of Technology, Guangzhou, 2014)
4. Q.L. Li, C.H. Huang, Heilongjiang Water Conservancy Science and Technology 28, 1 (2016)
5. X.X. Ren, W.Z. Tang, Sponge city runoff control rate and other indicators application (China Water Resources Press, Beijing, 2015)
6. X.H. Wang, Y.C. Zhang, P. Zhang, Water Resources Protection 12, 1 (2016)
7. Q. Wang, Research on the planning and design of urban green space in the Wenhuang Plain based on the concept of disaster reduction (Huazhong Agricultural University, Wuhan, 2017)

Research on the Coordination of Protection Coupling of Ecological Environment and Tourism Development

Li Yang[1,*],Pengcheng Su[2]

[1]Guilin University of Technology, Tourism Management, Guilin 541006, China
[2]Xi'An Jiaotong University, Business Administration, Xi'an 710048, China

Abstract. While tourism has become a strategic pillar industry, adding vitality to regional development, ecological and environmental problems have become increasingly prominent. Based on the evaluation index system of ecological environment and tourism development, the paper takes Guangxi Zhuang Autonomous Region as an example and uses entropy weighting method to reduce the dimension of the original data. Then, the coupled coordination model is used to compare and analyse the comprehensive development index of the two systems from 2009 to 2018, as well as the development level of the coupling coordination degree between the systems. The results are as follows: (1) Ecological environment and tourism development are interrelated and mutually restricted, with positive and negative feedback. (2) The comprehensive development level of ecological environment system and tourism development system is generally on the rise. The coupling degree fluctuates between the two systems, but the coordination degree and the comprehensive development index show a trend of steady growth on the whole, which accords with the characteristics of high coupling and not high coordination. (3) Finally, according to the development of the coupled coordination model, some suggestions on future ecological environment and tourism development are put forward.

1 Introduction

The ecological environment is composed of land resources, water resources, biological resources, climate resources and other elements, and is a composite ecosystem related to the sustainable development of society and economy. A benign ecological environment is the cornerstone to support the healthy development of tourism and an important part of increasing tourism attractiveness. The development of tourism also provides a stable financial guarantee for the protection of the ecological environment. However, with tourism in full swing, it will inevitably cause damage to the ecological environment, and the environmental pollution and waste of resources in turn restrict the development of tourism. The two systems influence and restrict each other. As a famous karst topography in Guangxi, the contradiction between ecological environment protection and tourism development is particularly prominent because of the destructive nature of karst topography. Therefore, the tourism industry must be developed within the scope of the ecological environment, so as to realize the benign interaction and coordinated development of the tourism industry and the ecological environment.

2 Domestic development

The relationship between tourism development and ecological environment has always been a research hot topic of scholars at home and abroad. Domestic scholars mainly use a combination of quantitative and qualitative methods, and study the relationship between the two systems through the gray GM model, PSR model, and coupling coordination model. Wang(2009) explores the relationship between ecological environment and tourism development from the perspective of coupled tourism, and proposes a sustainable development feedback mechanism for Kanas scenic spot[1]. Chen(2019) studies the mechanism of ecotourism in poverty alleviation from the perspective of ecological anthropology[2]. Although there are a lot of studies on the relationship between ecological environment and tourism development, there are few studies from the perspective of ethnic minorities in Guangxi. Therefore, the study of the relationship and development level between the ecological environment and tourism development in minority areas also provides suggestions for ecological protection and low-carbon development in the area.

3 Data collection and analysis

3.1 Data sources and processing

Constructing an evaluation index system is an important foundation for studying the level of coupling and coordination between the ecological environment and tourism development in Guangxi Zhuang Autonomous Region. The ecological environment and tourism

* Corresponding author: 1033323577@qq.com

development are two relatively complex systems, so the article selects the relevant indicators of the two systems based on the PRS model, the research results of relevant scholars and the existing mature research results[3-4]. The ecological environmental system index consists of two dimensions: the current situation of environmental pollution and the results of environmental governance. It covers 11 specific indicators and mainly reflects the environmental quality, pollution and governance of a region. The index of tourism development system is composed of two dimensions: tourism economy and tourism service, including eight specific indicators that mainly reflect the development advantage, scale and level of tourism. In order to ensure the authenticity, completeness and scientificity of the data, the data required by the two systems in the past ten years are selected from the 2009-2018 China Statistical Yearbook and the Guangxi Statistical Yearbook as samples. The coupling coordination model is used to further analyze the relationship between Guangxi's ecological environment and tourism industry.

In order to eliminate the difference in magnitude and dimension of the original data of the indicator, firstly, it is necessary to use the range method to perform dimensionless processing on each indicator data into categories, and then calculate the weight of the indicator according to the entropy weighting method and the analytic hierarchy process.

$$X_{ij}^{'} = \frac{X_{ij} - \min(X_{ij})}{\max(X_{ij}) - \min(X_{ij})} + 0.001 \quad (1)$$

$$X_{ij}^{'} = \frac{\max(X_{ij}) - X_{ij}}{\max(X_{ij}) - \min(X_{ij})} + 0.001 \quad (2)$$

$$H_j = -k(\sum_{i=1}^{m} P_{ij} \ln P_{ij}), k = \frac{1}{\ln m} \quad (3)$$

$$P_{ij} = (1 + X_{ij}^{'}) / \left(\sum_{i=z}^{m} (1 + X_{ij}^{'}) \right) \quad (4)$$

$$W_j = (1 - H_j) / (\sum_{j=1}^{n} (1 - H_j)) \quad (5)$$

Among these formulas, X_{ij} represents standardized data; P_{ij} represents the proportion of the numerical value; W_j represents the statistical weight of the jth index.

Table1. Evaluation Index System

System	Element	Index	Nature	Code	Statistical weight
Ecological environment system	The current situation of environmental pollution (0.2605)	Volume of industrial solid waste product/10,000 tons	-	X_1	0.0570
		Volume of industrial waste water discharged/10,000 tons	-	X_2	0.0501
		Volume of industrial waste gas discharged/100 million cu.m	-	X_3	0.0478
		Volume of sulfur dioxide discharged/10,000 tons	-	X_4	0.0512
		Total energy consumption/10,000 tce	-	X_5	0.0557
	The results of environmental governance (0.2921)	Public green space of parks per population/m^2	+	X_6	0.0540
		Coverage area of forestation of developed area/%	+	X_7	0.0489
		Rate of garbage no harmful disposal/%	+	X_8	0.0443
		Treatment rate of polluted water/%	+	X_9	0.0452
		Volume of comprehensive utilization of industrial solid waste/10,000 tons	+	X_{10}	0.0463
		Energy saving and environmental protection capital investment / 100 million yuan	+	X_{11}	0.0553
Tourism development system	Tourism economy (0.2367)	Number of inbound tourists/10,000 people	+	X_{12}	0.0531
		International tourism/foreign exchange income/ten-thousand dollars	+	X_{13}	0.0552
		Number of domestic tourists/10,000 person-times	+	X_{14}	0.0616
		Domestic tourism revenue/100 million yuan	+	X_{15}	0.0624
	Tourism service (0.2107)	Total number of travel agencies	+	X_{16}	0.0555
		Total number of star-rated hotels	+	X_{17}	0.0487
		Business revenue of enterprises above designated size of catering services/10,000 yuan	+	X_{18}	0.0546
		Business revenue of enterprises above designated size of hotels/10,000 yuan	+	X_{19}	0.0532

3.2 Research Methods

$$U_{i=1,2} = \sum_{j=1}^{n} W_j X_{ij} \quad (6)$$

$$\lambda = U_1 / U_2 \quad (7)$$

$$C = \sqrt{\frac{U_1 * U_2}{((U_1 + U_2)/2)^2}} \quad (8)$$

$$T = \alpha U_1 + \beta U_2 \quad (9)$$

$$D = \sqrt{C * T} \quad (10)$$

C is the coupling degree, the value range is (0,1), the closer

https://doi.org/10.1051/e3sconf/202123704014

the C value is to 1, the better the coupling degree of the two systems. T is a comprehensive evaluation index; α, β are both undetermined coefficient, considering that tourism development and ecological environment are mutually complementary and equally important, and α+β=1, so both take the value 0.5.

3.3 Data analysis

3.3.1 Overall analysis

According to the correlation model of coupling degree, the comprehensive evaluation indexes U_1 and U_2 of ecological environment and tourism development of Guangxi Zhuang Autonomous Region from 2009 to 2018, as well as the coupling degree C, system comprehensive evaluation index T and coupling coordination degree D between the two systems are obtained respectively. The coupling coordination level of each year is divided according to the "tenths method" proposed by Liao[5].

Table2. Comprehensive development index and coupling coordination

Year	U_1	U_2	C	T	D	Grade
2009	0.1628	0.0385	0.7863	0.1006	0.2813	mild disorder
2010	0.1334	0.1111	0.9958	0.1222	0.3489	moderate disorder
2011	0.1008	0.2239	0.9254	0.1623	0.3876	moderate disorder
2012	0.0851	0.3533	0.7911	0.2192	0.4164	near disorder
2013	0.0908	0.3856	0.7855	0.2382	0.4325	near disorder
2014	0.1131	0.4205	0.8174	0.2668	0.4670	near disorder
2015	0.1466	0.4810	0.8463	0.3138	0.5153	barely disorder
2016	0.1941	0.5258	0.8875	0.3600	0.5652	barely disorder
2017	0.1861	0.5961	0.8516	0.3911	0.5771	barely coordination
2018	0.1506	0.6835	0.7693	0.4171	0.5664	barely coordination

According to Table 2 and Figure 1, it can be found that the development trend of the two systems can be roughly divided into two stages. The first stage: ecological environment composite index from 2009 to 2010 is higher than the tourism industry. This shows that at this stage, Guangxi Zhuang Autonomous Region has a stable and good ecological environment and a thriving tourism industry, which can fully rely on a good natural ecological environment and strong tourist attraction to win the favor of a large number of tourists. The high ecological environment index is because Guangxi's tourism industry mainly relies on the region's rich tourism resources and cultural ecological environment, which is an important basis for realizing tourism as a strategic pillar industry. The relatively sluggish tourism industry was mainly affected by the international financial crisis in 2008 and the unexpected events. In this period, the government and the society are more inclined to protect Guangxi's green wealth than to pursue the economic benefits brought by tourism.

In the second stage, the tourism industry composite index from 2011 to 2018 is significantly higher than the ecological environment composite index, and there is a gap between the comprehensive evaluation values of the two systems. The tourism development index has increased rapidly from 0.2239 to 0.6835. With the support of local governments' policies to promote tourism development, the tourism industry has witnessed a spurt of development. The ecological environment composite index drops from 0.1008 in 2011 to 0.0908 in 2013, which indicates that the development of tourism in the past three years has caused damage to the ecological environment and brought about negative effects. The ecological environment index in 2014 begins to rise again, but in the following two years it begins to decline again. Generally speaking, the comprehensive evaluation index of ecological environment is always less than 0.5, and the overall ecological environment quality is not high, so its protection needs to be strengthened. Moreover, in the second stage of development, it can be found that the growth rate of the tourism industry is significantly higher than that of the ecological environment. Therefore, in the following regional development, attention should be paid to the actual situation and existing problems of the ecological environment, as well as the restoration and protection of ecological resources.

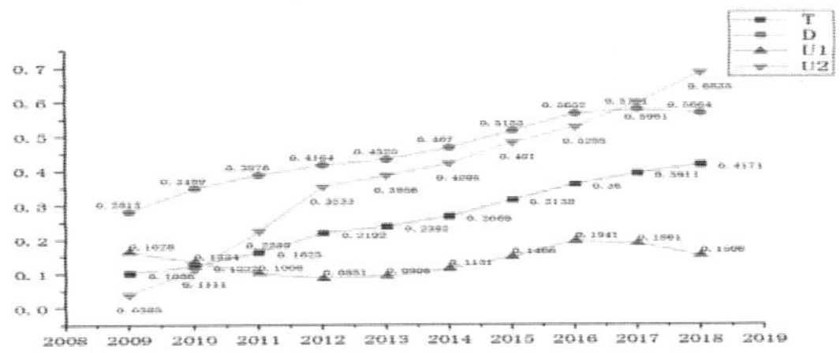

Fig1. Index Development Trend Chart

Table3. Comprehensive development index and coupling coordination

Year	λ	Type characteristics
2009	4.2286	Ecological environment is ahead of tourism development, the two are at low level coupling stage
2010	1.2007	Ecological environment is ahead of tourism development, the two are highly antagonistic.
2011	0.4502	Tourism development is ahead of ecological environment, the two are highly antagonistic.
2012	0.2409	Tourism development is ahead of ecological environment, and the two are at a low level.
2013	0.2355	Tourism development is ahead of ecological environment, and the two are at a low level.
2014	0.2690	Tourism development is ahead of ecological environment, and the two are at a low level.

2015	0.3048	Tourism development is ahead of ecological environment, and the two are in the run-in stage.
2016	0.3692	Tourism development is ahead of ecological environment, and the two are in the run-in stage.
2017	0.3122	Tourism development is ahead of ecological environment, and the two are in the run-in stage.
2018	0.2203	Tourism development is ahead of ecological environment, and the two are in the run-in stage.

According to Table 3, the ecological environment development system in Guangxi from 2009 to 2010 is ahead of the tourism industry system, and the two systems develop from low-level coupling stage to run-in stage. From 2011 to 2018, the tourism industry development system is ahead of the ecological environment system. The two systems transition from the antagonistic stage to the run-in stage, and there is still a great space for improvement from the coordinated development level. This also shows that the ecological environment of Guangxi cannot support the development of tourism industry well.

4 Conclusion and suggestions

4.1 Conclusion

In this paper, the coupled coordination degree model is used to analyze the evolution of the ecological environment and the coordination level of tourism development in Guangxi Zhuang Autonomous Region, so as to analyze the coordination relationship between the two systems.

(1) From the perspective of the coupling level, the two systems of ecological environment and tourism development are closely linked and have strong mutual coupling. The ecological environment lays a rich resource and material foundation for the development of tourism. While the tourism industry takes advantage of the development of ecological resources, it will also profoundly affect the ecological environment. The ecological environment and tourism development in Guangxi Zhuang Autonomous Region are on an upward trend. The comprehensive environment of the ecological environment is relatively good and the development level is relatively high. With the continuous development of the local unique ecological environment advantages, the tourism industry is developing rapidly.

(2) From the perspective of the coordination stage, the coupling development stage of the ecological environment and tourism development in Guangxi from 2009 to 2018 has improved from the initial moderate disorder to primary coordination. And the development type passes through antagonistic stage to running-in stage. However, the development of the two systems accords with the characteristics of high coupling and low coordination. Although the coupling degree is stable and high, the overall coupling coordination degree is low, failing to reach the ideal state, and there is still a large space for high-quality coordinated development.

4.2 Suggestions

4.2.1 Improving the carrying capacity of the ecological environment

The healthy and sustainable development of tourism depends on a good regional ecological environment, which is diverse and dynamic. Therefore, it is necessary to increase the carrying capacity within a reasonable range according to the specific conditions of different cities in Guangxi Zhuang Autonomous Region. Taking sustainable development and the "Two Mountains Theory" as the guiding concept, it adopts differentiated and long-term solutions, adheres to the coordinated development of the ecological environment and tourism, and forms a complex ecosystem of economic and natural ecological virtuous cycles.

4.2.2 Improve the quality of coupling coordination

When developing the tourism industry in Guangxi, it should not only be based on the favorable ecological advantages in the region, accelerate the green transformation and development of the tourism industry, but also pay attention to the steady advancement of the development of the ecological environment. The region must adhere to the coordinated development of the ecological environment and tourism.

References

1. Wang H, Yang Z P, Hang F. Research on Kanasi Lake's "Tourism-environment" Coupling Pattern Under Water Environment Stress[J]. Journal of Huazhong Normal University,43(2009):345-349.
2. Chen H P. Study on the Eco-tourism Poverty Alleviation in Minority Areas from the Perspective of Ecological Anthropology: Taking Zhongdong Miao Village in Guizhou Province as an Example[J]. Guizhou Ethnic Studies, 40(2019):139-145.
3. Gao N, Ma Y F, Li T S. Study on the Coordinative Development between Tourism Industry and Urbanization Based on Coupling Model: A Case Study of Xi'an[J]. Tourism Tribune, (2013):62-68.
4. Fang Y L, Huang Z F, Duan Z X. Coupling and Coordinating about Chinese Tourism Developing and Eco-Environment[J]. Economic Geography, 33(2013):195-201.
5. Liao C B. Quantitative Judgement and Classification System for Coordinated Development of Environment and Economy--A Case Study of the City Group in the Pearl River Delta[J]. Tropical Geography,3(1996):3-5.

The Application and Analysis of Opera Costume Elements in the Modern Costume Design

Zhang Jinbin

International Design Art College, Inner Mongolia Normal University.
Hohhot, Inner Mongolia, 010022, China

Abstract: This paper analyzes the application of opera costume elements in modern costume design, combines with the application of ideas in the paper, and applies opera costume elements to modern costume. Under this concept, four kinds of modern ready-to-wear clothes with opera costume elements are designed. Taking the works as examples, the whole process of the application design of opera costume elements in modern costume design is presented, which provides ideas and directions for the modern application design of opera costume elements in the future. It is also hoped that the research findings of this paper can give reference to designers.

1. An Analysis of the Application Design Ideas of Traditional Opera Costume Elements in Modern Costume Design

With Chinese national attention to its traditional culture, people from all walks of life pay more attention to its protection and inheritance, and clothing industry is no exception. Opera is an important part of Chinese traditional culture. If cultural elements can be used into fashion design, works made by designers will receive unexpected effects, which not only have aesthetic characteristics, but Chinese traditional culture can be inherited in a new way.

1.1. The Extraction and Application of the Style Elements of Opera Costume

Traditional Chinese opera costumes are diversified in modelings with excellent styles, fabrics, colors and patterns. As a result, opera costumes have become one of the design materials favored by modern fashion designers. As shown in Figure 1, the designer designs the work inspired from the opera "*Peony Pavilion*" (A play written by Tang Xianzu, a dramatist of the Ming Dynasty) with the theme of "Dream in the Deserted Garden". In the dream, "Regardless of her parents' detention, Du Liniang (the heroine of the *Peony Pavilion*) bravely rushed out of her boudoir and went to the garden with Chunxiang (Du's servant girl) to enjoy the scenery of spring. Du Liniang longs for love and she dreams of meeting a Xiucai (Xiucai, one who passed the imperial examination at the county level in the Ming and Qing dynasties) in front of the Peony Pavilion." The story is the starting point for the design. The designer chooses the split fishtail dress style to highlight the identity and status of Du Liniang, and at the same time, emphasizes her unusual act of rushing out of her boudoir to visit the garden. The design elements are from the collar type, the decoration, the layout and position of the pattern in the opera costume, and then those elements are combined with the design of modern dress style to reflect Du Liniang's shyness when she first dreamed about Xiucai. As shown in Figure 2, the theme design of Peking Opera costumes, the designer extracts the typical visual elements of opera costume style in "Mu Guiying Command" (a Peking opera) which are shawl and tassels (from design work 1, and design work 2), integrates them into the modern clothing, and designs the length of the coat according to the "Mang Pao" (a costume of Peking Opera that embroidered Python pattern) (design work 2). The two sets of costume design works collide with each other, sending out a strong cultural atmosphere of Peking Opera and a sense of modern fashion, which makes the costume of Peking Opera show great value that can be explored.

 https://doi.org/10.1051/e3sconf/202123704015

Figure 1 opera costume design of "*Peony Pavilion*"

Peking Opera Costume (Design Material)	Work 1	Work 2

Figure 2 the theme design of Peking Opera costume

1.2. The Collection and Application of Opera Costume Colors

Color is an important element of clothing. For clothing, there is a saying that "Color from far look, style form near look". Clothing color has a feature of strong visual, and different colors give people different visual feelings, with different symbolic meanings. For instance, green makes people feel fresh and elegant, with a sense of peace, comfort and growth; red is bold, warm, but with a sense of impatience. Color matching plays an important role in modern clothing design.

Traditional Chinese opera costumes are colorful, and their color matching is different due to various types of plays, roles, plots and so on. For instance, the younger characters such as Xiao Dan (Dan is one of the main characters of Peking Opera, mainly referring to the image of women, Xiao Dan means young women) and Xiao Sheng (Xiao Sheng is one of the roles of traditional opera, which refers to young men), etc., their costumes of traditional Chinese opera mostly use bright colors to reflect the young people's vigorous and passionate. But under the different characteristics of young people, the choice of clothing color will be different. For instance, in the opera "Dream in the Deserted Garden", in order to show Du Liniang's talent and her official identity, her costume color is usually elegant and luxurious. [1]. Costume color in Chinese opera contains the wisdom and emotion of ancient people, and complements the hierarchical system in the plot, the psychological temperament of characters in the opera, the customs and habits of the birthplace of the opera. It is also the source of modern closing color matching by designers. Generally speaking, there are two ways for designers to design the modern costumes: one is to directly copy and use traditional color matching in opera costumes; the other is to refine the traditional color of opera costumes and reconstruct the area of color matching. As shown in Figure 3, this series of costume colors designed by the designer come from the Peking Opera -- "Mu Guiying Command" of costume colors. The designer did not copy the color relations, but changed the area ratio of traditional costume colors, without losing the traditional charm, and integrating designer's innovation. In the meantime, for designers, the choice of cold tones or warm tones, bright color or dark color, and issues affected by wearing time, occasions and other factors should be taken into account when doing color matching.

Figure 3 modern application of Peking Opera clothing color matching

1.3. Costume Elements of Opera in Costume Design

In modern fashion design, pattern decoration is the design method that many designers like to adopt. Through the use of embroidery or printing and dyeing of patterns, clothing style visually gives viewers a kind of aesthetic enjoyment.

With the theme of "Sceneries of Time", Shiatzy Chen's 2014 spring/summer womenswear collection deconstructed the facial features of Peking Opera and drew them on the clothes with relatively easy strokes. Only a few lines create a pattern image that conforms to the modern aesthetic, and shows Oriental elements (Figure 4). The design inspiration of MUKZIN 2017 autumn/winter collection went back to the little-known "Maoer Opera" (A troupe or play formerly composed entirely of young actresses) in traditional Peking Opera. In the name of "Maoer", traditional design awakens a new era of women's consciousness. While retaining the

 https://doi.org/10.1051/e3sconf/202123704015

traditional image of "Maoer Opera", the designer endows the innovation in modeling, reconstructs the modern clothing in the form of pattern, and perfectly deduces the new era aesthetic characteristics of opera clothing (Figure 5).

Figure 4 "Sceneries of Time", Shiatzy Chen's 2014 spring/summer womenswear collection

Figure 5 MUKZIN 2017 autumn/winter work

1.4. Costume Elements of Opera in Costume Accessories

Accessories are the vital part of costume design. Opera costume elements are often applied in costume design. The term "element" not only refers to style modeling elements, colors, fabrics, etc., but also refers to costume accessories that play a role in setting off opera costume itself, such as helmets, crowns, hats, shoes and so on.

In Valentin Yudashkin's 2013 autumn/winter RTW fashion show works, the costume hat became the inspiration for designers, and they used design methods like modeling simplification, material transformation, color change, etc. to the headdress of autumn and winter works, which not only had the charm of traditional opera costumes, but also kept the modern fashion sense. The combination of costume hats and modern hats has no conflict (Figure 6). In Figure 7, the designer extracts the headscarf of the "*Peony Pavilion*" costume, retains its modeling and fabric texture, and applies it to the design of modern dress rather than its original color relations and pattern, which is very personalized and also in line with the aesthetic taste of modern consumers. The design of these two series makes the value of opera costumes to be extended and deepened, which shows the reference value and practical significance of design promotion.

Figure 6 the Valentin Yudashkin's 2013 autumn/winter RTW fashion show

Figure 7

2. The Application Design Process of Costume Elements in Modern Costume Design

As a treasure of Chinese traditional culture, opera originated from primitive song and dance, which is a comprehensive stage art with a long history. After the Han, Tang, Song and Jin Dynasties, a relatively complete opera art was formed. It is composed of literature, music, dance, art, martial arts, acrobatics and performing arts, with more than 360 types. However, with the emergence of new concepts and ideas, the development of opera is seriously restricted due to the decrease of audience and the shrinkage of performance market. Therefore, it is urgent to solve the problem of sustainability of opera culture.

Nowadays, in the new situation of advocating national cultural self-confidence, Chinese culture not only should be perceived by people in the traditional way, but also appears in people's daily life in a more diversified form. Thus, modern clothing is the best medium for cultural communication. However, when designers design and apply the elements of opera costumes, what are the steps in the specific process? How to combine the opera costume elements with modern costume design? In view of these problems above and combining many years of teaching experience and practice, the author puts forward the "three steps" in the application process, including the determination of theme, the selection and analysis of inspiration materials, and the extraction and transformation of inspiration elements.

2.1. Determination of Theme

When applying opera costume elements in modern costume design, first of all, designers should determine the theme in the form of words, such as Kunqu Opera, Sichuan Opera, Tibetan opera, or something else. Then, collect the pictures according to the theme and establish the database.

2.2. The Selection and Analysis of the Inspiration Materials

After the collection of pictures and the establishment of database, the designer should screen the theme data, select what he/she has interested in as the theme inspiration material, and analyze it (by analyzing its history and culture, summarizing its characteristics and value, etc.) to reserve for the next step of design. As shown in Figure 8, the designer chooses Peking Opera as the theme. Under this theme, the designer collects relevant materials, then classifies, selects and analyzes them, and extracts the main elements of Peking Opera Costumes which are headdress,

 https://doi.org/10.1051/e3sconf/202123704015

armour flags (the four triangular flags, made of silk and embroidered, strapped over the back of a warrior outside of his coat of mail) and facial makeup (a type of painting on the opera actors' faces). Then, the designer chooses facial makeup of Peking Opera as the main element of the design, and displays it with pictures and texts, so that the viewers can understand it at a glance.

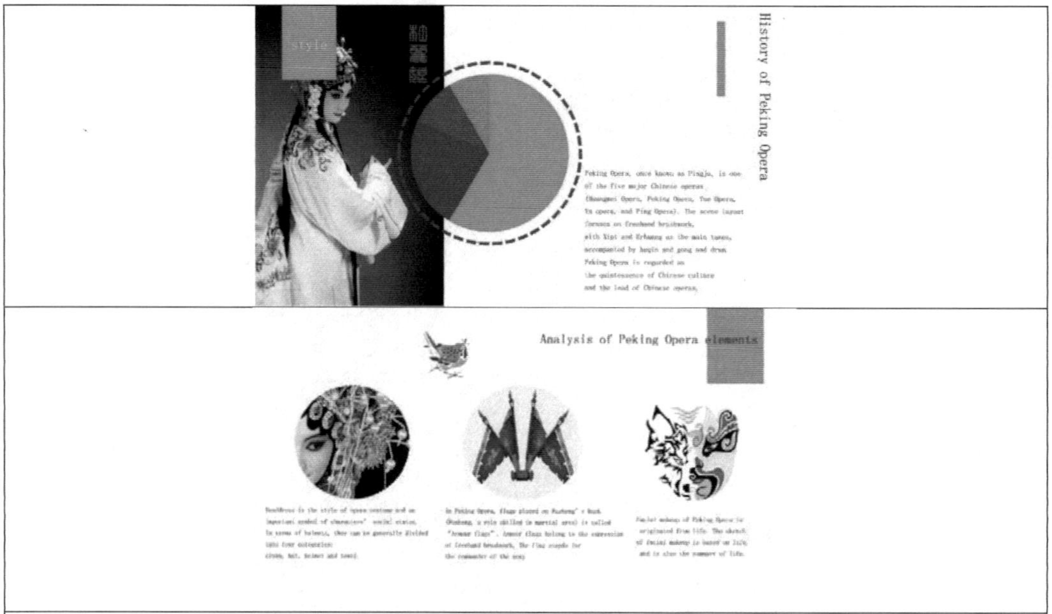

Figure 8 analysis of costume elements of Peking Opera

2.3. Elements Extraction and Application

After the steps mentioned above, next part will be the elements extraction and application. As shown in Figure 9, under the theme of Peking Opera, the designer extracts the elements of facial makeup (Peking Opera facial makeup, which goes beyond the concept of characters' makeup in opera and sets costume, is part of the costume of Peking Opera) and combines them with modern costume design. The designer adopts the method of patterning the Peking Opera facial makeup to simplify its original shape, remove the original color, and make it integrate with the fabric color. Then, design the layout according to the clothing style, to enlarge or reduce the facial makeup elements, which sets off the clothing style and has unique decoration.

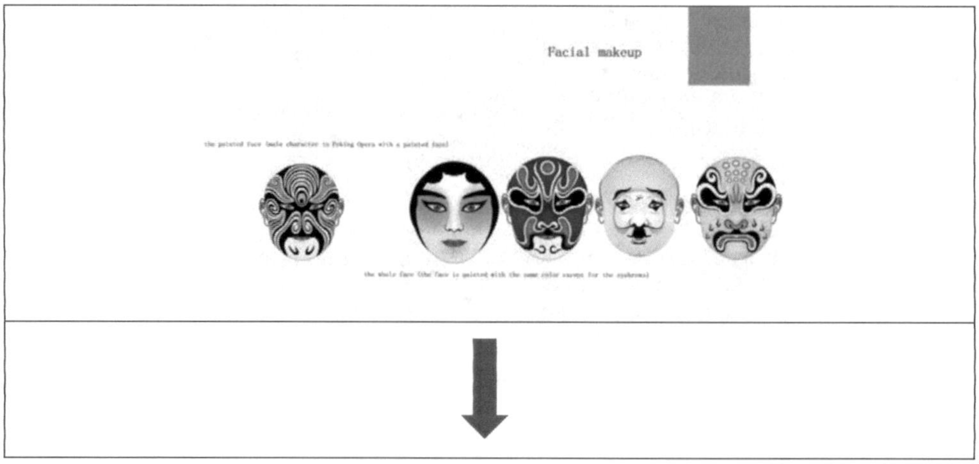

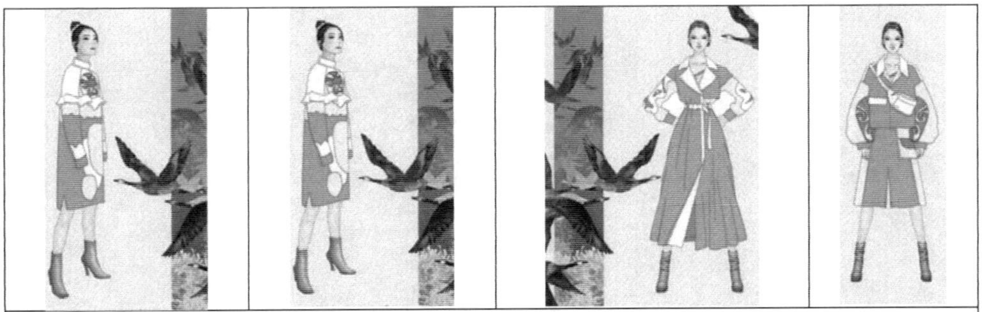

Figure 9 Four design examples

From the perspective of pattern application, the design ingeniously integrates the idea and process, and designs the clothing with the traditional Chinese opera culture in line with the modern aesthetics. It can be seen from the works that the application of opera costume elements in modern costume design is not completed in one go, but in steps. As a designer, we should know what to do and how to complete each step. In the process of modern fashion design, rationality and sensibility are mixed. If designers combine rationality and sensibility skillfully, they will be able to design works accepted by most people.

3 Conclusion

In view of the application of opera costume elements in modern costume design, the application of design process and ideas complement each other. Analyzing other designers' works and gaining inspiration from them will be the first step of application design; secondly, designers should make clear each step of design and solve problems they meet during the design process. The "three steps" of design and application proposed in this paper is bound to bring reference to relevant designers. The analysis of traditional opera costume elements in modern costume design is of great significance. It is also of great value for the inheritance of traditional culture and the development of modern costume design.

Reference

[1] Gao Shanhu. On the overseas communication of Sichuan Opera from the perspective of communication [J]. Journal of Qujing Teachers' College, 2017(4):117-120.

[2] Wang Peixi. On the stage costumes of the Peony Pavilion in Kunqu Opera [J]. Journal of National Academy of Chinese Theatre Arts, 2008, (04):55-59.

[3] Zhou Juan. Floating as duckweed, tough as grass--Brief analysis on the image of Nixi in Zhang Ailing's *Concatenate Traps*, 2014, (06):79-82.

[4] Qian Xulin. "Chinese style" in modern fashion design [J]. Journal of Clothing Research, 2018(10):416-421.

[5] Wang Qiming, & Lin Yaoyao. A study on the sleeve style of women's dress in Tang Dynasty [J]. Journal of Clothing Research, 2017(01):61-72.

[6] Li Yufei, & Liang Huie. Research Status and Trend of Qilu Ceremonial Costumes in Ming Dynasty [J]. Journal of Clothing Research, 2018(044):147-152.

[7] Zhu Wei. The cultural connotation of Beijing Opera Costume [J]. Home Drama, 2016(03):18-19

[8] Zhang Li, & Cheng Ganlin. The presentation and aesthetic features of opera symbols in Beijing Opera Costume [J]. Xi'an Polytechnic University, 2015(05):631-635.

[9] Tang Xiangning, Xu Liping, Ma Yuqing, Duan Xuwang, Han Chi & Wu Qiaoyan. Innovative design of silk scarves with Beijing Opera costume elements, [J]. Journal of Silk,2015(09):41-45.

[10] Yu Haiwei. Research on the application of traditional Beijing Opera dress pattern in modern Women's dress design under the market background, [J]. China Market, 2017(36):83-84.

[11] Lin Yanping. A study of costume patterns in traditional Chinese opera, [J]. Melliand China, 2017(08):52-54.

[12] Lin Yanping. Research on costume color culture in Chinese opera, [J]. Melliand China, 2016(04):52-55.

[13] Chen Mei. Analysis of the dress form of the traditional opera of Kun Opera, [J]. Donghua University, 2010.

[14] Pan Ting. A study of folk culture in the Peony Pavilion, [J]. Gannan Normal University, 2013.

Visual Analysis of Construction Waste Research Based on VOSviewer

HAN Lihua [1], GONG Zhiqi[1,2*]

[1] School of Civil Engineering, Qinghai University,810016,Xining,Qinghai,China.
[2] Qinghai Provincial Key Laboratory of Energy-saving Building Materials and Engineering Safety, 810016,Xining,Qinghai,China.

Abstract. Large-scale engineering construction projects have irreversibly damaged the ecological environment while promoting economic development and improving living standards. As the demand for realizing a resource-conscious and environment-friendly society continues to grow, people are increasingly interested in the sustainable development of construction waste. In this paper, the relevant literature in the field of construction waste is retrieved through the Web of Science database as the research object, and the VOSviewer measurement analysis tool is used to analyze the existing international research results. According to the research results, the research on construction waste in the field of construction research has attracted much attention in recent years. At the same time, domestic scholars have made outstanding contributions to this field, but they have the disadvantages of insufficient research depth and low citation rate. The research hotspots in the field of construction waste mainly focus on the construction waste recycling aggregate, construction waste management, and construction waste resource utilization strategy.

1 Introduction

Construction waste is an important part of municipal solid waste, it is wastes such as spoil and materials generated in the process of new construction, reconstruction, expansion and demolition of buildings, structures, pipe networks,etc.,and the process of house decoration. According to the stage of waste generation, construction waste can be divided into construction waste and demolished construction waste[1]. With the development of modern cities, people's requirements for infrastructure construction continue to increase, which has led to the rapid development of the construction industry, as well as the reconstruction of a large number of commercial and residential buildings, the amount of urban construction waste has continued to rise. According to statistics, the newly built building area in China is increasing at a rate of 2 billion square meters each year, and it will reach 30 billion square meters by 2020[2-3]. The continuous increase of construction waste and improper treatment methods have an irreversible impact on the ecological environment, for example, illegal dumping and landfilling have caused biodiversity and soil loss, and to a certain extent also polluted water resources [4].

In recent years, sustainable development based on green ecology has become an issue of increasing concern worldwide. Scholars are also increasing their research on construction waste, faced with massive research literature from different fields, traditional literature research methods are limited by time and have certain limitations, it is difficult to fully and accurately grasp the overall characteristics, frontier dynamics and evolutionary laws and trends of construction waste research from a large amount of research data. With the rapid development of information visualization technology, scientific knowledge graph analysis has become a new type of data mining and analysis technical method, it takes scientific knowledge as the research object, and directly and clearly expresses the internal knowledge structure and development process of a specific research field through graphical relationships.

Based on the status of publications in the field of architecture in the Web of Science database, this article statistically analyzes the number of publications in the field of architecture at home and abroad in recent years, countries with high production of documents, and research topics, use the bibliometric software VOSviewer to perform visual analysis of author co-occurrence and keyword co-occurrence on the documents in the database. the purpose of the article is to allow scholars in this field to quickly understand and master the development trend of construction waste, and to provide references for the research directions and hot spots of construction waste in my country.

2 Data sources and research methods

2.1 Data Sources

As the emphasis on the ecological environment has gradually increased in recent years, the research on construction waste is also developing rapidly and widely

* Corresponding author E-mail address:442953016@qq.com

used, and related documents are increasing year by year. In order to fully understand the global research status in the field of construction waste, this article intends to collect statistics in the core collection of Web of Science database for nearly 10 years, that is, the documents included in the database from January 2010 to February 2020, the title was set as the search field to conduct a literature search with the keywords "construction waste" and "Construction and demolition waste", after screening, 830 records were finally obtained as research texts, each record contains full-text record information such as the author, publishing organization, abstract, keywords, publication year, issue (volume), and references, and the above information is imported into the VOSviewer software to analyze its research development trend.

2.2 Research method

Bibliometric analysis is a method of statistical analysis of documents using the principles of bibliometrics. It mainly analyzes the structure and dynamic aspects of visualized scientific research. Bibliometrics mapping is an important technology in the field of bibliometrics, which can visualize knowledge fields and Relationship between articles, journals,etc[5]. This research uses bibliometric diagrams to determine the knowledge fields and research trends of prefabricated buildings based on existing literature.VOSviewer is a free document analysis software, which is widely used in the development of bibliometric drawing.

There are many bibliometric softwares, such as CiteSpace software[6] and CitNetExplorer[7]. Compared with these two, VOSviewer can generate distance-based maps to show co-authors, co-occurrence of keywords, citations, bibliographic coupling, and co-citation graphs. And other factors, its biggest advantage is that its visualization effect is better than other bibliometric analysis software. This method has also been widely used in previous literature review studies, so this article uses VOSviewer to analyze and identify the research focus in the field of construction waste, which lays the foundation for the analysis of hot spots in construction waste research.

This paper uses the bibliometric visualization software VOSviewer to analyze the co-occurrence of keywords and authors' co-occurrence in the field of construction waste in various countries in the past 10 years, It provides a reference for relevant researchers in the field of construction waste in china to choose research directions and grasp research hotspots.

3 Overview of Construction Waste Research

3.1 Literature publication year analysis

According to Fig.1, from the perspective of the distribution of the literature, the research interest in the field of construction waste at home and abroad has continued to rise. The number of articles published in 2020 rose from 28 in 2010 to 173 in 2019. The reason for the small number of articles published in 2020 is mainly due to the search time when selecting data. But on the whole, the continuous increase in the number of papers published indicates that research in the field of construction waste is becoming a research hotspot for scholars.

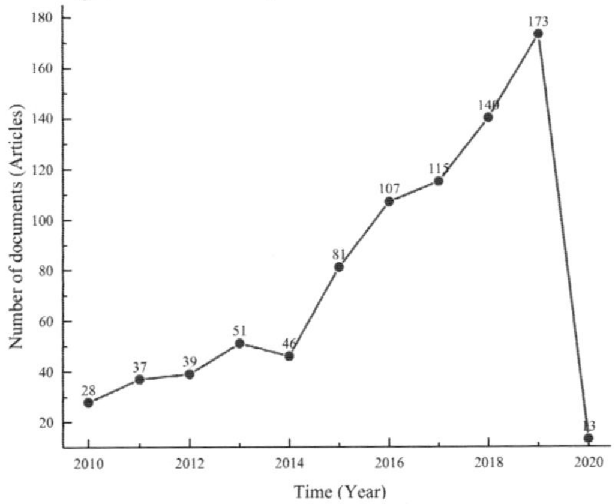

Fig1. Annual papers published in the field of construction waste.

3.2 Literature publication country analysis

Import the document records obtained from the Web of Science core collection database into VOSviewer, and set the cooperative relationship analysis unit to the country to draw a country co-occurrence map (Fig.2), a total of 34 key nodes and 98 leads were obtained. According to the color identification of the timeline, my country has conducted extensive research in this field in recent years,and our country has the largest research weight, with a total of 199 articles issued, accounting for 23.98% of the total number of articles issued. Followed by Brazil (89 articles accounted for 10.73%), Spain (89 articles accounted for 10.73%), Australia (58 articles accounted for 6.99%),the United Kingdom (48 articles accounted for 5.78%),the United States (39 articles accounted for 4.70%).

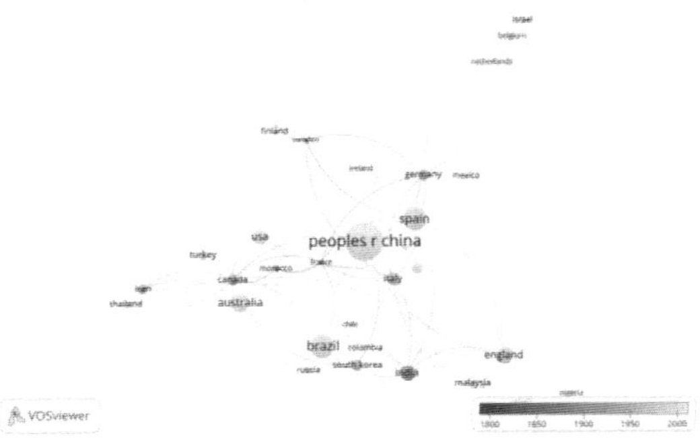

Fig2. National Co-occurrence Map of Construction Waste Research.

3.3 Representative scholars and articles

According to the full-text records imported into VOSviewer, set the analysis type to authors under the co-authorship menu, And set the minimum number of articles for each author to 3. According to the cluster view (Fig.3), a total of 5 clusters, 21 key nodes, and 128 leads are obtained. In the density view of Fig.4, the color of each key point is determined by the research density of that point. If the research contribution of this point is large, the number of citations will tend to dark yellow, otherwise it will appear blue. Through the density view, the authors of key research in this field can be judged [8]. By combining Fig.3 and Fig.2 to analyze, among the representative authors in the field of construction waste, Yuan hongping, Zuo jian and Wang jiayuan are cited more frequently, and the co-citation frequency is more than 24. In addition, other authors with higher weights are Wu huanyu, Lu weisheng, Zillante george, Poon chi sun, etc.

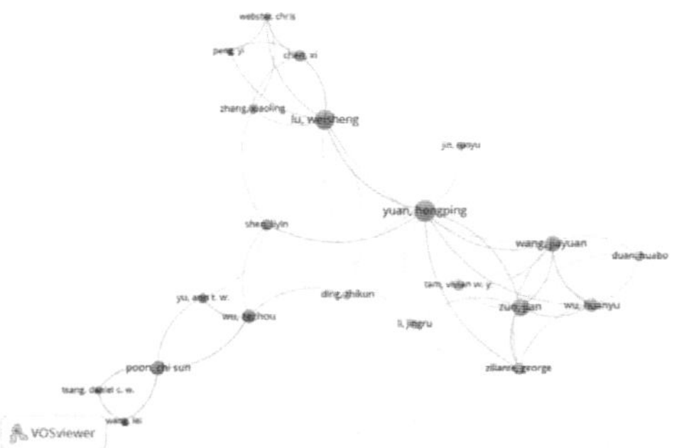

Fig3. Author cluster view of construction waste research.

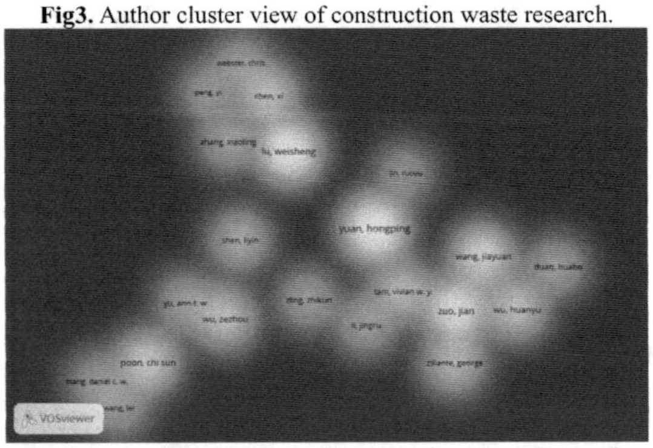

Fig4. Author Density View of Construction Waste Research.

Table 1 shows the information of papers that have been cited more than 100 times in the construction waste-related articles in the Web of Science core collection database. According to the information in the table, it can be concluded that the research content of highly cited papers mainly involves the following aspects: (1) Application of recycled aggregates: Silva RV[9] studied the main physical properties and composition of recycled aggregates used in concrete, and carried out statistical analysis on the data in the literature. A performance-based classification for the use of recycled aggregates in concrete construction; Behera M[10]introduced the production and use of recycled aggregates in concrete, and summarized their influence on the different properties of recycled concrete. The study found that the mechanical properties and durability of recycled concrete are generally lower than conventional concrete, however, recent studies have shown that the mechanical properties and durability of recycled concrete have received increasing attention. (2) Waste management: Yuan, H.[11]through literature research, concluded that the direction of construction waste management is mainly: from the traditional "cost-benefit analysis" to a broader view of sustainability "economy, environment and Social benefits", how to formulate a reasonable construction waste management fee, and how to formulate construction waste management based on the relationship between stakeholders; Yeheyis M[12] proposed a conceptual construction waste management framework to maximize 3R (reduce, reuse and recycle) and minimize construction by implementing a sustainable and comprehensive strategy throughout the life cycle of a construction project Waste disposal, and establishes life cycle-based sustainability indicators for construction waste, which are used to make decisions related to material selection, classification, recycling/reuse of construction waste, and treatment or disposal options. (3) Use waste to develop sustainable materials: Raut SP[13]describes various types of waste materials with different compositions, adding them to raw materials in different amounts, developing waste bricks, and studying bricks mixed with different waste materials Various physical, mechanical and thermal properties of the product, put forward application suggestions; Madurwar MV[14]discussed the potential application of agricultural waste as an alternative to sustainable building materials to reduce the use of natural resources and energy; Bravo M[15] analyzed the mechanical properties of concrete made from construction waste and recycled aggregates from all over Portugal. It was believed that when excessive recycled aggregates were added, the material properties would be reduced. Therefore, from the perspective of mechanical properties, a reasonable amount was added. This type of aggregate is feasible. (4) Key factors for on-site classification of construction waste: Wang J[16]adopted CSF methods, pilot studies, questionnaire surveys and face-to-face interviews, and concluded that six key factors are manpower, market for recycled materials, and waste classification capabilities, Management level, site space, construction waste classification equipment..

Table1. Statistics of citation information of construction waste related literature

Author	Time	Title	Cited frequency
Silva, R. V.	2014	Properties and composition of recycled aggregates from construction and demolition waste suitable for concrete production	275
Behera, Monalisa	2014	Recycled aggregate from C&D waste & its use in concrete - A breakthrough towards sustainability in construction sector: A review	223
Yuan, Hongping	2011	Trend of the research on construction and demolition waste management	169
Raut, S. P.	2011	Development of sustainable construction material using industrial and agricultural solid waste: A review of waste-create bricks	139
Madurwar, Mangesh V.	2013	Application of agro-waste for sustainable construction materials: A review	136
Wang, Jiayuan	2010	Critical success factors for on-site sorting of construction waste: A china study	114

 https://doi.org/10.1051/e3sconf/202123704016

Yeheyis, Muluken	2013	An overview of construction and demolition waste management in Canada: a lifecycle analysis approach to sustainability	106
Bravo, Miguel	2015	Mechanical performance of concrete made with aggregates from construction and demolition waste recycling plants	102

4 Construction waste research trends

4.1 Research hotspot

Based on the above analysis, set the data analysis type in the VOSviewer software to all keywords under the co-occurrence menu, and set the minimum frequency of keyword occurrences to 4, resulting in 5 clusters, 226 There are 7 key nodes and 7813 total leads. Table 2 shows the high-frequency keywords whose key frequency is greater than or equal to 30 times calculated by VOSviewer.

Table2. Keywords for construction waste research.

Number	Key words	Frequency	Co-occurrence frequency
1	Management	130	704
2	Concrete	111	571
3	Construction and demolition waste	115	567
4	Demolition waste	80	467
5	Performance	57	357
6	Waste management	58	325
7	Construction waste	64	305
8	Recycling	56	278
9	Model	45	275
10	Generation	45	267
11	behavior	46	261
12	mechanical-properties	39	259
13	china	39	256
14	fly-ash	40	249
15	sustainability	44	233
16	strength	39	230
17	system	32	211
18	cement	34	195
19	Compressive strength	35	189
20	aggregate	32	181

The keyword co-occurrence cluster view (Fig.5) and keyword co-occurrence density view (Fig.6) are obtained through keyword cluster analysis. In the cluster view, the same research field is displayed in the same color, and the size of the label indicates the research weight of the item. It can be seen from Fig.5 that there are 5 different research fields in the scope of construction waste research, combined with the density view to summarize and summarize,the 5 clustering themes are: the application of recycled aggregate(red cluster),construction waste management (Green cluster),construction waste recycling(blue cluster),construction waste resource utilization strategy (yellow cluster),development of recycled aggregate (purple cluster).

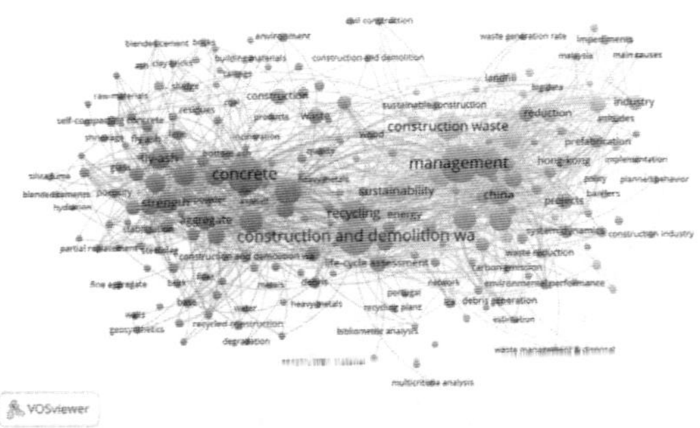

Fig5. Keyword co-occurrence cluster view.

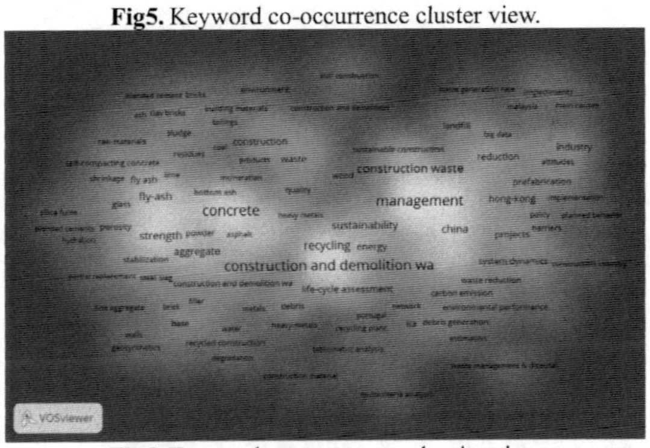

Fig6. Keyword co-occurrence density view.

4.2 Research areas

According to the construction waste research timeline in the five clustering themes and keyword co-occurrence tag view (Fig.7), the research fields are summarized as: construction waste recycled aggregate, construction waste management, and construction waste resource utilization strategies.

4.2.1 Construction waste recycled aggregate

With the accelerated development of urbanization and infrastructure, the extraction of raw materials required for construction projects has increased, and the demolition of old buildings and municipal facilities has produced a series of construction waste. At present, the domestic treatment and disposal methods of construction waste are mainly open-air stacking and landfilling. A small number of materials used for filling roads and foundations or antifreeze layers are recycled and reused. Construction projects have led to the shortage of natural resources and increasingly serious environmental damage. Therefore, the regeneration of construction waste as a resource is currently a hot area of research at home and abroad[17].

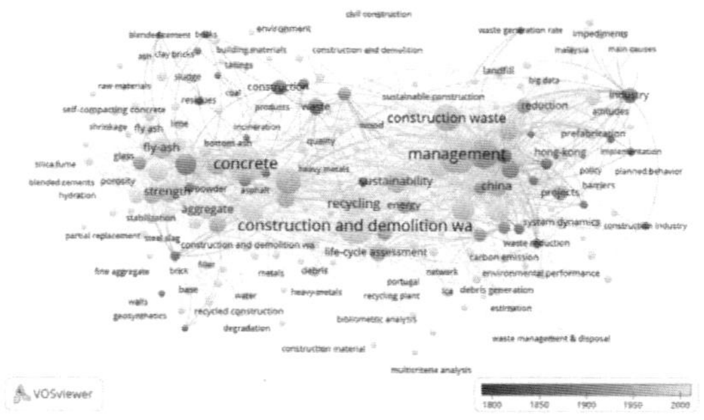

Fig7. Keyword co-occurrence tab view.

In 1946,Russian scientist Glushge proposed the concept of construction waste recycling and studied the basic properties of construction waste recycled concrete[18].In 1977,Japan formulated the specifications for the use of recycled aggregates and recycled concrete. The United States stipulated in the concrete aggregate standard in 1982 that waste concrete can be used as coarse and fine aggregates after treatment.my country only began to use construction waste as a resource in the 1980s[3].

The main components of construction waste are waste concrete and masonry, so in recent years, the above two are used as the main preparation materials of recycled aggregates for research at home and abroad. In 2014,Behera M[10]applied recycled aggregates with different properties to concrete for testing, and found that the mechanical properties and durability of recycled concrete are generally lower than conventional concrete, which provides a way to improve the mechanical properties and durability of recycled concrete in the future. research direction. In 2011, Raut SP[13]and others developed recycled bricks and stones by adding different amounts of construction waste to raw materials, and put forward corresponding opinions and measures based on different physical and chemical properties.

The development and application of recycled aggregates, on the one hand, solves the difficulty in the treatment of construction waste and the resulting environmental pollution. On the other hand, the reuse of construction waste, screening and re-creating the required construction materials can reduce construction materials consumption, thereby reducing the damage to the ecological environment, and achieving the purpose of protecting the earth's homeland on which mankind depends.

4.2.2 Construction waste management

The object of construction waste management is mainly the waste generated during the construction, decoration and demolition of the project, and its purpose is to reduce the generation of construction waste as much as possible at the source.Other construction waste generated is reused or recycled in a specific way[19].

In 1962, American economist Balding proposed the concept of circular economy[20], pointing out that human beings should transform economic activities from the traditional "resources-products-waste-end management"model to "resources-products-renewable resources".Cycle mode. As a new type of economic production mode, the development principle of circular economy is "reduction, resource utilization, and reuse" to maximize resource utilization, reduce waste discharge as much as possible, and achieve sustainable economic and social development. Therefore, the hierarchical model of construction waste management can be divided into four stages: reduction,reuse,recycling, and disposal of construction waste[11].

In the circular economy 3R (reduce, reuse, recycle) principle, construction waste reduction is the most effective way to manage construction waste, that is, to carry out cleaner production. Using this method can control the amount of construction waste generated from the source comprehensive management of the quantity, volume, type, and harmful substances of construction waste can achieve the dual goals of saving natural resources and reducing pollution[21].

4.2.3 Resource utilization strategy

According to the management level of construction waste, at this stage, construction waste management has changed from "end treatment" to "source reduction management off-site management"to realize the entire process of"resource management".In 2012, Yuan H[22]established a quantitative model of social benefits of construction waste resource management based on system dynamics theory,literature research and questionnaire surveys,and concluded that the resource management of construction waste can increase jobs and improve environmental pollution,Improve social satisfaction and other benefits.

At present, the utilization rate of construction waste in Japan and South Korea has reached over 97% each year, and over 90% in EU countries, as a large manufacturing country, China's resource utilization rate is only about 5%[23], which is far from other countries. It shows that China's construction waste resource utilization in the whole process management not only requires all stakeholders to give full play to their own advantages and capabilities. A systematic and flexible legal guarantee system for the recycling of construction waste is needed to guide the smooth progress of the recycling process of construction waste.

In recent years, various parts of the country have successively introduced construction waste resource management policies, which have put forward clear requirements in terms of source reduction management of construction waste, disposal and disposal fees, etc. Among them, although my country has achieved certain results in the construction waste disposal fee system, the issue of fee collection standards has not yet been resolved. If the fee is too low, producers will give priority to off-site disposal, which is not conducive to resource utilization. Excessive charges will lead to more illegal dumping activities, which will increase environmental pollution. According to the current domestic construction waste fees, it is basically unable to meet the operating costs of recycling, transportation, resource utilization and production, and infrastructure costs for treatment and disposal. In addition, when formulating a fee system, the environmental losses caused by the discharge of construction waste should be included. Therefore, the construction waste management departments in various places should hold a seminar on the construction waste treatment charging system in a timely manner to jointly discuss the reasonable construction waste treatment charging standards in the region, the fee collection process, and the usage of the fee. In addition, in order to facilitate the recycling and utilization of construction waste, the construction waste disposal fee system reflects the encouragement of the source classification collection of construction waste. For example, the collection price of mixed construction waste is much higher than that of classified construction waste, thereby stimulating the source classification of construction waste collect.

5 Conclusion

This study uses visualization tools to excavate and analyze 830 effective records in the Web of Science core collection database. It is found that the amount of publications in the field of construction waste has increased year by year. And Chinese scholars have made more prominent contributions to this field, and domestic scholars account for a large proportion of core authors in this field. The research hotspots in this field focus on construction waste recycled aggregates, construction waste management, and construction waste resource utilization strategies. Construction waste recycled aggregates and resource utilization strategies are research fields that have gradually emerged in recent years, and there are still many problems to be solved urgently.

Acknowledgements

This study was supported by the Qinghai Provincial Applied Basic Research Project Fund (Grant No.2018-ZJ-734),the National Natural Science Foundation of China (Grant No.71463047),and the Qinghai Provincial Key Laboratory of Energy-saving Building Materials and Engineering Safety (Grant No. 2018-ZJ-T01).

References

1. GB/T50743-2012. Technical Specification for Recycling of Engineering Construction Waste [S].
2. Tam, V.W., Fung, I.W., Sing, M.C.,et al. J.Cleaner Production,216-231, 109(2015).
3. Zhu Dongfeng. South China University of Technology (2010).
4. L Jaillon,C Poon. Construction Management and Economics, 953-966, 26 (9),(2008).
5. van Eck N, Waltman L. Scientometrics, 523-538. 84(2),(2009).
6. Chen C, Ibekwe SanJuan F, Hou J.Journal of the American Society for information Science and Technology, 1386-1409, 61(7),(2010).
7. Van Eck N J, Waltman L. Journal of Informetrics, 802-823, 8(4),(2014).
8. Gao Kai. Science and Technology Information Development and Economy, 95-98,12,(2015).
9. Silva R V , De Brito J , Dhir R K .Construction and Building Materials, 201-217, 65,(2014).
10. Behera M,Bhattacharyya S K,Minocha A K,et al. Construction & Building Materials, 501-516, 68,2014.
11. Yuan,H.,Shen,L.WasteManagement,670-679,31(4), (2011).
12. Yeheyis M,Hewage K,Alam M S, et al.Clean technologies and environmental policy,81-91,15(1),(2013).
13. Raut S P , Ralegaonkar R V ,Construction & Building Materials,4037-4042,25(10),(2011).
14. Madurwar M V, Ralegaonkar R V ,Mandavgane S A.. Construction & Building Materials,872-878,38,(2013).
15. Bravo M,Brito J D,Pontes J,et al.Journal of Cleaner Production,59-74,99(1),(2015).
16. Wang J,Yuan H,Kang X,et al.Resources Conservation & Recycling,931-936,54(11),(2010).
17. Xiong Feng. Chongqing University,(2017).
18. Wang Bo.HuazhongUniversity of Science and Technology,(2012).
19. Wu Zezhou. Chongqing University,(2012).
20. Zhang Jiayi. Bohai University,(2016).
21. Yang Qingdong.Inner Mongolia University of Science and Technology,(2012).
22. Yuan H.Waste Management, 0-1228,32(6),(2012).
23. Qin Yuebo.Nanjing Forestry University,(2009).

https://doi.org/10.1051/e3sconf/202123704017

Research Progress on Environmental Impact Assessment of Urban Construction Activities

CUI Songxu [1], GONG Zhiqi[1,2*]

[1] School of Civil Engineering, Qinghai University,810016, Xining, Qinghai, China.
[2] Qinghai Provincial Key Laboratory of Energy-saving Building Materials and Engineering Safety, 810016, Xining, Qinghai, China.

Abstract. The process of building new construction, reconstruction, and demolition has a very significant environmental impact. As the pillar industry of the national economy, the construction industry has put tremendous pressure on the environment in its production process. This pressure is not only manifested in the consumption of resources and energy, but also in the discharge of pollutants in the three states of solid, liquid and gas. This article aims to sort out the environmental impact assessment methods of construction activities at home and abroad, and sort out the research progress of environmental impact assessment. First, start from the macro background to explain the necessity of systematic analysis and evaluation of the environmental impact in the construction field, and then briefly comb the environmental impact assessment methods in the construction field at home and abroad from the aspects of applied research, advantages and limitations, and finally summarize and look forward to the article. It can provide scholars with a little reference and reference, and promote the further expansion and improvement of environmental impact assessment methods in the construction field.

1 Introduction

As an important industry of the national economy, the construction industry has a very important impact on the sustainable development of the environment, society and economy. Every year, large-scale urban construction activities bring convenience to people, but also produce relatively serious environmental impacts and a large amount of resource and energy consumption, which to a certain extent hinders the realization of sustainable development. If a building is regarded as a product, then the whole life cycle of the building is from raw material mining-transportation-processing-transportation-construction-operation and maintenance-demolition and disposal. The ecological environment brings greater pressure. This pressure is not only reflected in the supply of resources and energy, but also in the absorption of pollutants produced in each of the above-mentioned links. These pollutants may be discharged during energy production. The various types of greenhouse gases, toxic and harmful gases, sewage and solid waste, may also be solid, liquid, and gas three types of pollutants produced during the production of construction materials, building construction, renovation, and dismantling. According to the "China Building Energy Consumption Research Report" issued by the China Building Energy Conservation Association in 2017, China's building energy consumption was 947 million tons of standard coal, accounting for 21.10% of the country's total energy consumption, and building carbon emissions were 1.96 billion tons, accounting for 19.5% of the national energy carbon emissions [1]. The European Commission mentioned in the Building Energy Performance Directive in 2010 that the construction industry accounts for about 40% of the EU's total energy consumption [2]. In addition, studies have shown that as of 2015, the total amount of construction waste in my country has reached 7 billion tons, and the new construction waste generated annually exceeds 400 million tons [3]. From these data, the construction industry is a major challenge for us to reduce environmental pollution, achieve sustainable development, and move toward a low-carbon economy. Therefore, a systematic assessment of the environmental impact of urban construction activities is very necessary.

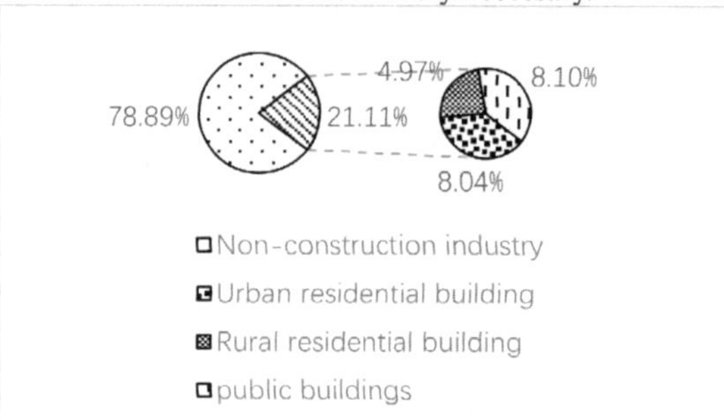

Fig1. The proportion of China's building energy consumption in 2017(The data are from China building Energy Consumption research report)

* Corresponding author E-mail address: 442953016@qq.com

2 Status of research on environmental impact assessment of construction activities

In response to the inevitable environmental problems caused by the construction industry, scholars have established corresponding associations, scientific research groups, and released relevant research methods and evaluation tools to alleviate or even eliminate this environmental problem. According to the literature review, the environmental evaluation system can be divided into: clause-based green building evaluation system, material flow analysis method, LCA and integrated evaluation method based on LCA idea. Next, we will briefly sort out the environmental impact assessment methods in the construction field at home and abroad from the aspects of applied research, advantages and limitations.

2.1 Ecological Footprint

The ecological footprint (EF) refers to the biologically productive area that is necessary to maintain a person, region, or country for survival or that can accommodate human wastes [4,5]. By comparing the ecological footprint demand with the carrying capacity of the natural ecosystem (also known as the ecological footprint supply), the sustainable development status of a country or region can be quantitatively judged, so as to make scientific plans and plans for future human survival and social and economic development. Suggest.

The calculation of the ecological footprint is based on two simple facts: 1. We can retain most of the resources consumed and most of the waste generated; 2. Most of these resources and waste can be converted into biological production that can provide these functions Sex land. First, the type of biological production area and its balancing treatment are carried out. The purpose of this stage is to convert these biological production areas with different ecological productivity into areas with the same ecological productivity, so as to summarize the ecological footprint and ecological carrying capacity; then proceed to the ecological per capita Footprint component calculation. This stage is to convert different energy consumption into a certain type of energy land area, that is, to estimate the land area required by natural assets based on the consumption rate of one of the energy sources; then summarize the ecological footprint; secondly, calculate the ecological Carrying capacity; the last is to judge whether it is an ecological deficit or an ecological surplus by the magnitude of the two values.

Domestic scholars have conducted various applied researches based on the ecological footprint method. Xu Zhongmin et al. [6] conducted a dynamic analysis of the ecological footprint, calculated the ecological footprint of Gansu Province in 1998, and quantitatively judged whether it was within the range of ecological carrying capacity; Liu Miao [7] based on the ecological footprint, analysed the ecological footprint of the upper reaches of the Minjiang River. Carrying capacity and landscape pattern changes were evaluated and predicted; Li Dan et al. [8] evaluated the ecological carrying capacity of the Yalong River source area, replaced the fossil energy land with woodland and grassland, and replaced it with a certain proportion. The final calculation More similar to the source area. Zhang Sheng et al. [9] established a multi-drive evaluation model to evaluate the water ecological carrying capacity of the Beijing-Tianjin-Hebei region; Gao Wei et al. [10] optimized the water ecological carrying capacity model of the basin based on the coupling process of water quantity and water quality. And studied the water ecological carrying capacity of the Panlong River Basin.

The ecological footprint method analysis index provides a "fair stalk" for measuring the degree of sustainability. It can objectively measure and compare the degree of sustainability in two dimensions in time and space, so that people can clearly know the sustainability of reality. How far is the goal, so as to help monitor the effectiveness of the implementation of sustainable programs. However, the calculation results can only reflect the impact of economic decisions on the environment, without paying attention to the energy consumption of ecological products and ecological services. At the same time, it also ignores other important factors in the development and utilization of resources, such as the promotion of industrial urbanization and the occupation of cultivated land due to pollution Land degradation caused by, erosion, etc. [11,12].

2.2 Material flow analysis

Material flow analysis refers to the systematic analysis of material flow and storage in a specific system within a certain time and space. It mainly involves the source, path and sink of material flow. According to the law of conservation of mass, the results of material flow analysis can always achieve the final material balance through all its input, storage and output processes [13].

The basic idea of the material flow analysis method can be traced back to many different research fields more than 100 years ago [13]. In the field of economics, Leontief proposed the input-output balance sheet in the 1930s [14,15]. Research on resource protection and environmental management first appeared in the 1970s, and these two initial application areas were: (1) urban metabolism; (2) analysis of pollutant migration paths in watersheds or urban areas [13]. Later, the theory of material balance, industrial metabolism, etc. laid a theoretical foundation for the application of material flow analysis. Austria, Japan, and Germany were the first to use material flow analysis to analyse the natural resources and material flow in their own economic systems [16], which marked the beginning of large-scale application research of material flow analysis worldwide. Since 1997, the World Resources Institute has conducted a comprehensive analysis of the material flow in the economic systems of five countries (the United States, Japan, Austria, Germany, and the Netherlands). At the same time, more and more countries have begun to use material flow analysis to conduct a comprehensive analysis and evaluation of their own economic systems, such as China, the United

Kingdom, Sweden, and Italy. In 2004, Paul H. and others co-authored the book Practical Handbook of Material Flow Analysis, which systematically introduced the concept, process, application scope and goals of material flow analysis, and stated the specific operation methods of its database management and software application. And listed many classic cases on environmental management, resource protection, waste management and regional material flow analysis and management.

Because the environment in which the economic system is located is a complex nonlinear system, and its behaviour is difficult to predict, it is difficult to establish a model to analyse it to find the results and mechanism of its change. The material flow analysis method can effectively analyse the material flow of the system from the perspective of the input-output balance sheet. Through analysis, we can get concise environmental pressure and sustainable development indicators. Of course, the material flow analysis method also has some shortcomings. The current material flow analysis only considers the quality of the material and ignores the different environmental impacts that different material flows may bring, which weakens the link between the material flow index and the environmental impact caused by the material flow. The material flow analysis method cannot determine what should be reduced in order to achieve sustainable development? This key issue. Furthermore, the purpose of the material flow analysis method is to quantitatively study the physical changes of the total material mass passing through the interface of the economy and the environment, and does not include the material flow in the economic system such as the transfer of products in different sectors, and the economic The system is treated as a "black box" [16, 17]. In addition, the data needed for material flow analysis in the economic system can almost be found at the national scale, but the relevant data at the small regional scale is often due to some objective reasons (such as the limitation of statistical content, the monopoly of large local companies and enterprises, etc.) It is difficult to obtain, and this also limits the application of material flow analysis methods in economic systems.

2.3 Green Building Evaluation System

Green building evaluation refers to the evaluation of the green environmental protection of the building through a series of standardized measures and evaluation processes to achieve the goal of green building. Since the 1990s, many countries in the world have established their own green building evaluation systems. However, due to differences in the level of economic development and construction industry development in different countries, as well as differences in geographic location, topography, history and culture, different countries When formulating a green building evaluation system, specific details such as the importance of evaluation indicators, the level of detail of evaluation provisions, and the stage integrity of the evaluation system are not the same [18]. The current green building evaluation system is as follows:

(1) Building Research Establishment Environmental Assessment Method. This method was proposed in 1990 by the British Institute of Architecture. The main evaluation objects are the new construction of residential buildings, ecological buildings, public buildings, industrial buildings, and the reconstruction of existing buildings. The evaluation content covers management, energy, water, materials, ecology, land and ecology, transportation, pollution, building comfort, innovation, etc. There are four evaluation levels: Excellent, Excellent, Good, and Passed.

(2) Green Building Challenge. In 1998, scholars from more than ten countries including Canada researched and proposed. The main assessment objects are residential buildings, office buildings, and school buildings that are newly built and rebuilt from existing buildings. The evaluation content involves resource consumption, indoor air quality, environmental load, economy, environmental sustainability, maintainability, and operation management. A five-level scoring system has been established.

(3) Leadership in Energy and Environmental Design. The US Green Building Council proposed the LEED system in 1998. The objects to be assessed are residential buildings and public buildings that are newly built and rebuilt from existing buildings. The evaluation content involves sustainable sites, water conservation, energy and atmosphere, materials and resources, indoor environmental quality, location and transportation, innovative design, and regional priority. The total number of projects is 69, 26-32 points are certified, 33-38 points are silver, 39-51 points are gold, and 52-69 points are platinum.

(4) Comprehensive Assessment System for Building Environmental Efficiency. In the context of the general trend of the sustainable development concept, in April 2001, the "Comprehensive Building Environmental Assessment Research Committee" was jointly established by industry (enterprises), politics (government), and academics (scholars) in Japan, and jointly launched the project research, and finally developed "building comprehensive environmental performance evaluation system"-CASBEE. The assessment objects are buildings of various purposes and scales that are newly built and rebuilt from existing buildings. The assessment content involves the performance and quality of the building environment (including indoor environment, service facilities and outdoor environment) and the load of the building environment (including resources, energy, materials, and off-site environment). A total of 1 to 5-star ratings.

In addition, there are the Canadian system, the Chinese system, the Australian and German systems and so on.

2.4 LCA

Life Cycle Assessment, referred to as LCA, is aimed at the entire life cycle of a product (building)—that is, the entire process from the extraction and transportation of raw materials to the design, manufacture, use, recycling and final disposal of the product, quantitative calculation and analysis of the product Actual and potential energy and resources consumed and environmental management tools corresponding to pollutants discharged [19]. The earliest use of LCA ideas for environmental impact assessment began in the United States and other Western countries in the 1970s. For example, the well-known Coca-Cola Company conducted quantitative analysis of the environmental resource utilization and environmental emissions of its packaging products. After the efforts of many scholars, organizations, and scientific research groups, some basic frameworks, databases, and the more famous evaluation software SimPro appeared one after another. At this time, people realized that a unified international name and standard was needed for better research and communication. The name LCA has been fixed internationally. In 1997, with the joint efforts of international scholars, the International Organization for Standardization issued the first standard of life cycle theory-ISO14040 "Life Cycle Assessment-Principles and Framework", and then successively issued ISO 14041 "Life Cycle Assessment- Purpose and scope determination, inventory analysis, ISO14042 "Life Cycle Assessment-Impact Assessment", ISO14043 "Life Cycle Assessment-Interpretation of Results", etc. International standards are of great significance to the research and application of LCA theory. Before and after the standards were published one after another, scholars have achieved positive results for LCA research.

LCA is composed of 4 interrelated parts, namely: purpose definition and scope determination, inventory analysis, impact evaluation, and result interpretation. Purpose definition and scope determination are the first steps in life cycle assessment. It is the starting point and foothold of all subsequent steps. At this stage, it is necessary to clearly put forward the purpose and background of the analysis, as well as point out the assumptions and constraints involved in the analysis, and set functional units [19].

The inventory analysis stage is usually the most time- and labour-intensive stage in life cycle assessment [20,21], because of the lack of data related to the research object. In general, it is easier to collect data than reference search data [22]. Therefore, many databases have been developed in recent years. Foreign countries are particularly prominent in this regard. Related software and databases have been developed towards commercialization, such as SimPro, GaBi, etc. Research in my country started relatively late, in which Xu Xiaoming [23] introduced a logical hierarchical connection environmental engineering database design method, that is, based on hierarchical analysis, based on connection relations, combined with object-oriented technology methods for environmental engineering database design; Liu Xialu [24] summarized and put forward the basic method of establishing the life cycle reference database in my country; Gong Xianzheng, Li Xiaoqing and others [25-27] developed the China Material Environment Database (SinoCenter) using ASP.NET and SQL Server, and established the China Material Life Cycle Analysis Database platform. Life cycle analysis is a widely used environmental management tool. Its specific application requires the support of relevant environmental load data and evaluation method system data at different levels, regions and different technical levels. The reliability of data directly affects the results of product life cycle assessment Credibility and application performance. Therefore, all countries in the world are actively constructing LCA databases in line with their own actual conditions, and have achieved good results. In recent years, the construction of LCA databases in my country has slowly caught up.

Impact evaluation is the more difficult and technical part of LCA, and its methodology and evaluation model are one of the focuses of scholars in this field [19]. The general steps of impact evaluation are: classification, characterization, standardization, and weighting. The purpose of life cycle impact assessment (LCIA) is to understand and evaluate potential environmental impacts, including their scale and importance, and result interpretation is to summarize the results of inventory analysis and impact assessment to form corresponding conclusions and recommendations. In the life cycle assessment, if the scope of the investigation, the definition and distribution of the system boundary in the inventory analysis are different, the data sources are different, the impact evaluation is characterized, and the weighting method is different, different conclusions will be produced. Therefore, it is necessary to explain the necessary results after the evaluation.

2.5 Integrated evaluation method based on LCA idea

Although LCA is currently a widely used environmental management tool, it still has some limitations. For LCA to obtain reliable and application-performing evaluation results, it needs massive and high-quality data support; the endpoints of the LCA evaluation system are a regional (or local) indicator, which requires corresponding geographic information (population). Density, economic level, geology, landform, etc.) cooperate with the interpretation results [28], so that the interpretation of the impact evaluation results has practical significance. Therefore, scholars have begun to try to combine LCA with information technology, and make up for the shortcomings of the LCA method by incorporating the element of GIS. Domestic scholar Liu Fang et al. [29,30] believe that to solve the problem of the lack of traditional LCA spatial location data, geographic information systems should be introduced to achieve the acquisition of spatial location data. He successively published articles discussing the necessity and possibility of introducing spatial location data into LCA, and proposed a new functional structure and database framework of product material management information system. Tian Yajun et al. [28] reviewed the progress of the regionalization of life cycle assessment, discussed the necessity and feasibility of introducing spatial information data into LCA, and the important role of GIS in the regionalization of LCA. Wang Xue [31] believes that the coupling of GIS and LCA can well supplement, improve and manage traditional data, help explore the spatial characteristics of the environmental impact of products, activities or processes, and conduct environmental impact assessments related to land use.

Some foreign scholars also believe that GIS should be integrated. Zea Escamilla E [32] researched that the use of geographic information system makes the development of building materials and LCIA characteristic data of buildings highly consistent. Geyer et al. [33] believe that geographically related factors play a very important role in specific cases of building life cycle assessment, but they are difficult to solve in life cycle assessment. When using geographic information system technology, factors such as land use, seismic hazard areas, and material transportation distances are very straightforward. Gao Yu [34] studied the use of BIM model construction, the application of LCA method to calculate the environmental impact of building materials or components, and the spatial analysis of the obtained data through buffer analysis in GIS and network analysis technology; Lu Lu et al. [35] proposed a technical method of coupling geographic information system (GIS) spatial analysis and life cycle analysis (LCA) to assess the development and utilization potential of energy plants, and found that the spatial grid evaluation method can provide more refined data support for scientific decision-making. Foreign scholars Hiloidhari et al. [36] combined GIS and LCA research methods to demonstrate the role of spatial LCA in agricultural waste residue bioenergy planning. Saad et al. [37] established regionalized environmental impact characteristic factors for soil ecosystems in different ecological regions of Canada. Mutel et al. [38] calculated the difference in ecotoxicity effects produced by power plants in different locations in the United States based on GIS tools. Österbring M et al. [39] constructed a geographic information system model based on a specific building for a multi-family residential building in Gothenburg, and used an exploratory LCA method to evaluate the environmental impact of the renovation. S. GARCÍA-PÉREZ et al. [40] used geographic information systems combined with life cycle assessment methods to characterize and evaluate urban areas from the perspective of energy efficiency.

GIS is a technical system composed of three parts: data input, storage, editing, spatial analysis and spatial query, visual expression and output. Using GIS technology to analyse the life cycle of a product can describe complex spatial and attribute data in the form of a map, and realize spatial visualization of environmental impact. In addition, it can further demonstrate the temporal and spatial distribution of environmental impacts of urban construction activities in different research periods, which can provide management methods for environmental quality management in the construction industry and realize refined management of the built environment.

3 Conclusion

There are many environmental impact assessment tools and methods. The clause-based green building evaluation system has obvious advantages in evaluating some indicators that are difficult to quantify; the material flow analysis method is aimed at the total material quality at the interface between the economy and the environment in a larger area Quantitative research is carried out on the physical changes of the country, but it cannot determine what should be reduced in order to achieve sustainable development? Moreover, the current analysis is mostly based on the national scale, and it is impossible to obtain sufficient statistical data at the urban scale due to objective reasons; the ecological footprint method only calculates the supply balance of land resources and petrochemical energy materials, and does not involve other environmental impacts. Therefore, it is not dominant in many environmental impact studies; the evaluation system based on the LCA idea can not only perform effective quantitative analysis and evaluation of the current environmental impact, but also provide a comprehensive evaluation of the building and its "cradle to grave". The environmental issues involved in the process are evaluated, but the data requirements are particularly high and the necessary geographic information is lacking. Therefore, further research on the integrated application of the two methods of coupling LCA and GIS is needed.

Acknowledgements

This study was supported by the Qinghai Provincial Applied Basic Research Project Fund (Grant No.2018-ZJ-734) , the National Natural Science Foundation of China (Grant No.71463047) and the Qinghai Provincial Key Laboratory of Energy-saving Building Materials and Engineering Safety (Grant No. 2018-ZJ-T01).

References

1. China Building Energy Association, Architecture.J.7 30 (2020)
2. Parliament, European. Official journal of the European union 13 (2010)
3. Xiao Xuwen,Feng Dakuo,Tian Wei,Const.Tech.J. 44 6-8 (2015)
4. Ress W E. Focus 6 121-130 (1996)
5. China Commission for International Cooperation on Environment and Development, World Environment 52-57 (2008)
6. Xu Zhongmin, Zhang Zhiqiang, Cheng Guodong. Acta Geographica Sinica 55 607-616 (2000)
7. Liu Miao. Shenyang Institute of Applied Ecology, Chinese Academy of Sciences,(2007)
8. Li Dan, Huang Chuanyou, Yin Tong, Hydropower Energy Science 36 38-41 (2018)
9. Zhang Sheng, Wang Tieyu, Zhang Hong, Acta Ecologica Sinica 37 4159-4168 (2017)
10. Gao Wei, Yan Chang'an, Li Jincheng, Acta Scientiae Circumstantiae,37 755-762 (2017)
11. Qu Xiuqi, Liu Miao, Li Chunlin, Hu Yuanman, Journal of Meteorology and Environment,35 113-119 (2019)
12. Li Tao, Gong Yabo Journal of Nanhua University (Natural Science Edition) 34 60-67+96 (2020)
13. Paul H, Helmut R Practical Handbook of Material Flow Analysis (2004)

14. Leontief W. International Arts and Sciences Press(1977)
15. Leontief W. Input-Output Economics Oxford University Press, (1966)
16. Huang Heping, Bi Jun, Zhang Bing, Acta Ecologica Sinica,27 368-379 (2007)
17. Zhang Ling, Yuan Zengwei, Bi Jun. Acta Ecologica Sinica 29 6189-6198 (2009)
18. Zhao Hongjia. Jilin University of Architecture, 2019.
19. Zheng Xiujun, Hu Bin. Progress in Science and Technology 30 155-160 (2013)
20. Rebitzer, G., Ekvall, T., Frischknecht, R. Environmental International 30 701-720 (2004)
21. Finnveden, G., Hauschild, M. Z., Ekvall, Journal of Environmental Management 91 1-21 (2009)
22. Baumann, H. , Tillman, A-M Lund University (2004)
23. Xu Xiaoming, Mao Zhiwei, Rong Kangquan, Mechanical and Electrical Engineering Technology 36 15-17 (2010)
24. Liu Xialu, Wang Hongtao, Chen Jian, Journal of Environmental Sciences 30 2136-2144 (2010)
25. Gong Xianzheng, Zhang Qun, Liu Yu, Journal of Beijing University of Technology 35 997-1001 (2009)
26. Gong Xianzheng, Nie Zuoren, Wang Zhihong, Progress in China Materials 30 1-7 (2011)
27. Li Xiaoqing, Gong Xianzheng, Nie Zuoren, Progress in Chinese Materials 35 171-178 (2016)
28. Tian Yajun, Deng Yelin, Zhang Yueling, Xie Kechang, CIESC Journal 37 2195-2201 (2016)
29. Liu Fang, Shi Jinfa, Lu Changde, Journal of Xinyang Normal University 23 467-469 (2010)
30. Liu Fang, Shi Jinfa, Lu Changde, Journal of Nanjing University of Aeronautics and Astronautics 43 91-94 (2011)
31. Wang Xue, Shi Xiaoqing, Acta Ecologica Sinica 37 1346-1357 (2017)
32. Edwin Zea Escamilla, Guillaume Habert, The International Journal of Life Cycle Assessment 22 1210-1219 (2017)
33. Geyer R, Stoms DM, Lindner JP, International Journal of Life Cycle Assessment 15 454-467 (2010)
34. Gao Yu, Shenzhen University, 2018.
35. Lu Lu, Fu Xinyu, Jiang Dong, Journal of Geo-Information Science 16 328-334 (2014)
36. Hiloidhari, M.; Baruah, D. C.; Singh, A.; Kataki, S.; Medhi, K.; Kumari, S.; Ramachandra, T. V.; Jenkins, B. M.; Thakur, I. S., Bioresource Technology 242 (2017)
37. Saad R, Margni M, Koellner T, Int. J. LCA 16 198-211 (2011)
38. Mutel L, Pfister S, Hellweg S. Environ. Sci. Technol.,46 1096-1103 (2011)
39. Magnus Österbring, Érika Mata, Liane Thuvander, Building and Environment 165 1-8 (2009)
40. S. García-Pérez, J. Sierra-Pérez, J. Boschmonart-Rives, International Journal of Sustainable Development and Planning 2 294-303 (2017)

Discussion on the Sustainable Development Pattern of Road Network Structure of Cities with a Population of 1-2 Million in China

Zhengmin Wen*, Zhenqiang Li, Wenshuo Luo, Yuxin FU, Juan Quan, and Guohua Zhou

Guilin University of Technology, Guilin, 541004, China

Abstract. The urban road network system is the main carrier of urban traffic. The constraint factor that urban road network has on traffic leads to major urban traffic problems. In this paper, a questionnaire survey and the Delphi method are determined to determine the case cities with a population of 1-2 million at home and abroad. Literature research and comparative, analytical, and inductive research methods are used to compare, analyse, and evaluate the advantages and disadvantages of the road network structure of the case cities, to summarize the development law of the urban road network, and to propose a sustainable development structure model for each development stage of urban road network: the mode made of ring freeway + trunk street (the embryonic stage), the mode made of ring freeway + expressway + trunk street (the incubation stage), the mode made of outer ring freeway + outer ring expressway and radial expressway + trunk street (the mature stage). The innovative point of this paper is to put forward a sustainable development pattern for the road network structure of cities with a population of 1-2 million in China.

1 Introduction

Sustainable development of urban transportation has become one of the key elements of the overall sustainable development of the city[1], and road traffic is the most important way of urban transportation. Many studies have shown that the development law of urban road network structure is closely related to urban scale, urban nature, social and economic development level, and reflects the development stage of the city to a certain extent[2]. With the change of urban development direction, urban structure changes accordingly. The structure of the urban road network is related to the smoothness and convenience of people flow and logistics in each functional area outside and inside the city, and it has a direct impact on people's daily work, life and many other aspects[3]. Whether the structure is reasonable and in line with the development requirements of the times is related to the sustainable development of the city[4], and more directly related to the convenience of our work and life. Therefore, it is of great significance to study the sustainable development pattern of urban road network structure.

At present, there are 656 cities in China, including 157 cities with a permanent population of less than 0.5 million, 256 cities with a population of 0.5-1 million, 159 cities with a population of 1-2 million, 31 cities with a population of 2-3 million, and 53 cities with a population of more than 3 million[5,6,7]. In the next 10 years, China will be in the rapid development period of urbanization, including the extension type and connotation type. On the one hand, the scale of the urban population will expand rapidly; on the other hand, the internal structure of the city will evolve rapidly. It is predicted that there will be more than 400 cities with 1-2 million people in China by 2030, accounting for more than 60% of the total number of cities in China; meanwhile, the development of each city is basically mature, the internal structure is basically finalized, and the road network structure is also basically finalized. After 2030, the road network structure is basically stable. Therefore, it is of great theoretical significance to explore the road network structure characteristics of 1 to 2 million-population cities for guiding the planning and construction of Chinese cities.

2 Research methods

2.1 Research object selection requirements and methods

From the perspective of the formation of the existing road network of Chinese cities and the development trend of the future urban scale, the selected research objects meet the following requirements: firstly, all the cities are in the Middle and Eastern China with relatively higher economic development level and relatively perfect road system, therefore, these cities are of relatively typical significance for the study of the urban road network structure of China; secondly, the urban population scales are 1-2 million.

In the selection of research objects, the questionnaire method and Delphi method are used. First of all, according to the above requirements, 50 cities with a population of

* Corresponding author: 812475701@qq.com

1-2 million are selected from the cities in East China, Central China, South China, North China, Southwest, Northwest, and Northeast China. The urban population, built-up area, GDP, and other indicators are listed in the form of a questionnaire. Then, 16 research objects are finally determined by the Delphi method. The results are as follows:

East China: Yangzhou (2.12 million), Fuzhou (2.1 million), Xiamen (1.55 million), Hefei (2 million)[5, 6, 7].

Central China: Luoyang (1.8 million), Zhuzhou (1.21 million)[5, 6, 7].

Northeast China: Jilin (1.5 million), Anshan (1.73 million)[5, 6, 7].

South China: Zhuhai (1.12 million), Shantou (1.98 million), Huizhou (1.32 million), Haikou (1.54 million), Guilin (1.11 million) [5, 6, 7].

North China: Handan (1.63 million), Baotou (1.7 million), Hohhot (1.32 million) [5, 6, 7].

2.2 Data acquisition method

After the research objects are determined, the basic information of urban traffic status, including the main transport mode, the road section, the road grade structure, and the other contents, will be sorted out first through the access to relevant urban literature and network search. Secondly, the traffic data of the urban road network map, the latest urban map, or urban traffic map in different periods of each city will be collected to compare the urban road network development laws. At the same time, combined with the methods of a field survey on some cities, the field survey is carried out on the main areas and road sections of the city to confirm the urban road network and public service facilities.

2.3 Compilation method of working base map

In this paper, the urban maps or urban traffic maps of the selected cities are taken as the base maps, and the external freeway and main urban roads (expressways and trunk streets) of the city are shown in the figures, and the external traffic facilities such as bus stations and railway stations, and tourist attractions, as well as the related elements of the public service facilities such as shopping malls, stadiums, squares, cinemas, libraries and exhibition halls are shown in the figures in the form of lines and points[8], and finally form the work base maps of the research.

2.4 Research methods

This paper mainly uses the literature research method to study the theory and the latest literature about the road network structure. Combined with the research methods of comparison, analysis, and induction, it compares and analyzes the road network of different cities at home and abroad, evaluating the advantages and disadvantages of the road network structure, summarizing the development laws of the urban road network, and puts forward the sustainable development pattern of China's cities' reasonable road network structure according to the actual conditions.

3 Research process and results

3.1 Current mode of the urban road network with a population-scale of 1-2 million in China

From the perspectives of urban external traffic, traffic between different urban functional areas, the connection between urban functional areas and major large-scale public service facilities, study the above urban road network structure, and the sustainable development pattern can be summarized as the following types (Table 1):

1) Ring freeway + ring and radial trunk streets. The city roads have formed a relatively perfect ring plus radial road network structure, and the outer-ring freeway has fully formed. Different urban functional areas are connected by ring trunk streets, and the large-scale public service facilities such as shopping malls, gymnasiums, cinemas, exhibition centers are connected by urban trunk streets. The city interior areas are connected with the entrance and exit of the freeway through the radial trunk streets. Through research, it is found that the roads between different functional areas and from the functional areas to large-scale public facilities are blocked, the burden of the main roads is heavy, and there is no connection between the expressway and the urban main roads; the connection between the urban main roads and the external traffic facilities is not enough, and there is no connection by the expressways, so it takes a lot of time to arrive at the railway station and the bus station in each area of the city; and the expressway system connecting the entrances and exits of the ring freeway has not been established yet. Fuzhou, Hefei, Handan, Luoyang, Guilin are the type of cities.

2) Semi-ring freeway + ring and radial trunk streets. The rudiment of the ring plus the radial road network has gradually appeared. The different urban functional areas are connected by weak ring trunk streets with each other, and the large-scale urban public facilities are interpenetrated on both sides of the main roads in the central area of the city. In this kind of city, the freeway appears semi-ring form in the periphery of the city. The trunk street extending from the urban center to both sides is connected with the entrance and exit of the freeway, and the urban road in the open area enters the freeway through the ring main road. For the urban road network of semi-ring freeways, the convergence of the inside and outside urban transport has increased the burden of the urban ring roads, and there is no necessary expressway connection between urban roads and freeways; the ring trunk street connection between different urban functional areas is weak, and the road connection between urban large public facilities and different urban functional areas is blocked, and there is no expressway connection with each other. Zhuzhou, Huizhou, Anshan are the type of cities.

Table1. Classification of urban road network structure in China

<table>
<tr><th>City</th><th>Road Network of Region</th><th>Road Network of City</th></tr>
<tr><td colspan="3">Ring Freeway + Ring and Radial Trunk Streets</td></tr>
<tr><td>Fuzhou</td><td>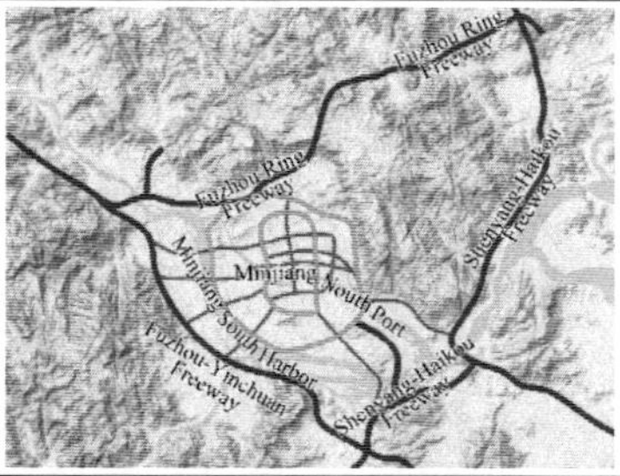
</td><td>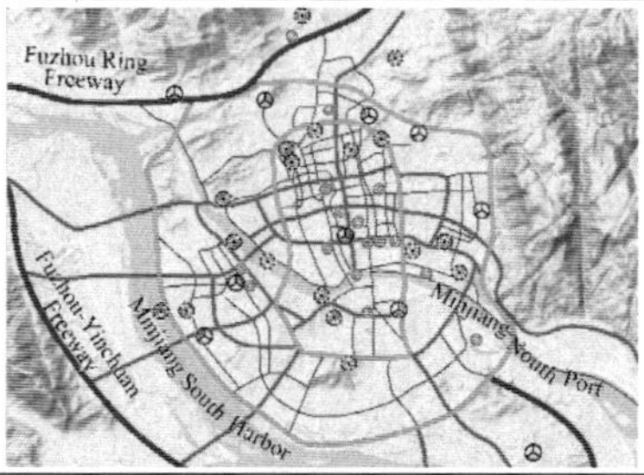
</td></tr>
<tr><td>Hefei</td><td>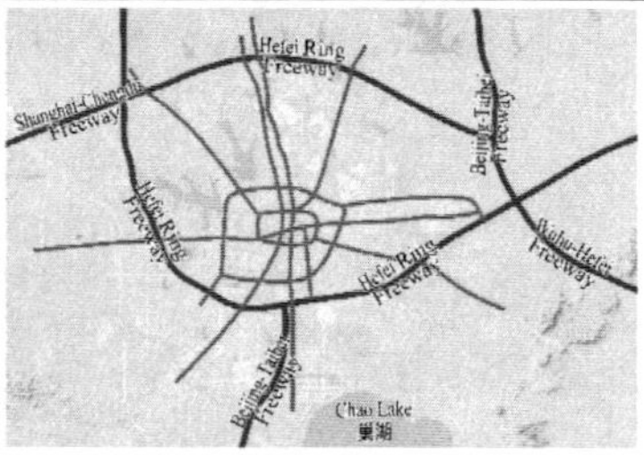
</td><td>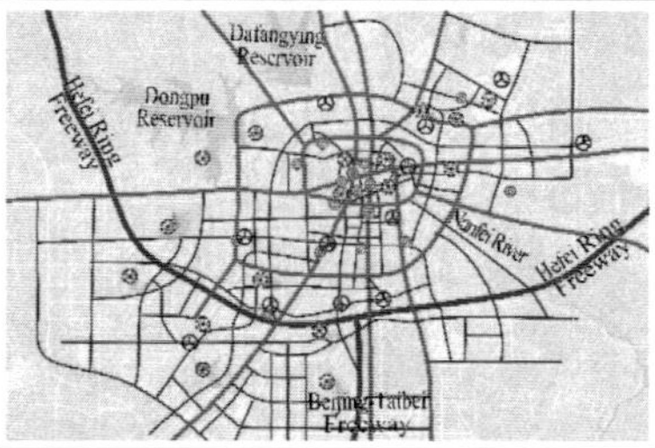
</td></tr>
<tr><td colspan="3">Semi- ring Freeway + Ring and Radial Trunk Streets</td></tr>
<tr><td>Zhuzhou</td><td>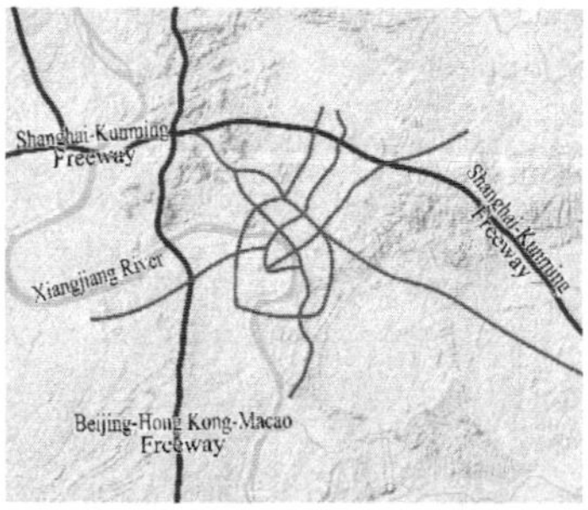
</td><td>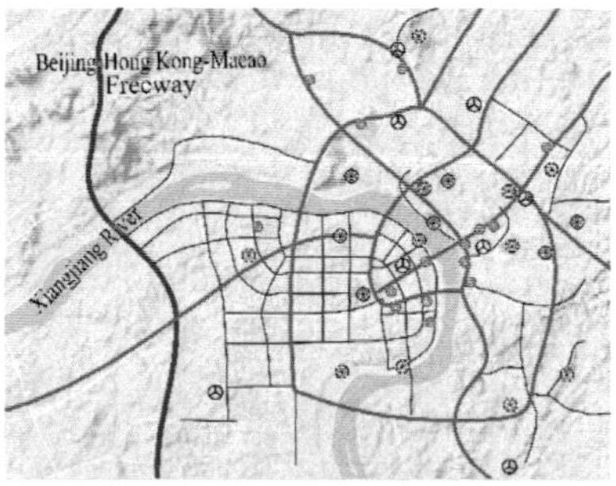
</td></tr>
<tr><td>Huizhou</td><td>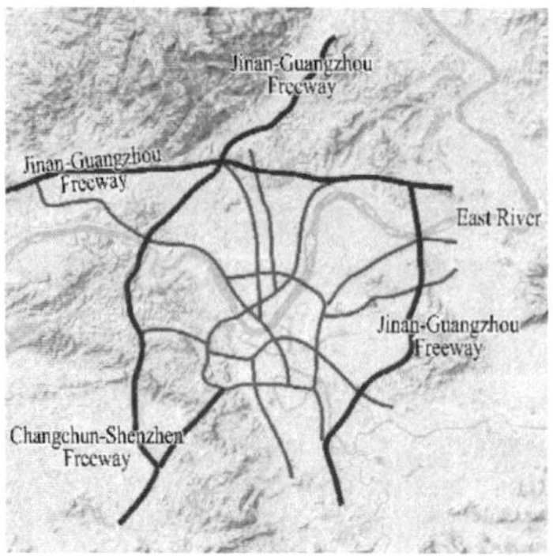
</td><td>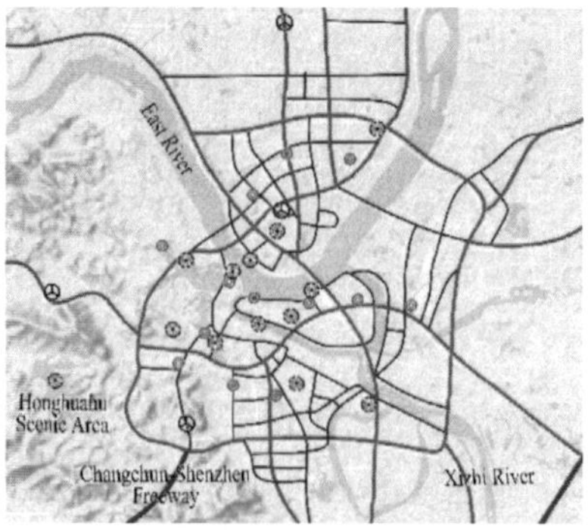
</td></tr>
</table>

Continued **Table 1.**

City	Road Network of Region	Road Network of City
Circumscribed Freeway + Ring and Radial Trunk Streets		

https://doi.org/10.1051/e3sconf/202123704018

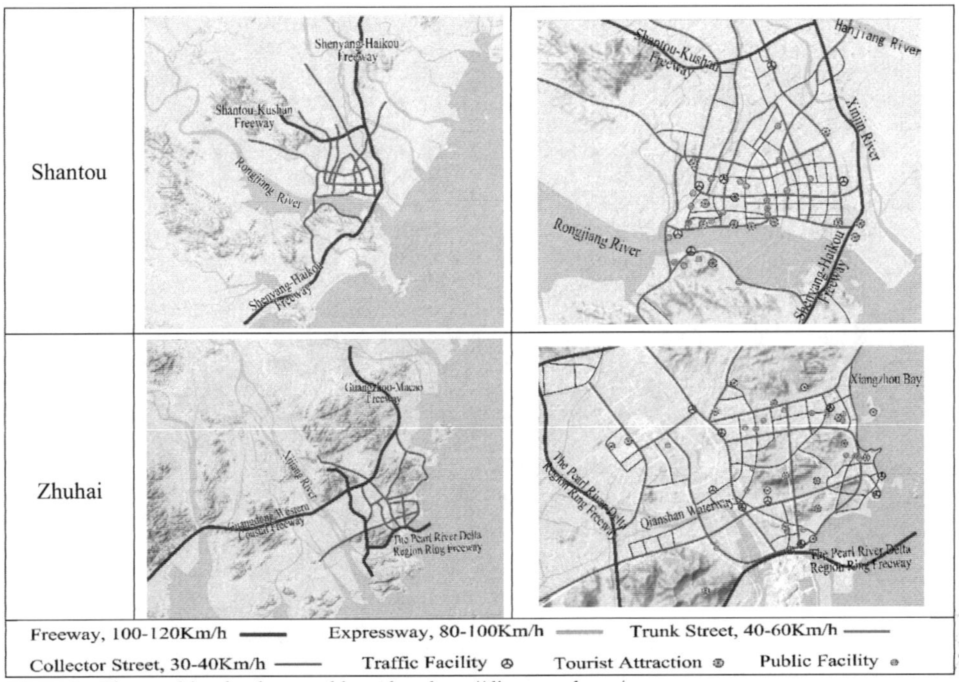

Sources: The working background based on http://ditu.google.cn/

3) Circumscribed freeway + ring and radial trunk streets. In this kind of coastal city, due to the influence of geographical locations, the freeways are all tangent to one side of a specific city. For coastal cities, the design standard of urban trunk streets is high, and urban public service facilities are organically distributed in the urban main roads network, and the roads between the different urban functional areas are generated in accordance with the city's shape. In such coastal cities, the connection of urban external traffic is only realized through the connection between urban roads and freeways on one side. Therefore, there is a lack of necessary expressway connection between them. The connections of trunk streets between different urban functional areas, and between large-scale urban public service facilities, are not enough, and there is a lack of expressway connection. In addition, considering that there should be a good connection between roads and ports, we should establish an expressway system to improve intermodal transport efficiency, such as Shantou, Zhuhai, Haikou, etc.

Based on the study of the urban road network structure characteristics with a population scale of 1-2 million in China, it is found that the road network structure types of cities with similar scale are quite different, while the road network structures of cities with different scales also have certain commonalities. From the perspective of the evolution of urban spatial structure and the development of urban road system, this phenomenon is closely related to three factors that affect the development and evolution of urban road system: the location of the city in the region, the layout of urban land, and the urban transport system. For example, Fuzhou, as the political, economic and cultural center of Fujian Province, as well as an important transport hub city in the province, has formed a road network structure of ring freeway + ring and radial trunk streets; while Shantou, a coastal city with a similar scale, has a fan-shaped urban land layout, thus forming an urban road network with fan-shaped roads as the skeleton.

3.2 Analysis on the road network structure of cities with a population of 1-2 million in China

Combined with the above research on the classification of the road network structure of cities with a population-scale of 1-2 million in China, the paper summarizes the characteristics of the urban road network structure of China's cities with a population-scale of 1 million, 1.5 million and 2 million respectively.

For a city with a population of about 1 million, due to the influence of the city scale and layout, the urban roads are arranged along with the public facilities, and the urban road network is relatively simple, which mainly presents a single ring plus cross-road network structure. The cross-roads converge in the urban center area internally, and are connected with the freeway entrance and exit externally. This road network mode has certain limitations. The cross main road not only bears the internal traffic of the city, but also bears the connecting function between the internal and external traffic of the city, which is not conducive to the development of urban traffic.

The city with a population of about 1.5 million, with the expansion of the city scale, the urban road network gradually evolved into a ring plus radial structure, and the ring freeway gradually took shape; the radial main roads connect the ring road and radiate outward to connect the ring freeway. Urban trunk streets and collector streets form the main framework of urban roads. This kind of road network structure has been gradually improved. The inner ring evacuates the traffic in the central area of the city, and the outer ring separates the internal and external traffic, and reaches the entrance and exit of the freeway through the main roads between the ring roads. However, the ring roads and the main roads are mainly the trunk streets and the collector streets, the level of the roads was relatively low, resulting in the insufficient connection of the roads.

Urban roads of cities with a population of about 2 million, have established a more mature road network system. The road network structure of the ring plus the radial expressways will be gradually formed. Through the ring plus the radial expressways, the traffic flow in the central area of the city can quickly enter and leave the city, and each area can reach the external transportation facilities and large-scale public service facilities, such as railway station and bus station, and the different functional areas can also quickly contact with each other. This kind of road network structure mode has been comparatively mature. The establishment of expressway system not only meets the rapid connection of urban internal traffic, but also makes the internal and external traffic separate from each other, and gradually builds a convenient, fast, efficient and safe urban traffic system[9].

The road network structure of cities with a population of 1-2 million in China has begun to take shape in a ring and radial way. However, because of some problems existing in the road network structure of most cities, it is proposed that in the planning stage of urban road network adjustment, the layout principles of urban expressways, trunk streets, collector streets, traffic and living roads, should be defined, and the quick connection between the urban functional groups as well as between the urban public service facilities should be considered, and the fast tracks inside and outside cities should be established. Besides, investigate the current situation, and forecast & analyse the future of the passenger flow, vehicle flow, passenger flow and cargo flow of each main road, and reserve the space of urban fast channel in combination with the actual situation and development trend of the roads and gradually build the ring roads according to the design standard of the fast tracks, and establish the fast connecting lines such as the fast tracks or the trunk streets between the rings, to play the role of "connecting" and "reaching" of the urban main roads[8,10].

3.3 Analysis of road network structure mode of foreign cities with a population-scale of 1-2 million

Foreign cities generally have large-scale freeway networks, and basically have freeways around the city periphery, such as Toulouse (935,400, 2013)[11] in the French Republic, Munich (1,300,000, 2010)[12] in the Federal Republic of Germany, Detroit (1,850,000, 1950; 880,000, 2013)[13] in the United States of America, the urban road network structure presents a double ring plus radiation shape, and the urban main roads form a ring around the city center, roads radiating outward from the center of the ring road, forming the inner-ring radial road network structure; the freeway forms a loop around the city periphery, and there are radial freeways and expressways connected with the urban main road through the ring freeway, forming the outer-ring radial freeway network structure. Except for some cities such as Detroit, due to the impact of the city's geographical location, the city center is located along the banks of the Detroit River, and the urban road network structure is in the form of fan-shaped radiation. The freeway and expressway form a ring around the central area of the city, together with the urban main roads, forming a road network structure radiating outward from the center of the city (Figure 1).

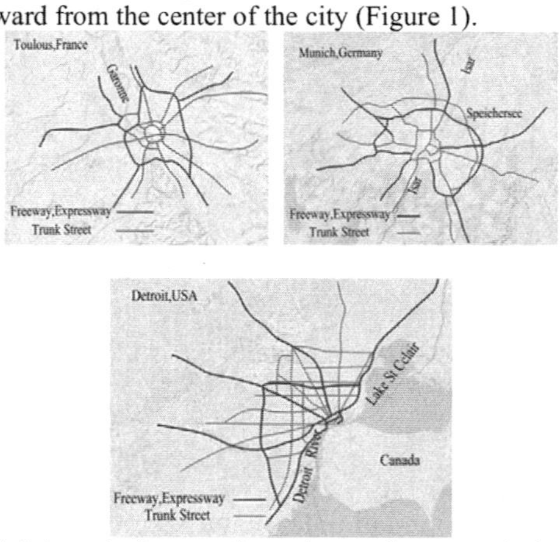

Fig1. Schematic diagram of road network structure in foreign cities. Sources: Based on the working base map provided by http://ditu.google.cn/

• The structural framework of the foreign urban road network has basically formed. Through the analysis of the structural characteristics of these mature road networks of the above cities, it is found that the road network presents certain laws.

• The freeway is closely connected with the city, and the freeways in most developed cities directly pass through the Central Business District area.

• The radial road network is relatively developed, with a large number of freeways and urban main roads radiating outwards.

• The freeway is connected unobstructedly with the urban main road, and the internal and external transport can be connected quickly.

• Shopping malls, gymnasiums and other public service facilities are also arranged on both sides of the freeway and expressway, but they are all connected through the construction of auxiliary roads to avoid interference with the freeway.

In addition, the internal road networks of city centre in western countries are mainly grid network, with uniform urban road spacing and dense distribution.

To sum up, this paper analyzes the characteristics of the foreign urban road network, and concludes that foreign urban roads have a relatively complete road grade structure, that is, the road grade system of the freeway -- expressway--urban main road[14]. At the same time, the layout of the freeway, expressway, and other high-level

https://doi.org/10.1051/e3sconf/202123704018

roads, is more flexible and occupies a higher proportion. In terms of road network structure, it combines the structure and characteristics of the city itself. Foreign cities have formed a road network structure mode with certain characteristics, for example, ring freeway + expressway + trunk street, outer-ring freeway + outer-ring expressway, and radial expressway + urban trunk street. When the city scale and urban economy develop to a certain stage, the characteristics of the urban road network layout will be more obvious, and the road network structure system will be more perfect.

3.4 Conception of the road network structure mode of cities with a 1-2 million population- scale in China

The research on the urban road network structure mode at home and abroad shows that "the ring + radial road network structure" has adapted to the development of the city, and become the inevitable trend of the sustainable development of the urban road network structure in the future. Based on the analysis of the road network structure of domestic research objects, three structure types of the urban road networks in China are found: ring freeway + ring and radial trunk street, semi-ring freeway + ring and radial trunk street, and circumscribed freeway + ring and radial main road. Chinese cities have already presented the ring and radial trunk street network, and there are freeways to establish connection with cities. However, compared with the foreign urban road network structure mode, the construction pace of urban expressways is slow in China, and the urban road grade structure is not intact; the city has built peripheral freeways, yet has not formed peripheral ring freeways.

Referring to the development characteristics of foreign urban road network structure, this paper summarizes the urban road network structure of the city with 1-2 million people, which has gone through three development stages: the embryonic stage, the incubation stage, and the mature stage, namely ring freeway + trunk street, ring freeway + expressway + trunk street, outer-ring freeway + outer-ring expressway, and radial expressway + trunk street. Combined with the analysis of the characteristics of urban road network structure at home and abroad, it is concluded that the establishment of peripheral ring freeway has become an important link to support the development of urban road network structure; the expressway is the inevitable choice of urban road development. In the process of the development of urban road network in China, the construction of urban peripheral fast tracks should take the construction of peripheral ring freeway as the goal, while the urban internal fast tracks should build the expressway system, and reserve the land for expressway development, and finally form the ring and radial expressway system. When the urban scale develops to a specific stage, the urban road network structure will correspond to the characteristics of the urban road network at a certain stage, and gradually form the corresponding road network structure mode.

4 Conclusions

Base on summarizing and analysing the characteristics of the road network structure of cities with 1-2 million people at home and abroad, this paper puts forward that with the expansion of the urban population scale, the sustainable development pattern of the urban road network structure has developed from the mode of ring freeway + trunk street (the embryonic stage) to the mode of ring freeway + expressway + trunk street (the incubation stage), and then to the mode of outer-ring freeway + outer-ring expressway and radial expressway + trunk street (the mature stage). The selected case cities have certain representativeness, and the research conclusions also have certain representativeness. The conclusion of the study has certain theoretical significance for guiding the planning and construction of the road networks of 1 to 2 million-population cities.

However, the urban road network structure is also related to many factors such as topography, socio-economic development conditions, and the number of cases is limited. The above conclusions cannot be applied to all cities.

There will be many problems in the process of development from the embryonic stage, the incubation stage, and then to the mature stage. What problems will be encountered and how to solve these problems will be the subject that the author plans to study next.

Acknowledgements

This topic comes from the scientific research project of Guilin science and Technology Bureau "Guilin modern transportation system construction", project numbered 20080612.

References

1. Pengjun Zhao, The Impacts of Land Use Intensification on Urban Transport Sustainability: Theoretical Thinking from Literature Review, Urban Development Studies, 2018, 25 (9): 108-116
2. He Gao, Shumin Feng, Caixiang Guo, Characteristic Analysis of Structural Types of Urban Road Network, Forest Engineering, 2006, 22(5): 28-31
3. Fei Shi, Wei Wang, Urban Road Network Structure Analysis, Urban Planning, 2007, 31(8): 68-73
4. Yosef Rafeq Jabareen, Sustainable Urban Forms: Their Typologies, Models, and Concepts, Journal of Planning Education and Research, 2006, 26: 38-52
5. 2015 China Urban Statistical Yearbook, 2015
6. China's top 500 cities with urban population, database of China Business Industry Research Institute[DB/OL], http://www.askci.com/news/paihang/20161208/15072081751.html, /2016-05-15
7. Ranking of population in main urban areas of China in 2014, ranking of China's urban population (top 50) by permanent population in central urban areas

[DB/OL], http://www.phbang.cn/city/145656.html, /2016-05-15

8. Xunchu Xu. Planning Thought of Urban Road Network System and Examples Analysis, Urban Transport of China, 2006, 4(1): 30-35
9. Pengfei Yang, Haodong Xie. The Optimization of Changsha Road Transportation Network Featured by Urban Space, Huazhong Architecture, 2018, 36(04):102-105
10. Lvmin Shao, Research on Xuzhou Expressway Network Planning, Proceedings of the 2017 China Urban Transport Planning Annual Conference, Urban Transport Planning Academic Committee of China Urban Planning Society, China Academy Of Urban Planning and Design, 2017: 1-8
11. Toulous [DB/OL], https://baike.baidu.com/item/%E5%9B%BE%E5%8D%A2%E5%85%B9/1568769, /2018-09-15
12. Munich [DB/OL], https://baike.baidu.com/item/%E6%85%95%E5%B0%BC%E9%BB%91/855, /2018-09-15
13. Urban culture, Detroit [DB/OL], https://baike.baidu.com/item/%E5%BA%95%E7%89%B9%E5%BE%8B/1660180, /2018-09-15
14. America Planning Association. Hierarchy of Street and Roads, Planning and Urban Design Standards, John Wiley & Sons, Inc., 2005: 226

Prediction of Underground Space Development Function of Existing Industrial District in City Based on CA-Markov Model

Xiaochun Hong [1,*], Xiang Ji [1,2,*]

[1]School of mechanics and civil engineering, China university of mining and technology, Xuzhou 211116, China
[2]Jiangsu collaborative innovation center for building energy saving and construction technology, Xuzhou 211116, China

Abstract. Scientific analysis of the spatial evolution of existing industrial areas in cities and prediction of future development needs will help to rationally allocate land resources in existing industrial areas in urban renewal, scientifically and rationally develop underground space, and promote the sustainable development of existing industrial areas. First of all, the development mode and leading function type of the existing industrial zone in the city are sorted out, and its corresponding underground space development function is further sorted out. It is found that the underground space development of the existing industrial zones in the city is closely related to the dominant functions and location of the ground renewal. To scientifically guide the development of underground space in existing industrial areas in cities, this study proposes a method based on the dynamics model and the CA-Markov model to predict the functions of underground space development in existing industrial areas in cities, which will improve the efficiency and Benefits to promote the rational allocation of land resources is of great significance.

1 Introduction

With the expansion of the city, some industrial relics with a high cultural value that were originally located at the edge of the city and gradually in the center of the city have gradually become an important part of the construction of urban characteristics. As a product of urban modernization, urban underground space plays a vital role in urban expansion, traffic pressure relief, and urban environment improvement. Existing urban industrial areas, as key objects of urban renewal, provide valuable land resources for the expansion of urban space. Through the development and utilization of the underground space of existing urban industrial areas, not only the dual goals of industrial land renewal and industrial historical heritage protection can be achieved, but also the urban space capacity will be improved and the conflict between land resource shortage and urban development will be alleviated. The establishment of a scientific method for predicting the functions of underground space in existing urban industrial zones is the key basis.

2 Literature review

In the renewal of existing industrial areas, the development of underground space plays a significant role in expanding urban land capacity, improving the urban functions of industrial relics, and promoting the integration of updated plot functions into the city [1]. Zhang Yue (2005) analyzed the effect of underground space development on the historical and cultural protection areas of Beijing's old city, and proposed multiple renewal models of historic districts based on underground space development [2], but did not verify the specific effects of these renewal models. Huang Leilei (2010) analyzed the principles, functions, development models, and spatial layout of underground space development in the renewal of the old city's industrial heritage sites based on the underground space development behavior in the renewal of the Sulun site in Suzhou, and found that underground space development was effectively promoted The integrity of the surface industrial heritage and the improvement of the vitality of the area [3], and then proposed the development guidelines for the development focus, development scope, and development intensity of the existing industrial zone in the context of protection [4,5]. Liu Jieyi et al. (2012) proposed a variety of underground space development strategies combined with heritage protection based on the concept of industrial heritage reuse. This industrial heritage protection method can increase land value and protect industrial heritage, and improve the sustainable development of urban industrial heritage and economic benefits. Obvious advantages in other aspects [6]. Zhao Yi (2015) arranged commercial, entertainment, warehousing, and municipal functions on the first basement floor, some commercial, parking, rail transit stations, air defenses, etc. on the second floor underground, parking, rail transit lines, and equipment rooms. The underground space development method set in the three underground floors promotes urban renewal, and the underground space should be reasonably developed under the intensity of ground development [7].

* Corresponding author: hongxiaochun00@163.com

Table1. Summary of main cases of renewal of existing urban industrial zones in China

Cases	**Long Museum**	**Oil Tank**	**B-LINK**	**MIFA 1862**	**Elite Valley**	**Hero World**	**Olympics Plaza**	**Su Lun Factory**	**Yuzhou Factory Site**	**Art Porcelain Factory**
Location	Shanghai	Shanghai	Shanghai	Shanghai	Shanghai	Shanghai	Beijing	Suzhou	Jingdezhen	Jingdezhen
Spatial location	City Center	City Center	City Center	City Center	City Center	Suburbs	Suburbs	City Center	City Center	City Center
Development model	Public space	Public space	Comprehensive development	Public space	Public space	Comprehensive development	Comprehensive development	Business service	Business service	Business service
The extent of historical features	fine	fine	fine	fine	fine	fine	fine	fine	fine	fine
Function matching situation	fine	fine	good	ordinary	fine	ordinary	good	good	fine	fine
Landscape environment	fine	fine	ordinary	fine	fine	ordinary	good	ordinary	good	good
Transportation Facilities	ordinary	ordinary	good	fine	fine	fine	fine	fine	fine	fine
Overall effect	ordinary	fine	fine	ordinary	ordinary	ordinary	good	good	fine	fine

Table2. Functional types of underground space in existing urban industrial areas

Cases	**Form of traffic**			**Exhibition space**	**Commercial space**	**Other public spaces**
	Rail transit	**Underground parking**	**Underground lane**			
Long Museum		√				
Oil Tank Art Center				√	√	√
B-LINK		√			√	√
MIFA 1862				√		
Elite Valley		√				
Hero World		√				
The 2022 Winter Olympics Plaza		√		√	√	
Su Lun Factory	√	√			√	
Yuzhou Factory Site		√			√	√
Art Porcelain Factory		√				

However, the underground space development in the renewal of existing industrial areas is not only affected by its own needs, but also by the development needs of the surrounding cities, and its underground space functions are also closely related to the ground renewal.

3 Existing urban industrial zone development model and leading functions of underground space

The main purpose of underground space development in existing urban industrial areas is to alleviate the contradiction between protection and renewal and the shortage of land resources, increase space capacity, improve regional functions, and improve environmental quality. Table 1 summarizes several representative cases in which my country has expanded the capacity of existing industrial areas through the development of underground space. It is clearly seen that the current renewal of existing industrial areas in China is mainly based on industrial tourism and creative industrial parks.

Existing urban industrial areas, as industrial historical relics, have distinctive cultural and historical resources, and are an important part of the city or region. The development of underground spaces has added some functions to better highlight the location and urban image, Bringing many opportunities for differentiated development of cities. Through the development and utilization of underground space, while protecting the urban texture on the ground, it also makes up for the lack of space, expands the urban capacity, drives regional development, and promotes better realization of renewal. Table 2 lists the types of underground space functions in existing urban industrial areas. The most common functions for developing underground spaces in existing urban industrial areas are underground parking and underground commerce. This aspect is driven by the protection of the surface features of existing urban industrial areas. Proper commercial development has also promoted the improvement of the economic benefits of existing urban industrial zones. Combined with the case situation in Table 1, it is seen that multi-functional development of underground space development functions based on meeting parking and commerce is an important way to improve the overall effectiveness of existing urban industrial areas.

Although underground parking and underground commercial development are common development methods in existing urban industrial areas, it is not difficult to find that there is an obvious alignment relationship between underground space and ground space, as shown in Figure 1. The selection of underground space functions corresponding to these different modes is affected by many factors. For example, from the perspective of regional development, when existing urban industrial areas that have been renewed are at the center of future urban development, they are simply from the perspective of space demand at this stage Setting out for underground space planning and design is difficult to meet the needs of future development. Urban traffic development trends corresponding to different traffic conditions also have a significant impact on the underground space development of existing urban

* Corresponding author: hongxiaochun00@163.com

industrial areas. Therefore, the use of scientific methods to simulate the evolution of the spatial structure of existing urban industrial areas is to judge the current and future development of existing urban industrial areas. Need and determine the scientific basis for the development of underground space in existing urban industrial zones.

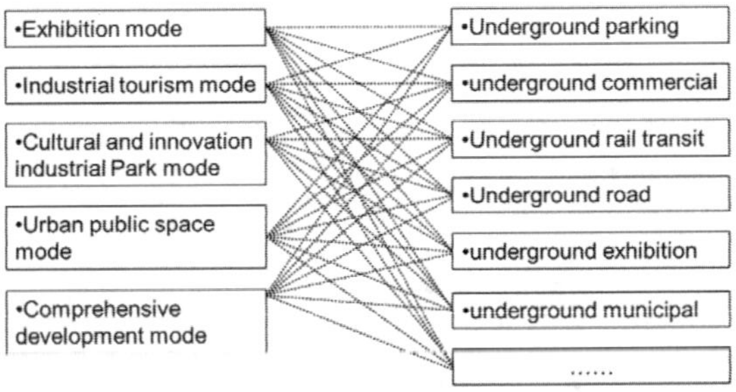

Fig1. Correspondence between ground development model and underground space.

4 Prediction of underground space development function based on the dynamic model and CA-Markov model

Based on the historical and current data of the existing urban industrial zone, using GIS technology, dynamic model and Markov model, analyze the evolution of the territorial spatial structure of the existing urban industrial zone, and discuss its changing laws; use the CA-Markov model to predict the existing The spatial pattern of land in urban industrial zones can reveal the Spatiotemporal evolution characteristics of the land in existing urban industrial zones. This Spatiotemporal evolution feature is the basis for the prediction of underground space development functions.

4.1 Land and space identification

At present, there are two common spatial identification methods, namely, merge classification method and quantitative measurement algorithm. The former is based on the current land use data, and the current data is merged and classified. The advantage is that the land function classification can be related to the current research area planning. The latter needs to construct an index system that meets the actual situation of the study area, and combine the index evaluation method to classify the land space type [8]. Due to the quantification of indicators and the differences between regions, comprehensive factors make this method not universal. Comprehensive comparison, merging The classification method is less difficult to apply, and it can be connected with the land classification system of the study area. Therefore, this study adopts the merged classification method.

4.2 Dynamic model of territorial space

The dynamic degree of land space is a reflection of the degree of changes in land space in different periods. The speed and amplitude of various spatial changes are usually expressed in a single-dynamic degree within a region, which can be reflected in a certain category during the research period. Other land spaces are transformed into other types of spatial activity[9], and the expression is:

$$K = \frac{S_i - S_{ii}}{S_{ii}} \times \frac{1}{\gamma} \times 100\% \qquad (1)$$

In the formula, K is the single dynamic index of a certain land space within a certain period time; Si is the area after a certain type of land space has been transferred, and Sii is the area when this type of land space has not been transferred; T is the study duration, this study Is the number of years. The comprehensive dynamic index is used to indicate the strength of the land space change in the study area, reflecting the comprehensive influence of people's development of space and other activities on the change of the land space structure. The expression is:

$$G = \frac{\sum_{i=1}^{n} \Delta S_{i-j}}{2\sum_{i=1}^{n} Si} \times \frac{1}{T} \times 100\% \qquad (2)$$

In the formula, G is the comprehensive dynamic index of the territorial space within a certain period of time; △ Si-j is the absolute value of the area of the i-th type of territorial space utilization type converted to other territorial space types during the study period, and Si is the absolute value of the area at the beginning of the study period. Type i area of land space utilization type.

Table3. Appropriate functions of underground spaces in existing urban industrial zones of different development types

Development model	Specific mode	Function type
Cultural development	Museum, Cultural and Creative Park	Underground exhibition hall, underground garage, underground commerce, underground storage, sunken space for public activities
Residential development	Residential community	Underground parking, underground storage
Business service development	Commercial streets, shopping centers, etc.	Comprehensive consideration of functions such as underground commerce and transportation facilities
Business office development	Urban industrial park, high-tech industrial park	Mainly underground garage, taking into account other supporting facilities
Leisure tourism development	City parks, industrial tourism	Mainly underground garage, taking into account commercial and municipal facilities; sunken space for public activities
Comprehensive development	Urban RBD, mixed function park	Coordinate the development of underground space, including: underground parking, cultural, commercial and other facilities
Other public facilities development	School, medical, sports, etc.	Mainly underground garage, taking into account other supporting facilities

Table4. Suggestions on the future development function of underground space in existing urban industrial areas

Cases	Form of traffic			Exhibition space	Commercial space	Other public spaces
	Rail transit	Underground parking	Underground lane			
Long Museum		√	○			
Oil Tank Art Center				√	√	√
B-LINK		√	○		√	√
MIFA 1862				√		○
Elite Valley	○	√			○	○
Hero World	○	√	○		○	○
The 2022 Winter Olympics Plaza	○	√		√	√	○
Su Lun Factory	√	√	○		√	○
Yuzhou Factory Site		√			√	√
Art Porcelain Factory		√	○		○	○

Remark: "√"Indicates the existing underground space function of the existing urban industrial zone,"○"Indicates new features in the future

4.3 CA -Markov model

4.3.1 Principle of the CA model

CA (Cellular Automata) consists of 4 parts: cell and cell state, neighborhood, and conversion rules.

The smallest unit of calculation; the state refers to the cell category; the neighborhood structure determines the conversion state of the current cell; the conversion rule is the cell conversion method. CA is a spatiotemporal dynamic simulation model in which space, time, and state are discretized. The general CA model formula is as follows [10]:

$$S_{(t+1)} = f(S_{(t)}, N) \quad (3)$$

In the formula, S is a finite and discrete state set of cells; N is the neighborhood of the cell; t and t+1 represent two different moments; f represents the state transition rule.

4.3.2 Markov transition matrix principle

The Markov model is studied by the method of probability theory. It is in a random time series, using a mathematical model to perform matrix analysis of time, and estimate the future possibility based on the current state and the trend of change. The Markov model can be expressed by the following formula [10]:

$$A_n = A_{n-1} X P_{ij} \quad (4)$$

In the formula, An and An-1 are the state of the land and space at two moments; Pij is the state transition probability matrix. Determining the state transition probability matrix is the key step of the model research, which can be expressed by the following formula [10]:

$$P = \begin{bmatrix} P_{11} & \cdots & P_{1n} \\ \vdots & \ddots & \vdots \\ P_{n1} & \cdots & P_{nn} \end{bmatrix} \quad (5)$$

In the formula, P is the probability of land space type conversion, each row in the matrix is the probability of land space type transferring to other types, and each column in the matrix is the probability of land space type being transferred from other types.

4.3.3 CA -Markov model

The CA-Markov model is a dynamic model with discrete-time, space, and state. Its birth combines the advantages of Markov model time dimension analysis and CA model space dimension analysis. It is an important research method of nonlinear science. [10].

4.3.4 Prediction of functions of underground space in existing industrial areas based on the CA-Markov model

Based on the CA and Markov model, we can intuitively analyze the development history, current situation, and future trends of existing industrial areas, and propose suitable functions for the development of underground space in existing urban industrial areas corresponding to different surface development models, see Table 3, combining surface development models and underground space development The function correspondence table, and the corresponding future underground space development functions of existing urban industrial parks are proposed in Table 4.

5 Conclusion

Urban renewal has become the main direction of urban development in China[11]. The renewal of existing urban industrial areas in busy cities has an indelible effect on the value enhancement of existing industrial areas and the enhancement of urban functions. This study builds a CA-Markov model to predict the development trend based on the objective laws of existing urban industrial zones and urban development. According to the development trend and the base characteristics and development suitability characteristics of existing urban industrial zones, the future development of underground space functions is proposed. Direction, to provide theoretical support for guiding the scientific development of existing urban industrial zones and the rational construction of underground space.

Acknowledgements

This research was funded by the National Key R&D

Program Projects of China, grant number 2018YFC0704903 and National Natural Science Foundation of China, grant number 51778611.We greatly thank all participating experts, the visually impaired and volunteers.

References

1. Huang Leilei, Cao Yuna. Development and Utilization of Underground Space in Old City Industrial Heritage Area[J]. Tianjin Construction Technology, 2010, 20(2):18- 19.
2. Zhang Yue. Research on the Development and Utilization of Underground Space in Beijing Old City Historical and Cultural Protection Area[D]. Beijing: Beijing University of Technology, 2005.
3. Huang Leilei. Research on the Development and Utilization of Underground Space in Old City Industrial Heritage Area[D]. Suzhou: Suzhou Institute of Science and Technology. 2010.
4. Xia Jian, Wang Yong, Huang Leilei. Research on Underground Space Development Planning Strategies of Old City Industrial Heritage in the Context of Protection[J]. Journal of Suzhou University of Science and Technology: Engineering and Technology Edition, 2015, 28(3): 65-69 .
5. Xia Jian, Wang Yong. Research on the integrated development of underground space in the industrial heritage area of the old city based on overall protection[J]. Huazhong Architecture, 2016, 34. (4):17-20.
6. Liu Jieyi, Xia Jian, Shen Qing. Discussion on the protection of industrial cultural heritage combined with the development of underground space[J]. Chinese Journal of Underground Space and Engineering, 2012(2):5-10+17.
7. Zhao Yi. Research on the renewal strategy of Ningbo urban center from the perspective of inventory planning[D]. Hangzhou: Zhejiang University, 2015.
8. Xue B, Xiao X, Li J. Identification method and empirical study of urban industrial spatial relationship based on POI big data: a case of Shenyang City, China[J]. Geography and Sustainability, 2020, 1(2): 152-162.
9. Adhikari S, Southworth J. Simulating forest cover changes of Bannerghatta National Park based on a CA-Markov model: a remote sensing approach[J]. Remote Sensing, 2012, 4(10): 3215-3243.
10. Fu X, Wang X, Yang Y J. Deriving suitability factors for CA-Markov land use simulation model based on local historical data[J]. Journal of environmental management, 2018, 206: 10-19.
11. Hong Xiaochun. Historical evolution, current situation and development trend of space planning [J]. Journal of xi 'an university of architecture and technology (social science edition),2020,39(02):20-30.

Slenderize Your Body: Application and Evaluation of Anatomy in Apparel Design

Yan Cui[1,2], Xiaogang Liu[1] *

[1] College of Fashion and Design, Donghua University, Shanghai 200051, China
[2] Cornell University, Fiber Science and Apparel Design, Ithaca, NY, USA

Abstract. The idea of thinness as beauty is still the dominant ideology of modern women, and body modification through clothing is one of the main motivations for women's clothing consumption. Denim, as a primary category of body shaping apparel, has been gaining attention from brands and developers in recent years. This paper investigates an ergonomics-based approach for slim-fit denim garments that combines aesthetic plastic and stereovision into the fashion design. Ninety-seven female subjects with a background in fashion design were studied through data visualization and statistical analysis. Women's perception of a denim clothing line with anatomy design factors (ADF) was studied. The results reveal that ADF has a positive effect on the visual perception of female consumers and can help subjects to visually optimize their body shape, increase their body satisfaction and purchase intent. This study provides theoretical and data support for the integration of fashion art and sustainable innovation.

1. Introduction

Women's quest for a slimmer figure continues to be a global trend. Nowadays, the average female weight has tended to be rising. However, the mainstream cultural expectations for ideal feminine beauty was contrary to the increasing trend of average female weight, and this confliction makes women keep struggling between their perceptions of actual-self and internal ideal-self [1]. A compelling example was that young females in Hong Kong pay more attention to their diets even though they have had slender body shape [2]. In the prior study about the female body image of Korea and Hong Kong, the perception of "ideal beauty" across different cultures has been entirely consistent at the present days has some cross-cultural consistency to a certain extent [3]. The ideal female image forms a cross-cultural standard due to the uninterrupted impact of Asian culture on western countries [4].

Female consumers try to pursuing a perfect body in various ways, and the same is true for clothing consumption [5]. Throughout history, clothing consumption has not only fulfilled a person's functional needs, such as keeping warm and covering the body. Another form of the functionality of clothing products is called soft skill [6]; it includes the emotional and aesthetic needs of consumers. The primary function of denim as a typical garment was to flatter the figure. The "China Jeans Market Research and Development Prospects (2018-2025)" report shows that 86.3% wear jeans regularly, and only 1.3% do not wear jeans. Denim products have become a fashion sign for young people. The data reveals that only 13.5% of people prefer to wear casual jeans, while 64.3% prefer skinny jeans. With the rise in aesthetic requirements, female consumers are becoming higher demanding of their body-shape. As a result, more and more jeans brands were taking fit and optimizing body shape as one of their essential design criteria. For example, Levi's "Lady Style" collection has a smaller back patch pocket with a heart-shaped back piece, which was higher than the rest of the brand's jeans and was cut in a princess cut to create a buttock lift. However, these design methods remain in the realm of traditional clothing design, where designers shape the body by changing the shape of cut pieces, raising the waistline, or even using highly stretchable fabrics. The denim garment used in this research explored a new design methodology, which combined a series of anatomical shadows of ideal body image with washing area and optical illusion patterns to optimize the wearer's body shape on visual perception [7]. Anatomy design method incorporates visual principles to alter the cut of the denim garment; all of these design elements help the wearer to optimize their figure. This paper examines the ability of new design methods to optimize stature and its impact on consumer preferences through statistical analysis and data visualization analysis.

2. Methodology

In order to understand Asian female's perceptions of denim wears in China, especially for Urban Households, multiple statistical analysis methods were used in this research. A professional online questionnaire collection

*Correspondence: liuxg@dhu.edu.cn

website issued the questionnaires. Participants can fill in the questionnaire either by the webpage or Mobile client. A total of 20 pre-tests were conducted to assess the efficacy and applicability of the questionnaire for denim wears. Minor revisions were made based on the observations and results of the pre-test. The final version of the questionnaire was designed and organized into eight sections [8, 9]: basic information, subjects' subjective perceptions of the new design; design evaluation of three types of clothing; the relationship between subjects' viewing perspectives and weight loss effects; the relationship between weight loss effects and willingness to purchase. Five-point Likert scale questions (anchored from strongly disagree=1, 2,3,4,5 to strongly agree) were used to this survey combining with the single choice question, multiple-choice questions, and open-ended questions. The selection of samples follows diversity and representative: the samples should contain primary types of denim wears; they should be represented the typical garments in the denim market. Thus, a denim garment collection (5 dress, 2 short jackets, and 15 jeans.) was selected from the current market in this study, which applied to the anatomy design method. The test garments use four major design factors [7, 10]: anatomy shading, anatomy patterning, and anatomy warping (Figure 1). The principle of the Anatomy design method shown in figure 1.

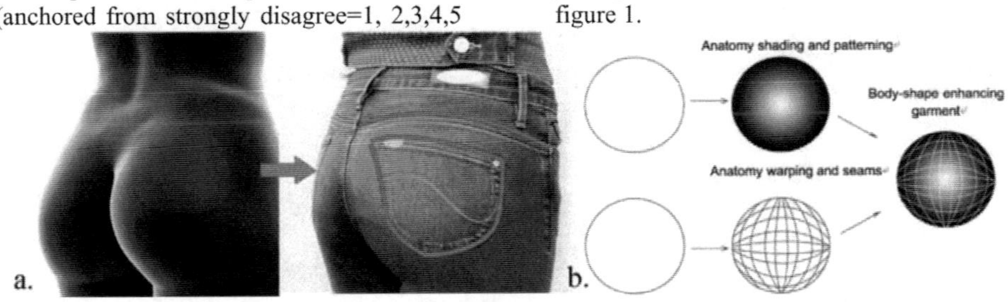

Figure 1. (a) left was a medically accurate illustration of female butt, the highlights area on this standard image were extracted (red area) and then applied it on jeans' design as washing patterns. In this way, the designer mapped the ideal body image onto the human body, creating a slim effect of garments. The effect helps the wearer improve body satisfaction. The red line corresponds to anatomy warping. **(b)** The design principle of the anatomy design method was to apply anatomy shading & patterning and anatomy warping & seams on the garment at the same time, which superimposed to form a strong three-dimensional visual perception [7, 11].

All participants were females and were recruited from a University in the Yangtze River delta area either at campuses or via personal contact by the researcher and research assistants, 96.91% of them came from Shanghai, based on the geographic analysis. Test garments were carried out by 100 female participants for the sake of result representativeness and authoritative, 96.9% of them in the age range of 20 to 30 years old with a standard deviation of 0.56(n=94). All of them were selected from design-related majors with relatively higher aesthetic sense and professional knowledge background. 97.9% of them were undergraduate students or above with a standard deviation of 0.57. During testing, 104 questionnaires were issued in total. Seven of them were excluded from the dataset due to inadequate responses or repeated submit, leaving a total of 97 valid questionnaires (93.27% return rate).

3. Results and Discussion

3.1. Subjective perceptions of the anatomy design method

An open-ended question was set in the test: please state the main reason why you choose this product. Then, with the online visualization tool Voyant Tools, the open-ended questions were analyzed to find out the testers' subjective feelings about the test garments. The corpus has 362 total words and 162 unique word forms, and vocabulary Density was 0.448. Most frequent words in the corpus were thin (11), good (10), jeans (9), looking (9), and comfortable (7). They were combining the results of cirrus, trends, and the original corpus (Figure 2a, b). The study found that subjects' subjective feelings about the test clothing were "thin, good-looking, and comfortable, " following with "longer leg" and "slim upper bod." Links' results reveal the relationships and distances between words. When "thin" was chosen as the core word, the words associated with it were "legs, looks, longer, beautiful, and design" In combination with the original corpus, it was found that when a garment flatters the wearer's figure, they perceive it as elegantly designed (Figure 2c). Furthermore, Bubblelines results indicate that the words thin and good-looking were highly overlapping. Garments' slim effect was directly related to consumers' subjective feelings of good looking, which means that "thinness as beauty" was still the mainstream thinking of young women.

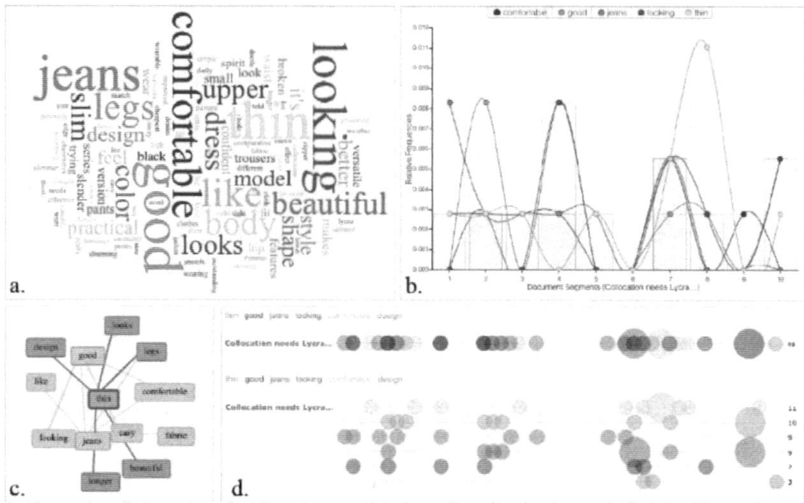

Figure 2. Data visualization and analysis results (Voyant Tools). **(a)** Cirrus analysis, **(b)** Trends analysis, **(c)** Links analysis, **(d)** Bubblelines analysis.

3.2. The primary site of action of body-enhancing effect for three types of garments

A multiple-choice analysis of where the three product categories specifically optimized body shape led to the following conclusions: jeans optimized the subjects' hips and Legs, visually curvy the hips, and slim the legs. The dress optimizes the subject's waist and abdomen, visually tapers the waist and conceals abdominal fat. The jacket optimized the subject's arms and abdomen, visually tapers the arms and also hides the belly fat. Overall, the jeans could optimize the whole body evenly, basically at 45% or more. On the other hand, Dresses and jackets can optimize mainly on individual parts. Particularly, 77.78% of the participants believe that the jacket modified their waist curve, while for the rest part was apparent. Generally, jeans were better than the remaining two types of clothing in terms of overall optimization.

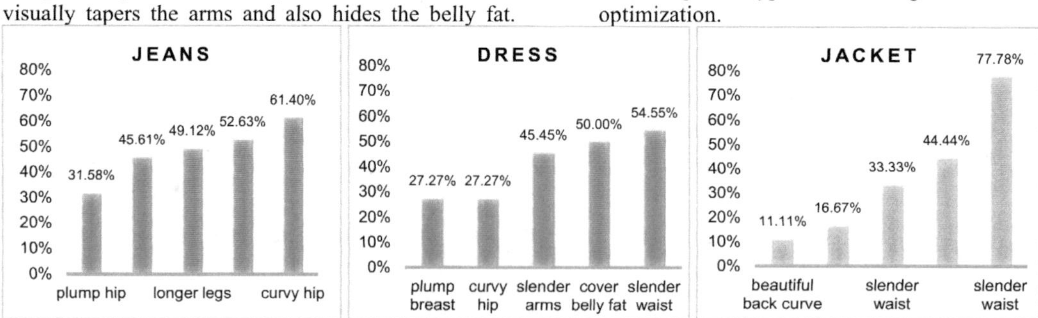

Figure 3. The primary site of action of body-enhancing effect for three types of garments.

3.3. Consumer's viewing-angle verse anatomy denim garments' body-shape enhancing ability

The test garments were designed with a series of anatomy design factors (ADF). All of them were two-dimensional patterns or seams before the garments were tried on human bodies. The original flat design factors would turn to three-dimensional elements after putting on the body. It was unclear whether the stereoscopic illusion function of ADF would increase or not due to garments' display state. Therefore, the main research purpose of this part was to explore whether the display mode of garments would affect the efficacy of ADF, and further to find out the best display mode for tested garments. It was known that Asian female consumers prefer a slimmer body-shape; we presumed that the weight-reduction effect of test garment was highly correlated with the body-shape enhancing ability. In this section, the visually weight-reduction effect was regarded as the body-shape enhancing ability. The result of this section showed that the weight reduction effect from mirror viewing angle was better than self-observation. 65.98% of females believe the test

Correspondence: liuxg@dhu.edu.cn

garments have 1kg to 3 kg weight-loss effect on visual perception by self-observation and 67.10% by mirror observation. In addition, the 3kg to 5 kg weight-loss effect increased from 4.12 %(self-observation) to 17.53 %(mirror). And 0kg weight-loss effect reduced from 29.9%(self-observation) to 14.43%(mirror). The paired samples T-test results shown the weight-loss effect from the mirror was greater than self-observation (t=5.57, P<0.01) (table 1)

Table 1. The paired samples T-test of weight-loss effect for different viewing angle.

		Paired Difference							
					95% Confidence Interval of the Difference				
		Mean	Std. Deviation	Std. Error Mean	Lower	Upper	t	df	Sig. (2-tailed)
Pair 1	Mirror – Self	.309	.547	.056	.199	.420	5.56996		.000

Besides, this section also tested the visual perception of other observers for tested garments, which would affect the consumer satisfaction assessment of garments. The comments or compliments from others would influence the self-satisfaction of social individuals. With the help of others opinion, subjects feel their actual-Self State were closer to ought-self state, and this would pass on this satisfaction to the clothing they wear, thereby consumers' satisfaction towards testing garments were increased. The result revealed that 73.2% of the subjects were told that they looked thinner than usual. The data were found to be unaligned by the tests of variance congruence of OTHERS(From others observation, how many people tell you that you have a better body size and shape?) and WTB (want to buy), so using SPSS Kruskal-Wallis H-test analysis of the relationship between OTHERS and WTB. Based on the hypothesis test summary, it can be assumed that the distribution of WTB was not identical across groups. The difference was statistically significant (H = 20.691, P = 0.002); The median of WTB was 3 in the 0 person group (n = 26); The median of WTB was 4 in 1 person group (n = 22) and 2-3 person group (n = 45); and the median of WTB was 5 in 4+ person group (n=3). Significance values have been adjusted by the Bonferroni correction for multiple tests. The distribution of WTB scores was statistically significant between the 0 and 2-3 person groups (Adj. P=0.001), 0 and 4+ person groups (Adj. P=0.008), with no difference between the other groups. This means that when less than one person tells the tester looks thinner, it does not affect the consumer' s desire to buy, but more than two people stating the same information will change the consumer' s purchasing willingness.

a.

Report

A_WTB

others	N	Median
1	26	3.00
2	23	4.00
3	45	4.00
4	3	5.00
Total	97	4.00

1= zero people told subjects she looks thinner
2=1 people told subjects she looks thinner
3=2-3 people told subjects she looks thinner
4=4 or more people told subjects she looks thinner

d. Pairwise Comparisons of others

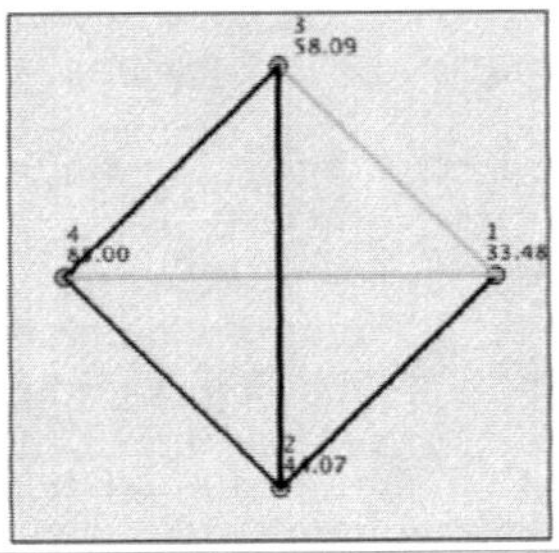

c. Each node shows the sample average rank of others.

Sample1-Samp...	Test Statistic	Std. Error	Std. Test Statistic	Sig.	Adj.Sig.
1-2	-10.584	7.559	-1.400	.161	.969
1-3	-24.608	6.505	-3.783	.000	.001
1-4	-51.519	16.101	-3.200	.001	.008
2-3	-14.024	6.768	-2.072	.038	.230
2-4	-40.935	16.209	-2.525	.012	.069
3-4	-26.911	15.746	-1.709	.087	.525

Each row tests the null hypothesis that the Sample 1 and Sample 2 distributions are the same.
Asymptotic significances (2-sided tests) are displayed. The significance level is .05.
Significance values have been adjusted by the Bonferroni correction for multiple tests.

Figure 4. (a) Median report between OTHERS and WTB, **(b)** & **(c)** pairwise comparisons of OTHERS.

The above results exposed anatomy design method was universal for typical clothing styles. This section further confirms the importance that young female consumers place on the slimming effect of clothing. At the same time, the anatomy design method visually gives the wearer a weight-loss effect. This effect has been verified not only by the wearer themselves but also by others who have observed wearers. The results show that the anatomy design method has a uniform and stable body modification effect on both others' perceptions and self-perceptions. Furthermore, tested garments had a better weight-reducing effect after trying on, which means the ADF played a better role in three-dimensional situations. In consequence, the best way to display tested garment was stereo exhibiting rather than hanging them up. Meanwhile, trying to ask consumers to wear such garments can promote their satisfaction. In order to identify the relationship between garment shaping ability and consumers' willingness to purchase, this section calculates the correlation between weight reduction effect and willingness to purchase using correlation analysis.

The correlation analysis indicated a direct correlation between body-shape enhancing ability and consumer behavior and preference (stage 2). All of the correlation coefficients were significant than zero, which means the body-shape enhancing ability of testing garments have a positive influence on female consumers' behavior and preference ($p < 0.01$). Specifically, female consumers' preference was directly associated with their mirror image improvement (Correlation coefficient=0.503, $p<0.01$). Among three viewing angles, mirror image has the most significant impact on consumer preference, following with friends' compliments and self-observation.

Table 2. Correlation analysis depicting body-shape enhancing ability influence on female consumer behavior and preference

	Self	Mirror	Others
Visual difference	0.417**	0.487**	0.451**
Ideal body shape	0.411**	0.495**	0.482**
WTB	0.416**	0.503**	0.447**
How much	0.225**	0.266**	0.313**

* p<0.05 ** p<0.01

4. Conclusion

This paper investigates a design method developed for slim denim clothing that incorporates human ergonomics, aesthetic plastic, and visual perception theory. Ninety-seven female subjects with fashion design or related background were invited to evaluate the products. A complete clothing collection was selected as test garment samples to study women's subjective perceptions of the anatomy design method through data visualization and statistical analysis. The relationship between the observation perspective and the garment shaping ability was calculated. Finally, the relationship between garment shaping ability and consumer purchase intention was analyzed. The study found that the participants' attitudes toward the anatomy design method were as follows Positive. They felt that tested garments left them with subjective feelings of thin, good-looking, and comfortable. The subject's perception depending on the observation angle and is better observed through a mirror than self-observation. Also, praise from others promotes consumers' willingness to purchase when more than two people have positive comments about the consumer. The willingness to buy changed from neutral to wanting to buy. At the same time, consumers perceived the test garments as visually different from other products, whether in a mirror view, self-observation, or evaluation by others, and It can help it create the perfect body. Moreover, the willingness to purchase most increases with the ability of the garment to shape the body. This study provides case studies and data to support the interdisciplinary application of apparel design for the development of slim-fit denim, and it also contributes to the fusion of fashion art and sustainable innovation.

Acknowledgment

The authors acknowledge the financial support of this work from the Discipline construction project (improving independent innovation and social service capabilities) under award number 107-08-0241028 from Shanghai Style Fashion Design &Value Creation Collaborative Innovation Center.

References

1. Saraceni R, Russell-Mayhew S. Cultural expectations of thinness in women: A partial replication and update of magazine content. Eating and Weight Disorders-Studies on Anorexia, Bulimia and Obesity 2007;12(3):e68-e74.
2. Lee S, Lee AM. Disordered eating in three communities of China: A comparative study of female high school students in Hong Kong, Shenzhen, and rural Hunan. International Journal of Eating Disorders. 2000;27(3):317-27.
3. Hsu H-J, Burns LD. Clothing evaluative criteria: A cross-national comparison of Taiwanese and United States consumers. Clothing and Textiles Research Journal. 2002;20(4):246-52.
4. DeLong M, LaBat K, Nelson N, Koh A, Kim Y. Global products, global markets: Jeans in Korea and the United States. Clothing and Textiles Research Journal. 2002;20(4):238-45.
5. Cui Y, Dai Y, Liu X, editors. The Influence of Evaluation Subject Visual Perception and Fitting Experience on Consumer Preference for the Body-Enhancing Garment. International Textile and Apparel Association Annual Conference Proceedings; 2019: Iowa State University Digital Press.
6. McDonagh-Philp D, Lebbon C. The emotional domain in product design. The Design Journal. 2000;3(1):31-43.
7. Peshek DJ, Mark JT, Marion B, Stephens K, Hoffman D, Zades SH, inventors; Google Patents, assignee. Body-enhancing garment and garment design2019.
8. Patterson PG, Spreng RA. Modelling the relationship between perceived value, satisfaction and repurchase intentions in a business-to-business, services context: an empirical examination. International Journal of service Industry management. 1997.
9. Alexander M, Jo Connell L, Beth Presley A. Clothing fit preferences of young female adult consumers. International Journal of Clothing Science and Technology. 2005;17(1):52-64.
10. Hoffman D, Peshek DJ, Dull SF, Zades SH, Fisher RON, Mark JT, et al., inventors; Google Patents, assignee. Anatomy shading for garments2019.
11. Cuenca-Guerra R, Quezada J. What makes buttocks beautiful? A review and classification of the

https://doi.org/10.1051/e3sconf/202123704020

determinants of gluteal beauty and the surgical techniques to achieve them. Aesthetic Plast Surg. 2004;28(5):340-7.

 https://doi.org/10.1051/e3sconf/202123704021

Research on the Urban Edge Space Form of Tanhualin Historical Living Block in Wuhan

Sang Xiangrui[1,*]

[1] College of Civil Engineering and Architecture, Wuhan University of Technology, Wuhan, Hubei, 430070, China

Abstract. Edge space is not only the main place for residents' public activities, but also the key component of urban public space. It promotes people's spontaneous activities and social activities. Reasonable urban fringe space design can create the possibility for people to transform the city from bottom to top. While preserving the cultural background, Tanhualin historical life block has formed a rich and varied space form. This paper analyses the composition of its marginal space to explore the importance of edge space to urban public activities.

1 Introduction

From the point of view of psychology, architecture and environment constitute the relationship between the figure and the ground in the urban plane form. In the process of their identity switching, the boundary space plays a defining role. Although the proportion of marginal space in the city is small, it provides convenient and rich public places for people. In recent years, more and more scholars began to pay attention to the topic of "urban acupuncture design", trying to stimulate the internal vitality of the city through small-scale transformation, so as to achieve urban renewal from the bottom to the top. In my opinion, to control the shape of the edge space is an important link to revitalize the city.

Kevin Lynch divides cognitive elements into five types in The Image of the City, and edge is one of them. It is a linear space independent of the road element, which is generally but not necessarily the boundary of the two regions. [1] From the macro point of view, the edge can be reflected in the form of riverside and expressway of the urban boundary; from the micro perspective, the edge space is the transition space between buildings and buildings, buildings and roads, roads and districts, or between multiple districts, and plays a role of limiting the adjacent two areas. Based on the actual situation of Tanhualin historical life block, this paper mainly explores the form and composition of its small-scale edge space from the micro level.

2 Research background

2.1 Overview of historical blocks

Tanhualin historical living block is located in the northeast of Wuchang old city. It is surrounded by Zhongshan Road in the East, Desheng Bridge and Fenghuang Mountain in the west, Liangdao Street in the south and Zhongshan Road in the north, with a total length of about 1.2km. As a relatively well-preserved historical block, it has gathered dozens of outstanding buildings with a century-old history, including churches, schools, medical buildings and many former residences of celebrities (Fig. 1), which vividly reflect the development of cultural elements such as religion, education and urban life of the old Wuchang. [2] In the process of urban modernization, Tanhualin block has been transforming the block in response to the needs of modern residents while protecting the historical context, forming a unique spatial form with the coexistence of the old and the new, diverse culture and rich levels. It has developed into an important cultural place for leisure activities of local residents and sightseeing for foreign tourists.

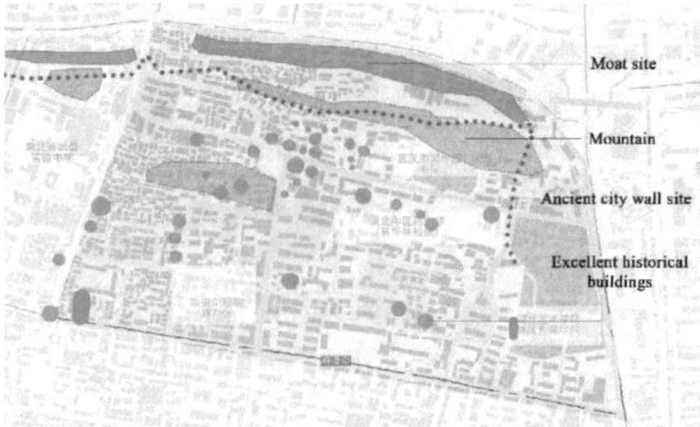

Fig. 1. Distribution map of historical and cultural elements in Tanhualin.

* Corresponding author: xiangruisang@whut.edu.cn

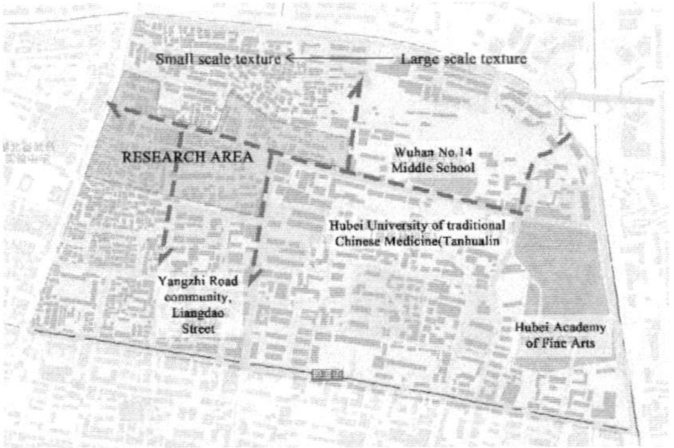

Fig. 2. Block texture and investigation route.

2.2. Spatial structure and research route

The whole Tanhualin block can be divided into three areas: commercial leisure, historical tour, and old city residence [3], but some of them interpenetrate with each other, forming a complex spatial form. The urban fabric transitions from the large-scale space to the small-scale living community from east to west, giving people a rich and varied space experience.

The survey time is selected in the time period from 2:00 pm to 5:30 pm on Wednesday to avoid the interference of the sudden increase of passenger flow on holidays to the normal state of life block. The survey route (Fig. 2) starts from Zhongshan Road in the East, and the entrance commercial area is dominated by large-scale space with pedestrian lines inside, echoing the texture of historical blocks, and realizing the transition from the main urban road to the Tanhualin historical block. In addition, subway stations and bus stops are located nearby, which is the main source of external passenger flow in the area. As a modern business district, small flower beds are often used to separate the building from the road in the edge of the building, or the corresponding pieces are placed according to the style of the store to attract tourists to stop, and various forms of seats are arranged to meet people's leisure needs (Fig. 3).

Fig. 3. Tanhualin business district.

The front section that officially enters the Tanhualin block will pass through Hubei University of Traditional Chinese Medicine and Wuhan No. 14 Middle School. Some modern buildings are preserved in this section, and now it has been transformed into functions of education, office or residence. The scale of the street is relatively large, and the vehicle and pedestrian areas are divided by paving transformation, planting of sidewalk trees and small road curbs, and mobile libraries and strip seats are installed at nodes to form a stay space (Fig. 4). The marginal space between the building and the road is raised according to the height difference. On the one hand, it can isolate the flow of people and traffic to protect the safety of pedestrians; on the other hand, it can be used as an extended outdoor space on the first floor of the building, with greenery in front of the door or leisure seats.

Fig. 4. Road and mobile library.

Fig. 5. Intersection of Tanhualin core area.

As we enter the core area of Tanhualin, the width of the road begins to shrink. At the same time, a liftable roadblock is set up at the entrance to block the passage of vehicles and define the scope of the pedestrian block (Fig. 5). Since then, it is the main research area of this paper. The core area of Tanhualin was originally dominated by a large number of small-scale residential areas and the remaining scattered religious, medical, educational and other types of historical preservation buildings. Later, with the main pedestrian street as the axis, the characteristic creative commerce was expanded to both sides. Combined with the terrain characteristics of the surrounding mountains, the horizontal street texture and scale change rich, and the vertical direction forms different views according to the terrain undulation. It is a special street with space experience. As the commercial pedestrian street on the central axis connects the residential areas on both sides of the north and south, which is one of the main routes for residents' daily life, so I also list it as a part of Tanhualin historical living block for investigation.

3 Investigation of Tanhualin living block

3.1 Central road

The pedestrian street in the center of Tanhualin is the main boundary dividing the north and south living communities, and the main place for people to go sightseeing and stroll. Its spatial form is embodied in a linear road connecting several node spaces of different scales (Fig. 6).

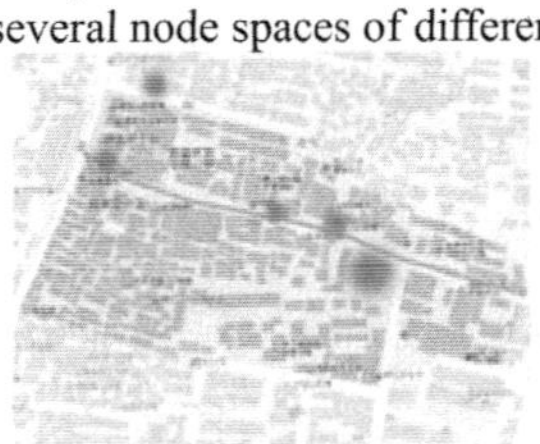

Fig. 6. Spatial form of central road.

Fig. 7. The central road.

3.1.1 Linear road

The construction of the marginal spaces on both sides of the linear road makes Tanhualin block form a harmonious style of petty bourgeoisie, culture and art, which is mainly reflected in three aspects: the street facade, plant decoration, and entrance step design (Fig. 7).

The change of pavement forms potentially divides the linear traffic in the center of the road and the edge space for people to stay at random on both sides of the road. Because most of the shops along the street are transformed from the first floor of the original residential buildings, and the building facades among different residential buildings are not forced to be unified. Therefore, diversity is created in the facade shape, material and color, which has the meaning of collage city. Some buildings have protruding balconies towards the street, which creates a sheltered gray space for the edge of the shop. Occasionally, residents sit on the doorstep to have a rest (Fig. 8). Some shops choose to use soft awning to delineate the entrance space for themselves as the transition to enter, or open windows beside the entrance to meet the needs of tourists (Fig. 9). They try to keep level with the facade of superstructure. Shops trying to keep level with the facade of the superstructure will give way to the entrance, forming an entrance platform that can put seats or decorative pieces, providing a place for people to stop when it rains.

Fig. 8. Residents sit on the steps to rest.

Fig. 9. Takeaway window.

In addition, the residents of Tanhualin district are very good at decorating the edge space with climbing plants such as Parthenocissus and small potted plants. The placement of some plants combined with the of entrance steps, with different heights, different leaf shapes and different colors, creates a small and delicate space atmosphere, attracting people to stay and watch at the edge of the road, and separates the noisy street from the quiet indoor environment of the shop. In addition to a few steps or a ramp at the entrance of some shops, sometimes there are narrow stairs along the edge of the building leading directly to the shops on the second floor. The side facade is usually decorated with color painting and other decorations. When there is a small flow of people, nearby residents sit here to rest (Fig. 10).

Fig. 10. People sitting on the steps to take pictures.

Fig. 11. The cafe behind Tanhualin Park.

3.1.2 Node space

Compared with the linear street, node space is a place for people to rest and communicate objectively. I regard it as the edge space of the whole street at a higher level. There is a large recreation square at the entrance of the core block, with high greening density and rest seats. The open private space is created through the shelter of trees. At the same time, it can lead to the building hidden behind the park (Fig. 11), which is the transition space between the road and the building. Walking along the pedestrian street, each section of the road will have a relatively open space for people to rest or used as a node for cultural promotion. According to the height difference, an elevated platform is formed and cultural sculptures are placed as symbolic meanings. This kind of edge space may be able to attract people's attention. However, if people sit down along the tree pools and flower beds for a rest, they may feel insecure about the view of the interior of the building behind. (Fig. 12) In contrast, it is more humanized to set wooden seats on the tree pool of arbors. The support on the back gives people a sense of psychological stability. People can rest and communicate at this node to alleviate the fatigue of walking. At the same time, they can have a positive sight communication with the surrounding shops and the crowd on the street. (Fig. 13)

Fig. 12. Cultural sculptures on the roadside

Fig. 13. Rest space at road node.

Fig. 14. People chat at the entrance of the community.

Fig. 15. Old people sit and rest at the entrance.

3.2 Community space

Compared with the central pedestrian street, the scale of community space is much narrower. For the closed community, residents seem to like to sit at the entrance of the lane for chatting (Fig. 14), perhaps because they can often meet acquaintances and can observe the movement of the outside crowd. As the old residential area did not consider providing special outdoor activity places for residents, the residents constantly spontaneously moved seats from their homes (Fig. 15), and reformed the relatively open edge space from bottom to top, which greatly improved the utilization rate of space.

For the open community formed by the intersection of large and small roadways in the south, the scale changes are more diverse and even slightly disordered. The influence factors of buildings on the edge space lie in two aspects: one is whether there is a balcony overhanging from the upper floor (Fig. 16), which gives people a sense of security due to being sheltered, which makes the external space penetrate inward, resulting in the trend of guiding people to enter the interior [4]; the second is whether there is a height difference between the building and the external road. At this time, steps and its attached platform become another element of the edge space.

If the building is higher than the road, the raised platform becomes the outward extension of the internal space. The elderly are glad to rest in front of their own house and sometimes talk to friends passing by. In summer, it may become a good place to eat, play chess and enjoy the cool (Fig. 17). If the building is basically level with the road, the edge space will often be transited by one or two steps or a gentle slope, and people's gathering behavior depends on whether it is open enough; if the building plane is lower than the road, considering the problem of rainwater backflow, this situation should be rare, and there is usually a circle of transitional space around the building to strengthen the sense of boundary of the building.

Fig. 16. The grey space formed by the balcony.

Fig. 17. The old sat on the platform and chatted.

In addition, the utilization rate of plants in the marginal space of the community is also very high. It fills the gap formed by the road in the plane view (Fig. 18), or decorates and guides the building entrance to enrich the street landscape and enhance the space vitality. During the investigation, we also saw interesting scenes. A resident standing outside communicated with the people inside directly through the window sill directly (Fig. 19). This kind of scene may be difficult to meet in the modern community, which fully reflects the humanity and community flavor of Tanhualin historical living block, and enriches the activity form of the crowd in the building edge space.

Fig. 18. Plant beside community Lane.

Fig. 19. The old woman communicated with the residents inside the house.

4. The relationship between marginal space and behavior

4.1 User perspective

According to the research of environmental psychology, compared with being in the center of outdoor space, the edge of building exterior wall, space boundary and other places can provide people with a sense of psychological protection, that is, the marginal tendency of behavior [5], which reflects in the living community is the marginal space mentioned in the article. Some scholars divide people's behavior activities in urban public space into three types, namely necessary activities, spontaneous activities and social activities. [6] In addition to the necessary activities may occur in specific places, which cannot be greatly affected by the environment, other forms of activities are subject to random changes due to the conditions of urban environment. From the survey, we can find that the edge space can provide a place for various forms of activities such as rest and conversation without disturbing other regional activities, which fully stimulates people's instinct to use public space and participate in public activities. Residents are willing to carry out spontaneous transformation of small space, so as to make the environment more adapt to the needs of their life. Edge space provides a foothold for people's fast-paced urban life, gathering emotions separated by reinforced concrete in every corner, trying to wake up the vitality of the city from the bottom up.

4.2 Urban design perspective

If the edge is defined as a clear boundary, the edge space is a more subjective area that spreads out to the surrounding area according to this line, and the scope is often between the architectural ancillary space and the urban functional space. Its use nature is not clearly defined to provide conditions for users to create diversified activities. From the negative aspect, the edge space does not play a major role in the composition of the city. In many cases, due to its vague intermediary, it is easily ignored by designers, resulting in the reduction of the utilization rate of public space; from the positive perspective, urban design can guide people's behavior and activities through the construction of edge space, and promote people to return to public life from closed rooms. The design has always been people-oriented. In the process of investigation, we have observed many spontaneous activities in the edge space. However, some small-scale roadways in the current living community may not be comfortable enough for residents. Narrow space and old facilities will bring bad space experience. Therefore, for the edge space of urban blocks, designers need to carry out comprehensive transformation from the aspects of scale, atmosphere and use experience, so that the urban environment and residents can better interact.

5. Conclusion

The development of modern cities provides convenience

for people's life, but the state of being accessible without leaving home makes people lose enthusiasm for public life. The development and reconstruction of Tanhualin historical life block symbolizes the vitality which is independent of the tide of urbanization. It maintains the vitality of the city from the inside to the outside with the distinct spirit of place and the close communication between people. The rich edge space provides an opportunity for people's emotional collision. As a small-scale urban design, it plays an important role in retaining urban characteristics and expanding public space.

References

1. K. Lynch. Trans. Xiang Bingren. The Image of the City [M]. China Construction Industry Press, Beijing. (1990)
2. Luo Caixia. Planning and design of public space in Tanhualin historical and Cultural District [D]. Huazhong Agricultural University. (2012)
3. Wang Wenhui, Yuan Liping. Discussion on the renewal and activation strategy of historical blocks under the concept of "urban repair" -- Taking Tanhualin historical block in Wuhan as an example [C]. Proceedings of 2019 China Urban Planning Annual Conference (02 urban renewal). China Urban Planning Society, Chongqing Municipal People's Government: China Urban Planning Society. (2019:2039-2049.)
4. Chen Wenjia, Zhu Chunyang, Zhou Yi. Study on the edge space form of urban life block -- Taking Tanhualin historical life block as an example [J]. Central China architecture. (2014,32 (04): 63-70)
5. Yao Xiaoyan. Study on the design of modern urban street edge space [D]. Hebei Agricultural University. (2007)
6. Jan Gail. Trans. He Renke. Life between buildings [M]. Beijing: China Construction Industry Press. (2002:13-17)

River corridor ventilation analysis and riverfront planning strategy in Tianjin's urban core area

Tong Ma[1], Tian Chen[1]*

[1] School of Architecture, Tianjin University, 92 Weijin Road, Nankai district, 300072 Tianjin, Peoples R China

Abstract. River corridors are important potential wind path in cities. This paper took the river in the core area of Tianjin as the research object. By using two ventilation indicators as frontal area index (FAI) and frontal area density (FAD), the overall ventilation and pedestrian level ventilation of riverfront area were quantitatively analyzed. The result showed that contrary to common belief, due to the high FAI of the riverfront area, river corridor in Tianjin's core area did not play the role of wind path but obstructed the wind flow. Also the low FAD indicator of the riverfront area lead to a better outdoor comfort in summer but worse comfort in winter. After verified the correlation between planning indicators (floor area ratio, building height and site coverage) and ventilation indicators, a cluster pattern urban riverfront development mode was proposed. Also a ventilation corridor planning methods based on the river direction and dominant wind direction was suggested. These planning strategy will be more conductive to the wind path and cooling island potential of the river corridors in urban core area.

1 Introduction

River corridor is precious natural landscape resource in urban area. Meanwhile, it is also urban cooling island and potential wind path. There are many existing studies on river microclimate effects, however, most of those studies focus on the cooling and wetting capacity of urban river corridors [1-4]. As an essential part of microclimate research, ventilation effects of river corridor are less considered. The river runs through cities is always considered as wind path as a matter of course due to its low surface roughness. However, few quantitative researches to support this assumption have been made.

Urban ventilation studies could be classified into three categories by scale: large scale (regional), mesoscale (city) and microscale (urban blacks) [5], as shown in figure 1. As the large-scale studies mainly focuses on regional atmospheric circulation, studies related to the ventilation effect of river corridors are mainly in mesoscale and microscale. Some mesoscale urban ventilation studies [6-8] point out urban river corridor as a potential main wind path to enhance ventilation in high dense urban area. However, no quantitative analysis is carried out in their studies. Zou et.al [9] conclude Haihe river which flows through the core area of Tianjin can be used as the secondary ventilation corridor to transform cooling air from suburban cool island by using remote sensing. However, those studies did not focus on rivers and mostly used 2-D analyzing tools as remoting sensing but not 3-D analyzing tools. This leads to the current lack of deepness on urban river ventilation research. Moreover, microscale studies, which mostly focus on the permeability of wet and cold air in wind leeward riverfront urban area [10-12], are difficult to systematically discuss the influence of the river on the urban ventilation due to its scale.

Another flaw in existing river ventilation studies is lack of cohesion between mesoscale and microscale [13]. Fortunately, an urban space analysis method based on three-dimensional model has been established in former urban ventilation studies to exam the correlation between urban spatial morphology and ventilation efficiency [14-15]. By using this methods, it is possible to establish the relationship between mesoscale and microscale. This method may also adopt in river corridor ventilation researches.

In summary, there are few studies that take rivers and riverfront areas as specific research objects and quantitative analyzed the ventilation effect of river corridors on urban areas in mesoscale. Therefore in this paper, taking Tianjin's core urban area as study area, the influence of river corridor and riverfront urban spatial morphology on urban ventilation will be discussed based on a three-dimension urban model. The findings presented in this work will fill in the research gap in urban river corridor ventilation and provide practical suggestions for urban planning.

* Corresponding author: chentian5561@vip.sina.com

https://doi.org/10.1051/e3sconf/202123704022

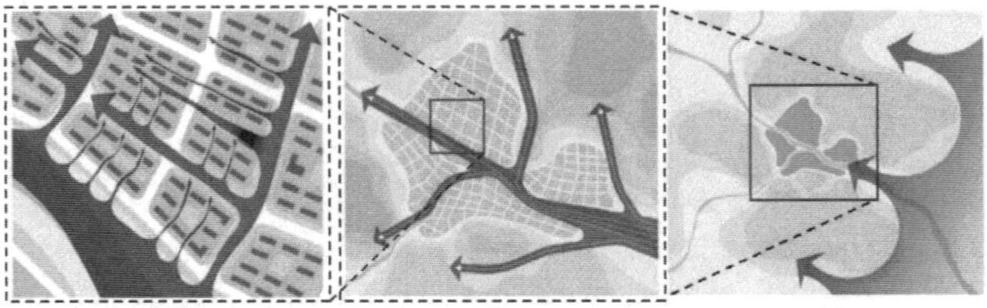

Fig1. Three scale in urban ventilation studies, microscale(a), mesoscale(b) and large scale(c), from Ren [5]

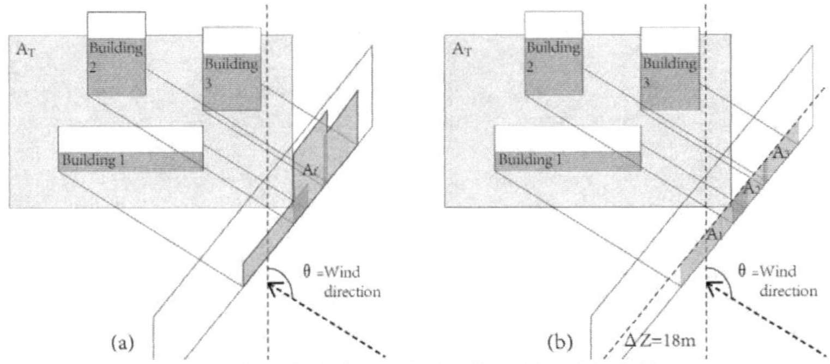

Fig2. The calculation methods of FAI(a) and FAD(b)

2 Research methods

2.1 Ventilation evaluation indicators

Wind speed is the fundamental index in urban ventilation studies. In this paper, two indicators as the overall wind speed and the pedestrian level wind speed will be selected as the essential index to analyze the effect of river corridor on urban ventilation.

The overall wind speed in urban area is determined by ground roughness when other conditions being equal. The higher the urban development intensity is, the more the windage resistance of buildings are, the higher the ground roughness is, and eventually lower the wind speed passing through the urban area [16]. Although a few indicators could be used to refer ground roughness, Frontal area index (FAI) is widely applied in urban planning field [9,14,15,17]. This indicator could accurately reflect the influence of urban canopy morphology and wind direction on the overall wind speed in high dense urban area. The higher the FAI is, the lower the overall wind speed. The formula is shown in eq.1, where $\lambda_{f(\theta)}$ represents the frontal area index of a certain area, A_F represents the front areas of buildings that face the wind direction of θ, and A_T represents the total lot area. Also figure 2a explain the calculation methods of FAI.

$$\lambda_{f(\theta)} = \frac{A_F}{A_T} \quad (1)$$

Wind speed on pedestrian level, which is another key indicator of urban ventilation, affects human outdoor comfort significantly. Past studies proved that Frontal area density (FAD) has the dominant influence on pedestrian level wind speed [16,20,21]. The higher FAD is, the lower wind speed is. Different from FAI, FAD only calculates the front areas of buildings within a certain height above the ground. The formula is shown in eq.2, $\lambda_{f(Z,\theta)}$ represents the frontal area density of a certain area, $A(\theta)$ represents the area of building surface that approach the wind direction of θ for a special height (18m in our study), and A_T represents the total lot area, $A(\theta)$=A1+A2+A3+...+An. Also figure 2b explain the calculation methods of FAD.

$$\lambda_{f(Z,\theta)} = \frac{A(\theta)_{proj(\Delta z)}}{A_T} \quad (2)$$

2.2 Urban riverfront 3-D model

The purpose of this paper is to study the influence of river corridor on the ventilation of urban core area. Therefore, three rivers running through Tianjin's core area (Haihe river, Xinkai river and South canal) were selected. The 1000-1500m range urban area on both side of the river was selected as study area, as shown in figure 3a. In order to quantitative evaluate the urban space, the study area was divided into grids to aggregate the spatial morphology index in each grid. There have been two aggregation methods used in past research. One is to aggregate buildings according to building blocks [20]. The other method involves manipulation of data using regular grids, which is commonly used in structural data related to urban ventilation [14,15]. Regular grids method was used in this paper due to river corridor cannot be divided into

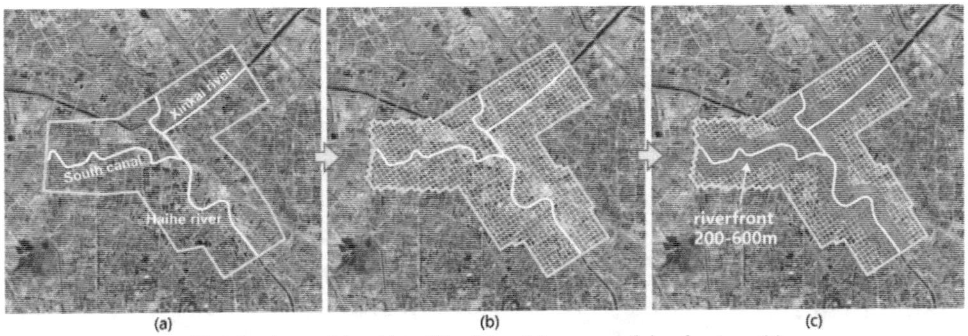

(a) (b) (c)

Fig3. Study area(a), grid partition(b) and the range of riverfront area(c)

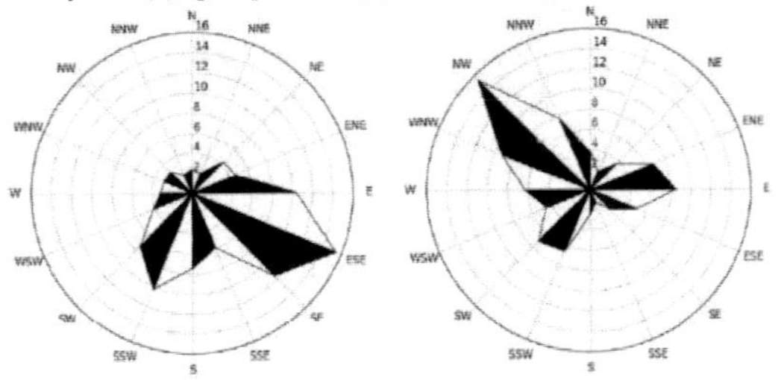

(b) (c)

Fig4. The wind direction of summer(a) and winter(b) in study area

any building blocks. As shown in figure 3b, study area was divided into 1130 regular grids with the size of 200m*200m. The FAI and FAD of each grid was calculated both in winter and summer. The overall FAI and FAD of urban riverfront area in 200m, 400m and 600m distant to the river (figure 3c) were calculated and compared with the whole study area. The spatial morphology difference between riverfront area and other urban area was compared to discussed the influence of river corridor on urban overall ventilation and pedestrian level ventilation. According to the meteorological data of Tianjin Meteorological Bureau in the past 15 years, Tianjin city's dominant wind direction is southeast by east in summer (figure 4a) and northwest in winter (figure 4b). Thus the FAI and FAD of study area in summer and winter were calculated respectively and then averaged.

3 Results and analysis

3.1 Overall ventilation analysis

The average FAI of every block in riverfront area within 200m, 400m and 600m range of the river corridor were calculated to identify if there are any spatial morphology difference between riverfront area and non-waterfront area, both in summer and winter, as shown in figure 5. The results showed that in summer, FAI of all study area (5a), riverfront 600m range (5b), riverfront 400m range (5c) and riverfront 200m range (5d) were respectively 0.131, 0.138, 0.137 and 0.140, and FAI beyond riverfront 600m range was 0.124. Meanwhile in winter, FAI of all study area (5e), riverfront 600m range (5f), riverfront 400m range (5g) and riverfront 200m range (5h) were respectively 0.129, 0.136, 0.135 and 0.140, and FAI beyond riverfront 600m range was 0.122. It could be seen that FAI in riverfront area are nearly 0.15 higher than non-riverfront area and FAI are highest in riverfront 200m range area, both in summer and winter.

Furthermore, the distribution of blocks that had a higher and lower FAI than the average within riverfront 600m range were mapped, as shown in figure 6. As the blocks with a lower FAI than the average value facilitating the natural ventilation while the blocks with a higher FAI than the average obstructing the natural ventilation, the relationship between the riverfront area and the ventilation corridor could be analyzed. It could be seen that the blocks obstructing ventilation were evenly distributed along the river in high dense riverfront areas. Meanwhile, the blocks facilitating ventilation showed little continuity. The only exception was the area around Tianjin railway station. The FAI of this area was obviously lower than surroundings and a wind corridor was formed due to the open space made by railways.

To sum up, the rivers in the urban core area of Tianjin didn't promoted urban ventilation due to the high FAI in riverfront area. In contrast, it slightly lowered the overall ventilation of urban core area. Moreover, the river corridor in Tianjin did not reach its potential in wind corridor as the high FAI blocks were linear distributed along the river, which was contrary to existing researches [8,9].

3.2 Pedestrian level ventilation analysis

The average FAD of every block in riverfront area within 200m, 400m and 600m range of the river corridor

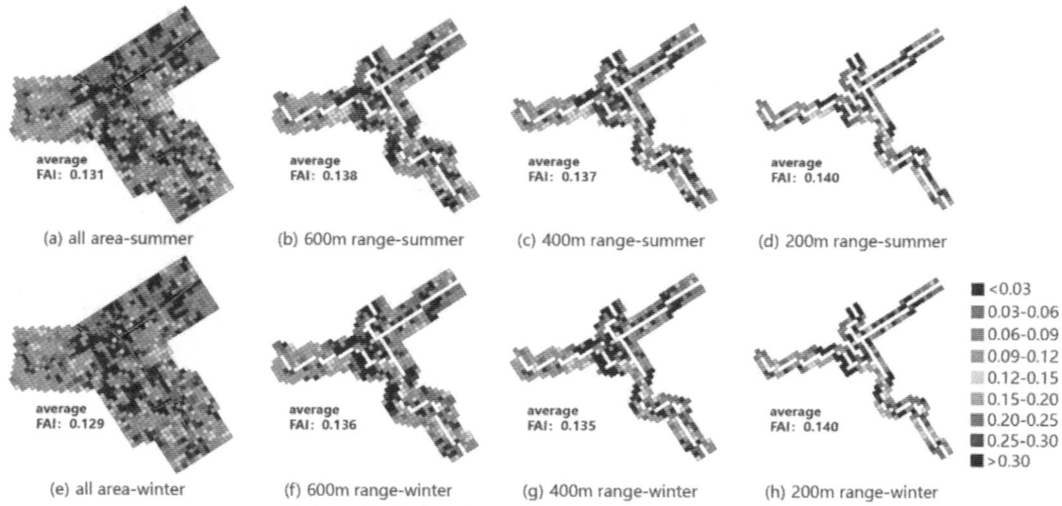

Fig5. FAI in whole study area and in different riverfront range

Fig6. In 600m range, FAI higher than average (a) and lower than average (b) in summer, and FAI higher than average (c) and lower than average (d) in winter

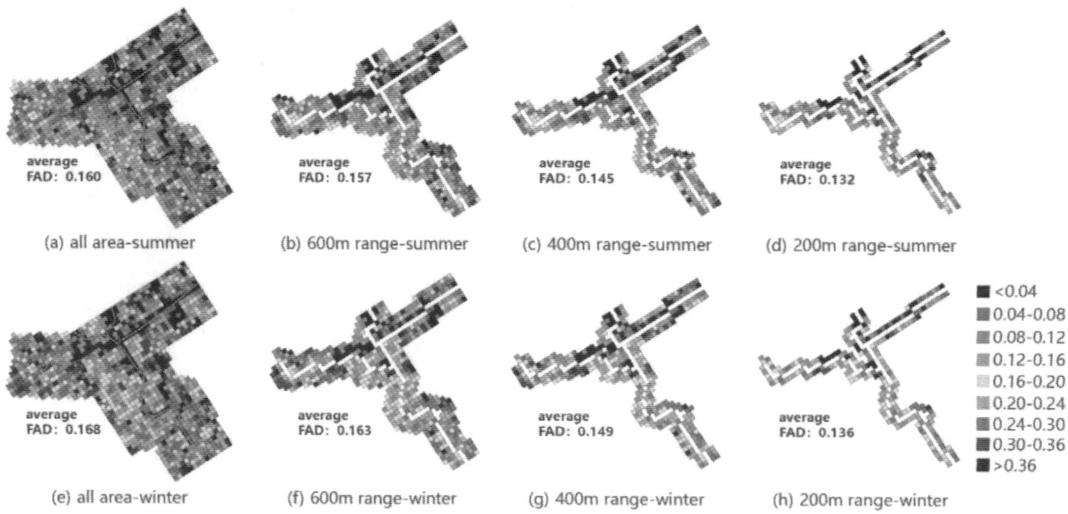

Fig7. FAD in whole study area and in different riverfront range

were calculated to analyse the effect of river corridor on pedestrian level wind condition in riverfront area, both in summer and winter, as shown in figure 7. The results showed that in summer, FAD of all study area (7a), riverfront 600m range (7b), riverfront 400m range (7c) and riverfront 200m range (7d) were respectively 0.160, 0.157, 0.145 and 0.132, and FAD beyond riverfront 600m range was 0.163. Meanwhile in winter, FAD of all study area (7e), riverfront 600m range (7f), riverfront 400m range (7g) and riverfront 200m range (7h) were

https://doi.org/10.1051/e3sconf/202123704022

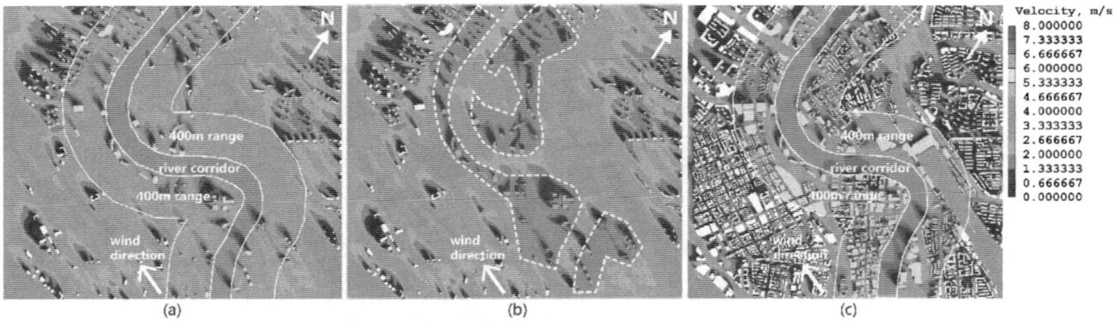

Fig8. Simulation result of wind speed at 27m height of 400m riverfront range(a), buildings obstruct ventilation(b) and wind speed at 1.5m height of 400m riverfront range(c)

respectively 0.168, 0.163, 0.149 and 0.136, and FAD beyond riverfront 600m range was 0.173. It was obvious that the FAD of riverfront area was significantly lower than non-riverfront area. The closer the area was to the river, the lower the FAD was. This caused the higher pedestrian level wind speed in riverfront area than non-riverfront area, especially in river banks. This enhanced the outdoor comfort on riverfront area in summer and reduced the comfort in winter, which was consistent to real feelings.

3.3 Numerical simulation analysis

To validate the ventilation efficiency indicators set in this paper, computational fluid dynamic numerical simulation was carried out to verify the relationship between spatial morphology and wind speed. K-ε turbulence model that commonly used in urban microclimate studies was selected. The simulation area was about 4000m*4000m within the study area. The initial wind speed at 10m was 2.8m/s, the wind direction was southeast to east, the model resolution was 5m and the urban roughness length was 0.75m.

Figure 8a shows the wind speed at 27m height of the simulation area. According to the statistics of the proportion of different wind speed intervals, the wind speed within river corridor was 2.55m/s, the wind speed within 400m range of riverfront area was 2.54m/s and the wind speed outside 400m range of riverfront area was 2.56m/s. As a result, river corridor had not promoted ventilation in high dense urban area as the wind speed in each area were almost the same. Figure 8b proved the area with high-rise buildings and high FAI caused the low wind speed zone within river corridor and wasted the ventilation potential of the river.

Figure 8c showed the wind speed at 2m height of simulation area. According to the statistics of the proportion of different wind speed intervals, the wind speed within river corridor was 2.12m/s, the wind speed within 400m range of riverfront area was 1.69m/s and the wind speed outside 400m range of riverfront area was 1.55m/s. This verified the higher pedestrian level wind speed in riverfront area than non-riverfront area due to its lower FAD.

4 Riverfront planning methods for ventilation optimization

By analyzing the value and distribution of FAI and FAD indicator of different range of riverfront area, this study figured out the influence of river corridor on the overall ventilation and pedestrian level ventilation for high dense urban area in Tianjin. The results are as below: 1) different from previous studies, instead of promoting ventilation, rivers in Tianjin's high dense urban area has a negative contribution on urban ventilation. 2) the potential of rivers as ventilation corridors and cold island effects cannot be realized due to the high FAI blocks along the river obstructed the air flow. 3) the low average FAD of riverfront area lead to higher pedestrian wind speed. As the pedestrian level wind speed is relatively high within river corridor than other urban area, the riverfront area is more comfortable in summer but less comfortable in winter.

4.1 Using planning indicator to optimize riverfront ventilation

The FAI and FAD could accurately reflect the effect of urban spatial morphology on ventilation. However, the calculation methods of those two indicators have to consider about the wind direction, building direction and building location in site. So it is difficult to use FAI and FAD in the early stage of urban planning practice. Therefore, the correlation between FAI/FAD and planning indicators as floor area ratio (FAR), building height (BH) and site coverage (SC) were established. The purpose is to reveal if FAI/FAD could be replaced by these planning indicators to optimize riverfront wind environment more convenient.

Figure 9a indicates that FAI is significantly positively correlated with FAR, with a Pearson correlation coefficient 0.886. Figure 9b indicates that FAD is weak positively correlated with FAR, with a Pearson correlation coefficient 0.489. The dispersion degree of FAD rapidly enhanced as FAR increase. Figure 9c indicates that FAI is significantly positively correlated with BH, with a Pearson correlation coefficient 0.793. The FAI of some blocks dispersed when BH is high.

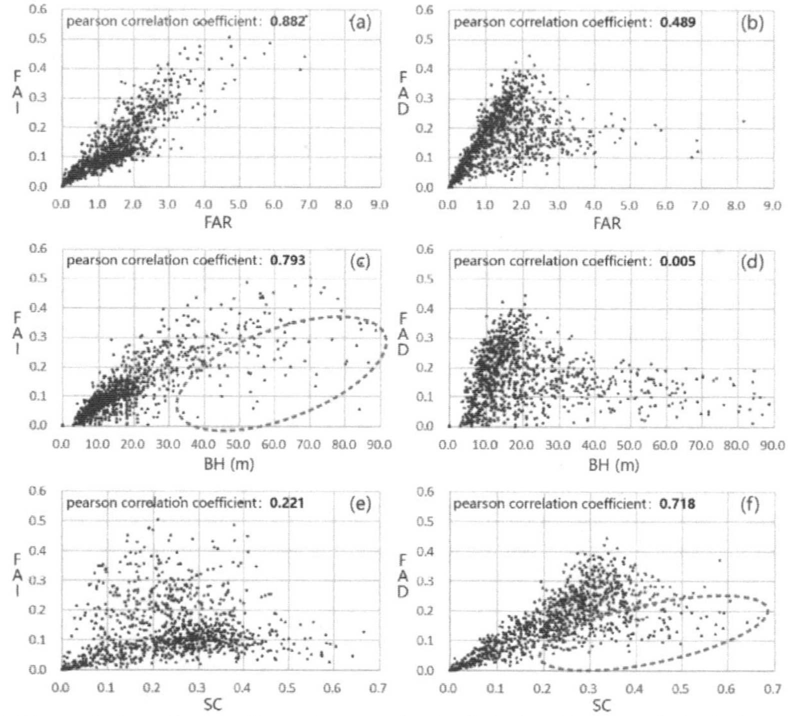

Fig9. The correlation and scatter diagram between planning index(x axis) and ventilation index(y axis)

These blocks have a common feature: small number of high-rise buildings. This caused these blocks with a high BH and a low FAI. This happens mainly because of the errors caused by the grid division, which is tolerable. Figure 9d indicates that FAD is not correlated with BH, with a Pearson correlation coefficient 0.005. Figure 9e indicates that FAI is not correlated with SC, with a Pearson correlation coefficient 0.258. Figure 9f indicates that FAD is significantly positively correlated with SC, with a Pearson correlation coefficient 0.740. The FAD of some blocks dispersed when SC is high. These blocks had the following features: 1) high site coverage with a low average building height and 2) a single building with a huge floor area. This two features will cause the low FAD with a high SC in blocks.

Therefore, in urban planning practice, FAR and BH could replace FAI to control the overall ventilation. In master planning, FAR is the essential indicator to controlling urban development intensity. So it is possible to optimize the riverfront wind environment by controlling the FAR of each urban block in the riverfront area in the early planning stage. BH could be used as a secondary indicator in skyline control or height limitation. The strong correlation between FAD and SC mean it is possible to effectively adjust the pedestrian level wind environment by controlling site coverage of a block. However, it should be noted that it is less useful to use this method when there are blocks with low-rise building clusters or buildings with huge floor plan area such as shopping malls on riverfront area. For these two kind of blocks, FAD need to be considered as a planning indicator directly.

4.2 Planning strategy of urban riverfront area

1) cluster riverfront layout mode. Many past studies suggest that low development intensity urban blocks with low building height and low site coverage should be located on both side of the wind corridors to help enhance the ventilation efficiency. Although this approach is rational theoretically, it is impractical in planning practice. Due to the urban development intensity is more determined by its land value, it is not possible to reduce the development intensity as the riverfront area has a more valuable natural landscape resources. The high natural landscape value causes the high FAR and further reduces the ventilation efficiency on riverfront area. Therefore, a cluster pattern of urban riverfront development is suggested in this study, as shown in figure 10. The blocks with high development intensity and building heights clustered as the overall FAR of the riverfront area stay the same. Meanwhile, the open space and block with low development intensity could be located between clusters. This could help the wind path cross the river and bring cold and humid air into the riverfront urban area to reduce urban heat island. For urban redevelopment area, FAR award and FAR transfer policy could be applied in blocks located in wind path to encourage a development mode with low building height and low density.

2) Planning urban wind corridor with river direction. River has a cooling and humidifying effect as well as its low surface roughness for ventilation. The microclimate effect of rivers can be maximized when cooling, humidification and ventilation effect combined.

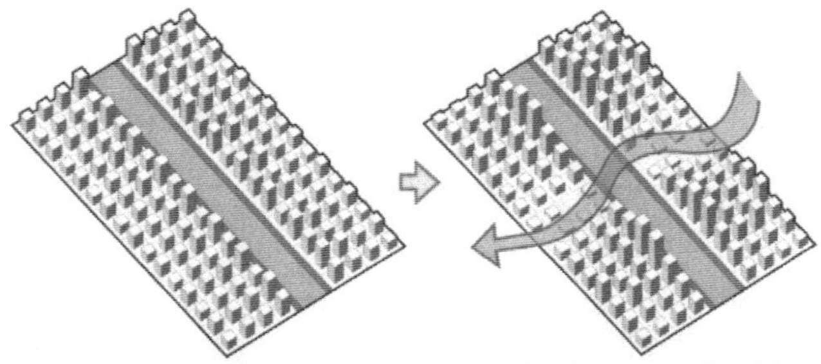

Fig10. From linear pattern to cluster pattern development on riverside area

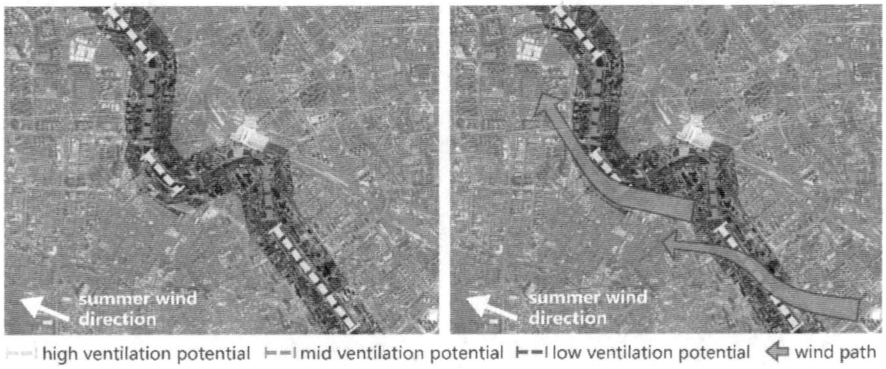

Fig11. Ventilation potential of river sections with different river-wind direction angle

However, as most rivers are natural formed, not all rivers have the potential to be wind path. Givoni [21] point out that the wind path is effective only if the angle between the wind path and wind direction less than 30°. Hence, when planning wind corridors, the rivers should be divided into several sections based on the its direction. The ventilation potential of each river section could be realized after figure out the angle between each river section and dominant wind direction, as shown in figure 11. Then the high ventilation potential section should be integrated into urban wind corridor system. As a result, the wind corridor will pass through more area of waterbody and bring cold and humid air into urban high dense area to alleviate urban heat island.

3) Different urban spatial morphology control on windward and leeward side of river. The riverfront area has a higher FAI and a lower FAD than other urban area, which means a low-dense and high-rise building arrangement is more popular in riverfront area. The reason for this condition is the maximum utilization of river landscape resources in urban development to enhance land value. However, the increase of building height will lead to the increase of wind dispersion and wind turbulence around the corner of tall buildings in pedestrian level [19]. Combined with the relatively high wind speed, this will cause the poor pedestrian level wind condition on riverfront area in winter. therefore, in planning practice, the windward and leeward riverfront area should set different planning strategy according to seasonal wind direction. Taking Tianjin as an example, it needs ventilation in summer and wind obstruction in winter. Thus in winter, the leeward riverfront should enhance the site coverage and lower the average building heights to reduce the pedestrian level wind speed and wind turbulence. A traditional development mode with high site coverage and narrow street is recommended and the podium of the high-rise building is a must. Meanwhile in summer, the leeward riverfront should reduce the site coverage to enhance the pedestrian level wind speed to allow cold and humid air penetrating into inner urban areas. Also the limitation for building height could be liberalized.

5 Conclusion

Taking Tianjin's urban core area as the research site, this study used two key indicators as FAI and FAD to analyze the effect of river corridor on urban ventilation. The following conclusions could help urban planners to understand the ventilation condition of riverfront area and develop targeted planning strategies. 1) different with past studies, this paper proved that river corridor in Tianjin's urban core area didn't promoting riverfront ventilation. On the contrary, it obstructed the wind path because of the high development intensity and high FAI on riverfront area. 2) the low FAD increased the pedestrian level wind speed in river banks which is conductive to summer ventilation but worsen the wind environment in winter. 3) FAR and building height were significantly positively correlated with FAI and site coverage was significantly positively correlated with FAD. These three planning indicators could be used to control riverfront wind environment in early planning stage. 4) in order to maximize the microclimate effect of the river, a cluster mode of riverfront urban development mode is proposed to promoting air flow through the river. 5) river sections that have a small angle with dominant wind direction should be considered as wind path and considered as part of the urban wind corridor system. 6) different spatial morphology planning methods should be applied in windward and leeward side of rivers according to the dominant wind direction in different seasons.

References

1. K. Yang, M. Tang, Y. Liu, et al. Analysis of microclimate effects around river and waterbody in Shanghai urban district. Journal of East China Normal University (Natural Science) **03:** 105-114 (2004)
2. E.A. Hathway, S. Sharples. The interaction of rivers and urban form in mitigating the Urban Heat Island effect: A UK case study. Building and Environment **58:** 14-22 (2012)
3. H. F. Li, Y. S. Li, Z. Lu, et al. Analysis on the thermal environment effect of river corridor landscape. Geography and Geo-information science **31(04)**: 51-54+133 (2015)
4. Z. Cai, G. F. Han, M. C. Chen. Do water bodies play an important role in the relationship between urban form and land surface temperature?. Sustainable Cities and Society **39:** 487-498 (2018)
5. C. Ren, R. Yang, C. Cheng, et al. Creating breathing cities by adopting urban ventilation assessment and wind corridor plan–The implementation in Chinese cities. Journal of Wind Engineering and Industrial Aerodynamics **182:** 170-188 (2018)
6. H. Z. Xi, S. Jiao, L. Y. Lu. Research on the mode of creating urban natural ventilation channel for the areas where its hot in summer and cool in winter-take Changsha as an example. Huazhong Architecture **28(06)**: 106-107 (2010)
7. J. Y. Li, X. H. Li, R. B. Xiao. Research on the planning and control method of urban ventilation corridors- a case study of regulatory planning of Baiyun new town, Guangdong. Landscape Architecture **05:** 92-96 (2014)
8. S. P. Zeng. Research on the theory of “source-flow-sink” ventilation corridor system construction and planning strategy-the case of the center districts of Tianjin. PHD thesis, Tianjin University (2016)
9. Z. Zou, J. Kang, Z. Y. Zhou, et al. Construction and control of ventilation corridors in the main urban area of megalopolis: a case of Tianjin. Building Science **35(06)**: 91-97+107 (2019)
10. J. Wang, Research on building layout strategy in riverfront of Shenzhen on wind environment. MSC thesis, Harbin Institute of Technology (2012)
11. X. C. Song, J. Liu, L. Yu. Impact analysis of riverfront residential building design in northern region on the thermal and humid environment in summer. Building Science **31(06)**: 14-19 (2015)
12. D. X. Zhang. Research on optimization strategy of wind environment in urban waterfront based on source and flow concept. MSC thesis, Shenyang Jianzhu University (2018)
13. W. W. Wang, F. N. Li, D. Wang. A review and trend of urban ventilation channel research. Urban Problems **02**: 36-40 (2018)
14. E. Ng, C. Yuan, L. Chen, et al. Improving the wind environment in high-density cities by understanding urban morphology and surface roughness: A study in Hong Kong. Landscape and Urban Planning **101(1)**: 0-74 (2011)
15. C. M. Hsieh, H. C. Huang. Mitigating urban heat islands: A method to identify potential wind corridor for cooling and ventilation. Computers Environment & Urban Systems **57**: 130-143 (2016)
16. T. T. Li. Research on the construction and planning of urban ventilation based on urban morphology and surface roughness – a case study of Shenzhen, China. MSC thesis, Shenzhen university (2017)
17. C. S. B. Grimmond, T. R. Oke. Aerodynamic properties of urban areas derived from analysis of surface form. J. Appl. Meteorol. **38**: 1262–1292 (1999)
18. C. Yuan, C. Ren, E. Ng. GIS-based surface roughness evaluation in the urban planning system to improve the wind environment – A study in Wuhan, China. Urban Climate **10**: 585-593 (2014)
19. T. Ma, T. Chen. Classification and pedestrian-level wind environment assessment among Tianjin’s residential area based on numerical simulation. Urban Climate **34:** 100719 (2020)
20. T. Gál, J. Unger. Detection of ventilation paths using high-resolution roughness parameter mapping in a large urban area. Building & Environment **44(1)**: 198-206 (2009)
21. B. Givoni. *Climate considerations in building and urban design*. Van Nostrand Reinhold (1998)

Current status and content of quantitative criteria in clinical trials of compression therapy

Congcong LEI[1] and Lu CHEN[2]

[1,2] Institute of textile and clothing, Shanghai University of Engineering Sciences, Shanghai 201600, China

Abstract. Objective: The scale of the treatment effect of lymphedema pressure has not been clarified, and there appears to be a lack of evidence for research in this area. There is no clear and uniform standard for measuring the experimental results of lymphedema treatment with pressure clothing, and there seems to be a lack of research on the experimental results measure. The purpose of this study was to summarize and quality evaluate the existing scientific evaluation criteria for the therapeutic measure of upper limb compression. To review the differences and similarities in efficacy evaluation criteria of pressure cuff, pressure bandage, CDT and other treatment methods. And some new ideas. Methods: Narrative review based on search in Pub-Med/Web of Science through keywords related to compression in lymphedema. A systematic review of published literature was performed. Results: In the review, 1716 articles were screened, of which 11 trials were included. The primary end point of the assessment and results was to measure changes in arm volume, secondary endpoints were the quality of life, patient compliance, range of motion of arm joints and other reason. Although the effectiveness of compression therapy can be obtained by comparing experiments, there is no uniform measurement of experimental results, so it is impossible to compare the effectiveness between different experiments. Conclusions: At present, in the treatment of lymphedema, the effect of the pressurized sleeve is recognized, and the accepted measure is the change in arm volume, and secondary criteria vary widely. The available literature lacks a unified measurement standard for experimental results and measurement items. Researchers should combine the actual situation of the local patients with lymphedema to construct a suitable measure of the effect of lymphedema compression treatment after breast cancer.

1 Introduction

Lymphedema is a chronic profressive disease without a cure in women, and lymphedema of upper limb following breast cancer treatment is one of the major long-term complications[1]. According to the most recent International Society of Lymphology (ISL) consensus document[2], We can learn about the current non-operative treatment of lymphedema (Figure 1). The generally accepted schemes of the combined treatment of lymphedema of upper is physical therapy and adjuvants.

Currently, the following methods are mainly used in the literature (Intermittent pneumatic compression, Complete Decongestive Therapy (CDT), manual lymphatic drainage (MLD), Compression garments). CDT is the most effective treatment which is a combination of four methods (MLD, compression therapy, a low-stretch elastic stocking or sleeve, continued exercises and skin care) as it reduces the symptoms of lymphedema and improves patients' quality of life and psychological states [3]. Multilayer low-stretch bandages or sleeve and elastic compression is the cornerstone of CDT[4]. Although CDT and other therapeutic techniques have demonstrated positive results, but the evidence base setting a scale in the treatment effect of lymphedema pressure has not been clarified.

Scientific norms and unified experimental quantitative standards can provide more targeted treatment programs for medical staff and patients, as well as provide strong support for scientific researchers' experiments.

In this article, we introduce the entity of lymphedema treatment, as well as medical treatment with Compression therapy. We also review therapies described in the literature. The status quo and main problems of the existing clinical trials were clarified to provide reference for domestic clinical trials.

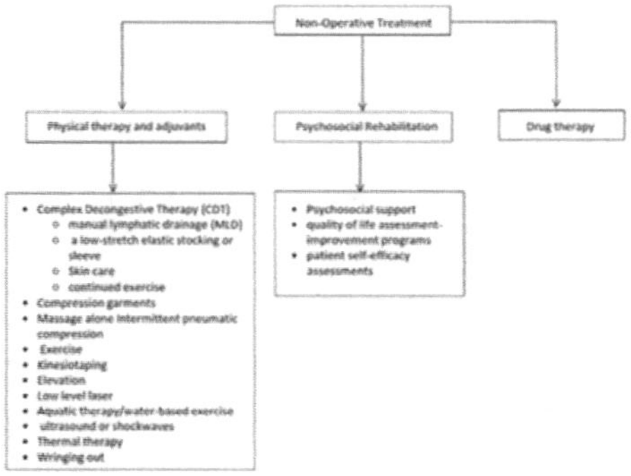

Figure 1. Current nonoperative methods of lymphedema.

2 Methods

Search strategy. A systematic search of the relevant guideline and systematic review databases was undertaken between 06 May 2020 and 05 June 2020. The following databases were searched: National Center for Biotechnology Information (NCBI), National Lymphedema Network (NLN) and Web of Science. Focusing on articles of the last 5 years and key search terms included 'Upper Extremity', 'elastic hosiery', 'pressure bandages', 'compression therapy', 'compression sleeves', 'lymph', 'lymphatic disease', 'nonsurgical therapy' and 'venous disease'. Intentionally no search was performed on lower extremity and intermittent pneumatic compression, which does not apply to this review. The 2020 International Society of Lymphology (ISL) consensus document is mainly harvested from the retrieval results of NLN. Initially, the searches produced 789 results in first search strategy and 272 in second one. (Figure 2).

Inclusion and exclusion criteria are as followed, Inclusion criteria: RCT within the last 5years, Outcome of Lymphedema, Lymphedema of upper limbs, Patients received compression sleeves or pressure bandages alone or in conjunction with MLD; exclusion criteria: without bandages or compression garments were used, without control test, preventive and therapeutic options, only discuss effectiveness and safety, discuss research method.

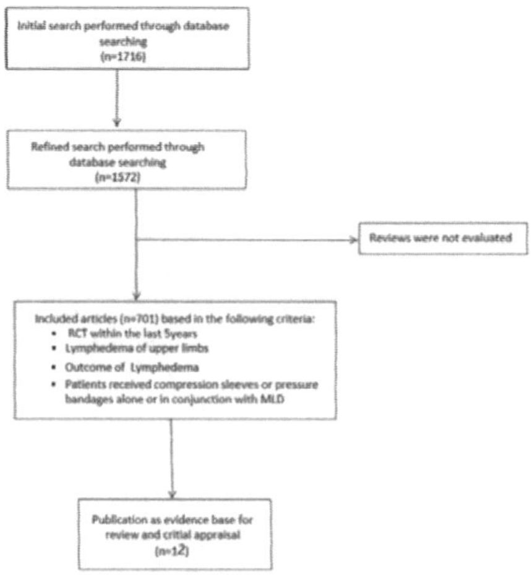

Figure 2. Summary of search strategy following Preferred Reporting Items for Systematic Reviews and Meta-Analyses (PRISMA) statement guidance. Lymphedema, Compression Bandages or stockings, CDT, randomized controlled trial, Upper Extremity.

3 Results subsection

The results of this search strategy were further refined by analyzing the review, a total of 56 potentially relevant publications were selected. After a decision to include only literature on compression stockings and upper extremity, a further review identified 11 relevant

 https://doi.org/10.1051/e3sconf/202123704023

publications as the evidence base for the measurement of experimental results. The characteristics of included study arms are displayed in Figure 3.

In 11 groups experiments, there are 7 groups of experimental time 3 months or less, 3 groups using CDT treatment, 5 groups used compression bandage or compressed cuff treatment. primary criterion for arm volumes or limb volume of 6 groups, and the criteria are different, Arm volume was measured by a plastic tape with light pressure (V = h (C2 + Cc + c2)/12p) in two article [5] [6]. In other article limb volume using a simplified frustum formula (summed truncated cone). The edema volume 1 group, lymphedema to cure the success rate of 3 groups, and a set of the velocity of blood flow as a primary standard.

In the secondary criteria, 8 groups contain quality of life, but the evaluation criteria are different, including EQ-5D questionnaire [7], LYMQOL arm questionnaire [8], validated questionnaire [6], QLQ-BR23 questionnaires [9], four questions[10], visual analogue scale [11], the Lymphedema Questionnaire [5]. There are 2 group that involve compliance [9] [8], the evaluation criteria is EORTC QLQ-C30 questionnaires [9]. It also takes Ranges of motion of the shoulder (flexion, abduction, and external rotation) as the detection standard. Usually they are measured with a goniometer [5] [10].

In actual treatment, patients often choose to despise or give up treatment due to comfort and other factors, and patients' compliance in treatment should also be included in the measurement criteria [8].

Author, year	Type of Physical therapy	Sample (n=)	Population (mean age)	stages of lymphedema	Group	Duration of compression treatment, mean (range)	Assessment and outcome measures: Primary endpoint	Assessment and outcome measures: Secondary endpoints
Ayhan, 2019	CDT	41	Female (56.05); as per the ISL classification system	Stage I-III	Before and after	3 weeks	Arm volumes	• the quality of life • neuropathic pain (NP)
Sezgin Ozcan, 2018	CDT	37	Female (53.6); as per the ISL classification system	BCRL diagnosis	Before and after	3 weeks	Arm volume measured by a plastic tape with light pressure (V = h (C2 + Cc + c2)/12p)	• Clinical symptoms of pain and heaviness with VAS • Ranges of motion of the shoulder (flexion, abduction, and external rotation) by a goniometer • musculoskeletal disorders of the upper extremity measured by DASH • health status measured by SF-36
Malloizel-Delaunay, 2019	IDT	93	Female ;as per the ISL classification system	Stage II	• Bandages +MLD • Bandages + MLD + Endermologie • Bandages + Endermologie	6 months	success rate (a reduction in upper limb excess volume ≥ 30% on day 5)	• Arm volume measured by a plastic tape with light pressure (V = h (C2 + Cc + c2)/12p) • quality of life measured by the EQ-5D questionnaire • adverse local and regional effects of IDT (bandages, Cellu M6, manual lymphatic drainage) • AE
Mestre, 2017	DLT	40	Female ;as per the ISL classification system	Stage II	• Day-time THUASNE lymphology hosiery + night-time MOBIDERM® Autofit device • Day-time THUASNE lymphology hosiery only	3 months	Arm volume measured by a plastic tape with light pressure (V = h (C2 + Cc + c2)/12p)	• Compliance • quality of life (LYMQOL arm questionnaire) • functional symptoms (heaviness, limb use lim-itation, pain) • sleep quality • Safety was assessed by numbers and percentages of patients with at least one reported AE
Miller, 2017	Compression garments	25	Male (n=19) and Female (n=114) (60.2)	primary lymphoedema	• previously worn compression garment • 3D flat knitted JOBST® Elvarex Plus compression garment	14 days	Lymphoedema symptoms	Patient' s opinion and quality of life measured by a validated questionnaire (Keeley et al, 2010
Ochalek, 2018	Compression sleeves	45	Female	breast cancer without distant metastases	• Compression group • No compression group	12 months	Arm volume was calculated by subtracting the volume of the normal limb from that of the affected limb	• physical activity (PA) by international Physical Activity Questionnaire (IPAQ) • Compliance by EORTC QLQ-C30 questionnaires • health-related quality of life (HRQOL) by QLQ-BR23 questionnaires
Pajero Otero, 2019	Kinesio taping and compression garments	30	Female	breast cancer–related lymphedema	• Kinesio taping • compression garment	71 days	lymphedema Relative Volume Change (RVC=((A2/U1)/(U2/A1))-1)	• range of motion of arm joints measured by a universal goniometer • self-perception of comfort was assessed via four questions • lymphedema-related symptoms (pain, tightness, heaviness, and hardness) measured by a verbal descriptive scale
Pujol-Blaya, 2019	Compression garments	42	Female (59.4)	A lymphoedema volume excess of 10% or more	• precast adjustable compression system group • multilayered compression bandages group	3 months	Excess volume ((excess volume pretreatment–excess volume posttreatment)× 100/excess volume pretreatment)	Patient' s opinion (pain, heaviness, tightness and hardness) measured by visual analogue scale
Rezende, 2017	Compression garments	20	Female (66.85)	breast cancer survivors	• Kinesiotherapy • FCB+kinesiotherapy • EC+kinesiotherapy	7 days	the maximum velocity of blood flow measured by Doppler SONARA/Tek	Limb volume, performed by an indirect method by sum of the approximate volume of the six truncated cones, formed by measure points of the circumferences of the seven points of the arm and forearm.
Sezgin Ozcan, 2018	CD and MLD	51	Female (66.85)	chronic Postmastectomy Upper Limb Lymphedema	• CD group • CDT group	6 months	limb volume using a simplified frustum formula (summed truncated cone)	• edema volume • limb-related volume change [(SL after treatment × NL before)/(NL after × SL before)]−1 • quality of life) from the Lymphedema Questionnaire
Suehiro, 2020	SEFS and CDT	51	SEFS(+): 71 SEFS(-):82	arm lymphedema treated for longer than 1 year using CDT	• SEFS was observed • SEFS was not observed	1 year	the edema ratio Edema ratio (%) = {(extremity volume in the affected limb) − (extremity volume in the contralateral limb)} × /(extremity volume in the contralateral limb) × 100	• Limb circumference • limb volume using the truncated cone method

Figure 3. Characteristics of the included studies

4 DISCUSSION

In this review, we aimed to summarize and quality assess the existing published evidence concerning The scale of the treatment effect of lymphedema pressure. Only tests carried out in the past 5 years were included because of the rapid development of Knit Garments, as well as significant changes in treatment strategies and surgical techniques for lymphedema pressure.

There are many methods to measure the excess volume, and since the limb volume changes with the increase/decrease of the patient's weight, and the different time of measurement also has an impact, a standard measurement scale should be established. However, when upper extremity lymphedema occurs after breast cancer treatment, it may affect other parts of the body and requires targeted treatment. Therefore, to compare and contrast the effectiveness of various treatment regimens, it is best to measure them in a standardized manner.

Although pressure cuff for the treatment of lymphedema international has been confirmed, there are few domestic studies of measure scale [12], and due to the individual specificity of lymphedema, often used in the treatment of CDT effect is best [13], and the CDT treatment professionals need to follow up in time, and the height of the patient to cooperate, so in formulating the unified measuring scale quite difficult. The optimal pressure values for lymphedema at different stages are different. In the actual treatment, it is necessary to timely track the patient's edema status and constantly adjust the pressure suit or pressure sleeve [14].

To improve the normality and accuracy of lymphedema stress therapy, more attention should be paid to the study of treatment outcome measurement scale in future studies.

References

1 Cal and Z. Bahar, "Women's Barriers to Prevention of Lymphedema After Breast Surgery and Home Care Needs: A Qualitative Study," (in eng), Cancer Nurs, vol. 39, no. 6, pp. E17-e25, Nov/Dec 2016.

2 "The diagnosis and treatment of peripheral lymphedema: 2020 Consensus Document of the International Society of Lymphology," (in eng), Lymphology, vol. 53, no. 1, pp. 3-19, 2020.

I. Tzani, M. Tsichlaki, E. Zerva, G. Papathanasiou, and E. Dimakakos, "Physiotherapeutic rehabilitation of lymphedema: state-of-the-art," (in eng), Lymphology, vol. 51, no. 1, pp. 1-12, 2018.

3 S. Vignes, "[Lymphedema: From diagnosis to treatment]," (in fre), Rev Med Interne, vol. 38, no. 2, pp. 97-105, Feb 2017. Les lymphœdèmes : du diagnostic au traitement.

4 D. Sezgin Ozcan, M. Dalyan, S. Unsal Delialioglu, U. Duzlu, C. S. Polat, and B. F. Koseoglu, "Complex Decongestive Therapy Enhances Upper Limb Functions in Patients with Breast Cancer-Related Lymphedema," (in eng), Lymphat Res Biol, vol. 16, no. 5, pp. 446-452, Oct 2018.

A. Miller, "Impact of seamless compression garments on limb functionality, comfort and quality of life," (in eng), Br J Community Nurs, vol. 22, no. Sup10, pp. S26-s37, Oct 1 2017.

5 J. Malloizel-Delaunay et al., "Endermology treatment for breast cancer related lymphedema (ELOCS): Protocol for a phase II randomized controlled trial," Eur J Obstet Gynecol Reprod Biol, vol. 241, pp. 35-41, Oct 2019.

6 S. Mestre et al., "Interest of an auto-adjustable nighttime compression sleeve (MOBIDERM® Autofit) in maintenance phase of upper limb lymphedema: the MARILYN pilot RCT," (in eng), Support Care Cancer, vol. 25, no. 8, pp. 2455-2462, Aug 2017.

7 K. Ochalek, T. Gradalski, Z. Szygula, and H. Partsch, "Physical Activity With and Without Arm Sleeves: Compliance and Quality of Life After Breast Cancer Surgery-A Randomized Controlled Trial," (in eng), Lymphat Res Biol, vol. 16, no. 3, pp. 294-299, Jun 2018.

8 V. Pajero Otero, E. García Delgado, C. Martín Cortijo, H. M. Romay Barrero, E. de Carlos Iriarte, and J. Avendaño-Coy, "Kinesio taping versus compression garments for treating breast cancer-related lymphedema: a randomized, cross-over, controlled trial," (in eng), Clin Rehabil, vol. 33, no. 12, pp. 1887-1897, Dec 2019.

9 V. Pujol-Blaya, S. Salinas-Huertas, M. L. Catasús, T. Pascual, and R. Belmonte, "Effectiveness of a precast adjustable compression system compared to multilayered compression bandages in the treatment of breast cancer-related lymphoedema: a randomized, single-blind clinical trial," (in eng), Clin Rehabil, vol. 33, no. 4, pp. 631-641, Apr 2019.

10 R. Hansdorfer-Korzon, J. Teodorczyk, A. Gruszecka, J. Wydra, and P. Lass, "Relevance of low-pressure compression corsets in physiotherapeutic treatment of patients after mastectomy and lymphadenectomy," (in eng), Patient Prefer Adherence, vol. 10, pp. 1177-87, 2016.

11 T. I. Grushina, "[What physiotherapeutic method for the treatment of post-mastectomy lymphedema is the most effective?]," (in rus), Vopr Kurortol Fizioter Lech Fiz Kult, vol. 94, no. 4, pp. 59-66, 2017. Kakoĭ metod fizioterapii postmastéktomicheskoĭ limfedemy naibolee éffektiven?

12 M. Karafa, A. Karafova, and A. Szuba, "The effect of different compression pressure in therapy of secondary upper extremity lymphedema in women after breast cancer surgery," (in eng), Lymphology, vol. 51, no. 1, pp. 28-37, 2018.

Design of Pedestrian Landscape Flying-Bird Arch Bridge with Cable-Stayed Cable and Lissajous Curved Arch Rib

Guangchuan Zhu[1*], Wenping Xu[1] and Jinfa Xu[2]

[1]College of Civil Engineering, Southeast University Nanjing, China,210096
[2]Jiangsu traffic engineering consulting and Supervision Co., Ltd, Nanjing, China,211800

Abstract. In view of the demand of super long span pedestrian landscape bridge with curved deck, this paper proposes a kind of cable-stayed and flying-brid arch bridge with Lissajous curved arch rib and double curved bridge deck. Its plane positive projection is Lissajous curve figure, the side elevation of the arch rib is the shape of the cable-stayed and flying-brid arch, and the arch foot cable and the upper space cable between the arch ribs are installed on the Lissajous spatial curved arch rib with concrete-filled steel tube to form a self balanced structure system, which floats and drags across the river. The main arch sling and the tail cable are installed, the double curved bridge deck and the central circular sightseeing platform are suspended, and multiple groups of spatial cables work well together to complement each other, with having complementary advantages. Combined with the actual project, the parameter design is carried out, the Midas finite element model is established, the internal force of the structure is calculated, the structural dynamic mode and stability analysis are carried out , the rationality of its structural performance is verified by analysis and research.

1 Introduction

With the development of the times, the spans of pedestrian landscape bridges are getting larger and larger. At the same time, more and more landscape bridges require the use of beautiful curved bridge decks. How to balance the mechanical performance requirements of super large-span structures and the aesthetic requirements of curved bridges is faced by bridge engineers [1-4].

When the curved bridge of the landscape river park has a large span, the beam bridge will encounter considerable difficulties in structure and construction [5-7], if a large-span concrete-filled steel tube flying swallow arch rib is used to suspend the double-span curve bridge deck, it is very suitable. A bridge across the river has a special and beautiful shape in aesthetics.

The cable-stayed flying swallow arch is a combination of a cable-stayed structure and a flying swallow arch bridge. The flying swallow arch bridge is the mainstay, and the cable-stayed structure is auxiliary. It is the product of the brave exploration of bridge design and construction technology. The cable-stayed flying-swallow arch has a novel structure and beautiful appearance, showing the characteristics of the flying swallow arch bridge and the cable-stayed bridge [8-10], the two bridge types work together and have complementary advantages.

In this paper, combined with a pedestrian landscape bridge in an ecological river, the geometric configuration research of the flying-swallow type cable-stayed arch bridge based on Lissajous's curved arch ribs is carried out, engineering parameter design is carried out, and the Midas finite element model is established for internal force analysis, modal analysis and stability analysis research to verify the rationality of its structure.

2 Geometric configuration study

The width of an ecological landscape river is 150 meters. Taking into account the requirements of the landscape, it is necessary to construct a double-span curve-shaped pedestrian landscape steel tube concrete flying swallow arch bridge, hoping to enhance the cultural landscape effect of the river park.

According to the smooth and graceful Lissajous curve graphics, the steel pipe is bent into Lissajous curve-shaped swift-stayed arch ribs by the pyrotechnic simmering process, and the lower tie rods between the arch ribs and the upper space cables between the arch ribs are set, to form a self-balanced structural system, reduce the unbalanced internal force of the support, and improve the bearing performance of the Lissajous figure steel tube concrete space arch rib. The space arch rib is equipped with slings and cable-stayed cables to suspend the double-width curved bridge deck, forming a flying-swallow type cable-stayed arch bridge with a double curved deck with Lissajous figure.

The Lissajous figure is a figure formed by using an oscilloscope to input different sinusoidal signals on the X-axis and Y-axis and superimposing them. The Lissajous figure is a regular and stable closed curve composed of simple harmonic vibrations of simple integer ratios, formed by two frequencies in mutually perpendicular directions.

*Corresponding author: Zhu Guangchuan: 857835935@qq.com

The parameter equation of Lissajous figure is:

$$x(\theta) = a\sin(\theta)$$
$$y(\theta) = b\sin(n\theta + \varphi)$$

in which $n \geq 1$ and $0 \leq \varphi \leq \frac{\pi}{2}$.

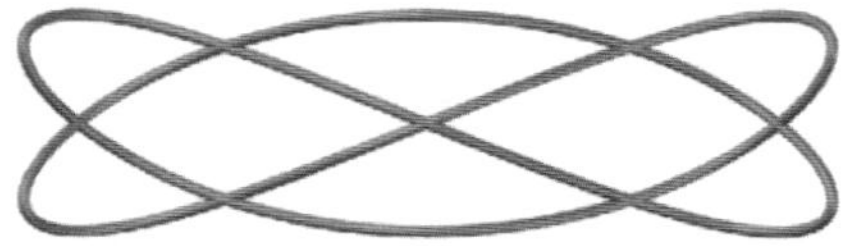

Fig1. The Lissajous figure

According to the structural characteristics of the flying -swallow type cable-stayed arch bridge, the spatial cable structure system of this bridge is composed of four parts: the lower tie bar between the arch ribs, the upper space cable between the arch ribs, the main arch sling and the rear stay cable. The spatial cables work together and complement each others' advantages, greatly improving the mechanical performance of the Lissajous-shaped concrete-filled steel tube flying swallow arch bridge structure.

The bridge adopts the construction technology of floating and dragging across the river. The lower tie rods between the arch ribs and the upper space cables between the arch ribs are installed, and the lower temporary cross cables between the arch ribs and the lower temporary transverse cables are installed to form a self-balanced structural system. Drag the Lissajous curve-shaped swallow arch across the river and place it on the foundation of the arch bridge smoothly. After the bridge is completed, the lower temporary cross cable and the lower temporary horizontal cable are removed.

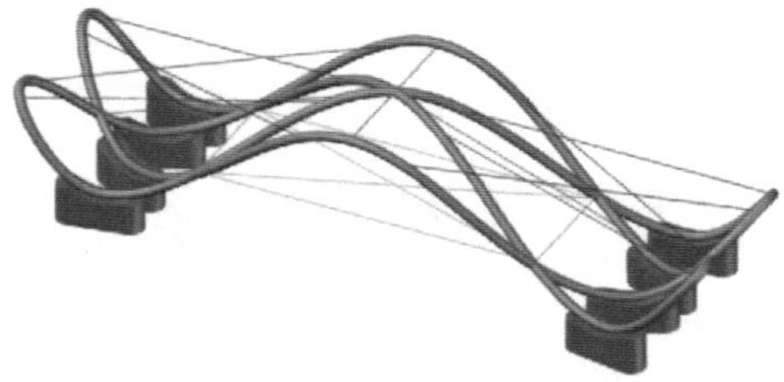

Fig2. Inter-arch space cable

In terms of structural mechanics, the flying-swallow type cable-stayed arch bridge structure is like a giant cantilever beam structure, which can be divided into two stress areas: the cantilever part and the mid-span part. The cantilever part needs to be equipped with upper cables, and the mid-span part needs to be configured with lower cables.

Among them, the lower tie rod cable between the arch ribs is composed of two permanent cables on the left and right, a temporary cross cable, and a temporary transverse cable. The lower tie rod cable of this bridge is mid-through, and the lower tie rod cable between the arch ribs is anchored at the X-shaped cross node of the Lissajous-shaped flying swallow arch, which is convenient for construction, convenient anchoring, and reasonable stress. The lower space between the arch ribs is like a wire cable with a positive bending moment, which can improve the flying swallow's bearing capacity and structural rigidity of the mid-span part of the arch structure.

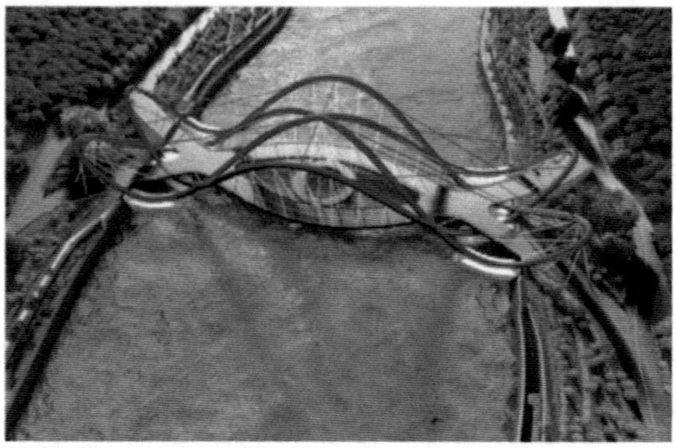

Fig3. Effect picture of flying-swallow style cable-stayed arch bridge with Lissajous curve arch rib

Install the main arch sling in the middle section of the Lissajous curve-shaped flying swallow arch rib, and install the oblique sling in the tail section of the arch rib. Lissajous's curve-shaped flying swallow-style cable-stayed arch rib structure suspends the double-width curve-shaped bridge deck, and the construction of the bridge deck stiffening beam is convenient.

The double-width curved bridge deck is composed of double-width cross-curved main beams and circular ring beams. The circular ring beam is located at the center of the double-width cross-curved main beam, which limits the relative displacement and deformation of the left and right double-width cross-curved main beams, can greatly improve the lateral structural force of the double-width cross-curved main beam, and can double as a sightseeing platform.

The flying-swallow type cable-stayed arch bridge with a double-span curve deck has the characteristics of a flying swallow-style cable-stayed arch bridge in the longitudinal direction, and its transverse direction is like a cable-stayed bridge system with inclined arch towers. Lissajous curve space inclined arch rib design is ingenious, with beautiful shape, rich facade, convenient construction of replacing cables, large spanning ability and good durability.

The cable-stayed` flying swallow arch pedestrian landscape bridge with curved arch ribs of Lissajous has a beautiful shape like a three-dimensional sculpture, like a dancing butterfly. A variety of curved design elements are integrated into the design, which is a perfect combination of geometry, architecture and structure, becoming a three-dimensional space curve pedestrian landscape bridge.

3 Design parameters

An ecological river has a landscape belt with parallel rivers built on both banks. The width of the river is 150 meters. The Lissajous curve-shaped flying swallow-style cable-stayed arch bridge with double-curved bridge decks spans the landscape river. The total length of the main span bridge is 190 meters, and the span is 40m + 110m + 40m. The seismic fortification of this bridge is 6 degrees.

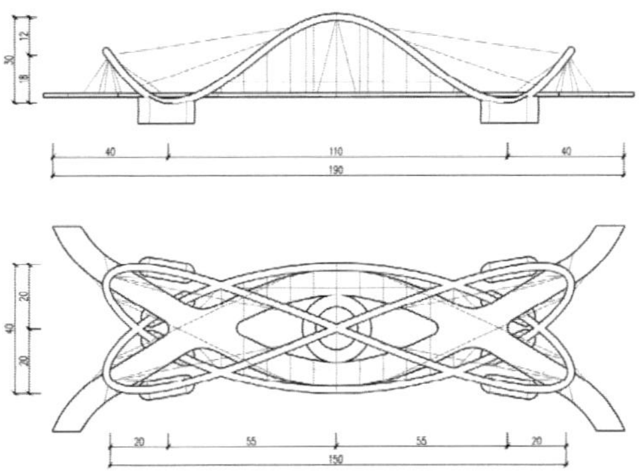

Fig4. General layout

Lissajous curve-shaped flying swallow arch rib is composed of three components connected end-to-end butterfly wing inclined arch rib, X-shaped cross-shaped arch rib and flying swallow inclined tower arch rib. The three steel pipe members are welded and connected as a whole. The plane orthographic projection is a three-leaf rose line figure, and the side elevation projection of the arch rib is a flying-swallow type cable-stayed arch bridge structure.

The horizontal projection length of the Lissajous graphic steel tube arch rib is 150 meters, the horizontal projection width is 40 meters, and the distance between the supports of the spatial arch rib is 110 meters. The diameter of the circular steel pipe of the arch rib is 2 meters, the wall thickness of the steel pipe is 20mm, and the inside of the steel pipe is filled with C50 concrete.

Lissajous curved flying-swallow arch ribs are installed with lower cables between the arch ribs and cables for the upper spaces between the arch ribs to form a self-balanced structural system. The floating plan is used to drag across the river, lift and translate, and place the flying swallow-style cable-stayed arch ribs on the foundation of the arch bridge smoothly.

The lower tie rod cable between the arch ribs is anchored at the X-shaped cross node of the Lissajous figure flying-swallow arch, forming a mid-through flying swallow arch bridge structure. There are 6 sets of arch-foot cables, of which the left and right permanent cables use 1770MPa galvanized high-strength steel wire, and each permanent arch-foot cable has a diameter of 0.4 meters. Besides, two high-strength steel wires of temporary cross cables with a diameter of 0.2 meters and a high-strength steel wire of temporary cross cables with a diameter of 0.1 meters are set up. After being dragged across the river by floating, 4 sets of temporary cables are removed.

The cable system of the upper space between the arch ribs is composed of three parts: the upper stay cable between the arch ribs, the longitudinal cable and the horizontal cable. The upper space cables between the arch ribs are made of 1770MPa galvanized high-strength steel wire, and the arch rib space cable system is like a negative bending moment wire cable, which can improve the bearing capacity and structural rigidity of the flying-swallow type cable-stayed arch structure.

Fig5. Effective picture of the completed bridge

The horizontal stay cables are anchored on the tops of the two left and right wing-shaped inclined arch ribs. The horizontal stay cable adopts high-strength steel wire cables with a diameter of 0.2 meters. The horizontal stay cables can ensure the overturning stability of the wing-shaped inclined arch ribs;

Longitudinal cable anchoring starts at the intersection of two swallow-like inclined tower arch ribs at one end of the arch bridge, spans the apex of the X-shaped cross-shaped arch rib, and anchors at the intersection of two swallow-like inclined tower arch ribs at the other end of the arch bridge. The longitudinal cable adopts a high-strength steel wire cable with a diameter of 0.2 meters;

The upper cable between the arch ribs are set between the butterfly wing-shaped inclined arch rib and the flying swallow-shaped inclined tower arch rib. The stay cable adopts a high-strength steel wire cable with a diameter of 0.15 meters. There are four cables on each side, and there is a total of eight cables between the arch ribs.

The double-width cross-curved main beam is a flat steel box beam structure with a beam height of 1.3 meters and a beam width of 9 meters. The circular ring beam is a steel box beam structure with a beam height of 1.2 meters and a beam width of 4 meters.

The sling system consists of three parts: the main arch herringbone sling, the circular central sling and the tail stay cable. The sling system suspends the double-width curved bridge deck and the central circular ring beam, the upper part of the herringbone sling is anchored on the butterfly wing-shaped inclined arch rib, and the lower part of the herringbone sling suspends the middle beam section of the double-width cross-curved main beam. The herringbone sling adopts a 1770MPa high-strength steel wire cable with a diameter of 0.10 meters, the spacing of the herringbone slings is 7.5 meters, each of 9 pairs of herringbone cables, and a total of 18 pairs of herringbone cables for the whole bridge.

The upper part of the tail stay cable is anchored on the arch rib of the flying-swallow type inclined tower, and the lower part of the tail stay cable is inclined to lift the two side span sections of the double-width cross-curved main beam. The rear stay cable adopts 1670MPa high-strength steel wire cable with a diameter of 0.10 meters, and there is a total of 24 rear stay cables for the whole bridge.

The upper part of the central sling of the ring is anchored on the top of the X-shaped cross-shaped arch rib, and the lower part of the central sling of the ring is suspended from the circular ring beam. The circular ring beam has 6 central slings in the inner ring and 4 central slings in the outer ring. The central sling adopts a 1770MPa high-strength steel wire cable with a diameter of 0.05 meters.

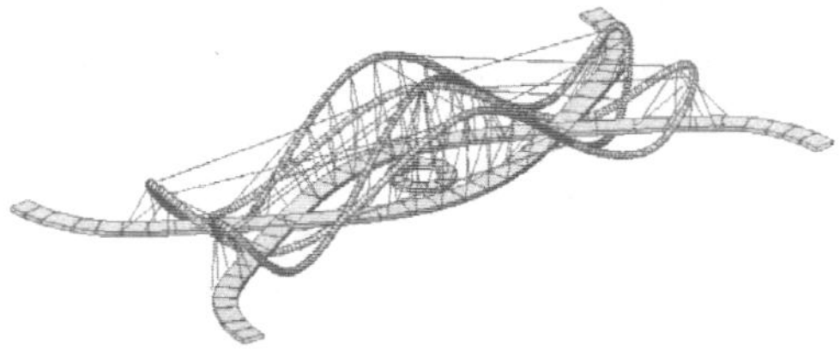

Fig6. Midas model

4 Calculation results under vertical load

Considering bridge deck paving and municipal pipelines, the standard value of uniformly distributed load is 6KN/m^2 for the additional dead load on bridge deck, and the standard value of uniformly distributed load is 5KN/m^2 for bridge deck live load. The calculation results under the vertical load are as follows:

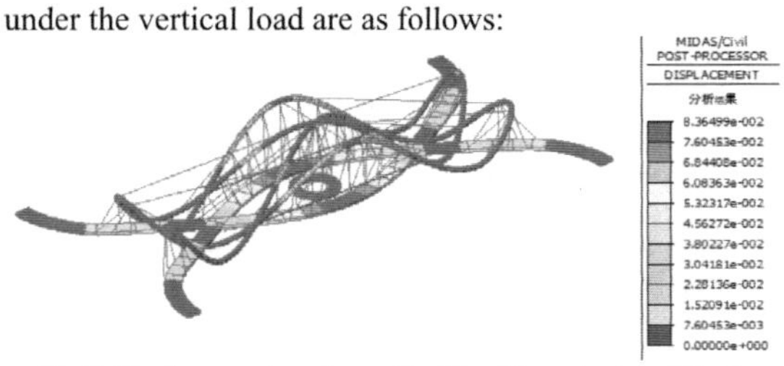

Fig7. Displacement under vertical load (constant and live)

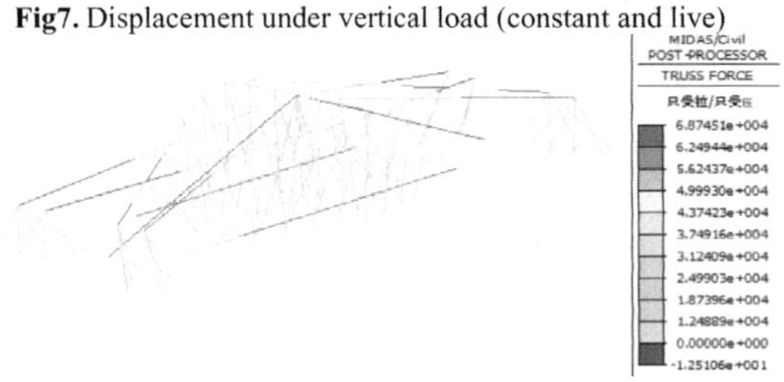

(a) Cable unit internal force

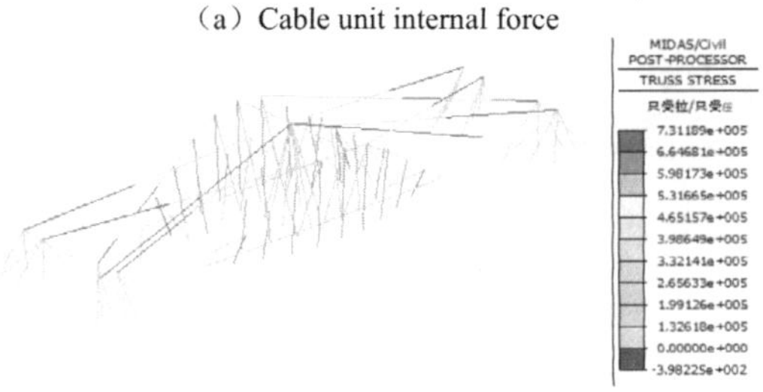

(b) Cable unit internal force

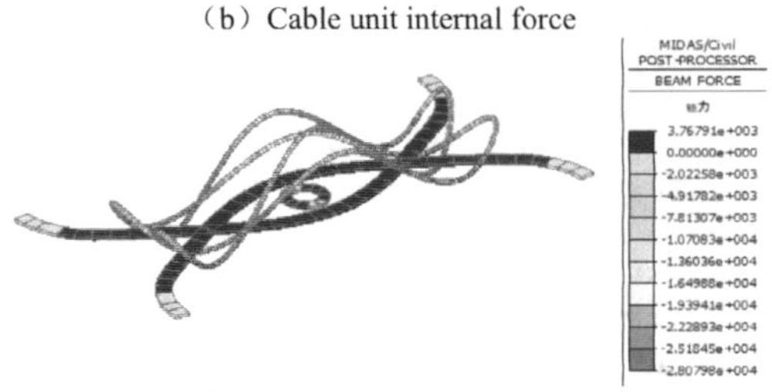

(c) Beam unit internal force

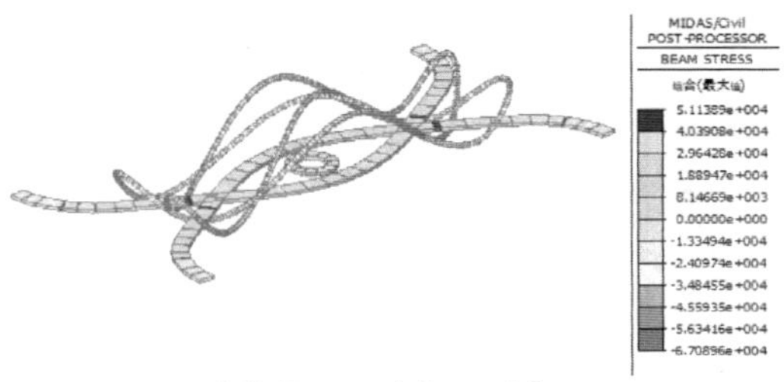

(d) Beam unit internal force

Fig8. Calculation results of internal force and stress under load

Under the action of the vertical live load, the maximum vertical displacement appears at the mid-span position, and the maximum deflection is only 0.084m, which meets the 1/500 limit specified in the code. The vertical structure of the bridge has a relatively large rigidity.

The maximum internal force of the main bridge sling is 8269.6KN, and the maximum stress of the sling is 731.2MPa, which meets the strength requirements.

The maximum internal force of the cable tie is 68745.1KN, and the maximum stress of the tie is 350.1MPa, which meets the strength requirements.

The maximum internal force of the bridge arch is 28079.8KN, and the maximum stress of the bridge arch is 67.1MPa, which meets the strength requirements. The maximum stress of the arch rib appears near the arch foot. The component material is steel tube concrete, the concrete is C50 concrete, and the steel is Q355 steel, which can meet the strength requirements.

5 Modal analysis results

In order not to miss any mode shapes, the sub-block method is used to solve the characteristic equations in the analysis process. In the analysis process, the sub-blocking method is used to solve the characteristic equations, the first 20 natural frequencies and the mode characteristics.

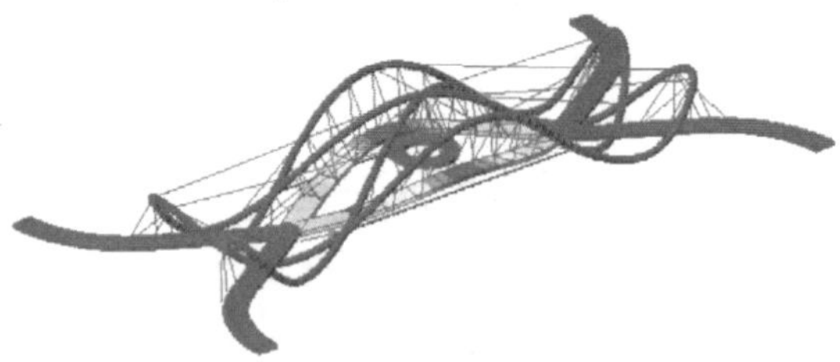

(a) 1st order mode (positive symmetrical vertical bending) (2.049Hz)

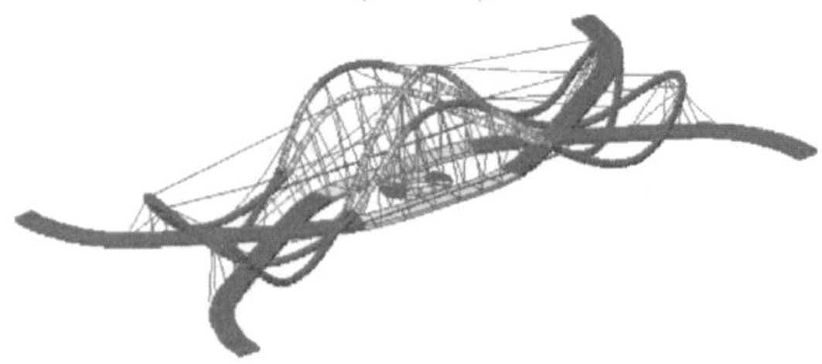

(b) 3rd order mode (positive symmetry side bending) (2.371Hz)

https://doi.org/10.1051/e3sconf/202123704024

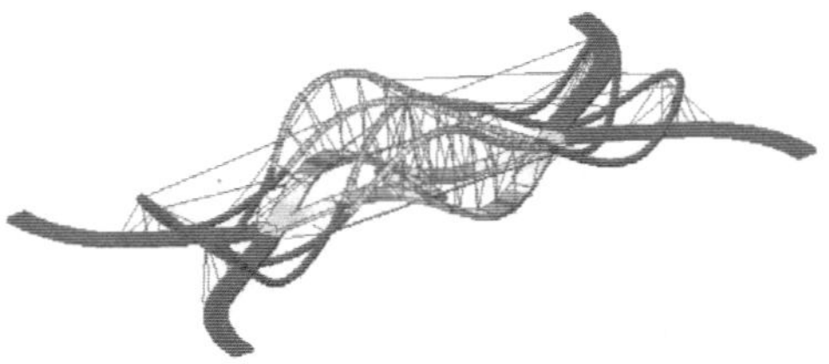

(c) 4th order mode (anti-symmetric vertical bending) (2.382Hz)

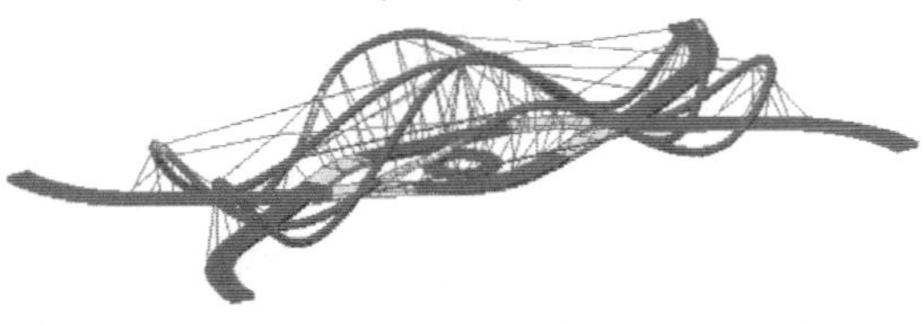

(d) 7th-order vibration mode (the arch is symmetrically bent) (3.209Hz)

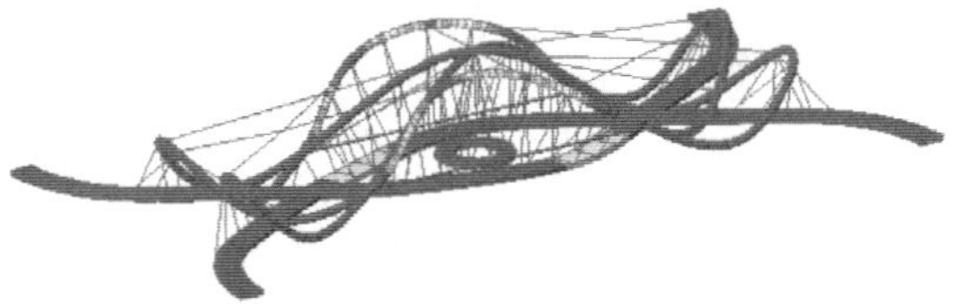

(e) 9th order mode (anti-symmetric side bending) (3.453Hz)

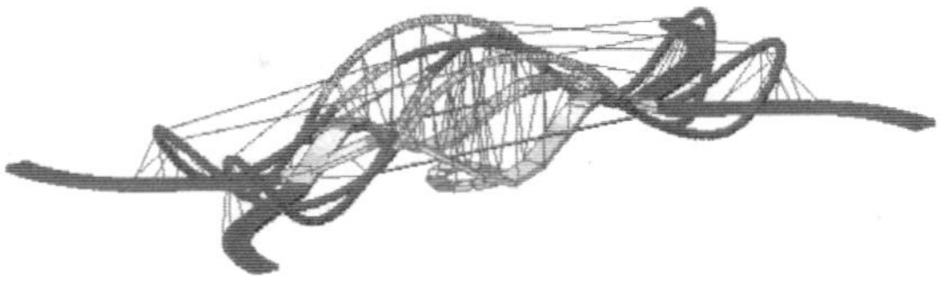

(f) 10th order mode (torsion of bridge deck) (3.987Hz)

Fig9. Typical modal

From the calculation results in the above figure, it can be seen that the first-order mode is a positive symmetric vertical bending with a frequency of 2.049 Hz; The third-order mode is positive symmetrical side bend, and the frequency is 2.371 Hz; The 10th order mode is a torsional mode with a frequency of 3.987 Hz.

On the whole, the vibration modes are dense and there is obvious grouping of vibration modes. The first 9 modes are dominated by lateral and vertical bending vibrations, and the torsional mode does not appear until the 10th order. The higher frequency of each order of the bridge structure indicates that the bridge structure has greater structural spatial rigidity.

6 Structural stability analysis

The buckling analysis of the designed bridge was carried out with Midas software, and the structural stability coefficient was obtained. In general, when the stability coefficient is the smallest, the possibility of structural instability is greatest. The following table lists the first 10 order stability factors.

Table1. Buckling mode calculation results

Mode	Eigenvalues	Allowable error
1	33.391	0.0000e+000
2	34.244	0.0000e+000
3	34.537	0.0000e+000
4	55.016	4.2530e-029
5	57.132	1.0357e-026
6	62.403	2.0364e-018
7	62.642	2.9969e-018
9	64.602	5.4086e-017
10	70.887	1.1194e-011

It can be seen from the above table that the minimum value of the stability safety factor of the bridge is 33.391, which fully meets the requirements of GB 50923-2013 "Technical Specification for Concrete Filled Steel Tubular Arch Bridges" greater than 4.0.

The analysis shows that the Lissajous curve-shaped flying-swallow type cable-stayed arch bridge connected end to end has good beam integrity, and the X-shaped cross-shaped arch rib can greatly improve the structural stability.

7 Conclusion

Based on the engineering background of a pedestrian landscape bridge with a main span of 110 meters , this paper conducts a research on the geometric configuration of a Lissajous curved flying swallow arch bridge with a double-span curve deck, establishes a Midas finite element model, and conducts the internal force analysis, dynamic mode and structural stability analysis under vertical load have drawn the following conclusions:

(1) According to the smooth and beautiful Lissajous figure, the steel pipe is bent into Lissajous curve-shaped flying swallow arch ribs by the pyrotechnic simmering process. The double-span curve bridge deck and the circular sightseeing platform are suspended to form a Lissajous curve arch rib flying-swallow type cable-stayed arch bridge. The shape is beautiful, like a dancing butterfly, with the beauty of three-dimensional sculpture.

(2) The curved steel tube arch ribs of the super-large-span Lissajous curved arch ribs and flying-swallow type cable-stayed arch bridges are raised and extended to form inclined arched towers, which is made in one go, smooth and continuous, the arch tower is integrated, and the structural integrity and stability are better.

(3) Lissajous curved swift-stayed arch ribs are installed with lower cables between the arch ribs and the upper space cables between the arch ribs to form a self-balancing structure system, which is dragged across the river by floating, and construction is quick and convenient.

(4) The space cable structure system is formed by the tie rod cable, the space cable between the arch ribs, the main arch sling and the tail stay cable. Multiple groups of space cables work together to complement each other, which greatly improves the mechanical performance of the Lissajous curve-shaped steel tube concrete flying swallow arch bridge structure.

(5) The Lissajous curve-shaped flying-swallow style cable-stayed arch bridge is like a giant cantilever beam structure, which can be divided into two stress areas: the

cantilever part and the mid-span part. The lower cable between the arch ribs bears a positive bending moment, and the upper space cable between the arch ribs bears a negative bending moment. The bridge has greater structural rigidity and bearing capacity. Under the action of vertical live load, the maximum deflection in the middle of the span is 0.084m, which meets the 1/500 limit specified in the code.

(6) Modal analysis shows that the frequency of each order of the bridge structure is relatively high, and there is an obvious mode grouping phenomenon, which indicates that the bridge structure has greater structural spatial rigidity. Among them, the first-order mode is positive symmetrical vertical bending, with a frequency of 2.049 Hz, which does not meet the 3 Hz specification. A mass frequency modulation damper can be added to control its dynamic response to solve the problem of pedestrian comfort.

(7) The Lissajous curve-shaped flying swallow-style cable-stayed arch bridge connected end to end has good structural integrity. Its X-shaped cross-shaped arch rib geometric configuration can greatly improve the structural stability, and the minimum stability safety factor is 33.391, which meets the requirements of the specification.

References

1. Kuo, M., & Xiao-Lan Meng. Towards a landscape-approach bridge design: Philosophy and practice[C]. *2011 International Conference on Electric Technology and Civil Engineering* (ICETCE), 2011.7102–7106.
2. B. Briseghella, L. Fenu, W. Huang, et al.Tensegrity footbridges with arch deck: static and dynamic behaviour[C].*6th International Conference on Arch Bridges*, Fuzhou, 2010.
3. Luigi Fenu ,Eleonor Congiua ,Davide Lavorato et al. Curved footbridges supported by a shell obtained through thrust network analysis[J]. *Journal of Traffic and Transportation Engineering*.2019,6(1):65-75.
4. Wang Tingzheng.Technical characteristics of cable-stayed flying swallow arch combination bridge[J]. *World Bridges*,2009(02):27-30. (In Chinese)
5. Zhao Yueyu,Yang Xiangzhan,Kang Houjun.Analysis of Dynamic Characteristics of Cable-stayed Arch Bridge[J]. *Highway*,2005,(11).36-39. (In Chinese)
6. Luo Shidong, Wang Xinguo, Wang Tingzheng, et al. Conception and Research on Innovative Technology of Long-span Cable-Stayed Arch Bridge[J]. *Bridge Construction*,2005,(6).31-33,51. (In Chinese)
7. shi-dong luo, Wang Xinguo ting-zheng wang, cassia Xing. Innovative Design and Research of long-span cable-stayed Arch Bridge [J]. Bridge Construction,2005(06):31-33+51.(In Chinese)
8. Lv Jiangen, Wang Ronghui. Comparison of mechanical properties between cable-stayed arch Composite bridge and ordinary arch Bridge [J]. *Highway*,2008(08):5-9.(In Chinese)
9. Wang Tingzheng. Technical Characteristics of Cable-stayed Flying-Arch Composite Bridge [J]. *World Bridge*,2009(02):27-30.(In Chinese)
10. Chen DE. Structural analysis of cable-stayed arch composite bridge [J]. *Journal of natural science of xiangtan university*,2009,31(03):123-127.(In Chinese)

Analysis on the Human Settlement Environment of Huizhou Ancient Villages Based on the Heritage of Ancient Roads -- a Case Study of Chengkan Village

Bi Zhongsong[1,2], *Chen Chun*[2], *Li Yunzhang*[2], and *Cheng Peng*[1,*]

[1*]College of Architectural and Engineering & Research Center of Huizhou Architecture, Huangshan University,245041 Huangshan, China
[2]College of Architecture and Environment, Sichuan University,610065 Chengdu, China

Abstract. The ancient road cultural route is an important part of Huizhou cultural heritage, which is closely related to Huizhou tradition. Huizhou ancient villages are important historical and cultural heritage. The important features of their locations and layouts are the high integration with nature, forming a traditional residential environment pattern of "backing mountain, surrounding water and facing screen". From the perspective of Huizhou Ancient Road, this paper selects Chengkan ancient village, which is an important cultural route of "Shuyuan Lingshan Chengkan ancient road", as the research object, and takes the construction of human settlements as the main research content. From the natural environment and cultural background, and the formed human environment as the starting point, the paper focuses on the village agglomeration spatial pattern, the village water system, the street space environment, the architectural environment and the Shuikou landscape environment to analyze the characteristics and concepts of traditional residential environment construction in Huizhou ancient villages represented by Chengkan village, so as to make a beneficial discussion on the regional rural revitalization, the protection and development of traditional village human settlement environment.

1 Introduction

China is a country with an ancient civilization mainly based on the agglomeration space pattern of villages, village water system, street space environment, building environment and Shuikou landscape environment. Under the special historical conditions of nature, culture and social system, China has formed its unique concept of human settlement environment[1]. Human settlement environment refers to the place where human beings live together, which is the surface space closely related to human survival activities, as a general term of natural environment, living environment and social and cultural environment. In a broad sense, human settlement environment is composed of natural environment, artificial environment, social and economic environment and cultural environment. In a narrow sense, human settlement environment only refers to people's living and living place is also a space for social activities integrating with nature[2].

Due to the influence of nature, society and history, the living environment of Huizhou is different from other regions[3]. Chengkan village is an important historical and cultural heritage of Huizhou ancient village. The main feature of its location and layout is the high integration with the natural environment. It skillfully inherits the traditional Chinese cultural thought of "unity of man and nature", forming the traditional village living environment of "backing mountain, surrounding water and facing screen".

2 Natural environment and historical and cultural background of Chengkan Village

2.1 Natural environment of ancient villages

Chengkan village is located in Chengkan Town, Huizhou District, Huangshan City, Anhui Province. It is surrounded by mountains on all sides, with Longshan and Changchun mountains in the north, Lingjin mountain in the east, Ge mountain in the west and xiajie mountain in the south. Its altitude is 180-350 meters, which makes the whole environment form the situation of "left green dragon, right white tiger, front rosefinch, and rear Xuanwu".

Chengkan village is located in a valley basin of about 80 hectares. Zongchuan river runs through the village from north to south, with a total length of about 19 kilometers and a drainage area of about 46 square kilometers. It is a tertiary tributary of Xin'an River, the main river in Huizhou. Zongchuan River enters Chengkan basin from Longshan mountain to Changchun mountain, flowing into fengle River in the south, and

* Corresponding author: cpaiwy@hsu.edu.cn

 https://doi.org/10.1051/e3sconf/202123704025

then injecting Lianjiang River, a tributary of Xin'an River in the East. In addition to the Zongchuan River, there are streams in the Chengkan area: Shikeng and huanlikeng in the north, dongshankeng and shibukeng in the southeast, Yaokeng and pengkeng in the West (Fig.1).

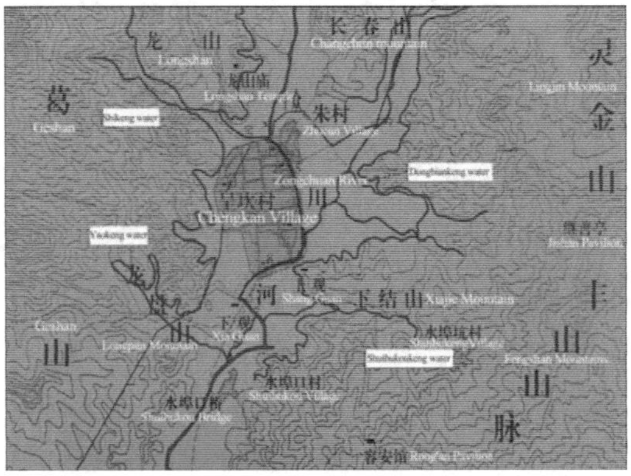

Fig. 1. Geographical environment and water system istribution map of Chengkan Village.

2.2 Historical and cultural background of ancient villages

The written history of Chengkan ancient village can be traced back to the end of the former Tang Dynasty, more than 1100 years ago. In order to avoid war and engage in business, Duke Luo qiuyin and his cousin Wenchang came to the place called Longxi at that time in order to avoid war and do business. According to historical records, the Luo brothers were familiar with Fengshui theory. They found that this place is surrounded by mountains on all sides and depressions in the middle. It is a typical place where "four rivers are contained in the village, and water is gathered as wealth". Therefore, they chose to live here.

Chengkan flourished in the song, yuan and Ming Dynasties. According to the genealogy of the Luo family in the Ming Dynasty, more than 110 people held official posts in the government. Among them, Luo Ruji, the Minister of bureaucrat in Song Dynasty, was more famous, and he was the chief judge of Yue Fei's unjust case. With the rise of Huizhou merchants, Luo's merchants spread all over the country. These officials and merchants returned to their hometown, built houses, ancestral temples and communities, which made Chengkan Village prosperous for a time.

2.3 Cultural environment of ancient villages formed under natural environment and cultural background

The geographical environment of "backing mountain, surrounding water and facing screen" and the development history of ancient villages spanning more than 1000 years have accumulated profound cultural heritages of Chengkan village(Fig.2). The development of Huizhou business economy has promoted the development of literature, art and architectural culture of Chengkan village, leaving a precious historical and cultural heritage for today's village.

Fig. 2. Composition of living environment in Chengkan Village.

Chengkan village is now a famous historical and cultural village in China. There are more than 130 cultural relics and historical buildings in different periods of Ming and Qing Dynasties, such as ancient ancestral halls, ancient dwellings, ancient community houses, ancient watchtowers, ancient pavilions, ancient wells, and ancient bridges. Chengkan ancient village has two national key cultural relics protection units: luodongshu temple and Chengkan village ancient building complex, including three ancestral halls, one community house, one school, two shops, three watchtowers, two ancient bridges, two ancient wells, one gate tower and thirty-four residences, with a total of 49.

Chengkan village is one of the most concentrated, preserved and characteristic ancient villages in southern Anhui Province. It contains a lot of historical information about the evolution and development of Huizhou traditional architecture from Ming Dynasty to Tsing Dynasty. It still retains some practices of song and Yuan Dynasties from the existing architectural remains of Ming Dynasty. It is a research on the historical evolution, structure and technology of Huizhou architecture. The scientific and technological development of Huizhou regional architecture in Ming and Tsing Dynasties is reflected from planning, site selection, environmental concept, spatial pattern, architectural form, architectural structure system, architectural modeling, etc.

3. The main modes to create the living environment in Chengkan village

3.1 Build living settlement of harmony between man and nature in the village

In the special geographical location and natural environment, people usually skillfully use the characteristics of this terrain to create a unique settlement pattern[4]. The traditional culture of Huizhou, which takes Chengzhu jurisprudence as the spiritual core, is closely combined with the unique geographical environment of Huizhou region, which has direct guidance and influence on the location, layout and construction of ancient village road.

Based on the traditional geomantic theory, Chengkan ancient village adjusts the village spatial form in the process of village site selection, construction and development, so as to achieve the perfect combination of spiritual pursuit and human settlement environment construction. The change of the original Zongchuan river channel in the ancient village is a typical example of the adjustment of village spatial form.

The existing Zongchuan river channel in Chengkan Village was reconstructed on a large scale in Hongzhi period of Ming Dynasty (1488-1506). After the diversion, the Zongchuan river is zigzag. In terms of village construction and development, the change of river course has two meanings[5] (Fig.3).

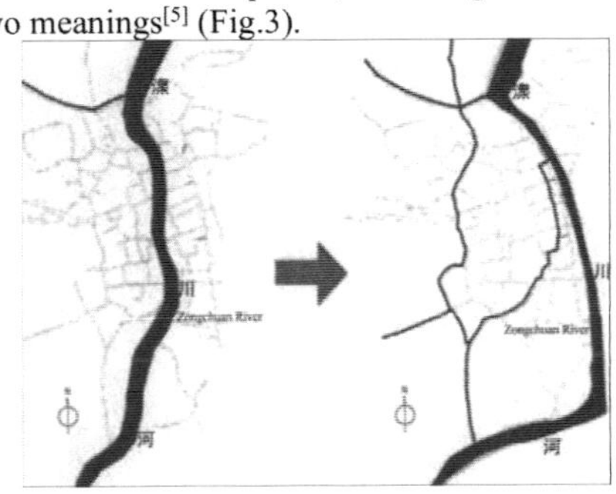

Fig. 3. Chengkan Village Zongchuan river channel change map (left: before change, right: after change).

On the one hand, with the rapid growth of the village population, the village construction also develops rapidly, and the Zongchuan river in the middle of the village hinders the development and construction of the village. The change of the river channel increases the ideal construction land of the village -- the land space on the west side of the Zongchuan river. On the other hand, it is close to the ideal geomantic state. The Chengkan Village before the river channel change met the Fengshui pattern of "back mountain and face screen" in ancient Huizhou area, but only "water ring" was not satisfing. Therefore, the change of "zigzag" river not only expanded the village construction land, but also formed the ideal geomantic pattern of "backing mountain, facing screen and surrounding water".

3.2 Build a benign development of the village water system by adapting to the terrain

Chengkan ancient village has a complete water conservancy system of "front River, middle ditch and back ditch"[6] (Fig.4). The "front River" refers to the Zongchuan river, which turns from the north to the southeast of the village, and then runs to the south, forming a good landscape of Shuikou Garden in the south of the village.

Fig. 4. Distribution of water system in Chengkan Village.

In order to solve the discharge of villagers' living, fire fighting, irrigation water and domestic sewage, two water channels were excavated manually: the "middle channel" in the East and the "back ditch" in the back street. The two will gather as the first-class in the south of the village. "Middle channel" is also called the east channel, which leads the river of Zongchuan River to Yongshun, Houluo general temple, Zhici, Luohuizheng mansion and Luohuitai mansion, through Shuizhen lane to the south of the village, and into the canal of the river. "Back ditch" means the small channel which leads to the west side of the village along the back street, and then flows into the middle channel.

In the 99th lane of the fifth street of Chengkan village, almost all households have water flowing in front of their doors (Fig.5), which shows that the water diversion and drainage facilities are not only reasonable, but also integrate the beauty of space form and environment. At the same time, the construction of this kind of facilities is not a single function consideration, but also a concept embodiment full of traditional cultural connotation. "Gathering water is like gathering money, taking four water in the village", the continuous flow of water expresses the long history and prosperous financial resources.

Fig. 5. Drainage system of Chengkan Village.

The water conservancy system of "Front river, middle ditch and back ditch" in Chengkan village and the unique street and lane pattern of "five streets and ninety-nine lanes" were formed by the ancients' careful planning and design based on a series of geomantic theories such as "pillow mountain, ring water, face screen", etc. ancient people in Huizhou not only respected nature and environment, but also successfully transformed nature and business Creating the environment, maintaining and expanding a balanced ecological environment.

3.3 Rich and varied street space environment

The main vein of the village extends along the Zongchuan River, which divides the village boundary into two parts: the East and the West. The two roads, called Daba and Xiaoba, enter the village along the river. The two roads meet at the head of Longxing bridge and are divided into front street and back street. These two streets are called "five streets" (Table 1), together with Zhongying street, Zhonger street and Xidong street in Hedong parallel. Vertical to the Zongchuan river, it is mostly alleys, known as ninety-nine lanes. Most of the streets and alleys in the village are very zigzag and changeable, ranging from less than 1 meter in width to about 3 meters in width(Fig.6).

Table 1. List of main streets and alleys in Chengkan Village.

Street Name	Length (m)	Width (m)
Xidong Street	231	2.0
Front Street	300	1.5-3.0
Zhong Ying Street	191	1.5-2.5
Zhong Er Street	127	1.5-2.0
Back Street	294	3.0

Fig. 6. The street space of Chengkan Village (from left to right is Xidong Street, Front Street, Zhong Ying Street)

3.4 Create a unique architectural space environment with Huizhou characteristics by terrain features

The design wisdom of living environment in Chengkan Village is also reflected in the construction of building space according to the terrain, mainly reflected in the building plane space and facade space.

The building characteristics of the plane layout in Chengkan village are mainly reflected in the natural layout. With the help of natural forms of mountains and rivers, there is no emphasis on certain geometric forms, let alone fixed patterns. The main house pursues the central axis as much as possible, but its wing rooms and other attached houses are flexible and changeable in the plane layout. The houses can be connected vertically and horizontally, and the deep courtyard is vertically and horizontally combined. The appearance is not required to be regular and symmetrical (Fig.7).

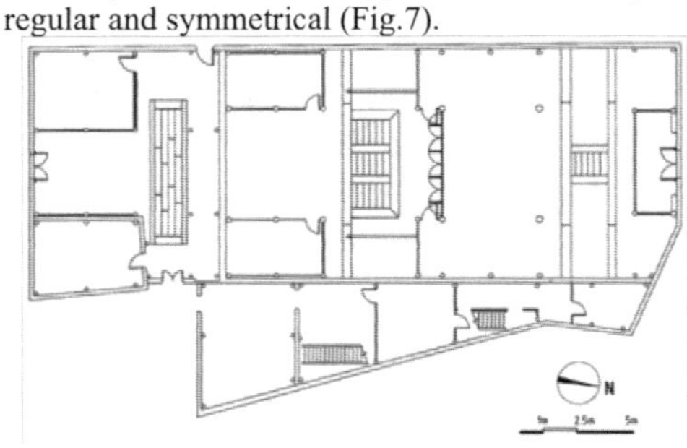

Fig. 7. Plan of Shangwu, Chengkan ancient village.

The ancient buildings in Chengkan village also show great flexibility in three-dimensional space. Because the number of main rooms and auxiliary buildings varies, the height of each house is different. At the same time, combined with the topography and topography of high / low changes, the mountain walls with different heights are formed. The whole ancient village has formed a rich and colorful group space image.

3.5 Keeping Shuikou, as a pioneer in Chinese public garden landscape construction

Shuikou is an indispensable and important part of the spatial form of Huizhou ancient villages, which forms Shuikou garden with regional cultural characteristics and humanistic spirit, and is a typical representative form of Huizhou regional traditional garden landscape. The appearance of Shuikou garden made up for the blank of Chinese ancient private garden, royal garden and temple garden, but no public garden.

In Huizhou ancient villages, combined with the natural relationship between the water system and the villages flowing through, water outlets are generally arranged at the head and end of the village. The combination of artificial construction and natural landscape of the water outlet not only respects and utilizes the natural form, but also creates artificially.

The Shuikou construction in Chengkan ancient village can also be divided into upper and lower water outlets. In the northern part of the village, Luodongshu Ancestral Hall is the Shuikou, also known as "Tianmen". According to local historical records, there are buildings such as Longshan Temple in the Shuikou area, surrounded by natural forest, which has basically no trace today. The Shuikou in longxingqiao, the southern head of the village, is also known as "Dihu". According to the theory of geomancy, it is an important part of storing wind and gathering gas. It is reported that there were upper and lower view gardens and upper and lower gardens in this area. Among them, there are Longxing bridge, Longxing Temple, Dutian temple, Guanyin temple and Baibu ladder in Shangguan, while those in Xiaguan include emperor temple, Wenhui hall, Wenchang Pavilion, Leji bridge, Kuang ruting and stone archway (Fig.8). Between the upper and lower views are the dam and the small dam, which makes the Zongchuan River turn westward from north to South at the foot of the mountain, and then to the South after the Longpan mountain, which makes the water flow at the water outlet bend, which is more in line with the traditional geomantic theory. Today's Xiashuikou, Longxing bridge is still towering on the Zongchuan river (Fig. 9,10), and you can feel the rich natural and cultural environment of Shuikou.

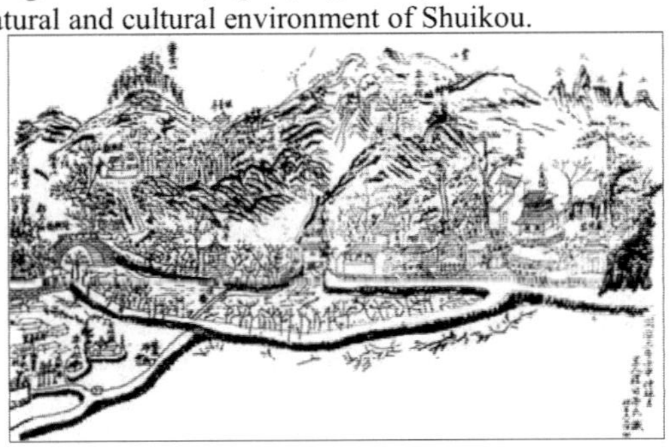

Fig. 8. Historical map of nanshuikou, Chengkan Village.

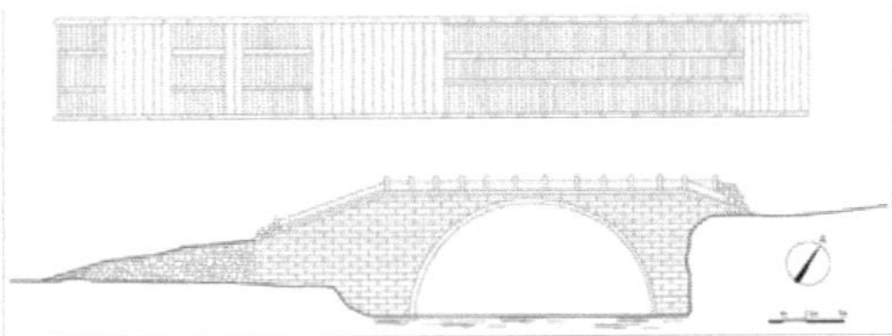

Fig. 9. Surveying and mapping drawings of Longxing bridge (top is plan and the bottom is elevation of Longxing bridge)

Fig. 10. Longxing bridge Conservation status in the south of Chengkan village

4 Conclusion

The geomantic environment surrounded by mountains and water reflects the scientific nature of ancient Huizhou ancestors in the selection of residential environment and the comprehensive application of simple traditional geomantic theory. With the development of science and the continuous improvement of human requirements for the quality of living environment, the construction of human settlement environment has gradually developed into an independent discipline, which has been widely concerned by professionals and ordinary people. Our ancestors have studied and practiced the construction of human settlement environment thousands of years ago. Chengkan village, as one of the typical representatives of ancient villages in Huizhou, not only has the general spatial pattern of human settlements in Huizhou traditional villages, such as "back mountain, surrounding water and screen", but also has the typical characteristics of its unique culture, nature, architecture and so on. It provides an important example for our research, discussion and practice of human settlement environment. Through the discussion and research of Huizhou ancient village human settlement environment, it is helpful for us to further understand and understand the construction concept of ancient human settlement environment, so as to provide practical research ideas and research methods for Rural Revitalization under the background of cultural self-confidence, initially solve the problem of "what to learn", and protect the traditional human settlement environment in Rural Revitalization Protection and inheritance need further discussion and practice, and then solve the problem of "how to learn", which is the main purpose and important practical goal of traditional human settlements.

Acknowledgements

This paper has been supported by the following fund projects. MOE (Ministry of Education of the People's Republic of China) Youth Fund Project of Humanities and Social Sciences (Project No.18YJC850002). Bidding Project of Key Research Base of Humanities and Social Sciences in Colleges and Universities of Anhui Province (Project No. SK2015A170). Special Funding Project of Innovatively Provincial Construction in Anhui Province (Project No.2020XZX001) . "10303 plan of Huizhou research talents training" project of Huangshan Social Science Federation. Natural Science Research Project of Anhui Education Department(Project No.KJHS2016B13). Anhui Education of Department Humanity and Social Science Research Project (Project No. SK2020A0468). National Innovation and Entrepreneurship Training Program for College Students(Project No. 201910375007X).

Thanks to Mr. Luo huiding of Chengkan Village and my tutor Professor Liu Boshan of Anhui University for their help, Anhui Huizhou Cultural Relics Engineering Survey and Design Co., Ltd. and Huangshan Social Science Federation for their support. At the same time, I would also like to thank other literature authors who have provided important references but failed to mark them one by one.

References

1. Wang Shusheng. Study on the construction of human settlements in historical cities along the Yellow River, Shanxi and Shanxi. Beijing: China Construction Industry Press, 2009
2. Wu Liangyong. Introduction to human settlements. Beijing: China Construction Industry Press, 2003

3. Hu Minxian. Spatial study on human settlements in Huizhou ancient villages. Beijing Forestry University, 2007
4. Gao Qian. Analysis of human settlements in Shijiagou ancient village. Journal of Xi'an University of science and technology, 2013,33 (4): 490-493
5. Ni Qi, Wang Yu, et al. Spatial composition principle of traditional villages with water system as the core in Huizhou area of China: investigation report of Chengkan Village in Huizhou area of Huangshan City. Urban architecture, 2008 (11): 107-108
6. Bi Zhongsong, Li Yunzhang, et al. Protection object composition and protection strategy of "ancient village architectural complex" cultural relics. Architecture and culture, 2015 (06): 200-204

Control Performance of Different Bra Designs and Materials during Physical Activity

Y Sun[1], KL Yick[1], P L Li[1], M Tang[1], A Yu[1], J Yip[1] and S P Ng[1]

[1] Institute of Textiles and Clothing, The Hong Kong Polytechnic University, Hong Kong

Abstract. Bras give support to the breasts and prevent damage to their different components during daily activities. Previous studies have however merely focused on specific types of bras and/or shoulder straps but failed to provide specific guidelines on bra features and materials that can be used to improve the support different functions of bras. The purpose of this study is to evaluate the breast control performance of various types and designs of bra under the conditions of walking and running. Two younger women and 1 older woman (all with a bra size of 36C/80C) have participated in a wear trial to don a total of 12 different commercial bras with different design features, and a changeable bra in which 12 different variables are examined, including the materials, strap width and type of underwire. The 3D movement of the breasts in the various bra conditions during walking at a speed of 4 km/h and running at 6 km/h is recorded and then compared with the braless condition by using an 11-camera motion capture system. The percentage of change in the nipple displacement in the vertical direction relative to the clavicle displacement is used to quantify and compare the breast control performance of the bras. A significant difference ($p<0.01$) in the ability of the bras to control movement is observed between the group of younger subjects and the mature subject (left breast: 11.49% vs. 35.42%, and right breast: 18% vs. 32.47%, respectively). The control performance of the bras is significantly different ($p<0.01$) during running and walking (left breast: 26.39% vs. 12.54%, and right breast: 30.04% vs. 15.61%, respectively). The correlation analysis showed that a small bra cup volume (length across the bra cup), taller gore, elongation of the bra strap and bra band, and rigid bra cup fabric allow good control of vertical breast displacement.

1. Introduction

In their natural state, breasts are likely to deform and sag under gravity due to the lack of anatomical support. Large inertial forces are exerted onto the breasts, especially during when the body moves to do daily activities or perform intensive exercise, which will cause stretching of the Cooper's ligaments in the breasts and lead to discomfort [1]. Steele [2] reported that some women can experience breast pain even with a mere 2 cm of breast displacement in the vertical direction. A good bra is therefore important to provide sufficient support of the breasts against gravity and control their excessive movement during daily activities [3]. To control movement of the breasts, bra performance is important and related to the design features and materials used [4]. In order to investigate the performance of a bra in preventing breast displacement during activity, objective measurements of the breast kinematics have been conducted by monitoring the three-dimensional (3D) trajectories of markers attached to certain points on a bra [5][6][7]. Starr et al. [8] compared two major types of sports bras (compression versus encapsulation bras) through a breast motion analysis. The compression sports bra caused greater vertical breast displacement in comparison to the encapsulation sports bra. Lu et al. [9] examined the material properties of sports bras and addressed the influence of type of material used in the sports bra design on the interaction between the bra and breasts. They concluded that both biaxial tensile and compression properties of bra materials should be considered in sports bra designs. Research studies have also focused on the evaluation of breast motion with different levels of bra support (high, medium, or low level of support) or bra type (everyday bra, or compression or encapsulation sports bra) [5][10]. However, the bra styles are usually limited in these studies which fail to address the large differences in bra features and their influence on controlling breast motion with different levels of activity. The purpose of this study is to therefore propose a systematic evaluation of bra features in controlling breast motion under the conditions of walking and running.

2. Methods

2.1. Participants

Two younger females (S1: 26 years old, and S2: 29 years old) and one older female (S3: 62 years old) were recruited for participation in the study. They all have a bra size of 36C/80C. They have not undergone any breast surgery and are not pregnant or breast feeding which may cause changes to the properties of the breast tissues or affect natural breast motion. The project was approved by the University Human Subjects Ethics Committee. Before taking part in the wear trial experiment, informed consent was received from all of the subjects.

2.2. Bra samples

A total of 13 bras were used for the wear trial, which included 12 commercial bras with different design features and 1 changeable bra which allows the replacement of bra components such as the materials, strap (different widths) and underwire (different types). As such, 12 different variables were examined, which provided 12 different bra conditions. The details of the bras are listed in Table 1 and shown in Figure 1.

Table 1. Details of commercial bra samples

	1	2	3	4
Bra style	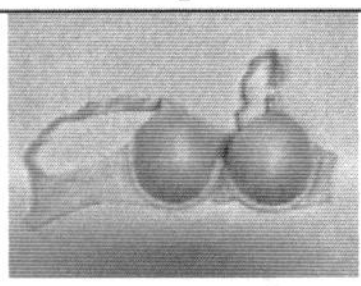	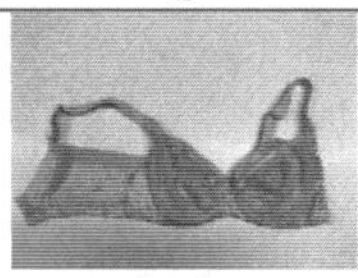	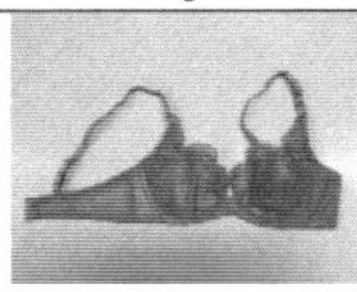	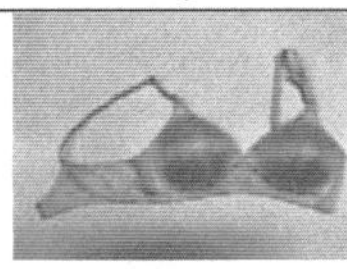
Design features	• Moulded foam cup • Hard underwire • No cradle • U-back • Hook and eye fastener	• Cut and sewn elastic lace and lining • Hard underwire • Elastic cradle • U-back • Hook and eye fastener	• Cut and sewn rigid lace • Hard underwire • Elastic cradle • Straight back • Hook and eye fastener	• Moulded foam cup • No underwire • Elastic cradle • U-back • Hook and eye fastener
	5	6	7	8
Bra style	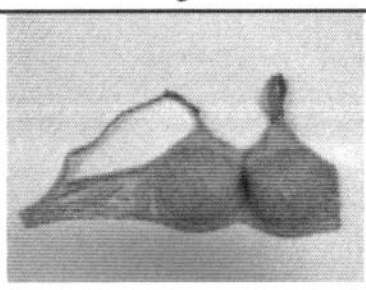	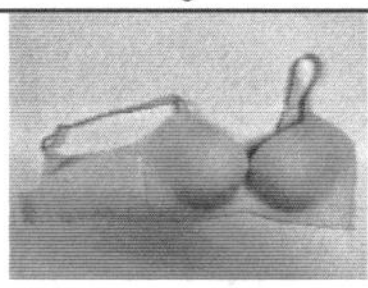	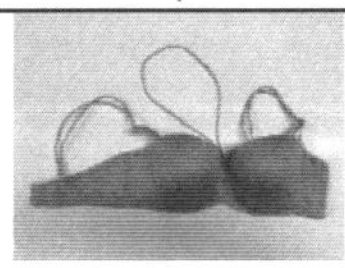	
Design features	• Cut and sewn foam cup • Soft underwire • Elastic cradle • U-back • Hook and eye fastener	• Moulded foam cup • Soft underwire • One piece cradle (moulded) • U-back • Hook and eye fastener	• Moulded foam cup • Hard underwire • Elastic cradle • U-back • Hook and eye fastener	• Elastic fabric inserted with moulded foam padding • No underwire • Stretch lace cradle • No straps • No closure
	9	10	11	12
Bra style	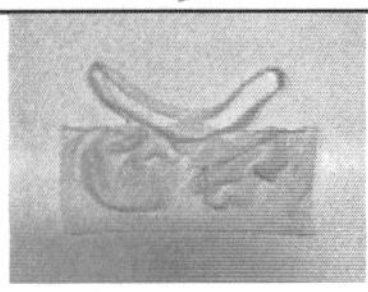	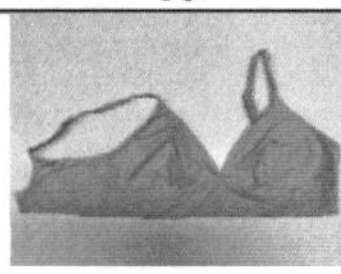	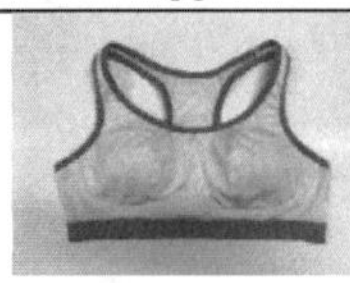	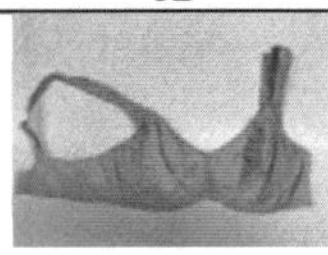
Design features	• Seamless knitted fabric • No underwire • Seamless cradle (knit) • U-back • No closure	• Cut and sewn elastic fabric • No underwire • Elastic cradle • U-back • Hook and eye fastener	• Elastic fabric inserted with moulded foam padding • No underwire • Elastic cradle • Racer back • No closure	• Cut and sewn elastic fabric • Hard underwire • Stretch lace cradle • U-back • Hook and eye fastener

https://doi.org/10.1051/e3sconf/202123704026

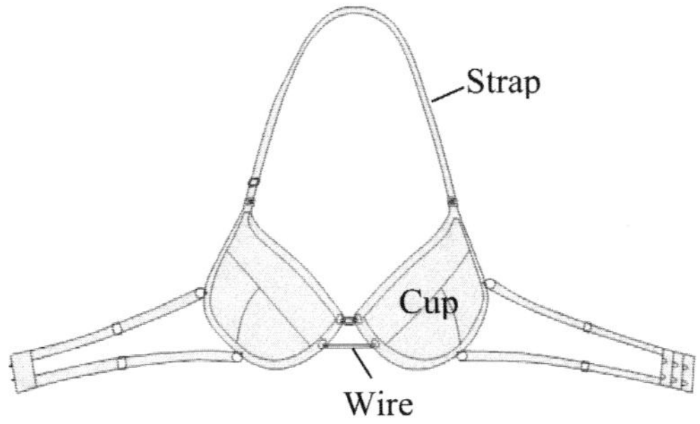

Changeable bra styles	Cup	Strap width (mm)	Wire
21	Moulded	12	2460 (light)
22	Moulded	12	2860 (heavy)
23	Moulded	16	2460
24	Moulded	16	2860
25	Woven	12	2460
26	Woven	12	2860
27	Woven	16	2460
28	Woven	16	2860
29	Knitted	12	2460
30	Knitted	12	2860
31	Knitted	16	2460
32	Knitted	16	2860

Figure 1. Changeable bra sample and different bra conditions

2.3. Motion capture experiment

In order to examine breast motion during activity, a Vicon motion capture system with 8 Vantage cameras and 3 Vero cameras was used to record the dynamic breast displacement. The subjects underwent 3 walking trials at a rate of 4 km/h and 3 running trials at 6 km/h in random order on a treadmill with 0° slope (braless, and wearing the 12 commercial bras and changeable bra with the 12 bra conditions). Each trial was recorded for 20 seconds with the subject standing, 60 seconds while she was walking/running and then 20 seconds with the subject standing again. Since displacement in the vertical direction contributes to most of the 3D dynamic movement of the breasts [11], only vertical displacement is examined in this study. The coordinates in the vertical direction (z axis in the Vicon system) of the markers on the two nipple points and clavicle were extracted for further analysis.

A well-designed bra can reduce excessive breast displacement during dynamic movement. In this study, the percentage of change in the nipple displacement (PND) in the vertical direction relative to the clavicle displacement in comparison to the braless condition is used as a criterion of bra performance in controlling breast displacement. The change is calculated by using equation (1). For each bra condition, 20 cycles of stride were extracted to obtain the average value of the relative peak/valley of the nipple displacement. The Figure 2 shows the relative nipple displacement (RND) in the condition of braless and wearing bra condition, respectively.

$$PND = \frac{A_n - A_b}{A_n} \times 100\% \quad (1)$$

where A_n represents the relative nipple displacement (RND) in the braless condition, and A_b is the RND in the bra wearing condition.

3. Results and discussion

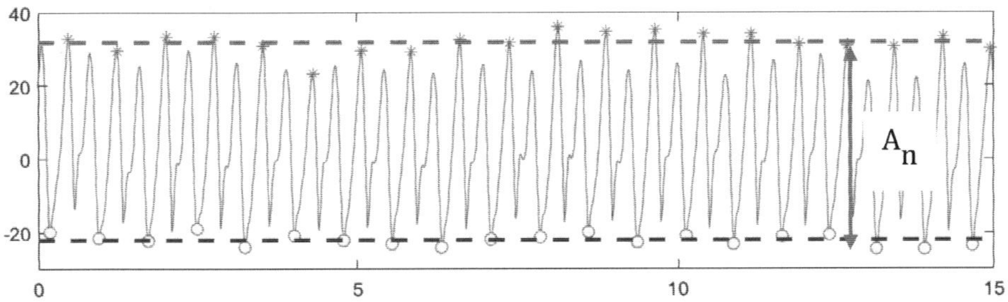

(a) Braless condition

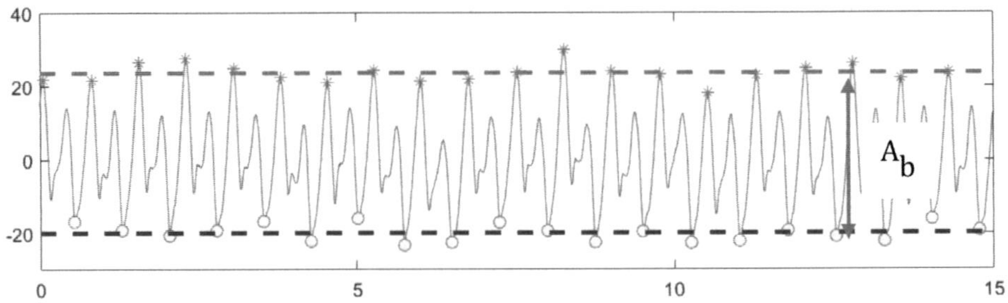

(b) Wearing bra condition

Figure 2. Relative nipple displacement (cm) in the condition of braless and wearing bra condition

The overall RND in the vertical direction obtained from Commercial Bras 1 to 12 and Changeable Bra Conditions 21 to 32 as compared to the braless (nude) state during running and walking is presented in Figures 3(a) and 3(b), respectively. The results indicate that the older subject (S3) shows a lower RND and higher PND than the younger subjects (S1 and S2) in all of the bra conditions during running and walking. A significant difference ($p<0.01$) of the ability of the bra to control movement can be observed between the group of younger subjects (PND of left breast: 11.49%, and PND of right breast: 18%) and older subject (PND of left breast: 35.42%, and PND of right breast: 32.47%). However, no significant difference can be found between the two younger subjects themselves ($p>0.01$). The control performance of all of the bras when donned by the subjects showed a significant difference ($p<0.01$) between running and walking (PND of left breast: 26.39% when running vs. 12.54% when walking, and PND of right breast: 30.04% when running vs. 15.61% when walking).

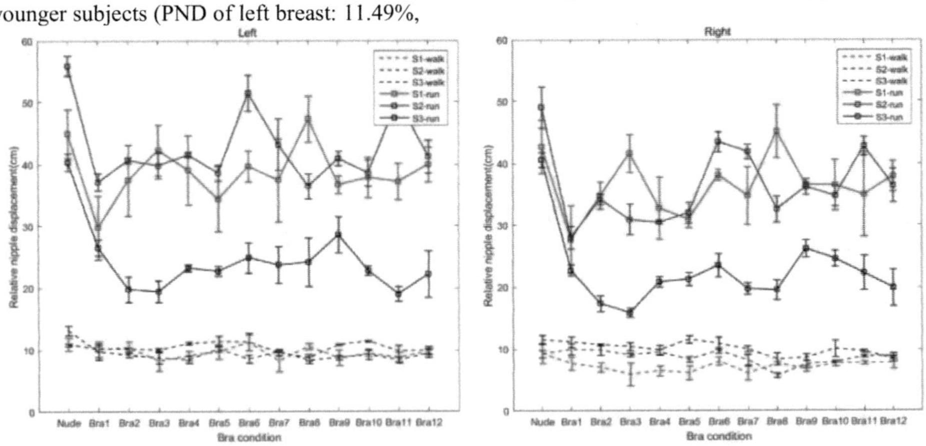

Figure 3(a). RND in vertical direction: Commercial bras and braless condition

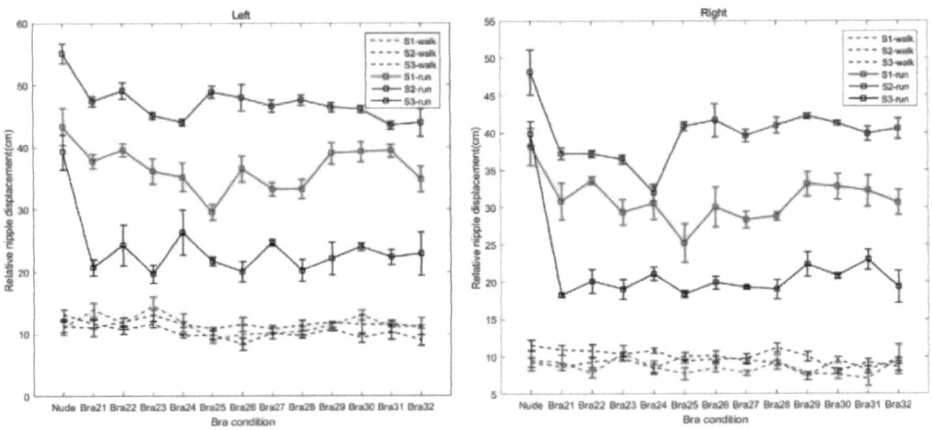

Figure 3(b). RND in vertical direction: Changeable bra and braless conditions

In order to evaluate the influence of the bra features on the control performance of the bra during walking and running, 11 measurements were taken for the commercial bras (including length of the neckline, length across the bra cup, height of the bra cup, length from the top to bottom of the gore, height of the gore, length of the wire channel, side seam, bra band, and length and width of the shoulder strap) and 6 material tests were carried out (including the elongation of the bra band, modulus and thickness of the shoulder strap, and thickness of the cup and its modulus in the course and wale directions) for the correlation analysis. No relationship was found between the bra measurements and percentage of control of breast displacement in the running condition. However, a negative relationship was found during walking for the measurement of the length across the bra cup (left breast: $r=-0.380$, $p=0.038$, and right breast: $r=-0.464$, $p=0.01$) and the PND. In addition, the height of the gore (left breast: $r=-0.330$, $p=0.049$, and right breast: $r=-0.352$, $p=0.035$) and elongation of the shoulder strap (left breast: $r=-0.442$, $p=0.01$, and right breast: $r=-0.368$, $p=0.035$) have a negative relationship with the RND. A smaller distance across the bra cup, taller gore and increased elongation of the bra strap could possibly enhance the control performance of a bra. The elongation of the bra band has a negative relationship with the RND (left breast: $r=-0.404$, $p=0.015$) and positive relationship with the PND (left breast: $r=0.426$, $p=0.01$), which shows that increased elongation of the bra band can enhance control over breast motion.

As for the changeable bras, the effects of the 12 variables, which include the bra cup materials, width of the shoulder strap and type of underwire on the control performance, were investigated. The correlation analysis revealed that there is no relationship between any of the variables and control performance in the running condition. However, the Young's modulus of the changeable bra cup material has a positive relationship ($r=0.343$, $p=0.041$) with the control performance of the left breast during walking. A T-test and one-way analysis of variance (ANOVA) were conducted to determine whether the RND and PND for the different bra conditions would have significant differences between/ among the different variables. It was found that there is no significant difference between the different strap widths and types of underwire (T-test, $p>0.05$). However, the results for the bra cup material showed a significant difference between the knitted and woven materials ($p<0.01$) in controlling the displacement of the left breast.

4. Conclusion

This study evaluates the control performance of various types and designs of bras during different activities. The PND in terms of the vertical displacement of the breasts under different bra conditions is determined and the results indicate that mature women who wear a bra can better control the displacement of their breasts in the vertical direction. During walking, the correlation analysis shows that a small cup volume (smaller distance across the length of the bra cup), taller gore, increased elongation of the bra strap & band and the use of rigid fabric for the bra cup can enhance the control of vertical breast displacement. However, there are some limitations in this study that prevent the generalization of the results. The sample size is small and the different bra features of the changeable bra do not shown any significant differences during running. In future studies, it is recommended that a larger sample size is used which could provide more data to carry out a statistical analysis.

Acknowledgment

We acknowledge financial support from ITF grant for project ITS/276/18.

Reference

1. Yu W and Zhou J 2016 Sports bras and breast kinetics *Advances in women's intimate apparel technology* (UK: Woodhead Publishing) pp 135-146
2. Steele J R 2013 The biomechanics of better bras: improving support and comfort during exercise *ISBS-Conference Proceedings Archive*
3. Liu K, Zhang L, Zhu C, Zhao X, Lu W, Li M and Wang J 2019 An analysis of influence factors of sports bra comfort evaluation based on different sizes *J. Text I.* **110** 1792-1799
4. Lawson L J and Lorentzen D 1990 Selected Sports Bras: Comparisons of Comfort and Support *Text. Res. J.* **8** 55-60
5. White J L, Scurr J C and Smith N A 2009 The effect of breast support on kinetics during overground running performance *Ergonomics* **52** 492-498
6. Zhou J, Yu W and Ng S P 2013 Identifying effective design features of commercial sports bras *Text. Res. J.* **83** 1500-1513
7. Zhou J, Yu W and Ng S P 2012 Studies of three-dimensional trajectories of breast movement for better bra design *Text. Res. J.* **82** 242-254
8. Starr C, Branson D, Shehab R, Farr C, Ownbey S and Swinney J 2005 Biomechanical analysis of a prototype sports bra *J. Text. Appar. Technol. Manag.* **4** 1-14
9. Lu M, Qiu J, Wang G and Dai X 2016 Mechanical analysis of breast-bra interaction for sports bra design *Mater. Today Commun.* **6** 28-36

https://doi.org/10.1051/e3sconf/202123704026

10. Scurr J C, White J L and Hedger W 2011 Supported and unsupported breast displacement in three dimensions across treadmill activity levels *J. Sports Sci.* **29** 55-61
11. Mason B R, Page K A and Fallon K 1999 An analysis of movement and discomfort of the female breast during exercise and the effects of breast support in three cases *J Sci Med Sport* **2** 134-144

On the Landscape Composition of Traditional Village in Huizhou District - A Case Study in Qizili Village

Cheng Peng[1], *Xiao Siyi*[2], *Li Yunzhang*[2], and *Bi Zhongsong*[1, 2*]

[1*]College of Architectural and Engineering & Research Centre of Huizhou Architecture, Huangshan University, 245041 Huangshan, China
[2]College of Architecture and Environment, Sichuan University, 610065 Chengdu, China

Abstract. The landscape features of traditional villages displayed in the Huizhou District were formed in a specific environment gradually. It was the result of the comprehensive effects of natural, historical, social, and cultural factors. This study took Qizili Village as an example and sorted out the composition and current situation of its natural landscape and human landscape. To refine the proper composition model of Qizili Village, we analyzed the landscape features from the perspectives of the natural environment, layout planning, traditional architecture, environmental elements, and traditional culture. Finally, we concluded the landscape connotation of Qizili Village.

1 Introduction

Traditional villages refer to ancient villages that have existed for a long period. Since these villages have strong social and economic values due to their enrichment of natural and cultural resources, they need to be protected and maintained [1]. China has attached more and more importance to the protection of traditional villages since the 21st century. However, traditional villages are still in danger of vanishment caused by continuous damages during the process of Chinese modernization, urbanization, and new rural construction [2].

Huizhou was an administrative division concept. It was once named Xin'an County in the Han Dynasty, and Kuanzhou in the Tang Dynasty. It was not until the Northern Song Dynasty, the emperor Huizong renamed Xin'an as Huizhou under his emperor title. The name Huizhou was used ever since. Ancient Huizhou governed six counties: She County, Yi County, Jixi, Xiuning, Qimen, and Wuyuan. The early villages were mainly formed by scattered settlements of local Shanyue indigenous peoples. The villages were initially formed due to the clan's migration of the Central Plains since Eastern Jin, late Tang, and Southern Song Dynasties. The villages came to a large scale at the beginning of the Ming Dynasty and reached the peak of prosperity in the middle of the Qing Dynasty [3]. It decayed gradually at the end of the Qing Dynasty.

The traditional villages in Huizhou are one of the typical representatives of ancient Chinese villages [4]. Historically, the indigenous inhabitants were mainly of the same clan or the same ethnic group. Additionally, the inconvenient transportation based on the mountainous geographic feature in southern Anhui led to the slow economic development in Ancient Huizhou. This economic disadvantage prevented the village from linking with the outside world. However, in the era without the advocation of active protection of local culture, the lagging development helped the inheritance of the local cultural resources. Therefore, the ancient villages were almost unchanged. In recent years, to the intense attention from the local governments, a wealth of traditional village resources remains the original appearance in Huizhou District with the rapid local economic progress. So far, most of these retained traditional villages are in living status. The landscape of traditional villages is preserved intact with a rich composition content (Fig.1).

When looking at the traditional villages, there are three layers: the outer layer refers to the natural mountains and rivers surrounding, the middle layer refers to the village and its spatial environment, and the inner layer refers to the village architectural details, such as buildings, bridges, pavilions, etc. All three layers are of strong Hui-style characteristics. The purpose of this study is to find the proper landscape composition mode of a traditional village based on the factors of nature, history, society, and culture [5]. To do so, we selected a representative village Qizili in Huizhou District. Qizili Village is located in the northeast of Shexian County, Anhui Province. It stays between the subtropical warm and humid monsoon climate zone. Thus, the climate conditions with four distinct seasons contribute to the lush vegetation outside the village, and intact mountain forest vegetation. 846 households with a population of 2631 people live in the village. Since Qizili Village was the transportation hub on the road of Ancient HuiHang Road, it linked Shexian County in Ancient Anhui Province and Chunan County in Ancient Zhejiang Province. Thus, Qizili Village played a crucial role in the

* Corresponding author: bzs@stu.scu.edu.cn

trade between Ancient Anhui Province and Ancient Zhejiang Province. It contributed to the economic development of both places [6]. Meanwhile, the above natural and social factors promoted the featured characteristics of the landscape in Qizili Village.

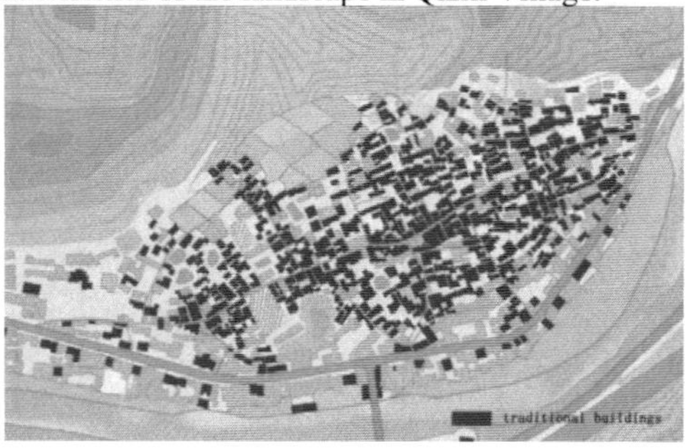

Fig. 1 Distribution map of traditional buildings

The research on the composition and current situation of Qizili Village helps to deepen recognization of the natural and human landscape of Huizhou traditional villages. This study provides a basis for how to preserve the natural environment, village pattern, architectural style, and traditional culture in traditional villages. The continuation of traditional village helps to improve the infrastructures and public facilities. By providing new ideas for the protection and development of traditional villages, it is beneficial for the living standards and the sustainable development of traditional villages.

2 Landscape composition of traditional villages

The precondition of the recognition of landscape features is to determine a reasonable classification of landscape types. This paper draws on the related content of village characteristic analysis and value evaluation in the *Basic Requirements for the Preparation of Traditional Village Protection and Development Planning* issued by the Ministry of Housing and Urban-Rural Development of the People's Republic of China to classifies the site investigation results of Qizili Village. We established two major aspects of a traditional village: nature and humanity. Specifically, it includes five categories of the natural environment, planning layout, traditional architecture, environmental elements, and traditional culture. The subcategories of the natural environment include village terrain, landform, climate, and vegetation. The subcategories of planning layout include village form, scale, structure, and layout. The subcategories of traditional buildings include building types, plane forms, structural types, materials, roofs, gables, Decorations, and other small items. The subcategories of environmental elements include street paving, pavilion walls, water wells, ponds, bridges, and other small items. The subcategories of traditional culture include surname beliefs, language customs, opera crafts, and other small items [7]. Based on the above classification, the landscape system that constitutes the traditional villages in Huizhou is complex and diverse, covering all aspects of nature and humanities, material, and non-material. The landscape system contains the characteristics of the traditional village landscape system in Huizhou completely.

3 Landscape characteristics of Qizili Village

3.1 Natural environment characteristics

Qizili Village, located in the eastern mountainous area of Shexian County, sits in the mountain valleys with distinct four seasons. The geographic feature of lush forest vegetation is favorable to form an excellent ecological environment. The types of landforms are complex and diverse. The local geological structure shows obvious folds and fractures. The soil is mainly mountainous red and yellow soil, suitable for the growth of various economic crops such as tea, mulberry, and chrysanthemum. Surrounded by mountains and rivers, Qizili Village maintains a relatively primitive ecological environment. The four mountains, including Hengyan, Wangyan, Ma'an, and Tanghou, Qizili, surround the village. The Changyuan River flows from east to west through the south of Qizili Village with an outer arc shape. Three smaller rivers in the village merge into the Changyuan River (Fig.2). It is a typical Taoyuan-style village in Huizhou District. The site selection of Qizili Village follows a relatively closed spatial pattern with mountains, forests, land, and water. The terrain is high in the north and low in the south. It is a true portrayal of the ideal landscape pattern consistent with the ancient cultural connotation that people should choose a lucky place to live [8]. The landscape of "mountain-water-field-dwelling" also highlights the characteristics of the overall environment of Qizili Village.

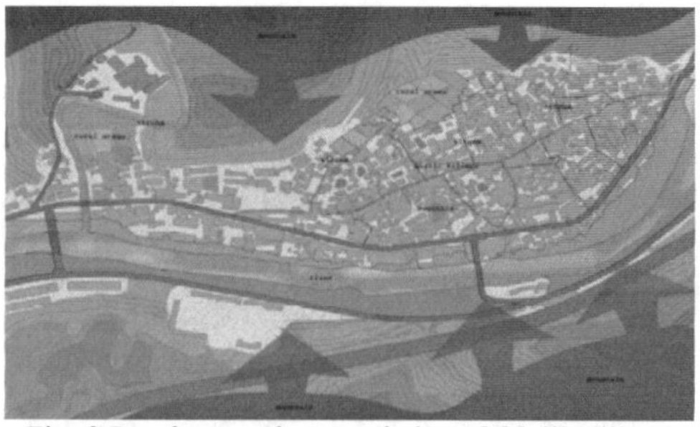

Fig. 2 Landscape characteristics of Qizili Village

3.2 Layout planning characteristics

Qizili Village has experienced a long history with an ancient name of Xizili. It was not until the Song Dynasty that the name had been changed to Qizili. Qizili Village reached its development peak during the Shaoxing period of the Southern Song Dynasty. Due to the increasingly dense population of the village and the limited space for survival and development, villagers began to move out to build new ones. Thus, the size of the village finally stabilized. Ancient Qizili Village had three parts: the village head, the middle street, and the village end. The village roads were tortuous, the steps were undulating, and the vertical space was full of changes. The terrain dropped from the east of the village to the west is nearly 10 meters with four-level stacking streets. The streets were flat, and people walked on them without the feeling of heights. The ancient village was built following the situation, using blocks to adjust the height of the foundation, forming a patchwork skyline. When entering the village from Shuikou, guided by roads, space formed three stages of semi-closed-open-closed. The ancient village took Dalu Street as the axis, with the alleys evenly distributed across the north and south as the front, and the traditional buildings as the center. The whole village was the layout as a fishbone, forming a village pattern of three streets and twenty-four alleys (Fig.3). Road pavilions and public activity spaces met the needs of transportation and communication. Nowadays, Qizili Village has a clear structure, rigorous overall layout, and reasonable functional layout, reflecting the form and development mode of traditional Chinese ancient village planning and construction [9].

Fig. 3 Layout planning characteristics

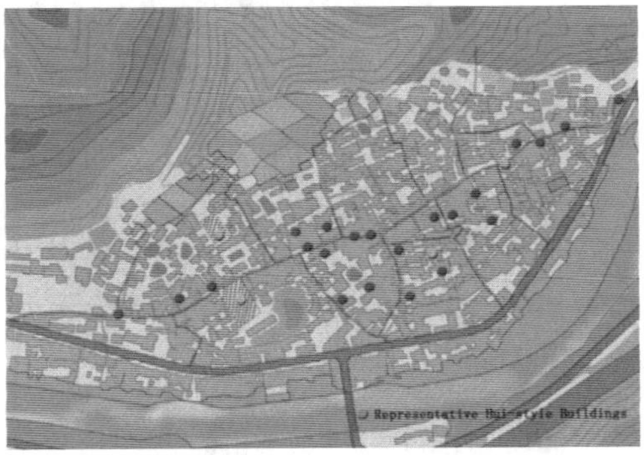

Fig. 4 Important traditional building

3.3 Traditional architectural characteristics

The gathering of ethnic groups in Qizili Village decided local architectural distribution. Ancestral halls and residential buildings were typically traditional buildings in Qizili Village were. There were 22 representative buildings, such as Huaiyuan Temple, Mingyi Hall, Zhisan Pavilion, Funding Hall, and others (Fig.4) in the village.

The characteristics of these buildings included harmonious size, a suitable scale, and the elegant color of the walls and tiles. Most of these buildings faced southwest. A typical Huizhou architectural form tended to be symmetrical layout with three rooms taking the patio as the centre. The centre was used as the main hall, with rooms on both sides, and the stairs were around the main hall (Fig.5). The structure adopted the beam-lifting type, with wooden pillars and wooden beams inside for load-bearing, brick exterior walls, and horsehead gables for isolation. The roofs of Huizhou architectural buildings were usually double-sloping or partial single-sloping with long and protruding eaves to prevent rainwater from splashing into the house as well as the "sit-bucket style" horse head gables to isolate fires and theft. Three kinds of carving with landscape themes: stone carving, wood carving, and brick carving were integrated naturally to highlight Hui-style architecture (Fig.6).

Compared with the square bricks mostly used in other areas of Huizhou, the buildings mostly used bluestone slabs in Qizili Village. The villagers made full use of bluestone slabs as most materials in the construction process due to the abundant stones in the local area.

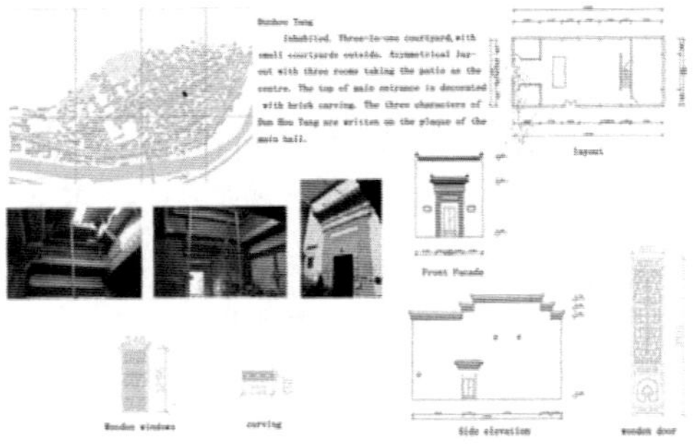

Fig. 5 Mapping of typical Hui-style building Dunhoutang

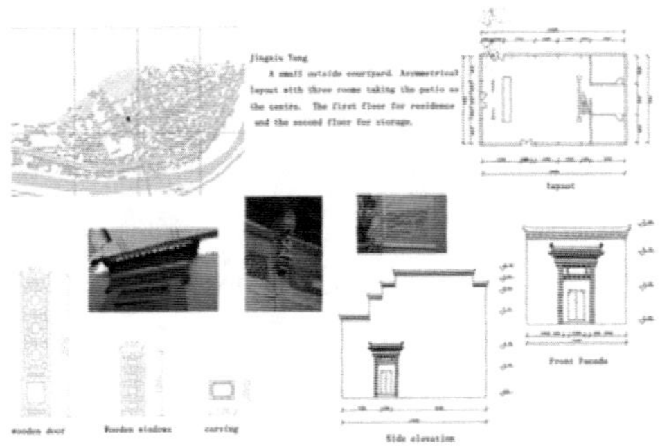

Fig. 6 Mapping of typical Hui-style building Jingxiutang

3.4 Environmental elements characteristics

Over 40 historic streets and alleys paved by bluestone slabs were built in Qizili Village during the Ming and Qing Dynasties (Fig.7). The main street was the official road of Yulingguan Ancient linking Anhui and Zhejiang. Hence, it was once the important channel for the exchange of materials and business of two commercial centres: Huizhou and Jiangnan. The Qixi Ancient Road in the north of the village led to Longchuan, Jixi, which was historically the main road for outsourcing the villagers' living materials. The Siyongtang Bridge was five miles upstream of the village. The south village saw the other six stone bridges, such as Shixiang Bridge and Huisheng Bridge. Since most villagers are mainly named Wang, three locust trees planted outside the Shuikou Pavilion in the west of the village imply that the officials named Wang were major roles in the officialdom. Wells and ponds scattered in the village. The existing Sangen Well, Jianggong Well, Houshan Well, Xitan Pond, Xujia Pond, Qianjie Pond, Gecko Pond are still providing villagers with drinking water and fire- fighting water nowadays.

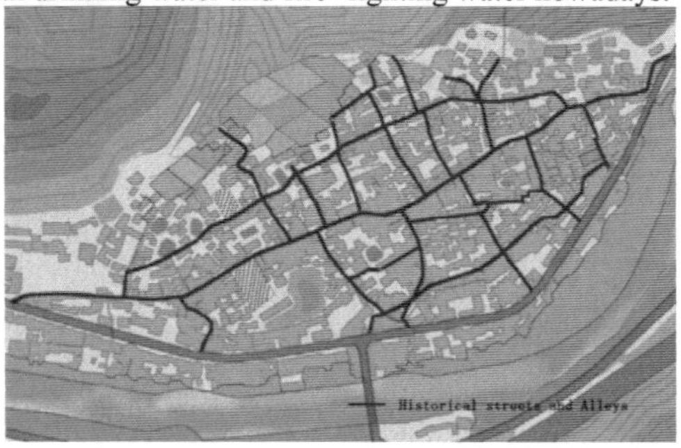

Fig. 7 Distribution map of streets and alleys

Historical environmental elements in Qizili Village were various (Fig.8), most of which were based on actual functions and uses. It fully reflected the simple life concept of the working people in the ancient mountainous areas. It showed the ecological concept of adapting measures to local conditions and making full use of the natural environment.

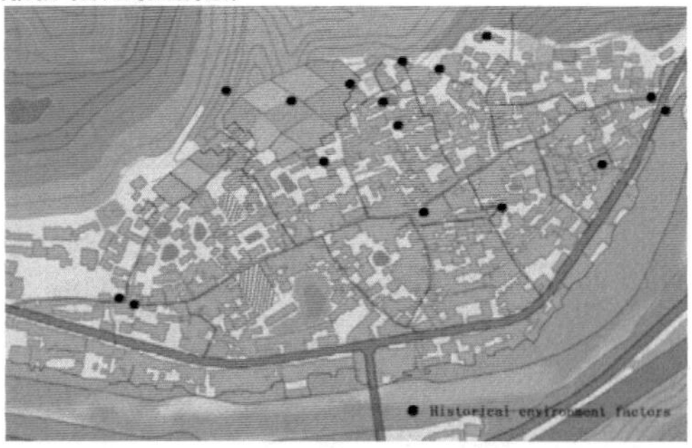

Fig. 8 Distribution map of other environmental elements

3.5 Traditional cultural characteristics

In the historical process of the development of Huizhou, each village formed a unique village culture manifested in clan etiquette, opera and ballads, and celebration customs [10]. Connecting the ancient Shexian County in Anhui Province to the west with Chun'an County in Zhejiang Province to the east, Qizili Village became a critical node of Huizhou's external relations. The combined effect of Huizhou and Jiangnan culture promoted the formation of the distinctive landscape in the long history of development as well as the unique production lifestyles and customs. These factors encouraged the village to form its unique cultural characteristics. With a population of over 80%, villagers in Qizili Village were named Wang. The inheritance of clans from generation to generation by the gathering of clans made it possible for the production and preservation of a large number of Huizhou documents. The representative documents included "Wang Family Genealogy", "Six Dragon Map", various medical books, and ancient books on family genealogy (Fig.9). The interweaving of culture led to a variety of custom activities in Qizili Village. For instance, measles lantern was a unique local traditional custom that was a lantern festival based on children's local operas. Any child who participated in this event would not suffer from measles for one year. Hence, the main participants were children (Fig.10). The Huizhou folk songs handed down in the village are rich in content and diverse in genres, such as chants, folk songs, minors, Buddhist and Taoist songs. 32 folk songs were written. Golden silk amber candied dates, a famous spot in Shexian County since the Ming and Qing Dynasties, still has production in Qizili Village. The ancient craftsmanship of candied dates has successfully passed on in Qizili Village. Other traditional festivals, such as welcoming the dragon lantern in the first month of the lunar calendar, sending the plague God in February, and celebrating the harvest on October 15, also spread to today.

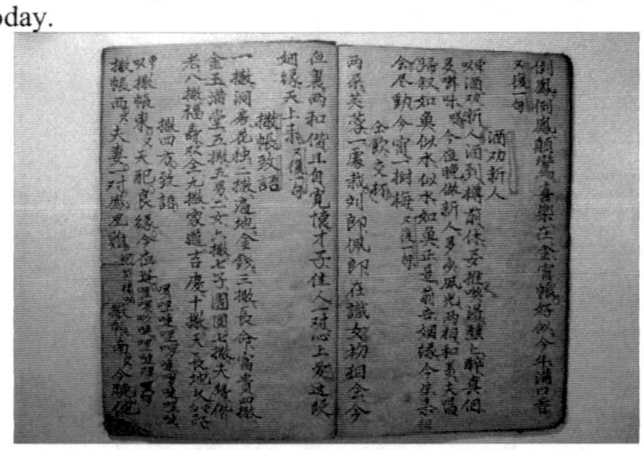

Fig. 9 Folklore activities notebook

Fig. 10 Measles lantern activities

4 Conclusion

This study took Qizili Village as an example and sorted out the composition and current situation of its natural landscape and human landscape. We focused on the characteristics of the Qizili Village landscape from the five major aspects of the natural environment, planning layout, traditional architecture, environmental elements, and traditional culture that constitute the landscape. The excellent natural environment of Qizili Village thanked the surrounding mountains and water. The entire village presented a reasonable layout and clear structure showing a high level of planning and construction. The village preserved a relatively complete and unique Huizhou traditional style with a relatively perfect layout and ancient buildings. The environmental elements with strong regional characteristics were diverse in forms as well as abounding in Qizili Village. Thus, the traditional villages in Huizhou represented by Qizili Village presented unique landscape characteristics in terms of village site selection, spatial layout, streets and alleys, individual buildings, and village culture.

Qizili Village provides a real historical example for understanding and studying the landscape features of Huizhou villages. Hence, this study contributes to the extant research by a distinguished example. The proper landscape composition of Qizili helps to improve the local environment, increase local infrastructure and public facilities, improve the living standards of villagers, and enable the sustainable development of traditional villages.

Acknowledgements

This study was funded by Special Funding Project of Innovatively Provincial Construction in Anhui Province (Project No.2020XZX001). Anhui Education of Department Humanity and Social Science Research Project "Research on the Landscape Features of Traditional Villages along Huizhou Ancient Road from the Perspective of Cultural Routes" (Project No. SK2020A0468) . MOE (Ministry of Education of the People's Republic of China) Youth Fund Project of Humanities and Social Sciences (Project No.18YJC850002). Bidding Project of Key Research Base of Humanities and Social Sciences in Colleges and Universities of Anhui Province (Project No. SK2015A170). "10303 Plan of Huizhou Research Talents Training" Project of Huangshan Social Science Federation.

References

1. Huang Chenglin, Preliminary Research on Huizhou Cultural Landscape.Geographical Research. 2000,19(3):257-263
2. Wu Xiaolin, Chen Ansheng, Wan Guoqing. World Cultural Heritage—Discussion on the Characteristics of Ancient Villages in Southern Anhui. Journal of Architecture. 2001(8):59-61.
3. Lu lin, Jiao Huafu, Cultural content of Huizhou architecture. Journal of Nanjing University. 1995(2):163-171.
4. Liu Peilin. Conservation and development of traditional settlements based on the concept of landscape gene integrity. Economic Geography. 2009,9(10):731-736.
5. Zhang Xichen, Hao Jingxin. Research on Laneway Landscape of Traditional Settlement in Southern Anhui.Journal of Jiangnan University.2002,1(2):179-183.
6. Chen Wei. The formation, development, and evolution of Huizhou vernacular architecture and traditional settlements. Huazhong Architecture. 2000,18(3):126-127.
7. Fang Xiansong. On the effect of ancient village landscape environment on the development of rural tourism. Shanxi Architecture.2010, (6):24-25.
8. Hu Zui, Liu Peilin. Spatial characteristics of landscape genes of traditional settlements in Hunan Province. Acta Geographica Sinica. 2013,68(2):219-231.
9. Li Guo. Research on the identification of traditional settlement landscape characteristics in Wuling Mountain based on integrated protection. Huazhong Agricultural University, 2017.
10. Zou Weihan. Analysis of landscape genes of traditional settlements along the Shu Road (South Shaanxi). New architecture, 2018(6):38-41

https://doi.org/10.1051/e3sconf/202123704028

The culture and inheritance of Huizhou architecture: taking the design practice of INK Weiping tourism cultural street as an example

Peng Zhiming[1], *Yang Yang*[2], *Xiao Hanyue*[3], *Li Yunzhang*[3], and *Bi Zhongsong*[1,3*]

[1*]College of Architectural and Engineering & Research Center of Huizhou Architecture, Huangshan University,245041 Huangshan, China[3]
[2]School of Architecture, University of Sheffield, Sheffield S10 2TN, UK
[3]College of Architecture and Environment, Sichuan University,610065 Chengdu, China

Abstract. This article uses the design analysis and insights of the Weiping Tourism Cultural Street, which the author participated in and practice to discuss how Huizhou culture can be protected and inherited in tourism cultural projects. Through the planning of the buildings of the project and the site planning of the surrounding environment, the formation of public spaces such as traditional streets, traditional theaters, as well as the influence of Huizhou culture on the space and architectural style and details of the building, the article interprets in detail the application of Huizhou architectural design and its traditional cultural connotations in cultural tourism development projects. At the same time, the article also discusses and summarizes the problems encountered in the protection and inheritance of Huizhou culture in the design and development process of the project, which has an enlightening effect on the better integration of the current practice of new Huizhou architecture with the current social and economic environment.

1 Research Background

Huizhou culture, Dunhuang culture, and Tibetan culture are listed as China's three major regional cultures. As the birthplace of Huizhou culture, Huizhou has unique natural geography and humanistic environment, which has not only cultivated the society, economy, art, and technology with distinctive local characteristics but also affected the local physical and spatial environment and the social relations between people closely related to it. In Huizhou, there are such a series of traditional Huizhou traditional buildings scattered in no matter of large and small villages; at different times of the year, villagers also hold various traditional festivals with local Huizhou cultural charm. And activities, such as *"Tai Wang Gong"* (celebrate the local protector of Huizhou), *"Tai She Gong"* (commemorate the god of land), the dance of grass dragon, *Mulian* Opera, etc. Huizhou's buildings and village spaces have an inseparable and close relationship with the local people because of these collective interactions.

Since Deng Xiaoping's reform and opening up the country in 1978, many foreign-funded enterprises and private capital have brought about significant changes in the appearance of China's cities. As a demonstration point of reform and opening up, many new cities, such as Shenzhen, have developed from a small village to a modern city full of tall buildings. In rural areas, the government has also implemented reforms. Scientific and technological agricultural methods have liberated a large amount of agricultural productivity, promoting drastic urbanization, modernization of life, and a relatively comfortable income level. These rural surplus labor forces have also migrated to the country's new round of urban strategies to promote the development of industry and service industries. Taking Huangshan City as an example, the rural population has continued to decrease this year, while the urban population has increased.[1] The older people stay in the villages working in agriculture, while young people move to the city for better job opportunities. This phenomenon has changed the proportion of the rural population structure. The rural area has been hollowed out. What is lost is the influence of the material and cultural environment of the next generation and the inheritance of intangible cultural heritage.

To protect Huizhou's culture and promote its inheritance and development, the Huangshan City Government has adopted a series of measures to increase the protection of ancient Huizhou buildings and villages as cultural carriers, such as the "Hundred Villages and Thousand Buildings" project. By the end of August 2012, the planning of 101 ancient villages has been completed, and the protection and utilization of 83 ancient villages and 1051 old houses have been implemented.[2] In addition to the government's policy and other macro-level planning, Huizhou villages are also actively participating in market operations, attracting social and private capitals to develop, protect, and promote rural tourism development.[3] In the current traditional Huizhou

* Corresponding author: bzs@stu.scu.edu.cn

area, traditional buildings and villages have formed a two-way pattern in which the government leads in protection and market investment for development. In academia, researches showed interests and dedications to traditional village culture protection and renewal,[4-6] restoration and protection of ancient buildings,[7] the inheritance of tangible and intangible Hui culture heritage, the creation of new Huizhou style architectural design techniques.[8-9] At the same time, it has also sparked intense discussions on whether tourism development has "prosperity" or "decline" for rural culture.[10-12]

As a practicing designer of Huizhou-style architecture, the author discovered through continuous practice and observation in daily life that the culture and inheritance of Hui-style architecture need to consider economic factors and the social environment. Simply treating buildings as cultural relics without intersections with local people's daily lives, this kind of culture is static and is for viewing, rather than blending into the times and being understood by people. With the rise of China's real estate market, many residential buildings have emerged as symbols of commodities and capital. The government's current primary efforts are put on protecting traditional Huizhou architecture; innovations in designs of contemporary Huizhou architecture are left as a commercial project in most of the cultural and tourism projects invested by private owners. Private sectors have an inestimable role in promoting the culture and inheritance of Huizhou architecture's development. The entrusted Chinese independent architects thus also played a crucial role in the process. They need to ensure the interests of commercial capital and promote the expression of culture and art through their building design and further balance the two. The construction of Huizhou's traditional material and cultural environment, the elements, and the modern expression of traditional craftsmanship are conducive to evoking *genius loci* [13] rooted in the site. At the same time, through local cultural interpretation and experience in tourism projects, it can be better for local people as well as visitors to have a deeper understanding and recognition of the traditionally inherited customs, ethics, and morals behind the architecture that constructs the cultural space of Huizhou.[14] This article combines the design process of the INK Weiping Tourism Cultural Street that the author participated in as an example and used the architect's perspective to show how Huizhou's traditional culture is expressed in the design of the new Hui-style architecture through a comprehensive analysis. Additionally, the article will analyse how to better promote Huizhou culture's inheritance and development through cultural tourism projects from design practice experiences.

2 Project introduction

The project is located in Weiping Town, Chun'an County, Zhejiang Province, west of Beihong Bridge Head (Figure 1). The west side of the plot is Weiping Plaza (Figure 2), the south side is Xin'an River Reservoir, the east side is Hongqiao Street and commercial and residential buildings, and the north side is adjacent to Hubin Road. The project is a comprehensive cultural tourism real estate project integrating cultural real estate, tourism real estate, and commercial real estate with the spiritual connotation of "inheriting Huizhou, creative reproduction, cultural platform." Old Weiping was initially affiliated with ancient Huizhou. Due to the Xin'an River water transport connected to the Huizhou area, the Huizhou culture was much influenced. The project aims to regenerate the historical and cultural district of old Weiping and brings back traditionally collective living sceneries and connections between people. By setting up cultural activities provided by the spatial design as a platform, local people and visitors can better understand and inherit the lost local living and endangered local culture.

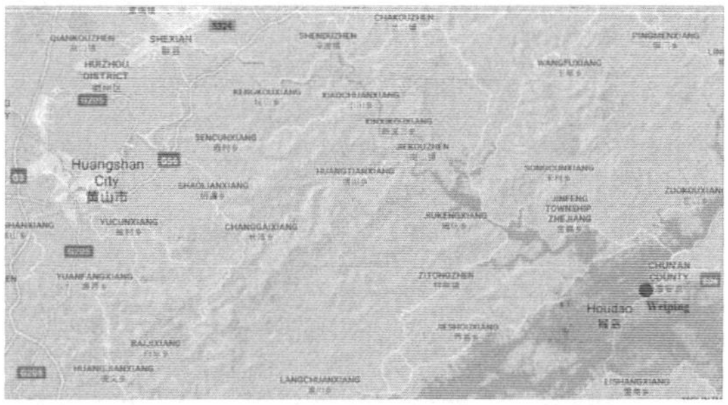

Fig. 1. The location of the project. Sources from the author.

Fig. 2. Left: site plan(red) and Weiping Plaza (green). Right: Weiping Plaza. Sources from the author.

3 Planning content and layout

The project is divided into four functional areas: entrance area, Weiping old street area, artist club area, and lakeside landscape area (Figure 3).

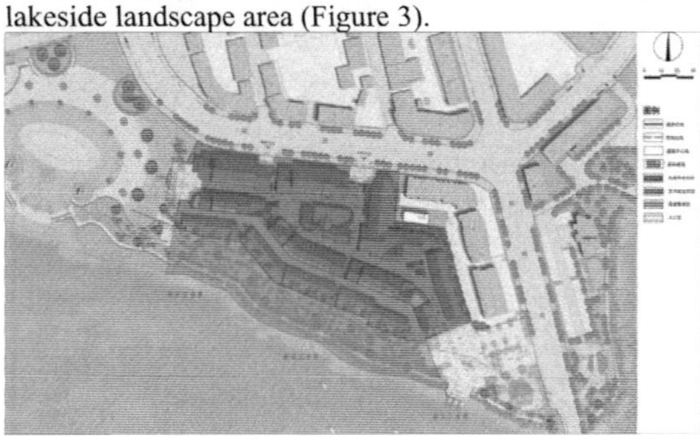

Fig. 3. Functional areas in the project: entrance area (yellow), Weiping old street area (red), artist studios (pink), lakeside landscape area (green). Sources from the author.

3.1 Entrance area

https://doi.org/10.1051/e3sconf/202123704028

According to the surrounding circulation and road traffic conditions, the INK Weiping Tourist Cultural project is divided into three main pedestrian entrances and an underground parking garage entrance, which implements the separation of pedestrians and vehicles. It inherits the open and suitable small-scale traditional block. Memorial arches and squares as symbols marking Huizhou village's main entrance as significant spatial nodes are designed in detail. For example, the west entrance mainly accepts the pedestrian flow of Weiping Square, and there is a "Ke Jia Shi Cheng" arch in front of the entrance axis (Figure 4), responding to the east-west axis of Weiping Square. At the east entrance, the commercial street buildings retreated, and Huizhou's representative *xiulou* (a loggia-like place in upper floors for Chinese women to do the needlework), *jiaolou* (corner building, see Figure 5) and arch was set up to strengthen the entrance space. Secondly, the entire Shui Mo Weiping Tourist Cultural Street is compared to a mandarin fish, so the east entrance is set up as an antique port at its original place to create the thriving river transport scene of old Weiping in the past. The stress of the entire project a pagoda named "Wannian Ge" was built at the fisheye position, and the base is the tea garden stone "Tianziji." Thirdly, the *zhaobi* (spirit screen) and other landscape architecture in traditional Hui-style garden are set up in the front square to strengthen the position of the entrance square. The west entrance mainly draws the pedestrian flow from Hubin Road. Through set-back commercial street buildings and balconies on the exterior walls to strengthen the link between buildings and the entrance space.

Fig. 4. Entrance arch on the axis formed by Weiping plaza and the street of the project. Sources from the author.

Fig. 5. Corner building marking the entrance at the east plaza. Sources from the author.

3.2 Weiping old street area

As the core area of the INK Weiping Tourist Cultural Street, Weiping old Street is divided into categories of the Front Street, the Back Street, the Horizontal Street, and the Lower Street. The buildings are arranged along the above-categorized streets in eight parts: (1) Leisure and cultural area of the waterfront area near the entrance square on the east side. There are open-air tea houses, cafes, bars, art creation rooms, and galleries. Such facilities are proposed to become an area with strong participation by tourists and rich cultural content of sightseeing, shopping, fashion, and leisure. (2) Local product exhibition area exhibits local products from the surrounding villages. Walnuts, tea, dried bamboo shoots, dried fish, seedless persimmon, and other native products and craft, at the same time, attract tourists by selling, introducing the history of unique products, allusions, exquisite display, performance, and personally making and tasting. (3) Traditional craft exhibition area arranges traditional craft workshops for teaching stone carving, brick carving, wood carving, bamboo carving, handicraft manufacture of the product like compasses, bamboo chimneys, and other small things. (4) Medical and health care area promotes Weiping Chinese medicine processing that has long been famous. (5) Particular food display area combines the existing Weiping annual fresh fish barbecue event. Exhibiting Weiping outstanding food such as hot sauce, dates, winter melon candy, peanut candy, moon cake, snow cake, etc. Allowing visitors to taste and participate in activities. (6) Museum of folk houses, celebrities, and specialties houses three reconstructed houses in Weiping area. Architectural models, measured drawings, large building samples, interior furniture, architectural components are displayed. (7) Set up some distinctive hotels along the lake or inside the courtyard building. (8) According to the religion, education, and culture of the old Weiping area, set up performance stage, academy, memorial hall, etc. Finally, in the Weiping Old Street area, some traditional shops are retained and restored according to the above functional areas, for examples, one hall, and one storehouse; pastry workshops, bucket rot workshops, soy sauce workshops, shoe shops, clothing shops, tea shops; rice shops, hotels, ceramic shops, wooden shops, iron shops, bamboo shops, medicine shops, gold and silver shops.

3.3 Artist studios and lakeside landscape area

INK Weiping Tourist Culture Street draws on the model of Beijing Songzhuang, Guanlan Printmaking Village, and Beijing 798 District, attracting 13 domestic art masters to settle in the 13 artists' studios are located in the project (Figure 6 and Figure 7). The belt-shaped areas along the river are all facing southeast. As studios for meeting between well-known domestic and foreign companies and artists, it provides a platform and opportunity for cultural exchanges and artistic creation. Lakeside landscape area is the Xin'an River waterfront greening area, featuring green land and tree planting. It provides a place where residents can walk and play, combining the natural landscape with Huizhou's cultural landscape.

Fig. 6. Artists' studios and lakeside landscape area. Sources from the author.

Fig. 7. Interior of Artist' studio. Sources from the author.

4 Analysis of building design

4.1 Building plan

As an essential carrier for inheriting Hui culture in Old Weiping, residential house design is set up based on the traditional shop mode in front and residence at the rear. The architectural scale and living facilities are adjusted and improved according to the modern lifestyle. The details adopt standard Huizhou architectural features such as window lattices, *queti* (bracket support), *mei ren'kao* (seating around the loggia), wood carvings, and stone carvings as decorations, installed with antique furniture. The building design plan can be divided or combined according to function through convenient and flexible dividing walls. The architectural plan of the artist club area adopts a modern architectural plan. The first floor is designed for reception and exhibition, and the second floor is for artists' rest and independent working studio. Simultaneously, a large area of terraces is set along the lake to conform to the contour of the terrain along the lake. It is an extended area convenient for artists to carry out various creative events concerning their studios.

4.2 Building façade

Hui-style architecture is an essential part of Weiping's regional environment, and it is the material carrier of Huizhou culture. The "white wall, grey tile, and horse head wall" constitutes the aesthetic theme of ancient Hui-style architecture. The project's architectural form and facade design are based on the inheritance of Hui-style architectural aesthetics in old Weiping Town. In order to echo the modern street facade of Hubin Road, the building facade adopts an improved form of traditional Hui-style commercial storefronts with two or three floors. The facade adopts a three-stage treatment method. The lower part is from the ground to the advertising plaque, and the store number in the first floor, from the lower edge of the balcony railing to the cornice is the middle part; and from the cornice to the roof ridge is the upper part. The lower part is as much open as possible to the street, and structural columns are set to the inner side of the exterior walls.

The door lintel is made of 30-40cm beams. The beam to the second-floor protruding part is connected by *que ti* and *xuan* (small landscape architecture for beautiful sceneries) components. The store plaque hangs on the protruding middle part below the floor railing. The middle protruding part adopts the method of creating voids to hide the modern air-conditioning behind louvers. The upper part from the cornice to the roof ridge uses small grey tiles, ending with dripping and hooks. Storefronts, on the one hand, are designed based on the traditional standard in detail. On the other hand, modern elements are applied to the building part like doors and windows and facade to meet contemporary business requirements. Horse headwalls on both sides gradually protrude according to the raising curvature of eaves and forming a varied and lively facade outline. The façade of the artist club area mainly draws on the traditional houses of Weiping; it adds modern architectural language such as large-area glass windows on the ground floor and an

https://doi.org/10.1051/e3sconf/202123704028

open terrace on the second floor to better display the internal creative activities in art production.

4.3 Spatial design in architecture

Spatial design in architecture continues and inherits the characteristics of Huizhou villages' external space. It underlines linear street space, converged node space (such as junctional block space), squares, etc (Figure 8). For street space, the facades of the storefronts in the blocks vary and show different expressions. Shops, houses, and theatres are laid out in a staggered arrangement. The spatial opening and closing, the width, the spatial form is always changing and dynamic. The primary node spaces include the east and west squares, the performance stage, the enlarged area inside the block, pagoda, etc. They are the highlights of the project and attractions for the users (Figure 9).

Fig. 8. The master plan of the project. Sources from the author.

Fig. 9. Aerial view of the project. Sources from the author.

5 Discussion and conclusion

In the project's design process, the reference and inheritance of Huizhou culture from large to small scales are gradually expanded. First, it is the consideration of the Chinese Fengshui or geomantic omen for the layout of the buildings in the Huizhou settlement, such as the back of the mountain and the water. The east square is the fisheye, and the west square is the fishtail, abstracting the triangular site into the image of "fish" in the project. Besides, during the construction of the site, the project is integrated into the surrounding urban space through urban design. For example, the East Square adjacent to Hongqiao street is the entrance of the ancient building complex, and it is also an essential node for guiding people to the waterfront space where old water port located and lakeside landscape area with collective memory left on the site. The West Plaza is connected to the city plaza-Weiping Plaza. It is connected with the ancient stage in the center of the building complex in the line of sight, creating another highly accessible urban space in the site. The project actively creates outdoor space in the site, transforms the traffic space into a suitable street space in the traditional Huizhou area, and connects the above-mentioned public spaces to improve the project's overall permeability and mobility. In addition to the traditional interpretation of architectural space, architectural elements such as *paifang* (entrance arch), *zhaobi* (entrance walls), *matou qiang* (horse headwalls), grey tiled sloping roofs, etc. have all become symbols in the representation of Huizhou culture embedded within. The project concept does not stop at the construction of the physical environment but also considers how to better integrate the later use with the building by considering its programs. The atmosphere of the traditional streets is inseparable from the traditional commercial streetscape. The handicrafts, the display of specialty food workshops, and the performances of singing troupes contribute to the inheritance of Huizhou culture together with the architectural design as a backdrop.

In the project's design process, the author also encountered unsatisfactory and unexpected issues that need to be improved in future practice. The shoppers did not organize as a community to standardize the design. As a result, shops in the project did not have a good uniform style in creating the signs. This issue resulted in a strong contrast between the white wall's Hui-style and the grey tiles and the modern signs and red banners (Figure 10).

Fig. 10. Signboards and banners contrast with traditional Hui-style architecture. Sources from the author.

In addition, the details of the design were not well grasped by the constructors who usually practice in the construction of modern architecture in the city area. It results in the low level of completion of the building, and the artistic conception of the design was not fully expressed. It can be seen from this project that the integration of design and construction in the traditional Huizhou architecture has shifted to the division of design and construction in modern architectural practice and cause further issues. The traditional architectural skills play a decisive role in realizing and whether the original design concepts can be fully expressed. In future design and planning, it is also necessary to consider integrating store signs and interior furniture into the building's overall design. The culture and inheritance in Huizhou through architectural design still needs to be continuously improved in future practice.

Acknowledgements

This paper has been supported by the following fund projects. MOE (Ministry of Education of the People's Republic of China) Youth Fund Project of Humanities and Social Sciences (Project No.18YJC850002). Bidding Project of Key Research Base of Humanities and Social Sciences in Colleges and Universities of Anhui Province (Project No. SK2015A170). "10303 plan of Huizhou research talents training" project of Huangshan Social Science Federation.

References

1. Huangshan City Statistics Bureau, *Huangshan City Statistical Yearbook*, 97(Huangshan, Huangshan City Statistics Bureau, 2020)
2. Huangshan City Cultural Committee, Decision **9**, 10-12(2012)
3. X. Yin et al., Anhui Daily 10, (2017)
4. Y. Yang, W. Huang, The Science Educaton Article Collects, 12, 205-206(2013)
5. X. Guan, X.Y, Wang, J. Zheng, Journal of Hefei University of Technology, **27(5)**, 98-102(2013)
6. J.H. Liao, Reform **290(4)**,130-139(2018)
7. Z.M, Peng, Anhui Jianzhu University master degree thesis, (2017)
8. D.Q, Shan, X.M, Li, China Engineering & Consulting, **03**, 30-33(2008)
9. W. Nie, Anhui Architecture 176(1),28-30(2011)
10. J. LI, Market Weekly, 4, 106-108(2016)
11. X. Kong, F.Y, Zhuo, C.S. Miao, Tropical Geography 36(2), 216-224(2016)
12. Z.F. Huang, R. Huang, Geographical Research **37(2)**, 233-249(2018)
13. C.N. Schulz, *Genius Loci: Towards a phenomenology of architecture*, (New York, Rizzoli, 1979)
14. A. Rossi, *The architecture of the city* (Cambridge, The MIT Press, 1984)

E3S Web of Conferences **237**, 04029 (2021) https://doi.org/10.1051/e3sconf/202123704029
ARFEE 2020

Study on Evaluation Index System of Urban Traffic Improvement Based on Analytic Hierarchy Process

Xiaolei Zheng[1]*,Yong Yin[1],Tao Yang[2]

[1] Chongqing Jianzhu college, Associate professor, Chongqing,China, 400039
[2] Senior engineer Doctor of Tianjin University, Chongqing Fuling District Xincheng District Management Committee, Chongqing,China, 408000

Abstract. The double development of urbanization and motorization has made the traffic demand of our country grow rapidly and diversified. the improvement of urban traffic has become an important condition for the sustainable development of the city. Based on the analytic hierarchy process (AHP), this paper establishes the evaluation index system of urban traffic improvement by comparing the importance of indexing by experts. It is found that improvement of road network, optimization of public traffic and improvement of intersection are the most important in urban traffic improvement. the intersection of the most important level of service. By using qualitative and quantitative methods, the Likert scale was used to classify the results and the four-quadrant method was used to analyze the urban traffic improvement evaluation index system.

1 Introduction

The pace of urbanization in China is accelerating and the economy is developing rapidly. The traffic demand is also growing rapidly. At the same time, traffic congestion has become a prominent problem in most cities in China[1] . At present, China is in the stage of rapid motorization, which makes it difficult to achieve a fundamental relief of traffic congestion in the central city, seriously restricting the healthy development of the city and reducing the quality of life of the residents.

Urban traffic improvement has become an important issue for sustainable urban development. Causes of urban traffic congestion: On the one hand, the increase is in traffic demand, on the other hand, the shortage of supply is caused by the limited urban land. Therefore, the improvement of urban traffic is a systematic issue[2]-[3].

The establishment of a scientific and objective evaluation system for urban traffic improvement results can help us better understand the causes of urban congestion and provide more diversified and intuitive methods for solving problems. In this paper, the evaluation index system for urban traffic improvement is established based on the principle of analytic hierarchy process (AHP) by experts, and the evaluation system is tested according to the Fuling District of Chongqing,China.

2 Construction of evaluation index system for urban traffic improvement

2.1 Principles of establishment of indicator system

The urban traffic evaluation index system should be able to represent the current status of various aspects of the urban transport system, and it can reflect the changing trends of various aspects of the urban transport system, and should be able to reflect the degree of coordination of the various subsystems of urban transport[4]. When constructing the urban traffic evaluation index system, it should meet the following principles:

Systematic: Urban traffic itself is a complex system project, and it should be evaluated from the perspective of system engineering. The subsystems should be divided into different levels according to the basic principles of system engineering. At the same time, we should pay attention to the correlation between the indicators to avoid partiality.

Scientific and objective: The selected indicators and the evaluation criteria must be scientific, objective, and consistent with the facts.

Reliable and independent: The selected indicators should reflect the idea of system and maintain independence between the indicators.

Combinations of qualitative and quantitative analysis: on the basis of qualitative index analysis, the method of combining qualitative and quantitative analysis can carry out quantitative treatment of indicators, so as to make them better measurable.

* Corresponding author: xiaoleitaotao@126.com

2.2 Selection of evaluation indicators

This paper adopts AHP to select the index of evaluation system[5]. The entire evaluation index system consists of three levels: evaluation target level A, criterion level C, and indicator level P. The target layer A is an evaluation of urban traffic improvement. According to the composition of the urban traffic system, the criteria layer C is composed of six subsystems: road network improvement, public traffic optimization, intersection improvement, parking facility improvement, slow-motion system optimization, and traffic management optimization. The first-level indicator layer P contains a total of 18 first-level evaluation indicators. The structure of each evaluation indicator is shown in the table:

Table 1. Indicator system for urban traffic improvement.

The target layer A	The criterion level C	The indicator level P	Remark	Evaluation method	coding
Urban Traffic Improvement Evaluation System	Road network improvement	Road network structure	road network layout, grade is reasonable	Combine	P11
		Road network accessibility	cut - off ratio	Quantitative	P12
		Road network density	road network density has improved	Quantitative	P13
	Public transportation optimization	Bus network structure	line network level, layout optimization	Combine	P21
		Road and bus facilities	Optimization of bus lanes, stations, etc	Combine	P22
		Public transport hub construction	, transfer system construction	Qualitative	P23
	Intersection improvement	Intersection spacing		Combined	P31
		Intersection control mode	Whether the type of control method is reasonable	Qualitative	P32
		Intersection service level		Quantitative	P33
	Improvement of parking facilities	Parking facilities supply and demand ratio		Quantitative	P41
		Parking guidance system		Combined	P42
		Parking policy and management	Whether or not there are different control strategies for partitions	Qualitative	P43
	Slow system optimization	Slow - running system network	network layout and hierarchical structure optimization	Combine	P51
		Street crossing facilities		Combine	P52
		Slow space facilities	non - separation and environment optimization	Combine	P53
	Traffic Management Optimization	Traffic information modernization		Combine	P61
		Traffic system management	traffic demand and supply management	Qualitative	P62
		Transportation policy guarantee	project implementation completion estimated	Qualitative	P63

3 Standards for the evaluation model system of urban traffic improvement

This paper uses the method of questionnaire survey on the Internet platform to take an anonymous survey of experts, so that they are ranked according to the importance of indicators, and then use the principle of the analytic hierarchy process as the principle of urban traffic improvement evaluation index system model calibration.

3.1 Principles of AHP Model

American Operations Research Specialist T. L. Saaty proposed AHP in the mid-1970s to decompose complex problems into hierarchical structures, and then juxtapose each factor in the next level with each factor in the previous level to construct a judgment matrix[6-7]. Through the calculation and analysis of the judgment matrix, the importance of each factor is ranked, so as to achieve the purpose of solving the problem.General AHP is carried out according to four steps, namely, establishing hierarchical structure, constructing judgment matrix, calculating relative weight of elements under a single criterion, and carrying out consistency test, total hierarchy sort and consistency test. This section is limited to space and will not be introduced in detail.

3.2 Standards for indicator weights

This is limited to space, only the criteria layer C calculation process of a questionnaire is displayed. First, a comparison proof C of the criterion layer for the target layer A is constructed.

$$U=\begin{bmatrix} 1 & 2 & 3 & 5 & 5 & 7 \\ \frac{1}{2} & 1 & 2 & 4 & 4 & 6 \\ \frac{1}{3} & \frac{1}{2} & 1 & 4 & 5 & 6 \\ \frac{1}{5} & \frac{1}{4} & \frac{1}{4} & 1 & 3 & 5 \\ \frac{1}{5} & \frac{1}{4} & \frac{1}{5} & \frac{1}{3} & 1 & 5 \\ \frac{1}{7} & \frac{1}{6} & \frac{1}{6} & \frac{1}{5} & \frac{1}{5} & 1 \end{bmatrix}$$

The normalized matrix of matrix U is:

$$U=\begin{bmatrix} 0.421 & 0.480 & 0.453 & 0.344 & 0.275 & 0.233 \\ 0.210 & 0.240 & 0.302 & 0.275 & 0.220 & 0.200 \\ 0.140 & 0.120 & 0.151 & 0.275 & 0.275 & 0.200 \\ 0.084 & 0.060 & 0.038 & 0.069 & 0.165 & 0.167 \\ 0.084 & 0.060 & 0.030 & 0.023 & 0.055 & 0.167 \\ 0.060 & 0.040 & 0.025 & 0.014 & 0.011 & 0.013 \end{bmatrix}$$

Determine the weight $\omega =$ (0.368 0.241 0.194 0.097 0.070 0.031) CR is 0.0076, it meet the consistency test requirements.

3.3 Model calibration results

Matlab was used to calibrate the index system, and after obtaining the weight of the judgment matrix of each class of the questionnaire, the arithmetic average method was used to calculate the value. The weight of each index of each layer is shown in table 2.

It can be seen from table 2 that in the urban traffic improvement evaluation system, the road network improvement plays the biggest role in the criterion layer, while the public transportation optimization and the intersection improvement play the biggest role.

Table 2. The weight of each index.

The target layer A	The criterion level C	Criterion layer weight	The indicator level P	The indicator level weight	The total weight	The sorting
Urban Traffic Improvement Evaluation System	Road network improvement	0.405	P11	0.473	0.191	1
			P12	0.129	0.052	7
			P13	0.398	0.161	2
	Public transportation optimization	0.203	P21	0.553	0.112	4
			P22	0.261	0.053	5
			P23	0.186	0.038	9
	Intersection improvement	0.195	P31	0.587	0.114	3
			P32	0.232	0.045	8
			P33	0.181	0.035	10
	Improvement of parking facilities	0.092	P41	0.575	0.053	6
			P42	0.157	0.014	17
			P43	0.268	0.025	12
	Slow system optimization	0.057	P51	0.547	0.031	11
			P52	0.197	0.011	18
			P53	0.256	0.015	15
	Traffic Management Optimization	0.048	P61	0.340	0.016	14
			P62	0.301	0.014	16
			P63	0.359	0.017	13

Improvements in parking facilities, optimization of urban slow-motion systems, and optimization of traffic management have played a minor role. Among the 18 index layers, only the road network structure, road network density, bus line network structure, and intersection service level weight values are greater than the average value, indicating that the weight distributions of the indicators in the indicator layer are more dispersed and the variance is larger.

4 Application of Evaluation Index System for Urban Traffic Improvement

Based on the results of model calibration and the results of the "Fengling District Jiangnan District Traffic Improvement Plan" (hereinafter referred to as the "planning"), this paper applies the urban traffic improvement evaluation index system. By inviting the digital experts to take an anonymous form and using the planning results as a reference, the Likert scale is used as a principle to combine the qualitative and quantitative methods to score the indicator items[8].

4.1 Introduction to the traffic improvement plan in Jiangnan District, Fuling District

The Jiangnan District of Fuling District is surrounded by the Yangtze River and Wujiang River and covers an area of about 17 square kilometers. It is the core urban area of Fuling District and has a planned population of about 400,000.

Based on the full investigation of the current situation, this plan has systematically planned the road network, public transportation, intersections, parking, slow moving systems, and traffic management.

The plan will adjust the grade and layout of the trunk line network in Fuling District, adjust the local road network, open up the broken road, and encrypt the entire road network. The planned density will be increased from 5.9km/km2 to 6.555.9km/km^2; The network has undergone major adjustments to alleviate the problem of serious overlapping of the original line network and a small radiation area. A total of 5 bus stations and 7 bus terminals have been arranged; reasonable improvements have been made to the important blocked intersections of the current situation. Its service level can reach Class C; a differentiated control strategy for parking is adopted, and parking spaces are added according to the needs of different communities; and the urban green space system,urban road traffic, and public service facility systems are combined to adjust the slow-moving system in Fuling. Optimize the slow space environment; traffic management is centered on "relaxation and diversion" to improve the efficiency and role of traffic management.

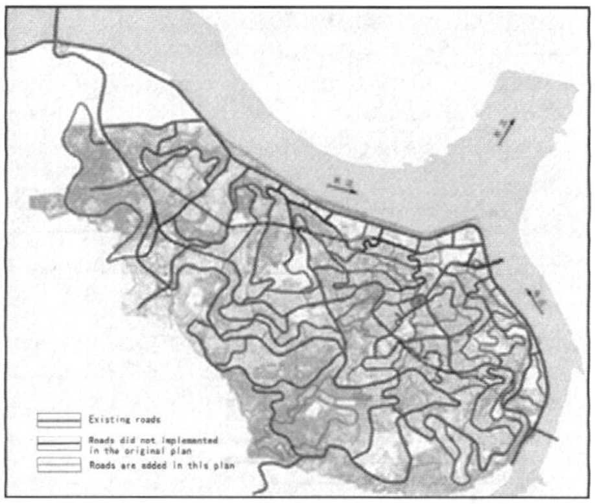

Fig. 1. The road network improvement scheme.

4.2 Results of traffic improvement evaluation in Fuling District

According to the Likert Scale, five divisions are very good, four divisions are better, three divisions are general, two divisions are worse, and one point is not good. The expert scores the average value, and then calculates the score of the criterion layer C and the score of the target layer A according to the index layer score.

Table 3 Each index score.

The target layer A	The criterion level C	Criterion score	The indicator level P	Index score
Urban Traffic Improvement Evaluation System	Road network improvement	4.5	P11	4.2
			P12	4
			P13	5
	Public transportation optimization	4.0	P21	4.1
			P22	3.5
			P23	4.2
	Intersection improvement	3.7	P31	4
			P32	3.7
			P33	2.6
	Improvement of parking facilities	4.5	P41	5
			P42	3.5
			P43	4.1
	Slow system optimization	3.6	P51	4.5
			P52	2.6
			P53	2.4
	Traffic Management Optimization	3.4	P61	2.5
			P62	4.1
			P63	3.6

From the table, the criterion level A score is 4.1, indicating that the results of the "Fengling District Jiangnan District Traffic Improvement Plan" are between good and excellent.

However, in the plan's forecast of the transportation plan, when the traffic demand increases by 80%, the road network service level of the improvement plan is not significantly reduced, and the various subsystems of the urban traffic can operate well.

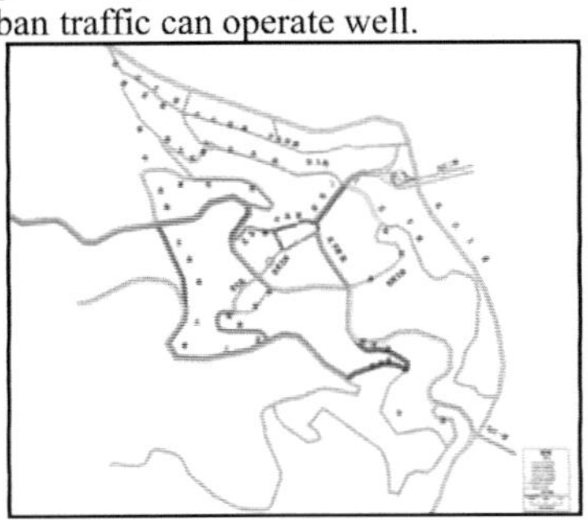

Fig. 2. Network operation analysis diagram.

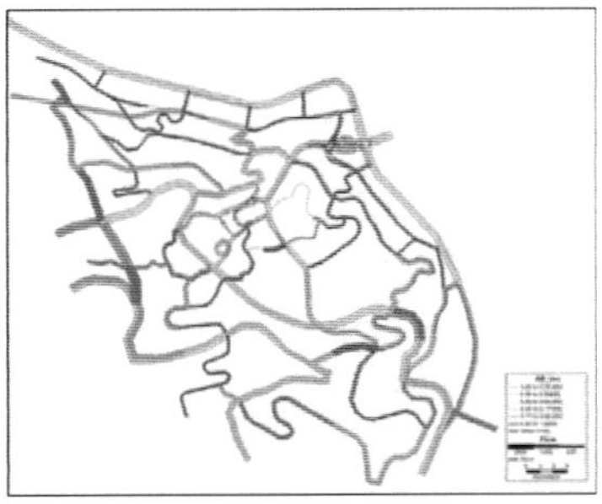

Fig. 3. Planning road network operation analysis diagram.

From this point of view, the improvement effect of the evaluation service system for this plan is 4.1, which is between good and excellent. This reflects the effect of the urban traffic improvement more accurately and the evaluation service system has higher reliability.

5 Conclusion

This paper establishes a comprehensive evaluation index system for urban traffic improvement through AHP and expert scoring. According to the principle of AHP, target layer A, criterion layer C, and indicator layer P are provided. Taking the urban traffic improvement evaluation system as the core, and using the six subsystems of roads, buses, intersections, parking, running, and management of the urban transportation system as the guideline, 18 evaluation indicators have been established. Through expert scoring, the importance of 22 evaluation indicators and 6 criteria subsystems is compared at the level of two or nine, and the weight of each index is obtained according to the principle of the analytic hierarchy process, so as to establish an urban traffic improvement evaluation system. It is found that road network optimization, optimization of public transportation, and improvement of intersections are the three most important indicators of urban traffic improvement, and they should also be the first issues to be noticed in the process of urban traffic improvement.

This paper uses the case of "Traffic Improvement Plan for Jiangnan Area of Fuling District" as a case to apply the evaluation system of urban traffic improvement. Through the combination of quantitative and qualitative methods, various indicators of each layer were scored according to the fifth scale of the Likert scale, and the overall evaluation of the plan was between good and excellent, which reflected this fact more realistically. The effect of urban traffic improvement and evaluation service system have high reliability.

Through the establishment and application of the urban traffic improvement evaluation index system, we can decompose the complex and diversified urban transportation problems and quickly find the advantages and disadvantages of the current traffic improvement programs. We have found positive solutions to our urban traffic problems. The role played a guiding role in the evaluation and optimization of urban transport improvement programs.

Acknowledgments：

Fund project: "Research on Evaluation Index System and Methods of Urban Traffic Improvement" Science and Technology Research project of Chongqing Education Commission in 2015（KJ1504208）

Fund Project: "Research on the Application of Prefabricated integrated pipe gallery in engineering" Science and Technology Research project of Chongqing Education Commission in 2018（KJQN201804307）

References:

1. Wenjuan Huang, Ximing Ye . Application of SWOT method in urban traffic improvement strategy formulation [J] . Traffic Standardization, **10,** (2012) : 116-119.
2. Bo Xue. Study on traffic improvement methods in urban centers of large cities [D] . Southwest Jiaotong University, (2014).
3. Qihang Yu. Study on the countermeasures of traffic improvement in urban center[D]. Nanjing Forestry University, (2013).
4. T L S. Transport planning with multiple criteria: The analytic hierarchy process applications and progress review[J]. Journal of Advanced Transportation, **29**(1995):81-126.
5. Li Q, Ping K. Application of Analytical Hierarchy Process in improving traffic and transport quality[J]. Journal of Hefei University of Technology, (2000).
6. Lianfen Wang , Shubai Xu . Introduction to analytic hierarchy process [M] . University press, (1990).
7. Yan Xu. Application of Qualitative and quantitative analysis methods in the research of Motorization process and Countermeasures in Beijing [J]. Road traffic and safety,**5**(2015):61-64.
8. Kulak O, Kahraman C. Fuzzy multi-attribute selection among transportation companies using axiomatic design and analytic hierarchy process[J]. Information Sciences, **170(**2–4) (2012):191-210.

Extraction and diversified application of embroidery insole fabric elements in Hongan, Eastern Hubei

Zhou Yuechao, Li Min, Chen Yu

College of industrial design, Hubei University of technology, Wuhan, China

Abstract. By the concept of redesign, the embroidery insole fabric elements of Hongan in Eastern Hubei Province is extracted and applied, and the application research of diversified creative products is carried out. On the premise of inheritance and innovation, the regional characteristics and cultural connotation of Hongan embroidered insole in Eastern Hubei Province are analysed. By understanding the connotation of the elements of Hongan embroidered insole fabric, the composition characteristics of common embroidered insole fabric is grasped, and the traditional fabric themes of Hongan embroidered insole is analysed and arranged; Extract the typical pattern fabric elements that can resonate with modern life, and extract, organize and redesign them; then through integration and reconstruction, carry out diversified application design of creative products. The fabric elements of Hongan embroidered insole are redesigned and applied in fashion, home textile products, clothing, interior decoration, etc. By redesigning, this paper analyses and arranges extracts and redesigns, integrates and reconstructs the elements of embroidered insole fabric in Hongan, Eastern Hubei Province, and studies the application of diversified creative product redesign. This modern process method and diversified application mode is not only the dynamic inheritance of traditional regional culture, but also the innovation and development of folk culture in Eastern Hubei Province. The inheritance and development of art traditional culture has important practical significance.

1. Introduction

With the development of human civilization, folk cloth art in Eastern Hubei [1], including Hongan embroidery, has undergone changes and is closely related to the quality of life. Hongan embroidery has distinctive artistic characteristics of folk cloth art in Eastern Hubei Province [2]. Hongan embroidery is diverse and rich in content [3]. Hongan embroidered insole is an indispensable daily necessities for people in Hongan. It not only has rich cultural connotation, but also has pleasant texture, excellent production, comfortable and durable [4]. Hongan embroidered insole contains feelings and principles, covering heaven, earth, customs and human feelings; it is widely used in many important moments of life, such as wedding, entrance examination, military service and so on. With the gradual improvement of modern people's living standards, the impact of the progress of contemporary civilization on traditional culture is also gradually obvious. As the production of Hongan embroidered insole is a form of oral transmission, the traditional manual production process cannot be mass produced, facing the situation of the loss of traditional handicraft [5]. Therefore, from the perspective of traditional culture living inheritance [6], this paper, based on the redesign concept [7-8], takes the embroidered insole in Hongan of Eastern Hubei as an example to study the extraction of fabric art elements and the design and application of diversified creative products.

2. Extraction of embroidery insole fabric elements from Hongan

2.1. Collation and analysis

Based on the concept of redesign, this paper sorts out and analyzes the embroidery insole fabric elements of Hongan, improves and uses the typical insole fabric elements, inherits and retains the profound connotation implied by the pattern theme of Hongan embroidered insole. The point, line and surface elements of Hongan embroidered insole pattern are embodied by different combinations.

The composition of the embroidered insole in Hongan is usually handed down by the embroidered women, which has a non-strict definition of pattern size and color matching. The composition and color matching of Hongan embroidered shoe mat are determined by the embroidery artists. The technique is at one's heart. It reflects the unique composition characteristics of Hongan embroidery and has typical regional characteristics of folk cloth art culture in Eastern Hubei. The inheritance of cultural connotation behind the pattern is the basis and core of redesign of

https://doi.org/10.1051/e3sconf/202123704030

traditional pattern elements of Hongan embroidered insole [9].

There are three types of composition of Hongan embroidered insole, as shown in Figure 1. One is realistic. It is beautified by the natural form of scenery. That is, on the basis of the prototype presented in life, it is processed by refining, strengthening, beautifying and layout. Realistic composition requires both the form and the spirit, and the form similarity reflects the spirit similarity. As shown in Figure 1 (a), "head loving fish" insole has complete fish shaped pattern, distinct gold carp, and both shape and spirit. The other is freehand brushwork. That is, on the basis of understanding the characteristics of natural scenery, embroidery women simplified the composition, generalized exaggeration, and expressed their ideas. The flower pattern of Hongan embroidered insole is just a kind of moral meaning. Some flower shapes are composed of geometric patterns, such as the insole of "blessing and longevity" in Figure 1 (b), which simplifies the abstract color matching pattern and expresses the meaning of mutual blessing between people. The third is the combination of realism and freehand brushwork. That is, realism and freehand brushwork are used in the composition of a pair of embroidered shoes. As shown in Figure 1 (c), "rich and noble" insole, copper coin realism and peony colorful Phoenix freehand brushwork complement each other [10].

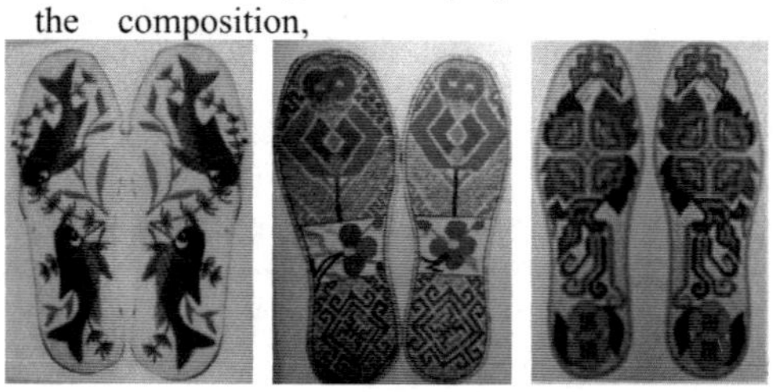

(a) (b) (c)

Figure 1. Embroidered insole pattern of Hongan.

2.2. Extraction and redesign

The extraction and redesign of embroidery insole fabric elements in Hongan mainly includes: selection of typical original insole pattern → extraction of typical pattern → arrangement of pattern → redesign of pattern, covering the implication, shape and color of typical insole elements [11]. The pattern features of Hongan embroidered insole are highly geometric and abstract. In the process of pattern extraction, arrangement and redesign of Hongan embroidered insole, the representative patterns and flower shape elements in the insole are described by computer simulation software; the patterns and flower shape elements are highly summarized and sorted; and then the redesign of various variations of patterns and flower shapes is carried out. In this way, the diversified creative product design idea not only retains the meaning of insole pattern and flower shape theme, but also conforms to the original regional style of Hongan embroidery pattern. Figure 2 lists the change graphics in the process of extraction, arrangement and redesign of the animal patterns of some Hongan embroidered insoles; Figure 3 lists the process change graphics of the extraction, arrangement and redesign process of the flower patterns of some Hongan embroidered insoles.

 https://doi.org/10.1051/e3sconf/202123704030

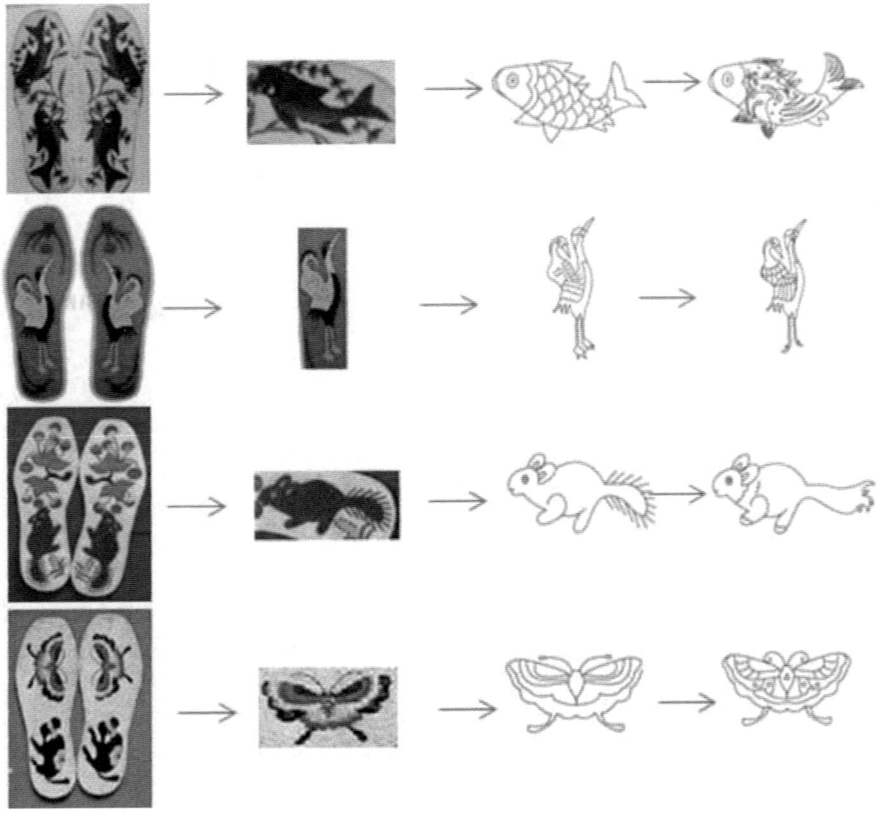

Insole pattern →pattern extraction → Pattern arrangement→ Pattern redesign

Figure 2. Extraction of animal pattern of Hongan embroidered insole.

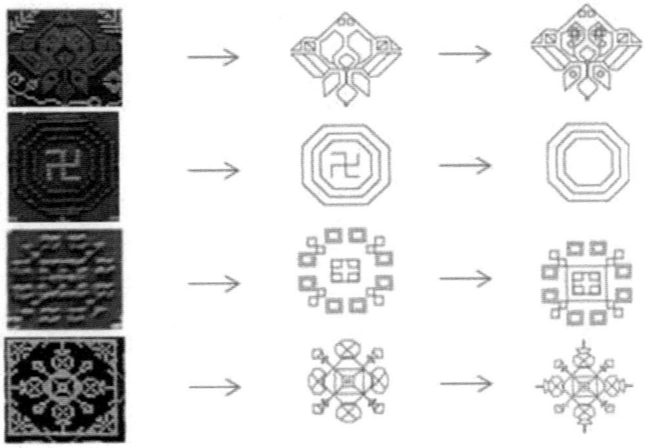

Pattern selection → Flower pattern arrangement → Redesign of flower pattern

Figure 3. Extraction of flower pattern of Hongan embroidered insole.

2.3. Integration and reconstruction

The integration and reconstruction of the typical patterns of Hongan embroidered insole includes the reconstruction of insole pattern and flower shape elements and color reconstruction. In other words, on the basis of retaining the refined and extracted pattern features and flower shape implication, the innovative combination of redesign elements is further carried out to form a transplantable multi creative combination pattern based on Hongan embroidered insole elements.

The integration and reconstruction of the typical patterns of Hongan embroidered insole includes the reconstruction of insole pattern and flower shape elements and colour reconstruction. In other words, on the basis of retaining the refined and extracted pattern features and flower shape implication, the innovative combination of redesign elements is further carried out to form a transplantable multi creative combination pattern based on Hongan embroidered insole elements.

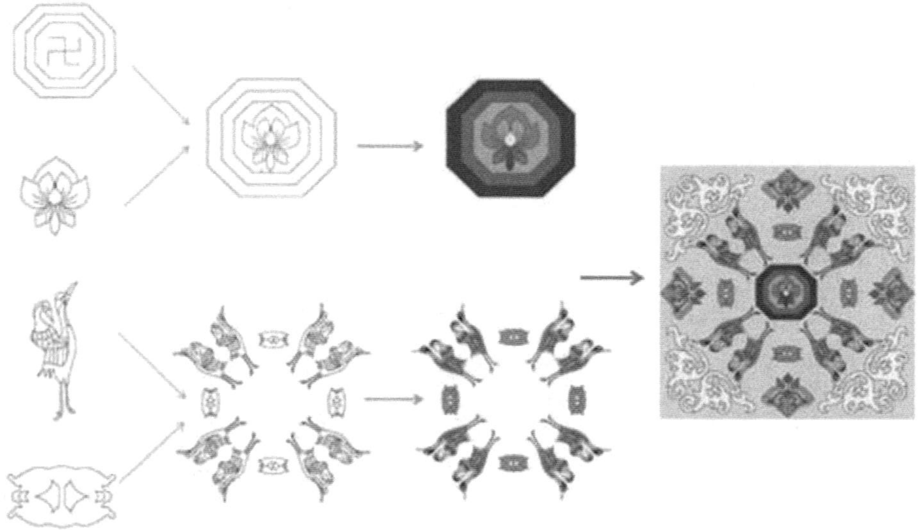

Redesign pattern -- Combinatorial reconfiguration --Color reconstruction--Combination design

Figure 4. Integration and reconstruction of typical pattern elements of "crane stepping on Lotus".

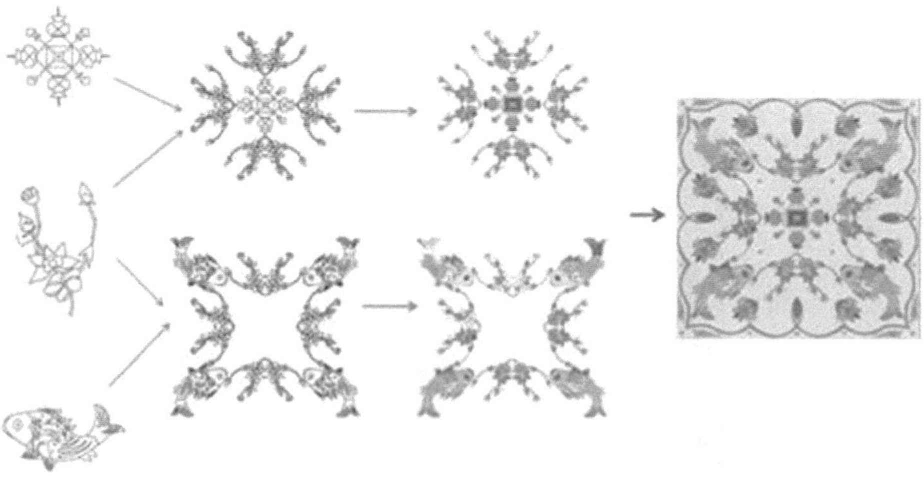

Redesign pattern -- Combinatorial reconfiguration --Color reconstruction--Combination design

Figure 5. Integration and reconstruction of typical pattern elements of "four happy head fish".

2.3.1. Reconstruction of insole pattern and pattern elements.

Reconstruction is not the reproduction and copy of embroidery patterns and flower shape elements [12], but based on the concept of redesign, through the use of digital simulation of computer simulation, the pattern and shape of embroidered insole in Hongan are highly abstract and internationalized, which not only retains the original meaning of embroidered insole, but also meets the needs of modern cultural life. The process of combined redesign is to combine the extracted, sorted and redesigned animal and lace patterns into symmetrical patterns or other ways. Figures 4 to Figures 6 show the process of re design combined pattern based on Hongan embroidered insole elements. Figure 4 shows the process of pattern reconstruction of embroidered insole of "crane treading Lotus"; Figure 5 shows the process of pattern reconstruction of embroidered insole of "four happy held fish"; and Figure 6 shows the process of pattern reconstruction of embroidered insole of "golden mouse and butterfly playing spring". Reconstruction of insole pattern and pattern elements. Reconstruction is not the reproduction and copy of embroidery patterns and flower shape elements [12], but based on the concept of redesign, through the use of digital simulation of computer simulation, the pattern and shape of embroidered insole in Hongan are highly abstract and internationalized, which not only retains the original meaning of embroidered insole, but also meets the needs of modern cultural life. The process of combined redesign is to combine the extracted, sorted and redesigned animal and lace patterns into symmetrical patterns or other ways. Figures 4 to Figures 6 show the process of re design combined pattern based on Hongan embroidered insole elements. Figure 4 shows the process of pattern reconstruction of embroidered insole of "crane treading Lotus"; Figure 5 shows the process of pattern reconstruction of embroidered insole of "four happy held fish"; and Figure 6 shows the process of pattern reconstruction of embroidered insole of "golden mouse and butterfly playing spring".

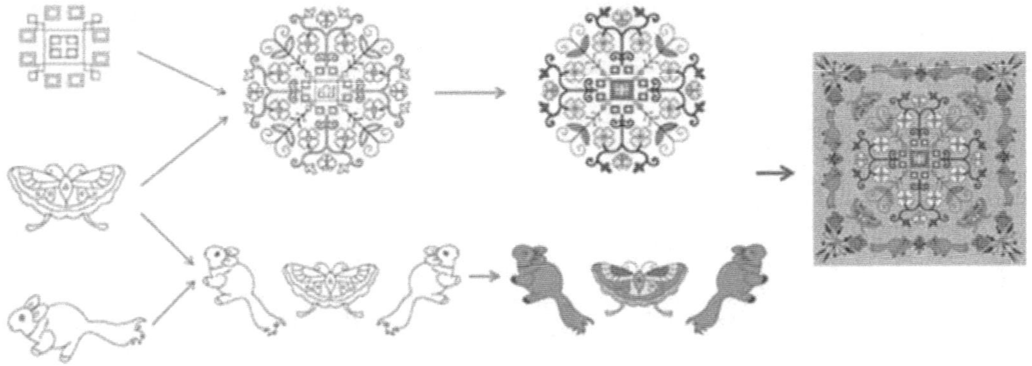

Redesign pattern -- Combinatorial reconfiguration --Colour reconstruction--Combination design

Figure 6. Integration and reconstruction of typical pattern elements of "golden mouse and butterfly make spring".

2.3.2. The pattern and pattern of insole are reconstructed.

The colour reconstruction should not only keep the inherent colour characteristics of Hongan embroidered insole pattern and flower shape elements, but also integrate into the popular colours of modern culture. The purpose of colouring is to recreate the diversified creative patterns of insoles on the basis of preserving regional characteristics. In the process of colour matching graphic redesign, the colour matching features of "five elements" of green, white, red, yellow and blue embroidered by Hongan are retained, and the harmony of colours is emphasized. In the process of colour reconstruction of the integrated and reconstructed multi insole elements combination pattern, the colour matching of combination pattern in Figure 6 is gradual harmony colour matching; the colour matching of combination pattern in Figure 4 and Figure 5 is comparative harmonious colour matching.

3. Hongan embroidered insole fabric elements diversified creative product design and Application

Hongan embroidered insole in the traditional folk cloth art in Eastern Hubei Province, the fabric design theme involves daily household, animals and plants, Buddhism and Zen, Chu culture, and words and Poems [13]. It seems that there are many types, but the production form is relatively single. Therefore, we can design and develop more fashionable and popular diversified creative products suitable for contemporary people's life by using the extracted insole fabric pattern and flower shape elements from the inheritance and diversity of product types.

3.1. Applied to fashion products

The re designed Hongan embroidered insole fabric element combination pattern, because of its portability, can be widely used, so modern fashion products, design and production of more practical value of creative products. Multiple insole fabric elements combination pattern can be applied to handbags, satchels, backpacks, shoulder bags and so on. As shown in Figure 7, it is a combination of multi Hongan embroidered insole fabric elements. The pattern is transplanted to the portable canvas bag to form a convenient and applicable household creative product with a certain cultural background.

https://doi.org/10.1051/e3sconf/202123704030

Figure 7. Redesigned handbag.

Figure 8. Redesigned pillow.

3.2. Applied to creative home textile products

The redesigned Hongan embroidered insole fabric elements can be combined with various home textile products to develop a series of creative home textile products. It can be applied to curtain, bed sheet and quilt cover, pillow and sofa set, etc. Single combination pattern can be applied to pillow, silk towel, and tablecloth and so on. As shown in Figure 8, a series of household pillow patterns are redesigned by using the single combination pattern of Hongan embroidered insole fabric elements. As shown in Figure 9, a series of silk scarves are redesigned by using the combination pattern of Hongan embroidered insole fabric elements.

Figure 9. Redesigned silk scarf.

3.3. Applied to modern clothing products

The combination pattern of Hongan embroidered insole fabric elements can also be used in modern clothing design. Through partial embellishment, animals, plants and flower patterns are partially applied to popular clothing, and can be selectively applied to T-shirts, coats, skirts, shoes and hats; through splicing or subsidies, it can be mixed and matched with modern clothing such as windbreaker, suit, dress, etc. As shown in Figure 10, the effect picture of red an embroidered insole combined with edge flower elements, as well as animal patterns such as golden mouse and crane, is applied to clothing. It is clear, concise and generous.

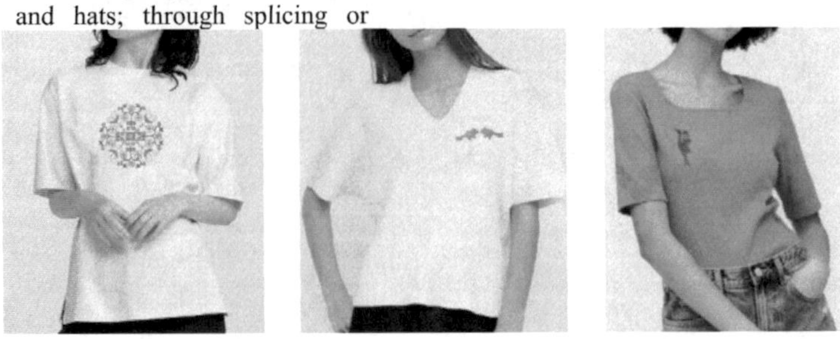

Figure 10. Redesigned clothing patterns.

3.4. Applied to interior decoration design products

The re designed Hongan embroidered insole fabric element combination pattern can also be used in interior decoration design [15]. Interior decoration design can directly transplant the combined pattern of redesign to decorative wallpaper, vase, screen and mural. For example, in Figure 11, the creative design of the combination pattern is applied to the interior

Vase Decoration design. The color is elegant and soft, elegant and generous.

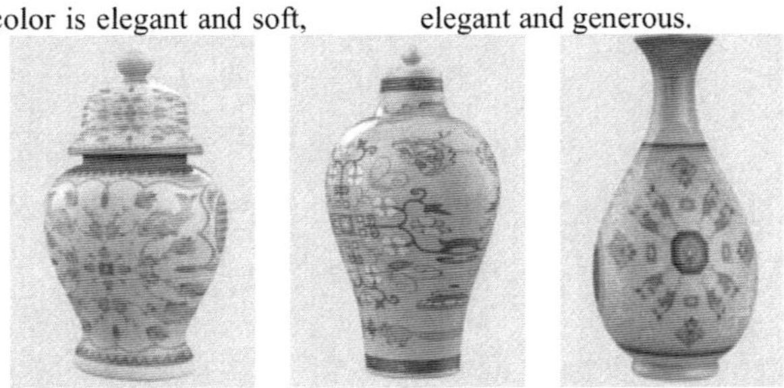

Figure 11. Redesigned decorative vase

4. Conclusions

Hongan embroidered insole fabric elements in the folk cloth art of Eastern Hubei Province are rich in patterns and patterns, which carry the emotional sustenance and changes of the times of Hongan people in Eastern Hubei Province. Through the sorting and analysis of the typical patterns and flower shaped cloth art elements of Hongan embroidered insole, extraction, arrangement and redesign, as well as integration and reconstruction, the multiple combination patterns formed from this have strong portability and can be widely used in all aspects of daily life, and then design diversified creative products with regional characteristics. This diversified creative product design idea not only retains the meaning of insole pattern and flower shape theme, but also conforms to the original regional style of Hongan embroidery pattern. Therefore, based on the concept of redesign, taking Hongan embroidered insole as the research object, it has important practical significance to extract the fabric art elements of Hongan embroidered insole and the application of diversified creative products.

Acknowledgments

This paper is supported by two foundation projects: Research on service design strategy for innovation of folk fabric products in Eastern Hubei Province (20YJA760036), supported by humanities and Social Sciences planning fund of the Ministry of education of China; and Research on collaborative service design based on innovation and development of folk handmade cloth art in Eastern Hubei Province (19D033).

References

1. Ding bin 2015 *Research on the folk cloth art of Eastern* (Hubei: Hubei Normal University)
2. Yuan Mengmeng 2019 The artistic characteristics of folk cloth art in Eastern Hubei and the origin of its style *Silk* vol 56 (04) pp 85-92
3. Huang Lin and Ye Hongguang 2017 Cultural characteristics and design application of Hongan embroidery *Tianjin Textile Technology* vol 05 pp 4-5
4. Wang Xia 2012 *Hongan xiuhuo: a series of intangible cultural heritage of Hubei Province* (Hubei: Hubei people's publishing press)
5. Wang Xia 2010 *Thinking about the protection and development of "Hongan embroidery activity" A preliminary study on the construction and development of the hometown of folk culture and art in China* (Beijing: Beijing shetu Culture Development Co., Ltd) pp 308-310
6. Chen Ruilian and Ye Hongguang 2016 Study on the protection and inheritance of Hongan embroidery Intangible Cultural Heritage *Tianjin textile branch* vol 4 pp 15-1
7. Jiang min 2013 Iinterpretation and Reflection on product redesign *Packaging engineering* vol 34 (18) pp 92-95
8. You Wenwen 2020 Study on the secondary design and reuse of garment waste fabric *Leather science and engineering* vol 30 (3) pp 88-92
9. Zhou Yang 2019 *Analysis and design of the pattern of yellow plum in Hubei Province* (Hubei: Hubei University of technology)
10. Lu Yanxin 2016 *Research on the characteristics of embroidery crafts in Hongan of Hubei Province* (Hubei: Wuhan University of technology)
11. Zhang Jingya 2015 *Research on Extraction of regional cultural elements and dynamic graphic design based on scene prototype* (Hunan: Hunan University)
12. Luoling 2017 *Research on the protection of folk embroidery from the perspective of the revitalization of traditional crafts* (Hubei: Wuhan Textile University)
13. Ma Jiao 2017 *Research on Hongan embroidery in Hubei* (Shanghai: Donghua University)
14. Chen Ruilian 2018 *Research on the application of folk embroidery in the design of modern women's wear* (Hubei: Wuhan Textile University)
15. Longzhou 2017 The application of "craft form qitheism" in the comparative study of Arts and crafts and decorative arts *Art observation* vol 2 pp 122-123

Social criteria to develop an in use holistic urban sustainable assessment tool: UHU2SAT

Antonio Sánchez Cordero[1*], Marta Videras Rodríguez[1], Sergio Gómez Melgar[2], and José Manuel Andujar Márquez[2]

[1]University of Huelva, *ETS Ingeniería. Programa de doctorado CYTIA.* e-mail: antonio.sanchez443@alu.uhu.es
[2] University of Huelva, *ETS Ingeniería, TEP192 Control y Robótica* web: http://www.controlyrobotica.com/

Abstract. Recently, several urban sustainability assessment tools (USAT) have appeared to reduce the human impact provided by the built environment. Few of them focus on the assessment of urban spaces like squares, streets, and parks, etc., but they don't operation and maintenance (OM) phase is not considered. It would be necessary to develop an in use holistic urban sustainable assessment tool (UHU2SAT) to specifically assess urban spaces under OM phase. This paper provides a qualitative research among 188 studies, with the aim to classify them according to the impacts they pursue: environmental (ENV), social (SOC) economic (ECO) and Others. Finally, the SOC criteria are reduced up to 21 and arranged into 4 different groups: Human comfort, Urban mobility, SOC Cohesion, and Health & safety. These SOC criteria have been discussed and revised according to the literature review to identify the most suitable indicators for the UHU2SAT. Finally, it can be concluded that this methodology could also be useful to obtain ENV and ECO criteria to provide a holistic assessment of the sustainability.

1 Introduction

The built environment constitutes and important part of the human footprint over our planet. Not only for the construction processes themselves, but also because the operation and maintenance (OM) and the recycling phases [1]. In fact, the impact of OM is more significant in developed countries [2].

According to the world's concerns, described in the United Nations' Sustainable Development Goals [3], many sustainability assessment tools have been recently appearing to reduce impacts from the built environment [4], [5]. At the beginning, these sustainability assessment tools (SAT) were focused on measuring the ENV impacts of new buildings [6]. As the time passed, they become more useful to improve buildings' sustainable design and construction through the inclusion of more factors of sustainability [7]. Then, researchers started to develop some more specific SAT adequate for different typologies, regional environment, and life cycle stages from design to operation and maintenance [8]. This provided a vast scenario with more than six hundred tools [5], [9], but focused on building' assessment.

In recent years, assessment of urban districts and cities become a reality with several SAT with specific features for that scale [10], but they are not as mature as building assessment, especially in case of urban spaces. For them, two examples of this are: CEEQUAL developed by Building Research Establishment (BRE) [11] and SITES developed by the US Green Building Council (USGBC) [12]. They both have the objective to assess urban infrastructure, landscaping and, works in public spaces, but without considering the OM phase of already in use urban spaces. Therefore, it is necessary to develop an in use holistic urban sustainable assessment tool (UHU2SAT), that can provide ground support for urban regeneration initiatives in urban spaces of developed cities. Also, it is necessary a driver for the uptake of sustainable pathways by recognizing excellence for sustainable urban spaces.

As some others declare [13], multi-criteria decision-making (MCDM) methodologies, and more specifically to ,multi-attribute decision-making (MADM) can be used to develop this tool [14]. The process includes the participation of different groups of experts and stakeholders that must evaluate priorities among different sustainable criteria [15], [16] [17]. Therefore, the accurate definition of the most appropriate evaluation criteria should be a fact of major importance in the creation and development of the UHU2SAT.

A truly holistic sustainability comes from the consideration of concerns like environmental (ENV) [7], social (SOC) [18], economic (ECO) [19], and others like, cultural [20], institutional [21], and technical [22]. All these impacts of sustainability must be evaluated to provide long term useful SAT. SOC impacts seems to be more difficult to evaluate according to the highest number of criteria found [23].

This research objective is to provide a critical review of 188 studies done in the recent years as well the most widely spread SAT for urban design, that establishes the foundations to define the most sustainable SOC criteria for the development of the future UHU2SAT.

* Corresponding author: antonio.sanchez443@alu.uhu.es

https://doi.org/10.1051/e3sconf/202123704031

2 Methodology

According to the objective of this paper, the methodology proposed as follows in Figure 2.1.

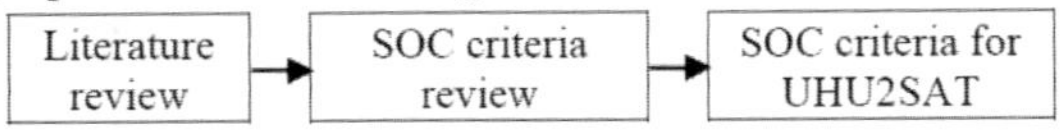

Fig. 2.1. Three steps methodology to obtain the SOC criteria.

2.1 Literature review

First, a literature review has been conducted simultaneously in Scopus and Web of Science, to be considered the most relevant scientific database for this purpose [24]. Three different lists of keywords have been used to proceed with the review:

- Type 1, group of words like urban sustainability assessment or indicators or criteria.
- Type 2, popular acronyms like Building Research Environmental Assessment Method for Communities (BREEAM C), Leadership in Energy & Environmental Design for Neighbourhood Development (LEED ND), *Deutsche Gesellschaft für Nachhaltiges Bauen* for Urban Districts (DGNB-UD), and The Comprehensive Assessment System for Built Environment Efficiency for Urban districts (CASBEE-UD) [5].
- Type 3, those more specific to the assessment of urban spaces, like CEEQUAL [11], and SITES [12].

Table 2.1. Most used USAT technical manual

USAT	Organization	Version	Ref.
SITES	GBCI	V2	[12]
CEEQUAL	BRE	V 6	[11]
BREEAM C	BRE	SD202 - 0.1:2012	[25]
LEED ND	USGBC	V 4	[26]
CASBEE-UD	JSBC	V 2014	[27]
DGNB-UD	DGNB	V 2016	[22]

Additionally, these database results have later been merged into database managed by Mendeley software™.

2.2 SOC criteria review

Second, the results obtained from the review have been organised according to the most relevant impacts: ENV, SOC, ECO and others, but only from 2010. Also, an additional review has been carried to discard those less relevant studies. Then, several international standards, providing the specific principles for sustainability within the construction sector, have been used to classify the most relevant impact of each criterion: The UN EN 15643 series, provided by the Technical Board 350 from the European Committee of Standardization, have been used to establish the basis of SOC criteria selection. Especially, those concerning to the general framework [28], and to the SOC impacts framework [29]. The 15643-3:2012 declares that accessibility, adaptability, health and comfort, neighbourhood stress, maintenance planning, security issues, public participation should be involved in SOC sustainability.

Additionally, several USAT, with a focus on SITES and CEEQUAL have been used to obtain basis criteria for SOC assessment, see Table 2.1. It must be mentioned that several of the criteria defined in any USAT cannot be classified into a specific sustainability impact, so they are placed according to their primary objective.

2.3 Most adequate SOC criteria for USAT

Finally, papers with criteria focused on SOC impacts have been carefully reviewed to keep only those adequate with: OM phase, specific for urban spaces.

Table 2.2. Evaluation methodology for criteria selection

Criteria	Objective	OM	Urban space	SOC
Criterion	Description	Y/N	Y/N	Y/N

3 Results

According to the methodology explained in Section 2, results from the literature research have been obtained and arranged into the following graphs and tables.

3.1 Literature review

Figure 3.1 shows the number of research items found in Scopus and Web of Science from 2010 using selected keywords established in Section 2.1.

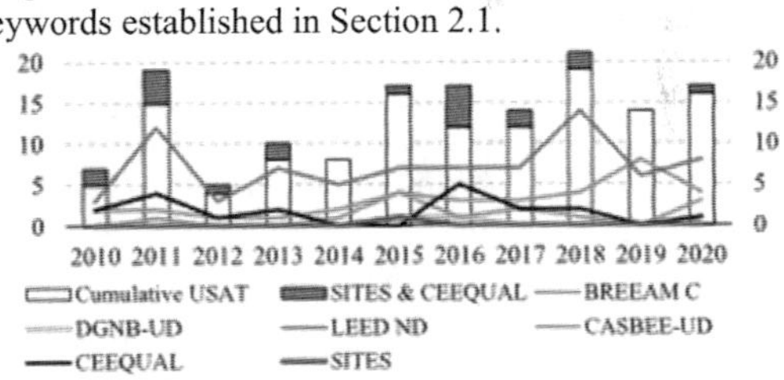

Fig. 3.1. Analysis of USATs' studies published from 2010.

White bars include BREEAM C, CASBEE UD, DGNB UD and LEED ND, while grey bars include those specific for urban spaces, CEEQUAL and SITES. It shows no clear trend in the yearly pattern. 2011 and 2018 were the more successful years, while 2012 had less publications than any other. Specifically, colour lines describe the trend in USAT's publications per year. There, LEED ND is more cited than any other USAT with 4 to 14 publications per year, while SITES and DGNB UD very few citations. Finally, publications with CEEQUEAL and SITES are less usual than all the others.

3.2 SOC criteria review

To provide a useful criteria analysis, the publications database has been reduced to consider only researches published in scientific journals, which later have been organised according the main impact of each of these ones, see Figure 3.2: ENV, SOC, ECO, or All. A growing pattern can be seen from 2010, except for the 2019, at All bars. For the rest of the bars in the figure, the trend is also increasing but with some irregularity. Considering any specific impact, the relevance ranking can be described as ENV is the first, while SOC and ECO present alternance. It is evident, that the research interests have been focused on ENV concern more than any other, but the distance is small, especially in 2016 when ENV and ECO showed the same relevance. For all years together, there are: 107 papers including ENV, 58 papers including SOC, and 68 papers including ECO, but these results may consider any combination.

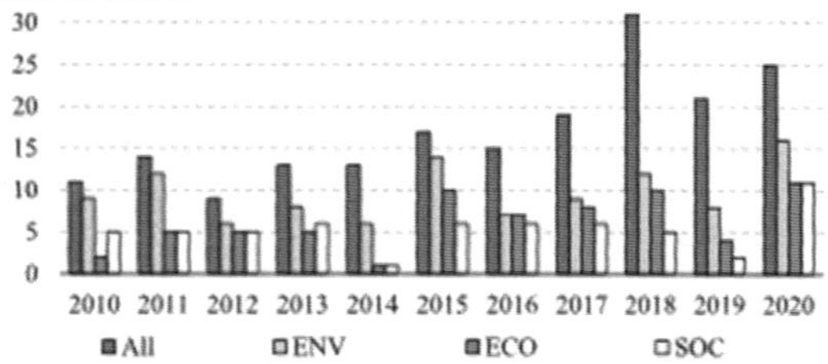

Fig. 3.2. Analysis of USATs' studies published from 2010.

To avoid this confusion, Figure 3.3. arranges the information to provide clear evidence of criteria focused profile. 188 journal papers from Figure 3.2, have been considered. It shows which research papers include only ENV, SOC, ECO and several combinations of them.

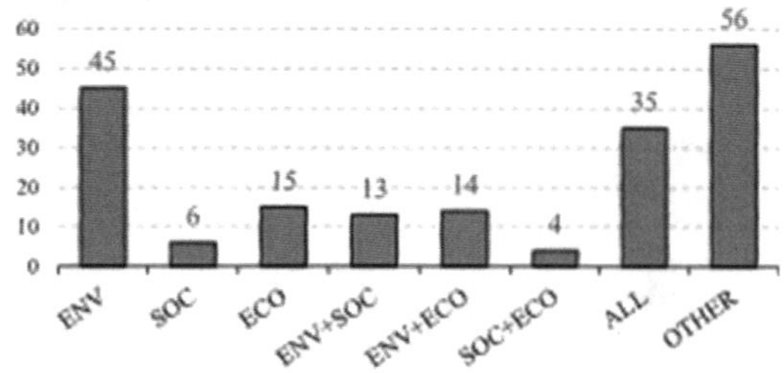

Fig. 3.3. Analysis of USATs' studies published from 2010.

According to Figure 3.3, there are 45 studies which contain ENV criteria, while there are only 6 about SOC criteria and 15 about ECO criteria. A combination of any of these options can be seen in the following options: ENV+SOC. ENV+ECO, SOC+ECO, and ALL together. Other contains those studies that do not clearly identify in which impact they focus.

Figure 3.3 shows that ENV criteria have been widely considered already, but ECO, and specially SOC criteria have not. Therefore, it could be necessary to provide complementary research about them.

3.3 Hierarchy Structure of the UUSSAT

Although several of the studies from Sections 3.1 and 3.2 have worked on the development of USAT adapted to regional specific aspects of sustainability, none of these studies consider the OM phase. It would be, then necessary to provide an adequate methodology for sustainability assessment analysis based on a proper criteria selection. According to that, the combination of both criteria, mainstream commercial and independent research USATs can provide a more comprehensive review of the current state of the art. Therefore, a more precise selection of criteria, which can be made with any of the variations of the Analytic Hierarchy Process defined by Saaty [30]. The structure here can be organised in two steps: category and criteria. Finally, and according to the structural decomposition suggested, the USATs analysed here can be summarised as follows:

Table 3.1. Evaluation methodology for criteria selection

USAT	Categories	Criteria	OM+ SOC
CEEQUAL	8	30 (Issues)	5 (16.66%)
SITES	10 (Section)	66 (Credit)	12 (18.18%)
BREEAM C	5	40 (Issues)	13 (32.50%)
CASBEE UD	4 (Topic)	35	9 (25.71%)
DGNB- UD	5 (Topic)	31	7 (22.58%)
LEED- ND	3	49 (Credit)	8 (16.32%)

CEEQUAL, now managed by BRE, is the most widely used USAT in civil engineering, infrastructure, landscaping and works in the public space [31]. Its scope includes construction or refurbishments, from strategy and design to construction, but it doesn´t include OM phase. It is structured in categories, issues, and assessment criteria which are evaluated through sets of indicators [11]. It includes 30 different criteria. According to the methodology showed in Table 2.2, 15 of these criteria could be adequate for the OM phase of an urban space assessment. Surprisingly, only 5 of them can consider SOC impact as the most important, which is the 16.66% of all criteria. See Table 3.1.

SITES is a rating system for sustainable land design and development managed by the Green Business Certification Inc (GBCI). It can apply to outside projects like: Open spaces (garden, parks, etc.), streetscapes and plazas, retail and office' outdoor areas, neighbourhood yards, public and private campuses; And also, any kind of infrastructure [12]. Its main goal is to enhance regenerative strategies for urban areas. As CEEQUAL SITES has been developed to assess construction or refurbishments for any of the urban places mentioned above, but it neither includes the OM phase. It is structured in sections, and credits which are evaluated through sets of indicators [12]. It includes 66 different credits, but they include 16 prerequisites that must be achieved prior to any further consideration. According to the methodology showed in Table 2.2, 33 of these criteria could be adequate for the OM phase of an urban space assessment, but finally, only 12 of them can considered SOC impact as the most important one, which is the 18.18% of all criteria. See Table 3.1.

According to their scope CEEQUAL, and SITES, are supposed to provide more adequate criteria for small scale urban spaces than BREEAM C, CASBEE-UD, DGNB-UD, and LEED-ND. However, they all have been analysed to obtain a more comprehensive result. Their hierarchy structure of these USATs can be seen in the Table 3.2, where the relevance of the OM+SOC criteria is: 32.50% in BREEAM C, 25.71% in CASBEE-UD, 22.58% in DGNB-UD, and 16.32% in LEED-ND.

In Figure 3.4. the graph bar shows only those criteria than can be adequate for urban spaces at the OM phase but divided into SOC or Other criteria. There, BREEAM C provides 13, and SITES provides 12, while the others provide from 9 to 5 SOC criteria. On the contrary, SITES provides 21, and LEED-ND provides 19, while the others provide from 15 to 10 criteria for any impact different than SOC. Some of these SOC criteria are very similar so they have been reduced to Table 3.2.

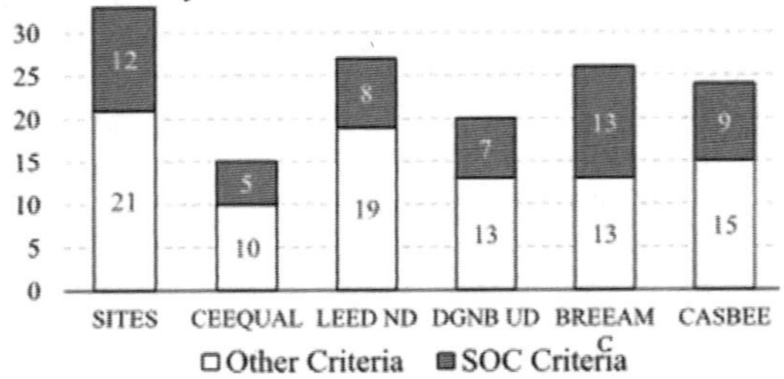

Fig. 3.4. OM urban space criteria comparison

Table 3.2. SOC criteria for the UHU2SAT

Group	Criteria	Aim	Reference
Human comfort	Micro-climate: Hygrothermal comfort on site	Promote a pleasant all year round comfort which meets the highest number of user's requirements	[25] [26] [22]
	Heat island effect: Hygrothermal comfort off site	Minimize the effects of urban heat island through the use of vegetation and reflective materials on surfaces	[12] [26]
	Lighting design	Provide a comfortable visual environment on-site that allows different age users to perform multiple outdoor tasks	[22]
	Lighting pollution control	Minimise, mitigate, and manage the negative effects of night-time lighting pollution for neighbours and astronomers	[11] [25] [26]
	Noise intrusion control	Minimise, mitigate, and manage the negative effects of noise pollution on different age users	[11] [25] [22]
Urban mobility	Connected pedestrian networks	Provide safe pedestrian ways not only on site, but also connected with walkways routes to the site surrounding facilities	[11] [25] [27]
	Emissions free vehicles network	Ensure facilities for bicycles and similar emissions free transports to enhance human health	[25] [26]
	Public transport network	Ensure public local transport networks on-site surroundings	[11] [25] [26] [27]
	Multi-modal transport networks	Improve human health and reduce pollution on site with networks that connects pedestrian, emissions free vehicles, and public lines	[11] [12] [25]
	Sustainable parking	Ensure the parking facilities have the precise size to provide service and it is not affecting any other urban mobility criterion	[25]
SOC Cohesion	Inclusive design	Guarantee the entire built environment accessible and usable for all, regardless of their physical status	[11] [25] [22]
	Commercial facilities nearby	Ensure accessible supplies nearby, that enhances usability of the urban space and integration into a wider area	[25] [26] [27][22]
	Green infrastructure and mental restoration	Guarantee inclusive high-quality green infrastructure space for all to provide visual and physical connections between users and nature	[12] [25] [26] [27]
	Recreation facilities	Provide physical activity and SOC networking facilities	[12] [26]
	Crime prevention	Ensure on-site 24-h safety conditions, through facilities like: night security lighting and security cameras among others	[27]
Health & safety	Hazardous materials	Assure no hazardous material is of claddings and coatings	[12]
	Plants management control	Guarantee human and plants health by promoting gardening practices with nearly zero use of pesticides and fertilizers	[12]
	Tobacco free space	Assure tobacco is ban within the urban space and its surroundings	[12]
	Maintenance protocol management	Minimization of air quality and noise disturbance through the proper use of maintenance management plan.	[11] [12] [22]
	Water disposal facilities	Provide free clean water disposal for all users	
	Evacuation plan	Evaluation of a disaster prevention plan and evacuation strategies	[27]

4 Discussion

From the information contained in Figure 3.2 can be seen how ENV, SOC and ECO criteria have been considered by the authors during the last 10 years. ENV appears in more studies than SOC and ECO, but the distance between them is not excessive. On the contrary, when data for all years are considered together, studies considering ENV criteria are much more than those considering SOC or ECO, even with any combinations between them. This is showing a lack of research interest when SOC criteria must be specifically developed. The interest of this study on developing SOC criteria for a new UHU2SAT has been justified.

Results shown in Table 3.1, provide the hierarchy structure of the most relevant USAT. According to the table, SITES and BREEAM C, contain the highest number of SOC criteria available for OM, with 12 and 13 each one. On the contrary, CEEQUAL has only 5 of them, which is inconsistent with the initial suggestion. If BREEAM-C, and CEEQUAL are both USAT from BRE, but with CEEQUAL developed to work at a smaller scale, how is it possible to have less SOC criteria adequate for the development of the UHU2SAT. First, the hierarchy structure of the USAT does not correspond with the same hierarchy. Second, BREEAM C includes more criteria than CEEQUAL [25] [11], so then the differences expressed in percentage are less important, 32.50 vs 16.60%. Third, the BREEAM-C, as CASBEE-UD, DGNB-UD and LEED-ND give more importance to SOC and ECO criteria than any other SAT [32]. Finally, BREEAM-C contains a higher number of SOC criteria related to urban mobility than CEEQUAL, but many of them have a very similar aim, so they can be merged into a few ones. Additionally, the scales are different [33].

Figure 3.4 shows those criteria obtained thorough the methodology explained in Table 2.2, where the criteria are divided into two groups: SOC versus Others. There, CEEQUAL seems less representative than SITES and any other USATs for the selection of useful criteria development of the UHU2SAT, see Figure 3.4.

The problem for many of these criteria is that they are not suitable for mall scale urban spaces as Medved, and Xu et al, suggest [19] [34]. Other authors suggest to assess sustainability, according to the size of the geographic scenario[35], or to use the proper criteria sets according to the scale to balance the SOC and the ECO impact [36]. Finally, others like Ameen et al proposed [17], a reduction of criteria from most relevant USATs, which has several similarities with the methodology sections 2.1, to 2.3.

Final results from Table 3.3. can be discussed according to be structure suggested into 4 groups that cover the most common aspects of SOC sustainability.

4.1 Human comfort criteria

This group includes 5 different criteria. Microclimate and Heat island effect aim to provide comfort to humans based on the hygrothermal conditions: temperature and relative humidity, which also are affected by solar radiation and wind. On site comfort can be measured through real-time monitoring devices and manage with adaptive comfort theories as other authors suggested [37], [38]. The percentage of the site where the comfort dissatisfaction is less than a defined limit can be used to guarantee minimum comfort conditions, according to a site's climate [25] [26] [22]. On the contrary, the heat island effect doesn´t have a direct impact on site, but it contributes to the district's warm up [12] [26]. It can be reduced through the proper use of reflective surfaces.

To enhance artificial lighting design on-site, best lighting design standards must be applied. It will guarantee a proper luminance level without glaring and appropriate for the tasks carried on-site [22]. On the contrary, these lighting devices should neither provide upwards lighting pollution, nor neighbours visual disturbance around the site. [11] [25] [26]. Both, criteria could be measured on site through proper electronic devices.

Finally, Acoustic comfort must be assessed through both, noise maps from reliable institutions that considers latest updates around-site and proper measuring devices to validate data from the acoustic map [11] [25] [22].

4.2 Urban mobility

Connected pedestrian [11] [25] [27]and Emissions-free vehicle network [25] [26] aim to assess the adequacy of the site to provide safe ways for alternative transport. The assessment would require the use of mobility maps of both on-site and the surrounding area.

The Public transport network can improve urban mobility through the presence of transport's lines to connect with the city, which is something considered in most of the USATs analysed in this study [11] [25] [26] [27] [17]. Additionally, these public transport lines could be connected to bicycle and/or pedestrian networks converting into multimodal networks [11] [12] [25], which introduce a higher benefit for user's mobility.

Finally, reduced parking facilities should be introduced, but only for institutional purposes, like police, firemen, or ambulance as well as handicapped private users [25].

4.3 Social Cohesion

Inclusive design aims the urban space to provide an integrative design to assure people of all ages, gender, and physical condition can use it [11] [25] [22]. It can be assessed through design plans and visual inspection on site according to best practices inclusive standards [39].

Commercial facilities around site enhance the usability, as well as provide a positive effect in other SOC and ECO criteria [25] [26] [27] [22]. It can be assessed with a district commerce on-site inspection, where variety is more important than repetition [25].

Green infrastructure is important to provide environments with visual and physical connections with nature [12] [25] [26] [27], where biophilic districts have shown a great potential on this concern. It can be assessed through ratios between green-soft versus grey-hard spaces [40], within the urban space or its surroundings. It can be used to comply with Recreation facilities criteria to provide more appealing urban spaces. Paved areas can be considered as recreation facilities if they allow physical activity and network facilities. Green infrastructure and Recreation facilities can both be assessed through as built plans and on-site verifications.

Crime prevention, which could be included in the Health and Safety group, aims to assure the site has been provided with security facilities: adequate lighting and cameras as well as vigilance plans [27]. It is, therefore, connected with lighting design criteria.

4.4 Health & Safety

Hazardous materials, Plants management control and Tobacco free space are willing to mitigate long term effects from the exposition to harmful chemicals on-site [12]. Hazardous materials can be assessed through on-sites report about cladding materials. Ecological gardening practices must follow ecological agriculture and can be assessed trough reviews of the gardening management plan [41]. Finally, Tobacco ban can be assessed through proper signage on-site inspection.

Water disposal facilities aim to enhance heath habits by providing free fresh water [42]. It should consider water quality, accessibility, proper sanitation plan, and infection control like Covid-19 and Legionella [43].

Maintenance protocol management aims to avoid environmental and human disturbance from noise and air pollution [11] [12] [22]. It can be properly assessed through the review of the maintenance plan, including gardening, cleaning, and occasional reparations.

Evacuation plan assesses the existence of a proper security plan in case of emergencies [27], with potential on-site risks and evacuation routes with safe meeting point.

In summary, several USATs have been used as sources to provide the most influencing SOC criteria to develop the UHU2SAT, where: Human comfort is more influenced by BREEAM-C, LEED-ND, and DGNB-UD, Urban mobility is more affected by BREEAM-C, SOC Cohesion is affected in the same way by BREEAM-C, LEED-ND, and CASBEE-UD., finally, Health & Safety group is widely affected by SITES.

5 Conclusion

The current state of the art about USAT has been proved insufficient to assess the holistic sustainability of urban spaces in use. This paper suggests new sets of SOC criteria classified into 4 groups, specifically useful to the develop a new in use holistic urban sustainable assessment tool UHU2SAT. Future research focused on ENV and ECO criteria should be necessary prior to the final development of the UHU2SAT.

References

1. L. Pérez-Lombard, J. Ortiz, and C. Pout, "A review on buildings energy consumption information," *Energy Build.*, vol. 40, no. 3, pp. 394–398, 2008.
2. I. Martínez-Zarzoso and A. Maruotti, "The impact of urbanization on CO2 emissions: Evidence from developing countries," *Ecol. Econ.*, vol. 70, no. 7, pp. 1344–1353, 2011.
3. UNECE, "SDG Priorities," 2019. [Online]. Available: https://www.unece.org/sustainable-development/sdg-priorities.html. [Accessed: 22-Oct-2019].
4. U. Berardi, "Clarifying the new interpretations of the concept of sustainable building," *Sustain. Cities Soc.*, vol. 8, pp. 72–78, 2013.
5. A. S. Cordero, S. G. Melgar, and J. M. A. Márquez, "Green building rating systems and the new framework level(s): A critical review of sustainability certification within Europe," *Energies*, vol. 13, no. 1, pp. 1–25, 2019.
6. D. T. Doan, A. A. Ghaffarianhoseini, N. Naismith, T. Zhang, A. A. Ghaffarianhoseini, and J. Tookey, "A critical comparison of green building rating systems," *Build. Environ.*, vol. 123, pp. 243–260, Oct. 2017.
7. B. R. Keeble, "The Brundtland Report: 'Our Common Future,'" *Med. War*, vol. 4, no. 1, pp. 17–25, 1988.
8. T. P. Obrecht, R. Kunic, S. Jordan, and M. Dovjak, "Comparison of Health and Well-Being Aspects in Building CertificationSchemes," *SUSTAINABILITY*, vol. 11, no. 9, May 2019.
9. S. Vierra, *Green Building Standards and certification systems*. Steven Winter Associates, 2014.
10. A. L. Chao, A. C. Gallego, V. Lopez-Chao, and A. Alvarellos, "Indicators framework for sustainable urban design," *Atmosphere (Basel).*, vol. 11, no. 11, pp. 1–18, 2020.
11. Building Research Establishment, "CEEQUAL Version 6 Technical Manual | International Projects SD6053:0.0." p. 197, 2019.
12. Sustinable Sites Initiative and Others, "SITES V2. Rating System for sustainable Land Design and Development," *Sustainable Site Initiative: Austin, TX, USA*. 2014.
13. S. Mahmoud, T. Zayed, and M. Fahmy, "Development of sustainability assessment tool for existing buildings," *Sustain. Cities Soc.*, vol. 44, pp. 99–119, Jan. 2019.
14. V. Penadés-Plà, T. García-Segura, J. V. Martí, and V. Yepes, "A review of multi-criteria decision-making methods applied to the sustainable bridge design," *Sustainability (Switzerland)*, vol. 8, no. 12. MDPI AG, 2016.
15. [15] development of sustainable assessment method for Saudi Arabia built environment: weighting system," *Sustain. Sci.*, vol. 10, no. 1, pp. 167–178, 2015.
16. H. Kang, Y. Lee, and S. Kim, "Sustainable building

assessment tool for project decision makers and its development process," *Environ. Impact Assess. Rev.*, vol. 58, pp. 34–47, 2016.

17. R. F. M. Ameen, M. Mourshed, and H. Li, "A critical review of environmental assessment tools for sustainable urban design," *Environ. Impact Assess. Rev.*, vol. 55, pp. 110–125, Nov. 2015.
18. K. Gompf, M. Traverso, and J. Hetterich, "Towards social life cycle assessment of mobility services: systematic literature review and the way forward," *Int. J. Life Cycle Assess.*, vol. 25, no. 10, pp. 1883–1909, Oct. 2020.
19. P. Medved, "A contribution to the structural model of autonomous sustainable neighbourhoods: New socio-economical basis for sustainable urban planning," *J. Clean. Prod.*, vol. 120, pp. 21–30, 2016.
20. L. Cities, U., & Governments, "Culture: fourth pillar of sustainable development.," *United Cities Local Gov. Policy Statement.*, 2010.
21. M. Holden, "Sustainability indicator systems within urban governance: Usability analysis of sustainability indicator systems as boundary objects," *Ecol. Indic.*, vol. 32, pp. 89–96, Sep. 2013.
22. Deutsche Gesellschaft für Nachhaltiges Bauen, "Overview of all criteria for quarters," 2020. [Online]. Available: https://www.dgnb-system.de/de/quartiere/kriterien/. [Accessed: 24-Nov-2020].
23. A. Merino-Saum, P. Halla, V. Superti, A. Boesch, and C. R. Binder, "Indicators for urban sustainability: Key lessons from a systematic analysis of 67 measurement initiatives," *Ecol. Indic.*, vol. 119, 2020.
24. A. Aghaei Chadegani *et al.*, "A comparison between two main academic literature collections: Web of science and scopus databases," *Asian Soc. Sci.*, vol. 9, no. 5, pp. 18–26, 2013.
25. Buiding Research Environmentl Assesment Method, "BREEAM Communities. Technical Manual SD202 - 0.1:2012." p. 175, 2012.
26. US Green Building Council, "LEED v 4 for Neigborhood development." 2018.
27. JSCB (Japan Sustainable Building Consortium) and Japan Sustainable Building Consortium, "CASBEE Certification System," 2019. [Online]. Available: http://www.ibec.or.jp/CASBEE/english/certification E.htm. [Accessed: 12-Sep-2019].
28. European Committee for Standardization, *EN 15643-1: 2010. Sustainability of construction works. Sustainability assessment of buildings. Part 1: General framework.* 2010.
29. European Committee for Standardization, *EN 15643-3: 2012. Sustainability of construction works. Assessment of buildings. Part 3: Framewoork for the assessment of social performance.* 2012.
30. M. Medineckiene, E. K. Zavadskas, F. Björk, and Z. Turskis, "Multi-criteria decision-making system for sustainable building assessment/certification," *Arch. Civ. Mech. Eng.*, vol. 15, no. 1, pp. 11–18, Jan. 2015.
31. I. Čadež and S. Hofmann, "Sustainability assessment scheme for transport infrastructure projects [Anforderungen an die Bewertung der Nachhaltigkeit von Straß enbauwerken]," *Bautechnik*, vol. 90, no. 10, pp. 609–613, 2013.
32. P. Wu, Y. Song, X. Hu, and X. Wang, "A preliminary investigation of the transition from green building to green community: Insights from LEED ND," *Sustain.*, vol. 10, no. 6, 2018.
33. D. Dizdaroglu and T. Yigitcanlar, "A parcel-scale assessment tool to measure sustainability through urban ecosystem components: The MUSIX model," *Ecol. Indic.*, vol. 41, pp. 115–130, Jun. 2014.
34. Y. Xu, R. Keivani, and A. J. Cao, "Urban sustainability indicators re-visited: lessons from property-led urban development in China," *Impact Assess. Proj. Apprais.*, vol. 36, no. 4, pp. 308–322, 2018.
35. W. W. Kropp and J. K. Lein, "Scenario Analysis for Urban Sustainability Assessment: A Spatial Multicriteria Decision-Analysis Approach," *Environ. Pract.*, vol. 15, no. 2, pp. 133–146, Jun. 2013.
36. C. Li and J. Li, "Assessing urban sustainability using a multi-scale, theme-based indicator framework: A case study of the Yangtze River Delta region, China," *Sustain.*, vol. 9, no. 11, 2017.
37. J. F. Nicol and M. A. Humphreys, "Adaptive thermal comfort and sustainable thermal standards for buildings," in *Energy and Buildings*, 2002, vol. 34, no. 6.
38. C.F.R.E.S., "RUROS home page. Rediscovering the Urban Realm and Open Spaces." [Online]. Available: http://alpha.cres.gr/ruros/. [Accessed: 02-Dec-2020].
39. K. C. Wu and L. Y. Song, "A case for inclusive design: Analyzing the needs of those who frequent Taiwan's urban parks," *Appl. Ergon.*, vol. 58, pp. 254–264, 2017.
40. T. Beatley and P. Newman, "Biophilic cities are sustainable, resilient cities," *Sustain.*, vol. 5, no. 8, pp. 3328–3345, 2013.
41. H. Eckert, G. Breitschuh, and D. R. Sauerbeck, "Criteria and standards for sustainable agriculture," *J. Plant Nutr. Soil Sci.*, vol. 163, no. 4, pp. 337–351, 2000.
42. S. Naji and J. Gwilliam, "Neighborhood sustainability assessment tools and water system adaptation: A framework to analyse the adaptive capacity in the physical-social context," *Int. J. Sustain. Dev. Plan.*, vol. 11, no. 6, pp. 907–916, 2016.
43. International WELL Building Institute (IWBI), "The WELL Community Standard." p. 148, 2017.

E3S Web of Conferences **237**, 04032 (2021)
ARFEE 2020 https://doi.org/10.1051/e3sconf/202123704032

Research on outdoor environment renewal for the aged in residential area——A case study of a residential area in Jinan

Yan Wang[1], Xiao Wang[1*]

[1]School of Civil Engineering and Architecture,University of Jinan,Jian 250022,China

Abstract: The model of home-based care in residential areas has gradually become an important way of providing for the aged in the age of aging in China, and more attention has been paid to the aging and renewal of existing residential areas.Taking the outdoor environment of residential areas as the research object, this paper analyses the environmental status quo of a residential area in Jinan through field investigation, and puts forward suggestions for aging renovation from the aspects of outdoor landscape system, road traffic system and supporting facilities of residential areas, so as to provide references for residential renovation.

1 The Introduction

According to statistics, by the end of 2017, the number of people aged 60 or above in China had reached 240 million, accounting for 17.3% of the total population.[1]The characteristics of "getting old before getting rich" are remarkable[2], and the elderly tend to "acquaintance society"[3].With the increasingly severe aging, China has put forward the "9073" pension model in combination with the actual conditions, which mentioned that 90% of the elderly are supported by social services through home-based nursing service.It can be seen that aging in place model[4] in residential areas will become the mainstream way for urban residents to provide for the aged.However, the current situation of existing residential areas in China is worrying. Due to the lack of timely management and maintenance, serious aging of buildings, serious lack of infrastructure in residential areas, and deterioration of outdoor public environment, it is difficult to provide the conditions required for home-based care model.Based on this, this paper attempts to conduct field research on the outdoor environment of Qinglong Backstreet community, analyze and summarize its current problems, and put forward appropriate suggestions for the renewal of aging, so as to provide some references for the renewal of residential environment of China's home-based care model.

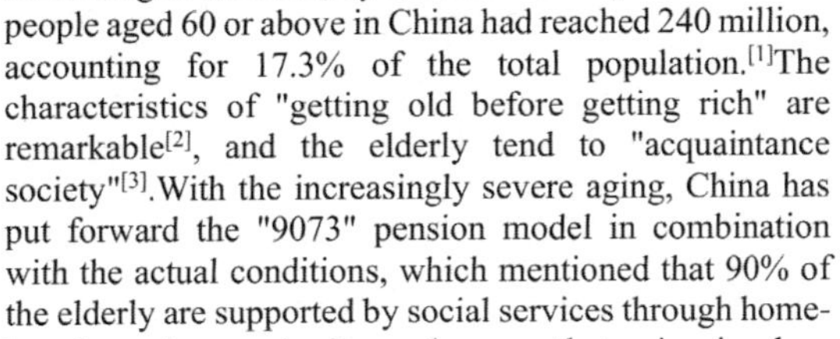

Fig. 1.Location map of Qinglong Community and the current situation of the environment outside the community.（Picture source, drawn by the author）

2 Investigation on the current situation of outdoor environment in residential areas

Taking Jinan city as the research scope, Qinglong Backstreet District, an existing residential area with a prominent aging problem, was selected as the research object.The research methods include field survey, questionnaire survey and online survey. The research content includes outdoor landscape system, road traffic system and supporting facilities of existing residential areas.

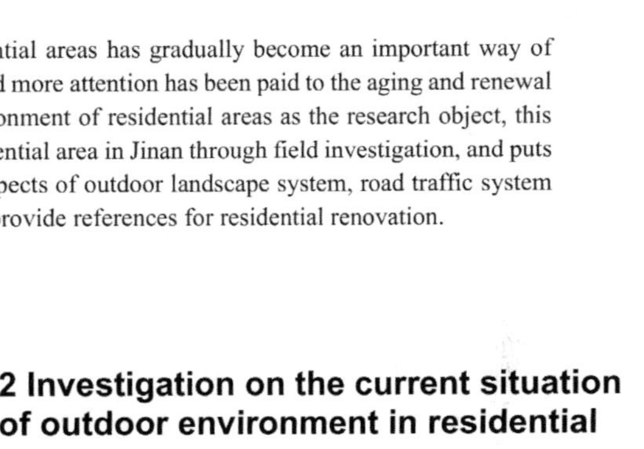

Fig. 2. The current situation of the community environment. (photo sources, taken by the author and Baidu Map)

*xiaowang2010@126.com

2.1 Residential Situation

Fig. 3. House analysis diagram. （Picture source, drawn by the author）

The residential buildings in The Qinglong Back Street community of Jinan were built in batches from the 1980s to the early 21st century, which was a relatively early residential community in Jinan.This community has a favorable geographical location. It is adjacent to the Peace Campus of Shandong Jianzhu University in the south, the central campus of Shandong University in the northeast, Daming Lake Scenic area in the west, Jinan central Hospital in the east, and several shopping centers. There are many bus stops and convenient transportation.The whole residential area is an open determinant layout, with traffic for people and cars.There are many original residents in the residential area, and the proportion of middle-aged and elderly residents is large.

2.2 Tight road and parking area

Fig. 4. Road analysis diagram. （Picture source, drawn by the author）

Due to the increase in the number of vehicles and the number of shopping malls nearby, especially the rush hour, the ground parking in the community is relatively tight, and the regional traffic pressure is great.The community is close to the Daming Lake scenic area. On weekends and holidays, there are a lot of external vehicles to visit the scenic area. The traffic pressure around the community is heavy, causing congestion and making it difficult for residents to travel.

The roads inside the community are circular, with many end-end roads and low land utilization, which affects the traffic convenience.The lack of activity space in the open community and the random parking of vehicles from the front to the back of the building to the road have a great impact on residents' travel, and the elderly are not safe to travel.People and cars are mixed on the road, and barrier-free road debris is piled everywhere, which is difficult to meet the travel needs of the elderly.It is easy to cause congestion because of the non-motor vehicle sheds built in disorder.Poor lighting of street lamps, serious damage to the road surface at the well cover of engineering pipelines, uneven road surface, not only affect the appearance of the community, but also difficult to ensure the safety of the elderly travel.

2.3 Lack of landscape and activity space

The squares are mainly scattered in the open space between residential buildings, with low and uneven coverage, low service level and simplistic functions.Greening is arranged at the edge of the square to define the boundary of the square. Due to the lack of regular management, the green land is piled up by residents with garbage and sundry things, and the residential area presents an aging and disordered state.
There is no clear landscape center, and the residential identification is not strong, which affects the elderly's environmental quality of natural life, psychological security, happiness, and serious shortage of public green space and activity space.The residential environment is short of pavilions, seats and other facilities for residents to gather and rest. The pipeline laying conflicts with the greening land, which affects the viewing degree of the greening and has potential safety hazards.

Fig. 5. Landscape analysis diagram. （Picture source, drawn by the author）

2.4 Aging of buildings and supporting facilities

The residential area is mainly composed of 6 floors, brick and concrete structure, red brick exterior wall, and most residential buildings are externally laid with insulation layer.The space between buildings is 10 meters, the walls are dirty, the pipeline facilities are exposed, and the local wall tiles are missing.

There is a lack of barrier-free facilities in the residential area, lack of management of entrances and exits, serious sanitation problems, garbage cans are left in disorder, unclear door signs and lack of identification.The public service facilities are mainly commercial and social service facilities, while the elderly service facilities and social welfare facilities are insufficient, which need to be solved and optimized urgently.

3 Outdoor environment suitable for aging and renewal strategy

Based on the three needs of the elderly: psychological needs, physiological needs, and behavioral activity needs, the existing strategies of community age-appropriate external environment were put forward -- material age-appropriate construction, spiritual age-appropriate construction.The Five principles of old-age care developed in Sweden - generalization, day-by-day, treating people as a whole, self-determination, influence

https://doi.org/10.1051/e3sconf/202123704032

and participation, and appropriately managed actions.From the three aspects of physical weakness, should increase the rest facilities, reduce the vertical traffic frequency, from the hardware facilities and software supporting comprehensive consideration, easy to identify, easy to reach, easy to communicate, barrier-free, safety and meet the physiological and psychological needs of the elderly.

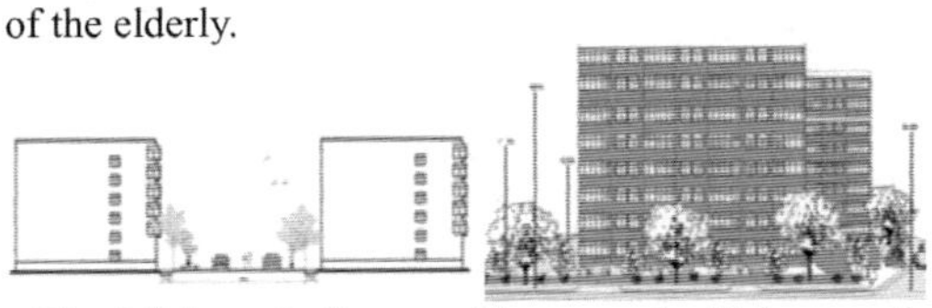

Fig. 6. Schematic diagram of human and vehicle separation.
Fig. 7..Schematic diagram of human and vehicle separation. (Picture source, drawn by the author)

3.1 Updating roads and parking Spaces

To strengthen the existing phenomenon of disorderly parking treatment, the demolition of the indiscriminate construction of car sheds, residential groups centralized layout of non-motor vehicle parking Spaces.The separation of people and vehicles is adopted to reduce end-end roads, improve land utilization and travel convenience, and meet the emergency response of ambulances and fire trucks.Clear parking space to make the road as smooth and straight as possible.Adding vertical parking spaces to solve the problem of "less land and more cars" and control the inflow of external vehicles, and relieve the traffic pressure inside and outside the community.The road is repaired to make it smooth, and the ground is paved with non-slip materials to meet the behavioral needs of the elderly.

3.2 Enrich the greening and outdoor activity space

Outdoor activities of the elderly can be divided into three aspects: exercise, rest and special activities.The required functional space can be added around the original square or the reused abandoned building (boiler, workshop).

For the reuse of open Spaces of different areas and terrains, different levels of "pocket parks" are arranged accordingly to improve the coverage rate of greening and square and expand the service scope."Pocket Park" sets up activity venues with different private densities: semi-private, semi-public and public places.The environmental design experiences the regional cultural characteristics, retains the original elements that reflect the regional characteristics, keeps the familiarity of the living community, gives the elderly a sense of belonging, and creates a "natural retirement community" [5]to increase the variety of green plants satisfying the northern climate, improve the sense of landscape design, and enrich the paving way.Adopt bright and concise patterns, colors and sounds, lighting to shape landscape nodes, entrances and exits, etc. for easy identification design.Lack of pavilions, seats and other facilities, piping and greening land conflict.Expand the general scale of the site, take into account the use of other groups, improve the utilization of land.Barrier-free treatment should be carried out in the main areas of elderly activities.The steps are equipped with a single handrail, which focuses on the junction between the platform and the ramp to meet the turning radius of the wheelchair.

3.3 Upgrade supporting facilities for aging

Fig.8. Schematic diagram of landscape renovation（Picture source, drawn by the author）

Improve the property management service system, focusing on three aspects: garbage collection, aging of water and electricity pipelines, and access control for internal and external personnel.In order to facilitate the elderly upstairs and downstairs, according to its existing structure, can consider the external elevator frame installation.Replace the door and window components, reduce the heat conduction speed in winter, improve the capacity of heat preservation, heat insulation and sound insulation.The supporting endowment system provides exchange activities such as shared garden and kitchen according to the level configuration of residential area, residential area and residential group.Housing for elderly care, living services, health care and rehabilitation, and recreational facilities should be added to promote the participation of the elderly, so as to achieve the existing community state of sharing, co-construction and symbiosis.

4 Conclusion

This paper takes the present situation of existing communities in Jinan city as the object of study and focuses on the design of age-appropriate transformation of the external environment of public space.This paper deals with the lack of senility in square, road and afforestation, and puts forward the strategy of senility improvement in outer space environment from the aspects of psychology, physiology and behavior of the elderly.The content of the age-appropriate transformation of the outer space environment of the regional difference and the urban-rural difference between the old and the new needs further study.Making use of the original community to reform for the aged is not only beneficial to the vulnerable groups, improving the living environment, but also can relieve the pressure of China's old-age care problems.

Acknowledgement

This research was financially supported by Shandong University Students Innovation and Entrepreneurship training platform(S201910427013X).

References

1. Source: National Bureau of Statistics, PRC.
2. Lu Fei, Luo Jingjing , Kang Wen.Research on the Strategy of Aging Improvement based on universal Design of Cold Region Housing Environment -- A case study of Lilac Home in Harbin.
3. Jingjing.Children's Wonderlan--Stockholm Experience [M] Shanghai: Shanghai Far East Publishing House, 2016.
4. BEDNEY B J,GOLDBERG R B,JOSEPHSON K.Aging in place in naturally occurring retirement communities:transforming aging through supportive service programs[J].Journal of Hosing for the Elderly,2010,24(3-4):304-321.
5. HUNT M E,GUNTER-HUNT G.Naturally occurring retirement communities[J].Journal of Housing for the Elderly,1986,3(3-4):3-2.

Comparison of Vertical Breast Displacement during Rope Skipping, Walking, Jogging and Running

X Sheng[1], X Chen[1,2,3*], Y Li[1,2,3], G Sun[1,2,3], H Xie[1,2,3]

[1]School of Textiles and Fashion, Shanghai University of Engineering Science, Shanghai, 201620, P.R. China
[2]Textile Industrial Key Laboratory of Ergonomics and Functional Clothing, Shanghai University of Engineering Science, Shanghai, 201620, P.R. China
[3]Sino-British Joint Lab for Smart Sportswear, Shanghai University of Engineering Science, P.R. China

Abstract. In order to investigate the effect of exercise modality and bra type on vertical breast displacement, seven participants with similar ages and body types were selected to take part in this study. Three-dimensional motion capture system (NDI Optotrak Certus) was used to collect the data of vertical breast displacement when participants were walking at the speed of 5 km/h, jogging at the speed of 7.5 km/h, running at the speed of 10 km/h on a treadmill, and rope skipping at the frequency of twice per second. It was found that there was no significant difference in vertical breast displacement between rope skipping, jogging and running. And the vertical breast displacement when wearing sports bra is significantly less than everyday bra in four exercise modalities. The results of this study might be useful for designing special sports bra for rope skipping modality.

1 Introduction

As a form of moderate-to-high intensity aerobic exercise,[1, 2] rope skipping is an effective way to lose weight, which is accepted and used by the majority of people.[3-5] Besides, previous researches showed that rope skipping exercise made a positive effect on vascular compliance, cardiopulmonary function, oxidative stress and antioxidant capacity.[5] Meanwhile, because of its low cost, rope skipping could be promoted to people from lower socioeconomic classes. Also, rope skipping is an activity which does not require a lot of space, hence is suitable for densely populated environments like school.[1] However, rope skipping includes a jumping component, which may cause excessive breast displacement. It is acknowledged that excessive breast displacement is related to breast discomfort even pain, which troubles many women, especially women with large breasts.[6, 7] Fortunately, breast displacement could be reduced by wearing sports bras.[8-10] Previous researches have investigated breast biomechanics during a range of exercise modalities, including treadmill walking[11], running[12] and swimming,[13] and found that increases in breast support (wearing bra) caused decreases in breast displacement.[13] But up to now, much less exploration on the vertical breast displacement during rope skipping was found.

Therefore, the first aim of this study was to assess vertical breast displacement during rope skipping, walking, jogging and running. The second aim of this study was to evaluate the effect of breast support on vertical breast displacement. Firstly, it was hypothesized that there would be significant differences in vertical breast displacement during rope skipping, walking, jogging and running. Secondly, it was hypothesized that breast motion would be significantly reduced as breast support increased from everyday bra to sports bra.

2 Methods

2.1 Participants

Seven healthy active women (mean ± SD (Standard Deviation): age = 20 ± 1 yr; breast circumference = 82 ± 8 cm; upper-breast circumference = 86 ± 3 cm; under-breast circumference = 73 ± 2 cm) who had no history of pregnancy, breast cancer or surgery were recruited for this study. According to the professional bra fitting criteria, each participant was fitted by an experienced bra fit assessor.[14] Four of participants were cup A (breast circumference minus under-breast circumference was 12-14cm), four of them were cup B (breast circumference minus under-breast circumference was 14-16cm), two of them were cup C (breast circumference minus under-breast circumference was 16-18 cm), and one of them was cup D (breast circumference minus under-breast circumference was 18-20 cm).

2.2 Experimental bras

Everyday bra and sports bra used in the experiment were both common bras. The experimental sports bra used in this study was one of the basic styles of sports bra (NIKE^{TM}, China) (Figure 1), which was designed for

* Corresponding author: chenxn@sues.edu.cn

medium intensity sports. Two different bra sizes (M and L) of this style were used. The fabric of the experimental sports bra is made of 88% recycled polyester and 12% spandex blended. The mesh fabric is made of 80% polyester and 20% spandex. Both the front side and back side of the liner are made of 100% polyester. Everyday bra used in this study was a common style (COSMO LADY, China), seen in Figure 2. The cup lining was made of 65% polyester and 35% cotton. The side lining was made of 84% polyamide and 16% spandex.

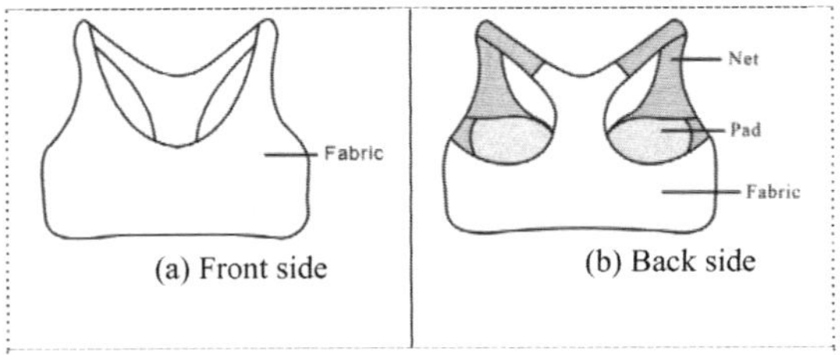

Figure 1. Structure of sports bra, the front side (a) and the back side (b).

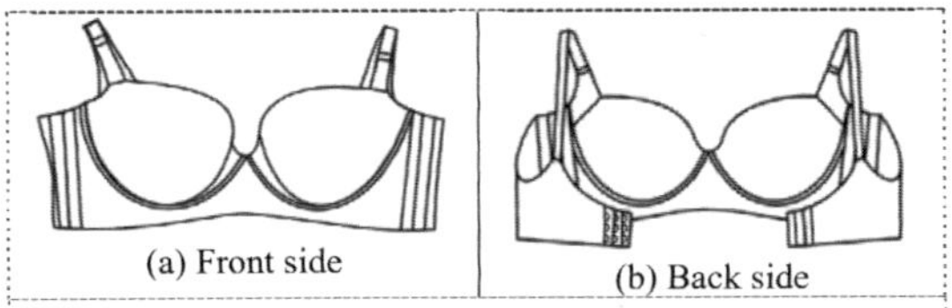

Figure 2. Structure of everyday bra, the front side (a) and the back side (b).

2.3 Measurement process

This experiment was conducted in the artificial climate chamber of the human motion and ergonomics laboratory of the institute. Three-dimensional motion capture systems (NDI Optotrak Certus) with three infrared cameras were employed to collect the motion of the markers pasted on body and bra. The whole testing process included the following steps: warm-up, markers pasting and data collection. The laboratory coordinate system was established with the X-axis positive rearward, Y-axis positive to the right, and Z-axis positive upward.

2.3.1 Warm-up. In order to prevent experiment accidents, all participants were provided with familiarization trials on the laboratory treadmill. During the familiarization period, they were allowed sufficient practice at a self-selected gait speed.

2.3.2 Markers Pasting. After warming up, participants were asked to put on the experimental bra. To quantify breast motion relative to trunk motion, one marker presenting the trunk movement was fixed on the suprasternal notch, another marker was positioned at left breast nipple to capture breast motion.

2.3.3 Data Collection. All data were collected for each participant, with participants required to complete four exercise trials in each bra condition. First, participants walked at the speed of 5 km/h on the treadmill in the everyday bra condition. Data was collected 30 s after the start of exercise to ensure steady-state breast motion had been achieved. The data collection duration was 60 s. Then, the same procedures were used in collecting breast motion data during treadmill jogging (7.5 km/h), treadmill running (10 km/h) and rope skipping (twice per second). Participants were given a sufficient rest period between trials. After all data collection in the everyday bra condition was completed, data in the sports bra condition was collected in the same procedures.

2.4 Statistical analysis

Assuming that the breast is symmetrical, the left breast was selected as the research object. That is, the data used in this study were all left breast data. Vertical breast displacement was the difference between the vertical coordinates of left breast nipple and suprasternal notch. The mean and SD values were calculated for each of the four exercise modalities for the average vertical breast displacement in the two bra conditions (everyday bra and sports bra).

In this study, all data of vertical breast displacement were checked for normality using a Kolmogorov-Smirnov test. A two-way ANOVA design with two within factors (exercise modality and bra type) was then conducted to reveal whether there was a significant effect of exercise modality and bra type on vertical breast displacement or interactions of exercise modality (walking, jogging, running, rope skipping) or bra type (everyday bra, sports bra). Bonferroni post hoc analyses were used to find where any significant difference exists. Statistical analyses were conducted in the Statistical Package for the Social Sciences (Version 25.0; SPSS Inc., Chicago, IL). The significance level was set at 0.05.

3 Results

3.1 Effect of exercise modality on vertical breast displacement

Mean and SD of vertical breast displacement was calculated for each of four exercise modalities in the two bra conditions (Figure 3). During the trials, vertical breast displacement ranged from 3.957 mm to 13.360 mm whilst walking at the speed of 5 km/h, and ranged from 6.579 mm to 35.143 mm whilst jogging at a speed of 7.5 km/h. It was showed in Figure 3 that there was a dramatically increase of vertical breast displacement from walking to jogging. However, only slight changes were found between jogging, running and rope skipping. Meanwhile, the results of two-way ANOVA revealed that there was a significant effect of exercise modality on vertical breast displacement ($p < 0.05$). The statistic results showed that there was no significant exercise modality-bra type interaction ($p = 0.403$). Bonferroni post-hoc analyses determined that there were significant differences in vertical breast displacement between walking and running, walking and jogging, walking and rope skipping ($p < 0.05$).

Nevertheless, there was no significant difference in vertical breast displacement between running, jogging and rope skipping.

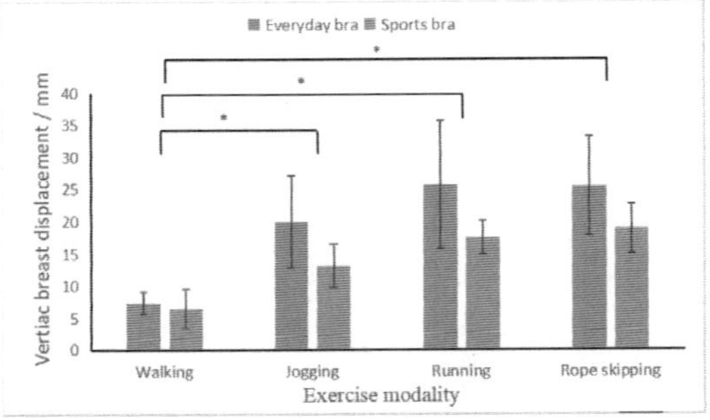

Figure 3. Mean (± SD) vertical breast displacement at four exercise modalities in each breast support condition (n=7). *Significant difference in vertical breast displacement during exercise modalities ($p < 0.05$).

3.2 Effect of bra type on vertical breast displacement

During the trials, vertical breast displacement ranged from 7.359 mm to 25.702 mm whilst the participants wearing everyday bra, and ranged from 6.490 mm to 18.832 mm whilst the participants wearing sports bra. The results of two-way ANOVA revealed that there was a significant effect of bra type on vertical breast displacement under each exercise modality ($p < 0.05$), whereby the vertical breast displacement in the everyday bra condition was significantly greater than when wearing the sports bra.

4 Discussion

This study investigated the effect of exercise modality and bra type on vertical breast displacement. It was found that there were significant differences between walking and running, walking and jogging, walking and rope skipping. However, no significant difference was found between rope skipping and jogging, rope skipping and running. It was also found that there was a significant reduction on vertical breast displacement whilst wearing a sports bra, compared to wearing an everyday bra. The details of these findings are as below.

4.1 Effect of exercise modality on vertical breast displacement

The first hypothesis could be partially accepted as the results of this study showed that vertical breast displacement differed significantly between walking and other three exercise modalities, confirming findings by Scurr et al.'s and Mason et al.'s research.[15-17] However, no significant difference was found in vertical breast displacement between rope skipping, jogging and running. These findings might be explained by the different intensity of these four modalities. Walking is a low-intensity exercise while running, jogging and rope skipping are all high-intensity exercises. The trunk movements of participants in running, jogging and rope skipping are similar in the vertical direction, with obvious up and down trunk movement. And it was speculated that such trunk movements would drive the breasts to do similar up and down movements. The findings suggested that breast motion research that analyzes treadmill jogging and running may be applicable to rope skipping, which might help to inform the requirements of breast support garments as a means to reduce breast pain for rope skipping.

4.2. Effect of bra type on vertical breast displacement

The results of this study confirmed that the vertical breast displacement in the sports bra condition was significantly reduced than when wearing the everyday bra, whatever the exercises modalities. Namely, sports bra is more effective in reducing vertical breast displacement than everyday bra. This corresponded with previous research[18, 19] and supported the second hypothesis of this study. According to Lu et al.'s research, the superiority of sports bra could be attributed to the tension caused by stretched bra fabrics and the compressed cup pad.[10] During exercise, the breast kinetic energy converts to the tensile energy of the fabric and pad material, and the compression energy of the pad material.[10] This may explain how vertical breast displacement was reduced in the sports bra condition. The findings of this study also supported the notion that women should be encouraged to wear sports bra during exercise to reduce the breast displacement and relieve the discomfort caused by excessive displacement.[7, 17, 20]

4.3. Strengths and limitations

A limitation of this study was that we only compared the vertical displacement between different exercise modalities. Previous studies had established that velocity and acceleration of movement during exercise and ground reaction forces that result from different breast support conditions were also factors affecting breast discomfort.[21-23] Therefore, future research might look to the comparisons of these three factors between rope skipping and other exercise modalities in different breast support conditions. There was also a limitation in the data processing. The marker representing breast motion was placed on the surface of bra instead of on the breast. The results might be biased because of possible relative displacement between bra and breast caused by rubbing during exercise. Furthermore, it was also acknowledged that the present study was limited in the small number of participants. Consequently, the results require further verification with more participants in the future.

Despite these limitations, to our knowledge, this study is the first to investigate the vertical breast displacement during rope skipping, and to compare the results during rope skipping with that during walking, jogging and running. Therefore, this information is important for designing special sports bras for rope skipping, which is expected to reduce excessive breast displacement and bra discomfort.

5 Conclusion

This study found that there was no significant difference in vertical breast displacement between rope skipping and jogging, and between rope skipping and running. Therefore, findings in previous literature in breast motion during treadmill jogging (7.5 km/h) or running (10 km/h) could be applicable to rope skipping. Future research should consider comparing the velocity and acceleration of movement and ground reaction forces exerted on breast between rope skipping and other exercise modalities in different breast support conditions.

Acknowledgement

This study was funded by the National Natural Science Foundation of China (Grant No.11802171), the open fund of Key Laboratory of Clothing Design and Technology (KLCDT2020-01), Teacher Training Programs on Human Factor and Ergonomics (201901024017) and National Key Research and Development Program of China (2018YFC2000905).

References

1. Ha A S, Lonsdale C, Ng J Y Y and Lubans D R 2014 *BMC Public Health* **14** 535
2. Ridley K and Olds T S 2008 *Medicine and science in sports and exercise* **40** 1439-46
3. Kang J, Zhou Y and He X 2008 China Sport Science and Technology 99-102
4. Hao M, Han W and Yamauchi T 2019 *Asia-Pacific journal of public health* **31** 348-58
5. Zhang Y 2014 The effects of rope skipping on cardiopulmonary function in adolescents: an experimental study (Chengdu: Chengdu sport University)
6. Jie Z, Winnie Y and Ng S P 2011 *Textile Research Journal* **81** 1234-48
7. Lorentzen, Deana, Lawson and LaJean 1987 *Physician & Sportsmedicine*
8. Steve H and Joanna Scurr 2011 *Sports Engineering* **14**
9. Maha S A A H M D and Frcsi 2000 *The Breast Journal* **6**
10. Lu M, Qiu J, Wang G and Dai X 2016 *Materials Today Communications* **6** 28-36
11. McGhee D E, Steele J R, Zealey W J and Takacs G J 2013 *Appl Ergon* **44** 112-8
12. Steve H, Alex M and Joanna Scurr 2013 *Proceedings of the Institution of Mechanical Engineers, Part P: Journal of Sports Engineering and Technology* **227**
13. Mills, Lomax, Ayres and Scurr 2015 *Journal of Sports Sciences* **33**
14. McGhee D and J. Steele 2010 *Journal of Science and Medicine in Sport* **13**
15. Scurr J C, Hedger J L W and Wendy 2011 *Journal of Sports ences* **29** 55-61
16. Scurr J, White J and Hedger W 2009 *Journal of Applied Biomechanics* **25** 322-9
17. Mason B R, Page K A and Fallon K 1999 *Journal of Science & Medicine in Sport* **2** 134-44
18. Zhou J, Yu W, Ng Z and Hale J 2009 *Journal of Fiber Bioengineering and Informatics* **2** 108-13
19. Bowles K A, Steele J R and Munro B 2008 *Br J Sports Med* **42** 670-3
20. White J, Scurr J and Hedger W 2011 *J Appl Biomech* **27** 47-53
21. White J L, Scurr J C and Smith N A 2009 *Ergonomics* **52** 492-8
22. Deirdre E M 2007 *British journal of sports medicine* **12**
23. Scurr J C, Hedger J L W and Wendy 2010 *Journal of Sports ences* **28** 1103-9

Research on Landscape Architecture Design Based on Corporate Social Responsibilities

Xinying Zhang[1,*], Wenjie Chen[2]

[1]College of Landscape Architecture, Sichuan Agricultural University, Chengdu 611130, P.R. China
[2]School of Economics and Management, Chongqing University of Posts and Telecommunications, Chongqing 400065, P.R. China

Abstract. As the natural environment becomes growingly deteriorated, environmental protection becomes a global common concern. Landscape architecture (LA) is concerned about the relationship among humans, the built, and natural environments, so it is of special significance to study corporate social responsibilities (CSR) of LA firms. This paper studied CSR of LA firms from the perspective of landscape architects, i.e., this paper explored LA design based on corporate social responsibilities. CSR of LA firms was classified into economic responsibilities and green responsibilities, so this paper studied LA design based on economic responsibilities and green responsibilities respectively. This paper contributes to the existing study of LA design by innovatively adding corporate social responsibilities into the LA design philosophies.

1 Introduction

With the rapid economic development, eco-environmental problems such as resource shortage, global warming, urban heat island effect, and ecological pollution occur frequently. Under this background, environmental protection becomes a global common concern, and a good ecological environment becomes the basic conditions and common demands for the economic and social development of all countries, so green development and protection of the eco-environment become the common objective for all countries. Landscape architecture (hereinafter abbreviated as LA) deals with the relationship among humans, the built and natural environments, so it plays an especially important role in protecting the eco-environment and facilitating green development in ways such as improving human living spaces, saving energies, enriching biodiversity, and preserving historic and cultural relics. Due to the inherently distinctive nature of LA, LA firms shoulder greater social responsibilities than any other firms to protect the ecological environment. As employees, landscape architects are assumed to take active initiative to help the firm they serve to perform its social responsibilities through the LA design.

2 Literature review

The concept of green responsibility evolves from corporate social responsibility. Since the concept of corporate social responsibility (hereinafter abbreviated as CSR) was first proposed by Oliver Sheldon in 1923, numerous scholars have done extensive research on CSR. Different scholars have made different descriptions of CSR from different perspectives. Carroll (1991) proposed a classic four-level pyramid model of CSR, as shown in Fig.1, indicating that economic responsibilities at the bottom of the pyramid are the required basic social responsibility for all firms.

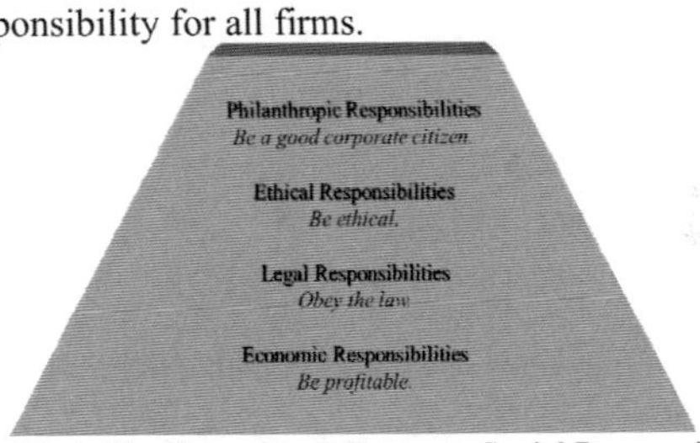

Fig.1. The Pyramid of Corporate Social Responsibility

Source: A. B. Carroll. The Pyramid of Corporate Social Responsibility: Toward the Moral Management of Organizational Stakeholders [J]. Business Horizons, **July/August**: 39-48 (1991).

CSR means that a firm should not only act in the best interests of its shareholders but also be responsible for all of its stakeholders, including employees, society, and environment. Accordingly, the concept of green CSR or green responsibility was proposed. A firm's green responsibility or green CSR refers to the firms' social responsibility in environmental protection and resource utilization. Specifically speaking, green responsibility involves that a firm should strive to avoid resource waste and minimize its destruction to the environment during its operation process, i.e., be responsible for the eco-environmental protection. To maintain harmonious co-existence among humans, nature, and society and facilitate the sustainable development of humans, nature,

* Corresponding author: la2017zxy@163.com

and society, firms are assumed to undertake green responsibilities. Green responsibilities are one domain of CSR, layered upon economic responsibilities.

Only two papers in Chinese are available on http://www.cnki.net that are concerned about the corporate social responsibilities of LA firms. Zhang et al. (2010) started with the corporate social responsibility (CSR), described its basic concept that contains various corporate activities, and discussed the way Japanese LA firms perform their corporate social responsibilities with three related examples of Japanese landscape. Wu (2009) suggested that LA firms should undertake the social responsibility of being a green firm and argued that the social responsibility of LA firms include 6 dimensions: commercial civilization, science and technology innovation, corporate culture, green practice, client support, and co-construction with other stakeholders.

LA firms plays a particularly important role in reducing resource waste and minimizing destruction to the environment, so it is of special significance to study CSR of LA firms. Without a doubt, LA firms should be environmentally responsible. As employees of a LA firm, landscape architects should also be environmentally responsible through their LA design works. Thus, this paper attempts to analyse CSR of LA firms from the perspective of landscape architects. In other words, this paper intends to explore LA design based on corporate social responsibilities.

3 LA design based on the economic responsibilities

A firm's primary responsibility is to make profits, i.e., the economic responsibility is the basic responsibility upon which all other responsibilities are based (Carroll,1991). Friedman (1970) also maintained that the social responsibility of business is to increase its profits. No firm can survive or develop on the market without profits over the time. Thus, landscape architects should endeavor to meet the economic responsibilities of both the firm and the society.

3.1 The economic responsibilities of LA design for the firm

To help the firm they serve to perform its economic responsibilities, i.e., maximize its profits, landscape architects are supposed to think strategically and act economically. They should make efforts to help the firm to make profits through their design works, i.e., to produce LA design products which enable the firm to obtain greater corporate performance with less costs. Put it another way, LA design products are supposed to serve the competitive strategy the firm has formulated. As for the firm adopting a cost leadership strategy, landscape architects are assumed to produce competitive LA products with relatively low costs, e.g., by adopting less complicated design processes or selecting low-cost building materials. In terms of the firm with a differentiation strategy, landscape architects should make efforts to produce distinctive LA products with various meaningful and aesthetic functions, say, by using unique design processes and selecting diversified building materials. In a word, it is one of the social responsibilities of landscape architects to help the firm in which they work to meet its economic responsibilities.

3.2 The economic responsibilities of LA design for the society

As environmentally responsible employees, landscape architects are also assumed to shoulder the social responsibility for the society. They should try to produce LA design products capable of generating social benefits and improving social well-being. Landscape architects are supposed to help to relieve the tensions between humans and the natural environments with their design works. For example, landscape architects may in some way ease the contradictions between LA and human demands for water by designing water-saving ecological parks. Besides, landscape architects should be conscious of designing various works that can protect and improve the urban environments which in turn facilitate the economic development of the surrounding areas. Put it another way, landscape architects are assumed to shoulder the responsibility of boosting the economic and social development or improve the total social well-being of the local area with their design works.

In the past decades, many local Chinese governments have paid increased attentions to LA design with an attempt to attract tourists and thus foster the development of the local tourism industry. Numerous successful cases of LA design are available throughout China, among which Chongqing may probably be recognized as the most typical one. Since 1990s, the Chongqing government have invested much in LA design according to the distinctive topographical and geomorphological conditions featured by a combination of rivers and hills with various altitudes. Chongqing is currently well-known for its unique nightscape lighting due to the local government's continuous efforts in LA design in the past years. Millions of tourists rush to Chongqing from different parts of China and even the world for sightseeing, bringing significant economic benefits to Chongqing. Landscape architects have undoubtedly made a great contribution to the economic and social development of Chongqing.

4 LA design based on the green responsibilities

Economic responsibilities are the basic responsibility for all firms. As for LA firms, the most important responsibility is the green responsibilities. Thus, the following sections focuses on the way LA firms perform their green responsibilities from the perspective of landscape architects. The work of landscape architects is always involved with the three types of relationship: relationship between humans and the built environment, relationship between humans and the natural environments, and relationship between the built and natural environments. Thus, the following sections attempt to explore the way in which landscape architects deal with the three types of relationship based on the green responsibilities.

4.1 Trying to achieve an effective unification of meaningful and aesthetic LA with human demands for better living spaces

LA design based on environmental responsibility focuses on the sustainable development of the ecological environment, but it never overlooks the social nature and aesthetic characteristics of LA. To coordinate the relationship between humans and the built environment, landscape architects should endeavor to design LA that is both meaningful and aesthetic so as to provide pleasing relaxation, leisure, and recreation places for humans. For example, when landscape architects are designing water landscape like fountains, they should first consider if the water landscape can bring an extraordinary visual effect for viewers, and meanwhile they should also value if the water landscape can be reached by viewers at short range. For another example, when landscape architects are designing park roads, they should consider both the roads' function to evacuate sight viewers and the sight viewers' aesthetic demands. To do so, landscape architects may design an over 100 meters long road covered with cobblestone that is wide enough to allow at least two pedestrians to walk simultaneously in parallel. In this way, sight viewers can experience a soothing feeling in their feet soles touching cobblestone and meanwhile watch the enjoyable green scenery along the road (Hu, 2020).

4.2 Trying to realize an organic combination of human demands for a better living environment with respect for the natural environment

With the economic and social development, people have greater demands for a better life quality than ever before, including a better living environment. To achieve the purpose of living in a better environment, people have been used to overexploitation of the natural resources, resulting in destruction and deterioration of the present natural environment. LA design is involved with the reshaping and optimization of the natural environment, so it may inevitably do some destruction to the existing natural environment. Thus, landscape architects should try every means to minimize the destruction of their design works to the natural environment and meanwhile meet humans' demands for pleasing living spaces. To achieve the above dual objective, landscape architects may conduct LA design according to the following steps (Wu, 2019).

(1) Before the LA design, landscape architects should make an in-depth investigation of the existing natural conditions in the landscape, and then propose a design scheme proper for the local conditions. Information necessary for the investigation usually include climate features, topographical and geomorphological conditions, hydrology, municipal facilities, traffic conditions, and plants.

(2) In terms of the selection of plants, local plants are preferred. If landscape architects find it necessary to introduce alien species of plants from other places, they must select those species adaptable to the local conditions.

(3) During the construction process of LA, landscape architects should stick to the belief of minimal destruction to or minimal reshaping of the natural environment in order to demonstrate the LA characteristics based on the protection of the local environment.

4.3 Trying to make a good coordination of the construction of modern LA with the preservation of historic and cultural relics

Areas vary in geography, history, culture, and economic development. Each area has its own geographical, historical, and cultural characteristics and is at a certain economic development stage. Before the LA design, landscape architects need to have a better understanding of the unique historical and cultural characteristics and the economic development level of the aera involved. They should incorporate the development characteristics into the historical tradition of the local area. Their LA design should show respect for the history and meanwhile be oriented toward the future. In other words, LA design should be intended to preserve historic and cultural sites whenever possible for the purpose of preventing the natural environment form being destroyed. Besides, LA design should also be aimed to demonstrate the economic and technological development characteristics of the local area to enable people to feel the change of time.

East Chendu Memory is a successful LA design case combing the construction of modern LA with the preservation of historic and cultural relics. *East Chendu Memory* was built at the old site of a former state-run electronic tube factory founded in the 1950s. At the old site, some of the factory areas with typical industrial characteristics were preserved as an industrial civilization relic, and the remaining areas were reconstructed into a cultural and recreational base. In terms of architectural style, *East Chendu Memory* combined the industrial tradition with the music fashion. Here, people can feel the fascination of an industrial relic, experience the evolution of the local city, and recall the past happy memories[1].

5 Conclusions

We live in an era when the natural environment becomes increasingly deteriorated. Any individual and organization has the social responsibility to protect the natural environment from being destroyed in order to achieve a sustainable harmonious development between humans and the nature. LA is concerned about the

[1] Translation of excerpts from the introduction to *East Chendu Memory* at https://baike.baidu.com/item/东郊记忆/6249828?fr=aladdin

relationship among humans, the built, and natural environments. Thus, it is of special significance to study CSR of LA firms. This paper has studied CSR of LA firms from the perspective of LA design. CSR of LA firms was classified into economic responsibilities and green responsibilities. Accordingly, this paper has discussed in detail LA design based on economic responsibilities and green responsibilities respectively.

Acknowledgment

We would like to thank the anonymous reviewers for their valuable suggestions. Financial support from Chongqing Social Science Found (Project#2018YBGL079) is also acknowledged.

References

1. A. B. Carroll. The Pyramid of Corporate Social Responsibility: Toward the Moral Management of Organizational Stakeholders [J]. Business Horizons, **July/August**: 39-48 (1991)
2. A. Zhang, Y. l. Zhang, M.L. Kong, J.H. Zhang. Corporate Social Responsibility (CSR) and Landscape architecture—In Order to Achieve Low-carbon Society [J]. Chinese Landscape Architecture, **6**:31-34 (2010) (in Chinese)
3. M. Friedman. The Social Responsibility of Business is to Increase Its Profits [N]. New York Times, **September 13**: 122-126 (1970)
4. G. C. Wu. Green Landscape Enterprises Should Take on Social Responsibilities [J]. Chinese Landscape Architecture, **9**:78-79 (2009) (in Chinese)
5. X.T. Wu. Analysis of the Application of Landscape Architecture Design to Environmental Protection [J]. Intelligent City, **19:**185-186(2019) (in Chinese)
6. C. R. Hu. Discussion of the Application of Humanized Design Belief in Landscape Architecture Design [J]. Housing and Real Estate, **18:**54 (2020) (in Chinese)

Towards "Autonomous": A review of the "stereotypical" behaviors in Robotic construction

Boyuan Yu[1] , Lei Zhang[2,3,*], Haoxing Zhang [1]
[1]The Bartlett school of Architecture, University College London, WC1E 6BT, London, United kingdom
[2]Post-doctoral Research Station for Landscape Architecture, Southeast University,Architecture School, 2 Si pai Lou, Xuanwu District, Nanjing, Jiangsu Province, China.
[3]Department of Art , Harbin university of science and technology, 52 Xuefu St., Nangang District, Harbin, Heilongjiang Province, China

Abstract. With the development of digitalization in the architecture industry, robotic construction is now attracting attention as a new technology in the architecture industry. As the number of research cases increases, some people question whether robotic construction is really an efficient and autonomous construction process. The purpose of this paper is to point out the "stereotypical" behavior of robots in autonomous construction in the architecture industry and to question the repetition of a single action by robots in terms of architectural theory and its history. It attempts to outline an objective relationship between robotic construction and architecture and to give architects a warning and advice. It is hoped that in the future, architects will be able to realize autonomous from the perspective of architectural theory and history, rather than simply being limited by the technical constraints of computer language translation and programming.

1 Introduction

Over the past few decades, the construction industry has suffered from poor efficiency, labor shortages and high accident rates at construction sites [1, 2]. In the context of industrial automation, robotic construction has received much attention as a potential solution to improve construction efficiency, reduce accident rates, perform complex tasks, and increase worker safety [3] . Institutions such as ETH z have set up digital construction labs for a lot of experimentation and research, and believe that robots will be a key element in the future of construction [4, 5]. At this stage, the use of robots in the construction industry has been demonstrated to contribute to a more thriving automation of the construction industry [5]. However, at the same time architects are constrained by the technology and mindset of construction, which is technically programmed to input commands to the robot, where the robot's job is only to materiality the output through code, essentially repeating the work and actions that humans have already done, i.e., performing "stereotypical" behaviors and repeating a single action [6]. This leads directly to the loss of traditional architectural language and craftsmanship, and in the long run to a lack of reflection by architects on architectural design and autonomous construction [4, 7]. In this way, we are closer to buildings that need more attention and research effort for autonomous robotic construction[8]. This article reviews and analyses the stereotyped behavior of autonomous execution in digital and robotic construction. Afterward, a dialogue on the potential problems of robotic construction based on architectural theories and historical perspectives will be organized, then a discussion and hypothesis on the orientation of the improvement of autonomous construction will be presented. In this article, we aim to provide some theoretical support for the design theory of autonomous construction and the direction of its technological advancements [8].

2 Autonomous Robotics and Architecture

2.1 Autonomous in Architecture

With the objective needs of industrialization, automati on, and industrialization, the application of numerical control machines in the field of architecture is emergi ng [5]. The 1990s attempted to introduce robotics into the construction industry, but the limitations of contr ol technology led to highly specialized standardized m anufacturing that was very expensive and inflexible, u ltimately leading to the failure of traditional industrial production techniques to digital manufacturing and th e first attempt in history for robotics to cross over int o the construction sector [4, 9]. In 2005, ETH z estab lished the world's first robotics laboratory dedicated to architectural design and manufacturing, reintroducing robotics in the field of architecture and signalling the beginning of architectural autonomy[9]. Over the next 10 years, the ETH has conducted a number of constr uction experiments to convert standardized industrial manufacturing techniques into unstandardized digital manufacturing techniques, making significant contribut

* Corresponding author: 1034571707@qq.com

https://doi.org/10.1051/e3sconf/202123704035

ions such as Gramazio, Fabio, Yuan, Phillip F., or M ario Carpo[8, 10]. From wavy silk walls to complex p avilions, robotic fabrication technology excels in conne cting elements and polymerizing generative materials, depicting the inherent stability of architecture and con struction through the means of technical construction （Fig. 1） [11, 12].Demonstrating the crucial importan ce of robotics in the architectural construction process, robotic construction reveals its hidden "constructivene ss" and establishes a direct link between design and c onstruction, showing the most relevant constructive ex pression of our time [13].However, in the meantime, a large number of scholars have argued that robotic co nstruction is merely an "Exotic" machine, a temporary mode [8].Others interpret it as a superficial investigat ion, based on a "decorative" study, especially the case of Gantenbein Vineyard Facade（Fig. 2） in Gramazi o's earlier book, The Robotic Touch (Foged 2016). Th ese questions concern purely superficial properties of t he discussion, subjective assertions and comments und er visual feedback [14]. Originally, objectively analyze d from the perspective of architectural theory and its history, the practice of architecture has been transform ed at a rapid pace with the development of digital tec hnologies in the field of architecture, but the definitio n of this direction, the original vision that introduced the discipline to the discipline, has been partially lost, resulting in cases of automated construction that indu lge in a strange kind of "Rushinianism," where archite cts are essentially using robots to complete the constr uction process taking the place of and replacing forme r craftsmen[8].As robots follow the instructions and g enerate data, they have no process knowledge, experie nced judgement or creativity [8]. They are developed for specific construction processes such as picking up, placing... Ultimately, they are rigid, developed as too ls for executing commands to increase the productivit y of "humans" instead of craftsmen [15].Consequently, as a means to achieve design and construction, robot s do not perfectly fit the vision of integrating design, the construction and materials for the autonomous con struction that was initially proposed at the introductio n of architecture （Fig. 3）[13].

Fig. 1. Brick Labyrinth, Zurich, 2017

Fig. 2. Gantenbein Vineyard Facade, 2006

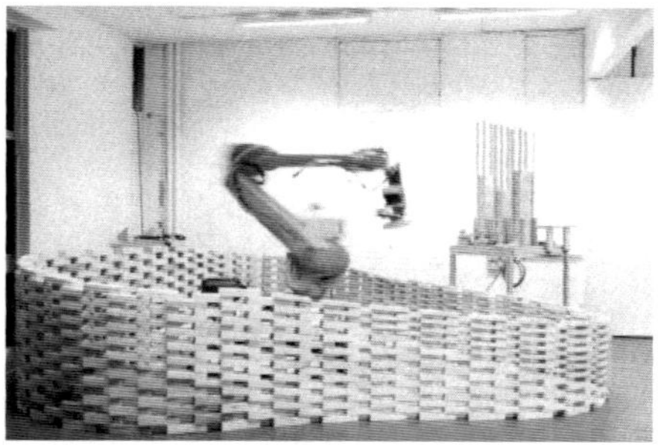

Fig. 3. ETH Zurich， Fabio Gramazio and Matthias Kohler

2.2 Autonomous Robotics

Autonomous construction is not a single act accomplished through the repetition of an action that mimics a human action [9]. On the one hand, such "stereotypical" behavior creates ambiguity about autonomous construction, and on the other hand, traditional building techniques lose their oral tradition of craftsmanship, forgetting to update craftsmanship techniques, which gradually results in the disappearance of traditional craftsmanship, and may result in a uniformity of architecture without individual expression [16].This lack of spirituality in the architecture of the era had occurred in the late stages of modernization and industrialization, but was noted by Ruskin [17].But now, in the era of digitalization and autonomy, we are faced with the same problem: architecture, as a medium that has been handed down for thousands of years and that records history, cannot lose its soul because of innovations in construction technology, a development that is counterintuitive in the disciplinary dimension [8].Objectively, the demands of construction technology go far beyond imitating existing construction technologies [18]. But robotic construction is not about imposing barriers to the building process, but rather about becoming more self-aware and contributing to the autonomous and responsive performance of the architectural discipline[8, 19].To achieve autonomous robotic construction should not be obsessed with the output of complex geometries and mindless repetition of single actions, repeating "stereotypical" behaviors （Fig. 4）[2, 16].

Fig. 4. UCL The Bartlett school.

In the field of architecture, robots should be based on a rational consideration of architecture itself in the process of "materiality" [8].Robotic construction technology is a powerful tool in building development, and architects should consider construction, materials, technology, processes and design holistically [2, 18].In this process it is necessary to explore the theoretical essence of architecture in response to the contribution of the "heritage" of architectural history, and the most important contribution of both to the discipline is the capacity for rational meditation that they cultivate. At the same time, robotic fabrication and robotic architecture are, in fact, linked by these intangible "nostalgic" (theoretical and historical) connections [8, 11].The best way to achieve a truly autonomous construction based on robots in this context is to think according to the logic of the discipline. In the future, robotics in architecture will require "fundamental changes in the early design stages as well as during construction that go well beyond imitating existing building technologies" [6]. Autonomous construction based on robotics should also reflect on the output process, possibly through innovations in machine learning, human-machine interaction, and artificial intelligence, and other technologies that attempt to go beyond the ability to mimic and repeat a single human action or task to achieve "autonomous" construction, exploring more possibilities for architectural design that are not available to humans. many possibilities to truly complete autonomous construction in the field of architecture（Fig. 5）.

Fig. 5. Human- machine interaction

3 Conclusion

The purpose of this review is to provide an objective architectural theoretical and historical perspective on robotics and its achievements and to point out that its current repetitive and uniform behavior, the performance of "rigid" tasks, and the autonomy it was intended to achieve when it was introduced to the discipline are at odds with each other. At the same time, the limitations of this paper are that it does not take into account the limitations of today's robotic technologies, and it also incorporates research that focuses solely on mechanical and computer technology in the perspective of architectural science. Robotic construction is the foundation of research on autonomous processes. It is hoped that in the future, architects will take into consideration the process, design, construction, materials and other factors related to the origins of architecture through robotics. Robots are more than capable of imitating human behavior and doing more than humans have yet to build, a truly autonomous construction.

Figure resources:

Fig 1:
https://gramaziokohler.arch.ethz.ch/web/e/lehre/327.html
Fig 2:
https://divisare.com/projects/260919-gramazio-kohler-ralph-feiner-gantenbein-vineyard-facade
Fig 3:
https://divisare.com/projects/260919-gramazio-kohler-ralph-feiner-gantenbein-vineyard-facade

Fig 4:
https://www.ucl.ac.uk/bartlett/about-us/our-resources/b-made-bartlett-workshops/b-made-robotics
Fig 5:
https://iaac.net/educational-programmes/masters-programmes/master-in-robotics-and-advanced-construction-mrac/

References

1. T. Bock , D.S., J. Fliedner, T. Huynh *Automatlic generation of the controlling-system for a wall construction robot* Automation in Construction, 1996(5（1）): p. 15–21.
2. Tibbits, S., *From Automated to Autonomous Assembly*. Architectural Design, 2017. **87**(4): p. 6-15.
3. Wu, Y., et al., *CU-brick cable-driven robot for automated construction of complex brick structures: From simulation to hardware realisation*, in *2018 IEEE International Conference on Simulation, Modeling, and Programming for Autonomous Robots (SIMPAR)*. 2018. p. 166-173.
4. Picon, A., *Robots and architecture: Experiments, fiction, epistemology*. Architectural Design, 2014(84(3)): p. 54-59.
5. Gramazio, F., M. Kohler, and J. Willmann, *Thriving digital materiality*, in *The robotic touch How robots change architecture*. 2014, Park Books. p. 15-22.
6. Buchli, J., et al., *Digital in situ fabrication - Challenges and opportunities for robotic in situ*

fabrication in architecture, construction, and beyond. Cement and Concrete Research, 2018. **112**: p. 66-75.
7. Lloret, E., et al., *Complex concrete structures.* Computer-Aided Design, 2015. **60**: p. 40-49.
8. Picon, A., *Digital fabrication, between disruption and nostalgia.* Instabilities and Potentialities, 2019: p. 223-238.
9. Gramazio, F., M. Kohler, and J. Willmann, *The robot*, in *The robotic touch How robots change architecture*. 2014, Park Books: Zurich. p. 103-115.
10. Gramazio, F. and J.M. Yoon, *Digital materiality in architecture*, in *The William Cooper Mack Thesis Fellowship Lecture (2018)*. 2018: New York.
11. Yuan, P.F. and K. Li, *Novel Bricks: A Scenario of Human–Machine Collaboration.* Architectural Design, 2020. **90**(5): p. 22-29.
12. Picon, A., *Afterword to Other Computations.* Dearq, 2020(27): p. 139-140.
13. Gramazio, F., M. Kohler, and J. Willmann, *A new Physis of construction*, in *The robotic touch How robots change architecture*. 2014, Park Books. p. 181-192.
14. Foged, I.W., *Bricks / Systems*, ed. I.W. Foged. 2016.
15. Peter, Z., *Hybrid space : new forms in digital architecture*. 1999, Thames & Hudson, London.
16. Yuan, P.F. and C. Yan, *Collaborative Networks of Robotic Construction.* Architectural Design, 2020. **90**(2): p. 74-81.
17. Richard, S., *The Craftsman*. 2008: Yale University Press.
18. Bock, T. and S. Langenberg, *Changing Building Sites: Industrialisation and Automation of the Building Process.* Architectural Design, 2014. **84**(3): p. 88-99.
19. Bauer, A., D. Wollherr, and M. Buss, *Human–Robot Collaboration: A Survey.* International Journal of Humanoid Robotics, 2011. **05**(01): p. 47-66.

Optimization of urban scenic spot planning based on ecological chain and reversed-planning theory—Using Yunlong Lake in Xuzhou city as an example

Lan Zhang [1], Difei Zhao [2,*], Rui Li [3], and Yingxing Chen[4]

[1] School of Architecture and Design, China University of Mining and Technology, Xuzhou, Jiangsu 221116, China
[2] Artificial Intelligence Research Institute, China University of Mining and Technology, Xuzhou, Jiangsu 221116, China
[3] School of Foreign Studies, China University of Mining and Technology, Xuzhou, Jiangsu 221116, China
[4] School of History, Culture&Tourism, Huaiyin Normal University, Huaian, Jiangsu 223300, China

Abstract. Modernized urban organization relies more on the foundation of ecological infrastructure (EI) with the planning objectives based more on synthetical benefits of economy, society and ecological environment. The effect of multiple factors ought to be taken into consideration in the urban scenic spot planning and and the process of formulating the related plan faces a challenge of how to organically bound environmental factors to tour function of urban scenic spot. This thesis, taking YunLong Lake, a scenic spot in XuZhou, JiangSu province as an example, studies how to ameliorate the planning of urban scenic spot combined with the ecological chain and reversed-planning theory. The result shows that utilizing the theory flexibly when analyzing the urban scenic spot planning design can ensure the existing ecological connection between each system and maintain the ecological stability of urban scenic spot. Thus, with the basis of ecological chain and reversed-planning theory, this paper puts forward a new planning concept and work flow, in order to provide new ideas for modernized urban scenic spot planning.

1 Introduction

Sustainable development is the coordinated development of economy, society and ecological environment [1]. Whereas, in the current urban planning and design, the emphasis on economic factors evidently higher than that on social and ecological one, which leads to a series of natural and social environment problems. Lake or wetland scenic spot, one part of modern cities, is a vital button to ameliorate the urban ecological environment [2-4]. How to unify the environmental benefits into tourism functions in urban scenic spots has become a puzzle faced by urban scenic spots planning. In recent years, the scientific idea of ecological environment has also been applied in the planning of urban scenic spots. Environmental design concepts such as imitating natural community design, artificially assisting natural succession, and regulating biological chains through biological control have also been utilized in the design and planning of urban scenic spots so as to optimize and improve the environmental functions of urban scenic spots. Accordingly, only when all function of urban scenic spots obtains the specific design in planning does each corresponding function can be ensured in operation after the completion of the design. This paper, taking YunLong Lake, a scenic spot in XuZhou, JiangSu province as an example, by synthetically considering the impact of natural environment on scenic spot planning, raises a new approach of connecting ecological chain and reversed-planning concept in modernized urban scenic spots.

2 Research Objects and Theoretical Basis

2.1 Overview of Yunlong Lake Scenic Spot in Xuzhou

Yunlong Lake Scenic Spot is located in the southern part of Xuzhou City (Fig. 1), covering an area of 44.7 square kilometers. It was approved as a national AAAAA-level scenic spot in August 2016. Yunlong Lake featuring the natural landscape, is a top urban scenic spot surrounded by mountains on three sides and a city on one side (Fig. 2), with Han culture, celebrity culture, and religious culture as the main content and also equipped with comprehensive functions such as sightseeing and leisure.

2.2 Ecological chain

Ecosystem refers to a unified whole in which all creatures inhabiting together in a certain area (i.e. biotic community) interact mutually and interdependently with their non-biological environment through continuous material circulation and energy flow process [5-6]. In order to sustain the ecological balance in system by employing a

* Corresponding author: diffidiffi@126.com

https://doi.org/10.1051/e3sconf/202123704036

self-regulating mechanism, the biological components in the ecosystem are directly or indirectly connected through the food web, maintaining the relative stability of the structure and function of the ecosystem. Whereas, the self-regulation capacity of the ecosystem is limited and the adjustment wouldn't take effect well once beyond the limitation. Furthermore, the ecological system of urban lake scenic spots is greatly affected by the city planning, mainly showing the marked effect on the ecological chain of lake scenic spots caused by artificial modification.

Fig. 1. Yunlong lake scenic spot in Xuzhou city

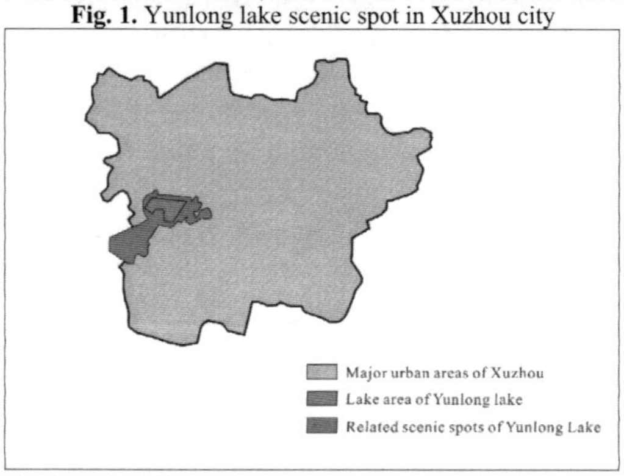

Fig. 2. The relationship between Yunlong lake, related scenic spots and Xuzhou city

2.3 Reversed-planning theory

Early planning, influenced by modernist architectural thoughts born in the industrial age which pursues function and efficiency, tend to consider economic benefits one-sidedly and ignore other factors such as ecology and humanity [7]. It is "reversed-planning " theory that criticized and reflected on traditional planning lacking of ecological consciousness [1]. Prior to the traditional planning due to its comprehensive consideration of various ecological factors, " Reversed-planning " refers to the first planning formulated for non-constructible land and restricted exploitation land on the basis of protecting the original ecological pattern or enabling it to display greater ecological benefits and as a result, forming a mandatory guidance for usable area. Traditional planning regards usable area as figure and non-constructive land as background in most cases, thus neglecting the significance of non-constructive land and even making it a negative element. While "reversed-planning "can coordinate the relationship between ecological environment and construction in the condition of not destroying the original ecological pattern and even generating its greater benefits.

3 New approach of Urban Scenic Spots Planning Based on Ecological Chain and Reversed-planning Theory

There exists discrepancy in the ecological chain between urban scenic spots and the natural environment. The remarkable impact of human activities in city as well as artificial buildings in scenic spots on the natural ecological environment of urban scenic spots weaken the already fragile ecological chain. Yet as for modern cities, urban scenic spots are still a vital functional unit of ecological regulation, and also a crucial human environment regulation unit. Consequently, the urban lake area landscape is able to function well in tourism economy, ecological and humanistic environment, only within the special ecological environment marked with urban human behaviours, that is, maintaining the stability of ecological environment function can maximize its tourism economic and humanistic environment effect. To sum up, the result of this research suggests that when planning and designing urban scenic spots, a sustainable and stable ecological chain ought to be fully understood and created first with the basis of sustaining the stability of urban ecological environment. What's more, gaining the environmental and ecological benefits of scenic spots planning requires further synthetically taking the impact of human factors into account while designing the ecological chain [8]. In the perspective of the ecological benefits of scenic spots, the research proposes a new design workflow as follows by the combination of the ecological chain and reversed-planning concepts:

Table 1. Work flow of urban scenic spots planning based on ecological chain and reversed-planning theory

Step	Function
Ecological chain study	To understand the ecosystem and key influencing factors of scenic spots
Planning of non-constructive land and restricted construction land	To ensure the urban lake area can function well in ecological level
Planning of constructive land	
Synthetically ameliorate the planning and design planning	To ensure that the scheme takes into account the environmental function and the scenic spot tour function

1) In the first place, we should investigate the urban scenic spots and surrounding environment by combining the regional characteristics of ecological environment so as to complete the design of ecological chain. What's more, the impact of man-made regulation factors on the urban lake landscape in the design of ecological chain demands to be taken into account;

2) On the basis of the ecological chain design with reversed-planning theory, we are supposed to plan for non-constructive land and restricted construction land by introducing concepts such as the connectivity of natural ecological units to ensure that the urban lake area can

function well in ecological level;

3) To carry out the planning of constructive land based on the project of non-constructive land and restricted construction land;

4) In the end, to synthetically ameliorate the established planning and design planning for the sake of forming the ultimate plan.

4 A design case study using Yunlong Lake

Taking Yunlong Lake as an instance, the planning and design are carried out in accordance with the workflow that combines ecological chain and reversed-planning theory (Fig. 3). To start with, the research analyses and classifies the environmental ecological chain of Yunlong Lake and its surrounding areas, and then divide the area into four ecological zones of lake water system, lake peripheral wetland, peripheral mountain forest, and cities in the boundary of ecological chain. These four ecological zones take the lake water system as the core zone which contains four levels, including the bottom biological layer, the middle biological layer, the surface biological layer and the lakeshore biological layer. The ecological chain, composed of the four biological layers, sustains the natural ecology of the lake area and the layers in the ecological chain influence each other [9-10]. Diverse manual maintenance measures need to be taken before planning and designing the scenic area. For example, the bottom biological layer of the water system demands taking lake dredging measures, the middle and surface biological layers calls for the monitoring towards lake water quality, and the lakeshore biological layer needs to maintain a certain connectivity on environment of lake and shore to reduce the over-segmentation of artificial buildings in the natural environment. When planning a scenic spot, laying out with the goal of keeping great biodiversity can drastically level up the quality of ecological environment in scenic spot.

After the research and design of dividing functional zones of lake natural environment based on the ecological chain, according to the reversed-planning theory, non-constructive land and restricted constructive land in scenic spots are obtained, and suitable constructive land is also gotten through reversing the background. In order to further consolidate the stable ecological pattern, each land and the connection space between them have been sorted out and transformed: biological corridors have been put into place, the division of artificial buildings on the lake shore to the natural environment has been decreased, and a considerable number of lake-land connections have been remained. In addition, by designing the island in the center of the lake, environmental functional segments such as the migratory bird conservation area and the amphibian breeding area have been added. In the area far away from city, the ecological zone of lake water system, the lake periphery wetland ecological zone and the peripheral mountain forest ecological zone are connected as one. The various ecological functions of urban lake scenic spots are all based on their original normal ecological environment. Hereby in designing, taking the natural environment as the first active element to design can better guarantee that the scenic spot achieves its diverse functions.

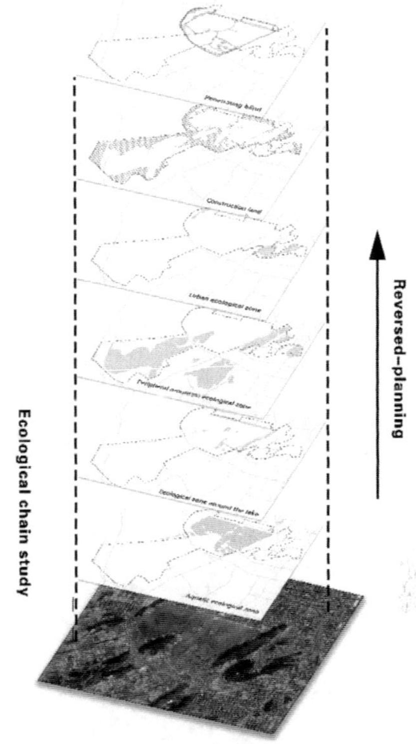

Fig. 3. Work flow of landscape design combining Reversed-planning and ecological chain

In the end, owing to perfecting the modification of Yunlong Lake's ecological function, the non-constructive land in the scenic land planning transforms to a lake ecological maintenance function zone, and the constructive land converts to a lake tourism and economic function zone. The latter can be further planned and divided into natural landscape sightseeing land and tourism expansion function land. In the specific design, the former mainly focuses on structures such as observation decks and submerged corridors, while the latter can be divided into cultural areas, catering area and retail area, etc. As for the next step of design, each structure should avoid cutting apart the natural environment, and the buildings ought to be concentrated as much as possible in their respective land for the sake of lessening the effect on the natural ecology of the lake. After obtaining the specific segmentations, the road network and the functional divisions are appropriately organized and regulated to form an integrated scenic spot planning and design. In addition, the introduction of artificial intelligence technology will improve the planning effect [12-13].

5 Conclusions and Prospects

Based on the design concept of ecological chain and reverse-planning, three functions of tourism service, economy and ecology can be unified under the condition of ensuring the excellent ecological function of urban lake landscape area, which is conducive to the improvement of urban natural environment and social environment. After all, it is of great significance to maintain the ecological stability of scenic spots in the city. By this way, this paper proposes a new design process that combines ecological chain and reverse-planning. That is, through design and planning measures such as multi-ecological function zoning, impeccable ecological chain design, as well as a series of design and planning measures of ecological planning first and architectural planning second, the optimized function of lake scenic spots aims to provide scientific basis and reference for landscape planning, tourism development and environment protection.

With the continuous expansion of urban space, the destruction of urban landscape environment is increasingly serious. Once the scenic area is built, its impact on the environment will begin to play a role, resulting in ecological loss which is hard to recover. Therefore, it is extremely crucial to fully incorporate environmental factors into the design concept. Comprehensively studying the ecological chain of scenic spots is the basis of environmental planning of scenic spots, which provides scientific basis for avoiding environmental impact caused by planning. The reversed-planning theory ensures that environmental factors are given priority by changing the workflow. The combination of ecological chain and reversed-planning theory ensures both theoretical advantages and practical feasibility of the planning. In the future, we plan to apply the new workflow in more cases of urban scenic spot planning to further promote the new workflow. The application of artificial intelligence technology in this process is also an exciting research direction. Collecting the information of the original topography, biology and ecological function of the scenic spot, intelligently analysing the ecological function and relationship of the scenic spot, and carrying out the fine design and planning, will be able to protect the sustainable development of the environment to the greatest extent. Intelligent analysis based on experience or algorithm can also dramatically reduce the human cost of planning and design. The results of intelligent analysis based on the same algorithm is instrumental in enhancing the comparability of planning schemes. Moreover , the application of artificial intelligence technology in this process is an stimulating research direction. In the future ecological chain design, we should refine the regional bionic community design, including animal habitat design and planting design, to enrich species diversity and build a stable cycle of the system. Above all, the results of intelligent analysis based on the same algorithm can be beneficial to intensify the comparability of planning schemes.

Acknowledgement

The research is sponsored by the Innovation Training Program for Artificial Intelligence Research Institute College Students, China University of Mining and Technology (No. CSA2020-02), the National innovation training program for College Students of China University of Mining and Technology (No. 201910290034Z) and Liu Baojun Science Foundation for Geoscience Young Scholars (No. DMSMX2019002). The author Difei Zhao also thanks to the China Scholarship Council (CSC) for providing the CSC scholarship for occupational training at the University of Queensland.

References

1. L Zhang, C Huang, D Zhao, H Du, Q Lu, H Yu, Y Yu, Y Chen, M Zhang. *Fresen Environ Bull* **29**, 10: 9095-9102 (2020) .
2. E Mertens. *Atmos Environ*. **33**, 24: 4115-4123(1999).
3. C Nilon, M Aronson, S Cilliers, D Cynnamon, L Frazee, M Goddard, K Neill, D Roberts, K Stander, E Stander, P Werner, M Winter, K Yocom.. *Bioence*. **67**, 4: 332-342(2017).
4. X Zhang, B Li, H Xu, M Wells, B Tefsen, Qin B.. *Water Research*. **151**, 3: 500-514(2019).
5. Yan, D., Xia, J. *Hube Forestry ence and Technology*. **153**, 5: 12-17(2008).
6. P Macnaghten, R Owen. *Nature*. **479**, 11: 293-293(2011).
7. A Tamguicht. *Economics Bulletin*, **15**, 9: 1-10(2007).
8. R King, C Richardson, U Romanowicz. *Ecosystems*. **7**, 1: 75-97(2004).
9. E Halfon, N Schito, R Ulanowicz. *Ecological Modelling*. **86**, 1: 0-36(1996).
10. A Neumann, D Kim, G Perhar, G Arhonditsis. *J Environ Manage*. **188**, 3: 308-321(2017).
11. Y Kongjian, L Dihua. *Annual meeting of Chinese Association for Science and Technology*. 26-37(2002).
12. S Laura, G Inma, O Eva, G Cesar. *Int J Artif Intell T*. **18**, 11: 717-738(2009).
13. J Fink, A Kobsa. *Artif Intell Rev*. **18**, 9: 33-74(2002)

 https://doi.org/10.1051/e3sconf/202123704037

Study on the Wearing Performance of Compression Stockings and Its Influence on the Blood Flow Velocity of Lower Limbs

Hainan Liu[1], Ling Cheng[2,*]

[1] School of Textile Science and Engineering, Tiangong University, 399 Binshui West Road, Tianjin, China
[2] School of Physical Science and Technology, Tiangong University, 399 Binshui West Road, Tianjin, China

Abstract. Compression stockings can reduce venous pressure and improve venous return by applying progressive pressure on the lower limbs from bottom to top, thereby preventing and treating venous diseases of the lower limbs. The purpose of this research is to conduct a preliminary understanding and analysis of compression stockings through various mechanical properties and feel styles of compression stockings, and to explore the effect of compression stockings on the blood flow of lower limbs by using blood flow speed as an experimental index. The results show that low-grade compression stockings are more effective in accelerating blood flow in healthy people under sedentary conditions.

1 Introduction

Chronic lower extremity venous disease is a common vascular disease and frequently-occurring disease. Its incidence increases with age. The incidence of women is higher than that of men. Chronic venous disease is a syndrome of a series of symptoms caused by poor venous blood return and high venous pressure due to abnormal structure or function of the vein[1-2]. Venous diseases account for about 60% of vascular surgical diseases and often occur in the lower extremities. Compared with medication and surgery, compression therapy is a scientific, simple and effective treatment method[3]. Studies have shown that compression therapy can increase the action of the calf muscle pump to reduce venous pressure and improve venous return. Appropriate compression can reduce the diameter of the main vein, thereby increasing blood flow velocity[4].

Compression stockings, a medical textile that can promote venous blood return to the heart and relieve symptoms related to venous diseases of the lower extremities, have been proven to be an effective non-surgical treatment[5-6].This pressure difference can speed up the blood flow, reduce or drive away the stagnation of venous blood at the distal end of the limb, and protect the venous valve to achieve the effect of prevention and treatment[7].

Scientific pressure levels and gradient distribution are the key to the medical performance of compression stockings, but the progressive pressure it presents is the result of the comprehensive effect of various mechanical properties of the fabric. Studies have shown that the production parameters of compression stockings have a significant impact on their pressure characteristics. Similarly, the wearing and medical performance of compression stockings are the comprehensive results of various mechanical properties of the fabric[8-9]. Therefore, the purpose of this study is to get a preliminary understanding of compression stockings from the perspective of evaluating fabric styles and analyzing the mechanical properties of materials, and use ultrasonic Doppler blood flow detectors to monitor the skin blood flow changes of healthy adults wearing different grades of compression stockings.

2 Materials and methods

2.1 The basic characteristics of compression stockings

This study tested and analyzed 3 grades of elastic compression stockings produced in Taiwan. According to the manufacturer's statement, their compression socks have passed the ISO 9002 quality system certification. Compression stockings exert the greatest pressure on the ankle and show a gradually decreasing pressure distribution up the leg. Therefore, the compression stockings of each grade are divided into three parts for analysis, as shown in **Figure 1**, and marked a, b and c respectively from top to bottom. Combined with the basic structure test of the sample, the basic characteristics of the tested sample are shown in **Table 1**.

*chengling@tiangong.edu.cn

https://doi.org/10.1051/e3sconf/202123704037

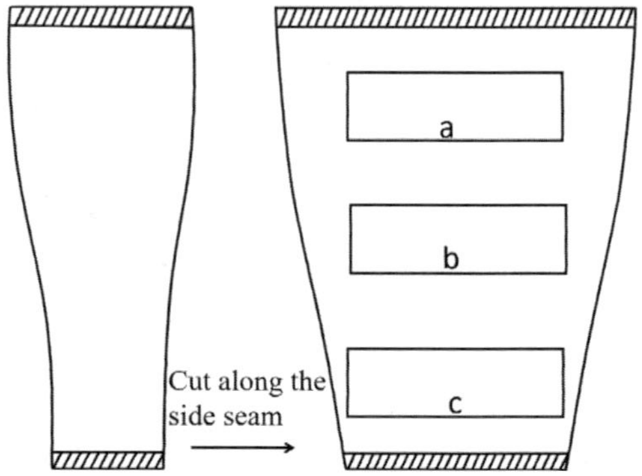

Figure 1. Schematic diagram of compression stockings sampling.

Table 1. The basic parameters of the sample.

Pressure rating		Thickness (cm)	Weight (g)	Transverse density (row/5cm)	Longitudinal density (row/5cm)	Recommended medical functions
Light	a	0.706	1.41	59	119	Prevent varicose vein and thrombosis, reduce heavy and fatigue.
	b	0.707	1.45	72	116	
	c	0.712	1.54	88	111	
Moderate	a	0.705	1.44	60	110	For the treatment of mild varicose veins, pain, swelling.
	b	0.720	1.41	76	106	
	c	0.736	1.43	94	101	
Strong	a	0.735	1.55	58	110	For the treatment of severe varicose veins, severe edema, leg ulcers, lymphedema.
	b	0.738	1.60	72	107	
	c	0.744	1.62	89	103	

2.2 Assessment of material performance

2.2.1 INSTRON-3300 Fabric universal strength machin. This instrument can be used to test textile materials, rubber and other materials for tensile, compression, tearing and other experiments. The latest technology is used in measurement, control, recording, etc., and an automatic control and data processing system is formed with a microprocessor as the core. It has the advantages of high measurement accuracy, wide measurement range, stable clamping force, and no damage to the sample.

2.2.2 PhabrOmeter Fabric feel evaluation system. The feel style of fabric is one of the important quality attributes of textile products, which affects the wear performance and comfort of clothing. PhabrOmeter fabric evaluation system is a new kind of measuring instrument system developed by American Xinsaibao Technology Company for fabric sensory performance evaluation. It can realize the fast, objective and high-throughput testing of multiple indicators of fabric handle style. A single test can obtain multiple indexes such as softness, smoothness, rigidity, drape, and wrinkle recovery at the same time.

2.2.3 Skin blood flow velocity measurement. The experimental equipment is a portable ultrasound Doppler blood flow detector, which has the advantage of non-invasive and continuous measurement of skin blood flow velocity. The device probe emits a certain frequency of ultrasound, and it will reflect when it encounters the red blood cells moving in the blood vessel. The movement speed of the red blood cells can be obtained by the frequency difference between the incident wave and the reflected wave, which is the blood flow speed.

3 Experimental protocol

The whole process of the test is carried out in a laboratory with an ambient temperature of (23±2) °C and a relative humidity of (65%±2%).Before testing, all fabric samples were allowed to stand for 24 hours in the aforementioned standard laboratory to minimize the instability of the knitted fabric.

3.1 Tensile performance test

To test the tensile properties of fabrics, most of them need to test the breaking strength of the fabric, that is, the maximum load when the sample is stretched and broken at a certain tensile speed. The sample for this test is compression stockings. In daily life, the fabric can rarely be pulled to fracture, testing its breaking strength is of little help to this experiment, so this test uses the constant elongation method Stretch the sample and analyze the stress-strain curve when the sample is stretched to the corresponding elongation. When wearing compression stockings, transverse stretching is produced in order to adapt to the shape of human legs, and the longitudinal deformation is small. Therefore, the deformation of pressure socks can be viewed approximately as a simplified problem of lateral stretching. In this study, only transverse stretching is carried out on fabrics to test the transverse tensile properties of fabrics, and the test indicators are shown in **Table 2**.

In the experiment, the sample was selected to be stretched to 15cm in one direction and the stretching rate was 100mm/min. The tested sample size is 5cm 15cm, and the effective size is 5cm 10cm.

Table 2. Tensile property test index.

Test index	Index introduction
Initial modulus	The initial modulus is used to describe the difficulty of the knitted fabric to deform under a small external force, and it is also one of the indicators that characterize the elasticity of the knitted fabric. The size of the initial modulus indicates how easy it is for the fiber to deform under a small load.
Elastic recovery rate	Elastic deformation refers to the recoverable deformation of the fabric after the external force is removed. For knitted fabrics, it is expressed by the tensile elastic recovery rate.
Plastic deformation rate	Plastic deformation refers to the deformation that the fabric cannot recover after the external force is removed under the specified conditions. It is expressed by the plastic deformation rate.

3.2 Evaluation of fabric handle style

In the evaluation of fabric handle style, the tested sample is a circle with a radius of 10 cm, and the test indexes are shown in **Table 3**.

Table 3. Test index of fabric hand style.

Test index	Index introduction
Stiffness	Stiffness is related to fabric flexibility. The higher the stiffness value is, the more crisp the fabric is.
Softness	The compression resistance of fabric is related to its softness. Low compression resistance can be considered as high softness. The higher the softness value is, the softer the fabric is.
Smoothness	Surface friction is a kind of surface resistance due to sliding. The greater the smoothness value, the smoother the fabric surface.
Drape	The smaller the drape index is, the better the drape of the fabric is.
Wrinkle Recovery Rate	The fabric was measured twice (with an interval of 5 minutes). The difference between the two test values is the index value of wrinkle recovery ability of the fabric. The higher the wrinkle recovery rate is, the better the wrinkle recovery performance is.
Stretch Index	Describe the elasticity of the fabric. The higher the elastic index is, the better the elasticity of the fabric is.

3.3 Blood flow velocity of skin was measured

The blood flow measurement experiment requires that the subjects do not exercise vigorously within 24 hours before the experiment, and do not eat spicy or alcohol-containing irritating food within 17 hours. After entering the laboratory, the subjects had a quiet rest for 30 minutes, so that they could adapt their mental and physical conditions to the current environment. Since the blood flow changes of the human body are affected by many minor factors, even if the same person is in different environment and mood, the blood flow changes are also different every day, so the test time is selected as 13:30-18:30 every day to ensure the accuracy of the experiment.

Immobilisation, and prolonged standing and sitting have been suggested to be the principal aetiological factors in the pathogenesis of such venous disorders. Therefore, a point above the knee in the sitting position is selected as the test point, and no pressure contact is guaranteed during the test.

4 Results and Discussion

4.1 Test results and analysis of tensile properties

It can be seen from **Figure 2(a)** that as the pressure level increases, the initial modulus value increases; the same level, the initial modulus value of the fabric at different positions from top to bottom is also increasing. Which indicates that the higher the pressure class is, the greater the pressure value is, the more difficult it is for the pressure socks to deform under the action of less external forces.

From **Figure 2 (b, c)**, it can be seen that as the pressure level increases, the elastic recovery rate decreases while the plastic deformation rate increases. That is to say, the elastic recovery rate of the fabric has a negative correlation with the clothing pressure. In other words, the better the elasticity of the compression stocking under the same elongation condition is, the lower the clothing pressure in this part is. This is because a knitted fabric with good elastic recovery requires less recovery force, and the pressure perpendicular to the surface of the human body is less, and the corresponding clothing pressure is also less. From the perspective of pressure comfort of tight-fitting knitted garments, under the same clothing allowance, garments made of knitted fabrics with good elasticity not only fit tightly, but also do not restrict the movement of the human body, and have better sports comfort. On the other hand, in the same grade of compression stockings, the elastic recovery rate of the sample position from top to bottom is increasing, and the plastic deformation rate is decreasing. It is known that the greater the elastic recovery rate of the knitted fabric is, the smaller the plastic deformation rate is, and the better its elasticity is. It indicates that the closer the sample position is to the ankle, the better its elasticity is.

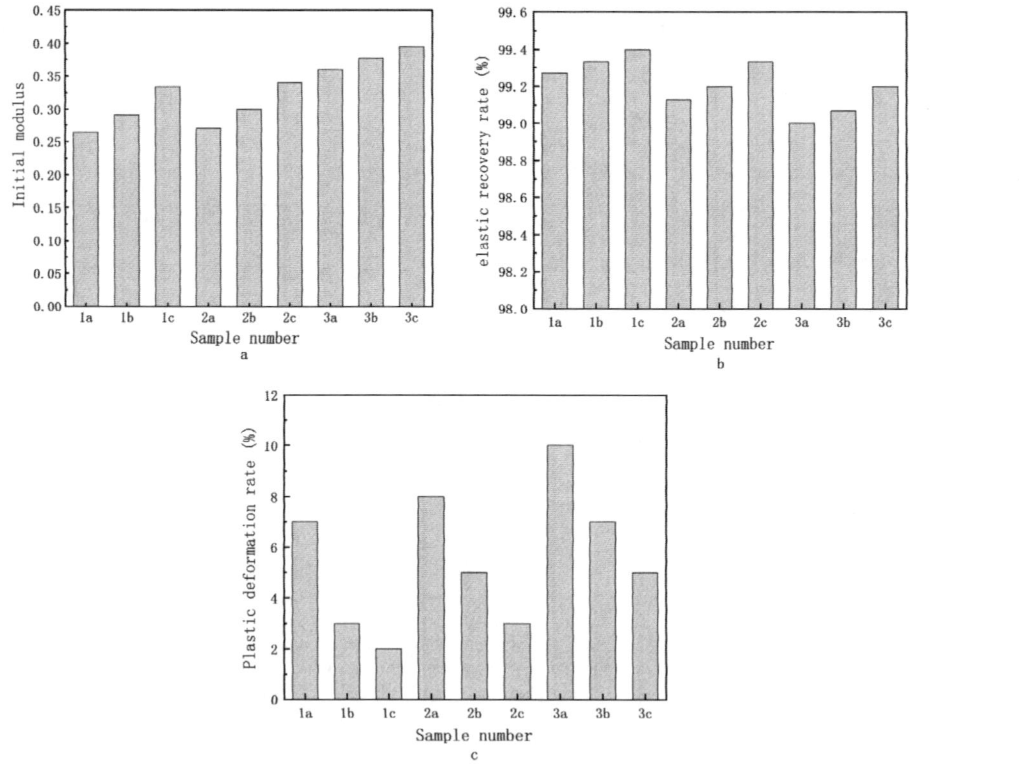

Figure 2. Histogram of tensile performance indicators.

4.1.1 Correlation analysis between sample parameters and performance indexes. Correlation analysis is to study whether there is a certain dependence relationship between phenomena, and to explore the related direction and degree of correlation between specific dependent phenomena. It is a statistical method to study the correlation between random variables. The correlation coefficient is a measure of the closeness of the relationship between two or more variables. The correlation coefficient calculated based on the sample data is called the sample correlation coefficient and is denoted as r. The value of the correlation coefficient r is between -1 and 1, but can be any value within this range. $0.8<|r|<1$ is high correlation, $0.5<|r|<0.8$ is moderate correlation, $0.3<|r|<0.5$ is low correlation, and $|r|<0.3$ can be regarded as irrelevant. In the case of positive correlation, the two variables change in the same direction; in the case of negative correlation, the two variables change in the opposite direction.

The spss software is used to analyze the correlation between the structural parameters of the sample and the tensile performance test indicators. The analysis results are shown in **Table 4**.

 https://doi.org/10.1051/e3sconf/202123704037

Table 4. Correlation between structural parameters and performance indexes.

	-	initial modulus	correlation	elastic recovery rate	correlation	Plastic deformation rate	correlation
Pearson correlation coefficient	Weight	0.923	Highly positive correlation	-0.365	Low negative correlation	0.135	irrelevant
	Thickness	0.848	Highly positive correlation	-0.412	Low negative correlation	0.13	irrelevant
	Transverse density	0.358	Low positive correlation	0.588	Moderate positive correlation	-0.807	Highly negative correlation
	Longitudin-al density	-0.483	Low negative correlation	0.153	irrelevant	0.154	irrelevant

It can be seen from **Table 4** that the correlation between the thickness and grammage of the sample and the initial modulus is positive and the correlation is strong, that is, the thickness and grammage have a significant effect on the initial modulus of the fabric. The transverse density of the sample had a moderate positive correlation with the elastic recovery rate of the fabric and a high negative correlation with the plastic deformation rate of the fabric, that is, increasing the transverse density of the fabric appropriately would increase its elastic property.

4.1.2 Linear regression analysis of stretch rate and load. Linear regression refers to a method of establishing a linear regression model for objects with correlations, and inferring another unknown quantity through a known quantity. As shown in Figure 3, the elongation and the corresponding load value of the 9 samples are distributed around a straight line, and the regression function between the two sets of data is considered to be a linear function. Using Spss software to curve fitting the relationship between the tensile rate of the sample and the tensile load, **Table 5** can be obtained.

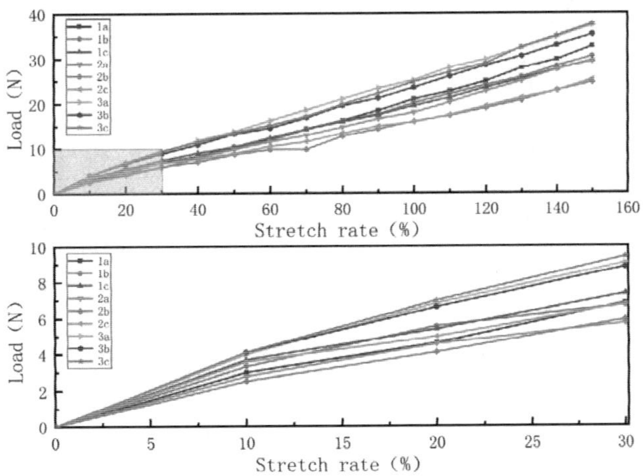

Figure 3. Sample tensile curve.

Table 5. Fitting equation of specimen tensile curve.

The sample number	Regression equation	Correlation coefficient
1a	y=0.21x+0.009	0.997
1b	y=0.194x+0.656	0.997
1c	y=0.186x+1.19	0.997
2a	y=0.188x+0.178	0.996
2b	y=0.155x+0.515	0.994

https://doi.org/10.1051/e3sconf/202123704037

2c	y=0.153x+1.275	0.992
3a	y=0.238x+1.636	0.998
3b	y=0.224x+1.474	0.997
3c	y=0.236x+1.339	0.996

The correlation coefficients of the curve fitting equations are all above 0.99, and the fitting effect is good, so the linear equation can be used to fit the stretching curve equation. The extensibility of fabric can be expressed by the tensile force under a certain elongation rate. The lower the tensile force under the same elongation rate is, the better the extensibility of fabric is. It can be known from the curve fitting equation of the sample that the tensile elasticity of the tested sample is very well, and a small load can produce a large elongation.

Figure 4 is the linear fitting equation diagram corresponding to three samples in different grades of compression stockings. It can be seen that when the sample at different positions has the same elongation, the required load value has little difference, especially at the two positions b and c. Among the three grades, sample a requires the largest load when large elongation occurs, that is, the compression stocking has the worst extensibility near the knee.

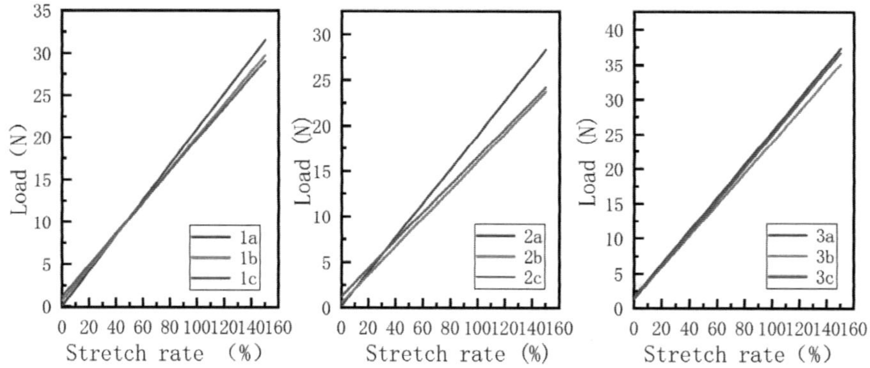

Figure 4. Linear fitting equation of stretch of socks with different grades of pressure.

Figure 5 is the linear fitting equation diagram of samples with the same position but corresponding to different grades. When the same elongation occurs, the maximum load is required for the three-stage pressure socks, and this is true regardless of whether the elongation is small or large. But for the first-stage and second-stage compression stockings, the required load value is not much different when the elongation is small; and when the greater elongation occurs, the load value required for the first-stage is significantly greater than the load value required for the second-stage. This may be caused by other properties of the fabric.

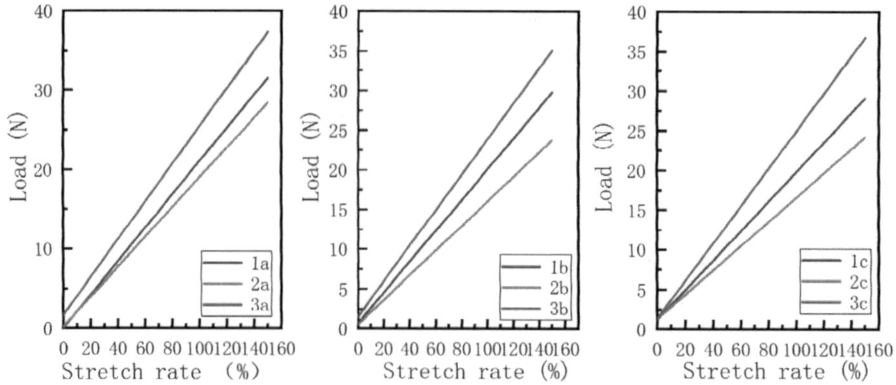

Figure 5. Drawing of linear fitting equation for different samples.

4.2 The result and analysis of fabric hand feeling style evaluation

The PhabrOmeter fabric feel evaluation system can simulate human hand touching the fabric and give sensory performance evaluation. Through the data, an objective understanding of the style characteristics and feel performance of compression stockings can be realized.

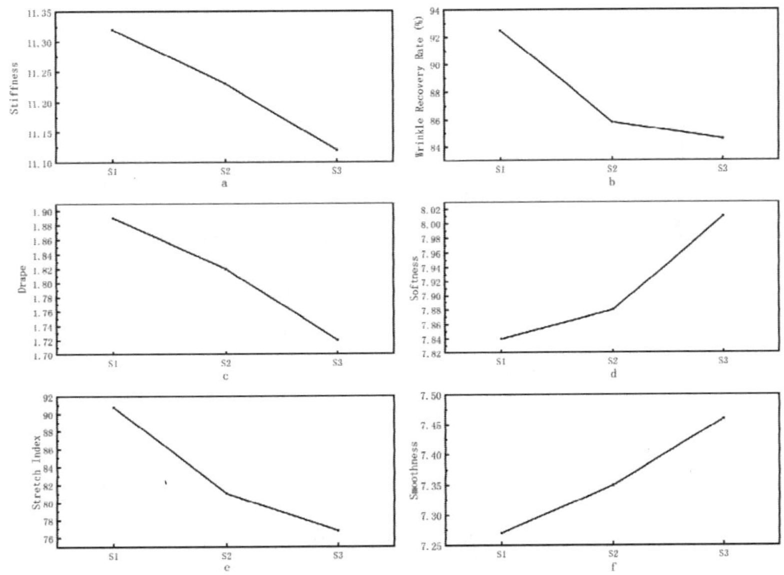

Figure 6. Hand feel style index.

Generally speaking, the different style indexes of three grades of pressure socks have little difference. The higher the pressure level is, the greater the softness index and smoothness index are. The higher the softness is, the softer the cloth body is, and the higher the smoothness is, the smoother the surface of the fabric is, and the lower the friction force is. If the pressure level is too high, the stronger the restraint to the legs, the wearing comfort will be lower than that of the pressure socks with weak pressure level. Therefore, reducing the friction on the legs and improving the softness of the socks will improve the wearing comfort of high-grade pressure socks. On the other hand, the lower the pressure level is, the weaker the leg restraint is. It can be seen from **Figure 6 (e)** that the drape coefficient of low-grade pressure socks is higher, which indicates that the effect of close fitting is not as good as that of high-grade pressure socks. In daily wearing, in order to ensure the accurate pressure value and gradient distribution of pressure socks, it is necessary to consider the fitness and shape retention of pressure socks. The stiffness, elasticity index and wrinkle recovery rate of low-grade pressure socks are higher than those of high-grade pressure socks. The higher the stiffness and wrinkle recovery are, the better the bending resistance of the fabric is, and the higher the elastic index is, the better the elasticity and relaxation resistance of the fabric are. On the whole, it can improve the weak binding and pressure attenuation.

4.3 Test results and analysis of skin blood flow velocity

When the compression stockings were not worn, the blood flow velocity of the subjects did not fluctuate much. An independent sample t test was performed on the blood flow speed when wearing different grades of compression stockings and when not wearing it, and it was concluded that the blood flow speed in each group of data was significantly different from the blood flow speed in the unbound state($p<0.01$).

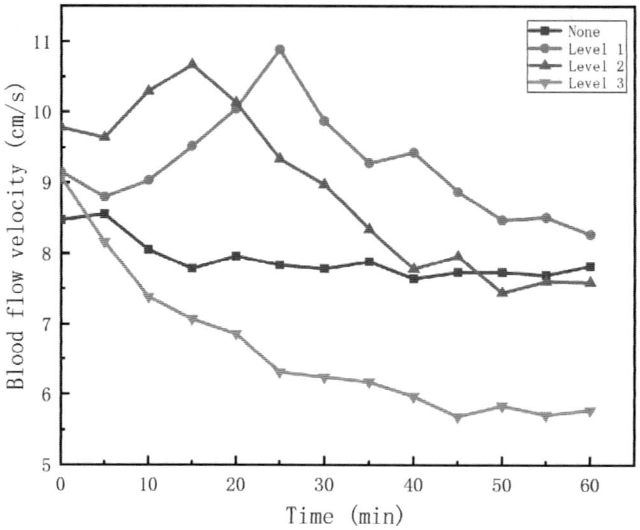

Figure 7. Diagram of blood flow velocity under different pressure states.

It can be seen from the figure that the primary and secondary pressure stockings can accelerate the blood flow speed in a period of time after wearing. But the difference is that the blood flow velocity of grade I pressure socks reaches the peak value after wearing them for about 25 minutes, which is 37% faster than that when they are not wearing. After that, the blood flow velocity began to slow down gradually in the remaining 35 minutes, but it was always higher than that when the pressure socks were not worn. On the other hand, the blood flow velocity of the secondary pressure socks reached the peak after 15 minutes of wearing, which was 35% faster than that without wearing. Then the blood flow velocity began to slow down gradually in the remaining 45 minutes, and the blood flow velocity value was slightly lower than that without pressure socks after wearing 45 minutes.

Compared with the first level compression stockings, the peak blood flow velocity of the two is almost the same. At the same time, the average blood flow velocity within the first 20 minutes of the second level compression stockings is significantly greater than that of the first level. This not only reflects the promoting effect of pressure on blood flow velocity, but also within a certain period of time, the greater the pressure value is, the greater the blood flow velocity is.

Compared with the second-level compression stockings, the third-level promotion of blood flow is only reflected in the instantaneous state just after wearing it. Within 60 minutes of the test, the blood flow rate has been in a state of decline. This means that excessive pressure can inhibit the blood flow rate of normal people.

4.3.1Correlation analysis. From the previous section, it can be seen that the pressure grade and wearing time of pressure socks have certain influence on blood flow velocity of lower limbs. Therefore, the correlation analysis of blood flow velocity, pressure level and time is made, as shown in **Table 6**.

Table 6. Correlation analysis table.

		Blood flow	Pressure rating	Time
	Blood flow	1.000	-.413**	-.450**
Pearson correlation	Pressure rating	-.413**	1.000	.054
	Time	-.450**	.054	1.000
	Blood flow		.002	.001
Sig. （bilateral）	Pressure rating	.002		.702
	Time	.001	.702	
	Blood flow	52	52	52
N	Pressure rating	52	52	52
	Time	52	52	52

According to the results of table correlation analysis, the correlation coefficient between blood flow velocity and pressure grade is -0.413, and the correlation coefficient between blood flow velocity and wearing time is -0.45, and the significance level is less than 0.01. Therefore, the correlation between blood flow velocity and pressure grade and wearing time is negative and strong.

5 Conclusion

This study mainly evaluated the wearing performance of compression stockings and their effect on human lower limbs from the perspective of fabric performance and physiology. In terms of pressure rating and mechanical properties, studies have shown that the structural parameters, tensile properties and overall effects of fabrics are closely related to the pressure rating of compression socks. For example, as the pressure level increases, the heavier the fabric is, the greater the initial modulus of the fabric is, that is, the stronger the fabric's resistance to low stress deformation.

It is known that the decrease in the venous blood flow velocity of the lower extremities is generally considered to be one of the main reasons for the formation of venous diseases of the lower extremities[10].Therefore, blood flow velocity can be an effective response to venous function. In this study, blood flow velocity was measured in healthy women, and it was found that there was little fluctuation in blood flow velocity without pressure. On the other hand, within 1 hour of wearing different grades of compression stockings, the first-level and second-level compression stockings increased the blood flow speed by 37% and 35%, respectively, while the third-level compression stockings had an inhibitory effect on the blood flow rate. The results show that compression stockings with lower pressure levels are more suitable for people who are sedentary in daily life, while those with higher pressure may be more suitable for people with severe venous diseases of the lower extremities.

The limitation of this study is that the subjects were female subjects with healthy vein function and leg shape, and blood flow velocity was not studied in patients with various degrees of venous diseases. Therefore, in order to make compression stockings play a greater role in the prevention and treatment of venous diseases, further research is still needed to investigate the effect of high-grade compression stockings on lower extremity venous diseases.

References

1.Hamdan and Allen 2012 Jama. 308(24) 2612-21.

2.Born D P, Sperlich B, Holmberg H C 2013 Int J Sport Physiol. 8(1) 4-18.

3.Fu W, Liu Y and Fang Y 2013 Int J Adv Robot Syst. 10(1) 66-.

4.W Blättler and Zimmet S E 2008 Phlebology. 23(5) 203-205.

5.Partsch H 2005 Dermatol Surg. 31:625–630.

6.Liu R, Kwok YL, Li Y, Lao TT, Zhang X and Dai XQ 2005 Dermatol Surg. 31:615–624.

7.Mosti G and Partsch H 2011 Eur J Vasc Endovasc Surg. 42 261-6.

8.Burak S 2016 J Ind Tex. 47(6) 1083-1096.

9.Liu R, Kwok Y L, Li Y, et al 2010 J Appl Polym Sci. 104(1) 601-610.

10.Sigel B, Edelstein AL, Felix RF and Memhardt CR 1973 Arch Surg. 106 38–43.

Landscape ecological risk assessment and spatiotemporal change analysis in Yonghe County

Jixuan Yan [1,2], Guang Li[2]*, Haiyan Wang[2], Meihua Zhang[1], Dongyuan Sun[1] and Xia Kang[1]

[1] College of Water Resources and Hydropower Engineering, Gansu Agricultural University, Lanzhou 730070, China
[2] College of Forestry, Gansu Agricultural University, Lanzhou 730070, China

Abstract. On the basis of land use data, combined with ArcGIS and fragstats4.2, the landscape ecological risk spatial mode and process feature of Yonghe county from 1980 to 2018 were analyzed. The results showed that: (1) from 1980 to 2018, the landscape pattern of Yonghe county changed obviously, the area of arable land and holt decreased, while the area of grassland, water and buildings grow a number. The conversion area between arable land and grassland is the largest. In Yonghe County, the fragmentation degree of landscape is on the rise, and the separation degree is also on the rise. The overall dominance of construction land is the largest, and the dominance of grassland is the smallest. (2) The landscape ecological risk level gradually subsided, the area of middle risk area decreased, and moderate risk areas are falling . From the spatial distribution analysis, the risk types of Yonghe County subsided from southeast to northwest. The high-risk areas were principally scattered in sangbi town and Jiaokou township. The main landscape type in this area was grassland, which was easily disturbed by human activities; the low-risk areas were principally scattered in Potou Township in the north and Yonghe County in the middle of the study area, and Woodland and buildings are the main landscape types land have strong anti-interference ability and low risk value.

1 Introduction

With the influence of the change of environmental management objectives and concepts, landscape ecological risk evaluate has gradually become one of the hot spots in the field of geography, ecology and other disciplines. With development of risk evaluate, asses the scale is constantly expanding, from the initial ecosystem to the present landscape pattern and regional ecological security [1].

In recent years, a large number of domestic scholars use different index construction methods to study regional ecological risk evaluate[2]. For example, Xu LAN, Zhou Rujia and other scholars have evaluated the regional ecological risk from landscape structure change, and carried out spatial autocorrelation analysis of regional ecological risk to study the spatial characteristics of ecological risk[3]. Therefore, this paper takes Yonghe County of Shanxi Province as an example to evaluate the landscape ecological risk of the water and soil conservation function ecological protection area in time series, so as to understand the landscape ecological risk status of the soil and water conservation area, so as to provide reference basis for land ecological security and regional ecological risk prevention and control in the study area.

2 Research area

Yonghe County, subordinate to Linfen City, Shanxi Province, is located in the south of Luliang mountains and the East Bank of Shanxi Shaanxi Grand Canyon inter reaches of the Yellow River, with a overall area of 1212 square kilometers (Fig. 1). It is a part of the Western Shanxi Plateau, with overlapping hills and ridges, vertical and horizontal gullies, serious soil erosion and extremely fragile ecological environment.

3 3 Data origin and research methods

3.1 Data origin and processing

This paper uses the land use data of resource and environment center of Chinese Academy of Sciences as the data source, which are respectively in 1980, 2000 and 2018, with a resolution of 30m × 30m. Yonghe county is divided into five categories of arable land, woodland, meadow, water area and construction land as risk receptors.

*corresponding author's e-mail:lig@gsau.edu.cn

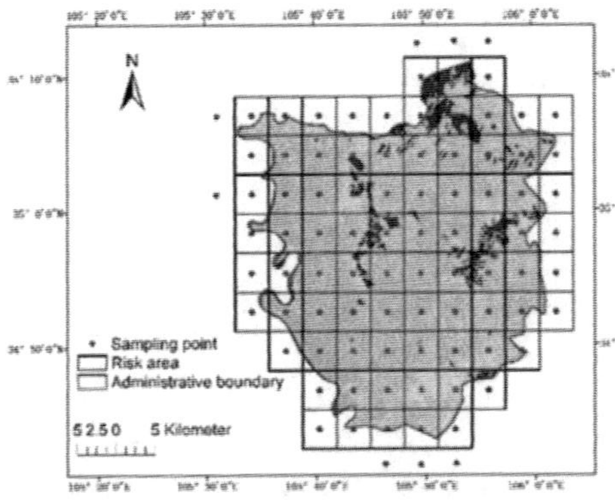

Fig. 1. Caption of the Figure 1. Below the figure.

3.2 Division of risk area

In order to collect the ecological risk evaluate unit and display the ecological risk index in the study area, arcgis10.5 was used to grid Yonghe county.This paper adopts a 1km × 1km square grid for equal interval sampling, and divides 82 evaluation units (Fig. 1). Ecological risk index of each risk area is calculated by using the ecological risk evaluate model, and the ecological risk value is assigned as the value of the sample plot center point, as the sample of spatial interpolation analysis.

3.3 Research method

3.3.1 Landscape disturbance index(E_i).

The index of landscape disturbance degree indicates the level of external disturbance to various landscapes[5]. The expression is as follows: $E_i = aC_i + bN_i + cD_i$

Where C_i is landscape fragmentation index; N_i is landscape separation index; D_i is landscape separation index. and a, b and c are the corresponding weights. Referring to the existing achievements, the indexes are 0.5, 0.3 and 0.2 respectively.

3.3.2 Landscape fragility index[6] (F_i).

The higher the value, the more unstable the ecosystem is. According to the existing results, the six landscape types were classified according to the ability of resisting external influence: water area = 5, unused land = 6, arable land = 4, woodland = 2, meadow = 3, buildings = 1, and the landscape vulnerability index was calculated by normalization.

3.3.3 Landscape ecological loss degree(R_i).

The landscape index method can clearly reflect the dynamic changes of landscape pattern by quantifying landscape indicators [7] . It reflects the loss degree of different landscapes when they are affected by the outside world. The formula is as follows: $R_i = E_i \times F_i$

Where E_i is landscape disturbance index; F_i is Landscape fragility index.

3.3.4 Ecological risk index(ERI_i).

The expression is as follows: $ERI_i = \sum_i^N \frac{A_{ki}}{A_k} R_i$,Where A_{ki} is the acreage for landscape i in the k sampling area; n is the number of landscape types;and A_k is the total acreage in the k sample [8].

3.3.5 Spatial autocorrelation analysis.

In the software arcgis10.5, the landscape ecological hazard value is assigned to the center points of 82 risk areas, and the data is optimized by using the semi variance variation function in statistics, and the ecological risk map is generated by using Kriging interpolation. In order to better identify the landscape ecological hazard changes in each period[9,10]. the risk value is classified into five ecological hazard levels, very high ecological wind Risk areas (ERI > 0.05), high ecological risk areas (0.035 ⩽ ERI < 0.05), moderate ecological hazard areas (0.020 ⩽ ERI < 0.035), low ecological risk areas (0.014 ⩽ ERI < 0.020), very low ecological hazard areas (ERI < 0.014).

4 Results

4.1 Land use change in Yonghe County

Based on ArcGIS, three land use maps (Figure 2) and transfer matrix of 1980-2000, 2000-2018 and 2000-2018 are obtained. In recent 40 years, the grow in number land types of Yonghe County include meadow, construction land and water area, which increase by 10.92 km^2 and 2.17 km^2 and 0.74 respectively, and the reduced land types are arable land and forest land, which are reduced by 11.3 km^2 and 2.6 km^2 respectively.

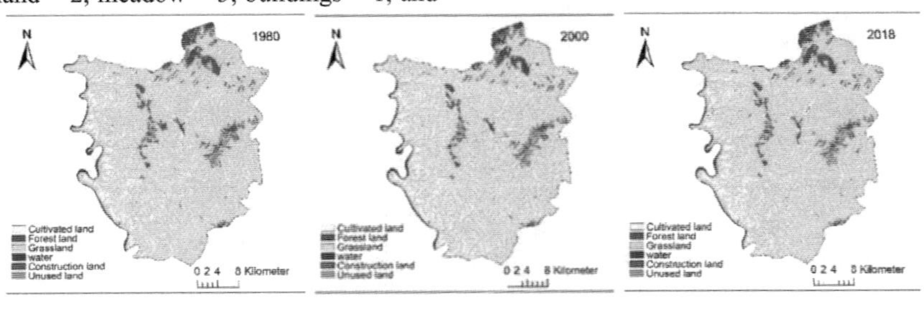

Figure 2. Land use change map in 1980, 2000 and 2018

 https://doi.org/10.1051/e3sconf/202123704038

Table 1. Land use change matrix from 1980 to 2000. km²

	Cultivatedland	forestland	grassland	Water	constructionland	2000Total
Cultivatedland	401.18	0.13	1.06	0.00	0.00	402.37
forestland	0.01	74.49	0.03	0.00	0.00	74.52
grassland	0.28	2.77	723.14	0.00	0.00	726.20
Water	0.00	0.00	0.00	13.36	0.00	13.37
constructionland	0.00	0.00	0.00	0.00	1.79	1.79
1980Total	401.48	77.39	724.23	13.37	1.79	0.00

Table 2. Land use change matrix from 2000 to 2018. Km²

	Cultivatedland	forestland	grassland	Water	constructionland	2018Total
Cultivatedland	342.81	0.76	46.37	0.09	0.11	390.17
forestland	1.07	70.27	3.33	0.02	0.02	74.72
grassland	56.75	3.39	674.86	0.10	0.06	735.16
Water	0.11	0.09	0.86	13.04	0.02	14.11
constructionland	1.61	0.01	0.77	0.00	1.58	3.96
2000Total	402.37	74.52	726.20	13.37	1.79	0.00

Table 3. Land use change matrix from 1980 to 2018. Km²

	Cultivatedland	forestland	grassland	Water	constructionland	2018Total
Cultivatedland	342.15	0.86	46.95	0.09	0.11	390.17
forestland	1.07	70.29	3.33	0.02	0.02	74.72
grassland	56.53	6.14	672.32	0.10	0.06	735.16
Water	0.11	0.09	0.85	13.04	0.02	14.11
constructionland	1.61	0.01	0.76	0.00	1.58	3.96
1980Total	401.48	77.39	724.23	13.37	1.79	0.00

From the perspective of time, the changes of land use types from 1980 to 2000 were relatively small, including arable land, meadow and buildings, which increased by 0.89 km², 1.96 km² and 0.000 9 km², respectively; the land use type with relatively large reduction was woodland, with a decrease of 2.86 km², and the water body was relatively stable, with a decrease of 0.0018 km² (Table 1).

However, from 2000 to 2018, the change of land use types was relatively large, meadow and construction land a grow in number by 8.96 km² and 2.17 km², arable land reduced by 12.19 km², and the increased area of forest land and water body was smaller, 0.20 km²and 0.74 km², respectively (Table 2). During the study period, the landscape pattern of Yonghe county was greatly affected by the rapid development of urbanization, especially during the period from 2000 to 2018, Buildings is on the increase trend, and the expansion of construction land and water area is inseparable from the occupation of arable land and meadow,which was also the reason for the large reduction of arable land and meadow area. In terms of the types of land use transferred in and transferred out, the area of grassland, water area and buildings has continued to increase in recent 40 years. The main supply sources of water area are meadow, arable land and a small amount of woodland. Among them, meadow is transferred 0.1km²to water area, and arable land is 0.09km² to water area; the main supply source of construction land is arable land, with a total of 0.11km²to buildings （ Table 3).

4.2 Analysis of landscape index

Using the software fragstats4.2 and excel, referring to the formula listed, the landscape pattern index of five types of landscape in the study area in 1980, 2000 and 2018 was calculated (Table 4).

Table 4. Landscape pattern index of Yonghe County

Landscape type	Time	Area /km²	Number	Fragmentation index	Abruption index	Predominance index	Obstruction index	Fragility index	Damnify index
Cultivated land	1980	401.44	1232	3.07	1.53	2.50	2.49	0.19	0.47
	2000	402.33	1235	3.07	1.52	2.50	2.49	0.19	0.47

	2018	390.14	1252	3.21	1.58	2.49	2.58	0.19	0.49
Woodland	1980	77.38	117	1.51	2.44	2.89	2.07	0.10	0.20
	2000	74.51	113	1.52	2.49	2.40	1.99	0.10	0.19
	2018	74.72	121	1.62	2.57	2.40	2.06	0.10	0.20
Grassland	1980	724.17	135	0.19	0.28	2.30	0.64	0.14	0.09
	2000	726.14	138	0.19	0.28	2.30	0.64	0.14	0.09
	2018	735.10	157	0.21	0.30	2.28	0.65	0.14	0.09
Water area	1980	13.37	21	1.57	5.98	2.88	3.16	0.24	0.75
	2000	13.37	23	1.72	6.26	2.88	3.32	0.24	0.79
	2018	14.11	14	0.99	4.63	2.85	2.45	0.24	0.58
Construction land	1980	1.79	18	10.08	41.47	6.78	18.84	0.05	0.90
	2000	1.79	18	10.08	41.45	6.78	18.83	0.05	0.90
	2018	3.96	30	7.57	24.11	4.02	11.82	0.05	0.56

From the results, it can be seen that with the increase of meadow and water area, the dominance degree is also rising. During the study period, the loss degree of arable land and grassland changed the most and showed an upward trend. The reason is that the land type dominated by plateau in Yonghe county has overlapping hills and ridges and ravines. Arable land and meadow are principally distributed, while the meadow area decreases, its fragmentation also increases. In the study area, the meadow vulnerability is relatively high, which leads to an upward trend of grassland loss; while the water area gradually reduces, leading to the decline of its fragmentation, at the same time, the vulnerability of water area is also relatively high, which leads to the increase of its loss degree.

4.3 Ecological risk changes

The spatial difference analysis of landscape ecological hazard in Yonghe county shows that the spatial arrangement of landscape ecological hazard in Yonghe county has obvious location and heterogeneity characteristics (Fig. 3). The risk type principally scattered in the northern part of the paper is low ecological hazard, with obvious spatial aggregation. This area is a contiguous forest land with obvious landscape advantages. Over the years, it has formed a relatively stable landscape structure with low landscape fragmentation and separation; The areas with low ecological risk are principally scattered in scattered in Yonghe county with complete landscape structure and strong anti-interference ability;the middle ecological risk is principally scattered in scattered in Yonghe county To be scattered in the area around the county seat of the study area, there are more fragmented woodlands and meadows in this area, which have poor stability and relatively fragile ecosystem; the higher ecological hazard is principally scattered in scattered in the southern part, including sangbi Town, Gedi Township and Jiaokou Township, where the terrain is low and there is a large range of meadow, and the human disturbance intensity is large; the ecological risk is high The area is principally located in the southwest of China, with more fragmented arable land and the highest potential ecological hazard of landscape.

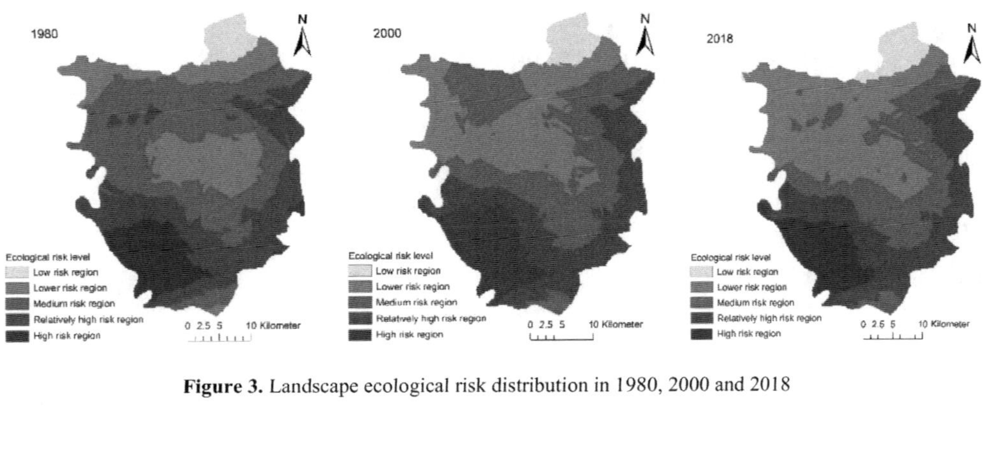

Figure 3. Landscape ecological risk distribution in 1980, 2000 and 2018

5 Discussion

From the core connotation of landscape ecological hazard, the main risk sources of landscape ecological hazard are

https://doi.org/10.1051/e3sconf/202123704038

the increase of man-made landscape and the decrease of natural landscape. Human activities change the pattern and composition structure of surface ecosystem, thus affecting the service function and spatial distribution of ecosystem, leading to the imbalance of ecosystem and the increase of ecological risk. The pattern and evolution characteristics of landscape ecological hazard not only directly reflect the characteristics of ecological environment, but also the ability of ecosystem to resist external interference and disturbance. In order to clearly express the spatial distribution characteristics of landscape ecological hazard degree in Yonghe county. The results show that:

From 2010 to 2015, the landscape pattern of Yonghe county has changed obviously, the area of woodland and arable land subsided significantly, meadow, water body and buildings expanded significantly, and the expansion area principally came from the transformation of arable land and grassland area. From 1980 to 2018, the landscape fragmentation with a grow in number as a whole. In 1980, the fragmentation degree of buildings was the highest, followed by arable land, woodland, water area and meadow. After nearly 40 years of change, except for buildings and water area, the fragmentation of other landscape types has grow in number. From 1980 to 2018, the level of separation of Yonghe county also changed significantly. Except for the water area and buildings, the level of separation of other landscape types growed to varying degrees. From 1980 to 2018, the overall landscape dominance of Yonghe county did not change much. The landscape dominance of buildings was the largest, followed by woodland and water area, and farmland and meadow were the smallest.

In 2015, the area of moderate risk area reduced significantly, while the area of mild risk area increased, and the area of high risk area subsided importantly.

In view of the analysis of spatial arrangement, the landscape risk types of Yonghe county showed a decreasing trend from south to north. The area with high risk is principally arable land, which has mostly rural residential areas, and is vulnerable to human activities, and the ecological environment is fragile; the northwest is principally the type of low-risk areas; most of the low-risk areas are in the north of the study area, and the main landscape type is woodland, with low degree of fragmentation, strong anti-interference ability and lowest risk value.

Acknowledgment

Fund Project: Gansu Provincial Special Project(SSCZZ-20160909)

References

1. Xu Yan, Gao Junfeng, Gao Yongnian. Landscape ecological risk assessment of Taihu Lake area based on dynamic change of land use. Lakes Science ,2011,23(4):642-648.
2. Gong Jie, Zhao Caixia, Xie Yuchu, Gao Yanjing. Ecological risk assessment and management of Bailong River Basin in Gansu Province based on landscape pattern. Journal of Applied Ecology 25(7):2041-2048.
3. Gao Bin, Li Xiaoyu, Li Zhigang, Chen Wei, he Xingyuan, Qi Shanzhong. Ecological risk analysis of Jinzhou Bay Coastal Economic Development Zone based on landscape pattern. Journal of Ecology ,2011,31(12):3441-3450.
4. Gao Yongnian, Gao Junfeng, Xu Yan. Landscape ecological risk effect of land use change in water ecological function area of Taihu Lake Basin. Journal of Natural Resources,2010,25(7):1088-1096.
5. Gao Bin, Li Xiaoyu, Li Zhigang, et al. Ecological Risk Analysis of Jinzhou Bay Coastal Economic Development Zone Based on Landscape Pattern [J]. Ecologica Sinica ,2011,31(12):3441-3450
6. Zhang Yue, Zhang Fei, Wang Juan, et al. Study on Ecological Risk Assessment and Prediction of Lake Abby Region Based on LUCC [J].]; and China Environmental Science 36(11):3465-3474.
7. Wang J, Cui B S, Liu J, et al. The effect of land use and its change on ecological risk in the Lancang River watershed of Yunnan Province at the landscape scale[J]. Acta Scientiae Circumstantiae, 2008, 28(2): 269-277.
8. Zhang Y, Zhang F, Zhou M, et al. Landscape ecological risk assessment and its spatiotemporal variations in Ebinur Lake region of inland arid area[J]. Chinese Journal of Applied Ecology, 2016, 27(1): 233-242.
9. Tian Y, Li B, Wang S. Study on landscape ecological risk of the coastal areas of Jiangsu Province[J]. Research of Soil and Water Conservation, 2015, 22(1): 241-245, 251.
10. Xu Y, Gao J F, Gao Y N. Landscape ecological risk assessment in the Taihu region based on land use change[J]. Journal of Lake Sciences, 2011, 23(4): 642-648.

Taiwan's Ecological Footprint, 2012-2018

Yung-Jaan Lee[1,*], *Po-Shu* Wu[1], and *Lei* Chai[1]

[1]Chung-hua Institution for Economic Research, Taiwan

Abstract. The Ecological Footprint (EF) is a measurement broadly adopted by the international community to measure the progress toward sustainability. Taiwan's EF methods refer to the annual reports of the Global Footprint Network (GFN). Therefore, the calculation method closely follows international trends and is updated accordingly. Since the first calculation of Taiwan's EF in 1998, Taiwan's EF has been revised several times. At present, the EF from 1994 to 2011 can be obtained. The purpose of this study is to update Taiwan's EF from 2012 to 2018. This study divides the biologically productive lands into six categories. Since there are two different data sources for fishing grounds and carbon emissions, Taiwan's EF can be calculated with four different results. Overall, Taiwan's EF shows a slow downward trend from 2012 to 2018. Furthermore, Taiwan's carbon footprint accounted for about 70% of the EF, followed by the cropland footprint, which accounted for about 20% of the EF. Compared with global trends, Taiwan's carbon footprint is about 10% higher than the global carbon footprint, indicating that Taiwan's carbon emissions are higher than the global average. With the global emphasis on carbon reduction, Taiwan needs to focus on improving carbon emissions.

1 Introduction

In 2020, the COVID-19 swept the world. Cities around the world, in the face of the pandemic, adopted shutdown strategies, including restrictions on going out, working from home, self-control and other outbreak control strategies. For the human society, the COVID-19 is an unprecedented disaster. However, from the perspective of nature, it is a rare respite. For example, the water quality of Venice's Grand Canal becomes clear and fish swims freely [1]. Many wild animals suddenly appeared in cities. Because of the shutdown of cities, it is not humans who walk down the street, but a variety of wild animals leaving their footprints and asking: where have humans gone?

When the COVID-19 arrived, there was a dead silence. Economic activities have slowed down as a result of human dormant. According to the latest report of the Global Footprint Network [2], as a result of the COVID-19 outbreak, the ecological footprint (EF) of human activities declined significantly in 2020, with data showing an 8.4% decline in forest footprints and a significant 14.5% decline in carbon footprints. This is the first time since the global financial crisis in 2008 that the carbon footprint has declined. Does this mean that the pandemic is the Earth's febrifuge?

Using the Ecological Footprint (EF) as a measurement, we can find that the world's greenhouse gas emissions have declined due to the pandemic. Meanwhile, slowing production of forest products due to slowing economic activities indicates that the carbon sequestration rate of forest increases. The above-mentioned EF data suggest that the pandemic helps make the environment move toward sustainability, which is a rare opportunity. Looking back at Taiwan's response to the 2020 pandemic, the performance was positive and optimistic. As a result, the measurement of Taiwan's EF becomes an important research agenda, which is the main purpose of this study.

Since Taiwan is not included in the Global Footprint Network (GFN), Taiwan's EF has long been calculated by Taiwan's Forestry Bureau and the Chung-Hua Institution for Economic Research [3-5]. At present, Taiwan' s EF had been calculated from 1994 to 2011. After 2012, Taiwan's EF has not been studied for systematic calculation, which is the main research goal of this study. First, this study introduces the concept of EF, the trend of research on EFs abroad and the study of Taiwan's EF in the past. Second, this study presents the results of Taiwan's EF from 2012 to 2018. This study then discusses the differences between Taiwan's EF trends and global trends, and puts forward suggestions for Taiwan's future research and related policy applications.

2 Literature Review

2.1 Ecological Footprint: Concept and Calculation

2.1.1 Concept of EF: measurement of sustainability

The ecological footprint is a sustainability measurement widely adopted by the international community in recent years, and its novelty lies in the transformation of sustainable development from a narrative definition to a computable indicator. The characteristics of this indicator transform complex statistical items into a clear single value - global hectares (gha) - which makes sustainability

* Corresponding author: yungjaanlee@gmail.com

a comprehensive environmental measurement that can be tracked over a long period of time and can be communicated in dialogue with other indicators.

The term ecological footprint consists of "ecological" and "footprint". Ecology refers to the ecological environment with the ability to survive and maintain (i.e., biological productivity). The survival and maintenance capacity refers to the ability of ecological environment to contain assignation and self-purification. Footprints are traces left by human activities and can be referred to as a portion of the lands occupied by humans. Thus, combining these two words, the ecological footprint can be understood as the areas occupied by humans for their survival [6]. In practice, the EF is based on the "occupied areas" perspective to calculate the sum of human consumption in a particular region (usually a country). The sum is then divided by the total population and the "environmental areas occupied per capita" can be obtained - that is, the “ecological footprint”.

The EF is an operational tool for measuring sustainable development. The most widely cited definition of sustainable development is the Brundtland Report's "development that meets the needs of the present without compromising the ability of future generations to meet their own needs " [7]. In order to develop sustainably, Daly [8] argued that the concept should be a transition from sustainable “growth” of quantitative expansion to sustainable “development” of qualitative improvement. Growth is a quantitative increase in scale, and development is a qualitative improvement or a release of potential.

Daly [8] suggested the need to establish an operational principle of sustainable development, which is the establishment of an "environmental carrying capacity" - "the maximum number of populations that the habitat can support without permanently detracting from the productivity of the ecosystem of a particular habitat". Furthermore, the Agenda 21, issued by the United Nations at the Earth Summit in Rio in 1992, advocated that "data-based information improves the quality of decision-making” provided a strong basis for the establishment of environmental monitoring indicators.

The EF can be regarded as a paradigm shift in the concept of "environmental carrying capacity". In the past, "environmental carrying capacity" was regarded as the feeding capacity of a piece of land (including food, water, energy, social welfare, etc.). The calculation of carrying capacity needed to consider regional characteristics and the vague definition of computation. Thus, Rees [9] responded to Daly's [8] viewpoint and proposed the ecological footprint concept.

Subsequently, Wackernagel and Rees [6] transformed the carrying capacity concept into the EF concept and proposed four types of biologically productive land, establishing the methodology of the EF as an indicator for measuring sustainable development. This set of calculations seeks to be comparable and traceable, and has the potential to correct comprehensive metrics. As data permit, the Global Footprint Network (GFN), a cross-border EF database, was established. On a two-year base, the National Footprint Account (NFA) was published, with a reference manual on trends and calculation methods for EF changes in 232 regions/countries worldwide.

2.1.2 Basic Formula for Calculating the EF

The purpose of calculating the EF is to respond to the following question: How much biosphere's regenerative capacity is needed to support human activities? More precisely, how much regenerative capacity does a particular human activity (food, forestry, etc.) require? In response to the above questions, the following assumptions are proposed for the calculation of the EF (Wackernagel and Rees, 1996):

(1). The earth's biological productivity to support the survival of all living things is limited, and human activities can be absorbed and assimilated by the regenerative capacity of the biosphere.

(2). The regenerative capacity is measured as ecological capacity. It is assumed that biological productivity is sustainable.

In practice, only the basic services of human activities need to be calculated, without double counting more than two services (the finer it is, the more natural functions it contains). A total of six biologically productive lands can be obtained (Table 1) (NFA, 2018).

Table 1. Six types of biologically productive lands.

Cropland: Cereals, nuts, potatoes, etc.	Grazing land: Meat.	Forest land: Wood.
Fishing ground: Aquatic.	Built-up land Land for housing and facilities.	Carbon footprint: Land for converting carbon dioxide.

Source: [10]

According to the latest reference manual published by the NFA [10], the six biologically productive lands include: cropland, grazing land, forest land, fishing ground, built-up land and carbon footprint. Each type of land is calculated based on a classified item, i.e., grazing land calculates the total consumption of meat. Therefore, this method of calculation is called the enumeration calculation.

The EF calculates the average consumption per person (Ci) for the six consumption items (i). To convert the six types of footprints into a uniform value, the per capita consumption (Ci) is divided by the average productivity (Pi) of that item (i), and the six values are added up to obtain the EF:

$$EF\ (gha/per\ capita) = Ci/Pi \qquad (1)$$

where,

Annual consumption per capita (Ci): total regional or national consumption divided by the population for a particular item.

Average biological productivity for a specific item (Pi): regional production divided by global average yield (i.e., conversion value).

The biologically productive lands can be simplified to global hectares (gha).

The above EF value is the most basic calculation concept. In practice, the EF should be justified through the correction of international trade and equivalent productivity, reflecting: (1) The international trades (import and export of consumer goods) are frequent and there is a need to consider the international trade correction. (2) The land productivity of different consumer goods is not the same. For example, croplands are highly intensively operated and therefore highly productive; and grazing lands are highly extensively operated and therefore less productive. Consequently, data correction through the "equivalent factor (EQF)" is required. The calculation formula for the actual consumer goods (i) is as follows.

EF (i) = (Production + Imports - Exports) / Total population X Equivalent Factor (EQF) (2)

The above formula reflects the concept of EF, which is calculated considering: (1) the adjustment of the enumerated items, (2) the relevant parameters containing corrections to the equivalent factors, and (3) changes in the way global production is calculated. Changes in these factors are detailed in the NFA's annual report.

2.2 Previous Research on Taiwan's EF

Since the Global Footprint Network (GFN) does not include Taiwan as an area for calculating EFs, the calculation of Taiwan's EFs has mainly studied and analyzed by scholars in Taiwan. The EF research is divided into two parts: First, the national EF account is carried out with a specific period of time as the target of long-term tracking. Second, relevant urban/regional development research and economic externality research use the EF as an assessment indicator of sustainable development, because of its mature, stable and traceable characteristics. Taiwan's EF accounts mainly refer to the annual reports of the Global Footprint Network (GFN) and the National Footprint Account (NFA), which detail the calculation of EF and include algorithmic changes, such as the adjustments of enumerated items and updates to balance parameters. Therefore, Taiwan's EF is calculated in keeping with international trends and updated simultaneously during major algorithmic changes. Taiwan's EF accounts are mainly conducted by Lee and his colleagues [3-5, 11]. Since the first calculation of Taiwan's EF in 1998, Taiwan's EF has been recalculated several times and the previous footprint data have been updated simultaneously. At present, there are studies covering Taiwan's EFs from 1994 to 2011. Therefore, the main purpose of this study is to update Taiwan's EF accounts from 2012 to 2018.

Taiwan's first EF account in 1998, referred to Wackernagel and Rees [6], divided the bioproductive land into six categories (cropland, grazing land, fishing ground, forest land, energy land and built-up land) and calculated that Taiwan's EF in 1996 was 4.67 hectares per person [4]. However, the EF value did not consider the equivalent factor (EQF) and would significantly underestimate the footprint.

Lee and his colleagues were commissioned by the Council of Agriculture in 2005-2006 to recalculate the EF after including equivalent factors, and to calculate the change in EFs over a 10-year period from 1994 to 2003. Their research found that the EF increased from 5.07 global hectares per person in 1994 to 5.14 global hectares per person in 2003 [3]. It can be found that after the inclusion of equivalent factors, the EF value grew from 4.67 to 5.07 global hectares/person, which accurately reflected the differences in biological productivity of different types of land. Moreover, it showed that the previous research underestimated Taiwan's EF, the difference between the two account is about 1.25 times.

As the global warming effect from increased greenhouse gas emissions threatens global sustainable development, the Global Footprint Network (GFN) and the National Footprint Account (NFA) replaced the "energy land" with the "carbon footprint" to address that the increase in carbon emissions will significantly enlarge the EF account. Lee [3] then recalculated Taiwan's EF in 2004 and found that Taiwan's EF had grown significantly from 5.14 global hectares/person to 6.72 global hectares/person (of which the carbon footprint accounted for 2.23 global hectares/person), which is equivalent to 42 Taiwans. The amount of CO2 emitted requires 14 Taiwans to absorb. The change of enumerated items allowed Taiwan's EF to grow by about 1.3 times.

The EF is an important indicator for measuring sustainable development. Its innovative enumerated calculation methods can estimate national scale footprint account and local scale footprint account. Lee [3] adopted a local-scale approach to analyze Taiwan's EF at the county and city level. The results found that consumption patterns and regional characteristics exhibit a significant relationship with Taiwan's EF. Through the comparison of EF, we can see that good-quality public services and effective land use are important pathway to reduce the footprint.

Wang et al. [11] referred to the research methods of Lee [6] and carried out a retrospective calculation of Taiwan's EF from 1994 to 2007. Their study results show that the EF of 2007 was 6.54 global hectares/person. The most important algorithmic changes are mainly in the built-up land and energy land. First, because hydropower data are not easy to obtain, so they are excluded from the calculation, the built-up land footprint only calculated the built-up area. The energy land was originally divided into "consumer goods" and "carbon footprint." However, because of the complexity of the item of consumer goods and calculation difficulties of data variation, the item of consumer goods was excluded from the calculation of energy lands. The energy lands only calculated the carbon footprint.

In response to the global trends in EF calculation and to continuously track changes in Taiwan's EF, Lee and Peng [5] recalculated Taiwan's EF. According to their research findings, Taiwan's EF in 2011 was 9.43 global hectares per person, of which carbon emissions amounted to 5.94 global hectares per person, accounting for more than 60%. The significant growth in footprint caused by carbon emissions is in line with global trends.

 https://doi.org/10.1051/e3sconf/202123704039

3 Results

Using the classification method of NFA [10], this study divides the EF into six categories of bioproductive lands: cropland, grazing land, fishing ground, forest land, built-up land and carbon footprint. There are two sources of data for fishing ground, with differences in production, imports and exports from different sources. In addition, the sources of carbon emissions are divided into domestic statistics and IEA statistics. Since there are two sources of fishing ground and carbon emissions, four different results can be calculated for Taiwan's EF (Fig. 1). Although Taiwan's EF fluctuated from 2012 to 2018, there was a slow downward trend. The four different colored lines represent the EF of the different data source combinations, with the differences caused by different sources of fishing ground data less than 0.1 gha. Conversely, different sources of carbon emissions data have a larger impact, with differences between about 0.1 gha and 0.2 gha.

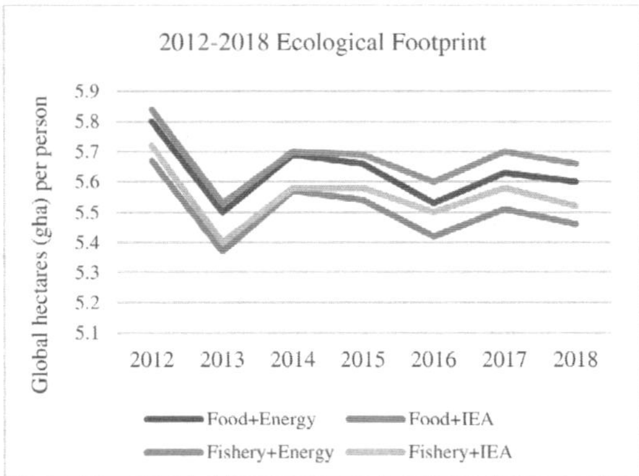

Fig. 1. Changes in Taiwan's EF from 2012 to 2018.

The proportion of carbon footprint and cropland footprint were high, and the impact on Taiwan's EF was relatively large, of which the carbon footprint is the most significant, and the footprint of the four items is less than 0.5 gha (Fig. 2). From 2012 to 2018, Taiwan's EF tended to stabilize to a slow decline, the decline of cropland footprint was more obvious, but the carbon footprint showed a slow upward trend. The proportion of carbon footprint to Taiwan's EF had gradually increased. At a time when the world attaches great importance to carbon emissions and advocates carbon reduction, Taiwan's carbon footprint has increase. Taiwan needs to focus on how to reduce its carbon footprint.

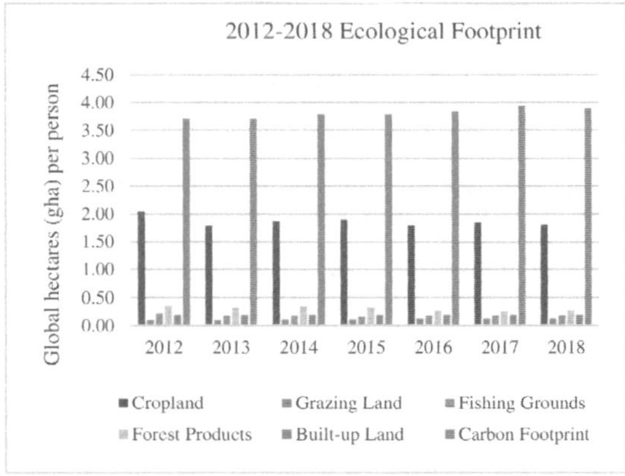

Fig. 2. Changes in Taiwan's Six Footprints from 2012 to 2018.

In the changing trend of various footprints (Fig. 3), carbon footprint accounted for most of the proportion, followed by the cropland footprint. Taiwan's EF remained stable to a slow decline from 2012 to 2018, with no significant change in footprints during that period, with cropland footprint ranging from about 1.5 to 2.0 gha and carbon footprint falling around 4.5 gha. In terms of the proportion of footprints, the proportion of cropland footprint remained at 20%, while the carbon footprint accounted for 70% of Taiwan's EF. As for the other four footprints, their proportions were low and had no significant changes.

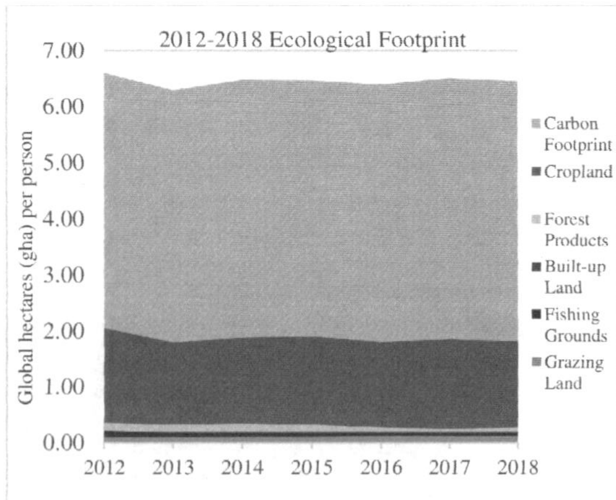

Fig. 3. Trend of Changes in Taiwan's Six Footprints from 2012 to 2018.

4 Conclusions and Suggestions

This study calculates the changes of Taiwan's EF from 2012 to 2018, and revises Taiwan's research methods according to the latest global EF research, of which the calculation of carbon footprint is the most important change.

Although Taiwan's EF fluctuated from 2012 to 2018, it showed a slow downward trend overall, indicating that the per capita consumption of resources decreased but not significantly. In addition, Taiwan's carbon footprint accounted for about 70% of the EF, while the global carbon footprint accounted for about 60%. At present, carbon reduction is a global goal. Taiwan's carbon footprint continues to increase, and the proportion is gradually rising. In order to be in line with international standards, with the global carbon reduction as the goal, and continue to reduce Taiwan's EF, carbon footprint reduction is one of the key items in urgent need of review and improvement.

Compared with the trend of global EF, the trend of Taiwan's EF is consistent with that of the world. Whether the trend of Taiwan's EF is caused by population growth still needs more studies to confirm. Furthermore, influenced by domestic land use patterns and forestry policies, Taiwan's grazelands are small and the forest products are low, therefore the footprints of both items as a proportion of the EF are lower than the global average.

This study only calculates Taiwan's EF of 2012-2018. The follow-up studies are encouraged to calculate Taiwan's ecological benchmark to examine whether there is an ecological deficit in Taiwan. Furthermore, it is recommended to use the latest calculation methods to trace back Taiwan's EF over the years to examine whether Taiwan is moving toward or away from sustainability.

References

1. NASA. NASA Probes Environment, COVID-19 Impacts, Possible Links (April 30, 2020) (Retrieved on December 7, 2020)
2. Earth Overshoot Day. (2020). Earth Overshoot Day 2020 (Global Footprint Network, https://www.overshootday.org/; Retrieved on December 7, 2020)
3. Y.J. Lee. *Analysis and Comparison of Taiwan's Ecological Footprint Trend* (Forestry Bureau, Council of Agriculture, Taiwan, 2006)
4. YJ. Lee, A.C. Chen, Socioeconomic Law and Institution Review **22** (in Chinese) (1998)
5. YJ. Lee, L.P. Peng, Sustainability **6**, 6170-6187 (2014)
6. M. Wackernagel, W. Rees. *Our Ecological Footprint: Reducing Human Impact on the Earth* (New Society Publishers, Canada, 1996)
7. WCED, *Our Common Future* (WCED, 1987)
8. H. Daly, Ecol. Econ. **2** (1990)
9. W. Rees, Environ. Urban. **4**, 2 (1992)
10. D. Lin, L. Hanscom, J. Martindill, M. Borucke, L. Cohen, A. Galli, E. Lazarus, G. Zokai, K. Iha, D. Eaton, M. Wackernagel. *Working Guidebook to the National Footprint Accounts* (Global Footprint Network, 2018)
11. B.C. Wang, F.Y. Chou, Y.J. Lee, Comput. Environ. Urban **36**, 4 (2012)

AUTHOR INDEX

AUTHOR INDEX

AUTHOR INDEX

AUTHOR INDEX

AUTHOR INDEX